D1241338

Practical Handbook of

MICROBIOLOGY

Second Edition

Practical Handbook of MICROBIOLOGY

Second Edition

Edited by

Emanuel Goldman
Lorrence H. Green

CRC Press is an imprint of the
Taylor & Francis Group, an **informa** business

CRC Press
Taylor & Francis Group
6000 Broken Sound Parkway NW, Suite 300
Boca Raton, FL 33487-2742

Printed in the United States of America on acid-free paper
10 9 8 7 6 5 4 3 2 1

International Standard Book Number-13: 978-0-8493-9365-5 (Hardcover)

Library of Congress Cataloging-in-Publication Data

Practical handbook of microbiology / editors, Emanuel Goldman and Lorrence H. Green. -- 2nd ed.
p. ; cm.
"A CRC title."
Includes bibliographical references and index.
ISBN 978-0-8493-9365-5 (hardcover : alk. paper) 1. Microbiology--Handbooks, manuals, etc. I. Goldman, Emanuel. II. Green, Lorrence H. III. Title.
[DNLM: 1. Microbiology. QW 52 P895 2008]

QR72.5.P73 2008
579--dc22 2008011858

Visit the Taylor & Francis Web site at
http://www.taylorandfrancis.com

and the CRC Press Web site at
http://www.crcpress.com

Dedication

To
Rosalind and Susan Green, Emily and Adrian King,
Michele and Matthew and Jordan Green.

Lorrence H. Green

To the memory of Salvador Luria.

Emanuel Goldman

Contents

Preface

The first edition of the *Practical Handbook of Microbiology* was published in 1989. Since that time, the field of microbiology has undergone many changes and has grown to encompass other disciplines as well. New chapters have been added and a number of chapters from the first edition were dropped. Tables in the first edition that were outdated have been replaced by tables in the individual chapters. This edition also contains a new broad and concise survey table of selected eubacteria. Areas generally considered part of microbiology that were not covered or covered only briefly in the first edition are now included with comprehensive introductory chapters.

This book was written to provide basic knowledge and practical information about working with microorganisms, in a clear and concise form. Although of use to anyone interested in the subject matter, the book is intended to especially benefit two groups of microbiologists: people trained as microbiologists who are highly specialized in working with one specific area of microbiology; and people who have been trained in other disciplines and use microorganisms as simply another tool or chemical reagent.

We hope our readers will share our enthusiasm for microbiology and find this book to be useful.

Acknowledgments

This generation of the *Practical Handbook of Microbiology* is due in large part to the dedication and perseverance of a former editor at CRC Press, Judith Spiegel. She saw the need for the book, cultivated a relationship with the scientific editors, and facilitated the process of approvals and preparation of this large project. Most of the contributors to the first edition were no longer available to update their chapters, although we were able to locate a few who were still active in the field and were willing to contribute to the second edition, for which we are most grateful. We would also like to acknowledge the able assistance of CRC staff in the production of the book, especially Jill Jurgensen, as well as Theresa Delforn, Karen Simon, and Barbara Norwitz. Finally, this book would not have been possible without the great efforts of so many contributors. We salute them all.

Emanuel Goldman
Lorrence H. Green

Editors

Emanuel Goldman is a professor in the Department of Microbiology and Molecular Genetics of New Jersey Medical School (NJMS), University of Medicine & Dentistry of New Jersey (UMDNJ). He graduated with honors from the Bronx High School of Science in 1962, received a B.A. cum laude from Brandeis University in 1966, where he was a chemistry major, and completed his Ph.D. in biochemistry at M.I.T. in 1972. He performed postdoctoral research at Harvard Medical School and at the University of California, Irvine, before joining the faculty of New Jersey Medical School in 1979, where he rose through the ranks to Professor in 1993. Among his awards and honors, Dr. Goldman was a Damon Runyon Fellow, a Lievre Senior Fellow of the California Division, American Cancer Society, and a recipient of a Research Career Development Award from the National Cancer Institute. Among his service activities, he was an officer and organizer of the New York–New Jersey Molecular Biology Club, served as a full member of an American Cancer Society Study Section, and continues to serve on the editorial board of *Protein Expression and Purification*. He was also twice elected by his colleagues to serve as the president of the UMDNJ chapter of the American Association of University Professors, and he was elected to serve as president of the Faculty Organization of NJMS. Among several areas of research activity, he has focused on the role of tRNA in elongation of bacterial protein synthesis, including uncharged tRNA, codon bias, and programmed translational frameshifts. In addition to numerous scientific peer-reviewed publications and publications in the lay press, he has contributed a chapter to *Zubay's Biochemistry* textbook, three chapters to the *Encyclopedia of Life Sciences*, and a chapter to the *Encyclopedia of the Human Genome*.

Lorrence H. Green, president of Westbury Diagnostics, Inc., earned a Ph.D. in cell and molecular biology from Indiana University in 1978. He followed this with three years of recombinant DNA and genetic research at Harvard University. In 1981, Dr. Green moved into industry by joining Analytab Products Inc., a major manufacturer of *in vitro* diagnostic test kits. During the next twelve years he helped to invent and manufacture over 40 diagnostic test kits, and rose to become the director of New Product Development and Product Support. In 1993, Dr. Green founded Westbury Diagnostics, Inc., a contract research and development laboratory also offering consulting services. In addition to his work at Westbury, Dr. Green has also served as an adjunct associate professor of microbiology at the New York College of Osteopathic Medicine, and as an adjunct assistant professor of biology at Farmingdale State College of the State University of New York. He has also lectured to microbiology students at the New York Institute of Technology and to business students at the State University of New York at Stony Brook. Dr. Green is the former chairman of the Microbiology Section of the New York Academy of Sciences, as well as the membership chairman and treasurer of the New York City branch of the American Society for Microbiology. From 2001 until 2004 he was a member of the Advisory Committee on Emerging Pathogens and Bioterrorism to the New York City Commissioner of Health. His long-term interest is the use of technology in the development of commercial products. He also enjoys providing mentorship and career advice to students at all levels. He has spoken at many career day events, and judged many regional science fairs.

Editors

Contributors

Sharon L. Abbott
Microbial Diseases Laboratory
Division of Communicable Disease Control
California Department of Public Health
Richmond, California

Elisabeth E. Adderson
Department of Infectious Diseases
St. Jude Children's Research Hospital
Memphis, Tennessee

Abdu F. Azad
Department of Microbiology and Immunology
University of Maryland School of Medicine
Baltimore, Maryland

Lourdes G. Bahamonde
Department of Internal Medicine
Long Island Jewish Health System
North Shore University Hospital
Manhasset, New York

Matthew E. Bahamonde
Department of Biology
Farmingdale State College
Farmingdale, New York

Sukhadeo Barbuddhe
Institute for Medical Microbiology
Justus-Liebig University
Giessen, Germany
Present address:
ICAR Research Complex for Goa
Old Goa, India

Subit Barua
Department of Biological Sciences
St. John's University
Jamaica, New York

Paramita Basu
Department of Biological Sciences
St. John's University
Jamaica, New York

Magda Beier-Sexton
Department of Microbiology and Immunology
University of Maryland
Baltimore, Maryland

Nagamani Bora
School of Biology
Newcastle University
Newcastle upon Tyne, United Kingdom

Michael Bott
Institute for Biotechnology
Research Centre Juelich
Juelich, Germany

Daniel R. Brown
Department of Infectious Diseases and Pathology
College of Veterinary Medicine
University of Florida
Gainesville, Florida

Eva M. Campodonico
Boyer Center for Molecular Medicine
Yale University School of Medicine
New Haven, Connecticut

Trinad Chakraborty
Institute for Medical Microbiology
Justus-Liebig University
Giessen, Germany

Judith A. Challis
Independent Scholar
Brewster, New York

Stuart Chaskes
Department of Biology
Farmingdale State College
Farmingdale, New York

Jongsik Chun
School of Biological Sciences
Seoul National University
Seoul, Republic of Korea

Peter M. Colaninno
ICON Central Laboratories
Farmingdale, New York

Rita R. Colwell
University of Maryland
College Park, Maryland
and
Bloomberg School of Public Health
Johns Hopkins University
Baltimore, Maryland

Carmen E. DeMarco
Oakwood Hospital System
Dearborn, Michigan

Nellie B. Dumas
Wadsworth Center
New York State Department of Health
Albany, New York

Peter Dürre
Institut für Mikrobiologie und Biotechnologie
Universität Ulm
Ulm, Germany

Lothar Eggeling
Institute for Biotechnology
Research Centre Juelich
Juelich, Germany

Patricia I. Fields
Enteric Diseases Laboratory Branch
Centers for Disease Control and Prevention
Atlanta, Georgia

Vincent A. Fischetti
Laboratory of Bacterial Pathogenesis and Immunology
Rockefeller University
New York, New York

Jed F. Fisher
Department of Chemistry and Biochemistry
University of Notre Dame
Notre Dame, Indiana

Collette Fitzgerald
Enteric Diseases Laboratory Branch
Centers for Disease Control and Prevention
Atlanta, Georgia

Steven Foley
Infectious Disease Laboratory
National Farm Medicine Center
Marshfield Clinic Research Foundation
Marshfield, Wisconsin

Joseph J. Gillespie
Department of Microbiology and Immunology
University of Maryland School of Medicine
Baltimore, Maryland
and
Virginia Bioinformatics Institute
Virginia Polytechnic Institute
Blacksburg, Virginia

Edmund R. Giugliano
North Shore Long Island Jewish Health System Laboratories
Lake Success, New York

Emanuel Goldman
Department of Microbiology and Molecular Genetics
New Jersey Medical School
University of Medicine and Dentistry
Newark, New Jersey

Lorrence H. Green
Westbury Diagnostics, Inc.
Farmingdale, New York

Sarah T. Gross
Department of Biology
Farmingdale State College
Farmingdale, New York

Torsten Hain
Institute for Medical Microbiology
Justus-Liebig University
Giessen, Germany

Violet I. Haraszthy
Department of Restorative Dentistry
School of Dental Medicine
State Universtiy of New York at Buffalo
Buffalo, New York

Irvin N. Hirshfield
Department of Biological Sciences
St. John's University
Jamaica, New York

J. Michael Janda
Microbial Diseases Laboratory
Division of Communicable Disease Control
California Department of Public Health
Richmond, California

Donna J. Kohlerschmidt
Wadsworth Center
New York State Department of Health
Albany, New York

Elizabeth Kutter
Evergreen State College
Olympia, Washington

Vincent J. LaBombardi
St. Vincent's Hospital
New York, New York

Peter S. Lee
Amgen, Inc.
Thousand Oaks, California

Stephen A. Lerner
Department of Medicine
Wayne State University School of Medicine
Detroit, Michigan

Martin J. Loessner
Institute of Food Science and Nutrition
Swiss Federal Institute of Technology
Zurich, Switzerland

Yvonne A. Lue
siParadigm Diagnostic Informatics
Oradell, New Jersey

Wlodek Mandecki
Department of Microbiology and Molecular Genetics
New Jersey Medical School
University of Medicine and Dentistry
Newark, New Jersey
and
PharmaSeq, Inc.
Monmouth Junction, New Jersey

Meghan May
Department of Infectious Diseases and Pathology
College of Veterinary Medicine
University of Florida
Gainesville, Florida

Dominique Missiakas
Department of Microbiology
University of Chicago
Chicago, Illinois

Shahriar Mobashery
Department of Chemistry and Biochemistry
University of Notre Dame
Notre Dame, Indiana

Kimberlee A. Musser
Wadsworth Center
New York State Department of Health
Albany, New York

Norberto J. Palleroni
Department of Biochemistry and Microbiology
Cook College
Rutgers University
New Brunswick, New Jersey

Charles S. Pavia
Department of Biomedical Sciences
New York College of Osteopathic Medicine
New York Institute of Technology
Old Westbury, New York

Mark S. Peppler
Department of Medical Microbiology and Immunology
University of Alberta
Edmonton, Canada

John B. Perkins
DSM Anti-Infectives B.V.
DAI Innovation
Delft, the Netherlands

David H. Pincus
R&D Microbiology
bioMérieux, Inc.
Hazelwood, Missouri

Catherine E.D. Rees
School of Biosciences
University of Nottingham, Sutton Bonington
Sutton Bonington, United Kingdom

D. Ashley Robinson
Department of Microbiology and Immunology
New York Medical College
Valhalla, New York

Ken S. Rosenthal
Department of Microbiology, Immunology, and Biochemistry
Northeastern Ohio Universities Colleges of Medicine and Pharmacy
Rootstown, Ohio

Craig R. Roy
Boyer Center for Molecular Medicine
Yale University School of Medicine
New Haven, Connecticut

Patricia Ryan
Rockefeller University Hospital
Rockefeller University
New York, New York

Olaf Schneewind
Department of Microbiology
University of Chicago
Chicago, Illinois

Frederick L. Schuster
Viral and Rickettsial Disease Laboratory
California Department of Public Health
Richmond, California

Susan E. Sharp
Kaiser Permanente – Northwest
Portland, Oregon

Sanjay K. Shukla
Molecular Microciology Laboratory
Clinical Research Center
Marshfield Clinic Research Foundation
Marshfield, Wisconsin

Trevor H. Stenson
Department of Medical Microbiology and Immunology
University of Alberta
Edmonton, Canada

Maxim Suvorov
Department of Chemistry and Biochemistry
University of Notre Dame
Notre Dame, Indiana

Ernestine M. Vellozzi
Department of Biomedical Sciences
Long Island University, C.W. Post
Brookville, New York

Audrey Wanger
University of Texas Medical School
Houston, Texas

Alan C. Ward
School of Biology
Newcastle University
Newcastle upon Tyne, United Kingdom

Jean Whichard
Enteric Diseases Laboratory Branch
Centers for Disease Control and Prevention
Atlanta, Georgia

Robert F. Whitcomb (Deceased)
Beltsville Agricultural Research Center
U.S. Department of Agriculture
Beltsville, Maryland

Joseph J. Zambon
Department of Periodontics and Endodontics
School of Dental Medicine
State University of New York at Buffalo
Buffalo, New York

Daniel R. Zeigler
Department of Biochemistry
The Ohio State University
Columbus, Ohio

Part I

Practical Information and Procedures for Bacteriology

1 Sterilization, Disinfection, and Antisepsis

Judith A. Challis

CONTENTS

BACKGROUND

The need for techniques to control the growth of microorganisms has been shaped by both cultural and scientific advances. From the days of early food preservation using fermentation of milk products and smoking of meats to extend the shelf life of foods, practical needs have contributed to the development of these techniques.

Formal development of clinical medical settings led to an awareness of the cause and effects of disease. These observations contributed to the techniques developed by Joseph Lister, Oliver Wendell Holmes, and Ignaz Semmelweis that prevented disease transmission between patients.

Society continues to shape our needs today, with the rampant evolution of drug-resistant microbes, a high nosocomial-infection rate in patient-care facilities, and the development of exploratory and permanently inserted medical devices. In addition, the threat of bioterrorism introduces the potential for the introduction of pathogenic microbes into both a type and breadth of environment not considered in the past.

Our requirements for sterilization, antisepsis, and sanitizing thus go beyond the historical needs of the research and clinical laboratories and commercial production requirements. This chapter not only reviews commonly used laboratory techniques, but also specifically addresses their limitation when used to reduce or eliminate bacteria on materials that were not originally designed to be sterilized or disinfected.

METHODS OF STERILIZATION

Sterilization is the removal of all microbes, including endospores, and can be achieved by mechanical means, heat, chemicals, or radiation (Table 1.1). When using heat, it may be either dry heat or moist heat. Traditionally, moist heat under pressure is provided by autoclaving and dry heat by ovens (see Table 1.1).

Disinfection is the process that eliminates most or all microorganisms, with the exception of endospores. Disinfectants can be further subcategorized as high-level disinfectants, which kill all microorganisms with the exception of large numbers of endospores with an exposure time of less

TABLE 1.1

Routine Methods of Sterilization[1]

Method [Ref.]	Temperature	Pressure	Time	Radiation (Mrad)
Dry heat [24]	150–160°C 302–320°F		>3 h	
Dry heat [24]	160–170°C 320–338°F		2–3 h	
Dry heat [24]	170–180°C 338–356°F		1–2 h	
Moist heat	135°C 275°F	31.5 psig	40 min	
Boiling—indirect [24]			1 hr	
Boiling—direct[2]			2 min	
Radiation—cobalt 60[3] [24]	Ambient		hours	2–3
Radiation—cesium 137[3] [24]	Ambient		hours	2–3
Electronic accelerators[4] [24, 25]			<1 sec	2.5
Ozone [26, 27]		See cited references	See cited references	

[1] Steam under pressure (i.e., autoclave).

[2] Boiling point of liquids- e.g., cumene (isopropylbenzene) 152°C (306 °F).

[3] Dependent on curies in source.

[4] Electrostatic (Van de Graaff), electromagnetic (Linac), direct current, pulsed transformer.

than 45 min; intermediate-level disinfectants, which kill most microorganisms and viruses but not endospores; and low-level disinfectants, which kill most vegetative bacteria, some fungi, and some viruses with exposure times of less than 10 min [1].

Antiseptics destroy or inhibit the growth of microorganisms in or on living tissues and can also be referred to as biocides [2]. Disinfectants are used on inanimate objects and can be sporostatic but are not usually sporocidal [2].

Steam sterilization or dry heat can be monitored by the use of biological indicators or by chemical test strips that turn color upon having met satisfactory conditions. These indicators are widely available. Usually the spores of species of *Geobacillus* or *Bacillus* spp. are used in either a test strip or suspension, as these organisms are more difficult to kill than most organisms of clinical interest [3]. Growth of the spores in liquid media after the cycle of sterilization is complete indicates the load was not successfully sterilized.

In the days following September 11, 2001, letters containing the spores and cells of *Bacillus anthracis* were mailed to several news media offices and to two U.S. senators. This introduced a pathogen into sites where the methods of safe handling or eradication of the pathogen were not in existence. A number of laboratories conducted experiments on the potential dispersal, detection, and eradication of organisms spread by such an event.

For example, Lemieux et al. [4] conducted experiments with simulated building decontamination residue (BDR) contaminated with 10^6 spores of *Geobacillus sterothermophilus* to simulate *Bacillus anthracis* contamination.

A single cycle did not effectively decontaminate the BDR. Only autoclave cycles of 120 min at 31.5 psig/275°F and 75 min at 45 psig/292°F effectively decontaminated the BDR. Two standard cycles at 40 min and 31.5 psig/275°F run in sequence were even more effective. The authors state that the second cycle's evacuation step probably pulled condensed water out of the pores of the materials, allowing better steam penetration. It was found that both the packing density and material type of the BDR significantly impacted the effectiveness of the decontamination process.

Therefore, when standard laboratory methods are used for materials made of unusual substances and density, not normally used in a laboratory, the method of sterilization must be monitored and tailored independently of that substance. Standard laboratory methods based on different and lower autoclave packing levels and lower density materials may not be effective.

Other common methods of sterilization include gases such as ozone, radiation, or less commonly electronic accelerators (see Table 1.1).

Solutions containing heat-labile components require a different approach. Filtration is generally the most accepted and easiest method. The FDA and industry consider 0.22-μ filters sterilization grade based on logarithmic reductions of one of the smaller bacteria *Brevundimonas diminuta* [5]. Usually a delay of 48 hours (hr) is required for the colony development of *Br. diminuta* before the colonies are large enough to be viewed for counting. Griffiths et al. [6] describe a Tn5 recombinant method in which organisms were cloned for bacterial bioluminescence (luxABCDE) and fluorescence (*gfp*). These organisms were detected after only 24 hr of incubation by either fluorescence or bioluminescence. They state that this method may aid in preventing quality control backlogs during filter manufacturing processes.

Filter integrity for sterilization is usually done by a bubble test to confirm the pore size of the manufactured filter [5, 7]. The bubble point is based on the fact that liquid is held in the pores of the filter (usually membrane) by surface tension and capillary forces, and the bubble pressure detects the least amount of pressure than can displace the liquid out of the pores of the filter.

While most solutions used in molecular biology and microbiology laboratories will be adequately sterilized with a 0.22-μ filter, those for tissue culture might also need to remove potential mycoplasma contamination. A 0.1-μ filter should be used to remove mycoplasma from tissue-culture solutions [8] (Table 1.2).

The removal of contaminants from air may be necessary in the case of fermentation or drug manufacturing in chemical reactions requiring some form of gas. Various types of filters are available for the removal of organisms from air. One such filter, as an example, the Aerex 2 (Millipore), can withstand 200 steam-in-place cycles at 293°F (145°C), resists hydraulic pressure at 4.1 bard (60 psid), and has the ability to retain all phages when challenged with the ΦX-174 bacteriophage at 29 nm in diameter and 10^7 to 10^{10} bacteriophages per cartridge [9]. In this particular scenario, the operator must assess their needs for how critical the product sterility (i.e., drugs versus topical cosmetics) must be, as well as the ability of the product to withstand pressure, heat, flow-rate, and other parameters of the manufacturing process, and then choose a filter accordingly.

In some cases, a filter is used to recover bacteria from dilute solutions of bacteria such as environmental water samples. The organism of significance whose retention on the filter is most important should determine the filter pore size. The smaller the filter pore size, the slower the sample flow rate and throughput. Standard methods define an acceptable recovery rate for filter-recovery of bacteria as being 90% of the number of bacteria (colony-forming units, CFU) recovered

TABLE 1.2
Filter-Sterilization

Size(μ)	Purpose
0.1	Mycoplasmal removal
0.22	Routine bacterial removal
0.45	Plate counts water samples
>0.45	Removal of particulates, some bacteria, yeast, and filamentous fungi

from the same sample by the spread plate method. In one study performed by Millipore, filtration with a 0.45-µ pore-size filter provided 90% recovery of 12 microorganisms used, ranging in size from *Br. diminuta* to *Candida albicans*. In some cases, the use of the larger 0.45-µ pore-size filter instead of the 0.22-µ filter also increased the size of the colony, facilitating counting of the CFUs [10] (Table 1.2).

It is important to note that all filter sterilization is relative. While a 0.22-µ filter will remove most bacteria, it will not remove viruses, mycoplasma, prions, and other small contaminants. Every sample type and the level of permissible substances in the filtrate must be assessed on a case-by-case basis. For example, filter-sterilization of water or solutions used in atomic force microscopy might still allow enough small viruses and other particulates through to interfere with sample interpretation, and therefore ultra-pure water may be needed.

DISINFECTANTS AND ANTISEPTICS

The regulation of disinfectants and antiseptics falls under the jurisdiction of different agencies, depending on where the chemical will be used. In the United States, the EPA regulates disinfectants that are used on environmental surfaces (e.g., floors, laboratory surfaces, patient bathrooms), and the FDA regulates liquid high-level sterilants used on critical and semi-critical patient care devices.

Labeling terminology can also be confusing, as the FDA uses the same terminology as the CDC [critical (sterilization), semi-critical (high and intermediate), and non-critical (low-level disinfection)], while the EPA registers environmental disinfectants based on the claims of the manufacturer at the time of registration (i.e., EPA hospital disinfectant with tuberculocidal claim = CDC intermediate level disinfectant) [11]. In addition, the use of disinfectants varies globally. For example, surface disinfectants containing aldehydes or quaternary ammonium compounds are less often used in American, British, and Italian hospitals than in German ones [12], thus making it difficult to define overall "best methods or practices." Whatever method is employed — mechanical, heat, chemical, or radiation — a method of assessment must be in place to test the initial effectiveness of the method on either the organism of the highest level of resistance to the method or a surrogate; and a standard must be included to monitor the effectiveness of the method for its regular use by general staff. While these practices are common for daily runs of the autoclave, this is used less often when general low-level disinfection occurs on non-critical surfaces. Table 1.3 lists a number of commonly used disinfectants, along with some of the organisms against which they have activity.

Disinfectants are often tested against cultures of the following bacteria: *Pseudomonas aeruginosa*, *Staphylococcus aureus*, *Salmonella typhuimuium*, *Mycobacterium smegmatis*, *Pevotella intermeida*, *Streptococcus mutans*, *Actinobacillus actinomycetemcomiticans*, *Bacteriodies fragilis*, and *Escherichia coli* [13]. One common approach is to use the broth dilution method, wherein a standard concentration of the organism is tested against increasing dilutions of the disinfectant. The minimum inhibitory concentration (MIC) of the organism is then determined.

While there are many reports of liquid disinfectant activity against liquid cultures, biofilms of the aforementioned organisms have survived when the liquid culture of the same organism has been killed [13]. Therefore, biofilm disinfection must be evaluated separately. Better efficacy in biofilm prevention and removal has been demonstrated by the use anti-biofilm products compared to detergent disinfectants containing quaternary ammonium compounds [14].

One problem with most disinfectants and antiseptics is their short effective life span. Hospitals and laboratories today are challenged with multiple drug-resistant organisms that may be transmitted from surfaces. These surfaces may be routinely recontaminated by either new sample processing or patient/visitor/staff activity. A new agent, a combination of antimicrobial silver iodine and a surface-immobilized coating (polyhexamethylenebiguanide), Surfacine™ is both immediately active and has residual disinfectant activity [15]. Additionally, it can be used on animate or inanimate objects. In tests on formica with vancomycin-resistant *Enterococcus* provided 100% effective on surface challenge levels of 100 CFU per square inch for 13 days [16]. Rutala and Weber [15]

TABLE 1.3
Partial Listing of Some Useful Disinfectants

Substance	Concentration	Time	Types of Organisms Killed	Ref.
Inorganic hypochlorite (bleach)	6000 ppm	5 min at 22°C	Adenovirus 8	[28]
	5000 ppm	10 min at 23–27°C	Bacteria, viruses, and fungi, hepatitis A and B, HSV-1 and 2	[1]
	2000 ppm	pH 7, 5°C, 102 min	*Bacillus anthracis* Ames (spores)[2]	[29]
		pH 7, 5°C, 68 min	*Bacillus anthracis* Sterne (spores)[2]	[30]
	1000 ppm	10 min at 20 ± 2°C	*Feline calicvirus* (norovirus surrogate)	[31]
Peracetic acid	0.3%		Sporocidal, bactericidal, viricidal, and fungicidal	[2]
Glutaraldehyde	2.65% in hard water (380–420 ppm calcium carbonate)	5 minutes at 22°C	Adenovirus 8	[28]
Ethyl Alcohol	70% undiluted	5 min at 22°C	Disinfectant	[28]
Surfacine	Manufacturer's instructions		VRE	[15, 16]
R-82 (quaternary ammonium compound)	856 ppm (1:256 dilution)	10 min/20°C ± 2°C	Feline calicvirus (norovirus surrogate)	[31]
Sterilox (superoxidized water)	Chlorine (170 ppm and 640 ppm)		High-level disinfectant	[28]
Ozone disinfection of drinking water	0.37 mg/liter	Contact-time 5 min at 5°C	Noroviruses	[26]
Acidic sodium dodecyl sulfate	1% SDS plus either 0.5% acetic acid and 50 mM glycine (pH 3.7) or 0.2% peracetic acid	30 min at 37°C	Prions: specifically PrP^{Sc}, see Ref. for details	[32]

[1] Highest concentration given for use with most resistant organism tested. See cited reference for refined values for each organism.

[2] The authors observed $\log_{10}$ inactivation of populations of spores, the inactivation increasing with contact time, but not sterilization.

also state that the manufacturer's test data demonstrated inactivation of bacteria, yeast, fungi, and viruses with inactivation time varying by organism.

DISINFECTION OF HANDS

One of the most important areas for infection control is the degerming and disinfection of the hands. Usually, mechanical methods such as scrubbing with a brush are most commonly used, but alternatively alcohol-based rubs are also used for hand disinfection.

Widmer [17] states that compliance for handwashing is less than 41% and, additionally, understaffing may further complicate the problem. In a survey of literature on handwashing by Widmer, the following compliance values were found:

- ICU (intensive care unit): 9–41%
- Ward/ICU: 32–48%
- Pediatric: 37%
- Surgical ICU: 38%
- All ICUs: 32%

Handwashing includes wetting the hands and wrists with water, taking a dose of soap with the forearm or elbow, rubbing the hands and wrists with the soap for 10 to 15 seconds, rinsing the hands, and then drying the hands without rubbing. Additionally, the faucet is turned off while holding the paper towel, so as not to recontaminate the hands [17].

Many authors now advocate waterless alcohol rubs. Kampf et al. [18] advocate a 1-minute handwashing at the beginning of the day, a 10-minute drying time, and then a rub in alcohol-based hand disinfectant. Gupta et al. [19] compared an alcohol-based waterless and a water-based, alcohol-based, water-aided scrub solution against a brush-based iodine solution under conditions encountered in community hospital operating rooms. The brush-based iodine solution performed best on the first day of the study; but when colony-count reductions were studied between the three over the course of 5 days, no significant difference was found. The participants found the alcohol-based waterless solution the easiest of the three to use with the lowest complaints of skin irritation. Another reason to advocate non-abrasive hand sanitizing methods is the fact that brush-based systems can damage skin and cause microscopic cuts that can harbor microbes [19].

Cooper et al. [20] found that even small increases in the frequency of effective hand washes were enough to bring endemic nosocomial infective organisms under control in a computer model. Raboud et al. [21], using a Monte-Carlo simulation, determined factors that would reduce transmission of nosocomial infection. The factors that were relevant included reducing the nursing patient load from 4.3 (day) and 6.8 (night) to 3.8 (day) and 5.7 (night), increasing the hand washing rates for visitors, and screening patients for methicillin-resistant *Staphylococcus aureus* (MRSA) upon admission.

HAZARD ANALYSIS CRITICAL CONTROL POINT AND RISK ASSESSMENT METHODS

Food-related industries have used Hazard Analysis Critical Control Point (HACCP) as a method to identify critical control points where lack of compliance or failure to meet standards are points where food-borne illness or contamination of a manufactured product may occur.

The HACCP can be used to promote two positive outcomes. First, a well-designed HACCP can be used as a diagnostic tool to find points in a system where failure might occur. For example, if a certain floor disinfectant has a maximum disinfectant activity at 80°F, a critical control point on the HACCP would be that the temperature was measured and remained at that temperature throughout use. The second area of use is in education, where those using the disinfectant could be asked during regular and routine surveys if they were measuring and using the disinfectant at that temperature. Meeting compliance standards by writing elegant in-house documents of required methods and actually monitoring that those standards are maintained by all staff, regardless of station, are two different goals — both of which have equal importance.

The HACCP is a qualitative rather than a quantitative tool. Larson and Aiello [22] list the seven steps of HACCP as follows:

1. Analyze the hazards.
2. Identify the critical control points.
3. Establish preventative measures with critical limits for each control point.
4. Establish procedures to monitor the critical control points.

5. Establish corrective actions to be taken when monitoring shows that a critical limit has not been met.
6. Establish procedures to verify the system is working properly.
7. Establish effective record-keeping for documentation.

When more quantitative measures are required, a microbial risk assessment model may be the method of choice. Larson and Aiello [22] use four categories of environmental site-associated contamination risks and need for contamination based on the work of Corvelo and Merkhoffer [23]. These include:

1. Reservoirs (e.g., wet sites: humidifiers, ventilators, sinks) with a high probability (80 to 100%) of significant contamination and an occasional risk of contamination
2. Reservoir/disseminators (e.g., mops, sponges, cleaning materials) with a medium risk (24 to 40%) of significant contamination and a constant risk of contamination transfer
3. Hand and food contact surfaces (e.g., chopping boards, kitchen surfaces, cutlery, cooking utensils) with a medium risk (24 to 40%) of significant contamination and a constant risk of contamination transfer
4. Other sites (e.g., environmental surfaces, curtains, floors) with a low risk (3 to 40%) of significant contamination and an occasional risk of contamination transfer

Corvelo and Merkhoffer [23] use six criteria in the evaluation of a risk-assessment-based framework: (1) logical soundness, (2) completeness, (3) accuracy, (4) acceptability, (5) practicality, and (6) effectiveness.

Whether a formal HACCP or risk-assessment or a site-specific method/process-specific model developed by the operator is used, a formal method of assessment of structure, appropriate use, compliance, education, and record-keeping is needed. The methods must be kept current with the practices of the laboratory, hospital, or manufacturing entity, and also with the evolution of resistance in microbes and introduction of new microbes into the environment.

The tables in this chapter represent a selective starting point for choosing a method of sterilization or disinfection. Many of the cited articles contain in-depth information and should be consulted for further details, as there are space limitations herein. New methods are also being developed on a daily basis.

REFERENCES

1. Rutula, W.A. and Weber, D.J. Uses of Inorganic Hypochlorite (Bleach) in Health-Care Facilities, *Clin. Microbiolog. Rev.*, 10, 597, 1997.
2. McDonnell, G. and Russell, A.D. Antiseptics and Disinfectants: Activity, Action, and Resistance, *Clin. Microbiolog. Rev.*, 12, 147, 1999.
3. Bond, W.W., Ott, B.J., Franke, K., and McCracken, J.E. Effective Use of Liquid Chemical Germicides on Medical Devices, Instrument Design Problems, in *Disinfection, Sterilization and Preservation*, 4th ed. Block, S.S., Ed., Lea and Febiger, Philadelphia, PA, 1991.
4. Lemieux, P., Sieber, R., Oasborne, A., and Woodard, A. Destruction of Spores on Building Decontamination Residue in a Commercial Autoclave, *Appl. Environ. Microbiol.*, 72, 7687, 2006.
5. Millipore, Personal Communication with Lisa Phillips, Millipore Technical Representative, 2006.
6. Griffiths, M.H., Andrew, P.W., Ball, P.R., and Hall, G. Rapid Methods for Testing the Efficacy of Sterilization-Grade Filter Membranes, *Appl. Environ. Microbiol.*, 66, 3432, 2000.
7. Millipore Technical Reference xitxsp121p24, http://www.millipore.com/userguides.nsf/docs/xitxsp121p24?open&lang=en
8. Millipore Technical Reference tn039, http://www.millipore.com/publications.nsf/docs/tn039
9. Millipore Technical Reference pf1080en00, http://www.millipore.com/publications.nsf/docs/pf1080en00

10. Millipore Technical Reference QA13, http://www.millipore.com/publications.nsf/docs/QA13
11. MMWR. Regulatory Framework for Disinfectants and Sterilants. Appendix A. 52 62, 2003.
12. Vitali, M. and Agolini, G. Prevention of Infection Spreading By Cleaning and Disinfecting: Different Approaches and Difficulties in Communicating, *Am. J. Infect. Control,* 34, 49, 2006.
13. Vieira, C.D., de Macedo Farias, Galuppo Diniz, C., Alvarez-Leite, M., L., da Silva Camargo, E., and Auxiliadora Roque de Carvalho, M. New Methods in the Evaluation of Chemical Disinfectants Use in Health Care Services, *Am. J. Infect. Control*, 33, 162, 2005.
14. Marion, K., Freny, J., Bergeron, E., Renaud, F.N., and Costerton, J.W. Using an Efficient Biofilm Detaching Agent: An Essential Step for the Improvement of Endoscope Reprocessing Protocols, *J. Hosp. Infect*., 62, 136, 2007.
15. Rutula, W.A. and Weber, D.J. New Disinfection and Sterilization Methods, *Emerg. Infect. Dis*., 7, 348, 2001.
16. Rutula, W.A., Gergen, M.F., and Weber, D.J. Evaluation of a New Surface Germicide (Surfacine™) with Antimicrobial Persistence, *Infect. Control Hosp. Epidemiol.,* 21, 103, 2000.
17. Widmer, A.F. Replace Hand Washing with the Use of a Waterless Alcohol Rub?, *Clin. Infect. Dis*., 31, 136, 2000.
18. Kampf, G., Kramer, A., Rotter, M., and Widmer, A. Optimizing Surgical Hand Disinfection, *Zentralbl. Chir*., 131, 322, 2006.
19. Gupta, C., Czubatyj, A.M., Briski, L.E., and Malani, A.K. Comparison of Two Alcohol-Based Surgical Scrub Solutions with an Iodine-Based Scrub Brush for Presurgical Antiseptic Effectiveness in a Community Hospital, *J. Hosp. Infection*, 65(1), 65, 2007.
20. Cooper, B.S., Medlye, G.F., and Scott, G.M. Preliminary Analysis of the Transmission Dynamics of Nosocomial Infections: Stochastic and Management Effects, *J. Hosp. Infect.,* 43, 131, 1999.
21. Raboud, J., Saskin, R., Sior, A., Loeb, M., Green, K., Low, D.E., and McGeer, A. Modeling Transmission of Methicillin-Resistant *Staphylococcus aureus* among Patients Admitted to a Hospital, *Infect. Control Hosp. Epidemiol.,* 26, 607, 2005.
22. Larson, E. and Aiello, E. Systematic Risk Assessment Methods for the Infection Control Professional, *Am. J. Infect. Control,* 34, 323, 2006.
23. Covello, V. and Merkhoffer, M. *Risk Assessment Methods: Approaches for Assessing Health and Environmental Risks,* Plenum Press, New York, 1993.
24. Gaughran, E.R.I. and Borick, P.R. Sterilization, Disinfection, and Antisepsis, in *Practical Handbook of Microbiology*. O'Leary, W., CRC Press, Boca Raton, FL, 1989, p. 297.
25. Helfinstine S.L., Vargas-Aburto, C., Uribe, R.M., and Woolverton, C.J. Inactivation of Bacillus Endospores in Envelopes by Electron Beam Irradiation, *Appl. Environ. Microbiol.,* 71, 7029, 2005.
26. Shin, G.-A. and Sobsey, M.D. Reduction of Norwalk Virus, Poliovirus 1, and Bacteriophage MS2 by Ozone Disinfection of Water, *Appl. Environ. Microbiol.,* 69, 3975, 2003.
27. Finch, G.R., Black, E.K., Labatiuk, C.W., Gyurek, L., and Belosevic, M. Comparison of *Giardia lamblia* and *Giardia muris* Cyst Inactivation by Ozone, *Appl. Environ. Microbiol.,* 59, 3674, 1993.
28. Rutula, W.A., Peacock, J.E., Gergen, M.F., Sobsey, M.D., and Weber, D.J. Efficacy of Hospital Germicides against Adenovirus 8, a Common Cause of Epidemic Keratoconjunctivitis in Health Care Facilities, *Antimicrob. Agents Chemother*., 50, 1419, 2006.
29. Rose, L.J., Rice, E.W., Jensen, B., Murga, R, Peterson, A., Donlan, R.M., and Arduino, M.J. Chlorine Inactivation of Bacterial Bioterrorism Agents, *Appl. Environ. Microbiol*., 71, 566, 2005.
30. Rice, E.W., Adcock, N.J., Sivaganesan, M., and Rose, L.J. Inactivation of Spores of *Bacillus anthracis* Sterne, *Bacillus cereus*, and *Bacillus thuringiensis* subsp, *Isrealensis* by Chlorination. *Appl. Environ. Microbiol.,* 71, 5587, 2005.
31. Jimenez, L. and Chiang, M. Virucidal Activity of a Quaternary Ammonium Compound Disinfectant against Feline Calcivirus: A Surrogate for Norovirus, *Am. J. Infect. Control*, 34, 269, 2006.
32. Peretz, D., Supattapone, S., Giles, K., Vergara, J., Freyman, Y., Lessard, P., Safar, J., Glidden, D., McCulloch, C., Nguyen, H., Scott, M., Dearmond, S., and Prusiner, S. Inactivation of Prions by Acidic Sodium Dodecyl Sulfate, *J. Virol.,* 80, 323, 2006.

2 Quantitation of Microorganisms

Peter S. Lee

CONTENTS

QUANTITATIVE MICROBIAL ENUMERATION

This chapter provides general information on the various methods used to estimate the number of microorganisms in a given sample. The scope of this chapter is limited to methods used for the enumeration of microbial cells and not the measurement of microbial cellular mass.

Most-Probable-Number (MPN) Method

General

The most probable number (MPN) method is a microbial estimate method used to enumerate viable cell counts by diluting the microorganisms, followed by growing the diluted microorganisms in replicate liquid medium dilution tubes. An example of an MPN dilution scheme used to prepare the various dilution test tube replicates is shown in Figure 2.1. After optimal microbial growth incubation, the positive and negative test results are based on the positive (visible turbidity) and negative (clear) replicate dilution tubes. The MPN method can also be referred to as the multiple tube fermentation method if fermentation tubes (Durham tubes) are also used inside the serial dilution tubes to measure gas production. The viable cell counts are then compared to a specific MPN statistical table to interpret the test results observed. There are also several versions of statistical tables used based on the sample matrix tested, test dilutions used, number of replicates per dilution used, and the statistical considerations used.[1–7] Tables 2.1 and 2.2 provide two examples of MPN statistical tables for three tenfold dilutions with three (3) or five (5) tubes at each dilution.[8] The confidence limits of the MPN are narrowed when a higher number of replicate dilution tubes is inoculated in the dilution series. A number of assumptions are also considered when using the MPN method of microbial estimation. The microorganisms to be estimated are distributed randomly and evenly separated within the samples tested in the liquid dilution tubes. The microorganisms to be estimated are also separated individually and are not clustered together, nor do they repel each other in the liquid dilution tubes. In addition, the utilization of optimal growth medium, incubation temperature, and incubation period is needed to allow any single viable cell to grow and become quantifiable in the liquid dilution tubes used. Primary equipment and materials used for this method are serial dilution

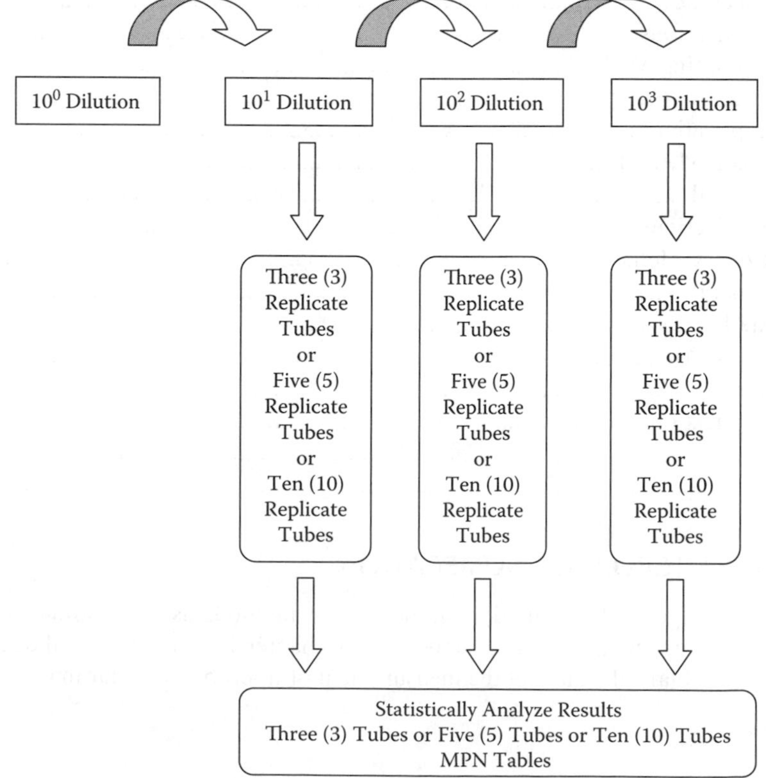

FIGURE 2.1 MPN serial dilution scheme.

TABLE 2.1

Three (3) Tubes Each at 0.1, 0.01, and 0.001 g Inocula and the MPNs per Gram with 95% Confidence Intervals[8]

Positive Tubes			MPN/g	Conf. Limit		Positive Tubes			MPN/g	Conf. Limit	
0.10	0.01	0.001		Low	High	0.10	0.01	0.001		Low	High
0	0	0	<3.0	—	9.5	2	2	0	21	4.5	42
0	0	1	3.0	0.15	9.6	2	2	1	28	8.7	94
0	1	0	3.0	0.15	11	2	2	2	35	8,7	94
0	1	1	6.1	1.2	18	2	3	0	29	8.7	94
0	2	0	6.2	1.2	18	2	3	1	36	8.7	94
0	3	0	9.4	3.6	38	3	0	0	23	4.6	94
1	0	0	3.6	0.17	18	3	0	1	38	8.7	110
1	0	1	7.2	1.3	18	3	0	2	64	17	180
1	0	2	11	3.6	38	3	1	0	43	9	180
1	1	0	7.4	1.3	20	3	1	1	75	17	200
1	1	1	11	3.6	38	3	1	2	120	37	420
1	2	0	11	3.6	42	3	1	3	160	40	420
1	2	1	15	4.5	42	3	2	0	93	18	420
1	3	0	16	4.5	42	3	2	1	150	37	420
2	0	0	9.2	1.4	38	3	2	2	210	40	430
2	0	1	14	3.6	42	3	2	3	290	90	1,000
2	0	2	20	4.5	42	3	3	0	240	42	1,000
2	1	0	15	3.7	42	3	3	1	460	90	2,000
2	1	1	20	4.5	42	3	3	2	1,100	180	4,100
2	1	2	27	8.7	94	3	3	3	>1,000	420	—

TABLE 2.2

Five (5) Tubes Each at 0.1, 0.01, and 0.001 g Inocula and the MPNs per Gram with 95% Confidence Intervals[8]

Positive Tubes			MPN/g	Conf. Limit		Positive Tubes			MPN/g	Conf. Limit	
0.10	0.01	0.001		Low	High	0.10	0.01	0.001		Low	High
0	0	0	<1.8	—	6.8	4	0	2	21	6.8	40
0	0	1	1.8	0.09	6.8	4	0	3	25	9.8	70
0	1	0	1.8	0.09	6.8	4	1	0	17	6	40
0	1	1	3.6	0.7	10	4	1	1	21	6.8	42
0	2	0	3.7	0.7	10	4	1	2	26	9.8	70
0	2	1	5.5	1.8	15	4	1	3	31	10	70
0	3	0	5.6	1.8	15	4	2	0	22	6.8	50
1	0	0	2	0.1	10	4	2	1	26	9.8	70
1	0	1	4	0.7	10	4	2	2	32	10	70
1	0	2	6	1.8	15	4	2	3	38	14	100
1	1	0	4	0.7	12	4	3	0	27	9.9	70
1	1	1	6.1	1.8	15	4	3	1	33	10	70
1	1	2	8.1	3.4	22	4	3	2	39	14	100
1	2	0	6.1	1.8	15	4	4	0	34	14	100
1	2	1	8.2	3.4	22	4	4	1	40	14	100
1	3	0	8.3	3.4	22	4	4	2	47	15	120
1	3	1	10	3.5	22	4	5	0	41	14	100
1	4	0	11	3.5	22	4	5	1	48	15	120
2	0	0	4.5	0.79	15	5	0	0	23	6.8	70
2	0	1	6.8	1.8	15	5	0	1	31	10	70
2	0	2	9.1	3.4	22	5	0	2	43	14	100
2	1	0	6.8	1.8	17	5	0	3	58	22	150
2	1	1	9.2	3.4	22	5	1	0	33	10	100
2	1	2	12	4.1	26	5	1	1	46	14	120
2	2	0	9.3	3.4	22	5	1	2	63	22	150

2	2	1	12	4.1	26	5	1	3	84	34	220
2	2	2	14	5.9	36	5	2	0	49	15	150
2	3	0	12	4.1	26	5	2	1	70	22	170
2	3	1	14	5.9	36	5	2	2	94	34	230
2	4	0	15	5.9	36	5	2	3	120	36	250
3	0	0	7.8	2.1	11	5	2	4	150	58	400
3	0	1	11	3.5	23	5	3	0	79	22	220
3	0	2	13	5.6	35	5	3	1	110	34	250
3	1	0	11	3.5	26	5	3	2	140	52	400
3	1	1	14	5.6	36	5	3	3	180	70	400
3	1	2	17	6	36	5	3	4	210	70	400
3	2	0	14	5.7	36	5	4	0	130	36	400
3	2	1	17	6.8	40	5	4	1	170	58	400
3	2	2	20	6.8	40	5	4	2	220	70	440
3	3	0	17	6.8	40	5	4	3	280	100	710
3	3	1	21	6.8	40	5	4	4	350	100	710
3	3	2	24	9.8	70	5	4	5	430	150	1100
3	4	0	21	6.8	40	5	5	0	240	70	710
3	4	1	24	9.8	70	5	5	1	350	100	1100
3	5	0	25	9.8	70	5	5	2	540	150	1700
4	0	0	13	4.1	35	5	5	3	920	220	2600
4	0	1	17	5.9	36	5	5	4	1600	400	4600
						5	5	5	>1600	700	—

tubes (may include use of Durham fermentation tubes), dilution replicate tubes, pipettes, specific growth medium and reagents, and incubator with appropriate optimal temperature setting. Sources of error using this method are improper or inadequate preparation of the test samples, serial dilution or dilution factor calculation error, and statistical interpretation error of observed results.

Applications

The MPN method is useful in the estimation of low microorganism counts where particulate matter or turbidity is present in the sample matrix, such as in milk, food, water, and soil. The MPN method is not particularly useful in the enumeration of fungi. It is mainly used in the enumeration of samples with low bacteria concentrations or bacteria that does not grow well on solid medium. MPN methods are used in the areas of food,[8–10] pharmaceutical,[11–13] environment,[14, 15] water,[16–19] agriculture,[20, 21] petroleum,[22, 23] and poultry[24] applications. Modified or alternate versions of the MPN method have also been developed to further expand the use of the standard MPN method to estimate bacteria by utilizing solid medium,[25] microtiter plates,[26–28] colometric reagents or defined substrate medium,[29–32] and polymerase chain reaction.[33]

Turbidimetric Method

General

The turbidimetric method is used to measure the turbidity in a suitable liquid medium inoculated with a given microorganism.[34–40] Turbidimetric measurement is a function of microorganism cell growth. This method of microbial enumeration is based on the correlation between the turbidity observed and changes in the microbial cell numbers. Standard optical density (O.D.) or turbidimetric curves constructed may be used to estimate the number of microbial cells for the observed turbidity values using a turbidimetric instrument. This is achieved by determining the turbidity of different concentrations of a given species of microorganism in a particular liquid medium using the standard plate count method to determine the number of viable microorganisms per millimeter of sample tested. The turbidimetric calibration standard curve generated is subsequently used as a visual comparison to any given suspension of that microorganism. The use of this method assumes that there is control of the physiological state of the microorganisms. It is based on the species of microorganism and the optimal condition of the turbidimetric instrument used. Primary equipment and materials used for this method are the turbidimetric reading instrument, turbidimetric calibration standards, optically matched tubes or cuvettes, appropriate recording instruments, dilution tubes, pipettes, specific growth medium and reagents, and incubator with appropriate optimal temperature setting. The turbidimetric reading instrument can be, for example, the spectrophotometer or turbidimeter (wavelength range of 420 to 615 nm shown in Figure 2.2). The typical turbidimetric standard used is the McFarland turbidimetric standard. These standards are

FIGURE 2.2 DR 5000 UV-VIS spectrophotometer. (Courtesy of Hach Co.)

used as a reference to adjust the turbidity of microbial suspensions to enable the number of microorganisms to be within a given quantitative range. These standards are made by mixing specified amounts of barium chloride and sulfuric acid to form a barium sulfate precipitate that causes turbidity in the solution mixture. If *Escherichia coli* is used, the 0.5 McFarland standard is equivalent to an approximate concentration of 1.5×10^8 colony forming units (or CFUs) per mL. The more commonly used standard is the 0.5 McFarland standard, which is prepared by mixing 0.05 mL of 1% barium chloride dehydrate with 9.95 mL of 1% sulfuric acid. There are other turbidimetric standards that can be prepared from suspensions of latex particles that can extend the expiration use and stability of these suspensions.[41, 42] Sources of error for this method include using a non-calibrated instrument, damaged or misaligned optical matched tubes or cuvettes, color or clarity of the liquid suspension medium used, condition of the filter or detector used, and quality of the turbidimetric instrument lamp output. Sources of error for this method can also be microbial related, such as clumping or settling of the microbial suspension, inadequate optimal growth condition used; or analyst related, such as lack of understanding the turbidimetric technique used, sample preparation error, and incorrect use of the turbidimetric standards.[39, 40]

Applications

Turbidimetric methods are used where a large number of microorganisms is to be enumerated or when the inoculum size of a specific microbial suspension is to be determined. These areas include antimicrobial, preservative, or biocide susceptibility assays of bacteria and fungi,[43–51] growth of microorganisms,[52–55] effects of chemicals on microorganisms,[56, 57] and control of biological production.[58]

Plate Count Method

General

The plate count method is based on viable cell counts. The plate count method is performed by diluting the original sample in serial dilution tubes, followed by the plating of aliquots of the prepared serial dilutions into appropriate plate count agar plates by the pour plate or spread plate technique. The pour plate technique utilizes tempered molten plate count agar poured into the respective plate and mixed with the diluted aliquot sample in the plate, whereas the spread plate technique utilizes the addition and spreading of the diluted aliquot sample on the surface of the preformed solid plate count agar in the respective plate.[59] An example of a plate count serial dilution scheme is shown in Figure 2.3. These prepared plate count agar plates are then optimally

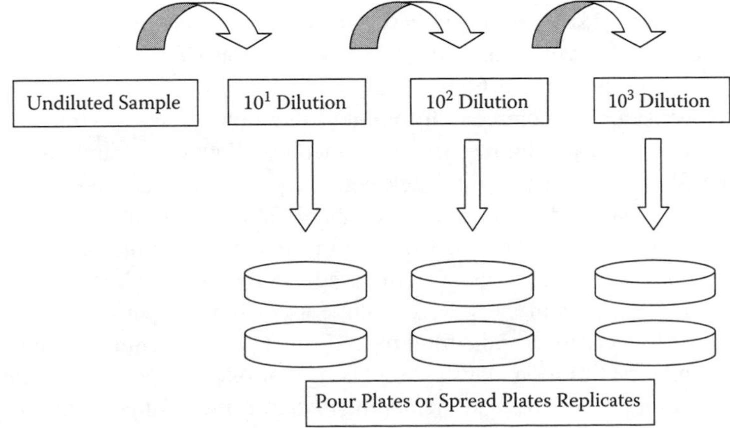

FIGURE 2.3 Plate count serial dilution scheme.

incubated and the colonies observed on these plate count agar plates are then counted as the number of colony forming units (CFUs). The counting of CFUs assumes that every colony is separate and founded by a single viable microbial cell. The total colony counts obtained in CFU from the incubated agar plates and the respective dilution factor used can then be combined to calculate the original number of microorganisms in the sample in CFU per mL. The typical counting ranges are 25 to 250 CFU or 30 to 300 CFU per standard plate count agar plate. Additional considerations for counting colony forming units (or CFUs) are counting of plate spreaders, too numerous to count (TNTC) reporting and statistics, rounding and averaging of observed plate counts, limit of detection, and limit of quantification of plate counts.[60–62] There are also optimal condition assumptions for the plate count method as changes to the plate count agar nutrient level or temperature can affect the surface growth of bacteria.[63, 64] Primary equipment and materials used for this method are serial dilution tubes (bottles); Petri plates or dishes; pipettes; specific growth medium, diluents, and reagents; incubator and water bath with appropriate optimal temperature setting; commercial colony counter (manual or automated); plate spreader or rod. Total bacteria and fungi can be enumerated separately using the plate count method based on the type of culture medium utilized.[65–69] Specific or selective culture medium can also be used in place of the standard plate count agar media more specific microbial enumeration.[70] Sources of error using this method are improper or inadequate preparation of the test samples, serial dilution error, suboptimal incubation conditions, undercounting due to cell aggregation or clumping, and analyst error in the colony counting or calculation of observed results.

Applications

The plate count method is used primarily in the enumeration of samples with high microorganism numbers or microorganisms that do not grow well in liquid media. Plate count methods are used in the areas of food,[65, 66, 71–74] pharmaceutical,[11–13] environmental including drinking water applications,[19, 75–77] and biofilm testing.[78] Modified or alternate versions of the plate count methods have also been developed to further enhance the use of the standard plate count method approach to estimate bacteria or fungi by utilizing the roll tube method,[79–82] drop plate method,[83, 84] spiral plate count method,[85–89] Petrifilm™,[90–94] SimPlate™,[31, 95–97] RODAC™ (replicate organism detection and counting plate) for environmental surface sampling,[98–102] dipslide or dipstick paddle method,[103, 104] and adhesive sheet method.[105]

Membrane Filtration Method

General

The membrane filtration (MF) method is based on viable cell enumeration. The MF method is conducted by filtering a known volume (typical volumes range from 100 to 1000 mL) of liquid sample through a sterile membrane filter (0.22 or 0.45 μm pore size). The filter membrane will retain microorganisms during membrane filtration of the liquid sample. After filtration, the membrane filter is placed onto a membrane filter agar medium plate or membrane filter medium pad. These membrane filter plates or pads are then optimally incubated and the colonies observed on these membrane filter plates or pads are counted as the number of colony forming units (CFUs). The total colony counts obtained in CFU from the incubated membrane filter plates and pads and the respective membrane filtered sample volume used can then be combined to calculate the original number of microorganisms in the sample volume used (in CFU per mL). The typical counting ranges are 20 to 80 CFU per membrane filter used. Primary equipment and materials used for this method are the membrane filtration manifold unit (single or multiple); specific membrane filter with an appropriate membrane pore size and membrane material; filter funnels; pipettes; serial dilution tubes; specific growth medium; sterile buffers or diluents, and reagents, incubator with appropriate optimal temperature setting; and sterile forceps. Sources of error using this method are improper or inadequate preparation of the test samples, sample dilution error, suboptimal incubation conditions, solid particles and membrane-adsorbed chemicals may interfere with microbial growth, and analyst error in the colony counting or calculation of observed membrane filtration results.

Applications

The membrane filtration method is used to concentrate the number of microorganisms from large volumes of liquid sample with low numbers of microorganisms. This flexible sample volume range allows for the use of larger sample volumes and increases in the sensitivity of microbial enumeration. The membrane filtration method is not particularly useful with turbid or highly particulate samples that may block the membrane pores. Membrane filtration methods are used in the areas of food,[106] pharmaceutical,[11–13, 107] and environmental including drinking and recreational water application.[19, 108–116] Modified or alternate versions of the membrane filtration estimate have also been developed to further enhance the use of this method in estimating bacteria by utilizing the defined substrate medium membrane filtration method (selective and/or differential medium)[117–120] and the higher microbial enumeration range hydrophobic grid-membrane filtration method.[121–126]

Direct Count Method

General

The number of microbial cells can be determined by direct microscopic examination of the sample within a demarcated region or field of the microscopic slide and counting the number of microbial cells observed per field under the microscope. The enumeration of microbial cells in the original sample can be rapidly calculated using an aliquot of known volume used for direct enumeration under the microscope slide (numbers of cells counted per field and number of microscopic fields counted) in number of microbial cells per milliliter or by using a counting standard.[127, 128] This direct count method requires a higher number of microorganisms per milliliter for enumeration, and there is the potential for counting both dead and living microbial cells. An extension of the direct count method is the use of fluorescent stains with the epifluorescence microscope or the direct epifluorescence filtration technique (DEFT).[129–131] Primary equipment and materials used for this method are the microscope (light or phase contrast or epifluorescent), microbial counting chambers (Petroff-Hauser or Helber or Nebauer) or membrane filter, microscope slide support, internal counting standards (latex particles), biological stains or fluorochromes for certain applications, and capillary tubes. Sources of error using this method include the experience of the analysts performing the microbial cell count observed under the microscope, cell aggregation or clumping, inadequate staining, and inadequate microscopic slide sample preservation or preparation.

Applications

Direct count methods are used in the areas of food,[132, 133] biocidal,[134, 135] metal working fluids,[136] and environmental including soil, air, water, or terrestrial ecological applications.[137–144] Modified or alternate versions of the direct count method have also been developed, such as the direct viable count method (DVC),[145–148] fluorescence *in situ* hybridization (FISH) method,[149, 150] and combined DVC-FISH method.[151]

Particle Counting Method

General

The coulter counting method, or electrical sensing zone (ESZ) method, is a particle count method in which microbial cells suspended in a weak electrolyte pass through an electric field of standard resistance within a small aperture. The measured resistance changes as the microbial cells pass through the aperture. The voltage applied across the aperture creates a sensing zone. If a constant voltage is maintained, a microbial cell passing through will cause a transient decrease in current as the resistance changes. A transient increase in voltage will occur if constant current is maintained. These transient impedance changes in the sensing zone can then be amplified and recorded electronically, thus allowing one to enumerate the number of microbial cells flowing through the aperture opening. The estimation of the total number of microbial cells in a given sample can be correlated with the microbial cell counts obtained for either a metered sample volume or a fixed time period for flow under constant pressure.[152–154]

The flow cytometry method is another particle count method that analyzes the light scatter and fluorescence emitted from individual microbial cells as these cells flow in a single-file manner through an intensely focused laser light source. The sample is injected into a pressurized fluid delivery system or sheath that forces the suspension of cells into laminar flow and aligned in single file along the central core through the sensing zone. As the cells pass through the sensing zone, a laser light source generates a light scatter pulse that is collected at small forward angles and right angles to the cell. The collected light pulse from the single cell is converted to an electrical signal (voltage) by either a photodiode or a photomultiplier. The captured data can then be analyzed as two-parameter histograms or by using multiparametric analysis. Fluorescent dyes or stain can be used to further enhance the use of the flow cytometry to discriminate between live or dead microbial cells. A combination of internal calibrating counting beads used per test, the number of events in a region containing microbial cell population and bead population, dilution factor used, and test volume will provide the estimated total number of microbial cells in a given sample.[155, 156]

Primary equipment and materials used for the particle count method are commercially available counters (coulter counters or flow cytometry counters), internal liquid bead counting standards (polystyrene beads with or without fluorophores), biological stains or fluorochromes for certain applications, specific reagents (electrolyte solutions or sheathing liquids or preservatives), disposable test tubes, micropipettes plus pipette tips, and vortex mixer. Sources of error using this method include the presence of bubbles or foreign particles while cell counting, which causes high background noise; partial blocking of the counting aperture; a higher number of cells, which may lead to higher probability of the coincident passage of two or more cells through the aperture; and background noise, which will become higher for enumerating low concentrations of cells.

Applications

Particle counting methods are used in the areas of food,[157, 158] pharmaceutical,[159, 160] and environmental applications.[161, 162] Flow cytometry methods can also be used for virus enumeration.[163] Modified or alternate versions of the standard particle counting methods have also been developed, including the solid state laser cytometry method.[164, 165]

Focal Lesion Method

General

The focal lesion method considers an infection of a suitable host by a specific virus to produce a focal lesion response that can be quantitated if it is within an appropriate range of the dilution of the viral sample used. The average number of focal lesions is a linear function of the concentration of virus employed. This method is used to enumerate the number of viruses because each lesion formed is focused at the site of the activity of an individual viral particle. The relationship between the lesion counts obtained and the virus dilution used is also statistically consistent. Primary equipment and materials used depend on the test system employed. Focal lesions can be generated on the skin of whole animals, on the chorioallantoic membrane of chicken embryo, or in cell culture monolayers. Focal lesions can also be produced in agar plate cultures of bacterial viruses and on the leaves of a plant rubbed with a mixture of virus plus abrasive. Sources of error using this method are nonspecific inhibition of focal lesion response by unknown factors, dilution or counting errors, overlapping of focal lesions, size of focal lesions and the numbers countable on the surface area tested, and insufficient viral replication to provide statistical significance.

Applications

The focal lesion method can be used for the enumeration of animal, plant, and bacterial viruses.[166–168]

Quantal Assay Method

General

The quantal assay method uses a statistical method with an all-or-none (cell culture mortality or cytopathic effect or death of experimental animals) response or outcome to enumerate the number of viruses in a given sample. Using a serial dilution approach, the endpoint of activity or infectivity is considered the highest dilution of the virus at which there is 50% or more positive response of the inoculated host. The exact endpoint is determined by interpolation from the cumulative frequencies of positive and negative responses observed at the various dilutions. The endpoint of the quantal titration is the viral dilution that has 50% positive and 50% negative responses. The reciprocal of the dilution yields an estimate of the number of viral units per inoculum volume of the undiluted sample and is expressed in multiples of the 50% endpoint. Primary equipment and materials used for the quantal assay method will depend on the test system employed either as groups of animals, use of chick embryo, or tubes/flasks of cell cultures. Sources of error using this method include dilution or counting errors and insufficient replication to provide statistical significance, and sample titration.

Applications

The quantal assay method is used in infectivity assays in animals or tissue-culture-infectious-dose assay applications in which the animal or cell culture is scored as either infected or not infected. The infectivity titer is proportionately measured by the animal or tissue culture infected. The main use of this method is in the enumeration of culturable waterborne viruses.[169, 170]

Plaque Assay Method

General

The plaque assay method utilizes serial dilution of a given viral suspension sample that is inoculated onto a confluent monolayer of a specific host cell culture. Following the adsorption or attachment step, the monolayer is overlaid with either a semisolid or solid medium to restrict the movement of the progeny virus in the vicinity of the infected host cells. After incubation, the infectious plaque plate is enumerated and examined for plaque formation. Each virus unit makes one plaque (an observed localized focus of infected and dead cells) after the cell monolayers are stained with general cellular stain or dye. The infectivity titer of the original viral suspension sample is recorded as plaque-forming units (PFUs) per milliliter. Primary equipment and materials used for this method include serial dilution tubes (bottles), Petri plates or dishes, pipettes, specific growth medium, specific cell culture or bacterial host, diluents, reagents, membrane filtration setup, stains/dyes as needed, and incubator and water bath with appropriate optimal temperature setting. Sources of error using this method include improper or inadequate preparation of the test samples, serial dilution error, suboptimal incubation conditions, and analyst error in the plaque counting or calculation of observed results.

Applications

The plaque assay method can be used for the enumeration of animal, plant, and bacterial viruses.[171–176]

SUMMARY

There are a number of test methods used for the quantitative enumeration of microorganisms. A more formal validation approach can also be used to compare new or alternate test methods for the quantitative enumeration of microorganisms.[177] This may include rapid test methods for the quantitative enumeration of microorganisms compared to the above standard microbial enumeration test methods.[178–181] Methods for quantitative enumeration of microorganisms will continue to be developed and validated to meet the various industry demands and the associated microbial quantitative enumeration applications, as evidenced by more recent statistical approaches to quantitative microbial estimation or counting methods and the uncertainty associated with microbiological testing.[182–185]

REFERENCES

1. Cochran, W.G., Estimation of bacterial densities by means of the "Most Probable Number," *Biometrics*, 6, 105, 1950.
2. Russek, E. and Colwell, R.R., Computation of most probable numbers, *Appl. Environ. Microbiol.*, 45, 1646, 1983.
3. de Man, J.C., MPN tables, corrected, *J. Appl. Microbiol. Biotechnol.*, 17, 301, 1983.
4. Tillett, H.E. and Coleman, R., Estimated numbers of bacteria in samples from non-homogeneous bodies of water, *J. Appl. Bacteriol.*, 59, 381, 1985.
5. Haas, C.N. and Heller, B., Test of the validity of the Poisson assumption for analysis of MPN results, *Appl. Environ. Microbiol.*, 54, 2996, 1988.
6. Hass, C.N., Estimation of microbial densities from dilution count experiments, *Appl. Environ. Microbiol.*, 55, 1934, 1989.
7. Garthright, W.E. and Blodgett, R.J., FDA's preferred MPN methods for standard, large or unusual tests, with a spreadsheet, *Food Microbiol.*, 20, 439, 2003.
8. U.S. Food and Drug Administration, Most probable number from serial dilutions, *Bacteriological Analytical Manual Online*, Department of Health and Human Services, 2006, Appendix 2.
9. Health Canada, Enumeration of coliforms, faecal coliforms and of *E. coli* in food using the MPN method in *MFHPB-19*, Health Product and Food Branch, 2002.
10. Health Canada, Enumeration of coliforms, faecal coliforms and of *E. coli* in water in sealed containers and prepackaged ice using the MPN method in *MFO-18*, Health Product and Food Branch, 2003.
11. USP, <61> Microbial Limit Tests, USP29-NF24, *The United States Pharmacopeial Convention*, 2006, 2503.
12. EP, 2.6.12 Microbiological examination of nonsterile products: total viable aerobic count, *European Pharmacopoeia, 5th ed.*, Supplement 5.6, 2006, 4398.
13. JP, 35. Microbial Limit Test, *Japanese Pharmacopoiea, 14th ed.*, 2001, 58.
14. Munoz, E.F. and Silverman, M.P., Rapid single-step most-probable-number method for enumerating fecal coliforms in effluents from sewage treatment plants, *Appl. Environ. Microbiol.*, 37, 527, 1979.
15. Guerinot, M.L. and Colwell, R.R., Enumeration, isolation, and characterization of N2-fixing bacteria from seawater, *Appl. Environ. Microbiol.*, 50, 350, 1985.
16. Maul, A., Block, J.C., and El-Shaarawi, A.H., Statistical approach for comparison between methods of bacterial enumeration in drinking water, *J. Microbiol. Meth.*, 4, 67, 1985.
17. Jagals, P. et al., Evaluation of selected membrane filtration and most probable number methods for the enumeration of faecal coliforms, *Escherichia coli* and Enterococci in environmental waters, *Quant. Microbiol.*, 2, 1388, 2000.
18. Environment Agency. The enumeration of coliform bacteria and *Escherichia coli* by a multiple tube most probable number technique, in *The Microbiology of Drinking Water—Part 4*, Environment Agency, U.K., 2002, 29.

19. APHA, *Standard Methods for the Examination of Water and Wastewater, 21st ed.*, American Public Health Association, Washington, D.C., 2005.
20. Dehority, B.A., Tirabasso, P.A., and Grifo Jr., A.P., Most-probable-number procedures for enumerating ruminal bacteria, including the simultaneous estimation of total and cellulolytic numbers in one medium, *Appl. Environ. Microbiol.*, 55, 2789, 1989.
21. Shockey, W.L. and Dehority, B.A., Comparison of two methods for enumeration of anaerobe numbers on forage and evaluation of ethylene oxide treatment for forage sterilization, *Appl. Environ. Microbiol.*, 55, 1766, 1989.
22. Fedorak, P.M, Semple, K.M., and Westlake, D.W.S., A statistical comparison of two culturing methods for enumerating sulfate-reducing bacteria, *J. Microbiol. Meth.*, 7, 19, 1987.
23. Eckford, R.E. and Fedorak, P.M., Applying a most probable number method for enumerating planktonic, dissimilatory, ammonium producing, nitrate reducing bacteria in oil field waters, *Can. J. Microbiol.*, 51, 725, 2005.
24. Capita, R. and Alonso-Calleja, C., Comparison of different most-probable-number methods for enumeration of *Listeria* spp. in poultry, *J. Food Prot.*, 66, 65, 2003.
25. Carvalhal, M.L.C., Oliveira, M.S., and Alterthum, F., An economical and time saving alternative to the most-probable-number method for the enumeration of microorganisms, *J. Microbiol. Meth.*, 14, 165, 1991.
26. Rowe, R., Todd, R., and Waide, J., Microtechnique for most-probable-number analysis, *Appl. Environ. Microbiol.*, 33, 675, 1977.
27. Gamo, M. and Shoji, T., A method of profiling microbial communities based on a most-probable-number assay that uses BIOLOG plates and multiple sole carbon sources, *Appl. Environ. Microbiol.*, 65, 4419, 1999.
28. Saitoh, S., Iwasaki, K., and Yagi, O., Development of a new most-probable-number method for enumerating methanotrophs, using 48-well microtiter plates, *Microbes & Environ*, 17, 191, 2002.
29. Edberg, S.C. et al., Enumeration of total coliform and *Escherichia coli* from source water by the defined substrate technology, *Appl. Environ. Microbiol.*, 56, 366, 1990.
30. Harriott, O.T. and Frazer, A.C., Enumeration of acetogens by a colometric most-probable-number assay, *Appl. Environ. Microbiol.*, 63, 296, 1997.
31. Jackson, R.W. et al., Multiregional evaluation of the SimPlate heterotrophic plate count method compared to the standard plate count agar pour plate method in water, *Appl. Environ. Microbiol.*, 66, 453, 2000.
32. Environment Agency, D. The enumeration of coliform bacteria and *Escherichia coli* by a defined substrate most probable number technique, in *The Microbiology of Drinking Water—Part 4*, Environment Agency, U.K., 2002, 44.
33. Inoue, D. et al., Comparative evaluation of quantitative polymerase chain reaction methods for routine enumeration of specific bacterial DNA in aquatic samples, *World J. Microbiol. Biotech.*, 21, 1029, 2005.
34. McFarland, J., The nephelometer: an instrument for estimating the number of bacteria in suspensions used for calculating the opsonic index for vaccines, *JAMA*, 49, 1176, 1907.
35. Kurokawa, M., A new method for the turbidimetric measurement of bacterial density, *J. Bacteriol.*, 83, 14, 1962.
36. Fujita, T. and Nunomura, K., New turbidimetric device for measuring cell concentrations in thick microbial suspensions, *Appl. Microbiol.*, 16, 212, 1968.
37. Koch, A.L., Turbidimetric measurements of bacterial cultures in some available commercial instruments, *Anal. Biochem.*, 38, 252, 1970.
38. Li, R.C., Nix, D.E., and Schentag, J.J., New turbidimetric assay for quantitation of viable bacteria densities, *Antimicrob. Agents Chemother.*, 37, 371, 1993.
39. Sadar, M.J., Turbidity science, *Technical Information Series—Booklet No.11*, Hach Company, 1998.
40. Sutton, S., Measurement of cell concentration in suspension by optical density, *Pharm. Microbiol. Forum Newslett.*, 12, 3, 2006.
41. Pugh, T.L. and Heller, W., Density of polystyrene and polyvinyl toluene latex particles, *J. Colloid. Sci.*, 12, 173, 1957.
42. Roessler, W.G. and Brewer, C.R., Permanent turbidity standards, *Appl. Microbiol.*, 15, 1114, 1967.
43. Kavanaugh, F., Turbidimetric assays: The antibiotic dose-response line, *Appl. Microbiol.*, 16, 777, 1968.

44. Galgiani, J.N. and Stevens, D.A., Turbidimetric studies of growth inhibition of yeasts with three drugs: inquiry into inoculum dependent susceptibility testing, time of onset, of drug effect, and implications for current and new methods, *Antimicrob. Agents Chemother.*, 13, 249, 1978.
45. Hughes, C.E., Bennett, R.L., and Beggs, W.H., Broth dilution testing of *Candida albicans* susceptibility to ketonazole, *Antimicrob. Agents Chemother.*, 31, 643, 1987.
46. Odds, F.C., Quantitative microculture system with standardize inocula for strain typing, susceptibility testing, and other physiologic measurement with *Candida albicans* and other yeasts, *J. Clin. Microbiol.*, 29, 2735, 1991.
47. Klepser, M.E. et al., Influence of test conditions on antifungal time-kill curve results: Proposal for standardized methods, *Antimicrob. Agents Chemother.*, 42, 1207, 1998.
48. Llop, C. et al., Comparison of three methods of determining MICs for filamentous fungi using different end point criteria and incubation periods, *Antimicrob. Agents Chemother.*, 44, 239, 2000.
49. USP, <51> Antimicrobial effectiveness testing, USP29-NF24, *The United States Pharmacopeial Convention*, 2006, 2499.
50. EP, 5.1.3 Efficacy of antimicrobial preservation, *European Pharmacopoeia, 5th ed.*, 2006, 447.
51. JP, 34. Microbial assay for antibiotics, *Japanese Pharmacopoiea, 14th ed.*, 2001, 54.
52. Meletiadis, J., teDorsthorst, D.T.A., and Verweij, P.E., Use of turbidimetric growth curves for early determination of antifungal drug resistance of filamentous fungi, *J. Clin. Microbiol.*, 41, 4718, 2003.
53. Begot, C., et al. Recommendation for calculating growth parameters by optical density measurements, *J. Microbiol. Meth.*, 25, 225, 1996.
54. Domínguez, M.C., de la Rosa, M., and Borobio, M.V., Application of a spectrophotometric method for the determination of post-antibiotic effect and comparison with viable counts in agar, *J. Antimicrobiol. Chemother.*, 47, 391, 2001.
55. Baty, F., Flandrois, J.P., and Delignette-Muller, M.L., Modeling the lag time of *Listeria monocytogenes* from viable count enumeration and optical density data, *Appl. Microbiol.*, 68, 5816, 2002.
56. Myllyniemi. A. et al., An automated turbidimetric method for the identification of certain antibiotic groups in incurred kidney samples, *Analyst*, 129, 265, 2004.
57. Pianetti, A. et al., Determination of the variability of *Aeromonas hydrophila* in different types of water by flow cytometry, and comparison with classical methods, *Appl. Environ. Microbiol.*, 71, 7948, 2005.
58. Vaishali, J.P., Tendulkar, S.R., and Chatoo, B.B., Bioprocessing development for the production of an antifungal molecule by *Bacillus lichenformis* BC98, *J. Biosci. Bioeng.*, 98, 231, 2004.
59. Clark, D.S., Comparison of pour and surface plate methods for determination of bacterial counts, *Can. J. Microbiol.*, 13, 1409, 1967.
60. Jennison, M.W. and Wadsworth, G.P., Evaluation of the errors involved in estimating bacterial numbers by the plating method, *J. Bacteriol.*, 39, 389, 1940.
61. Haas, C.H. and Heller, B., Averaging of TNTC counts, *Appl. Environ. Microbiol.*, 54, 2069, 1988.
62. Sutton, S., Counting colonies, *Pharm. Microbiol. Forum Newslett.*, 12, 2, 2006.
63. Salvesen, I. and Vadstein, O., Evaluation of plate count methods for determination of maximum specific growth rate in mixed microbial communities, and its possible application for diversity assessment, *J. Appl. Microbiol.*, 88, 442, 2000.
64. Fujikawa, H. and Morozumi, S., Modeling surface growth of *Escherichia coli* on agar plates, *Appl. Environ. Microbiol.*, 71, 7920, 2005.
65. U.S. Food and Drug Administration, Aerobic plate count, *Bacteriological Analytical Manual Online*, Department of Health and Human Services, 2001, chap. 3.
66. U.S. Food and Drug Administration, Yeast, mold and mycotoxins, *Bacteriological Analytical Manual Online*, Department of Health and Human Services, 2001, chap. 18.
67. Beuchat, L.R., Media for detecting and enumerating yeasts and moulds, *Int. J. Food. Microbiol.*, 17, 145, 1992.
68. Beuchat, L.R., Selective media for detecting and enumerating foodborne yeasts, *Int. J. Food. Microbiol.*, 19, 1, 1993.
69. Beuchat, L.R. et al., Performance of mycological media in enumerating dessicated food spoilage yeast: an interlaboratory study, *Int. J. Food. Microbiol.*, 70, 89, 2001.
70. EP, 2.6.13 Test for specified micro-organisms, *European Pharmacopoeia, 5th ed.*, Supplement 5.6, 2006, 4404.
71. Health Canada, Determination of the aerobic colony count in foods in *MFHPB-18*, Health Product and Food Branch, 2001.

72. Health Protection Agency, Aerobic plate count at 30°C: Surface plate method, *National Standard Method F10*, Issue 1, U.K., 2005.
73. Reasoner, D.J., Heterotrophic plate count methodology in the United States, *Int. J. Food Microbiol.*, 92, 307, 2004.
74. Gracias, K.S. and McKillip, J.L., A review of conventional detection and enumeration methods for pathogenic bacteria in food, *Can. J. Microbiol.*, 50, 883, 2004.
75. Pepper, I.L. et al., Tracking the concentration of heterotrophic plate count bacteria from the source to the consumer's tap, *Int. J. Food Microbiol.*, 92, 289, 2004.
76. Health Protection Agency, Colony count by the pour plate method, *National Standard Method W4*, Issue 4, U.K., 2005.
77. Health Canada, Heterotrophic plate count in *Guidelines for Canadian Drinking Water Quality — Guideline Technical Document*, Healthy Environments and Consumer Safety Branch, Ottawa, 2006.
78. Cerca, N. et al., Comparative assessment of antibiotic susceptibility of coagulase-negative *Staphylococci* in biofilm versus planktonic culture as assessed by bacterial enumeration or rapid XTT colorimetry, *J. Antimicrob. Chemother.*, 56, 331, 2005.
79. Vervaeke, I.J. and Van Nevel, C.J., Comparison of three techniques for the total count of anaerobics from intestinal contents of pigs, *Appl. Microbiol.*, 24, 513, 1972.
80. Grubb, J. A. and Dehority, B.A., Variation in colony counts of total viable anaerobic rumen bacteria as influenced by media and cultural methods, *Appl. Environ. Microbiol.*, 31, 262, 1976.
81. Moore, W.E.C. and Holdeman, L.V., Special problems associated with the isolation and identification of intestinal bacteria in fecal floral studies, *Am. J. Clin. Nutri.*, 27, 1450, 1974.
82. Nagamine, I. et al., A comparison of features and the microbial constitution of the fresh feces of pigs fed diets supplemented with or without dietary microbes, *J. Gen. Appl. Microbiol.*, 44, 375, 1998.
83. Neblett, T.R., Use of droplet plating method and cystine-lactose electrolyte-deficient medium in routine quantitative urine culturing procedure, *J. Clin. Microbiol.*, 4, 296, 1976.
84. Herigstad, B., Hamilton, M., and Heersink, J., How to optimize the drop plate method for enumerating bacteria, *J. Microbiol. Meth.*, 44, 121, 2001.
85. Gilchrist, J.E. et al., Spiral plate method for bacterial determination, *Appl. Microbiol.*, 25, 244, 1973.
86. Peeler, J.T. et al., A collaborative study of the spiral plate method for examining milk samples, *J. Food Prot.*, 40, 462, 1977.
87. Zipkes, M.R., Gilchrist, J.E., and Peeler, J.T., Comparison of yeast mold counts by spiral, pour, and streak plate methods, *J. AOAC Int*, 64, 1465, 1981.
88. Alonso-Calleja, C. et al., Evaluation of the spiral plating method for the enumeration of microorganisms throughout the manufacturing and ripening of a raw goat's milk cheese, *J. Food Prot.*, 65, 339, 2002.
89. Health Protection Agency, Aerobic plate count at 30°C: Spiral plate method, *National Standard Method F11*, Issue 1, U.K., 2004.
90. Knight, L.J. et al., Comparison of the Petrifilm dry rehydratable film and conventional methods for enumeration of yeasts and molds in foods: collaborative study, *J. AOAC Int.*, 80, 806, 1997.
91. Health Canada, Enumeration of total aerobic bacteria in food products and food ingredients using 3M™ Petrifilm™ aerobic plate counts in *MFHPB-33*, Health Product and Food Branch, 2001.
92. Ellis, P. and Meldrum, R., Comparison of the compact dry TC and 3M Petrifilm ACP dry sheet media methods with the spiral plate method for the examination of randomly selected foods for obtaining aerobic colony counts, *J. Food Prot.*, 65, 423, 2002.
93. Beloti, V. et al., Evaluation of Petrifilm™ EC and HS for total coliforms and *Escherichia coli* enumeration in water, *Braz. J. Microbiol.*, 34, 310, 2003.
94. Vail, J.H. et al., Enumeration of waterborne *Escherichia coli* with Petrifilm plates: Comparison to standard methods, *J. Environ. Qual.*, 32, 368, 2003.
95. Townsend, D.E. and Naqui, A., Comparison of SimPlate™ total plate count test with plate count agar method for detection and quantification of bacteria in food, *J. AOAC Int.*, 3, 563, 1998.
96. Feldsine, P.T. et al., Enumeration of total aerobic microorganisms in foods by SimPlate total count-color indicator methods and conventional culture methods: collaborative study, *J. AOAC Int.*, 86, 2573, 2003.
97. Ferrati, A.R. et al., A comparison of ready-to-use systems for evaluating the microbiological quality of acidic fruit juices using non-pasteurized orange juices as an experimental model, *Int. Microbiol.*, 8, 49, 2005.
98. Bruch, M.K. and Smith, F.W., Improved method for pouring Rodac plates, *Appl. Microbiol.*, 16, 1427, 1968.

99. Health Canada, Environmental sampling for the detection of microorganisms in *MFLP-41A*, Health Product and Food Branch, 1992.
100. Hacek, D.M. et al., Comparison of the Rodac imprint method to selective enrichment broth for recovery of vancomycin-resistant Enterococci and drug-resistant *Enterobacteriaceae* from environmental surfaces, *J. Clin. Microbiol.*, 38, 4646, 2000.
101. Collins, S.M. et al., Contamination of the clinical microbiological laboratory with vancomycin-resistant Enterococci and multidrug resistant *Enterobacteriaceae*: Implications for hospital and laboratory workers, *J. Clin. Microbiol.*, 39, 3772, 2001.
102. JP, 8. Microbiological evaluation of processing areas for sterile pharmaceutical products, *Japanese Pharmacopoiea, 14th ed.*, 2001, 1312.
103. Scarparo, C. et al., Evaluation of the DipStreak, a new device with an original streaking mechanism for detecting, counting, and presumptive identification of urinary tract pathogens, *J. Clin. Microbiol.*, 40, 2169, 2002.
104. Simpson, A.T. et al., Occupational exposure to metalworking fluid mist and sump fluid contaminants, *Am. Occup. Hyg.*, 47, 17, 2003.
105. Yamaguchi, N. et al., Development of an adhesive sheet for direct counting of bacteria on solid surfaces, *J. Microbiol. Meth.*, 53, 405, 2003.
106. U.S. Food and Drug Administration, Enumeration of *Escherichia coli* and total coliform bacteria, *Bacteriological Analytical Manual Online*, Department of Health and Human Services, 2002, chap. 4.
107. Bauters, T.G. and Nelis, H.J., Comparison of chromogenic and fluorogenic membrane filtration methods for detection of four *Candida* species, *J. Clin. Microbiol.*, 40, 1838, 2002.
108. USEPA, Improved enumeration method for recreational water quality indicators: Enterococci and *Escherichia coli*, U.S. Environmental Protection Agency Office of Science and Technology, Washington, D.C., 2000.
109. Environment Agency, A method for isolation and enumeration of Enterococci by membrane filtration, in *The Microbiology of Drinking Water — Part 5*, Environment Agency, U.K., 2002, 8.
110. USEPA, Method 1106.1 Enterococci in water by membrane filtration using a membrane-Enterococcus esculin iron agar (mE-EIA), U.S. Environmental Protection Agency Office of Water, Washington, D.C., 2002.
111. USEPA, Method 1600 Enterococci in water by membrane fitration using membrane-Enterococcus indoxyl-β-D-glucoside agar (mEI), U.S. Environmental Protection Agency Office of Water, Washington, D.C., 2002.
112. USEPA, Method 1604 Total coliforms and *Escherichia coli* in water by membrane filtration using a simultaneous detection technique (MI medium), U.S. Environmental Protection Agency Office of Water, Washington, D.C., 2002.
113. USEPA, Method 1603 *Escherichia coli* (E. coli) in water by membrane filtration using modified membrane-thermotolerant *Escherichia coli* agar (modified mTEC), U.S. Environmental Protection Agency Office of Water, Washington, D.C., 2002.
114. Health Protection Agency, Enumeration of Enterococci by membrane filtration, *National Standard Method W3*, Issue 31, U.K., 2006.
115. USEPA, Method 1103.1 *Escherichia coli* (E. coli) in water by membrane filtration using a membrane-thermotolerant *Escherichia coli* agar (mTEC), U.S. Environmental Protection Agency Office of Water, Washington, D.C., 2006.
116. Health Protection Agency, General techniques for the detection and enumeration of bacteria by negative pressure membrane filtration, *National Standard Method W1*, Issue 1, U.K., 2007.
117. Brenner, K.P. et al., A new medium for the simultaneous detection of total coliforms and *Escherichia coli* in water, *Appl. Environ. Microbiol.*, 59, 3534, 1993.
118. Ciebin, B.W. et al., Comparative evaluation modified m-FC and m-TEC media for membrane filter enumeration of *Escherichia coli* in water, *Appl. Microbiol.*, 61, 3940, 1995.
119. Rompré, A. et al., Detection and enumeration of coliforms in drinking water: current methods and emerging approaches, *J. Microbiol. Meth.,* 49, 31, 2002.
120. Grant, M.A., A new membrane filtration medium for the simultaneous detection of *Escherichia coli* and total coliforms, *Appl. Environ. Microbiol.*, 63, 3526, 1997.
121. Sharpe, A.N. and Michaud, G.L., Enumeration of high numbers of bacteria using hydrophobic grid-membrane filters, *Appl. Microbiol.*, 30, 519, 1975.
122. Tsuji, K. and Bussey, D.M., Automation of microbial enumeration: development of a disposable hydrophobic grid-membrane filter unit, *Appl. Environ. Microbiol.*, 52, 857, 1986.

123. Parrington, L.J., Sharpe, A.N., and Peterkin, P.I., Improved aerobic colony count technique for hydrophobic grid-membrane, *Appl. Environ. Microbiol.*, 59, 2784, 1993.
124. Health Canada, Enumeration of faecal coliforms in foods by hydrophobic grid-membrane filter (HGMF) method in *MFLP-55*, Health Protection Branch, 1998.
125. Health Canada, Enumeration of *Escherichia coli* in foods by hydrophobic grid-membrane filter (HGMF) method in *MFHPB-26*, Health Protection Branch, 2001.
126. Heikinheimo, A., Linström, M., and Korkeala, H., Enumeration and isolation of *cpe*-positive *Clostridium perfringens* spores from feces, *J. Clin. Microbiol.*, 42, 3992, 2004.
127. Kirchman, D. et al., Statistical analysis of the direct count method for enumerating bacteria, *Appl. Environ. Microbiol.*, 44, 376, 1982.
128. Humberd, C.M., Short report: enumeration of Leptospires using the coulter counter, *Am. J. Trop. Med. Hyg.*, 73, 962, 2005.
129. Hobbie, J.E., Daley, R.J., and Jasper, S., Use of nucleopore filters for counting bacteria by fluorescence microscopy, *Appl. Environ. Microbiol.*, 33, 1225, 1977.
130. Kepner Jr., R.L. and Pratt, J.R., Use of fluorochromes for direct enumeration of total bacteria in environmental samples: past and present, *Microbiol. Rev.*, 58, 603, 1994.
131. Brehm-Stecher, B.F. and Johnson, E.A., Single-cell microbiology: tools, technologies, and applications, *Microbiol. Molec. Biol. Rev.*, 68, 538, 2004.
132. Pettipher, G.L. and Rodrigues, U.M., Rapid enumeration of microorganisms in foods by the direct epifluorescent filter technique, *Appl. Environ. Microbiol.*, 44, 809, 1982.
133. Rapposch, S., Zangeri, P., and Ginzinger, W., Influence of fluorescence of bacteria stained with Acridine Orange on the enumeration of microorganisms in raw milk, *J. Dairy Sci.*, 83, 2753, 2000.
134. Matsunaga, T., Okochi, M., and Nakasono, S., Direct count of bacteria using fluorescent dyes: Application to assessment of electrochemical disinfection, *Anal. Chem.*, 67, 4487, 1995.
135. McBain, A.J. et al., Exposure of sink drain microcosms to Triclosan: Population dynamics and antimicrobial susceptibility, *Appl. Environ. Microbiol.*, 69, 5433, 2003.
136. Veillette, M. et al., Six months tracking of microbial growth in a metalworking fluid after system cleaning and recharging, *Ann. Occup. Hyg.*, 48, 541, 2004.
137. Bloem, J., Veninga, M., and Shepherd, J., Fully automated determination of soil bacterium numbers, cell volumes, and frequencies of dividing cells by confocal laser scanning microscopy and image analysis, *Appl. Environ. Microbiol.*, 61, 926, 1995.
138. Yu, W. et al., Optimal staining and sample storage time for direct microscopic enumeration of total and active bacteria in soil with two fluorescent dyes, *Appl. Environ. Microbiol.*, 61, 3367, 1995.
139. Bölter, M. et al., Enumeration and biovolume determination of microbial cells — A methodological review and recommendations for applications in ecological research, *Biol. Fertil. Soils*, 36, 249, 2002.
140. Terzieva, S. et al., Comparison of methods for detection and enumeration of airborne microorganisms collected by liquid impingement, *Appl. Environ. Microbiol.*, 62, 2264, 1996.
141. Ramalho, R. et al., Improved methods for the enumeration of heterotrophic bacteria in bottled waters, *J. Microbiol. Meth.*, 44, 97, 2001.
142. Lisle, J.T. et al., Comparison of fluorescence microscopy and solid-phase cytometry methods for counting bacteria in water, *Appl. Environ. Microbiol.*, 70, 5343, 2004.
143. Angenent, L.T. et al., Molecular identification of potential pathogens in water and air of a hospital therapy pool, *PNAS*, 102, 4860, 2005.
144. Helton, R.R., Liu, L., and Wommack. K.E., Assessment of factors influencing direct enumeration of viruses within estuarine sediments, *Appl. Environ. Microbiol.*, 72, 4767, 2006.
145. Kogure, K. et al., Correlation of direct viable count with heterotrophic activity for marine bacteria, *Appl. Environ. Microbiol.*, 53, 2332, 1987.
146. Roszak, D. B. and Colwell, R.R., Metabolic activity of bacterial cells enumerated by direct viable count, *Appl. Environ. Microbiol.*, 53, 2889, 1987.
147. Joux, F. and Lebaron, P., Ecological implication of an improved direct viable count method for aquatic bacteria, *Appl. Environ. Microbiol.*, 63, 3643, 1997.
148. Yokomaku, D., Yamaguchi, N., and Nasu, M., Improved direct viable count procedure for quantitative estimation of bacterial viability in freshwater environments, *Appl. Environ. Microbiol.*, 66, 5544, 2000.
149. Moter, A. and Göbel, U.B., Fluorescence in situ hybridization (FISH) for direct visualization of microorganisms, *J. Microbiol. Meth.*, 41, 85, 2000.

150. Kenzaka, T. et al., Rapid monitoring of *Escherichia coli* in Southeast Asian urban canals by fluorescent-bacteriophage assay, *J. Health Sci.*, 52, 666, 2006.
151. Baudart, J. et al., Rapid and sensitive enumeration of viable diluted cells of members of the family Enterobacteriaceae in freshwater and drinking water, *Appl. Environ. Microbiol.*, 68, 5057, 2002.
152. Shapiro, H.M., Microbial analysis at the single-cell level: tasks and techniques, *J. Microbiol. Meth.*, 42, 3, 2000.
153. Kubitschek, H.E. and Friske, J.A., Determination of bacterial cell volume with the coulter counter, *J. Bacteriol.*, 168, 1466, 1986.
154. Kogure, K. and Koike, I., Particle counter determination of bacterial biomass in seawater, *Appl. Environ. Microbiol.*, 53, 274, 1987.
155. Jepras, R.I. et al., Development of a robust flow cytometric assay for determining numbers of viable bacteria, *Appl. Environ. Microbiol.*, 61, 2969, 1995.
156. Davey, H.M. and Kell, D.B., Flow cytometry and cell sorting of heterogeneous microbial populations: the importance of single-cell analyses, *Microbiol. Rev.*, 60, 641, 1996.
157. Wang, X, and Slavik, M.F., Rapid detection of Salmonella in chicken washes by immunomagnetic separation and flow cytomtry, *J. Food Protect.*, 62, 717, 1999.
158. Gunasekera, T.S., Attfield, P.V., and Veal, D.A., A flow cytometry method for rapid detection and enumeration of total bacteria in milk, *Appl. Environ. Microbiol.*, 66, 1228, 2000.
159. Li, R.C., Lee, S.W., and Lam, J.S., Novel method for assessing postantibiotic effect by using the coulter counter, *Antimicrob. Agents Chemother.*, 40, 1751, 1996.
160. Okada, H. et al., Enumeration of bacterial cell numbers and detection of significant bacteriuria by use of a new flow cytometry-based device, *J. Clin. Microbiol.*, 44, 3596, 2006.
161. Lebaron, P., Parthuisot, N., and Catala, P., Comparison of blue nucleic acid dyes for flow cytometric enumeration of bacteria in aquatic systems, *Appl. Environ. Microbiol.*, 64, 1725, 1998.
162. Chen, P. and Li, C., Real-time quantitative PCR with gene probe, fluorochrome and flow cytometry for microorganisms, *J. Environ. Monit.*, 7, 257, 2005.
163. Marie, D. et al., Enumeration of marine virus in culture and natural samples by flow cytometry, *Appl. Environ. Microbiol.*, 65, 45, 1999.
164. Broadaway, S.C., Barton, S.A., and Pyle, B.H., Rapid staining and enumeration of small numbers of total bacteria in water by solid-phase laser cytometry, *Appl. Environ. Microbiol.*, 69, 4272, 2003.
165. Delgado-Viscogliosi, P. et al., Rapid method for enumeration of viable *Legionella pneumophila* and other *Legionella* spp. in water, *Appl. Environ. Microbiol.*, 71, 4086, 2005.
166. Marennikova, S.S., Gurvich, E.B., and Shelukhina, E.M., Comparison of the properties of five pox virus strains isolated from monkeys, *Arch. Virol.*, 33, 201, 1971.
167. Ajello, F., Massenti, M.F., and Brancato, P., Foci of degeneration produced by measles virus in cell cultures with antibody free liquid medium, *Med. Microbiol. Immunol.*, 159, 121, 1974.
168. Brick, D.C., Oh, J.O., and Sicher, S.E., Ocular lesions associated with dissemination of type 2 herpes simplex virus from skin infection in newborn rabbits, *Invest. Ophthalmol. Vis. Sci.*, 21, 681, 1981.
169. USEPA, Total culturable virus quantal assay in *USEPA Manual of Methods for Virology*, U.S. Environmental Protection Agency Office of Science and Technology, Washington, D.C., 2001, chap. 15.
170. Lee, H.K. and Jeong, Y.S., Comparison of total culturable virus assay and multiplex integrated cell culture-PCR for reliability of waterborne virus detection, *Appl. Environ. Microbiol.*, 70, 3632, 2004.
171. USEPA, Cell culture procedures for assaying plaque-forming viruses in *USEPA Manual of Methods for Virology*, U.S. Environmental Protection Agency Office of Science and Technology, Washington, D.C., 1987, chap. 10.
172. Cornax, R. et al., Application of direct plaque assay for detection and enumeration of bacteriophages for *Bacteroides fragilis* from contaminated-water samples, *Appl. Environ. Microbiol.*, 56, 3170, 1990.
173. USEPA, Procedures for detecting coliphages in *USEPA Manual of Methods for Virology*, U.S. Environmental Protection Agency Office of Science and Technology, Washington, D.C., 2001, chap. 16.
174. USEPA, Method 1602: Male-specific (F^+) and somatic coliphage in water by single agar layer (SAL) procedure, U.S. Environmental Protection Agency Office of Water, Washington, D.C., 2001.
175. Mocé-Llivina, L., Lucena, F., and Jofre, J., Double-layer plaque assay for quantification of Entroviruses, *Appl. Environ. Microbiol.*, 70, 2801, 2004.
176. Matrosovich, M. et al., New low-viscosity overall medium for viral plaque assay, *Virol. J.*, 3, 63, 2006.
177. EP, 5.1.6 Alternative methods for control of microbiological quality, *European Pharmacopoeia, 5th ed.*, Supplement 5.5, 2006, 4131.

178. de Boer, E. and Beumer, R.R., Methodology for detecting and typing foodborne microorganisms, *Int. J. Food Microbiol.*, 50, 119, 1999.
179. Fung, D.Y.C., Rapid methods and automation in microbiology, *Comp. Rev. Food Sci. Food Safety*, 1, 3, 2002.
180. Cundell, A.M., Opportunities for rapid microbial methods, *Eur. Pharm. Rev.*, 1, 64, 2006.
181. Hussong, D. and Mello, R., Alternative microbiology methods and pharmaceutical quality control, *Am. Pharm. Rev.*, 9, 62, 2006.
182. Feldsine, P, Abeyta, C., and Andrews, W.H., AOAC International methods committee guidelines for validation of qualitative and quantitative food microbiological official methods of analysis, *J. AOAC Int.*, 85, 1187, 2002.
183. Clough, H.E. et al., Quantifying uncertainty associated with microbial count data: a Bayesian approach, *Biometrics*, 61, 610, 2005.
184. Lombard, B. et al., Experimental evaluation of different precision criteria applicable to microbiological counting methods, *J. AOAC Int.*, 88, 830, 2005.
185. Health Protection Agency, Uncertainty of measurement in testing, *National Standard Method QSOP4*, Issue 5, U.K., 2005.

3 Culturing and Preserving Microorganisms

Lorrence H. Green

CONTENTS

INTRODUCTION

The use of culture as a diagnostic technique is described in the chapter entitled "Diagnostic Medical Microbiology." It is the purpose of this chapter to discuss only the principles of enrichment culture. This is a technique that has been in use for more than 100 years.[1] It consists of incubating a sample in a medium that encourages the growth of an organism of interest, while inhibiting the growth of others. In this way it can assist the technician in isolating pure colonies of microorganisms from mixtures in which the organisms represent only a very small percentage of the overall flora, and isolation by streaking may not be practical. In some cases, as in the isolation of salmonella, first growing a sample in an enrichment culture is a necessary first step to increase the small relative number of organisms usually found in the primary sample.[2] While the use of enrichment broths in food and environmental microbiology has been demonstrated, their use in clinical microbiology has not fully been established.[3]

DESIGNING AN ENRICHMENT MEDIUM

While the media used to enrich for specific microorganisms might differ greatly from each other, they should all contain an energy source, a carbon source, and a source of trace and major elements. In addition, the pH, temperature, and oxygen tension should be appropriate to the microorganism of interest.[1]

Energy Source

Most microorganisms are heterotrophs, and as a result obtain their energy by ingesting organic molecules. When these molecules are broken down, energy contained in the chemical bonds is liberated.[5] Some microorganisms are autotrophs and are capable of obtaining energy either by oxidizing inorganic chemical compounds (chemoautotrophs) or by directly utilizing light (photosynthetic).[5] Chemoautotrophs derive energy from reduced inorganic molecules or ions, which can include H_2, NH_4^+, Fe^{2+}, and S^0.[1,6] Photosynthetic microorganisms derive their energy directly from light sources, with bacteria and photosynthetic eukaryotes favoring light in the red portion of the spectrum, while Cyanobacteria favor light in the blue portion.[1]

Source of Carbon

Because autotrophic organisms derive their energy from light, this usually requires little more than a source of simple carbon, such as bubbling CO_2 gas into the media or by adding a solid carbonate source. Heterotrophic organisms can obtain their carbon from a wide variety of sources. The most common of these are carbohydrates. Many of these are actually used as the component parts of biochemical tests used to identify microorganisms. These include carbohydrates (i.e., glucose, maltose, lactose, sucrose, etc.); acids found in the TCA cycle (i.e., succinic, oxaloacetic, malic, α-ketogluaric, etc.); amino acids (i.e., alanine, arginine, glycine, serine, etc.); and fatty acids (i.e., lactic, pyruvic, butyric, propionic, etc.).

In the past few years, it has been realized that microorganisms exist that are capable of using toxic compounds as carbon sources. As such, research has focused on enriching samples for microorganisms that can be used for bioremediation of environmental spills. A recent article found an organism in a soil sample that was capable of using trichlorethene (TCE), recognized as the most commonly found contaminant at Superfund sites.[7]

Trace and Major Elements

All microorganisms also require varying amounts of certain elements. Major elements such as nitrogen, potassium, sodium, magnesium, and calcium can be found as salts. These can include chlorides (NH_4Cl, KCl, NaCl); sulfates ($MgSO_4 \cdot 7H_2O$, Na_2SO_4, $(NH_4)_2SO_4$); carbonates ($MgCO_3$, $CaCO_3$); nitrates ($Ca(NO_3)_2 \cdot 4H_2O$); and phosphates (K_2HPO_4, KH_2PO_4).[1] Other elements may also be required in trace amounts. These can include iron, zinc, manganese, copper, cobalt, boron, molybdenum, vanadium, strontium, aluminum, rubidium, lithium, iodine, and bromine.[1]

pH

While most organisms prefer to grow in a neutral pH environment, pH can be used as an enrichment method. Acidophiles can be enriched by lowering the pH in the media.[8]

Physical Components

Although researchers usually tend to focus on ingredients that can be added to a liquid medium, it should be remembered that manipulating the physical environment that the culture is placed in for temperature, oxygen tension, and even pressure may offer some microorganisms a competitive advantage in reproduction. The levels of these agents will be determined by which organism the technician is trying to isolate.[4]

Antimicrobial Agents

Enrichment for specific groups or species of microorganisms can not only be achieved by defining conditions that will preferentially allow them to grow, but can also be achieved through the use of agents that will inhibit the growth of competing organisms. In some instances, fast-growing organisms can be inhibited, to give fastidious or slower growing organisms an advantage in a particular sample. An interesting clinical consequence of the use of antimicrobial agents as a method of enrichment is the observation of resistant organisms in patients who have been hospitalized. Axon et al.[9] have found that hemodialysis patients, who usually have a greater chance of being exposed to antibiotics and of being hospitalized, are a reservoir of vancomycin-resistant Enterococcus infections in a nosocomial setting.

PRESERVATION OF MICROORGANISMS

INTRODUCTION

Many laboratory test procedures require the use of microorganisms as reagents, while advances in biotechnology have resulted in the creation of engineered microorganisms. In both instances, to obtain consistent results, these organisms must be preserved in a manner that will allow for their genetic stability and long-term survival. Preservation of microorganisms can be accomplished by a variety of methods. These can include subculturing them, reducing their metabolic rate, or putting them into a state of quasi-suspended animation.[10] The chosen method will usually depend on which organism one is trying to preserve. Despite the introduction of newer methodologies, many of the techniques that are used have not changed much since the previous edition of this book. Most still involve the use of drying, lyophilization, or storage in freezing or subfreezing temperatures.

METHODS

Serial Subculture

This is a simple method in which the cultures are periodically passed in liquid or agar media. Some cultures can be stored on agar media in sealed tubes and survive for as long as 10 years.[11] This method has been used extensively for fungi but usually requires storage under mineral oil.[10] Despite its simplicity and applicability for cultures that cannot survive harsher preservation methods, serial subculture is not a very satisfactory method. In our hands we have had problems with contamination, culture death, and the unintended selection of mutants. This has been particularly difficult in cases in which we were trying to develop new products for the identification of microorganisms. It was imperative to periodically confirm the identity of the organisms being used in the database.

Storage at Low Temperatures

Some species are capable of being stored for long periods at 4 to 8°C on agar plates or on slants. We also have been able to store cultures at –20°C for extended periods. We have found that this can be accomplished by growing the cultures in liquid media for 24 to 48 hours and then mixing one part sterile glycerol to three parts culture. While this does not work for all organisms, it allows the use of a standard refrigerator for preserving cultures. Many laboratories store organisms at –70 to –80°C by first mixing them with 10% glycerol.[12] Many cultures can be stored indefinitely in liquid nitrogen. A method for this can be found at www.cabri.org. While the use of liquid nitrogen is a good method for storing microorganisms, for most laboratories it requires the expense of constantly refilling a specialized thermos.

Freeze-Drying

This is a widely used method in which a suspension of microorganisms is frozen and then subjected to a vacuum to sublimate the liquid. The resulting dried powder is usually stored in vials sealed in a vacuum. Many factors can affect the stability of the culture. These include the growth media, the age of the cultures, the phase of the culture, and the concentration of the organisms in the suspension.[10, 13] The length of sample viability can vary substantially but cultures surviving for up to 20 years have been reported.[14] Freeze-drying requires a cryoprotective agent to provide maximal stability. Usually, the organisms are suspended in 10% skim milk.[15]

Storage in Distilled Water

This was cited as a method of preservation of *Pseudomonas* species and fungi in the first edition of this book.[10] Recent studies have found that it is still an effective method of preserving fungi,[16] as well as a wide variety of bacteria (including *Pseudomonas fluorescens, Erwinia* spp., *Xanthomonas campestris, Salmonella* spp., *Yersinia enterocolitica, Escherichia coli O157:H7, Listeria monocytogenes, and Staphylococcus aureus*.[17] Stationary-phase organisms grown on agar media and then suspended in 10 mL of sterile water were found to be stable when sealed with parafilm membranes and then stored at room temperature in the dark. Even greater stability could be obtained by suspending the organisms in a screw-capped tube with phosphate buffered saline (PBS) at pH 7.2, containing 15.44 μM KH_2PO_4, 1.55 mM NaCl, and 27.09 μM Na_2HPO_4.[17]

Drying

Sterile soil or sand has been used for preserving spore-forming organisms by adding suspensions and then drying at room temperature. In addition, bacterial suspensions have been mixed with melted gelatin and then dried in a desiccator. Both methods have produced samples that are stable for long periods of time.[10] Recently, a method has been developed in which a microliter quantity of a bacterial suspension is mixed with a pre-dried activated charcoal cloth based matrix contained within a resealable system that can then be stored. Experiments with *Escherichia coli* have found that viability of over a year at 4°C can be obtained.[18]

Recovery and Viability of Preserved Microorganisms

Several factors can affect the viability of microorganisms that have been stored by either freezing or drying. These can include the temperature at which microorganisms are recovered, the type and volume of the media used for recovery, and even how quickly microorganisms are solubilized in a recovery medium.[10]

Usually, the losses of preservation can be overcome by initially preserving large numbers of microorganisms. In doing this, one must be careful to avoid two problems. The first is that if only a very small fraction of organisms is recovered, this can result in the selection of biochemically distinct strains. The second is that if a culture has inadvertently been contaminated with even a few microorganisms, the preservation technique may lead to a selection for the contaminant.

Drying and freeze-drying have been known to cause changes in several characteristics of preserved microorganisms. These can include colonial appearance and pathogenicity.[10] After revival from a frozen or dried state, many protocols usually advise subculturing the organisms at least two or three times in an attempt to restore any characteristics that may have been lost.[10]

REFERENCES

1. Aaronson, S,. Enrichment culture, *CRC Practical Handbook of Microbiology*, O'Leary, W., Ed., CRC Press, Boca Raton, FL, 1989, 337.
2. Keen, J., Durso, I., and Meehan, L., Isolation of *Salmonella enterica* and Shiga-toxigenic *Escherichia coli* O157 from feces of animals in public contact areas of United States zoological parks, *Appl. Environ. Microbiol.*, 73, 362, 2007.
3. Miles, K.I., Wren, M.W.D., and Benson, S., Is enrichment culture necessary for clinical samples?, *Br. J. Biomed. Sci.*, 63, 87, 2006.

4. Ohmura, M., Kaji; S., Oomi, G., Miyaki, S., and Kanazawa, S., Effect of pressure on the viability of soil microorganisms, *High Pres. Res.,* 27, 129, 2007.
5. Kobayashi, G., Murray, P., Pfaller, M., and Rosenthal, K., *Medical Microbiology, 4th ed.*, Mosby, St. Louis, MO, 2002, 25.
6. Ohmura, N., Sasaki, K., Matsumoto, N., and Saiki, H., Anaerobic respiration using Fe^{3+}, S^0, and H_2 in the chemolithoautotrophic bacterium *Acidithiobacillus ferrooxidans, J. Bacteriol.*, 184, 2081, 2002.
7. Lee, S.-B., Strand, S., Stensel, H., and Herwig, R., *Pseudonocardia chloroethenivorans* sp. nov., a chloroethene-degrading actinomycete, *Int. J. Syst. Evol. Microbiol.,* 54, 131, 2004.
8. Johnson, D.B. and Hallberg, K.B., The microbiology of acidic mine waters, *Res. Microbiol.*, 154, 466, 2003.
9. Axon, R.N., Engemann, J.J., Butcher, J., Lockamy, K., and Kaye, K.S., Control of nosocomial acquisition of vancomycin-resistant Enterococcus through active surveillance of hemodialysis patients, *Infect .Control Hosp. Epidemiol.,* 25, 436, 2004.
10. Lapage, S., Redway, K., and Rudge, R., Preservation of microorganisms, *CRC Practical Handbook of Microbiology*, O'Leary, W., Ed., CRC Press, Boca Raton, FL, 1989, 321.
11. Antheunisse, J., Preservation of microorganisms, *Antonie van Leeuwenhoek, J. Microbiol. Serol.,* 38, 617, 1972.
12. Cameotra, S.S., Preservation of microorganisms as deposits for patent application, *Biochem. Biophys. Res. Commun.*, 353, 849, 2007.
13. Morgan, C.A., Herman, N., White, P.A., and Vesey, G., Preservation of micro-organisms by drying; a review, *J. Microbiol. Meth.*, 66, 183, 2006.
14. Miyamoto-Shinohara, Y., Sukenobe, J., Imaizumi, T., and Nakahara, T., Survival curves for microbial species stored by freeze-drying, *Cryobiology,* 52, 27, 2006.
15. Crespo, M.J., Abarca, M.L., and Cabañes, F.J., Evaluation of different preservation and storage methods for *Malassezia* spp., *J. Clin. Microbiol.,* 38, 3872, 2000.
16. Deshmukh, S.K., The maintenance and preservation of keratinophilic fungi and related dermatophytes, *Mycoses,* 46, 203, 2003.
17. Liao, C.-H. and Shollenberger, L.M., Survivability and long-term preservation of bacteria in water and in phosphate-buffered saline, *Lett. Appl. Microbiol.,* 37, 45, 2003.
18. Hays, H.C.W., Millner, P.A., Jones, J.K., and Rayner-Brandes, M.H., A novel and convenient self-drying system for bacterial preservation, *J. Microbiol. Methods,* 63, 29, 2005.

4 Stains for Light Microscopy

Stuart Chaskes

CONTENTS

The following is a selection of staining methods that may be of assistance to microbiologists.

GRAM STAIN[1, 2]

Background for the Gram Stain

The Gram stain was developed by Christian Gram in 1884 and modified by Hucker in 1921. The Gram stain separates bacteria into two groups: (1) Gram-positive microorganisms that retain the primary dye (Crystal violet) and (2) Gram-negative microorganisms that take the color of the counterstain (usually Safranin O). These results are due to differences in the structure of the cell wall. Crystal violet is attracted to both Gram-positive and Gram-negative microorganisms. The second step (Gram's Iodine, a mordant) stabilizes the Crystal violet into the peptidoglycan layer of the cell wall. The peptidoglycan layer is much thicker in Gram-positive bacteria than in Gram-negative bacteria; hence, the Crystal violet is more extensively entrapped in the peptidoglycan of Gram-positive bacteria. The third step (alcohol decolorization) dissolves lipids in the outer membrane of Gram-negative bacteria and removes the Crystal violet from the peptidoglycan layer. In contrast, the Crystal violet is relatively inaccessible in Gram-positive microorganisms and cannot readily be removed by alcohol in Gram-positive microorganisms. After the alcohol step, only the colorless Gram-negative microorganisms can accept the Safranin (counterstain). Carbol-fuchsin and Basic-fuchsin are sometimes employed in the counterstain to stain anaerobes and other weakly staining Gram-negative bacteria (including *Legionella* spp., *Campylobacter* spp., and *Brucella* spp.). The most frequent errors in the Gram stain often are associated with slide preparation. Thick slide preparations, excessive heat fixing that distorts the bacteria, and improper decolorizations are common problems encountered with the stain. Inexperience can lead to over-decolorization (too many Gram-negative bacteria) or under-decolorization (too many Gram-positive bacteria). The decolorization step is the most problematic part of the procedure. Bacteria are often called Gram variable if half are stained violet and the other half are pink. The details of the three-step Gram stain by Mesaros, Army, Strenkoski, and Leon (www.freepatentsonline.com/5827680.htlm) are available. The three-step method simultaneously decolorizes and counterstains Gram-negative bacteria. Most clinical laboratories are currently using the four-step Gram stain. A Modified Brown and Brenn Gram stain (*Surgical Pathology Staining Manual*, www.library.med.utah.edu/WebPath/HISTHTML/MANUALS/MANUALS.html) can be used to detect Gram-negative and -positive bacteria in tissue. In this method, Gram-positive bacteria stain blue, Gram-negative bacteria stain red, and the background color is yellow.

Standard Gram Stain Procedure

1. Fix the specimen with heat or use 95% methanol for 2 min.
2. Apply the primary stain (Crystal violet) for 1 min. Wash with tap water.
3. Apply the mordant (Gram's iodine) for 1 min. Wash with tap water.
4. Decolorize for 5 to 15 sec. Wash with tap water.
5. Counterstain with Safranin for 1 min. Wash with tap water.

The Gram Stain Reagents

1. Primary stain: 2 g Crystal violet, 20 mL 95% ethyl alcohol, 0.8 g ammonium oxalate, and 100 mL distilled water.
2. Gram's iodine: 2 g potassium iodide, 1 g iodine crystals, and 100 mL distilled water.
3. Decolorizer: 50 mL acetone and 50 mL ethanol.
4. Counterstain: 4.0 g Safranin, 200 mL 95% ethanol, and 800 mL distilled water.

Decolorizer/Counterstain for the Three-Step Gram Stain

1. Combine 0.40% Safranin, 0.30% Basic fuchsin, 90% ethanol, 10% distilled water, and acidify to pH 4.5.

Variations for Gram Stain Reagents

Many variations exist for the Gram stain. The Gram iodine mordant can be stabilized using a polyvinypyrrolidone-iodine complex. Slowing the decolorizing step is accomplished using only 95% ethanol. Some laboratories prefer to decolorize with isopropanol/acetone (3:1 vol:vol).

ACRIDINE ORANGE STAIN[1, 2, 10, 12]

Background for the Acridine Orange Stain

The Acridine orange stain is a sensitive method for detecting low numbers of organisms in cerebral spinal fluid (CSF), blood, buffy coat preparations from neonates, and tissue specimens. The Acridine orange stain can be employed in rapidly screening normal sterile specimens (blood and CSF) where low numbers of microorganisms usually exist. Acridine orange is a fluorochrome that binds to nucleic acids. Bacteria and yeasts stain bright orange/red, and tissue cells stain black to yellowish-green. The filter system used on a fluorescent microscope can affect the observed colors. Because red blood cells are not fluorescent, the Acridine orange stain can be useful for screening blood cultures for the growth of microorganisms. The stain maybe useful in interpreting thick purulent specimens that failed to give clear results with the Gram stain. Acridine orange may also be useful for staining miscellaneous microorganisms such as *Mycoplasma*, *Pneumocystis jiroveci* trophozoites, *Borrelia burgdorferi, Acanthamoeba, Leishmania,* and *Helicobacter pylori* purulent specimens. Additional information on the Acridine orange stain can be found at www.histonet.org.

Acridine Orange Standard Staining Procedure

1. Fix the smear in methanol for 2 min.
2. Stain with Acridine orange for 1 min.
3. Rinse with tap water.

Paraffin Acridine Orange Staining Procedure

1. Deparaffinize and hydrate to distilled water.
2. Stain sections with Acridine orange for 30 min.
3. Rinse sections in 0.5% acetic acid in 100% alcohol for about 1 min.
4. Rinse sections 2X in 100% alcohol.
5. Rinse sections 2X in xylene.
6. Mount sections in Fluoromount.

Reagents for Standard Stain

1. 0.2 M sodium acetate buffer (pH 3.75).
2. Add Acridine orange to buffer (0.02 g).

Reagents for Paraffin Procedure

1. Add 5 mL acetic acid to 500 mL distilled water.
2. Add 0.05 g Acridine orange to the diluted acetic acid.

MACHIAVELLO STAIN MODIFIED FOR *CHLAMYDIA* AND *RICKETTSIA*

Background for the Machiavello-Gimenez Stain

Chlamydia and *Rickettsia* will stain red against a blue background. The nuclei of *Rickettesia* may also stain blue. Pinkerton's adaptation of Machiavello's stain is used to detect *Rickettesia* in tissue samples. *Legionella* can also be stained by this method. Additional information on the Machiavello stain can be found at www.histonet.org.

Machiavello Stain Procedure (Modified) for *Chlamydia*

1. Heat fix the smear or air dry.
2. Stain the slide with Basic fuchsin for 5 min.
3. Wash in tap water and then place the slide in a Coplin jar that contains citric acid. The slide should remain in the citric acid for a few seconds. Decolorization of the *Chlamydia* will occur if the citric acid is left on too long.
4. Wash the slide thoroughly with tap water.
5. Stain the slide for 20 to 30 sec with 1% methylene blue.
6. Wash the slide with tap water and air dry.

Pinkerton's Adaptation of Machiavello's Stain for *Rickettesia* in Tissues

1. Deparaffinize and hydrate to distilled water.
2. Stain overnight with methylene blue (1%).
3. Decolorize with alcohol (95%).
4. Wash with distilled water.
5. Stain for 30 min with Basic fuchsin (0.25%).

6. Decolorize 1 to 2 sec in citric acid (0.5%).
7. Differentiate quickly with alcohol (100%).
8. Dehydrate with alcohol (95%), followed by 3X alcohol (100%).
9. Clear in xylene 3X and mount in Permount.

Reagents for the Machiavello Stain

1. **Basic fuchsin:** Dissolve 0.25 g Basic fuchsin in 100 mL distilled water.
2. **Citric acid:** Dissolve 0.5 g citric acid in 100 mL distilled water. Some procedures use 0.5 g in 200 mL distilled water. This reagent must always be fresh.
3. **Methylene blue:** Dissolve 1.0 g in 100 ml distilled water.

ACID-FAST STAINS[1, 2, 10, 12]

Background for the Acid-Fast Stain

The cell wall of Mycobacteria contains a large amount of lipid that makes it difficult for aqueous-based staining solution to enter the cell. Gram-staining is not acceptable because the results are often Gram variable, or beaded Gram-positive rods, or negatively stained images. In contrast, the acid-fast stains contain phenol, which allows Basic fuchsin (red) or auramine (fluorescent) to penetrate the cell wall. The primary stain will remain bound to the cell wall mycolic acid residues after acid alcohol is applied. The resistance of the Mycobacteria group to acid-alcohol decolorization has led to the designation of this group as acid-fast bacilli (AFB). A counterstain such as methylene blue is applied to contrast the Mycobacteria (red) from the non-AFB organisms (blue). Detailed information on acid-fast stains can be found at Centers for Disease Control and Prevention (CDC), Acid Fast Direct Smear Microscopy Manual (www.phppo.cdc.gov/dls/ila/acidfasttraining/participants.aspx). The CDC also supplies detailed information on the use of fluorochrome staining for the detection of acid-fast Mycobacteria.

Ziehl Neelsen Staining Procedure (Hot Method)

1. Fix the prepared slide with gentle heat.
2. Flood the slide with Carbol-fuchsin and heat to steaming only once.
3. Leave for 10 min.
4. Rinse with distilled water.
5. Decolorize with acid alcohol for 3 min.
6. Rinse with distilled water.
7. Counterstain with methylene blue for 1 min.
8. Rinse with distilled water, drain, and air dry.

Ziehl Neelsen Staining Reagents

1. **Primary Stain**: 0.3% Carbol-fuchsin. Dissolve 50 g phenol in 100 mL ethanol (95%) or methanol (95%). Dissolve 3 g Basic fuchsin in the mixture and add distilled water to bring the volume to 1 L.
2. **Decolorization Solution**: Add 30 mL hydrochloric acid to 1 L of 95% denatured alcohol. Cool and mix well before use. **Alternate decolorizing reagent** (without alcohol): Slowly add 250 mL sulfuric acid (at least 95%) to 750 mL distilled water. Cool and mix well before using.
3. **Counterstain:** 0.3% nethylene blue. Dissolve 3 g nethylene blue in 1 L distilled water.

TABLE 4.1
Reporting Systems for Acid-Fast Stains

Reporting System	AFB Seen
Negative	No AFB seen. Examine at least 100 fields
Report actual number	1–9 AFB per 100 fields
1+	10–99 AFB per 100 fields
2+	1–10 AFB per field. Examine at least 50 fields
3+	>10 AFB per field. Examine at least 20 fields

Reporting System for Acid-Fast Stains

Table 4.1 should be used to report the results of the Ziehl Neelsen or Kinyoun acid-fast stain.

Kinyoun Staining Procedure (Cold Method)

1. Fix and prepare the slide with gentle heat.
2. Stain with Kinyoun Carbol-fuchsin for 3 to 5 min.
3. Rinse with distilled water.
4. Decolorize with acid alcohol until the red color no longer appears in the washing (about 2 min).
5. Rinse with distilled water.
6. Counterstain with methylene blue for 30 to 60 sec.
7. Rinse with distilled water, drain, and air dry.

Kinyoun Staining Reagents

1. **Primary Stain:** Dissolve 40 g Basic fuchsin in 200 mL of 95% ethanol. Then add 1000 mL distilled water and 80 g liquefied phenol.
2. **Decolorizer:** 3% acid. Mix 970 mL of 95% ethanol and 30 mL concentrated hydrochloric acid.
3. **Counterstain:** 0.3% methylene blue. Dissolve 3 g methylene blue in 1 L distilled water.

Traunt Fluorochrome Staining Procedure

1. Air dry the smears and fix on a slide warmer at 70°C for at least 2 hr. A Bunsen burner may be substituted.
2. Stain with Auramine O-Rhodamine B solution for 15 min.
3. Rinse the slide with distilled water.
4. Decolorize with acid alcohol for approximately 2 min.
5. Rinse with distilled water and drain.
6. Counterstain with potassium permanganate for 2 min. Applying the counterstain for a longer time period may quench the fluorescence of the *Mycobacterium* species. Acridine orange is not recommended as a counterstain by the Centers for Disease Control and Prevention (CDC).
7. Smears are scanned with a 10X objective. It is sometimes necessary to use the 40X objective to confirm the bacterial morphology.
8. Examine at least 30 fields when viewing the slide under a magnification of 200X or 250X. Examine at least 55 fields when viewing the slide under a magnification of 400X. Examine at least 70 fields when viewing the slide under a magnification of 450X.

9. If only 1 or 2 acid-fast bacilli are observed in 70 fields, the results are reported as doubtful and should be repeated.

1+ = Observing either 1 to 9 AFB in 10 fields (magnification of 200–250X) or 2 to 18 AFB in 50 fields (magnification of 400–450X).

2+ = Observing either 1 to 9 AFB per field (magnification of 200–250X) or 4 to 36 AFB per 10 fields (magnification 400–450X).

3+ = Observing 10 to 90 AFB per field (magnification 200–250X) or 4 to 36 per field (magnification 400–450X).

4+ = Observing more than 90 AFB per field (magnification of 200–250X) or more than 36 AFB per field (magnification 400-450X).

Traunt Staining Reagents

1. **Auramine O-Rhodamine B:** Dissolve 0.75 g Rhodamine B and 1.5 g Auramine O in 75 mL glycerol.
2. **Decolorizer:** Add 0.5 mL concentrated hydrochloric acid to 100 mL of 70% ethanol.
3. **Counterstain:** 0.5% potassium permanganate. Dissolve 0.5 g potassium permanganate in 100 mL distilled water.

Modified Acid-Fast Stain for Detecting Aerobic Actinomycetes (including Nocardia), Rhodococcus, Gordonia, and Tsukamurella

The actinomycetes are a diverse group of Gram-positive rods, often with branching filamentous forms, and are partially acid-fast. A modified acid-fast stain using 1% sulfuric acid rather than the usual 3% hydrochloric acid can be used if Gram-positive branching or partially branching organisms are observed.

Modified Acid-Fast Stain for Detecting Cryptosporidium, Isospora, and Cyclospora

The modified acid-fast stain is used to identify the oocysts of the Coccidian species, which are difficult to detect with the trichrome stain. Fresh or formalin-preserved stools as well as duodenal fluid, bile, and pulmonary samples can be stained.

Modified Acid-Fast Staining Procedure

1. Prepare a thin smear by applying one or two drops of sample to a slide. The specimen is dried via a slide warmer at 60°C and then fixed with 100% methanol for approximately 30 sec.
2. Apply Kinyoun Carbol-fuchsin for 1 min and rinse the slide with distilled water.
3. Decolorize with acid alcohol for about 2 min.
4. Counterstain for 2 min with Malachite green. Rinse with distilled water.
5. Dry, a slide warmer can be used, and apply mounting media and a cover slip.
6. Examine several hundred fields under 40X magnification and confirm morphology under oil immersion.
7. The parasites will stain a pinkish-red color.

Modified Acid-Fast Staining Reagents

1. **Kinyoun Carbol-fuchsin:** see earlier procedure.
2. **Decolorizer:** 10 mL sulfuric acid and 90 mL of 100% ethanol.
3. **Counterstain:** 3% Malachite green. Dissolve 3 g Malachite green in 100 mL distilled water.

MYCOLOGY PREPARATIONS AND STAINS

Potassium Hydroxide (KOH) and Lactophenol Cotton Blue (LPCB) Wet Mounts[6, 7, 12]

Wet mounts using 10% KOH (potassium hydroxide) will dissolve keratin and cellular material and unmask fungal elements that may be difficult to observe. Lactophenol cotton blue preparations are useful because the phenol in the stain will kill the organisms and the lactic acid preserves fungal structures. Chitin in the fungal cell wall is stained by the Cotton blue. The two wet mount preparations (KOH and LPCB) can be combined. Additional variations include using KOH and Calcofluor white or KOH and dimethylsulfoxide for thicker specimens of skin or nail. An extensive online *Mycology Procedure Manual* (Toronto Medical Laboratories/Mount Sinai Hospital Microbiology Department) is available at http://microbiology.mtsinai.on.ca/manual/myc/index.shtml. Additional staining procedures for the fungi can be found at www.doctorfungus.org.

KOH Procedure

1. Thin smear scrapings from the margin of a lesion are placed on a slide.
2. Add 1 or 2 drops of KOH, place on a cover slip, and allow digestion to occur over 5 to 30 min. Alternately, the slide can be gently heated by passing it through a flame several times. Cool and examine under low power. The fungal cell wall is partially resistant to the effects of KOH. However, the fungi may eventually dissolve in the KOH if left in contact with the reagent for an excessive time period.

KOH Reagent

1. Dissolve 10 g KOH in 80 mL distilled water and then add 20 mL glycerol.

LPCB Procedure

1. Place a thin specimen on the slide and add 1 drop LPCB. Mix well and place a cover slip over the slide. Nail polish can be used to make a semi-permanent slide.
2. Observe under the microscope for fungal elements.

LPCB Reagent

1. Dissolve 0.5g Cotton blue in 20 mL distilled water and then add 20 mL lactic acid and 20 mL concentrated phenol. Mix after adding 40 mL glycerol.

10% KOH with LPCB Procedure

1. Place a thin specimen on a slide and add 1 or 2 drops KOH. Place a cover slip over the preparation and wait for 5 to 30 min at room temperature or gently heat for a few seconds.
2. Add 1 drop LPCB and add a cover slip; examine under the microscope.

India Ink or Nigrosin Wet Mount[6, 7, 12]

India ink or Nigrosin will outline the capsule of *Cryptococcus neoformans*. The stain is far less effective than the latex agglutination procedure when examining cerebral spinal fluid. In addition, human red or white blood cells can mimic the appearance of yeast cells. A drop of KOH can be added because human cells are disrupted and yeast cells remain intact. A positive India ink preparation is not always definitive for *Cryptococcus neoformans* because addition yeasts such as *Rhodotorula* may be encapsulated.

Capsule Detection with India Ink

1. Mix 1 drop of the specimen with 1 drop India ink or 1 drop Nigrosin on a slide. Place a cover slip over the preparation and let it rest for 10 min before examining under the microscope.

CALCOFLUOR WHITE STAIN (CFW)[6, 7]

Background for the Calcofluor White Stain

The CFW stain can be used for the direct examination of most fungal specimens. This stain binds to cellulose and chitin in the fungal cell wall. CFW is used in conjunction with KOH to enhance the visualization of the fungal cell wall. Positive results are indicated by a bright green to blue fluorescence using a fluorescent microscope. A non-specific fluorescence from human cellular materials sometimes occurs. A bright yellow-green fluorescence is observed when collagen or elastin is present. The CFW staining technique usually provides better contrast than Lactophenol aniline blue stains. The CFW stain can be used for the rapid screening of clinical specimens for fungal elements (www.microbiology.mtsinai.on.ca/manual/myc/index.shtml). Furthermore, a CFW preparation subsequently can be stained with Gomori methenamine silver and periodic acid-Schiff (PAS) stain without interference.

General CFW Staining Procedure

1. Mix equal volumes of 0.1% CFW and 15% KOH before staining.
2. Place the specimen on the slide and add a few drops of the mixed CFW-KOH solution. Place a cover slip over the material.
3. The slide can be warmed for a few minutes if the material does not clear at 25°C.
4. Observe the specimen under a fluorescence microscope that uses broadband or barrier filters between 300 and 412 nm. The maximum absorbance of CFW is at 347 nm.
5. A negative control consists of equal mixtures of CFW-KOH. A positive control consists of a recent *Candida albicans* culture.

CFW Staining Reagents

1. 0.1% CFW (wt/vol) solution. Store the solution in the dark. Gentle heating and/or filtration is necessary to eliminate precipitate formation. CFW is available as a cellufluor solution from Polysciences (Washington, PA) or as Fluorescent brightener from Sigma (St. Louis, MO).
2. 15% KOH. Dissolve 15 g KOH in 80 mL distilled water and add 20 mL glycerol. Store both reagents at 25°C.

CFW Stain for *Acanthamoeba* spp., *Pneumocystis jiroveci*, and *Microsporidium* spp.

This procedure is used only as a quick screening method and not for species identification. The CFW stain is not specific because many objects other than fungi and parasites will fluoresce. Fresh or preserved stool, urine, and other types of specimens can be used in the following procedure (www.dpd.cdc.gov/DPDx/HTML/DiagnosticProcedures.htm).

CFW Staining Procedure for *Acanthamoeba*, etc.

1. Use approximately 10 μL of preserved or fresh fecal, or urine, specimen to prepare a thin smear.
2. Fix the smear in 100% methanol for 30 sec.

3. Stain with CFW solution for 1 min. A 0.01% CFW solution in 0.1 M Tris-buffered saline with a final pH of 7.2 constitutes the staining reagent.
4. Rinse with distilled water, air dry the slide, and mount.
5. Examine under a UV fluorescence microscope using a wavelength at or below 400 nm.
6. The spores of *Microsporidium* will also exhibit a blue-white fluorescent color.

HISTOPATHOLOGIC STAIN FOR FUNGI: PERIODIC-ACID SCHIFF (PAS) STAIN[1, 2, 6, 7]

Background for the PAS Stain

Some laboratories prefer to use phase contrast microscopy in place of the PAS stain because similar results are often obtained. In the PAS stain, carbohydrates in the cell wall of the fungi and carbohydrates in human cells are oxidized by periodic acid to form aldehydes, which then react with the fuchsin-sulfurous acid to form the magenta color. Identification of fungal elements in tissue can be enhanced if a counterstain such as Light green is used. The background stains green and the yeast cells or hyphae will accumulate the magenta color. The online *Surgical Pathology-Histology Staining Manual* discusses several PAS methods (www.library.med.utah.edu/WebPath/HISTHTML/MANUALS/MANUALS.html).

PAS Staining Procedure

1. Deparaffinize and hydrate to distilled water.
2. Treat slides with 0.5% periodic acid for 5 min, followed by a rinse in distilled water.
3. Stain in Schiff's reagent for 15 to 30 min at room temperature. An alternate method is to microwave on high power for 45 to 60 sec. The solution should be a deep magenta color. **Schiff's reagent requires extreme caution because it is a known carcinogen.**
4. Wash in running water (about 5 min) to develop the pink color.
5. Counterstain in hematoxylin for 3 to 6 min. Light green can be substituted when fungi are suspected. Go to Step 8 if using Light green.
6. If using hematoxylin, wash in tap water, followed by a rinse in distilled water.
7. Place in alcohol to dehydrate and apply a cover slip and mount.
8. If using Light green, wash in tap water followed by 0.3% ammonia water.
9. Place in 95% alcohol and later in 100% alcohol (2X), followed by clear xylene (2X); then mount.

Reagents

1. **Periodic acid:** Dissolve 0.5 g periodic acid in 100 mL distilled water. The reagent is stable for 1 year.
2. **Harris' Hematoxylin:** Dissolve 2.5 g Hematoxylin in 25 mL of 100% ethanol. Dissolve 50 g potassium or ammonium alum in 0.5 L heated distilled water. Mix together the two solutions without heat. After mixing, the two solutions are boiled very rapidly with stirring (approximate time to reach a boil is 1 min).After removing from the heat, add 1.25 g mercury oxide (red). Simmer until the stain becomes dark purple in color. Immediately plunge the container into a vessel containing cold water. Add 2 to 4 mL glacial acetic acid to increase the staining efficiency for the nucleus. The stain must be filtered before use.
3. **Light green reagent:** Dissolve 0.2 g Light green in 100 mL distilled water. Dilute 1:5 with distilled water before use. The optimum concentration of Light green often requires several attempts to determine the correct concentration.

GROCOTT-GOMORI METHENAMINE SILVER (GMS) STAIN[12]

Background for the GMS Stain

Silver stains are useful in detecting fungal elements in tissues. The fungal cell wall contains mucopolysaccharides that are oxidized by GMS to release aldehyde groups, which later react with silver nitrate. Silver nitrate is converted to metallic silver, which becomes visible in the tissue. *Pneumocystis jiroveci* can be detected by several staining techniques, including Toluidine blue O, Calcofluor white, GMS, and Giemsa. Multiple studies have compared the ease and accuracy in detecting *Pneumocystis* with these stains. A single stain has not emerged as being consistently superior to the others. The procedure described is for the microwave method and employs 2% chromic acid. The conventional method employs 5% chromic acid. The chromic acid solution must be changed if it turns brown. Additional online information is available at www.histonet.org and www.library.med.utah.edu/WebPath/HISTHTML/MANUALS/MANUALS.html.

GMS Staining Procedure

1. Deparaffinize and hydrate to distilled water.
2. Oxidize with chromic acid (2%) in a microwave set at high power for 40 to 50 sec and wait an additional 5 min.
3. Wash in tap water for a few seconds and then wash in distilled water 3X.
4. Rinse the slide in sodium metabisulfate (1%) at room temperature for 1 min to remove the residual chromic acid.
5. Place the working methenamine silver solution in the microwave (high power) for 60 to 80 sec. Agitate the slide in the hot solution. The fungi should stain a light brown color at this stage.
6. Rinse the slide in distilled water 2X.
7. Tone in gold chloride (0.5%) for 1 min or until gray.
8. Rinse in distilled water 2X.
9. Remove the unreduced silver with sodium thiosulfate (2%) for 2 to 5 min.
10. Rinse in tap water, followed by distilled water.
11. Counterstain with diluted Light green (1:5) for 1 min. The optimal dilution of Light green can vary.
12. Rinse in distilled water.
13. Dehydrate, clear, and mount.

Reagents

1. **Chromic acid (2%):** Dissolve 2.0 g chromium trioxide in 100 mL distilled water.
2. **Sodium metabisulfate (1%):** Dissolve 1.0 g sodium metabisulfate in 100 mL distilled water.
3. **Borax (5%):** Dissolve 5.0 g sodium borate in 100 mL distilled water.
4. **Stock solution of methenamine silver:** Add 5 ml silver nitrate (5%) to 100 mL of methenamine (3%). Store the reagent in the refrigerator in a brown bottle.
5. **Working solution of methenamine silver:** Add 25 mL distilled water and 2.5 mL borax (5%) to 25 mL methenamine silver stock solution.
6. **Gold chloride (0.5%):** Dissolve 0.5 g gold chloride in 100 mL distilled water.
7. **Light green (0.2%):** Dissolve 0.2 g Light green in 100 mL distilled water and add 0.2 mL glacial acetic acid.

TRICHROME STAIN[3–5, 8, 11, 12]

Background for the Trichrome Stain

The Wheatley Trichrome stain is the last part of a complete fecal examination and it usually follows a direct iodine wet mount and or parasite concentration technique. The Wheatley procedure is a modification of the Gomori tissue stain. The procedure is the definitive method for the identification of protozoan parasites. Small protozoa that have been missed on wet mounts or concentration methods are often detected with the Trichrome stain. The Trichrome stain detects both trophozoites and cysts, and documents a permanent record for each observed parasite. The background material stains green; the cytoplasm of the cysts and trophozoites stain blue-green; and chromotoidal bodies (RNA), chromatin material, bacteria, and red blood cells stain red or purplish-red. Larvae or ova of some metazoans also stain red. Thin-shelled ova may collapse when mounting fluid is used. The fecal specimen may be fresh, fixed in polyvinyl alcohol (PVA), Schaudinn, or sodium acetate-acetic acid-formalin (SAF). The CDC has used various components of the Trichrome stain to detect *Microsporidium* spores from the fecal component. The complete Chromotrope staining procedures are available at the CDC's website (www.dpd.cdc.gov/DPDx/HTML/DiagnosticProcedures.htm).

Trichrome Staining Procedure

1. Place the PVA smears in 70% ethanol/iodine solution for 10 min. This step can be eliminated if the fixative does not contain mercuric chloride.
2. Transfer slide to 70% ethanol for 5 min. Repeat with a second 70% ethanol treatment for 3 to 5 min.
3. Stain in Trichrome for 10 min.
4. Rinse quickly in 90% ethanol, acidified with 1% acetic acid, for 2 or 3 sec.
5. Rinse quickly using multiple dips in 100% ethanol. Repeat the dipping in a fresh 100% ethanol. Make sure that each slide is destained separately using fresh alcohol. The most common mistake is excessive destaining.
6. Transfer slides into 100% ethanol for 3 to 5 min. Repeat a second time.
7. Transfer slide to xylene for 5 min. Repeat a second time.
8. Mount with Permount.
9. A 10X objective can be used to locate a good area of the smear. Then examine the smear under oil immersion and analyze 200 to 300 fields.

Reagents

1. **70% Ethanol with iodine:** Add sufficient iodine to 70% ethanol to turn the alcohol a dark tea color (reddish-brown). If the ethanol/iodine reagent is too weak, the mercuric chloride will not be extracted from the specimen. The end result will be the formation of a crystalline residue that will hamper the examination of the specimen.
2. **D'Antoni's iodine:** Dissolve 1 g potassium iodide and 1.5 g powdered iodine crystals in 100 mL distilled water.
3. **Trichrome stain:** Add 1 mL glacial acetic acid to the dry components of the stain, which are 0.60 g Chromotrope 2R, 0.15 g Light green SF, 0.15 g Fast green, and 0.70 g phosphotungstic acid. The color should be purple. Dissolve all the reagents in 100 mL distilled water.

IRON HEMATOXYLIN STAIN[3–5, 8, 9, 12]

Background for the Iron Hematoxylin Stain

The Iron Hematoxylin stain is used to identify the trophozoites and cysts of the Protozoa group. It is less commonly used than the Trichrome stain. The cysts and the trophozoites stain a blue-gray to black, and the background material stains blue or pale gray. Helminthes eggs and larvae are usually difficult to identify because of excessive stain retention. Yeasts and human cells (red blood cells, neutrophils, and macrophages) are also detected by the stain. The stain can be used with fixatives that include PVA, SAF, or Schaudinn's. Good fixation is the key in obtaining a well-stained fecal preparation. The simplest variation of the Iron Hematoxylin stain is the method of Tompkins and Miller. A modified Iron Hematoxylin stain that incorporates a Carbol-fuchsin step will detect some acid-fast parasites (including *Cryptosporidium parum* and *Isospora belli*).

Iron Hematoxylin Stain Procedure

1. Prepare a thin layer fecal smear and place it in fixative. If SAF is used, proceed to Step 4. Mayer's albumin can be used to ensure that the specimen will adhere to the slide.
2. Dehydrate the slide in ethanol (70%) for 5 min.
3. Transfer slide in iodine ethanol (70%) for 2 to 5 min. The solution should have a strong tea color.
4. Transfer the slide to 50% ethanol for 5 min.
5. Wash the slide in a constant stream of tap water for 3 min.
6. Transfer the slide into 4% ferric ammonium sulfate mordant for 5 min.
7. Wash the slide in a constant stream of tap water for 1 min.
8. Stain with Hematoxylin (0.5%) for 3 to 5 min.
9. Wash in tap water for about 1 min.
10. Destain the slide with 2% phosphotungstic acid for 2 min.
11. Wash the slide in a constant stream of tap water.
12. Transfer the slide to 70% ethanol that contains several drops of lithium carbonate (saturated).
13. Transfer the slide to 95% ethanol for 5 min.
14. Transfer the slide to 100% ethanol for 5 min. Repeat a second time.
15. Transfer slide to xylene for 5 min. Repeat a second time.
16. Mount with Permount.

Reagents

1. **70% Ethanol/iodine:** Add sufficient iodine to 70% ethanol to turn the alcohol a dark tea color (reddish-brown).
2. **D'Antoni's iodine:** Dissolve 1 g potassium iodide and 1.5 g powdered iodine crystals in 100 mL distilled water.
3. **Iron Hematoxylin stain:** Dissolve 10 g Hematoxylin in 1000 mL of 100% ethanol. Store the reagent at room temperature.
4. **Mordant:** Dissolve 10 g ferrous ammonium sulfate and 10 g ferric ammonium sulfate in 990 mL distilled water. Add 10 mL concentrated hydrochloric acid.
5. **Working Hematoxylin stain:** Mix (1:1) the mordant and the Hematoxylin stain.
6. **Saturated lithium carbonate:** Dissolve 1 g lithium carbonate in 100 mL distilled water.

PREPARATION OF BLOOD SMEARS FOR PARASITE EXAMINATION[3–5, 8, 12]

Trypanosoma, Babesia, Plasmodium, Leishmania, as well as most microfilariae, are detected from blood smears. Identification of these parasites is based on the examination of permanent blood films. The blood samples from malaria and babesia patients are best collected toward the end of a paroxysmal episode. Blood samples can be collected randomly from patients with trypanosomiasis. Blood samples should be collected after 10 p.m. on those microfilariae that exhibit nocturnal periodicity (*Wuchereria* and *Brugia).* Blood samples should be collected between 11 a.m. and 1 p.m. if *Loa Loa* (diurnal periodicity) is suspected. Thick blood smears are typically used as a screening tool, and thin blood smears are used to observe detailed parasite morphology. Blood parasites are usually identified from the thin films. The Giemsa, Wright, and Wright-Giemsa combination can be used to stain the blood smears. The Giemsa stains the cytoplasm of *Plasmodium* spp. blue, whereas the nuclear material stains red to purple, and Schuffners's dots stain red. The cytoplasm of *Trypanosomes, Leishmaniae*, and *Babesia* will also stain blue and the nucleus stains red to purple. The sheath of microfilariae often fails to stain but the nuclei will stain blue/purple.

Giemsa Staining Procedure for Thin Films

1. Fix the blood films in 100% methanol for 1 min.
2. Air dry the slides.
3. Place the slides into the working Giemsa solution. The working solution is 1 part Giemsa stock (commercial liquid stain) and 10 to 50 parts phosphate (pH 7.0 to 7.2) buffer. Experimentation with several different staining dilutions/times is often required to obtain optimum results.
4. Stain for 10 to 60 min.
5. Briefly wash under water or in phosphate buffer.
6. Wipe the stain off the bottom of the slide and air dry.

Reagents

1. **Stock Giemsa commercial liquid stain:** Dilute 1:10 with buffer for thin smears.
2. **Phosphate buffer, pH 7.0:** Dissolve 9.3 g Na_2HPO_4 in 1 L distilled water (Solution 1); and dissolve 9.2 g $NaH_2PO_4{\cdot}H_2O$ in 1 L distilled water (Solution 2). Add 900 mL distilled water to a beaker and 61.1 mL Solution 1 and 38.9 mL Solution 2. This reagent is used to dilute the Giemsa stain.
3. **Triton phosphate buffered wash:** Add 0.1 mL Triton X-100 to 1000 mL of the pH 7.0 phosphate buffer. This is the wash; tap water can also be used.

REFERENCES

1. Chapin, K.C., Principles of stains and media, in *Manual of Clinical Microbiology, 9th ed.*, Murray, P.R., Ed., ASM Press, Washington, DC, 2007, chap. 14.
2. Clarridge, J.E. and Mullins, J.M., Microscopy and staining, in *Clinical and Pathogenic Microbiology, 2nd ed.*, Howard, B.J., Ed., Mosby, St. Louis, MO, 1994, chap 6.
3. Garcia, L.S., Macroscopic and microscopic examination of fecal specimens, in *Diagnostic Medical Parasitology, 5th ed.*, ASM Press, Washington, DC, 2006, chap. 27.
4. Forbes, B.A., Saham, D.F., and Weissfeld, A.E., Laboratory methods for diagnosis of parasitic infections, in *Bailey & Scott's Diagnostic Microbiology, 12th ed.,* Mosby Elsevier, St. Louis, MO, 2007, chap. 49.
5. Heelan, J.S. and Ingersoll, F.W., Processing specimens for recovery of parasites, in *Essentials of Human Parasitology*, Delmar Thompson Learning, Albany, NY, 2002, chap. 2.
6. Kern, M.E. and Blevins, S.B., Laboratory procedures for fungal culture and isolation, in *Medical Mycology, 2nd ed.*, F.A. Davis, Philadelphia, PA, 1997, chap. 2.

7. Larone, D.H., Laboratory procedures; staining methods; media, in *Medically Important Fungi; A Guide to Identification, 4th ed.*, ASM Press, Washington, DC, 2002, part 3.
8. Leventhal, R. and Cheadle, R., Clinical laboratory procedures, in *Medical Parasitology, 5th ed.*, F.A. Davis, Philadelphia, PA, 2002, chap. 7.
9. Tompkins, V.N. and Miller, J.K., Staining intestinal protozoa with iron-hematoxylin-phophotungstic acid, *Am. J. Clin. Pathol.*, 17, 755, 1947.
10. Traunt, J.P., Brett, W.A., and Thomas, W., Jr., Fluorescence microscopy of tubercle bacilli stained with Auramine and Rhodamine, *Henry Ford Hosp. Med. Bull.*, 10, 287, 1962.
11. Wheatley, W.B., A rapid staining procedure for intestinal amoebae and flagellates, *Am. J. Clin. Pathol.*, 2, 990, 1951.
12. Woods, G.L. and Walker, H.W., Detection of infectious agents by use of cytological and histological stains; *Clin. Microbiolog. Rev.*, 9, 382, 1996.

5 Identification of Gram-Positive Organisms

Peter M. Colaninno

CONTENTS

INTRODUCTION

With human infections caused by Gram-positive organisms on the rise, it is imperative that these organisms get identified as expeditiously as possible. However, because of the sheer multitude of Gram-positive organisms that are implicated in disease, identification of these organisms can be challenging. What follows is a synopsis of identification methods for Gram-positive organisms that can hopefully make the process easier.

GRAM-POSITIVE COCCI

STAPHYLOCOCCUS

Colony Morphology: Colonies of *Staphylococcus* on sheep blood agar present themselves as smooth, yellow, white or off-white colonies somewhere in the area of 1 to 2 mm in diameter.[1] Colonies may exhibit β-hemolysis and may show varying degrees of growth. Sometimes, the β-hemolysis is not evident after 24 hr of incubation and requires further incubation. Due to the varying degrees of colony size and color that may mimic other Gram-positive organisms such as streptococci and micrococci, identification of *Staphylococcus* can sometimes be problematic. CHROMagar (CHROMagar Company) is a type of media that can alleviate this problem. This differential media contains chromogenic substrates that yield a certain colony color.[2]

Quick Tests: Perhaps the most common quick test employed to help identify colonies of *Staphylococcus* is the catalase test. This simple test can differentiate off-white or gray colonies of *Staphylococcus* from *Streptococcus* and is an invaluable tool. The modified oxidase test is another quick test that can differentiate *Staphylococcus* from *Micrococcus*, as is the lysostaphin test (Remel). Differentiation among the staphylococci can be achieved by the coagulase test, which tests for both bound and free coagulase. Alternatively, there are a multitude of latex agglutination tests available. Slidex Staph (bioMerieux), Bacti Staph (Remel), Staphytect (Oxoid), Staphylase (Oxoid), Staph Latex Slide Test (Arlington Scientific), Staphtex (Hardy), and Set-RPLA Latex Staph (Denka-Seiken) are all latex agglutination tests for the identification of *Staphylococcus aureus*. In addition, there are a few kits that utilize passive hemagglutination for the identification of *Staphylococcus aureus*, such as Staphyslide (bioMerieux), HemaStaph (Remel), and StaphyloSlide (Becton-Dickinson).[3, 4]

Conventional Methods: *Staphylococcus aureus* can further be identified by its ability to produce DNase and ferment mannitol. Both DNase agar and Mannitol Salt agar are readily available. When needed to speciate coagulase-negative staphylococci (CNS), there are a number of tests that can be employed. Carbohydrate utilization such as sucrose, xylose, trehalose, fructose, maltose, mannose, and lactose, as well as such tests as urease, nitrate reduction, and phosphatase, will all aid in the identification of CNS. *Staphylococcus saphrophyticus* is a frequent cause of urinary tract infections and can be identified by its resistance to novobiocin. Bacitracin, as well as acid production from glucose, are tests that can be employed to differentiate *Staphylococcus* from *Micrococcus*.[5]

Identification Strips: There are a number of manufacturers that have developed identification strips containing many of the aforementioned biochemicals. API ID 32 Staph (bioMerieux) and API STAPH (bioMerieux) utilize 10 and 19 tests, respectively, on their strips and are reliable in identifying most strains of CNS (coagulase-negative staphylococci).[6]

Automated Methods: The Vitek GPI card (bioMerieux), the MicroScan Rapid Pos Combo Panel and Pos ID 2 Panel (Dade/MicroScan), and the Phoenix Automated Microbiology System Panel (Becton Dickinson) are a few of the more common automated identification panels for staphylococcus as well as other Gram-positive organisms.[3, 7, 8]

Molecular Methods: Molecular methods have gained widespread popularity in the field of medical microbiology. GenProbe introduced its AccuProbe for *Staphylococcus aureus* culture confirmation a number of years ago, and Roche Molecular Systems has developed a RT-PCR test for the detection of the *mecA* gene for use on its LightCycler.[9, 10] Table 5.1 summarizes all the available tests for *Staphylococcus* species.

Micrococcus

Colony Morphology: Colonies of micrococcus present as dull, white colonies that may produce a tan, pink, or orange color. They are generally 1 to 2 mm in diameter and can stick to the surface of the agar plate.[1] A Gram stain will assist in the preliminary identification, as micrococci appear as larger Gram-positive cocci arranged in tetrads.

Quick Tests: The catalase test and the modified oxidase test are invaluable tools in helping to differentiate micrococci from staphylococci species.

Conventional Tests: Glucose fermentation; lysostaphin; furazolidone, and bacitracin can all be used to separate micrococci from staphylococci.[11]

Identification Strips: The API ID 32 Staph identification strip (bioMerieux) can be utilized to identify *Micrococcus* species.

TABLE 5.1 Identification Tests for *Staphylococcus*
Catalase test
Coagulase test
Lysostaphin test
Latex agglutination tests
Passive hemagglutination tests
DNase
Mannitol salt
Carbohydrate utilization
Novobiocin disk
Bacitracin disk
API ID 32 Staph
API STAPH
Vitek GPI
MicroScan Pos ID 2
BD Crystal Gram-Positive ID
AccuProbe
LightCycler

TABLE 5.2 Identification Tests for *Micrococcus*
Catalase test
Modified oxidase test
Lysostaphin
Furazolidone
Glucose fermentation
Bacitracin disk
API ID 32 Staph
API STAPH

Automated Methods: MicroScan has 2-hr Rapid Gram-positive panels that can identify most *Micrococcus* species.

Table 5.2 summarizes all of the available tests for *Micrococcus.*

Streptococcus, including *Enterococcus*

Colony Morphology: *Streptococcus* colonies are typically smaller on sheep blood agar than *Staphylococcus* colonies and measure about 1 mm in diameter. Agar such as Columbia CNA and PEA can enhance the growth of streptococci, resulting in larger colony size. Perhaps the one characteristic that can aid in the separation of *Streptococcus* species is hemolysis. β-Hemolysis, α-hemolysis, α-prime hemolysis, and γ-hemolysis are all produced by members of the genus.

Quick Tests: The catalase test is very useful in differentiating *Staphylococcus* colonies from *Streptococcus* colonies, especially γ-hemolytic streptococci. After the colony has been tentatively identified as strep, there are a number of latex agglutination and coagglutination tests that can provide definitive identification. Kits include Phadebact (Boule'), Streptex (Remel), Meritec-Strep (Utech), Patho-Dx (Remel), StrepPro (Hardy Diagnostics), StrepQuick (Hardy Diagnostics), Streptocard (BD), and Slidex Strep (bioMerieux).[3] Two useful tests to help aid in the identification of Enterococci are the pyrrolidonyl arylamidase (PYR) and leucine aminopeptidase (LAP) tests. These are available as disks (Remel) and can be inoculated directly with colonies for rapid identification. The DrySpot Pneumo (Oxoid) and the PneumoSlide (BD) are rapid tests for the identification of *Streptococcus pneumoniae*.

Conventional Tests: Because there are a multitude of *Streptococcus* species implicated in human disease, there also exist a number of conventional tests to aid in the identification of streptococci. Conventional tests that aid in the identification of β-hemolytic strep include the bacitracin disk, CAMP test, susceptibility to SXT, hippurate hydrolysis, Vogues Proskauer (VP), and PYR. α-Hemolytic strep can be identified using the optochin disk, bile solubility, esculin hydrolysis, 6.5% NaCl, arginine hydrolysis, and acid from mannitol. Enterococci can be identified utilizing bile esculin, 6.5% NaCl, SXT susceptibility, motility, pigment production, and VP.[11]

Identification Strips: The API Rapid ID 32 Strep and API 20 Strep (bioMerieux) and RapID STR strips (Remel) provide a number of biochemicals for the identification of streptococci, enterococci, and nutritionally variant strep.[12] The Remel BactiCard Strep is a card that has PYR, LAP, and ESC reactions encompassed on it.

Automated Methods: The MicroScan Pos ID panel and the Vitek GPI card can be utilized for identification of strep.

Molecular Methods: The Accu-Probe Enterococcus System is available. There are identification methods available on the LightCycler for Group A and Group B strep. In addition, the identification of the *van A* gene is available.[13]

Table 5.3 summarizes all of the available tests for *Streptococcus* and *Enterococcus* species.

Aerococcus

Colony Morphology: Aerococci appear as α-hemolytic colonies that closely resemble enterococci and the viridans streptococci.

Quick Tests: Two tests that aid in the identification of *Aerococcus* species are PYR and LAP.

Conventional Tests: Laboratory tests utilized for the identification and speciation of *Aerococcus* include bile esculin, 6.5% NaCl, hippurate hydrolysis, and acid production from sorbitol, lactose, maltose, and trehalose.[14]

ID Strips: The API Rapid ID 32 Strep strip contains biochemicals that can identify *Aerococcus*.[15]

Automated Methods: The Vitek GPI card can be used to identify *Aerococcus*.[11]

Leuconostoc

Colony Morphology: *Leuconostoc* are small, α-hemolytic colonies.

Quick Tests: The PYR and LAP disks can be utilized.

Conventional Tests: Arginine dihydrolase, gas from glucose, resistance to vancomycin, lack of growth at 45°C, and esculin hydrolysis are all useful tests in the identification of *Leuconostoc*.[11, 16, 17]

ID Strips: The API Rapid ID 32 Strep.

Automated Methods: Vitek GPI card.

Pediococcus

Colony Morphology: Pediococci are small, α-hemolytic colonies that resemble viridans strep.

Quick Tests: The PYR and LAP discs can be used.

TABLE 5.3
Identification Tests for *Streptococcus* and *Enterococcus*

Hemolysis
Catalase test
Latex agglutination tests
Coagglutination tests
PYR test
LAP test
Bacitracin disk
CAMP test
SXT disk
Hippurate hydrolysis
Voges Proskauer
Optochin disk
Bile solubility
Esculin hydrolysis
Arginine hydrolysis
Bile esculin
6.5% NaCl
Motility
Pigment production
Carbohydrate utilization
API 20 STREP
API Rapid ID 32 Strep
MicroScan Pos ID
Rapid ID STR
LightCycler

Conventional Tests: Conventional tests include acid production from glucose, arabinose, maltose, and xylose; lack of gas production in glucose, growth at 45°C, and resistance to vancomycin.[11, 16]

ID Strips: None available.

Automated Methods: Vitck GPI card.

Table 5.4 summarizes all of the available tests for *Aerococcus*, *Leuconostoc*, and *Pediococcus*.

Lactococcus

Colony Morphology: Lactococci are small, α-hemolytic colonies that resemble enterococci and viridans strep.

Quick Tests: LAP and PYR disks.

Conventional Tests: Bile esculin; 6.5% NaCl; arginine dihydrolase; hippurate hydrolysis and acid from glucose, maltose, lactose, sucrose, mannitol, raffinose, and trehalose are all conventional tests that can be used.[11, 16]

ID Strips: The API Rapid ID 32 Strep strip contains biochemicals for the identification of *Lactococcus* species.

Automated Methods: Vitek GPI card.

Gemella

Colony Morphology: Colonies of *Gemella* closely resemble colonies of viridans streptococci.

Quick Tests: The PYR disk can be used to aid in identification.

TABLE 5.4
Identification Tests for *Aerococcus, Leuconostoc* and *Pediococcus*

PYR test
LAP test
Bile esculin
6.5% NaCl
Hippurate hydrolysis
Arginine hydrolysis
Gas from glucose
Resistance to vancomycin
Growth/lack of growth at 45° C
Carbohydrate utilization
API 20STREP

Conventional Tests: Conventional tests include esculin hydrolysis, reduction of nitrite, arginine dihydrolase, and acid production from glucose, mannitol, and maltose.[11, 16]

ID Strips: The API Rapid ID 32 Strep, RapID STR, and API 20A (bioMerieux) can be used to identify *Gemella*.

Automated Methods: The Vitek GPI card.

Table 5.5 summarizes all the available tests for *Lactococcus* and *Gemella*.

MISCELLANEOUS GRAM-POSITIVE COCCI

There are a few organisms whose implication in human disease is questionable. *Alloiococcus* has been isolated from sputum, blood cultures, and tympanocentesis fluid; *Rothia* (*Stomatococcus*) has been isolated from blood cultures; *Globicatella* has been recovered from blood cultures; *Helcococcus* has been isolated from wounds; and *Vagococcus* is primarily zoonotic. Biochemical reactions for these organisms are shown in Table 5.6.[16, 18]

TABLE 5.5
Identification Tests for *Lactococcus* and *Gemella*

LAP test
PYR test
Bile esculin
6.5% NaCl
Arginine hydrolysis
Hippurate hydrolysis
Nitrate reduction
Carbohydrate utilization
API 20STREP
API RapidID 32Strep
API 20A
Vitek GPI

TABLE 5.6 Identification Tests for Miscellaneous Gram-Positive Cocci
α-Hemolysis
LAP test
PYR test
Bile Esculin
6.5% NaCl
Motility
Carbohydrate utilization
Arginine hydrolysis
Growth/lack of growth at 45°C

GRAM-POSITIVE BACILLI

CORYNEBACTERIUM

Colony Morphology: Colonies of *Corynebacterium* on 5% sheep blood agar appear as tiny, gray to white colonies that are generally non-hemolytic. Special media such as cysteine tellurite blood agar, and Tinsdale agar should be utilized whenever the *Corynebacterium diphtheriae* group is to be isolated.

Quick Tests: The Gram stain can be an invaluable tool for the identification of *Corynebacterium* as they can be differentiated from *Bacillus* by their coryneform shape and size, and lack of spores. The catalase test is useful in differentiating *Corynebacterium* from *Erysipelothrix* and *Lactobacillus*.

Conventional Tests: Useful tests for the identification of *Corynebacterium* include motility; lack of H_2S production in a TSI slant; esculin; nitrate reduction; urease; as well as the carbohydrates glucose, mannitol, maltose, sucrose, xylose, and salicin.[19, 20]

ID Strips: API Coryne Strip (bioMerieux)and RapID CB Plus (Remel) are identification strips that can be used for the identification of *Corynebacterium*.[21, 22]

Automated Methods: Vitek GPI Card.

Tablc 5.7 summarizes all of the available tests for *Corynebacterium*.

TABLE 5.7 Identification Tests for *Corynebacterium*
Catalase test
Motility
Esculin hydrolysis
Nitrate reduction
Lack of H_2S production
Urease
Carbohydrate utilization
API Coryne
RapID CB Plus
Vitek GPI

Bacillus

Colony Morphology: *Bacillus* species have many variations of colony morphology. For example, *Bacillus anthracis* appear as non-hemolytic medium to large gray, flat colonies with irregular swirls, while *Bacillus cereus* appear as large, spreading β-hemolytic colonies, and *Bacillus subtilis* appear as large, flat colonies that may be pigmented or β-hemolytic.

Quick Tests: The Gram stain and Spore stain are useful tools in identification.

Conventional Tests: Motility; gas from glucose; starch hydrolysis, growth in nutrient broth with 6.5% NaCl; indole; nitrate reduction;VP; citrate, gelatin hydrolysis; esculin hydrolysis; growth at 42°C; and various carbohydrate fermentations such as glucose, maltose, mannitol, xylose, and salicin. Susceptibility to penicillin can also help in the presumptive identification of *Bacillus anthracis*.[23]

ID Strips: Microgen Bacillus ID (Microgen Bioproducts) and the API CHB (bioMerieux).

Automated Methods: Vitek B card.

Table 5.8 summarizes all the available tests for *Bacillus.*

Listeria

Colony Morphology: Colonies appear on sheep blood agar as small, translucent, and gray with a small zone of β-hemolysis.

Quick Tests: Catalase and motility on a wet mount are useful.

Conventional Tests: Growth at 4°C; esculin hydrolysis; motility; fermentation of glucose, trehalose, and salicin; lack of H_2S production; CAMP test.[24]

ID Strips: API Listeria strip, API Rapid ID 32 Strep, API Coryne strip, and Microgen Listeria ID (Microgen Bioproducts).[25]

Automated Methods: Vitek GPI Card.

Molecular Methods: AccuProbe Listeria Identification Kit (GenProbe).

Table 5.9 summarizes all of the available tests for *Listeria.*

TABLE 5.8
Identification Tests for Bacillus

Identification Tests for Bacillus
Spore stain
Motility
Gas from glucose
Starch hydrolysis
6.5% NaCl
Indole
Nitrate reduction
Voges Proskauer
Citrate
Gelatin hydrolysis
Esculin hydrolysis
Growth at 42°C
Carbohydrate fermentation
Susceptibility to penicillin
API CHB
Microgen Bacillus
Vitek B

TABLE 5.9
Identification Tests for Listeria

β–Hemolysis
Catalase test
Motility (wet mount)
Growth at 4°C
Esculin hydrolysis
Motility media
Lack of H_2S production
CAMP test
Carbohydrate fermentation
API Listeria
API RapidID32 Strep
API Coryne
Microgen Listeria
Vitek GPI
AccuProbe ListerialD

Lactobacillus

Colony Morphology: *Lactobacillus* can have a variety of colony morphologies, ranging from tiny, pinpoint, α-hemolytic colonies resembling viridans strep, to small, rough, gray colonies.

Quick Tests: Gram stain and Catalases tests are useful in aiding in the identification.

Conventional Tests: Fermentation of glucose, maltose, and sucrose as well as urease and nitrate reduction tests are helpful.[3]

ID Strips: API 20A and API CHL (bioMerieux) can be used to identify some strains of *Lactobacillus.*

Automated Methods: Vitek ANI card.

Table 5.10 summarizes all the available tests for *Lactobacillus.*

Actinomyces, Nocardia, Streptomyces

Colony Morphology: Colonies of *Nocardia* can vary but are commonly dry, chalky white in appearance, and can sometimes be β-hemolytic on sheep blood agar with a late developing pigment. *Streptomyces* can appear as waxy, glabrous, heaped colonies. *Actinomyces* appear as white colonies with or without β-hemolysis.

TABLE 5.10
Identification Tests for *Lactobacillus*

Catalase test
Urease
Nitrate reduction
Carbohydrate fermentation
API CHL
API 20A
Vitek ANI card

Quick Tests: Modified acid-fast stain and catalase tests.

Conventional Tests: Caesin; tyrosine; xanthine; urease; gelatin hydrolysis; starch hydrolysis; nitrate reduction; lactose, xylose, rhamnose and arabinose fermentation. The BBL Nocardia Quad Plate (Becton-Dickinson) combines casein, tyrosine, xanthine and starch on one plate.[26–29]

ID Strips: *Actinomyces*: API Coryne, BBL Crystal, RapID ANA II (Remel), and Rap ID CB Plus.[30, 31]

Automated Methods: Vitek ANI card.

Table 5.11 summarizes all of the available tests for *Actinomyces, Nocardia*, and *Streptomyces*.

Gardnerella

Colony Morphology: Special media such as V agar and HBT agar must be used to isolate *Gardnerella vaginalis*. Colonies appear as tiny β-hemolytic after 48 hr.

Quick Tests: Gram stain; catalase; oxidase.

Conventional Tests: Hippurate hydrolysis; urease; nitrate reduction; a zone of inhibition with trimethoprim, sulfonamide, and metronidazole; fermentation of glucose, maltose, and sucrose.[32]

ID Strips: The API Rapid ID 32 Strep strip can be used to identify *Gardnerella*.

Automated Methods: MicroScan HNID panel (Dade/MicroScan).

Table 5.12 summarizes all of the available tests for *Gardnerella*.

MISCELLANEOUS GRAM-POSITIVE RODS

Arcanobacterium, *Erysipelothrix*, *Kurthia*, *Rhodococcus*, *Gordona*, *Oerskovia*, *Dermatophilus*, *Actinomadura*, *Nocardiopsis*, and *Tsukamurella* have all been implicated in human disease.[26, 33–38] Identification tests for these organisms appear in Table 5.13.

ANAEROBIC GRAM-POSITIVE ORGANISMS

Clostridium, *Bifidobacterium*, *Peptostreptococcus*, *Finegoldia*, *Peptococcus*, *Micromonas*, *Eggerthella*, *Eubacterium*, and *Propionibacterium* are all organisms that have been implicated in human disease.[39–45] Identification tests are listed in Table 5.14.

TABLE 5.11
Identification Tests for *Actinomyces*, *Nocardia*, and *Streptomyces*

Modified acid-fast stain
Casein
Tyrosine
Xanthinc
Urease
Gelatin hydrolysis
Nitrate reduction
Carbohydrate fermentation
API Coryne
BBL Crystal
RapID ANA II
Rap ID CB Plus
Vitek ANI card

TABLE 5.12
Identification Tests for *Gardnerella*

Catalase test

Oxidase test

Hippurate hydrolysis

Urease

Nitrate reduction

Carbohydrate fermentation

Zone of inhibition with trimethoprim, sulfonamide, and metronidazole

API Rapid ID 32 Strep

MicroScan HNID

TABLE 5.13
Identification Tests for Miscellaneous Gram-Positive Rods

Arcanobacterium, Erysipelothrix, Kurthia

Catalase test

Nitrate reduction

H_2S production

Motility

Urease

CAMP test

Carbohydrate fermentation

API Coryne

Vitek GPI card

Rhodococcus, Gordona, Oerskovia, Dermatophilus, Actinomadura, Nocardiopsis

Pigment production

Modified acid-fast stain

Urease

Nitrate reduction

Anaerobic growth

Motility

Aerial mycelium

API Coryne

TABLE 5.14
Identification Tests for Anaerobic Gram-Positive Organisms

Clostridium, Bifidobacterium, Eggerthella, Eubacterium, Propionibacterium
Anaerobic growth
β–Hemolysis
Kanamycin disk (1 mg)
Colistin disk (10 g)
Vancomycin disc (5 g)
Indole
Catalase
Lecithinase
Naegler test
Reverse CAMP test
Nitrate reduction
Urease
Arginine dihydrolase
API 20A
API Rapid ID 32A
API Coryne
API 50CH
API ZYM
BBL Crystal ANA ID
Vitek ANI card
Peptostreptococcus, Peptococcus, Finegoldia, Micromonas
Catalase
Indole
API 20A
API Rapid 32A
BBL Crystal ANA ID
Vitek ANI card

REFERENCES

1. Kloos, W.E. and Bannerman, T.L. Staphylococcus and Micrococcus, in *Manual of Clinical Microbiology, 7th ed.*, Murray, P.R., Baron E.J., Pfaller, M.A., Tenover F.C., and Yolken R.H., Eds. American Society for Microbiology, Washington, DC, 1999, 264.
2. Hedin, G. and Fang, H. Evaluation of two new chromogenic media, CHROMagar MRSA and *S. aureus* ID, for identifying *Staphylococcus aureus* and screening methicillin-resistant *S. aureus*. *J. Clin. Microbiol.*, 43, 4242, 2005.
3. Forbes, B.A., Sahm, D.F., and Weissfeld, A.S. Staphylococcus, Micrococcus and similar organisms, in *Bailey and Scott's Diagnostic Microbiology, Vol. 11.* Mosby, Inc., St. Louis, MO, 2002, p. 285.
4. Personne, P., Bes, M., Lina, G., Vandenesch, F., Brun, Y., and Etienne, J. Comparative performances of six agglutination kits assessed by using typical and atypical strains of *Staphylococcus aureus*. *J. Clin. Microbiol.*, 35, 1138, 1997.

5. Hansen-Gahrn, B., Heltberg, O., Rosdahl, V., and Sogaard, P. Evaluation of a conventional routine method for identification of clinical isolates of coagulase-negative *Staphylococcus* and *Micrococcus* species. Comparison with API-Staph and API Staph-Ident. *Acta Pathol. Microbiol. Immunol. Scand.*, 95, 283, 1987.
6. Layer, F., Ghebremedhin, B., Moder, K.A., Konig, W., and Konig, B. Comparative study using various methods for identification of *Staphylococcus* species in clinical specimens. *J. Clin. Microbiol.*, 44, 2824, 2006.
7. Bannerman, T.L., Kleeman, K.T., and Kloos, W. Evaluation of the Vitek systems Gram-positive identification card for species identification of coagulase-negative Staphylococci. *J. Clin. Microbiol.*, 31, 1322, 1993.
8. Salomon, J., Dunk, T., Yu, C., Pollitt, J., and Reuben, J. Rapid Automated Identification of Gram-Positive and Gram-Negative Bacteria in the Phoenix System. *Abstr. 99th General Meeting ASM*, 1999.
9. Chapin, K. and Musgnug, M. Evaluation of three rapid methods for the direct identification of *Staphylococcus aureus* from positive blood cultures. *J. Clin. Micrbiol.*, 41, 4324, 2003.
10. Shrestha, N.K., Tuohy, M.J., Padmanabhan, R.A., Hall, G.S., and Procop, G.W. Evaluation of the LightCycler Staphylococcus M grade kits on positive blood cultures that contained Gram-positive cocci in clusters. *J. Clin. Microbiol.*, 43, 6144, 2005.
11. Bascomb, S. and Manafi, M. Use of enzyme tests in characterization and identification of aerobic and faculatatively anaerobic Gram-positive cocci. *Clin. Microbiol. Rev.*, 11, 318, 1998.
12. Sader, H.S., Biedenbach, D., and Jones, R.N. Evaluation of Vitek and API 20S for species identification of enterococci. *Diagn. Microbiol. Infect. Dis.*, 22, 315, 1995.
13. Sloan, L.M., Uhl, J.R., Vetter, E.A., Schleck, C.D., Harmsen, W.S., Manahan, J., Thompson, R.L., Rosenblatt, J.E., and Cockerill III, F.R. Comparison of the Roche LightCycler vanA/vanB detection assay and culture for detection of vancomycin-resistant enterococci from perianal swabs. *J. Clin. Microbiol.*, 42, 2636, 2004.
14. Zhang, Q., Kwoh, C., Attorra, S., and Clarridge, III, J.E. *Aerococcus urinae* in urinary tract infections. *J. Clin. Microbiol.*, 38, 1703, 2000.
15. You, M.S. and Facklam, R.R. New test system for identification of aerococcus, enterococcus and streptococcus species. *J. Clin. Microbiol.*, 24, 607, 1986.
16. Facklam, R. and Elliott, J. Identification, classification and clinical relevance of catalase-negative, Gram-positive cocci, excluding the streptococci and enterococci. *Clin. Microbiol. Rev.*, 8, 479, 1995.
17. Bjorkroth, K.J., Vandamme, P., and Korkeala, H.J. Identification and characterization of *Leuconostoc canosum*, associated with production and spoilage of vacuum-packaged, sliced, cooked ham. *Appl. Environ. Microbiol.*, 64, 3313, 1998.
18. Clinical Microbiology Proficiency Testing Synopsis. Blood Culture Isolate: *Rothia mucilaginosa* (*Stomatococcus mucilaginosus*) — Nomenclature Change. M031-4, 2003.
19. von Graevenitz, A. and Funke, G. An identification scheme for rapidly and aerobically growing Gram-positive rods. *Zentralbl. Bakteriol.*, 284, 246, 1996.
20. Fruh, M., von Graevenitz, A., and Funke, G. Use of second-line biochemical and susceptibility tests for the differential identification of coryneform bacteria. *Clin. Microbiol. Infect.*, 4, 332, 1998.
21. Hudspeth, M.K., Gerardo, S.H., Citron, D.M., and Goldstein, E.J. Evaluation of the RapID CB Plus System for identification of *Corynebacterium* species and other Gram-positive rods. *J. Clin. Microbiol.*, 36, 543, 1998.
22. Gavin, S.E., Leonard, R.B., Briselden, A.M., and Coyle, M.B. Evaluation of the Rapid CORYNE Identification System for *Corynebacterium* species and other coryneforms. *J. Clin. Microbiol.*, 30, 1692, 1992.
23. Koneman, E., Allen, S., Janda, W., Schreckenberger, P., and Winn, W. 1997. *Color Atlas and Textbook of Diagnostic Microbiology, 5th ed.* Lippincott Williams & Wilkins, Philadelphia, PA.
24. Kerr, K.G. and Lacey, R.W. Isolation and identification of listeria monocytogenes. *J. Clin. Pathol.*, 44, 624, 1991.
25. Kerr, K.G., Hawkey, P.M., and Lacey, R.W. Evaluation of the API Coryne System for identification of *Listeria* species. *J. Clin. Microbiol.*, 31, 750, 1993.
26. Funke, G., von Graevenitz, A., Clarridge III, J.E., and Bernard, K.A. Clinical microbiology of *Coryneform* bacteria. *Clin. Microbiol Rev.*, 10, 125, 1997.
27. Kiska, D.L., Hicks, K., and Pettit, D.J. Identification of medically relevant *Nocardia* species with a battery of tests. *J. Clin. Microbiol.*, 40, 1346, 2002.

28. Wauters, G., Avesani, V., Charlier, J., Janssens, M., Vaneechoutte, M., and Delmee, M. Distribution of *Nocardia* species in clinical samples and their routine rapid identification in the laboratory. *J. Clin. Microbiol.,* 43, 2624, 2005.
29. Mossad, S., Tomford, J., Stewart, R., Ratliff, N., and Hall, G. Case report of Streptomyces endocarditis of a prosthetic aortic valve. *J. Clin. Microbiol.,* 33, 3335, 1995.
30. Santala, A., Sarkonen, N., Hall, V., Carlson P., Jousimies-Somer, H., and Kononen, E. Evaluation of four commercial test systems for identification of actinomyces and some closely related species. *J. Clin. Microbiol.,* 42, 418, 2004.
31. Morrison, J.R. and G.S. Tillotson, G.S. Identification of *Actinomyces* (*Corynebacterium*) *pyogenes* with the API 20 Strep System. *J. Clin. Microbiol.,* 26, 1865, 1988.
32. Piot, P., Van Dyck, E., Totten, P., and Holmes, K. Identification of *Gardnerella (haemophilus) vaginalis. J. Clin. Microbiol.,* 15, 19, 1982.
33. Dobinsky, S., Noesselt, T., Rucker, A., Maerker, J., and Mack, D. Three cases of *Arcanobacterium haemolyticum* associated with abscess formation and cellulitis. *Eur. J. Clin. Microbiol Inf. Dis.*, 18, 804, 1999.
34. Reboli, A.C. and Farrer, W.E. Erysipelothrix rhusiopathiae: an occupational pathogen. *Clin. Microbiol. Rev.,* 2, 354, 1989.
35. Lejbkowicz, F., Kudinsky, R., Samet, L., Belavsky, L., Barzilai, M., and Predescu, S. Identification of *Nocardiopsis dassonvillei* in a blood sample from a child. *Am. J. Infect. Dis.,* 1, 1, 2005.
36. Rihs, J.D., McNeil, M.M., Brown, J.M., and Yu, V.L. *Oerskovia xanthineolytica* implicated in peritonitis associated with peritoneal dialysis: case report and review of Oerskovia infections in humans. *J. Clin. Microbiol.,* 28, 1934, 1990.
37. Gil-Sande, E., Brun-Otewro, M., Campo-Cerecedo, F., Esteban, E., Aguilar, L., and Garcia-de-Lomas, J. Etiological misidentification by routine biochemical tests of bacteremia caused by *Gordonia terrea* infection in the course of an episode of acute cholecystitis. *J. Clin. Microbiol.,* 44, 2645, 2006.
38. Woo, P.C., Ngan, A.H., Lau, S.K., and Yuen, K.Y. Tsukamurella conjunctivitis: a novel clinical syndrome. *J. Clin. Microbiol.,* 41, 3368, 2003.
39. Brazier, J., Duerden, B., Hall, V., Salmon, J., Hood, J., Brett, M., McLauchlin, J., and George, R. Isolation and identification of *Clostridium* ssp. from infections associated with the injection of drugs: experiences of a microbiological investigation team. *J. Med Microbiol.,* 51, 985, 2002.
40. Ha, G.Y., Yang, C.H., Kim, H., and Chong, Y. Case of sepsis caused by *Bifidobacterium longum. J. Clin. Microbiol.*, 37, 1227, 1999.
41. Kageyama, A., Benno, Y., and T. Nakase, T. Phylogenic evidence for the transfer of *Eubacterium lentum* to the genus *Eggerthella* as *Eggerthella lenta* gen. nov., comb. nov. *Inter. J. Sys. Bacter.*, 49, 1725, 1999.
42. Bassetti, S., Laifer, G., Goy, G., Fluckiger, U., and Frei, R. Endocarditis caused by *Finegoldia magna* (formally *Peptostreptococcus magnus*): diagnosis depends on the blood culture system used. *Diag. Microbiol. Inf. Dis,*. 47, 359, 2003.
43. Collins, M., Lawson, P., Willems, A., Cordoba, J., Fernandez-Garayzabal, J., and Garcia, P. The phylogeny of the genus *Clostridium*: proposal of five new genera and eleven new species combinations. *Int. J. Syst. Bacteriol,*. 44, 812, 1994.
44. Jouseimies-Somer, H., Summanen, P., Citron, D., Baron, E., Wexler, H., and Finegold, S. 2002. *Wadsworth-KTL Anaerobic Bacteriology Manual, 6th edition,* Star Publishing, Belmont, CA.
45. Cavallaro, J., Wiggs, L., and Miller, M. Evaluation of the BBL crystal anaerobic identification system. *J. Clin. Microbiol.,* 35, 3186, 1997.

6 Identification of Aerobic Gram-Negative Bacteria

Donna J. Kohlerschmidt, Kimberlee A. Musser, and Nellie B. Dumas

CONTENTS

INTRODUCTION

Aerobic gram-negative bacteria are ubiquitous. Many are found throughout the environment and distributed worldwide. Other gram-negative bacteria are established as normal flora in human and animal mucosa, intestinal tracts, and skin. Many of these organisms are typically harmless, and several are even beneficial. Others account for a large percentage of food-borne illness, and some have been identified as potential weapons of bioterrorism. The identification of these organisms is necessary in many circumstances. The clinical bacteriologist identifying the pathogen responsible for the severe, bloody diarrhea of a hospitalized young child; an environmental laboratory identifying the bacterial contaminant that forced a recall of a commercial product; the researcher identifying and characterizing the bacteria that are degrading chemical pollutants in a river — all of these scientists find themselves faced with the task of identifying aerobic gram-negative bacteria. This can be accomplished using a variety of methods. It often involves a combination of conventional phenotypic tests based on biochemical reactions, commercial kits or systems, and/or molecular analysis.

This chapter addresses the Gram stain and cellular morphology of bacteria, which are key features for preliminary identification of the gram-negative genera. Descriptions are given of some fundamental phenotypic characteristics including growth on trypticase soy agar (TSA) with 5% sheep blood, growth on MacConkey agar, glucose fermentation, oxidase detection, and pigment production.

In general, gram-negative bacteria can be divided into two groups based on their ability to ferment glucose and produce acid in triple sugar iron (TSI) agar or Klingler's iron agar (KIA) as shown in Figure 6.1 and Figure 6.2. Descriptions of ten additional conventional biochemical tests that can be used to further characterize a gram-negative bacterium are included in this chapter: motility, acid from carbohydrates, catalase, citrate, decarboxylases, indole, methyl red, Voges-Proskauer, urease, and nitrate. However, there are many bacteria that are difficult to identify solely by conventional biochemical tests. Many reference laboratories use specific PCR or real-time PCR assays to assist in the identification of gram-negative bacteria, as well as 16S rRNA gene sequence analysis to identify unusual members of this group, when conventional methods give problematic results.

The identification algorithms in Figures 6.1 and 6.2 do not include those gram-negative bacteria that are incapable of growing on TSA with 5% sheep blood. For example, most *Haemophilus* species will grow only on chocolate agar, *Legionella* species require L-cysteine and will grow on buffered charcoal-yeast extract media, and growth of *Bordetella pertussis* requires the supplements contained in Bordet-Gengou or charcoal-horse blood agar. The reader can consult References 27 through 31 for identification algorithms for these organisms.

Laboratory protocols for the identification of unknown gram-negative bacteria should also include procedures for the safe handling of potential pathogens. If a Select Agent (potential bioterrorism threat agent), for example *Francisella tularensis, Burkholderia mallei, Burkholderia pseudomallei, Brucella* sp., or *Yersinia pestis*, is suspected at any point during the identification process, culture work on the open lab bench should cease immediately, and a biological safety cabinet should be used if further testing is necessary. An appropriate reference laboratory should be contacted to arrange the shipment of the specimen for complete or confirmatory identification.

CHARACTERISTICS FOR INITIAL IDENTIFICATION

Identification schemes for gram-negative organisms that grow on 5% sheep blood agar begin with performing a Gram stain, observing colony morphology, and assessing the organism's ability to ferment glucose and grow on MacConkey agar. Figures 6.1 and 6.2 show a strategy to divide gram-negative organisms based on Gram stain cellular morphology, TSI reaction, cytochrome oxidase production, pigment production, and growth on MacConkey agar. Detailed algorithms for the preliminary identification of gram-negative bacteria have been published by Schreckenberger and Wong;[8] by York, Schreckenberger, and Miller;[9] and by Weyant et al.[2]

Gram Stain

Performing a Gram stain of bacterial growth from isolated colonies should be one of the first steps in identifying an unknown bacterium. Gram-negative organisms have a cell wall composed of a single peptidoglycan layer attached to a lipopolysaccharide-phospholipid bilayer outer membrane with protein molecules interspersed in the lipids. Decolorizing with alcohol damages this outer membrane and allows the crystal violet-iodine complex to leak out. The organism then takes up the red color of the safranin counterstain.[1] Once proven to be gram-negative, bacteria can be further differentiated based on their cellular morphology. They may appear as pronounced rods, as seen in

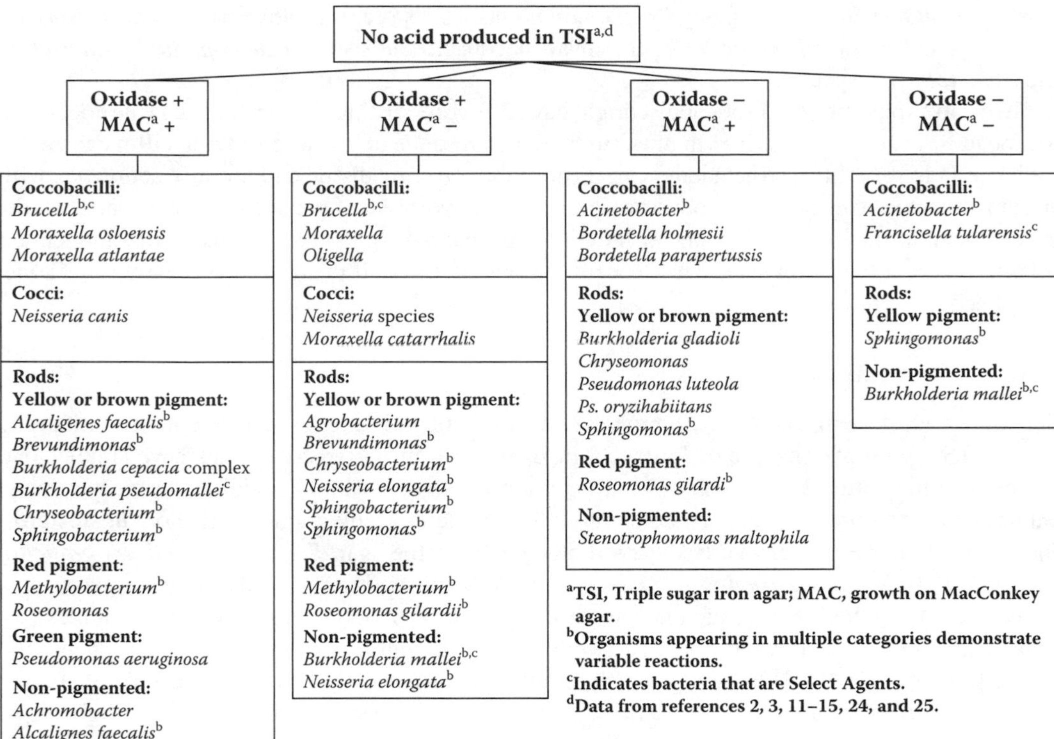

FIGURE 6.1 Gram-negative organisms that grow on TSA with 5% sheep blood and do not ferment glucose.

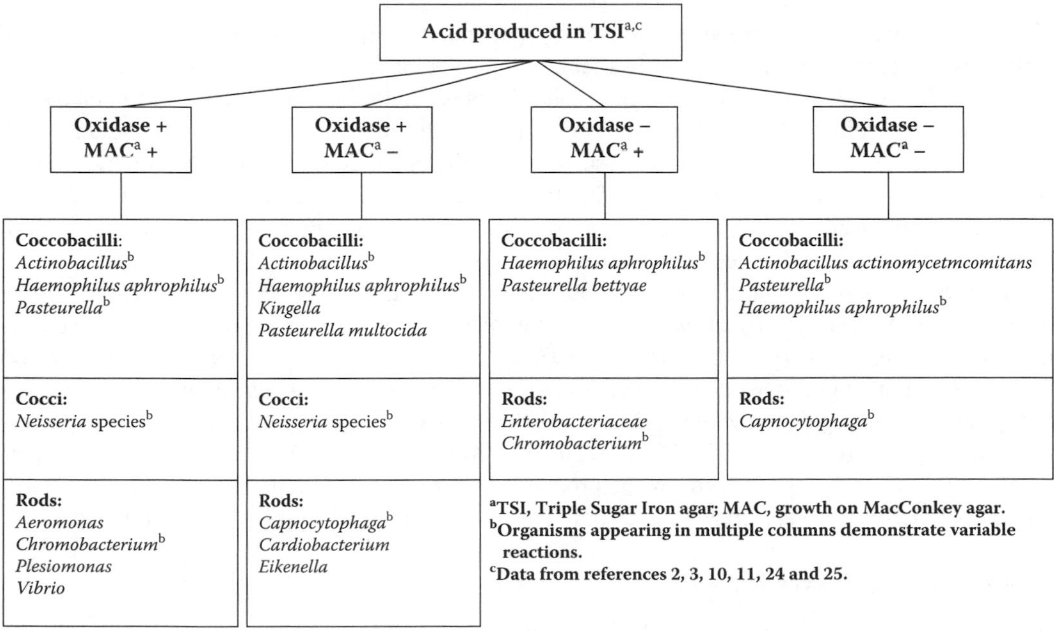

FIGURE 6.2 Gram-negative organisms that grow on TSA with 5% sheep blood and ferment glucose.

most *Enterobacteriaceae* and many *Pseudomonas* species; as gram-negative cocci, such as *Neisseria* species and *Moraxella catarrhalis*; or as oval coccobacilli, as seen in *Haemophilus*, *Francisella*, and *Acinetobacter* species.

Even an experienced microbiologist might have difficulty distinguishing between true cocci and coccobacilli in some cases. Growth of cultures in the presence of low levels of penicillin can cause the bacteria to reveal their true nature. Organisms that normally divide and produce short rods will instead continue to grow and produce filaments when exposed to penicillin. Gram stains made from growth at the edge of the zone of inhibition around a 10-μg penicillin disk will show either enlarged cocci if the organism is a true coccus, or long filaments if the organism is a gram-negative coccobacillus or rod.[2]

Colony Morphology

Colonies of most members of the *Enterobacteriaceae* family tend to be large and moist, and grow well on TSA with 5% sheep blood after 24 hr of incubation. *Neisseria* species have smaller but distinct colonies after 24 hr. Nonfermentative gram-negative rods and fastidious fermenters may require 48 hr or more of incubation before growth is evident. Many gram-negative organisms produce extracellular enzymes that lyse the red blood cells in the agar. *Escherichia coli*, *Morganella morganii*, *Pseudomonas aeruginosa*, *Aeromonas hydrophila*, and commensal *Neisseria* species can be recognized by the presence of complete clearing of the red blood cells around the colonies (β-hemolysis) on blood agar. Colonies of *Eikenella corrodens* and *Alcaligenes faecalis* produce only partial lysis of the cells, forming a greenish discoloration (α-hemolysis) around the colony.[3] Colonies of *Proteus mirabilis* and *Proteus vulgaris* are distinctive because of their "swarming" edges that can spread to cover the entire plate. *Photorhabdus* species, usually associated with nematodes but recently reported to also cause human infections, are bioluminescent: the colonies emit light when examined under conditions of total darkness.[4, 32]

Pigment Production

Gram-negative bacteria can produce a number of pigments that are helpful in establishing a species identification. Some pigments are water-insoluble and give color to the bacterial colonies themselves. These include carotenoids (yellow-orange), violacein pigments (violet or purple), and phenazines (red, maroon, or yellow). Other pigments such as pyocyanin and pyoverdin are water-soluble and diffusible. They confer color to the medium surrounding the colonies as well as to the colonies themselves. Media formulations developed specifically to detect pigment production contain special peptones and an increased concentration of magnesium and sulfate ions.[3] Pigment can also be observed in growth on TSA with 5% sheep blood, brain heart infusion agar (BHI), or Loeffler's agar.

Enterobacter species and the nonfermentative gram-negative rods *Chryseobacterium*, *Agrobacterium*, and *Sphingomonas* species can be distinguished from related organisms by their yellow pigment. *Burkholderia cepacia* complex or *Burkholderia gladioli* can be suspected when a colony of a nonfermentative gram-negative rod displays a yellow-brown pigment. *Chromobacterium violaceum* produces violet pigment while colonies of *Roseomonas*, *Methylobacterium,* and some species of *Serratia* appear pink to red. The fluorescent *Pseudomonas* group is distinguished from other nonfermentative gram-negative rods by the production of the fluorescent pigment pyoverdin. *Pseudomonas aeruginosa* isolates may produce both pyoverdin and the blue-green pigment pyocyanin.[3]

Growth on MacConkey Agar

The ability of an organism to grow and to ferment lactose on MacConkey agar offers clues to the organism's identification. MacConkey agar contains crystal violet and bile salts, which are inhibitory to most gram-positive organisms and certain gram-negatives.[5] *Enterobacteriaceae* will grow on MacConkey agar in 24 to 48 hr, while fastidious fermenters and nonfermentative gram-negative organisms may take up to 7 days for growth to appear. Most *Moraxella* species and most species of *Neisseria* can be distinguished from related organisms by their inability to grow on MacConkey agar. MacConkey agar also contains lactose and a neutral red indicator, which can further help characterize the organism. Gram-negative organisms that ferment lactose, such as *Klebsiella pneumoniae* and strains of *Escherichia coli*, will develop into purple colonies as the neutral red indicator changes due to the production of acids. Organisms that do not ferment lactose, such as *Proteus* or *Pseudomonas* species, appear colorless or transparent.

Triple Sugar Iron (TSI) Agar

TSI agar can be used to determine whether an organism is capable of fermenting glucose. This media contains glucose, lactose, and sucrose in a peptone and casein base with a phenol red pH indicator.[3, 6, 7] An agar slant is inoculated by first stabbing through the center of the medium and then streaking the surface of the slant, using an inoculum of 18- to 24-hr growth. If the organism is able to ferment glucose, acid will be produced in the agar, lowering the pH and changing the color of the medium from red to yellow. If gas is produced during fermentation, bubbles will form along the stab line, fracturing or displacing the medium. Most species of *Enterobacteriaceae* and *Aeromonas* will produce an acid reaction, often with the production of gas, after 24 hr of incubation at 35° to 37°C. *Pasteurella*, *Eikenella*, and other fastidious fermentative gram-negative rods will produce a weak acid reaction that may take up to 48 hr to appear, and typically will have no gas formation. *Pseudomonas* species, *Moraxella* species, and other nonfermentative gram-negative organisms will show no color change on TSI medium, or will produce a pink, alkaline reaction as the bacteria break down the peptones in the agar base.

Oxidase

The oxidase test determines the ability of an organism to produce cytochrome oxidase enzymes. A phenylenediamine reagent is used as an electron acceptor. In the presence of atmospheric oxygen and the enzyme cytochrome oxidase, the reagent is oxidized to form a deep blue compound, indol phenol. Two different reagents are commercially available: (1) tetramethyl-*p*-phenylenediamine dihydrochloride (Kovac's formulation) and (2) dimethyl-*p*-phenylenediamine dihydrochloride (Gordon and McLeod's reagent). The tetramethyl derivative is the more stable of the two, is less toxic, and is more sensitive in detecting weak results.[2, 3] Oxidase can be detected by adding drops of the test reagent directly onto 18- to 24-hr-old growth on a culture plate. Colonies will become blue if cytochrome oxidase is produced. An alternate, more sensitive oxidase test method uses filter paper saturated with reagent. A loopful of organism from an 18- to 24-hr-old culture is inoculated onto filter paper. Growth should be taken from a medium that contains no glucose. Development of a blue color at the site of inoculation within 10 to 30 sec indicates a positive reaction. Care must be taken to avoid using metal loops, which can give a false-positive reaction.[2] All members of the *Enterobacteriaceae* except *Plesiomonas shigelloides* are negative for oxidase. *Aeromonas*, *Vibrio*, *Neisseria*, *Moraxella*, and *Campylobacter* are all positive for oxidase, as are most *Pseudomonas* species. *Acinetobacter* species can be distinguished from other similar nonfermentative gram-negative rods by a negative oxidase reaction.

ADDITIONAL CONVENTIONAL BIOCHEMICAL TESTS

Gram-negative organisms can be further characterized phenotypically using a battery of media and substrates. The pattern of reactions is then compared with charts in the literature[2–4, 10–17, 24, 25] to arrive at a definitive identification. What follows is a brief description of some key tests used for identifying gram-negative organisms to the species level. More detailed descriptions of these test methods have been published elsewhere.[3, 6, 7, 18]

Acid from Carbohydrates

Many identification schemes for gram-negative organisms rely on the determination of the organism's ability to produce acid from a variety of carbohydrates. Acid can be produced by either fermentation or oxidation. Fermentative organisms are tested using a 1% solution of carbohydrate in peptone water containing a pH indicator. Different media formulations may incorporate bromcresol purple, phenol red, or Andrade's indicator with acid fuchsin.[2, 3, 7, 18] Ten to twelve different carbohydrates are inoculated with a suspension of the organism in saline or in broth without glucose. After 1 to 5 days of incubation at 35° to 37°C if the organism is capable of fermenting the carbohydrate, the resulting acid will change the color of the pH indicator. The tube will become yellow if bromcresol purple or phenol red is used as an indicator, or pink if Andrade's indicator is used.

Nonfermentative organisms may produce acid from carbohydrates by oxidation. This can be detected using Hugh and Leifson's Oxidation-Fermentation (OF) basal medium, a semi-solid peptone base with a bromthymol blue indicator and 1% concentration of carbohydrates. Two tubes of each carbohydrate are inoculated with the test organism by stabbing an inoculation needle into the center of the semi-solid medium. The medium in one tube of each set is overlaid with sterile mineral oil. The second tube is left with the cap loosened. Both tubes are incubated at 35° to 37°C for up to 10 days. If the isolate oxidizes the carbohydrate, acid will be produced in the presence of oxygen, causing a yellow color change in the tube with no oil. If the isolate ferments the carbohydrate, an acid change will occur in the tube overlaid with oil. An organism that is neither oxidative nor fermentative may produce an alkaline (blue) reaction in the open tube if the organism breaks down peptones in the medium base.[2, 3, 7, 18]

Catalase

The enzyme catalase breaks down hydrogen peroxide to form oxygen gas and water. When drops of 3% hydrogen peroxide are added to 18- to 24-hr growth of an organism, bubbles will form if the organism is producing the catalase enzyme. Any medium containing whole blood should not be used for catalase testing because red blood cells contain catalase and can cause a false-positive result. Although it contains blood, chocolate agar can be used for the catalase test because the red blood cells have been destroyed in preparing the media.[6] The catalase reaction is useful to differentiate *Kingella* species, which are catalase-negative, from most other fastidious fermenters and from *Moraxella* species, which are catalase-positive. *Neisseria gonorrhoeae* is strongly positive when 30% hydrogen peroxide (superoxol) is used. This characteristic is often included as part of a presumptive identification for *Neisseria gonorrhoeae* from selective media.[10, 18]

Citrate Utilization

The ability of a bacterium to use citrate as a sole source of carbon is an important characteristic in the identification of *Enterobacteriaceae*. Simmon's citrate medium formulation includes sodium citrate as the sole source of carbon, ammonium phosphate as the sole source of nitrogen, and the pH indicator bromthymol blue. The agar slant is inoculated lightly and incubated at 35° to 37°C for up to 4 days, keeping the cap loosened because the reaction requires oxygen. If the organism utilizes the citrate, it will grow on the agar slant and break down the ammonium phosphate to form ammonia. This will cause an alkaline pH change, turning the pH indicator from green to blue.[3, 6, 7] *Escherichia coli* and *Shigella* species are citrate negative, while many other *Enterobacteriaceae* are citrate positive. Among nonfermentative gram-negative rods, *Alcaligenes faecalis* and *Achromobacter xylosoxidans* are consistently positive for citrate utilization, while *Moraxella* species are consistently negative. A positive citrate reaction helps to distinguish *Roseomonas gilardii* from *Methylobacterium* species.

Decarboxylation of Lysine, Arginine, and Ornithine

This test determines whether an organism possesses decarboxylase enzymes specific for the amino acids lysine, arginine, and ornithine. Moeller's decarboxylase medium is commonly used, containing glucose, the pH indicators bromcresol purple and cresol red, and 1% of the L (levo) form of the amino acid to be tested. A control tube containing the decarboxylase base with no amino acid is inoculated in parallel. All tubes are overlaid with oil because decarboxylation requires anaerobic conditions. If the organism possesses the appropriate enzyme, hydrolyzation of the amino acid will occur, forming an amine and producing an alkaline pH change. A positive reaction will result in a purple color change. A negative reaction will retain the color of the base or become acidic (yellow) due to fermentation of glucose in the medium.[3, 6, 7, 18] Fermentative gram-negative rods require a light inoculum and 24 to 48 hr of incubation. Nonfermenters, however, may produce a weaker reaction and often need a heavy inoculum and prolonged incubation before a color change occurs. Decreasing the volume of substrate used can also enhance reactions in weakly positive organisms.[3]

Lysine, arginine, and ornithine reactions are key tests that help differentiate among the genera and species of the *Enterobacteriaceae*. For example, many species of *Citrobacter* and *Enterobacter* are positive for arginine and ornithine, while *Klebsiella* species are often only positive for lysine, and *Yersinia* species are typically positive for ornithine. *Lecleria adecarboxylata* is unable to decarboxylate any of the three amino acids, while *Plesiomonas shigelloides* is positive for all three. Among nonfermentative gram-negative rods, a positive arginine dihydrolase reaction is characteristic of *Pseudomonas aeruginosa* and the select agents *Burkholderia mallei* and *Burkholderia pseudomallei*. A positive lysine decarboxylase reaction is useful in distinguishing *Burkholderia cepacia* complex and *Stenotrophomonas maltophilia* from most other nonfermenting gram-negative rods.

Indole Production

The indole production test distinguishes organisms that produce tryptophanase, an enzyme capable of splitting the amino acid tryptophan to produce indole. Aldehyde in the test reagent reacts with indole to form a red-violet compound.[3, 18] The test can be performed in a variety of ways. A spot indole test is useful for rapid indole producers such as *Escherichia coli*. A drop of Kovac's reagent, containing 5% *p*-dimethylaminobenzaldehyde, is placed on the colony on an agar surface or on a colony picked up on a swab. The colony will form a red to pink color if positive.

Liquid or semisolid medium that has high tryptone content can be used for a tube test for indole. The medium is inoculated and incubated for 24 to 48 hr. A few drops of Kovac's reagent are added. If the organism produces tryptophanase, a pink color will develop at the meniscus.[3, 6, 18] *Escherichia coli* and several other *Enterobacteriaceae* are indole positive and can be tested using this method.

For weakly reacting nonfermenters and fastidious organisms, Ehrlich's reagent, containing 1% *p*-dimethylaminobenzaldehyde, can be used. In this method, 0.5 mL xylene is added to the organism growing in tryptone broth, in order to extract small quantities of indole, if present. The mixture is vortexed and allowed to settle. Six drops of Ehrlich's reagent are then added to the tube. If positive, a pink color will form below the xylene layer.[3, 6, 18] *Cardiobacterium hominis* and several nonfermentative gram-negative rods, including *Chryseobacterium indologenes,* will be indole positive by this method.

Motility

Motility is demonstrated macroscopically by making a straight stab of growth into semisolid medium, incubating for 24 to 48 hr at 35° to 37°C, and then observing whether the organism migrates out from the stab line, causing visible turbidity in the surrounding medium.[6] Various formulations of motility medium are available. *Enterobacteriaceae* can be tested in a 5% agar medium with a tryptose base. Incorporation of 2,3,5-triphenyltetrazolium chloride (TTC) into this medium can help in visualizing the bacterial growth. TTC causes the bacterial cells to develop a red pigment that can aid in the interpretation of the motility reaction. Nonfermentative gram-negative organisms that fail to grow in 5% agar medium can be tested using 3% agar in a base containing casein peptone and yeast extract. Fastidious organisms may demonstrate motility better in a 1% agar base containing heart infusion and tryptose.[5] *Shigella* species and *Klebsiella* species are always nonmotile. *Pseudomonas aeruginosa* and *Achromobacter xylosoxidans* are both motile. *Yersinia enterocolitica* is often nonmotile at 35° to 37°C but motile at 28°C.

Motility may be difficult to demonstrate in semi-solid medium if the organism grows slowly. Use of a wet mount made from a suspension of growth in water or broth may enable weak motility to be detected. The mount is observed with the 40X or 100X objective of a light microscope to determine whether the organisms change position.[2]

Methyl Red-Voges Proskauer (MR-VP)

The methyl red (MR) test is used to determine an organism's ability to both produce strong acid from glucose and to maintain a low pH after prolonged (48 to 72 hr) incubation. MR-VP medium formulated by Clark and Lubs is inoculated and incubated for at least 48 hr. Five drops of 0.02% methyl red reagent are added to a 1-mL aliquot of test broth. Production of a red color in the medium indicates a positive reaction.[3, 6, 18]

The Voges-Proskauer (VP) test determines an organism's ability to produce acetylmethylcarbinol (acetoin) from glucose metabolism. MR-VP broth is incubated for a minimum of 48 hr, followed by the addition of 0.6 mL of 5% α-naphthol and 0.2 mL of 40% potassium hydroxide to a 1-mL aliquot of test broth with gentle mixing. The 40% potassium hydroxide causes acetylmethylcarbinol, if present, to be converted to diacetyl. α-Naphthol causes the diacetyl to form a red color.[3, 6, 8]

The MR and VP tests are used as part of the identification of *Vibrio* and *Aeromonas* species and also help to differentiate among the genera and species of the *Enterobacteriaceae*. For example, *Escherichia coli*, *Citrobacter*, *Salmonella*, and *Shigella* species are positive for MR but negative for VP, while some species of *Enterobacter* are positive for VP but negative for MR. Several species of *Serratia* and *Klebsiella* can be distinguished from other *Enterobacteriaceae* because they are positive for both MR and VP. The tests are usually performed at 35° to 37°C, but *Yersinia enterocolitica* is VP-positive at 25°C and variable at 35° to 37°C.

Nitrate Reduction

The ability of an organism to reduce nitrate (NO_3) to nitrite (NO_2) or to nitrogen gas is an important characteristic used to identify and differentiate gram-negative bacteria. The organism is grown in a peptone or heart infusion broth containing potassium nitrate. An inverted Durham tube is included in the growth medium to detect the reduction of nitrites to nitrogen gas. After 2 to 5 days of incubation, 5 drops each of 0.8% sulfanilic acid and 0.5% *N,N*-dimethyl-naphthylamine are added to the test broth. If the organism has reduced the nitrate in the medium to nitrites, the sulfanilic acid will combine with the nitrite to form a diazonium salt. The dimethyl-naphthylamine couples with the diazonium compound to form a red, water-soluble azo dye. If there is no color development and no gas in the Durham tube, zinc dust can be used to determine whether nitrate has been reduced to nongaseous end products. Zinc dust will reduce any nitrate remaining in the medium to nitrite, and a red color will develop. If no red color develops after the addition of zinc, all the nitrate in the medium has been already reduced, indicating a positive nitrate reduction reaction. The development of a red color after the addition of zinc dust confirms a true negative reaction.[3, 6, 7, 18] All members of the *Enterobacteriaceae* except for certain biotypes of *Pantoea agglomerans* and certain species of *Serratia* and *Yersinia* are able to reduce nitrate. *Pseudomonas aeruginosa*, *Moraxella catarrhalis*, *Neisseria mucosa*, and *Kingella denitrificans* are characterized by a positive nitrate reaction. *Alcaligenes faecalis* can be distinguished from similar nonfermentative gram-negative organisms by its inability to reduce nitrate.

Urease

Detection of the ability of an organism to produce the enzyme urease can be accomplished using either broth or agar medium containing urea and the pH indicator phenol red. Urea medium is inoculated with isolated, 24-hr-old colonies and incubated at 35° to 37°C. If the organism produces urease, the urea will be split to produce ammonia and carbon dioxide. This creates an alkaline pH, causing the phenol red indicator to become bright pink.[7, 18] Some bacteria, such as *Proteus* species, *Brucella* species, and *Bordetella bronchiseptica,* are capable of producing a strongly positive urease reaction in less than 4 hr. A rapidly positive urease reaction is useful in differentiating the select agent *Brucella* species from other oxidase-positive, slowly growing, gram-negative coccobacilli. Other organisms must be incubated 24 hr or longer before a positive urease reaction becomes visible. Among the *Enterobacteriaceae*, *Proteus*, *Providencia*, and *Morganella* are urease positive, as are some *Citrobacter* and *Yersinia* species. A positive urease reaction helps distinguish *Psychrobacter phenylpyruvica* and *Oligella ureolytica* from related organisms. Urease is also a key test in the identification of *Helicobacter pylori* in gastric and duodenal specimens.

COMMERCIAL IDENTIFICATION SYSTEMS

There are several commercial systems available for the identification of gram-negative organisms. The Vitek® System (bioMerieux, Inc., Hazelwood, MO) and the Phoenix™ System (BD Biosciences, Sparks, MD) are miniaturized, sealed identification panels containing 30 to 51 dried biochemical substrates, depending on the particular test kit chosen. The systems incorporate modifications of conventional biochemicals and also test for hydrolysis of chromogenic and fluorogenic substrates, and for utilization of single carbon source substrates. A standardized inoculum is used, and the panels are continuously incubated in an automated reader. Reactions are detected by the instrument via various indicator systems. Results are compared with known bacteria in the system's database, to arrive at a final identification.[3, 20–23] API® kits (bioMerieux, Inc., Hazelwood, MO) and MicroScan® panels (Siemens Healthcare Diagnostics, Deerfeld, IL) also use miniaturized modifications of conventional biochemicals, but the systems are open, allowing reagents to be added by the technician, and giving the opportunity for visual confirmation of biochemical results.[3, 19, 22, 23] The Biolog MicroPlate™ system (Biolog, Inc., Hayward, CA) identifies gram-negative organisms by testing their ability to utilize 95 separate carbon sources. Reactions are based on the exchange of electrons generated during respiration, leading to a tetrazolium-based color change. The system produces a reaction pattern that can be classified using database software.[3, 19, 22]

Commercial systems are useful when a large number of isolates must be processed because such systems offer a decreased turn-around time and decreased hands-on time, as compared to conventional biochemical schemes. They perform with high levels of accuracy when identifying metabolically active organisms such as the *Enterobacteriaceae*, *Pseudomonas aeruginosa*, or *Stenotrophomonas maltophilia*. However, some systems are not designed to detect weak or delayed reactions, resulting in false-negative results. Some fastidious gram-negative organisms may not grow well in commercial systems and may thus require the use of additional testing methods. Also, commercial kits are limited by the range of bacteria included in the particular system's database. Organisms that are not included in the database will not be recognized and may be misidentified.

York,[26] Koneman et al.,[3] and O'Hara, Weinstein, and Miller.[22] provide detailed descriptions and evaluations of several commercial systems used for the identification of gram-negative organisms.

MOLECULAR ANALYSIS

Molecular analysis is becoming an increasingly important tool for the clinical bacteriologist, especially in the identification of aerobic gram-negative bacteria. PCR and real-time PCR assays offer specific and rapid identifications that can be key to the detection of food-borne pathogens in clinical specimens as well as in food and other environmental samples.[33, 34] Further, many gram-negative bacteria, such as *Legionella*, *Bordetella*, *Neisseria*, and *Haemophilus*,[35–38] are difficult to cultivate; PCR analyses offer superior sensitivity in identifying these organisms. The presence of certain specimen types, the initiation of antimicrobial therapy prior to specimen collection, or a delay in transport also can compromise the success of classical approaches to identification. Finally, molecular analyses are utilized for ruling in or out the presence of gram-negative bacterial select agents in samples; in such a situation, rapid issuance of test results by a reference laboratory can be crucial in detecting and averting the intentional dissemination of potentially deadly pathogens.[39]

Molecular analysis can quickly provide identification to the species level, or typing, for certain genera of gram-negative bacteria that cannot be differentiated by commercially available systems, or that would require the use of highly specialized protocols and reagents, and time-consuming, labor-intensive protocols. Examples include species determination for organisms of the *Burkholderia cepacia* complex[40] and the genera *Bordetella*,[41] *Brucella*,[42, 43] *Achromobacter*,[44] *Yersinia*,[45] and *Campylobacter*,[46] as well as the typing of *Escherichia coli*,[47] *Salmonella*,[48] and *Neisseria meningitidis*.[49] Additionally, the rapid characterization of antibiotic resistance, pathogenic plasmids, or toxin genes can be achieved. Examples include KPC-mediated carbapenem resistance in *Klebsiella pneumoniae*[50] and other *Enterobacteriaceae*,[51] the virulence plasmids of *Yersinia pestis*,[45] and the presence of Shiga toxins in *E. coli*.[34]

As previously mentioned, when conventional methods for gram-negative bacteria give inconclusive results, or when a specific PCR analysis or molecular test is not available, the use of 16S rRNA gene sequence analysis or other broad-range PCR may be successful.[52] Such tests are based on the fact that some genes present in all bacterial organisms contain both regions of highly conserved sequence and regions of variable sequence. The highly conserved regions can serve as primer binding sites for PCR, while the variable regions can provide unique identifiers enabling the precise identification of a particular strain or species. These sophisticated PCR applications require adequate laboratory facilities, trained and experienced staff with expertise in the use of molecular databases, and protocols in place to avoid contamination.[53, 54] Although such tests are time-consuming and costly, and may only be available at specialized laboratories such as reference laboratories, they can provide an answer when other modes of testing give problematic results or are not feasible.

ACKNOWLEDGMENTS

The authors thank Dr. Adriana Verschoor and Dr. Ronald J. Limberger, Wadsworth Center, New York State Department of Health, for critical readings of this chapter.

REFERENCES

1. York, M.K., Gram stain, in *Clinical Microbiology Procedures Handbook*, 2nd ed., Isenberg, H.D., Ed., ASM Press, Washington, DC, 2004, Vol. 1, Section 3.2.1.
2. Weyent, R.S., Moss, C.W., Weaver, R.E., Hollis, D.G., Jordan, J.G., Cook, E.C., and Daneshvar, M.I., *Identification of Unusual Pathogenic Gram-Negative Aerobic and Facultatively Anaerobic Bacteria*, 2nd ed., Williams & Wilkins, Baltimore, MD, 1995.
3. Koneman, E.W., Allen, S.D., Janda, W.M., Schreckenberger, P.C., and Winn, W.C., *Color Atlas and Textbook of Diagnostic Microbiology*, 5th ed., Lippincott Williams & Wilkins, Philadelphia, PA, 1997.
4. Farmer, J.J., *Enterobacteriaceae*: Introduction and Identification, in *Manual of Clinical Microbiology*, 8th ed., Murray, P.R. et al., Eds., ASM Press, Washington, DC, 2003, chap. 41.
5. Remel Product Technical Manual, Remel, Inc., Lenexa, KS, 1998.
6. MacFaddin, J.F., *Biochemical Tests for the Identification of Medical Bacteria*, 3rd ed., Lippincott, Williams & Wilkins Co., Philadelphia, PA, 2000.
7. Forbes, B.A., Sahm, D.F., and Weissfeld, A.S., *Bailey and Scott's Diagnostic Microbiology*, 10th ed., Mosby, St. Louis, MO, 1998.

8. Schreckenberger, P.C. and Wong, J.D., Algorithms for identification of aerobic gram-negative bacteria, in *Manual of Clinical Microbiology*, 8th ed., Murray, P.R. et al., Eds., ASM Press, Washington, DC, 2003, chap. 23.
9. York, M.K., Schreckenberger, P.C., and Miller, J.M., Identification of gram-negative bacteria, in *Clinical Microbiology Procedures Handbook*, 2nd ed., Isenberg, H.D., Ed., ASM Press, Washington, DC, 2004, Vol. 1, Section 3.18.2.
10. Janda, W.M. and Knapp, J.S., *Neisseria* and *Moraxella catarrhalis*, in *Manual of Clinical Microbiology*, 8th ed., Murray, P.R. et al., Eds., ASM Press, Washington, DC, 2003, chap. 38.
11. von Graevenitz, A., Zbinden, R., and Mutters, R., *Actinobacillus*, *Capnocytophaga*, *Eikenella*, *Kingella*, *Pasteurella*, and other fastidious or rarely encountered gram-negative rods, in *Manual of Clinical Microbiology*, 8th ed., Murray, P.R. et al., Eds., ASM Press, Washington, DC, 2003, chap. 39.
12. Kiska, D.L. and Gilligan, P.H., *Pseudomonas*, in *Manual of Clinical Microbiology*, 8th ed., Murray, P.R. et al., Eds., ASM Press, Washington, DC, 2003, chap. 47.
13. Gilligan, P.H., Lum, G., VanDamme, P.A.R., and Whittier, S., *Burkholderia*, *Stenotrophomonas*, *Ralstonia*, *Brevundimonas*, *Comamonas*, *Delftia*, *Pandoraea* and *Acidovorax*, in *Manual of Clinical Microbiology*, 8th ed., Murray, P.R. et al., Eds., ASM Press, Washington, DC, 2003, chap. 48.
14. Schreckenberger, P.C., Daneshvar, M.I., Weyant, R.S., and Hollis, D.G., *Acinetobacter*, *Achromobacter*, *Chryseobacterium*, *Moraxella* and other nonfermentative gram-negative rods, in *Manual of Clinical Microbiology*, 8th ed., Murray, P.R. et al., Eds., ASM Press, Washington, DC, 2003, chap. 49.
15. Chu, M.C. and Weyant, R.S., *Francisella* and *Brucella*, in *Manual of Clinical Microbiology*, 8th ed., Murray, P.R. et al., Eds., ASM Press, Washington, DC, 2003, chap. 51.
16. Ewing, W.H., *Edwards and Ewing's Identification of Enterobacteriaceae*, 4th ed., Elsevier Scientific, New York, 1986.
17. Janda, J.M. and Abbott, S.L., *The Enterobacteria*, 2nd ed., ASM Press, Washington, DC, 2006.
18. York, M.K., Traylor, M.M., Hardy, J., and Henry, M., Biochemical tests for the identification of aerobic bacteria, in *Clinical Microbiology Procedures Handbook,* 2nd ed., Isenberg, H.D., Ed., ASM Press, Washington, DC, 2004, Vol. 1, Section 3.17.
19. Truu, J., Talpsep, E., Heinaru, E., Stottmeister, U., Wand, H., and Heinaru, A., Comparison of API 20 NE and Biolog GN identification systems assessed by techniques of multivariate analyses, *J. Microbiol. Meth.*, 36, 193, 1999.
20. Salomon, J., Butterworth, A., Almog, V., Pollitt, J., Williams, W., and Dunk, T., Identification of gram-negative bacteria in the Phoenix™ system, presented at *9th European Congress of Clinical Microbiology and Infectious Diseases (ECCMID)*, Berlin, Germany, 1999.
21. Jorgensen, J.H., New phenotypic methods for the clinical microbiology laboratory. *Abstr. Intersci. Conf. Antimicrob. Agents Chemother.* 42:abstract 398, 2002.
22. O'Hara, C.M., Weinstein, M.P., and Miller, J.M., Manual and automated systems for detection and identification of microorganisms, in *Manual of Clinical Microbiology,* 8th ed., Murray, P.R. et al., Eds., ASM Press, Washington, DC, 2003, chap. 14.
23. O'Hara, C.M., Tenover, F.C., and Miller, J.M., Parallel comparison of accuracy of API 20E, Vitek GN, MicroScan Walk/Away Rapid ID and Becton Dickinson Cobas Micro ID-E/NF for identification of members of the family *Enterobacteriaceae* and common gram-negative, non-glucose-fermenting bacilli, *J. Clin. Microbiol.*, 31, 3165, 1993.
24. Garrity, G.M., Brenner, D.J., Krieg, N.R., and Staley, J.T., *Bergey's Manual of Systematic Bacteriology*, 2nd ed., Springer, New York, 2005.
25. Holt, J.G., Krieg, N.R., Sneath, P.H.A., Staley, J.T., and Williams, S.T., *Bergey's Manual of Determinative Bacteriology*, 9th ed., Lippincott, Williams & Wilkins, Philadelphia, PA, 2000.
26. York, M.K., Guidelines for identification of aerobic bacteria, in *Clinical Microbiology Procedures Handbook,* 2nd ed., Isenberg, H.D., Ed., ASM Press, Washington, DC, 2004, Vol. 1, Section 3.16.
27. Pasculle, A.W., *Legionella* cultures, in *Clinical Microbiology Procedures Handbook*, 2nd ed., Isenberg, H.D., Ed., ASM Press, Washington, DC, 2004, Vol. 1, Section 3.11.4.
28. McGowan, K.L., *Bordetella* cultures, in *Clinical Microbiology Procedures Handbook*, 2nd ed., Isenberg, H.D., Ed., ASM Press, Washington, DC, 2004, Vol. 1, Section 3.11.6.
29. Killian, M., *Haemophilus*, in *Manual of Clinical Microbiology*, 8th ed., Murray, P.R. et al., Eds., ASM Press, Washington, DC, 2003, chap. 40.
30. Loeffelholz, M.J., *Bordetella*, in *Manual of Clinical Microbiology,* 8th ed., Murray, P.R., et al., Eds., ASM Press, Washington, DC, 2003, chap. 50.

31. Stout, J.E., Rihs, J.D., and Yu, V.L., *Legionella*, in *Manual of Clinical Microbiology*, 8th ed., Murray, P.R., et al., Eds., ASM Press, Washington, DC, 2003, chap. 52.
32. Gerrard, J.G., McNevin, S., Alfredson, D., Forgan-Smith, R., and Fraser, N., *Photorhabdus* species: Bioluminescent bacteria as emerging human pathogens?, *Emerg. Infect. Dis.*, 9, 251, 2003.
33. Harwood, V.J., Gandhi, J.P., and Wright, A.C., Methods for isolation and confirmation of *Vibrio vulnificus* from oysters and environmental sources: a review, *J. Microbiol. Meth.*, 59, 301, 2004.
34. Bopp, D.J., Sauders, B.D., Waring, A.L., Ackelsberg, J., Dumas, N., Braun-Howland, E., Dziewulski, D., Wallace, B.J., Kelly, M., Halse, T., Musser, K.A., Smith, P.F., Morse, D.L., and Limberger, R.J., Detection, isolation, and molecular subtyping of *Escherichia coli* O157:H7 and *Campylobacter jejuni* associated with a large waterborne outbreak, *J. Clin. Microbiol.*, 41, 174, 2003.
35. She, R.C., Billetdeaux, E., Phansalkar, A.R., and Petti, C.A., Limited applicability of direct fluorescent-antibody testing for *Bordetella* sp. and *Legionella* sp. specimens for the clinical microbiology laboratory, *J. Clin. Microbiol.*, 45,2212, 2007.
36. Khanna, M., Fan, J., Pehler-Harrington, K., Waters, C, Douglass, P., Stallock, J., Kehl, S., Henrickson, K.J., The pneumoplex assays, a multiplex PCR-enzyme hybridization assay that allows simultaneous detection of five organisms, *Mycoplasma pneumoniae*, *Chlamydia (Chlamydophila) pneumoniae*, *Legionella pneumophila*, *Legionella micdadei*, and *Bordetella pertussis*, and its real-time counterpart, *J. Clin. Microbiol.,* 43, 565, 2005.
37. Hjelmevoll, S.O., Olsen, M.E., Sollid, J.U., Haaheim, H., Unemo, M., and Skogen, V., A fast real-time polymerase chain reaction method for sensitive and specific detection of the *Neisseria gonorrhoeae* porA pseudogene, *J. Mol. Diagn.*, 8,574, 2006.
38. Poppert, S., Essig, A., Stoehr, B., Steingruber, A., Wirths, B., Juretschko, S., Reischl, U., and Wellinghausen, N., Rapid diagnosis of bacterial meningitis by real-time PCR and fluorescence *in situ* hybridization, *J. Clin. Microbiol.*, 43, 3390, 2005.
39. Cirino, N.M., Musser, K.A., and Egan, C., Multiplex diagnostic platforms for detection of biothreat agents, *Expert Rev. Mol. Diagn.*, 4, 841, 2004. Review.
40. Mahenthiralingam, E., Bischof, J., Byrne, S.K., Radomski, C., Davies, J.E., Av-Gay, Y., and Vandamme, P., DNA-based diagnostic approaches for identification of *Burkholderia cepacia complex*, *Burkholderia vietnamiensis*, *Burkholderia multivorans*, *Burkholderia stabilis*, and *Burkholderia cepacia* genomovars I and III, *J. Clin. Microbiol.*, 38, 3165, 2000.
41. Koidl, C., Bozic, M., Burmeister, A., Hess, M., Marth, E., and Kessler, H.H., Detection and differentiation of *Bordetella* spp. by real-time PCR, *J. Clin. Microbiol.*, 45, 347, 2007.
42. Bricker, B.J., PCR as a diagnostic tool for brucellosis, *Vet. Microbiol.*, 90(1-4), 435, 2002.
43. Noviello, S., Gallo, R., Kelly, M., Limberger, R.J., DeAngelis, K., Cain, L., Wallace, B., Dumas, N., Laboratory-acquired brucellosis, *Emerg. Infect. Dis.*, 10, 1848, 2004.
44. Uy, H.S., Matias, R., de la Cruz, F., and Natividad, F., *Achromobacter xylosoxidans* endophthalmitis diagnosed by polymerase chain reaction and gene sequencing, *Ocul. Immunol. Inflamm.*, 13, 463, 2005.
45. Woron, A.M., Nazarian, E.J., Egan, C., McDonough, K.A., Cirino, N.M., Limberger, R.J., and Musser, K.A., Development and evaluation of a 4-target multiplex real-time polymerase chain reaction assay for the detection and characterization of *Yersinia pestis*, *Diagn. Microbiol. Infect. Dis.*, 56, 261, 2006.
46. LaGier, M.J., Joseph, L.A., Passaretti, T.V., Musser, K.A., and Cirino, N.M., A real-time multiplexed PCR assay for rapid detection and differentiation of *Campylobacter jejuni* and *Campylobacter coli*, *Mol. Cell. Probes.*, 18, 275, 2004.
47. Beutin, L. and Strauch, E., Identification of sequence diversity in the *Escherichia coli fliC* genes encoding flagellar types H8 and H40 and its use in typing of Shiga toxin-producing *E. coli* O8, O22, O111, O174, and O179 strains, *J. Clin. Microbiol.*, 45, 333, 2007.
48. Kim, S., Frye, J.G., Hu, J., Fedorka-Cray, P.J., Gautom, R., and Boyle, D.S., Multiplex PCR-based method for identification of common clinical serotypes of *Salmonella enterica* subsp. *enterica*, *J. Clin. Microbiol.*, 44, 3608, 2006.
49. Bennett, D.E. and Cafferkey, M.T., Consecutive use of two multiplex PCR-based assays for simultaneous identification and determination of capsular status of nine common *Neisseria meningitidis* serogroups associated with invasive disease, *J. Clin. Microbiol.*, 44, 1127, 2006.
50. Yigit, H., Queenan, A.M., Anderson, G.J., Domenech-Sanchez, A., Biddle, J.W., Steward, C.D., Alberti, S., Bush, K., and Tenover, F.C., Novel carbapenem-hydrolyzing beta-lactamase, KPC-1, from a carbapenem-resistant strain of *Klebsiella pneumoniae*, *Antimicrob. Agents Chemother.*, 45, 1151, 2001.

51. Bratu, S., Brooks, S., Burney, S., Kochar, S., Gupta, J., Landman, D., and Quale, J., Detection and spread of *Escherichia coli* possessing the plasmid-borne carbapenemase KPC-2 in Brooklyn, New York, *Clin. Infect. Dis.*, 44, 972, 2007.
52. Zbinden. A., Bottger, E.C., Bosshard, P.P., and Zbinden, R., Evaluation of the colorimetric VITEK 2 card for identification of gram-negative nonfermentative rods: comparison to 16S rRNA gene sequencing, *J. Clin. Microbiol.*, 45, 2270, 2007.
53. Maiwald, M., Broad-range PCR for detection and identification of bacteria, in *Molecular Microbiology: Diagnostic Principles and Practice*, Persing, D.H et al., Eds., ASM Press, Washington, DC, 2004, chap. 30.
54. Fredricks, D.N. and Relman, D.A., Application of polymerase chain reaction to the diagnosis of infectious diseases, *Clin. Infect. Dis.*, 29, 475, 1999.

7 Plaque Assay for Bacteriophage

Emanuel Goldman

CONTENTS

HISTORY

The Plaque Assay, an indispensable tool for the study of bacteriophage, was described in the earliest publications on the discovery of these viruses. Felix d'Hérelle, credited as co-discoverer of phage (along with Frederick Twort), described, in effect, the first plaque assay in 1917 [1] when he wrote the following:

> …if one adds to a culture of *Shiga* as little as a million-fold dilution of a previously lysed culture and if one spreads a droplet of this mixture on an agar slant, then one obtains after incubation a lawn of dysentery bacilli containing a certain number of circular areas of 1 mm diameter where there is no bacterial growth; these points cannot represent but colonies of the antagonistic microbes: a chemical substance cannot concentrate itself over definite points. (Translation from Stent [2].)

The "antagonistic microbes" were then named "bacteriophages."

Subsequently, the technique was refined by Gratia [3] and Hershey et al. [4], by pouring onto an agar plate a top layer of molten agar containing a phage-bacteria mixture. The molten agar solidified, fixing the phage particles in the semi-solid agar substrate; and as the bacteria grew during incubation to cover the plate as a lawn of bacterial growth, the embedded phage particles killed those bacteria in their vicinity, with the progeny phage infecting other bacteria in the vicinity to produce a zone of clearing, or a "plaque," on the lawn of bacterial growth. This method was described by Delbrück [5] as follows:

> In this technique a few milliliters of melted agar of low concentration, containing the bacteria, are mixed with the sample to be plated. The mixture is poured on the surface of an ordinary nutrient agar plate. The mixture distributes itself uniformly over the plate in a very thin layer and solidifies immediately. We used about 20 mL of 1.3 percent agar for the lower layer and 1.5 mL of 0.7 percent agar for the superimposed layer on petri plates of 7 cm diameter. With this technique 60 samples can be plated in about 20 minutes. The plaques are well-developed and can be counted after 4 hours' incubation.

The plaque assay was validated as showing that a single phage infecting a bacterium was sufficient to initiate formation of a plaque because there was linear proportionality between the plaques observed and the dilution of the phage sample [6]. Electron microscopy later indicated that under optimal conditions, there was a close approximation between physical phage particles and the presumptive phage particles that initiated plaque formation [7].

However, the "efficiency of plating" of a bacteriophage preparation can vary extensively, depending on conditions — especially the bacterial host strain. Varying susceptibility of different bacterial strains to different bacteriophage in fact was exploited for many years in the identification of bacteria, in a procedure called "phage typing." Although this procedure is not used as extensively today as in the past, a brief description of the principle is provided later in this chapter. Current use of phage for identification of bacteria, including phage typing, is discussed in Chapter 8.

METHODOLOGY

A suspension of bacteriophage of unknown titer is first subjected to serial dilution, as illustrated in Figure 7.1. A small portion (0.1 mL) of the suspension is added to 9.9 mL dilution fluid, for a 100-fold dilution. A small portion (0.1 mL) of this dilution, in turn, is added to a fresh tube containing 9.9 mL dilution fluid, for another 100-fold dilution. At this point, the original phage suspension has been diluted by a factor of 10,000. The process is repeated two more times, to a fourth tube where the dilution factor is 10^{-8}. Dilutions, of course, do not always have to be 1:100; other dilutions are perfectly acceptable, as long as the dilution factor is duly noted, as this is needed to obtain the titer of the original stock. In Figure 7.1, there is a fifth dilution tube with the addition of 1mL to 9 mL dilution fluid, to provide a 1:10 dilution, bringing the overall dilution factor to 10^{-9}. The need for large dilutions is because phage titers can be quite high.

Dilution fluid can be the growth medium for the host bacteria; but because this would be wasteful and expensive, in practice a simpler liquid diluent is used. Sterile saline solution can serve this purpose, or variations of saline with lower salt and a small addition of a growth nutrient, for example, sterile 0.25% (wt/vol) sodium chloride, 0.1% (wt/vol) tryptone.

Once the final dilution is made, 0.1 mL is added to tubes containing 2.5 mL soft agar that has been melted in a boiling water bath and equilibrated to approximately 50°C. Tubes can be kept in a water bath or, more commonly, a dry block heating unit. The dry block is preferable to avoid unwanted contamination of water-bath water from the outside of the tubes when the contents are poured onto plates. Tubes should be plated fairly soon after addition of the phage dilution, to avoid incubating the phage at the high temperature for an extended period of time. The soft agar contains the nutrients for growth of the bacteria (and can be a minimal media) with 0.65% (wt/vol) agar. Two drops of a fresh saturated bacterial suspension (about 0.1 mL, ~10^8 bacteria total) are added to the tubes, which are then mixed and poured onto nutrient agar (1 to 1.5% wt/vol) Petri plates (100 × 15 mm) that have been previously prepared and are at room temperature. The plate is briefly rocked to

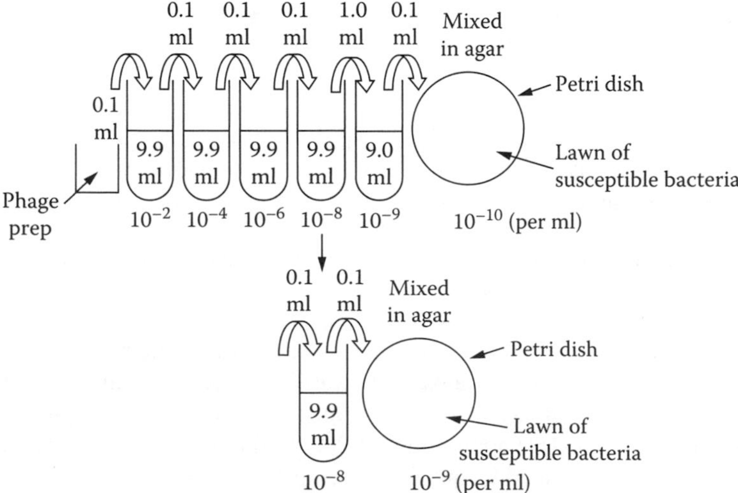

FIGURE 7.1 Serial dilutions.

facilitate uniform distribution of the top agar across the surface of the bottom agar, before the top agar solidifies. If the plates have been refrigerated and are not at room temperature, the top agar may solidify too quickly and not distribute uniformly, which will make plaque counting more difficult. The plates are then incubated at 37°C for several hours, usually overnight, to allow the bacteria to make a lawn and for the plaques to develop.

Because the titer is unknown at the outset, more than one dilution of the phage stock is plated in this way, so that a reasonable (countable) number of plaques will be obtained at least at one of the dilutions tested. Figure 7.1 illustrates two dilutions that are plated: 10^{-8} and 10^{-9}. If one obtained, for example, 15 plaques from the 10^{-9} dilution (and ~150 plaques from the 10^{-8} dilution), the titer of the original phage stock would be 1.5×10^{11} phage/mL in the original stock suspension. This number of phage is the reciprocal of the dilution factor and is derived by taking into account that only 0.1 mL of the dilution was added to the plate, which represents an additional dilution factor of 10^{-1} when considered on a per-milliliter basis. Hence, 15 plaques obtained from 0.1 mL of a 10^{-9} dilution $= 15 \times 10^{10}$ phage mL^{-1} in the original stock.

An example of actual plaques of a lytic phage grown on *Escherichia coli* is shown in Figure 7.2. The lawn is shown as a black background, although in an actual assay, it would likely be tan or brown, and the plaques are shown as clear white circular spots on the lawn. If this plate resulted from plating 0.1 mL of a 10^{-8} dilution (as in Figure 7.1, bottom), then the titer of the stock would be $\sim 1.5 \times 10^{11}$, because there are about 150 plaques on this plate. It is important to remember that each plaque represents the progeny phage from a single individual phage that was embedded in the top agar. The plaque as a whole may contain on the order of 10^6 phage. A good description of the preparation of phage and plaque assays can be found in a laboratory manual by Miller [8].

Different phage exhibit different morphologies in plaque assays. Some plaques are small and clear, with sharply defined borders. Some are large and can be more diffuse. Some phage exhibit turbid plaques, in which there is some bacterial growth within the plaque, as if those bacteria had developed resistance to the phage. This is the hallmark of a lysogenic phage, which at low frequency integrates its genome into the bacterial host and renders the host immune to reinfection by phage of the same type (see Chapter 43, "Introduction to Bacteriophage").

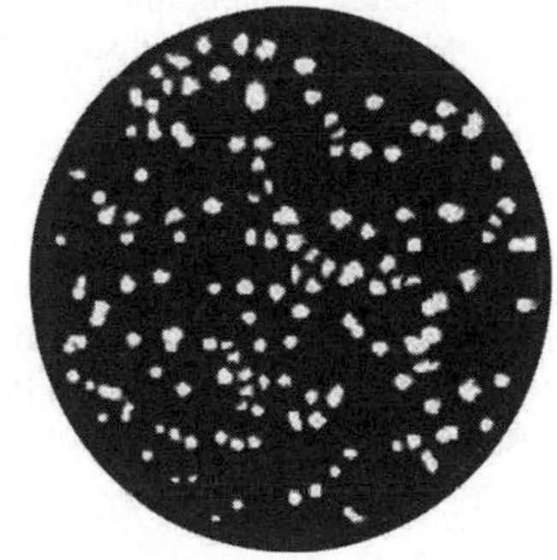

FIGURE 7.2 Bacteriophage plaques.

APPLICATION TO PHAGE TYPING

The ability of phage to form plaques on a lawn of susceptible bacteria was exploited to develop systematic methods for identifying strains of bacteria by their susceptibility to known stocks of individual phages, a technique known as "phage typing." A chapter in the first edition of this Handbook was devoted to this method [9], which was particularly useful for identifying strains of *Staphylococcus aureus* and *Salmonella typhi*, among others. An update on contemporary use of phage typing is included in the next chapter (Chapter 8).

The method requires a standard set of typing phages that are different for each species to be typed. A culture of the bacteria to be identified is spread over the surface of a nutrient agar plate. Drops from a set of typing phages are placed in a grid pattern on the surface of the plate, which is then incubated to allow the bacteria to form a lawn and for the phage to produce plaques. Following incubation, those phage in the set that can inhibit growth of the bacteria will form plaques. Because the phages were applied in a grid pattern, one can identify which phages the bacterial host is susceptible to, which will then be given a "phage-type" designation.

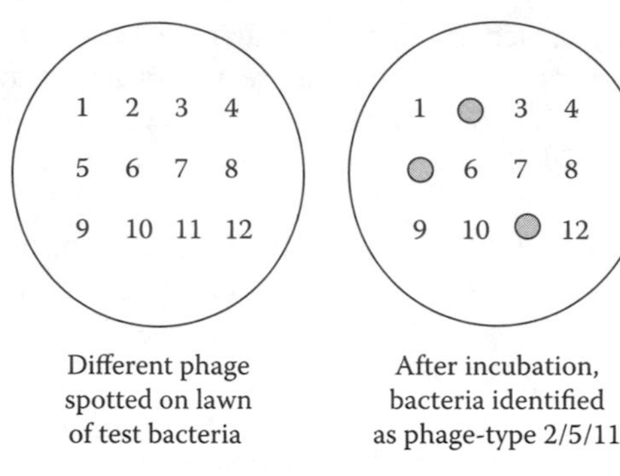

FIGURE 7.3 Phage typing.

This principle is illustrated in Figure 7.3. The plate on the left shows a grid pattern of 12 individual phages that are spotted on a surface containing the bacteria to be identified. The plate is incubated to allow the lawn (and phage) to grow. Following incubation, plaques are observed at the locations where phages 2, 5, and 11 were spotted, and not at the locations of the other nine phages in the set. If the bacteria to be identified had been a strain of *Neisseria meningitidis*, for example, the strain would thus be identified as *Neisseria meningitidis* phage type 2/5/11.

REFERENCES

1. d'Hérelle, F. Sur un microbe invisible antagoniste des bacilles dysentériques, *C. R. Acad. Sci. Paris,* 165, 373, 1917.
2. Stent, G.S. *Molecular Biology of Bacterial Viruses*, W.H. Freeman and Co., San Francisco, 1963, 5.
3. Gratia, A. Des relations numériques entre bactéries lysogènes et particules de bactériophage, *Ann. Inst. Pasteur,* 57, 652, 1936.
4. Hershey A. D., Kalmanson, G., and Bronfenbrenner, J. Quantitative methods in the study of the phage-antiphage reaction, *J. Immunol.*, 46, 267, 1943.
5. Delbrück, M. The burst size distribution in the growth of bacterial viruses (bacteriophages), *J. Bacteriol.*, 50, 131, 1945.
6. Ellis, E. L. and Delbrück, M. The growth of bacteriophage, *J. Gen. Physiol.*, 22, 365, 1939.
7. Luria, S.E., Williams, R.C. and Backus, R.C. Electron micrographic counts of bacteriophage particles, *J. Bacteriol.*, 61, 179, 1951.
8. Miller, J.H. Preparation and plaque assay of a phage stock, *Experiments in Molecular Genetics*, Cold Spring Harbor Laboratory, Cold Spring Harbor, 1972, 37.
9. Kasatiya, S.S. and Nicolle, P. Phage typing, *Practical Handbook of Microbiology*, 1st ed., O'Leary, W., Ed., CRC Press, Boca Raton, FL, 1989, 279.

8 Phage Identification of Bacteria

Catherine E.D. Rees and Martin J. Loessner

CONTENTS

INTRODUCTION

As the name suggests, bacteriophage (literally, "bacteria eating" from the Greek) were discovered by Edward Twort and Felix D'Herelle as lytic agents that destroyed bacterial cells (see Ref. 1 for a review of the history of their discovery and application). Bacteriophage* are, in fact, viruses that specifically infect members of the Bacterial kingdom. In the Eukaryotic kingdom, the visible diversity of biological forms makes it unsurprising that viruses have a specific host range, because it is clear that the organisms affected by the different viruses are very different. However, in the Bacterial kingdom, differences between bacterial genera are not as easy to detect and differentiation between members of a species is often only reliably determined at the molecular level. Hence, in the field of bacteriology, the fact that viruses have evolved to specifically infect only certain members of a genus or species seems surprising. However, as the molecular recognition events involved in eukaryotic virus infection are elucidated, it is clear that even subtle differences in cell surface proteins have profound effects on binding and infection of eukaryotic viruses (see Ref. 2 for a review), and this is also true for the bacteria-virus interaction.

Historically it was in 1926 that D'Herelle first reported the fact that not all members of a species could be infected by the same phage when he was investigating the use of bacteriophage to treat cholera.[3] He divided isolates of *Vibrio cholerae* into two groups on the basis of phage sensitivity. Until molecular methods of cell discrimination became more widely adopted, differentiation of cell type on the basis of phage sensitivity (or phage typing) was one of the best ways to discriminate between different isolates within a species, and much research was put into developing and refining the phage typing sets for a range of different bacteria. Antibody typing (serotyping) is also able to differentiate strains at the subspecies level but, as phage isolation and propagation is easier and cheaper than characterization of polyclonal or monoclonal antibodies, phage typing has often been

* There is ongoing debate about the plural form: bacteriophage or bacteriophages. For a discussion, see Ref. 4.

the first subspecies typing method developed for bacterial pathogens. Where robust and widely adopted phage typing schemes exist, their use has continued, and phage typing is still routinely used to sub-type isolates of both Gram-negative and Gram-positive bacteria (see section entitled "The Use of Phage Typing" for details). However, fewer new phage typing sets have been developed since DNA-based typing methods became more widely adopted. In contrast, methods for the detection of bacteria using bacteriophages are still being developed. This has resulted in the emergence of commercial tests, employing a variety of assay formats, for use in both medical and environmental applications. This chapter provides an overview of these technologies and describes some of the newer findings that may be developed into practical applications in the near future.

PHAGE TYPING

Choice of Phage

Bacteriophage can be divided into two broad groups on the basis of the biological outcome of infection. Virulent, or lytic, phages are those that always lyse the host cell at the end of the replication cycle. Temperate, or lysogenic, phages may cause lysis of the host cell after completing a lytic replication cycle, or alternatively they may enter a dormant phase of replication where their genome is maintained within the host cell without causing harm to that cell so that the virus is replicated as part of the cell during cell division. Because of this choice of replication pathway, infection with temperate phage will not always result in the formation of plaques (lytic areas of clearing in a lawn of bacteria) and therefore these are often not chosen for use in phage typing schemes. Much work has been put into the isolation of virulent phages that can be evaluated for use in phage typing sets, and these are often found in the natural environment of the target bacterium (for example, sewage is a good source of bacteria specific for enteric pathogens (see Ref. 5 through 12). However, other workers have chosen to try and identify phage variants or mutants (adapted phages) of previously isolated phage that have a new or altered host range. The fact that is was possible to select for such mutants was first shown by Craigie and Yen[13] as early as 1938. These workers identified the fact that phage could be selected from an infection that was now more lytic against that same strain on which they were last propagated, and the practice of phage adaptation is still used to improve the discriminatory power of phage sets.[11, 14] Perhaps more important for science as a whole, the observation led to the discovery of phage restriction barriers (see section entitled "Factors Affecting Phage Infection" and Ref. 15 and 16) and ultimately to restriction enzymes,[17–19] which then revolutionized the study of cellular molecular biology.

Factors Affecting Phage Infection

There are many factors that influence the ability of a phage to infect a particular cell type. For successful infection, the phage must recognize its host by binding to primary surface receptors. These are often molecules containing carbohydrates, such as lipopolysaccharides[20] (O antigen) in Gram-negative bacteria, teichoic acids[21, 22] (O antigen) in Gram-positive bacteria, or capsular polysaccharides[23] (K antigen). The fact that structural variations in these molecules occur with a relatively high frequency, which is used to distinguish cell types on the basis of serology, begins to explain why phages bind to different host cell types with variable efficiencies (see Ref. 24, 25). In addition to carbohydrates, cell surface proteins can be recognized, such as flagella or pili,[26–29] sugar uptake proteins,[30] membrane proteins[31, 32] or S-layer proteins.[33] Again, variations in these proteins occur that can be detected using specific antisera (for example, flagella represent H antigens); and, accordingly, phage binding occurs with a different affinity depending on the match between phage and host receptor molecule. Another level of variation can occur depending on the pattern of expression of the receptor by the host cell, which may be regulated by environmental conditions. This is particularly true for the expression of flagella and sugar transporters and therefore the particular culture conditions used can determine whether or not a phage can infect a cell.

Once the phage has penetrated the cell envelope, replication can be prevented by a variety of mechanisms. For example, if the cell contains a prophage (a dormant lysogenic phage), the regulatory proteins that suppress the lytic replication of the phage may also suppress replication of the newly infecting phage (see Ref. 34). Cells that possess DNA restriction-modification systems may reduce levels of infection by the destruction of incoming foreign DNA (the restriction barrier[35]) and host cell genes (abortive phage infection genes[36–38]) have been described that specifically block phage replication (for more details, see Ref. 39). In effect then, phage typing is an indirect way to investigate variation in cell structure and genetics using the ability of a phage to overcome these barriers to successfully infect a cell as a biological indicator of differences between cells.

Phage Typing Methods

Determining Phage Titer and RTD

To determine phage titer, tenfold serial dilutions of the phage are prepared in a buffer that contains divalent cations (often Mg^{2+}), to facilitate phage binding to the cell surface, and gelatin, to stabilize phage particle structure. Commonly used buffers are Lambda buffer (50 mM Tris-HCl [pH 7.5], 100 mM NaCl, 1 mM $MgSO_4$, and 0.2% gelatin)[40] or SM buffer (50 mM Tris-HCl [pH 7.5], 100 mM NaCl, 8 mM $MgSO_4$, and 0.01% gelatin).[40] Alternatively, the broth used for host cell infection can be used. Host cells are freshly grown in broth (usually overnight), diluted into fresh broth, and grown to mid-exponential phase of growth to approximately 1×10^8 cfu mL^{-1} (cfu = colony forming unit). This broth is often supplemented with Ca^{2+} to promote phage binding. The cells are then spread over an agar plate, allowed to air dry and then 10 to 20-µL samples[41] of each phage dilution spotted onto the lawn of the bacteria to produce discrete areas of infection (see Figure 8.1). Once the samples have dried, the plates are incubated overnight at the appropriate temperature and the numbers of plaques counted only in samples where discrete plaques are seen. This value is then used to either determine the titer of the phage stock, or the Routine Test Dilution (RTD; Ref. 42). The RTD is defined as the highest titer that just fails to produce confluent (complete) lysis of the cells within the sample area. The appearance of semi-confluent lysis indicates that the number of phage present is reaching a point of dilution where numbers are low enough to allow individual infection events to be detected and the growth of plaques from a point of infection can be visualized. This is an important point because the applications of high numbers of phage to a lawn of bacteria can produce confluent lysis although productive infection does not occur. This phenomenon can be due to the process of "lysis from without";[43] in essence, large numbers of phage binding to the cell surface damage the cell wall integrity and cells lyse before the phage have replicated. Alternatively, phage preparations may contain defective phage particles or bacteriocins[44] that again can cause complete cell lysis, or the phage suspension itself can contain chemical residues that inhibit lawn growth, giving the appearance of confluent lysis in the area where a sample has been applied. Any of these possibilities may lead to false areas of confluent lysis but are ruled out by dilution of the phage sample to the point where semi-confluent lysis occurs, which will not occur unless phage infect their host cells and grow.

The phage concentration to produce semi-confluent lysis depends on plaque morphology (phage producing fast-growing, large plaques will require fewer numbers to form semi-confluent lysis than phage that produce small discrete plaques) but is generally between 5×10^4 and 2×10^5 pfu mL^{-1} (Ref. 45) and should be two- to ten-fold lower than the concentration required to produce confluent lysis. The bacterial strain used to determine the RTD is always that used to propagate the phage so that host adaptation (see section entitled "Choice of Phage") does not affect the titer results.

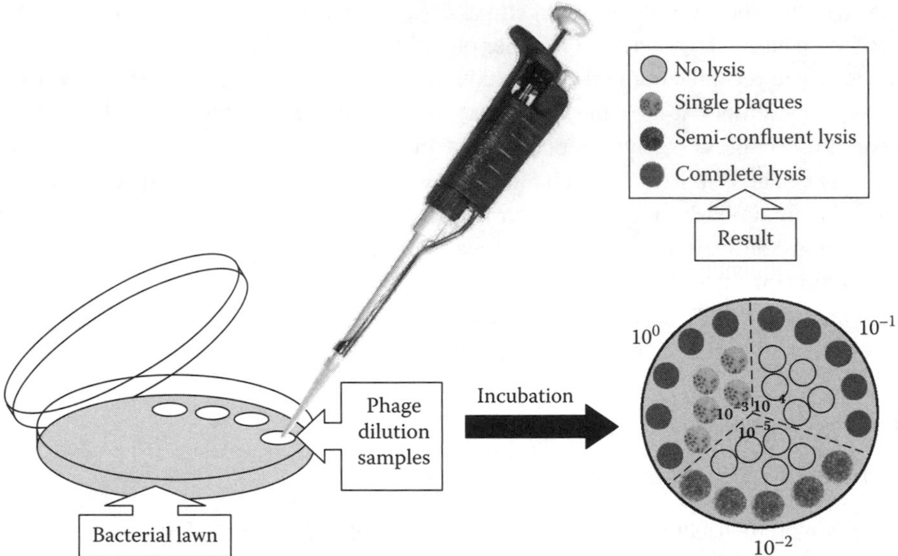

FIGURE 8.1 Phage titration and interpretation of results. Freshly grown host cells are spread over an agar plate to form a bacterial lawn. A tenfold serial dilution of the phage is prepared and then 10 to 20 μL samples of each phage dilution spotted onto the lawn and allowed to air dry. After incubation, numbers of plaques are counted only in samples where discrete plaques are formed. The Routine Test Dilution (RTD; Ref. 42) is defined as the highest titer that just fails to produce confluent (complete) lysis of the cells within the sample area.

Determining the Efficiency of Plating

Due to the subtle differences in cell surface components and cellular biochemistry that influence phage binding (described in section entitled "Factors Affecting Phage Infection"), even without host adaptation, the efficiency with which a particular phage will infect different strains of a bacterium will vary. This basically reflects the tightness of the interaction between the phage and host receptors, with stronger binding leading more often to productive infection events and weaker binding leading to fewer, and hence a lower titer. By comparing the titer of a phage preparation when tested against different strains of bacteria, a ratio of productive infection events can be determined. This value is termed the "efficiency of plating" (or EOP), with a value of 1 representing 100% efficiency (i.e., the titer on the two test strains is the same) and 0.9 representing 90% efficiency or a titer of one $\log_{10}$ lower. Because the binding of the phage to the cell can also be influenced by the differential expression of surface proteins and carbohydrates (see section entitled "Factors Affecting Phage Infection"), and this is influenced by the particular growth state of the cell, some variation in the titer of phage on the same strain can occur when the experiment is performed on different days. Therefore, a consistent difference of at least five- to ten-fold in titer is required before a reliable difference in EOP can be established. This degree of variability in expected titer is another reason for choosing semi-confluent lysis as the RTD when performing phage typing (see section entitled "Phage Typing Methods"). If a dilution was chosen that gave in the range of 10 to 20 discrete plaques, sometimes a result would be positive, and sometimes negative. The use of a higher phage dilution takes this inherent variation into account and allows a more robust typing result to be achieved.

Phage Typing Methods

For phage typing, a panel of phage characterized to have a discriminating host range is chosen and strains are infected with each phage at standard concentration (the RTD). The method is similar to that for determining phage titer, in that a bacterial lawn is prepared and samples of phages are applied to the surface (see Figure 8.2). Now samples of different phage from the phage typing set, each diluted to the RTD, are spotted onto the lawn in a grid pattern. In some typing schemes, both the RTD and 100X the RTD are tested. After incubation of the plates, infection is detected by the presence of plaques and patterns of susceptibility to individual phage are determined, leading to the determination of a phage type based on the pattern of sensitivity (see Figure 8.2). The results are not only scored as negative or positive, but are also given a semi-quantitative description that in some way reflects the EOP (i.e., at the RTD, if the phage infects with the same efficiency that it infects the host strain, semi-confluent lysis should occur. If confluent lysis is seen on the test strain, then the EOP is greater than 1; if single plaques appear, the EOP is less than 1). Different phage typing schemes use different conventions of scoring; commonly, letter descriptors are used for confluent lysis (CL), semi-confluent lysis (SCL), and opaque lysis (OL); this refers to the situation where growth of the lawn occurs within the zone of lysis so that the area is not cleared. These can be combined with < symbols to indicate intermediate degrees of lysis (e.g., <CL or <SCL), and then ± to +++ to indicate increasing numbers of discrete plaques. In typing schemes where more than one phage dilution is used, the scoring is more directly related to the EOP, where 5 = ++ reaction in the same dilution as on the propagating strain, 4 = ++ reaction in a dilution 10- to 100-fold more concentrated than that giving ++ on the propagating strain, 3 = ++ in a dilution 10^3- to 10^4-fold more concentrated than that giving ++ on the propagating strain, 2 = ++ in a dilution 10^5- to 10^6-fold more concentrated than that giving ++ on the propagating strain, and 1 indicates very weak lysis and a minus symbol (–) for no reaction. The number 0 is used to record evidence of lysis from without. The typical reaction patterns for a variety of phage can be found in the published literature and Public Health Agencies, and commercial companies still routinely use these methods for the sub-typing of a range of bacteria, such as *Salmonella enterica*,[46,47] *Escherichia coli* O157,[48,49] *Shigella*,[5,6] *Vibrio cholerae*,[50,51] *Staphylococcus aureus*,[45,52] *Listeria monocytogenes*,[53–55] and *Bacillus* spp.[56–58]

The Use of Phage Typing

These phage typing sets have been successfully used in epidemiological studies to monitor changes in the predominant organism causing disease in a population, and also the emergence of new

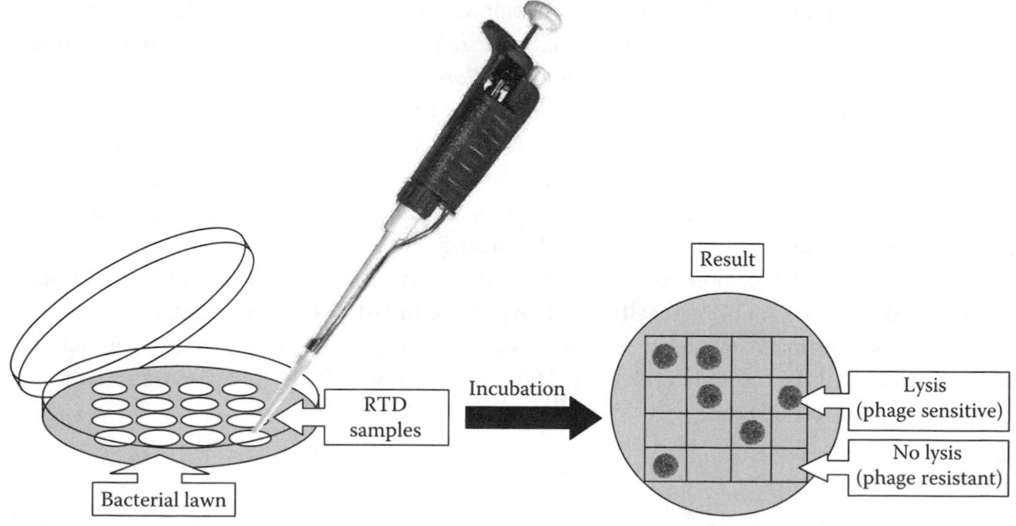

FIGURE 8.2 Phage typing. A panel of phage with a discriminating host range is each diluted to the RTD. 10 to 20 μL of each diluted phage are applied to the surface of a bacterial lawn of the test strain in a grid pattern. After incubation of the plates, infection is detected by the presence of plaques and patterns of susceptibility to individual phage are determined. See text for more details of qualitative scoring schemes used.

dominant clones. For *Salmonella enterica,* the phage typing system described by Callow[59] still provides a useful basis for the differentiation of strains from within the same serovar. This is particularly important for epidemiological studies due to the widespread distribution of some *Salmonella* serovars, and phage typing has been used worldwide to investigate a variety sources of bacterial contamination, environmental spread of the organism, and outbreaks of food poisoning.[60–63] These typing schemes have also been extended to include other members of the Enterobacteriacae,[64] which allows them to be used in a wider context for investigations of environmental sources of contamination. Similarly, studies of the spread of *Staphylococcus* have employed phage typing to monitor the spread of clones in hospital populations[65–67] and has been used to follow the dissemination of *Vibrio cholerae* strains in India.[11, 68] The effectiveness of these studies demonstrates the usefulness of this simple microbiological method; however, phage typing is not a rapid method and strains can only be typed after they have been isolated as pure cultures and identified.

Phage-Based Rapid Detection Methods

Traditional culture-based methods of detection exploit the specific growth requirements or cellular biochemistry of particular bacteria to achieve selective growth and diagnostic biochemical assays. Further identification can then be achieved using antibodies (or phage typing) to reveal more information about variation in cell surface structures. To achieve this, single cells present in a sample must be separated from the sample matrix and then grown until discrete colonies are formed. This is inevitably time consuming, and the drive of both the fields of medical microbiology and food microbiology is for rapid, more sensitive diagnostic tests. Using phage to detect the presence of their host cell is an obvious way to address this need. Phage-based tests use the specificity of the interaction between the virus and the host cell as a diagnostic tool, and in this sense is not really different from the use of phage in typing schemes. However, the specificity of interaction also means that the target cell can be detected in the presence of competitive flora, thus reducing the need for selective growth of the target bacterium before it is detected. Finally, the fact that the replication of the phage is so much faster than that of the host cell also allows methods to be developed that are more rapid than traditional culture.

Unlike phage selected for typing sets, phage with the widest host range are chosen for rapid detection methods, preferably those that can encompass all members of the genus or species to be detected. There are many examples of well-characterized, broad host range phage, but one that has been used more than most to develop these assays is *Salmonella* phage Felix 01; it has been shown that this phage can infect more than 95% of *Salmonella* isolates.[69] Similarly, two broad host range phage have been identified for *Listeria*: (1) A511,[70] which infects 95% of those serotypes most commonly associated with human disease (1/2a and 4b), and (2) P100,[71] which can infect and lyse more than 95% of serotypes 1/2 and 4 (*L. monocytogenes*) and 5 (*L. ivanovii*). Both of these phage have been used to develop rapid methods for the identification of *Listeria* in food samples. Broad host range phage specific for other bacteria can be readily isolated, often from those environments where the host cells are found naturally.[72–74] Using these broad host range phage, many different phage-based assay formats have been developed, but this section reviews the three approaches that have been developed into commercial assays. For an overview of other phage-based methods not described here, see Ref. 75.

Reporter Phage

In the reporter phage assay, phage are genetically engineered to deliver a reporter gene into the target bacterium. While the specificity of the assay comes from the phage host range, the speed and sensitivity of the assay come from the fact that the reporter genes are inserted into the phage so that they are expressed at high levels following phage infection, and sensitive instrumentation means that their synthesis can be detected at very low levels.

Reporter phage were first described in 1987[76] when the principle of reporter phage assays was demonstrated using a Lambda cloning vector containing the bacterial *lux* operon that was used for detection of *Escherichia coli* in milk samples. Even this simple model was able to detect as few as ten *E. coli* cells within 1 hr using a luminometer to detect the light produced by the infected cells.[77] This highlights one of the biggest advantages of the phage assays over many modern molecular tests: to generate the signal, the host cell must be viable to allow the reporter genes encoded by the phage to be expressed. This allows the same live/dead differentiation achieved using conventional culture techniques, and this is an important factor when carrying out analysis of processed foods. This group simply created wild-type *lux* phage by introducing the reporter genes by random mini Tn*10-luxAB* mutagenesis. These reporter phage were used to detect enteric pathogens on swab samples from abattoir, meat-processing factory surfaces, and animal carcasses.[78] In this study, light levels were compared with viable count and a good correspondence between the two measurements was found. The phage assay allowed detection of 10^4 cfu g^{-1} or cm^{-2} within 1 hr or <10 cfu g^{-1} or cm^{-2} following a 4-hr enrichment, demonstrating good sensitivity and that a far faster result could be achieved than by conventional culture. A similar level of sensitivity was also reported using a Lambda *lux* phage to detect enteric bacteria without enrichment.[79]

More recently, reporter phage have been created in a more directed manner, using homologous recombination to introduce the reporter gene into a precise location in the phage genome. For example, *luxAB* genes have been introduced into the *Salmonella* phages that infect groups B, D, and some group C serovars using this method.[80] A mixed phage preparation was used to detect as few as 10 cfu mL^{-1} of *Salmonella* cells with 6 hr of pre-incubation. Without pre-incubation, the limit of detection was 10^8 cfu mL^{-1} in 1 to 3 hr. The same method was used to introduce *luxAB* genes into the late gene region of the broad host range listeriaphage A511.[81] When this *lux* phage was tested in a range of different food types, detection levels of between 0.1 cells g^{-1} to 10 cells g^{-1} were achieved after a 20-hr selective enrichment (giving a total assay time of 24 hr), depending on the food type tested.[82] This work highlighted the fact that the nature of the matrix that contains the bacteria is an important factor in determining test sensitivity, as clearly the efficiency of phage infection was affected by either the food components or the high levels of competitive microflora present.

Phage structural genes were also targeted to introduce the fluorescent reporter gene *gfp* into the *Escherichia coli* O157:H7-specific phage PP01.[83] This reporter phage was used to specifically detect *E. coli* O157:H7 cells in the presence of other competitive *E. coli* cells, again demonstrating that phage specificity can be used to remove the need for selective removal of such competitors before identification is possible. The same group also introduced the *gfp* gene into phage T4, which was unable to lyse the *E. coli* host cells at the end of the replication cycle, leading to sustained expression of the *gfp* gene within the infected cells. These highly fluorescent cells could then be detected by automated methods.[84]

Another reporter phage assay for *Escherichia coli* O157:H7 has been developed by inserting the *lacZ* gene into phage T4.[85] In this case, the specificity of the assay does not come from the phage alone, but it has been combined in a three-step assay comprising enrichment, immunomagnetic separation (IMS), and the use of a colored enzyme substrate in a simple-to-use sample tube format (Phast Swab[86]). The advantage here is that by using traditional culture and separation methods, they have overcome the requirement to find a phage with a very defined specificity. A similar idea of IMS methods and reporter phage technology was also used to detect *Salmonella,* this time using phage P22 containing the ice nucleation reporter gene (*ina*) as a reporter.[87, 88] As few as 10 *Salmonella* Dublin cells mL^{-1} could be detected in the presence of a high level of competitor organisms using the phage alone, but the sensitivity of the assay for food samples was further increased by using it in combination with salmonellae-specific immunomagnetic bead separation.[89] This assay was developed into a commercial assay by the Idetek Corporation but is no longer available.

The advantage of the Phast Swab assay format is that the phage are packaged into a unit that does not require dispensing of liquids or opening once the sample has been added to the tube. This type of practical consideration makes the assay more accessible to workers in routine quality assurance labs. A second point in its favor is that it also achieves containment of the genetically engineered phage. In response to raised public concern about the release of GM organisms, nonreplicative forms of reporter phage have become favored because they minimize any possible risk of release of the GM phage. An example of this is the *lux* phage created by random mini Tn*10-luxAB* mutagenesis of phage NV10[90] that is specific for *Escherichia coli* O157:H7. In this case, the insertion of *lux* genes into the phage genome created a defective phage that could infect but not replicate.[91] Using this reporter phage, *E. coli* O157:H7 cells were detected within 1 hr of infection. Rather than relying on random mutagensis effects, Kuhn and co-workers[92] created a genetically "locked" phage that can express the luciferase gene after infecting any wild-type *Salmonella* cells but can only replicate in specially engineered host cells. However, this type of engineering program requires a considerable amount of genetic analysis of the phage to be used and is probably not practicable for a large number of different phage.

The most extensively studied reporter phage are those produced for the rapid detection of *Mycobacterium tuberculosis* since circumventing the need to culture this slow-growing organism reduces detection times from weeks to days. There is also less concern in the area of medical diagnosis about GM and, accordingly, the reporter phage have been extensively evaluated, although no commercial product has been produced. Reporter phage produced for the detection of *Mycobacteria* containing the reporter gene firefly luciferase (*luc* or *Fflux*)[85] have been produced using the lytic phage TM4[93, 94] and the temperate phage L5,[95] which was surprisingly effective at detecting cells despite the fact that it could form lysogens. However, the most effective *luc* reporter phage (or LRP) has been developed using phage TM4, and this has been extensively modified to optimize reporter gene expression.[94] Using these phage in combination with refinements of the sample prepation, mycobacteria were successfully detected in smear-positive human sputum samples within 24 to 48 hr[96] and, when compared with other standard microbiological testing methods, was found to perform favorably in the detection of positive clinical samples.[94]

Phage Amplification

In contrast to the reporter phage, this technology uses non-engineered phage and thus does not suffer from any of the complications surrounding the release of GM organisms. This also makes the development of new assays easier and less expensive, as demonstrated by the fact that one of the earliest publications describes the simultaneous development of tests for *Pseudomonas aeruginosa, Salmonella typhimurium,* and *Staphylococcus aureus.*[97] The other components of the assay are also standard microbiological materials and thus there is no need to purchase specialist equipment or for highly trained staff to carry out the assay. Although termed the "phage amplification assay," as growth of the phage is used to indicate the presence of the target bacterium, the test is effectively a *phage protection assay.* Samples to be tested for the presence of a target bacterium are mixed with phage and incubated to allow phage infection to occur. A virucide is then added to the sample to destroy any phage that have not infected their host cell. Next, the virucide is neutralized, the infected cells are mixed with more bacteria that will support phage replication (helper or sensor cells), and the entire sample is plated out using soft agar. During incubation of the plates, the phage finish their replication cycle, lyse the target cell, and the released phage go on to form plaques by infecting the helper cells present in the lawn. Each plaque represents a phage that was protected from the action of the virucide by infecting its specific target host (see Ref. 98). To develop an assay, all that is required is a phage with an appropriate host range and a virucide that will inactivate the phage while not damaging the host cell. This latter point is key to the assay; the phage must be protected within the cell, and the cell must be viable to allow phage replication to be completed.[99] Hence, like the reporter phage assays, phage amplification assays also provide live/dead differentiation.

The test still requires overnight incubation to allow growth of the helper cells to form a lawn; and because of this, although this assay method can be used to detect any bacterial cell, it is in the detection of slow-growing organisms — such as pathogenic mycobacteria — that the biggest advantage is gained. A phage amplification method for the detection of *Mycobacterium tuberculosis* in human sputum samples has been developed that takes only 2 days, compared to several weeks for other methods of detection, and is currently sold as the commercial test FAST*Plaque*TB®*.[100, 101] To achieve rapid detection using this assay, the phage chosen is D29, which has a broad host range and can infect both *M. tuberculosis* and also *M. smegmatis,* a member of the fast-growing group of *Mycobacteria* that can form colonies in 18 hr. *M. smegmatis* is used as the helper cells and lawn develops for plaque visualization after overnight incubation. The limitation of this assay is that, because of the broad host range of the phage, it will detect any type of *Mycobacterium*, including non-pathogens. However, when used to test human sputum samples, it is unlikely that high levels of non-pathogenic mycobacteria will be present and detection of high numbers of mycobacteria cells indicated that the patient requires treatment. A range of workers have carried out evaluations of this test and generally it performs as well as other culture-based tests.[102, 103] However, the main use of the assay has been in clinical settings in developing countries where it is difficult to afford the reagents and machines required for automated, culture-based, rapid methods, such as MGIT (Mycobacteria Growth Indicator Tube).[104]

The broad host range of this phage has made its application in this assay format problematical for samples where both pathogenic and non-pathogenic, environmental mycobacteria may be present. However, the broad host range of the phage means that it can be used to detect other mycobacteria if it is used in combination with other tests that improve the specificity of the assay. To this end, the FAST*Plaque*TB® reagents have been used for the detection of viable *Mycobacterium bovis* or *Mycobacterium avium* spp. *paratuberculosis* by combining the phage amplification assay with PCR confirmation of cell identity.[105] To do this, DNA is extracted from plaques at the end of the phage assay and then PCR is used to detect signature IS elements (IS900 for *M. paratuberculosis* and IS1081 for *M. bovis*). This combined phage amplification PCR assay was shown capable of detecting viable *M. paratuberculosis* in milk samples from infected cattle within 24 hr.[105]

Another combined assay that uses a variation of the phage amplification assay has been described for the detection of *Salmonella.*[106] Immunomagnetic separation is used to separate the target bacteria and the food sample, and the concentrated *Salmonella* cells are then infected with phage. Phage that have not infected a target cell are removed by washing, which also separates the *Salmonella* from the magnetic beads. These infected cells are incubated so that the phages can complete their replication cycle and then they are added to a culture of a phage propagating strain (signal amplifying cells or SACs). The optical density of the culture is monitored and cell lysis indicates that phages, protected within an infected cell, have been carried through the washing steps, indicating the presence of *Salmonella* in the original sample. Using this method, a detection limit of less than 10^4 cells mL^{-1} was achieved in 4 to 5 hr.

ATP Detection

ATP is generated in all living cells, and cells growing under given conditions contain constant levels of ATP, and therefore measuring ATP provides an indication of bacterial cells present in a sample. Dead cells rapidly deplete intracellular ATP levels due to the continued activity of intracellular ATPases; hence, direct measurement of ATP concentrations can be used to correlate with viable cell number and this is the basis of several commercially produced hygiene tests.[107] ATP is measured using firefly luciferase,[85] using a small portable luminometer.[108] Because a constant amount of light is produced per ATP molecule, there is a linear relationship between the number of photons of light and the number of ATP molecules (and hence cells) present in a sample. The reagents for this assay are the enzyme luciferase and the substrate luciferin, neither of which will freely permeate cell membranes. Therefore, to detect the ATP present within living cells, they must be first lysed open. In general hygiene tests, this is achieved using a chaotrophic lysing agent, but no information about the nature of the cells detected is given. To add specificity to these hygiene tests, phage[9] — or phage components[110–112] — have been used as specific lysing agents so ATP is only released from the target bacterial cells.

When carrying out hygiene tests, there is likely to be organic material present that will contribute to high background levels of ATP being present in samples, making the practical limit of detection for ATP assays of environmental surface swabs (usually a 100-cm^2 area) approximately 10^4 bacterial cells (there is approximately 10^{-15} g ATP per bacterial cell), and this is well above the levels of pathogenic bacteria expected to be found due to a cross-contamination event. To increase the sensitivity of the assay, the activity of the enzyme that converts ADP to ATP within the cell (adenylate kinase or AK) can be measured rather than ATP, because the levels of this enzyme per cell are also constant (see Ref. 113). Because the enzyme will continue to convert its substrate to ATP, which can then be detected by the luciferase/luciferin bioluminescence assay, overall sensitivity is approximately tenfold higher than measuring ATP alone. This AK assay has been used in combination with phage-mediated cell lysis and to add specificity, and the limit of detection can be reduced to less than 10^3 *Escherichia coli* or *Salmonella* cells,[114–116] which is a similar level of sensitivity achieved by other rapid methods of bacterial detection, such as immunoassays.[117]

The sensitivity of the pathogen-specific ATP assay has also been improved by combining it with other rapid methods of identification to form the commercially available *fastrAK*®* assay.[118] This system has been developed for rapid and sensitive detection of bacterial pathogens, such as *Salmonella*, *Escherichia coli* 0157 and *Listeria* in food samples. The assay has four stages that include an 8-hr pre-enrichment, immuno-magnetic concentration and separation of cells, specific phage-mediated lysis, and finally a luciferase AK assay to detect the presence of target cells. Each of these stages is designed to increase the specificity and sensitivity of the assay and it has been shown to be able to detect less than 10 bacterial cells in less than 11 hr, even in the presence of a competitive microflora at levels of 10^7 cfu g^{-1}. The level of sensitivity achieved is as good as conventional culture methods and is considerably faster than standard techniques.

OVERVIEW

The history of phage typing has been a little like phage therapy. Since its discovery, it has gone through different phases of popularity, but even today it continues to be a useful and sensitive tool for rapid discrimination of bacterial cell types. Phage typing *per se* has been superseded to some extent by molecular typing methods, but the technique remains a useful and sensitive method of investigating differences in cell structure. Rapid methods of detection have also shown promise but then fell out of favor, probably because no one phage can provide the "holy grail" of identification — that is, the one test that detects all target cells with 100% specificity and sensitivity. However, the recent trend of combining these with other rapid methods of bacterial detection to improve the speed and specificity of phage-based testing shows promise in producing assays that do indeed meet the requirements of the consumer; that is, simple, rapid tests with a high degree of discrimination. Phage identification methods have been in use for some 80 years now, and this recent revival of their fortunes in the form of commercial tests that are beginning to emerge suggests that they will continue to be used for some time to come.

REFERENCES

1. Summers, W.C., Bacteriophage research: early history, in *Bacteriophages: Biology and Applications,* Kutter E. and Sulakvelidze A., Eds., CRC Press, Boca Raton, FL, 2005, chap. 2.
2. Dimitrov, D.S., Virus entry: Molecular mechanisms and biomedical applications, Nat. Revs. Microbiol., 2, 109, 2004. Registered trademark of Biotec Laboratories Ltd, Ipswich, U.K
3. D'Herelle, F., *The Bacteriophage and Its Behaviour.* Williams and Wilkins Co., Baltimore, MD, 1926.
4. Ackermann, H.-W., Phage or Phages, *Bacteriophage Ecol. Group News*, 14, 2002. http://www.mansfield.ohio-state.edu/~sabedon/

5. Kallings, L.O., Lindberg, A.A., and Sjoberg. L., Phage typing of *Shigella sonnei, Arch. Immun. Ther. Exp.,* 16, 280, 1968.
6. Pruneda, R.C. and Farmer, J., Bacteriophage typing of *Shigella sonnei, J. Clin. Microbiol.,* 5, 66, 1977.
7. Adlakha, S., Sharma, K.B., and Prakash, K., Sewage as a source of phages for typing strains of *Salmonella typhimurium* isolated in India, *Indian J. Med. Res.,* 84, 1, 1986.
8. Baker, P.M. and Farmer. J.J., New bacteriophage typing system for *Yersinia enterocolitica, Yersinia kristensenii, Yersinia frederiksenii,* and *Yersinia intermedia*: correlation with serotyping, biotyping, and antibiotic susceptibility, *J. Clin. Microbiol.,* 15, 491, 1982.
9. Bhatia, T.R., Phage typing of *Escherichia coli* isolated from chickens, *Can. J. Microbiol.,* 23, 11, 1977.
10. Brown, D.R., Holt, J.G., and Pattee, P.A., Isolation and characterization of Arthrobacter bacteriophages and their application to phage typing of soil arthrobacters, *Appl. Environ. Microbiol.,* 35, 185, 1978.
11. Chakrabarti, A.K. et al. [Authors are Chakrabarti, Ghosh, Balakrish Nair, Niyogi, Bhattacharya, and Sarkar.], Development and evaluation of a phage typing scheme for *Vibrio cholerae* O139, *J. Clin. Microbiol.,* 38, 44, 2000.
12. Gaston, M.A., Isolation and selection of a bacteriophage-typing set for *Enterobacter cloaca, J. Med. Microbiol.,* 24, 285, 1987.
13. Craigie, J. and Yen, C.N., The demonstration of types of *B. typhosus* by means of preparations of type II Vi phage, *Can. Public Health J.,* 29, 448, 1938.
14. de Gialluly, C. et al. [Authors are de Gialluly, Loulergue, Bruant, Mereghetti, Massuard, van der Mee, Audurier, and Quentin.], Identification of new phages to type *Staphylococcus aureus* strains and comparison with a genotypic method, *J. Hosp. Infect.,* 55, 61, 2003.
15. Luria, S.E. and Human, M.L., A nonhereditary, host-induced variation of bacteria viruses, *J. Bacteriol.,* 64, 557, 1952.
16. Bertani, G. and Weigle, J.J., Host controlled variation in bacterial viruses, *J. Bacteriol*., 65, 113, 1953.
17. Linn, S. and Arber, S., A restriction enzyme from *Hemophilus influenzae*. Purification and general properties, *Proc. Natl. Acad. Sci. U.S.A.,* 59, 1300, 1968.
18. Meselson, M. and Yuan, R., DNA restriction enzyme from *E. coli., Nature,* 217, 1110, 1968.
19. Smith, H.O. and Wilcox, K.W., A restriction enzyme from *Hemophilus influenzae*: Purification and general properties, *J. Mol. Biol.,* 51, 379, 1970.
20. Eriksson, U., Adsorption of phage P22 to *Salmonella typhimurium, J. Gen. Virol.,* 34, 207, 1977.
21. Estrela, A.I. et al. [Authors are Estrela, Pooley, Delencastre, and Karamata.], Genetic and biochemical characterization of *Bacillus subtilis* 168 mutants specifically blocked in the synthesis of the teichoic acid poly(3-O-beta-D-glucopyranosyl-*N*-acetylgalactosamine 1-phosphate) — *gneA*, a new locus, is associated with UDP-*N*-acetylglucosamine 4-epimerase activity, *J. Gen. Microbiol.,* 137, 943, 1991.
22. Wendlinger, G., Loessner, M.J., and Scherer, S., Bacteriophage receptors on *Listeria monocytogenes* cells are the N-acetylglucosamine and rhamnose substituents of teichoic acids or the peptidoglycan itself, *Microbiology,* 142, 985, 1996.
23. Hung, C.H., Wu, H.C., and Tseng, Y.H., Mutation in the *Xanthomonas campestris xanA* gene required for synthesis of xanthan and lipopolysaccharide drastically reduces the efficiency of bacteriophage phi L7 adsorption, *Biochem. Biophys. Res. Commun.,* 291, 338, 2002.
24. Baggesen, D.L, Wegener, H.C., and Madsen, M., Correlation of conversion of *Salmonella enterica* serovar enteritidis phage type 1, 4, or 6 to phage type 7 with loss of lipopolysaccharide, *J. Clin. Microbiol.,* 35, 330, 1997.
25. Lawson, A.J. et al. [Authors are Lawson, Chart, Dassama, and Threlfall]. Heterogeneity in expression of lipopolysaccharide by strains of *Salmonella enterica* serotype Typhimurium definitive phage type 104 and related phage types, *Lett. Appl. Microbiol.,* 34, 428, 2002.
26. Joys, T.M., Correlation between susceptibility to bacteriophage PBS1 and motility in *Bacillus subtilis, J. Bacteriol*., 90, 1575, 1965.
27. Iino, T. and Mitani, M., Flagellar-shape mutants in *Salmonella, J. Gen. Microbiol*., 44, 27, 1966.
28. Merino, S., Camprubí, S., and Tomás, J.M., Isolation and characterization of bacteriophage PM3 from *Aeromonas hydrophila* the bacterial receptor for which is the monopolar flagellum, *FEMS Microbiol. Lett*., 69, 277, 1990.
29. Romantschuk, M. and Bamford, D.H., Function of pili in bacteriophage phi6 penetration, *J. Gen. Virol.,* 66, 2461, 1985.
30. Schwaartz, M., Phage Lambda receptor (LamB protein) in *Escherichia coli, Meth. Enzymol.,* 97, 100, 1983.

31. Heilpern, A.J. and Waldor, M.K., CTX phi, infection of *Vibrio cholerae* requires the *tolQRA* gene products, *J. Bacteriol.*, 182, 1739, 2000.
32. Sun, T.P. and Webster, R.E., Nucleotide sequence of a gene cluster involved in entry of E colicins and single-stranded DNA of infecting filamentous bacteriophages into *Escherichia coli*, *J. Bacteriol.*, 169, 2667, 1987.
33. Callegari, M.L. et al. [Authors are Callegari, Sechaud, Rousseau, Bottazzi, and Accolas.], Le recepteur du phage 832-B1 de *Lactobacillus helveticus* CNRZ 892 est une protéine, in *Proc. 23th Dairy Congress, Vol. I., Montreal,* 1990.
34. Harvey, D. et al. [Authors are Harvey, Harrington, Heuzenroeder, and Murray.], Lysogenic phage in *Salmonella enterica* serovar Heidelberg (*Salmonella* Heidelberg): implications for organism tracing, *FEMS Microbiol. Lett.*, 103, 291, 1993. Registered trademark of Alaska Food Diagnostics Ltd, Salisbury, U.K.
35. Frank, S.A., Polymorphism of bacterial restriction-modification systems — the advantage of diversity, *Evolution,* 48, 1470, 1994.
36. Chopin, M.C., Chopin, A., and Bidnenko, E., Phage abortive infection in lactococci: variations on a theme, *Curr. Opin. Microbiol.,* 8, 473, 2005.
37. Tangney, M. and Fitzgerald, G.F., AbiA, a lactococcal abortive infection mechanism functioning in *Streptococcus thermophilus*, *Appl. Environ. Microbiol.,* 68, 6388, 2002.
38. Tran, L.S.P. et al. [Authors are Tran, Szabo, Ponyi, Orosz, Sik, and Holczinger.], Phage abortive infection of *Bacillus licheniformis* ATCC 9800; identification of the *abi*BL11 gene and localisation and sequencing of its promoter region, *Appl. Microbiol. Biotechnol.,* 52, 845, 1999.
39. Petty, N.K. et al. [Authors are Petty, Evans, Fineran, and Salmond.], Biotechnological exploitation of bacteriophage research, *Trends Biotechnol.,* 25, 7, 2007.
40. Sambrook, J., Fritsch, E.F., and Maniatis, T., *Molecular Cloning: A Laboratory Manual,* 2nd ed., Cold Spring Harbor Laboratory Press, 1989.
41. Miles, A.A. and Misra, S.S., The estimation of the bactericidal power of the blood, *J. Hygiene*, 38, 732, 1938.
42. Adams, M.H., *Bacteriophages*, Interscience, New York, 1959.
43. Heagy, R.C., The effect of 2-4 DNP and phage T2 on *E. coli, J. Bacteriol.*, 59, 367, 1950.
44. Kekessy, D.A. and Piguet, J.D., New method for detecting bacteriocin production, *Appl. Microbiol.,* 20, 282, 1970.
45. Blair, J.E. and Williams, R.E.O., Phage typing of *Staphylococci, Bull. WHO,* 24, 771, 1961.
46. Anderson, E.S. et al. [Authors are Anderson, Ward, Saxe, and de Sa.], Bacteriophage-typing designations of *Salmonella typhimurium, J. Hyg. (London),* 78**,** 297, 1977.
47. Laszlo, V.G. and Csorian, E.S., Subdivision of common *Salmonella* serotypes — phage typing of *S. virchow, S. manhattan, S. thompson, S. oranienburg and S. bareilly*, *Acta Microbiol. Hung*., 35, 289, 1988.
48. Ahmed, R. et al. [Authors are Ahmed, Bopp, Borczyk, and Kasatiya.], Phage-typing scheme for *Escherichia coli* O157:H7, *J. Infect. Dis.,* 155, 806, 1987.
49. Khakhria, R., Duck, D., and Lior, H., Extended phage-typing scheme for *Escherichia coli* O157:H7, *Epidemiol. Infect*., 105, 511, 1990.
50. Basu, S. and Mukerjee, S., Bacteriophage typing of Vibrio ElTor, *Experimenta,* 24, 299, 1968.
51. Sarkar, B.L., Cholera bacteriophages revisited, *ICMR Bull.*, 32, No. 4, ICMR Offset Press, India, 2002.
52. Richardson, J.F. et al. [Authors are Richardson, Rosdahl,, Van Leeuwen, Vickery, Vindel, and Witte.], Phages for methicillin-resistant *Staphylococcus aureus*: an international trial, *Epidemiol. Infect.,* 122, 227, 1999.
53. Mclauchlin, J. et al. [Authors are Mclauchlin, Audurier, Frommelt, Gerner-Smidt, Jacquet, Loessner, Van Der Mee-Marquet, Rocourt, Shah, and Wilhelms.], WHO study on subtyping *Listeria monocytogenes*: results of phage-typing, *Int. J. Food Microbiol.,* 32, 289, 1996.
54. Rocourt, J., Taxonomy of the *Listeria* genus and typing of L. monocytogenes, *Pathologie Biologie.*, 44, 749, 1996.
55. Sword, C.P. and Pickett M.J., Isolation and characterization of bacteriophages from *Listeria monocytogenes, J. Gen. Microbiol.,* 25, 241, 1961.
56. Abshire, T.G., Brown, J.E., and Ezzell, J.W., Production and validation of the use of gamma phage for identification of *Bacillus anthracis*, *J. Clin. Microbiol.,* 43, 4780, 2005.

57. Brown, E.R. and Cherry, W.B., Specific identification of *Bacillus anthracis* by means of a variant bacteriophage, *J. Infect. Dis.*, 96, 34, 1955.
58. Ackermann, H.-W. et al. [Authors are Ackermann, Azizbekyan, Bernier, de Barjac, Saindoux, Valéro, and Yu], Phage typing of *Bacillus subtilis* and *B. thuringiensis, Res. Microbiol.*, 146, 643, 1995.
59. Callow, B.R., A new phage-typing scheme for *Salmonella typhimurium, J. Hyg. (London)*, 57, 346, 1959.
60. Cooke, F.J. et al. [Authors are Cooke, Day, Wain, Ward, and Threlfall], Cases of typhoid fever imported into England, Scotland and Wales (2000-2003), *Trans. Roy, Soc. Tropical Med. Hyg.,* 101, 398, 2007.
61. Michel, P. et al. [Authors are Michel, Martin, Tinga, and Dore], Regional, seasonal, and antimicrobial resistance distributions of *Salmonella typhimurium* in Canada — a multi-provincial study, *Can. J. Pub. Health,* 97, 470, 2006.
62. Matsui, T. et al. [Authors are Matsui, Suzuki, Takahashi, Ohyama, Kobayashi, Izumiya, Watanabe, Kasuga, Kijima, Shibata, and Okabe], *Salmonella enteritidis* outbreak associated with a school lunch dessert: cross-contamination and a long incubation period, Japan, 2001, *Epidemiol. Infect.,* 132, 873, 2004.
63. HPA, *Salmonella enteritidis* PT56 in Durham—final report, 2004, Comm. Dis. Rep. CDR Wkly 14(5) [serial online], http://www.hpa.org.uk/cdr/archives/2004/cdr0504.pdf
64. He, X. and Pan, R., Bacteriophage lytic patterns for identification of Salmonellae, Shigellae, Escherichia coli, Citrobacter freundii, and Enterobactera cloacae, *J. Clin. Microbiol.,* 30, 590, 1992.
65. Borer, A. et al. [Authors are Borer, Yagupsky, Peled, Porat, Trefler, Shprecher-Levy, Riesenberg, Shipman, and Schlaeffer], Community-acquired methicillin-resistant *Staphylococcus aureus* in institutionalized adults with developmental disabilities, *Emerg. Intect. Dis.,* 8, 966, 2002.
66. Gomes, A.R., Westh, H., and de Lencastre, H., Origins and evolution of methicillin-resistant *Staphylococcus aureus* clonal lineages, *Antimicrob. Agents Chemother.,* 50, 3237, 2006.
67. Salmenlinna, S., Lyytikainen, O, and Vuopio-Varkila, J., Community-acquired methicillin-resistant *Staphylococcus aureus,* Finland, *Emerg. Intect. Dis.,* 8, 602, 2002.
68. Sarkar, B.L. et al. [Authors are Sarkar, Roy, Chakrabarti and Niyogi], Distribution of phage type of *Vibrio cholerae* O1 biotype ElTor in Indian scenario (1991–98), *Indian. J. Med Res.,* 109, 204, 1999.
69. Cherry, W.B. et al. [Authors are Cherry, Davis, Edwards, and Hogan], A simple procedure for the identification of the genus *Salmonella* by means of a specific bacteriophage, *J. Lab. Clin. Med.*, 44, 51, 1954.
70. Loessner, M.J. and Busse, M., Bacteriophage typing of *Listeria* species, *Appl. Environ. Microbiol.,* 56, 1912, 1990.
71. Carlton, R.M. et al. [Authors are Carlton, Noordman, Biswas, de Meester, and Loessner], Bacteriophage P100 for control of *Listeria monocytogenes* in foods: genome sequence, bioinformatic analyses, oral toxicity study, and application, *Regul. Tox. Pharmacol.,* 43, 301, 2005.
72. Barman, S. and Majumdar, S., Intracellular replication of choleraphage phi 92, *Intervirology,* 42, 238, 1999.
73. Jensen, E.C. et al. [Authors are Jensen, Schrader, Rieland, Thompson, Lee, Nickerson, and Kokjohn], Prevalence of broad–host range lytic bacteriophages of *Sphaerotilus natans*, *Escherichia coli*, and *Pseudomonas aeruginosa*, *Appl. Environ. Microbiol.*, 64, 575, 1998.
74. Sullivan, M.B., Waterbury, J.B., and Chisholm, S.W., Cyanophages infecting the oceanic *Cyanobacterium prochlorococcus, Nature,* 424, 1047, 2003.
75. Rees, C.E.D. and Loessner, M.J., Phage for the detection of pathogenic bacteria, in *Bacteriophages: Biology and Applications,* Kutter, E. and Sulakvelidze, A., Eds., CRC Press, Boca Raton, FL, 2005, chap. 9.
76. Ulitzur, S. and Kuhn, J., Introduction of *lux* genes into bacteria; a new approach for specific determination of bacteria and their antibiotic susceptibility, in *Bioluminescence and Chemiluminescence: New Perspectives,* Slomerich, R. et al., Eds. (Editors are Slomerich, Andreesen, Kapp, Ernst, and Woods) John Wiley & Sons, New York, 1987, p. 463.
77. Ulitzur, S. and Kuhn, J., U.S. Patent 4,861,709, Detection and/or identification of microorganisms in a test sample using bioluminescence or other exogenous genetically introduced marker, 1989.
78. Kodikara, C.P., Crew, H.H., and Stewart, G.S.A.B., Near on-line detection of enteric bacteria using *lux* recombinant bacteriophage, *FEMS Microbiol. Lett.*, 83, 261, 1991.
79. Duzhii, DE. and Zavilgelskii, G.B., Bacteriophage *lambda:lux*: design and expression of bioluminescence in *E. coli* cells, *Molec. Gen. Mikrobiol. Virusol.,* 3, 36, 1994.

80. Chen, J. and Griffiths, M.W., *Salmonella* detection in eggs using *lux*+ bacteriophages, *J. Food Protect.*, 59, 908, 1996.
81. Loessner, M.J. et al. [Authors are Loessner, Rees, Stewart, and Scherer.], Construction of luciferase reporter bacteriophage A511::*luxAB* for rapid and sensitive detection of viable *Listeria* cells, *Appl. Environ. Microbiol.*, 62, 1133, 1996.
82. Loessner, M.J., Rudolf, M., and Scherer, S., Evaluation of luciferase reporter bacteriophage A511::*luxAB* for detection of *Listeria monocytogenes* in contaminated foods, *Appl. Environ. Microbiol.*, 63, 2961, 1997.
83. Oda, M. et al. [Authors are Oda, Morita, Unno, and Tanji.], Rapid detection of *Escherichia coli* O157:H7 by using green fluorescent protein-labelled PP01 bacteriophage, *Appl. Environ. Microbiol.*, 70, 527, 2004.
84. Tanji, Y. et al. [Authors are Tanji, Furukawa, Na, Hijikata, Miyanaga, and Unno.], *Escherichia coli* detection by GFP-labeled lysozyme-inactivated T4 bacteriophage, *J. Biotechnol.*, 114, 11, 2004.
85. Goodridge, L. and Griffiths, M., Reporter bacteriophage assays as a means to detect foodborne pathogenic bacteria, *Food Res. Int.*, 35, 863, 2002.
86. Goodridge, L., presented at *16th Evergreen International Phage Biology Meeting,* Washington, August 7–12, 2005.
87. Wolber, P.K. and Green, R.L., Detection of bacteria by transduction of Ice Nucleation genes, *Trends Biotechnol.*, 8, 276, 1990.
88. Wolber, P.K. and Green, R.L., New rapid method for the detection of *Salmonella* in foods, *Trends Food Sci. Technol.*, 1, 80, 1990.
89. Irwin, P. et al. [Authors are Irwin, Gehring, Tu, Brewster, Fanelli, and Ehrenfeld], Minimum detectable level of Salmonellae using a binomial-based bacterial ice nucleation detection assay (BIND), *J. AOAC Int.*, 83, 1087, 2000.
90. Khakhria, R., Duck, D., and Lior, H., Extended phage-typing scheme for *Escherichia coli* O157:H7, *Epidemiol. Infect.*, 105, 511, 1990.
91. Waddell, T.E. and Poppe, C., Construction of mini-Tn*10luxAB*cam/Ptac-ATS and its use for developing a bacteriophage that transduces bioluminescence to *Escherichia coli* O157:H7, *FEMS Microbiol. Lett.*, 182, 285, 2000.
92. Kuhn, J. et al. [Authors are Kuhn, Suissa, Wyse, Cohen, Weiser, Reznick, Lubinsky-Mink, Stewart, and Ulitzur.], Detection of bacteria using foreign DNA: the development of a bacteriophage reagent for *Salmonella*, *Int. J. Food Microbiol.*, 74, 229, 2002.
93. Jacobs, W.R. et al. [Authors are Jacobs, Barletta, Udani, Chan, Kalkut, Sosne, Kieser, Sarkis, Hatfull, and Bloom.], Rapid assessment of drug susceptibilities of *Mycobacterium tuberculosis* by means of Luciferase Reporter Phages, *Science,* 260, 819, 1993.
94. Bardarov, S. et al. [Authors are Bardarov, Dou, Eisenach, Banaiee, Ya, Chan, Jacobs, and Riska.], Detection and drug-susceptibility testing of M-tuberculosis from sputum samples using luciferase reporter phage: comparison with the Mycobacteria Growth Indicator Tube (MGIT) system, *Diag. Microbiol. Infect. Dis.*, 45, 53, 2003.
95. Sarkis, G.J., Jacobs, W.R., and Hatfull, G.F., L5 Luciferase reporter mycobacteriophages — a sensitive tool for the detection and assay of live *Mycobacteria*, *Molec. Microbiol.*, 15, 1055, 1995.
96. Riska, P.F. et al. [Authors are Riska, Jacobs, Bloom, McKitrick, and Chan.], Specific identification of *Mycobacterium tuberculosis* with the Luciferase Reporter Mycobacteriophage: use of *p*-nitro-α-actylamino-β-hydroxy propiophenone, *J. Clin. Microbiol.*, 35, 3225, 1997.
97. Stewart, G.S.A.B. et al. [Authors are Stewart, Jassim, Denyer, Newby, Linley, and Dhir.], The specific and sensitive detection of bacterial pathogens within 4 h using bacteriophage amplification, *J. Appl. Microbiol.*, 84, 777, 1998.
98. Mole, R.J. and Maskell, T.W.O'C., Phage as a diagnostic — the use of phage in TB diagnosis, *J. Chem. Technol. Biotechnol.*, 76, 683, 2001.
99. de Siqueira, R.S., Dodd, C.E.R., and Rees, C.E.D., Evaluation of the natural virucidal activity of teas for use in the phage amplification assay, *Int J. Food Microbiol.*, 111, 259, 2006.
100. McNerney, R. et al. [Authors are McNerney, Wilson, Sidhu, Harley, Al Suwaidi, Nye, Parish, and Stoker], Inactivation of mycobacteriophage D29 using ferrous ammonium sulphate as a tool for the detection of viable *Mycobacterium smegmatis* and *M. tuberculosis, Res. Microbiol.*, 149, 487, 1998.
101. Stewart, G.S.A.B. et al. [Authors are Stewart, Jassim, Denyer, Park, Rostas-Mulligan, and Rees.], Methods for rapid microbial detection. PCT Patent WO 92/02633, 1992.

102. Muzaffar, R. et al. [Authors are Muzaffar, Batool, Aziz, Naqvi, and Rizvi], Evaluation of the FAST-PlaqueTB assay for direct detection of *Mycobacterium tuberculosis* in sputum specimens, *Int J. Tubercul. Lung Dis.,* 6, 635, 2002.
103. Dinnes, J. et al. [Authors are Dinnes, Deeks, Kunst, Gibson, Cummins, Waugh, Drobniewski, and Lalvani], A systematic review of rapid diagnostic tests for the detection of tuberculosis infection, *Health Technol. Assess.,* 11, 314, 2007.
104. Reisner, B.S., Gatson, A.M., and Woods, G.L., Evaluation of Mycobacteria Growth Indicator Tubes for susceptibility testing *of Mycobacterium tuberculosis* to isoniazid and rifampin, *Diag. Microbiol. Infect. Dis.,* 22, 325, 1995.
105. Stanley, E.C. et al. [Authors are Stanley, Mole, Smith, Glenn, Barer, McGowan and Rees], Development of a new, combined rapid method using phage and PCR for detection and identification of viable *Mycobacterium paratuberculosis* bacteria within 48 hours, *Appl. Environ. Microbiol.,* 73, 1851, 2007.
106. Favrin, S.J., Jassim, S.A., and Griffiths, M.W., Development and optimisation of a novel immunomagnetic separation-bacteriophage assay for detection of *Salmonella enterica* serovar Enteritidis in broth, *Appl. Environ. Microbiol.*, 67, 217, 2001.
107. Baumgart. J., Rapid methods of process control of cleaning and disinfection procedures — their benefits and their limitations, *Zentralblat. Hyg. Umweltmed.*, 199, 366, 1996.
108. Stanley, P.E., A review of bioluminescence ATP techniques in rapid microbiology, *J. Biolumin. Chemilumin.,* 4, 375, 1989.
109. Sanders, M.F. A rapid bioluminescent technique for the detection and identification of *Listeria monocytogenes* in the presence of *Listeria innocua,* in *Bioluminescence and Chemiluminescence: Fundamental and Applied Aspects,* Campbell, A.K., Kricka, L.J., and Stanley, P.E., Eds., John Wiley & Sons, Chichester. 1995, p. 454.
110. Loessner, M.J, Bacteriophage endolysins — current state of research and applications, *Curr. Opin. Microbiol.*, 8, 480, 2005.
111. Stewart, G.S.A.B., Loessner, M.J., and Scherer, S., The bacterial *lux* gene bioluminescent biosensor revisited, *ASM News,* 62, 297, 1996.
112. Schuch, R., Nelson, D., and Fischetti, V.A., A bacteriolytic agent that detects and kills *Bacillus anthracis, Nature*, 418, 884, 2002.
113. Corbitt, A.J., Bennion, N., and Forsythe, S.J., Adenylate kinase amplification of ATP bioluminescence for hygiene monitoring in the food and beverage industry, *Lett. Appl. Microbiol.,* 6, 443, 2000.
114. Blasco, R. et al. [Authors are Blasco, Murphy, Sanders, and Squirrell], Specific assays for bacteria using phage-mediated release of adenylate kinase. *J. Appl. Microbiol.,* 84, 661, 1998.
115. Squirrel, D.J., Price, R.L., and Murphy, M.J., Rapid and specific detection of bacteria using bioluminescence, *Analyt. Chim. Acta,* 457, 109, 2002.
116. Wu, Y., Brovko, L., and Griffiths, M.W., Influence of phage population on the phage-mediated bioluminescent adenylate kinase (AK) assay for detection of bacteria, *Lett. Appl. Microbiol.,* 33, 311, 2001.
117. Nikitin, P.I., Vetoshko, P.M., and Ksenevich, T.I., Magnetic immunoassays, *Sensor Letts.,* 5, 296, 2007.
118. Alaska Food Diagnostics Ltd., *fastrAK* product information, www.alaskafooddiagnostics.com. (Accessed 12 January, 2007.)

9 Phage Display and Selection of Protein Ligands

Wlodek Mandecki and Emanuel Goldman

CONTENTS

INTRODUCTION

One of the principal modes of controlling biochemical processes in cells relies on protein binding to a ligand. The ligand could be another protein, a nucleic acid, or a small molecule. *Phage display* is a tool that makes possible the derivation of protein ligands that have never previously been known to exist in nature, and later, to use them in many types of experiments to study molecular interactions. The principle behind phage display is to make a protein or peptide in a bacterial cell, immobilize it on a phage particle, and tether to the peptide a molecular signature that fully describes the sequence of the peptide. This signature is a nucleic acid sequence that becomes part of the phage genome carried on the same phage particle as the displayed peptide and actually encodes the displayed protein. Different protein sequences can be displayed by simply modifying the sequence of the DNA through molecular engineering. The collective display of a large number of molecules on the phage is called a *phage library*.

The power of phage display is derived from the fact that in a single biochemical experiment, an extremely large number of proteins can be studied, up to 10^{11} and more. That is, this number of peptides can be made in association with the phage particles and used in an experiment in which the binding of all the peptides to a target molecule is tested for, after which the binders with the desired affinity to the target can be identified and characterized.

The type of phage most often used to display proteins is the filamentous phage of *Escherichia coli*, M13 (see also Chapter 43). Of particular significance for phage display are two coat proteins — minor protein pIII and major coat protein pVIII — as both of them have been appended with random peptides or engineered proteins to create phage libraries. George Smith, in 1985, was the first to demonstrate that a protein, a foreign antigen, can be displayed on the filamentous phage particle fused to the pIII protein [1] and selected for in an enrichment experiment from among a large excess of ordinary phage. Five years later, in 1990, he and co-worker Jamie Smith reported on the construction of the first phage library [2]. It was a library of six-amino-acid-long peptides. The authors derived and tested several peptide ligands to two monoclonal antibody targets. The significance of this discovery quickly became clear to the scientific community, and the new field was established with explosive growth following. Today, more than 350 articles on phage display are published yearly [3].

The field grew in several directions. Different types of protein scaffolds were displayed on the phage and subjected to randomization, creating many types of useful libraries. Of particular significance was the display of antibodies or antibody fragments, so-called single-chain antibodies on filamentous phage, first reported by the MRC group [4, 5] and the Scripps group [6, 7]. Today, phage display approaches are a significant source of therapeutically and diagnostically useful antibodies, with the complexity of libraries having increased from 10^6 clones early on to now over 10^{11} clones. Other phage proteins have been explored as targets for protein display; of particular interest is the display of peptides on pVIII [8–11] that, in principle, allows for a display of a large number of identical peptides, although the peptides are fairly short on the phage coat.

A large number of phage libraries have been constructed over the years, with a number of proteins having been displayed on the phage and having provided a basis for construction of new types of phage libraries. The proteins ranged from single-chain antibodies and F_{ab} fragments of antibodies, to protein fragments, enzymes (even as large as alkaline phosphatase), protein hormones, inhibitors, toxins, receptors, ligands, DNA and RNA binding proteins, cytokines, and others. Very different classes of target molecules were subjected to selection, including antibodies, enzymes, sugars, and small molecules. A number of applications were reported, ranging from epitope mapping, identification of new receptors and natural ligands, drug discovery, epitope discovery, and selection of new materials.

In addition, theoretical approaches to biopanning emerged [12–14], helping investigators implement the most efficient techniques for selecting binders. Most notably among the recent literature, a number of excellent books [15–17] and review articles [18–22] that focus on different aspects of phage display have been published.

FILAMENTOUS PHAGE BIOLOGY

The filamentous bacteriophages are a group of viruses that contain a circular single-stranded DNA genome carried in an elongated protein capsid cylinder. The Ff class of the filamentous phages, incorporating phage f1, fd, and M13, is of greatest significance to phage display, and phage M13 in particular (schematic diagram in Figure 9.1) has been used most often to construct phage libraries. The host of filamentous phage is *Escherichia coli*. A complete listing of phage proteins and the genes encoding them is provided in Table 9.1.

M13 Virion

The filamentous phage genome consists of 6407 nt single-stranded DNA and encodes 11 proteins. Five structural proteins include a minor coat protein pIII, the major coat protein pVIII, and three other proteins (pVI, pVII, and pIX) (Figure 9.2). Proteins pIII and pVI cluster on the adsorption end

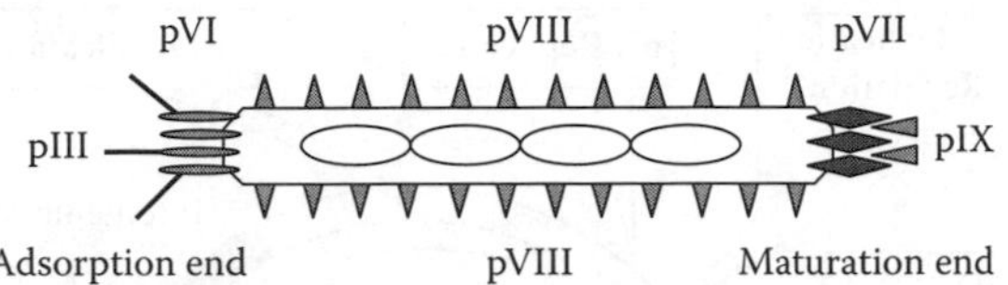

FIGURE 9.1 Schematic of coliphage M13 virion. (*Source:* Modified from Slide 1 (of 33) from http://www.niddk.nih.gov/fund/other/archived-conferences/2001/genoproteo/wang.pdf.)

TABLE 9.1
F-specific Filamentous Phage Genes/Proteins and Properties

Gene	Protein	Size (aa)	Function	Location	Used for Display?
I	I	348	Assembly	Inner membrane	
	XI	108	Assembly	Inner membrane	
II	II	409	Replication (nickase)	Cytoplasm	
	X	111	Replication	Cytoplasm	
III	III	406[a]	Virion component	Virion tip (end)	Yes (N-term)
IV	IV	405[a]	Assembly (exit channel)	Outer membrane	
V	V	87	Replication (ssDNA bp)	Cytoplasm	
VI	VI	112	Virion component	Virion tip (end)	Yes (C-term)
VII	VII	33	Virion component	Virion tip (start)	Yes (N-term)
VIII	VIII	50[a]	Virion component	Virion filament	Yes (N- and C-term)
IX	IX	32	Virion component	Virion tip (start)	Yes (N-term)

[a] Mature protein without signal sequence.

Source: With permission from Table 1 on p. 7 of Ref. 15 (© 2004, Oxford University Press).

of the virion, while proteins pVII and pIX cluster on the maturation end. These proteins are present in very few (three to five) copies in the virion, in sharp contrast with pVIII (2700 copies). The six non-structural proteins are replication proteins pII and pX, single-stranded DNA binding protein pV, and morphogenic proteins pI, pIV, and pXI (Arabic numerals, instead of Roman numerals, are sometimes used to designate the proteins and genes).

Information about the virion structure comes primarily from x-ray diffraction studies and electron micrography. The M13 genome is packaged into a phage capsid 65 Å in diameter and 9300 Å in length (Figure 9.3). The main building block for the capsid is the 50-amino-acid-residue-long protein pVIII, oriented at a 20° angle from the particle axis and assembled in a shingle-type array. The assembly of pVIII proteins leaves a cavity with a diameter of 25 to 35 Å, filled with phage DNA throughout the length of the virion. The phage DNA is oriented in the particle, determined by the packaging signal (PS), which forms an imperfect hairpin and is positioned at the maturation end of the phage particle. The PS is necessary and sufficient for incorporation of the DNA into the phage particle. The bulk of the pVIII protein (residues 6 to 50) forms an α-helix. Five N-terminal residues are surface-exposed, while residues near the C terminus interact with the genomic DNA. The amino terminus of pVIII has been the subject of extensive engineering in the process of creating phage libraries.

The ends of the particle have a different appearance in electron micrographs. The pointed end (also known as the adsorption, proximal, or infectious end) consists of about five copies each of pIII and pVI. The 406-residue pIII is, by far, the most commonly used protein in phage display. It consists of three domains (N1, N2, and CT) separated by two glycine-rich linkers (Figure 9.3B and C). Domains N1 and N2 appear in electron micrographs as knobby structures emanating from the pointed end of the particle into the medium and are key for infectivity. The crystal structure of

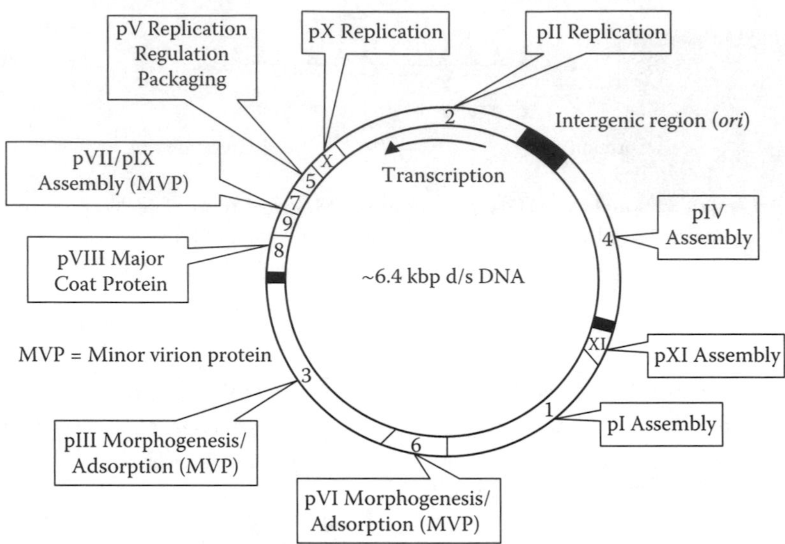

FIGURE 9.2 Phage M13 genome. The replicative form DNA (RF DNA) is depicted. The genes are designated by Arabic numerals, while the proteins they encode by Roman numerals. Proteins X and XI are not encoded by separate genes, but are derived from gene *2* and gene *1*, respectively, as discussed in the text. (*Source:* Modified from a GP Smith PowerPoint slide.)

a polypeptide comprising both domains has been determined (Figure 9.3C). The C-terminal 132 residues of the CT domain are proposed to interact with pVIII and form the proximal end of the particle; they are also thought to be buried. Most importantly, they are necessary and sufficient for pIII to be incorporated into the phage particle and for termination of assembly and release of phage from the cell.

The role of 112–amino-acid-long protein pVI is not fully understood. It is thought that the protein interacts with pVIII at the tip of the particle and is partially buried. The C-terminus of pVI has been used to display heterologous proteins [23], suggesting that it is accessible on the surface of the phage particle.

The other phage particle end, also known as the distal or maturation end, appears blunt in electron micrographs. It has approximately five copies each of two small proteins pVII (33 amino acids in length) and pIX (32 amino acids in length). While protein interactions at this particle end are not fully understood, it has been shown that the amino termini of both proteins can be used to display proteins (both light- and heavy-chain variable domains have been displayed [24]). In addition, a library of single-chain antibodies has been constructed as a fusion to the amino terminus of pIX [25].

M13 Phage Life Cycle

In describing the M13 life cycle, we focus on the aspects of direct significance to experimentation with phage display. More detailed descriptions are available [16, 26, 27].

To infect the bacterium, the bacteriophage uses the tip of the F conjugative pilus of *Escherichia coli* as the receptor. The three bacterial proteins required during phage infection are TolQ, TolR, and TolA, which appear to form a complex anchored in the cytoplasmic membrane; the complex protrudes into the periplasm of *E. coli*. Infection starts with the binding of the N2 domain of the phage pIII protein to the tip of the F pilus, followed by a retraction of the pilus, which brings the adsorption end of the particle to the periplasm. As a result of this binding, the N1 domain of pIII is released from the N2 domain and is allowed to interact with TolA. Thus, the tip of the pilus can

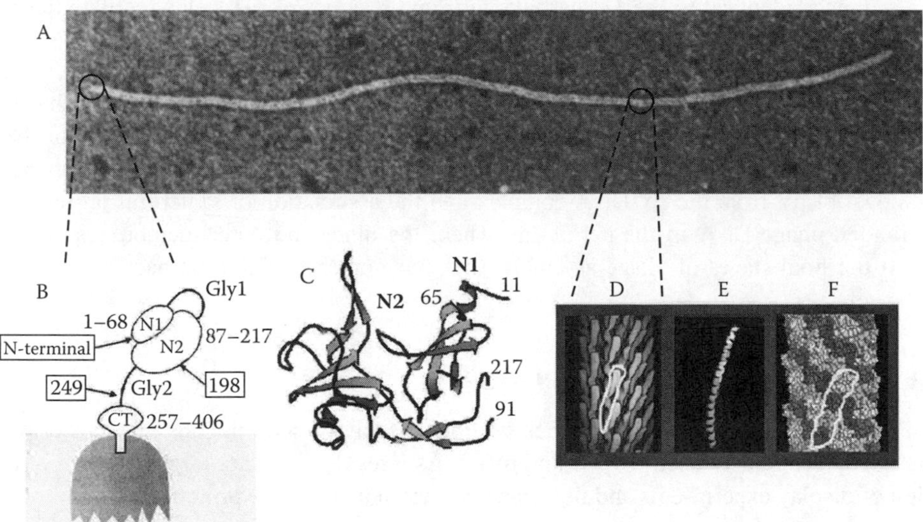

FIGURE 9.3 Panel A: Electron micrograph of a filamentous ssDNA phage. (Source: Courtesy of Professor R. Webster. GP Smith PowerPoint slide, with permission.) Panel B: Schematic of the pIII structure, including domains N1, N2, and CT, and glycine-rich linkers G1 and G2. Numbers indicate amino residues comprising the domains. (With permission, from Fig. 3 on p. 5 of Ref. 15; © 2004 Oxford University Press.) Panel C: Crystal structure of the N1 and N2 domains of pIII [27]. Image was generated using KiNG software and file 1gp3.pdb was provided by Protein Data Bank (www.rcsb.org). Numbering is according to the amino acid residues: 1 – amino terminus and the site for display of peptides and proteins fused to pIII; 1-65 - N1 domain; 91-217 – N2 domain. The structure of the 66–90 region, a glycine linker, has not been determined. Panel D: Computer rendering of the electron density map of M13 based on x-ray diffraction studies.[46] (With permission, from Lee Makowski and the publisher (© 1992 Elsevier)). The elongated rod-like structures are pVIII subunits forming the phage coat. The highlighted structure corresponds to a single pVIII subunit. Panel E: Model of the α-helical domain of pVIII.[47] (With permission from Valery Petrenko and the publisher (© 2007 Elsevier) based on x-ray fiber diffraction and solid-state NMR data.)[49] Four differently shaded areas are as follows (from top of figure to bottom): (1) the foreign peptide fused to the amino terminal region of pVIII; (2) amino acids 6–11 of pVIII; (3) amino acids 12–19 of the engineered pVIII; and (4) buried residues of pVIII (amino acids 20–50). Panel F: Space-filling model of the phage. About 1% of the phage length is shown. Shaded areas: same as in Panel E. The highlighted structure corresponds to a single pVIII subunit. Credits and permissions: same as in Panel E.

be considered a receptor, and TolA a coreceptor. The processes of phage particle disassembly and entrance of the single-stranded phage DNA into the cytoplasm that follow are not well understood. Phage pVIII protein is reused in the making of the new phage, and DNA, which is a (+)-strand (same orientation as phage mRNA), undergoes replication.

The (+)-strand becomes a template for synthesis of the complementary (–)-strand. The synthesis by bacterial enzymes and the introduction of negative supercoils by a gyrase produces a covalently closed, supercoiled, double-stranded DNA called the parental replicative form DNA (RF DNA). This RF DNA serves a triple role: initially, it is a template for transcription of mRNA and for synthesis of additional RF DNA, and after the sufficient amount of phage protein has accumulated, it serves as a template for synthesis of the (+)-strand to be packaged into new phages. Phage protein pII is necessary for nicking/closing the DNA to be replicated, while the accumulation of pV results in the formation of a complex between pV and (+)-strands, thereby preventing the conversion of the (+)-strand to RF DNA. The RF DNA/pV complex serves as a substrate for phage assembly. An additional protein, pX, created by an in-frame translational start within the p*II* mRNA and thus having a sequence identical to the 111 carboxy terminal residues of pII, helps regulate the replication of phage DNA.

Phage assembly is a complex process requiring five structural proteins (pIII, pVIII, pVI, pVII, and pIX); three assembly proteins (pI, pIV, and pXI); and at least one bacterial protein, thioredoxin. Protein pXI is created by an in-frame translational start within the gene *I* mRNA and thus having a sequence identical to the 108 carboxy terminal residues of pI. At the core of the assembly process is the removal of pV from the pV-DNA complex and the association of structural proteins around single-stranded phage DNA in the periplasm, where the inner and outer membranes are in close contact. At the final stages of phage assembly, the phage emerges from the bacterial cell without killing the cell.

PHAGE PROTEINS USED FOR DISPLAY

While all five phage coat proteins have been used for display of heterologous peptides or proteins, only two have been widely used — pIII and pVIII. As a result, there are large amounts of data on the design of display experiments and library construction using these proteins.

PROTEIN pIII

Of the two proteins pIII and pVIII, the former is by far the most commonly used. The insert location of choice is the amino terminus of pIII; that is, the heterologous gene is cloned between the gene *III* regions that correspond to the signal peptide of pIII and domain N1 of pIII. Upon secretion of pIII into the periplasmic space of the cell, the signal peptide is removed and the heterologous peptide ends up as a fusion to the amino terminus of the mature pIII protein.

There are several reasons for this cloning site preference. First, the replication processes of the phage seem to tolerate a wide range of insert sizes. This can be rationalized by realizing that insertion places the heterologous peptide at the adsorption end of the phage particle, which creates less steric hindrance during phage maturation; pIII emerges last during the transport of the particle from the bacterial cell into the medium. Proteins as large as alkaline phosphatase have been expressed as fusions to pIII [28]. On the other hand, many experiments with fusions of short peptides to pIII have also been quite successful.

Second, the structure of the amino-terminal half of pIII, comprising the N1 and N2 domains (Figure 9.3, panels B and C) is highly conducive to phage display experimentation. Both domains, but particularly the amino terminus, are exposed on the surface of the phage particle. Thus, one can expect the foreign peptide will be able to adopt its native fold and be exposed to the medium, which would allow for successful selection experiments.

Third, the purpose of many phage experiments is the selection of high-affinity binders to a biomolecule of choice. Under such conditions, a monovalent display is often desired, with expression on pIII being closer to this ideal (5 copies) than a clearly polyvalent (up to 2700 copies) display on pVIII. In real life, the valency of display is difficult to measure; it cannot be excluded that, due to unavoidable proteolysis, only a small fraction of phage coat proteins, whether pIII or pVIII, display a foreign peptide. Thus, pIII display might be closer to monovalent than the number of pIII molecules in the virion might indicate. Also, phagemid display (see below) will reduce the fraction of pIII fusion proteins to native proteins.

Alternative display locations in pIII, namely between the N1 and N2 domains and between the N1 and CT domains, also have been explored [29]; however, they have not been widely used.

PROTEIN pVIII

Display experiments on pVIII target the amino terminus of pVIII. That is, the heterologous gene is inserted between the region corresponding to the signal peptide and the mature pVIII protein.

Unlike pIII display, there is a significant length limit of six to eight amino acid residues on the size of the peptide displayed; this size limit seems to be connected to the interference of larger peptides with phage packaging and, in particular, with the transport of the phage particle through the pIV channel. Display of larger peptides fused to pVIII is only possible with hybrid phage display systems (see section entitled "Phagemid and Helper Phage" below).

Nomenclature of Phage Display Experiments

As alluded to above, it is desirable in many phage display experiments to reduce the number of foreign peptides displayed on the phage, relative to the native number of pIII or pVIII proteins (5 or 2700 each, respectively). This is done by introducing into the *Escherichia coli* cell additional copies of the wild-type gene *III* or gene *VIII*; the gene can reside on the same phage vector that harbors the heterologous gene, in which case the system is called 33 or 88, respectively. Alternatively, the gene can be harbored on another vector, and such systems are called 3+3 or 8+8. The single-gene systems are known as type 3 (p*III* only) or type 8 (p*VIII* only). The systems are depicted in Figure 9.4. Because the expression level of the wild-type genes can be controlled, the ratio of the heterologous to wild-type coat protein can be adjusted as needed.

Phagemid and Helper Phage

A popular implementation of the hybrid system involves the use of a phagemid, which is a cloning vector based on a plasmid (such as pBR322) and containing the replication origin of a filamentous phage as well as either gene *III* or gene *VIII* modified to accept gene inserts for display. The phagemid can be readily propagated in *Escherichia coli* leading to the accumulation of the pIII or pVIII fusion protein. To allow formation of viable phage, a co-infection with so-called M13 helper phage is required. Compared with the wild-type phage, the helper phage has a compromised origin of replication and is unable to be propagated efficiently under normal conditions. However, an *E. coli* cell harboring the phagemid and superinfected with helper phage will produce large amounts of a smaller version of the phage carrying phagemid DNA and displaying the heterologous proteins. Such particles are then used in affinity-selection experiments.

An often-used version of the M13 helper phage is M13K07 [30], available from New England Biolabs. It is an M13 derivative that carries the mutation Met40Ile in gII, with the origin of replication from P15A and the kanamycin resistance gene from Tn903 both inserted within the M13 origin of replication [30]. Other types of M13 helper phages are R408 and VCSM13 (Stratagene).

CONSIDERATIONS FOR SUCCESSFUL PHAGE DISPLAY EXPERIMENT

While a large number of applications of phage display have been reported, most phage display experiments fall into the following two categories based on the type of protein immobilized on the phage:

1. Preparation and screening of a random peptide library, with the goal of identifying a peptide capable of binding to a given protein under study
2. Cloning a protein, followed by randomization and selection, with the goal of finding a variant (mutant) of the protein possessing a desired property, typically a binding specificity to another molecule

Peptide Library

The first question to ask is whether an existing peptide library will be adequate for the particular experiment. A comprehensive list of random peptide libraries already in existence is provided in Ref. 18. In addition, New England Biolabs (www.neb.com) supplies the Ph.D. phage display system and

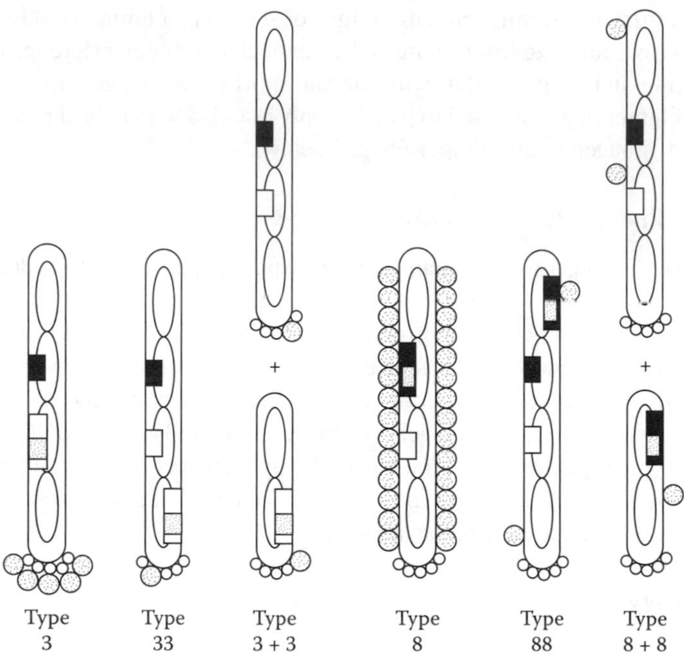

FIGURE 9.4 Types of phage vectors for display experiments. A Type 3 vector has a single recombinant gene *III* (open box) bearing a foreign DNA insert (hatched segment): all five copies of pIII are decorated with the foreign peptide encoded by the insert (hatched circles at the tip). The other phage genes, including *VIII* (black box), are normal. Type 33 vectors provide two genes *III*, one wild-type, the other an insert-bearing recombinant. The virions display a mixture of pIII molecules, only some of which are fused to the foreign peptide. Type 3+3 vectors resemble Type 33 in having two genes *III*; in this case, however, the recombinant gene is on a phagemid, a plasmid that contains, in addition to a plasmid replication origin and an antibiotic resistance gene, the filamentous phage "intergenic region." Type 8, 88, and 8+8 vectors are the gene *VIII* counterparts of the gene *III* vectors. (*Source:* With permission from GP Smith[48] and the publisher [© 1993 Elsevier]).

several types of libraries, in which seven or twelve amino acid random or disulfide-linked peptides are displayed on the phage. The diversity of the libraries is in the range of 1–3 × 10^9 different peptides.

If the decision is made to construct a new peptide library by cloning a random or partially random nucleic acid sequence within gene *III* or gene *VIII*, the desired length of the peptide must be determined, as well as the choice of cloning vector, cloning site, and signal sequence. The majority of peptide libraries constructed are of type 3 (see above) [18], and the gene insert is placed downstream of the signal sequence in front of gene *III*, often in a derivative of phage M13; this is a recommended (default) approach. It is more difficult to recommend the peptide length because while longer peptides provide a higher level of sequence heterogeneity (proportional to the peptide length) and more possibilities for formation of secondary/tertiary protein structures, these longer peptides might be more rapidly degraded proteolytically and the library more difficult to handle. Interpretation of the results might be more difficult as well [31].

Display of Proteins

Many different proteins have been displayed on the phage [15, 18], ranging from enzymes (alkaline phosphatase, trypsin, lysozyme, β-lactamase, glutathione transferase); hormones (human growth hormone, angiotensin); inhibitors (bovine pancreatic trypsin inhibitor, cystatin, tendamistan); toxins (ricin B chain); receptors (α subunit of IgE receptor, T cell receptor, B domain of protein A); ligands (substance P, neurokinin A); and cytokines (IL-3, IL-4); to different forms of antibodies

(scF_v, F_{ab}). The size of the protein, up to around 60,000 Da per subunit, does not seem to be of major concern, nor is the existence of a protein in a dimeric form, because a dimeric protein as large as alkaline phosphatase has been successfully displayed on the phage [32]. Similarly, a successful phage display experiment does not depend on whether or not the protein is secreted in its native form, because many intracellular proteins have been displayed on the phage (glutathione S-transferase, cytochrome b_{562}, FK506 binding protein, and others). The typical approach involves use of the 3+3 system (with an alternative being the single vector 3 system), with the gene cloned for expression on the amino terminus of pIII.

Antibody Libraries

Special attention should be given to the display of antibodies on filamentous phages due to their popularity and wide number of possible applications, but common difficulties are encountered with such libraries. Antibodies are disulfide-linked heterodimers composed of light and heavy chains, with their total molecular weight often exceeding 150,000 Da. The complexity of the antibody molecule in its native state renders it prohibitively difficult to display on the phage, which is why approaches have been developed to display a truncated single-chain version of the antibody molecule. The single-chain antibody (scAb) consists of the variable regions of the heavy and light chains connected with a short and flexible peptide linker (Figure 9.5A). Thus, the antigen-binding region of the antibody is preserved; however, functionalities associated with the remainder of both heavy and light chains, such as binding to various cell receptors and complement proteins, are removed upon the conversion to the single-chain form.

The construction of scAb libraries typically involves the joining of PCR amplified gene fragments corresponding to the F_v regions of the heavy and light chains through a glycine-rich linker, placing the signal sequence in front of the construct and followed by cloning between a suitable ribosome binding site/signal sequence and the 5′-terminal region of gene *III* (Figure 9.5B). Often, both mAb-tag, his-tag, and an amber stop codon are introduced into the construct as well.

CONSTRUCTION OF THE PHAGE LIBRARY

After choosing a phage vector system for library construction, a cloning strategy must be identified. A typical approach for inserting a peptide-encoding randomized sequence into a vector relies on the cloning of synthetic DNA. An example is presented in Figure 9.6, and the experiment is described in detail in Ref. 33.

The pCANTAB5E phage vector (GE Healthcare) was chosen for cloning, and its two unique restriction sites — *Sfi*I and *Not*I — were selected as the insertion sites for randomized DNA. As a result, the randomized peptide was expressed fused to the pIII protein and had on its amino terminus the FLAG peptide. The random insert sequence was of the type $(NNK)_m$ where N = A, C, G, or T; K = T or G; and m is the number of codons in the random sequence. In the example quoted, m = 40; however, m can vary greatly, depending on the experimental design, but it typically ranges from 8 to 30 codons. The purpose of the NNK codon design is to avoid two out of three stop codons. The third stop codon (TAG) can be suppressed if the phage is prepared in an *Escherichia coli* suppressor strain, such as TG1 [F' *traD36 lacI*^{q} *lacZM15 proA*$^+$*B*$^+$ / *supE* Δ(*hsdM-mcrB*)5(rk^-mk^- *mcrB*$^-$) *thi*, Δ(*lac-proAB*)], leading to incorporation of glutamine into the polypeptide chain in response to TAG codons. Despite this limitation, all amino acid residues can be incorporated into the polypeptide due to redundancy in the genetic code. In this example, the random sequence resided on a synthetic oligonucleotide (145 nt long) that was converted to double-stranded DNA in a PCR, digested with *Sfi*I and *Not*I, ligated into the pCANTAB5E vector, and electroporated into *E. coli*.

There are many methods to clone DNA into phage vectors. Other than the PCR-based cloning outlined above, approaches often used to clone randomized cDNAs that encode larger proteins

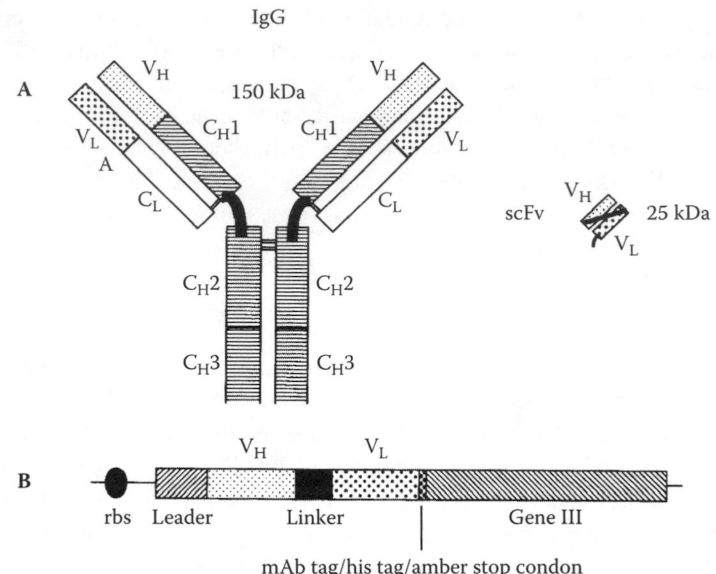

FIGURE 9.5 Antibody structure and recombinant forms used for phage display. The single-chain Fv (scFv) contain the antigen binding V_H and V_L domains from immunoglobulin G (IgG) and can be displayed on phages. In the case of the Fab and the diabody, one of the chains is secreted as a pIII fusion and the other is secreted as native non-fusion protein. Thick lines indicate peptide connections, and thin double lines indicate disulfide bonds. In the case of the diabody, the V_H and V_L encoding a single specificity are found on different chains, but come together at the protein level, as shown. (*Source:* With permission from p. 245 of Bradbury and Marks article in Ref. 15 [© 2004 Oxford University Press].)

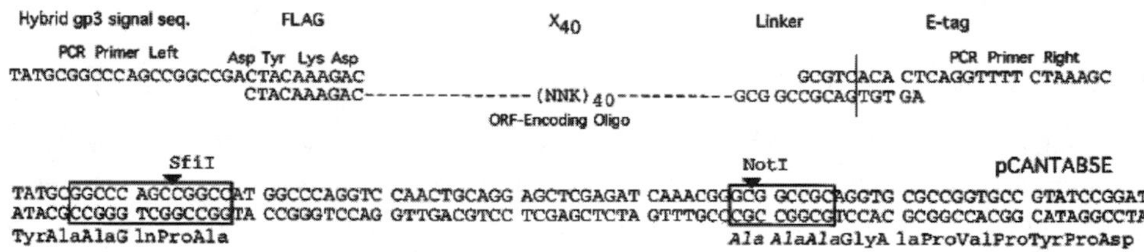

FIGURE 9.6 Cloning of a DNA insert carrying a randomized sequence into the pCANTAB5E phagemid. (Source: From Figure 1 on p. 1994 of Ref. 33.)

include Kunkel mutagenesis [19, 34], a variety of oligonucleotide-based mutagenesis methods, random mutagenesis, combinatorial infection and recombination, and DNA shuffling [17].

Library Diversity

The potential library diversity D goes up very rapidly with the number of randomized positions m, according to the formula $D = 20^m$ (assuming all 20 amino acids can be incorporated). For m = 8, the potential diversity exceeds 2×10^{10}, which is already larger than the number of independent clones in a typical phage library. This means that libraries in which more than eight positions are fully randomized will actually not contain a representation of all possible sequences — a reflection of the enormous diversity of protein sequences. It is reassuring for those who work with phage peptide libraries that binders to many (if not most) proteins are obtainable from such libraries despite a striking contrast between the number of clones in the phage library and the number of allowable sequences. The selection of the binder can be represented as a search for function in a multidimensional protein sequence space [18, 35, 36].

Achieving a library diversity of more than 10^{10} independent clones is not easy. While electroporation efficiency can be more than 10^{10} clones per 1 mL electrocompetent cells and the amount of DNA is typically not an issue, one needs to remember that the electroporation efficiency quoted applies to ideal conditions (i.e., native double-stranded supercoiled DNA), while the *in vitro* constructed DNAs typically transform with significantly lower efficiency. In the example quoted above [17], the peptide library of 1.5×10^{10} clones was constructed by combining five sublibraries, each of which was obtained from about 100 electroporations.

SELECTION (BIOPANNING) EXPERIMENT

The principle of phage selection (Figure 9.7) relies on exposing the target proteins immobilized on solid phase to the phage library, allowing peptides carried on phages to bind specifically to the target proteins and be retained on solid phase, and removing the excess phage by washing. Because this biopanning step will always result in an undesired retention of non-specifically bound phage, the step is repeated with the eluted and amplified phage. The total number of biopanning steps can vary from two to four or more, depending on the experimental strategy and conditions. The final product of the phage enrichment procedure is a group of phage clones that are considered potential binders to the target protein. Typically, the target-binding properties of the phage clone are confirmed in a phage ELISA assay, and the phage DNA is subjected to DNA sequencing, which identifies peptide sequences capable of binding to the target of interest. The biopanning cycle is schematically represented in Figure 9.8, and a comprehensive selection of biopanning protocols is presented in Ref. 16 and Ref. 37.

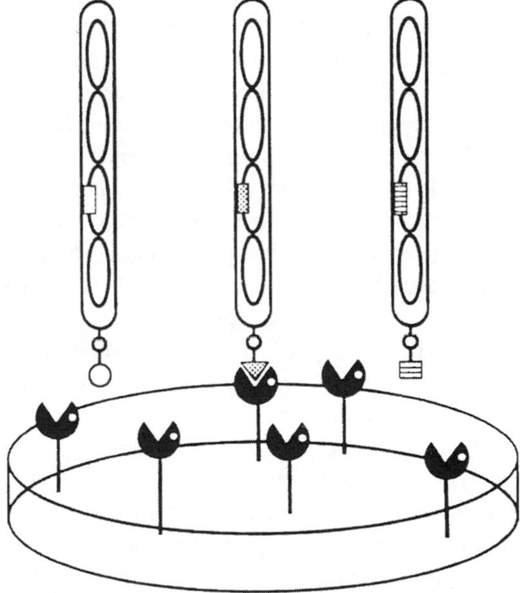

FIGURE 9.7 Phage display principle. Schematic diagram of a display library being "panned" on a Petri dish to which an open-jawed receptor has been tethered. Phages displaying peptides that bind the receptor are captured on the dish, while the remaining phages are washed away; the captured phages are eluted, cloned, and propagated by infecting fresh host cells, and the relevant part of their DNA sequenced to identify the peptide responsible for specific binding. (From Ref. 48, with permission from GP Smith and the publisher [© 1993 Elsevier].)

A great deal of variation is allowed in biopanning procedures. Different solid-phase materials have been used, such as polystyrene (the material of choice for making standard microtiter plates), nitrocellulose, or poly(vinylidene fluoride[PVDF]). The binding reaction can take place in solution; in this case, the target protein is biotinylated and the phage-target complex is harvested by binding it to streptavidin-coated microtiter plates [38]. The biopanning stringency also can be varied by adjusting the binding/elution buffer composition, time, temperature, etc. A general rule of thumb for protein targets known to bind to short linear epitopes (e.g., antibodies recognizing such epitopes) is to use more stringent conditions and fewer biopanning rounds, while target proteins that bind to more complex structures often require an optimization of the biopanning approach.

The analysis of phage clones after biopanning begins with an ELISA in which the peptide-bearing phage is immobilized in the well of a microtiter plate using an anti-phage antibody, and the presence and target-binding properties

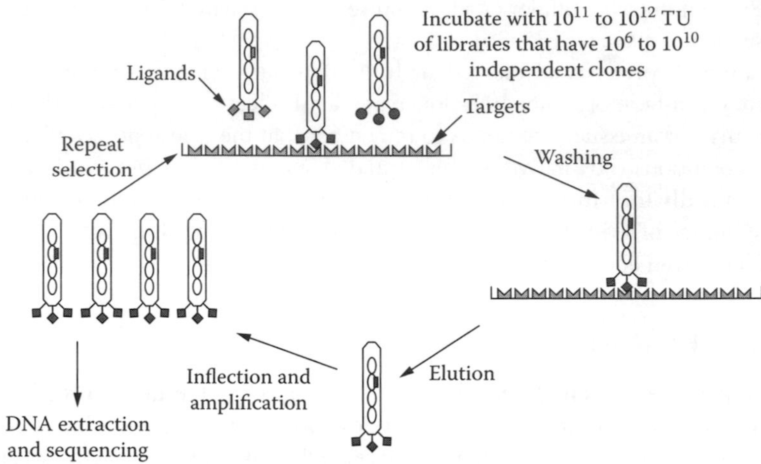

FIGURE 9.8 A schematic for the biopanning cycle. (Source: Modified from Slide 6 (of 33) from http://www.niddk.nih.gov/fund/other/archived-conferences/2001/genoproteo/wang.pdf.)

of the peptide are verified by binding of the target protein to the peptide. This is followed by a visualization of the target protein with an appropriate reagent, such as biotinylated HRPO (horseradish peroxidase) for the streptavidin-conjugated target protein.

In the next analysis step, several (often as many as 100 or more) ELISA-positive clones are subjected to DNA sequencing over the randomized region of the heterologous gene or cloned insert. The obtained DNA sequences are translated and the amino acid sequence of the peptide or the randomized region is defined. Often, a short linear region of homology can be readily spotted among the sequences and a consensus sequence can be derived.

Typical results from a simple biopanning experiment [39] are presented in Figure 9.9. In panel A, biopanning of a peptide library against an HCV monoclonal antibody (mAb) led to a simple linear sequence of six amino acid residues, RGRRQP, as a consensus. In panel B, with the target being an anti-HBsAg mAb, the consensus is still a short linear sequence, C*TC, where * is any amino acid residue. However, the presence of two cysteine residues strongly suggests that the binding form of the peptide adopts a looped conformation that is anchored by a disulfide bond formed by two cysteine residues. Nonetheless, the confirmation of such a secondary structure as a binding motif requires additional experimentation.

In view of the phage display complexities that can develop due to the high number of different molecules (more than 10^9) involved and largely unknown binding properties, it is wise to be guided by theoretical models and simulations of phage display [13, 41]. Complexities arise because randomized proteins displayed on the phage in the library have vastly differing binding affinities to the target. Depending on the target, the binding clones may or may not be abundant; and depending on the biopanning procedure, particularly its stringency and number of washes, different groups of clones can be retained or lost at varying rates, which can lead to uncertainties in obtaining a desired type of binder. Theoretical analyses as in Ref. 13 provide insight regarding the number of biopanning rounds and washes per round needed, while also helping to track the quantitative increase in the number of desired binding clones. Models can help one understand how the affinity distributions of phage populations change during panning.

EXPRESSION OF IDENTIFIED PROTEIN OR SYNTHESIS OF PEPTIDE

Obtaining consensus sequences can sometimes provide an important clue as to the significance of a particular region within a protein sequence, particularly when there is a match between the two

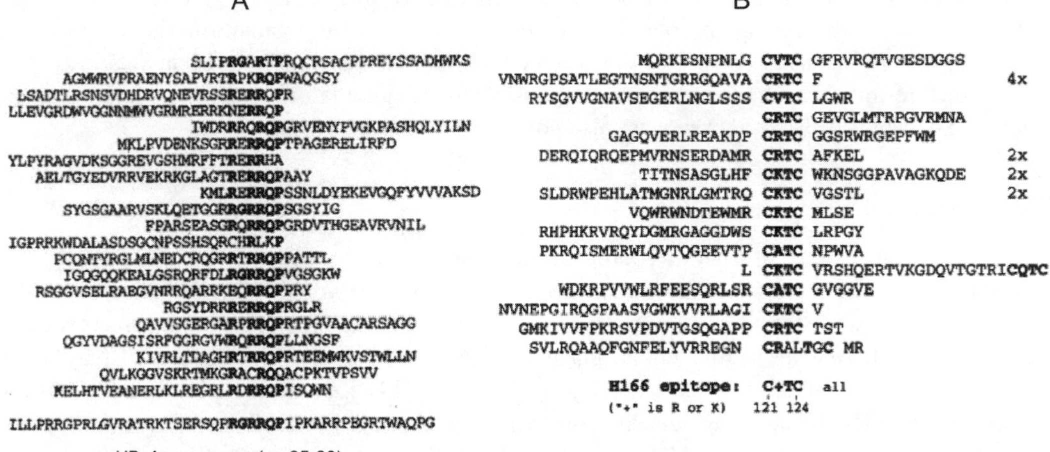

FIGURE 9.9 Examples of biopanning results: consensus sequences obtained from biopanning against two mAbs, against the HCV core protein (panel A), and HBV surface antigen (panel B). (*Source:* A: from Figure 5 on page 192 of Ref. 40; B: from Figure 1A on page 1999 of Ref. 39.)

sequences. But more often than not, further characterization of the binding clones is needed; and toward that goal, the protein displayed on the phage needs to be made in appropriate quantity and then purified and analyzed. Several approaches to this are possible.

First, certain phagemid vectors, such as those presented in Figure 9.5, have an amber stop codon (TAG) between the gene encoding the displayed protein and p*III*. A transfer of the phagemid to a non-suppressor strain of *Escherichia coli* can allow for expression of the protein as an entity not fused to pIII. Moreover, the vector may provide the his-tag or an antibody-tag, which is retained on the expressed protein in the non-suppressor strain; such a tag may facilitate protein purification and subsequent characterization by ELISA.

Second, the gene encoding the protein can be re-cloned into an expression system. A large number of expression systems are available, and a good choice to start with is the system typically used to make the native protein in quantity.

Third, for random peptide libraries on phage, peptide synthesis is a very good method to obtain the peptide in quantity (see, e.g., Ref. 31). Alternatively, the short genes encoding those peptides can be fused to a well-characterized affinity protein, such as maltose binding protein (MBP), and studies of the fusion protein can confirm the properties of the peptide from phage display.

COMMON DIFFICULTIES

While phage display is a powerful tool to select for a peptide or protein that binds specifically to a ligand, certain problems might arise due to the complexity of the approach. First, the custom library quality is difficult to judge and, in particular, the number of independent clones in the library and the valency of display (the actual number of displayed peptides/proteins). Second, certain clones could be under- or over-represented due to expression inefficiencies or toxicity. Third, procedural difficulties with selection (often, over-selection) may lead to the elimination of clones otherwise expected to be identified.

In addition, the genetic sequence encoding the displayed peptide has not always proven to be easily discerned. In our hands, screening a phage display random peptide library for binders to specific targets resulted in some instances with as many as 50% of all selected clones not having an obvious open reading frame; nevertheless, the peptide had to have been synthesized or the phage would not have bound to the target ligand [31].

In a series of experiments, we pursued the explanation for this anomaly for one specific clone and its derivatives by fusing the peptide-encoding sequence to a β-galactosidase reporter and expressing it in *Escherichia coli* [42–44]. Ultimately, our experiments indicated that an unexpected translational re-initiation phenomenon was responsible for the anomalous expression [44]; however, other unexpected translational events are also possible in such cases [45].

CONCLUSIONS

Phage display provides a myriad of approaches to identify protein ligands. In skilled hands, the procedures are quite rapid (a few days), and the results provide a wealth of information about the protein ligands and the target protein itself, as well as their interactions. Despite the emergence in recent years of alternatives to filamentous phage display, such as aptamers, ribosome display, yeast two-hybrid system, bacterial display, other phage systems (T4, T7, and lambda) (see Ref. 19 for references), the filamentous phage display nonetheless offers a very reliable set of tools and tested procedures and has been improved upon over many years by a large number of investigators. This method is an excellent starting point for the identification and activation of protein ligands.

REFERENCES

1. Smith, G.P. Filamentous fusion phage: novel expression vectors that display cloned antigens on the virion surface. *Science*, 228, 1315, 1985.
2. Scott, J.K. and Smith, G.P. Searching for peptide ligands with an epitope library. *Science*, 249, 386, 1990.
3. Petrenko, V. http://www.vetmed.auburn.edu/~petreva/Files/Lecture%201%20Introduction.pdf
4. McCafferty, J. et al., Phage antibodies: filamentous phage displaying antibody variable domains. *Nature*, 348, 552, 1990.
5. Griffiths, A.D. et. al. Human anti-self antibodies with high specificity from phage display libraries. EMBO J., 12, 725, 1993.
6. Kang, A.S. et al., Linkage of recognition and replication functions by assembling combinatorial antibody Fab libraries along phage surfaces. *Proc. Natl. Acad. Sci. U.S.A.*, 88, 4363, 1991.
7. Barbas, C.F. III et al., Assembly of combinatorial antibody libraries on phage surfaces: the gene *III* site. *Proc. Natl. Acad. Sci. U.S.A.*, 88, 7978, 1991.
8. Iannolo, G. et al. Modifying filamentous phage capsid: limits in the size of the major capsid protein, *J. Mol. Biol.*, 248, 5, 1995.
9. Malik, P., Marvin, D.A., and Perham, R.N. Role of capsid structure and membrane protein processing in determining the size and copy number of peptides displayed on the major coat protein of filamentous bacteriophage. *J. Mol. Biol.*, 260, 9, 1996.
10. Petrenko, V.A. et al. A library of organic landscapes on filamentous phage. *Protein Eng.*, 9, 797, 1996.
11. Ilyichev, A.A. et al. Construction of M13 viable bacteriophage with the insert of foreign peptides into the major coat protein. *Dokl Biochem. (Proc. Acad. Sci. USSR)* Engl. Transl., 307, 198, 1989.
12. Zhuang, G. et al. A kinetic model for a biopanning process considering antigen desorption and effective antigen concentration on a solid phase. *J. Biosci. Bioeng.*, 91, 474, 2001.
13. Mandecki, W., Chen, Y.C., and Grihalde, N. A mathematical model for biopanning (affinity selection) using peptide libraries on filamentous phage. *J. Theor. Biol.*, 176, 523, 1995.
14. Levitan, B. Stochastic modeling and optimization of phage display. *J. Mol. Biol.*, 10, 893, 1998.
15. Russel, M., Lowman, H.B., and Clackson, T. Introduction to phage biology and phage display. In *Phage Display: A Practical Approach* (The Practical Approach Series, 266), Clackson, T. and Lowman, H.B., Eds., Oxford University Press, 2004.
16. Barbas III, C.F., Burton, D.R., and Scott, J.K., Eds., *Phage Display: A Laboratory Manual*, Cold Spring Harbor Laboratory Press, 2001.
17. Sidhu, S.S., Ed. *Phage Display In Biotechnology and Drug Discovery*, Taylor & Francis, 2005.
18. Smith, G.P. and Petrenko, V.A. Phage display. *Chem. Rev.*, 97, 391, 1997.
19. Kehoe, J.W. and Kay, B.K. Filamentous phage display in the new millennium. *Chem. Rev.*, 105, 4056, 2005.

20. Bradbury, A.R. and Marks, J.D. Antibodies from phage antibody libraries. *J. Immunol. Meth.*, 290, 29, 2004.
21. Hoess, R.H. Protein design and phage display. *Chem Rev.*, 101, 3205, 2001.
22. Hoogenboom, H.R. Overview of antibody phage-display technology and its applications. *Methods Mol. Biol.*, 178, 1, 2002.
23. Fransen, M., Van Veldhoven, P.P., and Subramani, S. Identification of peroxisomal proteins by using M13 phage protein VI phage display: molecular evidence that mammalian peroxisomes contain a 2,4-dienoyl-CoA reductase, *Biochem. J.*, 340 (Pt 2), 561, 1999.
24. Gao, C. et al., A method for the generation of combinatorial antibody libraries using pIX phage display. *Proc. Natl. Acad. Sci. U.S.A.*, 99, 12612, 2002.
25. Gao, C. et al., A cell-penetrating peptide from a novel pVII-pIX phage-displayed random peptide library. *Bioorg. Med. Chem.*, 10, 4057, 2002.
26. Webster, E. Biology of filamentous bacteriophage. In *Phage Display of Peptides and Proteins* (Kay, B.K. et al., Ed.), Academic Press, San Diego, 1996.
27. Lubkowski, J. et al. The structural basis of phage display elucidated by the crystal structure of the N-terminal domains of g3p. *Nat. Struct. Biol.*, 5, 140, 1998.
28. McCafferty, J., Jackson, R.H., and Chiswell, D.J. Phage-enzymes: expression and affinity chromatography of functional alkaline phosphatase on the surface of bacteriophage. *Protein Eng.*, 4, 955, 1991.
29. Krebber, C. et al., Selectively-infective phage (SIP): a mechanistic dissection of a novel *in vivo* selection for protein-ligand interactions. *J. Mol. Biol.*, 268, 607, 1997.
30. Vieira, J. and Messing, J. Production of single-stranded plasmid DNA. *Methods Enzymol.*, 153, 3-11, 1987. San Diego: Academic Press.
31. Carcamo J. et al. Unexpected frameshifts from gene to expressed protein in a phage-displayed peptide library. *Proc. Natl. Acad. Sci. U.S.A.*, 95, 11146, 1998.
32. McCafferty, J., Jackson, R.H., and Chiswell, D.J. Phage-enzymes: expression and affinity chromatography of functional alkaline phosphatase on the surface of bacteriophage. *Protein Eng.*, 4, 955, 1991.
33. Ravera, M.W. et al. Identification of an allosteric binding site on the transcription factor p53 using a phage-displayed peptide library. *Oncogene*, 16, 1993, 1998.
34. Scholle, M.D., Kehoe, J.W., and Kay, B.K. Efficient construction of a large collection of phage-displayed combinatorial peptide libraries. *Comb. Chem. High Throughput Screen.*, 8, 545, 2005.
35. Mandecki, W. A method for construction of long randomized open reading frames and polypeptides. *Protein Eng.*, 3, 221, 1990.
36. Mandecki, W. The game of chess and searches in protein sequence space. *Tibtech*, 16, 200, 1998.
37. Dennis, M.S. and Lowman, H.B. Phage selection strategies for improved affinity and specificity of proteins and ligands. In *Phage Display, A Practical Approach* (The Practical Approach Series, No 266), Clackson, T. and Lowman, H.B., Eds., 2004, Oxford University Press.
38. Scholle, M.D., Collart, F.R., and Kay, B.K. *In vivo* biotinylated proteins as targets for phage-display selection experiments. *Protein Expr. Purif.*, 37, 243, 2004.
39. Chen, Y.C. et al. Discontinuous epitopes of hepatitis B surface antigen derived from a filamentous phage peptide library. *Proc. Natl. Acad. Sci. U.S.A.*, 93, 1997, 1996.
40. Grihalde, N.D. et al., Epitope mapping of anti-HIV and anti-HCV monoclonal antibodies and characterization of epitope mimics using a filamentous phage peptide library. *Gene*, 166, 187, 1995.
41. Levitan, B. Stochastic modeling and optimization of phage display. *J. Mol. Biol.*, 277, 893, 1998.
42. Goldman, E., Korus, M., and Mandecki, W. Efficiencies of translation in three reading frames of unusual non-ORF sequences isolated from phage display. *FASEB J.*, 14:603–611, 2000.
43. Zemsky, J, Mandecki, W, and Goldman, E. Genetic analysis of the basis of translation in the –1 frame of an unusual non-ORF sequence isolated from phage display. *Gene Expr.* 10, 109, 2002.
44. Song, L., Mandecki, W., and Goldman, E. Expression of non-open reading frames isolated from phage display due to translation reinitiation. *FASEB J.*, 17, 1674, 2003.
45. Shu, P., Dai, H., Mandecki, W., and Goldman, E. CCC CGA is a weak translational recoding site in *Escherichia coli*. *Gene,* 343, 127, 2004.
46. Glucksman, M.J., Bhattacharjee, S., and Makowski, L. Three-dimensional structure of a cloning vector. X-ray diffraction studies of filamentous bacteriophage M13 at 7 Å resolution. *J. Mol. Biol.,* 226, 455–470, 1992.
47. Petrenko, V.A. Landscape phage as a molecular recognition interface for detection devices, *Microelectronics J.,* 2007, doi:10.1016/j.mejo.2006.11.007, in press.
48. Smith, G.P. Surface display and peptide libraries. *Gene,* 128, 1–2, 1993.

49. Marvin, D.A., Welsh, L.C., Symmons, M.F., Scott, W.R., and Straus, S.K. Molecular structure of fd (f1, M13) filamentous bacteriophage refined with respect to x-ray fibre diffraction and solid-state NMR data supports specific models of phage assembly at the bacterial membrane. *J. Mol. Biol.*, 355, 294–309, 2006.

10 Diagnostic Medical Microbiology

Lorrence H. Green

CONTENTS

INTRODUCTION

Microbiologists have been identifying organisms since Pasteur first realized that in addition to producing the fine wines of France, microorganisms also caused disease. Since that time, it appears as if the chief goal of diagnostic microbiology has been to come up with the one test that would separate the helpful organisms from those that are harmful. The unfortunate thing is that every time someone developed this ultimate test, it was eventually discovered that there was always some overlap. The purpose of this chapter is to describe the major methods being used to identify microorganisms in the laboratory today.

CULTURE

This is the oldest technique used in the laboratory today and has a very simple premise. Organisms are grown on media and then identified using biochemical tests. Despite all the advances that have been made in the laboratory since this technique was first used by Pasteur and Koch more that 125 years ago, it remains the number-one method used in clinical laboratories today (personal observation). Many sources and formulations for media exist, and a description of them can be found in the *Difco and BBL Manual* (available in print from Fisher Scientific, Inc., Hampton, NH, or online at www.bd.com/ds/technicalCenter/inserts/difcoBblManual.asp).

Selecting the appropriate media seems like a daunting task but it can be made much easier if one realizes that all media can be categorized into one of three groups. To determine which media is best suited for a particular application, the technician should first determine which of the three groups the needs of the research fit into: (1) general, (2) selective, or (3) differential media.

Media Types

General Media: These have been developed to allow the widest variety of microorganisms to grow. Typically, they are used in applications in which the sample contains a mixture of microorganisms, and the specific one that is responsible for causing a particular medical or environmental effect is not known. As a result, the media that will support the growth of all of the organisms in a sample is needed. One of the most widely used general media is Trypticase Soy Blood Agar. It is usually prepared with 5% defibrinated sheep blood in a nutrient agar base.

Selective Media. These have been developed to allow only certain microorganisms to grow. Typically, they are used in applications in which the organism of interest makes up only a small portion of the organisms found in a sample. They are used when one wants to isolate specific pathogens in samples with an extensive normal flora. An example of a selective media is Thayer Martin Agar, which is used in the isolation of *Neisseria*.

Differential Media: These are a subset of selective media that were developed to not only select organisms, but also to help identify them. Not only will the organisms of interest grow on the plate, but they will also produce characteristically colored colonies. An example of differential media is MacConkey Agar, which is used to differentiate lactose-fermenting Enterobacteriacae from non-lactose fermenters. While both types of colonies will grow on the media, only the lactose-fermenting organisms will produce purple colonies.

Two other factors to consider when choosing growth media are whether the media is defined or undefined. Defined media consist of components that are all chemically defined. They are rarely encountered in the microbiology laboratory. Most media are undefined. Undefined media are usually prepared from digests of animal or plant proteins, and as such do not have a defined chemical formula. These media are usually tested for growth with a small, specific battery of microorganisms, prior to release for sale by the manufacturer. This can sometimes result in lot-to-lot variations in growth patterns for organisms that are not found on the manufacturer's quality control testing battery of microorganisms. If researchers are using an organism not being tested by the manufacturer, it is always a good idea to test any new lots of media with their own battery to determine if they are suitable.

COLONIAL MORPHOLOGY

Different species of organisms will usually have characteristic appearances following incubation on solid media. Characteristics used to describe colonies include:

Size
Shape
Rough vs. smooth
Mucoid

GRAM STAIN

Following isolation on a culture plate, the first test routinely performed is the Gram stain. A standard method is described in Chapter 4, *Stains for Light Microscopy* by Stuart Chaskes. After staining, organisms are examined under a microscope and characterized by reaction (+ = purple, − = pink); and by morphology (cocci = spheres, bacilli = rods, spirochete = helix). While this is the basic pattern, variations in shape do exist.

BIOCHEMICAL TESTS

Conventional biochemical testing is performed following the Gram stain. Organisms are suspended, usually in physiological saline, and then inoculated into test tubes containing chemicals and indicators, and incubated. The choice of chemicals depends on the results of the Gram stain. Many of these are described in the chapters dealing with specific organism groups in this book. Following incubation, the results are read and the identification of the organism is determined. These tests can consist of solutions of simple carbohydrates, which contain a dye that changes color with a change in pH; solutions of chromogenic substrates that change color when they are cleaved; or antibody tests that will specifically react to one organism. Tests can also consist of extraction and identification of cellular components by chromatography.

Most laboratories no longer routinely perform conventional testing but rely on commercial tests instead. The trend over the past few years has been to use single biochemical tests in instances where a clinician already suspects the identity of the microorganism. An example of this is the BBL™ MGIT™ used for the identification of mycobacteria. The clinical sample is added directly to a tube and monitored daily. The tube contains a fluorogenic substrate that will become visible in the presence of mycobacteria.

Most of the time, the clinician will use one of several multitest commercial kits. These kits are all similar in that they contain many biochemical tests packaged together. These tests are inoculated with a suspension of a single microorganism, and incubated. Following incubation, the results of the tests are read and the identity of the organism is determined. For example, the API 20 E® Strip (bioMerieux, Inc.) is used for identifying Gram-negative rods. In this product, the positive and negative results obtained for each test are used to generate a seven-digit code number. The company's database then uses this number to determine the identity of the microorganism. Table 10.1 lists several of these kits, as well as websites for the companies that supply them

While commercial manual tests are easier than conventional tests, they are still very labor intensive and cumbersome in high-volume laboratories. Since the mid-1980s, instruments have been introduced to perform as much of the work as is possible. Instruments are available that can perform the inoculation, incubation, reading, and determination steps automatically. Table 10.1 also lists some of the commercially available systems. O'Hara has written an excellent in-depth review of many of these systems.[1]

TABLE 10.1
List of Manual and Automated Identification Products

Company	Manual Product	Automated Product	Website
bioMerieux, Inc.	API Tests (20E®, Staph, Rapidec Staph)	Vitek 2	www.biomerieux-usa.com
MIDI, Inc.		The Sherlock® (Microbial Identification System (MIDI)	www.midi-inc.com
Remel	The RapID™, Micro-ID™		www.remelinc.com
Biolog	MicroLog	OmniLog® ID System	www.biolog.com
Trek Diagnostics Systems		Sensititre®	www.trekds.com
Becton Dickinson/BBL	Crystal™		www.bd.com
Becton Dickinson	Enterotube II™, Oxi/Ferm II™	The BD Phoenix™ 100	www.bd.com
Microscan		WalkAway®	www.dadebehring.com

This section concerns itself primarily with bacteria and fungi. Viruses, however, can also be identified using culture techniques. This is usually performed using specific cell lines instead of specific tests. The pathogen is inoculated into a tube containing a specific cell line, and incubated. Following this, the culture is examined for the presence of a distortion of the cell morphology called a cytopathic effect (CPE). This can be diagnostic for certain viruses. Alternatively, as described below, the inoculated cells can be stained using specific anti-viral immunofluorescent reagents and observed under the microscope. Specific cell cultures can be obtained from sources listed in Chapter 15, "Major Culture Collections and Sources."

ANTIBODY ASSAYS

Agglutination Assays: Antibodies or antigens, depending on the test, are attached to a solid support. The support can consist of latex, red blood cells, bacteria (co-agglutination), charcoal, liposomes, or colloidal gold. The sample is added and if the analyte is present, it binds to the corresponding molecule on the support. The support molecules then form a lattice, which is visible as a clump. The tests can be direct or indirect. A direct test is one in which the clinician is trying to detect a pathogen. An indirect test is one in which the clinician is looking for the body's response to a pathogen (e.g., an antibody). In a direct test, an antibody is bound to the support. In an indirect test, an antigen (or an antibody) is bound to the support.

Figure 10.1 shows an example of a direct and an indirect test. Tests for identifying common pathogens are available from a large number of commercial suppliers. There are also several suppliers of latex particles for researchers who wish to develop their own tests. Practical information for doing this can be found at the website of Bangs Laboratories (www.bangslabs.com).

Antibody Labeling Techniques: An antibody is labeled with a tag. This tag can be an enzyme, a fluorescent molecule, a radioactive element, or even a colored particle such as latex or gold. As above, tests can be direct or indirect. Figure 10.2 shows an example of an enzyme immunoassay

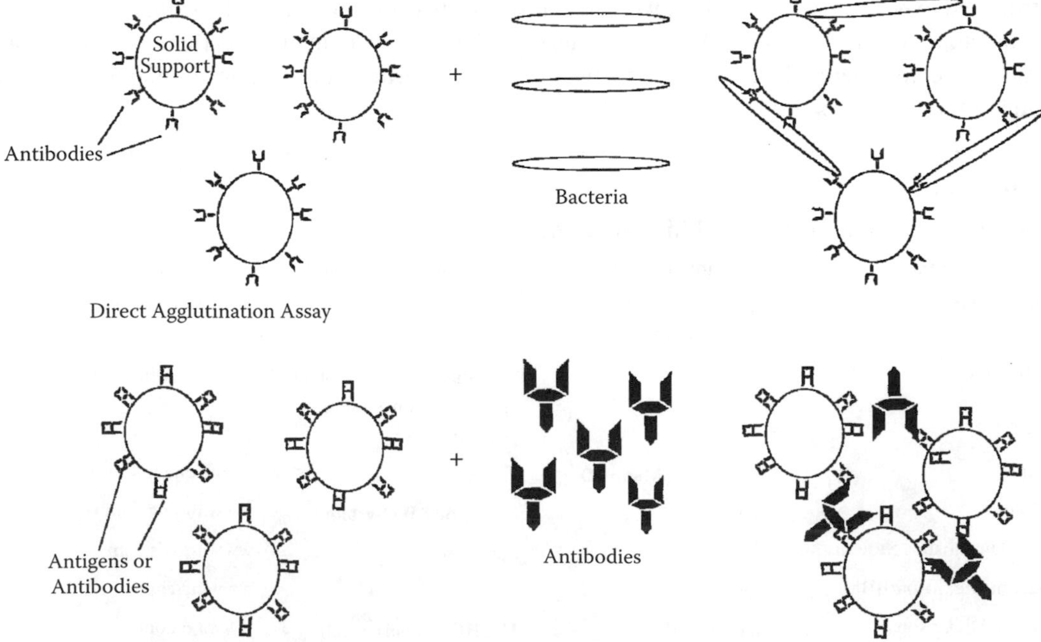

FIGURE 10.1 Agglutination assay.

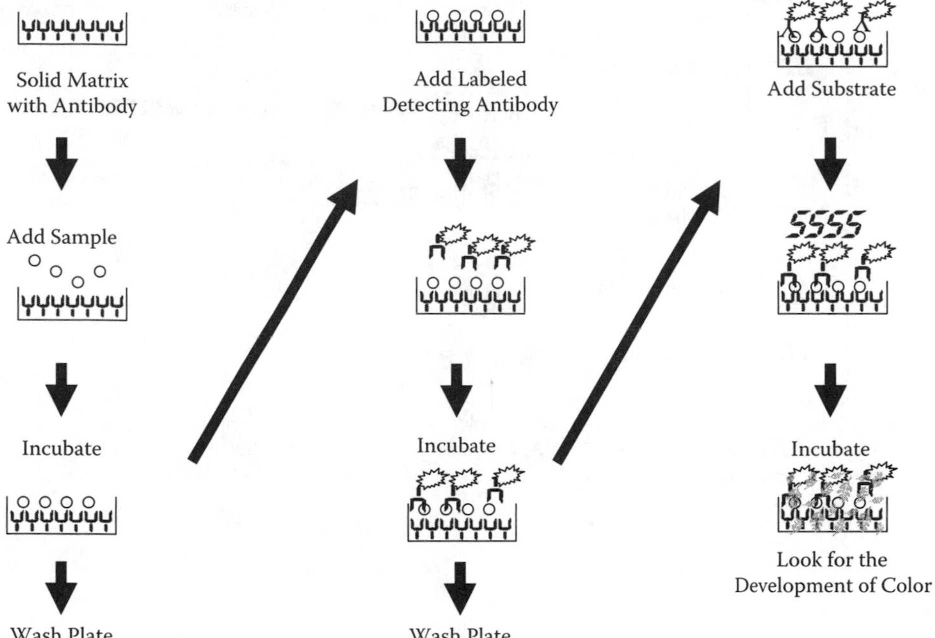

FIGURE 10.2. Steps in an enzyme immunoassay (EIA).

(EIA). To carry out this test, a solid matrix is required. The most frequently used is the plastic microtiter tray. The analyte is specifically trapped on the matrix using either an antibody (direct test) or an antigen (indirect test). The matrix is then washed and the labeled antibody is added. Following another washing step, the label is then detected. In EIAs or enzyme linked immunosorbent assays (ELISAs), the label is detected by adding a substrate and observing the formation of a color. In fluorescent immunoassays (FIAs), the label is detected using a fluorimeter. In radio immunoassays (RIAs), the label is detected using a Geiger or scintillation counter.

In recent years, labeled antibody assays have been improved with the introduction of lateral flow assays. As shown in Figurc 10.3, the test consists of a strip with several components printed on it. First, there is the conjugate pad, which contains the labeled antibody. There is also a line of trapping antibody, and finally a control line, which contains the analyte being assayed. The sample is added and capillary action moves it through the conjugate pad. If the sample contains the analyte, it will bind to the conjugate and pull it forward as it migrates down the strip. When this labeled complex encounters the trapping antibody, if the analyte is present, the complex will bind, forming a colored line. To allow for a positive control, these tests are designed to have an excess of labeled antibody. As the front continues to flow, this excess labeled antibody will bind to the analyte in the control line. The result is that if the analyte is present, two lines will be seen. If the analyte is absent, only the control line is observed. There are two major ways of visualizing the lines. In one type of test, the antibody is labeled with an enzyme and substrate must be added to visualize it. In another type of test, the antibody is labeled with colored markers (such as latex or gold particles) and can be viewed without any additional steps.

A recent improvement to the lateral flow assay is the Dual Path Platform (DPP™). As shown in Figure 10.4, the sample and conjugate have independent flow paths, which allows for greater sensitivity and also the ability to identify multiple analytes in a single sample.[2] (See www.chembio.com for full description of the technology.)

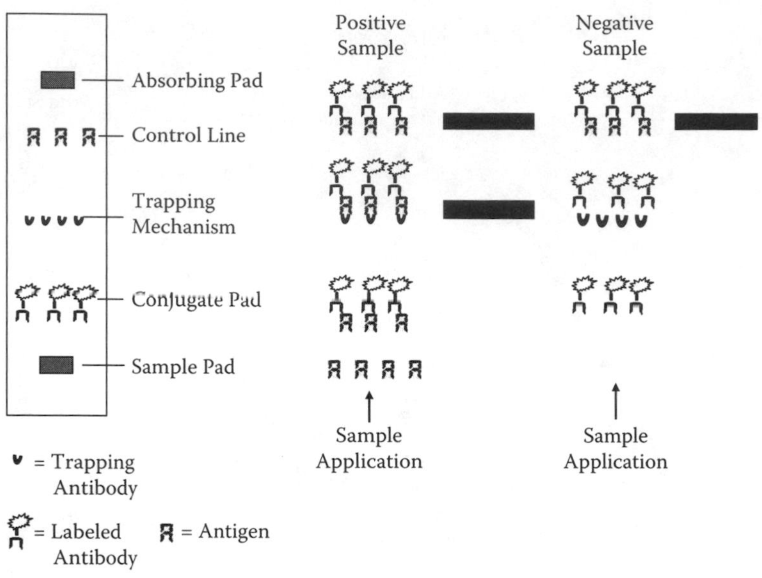

FIGURE 10.3. Lateral flow assay.

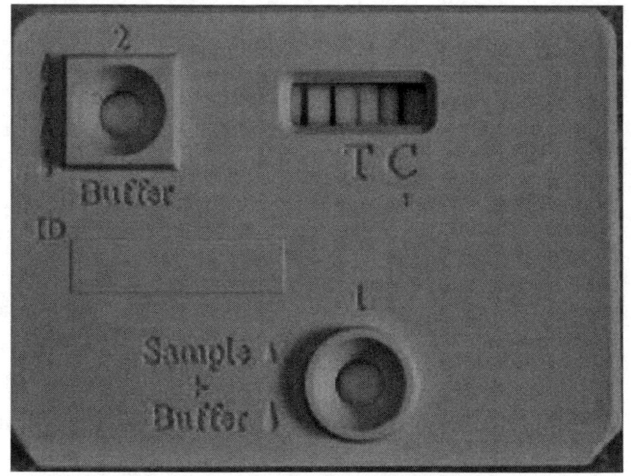

FIGURE 10.4. Dual path platform.

Antibody Labeling Technique Variations

Second Antibody: This procedure is used to increase sensitivity and to minimize the number of labeled antibodies that the lab must maintain. The steps are as described in Figure 10.2, except that the "detecting" antibody is unlabeled. After this antibody has bound to the analyte, it is detected using a generic labeled second antibody prepared in a different species. For example, individual unlabeled antibodies to separate microorganisms can be prepared in a rabbit. These rabbit antibodies can be used to bind the pathogen in the plate as in Figure 10.2. If one now adds a labeled anti-rabbit immunoglobulin antibody prepared in another animal, such as a goat, this will bind to the rabbit antibodies and be detected. Because there are many more sites to bind, this will increase the sensitivity of the assay.

Another advantage is that this single reagent can be used to detect any of the rabbit antibodies. As a result, a laboratory can produce rabbit antibodies to an entire array of pathogens, but only need produce the single labeled anti-rabbit immunoglobulin antibody to detect them.

Enhanced EIA: This is used to increase sensitivity. The steps are as described in Figure 10.2, except that the first antibody does not produce a colored end product; it produces a substrate for a second antibody. This second antibody now reacts with an increased amount of substrate and produces the colored end product.

Biotin/Avidin: Biotin is a vitamin that has a strong affinity for the glycoprotein avidin and can be bound to an antibody. After the antibody has bound to the analyte, an enzyme-labeled avidin is added. Upon adding the substrate, a colored end product is produced. This variation is used to increase sensitivity, and also to provide a system where multiple analytes can be detected using only a single labeled reagent.

Neutralization: This is a technique usually used to detect low levels of antibody in a patient's serum. A known amount of antigen is bound to a solid support. A labeled antibody is then added, and the amount of the end product is recorded. Following this, the procedure is repeated, but increasing amounts of the patient sample are mixed with the labeled antibody. If the patient sample contains an antibody to the antigen, it will compete with the labeled antibody and reduce (neutralize) the amount of end product detected. The clinician can calculate the level of antibody in the patient's blood by noting the quantities of patient sample and the decrease in end product.

Slide Tests: Labeled antibodies can also be used to specifically stain pathogens on a slide. One type of very sensitive assay is the direct fluoroimmuno assay. In this procedure, the sample is fixed to a slide and then the slide is incubated with a fluorescently labeled antibody. Following a washing step, the slide is viewed under a fluorescent microscope. If the analyte is present, fluorescence is observed. This test can be used to detect viruses growing in cell culture.

Many of the companies listed in Table 10.1 also supply antibody-based tests. Suppliers of specific antibodies as well as contracting services can be found at the website www.antibodyresource.com

NUCLEIC ACID PROBES

A labeled DNA probe that is complementary to a specific sequence in the organism of interest is synthesized. This probe is added to the sample and allowed to anneal. Following a washing step, the sample is assayed for the presence of the label. As with antibodies, the label can be a radioactive molecule, an enzyme, biotin/avidin, or a fluorescent molecule. Labeled probes have also been used to identify species on the bases of their ribosomal RNA sequences. Figure 10.5 shows an example of the use of a DNA probe.

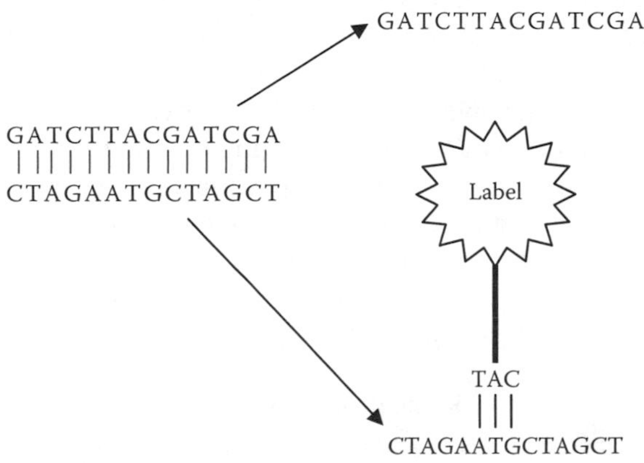

FIGURE 10.5. DNA probe.

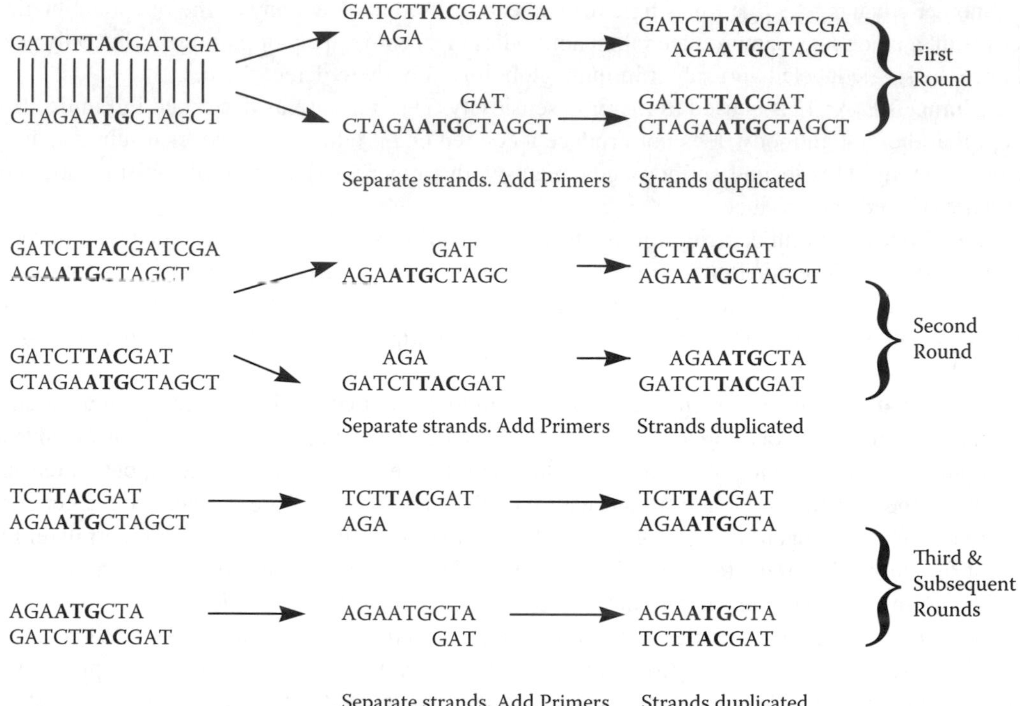

FIGURE 10.6. Polymerase chain reaction (PCR).

A disadvantage of this technique is that many DNA sequences used to identify an organism may only occur once in the genome. As a result, there are usually only limited molecules to detect. Advances in this field have relied on methods to either amplify the amount of material that can be detected, amplifying the label, or looking for other nucleic acids in the cell.

Sample Amplification Methods

Polymerase Chain Reaction: A specific region of DNA is copied exponentially. The DNA is amplified by supplying primers flanking the region of interest, nucleotides, and DNA polymerase, and then allowing them to undergo many rounds of replication. Figure 10.6 shows the basic steps. After the first round, there will be four strands. In addition to the original two strands, there will be one strand missing the 5′ end and one strand missing the 3′ end. After the second round of replication, strands that contain only the area of interest will have been produced. In subsequent rounds, these strands will be produced exponentially. Cycle times of 15 min are now possible, meaning that in 8 hr, the region of interest can be multiplied millions of times.[3] Once the DNA has been amplified, specific probes can be used to detect the pathogen. While the steps are fairly standard, different types of DNA may necessitate variations in the temperatures and in the times of incubation. In many instances, these parameters must be determined by the investigator.

Real-Time PCR: In recent years, the trend has been to detect the nucleic acid, as it is being multiplied in real-time. Two examples of commercial systems that are in use are the LightCycler® and Taqman®, both from Roche Diagnostics, Inc.

Taqman: Similar to conventional PCR, but in addition, a small probe is annealed to the amplicon. This probe has a fluorescent molecule at one end, and a fluorescent quencher at the other. As replication takes place, the taq polymerase in addition to adding bases to the primer also uses its nuclease activity to destroy the probe. This frees the fluorescent molecule from the quencher, and results in fluorescence that can be measured. A video representation of this technique can be viewed at www.med.unc.edu/anclinic/Tm.htm.

LightCycler: Primers are used as in regular PCR. In addition, two fluorescent probes are also included. Each fluorescent probe hybridizes to the same strand of DNA being copied, but contains a different fluorescent molecule. As the amplicon is synthesized, both fluorescent probes are incorporated. To detect them, light at a wavelength that excites the first fluorescent molecule is directed at the sample. When the first molecule is excited by this wavelength of light, it fluoresces a second wavelength. This wavelength will only excite the second molecule if it is on the same strand of the copy. When the second molecule is excited, it fluoresces at a third wavelength. The instrument then detects the third wavelength. Light at the third wavelength will only be detected if the amplicon contains both probes.

Information about both Taqman and LightCycler can be found at www.roche-diagnostics.com.

Variations of PCR:

Nested PCR amplification: This occurs in two steps, using two different primer pairs. It can increase the sensitivity and specificity of the PCR reaction.[4]

Booster PCR amplification: Occurs in two steps but differs from nested PCR in that the same primers are used for both steps.[5]

Hot Start PCR: PCR mixture, with the exception of one ingredient, is kept at temperatures above the nonspecific binding threshold. The missing component is added and cycling begins at an elevated temperature. This reduces nonspecific binding.[6]

Multiplex PCR: Several probes are used at one time. It can be used to distinguish multiple organisms in a single sample. This helps overcome the expense of examining a sample and looking for only one pathogen.[4]

Random Amplified Polymorphism DNA: Random primers are used for PCR in a sample. The amplified products are separated by size using gel electrophoresis. The patterns obtained from different samples are compared. This can be used to identify similar organism biotypes.[7]

Fluorescent Amplified-Fragment Length Polymorphism: Genomic DNA is specifically cut in a limited number of locations, using restriction endonucleases. These fragments are then ligated to double-stranded oligonucleotides. Fluorescene-labeled primers are then added, and several cycles of PCR are run. The labeled fragments are separated on a gel, by size. Different strains of an organism will give different patterns of fluorescent bands, which can be used as a fingerprint.[8]

Ligase Chain Reaction: The steps are similar to PCR. Oligonucleotide primers are constructed, but in LCR the primers actually flank the gene on both sides of each strand. The sample is heated to denature the DNA, and then cooled to allow the probes to anneal to their complementary strand. Ligase is added and fills in the gaps between the two primers, but only if they are aligned. As the reaction proceeds, the newly formed ligated primers as well as the initial DNA can now be used as templates. Again, this leads to an exponential replication of the DNA region of interest.[9]

Other Sample Amplification Methods: There are several other methods that use sophisticated molecular biology techniques primarily to amplify and quantify RNA viruses. Three of the most common are Amplicor™, commercially developed by Roche Molecular Systems; Transcription-Mediated Amplification, commercially developed by Gen-Probe; and Nucleic Acid Sequence Based Amplification (NASBA), commercially developed by bioMerieux Organon/Technika.

Amplicor: A specific DNA primer is annealed to a target RNA molecule. An enzyme with both DNA polymerase and reverse transcriptase activity is then used to complete the complementary DNA copy. Using this cDNA as a template, PCR is then carried out, using biotinylated primers. The biotinylated amplicon is then bound to a specific probe-coated microwell. The amplicon is detected using an avidin-HRP conjugate, which yields a colorimetric result.[10]

Transcription-Mediated Amplification (TMA). In TMA, RNA polymerase and reverse transcriptase are used to amplify RNA to produce a DNA copy. A double-stranded DNA molecule is then synthesized and used for transcription to produce new RNA copies of the original sequence. TMA is isothermal and does not require a thermal cycler. The most recent version of TMA can also amplify DNA. (See www.gen-probe.com for specific information.)

Nucleic Acid Sequence Based Amplification (NASBA). This technology is also isothermal and uses reverse transcriptase, RNA polymerase, and RNase H to amplify RNA molecules.[11] Recent articles have appeared, in which molecular beacons have been used to perform real-time NASBA.[12] A commercial real-time system, Nuclisens, is available from bioMerieux.

Reverse Hybridization: A series of probes are placed on an inert support. The organism DNA is labeled and hybridized to the support. Label will be visible in areas where there is homology between the organism and the probe. This information is used to identify the organism. This technique is especially useful in epidemiological studies in which one is trying to identify organism biotypes. In a recent article, PCR and reverse hybridization were combined to produce a test to identify the β-papilloma virus.[13]

Strand Displacement: This is a technology that involves using a restriction endonuclease to create a nick in a DNA molecule. Polymerases and primers are subsequently used to continuously produce new strands, downstream from the site of the nick. As each new strand is started, it replaces the one that had been previously created. This is an isothermal technique.[14]

Signal Amplification Methods

Branched Chain DNA: Unlike the technologies discussed above, this technology increases assay sensitivity by amplifying the signal molecules in a reaction. As described in Figure 10.7, a trapping sequence is immobilized on the bottom of a microtiter tray. The sample is added and hybridizes to it. An extender sequence, which is complementary to the sample and to an arm of the amplification

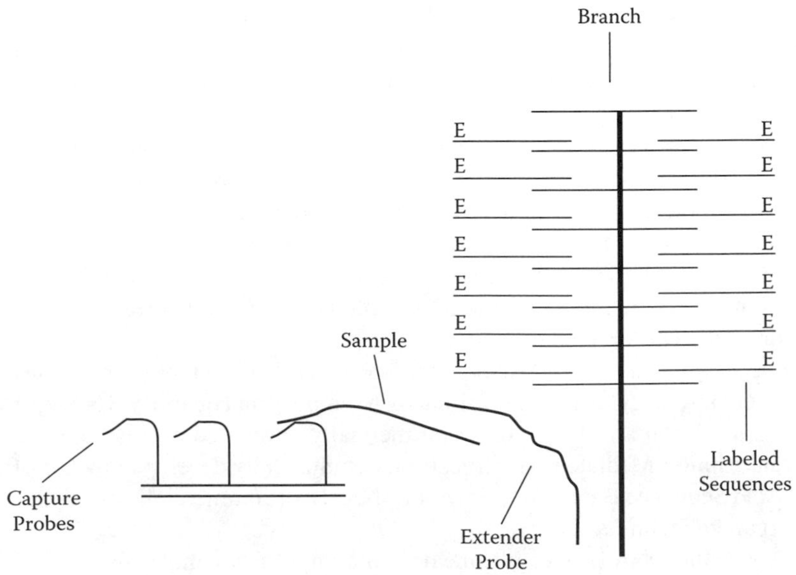

FIGURE 10.7. Branched chain reaction.

molecule, is then added and allowed to hybridize to the sample. An amplification molecule with many branches on it is then added and hybridizes to the other side of the extender sequence. In the next step, enzyme-labeled sequences that are complementary to the chains of the branch are added. Multiple numbers of these will hybridize with the amplification molecule and in the final step the label is detected. These assays are frequently used to estimate viral load in HIV and HCV[15] infection. Quantiplex is a branched-chain DNA system manufactured by Siemens Diagnostics.

Hybrid Capture: This technique was originally developed to test for the presence of DNA from Papilloma virus in cases of cervical dysplasia.[16] In this technique, RNA probes for the pathogen are added to the sample. If the pathogen is present, an RNA-DNA duplex forms. This duplex is then added to a microtiter plate with a trapping mechanism that captures RNA-DNA hybrids. Following this, a multiply chemiluminescently labeled antibody specific for RNA-DNA duplexes is added. Following the washing steps, the chemiluminescent reagent is added and the duplex is detected. Commercial systems are available from Digene, Inc. (www.digene.com).

DNA MICROARRAY TECHNOLOGY

DNA microarray technology refers to a technique in which hundreds or thousands of oligonucleotides are attached to a solid support (Figure 10.8). The sample DNA or RNA is then hybridized to the support. Following this, sites of binding are detected. Information can be found at www.genechips.com. This technology has been used in detecting genetic variations among a single species of microorganisms. In a recent study, Neverov et al.[17] used the technique to identify different strains of measles virus. The technique has also been used in trying to identify multiple pathogens from a single isolate.[18]

ELECTROPHORESIS

This is a method based on separating proteins, or nucleic acids, from microorganisms based upon their charge (Figure 10.9). A support matrix, usually a gel (acrylamide or agarose) or acetate strip, is prepared, and the sample, prepared in an appropriate buffer, is placed in or on the support. When a current is applied, charged molecules in the sample will migrate at a speed that is related to their charge. There are variations of this technique that create a chemical environment in which the molecules can also be separated by size or isoelectric point. Following electrophoresis, the support is stained and specific molecules can then be identified on the basis of how far they have migrated.

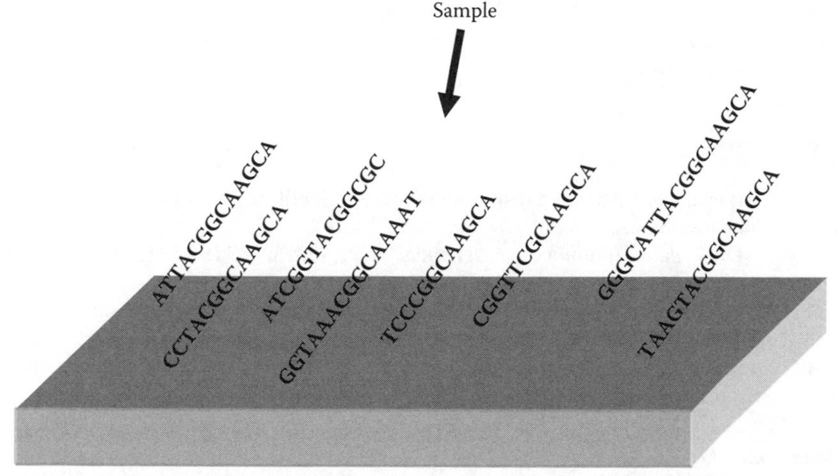

FIGURE 10.8 DNA microarray technology.

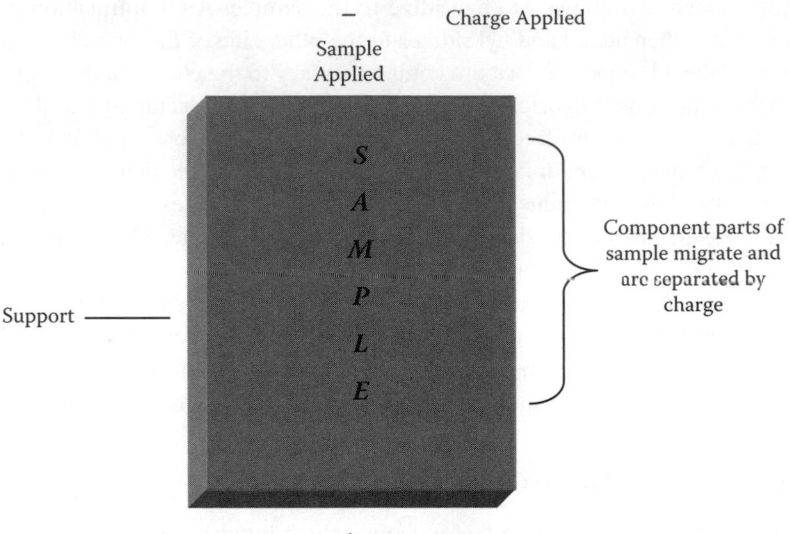

FIGURE 10.9 Electrophoresis.

Alternatively, the molecules can be removed from the support by blotting onto an absorbing media and used for further analysis.[19]

Polyacrylamide Gel Electrophoresis (PAGE): Microorganism proteins are run in the support and stained. Coomassie blue and methyl green are commonly used. Specific protein bands are identified.

Two-Dimensional Gels: The support matrix is subjected to an electric field and then rotated 90° and run again. This provides much greater resolution.

Pulse Field Gels: The support matrix is alternately exposed to electric fields in varying directions. The advantage of this technique is that it can be used for very large molecules. It is useful in identifying the sources of hospital-acquired infections.

Southern Blots: Microorganism DNA is digested using restriction endonucleases and is then run in the support. Following this, the material in the gel is transferred to either a nitrocellulose or nylon sheet. Staining the sheet with a DNA-specific stain or by hybridizing a labeled specific DNA probes, can then identify bands.

Northern Blots: Similar to above, except that the sample that is run is RNA. Bands can be identified by staining or by using labeled DNA probes.

Western Blots: Proteins are run and bands are stained using labeled antibodies. This is a confirmatory test in HIV testing.

REFERENCES

1. O'Hara, C.M. Manual and automated instrumentation for identification of Enterobacteriaceae and other aerobic Gram-negative bacilli. *Clin. Microbiol. Rev.,* 18, 147, 2005.
2. Esfandiari, J. Dual Path Immunoassay Device. U.S. Patent #7,189,522, and pending PCT/US #06099191.
3. Cheng, S., Fockler, C., Barnes, W., and Higuchi, R. Effective amplification of long targets from cloned inserts and human DNA. *Proc. Natl. Acad. Sci. U.S.A.,* 91, 5695, 1994.
4. Millar, B.C., Xu, J., and Moore J.E. Molecular diagnostics of medically important bacterial infections. *Curr. Issues Mol. Biol.,* 9, 21, 2007.
5. Ruano, G., Fenton, W., and Kidd, K.K. Biphasic amplification of very dilute DNA samples via 'booster' PCR. *Nucl. Acids Res.,* 17, 5407, 1989.
6. Chou, Q., Russell, M., Birch, D., Raymond, J., and Bloch, W. Prevention of pre-PCR mis-priming and primer dimerization improves low-copy-number amplifications. *Nucl. Acids Res.,* 20, 1717, 1992.

7. Van Leeuwen, W., Sijmons, M., Sluijs, J., Verbrugh, H., and van Belkum, A. On the nature and use of randomly amplified DNA from *Staphylococcus aureus*. *J. Clin. Microbiol.,* 34, 2770, 1996.
8. Kassama, Y., Rooney, P.J., and Goodacre, R. Fluorescent amplified fragment length polymorphism probabilistic database for identification of bacterial isolates from urinary tract infections. *J. Clin. Microbiol.,* 40, 2795, 2002.
9. Wiedmann, M., Wilson, W.J., Czajka, J., Luo, J., Barany, F., and Batt, C.A. Ligase chain reaction (LCR) — overview and applications. *PCR Methods Appl.*, 3, S51-64, 1994.
10. Revets, H., Marissens, D., De Wit, S., Lacor, P., Clumeck, N., Lauwers, S., and Zissis, G. Comparative evaluation of NASBA HIV-1 RNA QT, AMPLICOR-HIV Monitor, and QUANTIPLEX HIV RNA assay, three methods for quantification of human immunodeficiency virus type 1 RNA in plasma. *J. Clin. Microbiol.,* 34, 1058, 1996.
11. Compton, J. Nucleic acid sequence based amplification. *Nature,* 350, 91, 1991.
12. Nadal, A., Coll, A., Cook, N., and Pla, M. A molecular beacon-based real time NASBA assay for detection of *Listeria monocytogenes* in food products: role of target mRNA secondary structure on NASBA design. *J. Microbiol. Methods,* 68, 623, 2007.
13. De Koning, M., Quint, W., Struijk, L., Kleter, B., Wanningen, P., Van Doorn, L.-J., Weissenborn, S.J., Feltkamp, M., and Ter Schegget, J. Evaluation of a novel highly sensitive, broad-spectrum PCR-reverse hybridization assay for detection and identification of beta-papillomavirus DNA. *J. Clin. Microbiol.*, 44, 1792, 2006.
14. Abramson, R.D. and Myers, T.W. Nucleic acid amplification technologies. *Curr. Opin. Biotechnol.,* 4, 41, 1993.
15. Ross, R.S., Viazov, S., Sarr, S., Hoffmann, S., Kramer, A., and Roggendorf, M. Quantitation of hepatitis C virus RNA by third generation branched DNA-based signal amplification assay. *J Virol. Methods,* 101, 159, 2002.
16. Brown, D.R., Bryan, J.T., Cramer, H., and Fife, K.H. Analysis of human papillomavirus types in exophytic *Condylomata acuminata* by hybrid capture and Southern Blot techniques. *J. Clin. Microbiol.,* 31, 2667, 1993.
17. Neverov, A.A., Riddell, M.A., Moss, W.J., Volokhov, D.V., Rota, P.A., Lowe, L.E., Chibo, D., Smit, S.B, Griffin, D.E., Chumakov, K.M., and Chizhikov, V.E. Genotyping of measles virus in clinical specimens on the basis of oligonucleotide microarray hybridization patterns. *J. Clin. Microbiol.,* 44, 3752, 2006.
18. Li, Y., Liu, D., Cao, B., Han, W., Liu, Y., Liu, F., Guo, X., Bastin, D.A., Feng, L., and Wang, L. Development of a serotype-specific DNA microarray for identification of some *Shigella* and pathogenic *Escherichia coli* strains. *J. Clin. Microbiol.,* 44, 4376, 2006.
19. Lodish, H., Berk, A., Zipursky, L.S., Matsudaira, P., Baltimore, D., and Darnell, J. *Molecular Cell Biology, 4th ed.,* W.H. Freeman, New York, 2000, chap 3, chap 7.

11 Mechanisms of Action of Antimicrobial Agents

Carmen E. DeMarco and Stephen A. Lerner

CONTENTS

INTRODUCTION

Since the previous edition of this Handbook, many new antibiotics have been developed and are now on the market or undergoing clinical trials. There are many compounds with antimicrobial activity. This chapter, however, attempts to summarize current information about, and working models for, the mechanisms of action only of antimicrobial agents that are in clinical use or in advanced stages of investigation, or that illustrate important points or are used in laboratory research. This chapter also provides references to more extensive information.

The antimicrobial agents presented here will be grouped in three major classes, according to the target microbes — (1) antibacterials, (2) antimycobacterials, and (3) antifungals — and then, where applicable, by mechanisms of action into six major categories:

1. Inhibition of cell wall synthesis
2. Interference with membrane integrity
3. Inhibition of nucleic acid synthesis
4. Inhibition of protein synthesis
5. Inhibition of synthesis of essential small molecules
6. Miscellaneous (or unknown) effects

ANTIBACTERIAL AGENTS

Inhibition of Cell Wall Synthesis

The bacterial cell wall is a unique structure responsible for multiple functions: maintenance of cell shape, protection against osmotic pressure, attachment site for bacteriophage, or serving as a platform for bacterial appendages such as flagella, fimbriae, and pili.

The main component of the bacterial cell wall is the peptidoglycan layer. In Gram-positive bacteria, the cell wall is fifty to one hundred molecules thick, whereas in Gram-negative bacteria it is only one or two molecules thick. The Gram-negative cell wall is surrounded by an outer membrane, consisting of a phospholipid bilayer with incorporated protein molecules. Some of these proteins, the so-called porins, form aqueous channels through which hydrophilic molecules, including most antimicrobial agents, diffuse across the outer membrane to enter the periplasmic space, which lies between the outer membrane and inner (or cytoplasmic) membrane. The peptidoglycan lies in the periplasmic space but does not generally pose a physical barrier to penetration by most antimicrobials, because such small molecules easily pass through. Interference with cell wall biosynthesis generally has bactericidal consequences because cell wall lysis eventually ensues. In some cases, the interference with cell wall synthesis leads to liberation of lipoteichoic acids, which normally regulate the activity of hydrolytic enzymes that play roles in the dynamic changes of the cell wall in cell growth and division. When the regulation is lifted by the loss of the lipoteichoic acids, these enzymes may actively hydrolyze the cell wall, leading to its disintegration.

Drugs that Inhibit Cell Wall Biosynthetic Enzymes

Phosphonomycin (Fosfomycin). Phosphonomycin acts as an analog of phosphoenolpyruvate and binds covalently to a cysteinyl residue of the phosphoenolpyruvate: UDP-GlcNAc-3-enolpyruvyl transferase; this complex interferes with the formation of UDP-N-acetyl muramic acid (UDP-MurNAc) early in the synthesis of the bacterial cell wall.[48, 132]

Cycloserine. Cycloserine is an inhibitor of the cell wall peptidoglycan and is active against many Gram-positive and Gram-negative bacteria. However, it is mainly used as an antimycobacterial agent, so we discuss it more extensively in a subsequent subsection.

Drugs that Bind to Carrier Molecules

Bacitracin. In the presence of divalent cations, bacitracin complexes with the membrane-bound pyrophosphate form of the undecaprenyl (C_{55}-isoprenyl) carrier molecule that remains after the disaccharide-peptide unit is transferred from GlcNAc-MurNAc-pentapeptide pyrophospholipid to the nascent peptidoglycan chain. This binding inhibits the enzymatic dephosphorylation of this carrier to its monophosphate form, which is required for another round of synthesis and transfer of the disaccharide-peptide unit. Similar complex formation with intermediates of sterol biosynthesis in animal tissues (e.g., farnesyl pyrophosphate and mevalonic acid) may account in part for the toxicity of bacitracin in animal cells.[99, 103, 104]

Drugs that Combine with Cell Wall Precursors

Glycopeptides: Vancomycin and Ristocetin. Vancomycin and ristocetin are glycopeptide antibiotics that interfere with bacterial cell wall synthesis. Although the structure and action of vancomycin have been studied in greater detail, the mechanism of action of ristocetin is believed to be similar to that of vancomycin. However, due to its toxic effects, ristocetin is not used as an antimicrobial in clinical practice.

These drugs are bactericidal and may even kill cell wall-deficient bacteria (L forms), suggesting secondary effects not limited to cell wall synthesis.

Vancomycin is a 1450-Da (dalton) tricyclic glycopeptide in which are linked two chlorinated tyrosines, three substituted phenylglycines, glucose, vancosamine (a unique amino-sugar), *N*-methyl leucine, and aspartic acid amide. This structure can undergo hydrogen bonding to the acyl-D-alanyl-D-alanine terminus of the pentapeptide in various peptidoglycan precursors (e.g., UDP-MurNAc-pentapeptide, MurNAc-pentapeptide pyrophospholipid, and GlcNAc-MurNAc-pentapeptide pyrophospholipid). Binding of vancomycin occurs through a set of backbone contacts between the D-Ala-D-Ala dipeptide at the end of the pentapeptide chain and the amides that line a cleft formed by the cross-linked heptapeptide of the glycopeptide. The bound glycopeptide acts as a steric impediment, preventing lipid II and/or the nascent glycan from being processed further. Subsequently, the transglycosylation and/or transpeptidation steps of peptidoglycan synthesis are inhibited, weakening the peptidoglycan layers and exposing the cell to lysis.[41, 49, 81, 98, 131]

Lipoglycopeptides: Telavancin, Teicoplanin, Ramoplanin, Dalbavancin, and Oritavancin. Telavancin is a semisynthetic derivative of vancomycin, featuring the same glycopeptide core as vancomycin, but with the addition of an extended lipophilic (decylaminoethyl) side chain to the vancosamine sugar and a hydrophilic side chain represented by the negatively charged phosphonomethyl aminomethyl moiety to the resorcinol-like 4′-position of amino acid 7. In contrast to vancomycin, which binds to the cell wall, telavancin binds predominantly to the membrane. Telavancin thus has a dual mode of action. The lipophilic moiety promotes interaction with the cell membrane and anchors the molecule to the cell surface, providing improved binding affinity of the glycopeptide core for D-alanyl-D-alanine-containing peptidoglycan. The rapid bactericidal effect of telavancin is due to disruption of the barrier function of the membrane, leading to both concentration- and time-dependent loss of membrane potential and increased membrane permeability. The hydrophilic substituent increases the distribution in the body and promotes rapid clearance from the body, thus reducing nephrotoxicity.[43, 55, 79, 121]

Teicoplanin is a lipoglycopeptide antibiotic structurally resembling ristocetin, and approved for human use outside the United States. Teicoplanin differs from vancomycin in three ways:

1. The glycosylation number and placement are different from vancomycin.
2. There is a long-chain fatty acid substituent in the amide linkage to the GlcNAc sugar attached to PheGly.
3. The cross-linked heptapeptide scaffold is different at residues 1 and 3, allowing four side-chain cross links (1-3, 2-4, 4-6, 5-7), in contrast to the three in vancomycin.

Like vancomycin, teicoplanin acts by interacting and forming a complex with the substrate peptidoglycan units that have the pentapeptidyl tails terminating in D-Ala_4-D-Ala_5. This substrate sequestration effectively shuts down transpeptidation by making the N-acyl-D-Ala-D-Ala acceptor unavailable to the transpeptidases.[11, 122]

Ramoplanin is a novel lipoglycopeptide with a unique mechanism of action, causing alteration of cell wall peptidoglycan linkages and membrane permeability of Gram-positive bacteria, thereby disrupting the cell wall. Ramoplanin acts by inhibiting the late-stage assembly of the peptidoglycan monomer and its polymerization into mature peptidoglycan. It inhibits the ability of the MurG glycosyltransferase to link undecaprenyl-pyrophosphoryl-N-acetylmuramyl-pentapeptide (Lipid I) with uridyl-pyrophosphoryl-N-acetylglucosamine (UDP-GlcNAc) to release uridyl-diphosphate (UDP) and form the disaccharide undecaprenyl-pyrophosphoryl-N-acetylmuramyl-(N-acetylglucosamine)-pentapeptide (Lipid II), which forms the basis of the recurring disaccharides in the polysaccharide chain of the peptidoglycan. This mechanism of action might involve complexation of Lipid I by ramoplanin, because its antimicrobial activity was found to increase dramatically as levels of the antibiotic approached concentrations equivalent to predicted cellular pools of Lipid I.[38, 69, 121]

Dalbavancin is a semisynthetic lipoglycopeptide structurally related to teicoplanin, with an unusually long half-life of 6 to 10 days. Dalbavancin interrupts Gram-positive bacterial cell wall biosynthesis by binding, like vancomycin, to the terminal D-alanyl-D-alanine of the pentapeptide-glycosyl cell wall intermediate, thus inhibiting transpeptidation and leading to cell death. A unique property of dalbavancin is the ability to dimerize and anchor itself into bacterial membranes, improving its interaction with, and binding affinity to, D-alanyl-D-alanine residues of nascent peptidoglycan precursors and producing more rapid and potent bactericidal activity against methicillin-resistant *Staphylococcus aureus* (MRSA) and coagulase-negative staphylococci, as compared with vancomycin and teicoplanin.[60, 67, 21]

Oritavancin is a novel semisynthetic lipoglycopeptide antibiotic with concentration-dependent bactericidal activity against Gram-positive cocci, including glycopeptide-resistant enterococci and *Staphylococcus aureus* with intermediate susceptibility to glycopeptides. It also displays a prolonged postantibiotic effect and elevated serum protein binding as compared with vancomycin. The mechanism of action of oritavancin is similar to that of vancomycin, with some important differences that explain its activity against vancomycin-resistant pathogens: oritavancin can form dimers, which cooperatively bind to precursors ending in D-Ala-D-Lac, as well as those with the normal D-Ala-D-Ala, and the dimer is stabilized in the most favorable position by the hydrophobic interactions of the lipophilic side chain, anchoring oritavancin in the membrane; oritavancin has been shown to inhibit the transglycosylation step of cell-wall biosynthesis.[2, 13, 66, 121]

Drugs that Inhibit Enzymatic Polymerization and Attachment of New Peptidoglycan to Cell Wall (β-Lactams)

All β-lactam antibiotics contain the highly reactive β-lactam ring that binds to, and inactivates, various penicillin-binding proteins (PBPs), enzymes in the cytoplasmic membrane that catalyze various steps in the biosynthesis and reshaping of the cell wall peptidoglycan.

Penicillins and Cephalosporins. Penicillins and cephalosporins interfere with the terminal step in the cross-linking of peptidoglycan chains (i.e., transpeptidation of the N-terminus of the amino acid or peptide bridge from the pentapeptide onto the penultimate D-alanine of the pentapeptide chain, with release of the terminal D-alanine). It has been suggested that penicillin is a structural analog of a transition state of the D-alanyl-D-alanine terminus of the recipient peptidoglycan strand that is involved in transpeptidation. It thereby competes with the D-Ala-D-Ala bond in binding to the transpeptidase enzyme, compromising the formation of an acyl-enzyme intermediate with the penultimate D-alanine. The highly reactive CO-N bond of the β-lactam ring acylates the transpeptidase irreversibly, thereby inactivating it, so the cross-linking does not occur. Bacteria display in their cytoplasmic membrane a variety of penicillin-binding proteins (PBPs), which include other penicillin-sensitive enzymes in addition to the transpeptidases and D-alanine carboxypeptidases. The concept of multiple target sites for the actions of penicillins and cephalosporins is supported by the observations that various β-lactam antibiotics produce differential effects on the processes of cell division, elongation, and lysis. Specific β-lactam antibiotics bind different PBPs preferentially, and the differences in physiological and morphological effects of various β-lactams correlate with the affinity of their binding to different PBPs. The lytic and killing actions of penicillins and other β-lactams may not result directly from their inhibition of peptidoglycan synthesis, but may also involve the action of bacterial murein hydrolases. It has been observed that lipoteichoic acids, which inhibit these autolytic enzymes, are released from some bacteria upon treatment with penicillin *in vitro*. It has thus been postulated that the release of these endogenous inhibitors of murein hydrolase enzymes interferes with the regulation of these enzymes in the cell, so the unregulated hydrolases destroy the integrity of the cell wall and contribute to cell lysis and death. Modifications of the side chains of some penicillins and cephalosporins have permitted improved penetration through the porins in cell envelope structures, resulting in an enhanced antibacterial spectrum.[4, 10, 107, 127]

Carbapenems: Imipenem, Meropenem, Ertapenem, and Faropenem. Carbapenems diffuse more readily through the outer membrane of most Gram-negative bacteria, in part because of their zwitterionic characteristics. Furthermore, in *Pseudomonas aeruginosa*, carbapenems enter more rapidly through pores formed by a specific porin, outer membrane protein D2, whose mutational loss or alteration produces carbapenem resistance. Carbapenems are also generally less readily inactivated by bacterial β-lactamases than other β-lactams.[4, 116, 133]

Imipenem inhibits bacterial cell wall synthesis by binding to and inactivating specific PBPs. In *Escherichia coli* imipenem inhibits the transpeptidase activities of PBP-1A, -1B, and -2, and the D-alanine carboxypeptidase activities of PBP-4 and PBP-5. It also inhibits the transglycosylase activity of PBP-1A, and it is a weak inhibitor of the transpeptidase activity of PBP-3, in contrast to all other β-lactams and other carbapenems that preferentially bind to PBP-1 and -3. Imipenem induces sphere formation with subsequent cell rupture, and it does not inhibit septum formation by the cells, thus reducing the amount of lipopolysaccharide liberated during bacteriolysis. For human use, imipenem is administered in combination with cilastatin, an inhibitor of dehydropeptidase (DHP) I, an enzyme produced by the renal epithelial cells that hydrolyzes imipenem.[92, 133]

Meropenem contains a C-methyl substituent, and thus is not susceptible to the renal DHP-I.[127] Its bactericidal action is similar to that of imipenem, although various bacterial mechanisms of resistance may affect susceptibility to these two carbapenems differentially.

Ertapenem binds most strongly to PBP-2 of *Escherichia coli*, then PBP-3, and it also has good affinity for PBP-1A and -1B, achieving rapid bactericidal action without the prior filamentation that occurs with third-generation cephalosporins that bind primarily to PBP-3.[27, 62]

Faropenem is a new, orally administered penem antibiotic that displays both bactericidal and β-lactamase inhibitor effects and is less susceptible to the renal dipeptidase enzyme DHP-1 as compared with imipenem. It has preferential affinity for PBP-2 but also for PBP-1A, -1B, and -3. Faropenem exposure at concentrations below the minimum inhibitory concentration (MIC) for *Staphylococcus aureus* affected septum formation with a decrease in the number of viable cells at increasing antibiotic concentrations; exposure to concentrations equivalent to the MIC and above produces cell lysis. A similar effect is observed in *E. coli*, with changes in cell shape at concentrations below the MIC and cell lysis at or above the MIC.[27, 40]

Monobactams: Aztreonam. The binding of aztreonam to PBP-3 of Gram-negative aerobic organisms results in the formation of nonviable filamentous structures. Its bactericidal activity is limited to aerobic Gram-negative bacteria because it does not bind to PBPs in Gram-positive aerobic bacteria or in anaerobic bacteria.[110, 111]

Interference with Cytoplasmic Membrane Integrity

Drugs that Disorganize the Cytoplasmic Membrane

Polymyxins: Polymyxin B and Colistin (Polymyxin E). Polymyxin B is regarded as a model compound of polymyxins, which are polypeptides, and colistin (polymyxin E) is believed to have an identical mechanism of action.

Polymyxin B initiates disorganization of the cytoplasmic membrane and causes release of lipopolysaccharide and leakage of intracellular components, with an extremely rapid bactericidal effect. This effect is initiated by electrostatic interaction of polymyxin with the outer membrane of Gram-negative bacteria and competitive displacement of covalent cations (calcium and magnesium) that normally stabilize lipopolysaccharide molecules in the outer leaflet of the bacterial outer membrane. As a result of studies with liposome systems, it has been postulated that the mechanism of action of polymyxin involves an ionic interaction between the phospholipid phosphate and the ammonium group of polymyxin. Simultaneous proton transfers between these two groups, and also between an ammonium group on the phospholipid and a γ-amino group on polymyxin, result in neutralization of charge on the polar head of the phospholipid, which can alter the electrostatic and hydrophobic stabilizing forces of the membrane. Liposomes derived from methylated phospholipids are not sensitive to polymyxin; possible explanations are prevention of proper alignment of the phospholipid and antibiotic by the methyl groups, or an increase in the basicity of the phospholipid amino group, which would make proton transfer less favored. A unique and very beneficial property of colistin is its anti-endotoxin activity, resulting from its ability to neutralize bacterial lipopolysaccharides.[23, 34, 35, 46, 59, 63]

Drugs that Produce Pores in Membranes

Lipopeptides: Daptomycin. Lipopeptides are a class of antibiotics that are active against multidrug-resistant Gram-positive bacteria. Lipopeptides consist of a linear or cyclic peptide sequence, with either net positive or negative charge, to which a fatty acid moiety is covalently attached to its N-terminus.

Daptomycin is a cyclic lipopeptide with rapid bactericidal action that results from its binding to the cytoplasmic membrane in a calcium-dependent manner. The acyl chain of the daptomycin molecule partly inserts into the membrane. In the presence of Ca^{2+}, which bridges between the anionic daptomycin and the anionic headgroups of bacterial outer leaflet lipids, the daptomycin molecule is anchored deeper into the membrane and forms aggregates. Subsequently, the bacterial membrane is depolarized and perforated and cell death ensues; thus, daptomycin is rapidly bactericidal.[105]

Peptide Antibiotics: Gramicidin A and Gramicidin S. Gramicidin A is a linear pentadecapeptide with an alternating D- and L-amino acid structural sequence, which is active in a dimer form. In membranes, gramicidin A forms transient channels specific for monovalent cations.[31, 117–119, 126]

Gramicidin S is a cyclic decapeptide that interacts electrostatically with the membrane phospholipids, without incorporating into the lipid region of the membrane. It achieves its bactericidal action by promoting leakage of cytoplasmic solutes; it may also cause uncoupling of oxidative phosphorylation as a secondary effect. The interaction of this peptide with the phospholipid bilayer is non-specific; and in addition to the antimicrobial activity, it exhibits highly hemolytic activity by interacting with the lipid bilayer of erythrocytes, limiting its clinical use to topical preparations.[80, 88]

Inhibition of Nucleic Acid Synthesis

Inhibitors of DNA Replication

Quinolones. The principal targets for the antibacterial activity of quinolones are DNA gyrase and topoisomerase IV.

DNA gyrase is a tetramer enzyme with two GyrA and GyrB subunits (encoded by the *gyrA* and *gyrB* genes) that catalyzes the introduction of negative superhelical turns and elimination of positive supercoils at the replication fork in the double-stranded, covalently closed circular bacterial DNA. DNA gyrase acts before the replicating fork, preventing replication-induced structural changes.[30]

Topoisomerase IV is also an enzyme composed of pairs of two subunits, called ParC and ParE (homologous to GyrA and B but encoded by genes *parC/parE*) in *Escherichia coli* and with GrlA and GrlB (encoded by genes *grlA/grlB*) in *Staphylococcus aureus*. A *parF* locus has been identified in *Salmonella typhimurium* and suggested to be involved in topoisomerase IV activity. Topoisomerase IV acts behind the replicating fork and catalyzes removal of supercoils and decatenation (unlinking) of interlocked daughter DNA molecules.[57]

Both DNA gyrase and topoisomerase IV appear to be targeted by all quinolones, regardless of bacterial species. However, the preferred drug target depends on the organisms and also on the drug; for example, the preferred site of action for levofloxacin is topoisomerase IV, while the main target for moxifloxacin and gatifloxacin is DNA gyrase. In Gram-negative bacteria, DNA gyrase is generally bound more readily by quinolones; in Gram-positive bacteria, topoisomerase IV is generally preferred.[51, 71]

Nalidixic acid, the first quinolone discovered and used in clinical practice, has a narrow spectrum of activity.[12, 109] The fluorinated quinolones discovered in the mid-1980s have a much broader spectrum of activity and improved bioavailability. Several of these, such as norfloxacin, ciprofloxacin, levofloxacin, and moxifloxacin, have extensive usage in clinical practice for treatment of Gram-negative and some Gram-positive pathogens.[12, 30, 33, 51, 71, 102, 109]

Nitroimidazoles. Nitroimidazoles — with the class representative metronidazole — are cytotoxic to facultative anaerobic bacteria and obligate anaerobes through a four-step process.[32, 54, 114, 115]

1. Entry into the bacterial cell by diffusion across the cell membrane.
2. Reduction of the nitro group by the intracellular electron transport proteins. The nitro group of metronidazole acts as an electron recipient, capturing electrons that are usually transferred to hydrogen ions in the oxidative decarboxylation of pyruvate cycle; the concentration gradient created by this process promotes further uptake of metronidazole into the cell and formation of intermediate compounds and free radicals toxic to the cell.
3. Interaction of cytotoxic by-products with host cell DNA, resulting in DNA strand breakage, inhibited repair, and, ultimately, disrupted transcription and cell death.
4. The toxic intermediate compounds decay into inactive end-products.

Inhibitors of RNA Polymerase: Rifamycins (Rifampin and Rifaximin)

Rifamycins, with rifampin as a class representative, specifically inhibit bacterial DNA-dependent RNA polymerases by blocking the RNA chain initiation step. Rifampin is used as part of a combination regimen in the treatment of mycobacterial infections and some Gram-positive infections (e.g., *Staphylococcus aureus*).

These antibiotics bind very tightly to the β-subunit of the RNA polymerase in a 1:1 molar ratio. They do not bind well to β-subunits of RNA polymerase isolated from rifamycin-resistant mutants, and mutations are found in the β-subunits of RNA polymerase in such mutants. Mammalian enzymes are not affected by rifampin. Rifamycins also modify bacterial pathogenicity and may alter attachment and tissue toxicity.[1, 91, 128]

Rifaximin is a novel semisynthetic derivative of rifamycin, used for treatment of enteric infections due to the fact that it is virtually non-absorbable from the gastrointestinal tract and active against enteric infections. Like other rifamycins, the mechanism of action involves binding of rifaximin to the β-subunit of the bacterial DNA-dependent RNA polymerase, inhibiting initiation of chain formation in RNA synthesis.[1]

Inhibition of Ribosome Function and Protein Synthesis

Inhibitors of 30S Ribosomal Subunits

Aminoglycosides. The aminoglycoside antibiotics (also known as aminocyclitols) are hydrophilic sugars with multiple amino groups, protonated at physiological pH to function as polycations.

Two main categories of aminoglycosides are exemplified by the streptidine-containing class (primarily streptomycin) and the 2-deoxystreptamine-containing aminoglycosides that include neomycin, kanamycin (with its derivative amikacin), gentamicin, and tobramycin.

Aminoglycosides target accessible regions of polyanionic 16S rRNA on the 30S ribosomal subunit, decreasing the accuracy of the translation of mRNA into protein by contributing to the misreading of codons in the mRNA. As a result of the tight binding of 2-deoxystreptamine aminoglycosides in the major groove of helix H44 of 16S rRNA, the A site in the 30S ribosomal subunit undergoes a conformational change. Incorporation of non-cognate (or near-cognate) aminoacyl-tRNAs is allowed, and the error rate of the ribosome is increased. Elongation is also retarded but less drastically. As for the 2-doxystreptamine aminoglycosides, streptomycin binding also allows the binding of near-cognate tRNAs to the mRNA, so mistranslation ensues. Studies with ribosomes from streptomycin-resistant mutants have shown that protein S12 of the 30S ribosomal subunit also plays a role in the binding of streptomycin and the resulting effects. Studies have shown that aminoglycosides also have a role in the inhibition of assembly of the 30S ribosomal subunit.[18, 65, 70, 120]

Tetracycline, Doxycycline, and Minocycline. Tetracycline and the more recent members of the class, doxycycline and minocycline, exert their antibacterial effect by initially traversing the outer membrane of Gram-negative bacteria through the OmpF and OmpC porin channels as positively charged cation-tetracycline coordination complexes. The metal ion-tetracycline dissociates in the periplasm, and tetracycline diffuses through the inner (cytoplasmic) membrane and binds to the acceptor A site on the 30S ribosomal subunit, subsequently preventing the binding of the incoming aminoacyl-tRNA. Because this interference is reversible, tetracyclines are bacteriostatic. A specific major binding site and a lower affinity binding site of tetracycline have been identified by determining the crystal structure of the 30S ribosomal subunit with the bound drug.[14, 22, 84]

Glycylcyclines: Tigecycline. Tigecycline (GAR-936) is the 9-*t*-butylglycylamido derivative of minocycline. The central four-ring carbocyclic skeleton is essential for antibacterial activity, and the addition of the large *N*-alkyl-glycylamido group provides the drug with a broad spectrum of activity and allows evasion from two major mechanisms of tetracycline resistance: (1) tetracycline-specific efflux pump expulsion and (2) ribosomal protection by production of a protein that sterically blocks the binding of tetracycline but not of tigecycline to the 30S ribosomal binding site. Tigecycline exerts its effects by binding to the 30S ribosomal subunit of bacteria and blocking entry of aminoacyl transfer RNA into the A site of the ribosome; compared with tetracyclines, tigecycline binds five times more efficiently to this ribosomal target site and, subsequently, amino acid residues are prevented from incorporation into elongating peptide chains and protein synthesis is inhibited.[8, 77]

Inhibitors of 50S Ribosomal Subunits

Chloramphenicol. Chloramphenicol inhibits peptide bond formation on bacterial ribosomes by binding to the 50S ribosomal subunit at low- and high-affinity sites. This inhibition is explained by the binding of chloramphenicol to the peptidyltransferase center of the 50S ribosomal subunit in the A site, and subsequent blockage of the interaction of the aminoacyl-tRNA in the A site with the nascent polypeptide that is released from the tRNA in the P site.[47, 96]

Lincomycin and Clindamycin. Lincomycin and its chlorinated derivative, clindamycin, inhibit the peptidyl transferase function of the 50S ribosomal subunit, resulting in alteration of the P site to produce an enhanced binding capacity for the fMet-tRNA. At lower concentrations, these antibiotics also interact with the process of peptide chain initiation in cell-free systems.[20, 82, 90, 100]

Macrolides: Erythromycin, Azithromycin, and Clarithromycin. Erythromycin inhibits prokaryotic, but not eukaryotic, protein synthesis as a result of its binding to the 50S ribosomal subunit. Ribosomal protein L4 may be involved in this interaction, because some erythromycin-resistant bacterial mutants have exhibited an altered L4 protein. It has been postulated that erythromycin inhibits translocation by preventing the positioning of peptidyl-tRNA in the donor site. Other studies have indicated that the peptidyl transferase reaction may also be involved.[16, 20, 50, 86, 112]

Azithromycin and clarithromycin bind to the same receptor on the bacterial 50S ribosomal subunit and inhibit RNA-dependent protein synthesis by the same mechanism as erythromycin. Azithromycin appears to better penetrate through the outer envelope of Gram-negative bacteria such as *Mycoplasma catarrhalis* and *Haemophilus influenzae*.[50, 76, 85]

Ketolides: Telithromycin. Telithromycin's mechanism of action is essentially similar to that of erythromycin and other macrolides: by inhibiting bacterial protein synthesis through close interaction with the peptidyl transferase site of the 50S ribosomal subunit. Telithromycin has a higher binding affinity and mechanisms that may overcome the methylation of binding sites of the peptidyl transferase loop. At higher concentrations, telithromycin also is able to inhibit the formation of the 30S ribosomal subunit in addition to the inhibition of formation of the 50S ribosomal subunit.[50]

Streptogramins. Type A streptogramins, such as dalfopristin and virginiamycin M, inhibit the early stages of protein synthesis by inhibiting the elongation phase. Ribosomal initiation complexes are assembled in an apparently normal fashion, but are functionally inactive.

Type B streptogramins, such as quinupristin, inhibit late stages of protein synthesis by inhibiting the first two of the three elongation steps (AA-tRNA binding to the A site and peptidyl transfer from The P site) and do not affect the third step (translocation from the A site to the P site).

Type A and B streptogramins exert a synergistic inhibitory activity due to the conformational changes in the 50S subunit induced by the attachment of type A compounds resulting in greater affinity of binding of type B compounds.[24, 61, 72]

Fusidic Acid. Fusidic acid is a steroidal antibiotic that inhibits protein synthesis in prokaryotic and eukaryotic subcellular systems. The antibiotic binds to the ternary complex of prokaryotic EF-G (or EF-2 in eukaryotes) and GDP with the ribosome, preventing release of the elongation factor for further rounds of translocation. This sequestration of EF-G may not explain the inhibitory action of fusidic acid *in vivo* because growing cells appear to contain an excess of elongation factors; alternative explanations include (1) interference with subsequent complex formation between the ribosome and aminoacyl-tRNA-EF-Tu-GTP, and (2) possible effects of fusidic acid on both ribosomal A and P sites.[15, 26]

Oxazolidinones: Linezolid. Linezolid, the first oxazolidinone antibiotic approved for human use, inhibits bacterial growth by interfering with protein synthesis. Early results demonstrated binding to 50S but not 30S ribosomal subunits; however, the mechanism of action and the precise location of the drug binding site in the ribosome are not completely understood. Several ribosomal activities (such as formation of the initiation complex, synthesis of the first peptide bond, elongation factor G-dependent translocation and termination of translation) may be inhibited by oxazolidinones. Some of these activities depend on the functions of the ribosomal peptidyl transferase center. Interaction of oxazolidinones with the mitochondrial ribosomes provides a structural basis for the inhibition of mitochondrial protein synthesis, which correlates with clinical side-effects on the bone marrow that are associated with oxazolidinone therapy. Linezolid also exhibits antimycobacterial activity.[56, 72, 74]

Inhibition of Synthesis of Essential Small Molecules

Inhibitors of Dihydropteroate Synthetase

Sulfonamides. The sulfonamides are structural analogs of *p*-aminobenzoic acid (PABA) that compete with this natural substrate for dihydropteroate synthetases in the bacterial biosynthesis of dihydrofolate. In bacterial and mammalian cells, dihydrofolate is reduced by a reductase to tetrahydrofolate, which serves as a cofactor for one-carbon transfer in the biosynthesis of essential metabolites such as purines, thymidylate, glycine, and methionine. The susceptibility of bacterial dihydrofolate reductases to inhibition by trimethoprim thus permits sequential blockade of the biosynthesis of tetrahydrofolate in bacteria by the combination of a sulfonamide plus trimethoprim. Mammalian cells, which cannot synthesize dihydrofolate from PABA, derive preformed folates (folate, dihydrofolate, and also tetrahydrofolate) from the medium (or diet), whereas most bacteria are unable to take up these compounds and are thus dependent on the synthesis of tetrahydrofolate. Sulfonamides are generally used clinically in combination with trimethoprim.[44, 52]

Inhibitors of Dihydrofolate Reductase

Trimethoprim. Trimethoprim is a potent inhibitor of bacterial dihydrofolate reductases, which catalyze the reduction of dihydrofolate to the active tetrahydrofolate. Cofactors of tetrahydrofolate function as donors of one-carbon fragments in the synthesis of thymidylate and other essential compounds such as purines, methionine, panthotenate, and *N*-formyl methionyl-tRNA. The synthesis of thymidylate from uridylate involves not only carbon transfer, but also the oxidation of tetrahydrofolate to dihydrofolate, which must be reactivated by dihydrofolate reductase. The toxic effect of the limitation of tetrahydrofolate is primarily due to the deprivation of thymine, which results in thymine-less death. Mammalian cells are protected by the extraordinarily low affinity of their dihydrofolate reductase for trimethoprim.[44, 89]

ANTIMYCOBACTERIAL AGENTS

Inhibition of Cell Wall Synthesis

Isoniazid (Isonicotinic Acid Hydrazide, INH)

Isoniazid (INH) enters the mycobacterial cell by passive diffusion. INH is not toxic to the bacterial cell but acts as a prodrug and is activated by the mycobacterial enzyme KatG, a multifunctional catalase-peroxidase, in a process that generates reactive species that form adducts with NAD^+ and $NADP^+$. These compounds are potent inhibitors of critical enzymes in the biosynthesis of cell wall lipid and nucleic acid. Some INH-derived reactive species, such as nitric oxide, have a direct role in inhibiting mycobacterial metabolic enzymes.[113]

Cycloserine

D-Cycloserine, a structural analog of D-alanine, is a broad-spectrum antibiotic with activity against mycobacteria including *Mycobacterium tuberculosis* and also strains of other organisms, such as *Escherichia coli, Staphylococcus aureus, Nocardia,* and *Chlamydia* species. At concentrations of 100 to 200 μg mL^{-1}, D-cycloserine inhibits the synthesis of bacterial cell wall peptidoglycan by competitively inhibiting alanine racemase and D-alanyl-D-alanine synthetase. This prevents the formation of D-alanine from L-alanine and the synthesis of the D-alanyl-D-alanine dipeptide prior to its addition to the UDP-MurNAc tripeptide. The rigidly planar D-cycloserine has greater affinity for alanine racemase than does either D- or L-alanine.[93, 106]

Ethambutol

The mechanism of action of ethambutol is not completely understood. It appears to inhibit arabinosyl transferase enzymes that are involved in the biosynthesis of arabinoglycan and lipoarabinomannan which are essential elements within the mycobacterial cell wall.[7]

INTERFERENCE WITH CYTOPLASMIC MEMBRANE FUNCTION: PYRAZINAMIDE

Pyrazinamide is a prodrug that is converted into the active form, pyrazinoic acid, by bacterial nicotinamidase/pyrazinamidase. The exact mechanism of action is not well understood; however, it appears that the remarkable *in vitro* sterilizing effect of pyrazinamide is due to disruption of membrane energetics and inhibited membrane function in *Mycobacterium tuberculosis*.[134]

INHIBITION OF NUCLEIC ACID SYNTHESIS

Rifamycins: Rifampin, Rifabutin, and Rifapentine

The mechanism of action of rifamycins involves inhibition of DNA-dependent RNA synthesis caused by strong binding to the β-subunit of the DNA-dependent RNA polymerase of prokaryotes, with a binding constant in the range of 10^{-8} M. Eukaryotic enzymes are at least 10^2 to 10^4 times less sensitive to inhibition by rifampin. Rifabutin and rifapentine are semi-synthetic derivatives of rifampin. Rifabutin is active against a number of rifampin-resistant clinical pathogens, and rifapentine has a long half-life of approximately 12 hr, allowing its administration in once-weekly regimens.[36, 73, 75]

Fluoroquinolones as Antimycobacterial Agents

Fluorinated quinolones have been studied and used as second-line antituberculous drugs. Some fluoroquinolones, such as levofloxacin and moxifloxacin, are bactericidal against *Mycobacterium tuberculosis*, presumably by inhibition of its DNA gyrase at concentrations within achievable serum levels.[29, 53, 58, 97, 101]

INHIBITION OF PROTEIN SYNTHESIS

Aminoglycosides: Streptomycin and Amikacin

(See subsection entitled "Aminoglycosides")

Capreomycin

Capreomycin is a polypeptide antibiotic with a poorly understood mechanism of action, even active on non-replicating mycobacteria. It is thought to inhibit protein synthesis by binding to the 50S ribosomal subunit of the 70S ribosome, leading to abnormal protein synthesis that is ultimately fatal to the bacteria. Microarray studies have suggested that capreomycin also affects biochemical and metabolic pathways.[37]

ANTIFUNGAL AGENTS

INHIBITION OF CELL WALL SYNTHESIS

Echinocandins: Caspofungin, Micafungin, and Anidulafungin

Echinocandins are antifungal agents with a broad spectrum of activity, active against *Candida* and *Aspergillus* species.

Echinocandins impair formation of the fungal cell wall by inhibiting 1,3-β-D-glucan synthesis. They bind in a concentration-dependent manner to the 1,3-β-D-glucan synthase complex, a critical enzyme for the synthesis of 1,3-β-D-glucan polymers, which constitute a major component of the cell walls of many susceptible pathogenic fungi, providing the integrity and shape of the fungal cell wall, together with chitin.

In *Aspergillus* species, the echinocandins cause aberrant growth of hyphae at the apical tips but absolute killing does not occur. All echinocandins have similar activity *in vivo* and *in vitro* against *Aspergillus* species.

Against *Candida* species, micafungin and anidulafungin have similar minimum inhibitory concentrations (MICs) that are generally lower than the MIC of caspofungin. Echinocandins are highly active against *C. albicans, C. glabrata, C. tropicalis, C. dubliniensis,* and *C. krusei,* and less active against *C. parapsilosis, C. guillermondii,* and *C. lusitaniae.*

None of the echinocandins has activity against *Cryptococcus* species.[17, 19, 21, 64, 125, 129, 130]

Interference with Membrane Integrity

Polyenes: Amphotericin B and Nystatin

Amphotericin B is an amphipathic molecule produced by *Streptomyces nodosus.*

The integrity of sterol-containing membranes of eukaryotes is impaired by the binding of polyene antibiotics. At lower drug concentrations, potassium channel activity is increased. Studies with liposome systems have shown that interaction between these antibiotics and sterols causes formation of pores with radii of approximately 5 to 7 Å at higher drug concentrations, with resulting loss of intracellular potassium and various molecules, leading to impaired fungal viability.

The drug exists as aggregates in the cytoplasmic membrane and is anchored at the membrane-water interface by its polar end and extends its full length halfway through the lipid bilayer. The antibiotic molecules may orient themselves into a cylindrical half-pore with their polyhydroxylic surfaces interacting with each other and with water, and with their lipophilic surfaces facing outward. Molecular model building suggests that sterol molecules fit snugly between each pair of polyene molecules and stabilize these half-pores. Amphotericin B binds more tightly to ergosterol, the principal sterol in fungal membranes, than to cholesterol, the principal sterol in mammalian membranes. Pore formation results from hydrogen bonding between the rings of hydroxyl groups at the ends of two opposing half-pores.[3, 5, 45, 68, 78]

Azole Antifungal Drugs

First-generation azole antifungal drugs include fluconazole, miconazole, ketoconazole, and itraconazole. Newer azole antifungals are represented by voriconazole, ravuconazole, and posaconazole. The structural differences among these compounds dictate the antifungal potency, bioavailability, spectrum of activity, drug interaction, and toxic potential. The primary antifungal effect of the azoles occurs via inhibition of a fungal cytochrome P-450 mono-oxygenase enzyme (14α-demethylase) involved in the synthesis of ergosterol, the major sterol in the fungal cell membrane. The active site of this enzyme contains a protoporphyrin moiety that serves as a binding site for the antifungal azoles via a nitrogen atom in the imidazole or triazole ring. The remainder of the molecule binds to the apoprotein in a manner dependent on the structure of the individual azole. Imidazoles and triazoles inhibit the C-14α-demethylation of lanosterol in the ergosterol biosynthetic pathway; in some fungal species, they can also inhibit the Δ22-desaturase step. Once the membrane ergosterol is depleted and replaced with unusual 14α-methylsterols, alterations in the fungal membrane fluidity and permeability occur, and the enzymes involved in cell wall synthesis are affected with ultimate inhibition of cell growth. It has been shown that voriconazole, a second-generation triazole, exerts a postantifungal effect after exposure of yeasts to therapeutic concentrations.[25, 28, 39, 95, 108, 123]

Allylamines: Terbinafine

Terbinafine has a unique fungicidal mechanism of action in susceptible species, inhibiting squalene epoxidase, the enzyme that catalyzes the conversion of squalene to squalene 2,3 epoxide, a precursor of lanosterol, which is a direct precursor of ergosterol.[6, 9, 83, 94]

Inhibition of Nucleic Acid and Protein Synthesis

5-Fluorocytosine (5-FC) (Flucytosine)

5-Fluorocytosine is taken up into yeast cells by a cytosine permease and subsequently deaminated; 5-FC is specific for fungi because mammalian cells have very little cytosine deaminase. The resulting 5-fluorouracil is converted through several steps to 5-fluorodeoxyuridylic acid monophosphate, a noncompetitive inhibitor of thymidylate synthetase, which normally forms thymidylate from uridylate, and is involved in other steps in pyrimidine biosynthesis. The 5-fluorouracil also can be converted to 5-fluorouridine triphosphate, which, if incorporated into m-RNA, causes disruption in protein synthesis.[42, 78, 87, 124]

Miscellaneous (or Unknown) Effects: Griseofulvin

Griseofulvin inhibits only those fungi that possess chitin in their cell wall. However, long before the hyphae become noticeably distorted, demonstrable changes occur in the nuclei. Griseofulvin causes breakdown of spindles and development of irregular masses of chromatin. Although the exact mechanism of action of griseofulvin is not completely elucidated, it appears that its primary effect on fungi is due to its effect on the cell division process by interference with microtubule assembly and function.[78]

REFERENCES

1. Adachi, J.A. and H.L. DuPont. 2006. Rifaximin: a novel nonabsorbed rifamycin for gastrointestinal disorders. *Clin. Infect. Dis.*, 42:541–547.
2. Allen, N.E. and T.I. Nicas. 2003. Mechanism of action of oritavancin and related glycopeptide antibiotics. *FEMS Microbiol. Rev.*, 26:511–532.
3. Andreoli, T.E. 1974. The structure and function of amphotericin B-cholesterol pores in lipid bilayer membranes. *Ann. N.Y, Acad. Sci.*, 235:448–468.
4. Asbel, L.E. and M.E. Levison. 2000. Cephalosporins, carbapenems, and monobactams. *Infect. Dis. Clin. N. Am.*, 14:435–447, ix.
5. Balakrishnan, A.R. and K.R. Easwaran. 1993. Lipid-amphotericin B complex structure in solution: a possible first step in the aggregation process in cell membranes. *Biochemistry*, 32:4139–4144.
6. Balfour, J.A. and D. Faulds. 1992. Terbinafine. A review of its pharmacodynamic and pharmacokinetic properties, and therapeutic potential in superficial mycoses. *Drugs*, 43:259–284.
7. Belanger, A.E., G.S. Besra, M.E. Ford, K. Mikusova, J.T. Belisle, P.J. Brennan, and J.M. Inamine. 1996. The embAB genes of *Mycobacterium avium* encode an arabinosyl transferase involved in cell wall arabinan biosynthesis that is the target for the antimycobacterial drug ethambutol. *Proc. Natl. Acad. Sci. U.S.A.*, 93:11919–11924.
8. Bergeron, J., M. Ammirati, D. Danley, L. James, M. Norcia, J. Retsema, C.A. Strick, W.G. Su, J. Sutcliffe, and L. Wondrack. 1996. Glycylcyclines bind to the high-affinity tetracycline ribosomal binding site and evade Tet(M)- and Tet(O)-mediated ribosomal protection. *Antimicrob. Agents Chemother.*, 40:2226–2228.
9. Birnbaum, J.E. 1990. Pharmacology of the allylamines. *J. Am. Acad. Dermatol.*, 23:782–785.
10. Blumberg, P.M. and J.L. Strominger. 1974. Interaction of penicillin with the bacterial cell: penicillin-binding proteins and penicillin-sensitive enzymes. *Bacteriol. Rev.*, 38:291–335.
11. Borghi, A., C. Coronelli, L. Faniuolo, G. Allievi, R. Pallanza, and G.G. Gallo. 1984. Teichomycins, new antibiotics from *Actinoplanes teichomyceticus* nov. sp. IV. Separation and characterization of the components of teichomycin (teicoplanin). *J. Antibiot.* (Tokyo), 37:615–620.

12. Bourguignon, G.J., M. Levitt, and R. Sternglanz. 1973. Studies on the mechanism of action of nalidixic acid. *Antimicrob. Agents Chemother.*, 4:479–486.
13. Boylan, C.J., K. Campanale, P.W. Iversen, D.L. Phillips, M.L. Zeckel, and T.R. Parr, Jr. 2003. Pharmacodynamics of oritavancin (LY333328) in a neutropenic-mouse thigh model of Staphylococcus aureus infection. *Antimicrob. Agents Chemother.*, 47:1700–1706.
14. Brodersen, D.E., W.M. Clemons, Jr., A.P. Carter, R.J. Morgan-Warren, B.T. Wimberly, and V. Ramakrishnan. 2000. The structural basis for the action of the antibiotics tetracycline, pactamycin, and hygromycin B on the 30S ribosomal subunit. *Cell*, 103:1143–1154.
15. Burns, K., M. Cannon, and E. Cundliffe. 1974. A resolution of conflicting reports concerning the mode of action of fusidic acid. *FEBS Lett.*, 40:219–223.
16. Cannon, M. and K. Burns. 1971. Modes of action of erythromycin and thiostrepton as inhibitors of protein synthesis. *FEBS Lett.*, 18:1–5.
17. Cappelletty, D. and K. Eiselstein-McKitrick. 2007. The echinocandins. *Pharmacotherapy*, 27:369–388.
18. Carter, A.P., W.M. Clemons, D.E. Brodersen, R.J. Morgan-Warren, B.T. Wimberly, and V. Ramakrishnan. 2000. Functional insights from the structure of the 30S ribosomal subunit and its interactions with antibiotics. *Nature*, 407:340–348.
19. Carver, P.L. 2004. Micafungin. *Ann. Pharmacother.*, 38:1707–1721.
20. Champney, W.S. and C.L. Tober. 2000. Specific inhibition of 50S ribosomal subunit formation in *Staphylococcus aureus* cells by 16-membered macrolide, lincosamide, and streptogramin B antibiotics. *Curr. Microbiol.,* 41:126–135.
21. Chandrasekar, P.H. and J.D. Sobel. 2006. Micafungin: a new echinocandin. *Clin. Infect. Dis.,* 42:1171–1178.
22. Chopra, I. and M. Roberts. 2001. Tetracycline antibiotics: mode of action, applications, molecular biology, and epidemiology of bacterial resistance. *Microbiol. Mol. Biol. Rev.*, 65:232–260; second page, table of contents.
23. Clausell, A., M. Garcia-Subirats, M. Pujol, M.A. Busquets, F. Rabanal, and Y. Cajal. 2007. Gram-negative outer and inner membrane models: insertion of cyclic cationic lipopeptides. *J. Phys. Chem. B*, 111:551–563.
24. Cocito, C., M. Di Giambattista, E. Nyssen, and P. Vannuffel. 1997. Inhibition of protein synthesis by streptogramins and related antibiotics, *J. Antimicrob. Chemother.*, 39:7–13.
25. Como, J.A. and W.E. Dismukes. 1994. Oral azole drugs as systemic antifungal therapy. *N. Engl. J. Med.*, 330:263–272.
26. Cundliffe, E. and D.J. Burns. 1972. Long term effects of fusidic acid on bacterial protein synthesis *in vivo. Biochem. Biophys. Res. Commun.*, 49:766–774.
27. Dalhoff, A., T. Nasu, and K. Okamoto. 2003. Target affinities of faropenem to and its impact on the morphology of Gram-positive and Gram-negative bacteria. *Chemotherapy*, 49:172–183.
28. de Brabander, M., F. Aerts, J. van Cutsem, H. van den Bossche, and M. Borgers. 1980. The activity of ketoconazole in mixed cultures of leukocytes and *Candida albicans. Sabouraudia*, 18:197–210.
29. De Souza, M.V., T.R. Vasconcelos, M.V. de Almeida, and S.H. Cardoso. 2006. Fluoroquinolones: an important class of antibiotics against tuberculosis. *Curr. Med. Chem.*, 13:455–463.
30. Drlica, K. 1999. Mechanism of fluoroquinolone action. *Curr. Opin. Microbiol.*, 2:504–508.
31. Duax, W.L., J.F. Griffin, D.A. Langs, G.D. Smith, P. Grochulski, V. Pletnev, and V. Ivanov. 1996. Molecular structure and mechanisms of action of cyclic and linear ion transport antibiotics. *Biopolymers*, 40:141–155.
32. Edwards, D.I. 1993. Nitroimidazole drugs — action and resistance mechanisms. I. Mechanisms of action. *J. Antimicrob. Chemother.*, 31:9–20.
33. Emmerson, A.M. and A.M. Jones. 2003. The quinolones: decades of development and use. J. Antimicrob. *Chemother.*, 51(Suppl. 1):13–20.
34. Evans, M.E., D.J. Feola, and R.P. Rapp. 1999. Polymyxin B sulfate and colistin: old antibiotics for emerging multiresistant Gram-negative bacteria. *Ann. Pharmacother.*, 33:960–967.
35. Feingold, D.S., C.C. HsuChen, and I.J. Sud. 1974. Basis for the selectivity of action of the polymyxin antibiotics on cell membranes. *Ann. N.Y. Acad. Sci.*, 235:480–492.
36. Floss, H.G. and T.W. Yu. 2005. Rifamycin—mode of action, resistance, and biosynthesis. *Chem. Rev.*, 105:621–632.
37. Fu, L.M. and T.M. Shinnick. 2007. Genome-wide exploration of the drug action of capreomycin on Mycobacterium tuberculosis using Affymetrix oligonucleotide *GeneChips. J. Infect.*, 54:277–284.

38. Fulco, P. and R.P. Wenzel. 2006. Ramoplanin: a topical lipoglycodepsipeptide antibacterial agent. *Expert Rev. Anti Infect. Ther.*, 4:939–9945.
39. Garcia, M.T., M.T. Llorente, J.E. Lima, F. Minguez, F. Del Moral, and J. Prieto. 1999. Activity of voriconazole: post-antifungal effect, effects of low concentrations and of pretreatment on the susceptibility of *Candida albicans* to leucocytes. *Scand. J. Infect. Dis.*, 31:501–504.
40. Hamilton-Miller, J.M. 2003. Chemical and microbiologic aspects of penems, a distinct class of beta-lactams: focus on faropenem. *Pharmacotherapy*, 23:1497–1507.
41. Hammes, W.P. and F.C. Neuhaus. 1974. On the mechanism of action of vancomycin: inhibition of peptidoglycan synthesis in Gaffkya homari. *Antimicrob. Agents Chemother.*, 6:722–728.
42. Heidelberger, C. 1965. Fluorinated pyrimidines. *Prog. Nucleic Acid Res. Mol. Biol.*, 4:1–50.
43. Higgins, D.L., R. Chang, D.V. Debabov, J. Leung, T. Wu, K.M. Krause, E. Sandvik, J.M. Hubbard, K. Kaniga, D.E. Schmidt, Jr., Q. Gao, R.T. Cass, D.E. Karr, B.M. Benton, and P.P. Humphrey. 2005. Telavancin, a multifunctional lipoglycopeptide, disrupts both cell wall synthesis and cell membrane integrity in methicillin-resistant *Staphylococcus aureus. Antimicrob. Agents Chemother.*, 49:1127–1134.
44. Hitchings, G.H. 1973. Mechanism of action of trimethoprim-sulfamethoxazole. *I. J. Infect. Dis.*, 128 (Suppl.): 433–436.
45. Holz, R.W. 1974. The effects of the polyene antibiotics nystatin and amphotericin B on thin lipid membranes. *Ann. N.Y. Acad. Sci.*, 235:469–479.
46. HsuChen, C.C. and D.S. Feingold. 1973. The mechanism of polymyxin B action and selectivity toward biologic membranes. *Biochemistry*, 12:2105–2111.
47. Jardetzky, O. 1963. Studies on the mechanism of action of chloramphenicol. I. The conformation of chloramphenicol in solution. *J. Biol. Chem.*, 238:2498–508.
48. Kahan, F.M., J.S. Kahan, P.J. Cassidy, and H. Kropp. 1974. The mechanism of action of fosfomycin (phosphonomycin). *Ann. N.Y. Acad. Sci.*, 235:364–386.
49. Kahne, D., C. Leimkuhler, W. Lu, and C. Walsh. 2005. Glycopeptide and lipoglycopeptide antibiotics. *Chem. Rev.*, 105:425–448.
50. Katz, L. and G.W. Ashley. 2005. Translation and protein synthesis: macrolides. *Chem. Rev.*, 105:499–528.
51. Khodursky, A.B. and N.R. Cozzarelli. 1998. The mechanism of inhibition of topoisomerase IV by quinolone antibacterials. *J. Biol. Chem.*, 273:27668–27677.
52. Kompis, I.M., K. Islam, and R.L. Then. 2005. DNA and RNA synthesis: antifolates. *Chem. Rev.*, 105:593–620.
53. Kubendiran, G., C.N. Paramasivan, S. Sulochana, and D.A. Mitchison. 2006. Moxifloxacin and gatifloxacin in an acid model of persistent Mycobacterium tuberculosis. *J. Chemother.*, 18:617–623.
54. Lamp, K.C., C.D. Freeman, N.E. Klutman, and M.K. Lacy. 1999. Pharmacokinetics and pharmacodynamics of the nitroimidazole antimicrobials. *Clin. Pharmacokinet.*, 36:353–373.
55. Laohavaleeson, S., J.L. Kuti, and D.P. Nicolau. 2007. Telavancin: a novel lipoglycopeptide for serious Gram-positive infections. *Expert Opin. Investig. Drugs*, 16:347–357.
56. Leach, K.L., S.M. Swaney, J.R. Colca, W.G. McDonald, J.R. Blinn, L.M. Thomasco, R.C. Gadwood, D. Shinabarger, L. Xiong, and A.S. Mankin. 2007. The site of action of oxazolidinone antibiotics in living bacteria and in human mitochondria. *Mol. Cell,* 26:393–402.
57. Levine, C., H. Hiasa, and K.J. Marians. 1998. DNA gyrase and topoisomerase IV: biochemical activities, physiological roles during chromosome replication, and drug sensitivities. *Biochim. Biophys. Acta*, 1400:29–43.
58. Leysen, D.C., A. Haemers, and S.R. Pattyn. 1989. Mycobacteria and the new quinolones. *Antimicrob. Agents Chemother.*, 33:1–5.
59. Li, J., R.L. Nation, R.W. Milne, J.D. Turnidge, and K. Coulthard. 2005. Evaluation of colistin as an agent against multi-resistant Gram-negative bacteria. *Int. J. Antimicrob. Agents,* 25:11–25.
60. Lin, S.-W., P.L. Carver, and D.D. DePestel. 2006. Dalbavancin: A new option for the treatment of Gram-positive infections. *Ann. Pharmacother.*, 40:449–460.
61. Livermore, D.M. 2000. Quinupristin/dalfopristin and linezolid: where, when, which and whether to use? *J. Antimicrob. Chemother.*, 46:347–350.
62. Livermore, D.M., A.M. Sefton, and G.M. Scott. 2003. Properties and potential of ertapenem. *J. Antimicrob. Chemother.,* 52:331–344.
63. Lounatmaa, K., P.H. Makela, and M. Sarvas. 1976. Effect of polymyxin on the ultrastructure of the outer membrane of wild-type and polymyxin-resistant strain of Salmonella. *J. Bacteriol.,* 127:1400–1407.

64. Maesaki, S., M.A. Hossain, Y. Miyazaki, K. Tomono, T. Tashiro, and S. Kohno. 2000. Efficacy of FK463, a (1,3)-beta-D-glucan synthase inhibitor, in disseminated azole-resistant *Candida albicans* infection in mice. *Antimicrob. Agents Chemother.*, 44:1728–1730.
65. Magnet, S. and J.S. Blanchard. 2005. Molecular insights into aminoglycoside action and resistance. *Chem. Rev.*, 105:477–498.
66. Malabarba, A. and R. Ciabatti. 2001. Glycopeptide derivatives. *Curr. Med. Chem.*, 8:1759–1773.
67. Malabarba, A. and B.P. Goldstein. 2005. Origin, structure, and activity *in vitro* and *in vivo* of dalbavancin. *J. Antimicrob. Chemother.*, 55 (Suppl. 2):1115–1120.
68. Matsuoka, S. and M. Murata. 2003. Membrane permeabilizing activity of amphotericin B is affected by chain length of phosphatidylcholine added as minor constituent. *Biochim. Biophys. Acta*, 1617:109–115.
69. McCafferty, D.G., P. Cudic, B.A. Frankel, S. Barkallah, R.G. Kruger, and W. Li. 2002. Chemistry and biology of the ramoplanin family of peptide antibiotics. *Biopolymers*, 66:261–284.
70. Mehta, R. and W.S. Champney. 2002. 30S ribosomal subunit assembly is a target for inhibition by aminoglycosides in *Escherichia coli*. *Antimicrob. Agents Chemother.*, 46:1546–1549.
71. Mitscher, L.A. 2005. Bacterial topoisomerase inhibitors: quinolone and pyridone antibacterial agents. *Chem. Rev.*, 105:559–592.
72. Mukhtar, T.A. and G.D. Wright. 2005. Streptogramins, oxazolidinones, and other inhibitors of bacterial protein synthesis. *Chem. Rev.*, 105:529–542.
73. Munsiff, S.S., C. Kambili, and S.D. Ahuja. 2006. Rifapentine for the treatment of pulmonary tuberculosis. *Clin. Infect. Dis.*, 43:1468–1475.
74. Nagiec, E.E., L. Wu, S.M. Swaney, J.G. Chosay, D.E. Ross, J.K. Brieland, and K. L. Leach. 2005. Oxazolidinones inhibit cellular proliferation via inhibition of mitochondrial protein synthesis. *Antimicrob. Agents Chemother.*, 49:3896–3902.
75. Nahid, P., M. Pai, and P.C. Hopewell. 2006. Advances in the diagnosis and treatment of tuberculosis. *Proc. Am. Thorac. Soc.*, 3:103–110.
76. Neu, H.C. 1991. Clinical microbiology of azithromycin. *Am. J. Med.*, 91:12S–18S.
77. Noskin, G.A. 2005. Tigecycline: a new glycylcycline for treatment of serious infections. *Clin. Infect. Dis.*, 41 (Suppl. 5):S303–S314.
78. Odds, F.C., A.J. Brown, and N.A. Gow. 2003. Antifungal agents: mechanisms of action. *Trends Microbiol.*, 11:272–279.
79. Pace, J.L. and J.K. Judice. 2005. Telavancin (Theravance). *Curr. Opin. Investig. Drugs*, 6:216–225.
80. Pache, W., D. Chapman, and R. Hillaby. 1972. Interaction of antibiotics with membranes: polymyxin B and gramicidin S. *Biochim. Biophys. Acta*, 255:358–364.
81. Perkins, H.R. and M. Nieto. 1974. The chemical basis for the action of the vancomycin group of antibiotics. *Ann. N.Y. Acad. Sci.*, 235:348–363.
82. Pestka, S. 1971. Inhibitors of ribosome functions. *Annu. Rev. Microbiol.*, 25:487–562.
83. Petranyi, G., N.S. Ryder, and A. Stutz. 1984. Allylamine derivatives: new class of synthetic antifungal agents inhibiting fungal squalene epoxidase. *Science*, 224:1239–1241.
84. Pioletti, M., F. Schlunzen, J. Harms, R. Zarivach, M. Gluhmann, H. Avila, A. Bashan, H. Bartels, T. Auerbach, C. Jacobi, T. Hartsch, A. Yonath, and F. Franceschi. 2001. Crystal structures of complexes of the small ribosomal subunit with tetracycline, edeine and IF3. *EMBO. J.*, 20:1829–1839.
85. Piscitelli, S.C., L.H. Danziger, and K.A. Rodvold. 1992. Clarithromycin and azithromycin: new macrolide antibiotics. *Clin. Pharm.*, 11:137–152.
86. Poehlsgaard, J. and S. Douthwaite. 2005. The bacterial ribosome as a target for antibiotics. *Nat. Rev. Microbiol.*, 3:870–881.
87. Polak, A., and H. J. Scholer. 1975. Mode of action of 5-fluorocytosine and mechanisms of resistance. *Chemotherapy*, 21:113–130.
88. Prenner, E.J., R.N. Lewis, and R.N. McElhaney. 1999. The interaction of the antimicrobial peptide gramicidin S with lipid bilayer model and biological membranes. *Biochim. Biophys. Acta*, 1462:201–221.
89. Quinlivan, E.P., J. McPartlin, D.G. Weir, and J. Scott. 2000. Mechanism of the antimicrobial drug trimethoprim revisited. *Faseb J.*, 14:2519–2524.
90. Reusser, F. 1975. Effect of lincomycin and clindamycin on peptide chain initiation. *Antimicrob. Agents Chemother.*, 7:32–37.
91. Riva, S., A. Fietta, M. Berti, L.G. Silvestri, and E. Romero. 1973. Relationships between curing of the F episome by rifampin and by acridine orange in *Escherichia coli*. *Antimicrob. Agents Chemother.*, 3:456–462.

92. Rodloff, A.C., E.J. Goldstein, and A. Torres. 2006. Two decades of imipenem therapy. *J. Antimicrob. Chemother.,* 58:916–929.
93. Roze, U. and J.L. Strominger. 1966. *Alanine racemase from Staphylococcus aureus: Conformation of its substrates and its inhibitor, D-cycloserine*, pp. 92–94, Vol. 2.
94. Ryder, N.S. and M.C. Dupont. 1985. Inhibition of squalene epoxidase by allylamine antimycotic compounds. A comparative study of the fungal and mammalian enzymes. *Biochem. J.*, 230:765–770.
95. Sanati, H., P. Belanger, R. Fratti, and M. Ghannoum. 1997. A new triazole, voriconazole (UK-109,496), blocks sterol biosynthesis in *Candida albicans* and *Candida krusei. Antimicrob. Agents Chemother.*, 41:2492–2496.
96. Schlunzen, F., R. Zarivach, J. Harms, A. Bashan, A. Tocilj, R. Albrecht, A. Yonath, and F. Franceschi. 2001. Structural basis for the interaction of antibiotics with the peptidyl transferase centre in Eubacteria. *Nature*, 413:814–821.
97. Shandil, R.K., R. Jayaram, P. Kaur, S. Gaonkar, B.L. Suresh, B.N. Mahesh, R. Jayashree, V. Nandi, S. Bharath, and V. Balasubramanian. 2007. Moxifloxacin, ofloxacin, sparfloxacin, and ciprofloxacin against *Mycobacterium tuberculosis:* evaluation of *in vitro* and pharmacodynamic indices that best predict *in vivo* efficacy. *Antimicrob. Agents Chemother.,* 51:576–582.
98. Sheldrick, G.M., P.G. Jones, O. Kennard, D.H. Williams, and G.A. Smith. 1978. Structure of vancomycin and its complex with acetyl-D-alanyl-D-alanine. *Nature, 271:223–225.*
99. Siewert, G. and J.L. Strominger. 1967. Bacitracin: an inhibitor of the dephosphorylation of lipid pyrophosphate, an intermediate in the biosynthesis of the peptidoglycan of bacterial cell walls. *Proc. Natl. Acad. Sci. U.S.A.*, 57:767–773.
100. Spizek, J. and T. Rezanka. 2004. Lincomycin, clindamycin and their applications. *Appl. Microbiol. Biotechnol.*, 64:455–464.
101. Sriram, D., T.R. Bal, P. Yogeeswari, D.R. Radha, and V. Nagaraja. 2006. Evaluation of antimycobacterial and DNA gyrase inhibition of fluoroquinolone derivatives. *J. Gen. Appl. Microbiol.*, 52:195–200.
102. Stein, G.E. and E.J. Goldstein. 2006. Fluoroquinolones and anaerobes. *Clin. Infect. Dis.*, 42:1598–1607.
103. Stone, K.J. and J.L. Strominger. 1972. Inhibition of sterol biosynthesis by bacitracin. *Proc. Natl. Acad. Sci. U.S.A.*, 69:1287–1289.
104. Stone, K.J. and J.L. Strominger. 1971. Mechanism of action of bacitracin: complexation with metal ion and C 55-isoprenyl pyrophosphate. *Proc. Natl. Acad. Sci. U.S.A.*, 68:3223–3227.
105. Straus, S.K. and R.E. Hancock. 2006. Mode of action of the new antibiotic for Gram-positive pathogens daptomycin: comparison with cationic antimicrobial peptides and lipopeptides. *Biochim. Biophys. Acta,* 1758:1215–1223.
106. Strominger, J.L., E. Ito, and R.H. Threnn. 1960. Competitive inhibition of enzymatic reactions by oxamycin, *J. Am. Chem. Soc.,* 82:998–999.
107. Strominger, J.L., E. Willoughby, T. Kamiryo, P.M. Blumberg, and R.R. Yocum. 1974. Penicillin-sensitive enzymes and penicillin-binding components in bacterial cells. *Ann. N.Y. Acad. Sci.,* 235:210–224.
108. Sud, I.J., D.L. Chou, and D.S. Feingold. 1979. Effect of free fatty acids on liposome susceptibility to imidazole antifungals. *Antimicrob. Agents Chemother.,* 16:660–663.
109. Sugino, A., C.L. Peebles, K.N. Kreuzer, and N.R. Cozzarelli. 1977. Mechanism of action of nalidixic acid: purification of *Escherichia coli* nalA gene product and its relationship to DNA gyrase and a novel nicking-closing enzyme. *Proc. Natl. Acad. Sci. U.S.A.*, 74:4767–4771.
110. Sykes, R.B., and D.P. Bonner. 1985. Aztreonam: the first monobactam. *Am. J. Med.*, 78:2-10.
111. Sykes, R.B., D.P. Bonner, K. Bush, and N. . Georgopapadakou. 1982. Azthreonam (SQ 26,776), a synthetic monobactam specifically active against aerobic Gram-negative bacteria, *Antimicrob. Agents Chemother.*, 21:85–92.
112. Tanaka, S., T. Otaka, and A. Kaji. 1973. Further studies on the mechanism of erythromycin action. *Biochim. Biophys. Acta,* 331:128–140.
113. Timmins, G.S. and V. Deretic. 2006. Mechanisms of action of isoniazid. *Mol. Microbiol.,* 62:1220–1227.
114. Tocher, J.H. and D.I. Edwards. 1994. Evidence for the direct interaction of reduced metronidazole derivatives with DNA bases. *Biochem. Pharmacol.,* 48:1089–1094.
115. Tocher, J.H. and D.I. Edwards. 1995. The interaction of nitroaromatic drugs with aminothiols. *Biochem. Pharmacol.*, 50:1367–1371.

116. Trias, J. and H. Nikaido. 1990. Outer membrane protein D2 catalyzes facilitated diffusion of carbapenems and penems through the outer membrane of *Pseudomonas aeruginosa*. *Antimicrob. Agents Chemother.*, 34:52–57.
117. Urban, B.W., S.B. Hladky, and D.A. Haydon. 1980. Ion movements in gramicidin pores. An example of single-file transport. *Biochim. Biophys. Acta,* 602:331–354.
118. Urry, D.W. 1972. A molecular theory of ion-conductng channels: a field-dependent transition between conducting and nonconducting conformations. *Proc. Natl. Acad. Sci. U.S.A.,* 69:1610–1614.
119. Urry, D.W., M.C. Goodall, J.D. Glickson, and D.F. Mayers. 1971. The gramicidin A transmembrane channel: characteristics of head-to-head dimerized (L,D) helices. *Proc. Natl. Acad. Sci. U.S.A.*, 68:1907–1911.
120. Vakulenko, S.B. and S. Mobashery. 2003. Versatility of aminoglycosides and prospects for their future. *Clin. Microbiol. Rev.*, 16:430–450.
121. Van Bambeke, F. 2006. Glycopeptides and glycodepsipeptides in clinical development: a comparative review of their antibacterial spectrum, pharmacokinetics and clinical efficacy. *Curr. Opin. Investig. Drugs*, 7:740–749.
122. Van Bambeke, F., Y. Van Laethem, P. Courvalin, and P.M. Tulkens. 2004. Glycopeptide antibiotics: from conventional molecules to new derivatives. *Drugs*, 64:913–936.
123. Van den Bossche, H., G. Willemsens, W. Cools, P. Marichal, and W. Lauwers. 1983. Hypothesis on the molecular basis of the antifungal activity of N-substituted imidazoles and triazoles. *Biochem. Soc. Trans.*, 11:665–667.
124. Vanden Bossche, H., M. Engelen, and F. Rochette. 2003. Antifungal agents of use in animal health — chemical, biochemical and pharmacological aspects. *J. Vet. Pharmacol. Ther.,* 26:5–29.
125. Vazquez, J.A. and J.D. Sobel. 2006. Anidulafungin: a novel echinocandin. *Clin. Infect. Dis.*, 43:215–222.
126. Wallace, B.A. 1998. Recent advances in the high resolution structures of bacterial channels: Gramicidin A. J. Struct. *Biol.*, 121:123–141.
127. Walsh, C. 2003. *Antibiotics. Actions, Origins, Resistance.* ASM Press, Washington, DC.
128. Wehrli, W. 1983. Rifampin: mechanisms of action and resistance. *Rev. Infect. Dis.*, 5 (Suppl. 3): S407–S411.
129. Wiederhold, N.P. and J.S. Lewis, 2nd. 2007. The echinocandin micafungin: a review of the pharmacology, spectrum of activity, clinical efficacy and safety. *Expert Opin. Pharmacother.,* 8:1155–1166.
130. Wiederhold, N.P. and R.E. Lewis. 2003. The echinocandin antifungals: an overview of the pharmacology, spectrum and clinical efficacy. *Expert Opin. Investig. Drugs,* 12:1313–1333.
131. Williams, D.H. and J. Kalman. 1977. Structural and mode of action studies on the antibiotic vancomycin. Evidence from 270-MHz proton magnetic resonance. *J. Am. Chem. Soc.*, 99:2768–2774.
132. Wu, H.C. and P.S. Venkateswaran. 1974. Fosfomycin-resistant mutant of *Escherichia coli*. *Ann. N.Y. Acad. Sci.*, 235:587–592.
133. Zhanel, G.G., R. Wiebe, L. Dilay, K. Thomson, E. Rubinstein, D.J. Hoban, A.M. Noreddin, and J.A. Karlowsky. 2007. Comparative review of the carbapenems. *Drugs*, 67:1027–1052.
134. Zhang, Y., M.M. Wade, A. Scorpio, H. Zhang, and Z. Sun. 2003. Mode of action of pyrazinamide: disruption of *Mycobacterium tuberculosis* membrane transport and energetics by pyrazinoic acid. *J. Antimicrob. Chemother.,* 52:790–795.

12 Antibiotic Susceptibility Testing

Audrey Wanger

CONTENTS

BASIC PRINCIPLES OF ANTIMICROBIAL SUSCEPTIBILITY TESTING (AST)

The purpose of performing antimicrobial susceptibility testing (AST) is to provide *in vitro* data to help ensure that appropriate and adequate antimicrobial therapy is used to optimize treatment outcomes. In addition, the AST data generated daily can be statistically analyzed on an annual basis to generate an antibiogram that reflects the antimicrobial susceptibility and resistance patterns of important pathogens that prevail in a particular hospital. These hospital-specific AST epidemiologic data provide valuable guidance to the clinicians for the appropriate selection of empiric therapy, prior to the availability of culture and susceptibility results that often takes 2 to 3 days. The purpose of the AST of the culture pathogen is to help clinicians correct and/or modify empiric therapy as soon as the results become available.

Two basic methods of AST are available to laboratories: (1) qualitative and (2) quantitative. Qualitative methods, such as disk diffusion and abbreviated breakpoint dilution systems, are acceptable options for the testing of isolates from "healthy" patients with intact immune defenses and for less serious infections such as uncomplicated urinary tract infections. Both disk diffusion and breakpoint agar or broth dilution systems are considered satisfactory for predicting treatment outcome in these cases.[9] Quantitative systems that provide a minimum inhibitory concentration (MIC) value are more important in the treatment of serious infections such as endocarditis or osteomyelitis, and for infections in high-risk patient groups such as immunocompromised patients (for example, transplant patients) and those who are critically ill.

Depending on the size of the workloads, personnel resources available, and issues of convenience, many clinical laboratories prefer to use one or more automated systems for AST regardless of whether results obtained are qualitative or quantitative. These systems handle the majority of the higher volume specimens such as urines, and are usually supplemented with manual methods such as disk diffusion and other MIC tests to handle the limitations of automation as well as provide MIC results for different clinical situations.[14] Clinical laboratories should continue to develop and improve their AST algorithms to include different types of test systems in order to provide meaningful and accurate data depending on the type of patient, source of specimen, organism species and anticipated problems in detecting various types of resistance mechanisms.

Disk Diffusion

The principle behind the disk diffusion method, or "Kirby Bauer" test, is the use of a paper disk with a defined amount of antibiotic to generate a dynamically changing gradient of antibiotic concentrations in the agar in the vicinity of the disk.[2] The disk is applied to the surface of an agar plate inoculated with the test organism; and while the antibiotic diffuses out of the disk to form the gradient, the test organism starts to divide and grow and progresses toward a critical mass of cells. The so-called inhibition zone edge is formed at the critical time where a particular concentration of the antibiotic is just able to inhibit the organism before it reaches an overwhelming cell mass or critical mass.[1] At the zone demarcation point, the density of cells on the growth side is sufficiently large to absorb antibiotic in the immediate vicinity, thus maintaining the concentration at a sub-inhibitory level and enabling the test organism to grow. The critical times for the demarcation of the inhibition zone edge for most rapidly growing aerobic and facultative anaerobic bacteria vary between 3 and 6 hr, while critical times of fastidious organisms and anaerobic bacteria can vary from 6 to 12 hr or longer.

A density of approximately 10^8 CFU mL^{-1} (CFU = colony-forming unit) of the test organism in the inoculum suspension is used to obtain semi-confluent growth on the agar. The inoculum is prepared by suspending enough well-isolated colonies from an 18- to 24-hr agar plate in broth or physiologic saline (0.85%) to achieve a turbidity matching a 0.5 McFarland Standard. The Prompt™ (BD Diagnostic Systems, Sparks, MD) inoculum preparation system, which consists of a sampling wand and inoculum solution in a plastic bottle, can also be used to optimize workflow because the organisms are maintained at the same density for up to 6 hr. However, this system is not suitable for mucoid strains or fastidious organisms because the amount of cell mass absorbed onto the tip of the sampling wand may be insufficient to give the appropriate inoculum density. Furthermore, the inoculum solution contains Tween, a surfactant that can have potential antimicrobial effects on certain organism groups and can interfere with the antimicrobial activity of some classes of antibiotics. Colony counts of the inoculum suspension should be regularly performed to verify that the inoculum density is correct in terms of CFU mL^{-1}.

Because the disk diffusion is based on the use of a dynamically changing gradient, inoculum density variations can directly affect the critical times and influence the inhibition zone sizes, regardless of the susceptibility of the test organism. A heavy inoculum will shorten the critical time and lead to falsely smaller inhibition zone sizes, resulting in potentially false resistant results, while a light inoculum will cause the reverse effect and generate potentially false susceptible results.

The inoculum suspension should be used within 15 min of preparation. This is particularly important for fastidious organisms that lose their viability rapidly. A sterile cotton swab is dipped into the suspension and pressed firmly on the inside of the tube to remove excess liquid. The dried surface of the appropriate agar plate is inoculated by streaking the entire surface and then repeating this twice, rotating the plate 60° each time. This will result in an even distribution of the inoculum. The inoculated plate is then allowed to dry with the lid left ajar for no more than 15 min. Once the agar plate is completely dry, the different antibiotic disks are applied either manually or with a dispensing apparatus. In general, no more than twelve disks should be placed on a 150-mm agar plate or five disks on a 90-mm plate. Fewer disks are used when anticipating highly susceptible organisms.

Optimally, disks should be positioned at a distance of 30 mm apart and no closer than 24 mm apart when measured center to center, to minimize the overlap of inhibition zones. Most dispenser devices are self-tamping (disks are tapped or pressed onto the agar surface). If applied manually, the disk must be pressed down to make immediate and complete contact with the agar surface. Once in contact with the agar, the disk cannot be moved because of instantaneous diffusion of antibiotics from the disk to the agar.

Agar plates are incubated in an inverted position (agar side up) under conditions appropriate for the test organism. Plates should not be stacked more than five high to ensure that the plate in the middle reaches incubator temperature within the same time frame as the other plates. After 16 to 18 hr of incubation at 35°C for rapidly growing aerobic bacteria, or longer where appropriate for fastidious organisms or specific resistance detection conditions, the agar plate is examined to determine if a semi-confluent and even lawn of growth has been obtained before reading the plate. If individual colonies are seen, the inoculum is too light and the test should be repeated because zone sizes may be falsely larger. The same holds true for excessively heavy inoculum where zone edges may be very hazy and difficult to read and zone sizes may be falsely small.

If the lawn of growth is satisfactory, the zone diameter is read to the nearest millimeter using a ruler or sliding calipers. For Mueller Hinton agar (without blood supplements), zone diameters are read from the back of the plate. For blood containing agar, zone diameters are read from the surface of the agar. The zone margin, unless otherwise specified, is identified as the area in which no obvious visible growth is seen by the naked eye. Faint growth or micro-colonies detectable only with a magnifying glass or by tilting the plate should be ignored in disk diffusion tests.

Reading of zone sizes can be simplified using automated zone readers that use a camera-based image analysis system such as the BIOMIC (Giles Scientific, New York, NY); Aura (Oxoid, Basingstoke, U.K.); Protozone (Microbiology International, Frederick, MD); or SirScan (i2a., France).[15] The agar plate is placed into the zone reader instrument that is connected to a computer system. The system then reads the different zone sizes directly from the agar plate and converts the result to a susceptibility category interpretation using various zone-MIC interpretive guidelines in the software. Although zone sizes are automatically read, the user can intervene and adjust results. The BIOMIC system also converts zone sizes to MIC equivalents using a database of regression analyses based on the assumption of a linear correlation between the zone size and the MIC value This may not always be the case, however, particularly for fastidious, slow-growing organisms.

The main advantages of disk diffusion testing are the simplicity, lack of need for expensive equipment, and the cost-effective and flexible choice of antibiotics for testing. The main disadvantage of the disk diffusion method is that the information is qualitative, providing only susceptibility category results. It cannot be used to compare the potential efficacy of different agents or to fine-tune the therapy by making dosage adjustments. The other disadvantage is the amount of technologist time required for setting up the test and for manually reading the zone sizes and then consulting the interpretive standards prior to the interpretation of results. Certain antibiotics can be problematic to test with the disk diffusion method due to specific physicochemical properties of the molecule. Glycopeptides such as vancomycin have a high molecular weight (1450) and diffuse very slowly in agar. The slow diffusion results in a poorly resolved concentration gradient around the vancomycin disk, and an inhibition zone size that differs by only a few millimeters between susceptible and resistant strains.[4, 24]

Quantitative Testing (Minimum Inhibitory Concentration. MIC)

In critical infections such as endocarditis, meningitis, osteomyelitis, or in the immunocompromised host, accurate quantitative determination of the exact MIC value of the infecting organism is crucial to guide choices of optimal therapy. In these situations, an MIC value of 0.016 $\mu g\ mL^{-1}$ has significantly different therapeutic implications than does an MIC of 1 $\mu g\ mL^{-1}$ in influencing the antibiotic choice and dose relative to factors such as body weight, potential drug toxicity, and drug concentrations at the site of infection.[9]

Broth Macrodilution, Microdilution, or Agar Dilution

Broth macrodilution and agar dilution methods are infrequently performed today. These techniques are predominantly performed in reference or research laboratories because they are not easily automated and require that the laboratory acquire and prepare stock solutions, perform dilutions accurately, and dispense the aliquots either into tubes with broth or agar medium. Quality control of the broth and agar dilution plates also needs to be performed to verify the quality of the reagents before they can be used for testing the clinical isolates. The limited stability of many antibiotic agents in broth and/or agar also necessitates the preparation of fresh reagents, which makes these methods very cumbersome and time consuming. The setup and reading of these methods also require significant experience.

Semi-automated or automated broth microdilution provided by commercial manufacturers is most commonly used in clinical laboratories. The various systems and instruments currently available include Phoenix (Becton Dickinson, Sparks, MD); Vitek (bioMerieux, St. Louis, MO); Microscan (Dade Behring, Sacramento, CA); and Sensititre (Trek Diagnostic Systems, Westlake, OH). The advantage of these instrument-based systems includes automation in reading and interpretation. Some of these systems even include automated setup and inoculation. Most systems have both a species identification and AST function. In addition, they include computerized systems for the interpretation of results, and a database "expert" system to help look for inconsistencies between the organism identification and the anticipated susceptibility patterns against multiple antibiotics. Predetermined panels with different types of antibiotics, and to some extent preferred dilution ranges, are selected depending on the requirements of the particular hospital. Most of these systems work well for rapidly growing routine organisms that form the bulk of organisms that are routinely tested.[10]

Limitations may exist, however, that are associated with the resistance mechanisms. These include inducible and/or hetero-resistances, testing of mucoid organisms and slower growing and/or fastidious organisms. The primary underlying reason for these potential limitations is the combination of the use of a relatively small amount of the inoculum of the test organism and the shortened incubation time in the rapid same-day reading systems. The Vitek or Vitek 2 system is promoted as a rapid system with results available in as little as 4 to 6 hr for identification and slightly longer for susceptibility results. Inducible β-lactamases produced by Gram-negative aerobes such as *Enterobacter*, *Serratia,* and *Citrobacter* species can be difficult to detect, while resistance to cephalosporins may be falsely reported as susceptible in these automated systems. The Sensititre and MicroScan systems both use an "overnight" broth microtiter format. Results are read either automatically or manually after 18 to 24 hr of incubation. Both the Vitek and the Phoenix can only be read by the instrument. Sensititre can only be used for susceptibility testing and not for organism identification. It can be customized to provide specific panels for individual needs, however, particularly for small volume use.

Etest®

Etest is an agar-based predefined gradient method specifically developed for the determination of exact MIC values and is used to complement other routinely used AST methods. The product was approved by the U.S. FDA in 1991 and is currently cleared for clinical testing of Gram-positive and -negative aerobes, anaerobes, pneumococci, streptococci, haemophilus, gonococci, and yeasts (*Candida* species) with a wide variety of antimicrobial agents. Etest is a gradient technique that combines the principles of both the disk diffusion and agar dilution methods. A preformed and predefined gradient of antibiotic concentrations that spans across 15 dilutions of the reference

MIC method is immobilized in a dry format onto the surface of a plastic strip. When applied to the surface of an inoculated agar plate, the antibiotic on the plastic strip is instantaneously transferred to the agar in the form of a stable, continuous gradient directly beneath and in the immediate vicinity of the strip. The stability of the gradient is maintained for up to 18 to 20 hr, depending on the antibiotic. This covers the critical times of a wide range of pathogens, from rapid-growing aerobic bacteria to slow-growing fastidious organisms such as anaerobes and fungi as well as mucoid organisms. The predefined and stable gradient in Etest, unlike the unstable and dynamic gradient around a disk, provides inoculum tolerance. This means that a 100-fold variation in colony-forming units per milliliter (CFU mL^{-1}) has minimal effect on the results of homogeneously susceptible strains.[21]

A unique feature exclusive to Etest is the ability to use the method in a macro format through the use of a heavier inoculum. This helps optimize the detection of low-level resistance, inducible resistance, and heteroresistance and resistant subpopulations that may occur in varying frequencies. Important examples of heterogeneous resistance are oxacillin resistance in *Staphylococcus aureus* and hetero-vancomycin intermediate *S. aureus* (hVISA), both of which are well detected by Etest and the latter by the macro method format.[13, 23]

After incubation, a parabola-shaped inhibition zone centered alongside the test strip is seen. The MIC is read at the point where the growth/inhibition margin of the organism intersects the edge of the MIC calibrated strip. This can easily be seen with the naked eye for the majority of organisms. Tilting the plate or using a magnifying glass can be helpful to visualize microcolonies, hazes, or other colonies within the ellipse of inhibition when reading the endpoints for bactericidal drugs and for detecting different resistance phenomena such as heteroresistance to oxacillin in *Staphylococcus aureus.*

AST of Fastidious Organisms

Most fastidious organisms do not grow well enough in routine antimicrobial testing systems and require some type of supplementation — for example, blood. Disk diffusion was initially developed for susceptibility testing of rapidly growing aerobic bacteria, including enterococci, staphylococci, Enterobacteriaceae, and *Pseudomonas aeruginosa.* Modifications have been described by the CLSI (Clinical Laboratory Standards Institute, formerly NCCLS, United States) for certain fastidious organisms, including: *Haemophilus influenzae*, *Neisseria gonorrhoeae*, *Streptococcus pneumoniae,* and *Streptococcus* species.[7] Modifications for broth microdilution can also be used for testing fastidious organisms, although panels must be read manually and not by automated instruments. The MicroScan MicroStrep panel (Dade Behring, Sacramento, CA) is a freeze-dried panel containing Mueller Hinton broth with lysed horse blood for testing of *Streptococcus pneumoniae* and other *Streptococcus* species. Similar panels are available from Pasco (Becton Dickinson) and Sensititire (Trek Diagnostics). Vitek 2 (bioMerieux) also has a card specifically designed for susceptibility testing of *Streptococcus* species. Etest, being an agar-based system that can be easily adapted to various growth media, can be used for testing of fastidious organisms, including *Streptococcus* and *Haemophilus Neisseria* as well as many other fastidious and very slow-growing organisms.

The CLSI recently developed a document with guidelines for testing and interpretation of several groups of fastidious organisms such as *Corynebacterium*, *Listeria,* and *Bacillus* species.[5] Other unusual and/or rarely encountered opportunistic pathogens for which no CLSI guidelines are available can also be tested by an MIC method if the system supports good growth of these organisms and only the MIC value is reported without an interpretation. Scientific references can be used to guide the choice of appropriate media, incubation conditions, and antibiotics to test for these unusual organisms.

AST of Anaerobes

Antimicrobial resistance among many clinically important species of anaerobes has increased, which has made empiric therapy choices unpredictable. Few antimicrobials remain where these pathogens are expected to be uniformly susceptible.[11, 12] Although rare, metronidazole resistance has been found in *Bacteroides* species as well as other anaerobes. Equally important is the need to recognize inherent resistance to metronidazole in anaerobes such as *Propinoibacterium.* In 2004, the CLSI published a document describing agar dilution as the reference method for anaerobic susceptibility testing. Shortly afterward, a broth microdilution method for rapidly growing anaerobes (*Bacteroides* species only) was included in the reference standard; however, the method is only described for a limited number of antimicrobials and the panels are not commercially available.[6] Anaerobic susceptibility testing is indicated for the management of patients with serious infections who may require long-term therapy and for the guidance of appropriate therapy for organisms that are not predictably susceptible. The CLSI also recommends that clinical laboratories save significant anaerobes and perform AST at the end of the year in a batch format in order to create an anaerobic antibiogram to help guide empiric therapy. The media recommended by the CLSI is Brucella agar or broth supplemented with vitamin K and hemin. After inoculation in agar or broth, results can be read after 24 to 48 hr of incubation in an anaerobic environment.

AST of Fungi

In 1997, the CLSI published a standard method for antifungal susceptibility testing of yeasts, and later for filamentous fungi (molds) (2002).[17, 18] Although no MIC interpretive breakpoints exist for molds with any of the currently available antifungal agents, reproducible results have been achieved with the CLSI as well as alternative methods as documented in the literature. The CLSI reference methods for fungi include disk diffusion (yeast only), as well as broth macro and microdilution.[16] A commercially available broth microdilution method (Yeast One, Trek Diagnostics, Cleveland, OH), as well as an automated method (Vitek) and Etest, are FDA-cleared for antifungal testing for a limited set of organism and antifungal agent combinations. The recommended medium is RPMI 1640 buffered with MOPS in either a broth or agar format, the latter of which requires supplementation with 2% glucose for Etest. After incubating for 24 to 48 hr, the MIC results for *Candida* species can be read. Endpoints are determined based on the class of antifungal agent with fungicidal agents such as amphotericin being read at 100% inhibition and fungistatic agents such as the azoles read at 80% inhibition; or the first dilution showing a significant decrease in turbidity or growth. MIC breakpoints are published by the CLSI for fluconazole, itraconazole, voriconazole, and flucytocine for *Candida* species.

AST of Mycobacteria

The reference method for AST of slow-growing mycobacteria, in particular *Mycobacterium tuberculosis,* is the agar proportion method. This technique uses a single critical concentration of the antimycobacterial agent to define susceptibility/resistance. A standardized inoculum of the organisms is added to a Middlebrook 7H11 agar plate. The agent is then added to a single quadrant. Drugs tested include isoniazid, rifampin, ethambutol, and streptomycin. Colonies are counted after 3 weeks of incubation, and the number on the drug-containing quadrant is compared to the control quadrant, which does not contain any drug. Organisms are considered resistant to the drug if the number of colonies on the drug-containing quadrant of resistant colonies is greater than 1% of the control population.[8]

To provide more timely results, clinical laboratories use an automated or semi-automated broth based system for the detection of growth of *Mycobacterium tuberculosis* in the presence of a defined critical concentration of the drug in broth as described by the CLSI. The Bactec (Becton Dickinson, Sparks, MD) measures growth of *M. tuberculosis* at each critical drug concentration or, at the most, two concentrations. The system is considered a rapid method because it only requires approximately 1 week of incubation. If growth of the organism in the presence of the drug as measured by production of a radioactive carbon source in the headspace of the bottle is as good or better than the control, then it is considered resistant to that drug. Other non-radiometric automated systems are being evaluated for susceptibility testing of *M. tuberculosis*. These systems include Versa Trek (Trek Diagnostics); Mycobacteria Growth Indicator Tube (MGIT, Becton Dickinson, Sparks, MD); and MB/BacT ALERT 3D (bioMerieux, Durham, NC).[25]

Susceptibility testing of rapidly growing mycobacteria can be performed by either broth microdilution (the CLSI reference method) or by Etest. Broth dilution may require the laboratory to prepare its own trays because commercially available trays used for the testing of other organisms contain neither the necessary drugs nor the appropriate concentrations for mycobacterial testing.

SUSCEPTIBILITY INTERPRETATION

Information on drug levels, the site of infection, and MIC distributions of biological populations of bacteria and clinical outcome data from clinical trials are used to select MIC cutoffs or breakpoints to predict whether a pathogen with a certain zone-MIC correlate would respond to therapy with standard dosages. These breakpoints are published in the CLSI AST documents and are updated on a yearly basis as needed. The FDA also approves specific MIC breakpoints for antimicrobial agents when the drug is approved for clinical use, and these breakpoints are published in the package insert for the drug. Most clinical laboratories in the United States use the CLSI breakpoints; however, commercially available systems are required to use FDA breakpoints in their interpretive software. For several new agents, only FDA breakpoints may be available; and for some agents, FDA and CLSI breakpoints may be different, causing confusion for laboratorians.

Currently three categories of interpretation are used for reporting of AST results:

1. *Susceptible*. An infection caused by the organism will most probably respond to treatment with a standard dosage of the antibiotic as indicated for use for that infection.
2. *Intermediate*. These results are usually regarded as indicative of non-useful therapeutic options similar to the resistant category for treatment purposes. It also serves as a buffer zone to help prevent major categorical errors caused by slight changes in the zone sizes due to the influence of technical variables. However, infections at sites where the antibiotic is likely to be highly concentrated, such as β-lactams in the urinary tract, may respond to therapy.
3. *Resistant*. The infection is not likely to respond to therapy.

AGAR SCREEN METHODS

Agar-based methods are available for screening of certain resistance mechanisms.[22] The agar screen method uses a certain cutoff concentration of the antibiotic that is expected to inhibit the growth of normal susceptible strains and allow resistant phenotypes to grow. The test is performed by inoculating a defined volume of a suspension of the test strain from a pure culture onto the agar, followed by incubation for 18 to 24 hr. Examples of these agar screen methods include the use of Mueller Hinton agar supplemented with 4% NaCl and 6 μg mL^{-1} oxacillin to detect oxacillin resistance in *Staphylococcus aureus* (MRSA screen plate). The agar screen plate is inoculated with a 1-μL loop dipped into a bacterial suspension with a turbidity equivalent to 0.5 McFarland standard. After 24 hr of incubation at 35°C, the appearance of more than one colony is reported as MRSA screen positive and requires the laboratory to perform a confirmatory test to verify the resistance.

Brain Heart Infusion (BHI) agar supplemented with 6 μg mL^{-1} of vancomycin is used as a screen for vancomycin resistance in *Staphylococcus aureus* (VISA) or *Enterococcus* (VRE). The same procedure and interpretation as for the MRSA screen agar plates are used for these tests. Additional screening agars are available for detecting high-level aminoglycoside resistance (HLAR) in enterococci using BHI agar supplemented with 500 μg mL^{-1} of gentamicin and 2000 μg mL^{-1} of streptomycin. A positive HLAR screen result indicates that a synergistic effect will not be seen when the aminoglycoside is combined with either ampicillin, penicillin, or vancomycin. These screen agar plates are commercially available from several manufacturers.

Latex agglutination methods are also available as a direct method for detecting oxacillin resistance in *Staphylococcus aureus*. A suspension of pure colonies of the test organism is added to latex beads bound to antibody against the PBP2a antigen (marker for MRSA), and the presence of agglutination is interpreted as a positive MRSA result.[3] Due to the expense of this assay and the recent availability of molecular methods for detection of oxacillin resistance, these assays are not often used in clinical laboratories.

MOLECULAR METHODS

Molecular-based methods are becoming more common as adjuncts for specific testing needs in the clinical microbiology laboratory.[20] Several molecular tests are now packaged in an easy-to-use kit format where the amplification reaction takes place in a closed environment. This obviates the need for a separate room to perform the testing. Although most of the assays are more expensive than culture methods, the turnaround time of same-day results, theoretically available in several hours, is a major advantage for patient care and can contribute to cost savings to the hospital. Another advantage of these assays is that they do not rely on viable organisms for detection. The most commonly used molecular test for antimicrobial resistance detection is the polymerase chain reaction (PCR) method for the mecA gene to detect oxacillin resistance in *Staphylococcus aureus* (MRSA). Molecular assays that require a pure culture of the test organism are being replaced by newer assays that directly detect the presence of MRSA in patient specimens. These assays are currently approved by the FDA for use on nasal swabs and may likely be expanded to include skin and soft tissue as well as positive blood cultures in the near future.[19] Assays for the detection of other resistance mechanisms are also under evaluation. These include assays for the detection of resistance to vancomycin in *Enterococcus* and anti-mycobacterial agents, particularly rifampin, in *Mycobacterium tuberculosis*.

QUALITY CONTROL

The CLSI-recommended media for AST is cation-adjusted Mueller Hinton broth for broth methods or the standard Mueller Hinton agar for agar-based methods. The CLSI provides guidelines for the selection of quality control (QC) strains and the frequency with which QC should be performed for both disk diffusion and MIC methods. QC strains chosen for MIC testing should be such that the expected MIC value for the strain falls in the middle of the concentration range being tested for that antibiotic. However, this is seldom the case in MIC test systems that use abbreviated or limited dilution ranges of the drug. QC strains can be purchased in dehydrated form either as Culti-loops® (Remel), pellets, swabs (KWIK-STIK, Hardy Diagnostics), or lyophilized in multiple unit containers (LYFO-DISK, MicroBioLogics, St. Cloud, MN). After processing according to the manufacturer's instructions, organisms should be maintained on an agar slant, stored at 2 to 8°C, and subcultured weekly for up to 1 month, after which a new QC stock organism should be used. QC organisms can be stored for prolonged periods of time at –70°C in broth (such as Brain Heart Infusion or Brucella) or skim milk. Stock organisms should be subcultured at least twice when processed from the freezer storage before being used for quality control testing. Testing of QC strains should be done either on each day of clinical testing for infrequently performed tests or weekly after qualifying an initial 30-day QC validation. The CLSI provides flow diagrams for frequency of QC testing, troubleshooting, and corrective action for out-of-control results.

Antibiotic disks are available from several manufacturers (e.g., Becton Dickinson, Remel, and Hardy Diagnostics). They should be handled and stored according to the manufacturer's instructions. Disks can be stored at 4°C in disk dispensers containing desiccant for day-to-day use for up to 1 week for short-term purposes. Long-term storage of unopened cartridges should be at –20°C. Packages should be allowed to reach room temperature prior to opening to prevent moisture condensing on the outer surface from penetrating into the cartridge and deteriorating the antibiotic. The CLSI also provides recommendations for detecting unusual AST results and suggests that laboratories modify the list of agents to test based on the prevalence of resistance mechanisms in their institutions. Unusual or questionable results include carbapenem (imipenem, meropenem, ertapenem) resistance in strains belonging to the family Enterobacteriaceae that are susceptible to extended-spectrum cephalosporins (cefotaxime, ceftriaxone, or cefepime). Labile compounds such as imipenem may give unexpected resistant results due to degradation of the compound even with proper storage.

REFERENCES

1. Acar, J. and F.W. Goldstein. 1996. Disk susceptibility test, In V. Lorian (Ed.), *Antibiotics in Laboratory Medicine.* Williams and Wilkins, Baltimore, p. 1–51.
2. Bauer, A., W.M.M. Kirby, J.C. Sherris, and M. Turch. 1966. Antibiotic susceptibility testing by a standardized single disk method. *Am. J. Clin. Pathol.*, 45:493–496.
3. Cavassini, M., A. Wenger, K. Jaton, D.S. Blanc, and J. Bille. 1999. Evaluation of MRSA-Screen, a simple anti-PBP 2a slide latex agglutination kit, for rapid detection of methicillin resistance in *Staphylococcus aureus. J. Clin. Microbiol.*, 37:1591–1594.
4. CDC. 1997. Interim guidelines for prevention and control of staphylococcal infection associated with reduced susceptibility to vancomycin. *MMWR*, 46:626–635.
5. CLSI. 2006. *Methods for Antimicrobial Dilution and Disk Susceptibility Testing of Fastidious Bacteria, Approved Guideline.* CLSI, Wayne, Pennsylvania.
6. CLSI. 2007. *Methods for Antimicrobial Susceptibility Testing of Anaerobic Bacteria*, Approved Standard, 7th ed. CLSI, Wayne, Pennsylvania.
7. CLSI. 2007. *Performance Standards for Antimicrobial Susceptibility Testing*, Seventeenth Informational Supplement. CLSI, Wayne, Pennsylvania.
8. CLSI. 2003. *Susceptibility Testing of Mycobacteria, Nocardiae, and Other Aerobic Actinomycetes,* Approved Standard. CLSI, Wayne, Pennsylvania.
9. Craig, W. 1993. *Qualitative Susceptibility Tests versus Quantitative MIC Tests.* Elsevier Science Publishing Co., Inc., New York.
10. Evangelista, A.T. and A.L. Truant. 2002. *Rapid Systems and Instruments for Antimicrobial Susceptibility Testing of Bacteria.* ASM Press, Philadelphia.
11. Hecht, D.W. 2006. Anaerobes: antibiotic resistance, clinical significance, and the role of susceptibility testing. *Anaerobe*, 12:115–121.
12. Hecht, D.W. 2002. Evolution of anaerobe susceptibility testing in the United States. *Clin. Infect. Dis.*, 35:S28–S35.
13. Hubert, S.K., J.M. Mohammed, S.K. Fridkin, R.P. Gaynes, J.E. McGowan, Jr., and F.C. Tenover. 1999. Glycopeptide-intermediate *Staphylococcus aureus:* evaluation of a novel screening method and results of a survey of selected U.S. hospitals. *J. Clin. Microbiol.*, 37:3590–3593.
14. Jorgensen, J. and M.J. Ferraro. 1998. Antimicrobial susceptibility testing: general principles and contemporary practices. *Clin. Infect. Dis.*, 26: 973–980.
15. Korgenski, E. and J.A. Daly. 1998. Evaluation of the BIOMIC video reader system for determining interpretive categories of isolates on the basis of disk diffusion susceptibility results. *J. Clin. Microbiol.*, 36:302–304.
16. NCCLS. 2004. *Reference Method for Antifungal Disk Diffusion Susceptibility Testing of Yeasts,* Approved Standard. NCCLS, Wayne, PA.
17. NCCLS. 2002. *Reference Method for Broth Dilution Antifungal Susceptibility Testing of Filamentous Fungi,* Approved Standard. NCCLS, Wayne, PA.
18. NCCLS. 2002. *Reference Method for Broth Dilution Antifungal Susceptibility Testing of Yeasts.* Approved Standard, 2nd edition. NCCLS, Wayne, PA.

19. Paule, S.M., D.M. Hacek, B. Kufner, K. Truchon, R.B. Thomson, Jr., K.L. Kaul, A. Robicsek, and L.R. Peterson. 2007. Performance of the BD GeneOhm™ MRSA test before and during high-volume clinical use. *J. Clin. Microbiol.*, published online ahead of print, doi:10.1128/JCM.00670-07.
20. Rasheed, J.K., F. Cockerill, and F.C. Tenover. 2007. *Detection and Characterization of Antimicrobial Resistance Genes in Pathogenic Bacteria,* 9th ed. ASM Press, Washington, DC.
21. Sanchez, M.L. and R.N. Jones. 1992. Etest, an antimicrobial susceptibility testing method with broad clinical and epidemiologic application. *The Antimicrobic Newslett.*, 8:1–8.
22. Swenson, J.M., J.B. Patel, and J.H. Jorgensen. 2007. *Special Pheontypic Methods for Detecting Antibacterial Resistance*, 9th ed. ASM Press, Washington, DC.
23. Walsh, T.R., A. Bolmstrom, P. Ho, A. Qwarnstrom, M. Wootton, R.A. Howe, P.M. Bennett, and A.P. MacGowan. 2000. Evaluation of growth conditions in susceptibility testing to discriminate hetero-vancomycin intermediate *Staphylococcus aureus* (hVISA) and VISAs from methicillin resistant SA (MRSA). *Clin. Microbiol. Infect.*, 6:84.
24. Walsh, T.R., A. Bolmstrom, A. Qwarnstrom, P. Ho, M. Wootton, R.A. Howe, A.P. MacGowan, and D. Diekema. 2001. Evaluation of current methods for detection of staphylococci with reduced susceptibility to glycopeptides. *J. Clin. Microbiol.*, 39:2439–2444.
25. Woods, G L., N.G. Warren, and C.B. Inderlied. 2007. *Susceptibility Test Methods: Mycobacteria, Nocardia, and other Actinomycetes*, 9th ed. ASM Press, Washington, DC.

13 Bacterial Cell Wall Morphology and Biochemistry

Maxim Suvorov, Jed F. Fisher, and Shahriar Mobashery

CONTENTS

INTRODUCTION

The bacterial *cell wall* is a heteropolymer that is a structural component of the cell envelope of most bacteria, with the exception of the *Mycoplasma* and a small group of highly specialized prokaryotes.[94, 95, 119, 154, 162, 177] Although often used interchangeably, the terms "cell wall" and "cell envelope" are distinct. The *cell envelope* is the multilayer barrier that separates and protects the cytoplasm of the bacterium from its external environment. The cell envelope includes the peptidoglycan cell wall, the membrane(s), the periplasmic space, and the macromolecules that are associated with these structures. The integrity of the cell wall is necessary for bacterial viability. The cell wall heteropolymer is biosynthesized from both amino acid and saccharide precursors, and in final form consists of a three-dimensional array of glycan strands interconnected by peptide cross-links. In recognition of the presence of peptide cross-linking of the glycan strands, the term "peptidoglycan" is often used as a synonym for the cell wall, particularly where emphasis of the molecular basis for the cell wall structure is appropriate. The term "murein" is a synonym of peptidoglycan. Exhaustive hydrolysis of the murein, by glycosylases (such as lysozyme) and by cell wall peptidases, degrades the cell wall into its constituent muropeptides. The core structure of the intact muropeptide is a glycan strand. The structure of this glycan, that of a repeating [*N*-acetylglucosamine]-[*N*-acetylmuramic] (abbreviated as NAG-NAM) disaccharide, is highly conserved among the Gram-negative and Gram-positive bacteria. The structure of the peptide cross-links is, however, much more variable. In the Gram-positive and Gram-negative eubacteria, the cell wall is a single polymeric molecule that surrounds the entire bacterium. The term describing this single polymeric molecule is "sacculus." The sacculus of the Gram-negative bacterium is found in the periplasm, a space between the inner membrane and outer membrane. In the Gram-positive bacterium, the cell wall forms the exterior surface of the bacterium and overlays an inner (periplasm-like) space located above the single cell (plasma) membrane.

As stated above, the basic peptidoglycan chemical structure of the cell wall is conserved among the eubacteria. This strongly implies that the function of the cell wall — whether exposed to the environment (as is the case for Gram-positive bacteria) or protected from the environment (as is the case for Gram-negative bacteria) — is the same for both. The primary functions of the cell wall include preservation of the integrity of the bacterium,[160] the creation of cell shape,[6, 14, 35, 119, 121, 205, 213, 251] and as a scaffold for the attachment of the numerous macromolecular complexes. Examples of these complexes include porin proteins for the control of solute ingress and egress, complexes that enable adhesion,[233] complexes for control of transport and secretion, complexes that are used for environmental sensing, and complexes for propulsion (flagella). The cross-linked strands of the peptidoglycan form a three-dimensional barrier that segregates the cytoplasm of the bacterium from its environment, while containing the turgor pressure of the cytoplasm. The structure and integrity of the cell wall must nonetheless be dynamic, to enable the *simultaneous* preservation of structural integrity while enabling both cell growth and division. The ability of the cell wall to possess the dual properties of structural strength and rigidity to contain the turgor pressure and maintain cell shape, while remaining capable of facile restructuring during growth and division, only can be reconciled by the presence of extraordinarily sophisticated and highly regulated enzymatic systems for cell wall synthesis and remodeling. Those structures (including muropeptides) that are released during remodeling of the peptidoglycan are recognized, during bacterial infection, by recognition proteins that activate the innate immune response of the eukaryotic organism.[56, 57, 90, 134, 156, 195, 250] The structural basis for peptidoglycan recognition by the peptidoglycan recognition proteins (PGRPs) of innate immunity is being explored aggressively by structure-activity studies of peptidoglycan fragments, obtained by chemical synthesis or by peptidoglycan degradation,[68, 128, 135, 175, 216] and by crystallography.[41, 89] Co-crystallization of the directly bacteriocidal human PGRP with a synthetic fragment of the cell wall suggests that the bacteriocidal mechanism used by human PGRPs is similar to that of vancomycin (complexation of the peptide stems to prevent peptidoglycan cross-linking).[42]

The essentiality of an intact cell wall for bacterial viability, and the relative accessibility of the enzymes involved in cell wall synthesis and remodeling, render these enzymes as important targets for the action of many antimicrobial agents.[239] The search for new targets for new antimicrobials, for structural improvements of existing antimicrobials, and for the identification of new synergistic antimicrobial combinations are among the main moving forces for current research on the cell wall.

GRAM-POSITIVE AND GRAM-NEGATIVE BACTERIA

The technique invented by Danish bacteriologist Hans Christian Gram in 1884 categorizes most prokaryotes as either Gram-positive or Gram-negative on the basis of a standardized staining procedure. The Gram-staining procedure involves gentle drying of a bacterial suspension on the microscopy slide, addition of a solution of crystal violet (or methylene blue), rinsing with an iodine solution, decolorizing with ethanol, and final addition of safranin as a counterstain. Those bacteria that retain the crystal violet-iodine complex, and thus appear purple-brown colored under microscope, are referred to as Gram-positive. Those that do not retain the crystal violet stain appear red colored and are referred to as Gram-negative.[13, 15, 101] Although it was intuitively clear that Gram-positive and Gram-negative bacteria must have different cell envelopes, it took more than half a century to confirm this hypothesis. Developments in transmission electron microscopy (TEM) in the 1950s allowed the first images of bacterial cell envelopes. These images revealed that the Gram-negative bacteria have a complex cell envelope consisting of two membranes and the cell wall (Figure 13.1A). An inner plasma membrane surrounds the cytoplasm. The 5- to 7-nm-thick cell wall (in the Gram-negative *Escherichia coli*) is located within the periplasmic space between the inner membrane and the outer membrane.[16, 48, 99] The exterior surface of the Gram-negative bacterium — that is, the outer leaflet of the outer membrane bilayer — consists primarily of specialized

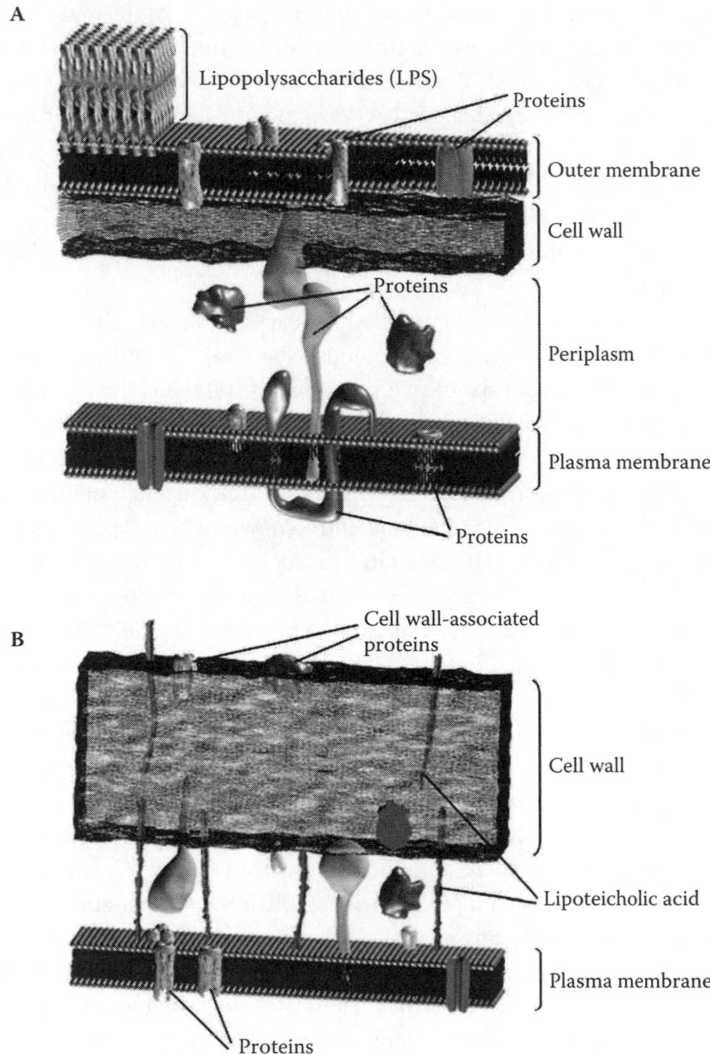

FIGURE 13.1 Cell envelope of Gram-positive and Gram-negative bacteria.

glycolipids referred to as lipopolysaccharide (LPS). In contrast to the Gram-negative microorganisms, Gram-positive bacteria have a single plasma membrane, with an approximately 20-nm-thick cell wall located above the membrane and forming the cell exterior[12, 16] (Figure 13.1B). The absence of the outer membrane, and the greater thickness of the Gram-positive cell wall compared to the Gram-positive bacteria, are believed to result in better retention of the iodine-dye complex, and to account for the characteristic color of the Gram-positive bacteria.

DIMENSIONS OF THE PEPTIDOGLYCAN

One of the most exciting developments with respect to the understanding of the organization of the bacterial cell envelope is the structural imaging of individual bacteria.[16, 107] Information on the spatial relationships of the components of the cell envelope, and their dimensions, has been obtained for both Gram-negative and -positive bacteria using cryo-transmission electron microscopy of cross-sectioned, vitreously frozen bacteria. These images show clearly the layered components of the cell envelope. The peptidoglycan of the Gram-negative bacterium *Escherichia coli* is found immediately underneath the outer membrane and above the periplasmic space.[151] At the mid-section of the bacterium, the *E. coli* peptidoglycan has a uniform thickness of 6.4 nm, consistent with earlier estimates obtained by atomic force microscopy.[248] The thickness of the mid-section peptidoglycan of the related Gram-negative bacterium (both are rod-shaped A1γ chemotypes) *Pseudomonas aeruginosa* is substantially thinner (2.4 nm).[149] In contrast, the thicknesses of the outer membranes (approximately 7 nm), plasma membranes (approximate 6 nm), and periplasmic spaces (approximately 22 nm) of the *E. coli* and *P. aeruginosa* cell envelopes are similar.[149] Tomography of cryo-TEM-sectioned *E. coli* suggests that the dimensions of the components of the cell envelope at the pole and at the mid-section are substantially similar.[145, 235]

Cryo-electron microscopy studies of Gram-positive bacteria indicate a greater complexity to the organization of their cell envelope than was previously believed.[17] Until recently, the Gram-positive cell envelope was thought to consist of two components: (1) the cell wall peptidoglycan as the exterior surface of the bacterium and (2) the plasma membrane immediately underneath the peptidoglycan. However, in both the Gram-positive rod (*Bacillus subtilis*)[151] and coccus (*Staphylococcus aureus*),[152] the cell envelope is seen to consist of at least three distinct components: (1) an outer-wall zone forming the bacterial surface (independently known to be the peptidoglycan) of variable thickness (*vide infra*), (2) an inner wall zone (16-nm thickness in *S. aureus*), and (3) the plasma membrane (5.4-nm thickness in *S. aureus*).[17, 152] The existence of the inner wall zone (and as well a possible fourth component within the inner wall zone, termed the "granular layer," and of unknown chemical composition) was confirmed in other Gram-positive bacteria by Zuber et al.[257]

The identity of the Gram-positive inner wall zone as the functional equivalent of the Gram-negative periplasmic space is strongly implied.[151, 152, 257] This assignment must be regarded as logical and appropriate, from a second perspective. The expectation of a periplasmic space[1] as a *necessary* space within the cell envelope — regardless of the Gram characterization — follows from the emerging structural studies of the biosynthetic enzymes that accomplish peptidoglycan biosynthesis.[140, 200] These multi-domain biosynthetic enzymes are localized within the outer leaflet of the plasma membrane and require space above this membrane to accomplish the sequential transglycosylation and transpeptidation events of peptidoglycan assembly (*vide infra*).[47, 138, 139, 161, 254] Moreover, cryo-electron microscopy of *Staphylococcus aureus* bacteria in the process of cell division shows the development of a mid-zone between the nascent cell walls that is logically suggested to contain the peptidoglycan hydrolytic enzymes — the autolysins — that act to accomplish daughter cell division.[150]

Three peptidoglycan zones, differentiated by their relative thickness, are observed in the dividing *Staphylococcus aureus*. Matias and Beveridge[150] observe that the thickness of the *S. aureus* peptidoglycan away from the septal region is approximately 19 nm, and increases to approximately 30 to 40 nm around the circumference of the septal region (termed the outer wall bridge zone). The sub-structure of the invaginating cell envelope contains a 10-nm-thick high-density zone, whose position is consistent with assignment as the nascent cell wall of the two daughter cells. An examination of the relative density of the Gram-positive *Bacillus subtilis* peptidoglycan shows greater density for the peptidoglycan near the periplasmic space, compared to the peptidoglycan at the exterior cell surface. This difference is expected, given the inside-to-outside mode of peptidoglycan biosynthesis used by this rod-shaped bacterium. As the cell wall grows outward, a progressive decrease in the extent of cross-linking is necessary to allow the corresponding increase in the cell diameter. In contrast, the *Escherichia coli* peptidoglycan is remarkably uniform in both dimension and density. All the cryo-TEM images show that the thickness of the peptidoglycan of intact cells is smaller than the thickness of lysed cell fragments. In the example of *S. aureus*, the difference in thickness is substantial (19 nm for intact cells compared to 33 nm for cell wall fragments).[152] This difference in thickness is interpreted as confirming the role of the peptidoglycan as the primary component of the cell envelope containing the turgor pressure of the bacterium.[151, 248]

CHEMICAL STRUCTURE OF MUREIN

In most eubacteria cell walls, the glycan strand component of the peptidoglycan is composed of a repeating β-1,4-linked disaccharide, interconnected by β-1,4-linkages. This core disaccharide is composed of the *N*-acetylglucosamine (abbreviated NAG or GlcNAc) and N-acetylmuramic acid (NAM or MurNAc) saccharides (Figure 13.2). In the mature murein, the glycan chains are cross-linked by means of the peptide cross-bridges that extend from the lactyl segment of the NAM saccharide. The NAG-NAM disaccharide core is identical for most Gram-positive and Gram-negative eubacteria. Exceptions do exist, however, such as pneumococcus and a small group of Gram-positive Archaebacteria of genera *Methanobacterium* and *Methanobrevibacterium*. These Archaea have a different peptidoglycan structure, termed "pseudomurein,"[112, 212] wherein the repeating disaccharide of the glycan strand is that of β-D-*N*-acetylglucosamine (NAG) β-1,3-linked to α-L-*N*-acetyltalosaminuronic acid (NAT).[201]

One fundamental property of the sacculus, that of the average length of its glycan strands, has proven to be an extraordinarily difficult property to measure. To ascertain the number of

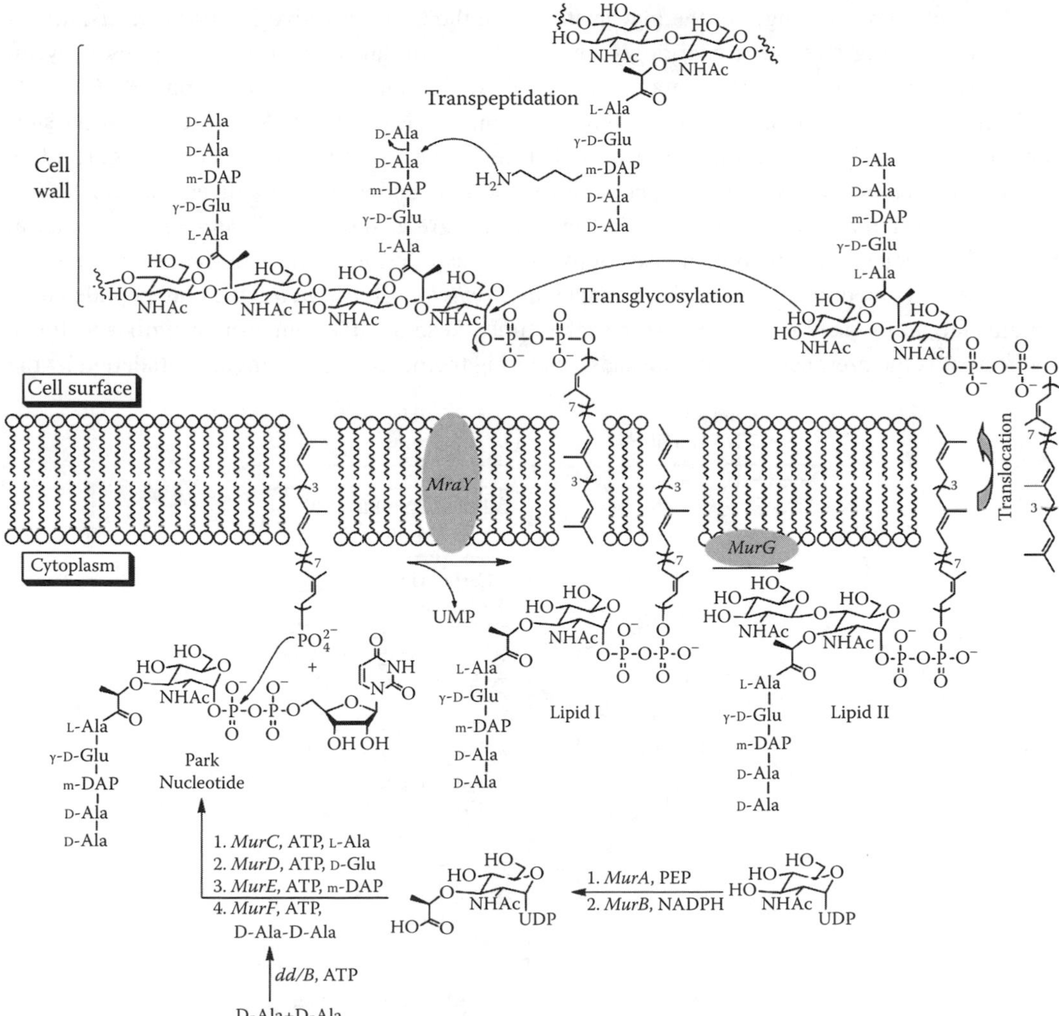

FIGURE 13.2 Assembly of the cell wall peptidoglycan. (Reproduced from *Curr. Opin. Chem. Biol.*, 6:786–793, 2002. With permission.)

NAG-NAM disaccharides in the glycan strand, exhaustive proteolytic degradation of the sacculus is necessary to cleave apart the peptide stems. The ensemble of discrete glycan strands thus obtained are separated by chromatography, and by rigorous mass spectral assignment of the separated peaks, are assigned a structure. Exhaustive glycosylase degradation, combined with proteolytic degradation, generates the ensemble of muropeptides. These are analyzed in analogous fashion. The distribution of structures, for both the glycan strand length and for the muropeptides, varies not only among the individual eubacteria, but moreover as a function of the stage of cell growth, the morphological transition, the presence of resistance determinants relating to the cell wall, and the presence of antibiotics. Moreover, there may exist distinct peptidoglycan domains within the sacculus (for example, a peptidoglycan that is different at the poles of a rod-shaped bacterium than that within the cylinder walls). The current experimental methodology cannot differentiate between such domains. Where an evaluation of the length distribution of the glycan strands has been made, a distribution of lengths is observed. In the Gram-negative *Escherichia coli*, the distribution of glycan lengths ranges from one to 80 NAG-NAM disaccharides, with the largest abundance of strands clustered at a length of five to ten disaccharide units.[93, 100, 133, 237] The distribution of the length of the glycan strands in the Gram-positive *Staphylococcus aureus* — notwithstanding its thicker peptidoglycan layer — is similar. In *S. aureus,* degree of glycan polymerization is from 3 to 10, with a maximum of 26 disaccharide units per strand.[29, 223]

Unlike the repeating disaccharide backbone, the amino acids of the NAM stem demonstrate substantial variability. This observation is especially true for the Gram-positive bacteria (Figure 13.3). The significance of these variations is not known, and may simply reflect a divergent evolution of the species. Differences in the length of the peptide stem are also seen. In Gram-negative bacteria, the murein oligopeptide stem is composed of five amino acid residues, while in some Gram-positive species, the oligopeptide is longer. For example, in *Staphylococcus aureus,* the oligopeptide chain contains ten amino acid residues.[201] At the same time, these peptide stems share similar features. One feature is the presence of D-amino acids, rendering the peptides resistant to degradation by the

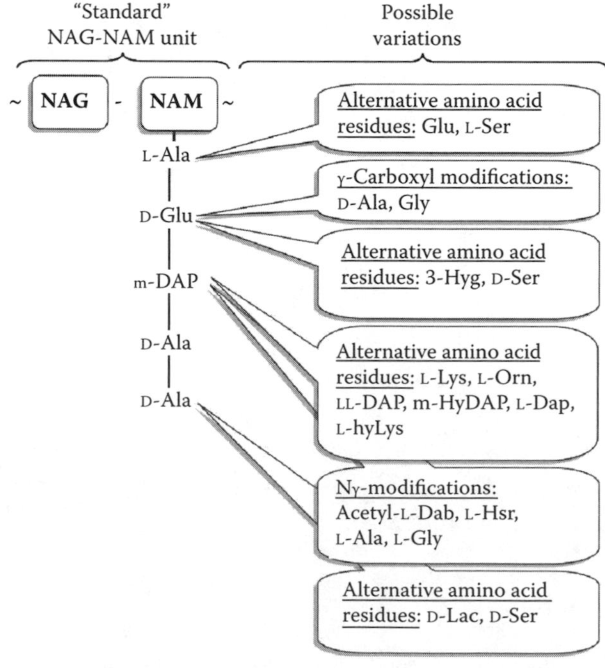

FIGURE 13.3 Diverse nature of murein.

majority of the peptidases. A second shared feature is the presence of a D-glutamic acid residue, linked in the chain through its γ-carboxyl rather than through its α-carboxyl.

The function of the stem peptides is to interconnect neighboring glycan chains to create a structurally sound cell wall. The extent of this cross-linking is variable among the bacteria. In *Escherichia coli,* the degree of cross-linking varies from 25 to 50%,[238] while the cross-linking value for *Staphylococcus aureus* is higher, estimated by Labischinski et al.[153, 193] as approximately 90%. The stem peptide also is an attachment point for all kinds of macromolecules. Among the attached macromolecules are proteins involved in the cell wall biosynthesis, transport of nutrients into the cell, and other processes. These proteins attach to the peptidoglycan by both covalent bonding and by electrostatic association.[200] In some areas of the cell wall, contact with membrane proteins can be seen on electron microscopy images as adhesion zones between the peptidoglycan and the plasma membrane.[48]

Not surprisingly, the cell wall peptidoglycan is firmly fixed to the adjacent membrane(s). In Gram-positive bacteria, the peptidoglycan layer is covalently linked to membrane-bound molecules of teichoic acid and teichuronic acid.[51, 52, 197, 224] These polymers anchor the cell wall to the plasma membrane. Teichoic acid and teichuronic acid are not the only molecules that covalently modify the peptidoglycan. Many proteins are attached to the peptidoglycan. Among the functions of these proteins are adherence, invasion, signaling, and interactions with the host immune system or the environment.[206] These proteins are attached to the peptidoglycan by a group of membrane-bound enzymes called sortases.[65, 132, 146] Sortase recognizes an LPXTG motif in the leader sequence of these proteins and forms covalent bonds between the threonine of this sequence and the cysteine residue of the active site. The leader sequence during this reaction undergoes cleavage between the threonine and glycine of the LPXTG motif. The protein/sortase acyl-enzyme complex is translocated across the plasma membrane. On the outer side of the plasma membrane, sortase undergoes deacylation, catalyzing formation of an amide bond between the protein and an amino group of the peptidoglycan stem peptide.[225]

In Gram-negative bacteria, the peptidoglycan is attached to both the inner and outer membranes. The most common process for attachment of the peptidoglycan to the outer membrane involves the Braun lipoprotein (Lpp), an abundant lipoprotein of the inner leaflet of the outer membrane.[32, 104, 167] The structure of the Braun lipoprotein is that of a coiled homotrimer of an α-helical peptide, approximately 76 Å in length.[22, 136, 209] In the Gram-negative Enterobacteriaceae, this lipoprotein has three fatty acids attached to its *N*-terminal cysteine, two as a diacylglyceride and one through an *N*-acyl link to the cysteine.[192] At the other end of the Lpp lipoprotein is a *C*-terminus lysine residue. The ε-amino group of this L-lysine is attached covalently to the L-center of the *meso*-diaminopimelic acid residue at position 3 of the peptide stem.[81, 142] Despite this apparent structural role, and the abundance of the Braun lipoprotein (as many as 5×10^5 and 2.5×10^5 for the free and the bound forms, respectively, copies per cell),[104] deletion of this lipoprotein is not lethal. In addition to this lipoprotein, the abundant *omp* porins of the outer membrane adhere to the peptidoglycan by non-covalent peptidoglycan recognition domains located at the *C*-terminus of the *omp* porin.[131] The structure of the peptidoglycan-binding motif, without peptidoglycan bound, of a *Neisseria meningitides* porin was determined by Grizot and Buchanan.[87] Deletion of the OmpA porin in *Escherichia coli* gives a viable but fragile phenotype. A third peptidoglycan-associated protein complex in Gram-negative bacteria is the Tol-Pal system.[30, 75, 185] This system comprises a five-protein complex that spans the periplasm and interconnects the outer membrane, the peptidoglycan, and the inner membrane. The function of the Tol-Pal system is maintenance of the cell envelope integrity (especially with respect to membrane invagination during cell division) and in macromolecule transport across the periplasm.[80] Each of the proteins of the Tol-Pal system interacts with each other, and with the peptidoglycan.[38, 39 46] NMR studies demonstrate that the basis for the interaction of the Pal protein with the peptidoglycan is non-covalent association, whereas an *N*-terminal lipid of the Pal protein holds this protein into the outer membrane.[46] The structure of a Pal-peptidoglycan complex was determined by NMR.[176] The stator proteins of the bacterial flagellar assembly have the dual function of ion transport and anchoring the flagellar assembly into the sacculus through peptidoglycan recognition domains of the stator.[24, 102, 124, 207, 218]

All of the examples given above represent a minute fraction of macromolecules found in the cell wall. As our knowledge about the bacterial cell wall increases, more molecules appear to interact with the peptidoglycan.

PEPTIDOGLYCAN BIOSYNTHESIS

Bacteria display an astonishing breadth of shape and morphological differences (as exemplified by the rod-shaped *Escherichia coli* Gram-negative and the coccus-shaped *Staphylococcus aureus* Gram-positive bacteria).[35, 213, 251] Nonetheless, the remarkable similarity of the chemical structures of the two peptidoglycans implies a fundamental similarity for their biosynthetic assembly. This similarity is indeed found.[100, 237] In the intensively studied Gram-negative bacterium *E. coli*, murein biosynthesis starts in the cytoplasm with the conversion of fructose-6-phosphate to UDP-*N*-acetylglucosamine by the Glm enzymes (Figure 13.4). Sequential addition of the stem peptide amino

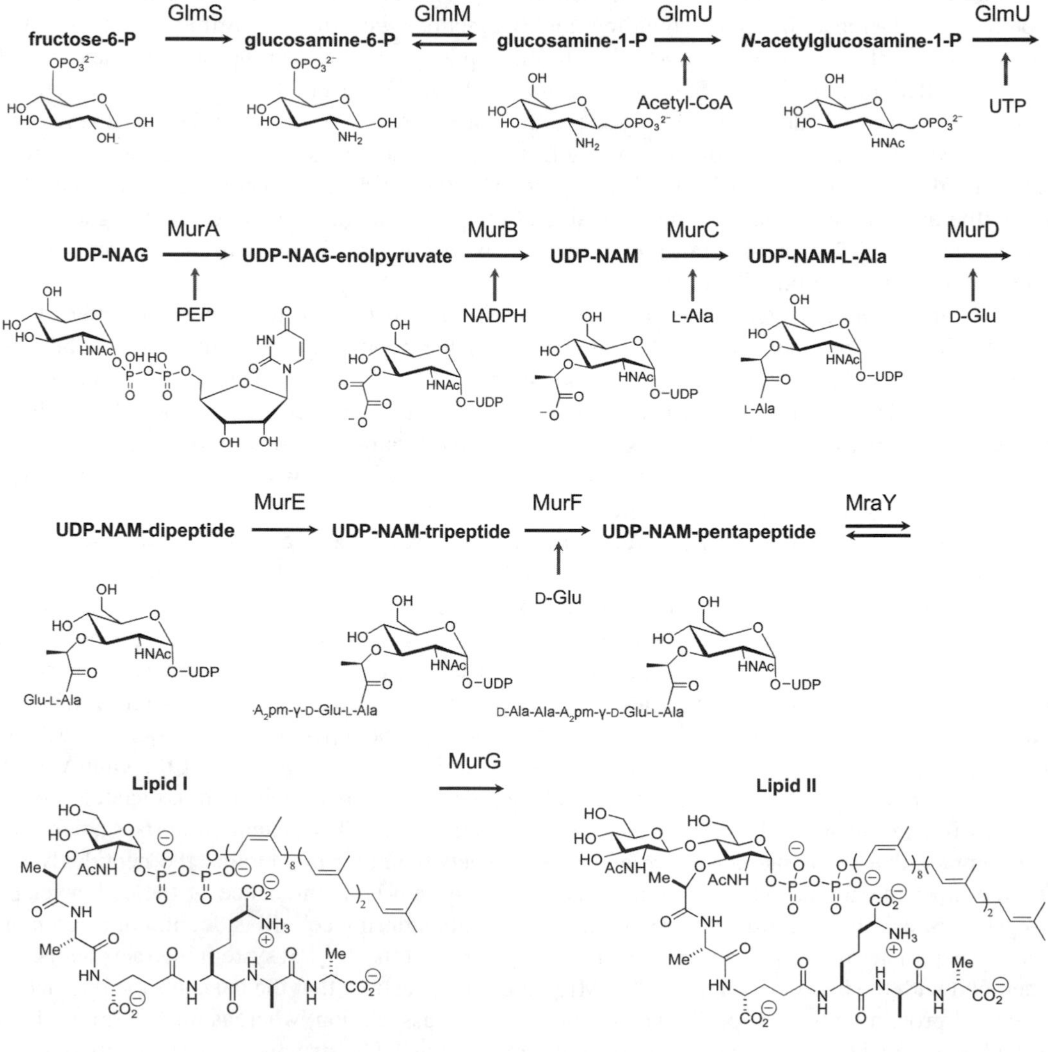

FIGURE 13.4 Synthesis of murein precursors in cytoplasm.

acid residues, catalyzed by the Mur proteins, follows. Concurrently, the MraY enzyme transfers *N*-acetylglucosamine from UDP-*N*-acetylglucosamine to the phosphate group of a C_{55}-isoprenoid alcohol (also termed undecaprenol) phosphate. The resulting glycolipid is called Lipid I. MraY catalysis uses the undecaprenol phosphate located on the inner surface of the plasma membrane, and newly synthesized Lipid I is incorporated into this inner surface. This is a membrane-anchoring event, and the ensuing reactions of cell wall biosynthesis involve membrane-bound enzyme catalysis of membrane-bound intermediates. The MurG enzyme adds *N*-acetylglucosamine to the C-4 position of the NAM, giving the Lipid II disaccharide intermediate. Newly biosynthesized Lipid II remains anchored to the inner side of plasma membrane by its undecaprenyl lipid moiety. Its transfer to the outer side of the plasma membrane — a necessary event to enable the final steps of cell wall synthesis — is accomplished by the presumed action of a lipid translocation protein, termed flippase.[53, 64] The mechanism of action of the flippase is poorly understood. Indeed, the Lipid II flippase[231] is yet to be identified (in contrast to the better-understood MsbA ABC transporter used in Gram-negative lipopolysaccharide translocation).[64] As soon as Lipid II is translocated across the plasma membrane, the periplasmic stage of murein biosynthesis begins.

This stage of the peptidoglycan synthesis (Figure 13.2) is much more complex than the cytoplasmic stage, as it involves the spatial and temporal coordination of multiple enzyme systems. It is also the least well-understood aspect of the cell wall biosynthesis, and yet an understanding of its mechanism is of the utmost importance for the design of antimicrobial agents targeting the cell wall machinery. The key enzymes of the periplasmic phase of cell wall assembly and remodeling are the penicillin-binding proteins (PBPs). As follows from their name, these enzymes are the target of the β-lactam class of antibiotics (including the penicillins, cephalosporins, and carbapenems). The majority of the PBPs are membrane-bound proteins, anchored to the outer surface of plasma membrane by a transmembrane anchor domain. The PBPs are a structurally related but mechanistically diverse ensemble of enzymes that catalyze the final transglycosylation, transpeptidation, and dd-carboxypeptidation reactions of cell wall synthesis.

The PBPs divide between two subclasses. The first class is the high-molecular-weight (HMW) PBPs. These are multidomain (containing membrane binding, transglycosylase, and transpeptidase domains) enzymes. Based on the particular catalytic activity of *N*-terminal domain, the HMW PBPs subdivide into two classes, A and B[77, 82] (Figure 13.5). Class A HMW PBPs, exemplified by PBP1a and PBP1b of *Escherichia coli*, have both transglycosylation and transpeptidation activities. These two PBPs catalyze sequential formation of the glycan strands (transglycosylation) of the peptidoglycan, followed by the cross-linking of these strands by transpeptidation of the stem peptide. Lipid II acts as both the glycosyl donor and the glycosyl acceptor for the progressive elongation of the glycan strands by transglycosylase domain catalysis.[256] Class B HMW PBPs, exemplified by the *E. coli* PBP2 and PBP3, possess transpeptidase activity while having a non-penicillin-binding domain with unknown function.

The LMW PBPs have DD-carboxypeptidase (*Escherichia coli* PBP5, PBP6, and PBP6b) and DD-endopeptidase (PBP4, PBP7, MepA) activities. The LMW PBPs control the degree of cross-linking of the peptidoglycan by hydrolysis of the D-Ala–D-Ala bond of the stem peptides, and participate in murein degradation and recycling by hydrolytic cleavage of the peptide cross-bridges. The number of PBPs and the relative distribution of these PBPs between the HMW and LMW classes are quite variable among bacteria. Nonetheless, the peptidase domains — whether domains having transpeptidase or carboxypeptidase activity — of all PBPs have an active site Ser-X-X-Lys (SXXK) tetrapeptide, the signature sequence of the serine acyl transferase superfamily. The serine is the catalytic residue used for acylation, and for acyl transfer to either water (carboxypeptidase, endopeptidase) or to a peptide amine (transpeptidase). This same serine is acylated by the β-lactam antibiotics during PBP inactivation by these antibiotics.[69, 118, 186] The lysine of this tetrapeptide is the catalytic base that assists serine engagement in these reactions (acylation by the peptide stem and by the β-lactam antibiotics, and acyl group transfer of the peptide stem).

Organism	Class	Gene	Protein	Function, expression	Localization (method[s])
B. subtilis	A	*ponA*	PBP1a/b	Cell division-specific TG/TPase veg	Septal (IF, GFP)
		pbpD	PBP4	Not known, veg	Distributed along membrane with distinct spots at periphery (GFP)
		pbpF	PBP2c	Synthesis of spore PG, veg, late stages of spo	Distributed along membrane, redistributed to prespore during sporulation (GFP)
		pbpG	PBP2d	Synthesis of spore PG, spo	Distributed along membrane (GFP); redistributed to prespore during sporulation
	B	*pbpA*	PBP2a	Synthesis of lateral wall, veg	Evenly distributed along the membrane (GFP)
		pbpH	PbpH	Synthesis of lateral wall, veg	Evenly distributed along the membrane (GFP)
		pbpB	PBP2b	Cell division-specific TPase, veg, spo	Septal (IF, GFP)
		pbpC	PBP3	Not known, veg, low expression during spo	Distinct foci and bands at cell periphery (GFP)
		spoVD	SpoVD	Synthesis of spore PG, spo	Not known
		pbpI	PBP4b	Not known, spo	Evenly distributed along the membrane (GFP)
	LMW carboxypeptidase	*dacA*	PBP5	Major DD-carboxypeptidase	Distributed along membrane with distinct spots at periphery (GFP)
		dacB	PBP5*	Control of peptide cross-linking in spore PG, spo	Not known
		dacC	PBP4a	Not known, late stationary phase	Distinct foci and bands at cell periphery (GFP)
		dacF	DacF	Control of peptide cross-linking in spore PG, spo	Not known
	LMW endopeptidase	*pbpE*	PBP4*	Not known, spo	Distinct foci and bands at cell periphery (GFP)
		pbpX	PbpX	Not known, veg	Septal, spiral outgrowth to both asymmetric septa during sporulation
E. coli	A	*ponA*	PBP1a	General PG synthesis	
		ponB	PBP1b (α, β, γ)	General PG synthesis	
		pbpC	PBP1c	Functions as a TGase only; binds to PBPs1B, −2, and −3 and MltA	
	B	*pbpA*	PBP2	Elongation-specific TPase	Spot-like pattern along lateral membrane, division site (GFP)
		pbpB or *ftsI*	PBP3	Cell division-specific TPase	Division septum (IF, GFP)
	LMW carboxypeptidase	*dacA*	PBP5	Control of cell shape	
		dacC	PBP6		
		dacD	PBP6b		
	LMW endopeptidase	*dacB*	PBP4	Control of cell shape in concert with PBP5	
		pbpG	PBP7	Control of cell shape in concert-with PBP5	
		mepA	MepA	Penicillin-insensitive endopeptidase	
	β-Lactamase	*ampC*	AmpC	Affects cell shape	
		ampH	AmpH	Affects cell shape	
S. aureus	A	*pbp2* or *pbpB*	PBP2	Cell division	Division septum (IF, GFP)
	B	*pbpA*	PBP1	Unknown	Not known
		pbpC	PBP3	Unknown	Not known
		mecA	PBP2A or PBP2′	Protein from extraspecies origin that confers beta-lactam resistance	Not known
	LMW transpeptidase	*pbpD* or *pbp4*	PBP4	Secondary cross-linking of PG	Not known
S. pneumoniae	A	*pbp1a*	PBP1a	Cell division	Septal
			PBP1b	Unknown	Septal or equatorial
		pbp2a	PBP2a	Peripheral cell wall synthesis	Equatorial
	B	*pbp2x* or *pbpX*	PBP2x	Cell division	Septal
		pbp2b	PBP2b	Peripheral cell wall synthesis	Equatorial
	LMW DD-carboxypeptidase	*dacA*	PBP3	Regulates cross-linking degree; coordination of the division process	Evenly distributed in both hemispheres and absent from the future division site

FIGURE 13.5 PBPs of gram-positive and gram-negative bacteria. (Reproduced from *Microbiol. Mol. Biol. Rev.*, 69:585–607, 2005. With permission.)

The probable specific functions of the individual PBPs have been assigned by gene knockout studies[6, 168] and *in vitro* experiments with homogeneous proteins.[97, 204] The loss of function in PBP knockouts, and in PBPs selectively inactivated by particular β-lactams, have also suggested function by correlation to altered bacterial shape. Although these methods are not uninformative, the interpretation of these observations has the serious limitation that most of the HMW PBPs are essential, and their loss of function is lethal. Therefore, the best method for the study of the function of these proteins is the use of chemically synthesized substrates, or substrates isolated from the bacterium, with the homogenous PBP.[10, 204]

Analysis of the reaction products is done using similar HPLC-MS methods, similar to those used for muropeptide analysis. A good example of this approach is the study of *Escherichia coli* PBP1b by Bertsche et al.[10] and Schwartz et al.[203, 204] Their experiments demonstrated that Lipid II first assembles into chains of various length by transglycosylase-catalyzed, Lipid II-dependent, β-1,4–linked glycan strand formation, which are subsequently cross-linked. The reason for transglycosylation preceding transpeptidation is suggested by the structure of the HMW PBP. It is probable that the rate of transpeptidation is regulated, perhaps in allosteric fashion, by the elongating peptidoglycan strand.[71]

The transpeptidation cross-linking reaction itself occurs in different patterns. The most common cross-linking found in *Escherichia coli* (and many other bacteria) is a 4→3 cross-linking[81] or D,D-transpeptidation,[100] wherein cross-linking occurs between the carboxy-terminus of the D-alanine in position-4 of one stem peptide and the lateral amino group at position-3 of the second strand (Figure 13.6A). Alternatively, the residue in position-3 (either *meso*-diaminopimelic acid or L-lysine) may cross-link. These D,D-transpeptidation reactions are blocked by β-lactam PBP inactivation. A second type of cross-linking is that of 3→3 interpeptide transpeptidation (Figure 13.6B),[81] referred to as L,D-transpeptidation.[100] Originally discovered in some *Streptomyces*, *Clostridium*, and *Mycobacterium* species, 3→3 cross-linking was later detected in *Escherichia coli*. This type of cross-bridge is more typical for *E. coli* treated with β-lactam antibiotics and multidrug resistant strains. The extent of 3→3 cross-linking also depends on the growth stage, changing in *E. coli* from 1% in log phase to 45% in late log phase. The insensitivity of the Enterobacteria to the β-lactam antibiotics correlates with their ability to construct their cell wall with 3→3 cross-linking.[18, 141, 143]

One of the most important unanswered questions is the reason for the existence of multiple PBPs with nearly identical enzymatic functions. A suggested explanation is specialization among the PBPs, wherein different PBPs are used at different stages of peptidoglycan synthesis, remodeling, and degradation that occur during the processes of cell growth, septation, and sporulation.[5] Fluorescent microscopy of labeled PBPs indicates that they localize in specific places,[59, 199] and interact with specific cytoskeleton proteins,[2, 3, 36, 37, 91, 181, 232, 234, 243] consistent with specialized roles for these PBPs.[198, 200, 236] The cytoskeleton machinery that transports the PBPs to "construction sites" can spatially separate PBPs with similar enzymatic functions.[182] These same kinetic experiments demonstrate different substrate preferences for different PBPs.[184] The growing evidence on the multienzymatic/multisubunit nature of peptidoglycan synthesis[100, 178, 200] is also consistent with the likelihood that PBPs possess different *in vivo* roles despite similar *in vitro* activities.

The likelihood that peptidoglycan biosynthesis occurs in concert with the creation of the membrane(s) of the cell envelope, and involves a multi-enzyme "hyperstructure"[36, 172] that includes the enzymes and proteins of the cytoskeleton and of Sec protein translocation,[33, 109, 208] is further suggested by a helical motif within the cell envelope of the rod-shaped bacterium. Direct evidence for helical domains to the peptidoglycan sacculus, corresponding to perhaps eight to ten turns over the length of the bacterium, was reported by Tiyanont et al.[222] in *Bacillus subtilis* using fluorescently labeled vancomycin and ramnoplanin, antibiotics that bind directly to peptidoglycan peptide stems.[66] Ghosh and Young, using fluorescent labeling of the outer membrane of *Escherichia coli* during cell growth, observed a helical disposition for the LPS and for the proteins of the outer membrane.[76] Helical patterns are seen in *E. coli* for chemoreceptor[219] location on the outer membrane.[208, 255] Whether these domains can mirror the organization of the cytoskeleton

Mycobacterium, Corynebacterium, Bacillus, E.coli

Streptomyces, Clostridium

E. faecium, E. hirae

S. pneumoniae, E. faecalis, S. aureus

Mycobacterium, E. coli

Streptomyces, C. perfringens

E. faecium

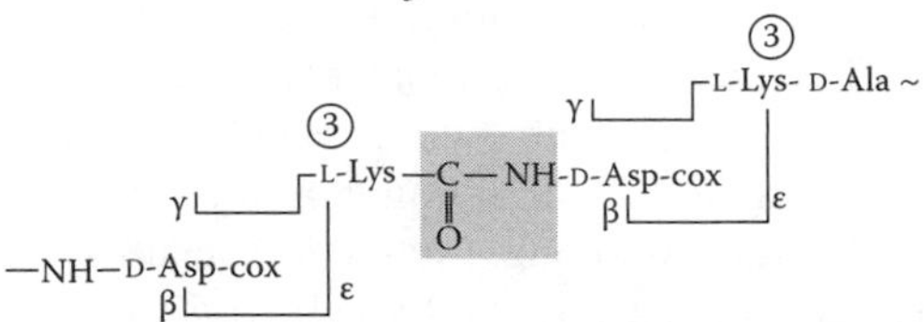

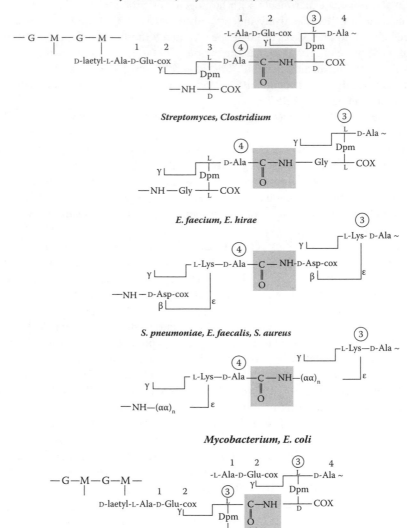

FIGURE 13.6 (A) Types of cross-linking in murein, and (B) types of cross-linking in murein. (Reproduced from *Microbiol. Mol. Biol. Rev.*, 66:702–738, 2002, with permission.)

remains an open and intriguing question.[180, 198, 213] While it is tempting to suggest, on the basis of the study by Tiyanont et al.,[222] that the boundaries of the helices correlate with a different pattern of peptidoglycan cross-linking, direct visualization of the sacculus has yet to provide evidence in support of domains, whether helical or otherwise.[145, 158]

THREE-DIMENSIONAL STRUCTURE OF PG

Although it is obvious that the cell wall peptidoglycan has a three-dimensional structure, there is no direct experimental data revealing the overall cross-linking pattern among the NAG-NAM glycan strands. Isolation of the bacterial peptidoglycan is done by the trivial operations of boiling the bacterial cell pellet in 4% sodium dodecyl sulfate, recovery of the murein sacculi by centrifugation, fractionation, and enzymatic removal of the nucleic acids and cell wall associated proteins.[55] Structural analysis of this polymer, however, is extraordinarily difficult. Degradation of the polymer indicates that despite large similarity in overall structure, the distribution of glycan strand length and pattern of stem cross-linking vary not just with the bacterium, but also as a function of cell growth and the nature of environmental stressors. As a consequence of its heterogeneity, the murein polymer is not crystalline and is therefore not capable of x-ray crystal structure determination. NMR analysis is similarly precluded. Atomic force microscopy (AFM) analysis, while consistent with the structural heterogeneity, is not informative as to molecular-level structure.[226, 248] Cell cryomicroscopy and tomography, although extremely useful for the determination of the sub-structure dimensions of the cell envelope, are likewise uninformative as to molecular structure. Four methods dominate current efforts to probe the detailed chemical structure of the peptidoglycan. The first is refinement of the classical method of radiolabeled biosynthetic precursor incorporation. The second is fluorescent imaging using specific probes of peptidoglycan structure. The third is the complete chemical synthesis of homogeneous peptidoglycan substructure, to ascertain higher-order structural patterns. Finally, the experimental data obtained from all these methods are used as constraints toward the computational simulation of the glycan and cross-linked stem arrangement of the peptidoglycan.

A critical value in the determination of the cell wall structure is the amount of the NAG-NAM in the cell wall. The answer was obtained by Wientjes et al.[246] by measurement of [^{3}H]-labeled *m*-diaminopimelic acid (*m*-Dap) incorporation into the *Escherichia coli* cell wall. Because diaminopimelic acid is found exclusively in the peptidoglycan, the number of *m*-Dap molecules incorporated equals the number of NAG-NAM disaccharide segments. The value obtained was 3.5×10^6 molecules, equivalent to a mass of 2.3×10^9 Da for the sacculus. Given the estimated value for the surface area of the *E. coli* bacterium of 10×10^6 nm^2, this value is consistent with a preponderance of a thin ("monolayered") cell wall. The conclusion about the relative thinness of the murein of the *E. coli* cell wall was confirmed by Labischinski et al. by small-angle neutron scattering.[130] According to these data, 75 to 80% of the peptidoglycan was interpreted as single layered, and 20 to 25% as triple layered.

Although the term "layer" is often used to describe the peptidoglycan, it is a term with uncertain definition. The common presumption is that the peptidoglycan layer is a single (NAG-NAM)$_n$ glycan strand located within the periplasm. A further presumption implicit in this structural formulation of the term "layer" is that the orientation of the glycan is parallel to the membrane surfaces. This presumption is possible but is by no means proven. The glycan chains may form planes ("layers") that are parallel (the classic peptidoglycan model), orthogonal, diagonal, or spiral with respect to the membrane surface. It is not possible to describe a peptidoglycan "layer," and without such a description it is not possible to formulate a description of the structural boundaries that define the cross-linked peptidoglycan.

The most common representation of the cell wall is the planar model.[31, 183] This model postulates the existence of layers that have adjacent glycan strands that run parallel to the cell surface. The stem peptides protrude above, below, and to the sides of the glycan strands, with some of these peptides forming cross-links connecting neighboring glycan strands.[31, 183] The peptides protruding above and below the plane of the glycan layer are mostly free and need not form cross-links, consistent with the experimentally measured cross-linking of less than or equal to 50%. Based on considerations of the possible growth mechanisms, the glycan strands in this model may be placed perpendicular to the long axis of the cell. Computer simulation of this type of cell wall also predicts a representative size of the pores in the peptidoglycan of approximately 2 by 6 nm.[183] This parameter corresponds well with the pore diameter estimated experimentally by Demchik et al.[58]

The planar model also explains the elasticity of the cell wall, which is mostly attributed to stretchable peptide cross-links. Another strong argument for this model is the fact that this model allows for the experimentally estimated 3.5×10^6 NAG-NAM units to cover the entire surface of the cell.[31, 238] Koch[119, 123] proposed a modification of this model. He believes that under mechanical stress, the peptidoglycan strand should look like a zigzag line. This type of structure would better explain such mechanical properties of sacculus as the elasticity in different directions that is observed experimentally.[248] Structural studies of bacterial peptidoglycan unquestionably demonstrate that the *in vivo* structure of peptidoglycan is not perfect in the sense that the degree of cross-linking and the length of the glycan strands have a high degree of variability. The planar model allows (at least in theory) for those imperfections in the cell wall structure without a significant loss of mechanical strength. The multilayer areas detected by the neutron scattering experiment in this case can be attributed to the cell wall growth areas. There are a few different mechanisms proposed for the growth of this type of cell wall, as discussed later.

An alternative model for the tertiary peptidoglycan structure was proposed by Dmitriev et al.[62, 63] They argue that a less than 100% degree of peptidoglycan cross-linking would render the planar structure dysfunctional. Based on computer simulation, they argue for a model where the glycan strands extend perpendicular to the plasma membrane. Peptide stem cross-linking creates a scaffold-like array, and hence their proposed arrangement is called the scaffold model. The peptidoglycan is suggested to have a fourfold helical conformation, with each turn of the helix being made of four disaccharide units and with the four peptide side chains oriented outward and perpendicular to each other. Each pair of peptides belonging to two adjacent strands may cross-link if they are located in the same plane. Each strand may cross-link with four others. The scaffold glycan strand arrangement allows even short glycan chains with less than complete cross-linking to productively incorporate into a stress-bearing zone of the murein matrix. Dmitriev et al.[62, 63] interpret the experimental data for the sacculus mass as being consistent with their scaffold model. A contrary opinion is given by Vollmer and Höltje.[238] They argue that in the maximally stretched scaffold arrangement, the peptidoglycan would cover no more than 40% of the cell surface area.

The latest perspective on the peptidoglycan structure was proposed by Mobashery et al.[70, 157] This model is based on the NMR structure of a synthetic $(NAG\text{-}NAM)_2$ muropeptide. The tertiary structure of this muropeptide is a right-handed helical glycan. This same conformation is preserved in the complex between this synthetic muropeptide and a human peptidoglycan recognition protein.[42] Extrapolation of this structure to a longer glycan strand indicated threefold (three NAG-NAM disaccharide units) periodicity to the helix. This value is different from the value — of four NAG-NAM units per helical turn — used for all previous peptidoglycan models.[122, 183] Given the necessity for the peptidoglycan to embed many macromolecules of the periplasm and the outer membrane,[169, 241] this model suggests the perpendicular orientation of the glycan strands to the membrane surface, as suggested by the scaffold model. Each peptidoglycan strand may connect with up to three neighboring strands, resulting in a honeycomb pattern having the intrinsic porosity necessary for protein entry.

All the peptidoglycan models have their own advantages and disadvantages.[252] No experimental data are as yet decisive to favor one model over another.

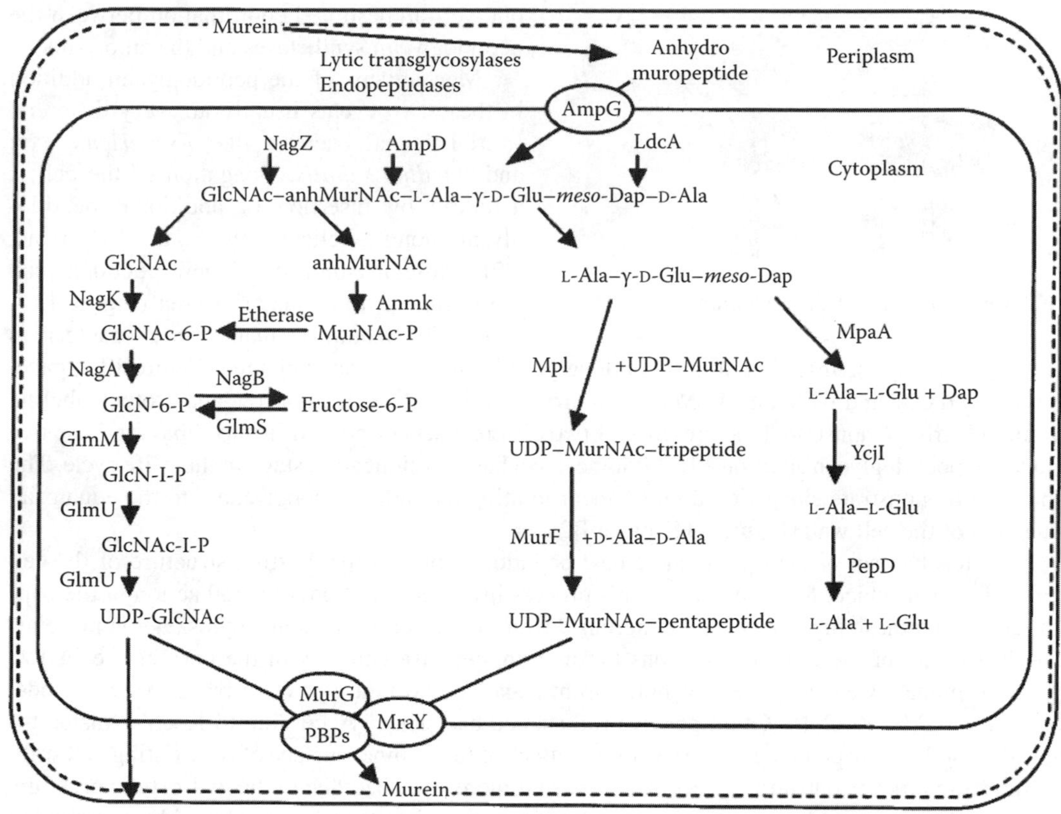

FIGURE 13.7 Murein turnover in Gram-negative bacteria. (Reproduced from *J. Bacteriol.*, 187:3643–3649, 2005. With permission.)

CELL WALL AS A DYNAMIC STRUCTURE

Constant remodeling of the bacterial cell wall is necessary to accommodate the processes of cell growth, division, and sporulation. Cell wall remodeling involves the controlled degradation and recycling of old murein, concurrent with the synthesis of new peptidoglycan. *Escherichia coli* is estimated to remodel 40 to 50% of its murein per generation.[84, 230] Because so much of the cell wall material undergoes remodeling, bacteria have developed very efficient systems for recycling the murein materials.[174] Approximately 90% of the turnover material is recovered and reinserted into the cell wall.[230] The main steps of this process are presented in Figure 13.7.

The process of murein remodeling requires specific murein hydrolyzing enzymes, called autolysins. This diverse group of enzymes includes lytic transglycosylase, DD-endopeptidase, carboxypeptidase, *N*-acetylmuramidase (lysozyme), acetylglucosaminidase, and *N*-acetylmuramyl-L-alanine amidase activities.[9, 21, 111] There are 18 known murein hydrolases in *Escherichia coli*.[242] The large number of hydrolases and the similarity of their enzymatic reactions result in redundancies of function.[96] As a result, the traditional method of gene knockout to assign function has experienced limited success. Moreover, the fact that the cell wall does not spontaneously undergo degradation by these enzymes implies that their activities are strictly regulated. This presumption has experimental evidence.[194] The soluble lytic transglycosylase Slt70 autolysin of *E. coli* selectively binds PBP3, PBP7, and PBP8. In this complex, the lytic transglycosylase activity of Slt70 is both stabilized and stimulated. Multiplicity of the murein hydrolases implies the presence of multiple regulatory mechanisms controlling the activity of these enzymes. Although not much information on this subject is available, processes such as cell growth and septum formation are unimaginable with-

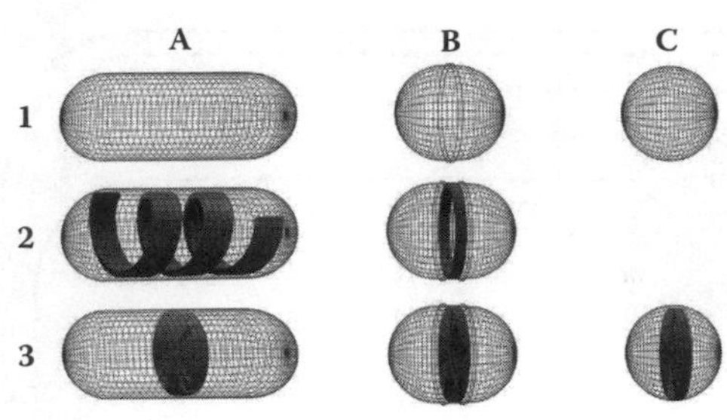

FIGURE 13.8 Growth and septation.

out well-orchestrated coordination between the peptidoglycan synthetases and the autolysins.

Mechanisms of the peptidoglycan addition in these two events usually are very different. In rod-shaped bacteria like *Escherichia coli* and *Bacillus subtilis,* elongation of the cell is achieved by insertion of additional peptidoglycan along a helical path (Figure 13.8, panel A2), while in some cocci new peptidoglycan is inserted in the equatorial area (Figure 13.8, panel B2). Septum formation in all bacteria is accomplished by the insertion of new peptidoglycan in the equatorial ring area (Figure 13.8, panel A3, B3, and C3), and AFM and TEM data confirm this.[34, 226, 257] Gene knockout and protein labeling studies clearly demonstrated the presence of two different sets of proteins responsible for these two modes of peptidoglycan addition.[6, 199, 211] Some cocci have no elongation stage in their life cycle. The ones that demonstrate elongation do it by incorporating the new peptidoglycan into the equatorial ring area of the cell wall (Figure 13.8, panel B2).

The mechanism of incorporation of new peptidoglycan into the existing structure of the cell wall remains a subject for speculation. This process involves a well-coordinated action of the peptidoglycan-synthesizing enzymes, the murein-degrading proteins, and some cytoskeleton proteins. The last group of the proteins positions the murein synthase complex in the correct site. A few different models were proposed for both Gram-negative and Gram-positive bacteria. One model was proposed by Koch for Gram-positive rod-shaped bacteria.[120] According to Koch's "inside-to-outside" model, new peptidoglycan strands are attached to the inner surface of the existing cell wall. As the strand moves outward, it becomes more and more stretched due to the hydrostatic pressure. Almost all the flexibility in murein comes from the peptides; therefore the peptides connecting neighboring strands of peptidoglycan stretch and become more susceptible to autolysins and the murein layer undergoes degradation.

Another growth model was proposed by Höltje[100] for *Escherichia coli*. According to his "three-for-one" model, new material is attached to the existing murein, allowing insertion of the new material into the stress-bearing layer without compromising its integrity. As follows from its name, for each old (docking) strand that is removed, three new strands are inserted (Figure 13.9). This pattern

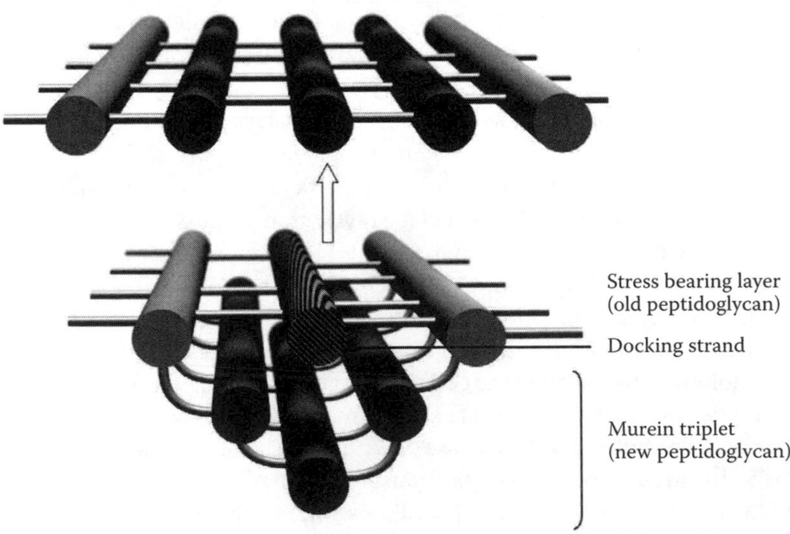

FIGURE 13.9 "Three-for-one" mechanism.

of murein growth necessarily requires coordination between the enzymes of peptidoglycan synthesis and those (the autolysins) of degradation. To accommodate this need, a hypothetical "murein synthase" multienzyme complex having both types of enzymes was proposed. The existence of such a complex was suggested by experimental data,[194] wherein interaction between the subunits was observed by plasmon resonance.[11] Possible enzymatic activities within such a complex might include those of a lytic transglycosylase, an endopeptidase, a transpeptidase, and a bifunctional transpeptidase-transglycosylase, transglycosylase. This synthase complex is anchored to both of the membranes as the HMW PBPs are bound to the outer side of the plasma membrane and the Mlt proteins are known to be anchored on the inner side of the outer membrane.[137] The complex is directed to the site of septum formation by the Z-ring formed by the tubulin-like cytoskeleton protein FtsZ.[2, 198] Z-ring formation involves multiple proteins that stabilize and anchor the ring to the plasma membrane. It is believed that the Z-ring pulls the plasma membrane inward. The cell wall enzymes would follow membrane invagination, eventually completing the cell wall septum necessary for daughter cell separation. Later, autolysin degradation through the middle layer of the septum would result in separation of the daughter cells.[150]

Cell growth and division are the major reasons for murein remodeling. In addition to these two events, the cell wall must accommodate other processes. Each also requires cell wall remodeling. One such event is sporulation. Bacteria from genera *Bacillus* and *Clostridium* form endospores in order to survive environmental stress. These endospores are among the most resilient of biological structures, able to withstand extreme temperatures, lack of water, and other environmental extremes. It is intuitively logical that the drastic difference in relative survivability, as exists between vegetative cells and the endospores, requires substantive changes in the murein structure.

Most of the information on endospore murein composition has been obtained with the Gram-positive bacterium *Bacillus subtilis*. The analytical method to evaluate the differences in the murein composition was HPLC isolation of muropeptides for MS evaluation of structure.[4] This analysis showed that the endospore murein is different from the vegetative cell murein in many ways. The most prominent changes are a reduction in the degree of cross-linking in the spore peptidoglycan due to removal of the peptide chains, and the absence of teichoic acid.[4, 187] The observed percentage of cross-linking in the endospores was only 2.9%, 11-fold less than that of the vegetative cell wall.[4] This reduction is attributed to complete removal of the peptide chains in 50% of the NAM saccharides, with their conversion to a muramic-δ-lactam structure.[25, 73, 74, 103, 187] Of the remaining NAM saccharides, 24% have their peptides truncated to single L-Ala residues, thus preventing peptide cross-link formation. Muramic-δ-lactam is a marker for germination lytic enzymes. The purpose of the other changes in the murein during sporulation is still debated.

The changes in the murein structure described above are not limited to sporulation. Throughout the life cycle of the cell, the peptidoglycan continuously undergoes modifications and alterations. Different characteristics of the cell wall can be attributed to age of the cell[79, 165] and stage of the life cycle.[28, 187] Some parasitic bacteria modify their cell wall to minimize or neutralize the immune response of the host organism during invasion. *Listeria monocytogenes*, one such organism, is resistant to the host bacteriolytic enzyme, lysozyme. Lysozyme is an *N*-acetyl-muramoylhydrolase that is found in many organisms. The ability of lysozyme to hydrolyze the glycosidic bond between the NAG and NAM saccharides of the peptidoglycan accounts for its bacteriocidal activity.[148] The structural basis for *Listeria* resistance to lysozyme was discovered upon analysis of the *Listeria* murein by Boneca et al.[28] Approximately one-half of the saccharides of the peptidoglycan of this pathogen are *N*-deacetylated. The *N*-deacetylase enzyme responsible for this peptidoglycan modification was identified. In addition to rendering *Listeria* resistant to lysozyme, this modified peptidoglycan is also poorly recognized by macrophages, the main producers of lysozyme.[227] Similar peptidoglycan modifications helping pathogenic bacteria to evade the immune system of the host organism are found in many other Gram-positive and Gram-negative microorganisms.[148]

Changes in the cell wall induced by the presence of antimicrobial agents in the environment are another example of the adaptation of this structure, as noted earlier in this chapter. Changes in physical factors, such as humidity or temperature, lead to changes in cell wall composition. Antibiotics that target the enzymes of cell wall biosynthesis also induce structural changes. One of the most recent examples of adaptability of bacterial cell wall biosynthesis machinery is the emergence of vancomycin-resistant *Staphylococcus aureus* (VRSA). The first *S. aureus* strain resistant to vancomycin was reported in 1996 in Japan; soon after, strains with the same antibiotic resistance profile were reported in many other countries. The danger of VRSA cannot be underestimated, as vancomycin is the antibiotic of choice for the treatment of the increasingly common nosocomial pathogen, methicillin-resistant *S. aureus* (MRSA). Emergence of a resistance mechanism to vancomycin would render vancomycin ineffective against these pathogens. Soon after the first reports of VRSA, microbiologists found that these strains had a thicker cell wall than vancomycin-susceptible *S. aureus* (VSSA). An average cell wall of 16 VRSA strains isolated from seven countries was 31.3 ± 2.6 nm, while in VSSA strains this parameter was 23.4 ± 1.9 nm.[50, 98] The increased thickness of the peptidoglycan layer could result from any number of events in cell wall biosynthesis. Although the genetic determinant for the VRSA phenotype is not yet elucidated, it is evident that the simple change to a thicker cell wall renders vancomycin ineffective against VRSA strains. Vancomycin inhibits cross-linking of bacterial peptidoglycan by binding D-alanyl-D-alanine residues of the peptidoglycan strands. In VRSA stains, an increase in the amount of peptidoglycan results in affinity trapping of vancomycin and clogging of the peptidoglycan mesh, preventing the antibiotic from penetrating deeper into the cell wall.

An altogether different basis for vancomycin resistance has emerged in *Enterococcus faecium* and *Enterococcus faecalis*.[49, 60] In vancomycin-resistant strains of these species, the terminal part of the murein peptide chains was modified. The terminal D-Ala-D-Ala sequence was replaced with D-Ala-D-Lac, or D-Ala-D-Ser, with a drastic reduction in the stability of the noncovalent vancomycin-peptidoglycan complex.[191] These modified peptidoglycans account for the self-resistance on the part of the microorganisms that biosynthesize these glycopeptides.[7] Indeed, an intimate connection between the auto-resistance mechanisms used by antibiotic-producing microorganisms, and the ensuing development of antibacterial resistance is probable.[170] The introduction of the D-Ala-D-Lac moiety into the cell wall is made possible by a cluster of five genes, with a key step performed by the product of the *vanA* gene. The lateral transfer of the *vanA* gene to *Staphylococcus aureus*, specifically to MRSA, took place in 2002 for the first time. Vancomycin-resistant MRSA is one of the most notorious pathogens, for which treatment options are very limited.

CELL WALL BIOSYNTHESIS ENZYMES AS TARGETS FOR ANTIBIOTICS

The cell wall continues as a target for antibiotic discovery. Because the integrity of the cell wall is vitally important to bacteria, many of the enzymes of murein biosynthesis are targets for the chemotherapeutic control of bacterial infection. Exhaustive efforts to identify these targets, and to optimize the efficacy of drugs targeted to these enzymes, initiated the antibacterial era in the middle of the 20th century. Hundreds of cell wall biosynthesis inhibitors now are used for antimicrobial chemotherapy. These antibiotics (when isolated from Nature) and antibacterials (when prepared by chemical synthesis) have saved millions of human lives. Currently, most of the antibiotics and antibacterials used in clinical practice target an aspect of cell wall biosynthesis. The main reasons for the diversity of antimicrobial agents targeting the murein, and the proteins involved in its biosynthesis, relate to clinically important differences relating to structure and target among the many Gram-negative, and Gram-positive, bacteria.

The antibiotics that target bacterial cell wall biosynthesis form two major groups: (1) transpeptidation inhibitors and (2) transglycosylation inhibitors. All β-lactam antibiotics (penicillins, cephalosporins, carbapenems, and monobactams) and some glycopeptide antibiotics (vancomycin) inhibit transpeptidation. The β-lactams and vancomycin inhibit transpeptidase-dependent stem cross-linking by different mechanisms. The β-lactams mimic the D-alanyl-D-alanine terminus of the peptide stem and acylate the active site serine.[81, 221] In contrast, vancomycin noncovalently binds to this same D-alanyl-D-alanine terminus. This intermolecular complex is sufficiently stable to prevent the use of the muropeptide as a substrate by the transpeptidase.[190] Bacitracin is another antibiotic that acts by inhibition of cell wall biosynthesis. It interferes with the dephosphorylation of the C_{55}-prenol pyrophosphate of Lipid II by forming a complex with this cell wall precursor,[129, 214] thus inhibiting murein biosynthesis at an early stage. In addition to these three classes of inhibitors of murein biosynthesis, there are many other inhibitors (and discovery strategies) being evaluated for antibacterial discovery.[23, 45, 155, 164, 202, 210, 235, 245] Because synthesis of the bacterial cell wall involves so many enzymes, there are many possible points for disruption of cell wall biosynthesis.

PEPTIDOGLYCAN INTEGRITY AS A RESPONSE STIMULUS

The perception of the peptidoglycan as a simple, albeit essential, structural polymer is now understood to be egregiously simplistic. Although the processes used by the bacterium to sense — and to respond to — compromise in the structural integrity of its peptidoglycan are only now being uncovered, it is already evident that peptidoglycan integrity directly correlates to gene expression and ultimately to mutation, horizontal gene transfer, and virulence factor expression. Not surprisingly, much of the evidence for these bacterial responses to compromise in the integrity of their peptidoglycan derives from observation of the bacterial response to antibiotic (notably β-lactam) exposure. These responses may be divided, for the purpose of this short discussion, between those relating to stress and those relating to resistance determinants. The fundamentals — although certainly not all the details — of the sensing process used by β-lactam-resistant Gram-positive bacteria are in place. Signal transduction is initiated by covalent acylation by the β-lactam[19, 83, 247] of cell surface sensor proteins.[8, 72, 114] Cross-membrane signal transduction is induced as a consequence of this acylation,[40, 147, 220] initiating expression of the genes for β-lactamase and/or the relatively β-lactam-insensitive PBP2a transpeptidase.[71, 92, 117, 196] In contrast to the Gram-positive bacteria, β-lactam sensing to initiate expression of resistance determinants (primarily β-lactamase expression) in Gram-negative bacteria occurs upon detection of an imbalance in muropeptide availability.[105, 106, 125, 126, 171, 215, 229, 244] Neither the molecular basis of the imbalance nor the rudiments of the signal transduction is well understood. Moreover, it is now evident that *both* Gram-positive and Gram-negative bacteria also respond to disrupted peptidoglycan integrity by yet other responses.

The first of these responses, the SOS response, is found in varying capacities among all bacteria (it is considered a robust response in *Escherichia coli*, *Pseudomonas aeruginosa*, and *Bacillus subtilis*, and as less robust in *Staphylococcus aureus*).[116] As the SOS response is regarded as primarily a response that enables DNA repair and damage tolerance to genotoxic agents such as fluoroquinolones,[20, 43, 44, 108, 228] the ability of β-lactams to induce the SOS response in bacteria may be regarded as surprising. In *S. aureus*, β-lactams induce the LexA/RecA proteins, trigger prophage induction, and effect high-frequency horizontal transfer of pathogenicity islands.[144, 228] Genetic exchange is also induced by aminoglycosides, fluoroquinolones, and mitomycin *c* in *Streptococcus pneumoniae* (a bacterium regarded as lacking the SOS response).[189] In Gram-negatives, the transcription profile resulting from ceftazidime inhibition of *P. aeruginosa* PBP3 was different, and having only partial overlap, from the transcription profile induced by ciprofloxacin.[26] PBP3 inactivation in *E. coli* results in DpiBA two-component signal transduction induction of both LexA/RecA and (error-prone) DNA polymerase IV activity, and the transient halt of cell division.[159] A complementary study in *E. coli* did not observe the LexA/RecA dependency but confirmed a DNA polymerase IV-dependent increase in the reversion of a +1 Lac frameshift mutation.[179]

Recognition that a second loss-of-peptidoglycan integrity-induced response is induced in many Gram-positive bacteria, and also in the mycobacteria, recently has emerged. Yeats et al. describe the sensors involved in this response as PASTA (for penicillin-binding protein and serine-threonine kinase associated domain)-containing proteins,[110, 249] exemplified by the mycobacteria PknB kinase.[27, 86, 88, 166, 173, 240, 253] The PknB kinase is found in a highly conserved actinobacteria cell wall and cell shape operon,[113] and its expression is essential for mycobacterial cell growth.[67] The PknB structure consists of an intracellular eukaryote-type ser/thr kinase domain, a transmembrane domain, and a fourfold repeat of the PASTA extracellular domains. The tertiary structure of these domains is known from the appearance of PASTA domains[110, 249] also in PBP2x, the penicillin-resistant transpeptidase from *Strepococcus pneumoniae*.[61, 85] Dasgupta et al.[54] report that the PknB kinase phosphorylates the *Mycobacterium tuberculosis* PBPA, a HMW penicillin binding protein that is recruited to the division site and is required for growth and septation. Jones and Young[110] argue that the evidence strongly suggests a role for the peptidoglycan-binding PASTA domain in control of peptidoglycan remodeling not just in cell growth, but also sporulation and biofilm formation. A recent study on the PrkC one-component kinase of penicillin-resistant *Enterococcus faecalis* has broadened the role of these kinases to include modulation of antibacterial resistance and intestinal persistence.[127] *E. faecalis* lacking this kinase, when grown under stress, shows significant morphological defects in their cell wall. Moreover, this same mutant is highly sensitized to certain β-lactams (such as cefuroxime) but is much less sensitized to other β-lactams (such as ampicillin) and to other antibiotics (such has bacitracin and vancomycin). Kristich et al.[127] suggest that this differential sensitivity indicates a specific role for one PBP (*E. faecalis* PBP5) in PrkC-dependent activation of the resistance response.

Finally, there is also emerging evidence that muropeptide generation, resulting from hydrolytic modification of the peptidoglycan, plays a significant role in the activation of mycobacteria from host-evading dormancy.[115, 163, 217]

CONCLUSION

All of the surmises, dating from recognition of the PBPs as the biological targets of the β-lactam antibiotics, as to the importance of the peptidoglycan sacculus as a critical structure of the bacterial cell envelope have proved correct. The experiments that have provided this proof, however, simultaneously have proved incorrect the assertion that the peptidoglycan cell wall should be defined as a mere structural edifice. Rather, the peptidoglycan is a dynamic structure, in states of constant remodeling and regeneration, and one that is intimately connected to all the events of bacterial reproduction. The research vistas that today surround the peptidoglycan include roles — as either a nucleation point, or as a scaffold — within the bacterial hyperstructures of cytoskeletal organization and cell envelope synthesis; as a stimulus for signal transduction within the bacterium, and by the host during bacterial invasion; and as a structure of opportunity for further antibiotic discovery. As is true of all vibrant fields of scientific study, experimental study yields far more new questions than it resolves old questions. So it is with the peptidoglycan. The allure of the peptidoglycan, whether measured by importance or by opportunity, is undiminished.

REFERENCES

1. M. Ehrmann, Ed. 2007. *The Periplasm*, 1st ed., ASM Press, Washington, DC.
2. Aaron, A., G. Charbon, H. Lam, H. Schwarz, W. Vollmer, and C. Jacobs-Wagner. 2007. The tubulin homologue FtsZ contributes to cell elongation by guiding cell wall precursor synthesis in Caulobacter crescentus. *Mol. Microbiol.*, 64:938–952.
3. Arends, S.J.R., K.B. Williams, R.J. Kustusch, and D.S. Weiss. 2007. Cell division. In M. Ehrmann (Ed.), *The Periplasm*. ASM Press, Washington, DC., p. 173–197.
4. Atrih, A., P. Zollner, G. Allmaier, and S.J. Foster. 1996. Structural analysis of *Bacillus subtilis* 168 endospore peptidoglycan and its role during differentiation. *J. Bacteriol.*, 178:6173–6183.
5. Barak, I. and A. J. Wilkinson. 2007. Division site recognition in *Escherichia coli* and *Bacillus subtilis*. *FEMS Microbiol. Rev.*, 31:311–326.
6. Begg, K.J. and W.D. Donachie. 1985. Cell shape and division in *Escherichia coli*: experiments with shape and division mutants. *J. Bacteriol.*, 163:615–622.

7. Beltrametti, F., A. Consolandi, L. Carrano, F. Bagatin, R. Rossi, L. Leoni, E. Zennaro, E. Selva, and F. Marinelli. 2007. Resistance to glycopeptide antibiotics in the teicoplanin producer is mediated by van gene homologue expression directing the synthesis of a modified cell wall peptidoglycan. *Antimicrob. Agents Chemother.*, 51:1135–1141.
8. Berger-Bachi, B. 1999. Genetic basis of methicillin resistance in *Staphylococcus aureus. Cell Mol. Life* Sci., 56:764–70.
9. Bernadsky, G., T.J. Beveridge, and A.J. Clarke. 1994. Analysis of the sodium dodecyl sulfate-stable peptidoglycan autolysins of select Gram-negative pathogens by using renaturing polyacrylamide gel electrophoresis. *J. Bacteriol.*, 176:5225–5232.
10. Bertsche, U., E. Breukink, T. Kast, and W. Vollmer. 2005. *In vitro* murein peptidoglycan synthesis by dimers of the bifunctional transglycosylase-transpeptidase PBP1B from *Escherichia coli. J. Biol. Chem.*, 280:38096–38101.
11. Bertsche, U., T. Kast, B. Wolf, C. Fraipont, M.E.G. Aarsman, K. Kannenberg, M. von Rechenberg, M. Nguyen-Disteche, T. den Blaauwen, J.V. Holtje, and W. Vollmer. 2006. Interaction between two murein (peptidoglycan) synthases, PBP3 and PBP1B, in *Escherichia coli. Mol. Microbiol.*, 61:675–690.
12. Beveridge, T.J. 2006. Cryo-transmission electron microscopy is enabling investigators to examine native, hydrated structures in bacteria and biofilms. *Microbe*, 1:279–284.
13. Beveridge, T.J. 2002. The structure of bacterial surfaces and its influence on stainability. *J. Histotechnol.*, 25:55–60.
14. Beveridge, T.J. 2006. Understanding the shapes of bacteria just got more complicated. *Mol. Microbiol.*, 62:1–4.
15. Beveridge, T.J. 2001. Use of the Gram stain in microbiology. *Biotech. Histochem.*, 76:111–118.
16. Beveridge, T.J. 2006. Visualizing bacterial cell walls and biofilms. *Microbe*, 1:279––284.
17. Beveridge, T.J. and V.R. Matias. 2006. Ultrastructure of Gram-positive cell walls, In V.A. Fischetti, R.P. Novick, J.J. Ferretti, D.A. Portnoy, and J.I. Rood (eds.), *Gram-Positive Pathogens*, second ed. ASM Press, Washington, DC, p. 3–11.
18. Biarrotte-Sorin, S., J.E. Hugonnet, V. Delfosse, J.L. Mainardi, L. Gutmann, M. Arthur, and C. Mayer. 2006. Crystal structure of a novel β-lactam-insensitive peptidoglycan transpeptidase. *J. Mol. Biol.*, 359:533–538.
19. Birck, C., J.Y. Cha, J. Cross, C. Schulze-Briese, S.O. Meroueh, H.B. Schlegel, S. Mobashery, and J.P. Samama. 2004. X-ray crystal structure of the acylated β-lactam sensor domain of BlaR1 from *Staphylococcus aureus* and the mechanism of receptor activation for signal transduction. *J. Am. Chem. Soc.*, 126:13945–13947.
20. Bisognano, C., W.L. Kelley, T. Estoppey, P. Francois, J. Schrenzel, D. Li, D.P. Lew, D.C. Hooper, A.L. Cheung, and P. Vaudaux. 2004. A recA-LexA-dependent pathway mediates ciprofloxacin-induced fibronectin binding in *Staphylococcus aureus. J. Biol. Chem.*, 279:9064–9071.
21. Biswas, R., L. Voggu, U.K. Simon, P. Hentschel, G. Thumm, and F. Gotz. 2006. Activity of the major staphylococcal autolysin Atl. *FEMS Microbiol. Lett.*, 259:260–268.
22. Bjelic, S., A. Karshikoff, and I. Jelesarov. 2006. Stability and folding/unfolding kinetics of the homotrimeric coiled coil Lpp-56. *Biochemistry*, 45:8931–8939.
23. Black, M.T. and J. Hodgson. 2005. Novel target sites in bacteria for overcoming antibiotic resistance. *Adv. Drug Deliver. Rev.*, 57:1528.
24. Blair, D. 2006. Fine structure of a fine machine. *J. Bacteriol.*, 188:7033–7035.
25. Blair, D.E. and D.M. van Aalten. 2004. Structures of *Bacillus subtilis* PdaA, a family 4 carbohydrate esterase, and a complex with N-acetyl-glucosamine. *FEBS Lett.*, 570:13–19.
26. Blazquez, J., J. Gomez-Gomez, A. Oliver, C. Juan, V. Kapur, and S. Martin. 2006. PBP3 inhibition elicits adaptive responses in *Pseudomonas aeruginosa. Mol. Microbiol.*, 62:84–99.
27. Boitel, B., M. Ortiz-Lombardia, R. Duran, F. Pompeo, S. Cole, C. Cervenansky, and P. Alzari. 2003. PknB kinase activity is regulated by phosphorylation in two Thr residues and dephosphorylation by PstP, the cognate phospho-Ser/Thr phosphatase, in *Mycobacterium tuberculosis. Mol. Microbiol.*, 49:1493–1508.
28. Boneca, I.G., O. Dussurget, D. Cabanes, M.A. Nahori, S. Sousa, M. Lecuit, E. Psylinakis, V. Bouriotis, J.P. Hugot, M. Giovannini, A. Coyle, J. Bertin, A. Namane, J.C. Rousselle, N. Cayet, M. Prevost, V. Balloy, M. Chignard, D.J. Philpott, P. Cossart, and S.E. Girardin. 2007. A critical role for peptidoglycan N-deacetylation in *Listeria* evasion from the host innate immune system. *Proc. Natl. Acad. Sci. U.S.A.*, 104:997–1002.

29. Boneca, I.G., Z.H. Huang, D.A. Gage, and A. Tomasz. 2000. Characterization of *Staphylococcus aureus* cell wall glycan strands, evidence for a new β-N-acetylglucosaminidase activity. *J. Biol. Chem.*, 275:9910–9918.
30. Bonsor, D.A., I. Grishkovskaya, E.J. Dodson, and C. Kleanthous. 2007. Molecular mimicry enables competitive recruitment by a natively disordered protein. *J. Am. Chem. Soc.*, 129:4800–4807.
31. Braun, V., H. Gnirke, U. Henning, and K. Rehn. 1973. Model for the structure of the shape-maintaining layer of the *Escherichia coli* cell envelope. *J. Bacteriol.*, 114:1264–1270.
32. Braun, V. and K. Rehn. 1969. Chemical characterization, spatial distribution and function of a lipoprotein (murein-lipoprotein) of the *E. coli* cell wall. The specific effect of trypsin on the membrane structure. *Eur. J. Biochem.*, 10:426–438.
33. Buist, G., A.N. Ridder, J. Kok, and O.P. Kuipers. 2006. Different subcellular locations of secretome components of Gram-positive bacteria. *Microbiology,* 152:2867–2874.
34. Burdett, I.D. and R.G. Murray. 1974. Electron microscope study of septum formation in *Escherichia coli* strains B and B-r during synchronous growth. *J. Bacteriol.*, 119:1039–1056.
35. Cabeen, M.T. and C. Jacobs-Wagner. 2005. Bacterial cell shape. *Nat. Rev. Microbiol.*, 3:601–610.
36. Carballido-Lopez, R. 2006. The bacterial actin-like cytoskeleton. *Microbiol. Mol. Biol. Rev.*, 70:888–909.
37. Carballido-Lopez, R. 2006. Orchestrating bacterial cell morphogenesis. *Mol. Microbiol.*, 60:815–819.
38. Cascales, E., A. Bernadac, M. Gavioli, J.C. Lazzaroni, and R. Lloubes. 2002. Pal lipoprotein of *Escherichia coli* plays a major role in outer membrane integrity. *J. Bacteriol.*, 184:754–759.
39. Cascales, E. and R. Lloubes. 2004. Deletion analyses of the peptidoglycan-associated lipoprotein Pal reveals three independent binding sequences including a TolA box. *Mol. Microbiol.*, 51:873–885.
40. Cha, J. and S. Mobashery. 2007. Lysine N(zeta)-decarboxylation in the BlaR1 protein from *Staphylococcus aureus* at the root of its function as an antibiotic sensor. *J. Am. Chem. Soc.*, 129:3834–3835.
41. Chang, C.I., Y. Chelliah, D. Borek, D. Mengin-Lecreulx, and J. Deisenhofer. 2006. Structure of tracheal cytotoxin in complex with a heterodimeric pattern-recognition receptor. *Science*, 311:1761–1764.
42. Cho, S., Q. Wang, C.P. Swaminathan, D. Hesek, M. Lee, G.J. Boons, S. Mobashery, and R.A. Mariuzza. 2007. Structural insights into the bacteriocidal mechanism of human peptidoglycan recognition proteins. *Proc. Natl. Acad. Sci. U.S.A.*, 104:8761–8766.
43. Cirz, R.T., M.B. Jones, N.A. Gingles, T.D. Minogue, B. Jarrahi, S.N. Peterson, and F.E. Romesberg. 2007. Complete and SOS-mediated response of *Staphylococcus aureus* to the antibiotic ciprofloxacin. *J. Bacteriol.*, 189:531–539.
44. Cirz, R.T., B.M. O'Neill, J.A. Hammond, S.R. Head, and F.E. Romesberg. 2006. Defining the *Pseudomonas aeruginosa* SOS response and its role in the global response to the antibiotic ciprofloxacin. *J. Bacteriol.*, 188:7101–7110.
45. Clardy, J., M.A. Fischbach, and C.T. Walsh. 2006. New antibiotics from bacterial natural products. *Nat. Biotechnol.*, 24:1541–50.
46. Clavel, T., P. Germon, A. Vianney, R. Portalier, and J.C. Lazzaroni. 1998. TolB protein of *Escherichia coli* K-12 interacts with the outer membrane peptidoglycan-associated proteins Pal, Lpp and OmpA. *Mol. Microbiol.*, 29:359–367.
47. Contreras-Martel, C., V. Job, A.M. Di Guilmi, T. Vernet, O. Dideberg, and A. Dessen. 2006. Crystal structure of penicillin-binding protein 1a (PBP1a) reveals a mutational hotspot implicated in β-lactam resistance in *Streptococcus pneumoniae. J. Mol. Biol.*, 355:684–696.
48. Costerto, J.W., J.M. Ingram, and K.J. Cheng. 1974. Structure and function of cell envelope of Gram-negative bacteria. *Bacteriol. Rev.*, 38:87–110.
49. Courvalin, P. 2006. Vancomycin resistance in Gram-positive cocci. *Clin. Infect. Dis.*, 42(Suppl. 1): S25–34.
50. Cui, L., A. Iwamoto, J.Q. Lian, H.M. Neoh, T. Maruyama, Y. Horikawa, and K. Hiramatsu. 2006. Novel mechanism of antibiotic resistance originating in vancomycin-intermediate *Staphylococcus aureus. Antimicrob. Agents Chemother.*, 50:428–38.
51. D'Elia, M.A., K.E. Millar, T.J. Beveridge, and E.D. Brown. 2006. Wall teichoic acid polymers are dispensable for cell viability in *Bacillus subtilis. J. Bacteriol.*, 188:8313–8316.
52. D'Elia, M.A., M.P. Pereira, Y.S. Chung, W. Zhao, A. Chau, T.J. Kenney, M.C. Sulavik, T.A. Black, and E.D. Brown. 2006. Lesions in teichoic acid biosynthesis in *S. aureus* lead to a lethal gain of function in the otherwise dispensable pathway. *J. Bacteriol.*, 188:4183–4189.
53. Daleke, D.L. 2007. Phospholipid flippases. *J. Biol. Chem.*, 282:821–825.

54. Dasgupta, A., P. Datta, M. Kundu, and J. Basu. 2006. The serine/threonine kinase PknB of *Mycobacterium tuberculosis* phosphorylates PBPA, a penicillin-binding protein required for cell division. *Microbiology*, 152:493–504.
55. de Pedro, M.A., J.C. Quintela, J.V. Holtje, and H. Schwarz. 1997. Murein segregation in *Escherichia coli*. *J. Bacteriol.*, 179:2823–2834.
56. Dekimpe, S.J., M. Kengatharan, C. Thiemermann, and J.R. Vane. 1995. The cell wall components peptidoglycan and lipoteichoic acid from *S. aureus* act in synergy to cause shock and multiple organ failure. *Proc. Natl. Acad. Sci. U.S.A.*, 92:10359–10363.
57. Delbridge, L.M. and M.X. O'Riordan. 2007. Innate recognition of intracellular bacteria. *Curr. Opin. Immunol.*, 19:10–16.
58. Demchick, P. and A.L. Koch. 1996. The permeability of the wall fabric of *Escherichia coli* and *Bacillus subtilis*. *J. Bacteriol.*, 178:768–773.
59. Den Blaauwen, T., M.E. Aarsman, N.O. Vischer, and N. Nanninga. 2003. Penicillin-binding protein PBP2 of *Escherichia coli* localizes preferentially in the lateral wall and at mid-cell in comparison with the old cell pole. *Mol. Microbiol.*, 47:539–547.
60. Depardieu, F., I. Podglajen, R. Leclercq, E. Collatz, and P. Courvalin. 2007. Modes and modulations of antibiotic resistance gene expression. *Clin. Microbiol. Rev.*, 20:79–114.
61. Dessen, A., N. Mouz, E. Gordon, J. Hopkins, and O. Dideberg. 2001. Crystal structure of PBP2x from a highly penicillin-resistant *S. pneumoniae* clinical isolate: a mosaic framework containing 83 mutations. *J. Biol. Chem.*, 276:45106–45112.
62. Dmitriev, B., F. Toukach, and S. Ehlers. 2005. Towards a comprehensive view of the bacterial cell wall. *Trends Microbiol.,* 13:569–574.
63. Dmitriev, B.A., F.V. Toukach, K.M. Schaper, O. Holst, E.T. Rietschel, and S. Ehlers. 2003. Tertiary structure of bacterial murein: The scaffold model. *J. Bacteriol.*, 185:3458–3468.
64. Doerrler, W.T. 2006. Lipid trafficking to the outer membrane of Gram-negative bacteria. *Mol. Microbiol.*, 60:542–552.
65. Dramsi, S., P. Trieu-Cuot, and H. Bierne. 2005. Sorting sortases: a nomenclature proposal for the various sortases of Gram-positive bacteria. *Res. Microbiol.* 156:289–297.
66. Fang, X., K. Tiyanont, Y. Zhang, J. Wanner, D. Boger, and S. Walker. 2006. The mechanism of action of ramoplanin and enduracidin. *Mol. BioSystems*, 2:69–76.
67. Fernandez, P., B. Saint-Joanis, N. Barilone, M. Jackson, B. Gicquel, S. Cole, and P. Alzari. 2006. The Ser/Thr protein kinase PknB is essential for sustaining mycobacterial growth. *J. Bacteriol.*, 188:7778–7784.
68. Filipe, S.R., A. Tomasz, and P. Ligoxygakis. 2005. Requirements of peptidoglycan structure that allow detection by the Drosophila Toll pathway. *EMBO Rep.*, 6:327–333.
69. Fisher, J.F., S.O. Merouheh, and S. Mobashery. 2005. Bacterial resistance to β-lactam antibiotics: compelling opportunism, compelling opportunity. *Chem. Rev.*, 105:395–424.
70. Fisher, J.F., S.O. Merouheh, and S. Mobashery. 2006. Nanomolecular and supramolecular paths toward peptidoglycan structure. *Microbe*, 1:420–427.
71. Fuda, C., D. Hesek, M. Lee, W. Heilmayer, R. Novak, S.B. Vakulenko, and S. Mobashery. 2006. Mechanistic basis for the action of new cephalosporin antibiotics effective against methicillin- and vancomycin-resistant *Staphylococcus aureus*. *J. Biol. Chem.*, 281:10035–10041.
72. Fuda, C.C., J.F. Fisher, and S. Mobashery. 2005. β-Lactam resistance in *Staphylococcus aureus*: the adaptive resistance of a plastic genome. *Cell Mol. Life Sci.*, 62:2617–2633.
73. Fukushima, T., T. Kitajima, and J. Sekiguchi. 2005. A polysaccharide deacetylase homologue, PdaA, in *Bacillus subtilis* acts as an N-acetylmuramic acid deacetylase *in vitro*. *J. Bacteriol.*, 187:1287–1292.
74. Fukushima, T., H. Yamamoto, A. Atrih, S.J. Foster, and J. Sekiguchi. 2002. A polysaccharide deacetylase gene (pdaA) is required for germination and for production of muramic delta-lactam residues in the spore cortex of *Bacillus subtilis*. *J. Bacteriol.*, 184:6007–6015.
75. Gerding, M.A., Y. Ogata, N.D. Pecora, H. Niki, and P.A. de Boer. 2007. The trans-envelope Tol-Pal complex is part of the cell division machinery and required for proper outer-membrane invagination during cell constriction in *E. coli*. *Mol. Microbiol.*, 63:1008–1025.
76. Ghosh, A.S. and K.D. Young. 2005. Helical disposition of proteins and lipopolysaccharide in the outer membrane of *Escherichia coli*. *J. Bacteriol.*, 187:1913–1922.
77. Ghuysen, J.M. 1991. Serine β-lactamases and penicillin-binding proteins. *Annu. Rev. Microbiol.*, 45:37–67.

78. Gilmore, M.E., D. Bandyopadhyay, A.M. Dean, S.D. Linnstaedt, and D.L. Popham. 2004. Production of muramic delta-lactam in *Bacillus subtilis* spore peptidoglycan. *J. Bacteriol.*, 186:80–89.
79. Glauner, B. and J.V. Holtje. 1990. Growth pattern of the murein sacculus of *Escherichia coli*. *J. Biol. Chem.*, 265:18988–18996.
80. Goemaere, E.L., E. Cascales, and R. Lloubes. 2007. Mutational analyses define helix organization and key residues of a bacterial membrane energy-transducing complex. *J. Mol. Biol.*, 366:1424–1436.
81. Goffin, C. and J.M. Ghuysen. 2002. Biochemistry and comparative genomics of SxxK superfamily acyltransferases offer a clue to the mycobacterial paradox: presence of penicillin-susceptible target proteins versus lack of efficiency of penicillin as therapeutic agent. *Microbiol. Mol. Biol. Rev.*, 66:702–738.
82. Goffin, C. and J.M. Ghuysen. 1998. Multimodular penicillin-binding proteins: an enigmatic family of orthologs and paralogs. *Microbiol. Mol. Biol. Rev.*, 62:1079–1093.
83. Golemi-Kotra, D., J.Y. Cha, S.O. Meroueh, S.B. Vakulenko, and S. Mobashery. 2003. Resistance to β-lactam antibiotics and its mediation by the sensor domain of the transmembrane BlaR signaling pathway in *Staphylococcus aureus*. *J. Biol. Chem.*, 278:18419–25.
84. Goodell, E.W. 1985. Recycling of murein by *Escherichia coli*. *J. Bacteriol.*, 163:305–310.
85. Gordon, E., N. Mouz, E. Duee, and O. Dideberg. 2000. Crystal structure of the PBP2x from *S. pneumoniae* and its acyl-enzyme form: Implication in drug resistance. *J. Mol. Biol.*, 299:477–485.
86. Greenstein, A.E., N. Echols, T.N. Lombana, D.S. King, and T. Alber. 2007. Allosteric activation by dimerization of the PknD receptor Ser/Thr protein kinase from *Mycobacterium tuberculosis*. *J. Biol. Chem.*, 282:11427–11435.
87. Grizot, S. and S. K. Buchanan. 2004. Structure of the OmpA-like domain of RmpM from *Neisseria meningitidis*. *Mol. Microbiol.*, 51:1027–1037.
88. Grundner, C., L. Gay, and T. Alber. 2005. *Mycobacterium tuberculosis* serine/threonine kinases PknB, PknD, PknE, and PknF phosphorylate multiple FHA domains. *Protein Sci.*, 14:1918–1921.
89. Guan, R., P.H. Brown, C.P. Swaminathan, A. Roychowdhury, G.J. Boons, and R.A. Mariuzza. 2006. Crystal structure of human peptidoglycan recognition protein I alpha bound to a muramyl pentapeptide from Gram-positive bacteria. *Protein Sci.*, 15:1199–1206.
90. Guan, R. and R.A. Mariuzza. 2007. Peptidoglycan recognition proteins of the innate immune system. *Trends Microbiol.*, 15:127–134.
91. Guzman, L.M., J.J. Barondess, and J. Beckwith. 1992. FtsL, an essential cytoplasmic membrane protein involved in cell division in *Escherichia coli*. *J. Bacteriol.*, 174:7716–7728.
92. Hanique, S., M.L. Colombo, E. Goormaghtigh, P. Soumillion, J.M. Frere, and B. Joris. 2004. Evidence of an intramolecular interaction between the two domains of the BlaR1 penicillin receptor during the signal transduction. *J. Biol. Chem.*, 279:14264–14272.
93. Harz, H., K. Burgdorf, and J.V. Holtje. 1990. Isolation and separation of the glycan strands from murein of *Escherichia coli* by reversed-phase high-performance liquid chromatography. *Anal. Biochem.*, 190:120–128.
94. Hasselbring, B.M. and D.C. Krause. 2007. Cytoskeletal protein p41 is required to anchor the terminal organelle of the wall-less prokaryote *Mycoplasma pneumoniae*. *Mol. Microbiol.*, 63:44–53.
95. Hatch, T.P. 1996. Disulfide cross-linked envelope proteins: the functional equivalent of peptidoglycan in chlamydiae?. *J. Bacteriol.*, 178:1–5.
96. Heidrich, C., A. Ursinus, J. Berger, H. Schwarz, and J.V. Holtje. 2002. Effects of multiple deletions of murein hydrolases on viability, septum cleavage, and sensitivity to large toxic molecules in *Escherichia coli*. *J. Bacteriol.*, 184:6093–6099.
97. Hesek, D., M. Suvorov, K. Morio, M. Lee, S. Brown, S.B. Vakulenko, and S. Mobashery. 2004. Synthetic peptidoglycan substrates for penicillin-binding protein 5 of Gram-negative bacteria. *J. Org. Chem.*, 69:778–784.
98. Hiramatsu, K. 2001. Vancomycin-resistant *Staphylococcus aureus*: a new model of antibiotic resistance. *Lancet Infect. Dis.*, 1:147–155.
99. Hobot, J.A., E. Carlemalm, W. Villiger, and E. Kellenberger. 1984. Periplasmic gel: new concept resulting from the reinvestigation of bacterial cell envelope ultrastructure by new methods. *J. Bacteriol.*, 160:143–152.
100. Holtje, J.V. 1998. Growth of the stress-bearing and shape-maintaining murein sacculus of *Escherichia coli*. *Microbiol. Mol. Biol. Rev.*, 62:181–203.
101. Horobin, R. 2002. Biological staining: mechanisms and theory. *Biotech. Histochem.*, 77:3–13.
102. Hosking, E.R., C. Vogt, E.P. Bakker, and M.D. Manson. 2006. The *Escherichia coli* MotAB proton channel unplugged. *J. Mol. Biol.*, 364:921–937.

103. Hu, K., H. Yang, G. Liu, and H. Tan. 2006. Identification and characterization of a polysaccharide deacetylase gene from *Bacillus thuringiensis*. *Can. J. Microbiol.*, 52:935–941.
104. Inouye, M. 1974. A three-dimensional molecular assembly model of a lipoprotein from the *Escherichia coli* outer membrane. *Proc. Natl. Acad. Sci. U.S.A.*, 71:2396–2400.
105. Jacobs, C., J.M. Frere, and S. Normark. 1997. Cytosolic intermediates for cell wall biosynthesis and degradation control inducible β-lactam resistance in Gram-negative bacteria. *Cell*, 88:823–832.
106. Jacobs, C., L.J. Huang, E. Bartowsky, S. Normark, and J.T. Park. 1994. Bacterial cell wall recycling provides cytosolic muropeptides as effectors for β-lactamase induction. *EMBO J.*, 13:4684–4694.
107. Jensen, G.J. and A. Briegel. 2007. How electron cryotomography is opening a new window onto prokaryotic ultrastructure. *Curr. Opin. Struct. Biol.*, 17:260–267.
108. Jerman, B., M. Butala, and D. Zgur-Bertok. 2005. Sublethal concentrations of ciprofloxacin induce bacteriocin synthesis in *Escherichia coli*. *Antimicrob. Agents Chemother.*, 49:3087–3090.
109. Jilaveanu, L.B. and D.B. Oliver. 2007. In vivo membrane topology of *Escherichia coli* SecA ATPase reveals extensive periplasmic exposure of multiple functionally important domains clustering on one face of SecA. *J. Biol. Chem.*, 282:4661–4668.
110. Jones, G. and P. Dyson. 2006. Evolution of transmembrane protein kinases implicated in coordinating remodeling of Gram-positive peptidoglycan: inside versus outside. *J. Bacteriol.*, 188:7470–7476.
111. Kajimura, J., T. Fujiwara, S. Yamada, Y. Suzawa, T. Nishida, Y. Oyamada, I. Hayashi, J. Yamagishi, H. Komatsuzawa, and M. Sugai. 2005. Identification and molecular characterization of an N-acetylmuramyl-L-alanine amidase Sle1 involved in cell separation of *Staphylococcus aureus*. *Mol. Microbiol.*, 58:1087–1101.
112. Kandler, O. and H. Konig. 1978. Chemical composition of the peptidoglycan-free cell walls of methanogenic bacteria. *Arch. Microbiol.*, 118:141–152.
113. Kang, C.M., D.W. Abbott, S.T. Park, C.C. Dascher, L.C. Cantley, and R.N. Husson. 2005. The *Mycobacterium tuberculosis* serine/threonine kinases PknA and PknB: substrate identification and regulation of cell shape. *Genes Dev.*, 19:1692–1704.
114. Katayama, Y., H.Z. Zhang, D. Hong, and H.F. Chambers. 2003. Jumping the barrier to β-lactam resistance in *Staphylococcus aureus*. *J. Bacteriol.*, 185:5465–5472.
115. Keep, N.H., J.M. Ward, M. Cohen-Gonsaud, and B. Henderson. 2006. Wake up! Peptidoglycan lysis and bacterial non-growth states. *Trends Microbiol.*, 14:271–276.
116. Kelley, W.L. 2006. Lex marks the spot: the virulent side of SOS and a closer look at the LexA regulon. *Mol. Microbiol.*, 62:1228–1238.
117. Kerff, F., P. Charlier, M.L. Colombo, E. Sauvage, A. Brans, J.M. Frere, B. Joris, and E. Fonze. 2003. Crystal structure of the sensor domain of the BlaR penicillin receptor from *Bacillus licheniformis*. *Biochemistry*, 42:12835–12843.
118. Kernodle, D.S. 2006. Mechanisms of Resistance to β-lactam antibiotics, *Gram-Positive Pathogens*, second ed. ASM Press, Washington, DC, p. 769–781.
119. Koch, A.L. 2006. *The Bacteria: Their Origin, Structure, Function and Antibiosis*. Springer, Dordrecht, The Netherlands.
120. Koch, A.L. 1985. Inside-to-outside growth and turnover of the wall of Gram-positive rods. *J. Theor. Biol.*, 117:137.
121. Koch, A.L. 2005. Shapes that *Escherichia coli* cells can achieve, as a paradigm for other bacteria. *Crit. Rev. Microbiol.*, 31:183–190.
122. Koch, A.L. 2000. Simulation of the conformation of the murein fabric: the oligoglycan, penta-muropeptide, and cross-linked nona-muropeptide. *Arch. Microbiol.*, 174:429–439.
123. Koch, A.L. 1998. The three-for-one model for gram-negative wall growth: a problem and a possible solution. *FEMS Microbiol. Lett.*, 162:127–134.
124. Kojima, S. and D.F. Blair. 2004. Solubilization and purification of the MotA/MotB complex of *Escherichia coli*. *Biochemistry*, 43:26–34.
125. Korsak, D., S. Liebscher, and W. Vollmer. 2005. Susceptibility to antibiotics and β-lactamase induction in murein hydrolase mutants of *Escherichia coli*. *Antimicrob. Agents Chemother.*, 49:1404–1409.
126. Kraft, A.R., J. Prabhu, A. Ursinus, and J.V. Holtje. 1999. Interference with murein turnover has no effect on growth but reduces β-lactamase induction in *Escherichia coli*. *J. Bacteriol.*, 181:7192–7198.
127. Kristich, C.J., C.L. Wells, and G.M. Dunny. 2007. A eukaryotic-type Ser/Thr kinase in *Enterococcus faecalis* mediates antimicrobial resistance and intestinal persistence. *Proc. Natl. Acad. Sci. U.S.A.*, 104:3508–3513.

128. Kumar, S., A. Roychowdhury, B. Ember, Q. Wang, R. Guan, R.A. Mariuzza, and G.J. Boons. 2005. Selective recognition of synthetic lysine and meso-diaminopimelic acid-type peptidoglycan fragments by human peptidoglycan recognition proteins Iα and S. *J. Biol. Chem.*, 280:37005–37012.
129. Kuroda, M., S. Nagasaki, and T. Ohta. 2007. Sesquiterpene farnesol inhibits recycling of the C55 lipid carrier of the murein monomer precursor contributing to increased susceptibility to β-lactams in methicillin-resistant *Staphylococcus aureus*. *J. Antimicrob. Chemother.*, 59:425–432.
130. Labischinski, H., E.W. Goodell, A. Goodell, and M.L. Hochberg. 1991. Direct proof of a "more-than-single-layered" peptidoglycan architecture of *Escherichia coli* W7: a neutron small-angle scattering study. *J. Bacteriol.*, 173:751–756.
131. Lazdunski, C.J., E. Bouveret, A. Rigal, L. Journet, R. Lloubes, and H. Benedetti. 1998. Colicin import into *Escherichia coli* cells. *J. Bacteriol.*, 180:4993–5002.
132. Lee, S.G. and V.A. Fischetti. 2006. Purification and characterization of LPXTGase from *Staphylococcus aureus*: the amino acid composition mirrors that found in the peptidoglycan. *J. Bacteriol.*, 188:389–398.
133. Li, S.-Y., J.-V. Holtje, and K.D. Young. 2004. Comparison of HPLC and fluorophore-assisted carbohydrate electrophoresis methods for analyzing peptidoglycan composition of *Escherichia coli*. *Anal. Biochem.*, 326:1–12.
134. Liang, M.D., A. Bagchi, H.S. Warren, M.M. Tehan, J.A. Trigilio, L.K. Beasley-Topliffe, B.L. Tesini, J.C. Lazzaroni, M.J. Fenton, and J. Hellman. 2005. Bacterial peptidoglycan-associated lipoprotein: a naturally occurring toll-like receptor 2 agonist that is shed into serum and has synergy with lipopolysaccharide. *J. Infect. Dis.*, 191:939–948.
135. Lim, J.H., M.S. Kim, H.E. Kim, T. Yano, Y. Oshima, K. Aggarwal, W.E. Goldman, N. Silverman, S. Kurata, and B.H. Oh. 2006. Structural basis for preferential recognition of diaminopimelic acid-type peptidoglycan by a subset of peptidoglycan recognition proteins. *J. Biol. Chem.*, 281:8286–8295.
136. Liu, J., W. Cao, and M. Lu. 2002. Core side-chain packing and backbone conformation in Lpp-56 coiled-coil mutants. *J. Mol. Biol.*, 318:877–888.
137. Lommatzsch, J., M.F. Templin, A.R. Kraft, W. Vollmer, and J.V. Holtje. 1997. Outer membrane localization of murein hydrolases: MltA, a third lipoprotein lytic transglycosylase in *Escherichia coli*. *J. Bacteriol.*, 179:5465–5470.
138. Lovering, A.L., L. De Castro, D. Lim, and N.C. Strynadka. 2006. Structural analysis of an "open" form of PBP1B from *Streptococcus pneumoniae*. *Protein Sci.*, 15:1701–1709.
139. Lovering, A.L., L.H. de Castro, D. Lim, and N.C. Strynadka. 2007. Structural insight into the transglycosylation step of bacterial cell-wall biosynthesis. *Science*, 315:1402–1405.
140. Macheboeuf, P., C. Contreras-Martel, V. Job, O. Dideberg, and A. Dessen. 2006. Penicillin binding proteins: key players in bacterial cell cycle and drug resistance processes. *FEMS Microbiol. Rev.*, 30:673–691.
141. Magnet, S., A. Arbeloa, J.L. Mainardi, J.E. Hugonnet, M. Fourgeaud, L. Dubost, A. Marie, V. Delfosse, C. Mayer, L.B. Rice, and M. Arthur. 2007. Specificity of L,D-transpeptidases from Gram-positive bacteria producing different peptidoglycan chemotypes. *J. Biol. Chem.*, 282:13151–13159.
142. Magnet, S., S. Bellais, L. Dubost, M. Fourgeaud, J. Mainardi, S. Petit-Frere, A. Marie, D. Mengin-Lecreulx, M. Arthur, and L. Gutmann. 2007. Identification of the L,D-transpeptidases responsible for attachment of the Braun lipoprotein to *Escherichia coli* peptidoglycan. *J. Bacteriol.*, 189:3927–3931.
143. Mainardi, J.L., M. Fourgeaud, J.E. Hugonnet, L. Dubost, J.P. Brouard, J. Ouazzani, L.B. Rice, L. Gutmann, and M. Arthur. 2005. A novel peptidoglycan cross-linking enzyme for a β-lactam-resistant transpeptidation pathway. *J. Biol. Chem.*, 280:38146–38152.
144. Maiques, E., C. Ubeda, S. Campoy, N. Salvador, I. Lasa, R.P. Novick, J. Barbe, and J.R. Penades. 2006. β-Lactam antibiotics induce the SOS response and horizontal transfer of virulence factors in *Staphylococcus aureus*. *J. Bacteriol.*, 188:2726–2729.
145. Marko, M., C. Hsieh, R. Schalek, J. Frank, and C. Mannella. 2007. Focused-ion-beam thinning of frozen-hydrated biological specimens for cryo-electron microscopy. *Nat. Methods,* 4:215–217.
146. Marraffini, L.A., A.C. Dedent, and O. Schneewind. 2006. Sortases and the art of anchoring proteins to the envelopes of Gram-positive bacteria. *Microbiol. Mol. Biol. Rev.,* 70:192–221.
147. Marrero, A., G. Mallorqui-Fernandez, T. Guevara, R. Garcia-Castellanos, and F.X. Gomis-Ruth. 2006. Unbound and acylated structures of the MecR1 extracellular antibiotic-sensor domain provide insights into the signal-transduction system that triggers methicillin resistance. *J. Mol. Biol.,* 361:506–521.
148. Masschalck, B. and C.W. Michiels. 2003. Antimicrobial properties of lysozyme in relation to foodborne vegetative bacteria. *Crit. Rev. Microbiol.*, 29:191–214.

149. Matias, V.R., A. Al-Amoudi, J. Dubochet, and T.J. Beveridge. 2003. Cryo-transmission electron microscopy of frozen-hydrated sections of *Escherichia coli* and *Pseudomonas aeruginosa*. *J. Bacteriol.*, 185:6112–6118.
150. Matias, V.R. and T.J. Beveridge. 2007. Cryo-electron microscopy of cell division in *Staphylococcus aureus* reveals a mid-zone between nascent cross walls. *Mol. Microbiol.*, 64:195–206.
151. Matias, V.R. and T.J. Beveridge. 2005. Cryo-electron microscopy reveals native polymeric cell wall structure in *Bacillus subtilis* 168 and the existence of a periplasmic space. *Mol. Microbiol.*, 56:240–251.
152. Matias, V.R.F. and T.J. Beveridge. 2006. Native cell wall organization shown by cryo-electron microscopy confirms the existence of a periplasmic space in *Staphylococcus aureus*. *J. Bacteriol.*, 188:1011–1021.
153. Matsunashi, M., C. Dietrich, and J.L. Strominger. 1967. Biosynthesis of the peptidoglycan of bacterial cell walls. *J. Biol. Chem.*, 242:3191–3206.
154. Mattman, L.H. 2000. *Cell Wall Deficient Forms: Stealth Pathogens*, 3rd ed. CRC Press, Inc., Boca Raton, FL.
155. McDevitt, D., D.J. Payne, D.J. Holmes, and M. Rosenberg. 2002. Novel targets for the future development of antibacterial agents. *J. Appl. Microbiol.*, 92 (Suppl.):28S–34S.
156. McDonald, C., N. Inohara, and G. Nunez. 2005. Peptidoglycan signaling in innate immunity and inflammatory disease. *J. Biol. Chem.*, 280:20177–20180.
157. Meroueh, S.O., K.Z. Bencze, D. Hesek, M. Lee, J.F. Fisher, T.L. Stemmler, and S. Mobashery. 2006. Three-dimensional structure of the bacterial cell wall peptidoglycan. *Proc. Natl. Acad. Sci. U.S.A.*, 103:4404–4409.
158. Meyer, P. and J. Dworkin. 2007. Applications of fluorescence microscopy to single bacterial cells. *Res. Microbiol.*, 158:187–194.
159. Miller, C., L.E. Thomsen, C. Gaggero, R. Mosseri, H. Ingmer, and S.N. Cohen. 2004. SOS response induction by β-lactams and bacterial defense against antibiotic lethality. *Science*, 305:1629–1631.
160. Mitchell, P. and J. Moyle. 1956. Osmotic function and structure of bacteria. *Symp. Soc. Gen. Microbiol.*, p. 150–180.
161. Morlot, C., L. Pernot, A. Le Gouellec, A.M. Di Guilmi, T. Vernet, O. Dideberg, and A. Dessen. 2005. Crystal structure of a peptidoglycan synthesis regulatory factor PBP3 from *Streptococcus pneumoniae*. *J. Biol. Chem.*, 280:15984–15991.
162. Moulder, J.W. 1993. Why is *Chlamydia* sensitive to penicillin in the absence of peptidoglycan? *Infect. Agents Dis.*, 2:87–99.
163. Mukamolova, G.V., A.G. Murzin, E.G. Salina, G.R. Demina, D.B. Kell, A.S. Kaprelyants, and M. Young. 2006. Muralytic activity of *Micrococcus luteus* Rpf and its relationship to physiological activity in promoting bacterial growth and resuscitation. *Mol. Microbiol.*, 59:84–98.
164. Mukhopadhyay, A. and R.T. Peterson. 2006. Fishing for new antimicrobials. *Curr. Opin. Chem. Biol.*, 10:327–333.
165. Nanninga, N. 1998. Morphogenesis of *Escherichia coli. Microbiol. Mol. Biol. Rev.*, 62:110–129.
166. Narayan, A., P. Sachdeva, K. Sharma, A.K. Saini, A.K. Tyagi, and Y. Singh. 2007. Serine threonine protein kinases of mycobacterial genus: phylogeny to function. *Physiol. Genomics,* 29:66–75.
167. Neilsen, P.O., G.A. Zimmerman, and T.M. McIntyre. 2001. *Escherichia coli* Braun lipoprotein induces a lipopolysaccharide-like endotoxic response from primary human endothelial cells. *J. Immunol.*, 167:5231–5239.
168. Nelson, D.E. and K.D. Young. 2001. Contributions of PBP 5 and DD-carboxypeptidase penicillin binding proteins to maintenance of cell shape in *Escherichia coli. J. Bacteriol.,* 183:3055–3064.
169. Nikaido, H. 2003. Molecular basis of bacterial outer membrane permeability revisited. *Microbiol. Mol. Biol.* Rev., 67:593–656.
170. Nodwell, J.R. 2007. Novel links between antibiotic resistance and antibiotic production. *J. Bacteriol.,* 189:3683–3685.
171. Normark, S. 1995. β-Lactamase induction in Gram-negative bacteria is intimately linked to peptidoglycan recycling. *Microb. Drug Resist.*, 1:111–114.
172. Norris, V., T. den Blaauwen, A. Cabin-Flaman, R. Doi, R. Harshey, L. Janniere, A. Jimenez-Sanchez, D. Jin, P. Levin, E. Mileykovskaya, A. Minsky, M.J. Saier, and K. Skarstad. 2007. Functional taxonomy of bacterial hyperstructures. *Microbiol. Mol. Biol. Rev.*, 71:230–253.
173. Ortiz-Lombardia, M., F. Pompeo, B. Boitel, and P.M. Alzari. 2003. Crystal structure of the catalytic domain of the PknB serine/threonine kinase from *Mycobacterium tuberculosis. J. Biol. Chem.*, 278:13094–13100.

174. Park, J.T. 1993. Turnover and recycling of the murein sacculus in oligopeptide permease-negative strains of *Escherichia coli*: indirect evidence for an alternative permease system and for a monolayered sacculus. *J. Bacteriol.*, 175:7–11.
175. Park, J.W., B.R. Je, S. Piao, S. Inamura, Y. Fujimoto, K. Fukase, S. Kusumoto, N.C. Ha, K. Soderhall, N.C. Ha, and B.L. Lee. 2006. A synthetic peptidoglycan fragment as a competitive inhibitor of the melanization cascade. *J. Biol. Chem.*, 281:7747–7755.
176. Parsons, L.M., F. Lin, and J. Orban. 2006. Peptidoglycan recognition by Pal, an outer membrane lipoprotein. *Biochemistry*, 45:2122–2128.
177. Pavelka, M.J. 2007. Another brick in the wall. *Trends Microbiol.*, 15:147–149.
178. Pereira, S.F., A.O. Henriques, M.G. Pinho, H. de Lencastre, and A. Tomasz. 2007. Role of PBP1 in cell division of *Staphylococcus aureus*. *J. Bacteriol.*, 189:3525–3531.
179. Perez-Capilla, T., M.R. Baquero, J.M. Gomez-Gomez, A. Ionel, S. Martin, and J. Blazquez. 2005. SOS-independent induction of dinB transcription by β-lactam-mediated inhibition of cell wall synthesis in *Escherichia coli*. *J. Bacteriol.*, 187:1515–1518.
180. Peters, P., M. Migocki, C. Thoni, and E. Harry. 2007. A new assembly pathway for the cytokinetic Z ring from a dynamical helical structure in vegetatively growing cells of *Bacillus subtilis*. *Mol. Microbiol.*, 64:487–499.
181. Pinho, M.G. and J. Errington. 2003. Dispersed mode of *Staphylococcus aureus* cell wall synthesis in the absence of the division machinery. *Mol. Microbiol.*, 50:871–881.
182. Pinho, M.G. and J. Errington. 2005. Recruitment of penicillin-binding protein PBP2 to the division site of *Staphylococcus aureus* is dependent on its transpeptidation substrates. *Mol. Microbiol.*, 55:799–807.
183. Pink, D., J. Moeller, B. Quinn, M. Jericho, and T.J. Beveridge. 2000. On the architecture of the Gram-negative bacterial murein sacculus. *J. Bacteriol.*, 182:5925–5930.
184. Pisabarro, A.G., R. Prats, D. Vaquez, and A. Rodriguez-Tebar. 1986. Activity of penicillin-binding protein 3 from *Escherichia coli*. *J. Bacteriol.*, 168:199–206.
185. Pommier, S., M. Gavioli, E. Cascales, and R. Lloubes. 2005. Tol-dependent macromolecule import through the *Escherichia coli* cell envelope requires the presence of an exposed TolA binding motif. *J. Bacteriol.*, 187:7526–7534.
186. Poole, K. 2004. Resistance to β–lactam antibiotics. *Cell Mol. Life Sci.*, 61:2200–2223.
187. Popham, D.L. 2002. Specialized peptidoglycan of the bacterial endospore: the inner wall of the lockbox. *Cell Mol. Life Sci.*, 59:426–433.
188. Popham, D.L. and K.D. Young. 2003. Role of penicillin-binding proteins in bacterial cell morphogenesis. *Curr. Opin. Microbiol.*, 6:594–599.
189. Prudhomme, M., L. Attaiech, G. Sanchez, B. Martin, and J.P. Claverys. 2006. Antibiotic stress induces genetic transformability in the human pathogen *Streptococcus pneumoniae*. *Science*, 313:89–92.
190. Reynolds, P.E. 1989. Structure, biochemistry and mechanism of action of glycopeptide antibiotics. Eur. *J. Clin. Microbiol. Infect. Dis.*, 8:943–50.
191. Reynolds, P.E. and P. Courvalin. 2005. Vancomycin resistance in *Enterococci* due to synthesis of precursors terminating in D-alanyl-D-serine. *Antimicrob. Agents Chemother.*, 49:21–25.
192. Robichon, C., D. Vidal-Ingigliardi, and A.P. Pugsley. 2005. Depletion of apolipoprotein N-acyltransferase causes mislocalization of outer membrane lipoproteins in *Escherichia coli*. *J. Biol. Chem.*, 280:974–983.
193. Rohrer, S., K. Ehlert, M. Tschierske, H. Labischinski, and B. Berger-Bachi. 1999. The essential *S. aureus* gene *fmhB* is involved in the first step of peptidoglycan pentapeptide interpeptide formation. *Proc. Natl. Acad. Sci. U.S.A.*, 96:9351–9356.
194. Romeis, T. and J.V. Holtje. 1994. Specific interaction of penicillin-binding proteins 3 and 7/8 with soluble lytic transglycosylase in *Escherichia coli*. *J. Biol. Chem.*, 269:21603–21607.
195. Royet, J. and R. Dziarski. 2007. Peptidoglycan recognition proteins: pleiotropic sensors and effectors of antimicrobial defences. *Nat. Rev. Microbiol.*, 5:264–277.
196. Safo, M.K., Q. Zhao, T.P. Ko, F.N. Musayev, H. Robinson, N. Scarsdale, A.H. Wang, and G.L. Archer. 2005. Crystal structures of the BlaI repressor from *Staphylococcus aureus* and its complex with DNA: insights into transcriptional regulation of the bla and mec operons. *J. Bacteriol.*, 187:1833–1844.
197. Schaffer, C. and P. Messner. 2005. The structure of secondary cell wall polymers: how Gram-positive bacteria stick their walls together. *Microbiology*, 151:643–651.
198. Scheffers, D. 2007. Cell wall growth during elongation and division: one ring to bind them?. *Mol. Microbiol.*, 64:877–880.

199. Scheffers, D.J., L.J. Jones, and J. Errington. 2004. Several distinct localization patterns for penicillin-binding proteins in *Bacillus subtilis. Mol. Microbiol.,* 51:749–764.
200. Scheffers, D.J. and M.G. Pinho. 2005. Bacterial cell wall synthesis: new insights from localization studies. *Microbiol. Mol. Biol. Rev.*, 69:585–607.
201. Schleifer, K.H. and O. Kandler. 1972. Peptidoglycan types of bacterial cell walls and their taxonomic implications. *Bacteriol. Rev.,* 36:407–477.
202. Schmid, M.B. 2006. Do targets limit antibiotic discovery?. *Nat. Biotechnol.,* 24:419–420.
203. Schwartz, B., J.A. Markwalder, S.P. Seitz, Y. Wang, and R.L. Stein. 2002. A kinetic characterization of the glycosyltransferase activity of *Eschericia coli* PBP1b and development of a continuous fluorescence assay. *Biochemistry*, 41:12552–12561.
204. Schwartz, B., J.A. Markwalder, and Y. Wang. 2001. Lipid II: total synthesis of the bacterial cell wall precursor and utilization as a substrate for glycosyltransfer and transpeptidation by penicillin binding protein (PBP) 1b of *Escherichia coli. J. Am. Chem. Soc.,* 123:11638–11643.
205. Schwarz, U., A. Asmus, and H. Frank. 1969. Autolytic enzymes and cell division of *Escherichia coli. J. Mol. Biol.,* 41:419–429.
206. Scott, J.R. and T.C. Barnett. 2006. Surface proteins of Gram-positive bacteria and how they get there. *Annu. Rev. Microbiol.*, 60:397–423.
207. Shinohara, A., M. Sakuma, T. Yakushi, S. Kojima, K. Namba, M. Homma, and K. Imada. 2007. Crystallization and preliminary X-ray analysis of MotY, a stator component of the *Vibrio alginolyticus* polar flagellar motor. *Acta Crystallograph. Sect. F Struct. Biol. Cryst. Commun.*, 63:89–92.
208. Shiomi, D., M. Yoshimoto, M. Homma, and I. Kawagishi. 2006. Helical distribution of the bacterial chemoreceptor via colocalization with the Sec protein translocation machinery. *Mol. Microbiol.*, 60:894–906.
209. Shu, W., J. Liu, H. Ji, and M. Lu. 2000. Core structure of the outer membrane lipoprotein from *Escherichia coli* at 1.9 A resolution. *J. Mol. Biol.*, 299:1101–1112.
210. Somner, E.A. and P.E. Reynolds. 1990. Inhibition of peptidoglycan biosynthesis by ramoplanin. *Antimicrob. Agents Chemother.*, 34:413–419.
211. Spratt, B.G. 1975. Distinct penicillin binding proteins involved in the division, elongation, and shape of *Escherichia coli* K12. *Proc. Natl. Acad. Sci. U.S.A.*, 72:2999–3003.
212. Steenbakkers, P.J., W.J. Geerts, N.A. Ayman-Oz, and J.T. Keltjens. 2006. Identification of pseudomurein cell wall binding domains. *Mol. Microbiol.*, 62:1618–1630.
213. Stewart, G. 2005. Taking shape: control of bacterial cell wall biosynthesis. *Mol. Microbiol.*, 57:1177–1181.
214. Stone, K.J. and J.L. Strominger. 1971. Mechanism of action of bacitracin: complexation with metal ion and C55-isoprenyl pyrophosphate. *Proc. Natl. Acad. Sci. U.S.A.,* 68:3223–3227.
215. Stubbs, K.A., M. Balcewich, B.L. Mark, and D.J. Vocadlo. 2007. Small molecule inhibitors of a glycoside hydrolase attenuate inducible AmpC mediated β-lactam resistance. *J. Biol. Chem.*, 282: 21382–21391.
216. Swaminathan, C.P., P.H. Brown, A. Roychowdhury, Q. Wang, R. Guan, N. Silverman, W.E. Goldman, G.J. Boons, and R. Mariuzza. 2006. Dual strategies for peptidoglycan discrimination by peptidoglycan recognition proteins (PGRPs). *Proc. Natl. Acad. Sci. U.S.A.*, 103:684–689.
217. Telkov, M.V., G.R. Demina, S.A. Voloshin, E.G. Salina, T.V. Dudik, T.N. Stekhanova, G.V. Mukamolova, K.A. Kazaryan, A.V. Goncharenko, M. Young, and A.S. Kaprelyants. 2006. Proteins of the Rpf (resuscitation promoting factor) family are peptidoglycan hydrolases. *Biochemistry* (Moscow), 71:414–422.
218. Terashima, H., H. Fukuoka, T. Yakushi, S. Kojima, and M. Homma. 2006. The Vibrio motor proteins, MotX and MotY, are associated with the basal body of Na-driven flagella and required for stator formation. *Mol. Microbiol.,* 62:1170–1180.
219. Thiem, S., D. Kentner, and V. Sourjik. 2007. Positioning of chemosensory clusters in *E. coli* and its relation to cell division. *EMBO J.*, 26:1615–1623.
220. Thumanu, K., J. Cha, J.F. Fisher, R. Perrins, S. Mobashery, and C. Wharton. 2006. Discrete steps in sensing of β-lactam antibiotics by the BlaR1 protein of the methicillin-resistant *Staphylococcus aureus* bacterium. *Proc. Natl. Acad. Sci. U.S.A.*, 103:10630–10635.
221. Tipper, D.J. and J.L. Strominger. 1965. Mechanism of action of penicillins: a proposal based on their structural similarity to acyl-D-alanyl-D-alanine. *Proc. Natl. Acad. Sci. U.S.A.*, 54:1133–1141.
222. Tiyanont, K., T. Doan, M.B. Lazarus, X. Fang, D.Z. Rudner, and S. Walker. 2006. Imaging peptidoglycan biosynthesis in *Bacillus subtilis* with fluorescent antibiotics. *Proc. Natl. Acad. Sci. U.S.A.,* 103:11033–11038.

223. Tomasz, A. 2006. The staphylococcal cell wall, p. 443–455. In V.A. Fischetti, R.P. Novick, J.J. Ferretti, D.A. Portnoy, and J.I. Rood (Eds.), *Gram-Positive Pathogens*, second ed. ASM Press, Washington, DC.
224. Tomasz, A. and W. Fischer. 2006. The cell wall of *Streptococcus pneumoniae*. In V.A. Fischetti, R.P. Novick, J.J. Ferretti, D.A. Portnoy, and J.I. Rood (Ed.), *Gram-Positive Pathogens*, second ed. ASM Press, Washington, DC, p. 230–240.
225. Ton-That, H., G. Liu, S.K. Mazmanian, K.F. Faull, and O. Schneewind. 1999. Purification and characterization of sortase, the transpeptidase that cleaves surface proteins of *Staphylococcus aureus* at the LPXTG motif. *Proc. Natl. Acad. Sci. U.S.A.*, 96:12424–12429.
226. Touhami, A., M.H. Jericho, and T.J. Beveridge. 2004. Atomic force microscopy of cell growth and division in *Staphylococcus aureus. J. Bacteriol.*, 186:3286–3295.
227. Travassos, L.H., S.E. Girardin, D.J. Philpott, D. Blanot, M.A. Nahori, C. Werts, and I.G. Boneca. 2004. Toll-like receptor 2-dependent bacterial sensing does not occur via peptidoglycan recognition. *EMBO, Rep.* 5:1000–1006.
228. Ubeda, C., E. Maiques, E. Knecht, I. Lasa, R.P. Novick, and J.R. Penades. 2005. Antibiotic-induced SOS response promotes horizontal dissemination of pathogenicity island-encoded virulence factors in *staphylococci. Mol. Microbiol.*, 56:836–844.
229. Uehara, T. and J.T. Park. 2002. Role of the murein precursor UDP-N-acetylmuramyl-L-Ala-gamma-D-Glu-meso-diaminopimelic acid-D-Ala-D-Ala in repression of β-lactamase induction in cell division mutants. *J. Bacteriol.*, 184:4233–4239.
230. Uehara, T., K. Suefuji, N. Valbuena, B. Meehan, M. Donegan, and J.T. Park. 2005. Recycling of the anhydro-N-acetylmuramic acid derived from cell wall murein involves a two-step conversion to N-acetylglucosamine-phosphate. *J. Bacteriol.*, 187:3643–3649.
231. van Dam, V., R. Sijbrandi, M. Kol, E. Swiezewska, B. de Kruijff, and E. Breukink. 2007. Transmembrane transport of peptidoglycan precursors across model and bacterial membranes. *Mol. Microbiol.*, 64:1105–1114.
232. van den Ent, F., M. Leaver, F. Bendezu, J. Errington, P. de Boer, and J. Lowe. 2006. Dimeric structure of the cell shape protein MreC and its functional implications. *Mol. Microbiol.*, 62:1631–1642.
233. Van Loosdrecht, M.C.M., J. Lyklema, W. Norde, G. Schraa, and A.J.B. Zehdner. 1987. The role of bacterial cell wall hydrophobicity in adhesion. *Appl. Environ. Microbiol.*, 53:1893–1897.
234. Varma, A. and K.D. Young. 2004. FtsZ collaborates with penicillin binding proteins to generate bacterial cell shape in *Escherichia coli. J. Bacteriol.*, 186:6768–6774.
235. Vicente, M., J. Hodgson, O. Massidda, T. Tonjum, B. Henriques-Normark, and E.Z. Ron. 2006. The fallacies of hope: will we discover new antibiotics to combat pathogenic bacteria in time?. *FEMS Microbiol. Rev.*, 30:841–852.
236. Vollmer, W. 2006. The prokaryotic cytoskeleton: a putative target for inhibitors and antibiotics?. *Appl. Microbiol. Biotechnol.*, 73:37–47.
237. Vollmer, W. 2007. Structure and biosynthesis of the murein (peptidoglycan) sacculus. In M. Ehrmann (Ed.), *The Periplasm*. ASM Press, Washington, DC, p. 198–213.
238. Vollmer, W. and J.V. Holtje. 2004. The architecture of the murein (peptidoglycan) in Gram-negative bacteria: vertical scaffold or horizontal layer(s)?. *J. Bacteriol.*, 186:5978–5987.
239. Walsh, C.T. 2003. *Antibiotics: Actions, Origins, Resistance*. ASM Press, Washington, DC.
240. Wehenkel, A., P. Fernandez, M. Bellinzoni, V. Catherinot, N. Barilone, G. Labesse, M. Jackson, and P.M. Alzari. 2006. The structure of PknB in complex with mitoxantrone, an ATP-competitive inhibitor, suggests a mode of protein kinase regulation in *Mycobacteria. FEBS Lett.*, 580:3018–3022.
241. Wei, C., J. Yang, J. Zhu, X. Zhang, W. Leng, J. Wang, Y. Xue, L. Sun, W. Li, J. Wang, and Q. Jin. 2006. Comprehensive proteomic analysis of S*higella flexneri* 2a membrane proteins. *J. Proteome Res.*, 5:1860-5.
242. Weiss, D.S. 2004. Bacterial cell division and the septal ring. *Mol. Microbiol.*, 54:588–597.
243. Weiss, D.S., J.C. Chen, J.M. Ghigo, D. Boyd, and J. Beckwith. 1999. Localization of FtsI (PBP3) to the septal ring requires its membrane anchor, the Z ring, FtsA, FtsQ, and FtsL. *J. Bacteriol.*, 181:508–520.
244. Wiedemann, B., D. Pfeifle, I. Wiegand, and E. Janas. 1998. β-Lactamase induction and cell wall recycling in Gram-negative bacteria. Drug Resist. Updat., 1:223–226.
245. Wiedemann, I., E. Breukink, C. van Kraaij, O.P. Kuipers, G. Bierbaum, B. de Kruijff, and H.G. Sahl. 2001. Specific binding of nisin to the peptidoglycan precursor lipid II combines pore formation and inhibition of cell wall biosynthesis for potent antibiotic activity. *J. Biol. Chem.*, 276:1772–1729.

246. Wientjes, F.B., C.L. Woldringh, and N. Nanninga. 1991. Amount of peptidoglycan in cell-walls of Gram-negative bacteria. *J. Bacteriol.*, 173:7684–7691.
247. Wilke, M.S., T.L. Hills, H.Z. Zhang, H.F. Chambers, and N.C. Strynadka. 2004. Crystal structures of the Apo and penicillin-acylated forms of the BlaR1 β-lactam sensor of *Staphylococcus aureus*. *J. Biol. Chem.*, 279:47278–47287.
248. Yao, X., M. Jericho, D. Pink, and T.J. Beveridge. 1999. Thickness and elasticity of Gram-negative murein sacculi measured by atomic force microscopy. *J. Bacteriol*., 181:6865–6875.
249. Yeats, C., R.D. Finn, and A. Bateman. 2002. The PASTA domain: a β-lactam-binding domain. *Trends Biochem. Sci.*, 27:438.
250. Yoshimura, A., E. Lien, R.R. Ingalls, E. Tuomanen, R. Dziarski, and D. Golenbock. 1999. Cutting edge: recognition of Gram-positive bacterial cell wall components by the innate immune system occurs via toll-like receptor 2. *J. Immunol.*, 163:1–5.
251. Young, K.D. 2006. The selective value of bacterial shape. *Microbiol. Mol. Biol. Rev.*, 70:660–703.
252. Young, K.D. 2006. Too many strictures on structure. *Trends Microbiol.,* 14:155–156.
253. Young, T.A., B. Delagoutte, J.A. Endrizzi, A.M. Falick, and T. Alber. 2003. Structure of *Mycobacterium tuberculosis* PknB supports a universal activation mechanism for Ser/Thr protein kinases. *Nat. Struct. Biol.*, 10:168–174.
254. Yuan, Y., D. Barrett, Y. Zhang, D. Kahne, P. Sliz, and S. Walker. 2007. Crystal structure of a peptidoglycan glycosyltransferase suggests a model for processive glycan chain synthesis. *Proc. Natl. Acad. Sci. U.S.A.*, 104:5348–5353.
255. Zhang, P., C.M. Khursigara, L.M. Hartnell, and S. Subramaniam. 2007. Direct visualization of *Escherichia coli* chemotaxis receptor arrays using cryo-electron microscopy. *Proc. Natl. Acad. Sci. U.S.A.*, 104:3777–3781.
256. Zhang, Y., E.J. Fechter, T. Wang, D. Barrett, S. Walker, and D.E. Kahne. 2007. Synthesis of heptaprenyl-Lipid IV to analyze peptidoglycan glycosyltransferases. *J. Am. Chem. Soc.*, 129:3080–3081.
257. Zuber, B., M. Haenni, T. Ribeiro, K. Minnig, F. Lopes, P. Moreillon, and J. Dubochet. 2006. Granular layer in the periplasmic space of Gram-positive bacteria and fine structures of *Enterococcus gallinarum* and *Streptococcus gordonii* septa revealed by cryo-electron microscopy of vitreous sections. *J. Bacteriol.*, 188:6652–6660.

14 Bacterial Cell Breakage or Lysis

Matthew E. Bahamonde

CONTENTS

INTRODUCTION

While bacterial cell lysis is a relatively straightforward process, there are significant variations in its practice. The type of lysis often depends on what subcellular component is to be purified, such as the isolation of DNA vs. protein, the scale of the project, miniprep vs. industrial-scale, and the type of microbe, Gram-negative vs. Gram-positive. In general, microbial lysis methods can be broken down into three categories based on the relative harshness/intensity of the method; Gentle, Moderate, or Vigorous (see Table 14.1. for examples of each).[1] Many of these methods give comparable results and only personal preference plays a role in selection. Others are better suited to specific tasks and should be looked on with preference.

The methods of lysis have been categorized by type of subcellular component to be isolated. The most common components isolated are nucleic acids, DNA (either genomic or plasmid) and RNA, proteins (either endogenous or recombinant), cell membranes (ghosts), and cell wall (liposaccharide). For both nucleic acids and proteins, there is also discussion of the scale of the project, with attention paid to the requirements of large-scale lysis. Differences in lysis between Gram-positive and Gram-negative bacteria as well as archaebacteria are pointed out where appropriate.

NUCLEIC ACID LYSIS

The two primary types of DNA (genomic and plasmid) extracted from bacteria require different lysis methods. The extraction of RNA requires additional care due to the fragility of the RNA molecule. For either type, a gentle method is best to prevent cleavage or denaturation of the DNA or degradation of the RNA.

TABLE 14.1
Various Lysis Methods Catergorized by Their Intensity

Gentle	Moderate	Vigorous
Alkaline vlysis	Mortar/pestle-French press	Manton-Gaulin Homogenizer
Triton X-100	Blender	Bead mill
Phenol		Sonication
Boiling		
Proteinase K		
Freeze/thaw		
Lysozyme/EDTA		

Plasmid DNA

During the preparation of plasmid DNA, lysis should allow for the plasmids to exit the cell without releasing any of the genomic DNA. This is commonly accomplished via alkaline/sodium dodecyl sulfate (SDS) lysis. During this procedure, the bacteria are exposed to a mild (0.2 N) sodium hydroxide (NaOH) solution in the presence of 1% SDS. Exposure to these chemicals is kept relatively short, usually around 5 minutes, and is often done on ice to further slow the reaction. The reason is to prevent complete loss of the cell membrane due to solubilization by the SDS. The SDS also begins to denature proteins. In the NaOH, both proteins and DNA are denatured and RNA is degraded. The genomic DNA is often sheared during the lysis process; and upon denaturation, complementary strands can drift away from each other. The plasmid DNA is also denatured but because it is relatively small and circular, the complementary strands will be joined like links in a chain. When reannealing occurs, the strands will basepair, properly reconstituting the double-stranded DNA plasmids. The genomic DNA strands will not be near their complement and will reanneal into a large mass of intertwined DNA. This reannealing occurs due to neutralization by the addition of acidic potassium or sodium acetate. The addition of these salts also causes proteins and SDS to precipitate. This precipitation traps the larger genomic DNA fragments. The preparation is cleared by centrifugation, leaving plasmid DNA, some RNA, and a few proteins in the supernatant. RNase treatment followed by extraction with organic solvents leaves a solution of nearly pure plasmid DNA that can be precipitated with alcohol.[2, 3] The alkaline lysis method has become the method of choice for the preparation of plasmid DNA and is the method found in the commercial plasmid DNA kits listed in Table 14.2.

A second common method of lysis uses heat (boiling) instead of high pH and a non-ionic detergent such as Triton X-100 or Tween-20 instead of SDS. Although these detergents are not as disruptive as SDS, the heat completely denatures both the proteins and the DNA. Upon cooling, coagulated proteins trap genomic DNA and both precipitate out of solution. The boiling method does contain more RNA as it is not destroyed by the heat as it is by high pH. This method is often augmented by the addition of lysozyme (activity explained below).[4, 5]

Prior to the lysis steps outlined above, there is usually the requirement of concentrating the bacterial cells via centrifugation and resuspension in a smaller volume. Usually the buffered solution used to resuspend contains a chelating agent, typically ethylenediaminetetraacetic acid (EDTA), and an osmotic regulator such as glucose. EDTA is a chelator of divalent cations and helps to inactivate DNases by removing Mg^{2+}. It also aids in disrupting the outer envelope of Gram-negative bacteria by removing Mg^{2+} from the lipopolysaccharide (LPS) layer. The osmotic regulator is important to prevent the bacterial cells from bursting in the hypotonic solution, which would defeat the purpose of the controlled lysis procedures.

TABLE 14.2
Commercial Suppliers of Bacterial Nucleic Acid Purification Kits

Supplier[1]	Plasmid DNA[2]	Genomic DNA	RNA
Qiagen Inc. www.qiagen.com	QIAprep Spin Miniprep Kit HiSpeed Plasmid Kit QIAfilter Plasmid Kit QIAGEN Plasmid Kit EndoFree Plasmid Kit	DNeasy Blood & Tissue Kit	RNeasy Protect Bacteria Mini Kit
Promega Corp. www.promega.com	Wizard® MagneSil® Plasmid DNA Purification System Wizard® Plus DNA Purification System Wizard® SV 9600 Plasmid DNA Purification System	Wizard® Genomic DNA Purification Kit	RNA PureYield™
Bio-Rad Laboratories www.biorad.com	Aurum Plasmid Kits	*AquaPure* Genomic DNA Isolation Kit	AquaPure RNA Isolation Kit Aurum Total RNA 96 Kit
Ambion Inc. www.ambion.com			RiboPure™-Bacteria Kit
Sigma-Aldrich www.sigmaaldrich.com	GenElute™ Plasmid Kit	GenElute™ Bacterial Genomic Kit	GenElute Bacterial Total RNA Purification Kit
Qbiogene Inc. www.qbiogene.com	PhoenIX™ Kit RapidPURE™ Plasmid Kit RPM® Kit	GNOME® DNA Kit FastDNA® Kit	FastRNA® Pro Soil-Direct Kit RNaid® Plus Kit
MoBio Laboratories www.mobio.com	UltraClean™ Standard Plasmid Prep Kit UltraClean-htp™ 96 Well Plasmid Prep Kit	UltraClean™ Microbial DNA Kit UltraClean™-htp 96 Well Microbial DNA Kit PowerMicrobial™ Midi DNA Isolation Kit PowerSoil™ DNA Isolation Kit PowerMax™ Soil DNA Isolation Kit UltraClean™ Water DNA Kit-	UltraClean™ Microbial RNA Kit RNA PowerSoil™ Total RNA Isolation Kit
EMD Chemicals Inc www.emdbiosciences.com	Mobius™ Plasmid Kits UltraMobius™ Plasmid Kits	MagPrep® Bacterial Genomic DNA Kit	
Epicentre Biotechnologies www.epibio.com	PlasmidMAX™ DNA Isolation Kit	MasterPure™ Gram Positive DNA Purification Kit SoilMaster™ DNA Extraction Kit WaterMaster™ DNA Purification Kit	Ready Lyse™ Lysozyme Solution
Invitrogen Corp. www.invitrogen.com	PureLink™ HiPure Plasmid Purification Kits	ChargeSwitch® gDNA Mini Bacteria Kit Easy-DNA™ Kit PureLink™ Genomic DNA Purification Kit	PureLink™ Micro-to-Midi™ Total RNA Purification System
Clontech Laboratories Inc. www.clontech.com	NucleoBond® Plasmid Purification Kits NucleoSpin Plasmid Purification Kits--	NucleoSpin® Tissue Kit	NucleoSpin® RNA II Kit

[1] The suppliers shown are the most common, but are by no means the only suppliers. The kits listed are representative of the suppliers product line and do not constitute their entire catalog.

[2] Plasmid DNA Kits are often available for varying yields of DNA: Mini up to 20 μg, Midi up to 100 μg, Maxi up to 500 μg, Mega up to 2500 μg, and Giga up to 10000 μg.

Genomic DNA

Unlike a plasmid DNA preparation, a genomic or chromosomal DNA preparation requires the complete degradation of the cell's peptidoglycan layer and cell membrane. Gram-negative bacteria also need their LPS layer disrupted. Digestion with the enzyme lysozyme is required for Gram-positive bacteria and is useful (but not required) for Gram-negative bacteria. Lysozyme hydrolyzes β(1-4)-linkages between N-acetylmuraminic acid and N-acetyl-D-glucosamine residues in peptidoglycans.[6] Other enzymes (Table 14.3) may also prove useful for specific bacterial types that may be resistant to lysozyme. The cells are then subjected to extended incubation (>15 minutes) in 1% SDS at elevated temperature (37° to 45°C). This is either followed or accompanied by proteolytic enzyme treatment to further remove proteins. Proteinase K is the enzyme of choice because it actually works better in the presence of SDS. Treatment with DNase-free RNase is also required to remove RNA contamination. As with plasmid preparations, chelation of divalent cations such as Mg^{2+} with EDTA is important for both lysis and for maintaining the integrity of the chromosomal DNA.[7] Finally, it is critical to treat the lysate gently to prevent shearing of the chromosomal DNA. The purification of genomic DNA, like plasmid DNA, has been reduced to a commercial kit. A list of suppliers is found in Table 14.2.

Mechanical disruption has been found to increase DNA yields when used in conjunction with enzymatic lysis. This is of particular concern when trying to increase the sensitivity of PCR reactions.[8] Some commercial kits, such as those from MoBio Laboratories (www.mobio.com), rely on mechanical, concussive disruption in the form of bead beating (see Table 14.2), which breaks up bacterial cell walls mechanically by vibrating bacteria with microbeads at high speed.[9] This method can be used with any bacterial strain and is particularly usefully for pathogenic bacteria as there are virtually no unlysed cells, a possibility with enzymatic disruption due to uneven mixing or inactivated enzyme.

RNA

During RNA isolation, the most important issue is stabilization of the RNA. RNases, unlike DNases, are very stable and are found throughout the lab. In addition, because bacterial mRNAs lack a 5′ cap and 3′ poly(A) tail, they are more unstable than their eukaryotic counterparts, with

TABLE 14.3
Enzymes Used During Bacterial Lysis

Enzyme	Mode of Action	Sensitive Microbe
Achromopeptidase	A lysyl endopeptidase	Gram-positive bacteria resistant to lysozyme
Labiase	Contains β-N-acetyl-D-glucosaminidase and lysozyme activity	Gram-positive bacteria such as *Lactobacillus*, *Aerococcus*, and *Streptococcus*
Lysostaphin	Cleaves polyglycine cross-links in peptidoglycan	*Staphylococcus* species
Lysozyme	Cleaves between N-acetylmuraminic acid and N-acetyl-D-glucosamine residues in peptidoglycans	Gram-positive cells, with Gram-negative cells being less susceptible
Mutanolysin	Cleaves N-acetylmuramyl-β(1-4)-N-acetylglucosamine linkages	Lyses *Listeria* and other Gram-positive bacteria such as *Lactobacillus* and *Lactococcus*

Source: Modified from: http://www.sigmaaldrich.com/Area_of_Interest/Biochemicals/Enzyme_Explorer/Key_Resources/Lysing_Enzymes.html

an average half-life of approximately 3 min.[10] Traditionally, chaotropic (chaos producing) agents such as guanidinium isothiocyanate or guanidinium hydrochloride are used to inactivate RNases and preserve RNA. However, there are proprietary reagents such as RNAprotect (by Qiagen Inc., www.qiagen.com) that can be added to a bacterial culture prior to lysis to protect RNA and reduce artifactual changes in expression that may occur during lysis. In addition to treating the culture, all solutions, glassware, and plasticware must be RNase-free. Solutions except those containing Tris must be treated with 0.1% diethylpyrocarbonate (DEPC) to inactivate RNases. Autoclaving of solutions after DEPC treatment destroys the DEPC. Tris solutions cannot be treated because Tris reacts with DEPC to inactivate it. Instead, Tris solutions are made using DEPC-treated water. Glassware must be baked at 300°C for at least 4 hr (autoclaving does not inactivate all RNases).[5] Plasticware should be bought already RNase free. Alternatively, plasticware and equipment can be cleaned with Ambion's RNase *Zap* (www.ambion.com).

The actual lysis can be done either enzymatically or mechanically. Gram-negative bacteria can be lysed with lysozyme alone or in combination with Proteinase K. Gram-positive bacteria should be lysed with both lysozyme and Proteinase K. Certain difficult-to-lyse Gram-positive bacteria may require mechanical disruption with bead beating or a tissue homogenizer in addition to enzymatic lysis. A tissue homogenizer, sold under various brand names, is a rapidly turning rotor that draws liquid-suspended cells into a core region. The cells are repeatedly cycled through narrow slits that mechanically shear them. Mechanical lysis on its own is suitable for a wide range of bacteria and is rapid, but overall RNA yields are generally lower than when used in conjunction with enzymatic methods.[11] Depending on the purification method, RNase-free DNase may be required to remove any contaminating DNA.

Large-Scale Lysis

Most bacterial lysis for plasmid DNA is done in the laboratory, with yields varying in size from 20 μg (minprep) to 10 mg (gigaprep). The advent of applications such as gene therapy and DNA vaccines will require the production of multigram to kilogram amounts of purified plasmid DNA in a cost-effective manner. Traditional protocols such as alkaline lysis do not scale up well due to complex mixing requirements that can leave local regions of extremely high pH, which can irreversibly damage plasmids. Use of lysozyme is also problematic due to cost. Large-scale lysis of bacterial filter cakes via fragilization followed by heat has been shown to be viable.[12]

PROTEIN LYSIS

Lysis for protein extraction, either endogenous or recombinant proteins, requires mechanical or chemical (enzymatic and mild detergent) disruption because protocols such as alkaline lysis deliberately destroy proteins. The addition of serine protease inhibitors such as phenylmethylsulfonyl fluoride (PMSF) or aprotinin is also strongly encouraged to improve yield. Lysis also needs to be carried out either on ice or at 4°C to protect the proteins from denaturing heat.

These methods are primarily for the recovery of exogenous, recombinant proteins but are also useful for the recovery of endogenous bacterial proteins.

Mechanical Lysis

Traditionally, bacteria are lysed through liquid homogenization. In this method, cells are sheared by forcing them through a small opening. With bacteria, this is often done with a handheld tissue homogenizer or the larger French press. A French press consists of a piston that applies high pressure to a sample, forcing it through a tiny hole in the press. Only two passes are required for efficient lysis due to the high pressures used with this process. The equipment is expensive but the French press is often the method of choice for breaking bacterial cells mechanically. Typical sample volumes range from 40 to 250 mL. Sonication or the use of high-frequency sound waves is commonly used for samples of less than 100 mL. A vibrating probe is immersed in a liquid cell suspension, and energy from the probe creates shock waves that reverberate through the sample. The primary disadvantage of mechanical disruption — in addition to the need for expensive specialized equipment — is the potential for sample overheating. Either generalized sample heating or localized regions of overheating within a sample can denature proteins, thus lowering the final yield. Lysis must not only be performed at low temperature, but particularly with sonication, the lysis needs to be done in short bursts with enough time in between to allow the sample to cool again.

Because mechanical lysis breaks cells completely, all intracellular components (including nucleic acids) are released. These can increase the viscosity of the lysate and complicate downstream processing. To resolve this, either DNase and RNase together are added to the sample or Benzonase® (Registered trademark of Merck KgaA, Darmstadt, Germany) alone. Benzonase is an endonuclease that degrades all forms of DNA and RNA (single stranded, double stranded, linear, and circular). One final method, freeze/thaw, can also be used to lyse bacterial cells for the release of recombinant proteins.[13] This technique involves repeated cycles of freezing a cell suspension and then thawing to room temperature or 37°C. Ice crystals that form during the freezing process cause the cells to swell and ultimately burst. The problem is that the process can consume a lot of time and is effective only on relatively small samples.

Chemical Lysis

For small-scale lysis, chemical methods are effective and economical. Many companies sell protein lysis kits based on this method (Table 14.4). These kits contain a non-ionic, detergent-based lysis buffer, lysozyme, and DNase/RNase or Benzonase. Non-ionic detergents are considered more gentle than SDS and less likely to denature proteins.[14] However, most purification schemes will require the extra step of removal of the detergent. The major disadvantage of these methods is their relative expense, due primarily to the enzymes involved.

Large-Scale Lysis

In commercial or very large-scale laboratory settings, it is common to require lysis of 10s of liters to 1000s of liters of cells. As stated above, chemical means are not viable at large-scale due to the expense, so mechanical lysis is the most common method of large-scale lysis. Of the methods discussed, neither the French press nor sonication is amenable to large-scale protein production. The most common methods are bead beating with a mechanical ball mill or the Manton-Gaulin Homogenizer. These are effective at lysing large volumes of cells. Continuously operating ball mills have the capability of lysing up to 40 L per hour. Due to the tremendous heat generated by the process, ball mills are jacket-cooled, some employing liquid nitrogen to prevent overheating of samples.[15]

BACTERIAL GHOSTS

Bacterial ghosts are empty cells with intact non-denatured envelopes.[16] They have potential uses as drug carriers and vaccine adjuvants.[16–18] They retain all the functional and antigenic determinants found in the living bacteria. Bacterial ghosts are generated through controlled expression of the PhiX174 gene *E* in Gram-negative bacteria. PhiX174 gene *E* encodes a membrane protein that fuses the inner and outer membranes, creating a tunnel through which the cytoplasmic content of the bacteria is expelled. Except for the tunnel generated by the *E* gene product, the morphology and cell surface structures are unaffected by the lysis. The lumen of the ghost can be filled with vaccine[17] or drugs.[16] In addition to the greater carrying capacity as compared to other delivery systems, ghosts also have the potential to target tissues such as the mucosal surfaces of the gastrointestinal and respiratory tracts, and their uptake by phagocytes and M cells.[16]

TABLE 14.4
Commercial Suppliers of Bacterial Protein Purification Kits

Supplier	Kits
Bio-Rad Laboratories www.biorad.com	MicroRotofor lysis kit ReadyPrep Protein Extraction Kit (Soluble/Insoluble)
EMD Chemicals Inc. www.emdbiosciences.com	BugBuster® 10X Protein Extraction Reagent BugBuster® HT Protein Extraction Reagent RoboPop™ Ni-NTA His•Bind® Purification Kit RoboPop™ GST•Bind™ Purification Kit
Epicentre Biotechnologies www.epibio.com	EasyLyse™ Bacterial Protein Extraction Solution Ready-Lyse™ Lysozyme Solution for Protein Extraction ReadyPreps™ Protein Preparation Kit (for total cellular proteins) PeriPreps™ Periplasting Kit (for periplasmic proteins)
Pierce Biotechnology http://www.piercenet.com	B-PER II Bacterial Protein Extraction Reagent B-PER Bacterial Protein Extraction Reagent
Promega Corp. www.promega.com	MagneHis™ Protein Purification System MagneGST™ Protein Purification System Maxwell® 16 Polyhistidine Protein Purification Kit
Qbiogene Inc. www.qbiogene.com	FastProtein Blue Matrix
Qiagen Inc. www.qiagen.com	Qproteome Bacterial Protein Prep Kit His-tag Purification and Detection Kits Two-Step Affinity Purification System
Sigma-Aldrich www.sigmaaldrich.com	CelLytic™ Express CelLytic B Plus Kit

LPS LYSIS

While not technically lysis, the liposaccharide cell envelope of Gram-positive bacteria can be removed and purified via extraction. The most common method uses phenol and hot water.[19] Alternatives include butanol and water, as well as detergent.[20]

ARCHAEBACTERIA

Because the cell walls of archaebacteria differ from those of eubacteria, enzymatic lysis is not the optimal method of disruption. Archaebacterial cell walls are composed of the polysaccharide pseudomurein and contain no peptidoglycan.[21] This precludes the use of lysozyme. However, most mechanical methods of lysis, such as the French press or bead beating, are effective and allow archaebacteria to be treated the same as eubacteria.

CONCLUSIONS

The lysis methods discussed above are the most popular but not necessarily the sole methods of lysis. For a more exhaustive list of lysis methods, see the prior edition of this Handbook.[22]

REFERENCES

1. Scopes, R.K., *Protein Purification: Principles and Practice*, 3rd ed. Springer-Verlag, New York, 1994.
2. Birnboim, H.C., A rapid alkaline extraction method for the isolation of plasmid DNA, *Methods Enzymol.,* 100, 243–255, 1983.
3. Birnboim, H.C. and Doly, J., A rapid alkaline extraction procedure for screening recombinant plasmid DNA, *Nucleic Acids Res.,* 7(6), 1513–1523, 1979.
4. Ausubel, F.M., *Current Protocols in Molecular Biology*, John Wiley & Sons, New York, 2001.
5. Sambrook, J. and Russell, D.W., *Molecular Cloning : A Laboratory Manual*, 3rd ed. Cold Spring Harbor Laboratory Press, Cold Spring Harbor, NY, 2001.
6. Feiner, R.R., Meyer, K., and Steinberg, A., Bacterial lysis by lysozyme, *J. Bacteriol.,* 52(3), 375–384, 1946.
7. Brown, W.C., Sandine, W.E., and Elliker, P.R., Lysis of lactic acid bacteria by lysozyme and ethylenediaminetetraacetic acid, *J. Bacteriol.,* 83, 697–698, 1962.
8. Zoetendal, E.G., Ben-Amor, K., Akkermans, A.D., Abee, T., and de Vos, W.M., DNA isolation protocols affect the detection limit of PCR approaches of bacteria in samples from the human gastrointestinal tract, *Syst. Appl. Microbiol.,* 2(3), 405–410, 2001.
9. Odumeru, J., Gao, A., Chen, S., Raymond, M., and Mutharia, L., Use of the bead beater for preparation of *Mycobacterium paratuberculosis* template DNA in milk, *Can. J. Vet. Res.,* 65(4), 201–205, 2001.
10. McGarvey, D.J. and Quandt, A., *Stabilization of RNA Prior to Isolation*, QIAGEN GmbH, 2003.
11. QIAGEN, I., RNAprotect® Bacteria Reagent Handbook, 2005.
12. O'Mahony, K., Freitag, R., Hilbrig, F., Muller, P., and Schumacher, I., Proposal for a better integration of bacterial lysis into the production of plasmid DNA at large scale, *J. Biotechnol.,* 119(2), 118–132, 2005.
13. Johnson, B.H. and Hecht, M.H., Recombinant proteins can be isolated from *E. coli* cells by repeated cycles of freezing and thawing, *Biotechnology (N.Y.),* 12(13), 1357–1360, 1994.
14. Godson, G.N. and Sinsheimer, R.L., Lysis of *Escherichia coli* with a neutral detergent, *Biochim. Biophys. Acta,* 149(2), 476–488, 1967.
15. Majors, R., Sample Preparation for Large-Scale Protein Purification, in *LC•GC Europe,* 2006.
16. Huter, V., Szostak, M.P., Gampfer, J., Prethaler, S., Wanner, G., Gabor, F., and Lubitz, W., Bacterial ghosts as drug carrier and targeting vehicles, *J. Control Release,* 61(1–2), 51–63, 1999.
17. Jalava, K., Hensel, A., Szostak, M., Resch, S., and Lubitz, W., Bacterial ghosts as vaccine candidates for veterinary applications, *J. Control Release,* 85(1–3), 17–25, 2002.
18. Lubitz, W., Witte, A., Eko, F. O., Kamal, M., Jechlinger, W., Brand, E., Marchart, J., Haidinger, W., Huter, V., Felnerova, D., Stralis-Alves, N., Lechleitner, S., Melzer, H., Szostak, M.P., Resch, S., Mader, H., Kuen, B., Mayr, B., Mayrhofer, P., Geretschlager, R., Haslberger, A., and Hensel, A., Extended recombinant bacterial ghost system, *J. Biotechnol.,* 73(2–3), 261–73, 1999.
19. Westphal, O. and Jann, K., Bacterial lipopolysaccharides. Extraction with phenol-water and further applications of the procedure., *Methods Carbohydr. Chem.,* 5, 83–93, 1965.
20. Joiner, K.A., McAdam, K.P., and Kasper, D.L., Lipopolysaccharides from *Bacteroides fragilis* are mitogenic for spleen cells from endotoxin responder and nonresponder mice, *Infect. Immun.,* 36(3), 1139–1145, 1982.
21. Balch, W.E., Fox, G.E., Magrum, L.J., Woese, C.R., and Wolfe, R.S., Methanogens: reevaluation of a unique biological group, *Microbiol. Rev.,* 43(2), 260–296, 1979.
22. O'Leary, W.M., *Practical Handbook of Microbiology,* CRC Press, Boca Raton, FL, 1989.

15 Major Culture Collections and Sources

Lorrence H. Green

Many sources exist for obtaining microbiological species in pure culture, which can be used as references. Foremost among these is the American Type Culture Collection (ATCC) (P.O. Box 1549, Manassas, VA 20108, USA) (www.ATCC.org).

Several companies such as Microbiologics (www.microbiologics.com) and Remel (www.remelinc.com) also supply ATCC microorganisms.

The World Data Centre for Microorganisms (http://wdcm.nig.ac.jp/hpcc.html) lists 526 culture collections in 67 countries in the world. Among these are:

Agricultural Research Service Culture Collection (NRRL)
(http://nrrl.ncaur.usda.gov/cgi-bin/usda/)

Australian Collection of Microorganisms:
(http://ilcfmp5.facbacs.uq.edu.au/smms/ACM/FMPro?-db=ACM.fp5&-format=ACM_info.html&-view)

Belgium Coordinated Collection of Microorganisms (BCCM)
(http://bccm.belspo.be/index.php)

Centraalbureau voor Schimmelcultures. An Institute of the Royal Netherlands Academy of Arts and Sciences
(www.cbs.knaw.nl/collections/index.htm)

Collection de l'Institut Pasteur (CIP)
(http://www.crbip.pasteur.fr/)

German Collection of Microorganisms and Cell Cultures
(www.dsmz.de).
This site also offers up-to-date information about nomenclature
(www.dsmz.de/microorganisms/bacterial_nomenclature.php)

Japan Collection of Microorganisms
(www.jcm.riken.go.jp)

The National Collection of Industrial, Marine and Food Bacteria (Scotland)
(www.ncimb.com/culture.html)

16 Epidemiological Methods in Microbiology

D. Ashley Robinson

CONTENTS

INTRODUCTION

Epidemiological methods in microbiology include laboratory and analytical tools that are used to study the microbial distributions and determinants of infectious disease in human populations. The scope of this chapter is limited to a discussion of the tools used to study the molecular epidemiology of bacterial infectious disease. Natural populations of named bacterial species can accumulate immense levels of biological variation, but only a portion of their variation will contribute to their ability to cause infectious disease. Increasingly powerful tools have been developed to probe bacterial chromosomes for the variation of greatest epidemiological relevance. We discuss the criteria used to evaluate different tools, the concepts used to interpret the biological variation revealed by these tools, and the epidemiological applications for which these tools are deployed.

OVERVIEW OF STRAIN TYPING

SOME DEFINITIONS AND ASSUMPTIONS

Although the fundamental concept of bacterial species is still subject to debate, some subspecies taxonomic units of epidemiological relevance have been defined [1, 2]. An isolate is defined as a group of bacterial cells that represent a pure culture obtained from a single colony grown on a solid medium [3, 4]. Isolates are collected from sources such as clinical specimens, fomites, and environmental samples. Genotypes and phenotypes refer to characteristics of genes and their expression products, respectively. A strain is defined as a group of isolates that are genotypically or phenotypically identical to each other and distinct from other such isolates [3, 4]. Strain typing tools are the laboratory techniques used to assess variation in genotypes and phenotypes. Markers refer to the specific genes and expression products assessed by strain typing. A clone is defined as a group of isolates genotypically identical to each other due to common ancestry [3, 4]. Multiple genotypic markers must be examined to classify isolates into a clone.

Isolates that are part of the same chain of transmission are defined as epidemiologically related. A fundamental assumption that underlies the epidemiological use of strain typing is that isolates that are epidemiologically related are also clonally related [5, 6]. This assumption is based on the reasoning that if isolates share a recent common ancestor, they are also likely to share a common set of virulence factors. Virulence factors refer to the bacterial products that enable bacteria to damage their host and result in infectious disease. It is reasoned that a common set of virulence factors will result in a common pattern of infectious disease. Thus, clones are actually proxies for virulence factors.

CRITERIA OF STRAIN TYPING TOOLS

Numerous strain typing tools are available to distinguish among isolates of a virulent bacterial species. However, the selection of a strain typing tool should be tailored to the particular epidemiological problem presented [7, 8]. We note that not all epidemiological problems require the use of strain typing. For example, an inadequately ventilated environment can be linked to an outbreak of pneumonia without any knowledge of the microscopic organism that is causing the disease. It has been emphasized that strain typing should be viewed as complementary to a thorough epidemiological investigation, and should be done as a means to achieve the humanitarian goal of epidemiology to prevent infectious disease [9, 10]. We also note that strain typing has applications outside the domain of public health research. The need to distinguish among bacterial isolates arises in systematics and taxonomy, population genetics, and other basic endeavors that intersect microbiology.

Several considerations must be made when selecting an appropriate strain typing tool for a particular epidemiological application [7, 8]. Not all biological variation provides a valid target for strain typing. The validity of a strain typing tool is based on its ability to distinguish between epidemiologically related and unrelated isolates, which defines the sensitivity and specificity of the tool, respectively [11]. A strain typing tool should be validated for a particular epidemiological application, which means that it should be able to confirm established epidemiological findings. No "gold standard" exists that can be used for validation. Validation of a strain typing tool is done empirically by comparison of the tool with previously validated tools and by demonstration that epidemiologically related and unrelated isolates can be distinguished [11].

Three basic performance criteria have been used to evaluate different strain typing tools [3, 10, 12]. These criteria include discriminatory ability, reproducibility, and typeability. Discriminatory ability is defined as the ability of a strain typing tool to detect differences between isolates. Simpson's index of diversity is widely used to quantify discriminatory ability and takes into account both the richness and abundance of types in a sample of isolates [13]. This index estimates the probability that two isolates randomly selected from a sample will have different types. A statistical test is available to compare the discriminatory ability of different strain typing tools applied to the same sample of isolates, or to compare different samples of isolates (e.g., from cases and controls) evaluated with the same strain typing tool [14]. Reproducibility is defined as the ability of a strain typing tool to yield the same results for the same isolates upon repeated testing. Both within- and between-laboratory reproducibility are evaluated. Typeability is defined as the ability of a strain typing tool to score a positive result for each isolate. Non-typeable isolates produce either null or ambiguous results. All three performance criteria are affected by experimental variation in laboratory techniques and by biological variation in the isolates [3, 10, 12].

Numerous convenience criteria are also used to evaluate different strain typing tools [15]. These criteria include the ease of performance (technical simplicity), the speed of performance (rapidity), the applicability to a wide taxonomic range of species (versatility), and cost. Additional criteria have been proposed, including epidemiological concordance, stability, and ease of interpretation [3, 12]. Epidemiological concordance is analogous to validity. Stability and ease of interpretation are discussed separately below.

INSIGHTS FROM BACTERIAL POPULATION GENETICS

Mutation and Recombination

To interpret the results of strain typing tools, the stability of the markers should be taken into account. Stability is defined as the constancy of the markers over time. Stability is best considered within a population genetics framework, which allows the causes of biological variation to be studied [16–18]. Bacteria are haploid, which means that they contain a single copy of their chromosome(s). Bacteria also reproduce asexually, which means that two identical daughter cells, or clones, are produced from a single parent cell. Mutations refer to the heritable genetic variations that occur within DNA. These variations include nucleotide substitutions, insertions, and deletions, which occur during DNA replication. The movements of mobile genetic elements (e.g., insertion sequences, larger transposons, lysogenic phage), which are not necessarily linked to replication, are also considered to be mutations. When the bacterial DNA mismatch-repair machinery undergoes mutations that impair its function, clones can experience a high rate of mutation. The frequency of mutations in a population can be altered by stochastic fluctuations in population size (genetic drift) and by natural selection. Some mutations can increase the fitness of a clone in a particular environment, and thus its frequency, while other mutations have no effect on fitness or are deleterious and are slowly replaced. Demographic factors that reduce genetic variation, such as population subdivisions and bottlenecks, also alter the frequency of mutations. These processes combine to cause bacteria to naturally form a clonal population genetic structure [16–18].

Horizontal genetic transfer and recombination between clones can occur if the clones have both a mechanism (genetic factors) and opportunity (ecological factors) to exchange genes. Parasexual processes such as conjugation, transduction, and transformation enable clones to horizontally transfer genetic material [19]. Recombination between clones can reduce genetic variation if it occurs frequently within a population. In this sense, recombination is a homogenizing force that can speed the replacement of less favorable mutations. Recombination between clones can preserve genetic variation if it occurs frequently between different populations. In this sense, recombination can allow rare alleles to escape extinction. At a molecular level, recombination can increase genetic variation by combining mutations, thereby creating new alleles. In population genetics parlance, a unique position on a chromosome is called a locus, and a unique combination of mutations at a locus is called an allele. The consequences of recombination between clones are profound for bacterial population genetic structure and infectious disease epidemiology. Recombination between clones can cause virulence factors to embark on independent chains of transmission. If recombination occurs frequently enough, clones can become too ephemeral to be epidemiologically relevant taxonomic units to identify [20].

Criteria of Clonality

Two basic criteria are used to determine whether the genetic structure of a bacterial population is more consistent with having been formed by clonal or recombinant processes [21]. These criteria include (1) the repeated isolation of a spatially and temporally stable clone and (2) the non-random association of alleles at different loci. It has been emphasized that a clone should occur more frequently than a random assortment of alleles would produce [20]. The repeated isolation of a clone in a bacterial population that experienced frequent recombination would occur only if natural selection favored a unique combination of alleles across loci. Such selection would be unlikely if the study examined genetic variation in multiple loci that encode products of unrelated physiological function. In this case, the simplest explanation is that recombination is rare and that periodic selection is influencing genetic variation in a relatively clonal population [22].

As clones propagate and experience mutations, clonal lineages develop. Because the chromosomes within a clonal lineage share a common ancestry, the alleles that occur at one locus will be statistically associated with the alleles that occur at other loci. This non-random occurrence of genetic variation is called linkage disequilibrium (LD) [21]. However, processes other than clonal population growth can cause LD. For example, bacterial populations that consist of genetically isolated subpopulations, where recombination occurs frequently within but not between subpopulations, can produce apparent LD if the strain typing data of the subpopulations are analyzed together. In this case, the LD that is detected is due to fixed mutations that distinguish the subpopulations. Such bacterial subpopulations have been called cryptic species [20]. Additionally, the spread of highly fit clones can produce temporary LD even if recombination occurs frequently within the bacterial population. These highly fit clones can achieve wide geographic distributions, but their genetic identity eventually will be eroded by recombination with other clones. This situation describes an epidemic population genetic structure and, interestingly, may have been first recognized by epidemiologists [20, 23].

The randomizing effects of recombination serve as null hypotheses for a variety of statistical tests that are used to detect LD. Some tests require strain typing tools that can identify distinct alleles at a locus [20, 24, 25], whereas other tests are more general and can be applied to a variety of strain typing tools [26]. New tests have been developed that offer increasingly powerful means to study bacterial population genetic structure [27]. These types of tests have demonstrated that virulent bacterial species display a spectrum of population genetic structures, ranging from the extremes of clonal to freely recombining (panmictic) [20, 25]. Even within the boundaries of a named bacterial species, there can be differences in the relative influence of recombination vs. mutation on population genetic structure [28, 29].

In addition to recombination, natural selection can also prevent or promote the accumulation of variation at a locus. Strong LD can cause the effects of natural selection at one locus to impact variation at nearby loci via a hitchhiking phenomenon [30]. Markers therefore differ in the rate at which they evolve and in the quality of their phylogenetic signal [16–18]. It is important to keep in mind that the stability of the markers that are assessed by strain typing is a reflection of the natural histories of both clones and their genes; thus, clone phylogeny should be distinguished from marker phylogeny [31].

DESCRIPTION OF STRAIN TYPING TOOLS

Phenotypes and Genotypes

Strain typing tools are commonly classified according to whether they assess phenotypes or genotypes. Historically, strain typing tools were based on phenotypes. While phenotypic characteristics can be highly correlated with epidemiological relationships, they are generally poor indicators of evolutionary relationships [32, 33]. The expression of phenotypes can be strongly influenced by the environment and other variable characteristics (e.g., temperature, nutrients, growth phase), so a common phenotype might reflect something other than a common ancestry in a group of isolates [32, 33]. Phenotyping tools have been used to assess variation in metabolic characteristics (biotyping), susceptibility to lytic bacteriophage (phage typing), susceptibility to antibiotics (resistotyping), immunological reactivity to antibodies (serotyping), and other characteristics. Multilocus enzyme electrophoresis (MLEE) is a phenotyping tool that was pivotal to the development of bacterial population genetics [20, 22]. MLEE indirectly assesses variation in gene sequence. It relies on differences in the electrostatic charge of different alleles at expressed enzyme loci to migrate at different rates during starch gel electrophoresis [34]. MLEE is the forerunner of a genotyping tool called multilocus sequence typing (MLST) that directly assesses variation in gene sequence from fragments of enzyme loci and other loci that encode products with cellular housekeeping functions [35]. Because of the relative evolutionary instability of phenotypic characteristics, phenotyping tools can exhibit problems with all three performance criteria [32, 33].

The MLEE-to-MLST transition mirrors the general trend of strain typing tools to develop from phenotypic to genotypic characteristics [36]. Genotyping tools have now been developed that allow many isolates to be typed at once (high throughput) and that allow much of the chromosome to be assessed for variation (high coverage) [37]. Unfortunately, the development of new genotyping tools has tended to outpace their evaluation for particular epidemiological applications [38]. Numerous reviews of genotyping tools have been published that provide a qualitative score of various performance criteria (e.g., low, moderate, or high discriminatory ability) [5, 10, 18, 39]. However, it is not obvious how well these scores reflect the criteria in different species. In addition to technical factors that can similarly influence these scores in all species, biological factors that can differ between species also influence these scores. For example, recombination between clones can increase the number of alleles per locus. Thus, a genotyping tool that can identify distinct alleles at a locus may show higher discriminatory ability in recombinant species when compared to more clonal species. Species-specific evaluations of genotyping tools may therefore be the most informative [40].

Band-Based Genotyping Tools

We find it convenient to classify genotyping tools according to whether the data they generate are electrophoretic banding patterns (band-based), nucleotide sequences (sequence-based), or the presence or absence of genes (binary-based). Band-based genotyping tools are also called imaged-based and molecular fingerprinting tools. In general, they indirectly assess variation in gene sequence and are based on the polymerase chain reaction (PCR) and restriction fragment length polymorphism (RFLP). PCR was a major innovation in molecular biology developed in the 1980s. A "gold rush" of genotyping tools making use of PCR occurred throughout the 1990s [33, 39]. PCR is used to exponentially and selectively replicate, or amplify, a DNA target through the design of primer sequences that flank the target. PCR reaction mixtures include a thermostable DNA polymerase and other reagents that amplify the DNA target through cycles of template denaturation, primer annealing, and primer extension. The amplified products, or amplicons, can be separated with agarose gel electrophoresis and visualized with a stain that binds DNA (e.g., ethidium bromide). PCR is a highly sensitive technique, and can be applied directly to clinical specimens.

The amplification of random sequences has been called randomly amplified polymorphic DNA (RAPD) and arbitrarily primed PCR (AP-PCR) [41, 42]. This technique uses short random primer sequences (~10 bp) and low stringency amplification conditions. The primers will bind random loci in the chromosome and will initiate replication if two primer sites are located sufficiently close to each other and oriented properly on complementary strands. The number and location of these random loci can differ among isolates. Repetitive sequences are found in all bacterial chromosomes, and have been frequently used as primer sites [43, 44]. Several families of taxonomically widely distributed repetitive sequences have been identified, including repetitive extragenic palindromic (REP) sequences, enterobacterial repetitive intergenic (ERIC) sequences, and mosaic repeat sequences called BOX elements. The number and location of these repetitive loci can differ among isolates. Another feature of bacterial chromosomes is short sequence repeats (SSR), also called variable number of tandem repeats (VNTR), that occur together and often within genes [45, 46]. Amplification of these sequences can reveal differences in copy number among isolates. An even more powerful extension of this technique is called multilocus VNTR analysis (MLVA), which amplifies different SSR sequences from multiple loci [47]. Finally, most bacterial chromosomes have multiple copies of their 16S and 23S ribosomal RNA (rRNA) operons. The homology of these sequences can result in intra-chromosomal recombination events that alter the length of the rRNA intergenic spacer regions among isolates [48].

RFLP refers to variation that is revealed through the digestion of DNA templates with restriction endonucleases [5, 10, 39]. These enzymes recognize short nucleotide sequences (~4 to 6 bp) called restriction sites with varying levels of specificity depending on the enzyme. If the restriction sites occur within a DNA template, then the enzymes will precisely cut the template at these sites, resulting in pieces of a template called restriction fragments. These fragments can be separated by agarose gel electrophoresis. The number and location of the restriction sites can differ among isolates. Banding patterns from RFLP analyses can be made less complex by Southern blotting and hybridization with probes of rRNA genes (ribotyping), insertion sequences (IS typing), and other loci [5, 10, 39]. One of the most widely used band-based genotyping tools is called pulsed field gel electrophoresis (PFGE) [49]. This technique uses restriction endonucleases that recognize relatively few restriction sites in a chromosome, resulting in a small number (~10 to 30) of large fragments (~10 to 800 kb) [4, 10]. These large restriction fragments are separated by special electrophoresis equipment that alternates the direction of current. PCR techniques have also been combined with RFLP techniques. For example, a band-based genotyping tool called amplified fragment length polymorphism (AFLP) is based on the digestion of chromosomal DNA followed by ligation of primer sites for PCR amplification onto the ends of the restriction fragments [50, 51]. Capillary electrophoresis can be used to separate the resulting complex mixture of AFLP amplicons.

While discriminatory ability is a strength of many band-based genotyping tools, reproducibility is problematic. Reproducibility is improved for PCR techniques when specificity is increased between the primer sequences and the DNA targets that they bind. For example, PCR based on REP sequences can lead to more reproducible results than RAPD [40, 44]. Standardized protocols aimed at improving reproducibility, and thus interpretation of data, have been developed for some band-based genotyping tools. For example, PFGE protocols have been standardized to the point that public health networks utilize this tool for national surveillance of foodborne bacterial pathogens [52]. Convenience criteria such as rapidity and cost are strengths of PCR techniques, but are problematic for some RFLP techniques. Despite their drawbacks, band-based genotyping tools have been very useful for local outbreak investigations [53].

Sequence-Based Genotyping Tools

Sequence-based genotyping tools assess variation directly from the nucleotide sequences of either a single locus or multiple loci. These tools provide a fundamental view of genetic variation [35]. The Sanger method of sequencing is currently used for strain typing. PCR templates are linearly replicated using a reaction mixture that includes fluorescently labeled dideoxynucleotides and other reagents. Incorporation of dideoxynucleotides interrupts replication, leading to a ladder of amplicons, each with its 3′-terminus marked by a fluorescent label that can be detected by an automated instrument. The development of new methods for sequencing based on coupled enzymatic reactions provides reason for optimism that larger portions of a bacterial chromosome may become economical to sequence in the future [54]. In the meantime, different regions of a chromosome must be selectively targeted for sequencing. Single locus sequence typing (SLST) has targeted highly variable sequences within a locus such as SSR sequences [55]. In contrast, MLST has targeted conserved loci such as the sort of housekeeping loci used for phenotyping by MLEE [56]. The

MLST technique utilizes partial nucleotide sequences (~450 to 500 bp) from a standard set (~7) of housekeeping loci. Unique sequences at each locus define alleles, and unique combinations of alleles across loci define sequence types (STs). MLST databases of allele and ST definitions are publicly accessible via the Internet for more than 30 microbial species, most of which are bacterial species [57].

Sequence-based genotyping tools achieve the highest levels of reproducibility. SLST and MLST have exhibited enough discriminatory ability to be utilized for local outbreak investigations [58, 59]. Public health networks have used MLST for international surveillance of respiratory bacterial pathogens [57]. Because MLST assesses variation in relatively stable genes, it has also been an important tool for studying bacterial population genetics [17, 25]. Convenience criteria such as rapidity and cost have probably been the major impediments to the wider application of sequence-based genotyping tools.

Binary-Based Genotyping Tools

Binary-based genotyping tools assess variation in gene content and are generally based on hybridization of labeled chromosomal DNA to an array of sequences that are fixed onto a surface such as a membrane or glass slide (microarray or gene-chip) [60]. The reverse of this technique can also be performed, whereby labeled sequences are hybridized to an array of chromosomal DNA preparations that are fixed onto a surface (library on a slide) [61]. The sequences can consist of parts of a chromosome, such as virulence factors and other markers, or they can be more representative of the entire contents of a reference strain's genome using arrayed amplicons or oligonucleotides as done with the binary-based genotyping tool called comparative genomic hybridization (CGH) [62–65]. A genome is defined as the entire complement of genetic material of an isolate, and includes both the chromosome and extrachromosomal plasmids. CGH studies have shown that up to a quarter of the contents of a genome can differ among isolates [60, 64, 65]. Suppressive subtractive hybridization (SSH) is a different type of binary-based genotyping tool that involves PCR-enriched cloning of the differences between two strains, and identifies sequences that are present in a tester strain's genome and absent from a reference strain's genome [66, 67]. SSH studies have revealed new candidate virulence factors for infectious diseases that have complicated etiologies [68–71]. Binary-based genotyping tools can exhibit high discriminatory ability but they can also present problems with reproducibility and they can be time-consuming and costly to use. These tools have most often been applied to studies of the association between virulence factors and infectious disease.

INTERPRETATION OF STRAIN TYPING DATA

The interpretation of strain typing data can be complicated even if the performance of the selected strain typing tool and the stability of the surveyed markers are well-characterized. In some cases, the fundamental assumption underlying the epidemiological use of strain typing is violated; epidemiologically related isolates will not always be clonally related. For example, contamination of food with sewage can result in diarrheal outbreaks due to multiple, unrelated clones [7]. Likewise, susceptible persons can become ill by simultaneous infection with multiple, unrelated clones [5, 11]. Furthermore, recombination between clones can spread virulence factors (e.g., antibiotic resistance genes) to multiple, unrelated clones [72]. For other reasons, clonally related isolates will not always be epidemiologically related. The selected strain typing tool can be too sensitive or not sensitive enough to identify the epidemiologically relevant taxonomic units. For example, undetected variations, such as virulence factors that are imported on mobile genetic elements or point mutations that slightly alter gene expression, may cause isolates that appear to be identical based on strain typing to behave with distinct epidemiological patterns [73, 74]. To reinforce the notion that strain typing tools typically do not detect all the variations that occur within a bacterial chromosome, it has been emphasized that the genotypically identical isolates of a given clone should be referred to as indistinguishable rather than identical [4, 10]. Finally, infectious disease is a particular outcome of an interaction that is influenced by both bacterial virulence factors and by host risk factors. Strain typing provides information about the bacterial determinants of infectious disease but, especially for facultative bacterial pathogens, the host's risk factors can decisively influence the outcome of the bacteria-host interaction.

Strain typing data is more straightforward to interpret when isolates are shown to be either identical or completely different. The major difficulty in interpretation arises when isolates are shown to be similar, but not identical, because the level of similarity that is epidemiologically relevant must be asserted [6, 8, 10, 18]. Interpretive criteria have been proposed but these criteria are specific for particular strain typing tools and particular epidemiological applications. For example, PFGE data are interpreted in local outbreak investigations according to the number of banding pattern differences that can arise per genetic event [4, 10]. The loss of a restriction site due to a single point mutation can result in a three-band difference. Thus, to allow for some variation to accumulate during the course of an outbreak, isolates that differ by one to three bands are interpreted as probably part of the same outbreak. Sequence-based strain typing tools allow genetic events to be more precisely discerned [58], but judgment calls regarding the epidemiological relevance of bacterial variation are made with all strain typing tools. Clearly, interpretive criteria depend on how rapidly clones and markers evolve and the elapsed time since the isolates shared a common ancestor [6, 18]. Typically, none of these parameters are known with precision. On the one hand, quantification of the rate of marker evolution in a laboratory may not reflect its rate of evolution in nature. On the other hand, the elapsed time since isolates shared a common ancestor is classified into the broad categories of short term or long term, depending on the particular epidemiological application.

The analysis of clonal relatedness can be greatly facilitated by phylogenetic clustering algorithms. These algorithms enable the tree-like (bifurcating) structures of clonal lineages to be reconstructed from strain typing data, and they enable the statistical support for these structures to be quantified [75, 76]. Moreover, algorithms are available to detect the network-like (reticulating) structures that result from recombination between clones [77, 78]. The biological variation that can accumulate within bacterial populations is immense [22]. Phylogenetic clustering provides a means to collapse the variation that is detected through strain typing into biological groups that might have some epidemiological relevance. The need to use sophisticated phylogenetic clustering algorithms depends on the particular epidemiological application [7]. However, we note that even the classification of genotypically indistinguishable isolates into clones follows a phylogenetic clustering algorithm, albeit a simple one. In summary, strain typing data are used to generate hypotheses about the clonal relatedness of isolates, which can be tested independently of any epidemiological information. Interpretive criteria are used to attempt to link these strain typing hypotheses to epidemiological hypotheses.

EPIDEMIOLOGICAL APPLICATIONS OF STRAIN TYPING

COMPARATIVE AND LIBRARY TYPING

We find it convenient to classify the epidemiological applications of strain typing according to the scale of inquiry. These applications include local outbreak investigations (comparative typing), global surveillance studies (library typing), and virulence factor discovery efforts (association mapping). The scale of inquiry influences the precision with which clones must be identified. The distinction between comparative and library typing has been reviewed [8]. However, a systematic approach to association mapping is a relatively new development [79].

Comparative typing refers to the use of strain typing tools to assess the clonal relatedness of epidemiologically related and unrelated isolates that are collected on a limited spatial and temporal scale [8]. Comparative typing can be applied to test prior epidemiological hypotheses that have identified a source as a possible cause of an outbreak of infectious disease. Strain typing is essentially confirmatory for this application because it is expected that the epidemiologically related isolates will also be clonally related; infection control decisions can be made without the strain typing data [7]. Comparative typing can also be applied to identify unsuspected sources in a spatial or temporal cluster of infectious disease. With this application, strain typing can point to a need for an epidemiological investigation or it can prevent a superfluous investigation [7, 11]. Finally, comparative typing can be applied to distinguish between reinfection of a person with a new clone vs. reactivation of a chronic infection (e.g., tuberculosis) [80]. Strain typing tools with high discriminatory ability are needed for these applications because the putative outbreak isolates must be distinguished from potentially closely related sporadic isolates that constitute the endemic bacterial population. PCR tools and SLST are examples of comparative typing systems [53, 59]. Comparative typing allows closely related isolates to be classified into clones, and their routes of transmission and reservoirs to be studied. This information can contribute to the development of infection control practices within an institution or within a community [7, 8, 53].

Library typing refers to the use of strain typing tools to assess the clonal relatedness of surveillance isolates collected on a broad spatial and temporal scale [8]. Library typing can be applied to monitor the global circulation of clones and their distributions in different host populations. These data allow the emergence or re-emergence of clones to be detected, and they can distinguish between the spread of clones vs. the spread of virulence factors [6, 7, 37]. Strain typing tools that exhibit high reproducibility and assess relatively stable markers are needed for this application. In addition, these tools should be high throughput, should have standardized protocols, and should provide a nomenclature that facilitates between-laboratory communication about the clones [8]. PFGE and MLST are examples of library typing systems [52, 57]. Library typing allows distantly related isolates to be classified into clonal lineages, and their patterns of spread and virulence trends to be studied [72, 81]. This information can contribute to the development of public health strategies at national and international levels [8, 37].

Association Mapping

Association mapping refers to the use of strain typing tools to identify the precise bacterial determinants (virulence factors) of infectious disease [79]. Association mapping can be used to identify virulence factors even when it is not clear whether clones differ in their virulence potential [68, 69, 71]. The virulence of a clone is often determined by its gene content, with more virulent clones differing from less virulent clones by the presence of conspicuous genetic elements. We note that the loss of genes can also lead to increased virulence [82]. Strain typing tools with high coverage of the chromosome, such as the binary-based CGH and SSH tools, are needed for this application [79]. Association mapping studies resemble epidemiological case-control studies, whereby isolates that are sampled from persons with infectious disease (cases) are compared for gene content with isolates that are sampled from persons without infectious disease (controls). Principles of sound epidemiological study design should be taken into account [7]. Moreover, because bacterial chromosomes can exhibit strong LD, it is necessary to subsequently assess whether the putative virulence factors that are identified have a causal association with infectious disease or are merely genetically linked to the authentic virulence factors [79, 83]. Experimental studies are needed for these subsequent steps.

Association mapping provides an observational approach to virulence factor discovery that is based on discerning bacterial exposures in human populations. In contrast, the experimental approach to virulence factor discovery that is often used in microbiology is based on assigning bacterial exposures in laboratory microcosm. The observational approach allows multiple bacterial exposures to be simultaneously evaluated in their natural setting, but its results can be confounded by imperfect study design. By comparison, the experimental approach addresses the biological plausibility of causation between bacterial exposures and infectious disease, but its results are often restricted to the simultaneous evaluation of a single bacterial exposure in a model of infectious disease that imperfectly represents the natural setting. Thus, the observational and experimental approaches to virulence factor discovery should be viewed as complementary to each other.

CONCLUDING REMARKS

Improvements in the laboratory tools used for strain typing and in the analytical tools used for interpreting strain typing data have led to substantial progress in identifying the bacterial determinants of infectious disease. It is now possible to identify these determinants on a global scale with molecular resolution. Knowledge of the characteristics of different strain typing tools, the population genetic structure of different bacterial species, and the peculiarities of different epidemiological applications should be integrated within an overarching conceptual framework. This need will become more acute as high-throughput, high-coverage genotyping tools become more routinely used to address epidemiological problems.

ACKNOWLEDGMENTS

We thank L. Riley, D. Smyth, and R. Willems for reviewing the manuscript. Work in our laboratory is supported in part by grants from the American Heart Association and the National Institute of General Medical Sciences (GM080602).

REFERENCES

1. Stackebrandt, E. et al., Report of the ad hoc committee for the re-evaluation of the species definition in bacteriology, *Int. J. Syst. Bacteriol.*, 52, 1043, 2002.
2. Dijkshoorn, L., Ursing, B.M., and Ursing, J.B., Strain, clone and species: comments on three basic concepts of bacteriology, *J. Med. Microbiol.*, 49, 397, 2000.
3. Struelens, M.J., Consensus guidelines for appropriate use and evaluation of microbial epidemiologic typing systems, *Clin. Microbiol. Infect.*, 2, 2, 1996.
4. Tenover, F.C. et al., Interpreting chromosomal DNA restriction patterns produced by pulsed-field gel electrophoresis: criteria for bacterial strain typing, *J. Clin. Microbiol.*, 33, 2233, 1995.
5. Arbeit, R.D., Laboratory procedures for the epidemiologic analysis of microorganisms, in *Manual of Clinical Microbiology*, 6th ed., Murray, P.M. et al., Eds., ASM Press, Washington, 1995, chap. 17.
6. Struelens, M., Molecular typing: a key tool for the surveillance and control of nosocomial infection, *Curr. Opin. Infect. Dis.*, 15, 383, 2002.
7. Foxman, B. et al., Choosing an appropriate bacterial typing technique for epidemiologic studies, *Epidemiol. Perspect. Innov.*, 2, 10, 2005.
8. Struelens, M.J., De Gheldre, Y., and Deplano, A., Comparative and library epidemiological typing systems: outbreak investigations versus surveillance systems, *Infect. Control Hosp. Epidemiol.*, 19, 565, 1998.
9. Foxman, B. and Riley, L., Molecular epidemiology: focus on infection, *Am. J. Epidemiol.*, 153, 1135, 2001.
10. Tenover, F.C., Arbeit, R.D., and Goering, R.V., How to select and interpret molecular strain typing methods for epidemiological studies of bacterial infections: a review for healthcare epidemiologists, *Infect. Control Hosp. Epidemiol.*, 18, 426, 1997.
11. Riley, L.W., *Molecular Epidemiology of Infectious Diseases: Principles and Practices,* ASM Press, Washington, DC, 2004, chap. 1.
12. Maslow, J. and Mulligan, M.E., Epidemiologic typing systems, *Infect. Control Hosp. Epidemiol.*, 17, 595, 1996.
13. Hunter, P.R. and Gaston, M.A., Numerical index of the discriminatory ability of typing systems: an application of Simpson's index of diversity, *J. Clin. Microbiol.*, 26, 2465, 1988.
14. Grundmann, H., Hori, S., and Tanner, G., Determining confidence intervals when measuring genetic diversity and the discriminatory abilities of typing methods for microorganisms, *J. Clin. Microbiol.*, 39, 4190, 2001.

15. Van Leeuwen, W.B., Molecular approaches for the epidemiological characterization of *Staphylococcus aureus* strains, in *MRSA: Current Perspectives*, Fluit, A.C. and Schmitz, F.J., Eds., Caister Academic Press, Norfolk, VA, 2003, chap. 5.
16. Tibayrenc, M., Toward an integrated genetic epidemiology of parasitic protozoa and other pathogens, *Annu. Rev. Genet.*, 33, 449, 1999.
17. Tibayrenc, M., Bridging the gap between molecular epidemiologists and evolutionists, *Trends. Microbiol.*, 13, 575, 2005.
18. Van Belkum, A. et al., Role of genomic typing in taxonomy, evolutionary genetics, and microbial epidemiology, *Clin. Microbiol. Rev.*, 14, 547, 2001.
19. Maynard Smith, J., Dowson, C.G., and Spratt, B.G., Localized sex in bacteria, *Nature*, 349, 29, 1991.
20. Maynard Smith, J. et al., How clonal are bacteria?, *Proc. Natl. Acad. Sci. U.S.A.*, 90, 4384, 1993.
21. Maynard Smith, J., Do bacteria have population genetics?, in *Population Genetics of Bacteria*, Baumberg, S. et al., Eds., Cambridge University Press, Cambridge, 1995, chap. 1.
22. Selander, R.K. and Levin, B.R., Genetic diversity and structure in *Escherichia coli* populations, *Science*, 210, 545, 1980.
23. Ørskov, F. and Ørskov, I., Summary of a workshop on the clone concept in the epidemiology, taxonomy, and evolution of the Enterobacteriaceae and other bacteria, *J. Infect. Dis.*, 148, 346, 1983.
24. Feil, E.J. et al., Estimating recombinational parameters in *Streptococcus pneumoniae* from multilocus sequence typing data, *Genetics*, 154, 1439, 2000.
25. Feil, E.J. et al., Recombination within natural populations of pathogenic bacteria: short-term empirical estimates and long-term phylogenetic consequences, *Proc. Natl. Acad. Sci. U.S.A.*, 98, 182, 2001.
26. Tibayrenc, M., Population genetics and strain typing of microorganisms: how to detect departures from panmixia without individualizing alleles and loci, *C. R. Acad. Sci. Paris*, 318, 135, 1995.
27. Didelot, X. and Falush, D., Inference of bacterial microevolution using multilocus sequence data, *Genetics*, 175, 1251, 2007.
28. Robinson, D.A. et al., Evolution and virulence of serogroup 6 pneumococci on a global scale, *J. Bacteriol.*, 184, 6367, 2002.
29. Robinson, D.A. et al., Evolution and global dissemination of macrolide-resistant group A streptococci, *Antimicrob. Agents Chemother.*, 50, 2903, 2006.
30. Maynard Smith, J. and Haigh, J., The hitch-hiking effect of a favourable gene, *Genet. Res.*, 23, 23, 1974.
31. Robinson, D.A. et al., Evolutionary genetics of the accessory gene regulator (*agr*) locus in *Staphylococcus aureus*, *J. Bacteriol.*, 187, 8312, 2005.
32. Tenover, F.C. et al., Comparison of traditional and molecular methods of typing isolates of *Staphylococcus aureus*, *J. Clin. Microbiol.*, 32, 407, 1994.
33. Zaidi, N., Konstantinou, K., and Zervos, M., The role of molecular biology and nucleic acid technology in the study of human infection and epidemiology, *Arch. Pathol. Lab. Med.*, 127, 1098, 2003.
34. Selander, R.K. et al., Methods of multilocus enzyme electrophoresis for bacterial population genetics and systematics, *Appl. Environ. Microbiol.*, 51, 873, 1986.
35. Maiden, M.C. et al., Multilocus sequence typing: a portable approach to the identification of clones within populations of pathogenic microorganisms, *Proc. Natl. Acad. Sci. U.S.A.*, 95, 3140, 1998.
36. Robertson, B.H. and Nicholson, J.K., New microbiology tools for public health and their implications, *Annu. Rev. Public. Health*, 26, 281, 2005.
37. Van Belkum, A., High-throughput epidemiologic typing in clinical microbiology, *Clin. Microbiol. Infect.*, 9, 86, 2003.
38. Achtman, M., A surfeit of YATMs, *J. Clin. Microbiol.*, 34, 1870, 1996.
39. Olive, D.M. and Bean, P., Principles and applications of methods for DNA-based typing of microbial organisms, *J. Clin. Microbiol.*, 37, 1661, 1999.
40. Witte, W., Strommenger, B., and Werner, G., Diagnostics, typing and taxonomy, in *Gram Positive Pathogens*, 2nd ed., Fishetti, V.A. et al., Eds., ASM Press, Washington, DC, 2006, chap. 31.
41. Williams, J.G. et al., DNA polymorphisms amplified by arbitrary primers are useful as genetic markers, *Nucleic Acids Res.*, 18, 6531, 1990.
42. Van Belkum, A. et al., Multicenter evaluation of arbitrarily primed PCR for typing of *Staphylococcus aureus* strains, *J. Clin. Microbiol.*, 33, 1537, 1995.
43. Versalovic, J., Koeuth, T., and Lupski, J.R., Distribution of repetitive DNA sequences in eubacteria and application to fingerprinting of bacterial genomes, *Nucleic Acids Res.*, 19, 6823, 1991.

44. Van Belkum, A. et al., PCR fingerprinting for epidemiological studies of *Staphylococcus aureus*, *J. Microbiol. Meth.*, 20, 235, 1994.
45. Van Belkum, A. et al., Short-sequence DNA repeats in prokaryotic genomes, *Microbiol. Mol. Biol. Rev.*, 62, 275, 1998.
46. Frenay, H.M. et al., Discrimination of epidemic and nonepidemic methicillin-resistant *Staphylococcus aureus* strains on the basis of protein A gene polymorphism, *J. Clin. Microbiol.*, 32, 846, 1994.
47. Lindstedt, B.A., Multiple-locus variable number tandem repeats analysis for genetic fingerprinting of pathogenic bacteria, *Electrophoresis*, 26, 2567, 2005.
48. Witte, W. et al., Increasing incidence and widespread dissemination of methicillin-resistant *Staphylococcus aureus* (MRSA) in hospitals in central Europe, with special reference to German hospitals, *Clin. Microbiol. Infect.*, 3, 414, 1997.
49. Snell, R.G. and Wilkins, R.J., Separation of chromosomal DNA molecules from *C. albicans* by pulsed field gel electrophoresis, *Nucleic Acids Res.*, 14, 4401, 1986.
50. Vos, P. et al., AFLP: a new technique for DNA fingerprinting, *Nucleic Acids Res.*, 23, 4407, 1995.
51. Grady, R. et al., Genotyping of epidemic methicillin-resistant *Staphylococcus aureus* phage type 15 isolates by fluorescent amplified length polymorphism analysis, *J. Clin. Microbiol.*, 37, 3198, 1999.
52. Swaminathan, B. et al., PulseNet: the molecular subtyping network for foodborne bacterial disease surveillance, United States, *Emerg. Infect. Dis.*, 7, 382, 2001.
53. Singh, A. et al., Application of molecular techniques to the study of hospital infection, *Clin. Microbiol. Rev.*, 19, 512, 2006.
54. Margulies, M. et al., Genome sequencing in microfabricated high-density picolitre reactors, *Nature*, 437, 376, 2005.
55. Shopsin, B. et al., Evaluation of protein A gene polymorphic region DNA sequencing for typing of *Staphylococcus aureus* strains, *J. Clin. Microbiol.*, 37, 3556, 1999.
56. Enright, M.C. et al., Multilocus sequence typing for characterization of methicillin-resistant and methicillin-susceptible clones of *Staphylococcus aureus*, *J. Clin. Microbiol.*, 38, 1008, 2000.
57. Maiden, M.C., Multilocus sequence typing of bacteria, *Annu. Rev. Microbiol.*, 60, 561, 2006.
58. Feavers, I.M. et al., Multilocus sequence typing and antigen gene sequencing in the investigation of a meningococcal disease outbreak, *J. Clin. Microbiol.*, 37, 3883, 1999.
59. Mellmann, A. et al., Automated DNA sequence-based early warning system for the detection of methicillin-resistant *Staphylococcus aureus* outbreaks, *PLoS Med.*, 3, e33, 2006.
60. Fitzgerald, J.R. et al., Evolutionary genomics of *Staphylococcus aureus*: insights into the origin of methicillin-resistant strains and the toxic shock syndrome epidemic. *Proc. Natl. Acad. Sci. U.S.A.*, 98, 8821, 2001.
61. Zhang, L. et al., Library on a slide for bacterial comparative genomics, *BMC Microbiol.*, 22, 4, 2004.
62. Saunders, N.A. et al., A virulence-associated gene microarray: a tool for investigation of the evolution and pathogenic potential of *Staphylococcus aureus*, *Microbiology*, 150, 3763, 2004.
63. Van Leeuwen, W., Validation of binary typing for *Staphylococcus aureus* strains, *J. Clin. Microbiol.*, 37, 664, 1999.
64. Dunman, P.M. et al., Uses of *Staphylococcus aureus* GeneChips in genotyping and genetic composition analysis, *J. Clin. Microbiol.*, 42, 4275, 2004.
65. Lindsay, J.A. et al., Microarrays reveal that each of the ten dominant lineages of *Staphylococcus aureus* has a unique combination of surface-associated and regulatory genes, *J. Bacteriol.*, 188, 669, 2006.
66. Agron, P.G. et al., Use of subtractive hybridization for comprehensive surveys of prokaryotic genome differences, *FEMS Microbiol. Lett.*, 211, 175, 2002.
67. Akopyants, N.S. et al., PCR-based subtractive hybridization and differences in gene content among strains of *Helicobacter pylori*, *Proc. Natl. Acad. Sci. U.S.A.*, 95, 13108, 1998.
68. Zhang, L. et al., Molecular epidemiologic approaches to urinary tract infection gene discovery in uropathogenic *Escherichia coli*, *Infect. Immun.*, 68, 2009, 2000.
69. Xie, J. et al., Molecular epidemiologic identification of *Escherichia coli* genes that are potentially involved in movement of the organism from the intestinal tract to the vagina and bladder, *J. Clin. Microbiol.*, 44, 2434, 2006.
70. Bille, E. et al., A chromosomally integrated bacteriophage in invasive meningococci, *J. Exp. Med.*, 201, 1905, 2005.
71. Oleastro, M. et al., Identification of markers for *Helicobacter pylori* strains isolated from children with peptic ulcer disease by suppressive subtractive hybridization, *Infect. Immun.*, 74, 4064, 2006.

72. Enright, M.C. et al., The evolutionary history of methicillin-resistant *Staphylococcus aureus* (MRSA), *Proc. Natl. Acad. Sci. U.S.A.*, 99, 7687, 2002.
73. Sandgren, A. et al., Effect of clonal and serotype-specific properties on the invasive capacity of *Streptococcus pneumoniae*, *J. Infect. Dis.*, 189, 785, 2004.
74. Silva, N.A. et al., Genomic diversity between strains of the same serotype and multilocus sequence type among pneumococcal clinical isolates, *Infect. Immun.*, 74, 3513, 2006.
75. Feil, E.J. et al., eBURST: inferring patterns of evolutionary descent among clusters of related bacterial genotypes from multilocus sequence typing data, *J. Bacteriol.*, 186, 1518, 2004.
76. Perez-Losada, M. et al., New methods for inferring population dynamics from microbial sequences, *Infect. Genet. Evol.*, 7, 24, 2007.
77. Posada, D. and Crandall, K.A., Evaluation of methods for detecting recombination from DNA sequences: computer simulations, *Proc. Natl. Acad. Sci. U.S.A.*, 98, 13757, 2001.
78. Cassens, I., Mardulyn, P., and Milinkovitch, M.C., Evaluating intraspecific "network" construction methods using simulated sequence data: do existing algorithms outperform the global maximum parsimony approach, *Syst. Biol.*, 54, 363, 2005.
79. Falush, D. and Bowden, R., Genome-wide association mapping in bacteria, *Trends Microbiol.*, 14, 353, 2006.
80. Murray, M. and Alland, D., Methodological problems in the molecular epidemiology of tuberculosis, *Am. J. Epidemiol.*, 155, 565, 2002.
81. Robinson, D.A. et al., Re-emergence of early pandemic *Staphylococcus aureus* as a community-acquired methicillin-resistant clone, *Lancet,* 365, 1256, 2005.
82. Maurelli, A.T. et al., "Black holes" and bacterial pathogenicity: a large genomic deletion that enhances the virulence of *Shigella* spp. and enteroinvasive *Escherichia coli*, *Proc. Natl. Acad. Sci. U.S.A.*, 95, 3943, 1998.
83. Johnson, J.R. et al., Experimental mouse lethality of *Escherichia coli* isolates, in relation to accessory traits, phylogenetic group, and ecological source, *J. Infect. Dis.*, 194, 1141, 2006.

Part II

Information on Individual Genus and Species, and Other Topics

17 The Family Enterobacteriaceae

J. Michael Janda and Sharon L. Abbott

CONTENTS

INTRODUCTION

There is no single family in the γ class of the phylum Proteobacteria that has had a greater impact on medicine, public health, molecular genetics and phylogeny, pathogenesis, gene structure, regulation, and function, or microbial ecology and physiology than the Enterobacteriaceae. The current edition of *Bergey's Manual of Systematic Bacteriology* lists 42 genera and over 140 validly published species in this family.[1] Current genera may contain as few as one (e.g., *Hafnia, Plesiomonas*) to as many as 12 distinct species (e.g., *Enterobacter*). The metamorphosis of the family over the past few decades and the explosion in the number of recognized taxa is a testimony to the early pioneering studies by the Centers for Disease Control and Prevention and the Pasteur Institute (French: Institut Pasteur) using DNA-DNA relatedness parameters and to more recent phylogenetic investigations employing 16S rRNA sequencing.

In contrast to the common misconception that enterobacteria are solely inhabitants of the gastrointestinal tract of vertebrates, the family Enterobacteriaceae is rather widely dispersed in nature and many species exist in free-living states in a variety of niches in the biosphere. These ecologic niches can be broadly broken down into four major groups, namely strains (1) principally associated in commensal or saprophytic states in the alimentary tract of humans and other vertebrates or at extraintestinal sites; (2) intimately associated with plants or plant diseases; (3) chiefly found in water, soil, invertebrate species, and industrial processes; and (4) obligate endosymbionts or commensals of insects.[2] Some species have developed such stable symbiotic relationships with their hosts that no genomic rearrangements or gene acquisitions have occurred over the past 50 million years.[3]

Due to the "molecular taxonomy" revolution, the classic biochemical features that previously identified membership in this family are no longer unequivocal, and their use has had to be redefined (Table 17.1). While still valid for most genera and species, an increasing number of exceptions continue to be reported. For bacteria that cannot be grown on routine media or that lack the required gene arrays to be cultured *in vitro*, inclusion in the family Enterobacteriaceae is based on a genetic rather than phenetic definition. Accurate biochemical identification of less commonly encountered groups of bacteria in this family requires reliance on a large battery of phenotypic tests, many of

TABLE 17.1
Defining Phenotypic Traits of the Family Enterobacteriaceae

Character	Marker for Family	Notable Exceptions
Metabolism:		
Facultatively anaerobic rods	+	*Alterococcus agarolyticus*
Morphology:		
Gram reaction	–	
Structure:		
Possession of ECA	+	*Dickeya chrysanthemi*
Spore formation	–	*Serratia marcescens (rare)*
Biochemical:		
Oxidase	–	*Plesiomonas*
Catalase	+	*Shigella dysenteriae 1, Xenorhabdus*
Nitrate reductase	+	*Erwinia and Yersinia (some)*
D-glucose	+	
D-xylose	+	*Cedecea (some), Edwardsiella, Morganella*

Note: ECA, enterobacterial common antigen.

which are not readily available on commercial panels. Several universal molecular probes targeting the small (16S) or large (23S) subunits of the ribosome have been developed that recognize almost all members of the enterobacterial family.[4,5] Although traditionally most members of this family have been susceptible to a wide variety of antimicrobial agents, including extended-spectrum cephalosporins, the exceptionally high usage of these drugs in hospital settings has led to the development of resistance to many β-lactam antibiotics via extended-spectrum β-lactamases (ESBLs) and carbapenemases.[6] Isolates of species commonly found to harbor such resistance factors include *Klebsiella pneumoniae, Enterobacter cloacae,* and *Escherichia coli,* and such emerging drug resistance is a major concern in healthcare settings.

Many species in the family Enterobacteriaceae serve as prototypes or models for studying global processes thought to play important roles in microbial communities, cell-to-cell communication, and gene expression with regard to virulence. *Escherichia coli* has long been an established vector system to clone and express genes and their gene products. Other high-profile processes that enterobacteria are thought to be intimately involved in include quorum sensing (*E. coli, Salmonella*)[7] and biofilm formation (*E. coli, Klebsiella*).

KLEBSIELLA AND *ENTEROBACTER*

KLEBSIELLA

Members of the genus *Klebsiella* are widely distributed in nature and are ubiquitous in forest environments, soil, vegetation, and water.[2] They can exist in a free-living state in the environment for prolonged periods of time. Klebsiellae are also found as normal commensals in the gastrointestinal tract of many vertebrates and mammals, including birds, reptiles, and even insects.

It is presently difficult attempting to define biochemical characteristics for inclusion in the genus *Klebsiella* because the genus has and is undergoing major taxonomic revisions. Most klebsiellae are non-motile, Voges-Proskauer (VP) positive (acetylmethycarbinol production), and indole-negative; however, there are exceptions to each of these traits. Other common traits that are traditionally associated with many/most *Klebsiella* species include urea hydrolysis and fermentation of *m*-inositol, a carbohydrate-like compound that is not commonly utilized by many other enterobacteria groups. Many klebsiellae isolates, particularly those of *Klebsiella pneumoniae* can produce copious capsular material (K antigen) that can often be observed on primary plating of isolates recovered from clinical specimens.

In the very broadest sense, up to 12 separate species have been listed in the genus *Klebsiella* as recently as 2000, including *K. mobilis* (=*Enterobacter aerogenes*).[1] Current taxonomic refinements suggest that three species (*K. ornithinolytica, K. planticola,* and *K. terrigena*) should be transferred to a new genus, *Raoultella,*[8] with two other previous species relegated to subspecies status within *K. pneumoniae* (ssp. *ozaenae* and ssp. *rhinoscleromatis*).[1] Further polyphasic taxonomic investigations will be required to clearly resolve these classification issues within the above-described proposed genera.

Klebsiella, and in particular *K. pneumoniae* and *K. oxytoca*, are extremely important human pathogens in the healthcare setting. These species are often responsible for devastating hospital-associated illnesses such as septicemia and endocarditis, pneumonia, and infections of the face (orbital cellulitis and endophthalmitis), bones (osteomyelitis), and urinary tract.[2] Of particular concern with regard to the treatment of such infections is the rapid emergence of ESBLs in this species and the difficulty in eradicating such systemic infections with the currently available arsenal of drugs.[9] Another trend with regard to *Klebsiella* infections involves the emergence of serious illnesses in Asia involving hepatic and lung abscesses apparently caused by *K. pneumoniae* strains carrying specific virulence determinants.[10, 11] Persons with underlying thalassemias in Asia appear to be especially prone to developing *Klebsiella* disease.[12] *K. pneumoniae* is also an important cause of animal infections, including mastitis in dairy cows.[13]

ENTEROBACTER

Most of the statements previously made regarding the genus *Klebsiella* equally apply to the genus *Enterobacter.* This includes their ubiquitous environmental distribution (trees, plants, crops, soil, water, foods), an incredibly complicated taxonomy, a lack of definitive phenotypic traits to characterize the genus, enhanced drug resistance mediated via ESBLs, and their overall importance as healthcare-associated pathogens.

This genus has numerous complex taxonomic issues, the most important of which concerns the preeminent pathogen of the genus, *Enterobacter cloacae. E. cloacae* as currently defined is genetically heterogeneous, being composed of from five to twelve distinct genomic groups within the complex that cannot be unambiguously separated from one another on a biochemical basis.[2] The second most common species in the genus, *E. aerogenes*, will most likely be eventually reclassified as a member of *Klebsiella* (see above), while another long-standing member (*E. agglomerans*) has already been reclassified into a new genus, *Pantoea.*[14] Most members of the genus are VP and ornithine decarboxylase-positive, utilize citrate, and ferment L-arabinose, L-rhamnose, raffinose, and D-xylose; exceptions to each of these characteristics exist.[2] Some species such as *E. sakazakii* produce yellow-pigmented colonies upon subculture.

Enterobacter cloacae causes many serious, life-threatening medical complications associated with hospitalization, among which are pediatric and adult bacteremia, central nervous system (CNS) disease and meningitis, and respiratory and urinary tract infections.[2, 15, 16] A less common but equally devastating pathogen is *E. sakazakii,* which causes serious invasive disease (meningitis, brain abscess) in infants and is linked to contaminated powered infant formula.[17, 18] *E. cloacae* and *E. asburiae* produce saxitoxins in the rumina of cattle associated with bovine paraplegic syndrome and can cause coliform mastitis in dairy cows.[2]

PROTEUS, MORGANELLA, AND *PROVIDENCIA*

In the 1980s, these three genera were grouped together in the tribe Proteeae, based on similar phenotypic and morphologic features. In many ways these associations still remain valid. Phylogenetically, these genera appear at the periphery of the Enterobacteriaceae based on 16S rRNA sequencing, and they are only 20% or less related to core members of this family (such as *Escherichia coli, Salmonella, Enterobacter,* and *Citrobacter*).

A number of biochemical features are almost exclusively associated with these three genera in the family Enterobacteriaceae and include production of phenylalanine deaminase, elaboration of a reddish-brown pigment on media containing DL-tryptophan, and the ability to degrade L-tryosine crystals via a tyrosine phenol-lyase. The most remarkable distinguishing feature within this tribe is the ability of many *Proteus* strains belonging to the species *Proteus mirabilis* and *P. vulgaris* to form concentric rings of outgrowth on solid media such as MacConkey agar.[2] This swarming phenomenon is due to conversion of swimmer cells in broth to swarmer cells in agar and is associated with cellular elongation and increased flagellin synthesis. Other features that help separate these three genera from one another include formation of H_2S on Triple Sugar Iron (TSI) agar slants and degradation of gelatin by *Proteus* and the fermentation of mannose by both *Morganella* and *Providencia.*

The singular disease principally associated with *Proteus, Morganella,* and *Providencia* is urinary tract infections (UTIs), and the preeminent pathogen of this group is *Proteus mirabilis* (up to 5% of all UTIs).[2] *P. mirabilis* typically produces uncomplicated UTIs in healthy women and less frequently in young children. Less often, *P. mirabilis* can cause complicated UTIs in persons with structural, physiologic, or neurologic disorders of the urinary tract necessitating catheterization. The virulence factors thought to be operative in *P. mirabilis* UTIs have been extensively studied and include urease, an IgA degrading protease, a hemolysin, and cell-associated fimbriae.[19] Proteeae are also common causes of bacteremia[20] and a variety of other less frequently encountered clinical conditions including CNS disease, wound infections, and rheumatoid arthritis. A recent finding has been the association of *Providencia* species and in particular *Providencia alcalifaciens* with episodes of travelers' diarrhea or outbreaks of gastroenteritis.[21, 22]

Members of the Proteeae are often isolated from homeothermic and poikilothermic species, including dogs, cows, birds, snakes, and fish.[2] *Morganella morganii,* a copious histamine producer, is associated with the spoilage of fish and the clinical condition known as scombroid poisoning.[23] A number of infections in animals have been attributed to these groups, including equine abortions, hoof canker, stomatitis, neonatal diarrhea in calves, and a large variety of additional systemic infections.

SALMONELLA

The genus *Salmonella* has two species, *S. bongori* (previously subspecies V) and *S. enterica*, with the latter species subdivided into six subspecies. All the commonly encountered serotypes belong in subspecies I of *S. enterica* while organisms previously known as *Arizona* fall within subspecies IIIA and IIIB. *S. bongori* and serotypes from all six subspecies of *S. enterica* have been isolated from humans although isolates of *S. bongori* and *S. enterica* subspecies VI are rare,[24] reptiles serve as the most common source for human infections caused by subspecies II, IIIA, IIIB, and IV.[25] Several serotypes are host-adapted (but can cause disease in other hosts, e.g., Dublin in cattle) or host-restricted (Gallinarum in poultry, Typhi in humans) but most serotypes cause disease in both warm- and cold-blooded animals. The gastrointestinal tract is the natural habitat for salmonellae but unlike *Escherichia coli* it can survive for weeks to years in water and soil.[26] Any food contaminated by fecal matter, including fresh fruits, vegetables, and shellfish, may serve as a source of salmonella infection; however, meats, raw milk, and eggs are the most frequently involved causes.[2] Infections may be acquired by direct transmission from both domesticated and wild animals and birds, although infrequently.

Non-typhoidal salmonellae (NTS) is a major cause of diarrheal disease in both industrialized and developing countries. In the United States between 1996 and 1999, the annual incidence rate was estimated to be 13.4 cases/100,000 population.[27] Rates for infants and children less than 6 years of age were even greater, with 117 and 34 cases/100,000 population, respectively. The mortality rate during this time was 0.6% overall, with estimates of the annual cost of salmonellae infections ranging from \$0.5 to \$3.2 billion. For typhoid fever (caused by *Salmonella enterica* serotype Typhi) the highest incidence rates occur in south-central and Southeast Asia (>100 cases/100,000 population), followed by other areas of Asia, Africa, Latin America, the Caribbean, and Oceania (exclusive of Australia and New Zealand) with rates of 10 to 100. In industrialized countries (Europe, North America, etc.), the rate is less than 10 per 100,000 population.[28] Fatality rates with typhoid have been reported to be less than 2% to 50% in endemic areas as opposed to less than 1% in the United States.[2]

Gastrointestinal illness, the most common infection caused by NTS, is characterized by abdominal pain and watery diarrhea (mild to cholera-like) with mucus and, in some cases, blood (occult or visible). Patients with NTS may develop a temporary carrier state usually lasting 1 to 2 months (occasionally up to a year with infants). Permanent carriers are rare (1%) with NTS infection; however, 2 to 5% of typhoid patients will become chronic carriers; 25% of typhoid carriers have no history of disease.[2] NTS cause a variety of extraintestinal illnesses, the most common and/or severe of which include sepsis, UTI, meningitis, and osteomyelitis. Enteric fever is caused by serotypes Typhi, Paratyphi A and C, all of which are host-adapted to humans, and serotype Paratyphi B. Untreated typhoid fever has three stages and persists approximately 3 to 4 weeks.[29] During the first week, symptoms are consistent with gastroenteritis, and stool cultures will be transiently positive while blood is usually negative. The second stage includes bacteremia and other extraintestinal symptoms, and cultures from blood, stool, bone marrow, and rose spots (if present) are generally positive. If complications arise, they usually occur in the third week; otherwise, the patient begins convalescence; cultures are not usually positive at this time, with the possible exception of stool.

Both NTS and serotypes causing enteric fever are easily isolated from stool in the acute phase of disease on traditional enteric isolation media such as MacConkey (MAC), xylose-lysine-desoxycholate (XLD), Hektoen (HE), and/or Salmonella-Shigella (SS) agars. Other media specific for salmonellae are also available (bismuth sulfite, brilliant green), as well as a variety of chromogenic agars. The sensitivity of the chromogenic agars vs. conventional media varies but these media have the advantage of fewer false positives. To detect carriers, pre-incubation in enrichment broth is recommended. Selenite F with cystine and Hajna's Gram-negative broth are generally used; tetrathionate does not recover serotype Typhi and should not be used for human specimens. Most standard blood culture methods are satisfactory for recovery of both NTS and agents of typhoid fever; however, for the latter group, use of 10% Oxgall in distilled water in lieu of tryptic soy broth and using a 1:10 dilution of blood (5 mL) to medium (25 mL), or culture of bone marrow, will increase the chance of recovery of these agents.[30, 31] Rapid diagnostic immunological assays have recently been developed for use in areas where blood or bone marrow culture is not readily available.[2]

All commercial systems approved for clinical diagnostic use in the United States identify *Salmonella* satisfactorily for phenotypically typical strains;[2] however, they fail to identify lactose-, sucrose-, or indol-positive strains of salmonellae. Strains of serotype Typhi are anaerogenic, produce little or no H_2S, and give negative reactions for citrate and ornithine. All subspecies of NTS, including serotype Typhi, are easily identified with a minimal number of conventional media, including lactose- and sucrose-positive strains that are not correctly identified using commercial systems. Most subgroup I strains of NTS are indol- and *o*-nitrophenyl-β-D-galactopyranoside (ONPG)-negative and H_2S-, LDC-, and citrate-positive and ferment dulcitol, a combination distinctive for salmonellae. Subgroup III strains are ONPG- and malonate-positive and dulcitol-negative. For epidemiological purposes, salmonellae are further characterized to serotype using the Kauffmann-White schema.[32] Cell surface (somatic) antigens place the organism within a group while the type is determined by a particular combination of flagellar antigens. There are more than 2000 serotypes of salmonellae distributed in more than 40 groups, although approximately 95% of salmonellae isolated from humans belong to groups A, B, C_1, C_2, D, and E. Typhi strains have a Vi (for virulence) capsule that masks the somatic antigen; full identification of these strains requires biochemical and serological testing, including testing for Vi antigen. Because of their frequent involvement in outbreaks, salmonellae are often subjected to additional fingerprinting methods, including phage typing or pulsed-field gel electrophoresis.

As with most enteric pathogens, salmonellae virulence is multifactorial, and both chromosomal and plasmid determinants are involved. Five pathogenicity islands (PIs) have been identified in salmonellae, two of which (*Salmonella* PI-1 and PI-2) contain type III secretion systems that are involved in invasion and intracellular multiplication. Other chromosomal genes are involved in adhesion and colonization of the bacterium to the intestine and resistance to gastric acidity. Determinants on virulence plasmids mediate colonization, resistance to complement-mediated lysis, and enhance the ability of strains to multiply in extraintestinal tissue.

SHIGELLA AND *ESCHERICHIA COLI*

The genera *Shigella* and *Escherichia* were found to belong to a single genus by DNA hybridization studies done in 1973.[33] The genus name *Shigella* has been retained because of the medical importance of these strains and the fact that most *Escherichia coli* strains are simple commensals of the intestinal tract of mammals and other animals. However, there are both diarrheagenic and extraintestinal pathogenic groups of *E. coli* and these are described in Table 17.2 (acronyms from this table are used throughout the text). *E. coli* can be serotyped, and specific types are associated with each of the above groups, the most notable of which is *E. coli* O157:H7 with the STEC group. Shigellae are broken down into four groups based on their cell surface (somatic) antigens and are designated as A (*Shigella dysenteriae*), B (*S. flexneri*), C (*S. boydii*), and D (*S. sonnei*). Three groups (A, B, and C) are divided into types and *S. flexneri* types are further subdivided into subtypes. There are several provisional serotypes of shigellae for which commercial antiserum is not available, and these strains can only be identified in reference laboratories.

Escherichia coli make their way into the food chain via fecal contamination of animal carcasses during slaughter, with subsequent improper handling (storage and/or cooking) of the product or by fecal adulteration of agricultural crops in the field or packing sheds. Testing of food products collected from commercial businesses confirms these routes of contamination with 72%, 55%, and 18% of meat, vegetables, and fish or shellfish samples, respectively, containing *E. coli* in one study.[34] Shigellae are host-adapted to the human intestine and, outside of humans, have only rarely been found to infect dogs or primates; isolations of shigellae from food and water only occur following contamination with human feces.[2]

Although reliable statistics for extraintestinal infections are not available, *Escherichia coli* can cause infection, and has been isolated from virtually every organ and anatomical site.[35] In 2000, in the United States alone, the estimate of mortality from *E. coli* bloodstream infections was 40,000 deaths, with $1.1 to $2.8 billion in healthcare costs.[35] Estimates for *E. coli* UTIs included 6 to 8 million cases of uncomplicated cystitis, 250,000 cases of pyelonephritis, and a quarter to half million cases of catheter-associated UTI.[35] The World Health Organization estimates there are more than 163 million cases of shigella per year in developing countries.[36] The mortality rate exceeds 1 million, and 61% of deaths occur in children under 5 years of age. In industrialized nations, the estimate of infections is 1.5 million per year with a 0.2% mortality rate. *Shigella flexneri* is the prevalent serotype seen in developing countries while *S. sonnei* predominates in developed countries.[2]

Cases of *Escherichia coli* sepsis usually arise following UTI and occur more frequently in women than in men.[37] Approximately 85 to 95% of patients who develop *E. coli* sepsis have underlying conditions with liver disease, hematologic malignancies, and solid tumors or other cancers being the most common.[2] *E. coli* meningitis can occur in any age group; neonates primarily acquire infection by vertical transmission from colonized mothers during or immediately following delivery and

TABLE 17.2
Pathogenic Groups of *Escherichia*

	Group Designation	Disease, Site of Infection	Pathogenic Mechanism
Diarrheagenic	Enteropathogenic (EPEC)	Watery diarrhea, small intestine	Adherence
	Enteroinvasive (EIEC)	Watery diarrhea followed by bloody diarrhea, colon	Invasion
	Enterotoxigenic (ETEC)	Watery diarrhea, small intestine	Heat-stable and/or heat-labile toxin
	Shiga-toxin producing (STEC) Also known as: Enterohemorrhagic (EHEC) Verotoxigenic (VTEC)	Watery diarrhea, sometimes bloody, especially for O157:H7, colon	Shiga toxins, type 1 and/or type 2
	Enteroaggregative (EaggEC)	Watery diarrhea, small intestine and colon	Adherence
	Diffuse adhering (DAEC)	Watery diarrhea, small intestine	Adherence
Extraintestinal (ExPEC)	Uropathogenic (UPEC)	Urinary tract	Adherence, toxins
	Bloodstream/meninges-associated (MNEC, BMEC)	Bloodstream, meninges	

less frequently via the nosocomial route. Approximately 90% of these strains carry a K1 capsule (which is identical to the type B polysaccharide of *Neisseria meningitidis*).[2] *E. coli* cystitis is characterized by acute dysuria and lower back pain, while pyelonephritis involves the upper urinary tract (kidneys and/or pelvis) and may be of a mild nature or cause vomiting and nausea with concomitant sepsis.[38] Gastrointestinal symptoms for the diarrheagenic groups of *E. coli* are given in Table 17.2. *Shigella* infections are typically characterized by generalized symptoms of fever, fatigue, anorexia, and malaise at onset, followed by watery diarrhea and abdominal pain.[2] Blood and mucus are present in approximately 40 and 50% of stools, respectively, with bloody stools occurring more commonly in children than in adults.[2] Although such isolations are infrequent, shigella can be isolated from sources other than stool, including urine, blood, wounds, sputum, gallbladder, and ears.[39] *Shigella* bacteremia is rare, occurring primarily in developing countries in malnourished children with decreased serum bactericidal activity.[2] Both STEC and shigellae, especially *S. dysenteriae* type 1, can cause hemolytic uremic syndrome (HUS), a disease characterized by microangiopathic hemolytic anemia, thrombocytopenia, and acute renal failure. *Shigella*- and *E. coli*-associated HUS is linked to the production of Shiga-toxin (Stx) by these organisms. In the United States, the most common serotype associated with HUS is O157:H7; however, other serotypes of STEC are capable of causing the disease.[40]

Escherichia coli can be easily isolated from blood by routine blood culture methods but if a single bottle is used, aerobic incubation recovers more isolates than anaerobic (non-vented) incubation.[2] Chromogenic agars are equal or superior to routine media (blood or MacConkey agars) used for UTI isolation in terms of numbers recovered, and have the added advantage that *E. coli* can be readily distinguished by their colony appearance and easily confirmed by a spot indole test.[2] Isolation of *E. coli* strains involved in diarrheal disease is a much more difficult task because most are biochemically indistinguishable from commensal strains. *E. coli* O157:H7 strains, which are delayed D-sorbitol fermenters, are the exception to this rule, and this trait is exploited in media used for their isolation. Sorbitol MacConkey (SMAC) agar contains D-sorbitol as a substitute for lactose present in regular MacConkey agar and is the primary isolation medium for O157 strains that present as colorless colonies.[41] Other formulations of SMAC have been developed but the most widely used is SMAC with cefixime and tellurite (CT-SMAC) incorporated to increase selectivity.[42] Immunomagnetic separation is a sensitive isolation technique for O157 and for a limited number of other serotypes of STEC, in which somatic antibody is coated on magnetic beads. Beads are commercially available (Dynal Biotech, Oslo, Norway) but are not FDA approved for use in clinical laboratories in the United States. After incubation, the beads, which are removed using a magnet, are washed and placed on appropriate media and any *E. coli* O157 bound to the beads produce colonies. Enzyme immunoassays (EIAs), which are commercially available and FDA approved, can be used to detect the somatic O157 antigen or Shiga toxin (Meridian Bioscience, Cincinnati, OH; Remel, Lenexa, KS). Stx can be detected directly from stool as well as from an isolate of STEC but there is a 3-log difference between EIA Stx detection from stool and the more sensitive tissue culture technique using Vero cells. A latex agglutination assay for Stx is also commercially available (Oxoid, Ogdensburg, NY). Both the latex assay and the EIA detect the presence of Stx but unlike the Vero cell assay with neutralizing antibody to type 1 and 2, neither can detect which toxin type(s), 1 or 2, are present. Testing serum for O157 antibody when culture is negative is a sensitive method to determine infection;[40] this assay is available from only a few reference laboratories and usually is restricted to patients involved in outbreaks. Isolates of *E. coli* from other types of diarrheal diseases are detected by determining the presence of the virulence factor(s) causing the infection, that is, detection of invasion by cell culture assay or detection of the *ipa* gene for invasion by polymerase chain reaction (PCR) for EIEC, etc. As with most specialized assays, these tests are available in a limited number of laboratories. Direct plating of stool is the method of choice for the isolation of shigellae; if culture is not performed within 2 to 4 hr of taking the stool, a transport medium must be used because shigellae are extremely sensitive to metabolic end products of other bacteria, phages, etc. present in stool. MAC, XLD, and HE agars work well for isolation of shigella, as does SS agar, but the latter may be too inhibitory for some strains; colonies of shigella will be colorless on these media (colonies on HE may appear green because of the color of the medium). *Shigella sonnei* is easily recognizable on plating media as it throws off large, rough colonies as well as smaller convex, smooth colonies; laboratories should select the smooth colonies for further testing.

Biochemically, typical strains of *Escherichia coli* present no challenge to commercial identification systems.[2] However, *E. coli* are phenotypically diverse organisms and may be biochemically inactive (indol-, lysine- or ornithine decarboxylase-, and ONPG-negative) or conversely can give positive reactions for tests (H_2S, urea) that are usually negative; such strains are not accurately identified by commercial kits.[2] Shigellae do not fair as well as *E. coli* in commercial identification systems. In data derived from various studies, of 198 strains tested collectively, 14 different systems correctly identified only 162 shigella.[2] There is considerable biochemical variation between species and types within species; distinctive traits include lack of mannitol fermentation for *S. dysenteriae* strains, ODC- and ONPG-positive reactions for *S. sonnei*, and gas-production from D-glucose for some strains of *S. flexneri* type 6 and *S. boydii* type 14.

Virulence characteristics are specific for each of the diarrheagenic *Escherichia coli* groups and for strains of ExPEC. Numerous genes, located on both the chromosome and on plasmids, are involved for each group but the primary virulence mechanism is given in Table 17.2. Shigellae are invasive pathogens; a 220-kb plasmid encodes the genes involved in invasion but genes that regulate plasmid-borne virulence factors are located on the chromosome along with other virulence genes.

YERSINIA

Of the twelve species within the genus *Yersinia,* only three are pathogenic for humans. While most strains of *Y. pestis* and *Y. pseudotuberculosis* are virulent, only certain *Y. enterocolitica* strains with specific serotype/biotype combinations are thought to cause disease. *Y. pestis* is a zoonotic organism causing epidemic plague in a number of rodents as well as humans.[43] Rats (*Rattus rattus* and *R. norvegicus*) play a major role in the spread of plague but are not significant in maintaining *Y. pestis* in the wild; enzootic rodents highly resistant to the organism (carriers) are believed to be the reservoir.[44] Unlike domesticated dogs, ferrets, coyotes, skunks, and raccoons that are highly resistant to infection, cats can develop clinical plague.[44] Lagomorphs, birds, and pigs, and rodents and aquatic environments are major reservoirs for *Y. pseudotuberculosis* and *Y. enterocolitica,* respectively, although both can be isolated from a wide variety of sources.[2]

It is now believed *Yersinia pestis* probably originated in Asia, not Africa,[45] and was responsible for three pandemics occurring in the 1st, 14th to 17th, and 19th centuries. Plague now primarily occurs as sporadic cases associated with rodent contact; however, since 1994, outbreaks have been identified in India, Zaire, and Mozambique.[44] In endemic areas, plague may occur as a mild or inapparent disease, detected primarily in seroprevalence studies.[44] In the United States, Perry and Fetherston[46] found that squirrels, rabbits, and prairie dogs were the most common vectors involved in plague cases but sources could not be identified for a third of the cases. Infected cats have transmitted plague to humans by scratches (bubonic) and aerosolized droplets (pneumonic).[47, 48] Large outbreaks of *Y. enterocolitica* and *Y. pseudotuberculosis* are generally infrequent and are exclusively community-acquired from consumption of contaminated foods; person-to-person transmission is rare. When yersiniae occur in nosocomial infections, an uncommon event, transmission is again via the foodborne route rather than person-to-person spread.

In plague, the organism migrates via the bloodstream to the lymphatic system, causing inflammation of the lymph nodes (buboes) in the groin, axilla, or neck[44] with an incubation period of 2 to 8 days following the fleabite. Plague is characterized by fever, chills, headache, and weakness, and, if present, gastrointestinal symptoms of abdominal pain, nausea, vomiting, and diarrhea. Secondary pneumonic plaque follows spread from the bloodstream to the respiratory tract and is distinguished by bronchopneumonia, cavitation, or consolidation with bloody or purulent sputum.[44] Patients with secondary pneumonic plague can infect close contacts with primary pneumonic plague in 1 to 3 days; these cases will be fatal if not treated within 1 day. In primary septicemic plague, which occurs in 25% of cases, symptoms resemble those of any Gram-negative infection (there are no buboes) but the mortality rate is 30 to 50%.[44, 46] *Yersinia pseudotuberculosis* is most commonly isolated from bloodstream infections and, as with *Y. enterocolitica,* occurs in patients with underlying medical conditions such as iron overload conditions.[2] *Y. enterocolitica* is also a major cause of transfusion-related sepsis.[49] Enteritis is the most frequently encountered gastrointestinal illness associated with *Y. enterocolitica* and *Y. pseudotuberculosis* but both can cause mesenteric lymphadenitis and terminal ileitis, conditions that mimic appendicitis. *Y. enterocolitica* serogroups of biotype 1A are regarded as nonpathogenic because they lack the virulence characteristics of pathogenic strains. Patients who carry the HLA-B27 allele are at increased risk of developing reactive arthritis (RA) following gastrointestinal illness; approximately 10% of patients with yersinia-associated RA develop chronic arthritis.[2]

Yersinia pestis present in blood, bubo aspirates, sputum, or cerebral spinal fluid exhibit a characteristic bipolar appearance when stained with Wayson's or Giemsa stain. Colonies on blood agar are opaque with a fried egg appearance appearing within 48 hr; however, plates should be held 7 days. *Y. pseudotuberculosis* is most often isolated from blood, and routine blood culture techniques are satisfactory. Like, *Y. enterocolitica*, *Y. pseudotuberculosis* can be isolated from MacConkey or Salmonella-Shigella agars although recovery on these media is variable for both organisms and colonies are often small (<1 mm) or pinpoint. On cefsulodin-Irgasan-novobiocin (CIN), a selective media designed specifically for isolation of *Y. enterocolitica*, colonies have a bull's-eye appearance with a red center and colorless apron. Both *Aeromonas* and *Plesiomonas shigelloides* also grow on CIN, appearing as bull's-eyes and colorless colonies, respectively, which allows this medium to be used to isolate pathogens other than yersinia. CIN, as prepared for *Y. enterocolitica*, may be inhibitory for strains of *Y. pestis* (unpublished data).

With respect to the identification of yersinia, the performance of commercial identification systems, overall, is mediocre, with *Yersinia pseudotuberculosis* faring the worst (correctly identified only 83% of the time).[2] Only three systems (API 20E, MicroScan Walk/Away, and VITEK) licensed for clinical use in the United States include *Y. pestis* in their database. All isolates of this organism should be sent to an appropriate public health reference laboratory immediately if the isolate is suspicious for plague. Most kits identify *Y. enterocolitica* as "*Y. enterocolitica* group," a designation that includes *Y. frederiksenii*, *Y. intermedia*, and *Y. kristensenii*, species generally considered non-pathogenic; systems that identify *Y. enterocolitica* as a species (API 20E and RapID) do not distinguish between virulent and non-virulent biotypes. Using conventional media, *Y. pestis* and *Y. pseudotuberculosis* are non-motile at 35°C, do not ferment sucrose, and are LDC-, ODC-, and ADH-negative; unlike *Y. pestis*, *Y. pseudotuberculosis* is urea-positive and L-rhamnose-negative. *Y. enterocolitica* hydrolyzes urea, ferments sucrose but not L-rhamnose, and is ODC-positive.

All three pathogenic species have a type III secretion system, also called the Yop virulon, located on a 70-kb plasmid. The Yop virulon is composed of a Ysc secretion apparatus (protein pump), Yop effector (causes pore formation in the host cell), and Yop translocator (blocks the host cell's ability to respond to infection) proteins.[50] Furthermore, a chromosomal high-pathogenicity island, containing genes encoding for a siderophore with a high affinity for iron, occurs in *Yersinia enterocolitica* pathogenic serogroup IB, *Y. pseudotuberculosis* serogroups I and III, and *Y. pestis*.[51]

ADDITIONAL MEMBERS OF THE FAMILY ENTEROBACTERIACEAE

There are many other genera and species in the family Enterobacteriaceae that cause serious diseases or illnesses in humans, animals, plants, and other living creatures. Many of these diseases are not specific to a single entity; that is, many different species can cause bloodborne disease in humans to varying degrees. However, a number of syndromes are either species-specific or are intimately linked to the genus and/or species, and some of these are presented in Table 17.3.

The genus *Citrobacter* consists of a large number of species (n = 11), most of which are lysine decarboxylase-negative (LDC-negative) and citrate-positive. Members of the *C. freundii* complex are also H_2S-positive on TSI. The most common species involved in human disease belong to the

TABLE 17.3
Selected Examples of Major Diseases Associated with Other Members of the Family Enterobacteriaceae

Category	Disease	Agent(s)
Humans	Gastroenteritis	*Edwardsiella tarda, Plesiomonas shigelloides*
	Neonatal meningitis, brain abscess	*Citrobacter koseri*
Animals	Murine colonic hyperplasia	*Citrobacter rodentium*
	Amber disease (grub)	*Serratia entomophila*
	Insect sepsis	*Photorhabdus*
Plants	Bark cankers	*Brenneria species*
	Stem canker, fire blight	*Erwinia species*
	Black spot necrosis, pink disease	*Pantoea species*
	Soft rot, black leg disease	*Pectobacterium carotovorum*
	Bark necrosis	*Samsonia*

C. freundii complex, where they primarily cause bacteremia and UTIs.[2] A less frequently isolated indole-positive species, *C. koseri* causes a variety of CNS diseases (ventriculitis, meningitis, brain abscesses) in neonates under 2 months of age. Risk factors for developing CNS disease caused by *C. koseri* include a gestation age of less than 37 weeks and low birth weight.[52] *C. rodentium*, a citrobacter species that does not cause disease in humans, can cause massive outbreaks of morbidity and mortality in mouse and gerbil colonies.[53]

Several other enterobacteria typically cause diarrhea as their most common clinical presentation. *Edwardsiella tarda*, an H_2S-, indole-, and LDC-positive species, causes watery to bloody diarrhea, particularly in subtropical regions of the world. Infection is often associated with close contact with reptiles. *Plesiomonas shigelloides*, the only oxidase-positive member of the family, causes enteritis and is associated with consumption of seafood or shellfish or contaminated water.

At least six genera, including the potential agent "*Phlomobacter*" that is associated with marginal chlorosis of strawberries, are considered phytopathogens. Not all isolates of these individual species cause disease under all circumstances, and many exist as commensals or part of the microbial rhizosphere. Many of these strains exhibit varying degrees of pigmentation (e.g., yellow) upon subculture in the laboratory. Specific tests have been designed to measure pathogenicity, such as the onion maceration test.

REFERENCES

1. Brenner, D.J., Krieg, N.R., and Staley, J.T. *Bergey's Manual of Systematic Bacteriology*, 2nd ed., Vol. 2, Part B. Springer, New York, 2005.
2. Janda, J.M. and Abbott, S.L. *The Enterobacteria*, 2nd ed., ASM Press, Washington, DC, 2005.
3. Tamas, I. et al. 50 million years of genomic stasis in endosymbiotic bacteria, *Science*, 296, 2376, 2002.
4. Ootsubo, M. et al. Oligonucleotide probe for detecting Enterobacteriaceae by *in situ* hybridization, *J. Appl. Microbiol.*, 93, 60, 2002.
5. Bohnert, J., Hübner, B., and Botzenhart, K. Rapid identification of *Enterobacteriaceae* using a novel 23S rRNA-targeted oligonucleotide probe, *Int. J. Hyg. Environ. Health,* 203, 77, 2000.
6. Livermore, D.M. and Woodford, N. The β-lactamase threat in *Enterobacteriaceae, Pseudomonas,* and *Acinetobacter, Trends Microbiol.*, 14, 413, 2006.
7. Walters, M. and Sperandio, V. Quorum sensing in *Escherichia coli* and *Salmonella*, *Int. J. Med. Microbiol.*, 296, 125, 2006.
8. Drancourt, M., et al. Phylogenetic analyses of *Klebsiella* species delineate *Klebsiella* and *Raoultella* gen. nov., with description of *Raoultella ornithinolytica* comb. nov., *Raoultella terrigena* comb. nov. and *Raoultella planticola* comb. nov., *Int. J. Syst. Evol. Microbiol.*, 51, 925, 2001.
9. Elliott, E. et al. *In vivo* development of ertapenem resistance in a patient with pneumonia caused by *Klebsiella pneumoniae* with an extended-spectrum β-lactamase, *Clin. Infect. Dis.*, 42, e95, 2006.
10. Chuang, Y.-P. et al. Genetic determinants of capsular serotype K1 of *Klebsiella pneumoniae* causing primary pyogenic liver abscess, *J. Infect. Dis.*, 193, 645, 2006.
11. Wang, J.L. et al. Changing bacteriology of adult community-acquired abscess in Taiwan: *Klebsiella pneumoniae* versus anaerobes, *Clin. Infect. Dis.*, 40, 915, 2005.
12. Vento, S., Cainelli, F., and Cesario, F. Infections and thalassaemia, *Lancet Infect. Dis.*, 6, 226, 2006.
13. Munoz, M.A. et al. Fecal shedding of *Klebsiella pneumoniae* by dairy cows, *J. Dairy Sci.*, 89, 3425, 2006.
14. Gavini, F. et al. Transfer of *Enterobacter agglomerans* (Beijerinck 1888) Ewing and Fife 1972 to *Pantoea* gen. nov. and description of *Pantoea dispersa* sp. nov., *Int. J. Syst. Bacteriol.*, 39, 337, 1989.
15. Kang, C.-I. et al. Bloodstream infections caused by *Enterobacter* species: predictors of 30-day mortality rate and impact of broad-spectrum cephalosporin resistance on outcome, *Clin. Infect. Dis.*, 39, 812, 2004.
16. Foster, D.R. and Rhoney, D.H. *Enterobacter* meningitis: organism susceptibilities, antimicrobial therapy and related outcomes, *Surg. Neurol.,* 63, 533, 2005.
17. Bowen, A.B. and Braden, C.R. Invasive *Enterobacter sakazakii* disease in infants, *Emerg. Infect. Dis.,* 12, 1185, 2006.

18. Drudy, D. et al. *Enterobacter sakazakii*: an emerging pathogen in powdered infant formula, *Clin. Infect. Dis.*, 42, 996, 2006.
19. Coker, C. et al. Pathogenesis of *Proteus mirabilis* urinary tract infection, *Microbes Infect.*, 2, 1497, 2000.
20. Kim, B.-N. et al. Bacteraemia due to tribe Proteeae: a review of 132 cases during a decade (1991–2000), *Scand. J. Infect. Dis.*, 35, 98, 2003.
21. Yoh, M. et al. Importance of *Providencia* species as a major cause of travellers' diarrhea, *J. Med. Microbiol.*, 54, 1077, 2005.
22. Murata, T. et al. A large outbreak of foodborne infection attributed to *Providencia alcalifaciens*, *J. Infect. Dis.*, 184, 1050, 2001.
23. Lorca, T.A. et al. Growth and histamine formation of *Morganella morganii* in determining the safety and quality of inoculated and uninoculated bluefish (*Pomatomus saltatrix*), *J. Food Prot.*, 64, 2015, 2001.
24. Aleksic, S., Heinzerling, F., and Bockemuhl, J. Human infection caused by salmonellae of subspecies II to VI in Germany, 1977–1992, *Zentbl. Bakteriol.*, 283, 391, 1996.
25. Mermin, J. et al. Reptiles, amphibians and human *Salmonella* infection: a population-based, case-control study, *Clin. Infect. Dis.*, 38 (Suppl. 3), S253, 2004.
26. LeMinor, L. The genus *Salmonella*, in M.P. Starr, H. Stolp, H.G. Truper, and H.G. Schlegel (Eds.), *The Prokaryotes: A Handbook on Habitats, Isolation, and Identification of Bacteria*. Springer-Verlag, Berlin, Germany, 1981, p. 2760–2774.
27. Voetsch, T.J. et al. FoodNet estimate of the burden of illness caused by non-typhoidal *Salmonella* infections in the United States, *Clin. Infect. Dis.*, 38 (Suppl. 3), S127, 2004.
28. Crump, J.A., Luby, S.P., and Mintz, E.D. The global burden of typhoid fever, *Bull. WHO*, 82, 346, 2004.
29. Goldberg, M.B. and Rubin, R.H. The spectrum of Salmonella infection, in R.C. Moellering and Gorbach, S.L. (Eds.), *Infectious Diarrhea*. The W.B. Saunders Co., Philadelphia, PA, 1988
30. Wain, J. et al. Quantitation of bacteria in bone marrow from patients with typhoid fever; relationships between counts and clinical features, *J. Clin. Microbiol.*, 39, 1571, 1998.
31. Escamilla, J., Florez-Ugarte, H., and Kilpatrick, M.E. Evaluation of blood clot cultures for isolation of *Salmonella typhi*, *Salmonella paratyphi-A*, and *Brucella melitensis*, *J. Clin. Microbiol.*, 24, 388, 1986.
32. Popoff, M.Y. *Antigenic Formulas of the Salmonella Serovars, 8th edition*. WHO Collaborating Centre for Reference and Research on *Salmonella*, Institut Pasteur, Paris, France, 2001.
33. Brenner, D.J. et al. Polynucelotide sequence relatedness among *Shigella* species, *Int. J. Syst. Bacteriol.*, 23, 1, 1973.
34. Venkateswaran, K., Murakoshi, A., and Satake, M. Comparison of commercially available kits with standard methods for the detection of coliforms and *Eschericha coli* in foods, *Appl. Environ. Microbiol.*, 62, 2236, 1996.
35. Russo, T.A. and Johnson, J.R. Medical and economic impact of extraintestinal infections due to *Escherichia coli*: focus on an increasingly important endemic problem, *Microbe Infection*, 5, 449, 2003.
36. Kotloff, K.L. et al. Global burden of *Shigella* infections: implications for vaccine development and implementation of control strategies, *Bull. WHO*, 77, 651, 1999.
37. Grandsen, W.R. et al. Bacteremia due to *Escherichia coli*: a study of 861 episodes, *Rev. Infect. Dis.*, 12, 1008, 1990.
38. Stamm, W.E. and Hooton, T.M. Management of urinary tract infections in adults, *N. Engl. J. Med.*, 329, 1328, 1993.
39. Gupta, A. et al. Laboratory-confirmed shigellosis in the United States, 1989–2002: epidemiologic trends and patterns, *Clin. Infect. Dis.*, 38, 1372, 2004.
40. Banatvala, N. et al. The United States National Prospective Hemolytic Uremic Syndrome Study: microbiologic, serologic, clinical, and epidemiologic findings, *J. Infect. Dis.*, 183, 1063, 2001.
41. March, S.B. and Ratnam, S. Sorbitol-MacConkey medium for detection of *Escherichia coli* O157:H7 associated with hemorrhagic colitis, *J. Clin. Microbiol.*, 23, 869, 1986.
42. Zadik, P. M., Chapman, P.A., and Siddons, C.A. Use of tellurite for the selection of verotoxigenic *Escherichia coli* O157, *J. Med. Microbiol.*, 39, 155, 1993.
43. Quan, T.J. *Yersinia pestis*, in A. Balows, H.G. Truper, M. Dworkin, W. Harder, and K.-H. Schleifer (Eds.). *The Prokaryotes, A Handbook on the Biology of Bacteria. Ecophysiology, Isolation, Identification, Applications*, 2nd ed. Springer-Verlag, Berlin, Germany, 1991, p. 2888–2989.

44. Smego, R.A., Frean, J., and Koornhof, H.J. Yersiniois I: microbiological and clinicoepidemiological aspects of plague and non-plague *Yersinia* infection, *Eur. J. Clin. Microbiol. Infect. Dis.,* 18, 1, 1991.
45. Achtman, M. et al. Microevolution and history of the plague bacillus, *Yersinia pestis*, *Proc. Natl. Acad. Sci. U.S.A.*, 101, 17837, 2004.
46. Perry, R.D. and Fetherson, J.D. *Yersinia pestis* — etiologic agent of plague, *Clin. Microbiol. Rev.*, 10, 35, 1997.
47. Doll, J.M. et al. Cat-transmitted fatal pneumonic plague in a person who traveled from Colorado to Arizona, *Am. J. Trop. Med. Hyg.*, 51, 109, 1994.
48. Weniger, B.G. et al. Human bubonic plague transmitted by a domestic cat scratch, *JAMA,* 251, 927, 1984.
49. Wagner, S.J., Friedman, L.I., and Dodd, R.Y., Transfusion-associated bacterial sepsis, *Clin. Microbiol. Rev.*, 7, 290, 1994.
50. Ruckdeschel, K. *Yersinia* species disrupt immune responses to subdue the host, *ASM News*, 66, 470, 2000.
51. Carniel. E. The *Yersinia* high-pathogenicity island: an iron-uptake island, *Microbes Infect.*, 3, 561, 2001.
52. Doran, T.I. The role of *Citrobacter* in clinical disease of children: review, *Clin. Infect. Dis.*, 28, 384, 1999.
53. Luperchio, S.A. et al. *Citrobacter rodentium*, the causative agent of transmissible murine colonic hyperplasia, exhibits clonality: synonymy of *C. rodentium* and mouse-pathogenic *Escherichia coli*, *J. Clin. Microbiol.*, 38, 4343, 2000.

18 The Genus *Pseudomonas*

Norberto J. Palleroni

CONTENTS

INTRODUCTION

Members of the genus *Pseudomonas* are very common in nature and can be isolated from a large variety of natural materials. A number of strains are notorious for their nutritional versatility toward organic compounds of low molecular weight in media totally devoid of organic growth factors; this capacity, combined with a fast growth rate, allows them to predominate in the microflora growing in natural media that have a reaction close to neutrality and some organic matter in solution.[2, 12, 21, 29]

The basic morphological features common to almost all species are the straight rod shape and the presence of one or several polar flagella. No spores are produced, and the Gram reaction is negative. These morphological attributes define the "pseudomonads" but admission to the genus *Pseudomonas* requires some additional physiological properties, such as an energy metabolism purely respiratory and nutrition of the chemoorganotrophic type. These properties, together with the widespread occurrence, determine the great importance of these organisms as participants in the carbon cycle in nature[2] and, therefore, in their inclusion in bioremediation projects.

Since the creation of the generic name *Pseudomonas* by Migula,[10] from the beginning of the 20th century many species names were assigned to the genus, but the number has undergone oscillations mainly due to the description of new species and to changes (sometimes quite profound) in the generic definition.[5, 21] At present, as we shall see, *Pseudomonas* is reduced to only a fraction of the species originally assigned to it, although the second edition of *Bergey's Manual of Systematic Bacteriology* includes the descriptions of 61 species, to which a few more have been added since January 2000, the final date of this survey.[16] A comprehensive chapter on the taxonomy of the genus, which includes a useful species table, has been included in a recently published three-volume treatise entitled *Pseudomonas*.[21]

The species have a DNA base composition ranging from 57 to 69 mol% guanine plus cytosine (G+C).[9] No internal discontinuities in this range are useful for the separation of subgeneric categories because, in fact, some of the natural groups within the genus cover a G+C span almost as wide as that of the whole genus.[19]

METHODOLOGY

Comprehensive descriptions of bacterial species should consist of a combination of phenotypic characteristics[29] and phylogenetic considerations based on comparative molecular studies.[11, 20] The main phenotypic characteristics of interest in the traditional characterization and in the taxonomy of *Pseudomonas* are presented in Table 18.1.

In the routine analysis of bacteria of this genus, not all these properties are investigated, and in some cases the salient ones are sufficient to orient the subsequent characterization. For example, the colony appearance of many strains of *Pseudomonas stutzeri*[8, 31] or the production of a diffusible greenish pigment by organisms of the fluorescent group,[6, 29] greatly simplify further analysis. However, for the purpose of publication of newly isolated strains, a balanced report of phenotypic characteristics and a phylogenetic analysis by the study of the nucleic acids are normally required.

What follows is a brief description of the main phenotypic characteristics of *Pseudomonas* strains.

TABLE 18.1
Characteristics of *Pseudomonas* Strains Used for the Identification Genus Species

A) Traditional Properties:
Morphology (shape, size of cells) and disposition (isolated cells, cells in pairs and/or chains)
Motility, and if this is present, flagella studies (number, insertion, length, wavelength)
Chemical properties of reserve materials, if present
Pigments (cell-associated or diffusible, appearance under low wavelength UV radiation)
Temperature relationships
Growth factor requirements
Oxidase reaction
Hydrolysis of gelatin, starch, Tween 80, lecithin (egg-yolk reaction)
Levan formation from sucrose
Arginine dihydrolase reaction
Ring fission mechanisms
Utilization of carbon sources
Utilization of nitrogen sources, nitrate reduction, denitrification
Metabolic pathways
Regulatory mechanisms
Immunological studies
Genetic studies
B) Molecular Characterization:
DNA base composition
Nucleic acid hybridizations (DNA-DNA, rRNA-DNA)
Nucleic acid sequences
Amino acid sequence of proteins

Pigments

In the early taxonomic treatments of *Pseudomonas*, pigment production was a characteristic of primary importance at the generic level. Later, many nonpigmented species were included in the genus but the character of pigmentation retained an important place among the diagnostic traits of some species.

The type species of the genus, *Pseudomonas aeruginosa,* is capable of producing a wide variety of pigments.[28, 29] Usually, pigment production can be induced or enhanced in special culture media, but repeated transfer of strains in the laboratory sometimes results in total loss of pigment production. In fact, pigmentation is one of the most erratic of all phenotypic traits.

Diffusible pigments that fluoresce under ultraviolet (UV) radiation of short wavelength (ca. 254 nm) are collectively called "fluorescent pigments" and characterize important species now placed in the so-called fluorescent group. Other common pigments (blue, red, yellow, or green) are those belonging to the chemical family of the phenazines; some of them are insoluble and others diffuse into the medium. Some soluble phenazine pigments fluoresce under long-wavelength UV radiation (ca. 350 nm), an important distinction with respect to the "true" fluorescent pigments described previously. Some species of *Pseudomonas* produce carotenoid pigments. The production of these compounds is not too useful for the circumscription of species groups within the genus; thus, among the many species of the fluorescent group, only *P. mendocina* and one strain of *P. alcaligenes* are known to produce carotenoids.[13]

Temperature Relationships

Growth at 4°C and at 41°C has been determined for a large number of strains. *Pseudomonas* species are typically mesophilic; some, such as *P. fluorescens*, can grow at 4°C and can occasionally be identified in the psychrophilic flora responsible for food spoilage. *Pseudomonas aeruginosa* does not grow at 4°C, but can grow at 41°C, and many strains still do well at 44°C. Determination of these extreme growth temperatures requires the use of well-regulated water baths; air incubators are very unreliable for this purpose.

Growth Requirements: Utilization of Carbon Compounds

All well-characterized species of the genus are independent of special nutritional requirements and can grow in simple media containing phosphate at a pH close to neutrality, magnesium sulfate, ammonium chloride, and trace amounts of iron and calcium, usually provided by a mixture of ferric ammonium citrate and calcium chloride. To this medium should be added a carbon source, ideally at a concentration of 0.1 to 0.2% (vol/wt).[18]

In the case of strains of the various pathovars of *Pseudomonas syringae*, which is a phytopathogenic species, growth in the minimal medium is usually very slow,[24] and it is stimulated by the addition of small amounts of complex organic compounds such as yeast extract. Some substrates are toxic at the above-mentioned concentration, but growth can be detected at concentrations ten or twenty times lower. Some substrates such as camphor, naphthalene, geraniol, and other aromatic, volatile compounds are placed in the lid of the Petri dish,[29] and the cultures should be incubated in a separate incubator or placed in tight containers so as not to contaminate the atmosphere with the vapors.

Nitrate Reduction and Denitrification

Ammonium salts are good nitrogen sources for growth, and in addition, most *Pseudomonas* species can utilize nitrate as a nitrogen source, and this implies the possession of a reductive system for the conversion of nitrate into amino compounds. The first step in the reduction probably gives nitrite in all cases; this step is the basis for the reaction called reduction of nitrate, which is assayed by testing for the appearance of nitrite in the nitrate medium. The reduction-of-nitrate reaction was performed in very few of the *Pseudomonas* strains of the Berkeley collection because it was thought that the reaction provided little taxonomically useful information; in addition, the accumulation of nitrite probably depends on the activity of the components of the reduction system, so that nitrite may be undetectable when its further reduction is not a limiting step of the process.

Some species of *Pseudomonas* can carry the reduction of nitrate to nitrogen gas or nitrous oxide, a process known as denitrification. Denitrifiers can grow anaerobically using nitrate in place of oxygen as the terminal electron acceptor. Nitrogen gas escapes to the atmosphere and represents a wasteful process in nature. Molecular nitrogen cannot be utilized as a nitrogen source by most strains, and it was thought that the property was negative for all species. However, fixation by some strains of *Pseudomonas stutzeri* has been reported.[8]

Extracellular Polysaccharides

Levan formation from sucrose is taxonomically important.[4] It can be determined in various media with a high concentration (4%) of the disaccharide. The cells use the glucose moiety and fructose is accumulated extracellularly in the form of a polysaccharide. The property is found in some of the biovars of *Pseudomonas fluorescens*.

Oxidase Reaction and Cytochrome Composition

The oxidase reaction[27] is positive for most *Pseudomonas* species but it is negative for the fluorescent plant pathogen *P. syringae*. A positive reaction is correlated with the presence of cytochrome c in the cells.

Production of Hydrolytic Enzymes

A number of substrates (gelatin, starch, poly-13-hydroxybutyrate, Tween 80, egg yolk) have been used for the detection of various specific hydrolytic activities. Gelatinases are of widespread occurrence in the genus; a large number of strains also hydrolyze sorbitan monooleate polyoxyethylene (Tween 80) and manifest strong lecithinase action in the egg yolk reaction. Starch and poly-13-hydroxybutyrate are far less commonly hydrolyzed (Table 18.2).

Arginine Dihydrolase Reaction

This reaction is carried out under anaerobic conditions by some *Pseudomonas* species, and it can be tested by direct chemical determination of arginine disappearance or ornithine appearance,[25, 26] and also, less specifically, by pH changes.[30] All strains that have a positive arginine dihydrolase reaction can grow with arginine as the sole carbon source; on the other hand, some strains capable of growth with arginine do not decompose the amino acid anaerobically.[29] As a taxonomic character, the reaction is very satisfactorily correlated with many other phenotypic properties, and therefore appears to be very valuable.

Production of Acids from Sugars and Sugar Alcohols

The tests for production of acids from various carbohydrates have been quite popular in bacteriology for determinative purposes.[27] In *Pseudomonas*, these characters have remained largely unexplored, due to ambiguities in the interpretation of the results and to the possibility of redundancy because several sugars may be oxidized by the same enzyme.[18]

Ring Fission Mechanisms

The mechanisms for the degradation of aromatic compounds have been studied extensively in some members of the genus. In these pathways, hydroxylation of the aromatic compound precedes its

TABLE 18.2
Selected Characters of *Pseudomonas* Species and of Related Organisms

Species	Number of flagella	PHB as reserve material	Fluorescent pigments	Phenazine pigments	Carotenoid pigments	Growth factors requirement	Denitrification	Oxidase reaction	Arginine dihydrolase	Growth at 4°C	Growth at 41°C	Gelatin liquefaction	PHB depolymerase	Starch hydrolysis	Cleavage of diphenols	Hydrogen/CO_2 utilization	G+C percent in DNA
A																	
P. aeruginosa	1	–	+	+	–	–	+	+	+	–	+	+	–	–	*o*	–	67
P. putida	>1	–	+	–	–	–	–	+	+	v	–	–	–	–	*o*	–	60-63
P. fluorescens	>1	–	+	–	–	–	v	+	+	+	–	+	–	–	*o*	–	59-61
P. chlororaphis	>1	–	+	+	–	–	v	+	+	+	–	+	–	–	*o*	–	63
P. syringae	>1	–	+	–	–	–	–	–	–	v	–	v	–	–	*o*	–	59-61
P. cichorii	>1	–	+	–	–	–	–	+	–		–	–	–	–	*o*	–	59
P. stutzeri	1	–	–	–	–	–	+	+	–	–	v	–	–	+	*o*	–	61-66
P. mendocina	1	–	–	–	+	–	+	+	+	–	+	–	–	–	*o*	–	63-64
P. alcaligenes	1	–	–	–	v	–	–	+	+	–	+	+	–	–			66-68
P. pseudoalcaligenes	1	v	–	–	–	–	–	+	+	–	+	v	–	–			62-64
B*																	
B. pseudomallei	>1	+	–	–	–	–	+	+	+	–	+	+	+	+	*o*	–	69
B. mallei	0	+	–	–	–	–	v	+	+	–	+	+	+	v	*o*	–	69
B. caryophylli	>1	+	–	–	–	–	v	+	+	–	+		–	–	*o*	–	65
B. cepacia	>1	+	–	v	–	–	–	+	–	–	v	+	–	–	*o*	–	67-68
B. gladioli	>1	+	–	–	–	–	–	+	–	–	+	+	–	–	*o*	–	68
Pa. lemoignei	1	+	–	–	–	–	–	+	–	–	+	–	+	–		–	58
C. testosteroni	>1	+	–	–	–	–	–	+	–	–	–	–		–	*m*	–	62
C. acidovorans	>1	+	–	–	–	–	–	+	–	–	–	–	–	–	*m*	–	67
A. delafieldii	1	+	–	–	–	–	–	+	–	–	–	–	+	–	*m*	–	65-66
R. pickettii	1	+	–	–	–	–	+	+	–	–	+	–	–	–	*o*	–	64
R. solanacearum	>1	+	–	–	–	–	v	+	–		–	–	–	–		–	66-68
A. facilis	1	+	–	–	–	–	–	+	–		–	+	+	–	*m*	+	62-64
Pe. saccharophila	1	+	–	–	–	–	–	+	–		–	–	–	+	*m*	+	69
H. flava	1	+	–	–	+	–	–	+	–		–	–	–	–	*m*	+	67
H. palleronii	1	+	–	–	+	–	–	+	–		–	–	–	–	*m*	+	67
S. maltophilia	>1	–	–	–	–	+	–	–	–	–	–	+	–	–		–	67
Br. vesicularis	1	+	–	–	+	+	–	+	–	–	–	–	–	–		–	66
Br. diminuta	1	+	–	–	–	+	–	+	–	–	v	–	–	–		–	66-67

Code: + = positive for all or most strains; – = negative for all or most strains; v = positive for a variable number of strains; *m* = *meta* type of ring cleaveage; *o* = *ortho* type of ring cleavage.

• *A: Acidovorax; B: Burkholderia; Br: Brevundimonas; C: Comamonas; H: Hydrogenophaga; Pa: Paucimonas; Pe: Pelomonas; R: Ralstonia; S: Stenotrophomonas.*

cleavage, and this last step is carried out in a manner that is characteristic of every group of organisms.[29] Taxonomic conclusions, however, should be drawn only when different organisms are compared under the same conditions of induction. In Table 18.2, the two main methods of cleavage are indicated as *o* and *m* (*ortho* and *meta*) and they are carried out by 1,2- or 2,3-dioxygenases, respectively.

Nutritional Properties

As discussed previously, the basic nutritional requirements for most *Pseudomonas* species are quite simple, because the strains can grow in mineral media supplemented with a single organic compound as the source of carbon and energy. The number of organic compounds that, when tested individually, can support growth varies within wide limits, and both the number and the types of utilizable compounds are important from the taxonomic point of view.[2, 29]

The ease with which the nutritional analysis can be performed and the wealth of taxonomic information that can be obtained have been the main reasons for the preferential attention given to this aspect of *Pseudomonas* physiology in modern taxonomic treatments of the genus. The mass of clear-cut nutritional data obtainable is ideally suited for numerical analysis. A methodology useful for the analysis of a large number of strains has been described, and it is a modification of the procedure described many years ago by den Dooren de Jong, who suggested that this analysis may facilitate the classification of *Pseudomonas* species.[2]

The nutritional analysis of *Pseudomonas* has been very valuable for the clear circumscription of several species and species groups. However, it is only fair to admit that in some cases this approach has been insufficient for unraveling the formidable taxonomic complexities of some groups. Thus, several biovars of the species *Pseudomonas fluorescens* are still in highly unsatisfactory taxonomic condition, and perhaps the number of internal subdivisions will increase after the analysis of a much larger collection of strains.

A similar situation holds for *Pseudomonas stutzeri*, a species of marked internal heterogeneity,[19] for which the nutritional analysis has done little to define clear internal subdivisions, but a recent review[8] shows that work mostly done at the Universidad de Islas Baleares, in Mallorca, has contributed to a better understanding of this fascinating taxon.

Table 18.2 summarizes data on some selected phenotypic properties diagnostically useful for the differentiation of species of the genus, as well as the characteristics of strains of other species that were classically assigned to *Pseudomonas,* and Table 18.3 summarizes data on nutritional characteristics of the two groups of species. The genera to which the second set of species belongs at present are indicated by the initial letter (genus) only, but the full generic name is added as a footnote (*) to the section B of the tables.

In Vitro Nucleic Acid Hybridization

Nucleic acid hybridization results have been particularly rewarding in understanding the internal complexity of the genus *Pseudomonas.* Deoxyribonucleic acid (DNA-DNA) hybridization experiments, as bases of homology studies, have generally supported the conclusions based on the phenotypic properties of the various species.[7, 14, 17, 23] In addition, the DNA-DNA hybridization studies also revealed internal heterogeneity in groups that appeared rather homogeneous on phenotypic grounds. DNA homology groups were therefore outlined, including species linked directly or indirectly to one another by some detectable level of DNA homology.

Curiously, the various DNA homology groups did not share any level of similarity among them, a situation that suggested that perhaps the distant relationships among the DNA homology groups could be estimated by resorting to the study of some genes of a conservative nature. In view of the fact that no prior experience on this approach for taxonomic studies was available from other working groups, a decision was made to attempt hybridization experiments between ribosomal RNA (rRNA) and total DNA. The approach led to the discovery of the fact that the genus *Pseudomonas* was, in fact, a complex of five RNA homology groups deserving separate generic designations.[20]

The significance of this work not only pointed the way toward a definitive taxonomy of *Pseudomonas* and the precise circumscription of the genus, but it also could be extended to general problems of phylogenetic relationships among species.[3] In the case of the *Pseudomonas* taxonomy, there are available a number of review treatments that will be very useful to the interested reader.[1, 7, 11, 14, 16, 20] Part B of Tables 18.2 and 18.3 includes the species assigned to new genera, which before the studies of rRNA/DNA hybridization were part of the genus *Pseudomonas*.

In the brief discussion below, some properties of the most important group (which retains the name *Pseudomonas*) will be presented. Tables 18.2 and 18.3 also include some properties of species previously assigned to *Pseudomonas* but now representing various other genera.

The *Pseudomonas fluorescens* RNA Homology Group

Pseudomonas fluorescens occupies a central position in this group, which includes both fluorescent and nonfluorescent pseudomonads. At present, the members of this group are *P. aeruginosa* (the type species of the genus), *P. putida* (two biovars), *P. fluorescens* (five biovars), *P. chlororaphis, P. syringae, P. cichorii, P. stutzeri, P. mendocina, P. alcaligenes,* and *P. pseudoalcaligenes*. A number of species (mostly fluorescent) of recent descriptions are not included in the tables, and they can be found in the last edition of *Bergey's Manual of Systematic Bacteriology*.[16]

The natural relationships of the species in the group are suggested by the results of nucleic acid homology and by the similarity in some phenotypic characters, biochemical pathways, immunological properties of enzymes, and regulatory mechanisms.

The different species within the group (A) show various degrees of uniformity. *Pseudomonas aeruginosa* and *P. mendocina* appear to be internally quite homogeneous, whereas *P. stutzeri* and *P. fluorescens* can be subdivided into subgroups or biotypes because they are clearly heterogeneous.

Pseudomonas aeruginosa has received considerable attention as an opportunistic pathogen. The interested reader can find excellent discussions on important aspects of the biology of this species in a number of treatises.[6, 22]

Pseudomonas aeruginosa is common in soils, from which it can be isolated by direct streaking on plates of nutrient agar, or, more conveniently, after enrichment under denitrification conditions with appropriate carbon sources. Several pigments can be produced by strains of the species; the blue pyocyanine is one of the most characteristic; and in addition, fluorescent pigment (an important siderophore) is produced by most strains.

The diagnosis of *Pseudomonas aeruginosa* is fairly simple: monotrichous flagellation, pigment production, capacity for growth at 41°C, ability to denitrify, and growth with geraniol and with acetamide as carbon sources. This constellation of characters is very useful although strains isolated from nature may lack one or several of the properties listed.

The species is phylogenetically somewhat isolated from others in the group. The DNA homology among strains is high (usually 90% or more by the competition technique) and the homology is considerably lower with other fluorescent pseudomonads.

Pseudomonas aeruginosa can be further subdivided into "types" by serological, phage, or pyocin typing. This internal classification of the species is useful for epidemiological purposes because it facilitates tracing the origin of infections. The typing methods are capable of detecting very small differences among strains, and even a homogeneous species such as this one presents a serious task to any single typing method. It is now accepted that typing gives the best results when at least two typing approaches are used simultaneously.

Pseudomonas fluorescens has been subdivided into a number of biotypes; two of these (D and E) have recently been restored to their original species level; but because of priority rules of nomenclature, only the name *P. chlororaphis* is valid.[7] The internal subdivision of *P. fluorescens* has been based on characters such as levan formation from sucrose, denitrification, pigment production, and

TABLE 18.3A
Some Diagnostically Important Nutritional Properties of *Pseudomonas* Species and of Species Previously Assigned to the Genus

Species	D-Fucose	Glucose	Trehalose	Cellobiose	Starch	*m*-Inositol	Geraniol	2-Keto-gluconate	Maleate	Glycollate	Lactate
A											
P. aeruginosa	–	+	–	–	–	–	+	+	–	–	+
P. putida	–	+	–	–	–	–	–	+	–	(–)	+
P. fluorescens	–	+	+	–	–	+	–	+	(–)	(–)	+
P. chlororaphis	–	(+)	(+)	–	–	+	–	v	–	–	+
P. syringae	–	+	–	–	–	(+)	–	–	–	–	+
P. cichorii	–	+	–	–	–	+	–	–	–	–	(–)
P. stutzeri	–	(+)	–	–	(+)	–	–	–	–	(+)	+
P. mendocina	–	+	–	–	–	–	+	–	–	+	+
P. alcaligenes	–	–	–	–	–	–	–	–	–	–	+
P. pseudoalcaligenes	–	–	–	–	–	–	–	–	–	–	+
B*											
B. pseudomallei	+	+	+	+	+	+	–	+	–	–	+
B. mallei	+	+	+	+	(+)	(+)	–	(+)	–	–	+
B. caryophylli	+	+	(–)	+	–	+	–	+	–	+	+
B. cepacia	+	+	(+)	+	–	+	–	+	–	(+)	+
B. gladioli	(+)	+	+	+	–	+	–	+	–	–	+
Pa. lemoignei	–	–	–	–	–	–	–	–	–	–	–
C. testosteroni	–	–	–	–	–	–	(–)	–	–	–	–
C. acidovorans	–	–	–	–	–	(–)	–	–	+	+	+
A. delafieldii	–	+	–	–	–	–	–	+	(+)	–	+
R. pickettii	–	+	–	–	–	–	(–)	+	(+)	+	+
R. solanacearum	–	+	(+)	–	–	(+)	–	–	–	(–)	(+)
A. facilis	–	+	–	–	–	–	–	–	–	–	+
Pe. saccharophila	–	+	+	+	+	–	–	–	–	–	+
H. flava	–	+	+	+	–	+		–	–	–	+
H. palleronii	–	+	–	–	–	+		–	–	–	+
S. maltophilia	–	+	+	+	–	–	–	–	–	–	+
Br. vesicularis	–	+	–	+	–	–	–	–	–	–	–
Br. diminuta	–	–	–	–	–	–	–	–	–	–	–

Code: + = positive for all strains; – = negative for all strains; (+) = positive for 50% or more of the strains; (–) = negative for 50% or more of the strains; (v) = positive for a variable number of strains (in Table 18.3A); (m) = positive after mutation (in Table 18.3B).

• *A: Acidovorax; B: Burkholderia; Br: Brevundimonas; C: Comamonas; H: Hydrogenophaga; Pa: Paucimonas; Pe: Pelomonas; R: Ralstonia; S: Stenotrophomonas*

TABLE 18.3B
Some Diagnostically Important Nutritional Properties of *Pseudomonas* Species and of Species Previously Assigned to the Genus

Species	PHB	Adipate	*m*-Hydroxybenzoate	Testosterone	Acetamide	Arginine	Valine	Norleucine	D-Tryptophan	Betaine	Pantothenate	Ethylene glycol
A												
P. aeruginosa	–	+	–	–	+	+	+	–	–	+	–	–
P. putida	–	–	(–)	(–)	(–)	+	+	–	–	+	(–)	(–)
P. fluorescens	–	(–)	(–)	(–)	–	+	+	–	–	+	(–)	–
P. chlororaphis	–	–	(–)	(–)	–	+	(+)	–	–	+	–	–
P. syringae	–	–	–	–	–	(+)	–	–	–	(+)	–	–
P. cichorii	–	–	–	–	–	+	–	–	–	+	–	–
P. stutzeri	–	(–)	–	–	–	–	(+)	–	–	(–)	–	+
P. mendocina	–	–	–	–	–	+	+	–	–	+	–	(+)
P. alcaligenes	–	–	–	–	–	(+)	–	–	–	–	–	–
P. pseudoalcaligenes	–	–	–	–	–	+	–	–	–	(+)	–	(–)
B*												
B. pseudomallei	+	+	–	–	–	+	+	–	–	+	–	(–)
B. mallei	+	(+)	–	–	–	+	(+)	–	–	+	–	–
B. caryophylli	–	–	–	–	–	+	(–)	–	–	+	–	–
B. cepacia	–	+	+	(+)	(+)	+	(+)	–	m	+	–	–
B. gladioli	–	+	–	–	–	+	+	(–)	–	+	–	–
Pa. lemoignei	+	–	–	–	–	–	–	–	–	–	–	–
C. testosteroni	(–)	+	+	+	–	–	–	+	–	–	–	–
C. acidovorans	–	+	+	–	+	–	–	+	+	–	–	–
A. delafieldii	+	+	–	–	–	–	–	–	–	–	–	–
R. pickettii	–	+	–	–	–	–	+	–	(+)	–	–	–
R. solanacearum	–	–	–	–	–	–	–	–	–	(–)	–	–
A. facilis	+	–	–	–	–	–	–	–	–	–	–	–
Pe. saccharophila	–	m	–	–	–	–	–	–	–	–	–	m
H. flava	–	–	+	–	–	–	–	–	–	–	–	–
H. palleronii	–	+	+	–	–	–	–	–	–	–	–	–
S. maltophilia	–	–	–	–	–	–	–	–	–	–	–	–
Br. vesicularis	–	–	–	–	–	–	–	–	–	–	–	–
Br. diminuta	–	–	–	–	–	–	–	–	–	–	+	–

Code: + = positive for all strains; – = negative for all strains; (+) = positive for 50% or more of the strains; (–) = negative for 50% or more of the strains; (v) = positive for a variable number of strains (in Table 18.3A); (m) = positive after mutation (in Table 18.3B).

• *A: Acidovorax; B: Burkholderia; Br: Brevundimonas; C: Comamonas; H: Hydrogenophaga; Pa: Paucimonas; Pe: Pelomonas; R: Ralstonia; S: Stenotrophomonas*

various nutritional characters. The neotype of the species, originally proposed by Rhodes, has been allocated by Stanier and associates[29] to biotype A, which is characterized by the capacity for levan formation and the inability to denitrify.

Pseudomonas putida can be differentiated from *P. aeruginosa* and *P. fluorescens* by means of a constellation of negative characters, namely, the inability to liquefy gelatin, to denitrify, and to produce levan from sucrose, the absence of lipase and lecithinase activity, and the incapacity for growth at 41°C and (with few exceptions) at 4°C. The strains of *P. putida* can be grouped into two biotypes but it is now considered that only biotype A can represent typical *P. putida* because the strains of biotype B appear to be more closely related to *P. fluorescens* than to biotype A. The strains of biotype A are rather homogeneous in phenotypic properties, although, surprisingly, they have considerable heterogeneity in DNA homology.

Many plant-pathogenic fluorescent pseudomonads can be assigned to one or two species — *Pseudomonas syringae* or *P. cichorii* — and these two species constitute a group that is considered to represent a phylogenetically separate branch within the *P. fluorescens* RNA homology group. There are other phytopathogenic fluorescent species outside the *syringae-cichorii* group; *P. aeruginosa* and some strains of *P. fluorescens* biotype B (*P. marginalis*) can also be agents of plant diseases.

The group constituted by *Pseudomonas syringae* and *P. cichorii* can be separated from other organisms of the *P. fluorescens* group on the basis of their negative arginine dihydrolase reaction (see Table 18.2). *Pseudomonas syringae* is a collective species including an enormous number of nomen species that have been named according to the host of origin. *Pseudomonas syringae* includes oxidase-negative strains, whereas *P. cichorii* is the name reserved for the oxidase-positive.

The present state of the nomenclature of the fluorescent phytopathogens is very confusing, and our knowledge of the phylogenetic relationships among the various types is still very limited. It is thought that some of the species now considered to be synonyms of *Pseudomonas syringae* deserve independent species rank.

The most important nonfluorescent members of the *Pseudomonas fluorescens* RNA homology group are *P. stutzeri, P. mendocina, P. alcaligenes*, and *P. pseudoalcaligenes. Pseudomonas stutzeri* is the species that has the broadest range of DNA base composition of the whole genus (from 61 to 66 mol% G+C), and, as expected, it is also heterogeneous in DNA homology. As mentioned before, an extensive review on this interesting species has been published recently.[8] The general phenotypic properties of the species are readily identifiable: wrinkled appearance of the colonies, capacity for denitrification, and use of starch and ethylene glycol for growth. The colonies may lose their wrinkled appearance after repeated transfers in laboratory media, but in some cases the original type can be recovered after sub-cultivation under conditions of denitrification.

Pseudomonas mendocina is a species somewhat similar to *P. stutzeri*; the strains are denitrifiers and can use ethylene glycol for growth. In contrast to most members of the *P. fluorescens* RNA group, they produce carotenoid pigments.

Pseudomonas alcaligenes and *P. pseudoalcaligenes* are nonfluorescent members of the *P. fluorescens* RNA homology group. *Pseudomonas alcaligenes* is still a poorly defined species, due to the low similarity among the three strains that have been described. One of these strains is particularly aberrant and produces carotenoid pigments. *Pseudomonas pseudoalcaligenes* strains are phenotypically heterogeneous but the DNA homology is relatively high among the 15 strains known at present. Outside of the species, the homology is higher with the strains of *P. mendocina* than with *P. alcaligenes.* Some strains of *P. pseudoalcaligenes* accumulate poly-β-hydroxybutyrate as carbon reserve material, a property uniformly negative in all other members of the RNA homology group.

Many review articles have been written on *Pseudomonas*, and the general impact of the studies on phylogenetic studies of prokaryotic organisms has been discussed.[15]

REFERENCES

1. De Ley, J. 1992. The Proteobacteria. Ribosomal RNA cistron similarities and bacterial taxonomy, p. 2111-2140. In A. Balows, H. G. Trüper, M. Dworkin, W. Harder, and K.-H. Schleifer (eds.), The Prokaryotes, a *Handbook on the Biology of Bacteria, Ecophysiology, Isolation, Identification and Applications*. Springer Verlag, New York.
2. den Dooren de Jong, L. E. 1926. *Bijdrage tot de kennis van het mineralisatieprocess*. Nijgh & Van Ditmar, Rotterdam.
3. Fox, G. E., E. Stackebrandt, R. B. Hespell, J. Gibson, J. Maniloff, T. A. Dyer, R. S. Wolfe, W. E. Balch, R. S. Tanner, L. J. Magrum, L. B. Zablen, R. Blakemore, R. Gupta, L. Bonen, B. J. Lewis, D. A. Stahl, K. R. Luehrsen, K. N. Chen, and C. R. Woese. 1980. The phylogeny of prokaryotes. *Science* 209:457-463.
4. Fuchs, A. On the synthesis and breakdown of levan by bacteria. 1-178. 1959. Rijks-Universitcit te Leidcn. 3-4-1959. Ref Type: Thesis/Dissertation
5. Haynes, W. C. and W. H. Burkholder. 1957. Genus I. *Pseudomonas* Migula, 1894, p. 89-152. In R. S. Breed, E. G. D. Murray, and N. R. Smith (eds.), *Bergey's Manual of Determinative Bacteriology*. The Williams & Wilkins Company, Baltimore.
6. Jessen, O. 1965. *Pseudomonas aeruginosa and other green Fluorescent Pseudomonads. A Taxonomic Study*. Munksgaard, Copenhagen.
7. Johnson, J. L. and N. J. Palleroni. 1989. Deoxyribonucleic acid similarities among *Pseudomonas* species. *Int. J. Syst. Bacterio*l. 39:230-235.
8. Lalucat, J., A. Bennasar, R. Bosch, E. García-Valdés, and N. J. Palleroni. 2006. Biology of *Pseudomonas stutzeri*. *Microbiol. Mol. Biol. Rev*. 70:510-547.
9. Mandel, M. 1966. Deoxyribonucleic acid base composition in the genus *Pseudomonas*. *J. Gen. Microbiol*. 43:273-292.
10. Migula, W. 1894. Ueber ein neues System der Bakterien. *Arbeiten aus dem bakteriologischen Institut der technischen Hochschule zu Karlsruhe* 1:235-238.
11. Moore, E. R. B., B. J. Tindall, V. A. P. Martins dos Santos, D. H. Pieper, J.-L. Ramos, and N. J. Palleroni. 2005. *Pseudomonas*: Nonmedical, p. 1-66. In M. Dworkin et al. (Eds.), *The Prokaryotes: An Evolving Electronic Resource for the Microbiological Community*, release 3.20. Springer-Verlag, New York.
12. Palleroni, N. J. 1978. *The Pseudomonas Group*. Meadowfield Press Ltd., Shildon.
13. Palleroni, N. J. 1984. Genus I. *Pseudomonas migula* 1894, p. 141-199. In N. R. Krieg and J. G. Holt (eds.), *Bergey's Manual of Systematic Bacteriology*. The Williams & Wilkins Co., Baltimore.
14. Palleroni, N. J. 1993. *Pseudomonas* classification. A new case history in the taxonomy of Gram-negative bacteria. *Antonie Van Leeuwenhoek - International Journal of Microbiology* 64:231-251.
15. Palleroni, N. J. 2003. Prokaryote taxonomy of the 20th century and the impact of studies on the genus *Pseudomonas*: a personal view. *Microbiol*. 149:1-7.
16. Palleroni, N. J. 2005. Genus I. *Pseudomonas*, p. 323-379. *Bergey's Manual of Systematic Bacteriology. Part B. The Gammaproteobacteria*. Springer, New York.
17. Palleroni, N. J., R. W. Ballard, E. Ralston, and M. Doudoroff. 1972. Deoxyribonucleic acid homologies among some *Pseudomonas* species. *J. Bacteriol*. 110:1-11.
18. Palleroni, N. J. and M. Doudoroff. 1972. Some properties and subdivisions of the genus *Pseudomonas*. *Annu. Rev. Phytopathol*. 10:73-100.
19. Palleroni, N. J., M. Doudoroff, R. Y. Stanier, R. E. Solanes, and M. Mandel. 1970. Taxonomy of the aerobic pseudomonads: The properties of the *Pseudomonas stutzeri* group. *J. Gen. Microbiol*. 60:215-231.
20. Palleroni, N. J., R. Kunisawa, R. Contopoulou, and M. Doudoroff. 1973. Nucleic acid homologies in the genus *Pseudomonas*. *Int. J. Syst. Bacteriol*. 23:333-339.
21. Palleroni, N. J. and E. R. B. Moore. 2004. Taxonomy of pseudomonads: experimental approaches, p. 3-44. In J.-L. Ramos (ed.), *Pseudomonas*. Kluwer Academic/Plenum Publishers, New York.
22. Pitt, T. L. 1998. *Pseudomonas*, *Burkholderia* and related genera, p. 1109-1138. In A. Balows and B. I. Duerden (eds.), *Systematic Bacteriology*. Arnold, London.
23. Ralston, E., N. J. Palleroni, and M. Doudoroff. 1976. Phenotypic characterization and deoxyribonucleic acid homologies of the *Pseudomonas alcaligenes* group. *Int. J. Syst. Bacteriol*. 26:421-426.
24. Sands, D. C., M. N. Schroth, and D. C. Hildebrand. 1970. Taxonomy of phytopathogenic pseudomonads. *J. Bacteriol*. 101:9-23.
25. Sherris, J. C., N. W. Preston, and J. G. Shoesmith. 1957. The influence of oxygen on the motility of a strain of *Pseudomonas* sp. *J. Gen. Microbiol*. 16:86-96.

26. Sherris, J. C., J. G. Shoesmith, M. T. Parker, and D. Breckon. 1959. Tests for the rapid breakdown of arginine by bacteria: their use in the identification of pseudomonads. *J. Gen. Microbiol.* 21:389-396.
27. Skerman, V. B. D. 1967. *A Guide to the Identification of the Genera of Bacteria*. Williams & Wilkins, Baltimore.
28. Sneath, P. H. A. 1960. A study of the bacterial genus *Chromobacterium. Iowa St. J. Sci.* 34:243-500.
29. Stanier, R. Y., N. J. Palleroni, and M. Doudoroff. 1966. The aerobic pseudomonads: A taxonomic study. *J. Gen. Microbiol.* 43:159-271.
30. Thornley, M. J. 1960. The differentiation of *Pseudomonas* from other Gram-negative bacteria on the basis of arginine metabolism. *J. Appl. Bacteriol.* 23:37-52.
31. van Niel, C. B. and M. Allen. 1952. A note on *Pseudomonas stutzeri. J. Bacteriol.* 64:413-422.

19 The Family Neisseriaceae

Yvonne A. Lue

CONTENTS

TAXONOMY

Bergey's Manual of Systematic Bacteriology defines the family of Neisseriaceae as consisting of two separate rRNA superfamilies that have no level of relatedness. The true Neisseriaceae and two *Kingella* ssp. belong to the β-subclass of Proteobacteria. False *Neisseria*, *Moraxella*, *Branhamella,* and *Acinetobacte*r genera belong to the γ-subclass of the Proteobacteria, and are removed from the family of Neisseriaceae. In addition, genetic studies have shown that *Eikenella*, *Simonsiella*, *Alyseilla* EF-4a, EF-4b, M-5, and M-6 (CDC groups) are related to the true Neisseriaceae and two *Kinegella* ssp. The present classification of the family of Neisseriaceae is *Neisseria, Kingella, Eikenella, Simonsiella Alyseilla* EF-4a, EF-4b, M-5 and M-6. *Neisseria gonorrhoeae* and *N. meningitidis* will be the only species discussed in this chapter because of pathogenicity in man. Species of the genus *Neisseria* found in the humans and animals inhabiting mucous membranes are listed in Table 19.1.

The *Neisseria* are Gram-negative diplococci, with adjacent sides that are flattened, giving the organisms a "coffee-bean" appearance. These organisms are aerobic, non-motile, grow best in an atmosphere of 5 to 10% CO_2, produce cytochrome oxidase, and do not form spores. Acid from carbohydrates, nitrate reduction, polysaccharide from sucrose, presence of deoxyribonuclease (DNase), superoxol (30% hydrogen peroxide), pigment production, and the resistance to Colistin (10 μg) are used to differentiate the species (Table 19.2). Non-growth-dependent rapid carbohydrate tests are available for the detection of acid production from carbohydrates for the speciation of *Neisseria*.

NEISSERIA GONORRHOEAE

Neisseria gonorrhoeae, the gonococcus, is the causative agent of gonorrhea, a sexually transmitted disease, and of several other conditions, which are sequelae to the infection. The organism infects only man, and transmission of the organism is almost always by person-to-person contact. It is the second most reported bacteria causing sexually transmitted disease.

Microscopically, the organism is indistinguishable from many other *Neisseria;* it is, however, the most fastidious in its growth requirements. The gonococcus requires the amino acid cysteine

TABLE 19.1
Host Range of *Neisseria* Species in Humans and Animals

	Species
Humans:	
Urogenital tract	*N. gonorrhoeae, N. meningitidis*
Oropharynx	*N. meningitidis, N. lactamica, N. cinerea, N. flavescens, N. subflava, N. sicca, N. mucosa, N. elongata*
Animals: Oropharynx:	
Cat	*N. canis*
Dog	*N. weaveri*
Guinea pig	*N. denitrificans and N. caviae*
Iguana	*N. iguanae*
Rabbit	*N. cuiniculi*
Monkey	*N. macacae*
Sheep and cattle	*N. ovis*

Source: Modified from Identification of *N. gonorrhoeae* and related species, Centers for Disease Control and Prevention, Sexually Transmitted Disease, 2004

and a usable energy source (i.e., glucose, pyruvate, or lactose), and will not grow on blood agar. It will grow on supplemented chocolate agar. The gonococcus ferments glucose but not maltose, fructose, or sucrose.

Laboratory Diagnosis

Diagnosis of gonococcal disease should be made at two levels: presumptive and confirmed. The isolation of the organism and the diagnosis of the disease must be reported to the public health authorities. There are serious medicolegal consequences for misdiagnosing gonorrhea or misidentifying strains of *Neisseria gonorrhoeae.*

The Gram stain showing Gram-negative intracellular microorganisms is presumptive for the diagnosis of gonorrhea. This test has a sensitivity of 50 to 70% and generally is used for symptomatic males. For symptomatic females, an endocervical smear has 40 to 60% sensitivity. A definitive diagnosis for females must be made by the recovery of *Neisseria gonorrhoeae* in culture from an endocervical specimen or by detection using a nucleic acid procedure. The Gram stain has low predictive value and is not useful for the asymptomatic female.

The site of specimen collection depends on the gender and sexual practices of the patient. Sites for obtaining specimens include urethra (males), endocervix (female), rectal canal, oropharynx, joint fluid, blood, skin lesions, and conjunctiva, depending on the clinical impression. The gonococcus is the most sensitive of the *Neisseria* species to adverse conditions, such as drying and extremes in temperature.

The correct transportation of the specimen is very important in maintaining the viability of the gonococcus for definitive identification. The specimen must be collected with a Dacron or rayon swab that must be transported in Stuart's or Amies buffered semisolid transport media at room temperature. Culture media, such as Martin-Lewis, Thayer-Martin, and New York City medium in JEMBEC plates, can be used to transport the specimens. The medium, at room temperature, is inoculated and the CO_2-generating tablet is placed in the area provided. The plate is then placed in a plastic bag, sealed, and sent to the laboratory at room temperature. Upon receipt, the inoculated plate is placed in the incubator at 35 ± 1°C with 5° to 10% CO_2.

TABLE 19.2
Differential Characteristics of Human *Neisseria* Species

Species	*Acid from					Nitrate Reduction	Polysaccharide from Sucrose	DNase	Superoxol	Pigment	Colistin Resistance 10 μg
	G	M	S	F	L						
N. gonorrhoeae	+	–	–	–	–	–	–	–	Strong (4+) positive (explosive)	–	R
N. gonorrhoeae ssp. kochii	+	–	–	–	–	–	–	–	Strong (4+) positive (explosive	–	R
N. meningitidis	+	+	–	–	–	–	–	–	Weak (1+) to strong (4+) positive	–	R
N. lactamica	+	+	–	–	+	–	–	–	Weak (1+) to strong (3+) positive	+	R
N. polysaccharea	+	+	–	–	–	–	+	–	Weak (1+) to strong (3+) positive	–	(R)
N. cinerea	–	–	–	–	–	–	–	–	Weak (2+) positive	–	(R)
N. flavescens	–	–	–	–	–	–	+	–	Weak (2+) positive	+	S
N. mucosa	+	+	+	+	–	+	+	–	Weak (2+) positive	d	S
N. subflava biovar subflava	+	+	–	–	–	–	–	–	Weak (2+) positive	+	S
N. subflava biovar flava	+	+	–	+	–	–	–	–	Weak (2+) positive	+	S
N. subflava biovar perflava	+	+	+	+	–	–	+	–	Weak (2+) positive	+	(R)
N. sicca	+	+	+	+	–	–	+	–	Weak (2+) positive	–	S
N. elongata	–	–	–	–	–	–	–	+	–	–	S

Abbreviations: G, glucose; M, maltose; L, lactose; S, sucrose; – most strains negative; + strains dependent (some strains positive, some strains negative); R most strains resistant; R most strains susceptible, some strains known to be resistant; S, strains suitable not known to be resistant.

Source: Modified from Identification of N. gonorrhoeae and related species Centers for Disease Control and Prevention, Sexually Transmitted Disease 2004.

*Cystine-tryptic digest semisolid agar containing 1% of carbohydrate with phenol red pH indicator.

Specimens submitted on swabs must be inoculated on an appropriate medium within 6 hr of collection. Specimens collected from sites with normal bacteria flora must be inoculated on a selection medium such as Thayer-Martin, Martin-Lewis, or New York City agar. Chocolate agar medium is used for specimens from normally sterile body sites. All media must be incubated in the atmosphere or 3 to 7% CO_2 at 35° to 37°C and examined every 24 hr for 72 hr. A higher concentration of CO_2 can inhibit the growth of some strains.

The colonies of *Neisseria gonorrhoeae* are about 0.5 mm in diameter, glistening, and raised. A pinkish-brown pigment is visible when sweeping the colonies with a cotton swab. For definitive identification, the organism must be a Gram-negative diplococci, oxidase positive, and ferment glucose, but not maltose, lactose, sucrose, or fructose. Chromogenic enzyme substrate tests based on the hydrolysis of biochemical substrate by bacterial enzyme can identify *N. gonorrhoeae* from other *Neisseria*. Definitive identification can also be made with direct fluorescent monoclonal antibodies, coagulation (Phadebact Monoclonal GC test, and GonoGen), and the DNA probe (Gen Probe) tests.

Nucleic acid tests do not require culture and therefore allow for the detection of the gonococcus in endocervical, urethral, and urine specimens. Nucleic acid hybridization tests for confirmation are available for endocervical and urethral specimens. Urine, endocervical, and urethral specimens can be used for nucleic acid amplified tests (NAATs). For the nucleic acid detection tests, each manufacturer has a unique system for the transportation of the specimens. The NAAT cannot be used for specimens from non-genital sites and should not be employed as a test of cure.

Specimens from patients suspected of sexual abuse must be submitted from the site of abuse for the recovery and definitive identification of *Neisseria gonorrhoeae*. Nucleic acid tests cannot be used for the specimens from these patients. The identification of *N. gonorrhoeae* in sexual abuse cases should be determined by at least two methods. The sensitivity pattern of the isolates from the alleged abuser and the patient is of value for the prosecution.

Gonococcal Disease

Acute anterior urethritis is the most common manifestation of gonorrhea. This usually appears in 90% of males 2 to 8 days after exposure and is manifested by a purulent discharge. Acute epididymitis is the most common local complication of the infection. In females, the endocervix is the primary site of infection. Urethral infection and infection of the Bartholin's glands are seen in women who have gonococcal cervicitis. Ascending infection occurs in 10 to 20% of females with a gonococcal infection. Asymptomatic infection can lead to pelvic inflammatory disease (PID), which can result in obstruction of the fallopian tubes.

Manifestation of disseminated gonococcal disease in males and females includes dermatitis, arthritis-tenosynovitis syndrome, monoarticular septic arthritis, perihepatitis, endocarditis, and meningitis. Ophthalmia neonatorum, a disease of newborns, can be prevented by prophylactic washing of the eyes of the infant with a solution of penicillin or silver nitrate immediately after birth.

When sulfonamides first became available, they were used successful in the treatment of gonorrhea. During World War II, strains of *Neisseria gonorrhoeae* resistant to sulfonamides appeared. Later, penicillin became the drug of choice; but with the acquisition of plasmid-mediated resistance, this treatment is now ineffective. The Centers for Disease Control and Prevention issued guidelines for treatment in 2006. For patients with uncomplicated gonococcal infections of the cervix, urethra, and rectum, the recommendations are for single dose of ceftriaxone (125 mg IM), cefixime (400 mg orally), ciprofloxacin (500 mg orally), ofloxacin (400 mg orally), or levofloxacin (250 mg orally). Quinolone resistance has been detected and should not be used for men who have sex with men, a history of recent travel, a partner's history of recent travel, or an infection acquired in California or Hawaii. There are specific treatment recommendations from the Centers for Disease Control and Prevention for patients with disseminated gonococcal infections.

Neisseria meningitidis

Colonies of this species grow to be 1 mm in diameter after 18 to 24 hr of incubation and are usually gray and mucoid on primary isolation media. The organism grows very well on sheep blood or chocolate agar at 37°C and is enhanced by incubation in an atmosphere of 5 to 10 % CO_2. Identification can be made by observing acid production from glucose and maltose, but not from sucrose, fructose, or lactose.

Meningococci are separated into 13 serologic groups (A, B, C, D, H, I, K, L, X, Y, Z, W135, and 29E) on the basis of capsular polysaccharide. Encapsulated cells of freshly isolated organisms demonstrate the quellung reaction when mixed with homologous antiserum. Serogroups A, B, and C are usually isolated from epidemic and endemic cases.

The pathogenicity of meningococcus is linked to a potent endotoxin. It comprises approximately 50% of the outer membrane of the organism. The endotoxin has been characterized as a lipo-*oligo*-saccharide because it lacks the multiple repeating sugars of a lipo*poly*saccharide. The endotoxin has been implicated as the cause of septic shock seen in patients with meningococcemia.

Meningococcal Disease

Man is the only natural host for *Neisseria meningitidis*, and the disease is maintained and spread by carriers harboring the organism in their nasopharynx. Infection commonly results from asymptomatic nasopharyngeal carriage. There is a 10 to 20% carriage rate in the population at any given time. The organism evades the immune system and enters the bloodstream to cause disease. Diseases caused by *N. meningitidis* include meningoencephalitis, meningitis with or without meningococcemia, meningococcemia without meningitis, and bacteremia without complications. Hematogenous spread of the organism can lead to endocarditis, percardititis, arthritis, endophthalmistis, osteomyelitis, and peritonitis. Risk factors of meningococcal disease include individuals with complement or properdin component deficiencies, hepatic failure, systemic lupus erythematous, multiple myeloma, and asplenia. Meningococci have been implicated as the etiological agent of approximately 2 to 4% of community-acquired pneumonia. Pharyngitis is associated with recent contact with carriers of *N. meningitidis*, and can be a prior symptom and sign of serious disease.

Diagnosis

For diagnosis of meningococcal disease, the organism should be isolated from CSF (cerebrospinal fluid), blood, or aspirates of petechiae. Synovial fluid, sputum, conjunctival and nasopharyngeal swabs should be obtained when clinically indicated. These specimens must be inoculated onto blood and chocolate media as quickly as possible. Gram-stained smears showing Gram-negative diplococci of CSF, aspirates, and other fluids can give the presumptive diagnosis. Latex agglutination and coagglutination tests are direct antigen tests to detect meningococcal capsular polysaccharide in CSF, urine, and serum.

Chemotherapy and Prophylaxis

With the introduction of sulfonamides, chemotherapy was the main method of treating meningococcal disease. The appearance of resistance to this agent in 1963, however, has made this treatment ineffective. Benzyl penicillin is the preferred agent for treatment of meningococcal meningitis, as long as the strain does not produce β-lactamase. Cefotaxime and cetriaxone are effective alternate agents. Vaccines have been successful in reducing the incidence of disease in the military, in containing epidemics caused by serogroup A, and in individuals with complement disease. This protection is not long-lasting. There is no vaccine for serogroup B. Currently there are two vaccines against *Neisseria meningitidis* available in the United States. Available since 1981, meningococcal polysaccharide vaccine (MPSV4 or Menomune®) and meningococcal conjugate vaccine (MCV4 or MenactraT) were licensed in 2005. Both vaccines can prevent four types of meningococcal disease.

These include two of the three types most common in the United States (serogroups C, Y, and W-135) and a type that causes epidemics in Africa (serogroup A). Meningococcal vaccines cannot prevent all types of the disease, but they can protect many people who might become sick if they do not receive the vaccine. Meningitis cases must be reported to state and/or local health departments to assure follow-up of close contacts and recognize outbreaks.

Laboratory Safety

Neisseria meningitidis is now classified as a biosafety level 2 organism, which requires that all manipulations of specimens that are suspected of containing this organism, and cultures of it, must be conducted in a biological safety cabinet. It is recommended that laboratory personnel be offered the meningococcal vaccine. Vaccine will not decrease the risk of infection but can reduce the potential risk of laboratory-acquired disease.

REFERENCES

1. O'Leary, W., *Practical Handbook of Microbiology*, first edition. CRC Press, Boca Raton, FL, 1989.
2. Amstrong, D., Cohen, J., et al. *Infectious Diseases*. Mosby, London, 1999.
3. Murray, P.R. et al. *Manual of Clinical Microbiology*, 8th edition. American Society for Microbiology, Washington, DC, 2003.
4. *Identification of N. Gonorrhoeae and Related Species*. Centers for Disease Control and Prevention, Sexually Transmitted Disease, Atlanta, GA, 2004.
5. Centers for Disease Control (CDC) / Division of Bacterial and Mycotic Diseases. Meningococcal Disease, 2005.
6. Centers for Disease Control (CDC), *Sexually Transmitted Diseases Treatment Guidelines 2006*. Morbidity and Mortality Weekly Report, Recommendation and Reports (MMWR), 55(RR11), 1–94, 2006.

20 Microbiological and Clinical Aspects of the Pathogenic Spirochetes

Charles S. Pavia

CONTENTS

SUMMARY

The spirochetes are a unique group of bacteria that can be distinguished morphologically from most other bacteria based on their large size and helical or corkscrew-shaped appearance. They also possess flagella that are internal rather than extracellular, which is more characteristic of most other motile organisms. Spirochetes usually cause diseases that are non-acute upon initial exposure of the host to the infectious agent, but they can have devastating consequences on the human body later on, in the absence of curative antibiotic therapy.

The sexually transmitted treponemal disease, syphilis, has been with us for many centuries, perhaps as far back as Biblical times, whereas the tick-borne borrelial-caused illness, Lyme disease, has only recently been described as a serious clinical entity. Both of these pathogens can cause a variety of multi-system disorders that can be easily confused with other infectious or non-infectious illnesses. Early diagnosis of syphilis and Lyme disease is essential for successful antibiotic treatment and prevention of chronic debilitating sequelae. Other spirochetal infections caused by various other *Borrelia, Leptospira,* and *Treponema* are uncommon in developed countries but still can cause significant morbidity in various other parts of the world.

All spirochetal infections rely heavily on serologic techniques for verifying or establishing the diagnosis. Non-medical preventive measures include avoiding contact with (1) the appropriate insect vector, (2) an infected sex partner or body fluid, and (3) contaminated fomites or skin lesions. Although vaccines for human use are currently unavailable, there is hope that continuing research along these lines will lead to the development of safe and effective ones.

INTRODUCTION

This chapter focuses on the pathogenic spirochetes, highlighting their unique basic biological properties and the important clinical, pathologic, and diagnostic entities and immune phenomena associated with the diseases caused by them. Major emphasis is placed on the causative agents of Lyme disease (*Borrelia burgdorferi* and related strains) and syphilis (*Treponema pallidum*), because these represent the most common spirochetal diseases worldwide, including North America, and have generated the most interest and discussions among clinicians, scientists, patients, and other members of the lay public. In addition, their social, behavioral, as well as biomedical implications have long been recognized and these have led to considerable scientific and political debate.

Basic Biology of the Spirochetal Bacteria

Spirochetes are a highly specialized group of motile Gram-negative spiral-shaped bacteria (Table 20.1), usually having a slender and tightly helically coiled structure. They range from 7 to 50 μm in length. One of the unique features of spirochetes is their motility by a rapidly drifting rotation, often associated with a flexing or undulating movement along the helical path. Such locomotion is due to the presence of axial fibrils, also known as flagella, that are wound around the

TABLE 20.1
Unusual Features of the Pathogenic Spirochetes

Large bacteria: up to 50 μ in length, but very thin in diameter compared with other bacteria (e.g., cocci and rods are 1 to 3 μ in length), red blood cell diameter is 6 to 8 μ; they also exhibit a unique spiral, helical shape (Figure 20.1).
Most of them require special staining techniques and microscopy for visualization, such as silver stain, fluorescence, or dark-field microscopy.
They are nutritionally fastidious and exhibit a slow rate of growth: estimated 24- to 33-hr division time *in vivo* and slightly less *in vitro*; compare with *Escherichia coli*: 20 min.
They are extremely sensitive to elevated temperatures (≥38°C). Pathogenic treponemes *cannot* be cultivated on artificial media; other spirochetes can be grown with some difficulty or with special media.
They cause chronic, stage-related, and sometimes extremely debilitating or crippling disease in the untreated host.
They do *not* seem to produce toxins.
The interplay or interrelationship between the invading spirochetes and the subsequent host response as factors in the disease process have yet to be clearly or fully defined.
They have endoflagella intertwined between the cell wall and protoplasmic cylinder — also called axial fibrils. Most bacterial flagella are extracellular.
Most pathogenic spirochetes (*Borrelia, Treponema)* are microaerophilic (once thought to be anaerobes).
Borrelia burgdorferi and its related strains are perhaps the most unique of the spirochetes, by having linear plasmids that code for outer-surface proteins.

main body (protoplasmic cylinder) and enclosed by the outer cell wall or sheath of these organisms. These bacteria belong to the order Spirochaetales, which includes two families: Spirochaetaceae and Leptospiraceae. Important members of these groups include the genera *Borrelia, Leptospira,* and *Treponema.*

The spirochetes generally comprise a fastidious group of bacteria; that is, they can be difficult to grow (and therefore to study) in the laboratory, often requiring highly specialized media and culture conditions (such as low oxygen tension) to optimize their replicating capabilities. Some, such as *Treponema pallidum*, can only be maintained consistently in a replicating state by *in vivo* passage in rabbits. The spirochetes live primarily as extracellular pathogens, rarely (if ever) growing within a host cell. A few reports have claimed finding spirochetes inside host cells but there is no evidence indicating that these are live or replicating organisms. Unlike most bacteria, spirochetes do not stain well with aniline dyes such as those used in the Gram stain procedure. However, their cell walls do resemble, both structurally and biochemically, those of other Gram-negative bacteria and are thus classified within this very large group of bacteria. Other staining procedures, such as silver, Giemsa, or the Wright stain, may be useful. The best way to visualize spirochetes, especially when they are alive and motile, is through the use of dark-field or phase-contrast microscopy or after staining with a fluorochrome dye, such as acridine orange,[1] and then viewing under a microscope equipped for fluorescence microscopy (Figure 20.1). When present, their appearance in tissue specimens can often be revealed best by the silver-staining technique.

The infections caused by the spirochetes are important public health problems throughout the world, leading to such diseases as Lyme and the relapsing fever borrelioses, syphilis and the other treponematoses, and leptospirosis (Table 20.2). A better understanding of the molecular biology, pathogenesis, and immunobiology of the disease-causing spirochetes has become crucial in efforts to develop effective vaccines because there has been no significant modification in excessive sexual activity, personal hygiene practices, or vector control. Further knowledge of immune responses to spirochetes is essential for their eventual control by immunization, and studies of the host-spirochete relationship have led to important new insights related to the immunobiology of these pathogens. Serologic techniques have long been used as indispensable diagnostic tools for the detection of many of the spirochetal diseases, especially Lyme disease and syphilis. Unfortunately, as may occur in other infectious processes, the host response to the spirochetes, as part of the normal protective mechanisms, may paradoxically cause an immunologically induced disease in the affected individual, leading to the complications of arthritis and the neuropathies of Lyme disease, as well as aortitis, immune-complex glomerulonephritis, and the gummatous lesions of syphilis.

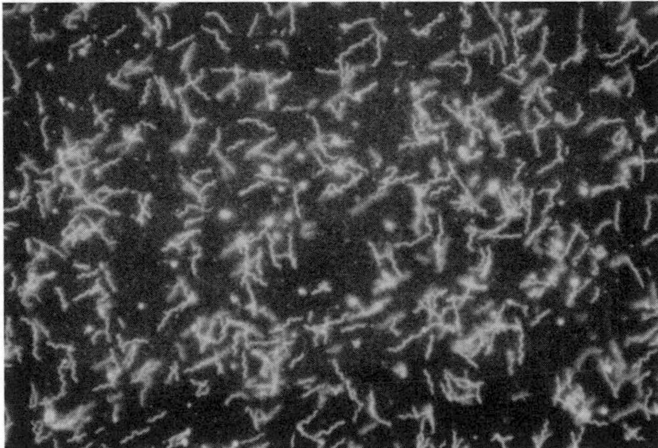

FIGURE 20.1 Photomicrograph of culture-grown *Borrelia burgdorferi* (sensu stricto) stained with the fluorochrome dye, acridine orange, and visualized through a microscope equipped for fluorescence microscopy; magnification 500X.

TABLE 20.2
Epidemiology of Spirochetal Infections

Pathogenic Spirochete	Human Disease	Vector or Source of Infection
Borrelia		
B. burgdorferi (sensu lato)*	Lyme disease or	Ixodid ticks
B. burgdorferi (sensu stricto)	Lyme borreliosis	*I. scapularis* or *pacificus*
B. afzelii		*I. ricinus*
B. garinii		
B. recurrentis	Epidemic relapsing fever	Body louse, ped. humans
B. hermsii		
B. turicatae	Endemic relapsing fever	Ornithodoros ticks
B. parkeri		
Leptospira		
Leptospira interrogens	Leptospirosis (Weil's Disease)	Exposure to contaminated animal urine
Treponema		
T. pallidum subspecies *pallidum*	Syphilis	Sexual contact, transplacental
T. pallidum ssp. *endemicum*	Bejel (endemic syphilis)	Direct contact with contaminated eating utensils
T. pallidum ssp. *pertenue*	Yaws	Direct contact with infected skin lesions
T. carateum	Pinta	

* Sensu stricto strains are found almost exclusively in North America, while *B. afzelii* and *B. garinii* are found only in Europe.

LYME DISEASE

Lyme Disease: General Features

In the mid-1970s, a geographic clustering of an unusual rheumatoid arthritis-like condition, involving mostly children and young adults, occurred in northeastern Connecticut. This condition proved to be a newly discovered disease named Lyme disease, after the town of its origin.[2] The arthritis was characterized by intermittent attacks of asymmetric pain and swelling, primarily in the large joints (especially the knees) over a period of a few years. Epidemiological and clinical research showed that the onset of symptoms was preceded by an insect bite and unique skin rash probably identical to that of an illness following a tick bite, first described in Europe at the turn of the century.[3]

The beneficial effects of penicillin or tetracycline in early cases suggested a microbial origin (likely bacterial) for what was initially called Lyme arthritis.

Lyme disease is now the most common tick-transmitted illness, and it has been reported in almost all of the contiguous states of the United States and in numerous countries worldwide. However, it occurs primarily in three geographic regions in North America: (1) the coastal areas of the Northeast from Maine to Maryland, (2) the Midwest in Wisconsin and Minnesota, and (3) the far West in parts of California and Oregon. These geographic areas parallel the location of the primary tick vector of Lyme disease in the United States — *Ixodes scapularis* (formerly called *I. dammini)* in the east and Midwest and *I. pacificus* in the far west. Outside North America, Lyme disease is most prevalent in western Europe, especially in Austria, Germany, and the Scandinavian countries, corresponding to the distribution of *I. ricinus* ticks. The greatest concentration of cases is in the northeastern United States, particularly in New York State, where the disease is endemic on eastern Long Island and just north of New York City in neighboring Westchester and Putnam counties.

In the early 1980s, spirochetal organisms were isolated and cultured from the midguts of *Ixodes* ticks taken from Shelter Island, NY (an endemic focus), and shortly thereafter they were cultured from the skin rash site, blood, and cerebrospinal fluid of patients with Lyme disease. This newly discovered spirochete called *Borrelia burgdorferi* is microaerophilic, resembles other spirochetes morphologically, and is slightly larger than the treponemes. Recently, the Lyme disease spirochetes have been given updated taxonomic designations under the broad category of *B. burgdorferi* (sensu lato) [Table 20.2]. Unlike the pathogenic treponemes, *B. burgdorferi* and related strains can be readily cultivated *in vitro* in a highly fortified growth media.[4]

Protection to *Borrelia burgdorferi* may develop slowly, and it is unclear whether resistance to reinfection occurs. Experimental animal studies have shown that immune sera can transfer protection to normal recipients challenged with *B. burgdorferi.*[5] Monoclonal antibodies to borrelial outer surface proteins are also protective,[6] and these were the major target antigens for the first vaccine that became available for human use. It should be noted that, for a variety of reasons, the vaccine was withdrawn from the market by the manufacturer in 2002.

Lyme Disease: Clinical Aspects

Lyme disease is an illness having protean manifestations with symptoms that include the following: (1) an erythematous-expanding red annular rash with central clearing; (2) fever, headache, stiff neck, nausea, and vomiting; (3) neurologic complications such as facial nerve (Bell's) palsy and meningitis; and (4) arthritis in about 50% of untreated patients.[7] These symptoms occur most frequently from May to November when ticks are active and numerous, and people are engaged in many outdoor activities. The most characteristic feature of early Lyme disease is a skin rash, often referred to as erythema migrans (EM), which appears shortly (3 to 32 days) after a bite from an infected tick. The lesion typically expands almost uniformly from the center of the bite and is usually flat or slightly indurated with central clearing and reddening at the periphery. It is noteworthy, however, that many Lyme disease victims either do not recall being bitten by a tick, or they do not notice a rash developing at an unrecognized tick bite, or possibly they do not go on to develop a classic EM at all. On the other hand, at various intervals after the initial rash, some patients develop similar but smaller multiple secondary annular skin lesions that last for several weeks to months. Due to the expanding nature of this rash and its originally described chronic form in European patients, it was initially referred to as erythema chronicum migrans. However, this term is rarely used now, with the term EM or multiple EM being the current descriptor. Biopsy of these skin lesions reveals a lymphocytic and plasmacytic infiltrate, and *Borrelia* can be cultured from them. Various flu-like symptoms such as malaise, fever, headache, stiff neck, and arthralgias are often associated with EM. The serious extracutaneous manifestations of Lyme disease may include migratory and polyarticular arthritis, neurologic and cardiac involvement with cranial nerve palsies and radiculopathy, myocarditis, and arrhythmias. Lyme arthritis typically involves a knee or other large joint. It may enter a chronic phase, leading to destruction of bone and joints if left untreated. Interestingly, Lyme arthritis is less common in Europe than in the United States but neurologic complications are more prevalent in Europe. Another geographic consideration involves the later-stage chronic skin condition known as acrodermitis chronicum atrophicans, which can occur in Europe but has rarely been observed in North America. Unique strain variations expressing antigenic sub-types between European and North American isolates of *Borrelia burgdorferi*[8] probably explain these dissimilarities, and this has led to additional species designations for other related Lyme disease-causing spirochetes, such as *B. afzelii* and *B. garinii,* which are not found in North America. Also, several other species within the senso-lato geno-complex, such as *B. lusitania* (found in Portugal) and *B.*

japonica (found in certain parts of the Far East), have been identified but their pathogenic capabilities, as possible causes of Lyme disease, have yet to be well characterized.

In most cases, humoral and cell-mediated immune responses are activated during borrelial infection.[7] Antibody, mostly of the IgM class, can be detected shortly after the appearance of EM; thereafter, a gradual increase in overall titer and a switch to a predominant IgG antibody response occurs for the duration of an untreated infection. Most notably, very high levels of antibody have been found in serum and joint fluid taken from patients with moderate to severe arthritis, making this the most actively invasive and debilitating form of Lyme disease. Although the presence of such high antibody titers against *Borrelia burgdorferi* may reduce the spirochete load somewhat, they appear not to ameliorate the disease process completely and, indeed, may actually contribute to some of the pathologic changes. These consistent serologic responses have led to the development of laboratory tests designed to aid in the diagnosis of Lyme borreliosis. For cellular immunity, lymphocyte transformation assays have shown that peripheral blood T cells from Lyme disease patients respond to borrelial antigens primarily after early infection and following successful treatment.[9, 10] Also, addition of antigens to synovial cells *in vitro* from infected patients triggers the production of interleukin-1, which could account for many of the harmful inflammatory reactions associated with this disease.[11] Other *in vitro* studies[12, 13] have shown that human mononuclear and polymorphonuclear phagocytes can both ingest and presumably destroy *Borrelia*. Thus, borrelial antigen-stimulated T cells or their products may activate macrophages, limiting dissemination and resulting in enhanced phagocytic activity and the eventual clearance of spirochetes from the primary lesion.

Lyme Disease: Diagnosis

Clinically, Lyme disease mimics other disorders, many of which are not infectious and therefore would ordinarily not be responsive to antibiotic therapy. Because *Borrelia burgdorferi* is the causative agent, finding the organism, or any of its key detectable components, in suspected cases is the most definitive diagnosis. However, in the vast majority of cases, the Lyme spirochete cannot be isolated or identified, and immune responses (antibody production) specific for *B. burgdorferi* must be used to confirm the diagnosis.[14] Unfortunately, the antibody response is often not detectable in the early treatable stage, and the clinical impression cannot always be confirmed.

Isolation of the spirochete unambiguously confirms the diagnosis of Lyme borreliosis. Recovery of *Borrelia burgdorferi* is possible but the frequency of isolation from the blood or other body fluids of acutely ill patients is very low.[1, 15, 16] Better success rates in isolating *Borrelia* have been achieved after culturing skin biopsy specimens of clear-cut EM rashes.[17] Despite such success, borrelial cultivation can usually only be done in a few laboratories or institutions because the medium is expensive and cultures can take up to 8 to 12 weeks of incubation before spirochetes are detected,[1] but most are positive within 4 weeks.

Visualization of the spirochetes in tissue or body fluids also has been used to diagnose Lyme borreliosis.[18] In the early stage of the disease, when EM is present, the Warthin-Starry or modified Dieterle silver stain can identify spirochetes in half or more of skin biopsies obtained from the outer portion of the lesion.[19] However, few microorganisms are present and they can be confused with normal skin structures or tissue breakdown products by inexperienced laboratory personnel. Immunohistologic examination of tissue has rarely been successful in determining the presence of *Borrelia* and, in chronic Lyme disease, spirochetes are rarely detectable by any microscopic technique.[20]

Serologic tests are, for all practical purposes, the only detection systems routinely available for the confirmation of Lyme borreliosis. One of the standard serological tests, either an enzyme-linked immunosorbent assay (ELISA) or an indirect immunofluorescence assay (IFA),[14] is available in many public and private laboratories. Blood samples obtained within 3 weeks of the onset of EM are frequently serologically negative in both assays.[21] In addition, these assays have not been standardized, with laboratories using different antigen preparations and "cut-off" values. Workers, using the same set of sera, have reported interlaboratory variation in results and interpretations.[22, 23] There is also considerable variability in the serologic response pattern of patients with Lyme disease. Finally, if antibiotics are administered during early illness, antibody production can be aborted or severely curtailed.[7, 14]

The existence of antigenically different strains of *Borrelia burgdorferi* sensu lato throughout the world[8] may account for some of the variability in antibody response patterns. In addition, assays currently either use the whole spirochete or a crude bacterial sonicate as antigen, or a mixture of recombinant-derived proteins. While the latter results in greater specificity, they tend to be less sensitive. With these assays, cross-reactions have been observed with other spirochetes, in particular *Treponema pallidum* and the relapsing fever *Borrelia* species.[24] For all of these assay systems, false-positive reactions are relatively rare except possibly in hyperendemic areas, where considerably more testing (often unnecessary) would be expected because of heightened awareness of Lyme disease within the general community. The cause of these positives is not always clear but they can occur if a patient has certain other disorders, such as syphilis (active or inactive), infectious mononucleosis, lupus, or rheumatoid arthritis. In the absence of a clear-cut EM rash, laboratory testing plays a vital role in establishing or confirming the diagnosis.

Attempts to improve antibody detection have used Western (immunoblot) analysis for the detection of IgM and IgG antibodies and have used purified flagellin antigen in the ELISA. Immunoblots are more sensitive and more specific than ELISAs. Although not standardized, commercial immunoblot test kits are now being offered to further verify a routine serologic test result, especially in troublesome cases, in conjunction with the two-tiered testing algorithm recommended by the CDC,[25] as well as for further confirmation of a true serologic response to specific borrelial antigens and possible infection. Some studies have shown, however, that immunoblotting could not overcome the inability to detect antibody during the first 3 weeks of infection.[26] The performance of the ELISA can also be improved using purified flagellin protein as antigen.[27] Antibodies to the 41-kDA, flagellum-associated component peak at 6 to 8 weeks, and often appear very early (<4 weeks) during an initial infection. Unfortunately, epitopes on this antigen are shared by many other spirochetes, and neither IgM nor IgG antibodies to this antigen are specific for *Borrelia burgdorferi*.[26] It is noteworthy, however, that recent development of an ELISA using recombinant VlsE (or C6) peptides showed considerable improvement in both specificity and sensitivity over prior assay systems.[28]

The prevailing sentiment within (as well as outside) the Lyme disease research and diagnostic community is that serological verification of this disorder is still fraught with difficulties. False-negative results are likely to occur if serum is obtained within 4 weeks of initial infection or if the patient has been treated with antibiotics. False-positives occur if large numbers of patients with a low *a priori* probability of having Lyme borreliosis are examined. Interlaboratory agreement on what constitutes a positive result varies, in part because of the diversity of detection systems being used by diagnostic testing facilities, which contain different antigen preparations and cut-off values developed by various manufacturers for serodiagnostic purposes. Compounding this problem, diagnosis of initial infection can be clinically difficult because only 60 to 75% of patients with Lyme borreliosis present with or recall EM, or have a clear and consistent epidemiologic history.[29]

Recently, attempts meant to address these apparent shortcomings in serologic testing have led to the development of non-serologic diagnostic procedures such as the polymerase chain reaction (PCR) and the lymphocyte proliferation assay. Using the PCR and selective probes, it is possible to detect a single organism in a serum or tissue sample, and such gene amplification procedures show great promise for the early detection of *Borrelia burgdorferi*.[30, 31] In this regard, using primers directed at the rRNA genes of *B. burgdorferi*, it is possible to find the Lyme disease spirochete directly from skin biopsy material[32] as well as from short-term cultures of tissue extracts.[33] From a realistic standpoint, however, and because it may be technically demanding and not cost-effective for most diagnostic labs handling just a few specimens, PCR may continue to be primarily a research tool rather than a routine diagnostic procedure.

While PCR-based procedures seem promising as an exquisite and novel diagnostic tool for selective stages of Lyme disease, serological testing, for a variety of reasons, will continue to be the mainstay for the laboratory detection of the vast majority of Lyme disease cases, as it currently is for syphilis (caused by a related spirochete) and for certain other infectious and non-infectious disorders. Nonetheless, continued refinements along these lines will be needed and should be geared toward developing as economical a system as possible combined with one having optimal sensitivity and specificity.

Finally, attention has recently focused on an assay system[9, 10] designed to measure past or current exposure to *Borrelia burgdorferi* by virtue of the patient's lymphocytes to respond *in vitro* to undergo DNA synthesis in the presence of specific borrelial antigens. This laboratory procedure is generally considered a good *in vitro* correlate of the classic DTH reaction,[34] analogous to what is found when measuring *in vivo* tuberculin sensitivity following a PPD skin test. For purposes related to Lyme disease, it was found that these lymphocyte proliferation assays have been helpful in identifying patients with active disease in the absence of detectable antibody (serologic) responses. Like serologic tests, however, this assay has yet to be standardized and there is growing concern over the evidence[9] for elevated responses occurring in some healthy controls, thereby limiting the usefulness of this technique.

Lyme Disease: Prophylactic Measures

Avoiding *Borrelia*-infected ticks or tick-infested areas will guarantee protection against Lyme infection. For those living in endemic areas, a few simple precautions will help minimize possible exposure. These include wearing clothing that fully protects the body and using repellents that contain DEET (diethyltoluamide). If a tick does attach to the skin, careful removal with tweezers shortly after it attaches, followed by application of alcohol or another suitable disinfectant, will make borrelial transmission unlikely.

In the early 1990s, considerable attention turned toward the development of a vaccine for Lyme disease. For a few years already, a canine vaccine consisting of whole inactivated organisms (Bacterin) was available for veterinary purposes.[35] Those being developed for humans consisted of recombinant outer surface proteins (Osp) of *Borrelia burgdorferi*. In 1998, the FDA gave final approval to the first Lyme disease vaccine (called LYMErix®), which was shown to be safe and effective in extensively conducted clinical trails. It consisted of DNA-derived OspA of B31 — a well-characterized tick isolate of *B. burgdorferi* — incorporated with an adjuvant.[36] However, in less than 4 years, the vaccine was withdrawn by the manufacturer, mostly due to concerns over its safety and unfavorable publicity resulting from claims of purported serious side effects in a select group of vaccinees. Although studies continue with other vaccine candidates (especially in Europe), it is unclear when they may be available for widespread use.

Lyme Disease: Treatment

In general, early Lyme disease is readily treatable with a 2- to 3-week course of antibiotics, such as amoxicillin and doxycycline.[37] Later complications, such as arthritis and the neuropathies, may require more intense and prolonged antibiotic therapy, including intravenous treatment, with ceftriaxone being the preferred parenteral drug. Based on a recent study showing favorable outcomes in treating patients with a single dose of doxycycline soon after a documented tick bite,[38] consideration should be given to following this treatment strategy, especially in highly endemic areas.

RELAPSING FEVER BORRELIOSIS

Relapsing fever is an acute febrile disease of worldwide distribution and is caused by arthropod-borne spirochetes belonging to the genus *Borrelia*.[39, 40] Two major forms of this illness are louse-borne relapsing fever (for which humans are the reservoir and the body louse *Pediculus humanus* is the vector) and tick-borne relapsing fever (for which rodents and other small animals are the major reservoirs and ticks of the genus *Ornithodoros* are the vectors). *Borrelia recurrentis* causes louse-borne relapsing fever and is transmitted from human to human following the ingestion of infected human blood by the louse and subsequent transmission of spirochetes onto the skin or mucous membranes of a new host when the body louse is crushed. The disease is endemic in parts of central and east Africa and South America. The causative organisms of tick-borne relapsing fever are numerous and include *B. hermsii, B. turicatae,* and *B. parkeri* in North America; *B. hispanica* in Spain; *B. duttonii* in eastern Africa; and *B. persica* in Asia. Ticks become infected by biting and sucking blood from a spirochetemic animal. The infection is transmitted to humans or animals when saliva is released by a feeding tick through bites or penetration of intact skin.

After an individual has been exposed to an infected louse or tick, *Borrelia* penetrate the skin and enter the bloodstream and lymphatic system. After a 1- to 3-week incubation period, spirochetes replicate in the blood and there is an acute onset of shaking chills, fever, headache, and fatigue. Concentrations of *Borrelia* can reach as high as 10^8 spirochetes per milliliter blood, and these are clearly visible after staining blood smears with Giemsa or Wright's stain. During the febrile disease, *Borrelia* are present in the patient's blood but disappear prior to afebrile episodes and subsequently return to the bloodstream during the next febrile period. Jaundice can develop in some severely ill patients as a result of intrahepatic obstruction of bile flow and hepatocellular inflammation; if left untreated, patients can die from damage to the liver, spleen, or brain. The majority of untreated patients, however, recover spontaneously. They produce borrelial antibodies that have agglutinating, complement-fixing, borreliacidal, and immobilizing capabilities and that render patients immune to reinfection with the same *Borrelia* serotype. Serologic tests designed to measure these antibodies are of limited diagnostic value because of antigenic variation among strains and the coexistence of mixed populations of *Borrelia* within a given host during the course of a single infection. Diagnosis in the majority of cases requires demonstration of spirochetemia in febrile patients.

LEPTOSPIROSIS (WEIL'S DISEASE)

Leptospiral infections are zoonoses widely distributed throughout the world. Human disease is somewhat more common in tropical than in temperate areas, and infection occurs more in the summer or autumn in temperate regions.[41] Most humans are infected by leptospires in water used for drinking, washing, bathing, or swimming primarily after contact with contaminated urine, from leptospiruric animals, that has entered the water supply. The most common sources of organisms are chronically infected rats and other rodents, cattle, horses, pigs, and dogs. Numerous wild animals and even reptiles and amphibians may be carriers. In North America, unvaccinated dogs are the major reservoir for exposure of humans to this disease. The routine vaccination of dogs against *Leptospira* is probably an important preventive measure. Leptospirosis is an occupational hazard for rice and sugar cane field workers, military personnel during field exercises, people who swim in contaminated ponds or streams, farmers, veterinarians, and sewer and abattoir workers. In tropical rainforests, infected rodents contaminate stream banks as well as the water. Rainstorms raise the level of the water, which washes leptospires into streams and makes swimming very hazardous. Related to this concern, large outbreaks of leptospirosis were reported among athletes participating in triathalon events held recently in both the United States in 1998[42] and Malaysia in 2000.[43]

Leptospirosis is an acute, febrile disease caused by various serotypes of *Leptospira* (there exist about 170 serovars in 20 serologic groups). Often referred to as Weil's disease, infection with *Leptospira interrogans* causes diseases that are extremely varied in their clinical presentations and that are also found in a variety of wild and domestic animals. After entering the body through the mucosal surface or breaks in the skin, leptospiral bacteria cause an acute illness characterized by fever, chills, myalgias, severe headaches, aseptic meningitis, rash, hemolytic anemia, uveitis, conjunctival suffuseness, and gastrointestinal problems. Most human infections are mild and anicteric, although in a small proportion of victims, severe icteric disease can occur and be fatal, primarily as a result of renal failure, myocarditis, and damage to small blood vessels. After infection of the kidneys, leptospiras are excreted in the urine. Liver dysfunction with hepatocellular damage and jaundice is common. Antibiotic treatment is curative if begun during early disease, but its value, thereafter, is questionable.

Diagnosis of leptospirosis depends on either seroconversion or the demonstration of spirochetes in clinical specimens. The macroscopic slide agglutination test, which uses formalized antigen, offers safe and rapid antibody screening. Measurement of antibody for a specific serotype, however, is performed with the very sensitive microscopic agglutination test involving live organisms. This method provides the most specific reaction with the highest titer and fewer cross-reactions. Agglutinating IgM-class-specific antibodies are produced during early infection and persist in high titers for many months. Protective and agglutinating antibodies often persist in sera of convalescent patients and may be associated with resistance to future infections.

SYPHILIS

Syphilis: General Features

The origins and history of syphilis are filled with many mysteries and hypotheses. Biblical references suggest its presence in early civilization. Other evidence points to its prevalence primarily after Columbus and his crew returned to Europe from the New World in 1493. From that point on, syphilis spread throughout Europe, affecting all levels of society including political and religious leaders. Indeed, the dementia and insanity associated with late-stage syphilis that may have occurred in certain afflicted rulers or monarchs probably changed the course of history during the 16th and 17th centuries. In the pre-antibiotic era, many toxic drugs were used for treatment; and as early as 1905, a blood test was developed for the diagnosis of syphilis — the so-called Wasserman test — the prototype for the current nonspecific serologic tests designed to measure antibodies to cardiolipin.

Syphilis is still a significant worldwide problem and, after gonorrhea and chlamydial infections, it is the third or fourth most commonly reported sexually transmitted disease in the United States. The most recent rise in both same-sex and heterosexual infection occurred in the mid-1980s to the early 1990s (Figure 20.2), which coincided and co-factored with the huge number of newly diagnosed HIV cases. At the same time, this also paralleled an alarming increase in congenital syphilis in many urban areas where drug abuse and the frequent exchange of sexual services for drugs were common practices among those using illicit drugs.

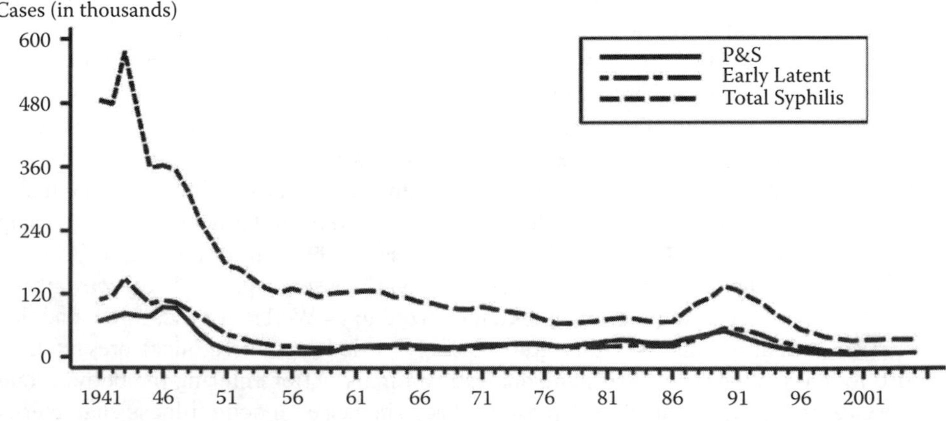

FIGURE 20.2 Syphilis: reported cases by stage of infection: United States, 1941–2005. (Source: From Department of Health and Human Services, CDC. *Sexually Transmitted Disease Surveillance*, p. 33; November 2006.)

Treponema pallidum is the spirochetal bacterium that causes syphilis — once referred to as "The Great Imitator" (or "Mimicker"), due primarily to the myriad of clinical presentations often associated with this disease.[44] If left untreated, syphilis can have severe pathologic effects, leading to irreversible damage to the cardiovascular, central nervous, and musculoskeletal systems.[45] This treponeme's genome was sequenced in 1998, yet it still cannot be grown on artificial media; it is highly motile, infectious, and replicates extracellularly and very slowly *in vivo.* Limited growth in tissue culture has been achieved but this pathogen must still be passaged *in vivo* using rabbits to maintain sufficient numbers of live organisms needed for conducting meaningful laboratory studies. Because of this problem, it has taken many years of research to acquire our current understanding of the treponemes and has probably delayed efforts toward any possible vaccine development. *T. pallidum*, unlike the other treponemes, is shorter and more tightly coiled than the *Borrelia* and *Leptospiras*.

With the institution of antibiotic therapy in the mid-1940s, the incidence of syphilis fell sharply from a high of 72 cases per 100,000 population in 1943 to about 4 per 100,000 in 1956. During the late 1970s and early 1980s, syphilis increased rapidly within the homosexual community. Such findings can be attributed, in part, to changing lifestyles, sexual practices, and other factors such as an unusually high prevalence and reduced efficacy of antibiotics in patients with acquired immunodeficiency syndrome (AIDS). Syphilis continues to rank annually as the third or fourth most frequently reported communicable disease in the United States, although its overall prevalence pattern has remained almost unchanged for the past several years (Figure 20.2).

The course of syphilis in humans is marked by several interesting phenomena. Without treatment, the disease will usually progress through several well-defined stages (somewhat resembling Lyme disease). This is unlike most other infectious diseases, which are ultimately eliminated by the host's immune system or, in severe cases, result in death. The relatively slow generation time of treponemes, which is estimated at 30 to 33 hr, contributes to this unique course. During the first two stages (primary and secondary syphilis), there is almost unimpeded rapid growth of *Treponema pallidum*, leading to an early infectious spirochetemic phase of the disease. The third stage (tertiary syphilis) occurs much later, following a prolonged latency period. Alterations in this stage are due primarily to tissue damaging immune responses elicited by small numbers of previously deposited or disseminated spirochetes.

Syphilis activates both humoral and cell-mediated immunity but this protection is only partial. The relative importance of each type of immune response is not fully known. Protective immunity against re-exposure is incomplete, especially during early stages when it develops relatively slowly.

Syphilis: Clinical Aspects

The severe late manifestations or complications of syphilis occur in the blood vessels and perivascular areas. However, sexual contact is the common mode of transmission with inoculation of infectious treponemes onto the mucous membranes of genital organs or of the oropharynx.

The first clinically apparent manifestation of syphilis (primary syphilis) is an indurated, circumscribed, relatively avascular and painless ulcer (chancre) at the site of treponemal inoculation. Spirochetemia with secondary metastatic distribution of microorganisms occurs within a few days after onset of local infection but clinically apparent secondary lesions may not be observed for 2 to 4 weeks. The chancre lasts 10 to 14 days before healing spontaneously.

The presence of metastatic infection (secondary syphilis) is manifested by highly infectious mucocutaneous lesions of extraordinarily diverse description as well as headache, low-grade fever, diffuse lymphadenopathy, and a variety of more sporadic phenomena. The lesions of secondary syphilis ordinarily go on to apparent spontaneous resolution in the absence of treatment. However, until solid immunity develops (a matter of about 4 years), 25% of untreated syphilitic patients may be susceptible to repeated episodes of spirochetemia and metastatic infection.

Following the resolution of secondary syphilis, the disease enters a period of latency with only abnormal serologic tests to indicate the presence of infection. During this time, persistent or progressive focal infection is presumably taking place but the precise site remains unknown in the absence of specific symptoms and signs. One site of potential latency, the central nervous system, can be evaluated by examining the cerebrospinal fluid in which pleocytosis, elevated protein levels, and a positive serologic test for syphilis are indicative of asymptomatic neurosyphilis.

Only about 15% of patients with untreated latent syphilis go on to develop symptomatic tertiary syphilis. Serious or fatal tertiary syphilis in adults is virtually limited to disease of the aorta (aortitis with aneurysm formation and secondary aortic valve insufficiency), the central nervous system (tabes dorsalis, general paresis), the eyes (interstitial keratitis), or the ears (nerve deafness). Less frequently, the disease becomes apparent as localized single or multiple granulomas known as "gummas." These lesions are typically found in the skin, bones, liver, testes, or larynx. The histopathologic features of the gumma resemble those of earlier syphilitic lesions, except that the vasculitis is associated with increased tissue necrosis and often frank caseation. With its myriad of organ system involvement and symptomatology, syphilis, not surprisingly, has long been called the "The Great Imitator."

Congenital syphilis is the direct result of treponemes crossing the placenta and fetal membranes, especially during mid-pregnancy, leading to spirochetemia and widespread dissemination after entering the fetal circulation. Fetal death and abortion can occur. Surviving babies with the disease have prominent early symptoms of hepatosplenomegaly, multiple long bone involvement, mouth and facial anomalies (e.g., saddle nose), and skin lesions. Treponemal antibodies (especially IgM) found in the newborn's blood is highly diagnostic.

Syphilis: Diagnosis

In its primary and secondary stages, syphilis can be diagnosed by dark-field microscopic examination of material from suspected lesions. Diagnostic serologic changes do not begin to occur until 14 to 21 days following acquisition of infection. Serologic tests provide important confirmatory evidence for secondary syphilis but are the only means of diagnosing latent infection. Many forms of tertiary syphilis can be suspected on clinical grounds but serologic tests are important in confirming the diagnosis. Spirochetes are notoriously difficult to demonstrate in the late stages of syphilis.

Two main categories of serologic tests for syphilis are available: (1) tests for nonspecific reaginic antibody and (2) tests for specific treponemal antibody.

Tests for Reaginic or Anti-Cardiolipin Antibody

This is an unfortunate and confusing designation; there is no relationship between this antibody and IgE reaginic antibody. Patients with syphilis develop an antibody response to a tissue-derived substance (from beef heart) that is thought to be a component of mitochondrial membranes and is called "cardiolipin." Antibody to cardiolipin antigen is known as Wassermann, or reaginic antibody. Numerous variations (and names) are associated with tests for this antigen. The simplest and most practical of these are the VDRL test (Venereal Disease Research Laboratory of the U.S. Public Health Service), which involves a slide microflocculation technique and can provide qualitative and quantitative data, and the rapid plasma reagin (RPR) circle card test. Positive tests are considered to be diagnostic of syphilis when there is a high or increasing titer or when the medical history is compatible with primary or secondary syphilis. The tests may also be of prognostic aid in following response to therapy because the antibody titer will revert to negative within 1 year of treatment for seropositive primary syphilis or within 2 years of that for secondary syphilis. Because cardiolipin antigen is found in the mitochondrial membranes of many mammalian tissues as well as in diverse microorganisms, it is not surprising that antibody to this antigen should appear during other diseases. A positive VDRL test may be encountered, for example, in patients with infectious mononucleosis, leprosy, hepatitis, or systemic lupus erythematosus. Although the VDRL test lacks specificity for syphilis, its great sensitivity makes it extremely useful.

Tests for Treponemal Antibody

The first test used for detecting specific antitreponemal antibody was the *Treponema pallidum* immunobilization (TPI) test. Although highly reliable, it proved too cumbersome for routine use. A major test used until recently was the fluorescent *T. pallidum* antibody (FTA) test. If virulent *T. pallidum* grown from an infected rabbit testicle is placed and fixed on a slide and then overlaid with serum from a patient with antibody to treponemes, an antigen-antibody reaction will occur. The bound antibody can then be detected by means of a fluoresceinated anti-human immunoglobulin antibody. The specificity of the test for *T. pallidum* is enhanced by first absorbing the serum with nonpathogenic treponemal strains. This modification is referred to as the FTA-ABS test. (If specific fluoresceinated anti-IgM antibody to human gamma globulin is used, the acuteness of the infection or the occurrence of congenital syphilis can be assessed. However, this test may sometimes be falsely positive or negative in babies born of mothers with syphilis.)

The FTA-ABS test is reactive in approximately 80% of patients with primary syphilis (vs. 50% for the VDRL test). Both tests are positive in virtually 100% of patients with secondary syphilis. Whereas the VDRL test shows a tendency to decline in titer after successful treatment, the FTA-ABS test may remain positive for years. It is especially useful in confirming or ruling out a diagnosis of syphilis in patients with suspected biologic false-positive reactions to the VDRL test. However, even the FTA-ABS test may be susceptible to false-positive reactions, especially in the presence of lupus erythematosus.

The microhemagglutination-*Treponema pallidum* (MHATP) test, a simple passive hemagglutination test, is a satisfactory substitute for the FTA-ABS test. Treponemal antigens are coupled onto animal erythrocytes and, in the presence of specific antibody, agglutination or clumping occurs, indicative of a positive reaction. Its principal advantages are economy of technician time and low cost. Its results correlate closely with those of the FTA-ABS test, except during primary and early secondary syphilis when both the VDRL and FTA-ABS are more likely to show reactivity. The VDRL test is the only one that can be used with reliability in the evaluation of cerebrospinal fluid.

The interpretation of serologic data from patients with syphilis may be extremely complex in some cases. For example, a prozone phenomenon may be encountered in secondary syphilis; serofastness may characterize late syphilis; and the VDRL test may be negative in up to a third of patients with late latent syphilis.

Syphilis: Prophylactic Measures

Prevention of syphilis requires the practice of safe sex techniques such as the use of condoms. These, if used properly, can be an effective barrier against the sexual transmission of *Treponema pallidum*. Early treatment with antibiotics is the only known way to prevent the later ravages of syphilis. Experimental vaccines have proven impractical or fail to afford complete protection.

Syphilis: Treatment

Penicillin is the drug of choice for syphilis in all its stages. Because the lesions of tertiary syphilis may be irreversible, it is crucial to identify and treat the disease before tertiary lesions begin. The AIDS patients with syphilis must be treated more intensively with penicillin.[46] This reinforces the notion that curing syphilis depends on interactions between an intact immune system and the treponemicidal effects of antibiotics.

NON-VENEREAL TREPONEMATOSES

The causes of yaws (*Treponema pallidum* ssp. *pertenue*), pinta (*T. carateum*), and bejel (*T. pallidum* and *endemicum*) are human pathogens responsible for this group of contagious diseases, which are endemic among rural populations in tropical and subtropical countries.[47] Unlike syphilis, these diseases are not transmitted by sexual activity but arise when treponemes are transmitted primarily by direct contact, mostly among children living under poor sanitary conditions. These three treponemal species are morphologically and antigenically similar to *T. pallidum* but give rise to slightly different disease manifestations. Pinta causes skin lesions only, yaws causes skin and bone lesions, and bejel (so-called endemic syphilis) affects the mucous membranes, skin, and bones. They do resemble venereal syphilis by virtue of the self-limiting primary and secondary lesions, a latency period with clinically dormant disease, and late lesions that are frequently highly destructive. The serologic responses for all three diseases are indistinguishable from one another and from that of venereal syphilis, and there is the same degree of slow development of protective immunity associated with prolonged untreated infection.

The spirochetes that cause yaws and pinta are transmitted through broken skin after close contact with infected people. The diseases are most prevalent in rural tropical areas where hygiene is poor and living conditions are crowded. Yaws occurs most often in 2- to 5-year-olds, and pinta occurs most often in older children and adolescents. During global campaigns in the 1950s and 1960s, both diseases were nearly eradicated, but their prevalence has increased in the absence of continuous control measures.

With yaws, there is usually an incubation period of 2 weeks to 3 months before a papular-type lesion appears on the face, arms, legs and persists for weeks or months. This papilloma expands slowly to form multiple papular or raspberry-like lesions that may ulcerate in the center. The primary papillomas heal spontaneously within months. Similar secondary lesions may erupt, heal, and spread in successive crops. Also, painful papillomas and hyperkeratosis may appear on the palms and soles, along with lymphadenopathy, and destructive osteitis and periosteitis of the long bones. A variety of skin lesions can appear in the late stage of the disease, with gummatous lesions at several sites. Yaws cripples and deforms but is rarely fatal.

With pinta, 1 to 8 weeks after infection, a scaling papule appears on the hands, legs, face, or dorsum of the feet. The papule may be accompanied by lymphadenitis. In 3 to 12 months, a maculopapular erythematous rash appears at the sites of the primary lesions. It may change color from blue to violet to brown and finally become depigmented. The rash and pigmented lesions can recur for years, and the depigmented lesions can cause considerable disfigurement.

Spirochetes can be found in early skin lesions in patients having either yaws or pinta. Serologic tests for syphilis become positive during the early stages of either disease. Small doses of an antibiotic such as penicillin cures the infection. Improved sanitation, including increased use of soap and water, will prevent transmission. Case monitoring and antibiotic treatment control these diseases.

SPIROCHETAL INFECTIONS DURING PREGNANCY

Except for syphilis, there is little evidence or few well-documented cases for other maternally acquired spirochetal infections that lead to congenital disease.[48] Nonetheless, syphilis was probably the first infectious disease where transplacental transmission was first recognized, and which could lead to serious outcomes to the mother or the developing fetus, involving miscarriages, stillbirths, or neonatal disease. Syphilis is now a relatively rare occurrence in the newborn, except in certain large urban settings such as New York City. *Treponema pallidum* bacteria are usually acquired transplacentally, although intra-partum acquisition is possible. A mother with untreated primary or secondary syphilis is unlikely to have normal children; about half will be premature or suffer perinatal death, and the remainder will have congenital disease. These proportions decrease to 10% perinatal death and 10% congenital syphilis when maternal infection is acquired during late gestation. About 50% of infected newborns are initially asymptomatic. Babies with congenital infection may have a vesicular or bullous rash that includes the palms and soles; they may have rhinitis (sniffles), a maculopapular rash, pneumonia, condylomata lata, hepatosplenomegaly, generalized lymphadenopathy, and/or abnormal CSF. These manifestations may not be apparent at birth but may develop during the first few weeks of life. Radiographs often reveal bony lesions at 1 to 3 months post-natally, and these are characteristic for symmetric metaphyseal involvement with elevation of the periosteum, and osteomyelitic lesions, most often involving the humerus and tibia. These bone changes occur in 90% of infants who manifest with congenital syphilis. The osteochondritis and periostitis may be painful and may be manifested by the pseudoparalysis of a limb due to pain. Diagnosis may be problematic and usually depends on serologic confirmation of clinically based suspicions.

Infected infants are often asymptomatic at birth; and if infection was late in the pregnancy, they may be seronegative. Either the VDRL/RPR tests or the FTA-ABS/MHA-TP tests may be used, but it should be noted that any maternal serologic reactivity, in the form of IgG antibodies, will be transmitted to the newborn. All these tests measure both IgG and IgM. However, for measuring IgM only (produced by the newborn), recently developed ELISAs and the FTA-IgM test can be used. This would be helpful in identifying an actual congenital infection because IgM is not transferred across the placenta. Problems exist, however, in the overall sensitivity and specificity of these newer assays. The infant's passively acquired VDRL reactivity gradually decreases by 3 months, and passive FTA-ABS/MHA-TP reactivity should decrease significantly by 6 months.

All infants suspected of having congenital syphilis on the basis of maternal history or physical findings should be carefully examined, investigated by serology, and have a lumbar puncture to collect spinal fluid (CSF) for routine analysis, a possible dark-field exam, and a VDRL/RPR test. The placenta should be examined for focal villitis and other possible abnormalities. If the serologically positive mother did not receive adequate therapy or received non-penicillin therapy, or if adequate follow-up is not assured, the child should be treated with penicillin at birth. Infants with neurosyphilis should be seen at follow-up at 3 months, then at 6-month intervals for repeat serology, CSF exam, and clinical reevaluation, until it is clear that VDRL/RPR titers are falling.

REFERENCES

1. Nadelman, R.B., Pavia C.S., et al. Isolation of *Borrelia burgdorferi* from the blood of seven patients with Lyme disease. *Am. J. Med.*, 88:21, 1990.
2. Steere, A.C. et al. Lyme arthritis: an epidemic of oligoarticular arthritis in children and adults in three Connecticut communities. *Arthritis Rheumat.*, 20:7, 1977.
3. Afzelius, A. Report to Verhandlungen der dermatologischen Gesellshaft zu Stockholm. *Acta Derm. Venereol.*, 2:120, 1921.
4. Barbour, A. Isolation and cultivation of Lyme disease spirochetes. *Yale J. Biol. Med.*, 57:521, 1984.
5. Pavia, C.S. et al. Activity of sera from patients with Lyme disease against *Borrelia burgdorferi. Clin Infect. Dis.*, 25(Suppl. 1):S25, 1997.
6. Simon, M.M. et al. Recombinant outer surface protein A from *Borrelia burgdorferi* induces antibodies protective against spirochetal infection in mice. *J. Infect. Dis.,* 164:123, 1991.
7. Steere, A.C. Lyme borreliosis in 2005, 30 years after initial observations in Lyme, CT. *Wien. Klin. Wochenschr.*, 118:625, 2006.
8. Barbour, A. et al. Heterogeneity of major proteins of Lyme disease borreliae: a molecular analysis of American and European isolates. *J. Infect. Dis.,* 152:478, 1985.
9. Horowitz, H.W., Pavia, C.S., et al. Sustained cellular immune responses to *Borrelia burgdorferi*: lack of correlation with clinical presentation and serology. *Clin. Diagn. Lab. Immunol.,* 1:373, 1994.
10. Dattwyler, R.J. et al. Seronegative Lyme disease: dissociation of specific T- and B-lymphocyte responses to *Borrelia burgdorferi. N. Engl J. Med.,* 319:1441, 1988.
11. Habicht, G.S. et al. Lyme disease spirochetes induce human and mouse interleukin-I production. *J. Immunol.*, 134:3147, 1985.
12. Benach, J.L. et al. Interaction of phagocytes with the Lyme disease spirochete; role of the Fc receptor. *J. Infect. Dis.,* 150:497, 1984.
13. Georgilis, K. et al. Infectivity of *Borrelia burgdorferi* correlates with resistance to elimination by phagocytic cells. *J. Infect. Dis.*, 163:150, 1991.
14. Aguero-Rosenfeld, M.E. et al. Diagnosis of Lyme borreliosis. *Clin. Microbiol. Rev.,* 18:484, 2005.
15. Steere, A.C. et al. The spirochetal etiology of Lyme disease. *N. Engl. J. Med.*, 308:733, 1983.
16. Wormser, G.P. et al. Improving the yield of blood cultures in early Lyme disease. *J. Clin. Microbiol.,* 38:1648, 1998.

17. Wormser, G.P. et al. Use of a novel technique of cutaneous lavage for diagnosis of Lyme disease associated with erythema migrans. *JAMA*, 268:1311, 1992.
18. Park, H.J. et al. *Erythema chronicum migrans* of Lyme disease: diagnosis by monoclonal antibodies. *J. Am. Acad. Dermatol.*, 15:111, 1986.
19. Duray, P.H. et al. Demonstration of the Lyme disease spirochetes by modified Dieterle stain method. *Lab. Med.*, 16:685, 1985.
20. Duray, P.H. and Steere, A.C. Clinical pathology correlates of Lyme disease by stage. *Ann. N.Y. Acad. Sci.*, 539:65, 1988.
21. Hedberg, C.M. et al. An interlaboratory study of antibody to *Borrelia burgdorferi. J. Infect. Dis.*, 155.1325, 1987.
22. Bakken, L.L. et al. Performance of 45 laboratories participating in a proficiency testing program for Lyme disease serology. *JAMA*, 268:891, 1992.
23. Schwartz, B.S. et al. Antibody testing in Lyme disease: a comparison of results in four laboratories. *JAMA*, 262:3431, 1989.
24. Grodzicki, R.L. and Steere, A.C. Comparison of immunoblotting and indirect ELISA using different antigen preparations for diagnosing early Lyme disease. *J. Infect. Dis.*, 157:790, 1988.
25. Centers for Disease Control and Prevention. Case definitions for infectious conditions under public health surveillance: Lyme disease. *MMWR,* 46:20, 1997.
26. Dattwyler, R.J. et al. Immunological aspects of Lyme Borreliosis. *Rev. Infect. Dis.*, 11(S6):1494, 1989.
27. Hanson, L. et al. Measurement of antibodies to the *Borrelia burgdorferi* flagellum improves serodiagnosis in Lyme disease. *J. Clin. Microbiol.*, 26:338, 1988.
28. Bacon, R.M. et al. Serodiagnosis of Lyme disease by kinetic enzyme-linked immunosorbent assay using recombinant VlsE1 or peptide antigens of *Borrelia burgdorferi* compared with 2-tiered testing using whole-cell lysates. *J. Infect. Dis.,* 187:1187, 2003.
29. Weinstein, A. and Bujak, D.I. Lyme disease: a review of its clinical features. *N.Y. J. Med.*, 89:566, 1989.
30. Rosa, P.A. and Schwann, T.G. A specific and sensitive assay for the Lyme disease spirochete *B. burgdorferi* using the polymerase chain reaction. *J. Infect. Dis.,* 160:1018, 1989.
31. Persing, D.H. et al. Detection of *B. burgdorferi* infection in *Ixodes dammini* ticks with the polymerase chain reaction. *J. Clin. Microbiol.,* 28:566, 1990.
32. Schwartz, I. et al. Diagnosis of early Lyme disease by polymerase chain reaction amplification or culture of skin biopsies from erythema migrans. *J. Clin. Microbiol.*, 30:3082, 1992.
33. Schwartz, I. et al. Polymerase chain reaction amplification of culture supernatants for rapid detection of *Borrelia burgdorferi. Eur. J. Clin. Microbiol. Infect. Dis.,* 12:879, 1993.
34. Oppenheim,, J. Relationship of *in vitro* lymphocyte transformation to delayed hypersensitivity in guinea pigs and man. *Fed. Proc.,* 27:21, 1968.
35. Chu, H.J. et al. Immunogenicity and efficacy study of a commercial *B. burgdorferi* bacterin. *J. Am. Vet. Med. Assoc.*, 201:403, 1992.
36. Steere, A.C. et al. Vaccination against Lyme disease with recombinant *Borrelia burgdorferi* outer surface lipoprotein A with adjuvant. *N. Engl. J. Med.,* 339:209, 1998.
37. Wormser, G.P. et al. The clinical assessment, treatment, and prevention of lyme disease, human granulocytic anaplasmosis and babesiosis: clinical practice guidelines by the Infectious Diseases Society of America. *Clin Infect. Dis.*, 43:1089, 2006.
38. Nadelman, R.B. et al. Prophylaxis with single-dose doxycycline for the prevention of Lyme disease after an *Ixodes scapularis* tick bite. *N. Engl. J. Med.,* 345:79, 2001.
39. Roscoe, C. and Epperly, T. Tick-borne relapsing fever. *Am. J. Physician*, 72:2039, 2005.
40. Rebaudat, S. and Parola, P. Epidemiology of relapsing fever borreliosis in Europe. *FEMS Immunol. Med. Microbiol.*, 48:11, 2006.
41. Lecour, H. et al. Human leptospirosis — a review of 50 cases. *Infection*, 17:8–12, 1989.
42. Morgan, J. et al. Outbreak of leptospirosis among triathlon participants and community residents in Springfield, Illinois, 1998. *Clin. Infect. Dis.*, 34:1593, 2002.
43. Sejuar, J. et al. Leptospirosis in "Eco-Challenge" athletes, Malaysian Borneo, 2000. *Emerg. Infect. Dis.*, 9:702, 2003.
44. Peeling, R,W. and Hook, E.W. The pathogenesis of syphilis: the Great Mimicker, revisited. *J. Pathol.*, 208:224, 2006.
45. Lafond, R.E. and Lukehart, S.A. Biological basis for syphilis. *Clin. Microbiol. Rev.,* 19:24, 2006.

46. Zellan, R.E. and Augenbraun, M. Syphilis in the HIV-infected patient: an update on epidemiology, diagnosis and treatment. *Curr. HIV/AIDS Rep.*, 1:142, 2004.
47. Antel, G.M. et al. The endemic treponematoses. *Microbes Infect.,* 4:83, 2002.
48. Remington, J.S. and Klein, J.O. *Infectious Diseases of the Fetus and Newborn Infant*, 6th edition, Saunders/Elsevier, Amsterdam, The Netherlands, 2005.

21 The Genus *Vibrio* and Related Genera

Rita R. Colwell and Jongsik Chun

CONTENTS

INTRODUCTION

Vibrios are short curved or straight cells, single or united into spirals, that grow well and rapidly on the surfaces of standard culture media, and can be readily isolated from estuarine, marine, and fresh water samples. These heterotrophic organisms vary in their nutritional requirements; some occur as parasites and pathogens for animals and for man. The short curved, asporogenous, Gram-negative rods that are members of the genus *Vibrio* are most commonly encountered in the estuarine, marine, or fresh water habitat. Distinguishing species of the genus *Vibrio* from other related genera can be presumptively accomplished by examining Gram stains of carefully prepared specimens, followed by electron microscopy to confirm morphology. In contrast, *Spirillum* isolates are frequently seen as rigid, helical cells with a single or several turns, motile by means of bipolar polytrichous flagella, whereas *Vibrio* species are short rods with a curved axis, motile by means of a single polar flagellum. However, *Vibrio* species may be short, straight rods (1.5 to 3.0 × 0.5 µm), or they may be S-shaped or spiral-shaped when individual cells are joined. Possession of two or more flagella in a polar tuft has also been demonstrated in *Vibrio* species, as have lateral flagella, notably *Vibrio parahaemolyticus*. Thus, to identify and classify *Vibrios*, physiological and biochemical taxonomic tests are made.[1] *Vibrio* species are facultatively anaerobic, with both a respiratory (oxygen utilizing) and a fermentative metabolism. Related genera may be aerobic or microaerophilic, with a strictly respiratory metabolism (oxygen is the terminal electron acceptor). Early on, Hylemon et al.[2] proposed a division of the genus *Spirillum* into three genera — *Spirillum*, *Aquaspirillum*, and *Oceanospirillum*; supported by the work of Carney et al.,[3] results based on molecular genetic studies have seen the reduction of *Spirillum* spp. to single species (i.e., *Spirillum volutans*), with other species moved to other genera. *Vibrios* require 1 to 3% NaCl for growth and the salt requirement is absolute.

In the 8th edition of *Bergey's Manual*,[1] published in 1974, the genus *Vibrio* was placed among the Gram-negative facultatively anaerobic rods in the family Vibrionaceae. The gray zone between the genera *Vibrio* and *Pseudomonas* has been considered in the past to be somewhat vague. However, evidence sharpened the demarcation between the two genera, and *Pseudomonas* species are recognized as oxidative, whereas *Vibrio* species are fermentative, producing acid in carbohydrate-containing media without formation of gas, and possess a DNA with G+C in the range from 39 to 49 mole %. Molecular genetic and genomic sequence data have proved useful in establishing relationships among the genera *Spirillum*, *Pseudomonas*, *Aeromonas,* and *Vibrio.*

GENUS *VIBRIO* PACINI 1854

The type species for the genus *Vibrio* is *Vibrio cholerae* Pacini 1854. The genus *Vibrio* has been expanded to include many more species than the five species of *Vibrio* listed in the earlier edition of *Bergey's Manual*.[1] The strictly anaerobic species have been removed from the genus, and no cultures of obligately anaerobic *Vibrio* species are extant. The microaerophilic species, including *Vibrio fetus* and *Vibrio bubulus*, have been transferred to the genus *Campylobacter.*

Of the species of *Vibrio* listed in the latest edition of *Bergey's Manual*,[4] those not attacking carbohydrates have been transferred to the genus *Commamonas*. Davis and Park[5] described *Commamonas* as Gram-negative rod-like bacteria giving no reaction on carbohydrate media and an alkaline reaction in the Hugh and Leifson test.[6] Other test results confirmed the observations of Davis and Park,[5] and Sebald and Véron[7] assigned *Vibrio percolans*, *Vibrio cyclosites*, *Vibrio neocistes*, and *Vibrio alcaligenes* to the genus *Commamonas*, which in the latest edition of *Bergey's Manual* is now listed among species of *Pseudomonas*. Historically, Colwell and Liston[8] observed that *Vibrio cuneatus* produced a green fluorescent pigment, and subsequent DNA studies[9] confirmed the conclusion that this species should be assigned to the genus *Pseudomonas*.

The C27 organisms of Ferguson and Henderson,[10] previously assigned to *Aeromonas* by Ewing et al.,[11] to *Plesiomonas* by Habs and Schubert[12] and by Eddy and Carpenter,[13] and to *Fergusonia* by Sebald and Véron,[7] were suggested to belong to the genus *Vibrio* by Hendrie et al.[14] However, in the latest edition of *Bergey's Manual*, these organisms are found with genus *Plesiomonas*. The number of characterized and defined species resident in the genus *Vibrio* has been significantly increased with many novel species having been described.

The description of the genus *Vibrio*, as amended by the Subcommittee on Taxonomy of Vibrios, International Committee on Nomenclature of Bacteria, is concise and remains a reasonably good working definition:

> Gram-negative, asporogenous rods which have a single, rigid curve or which are straight. Motile by means of a single, polar flagellum. Produce indophenol oxidase and catalase. Ferment glucose without gas production. Acidity is produced from glucose by the Embden-Meyerhof glycolytic pathway. The guanine plus cytosine in the DNA of *Vibrio* species is within the range of 40 to 50 moles percent.

The genus *Vibrio* has been the subject of many polyphasic studies and at present contains 68 species. The genus *Listonella* proposed by MacDonell and Colwell[15] is polyphyletic and belongs to the genus *Vibrio* on the basis of 16S rRNA phylogeny (Figure 21.1). Most of the present members of the genus *Vibrio* form a monophyletic group, except for *Vibrio calviensis,* which should be transferred to the genus *Enterovibrio.*[16]

The type species of the genus, *Vibrio cholerae* Pacini 1854, can be succinctly described as follows: producing L-lysine and L-ornithine decarboxylases; not producing L-arginine dihydrolase and hydrogen sulfide (Kligler iron agar). The G+C content in the DNA of *Vibrio cholerae* is approximately 48 ± 1%. *Vibrio cholerae* includes strains that may or may not elicit the cholera-red (nitroso-indole) reaction, may or may not be hemolytic, may or may not be agglutinated by Gardner and Venkatraman O group I antiserum, and may or may not be lysed by Mukerjee *Vibrio cholerae* bacteriophages I, II, III, IV, and V.[17]

Vibrio cholerae strains possess a common H antigen and can be serologically grouped into many serotypes according to their O antigens. Strains agglutinated by Gardner and Venkatraman O group I antiserum are in serotype I and were the principal cause of cholera in man until the emergence of *Vibrio cholerae* serotype O139. An M antigen can obscure the agglutinability of mucoid strains of *Vibrio cholerae*. General reviews on cholera have been published.

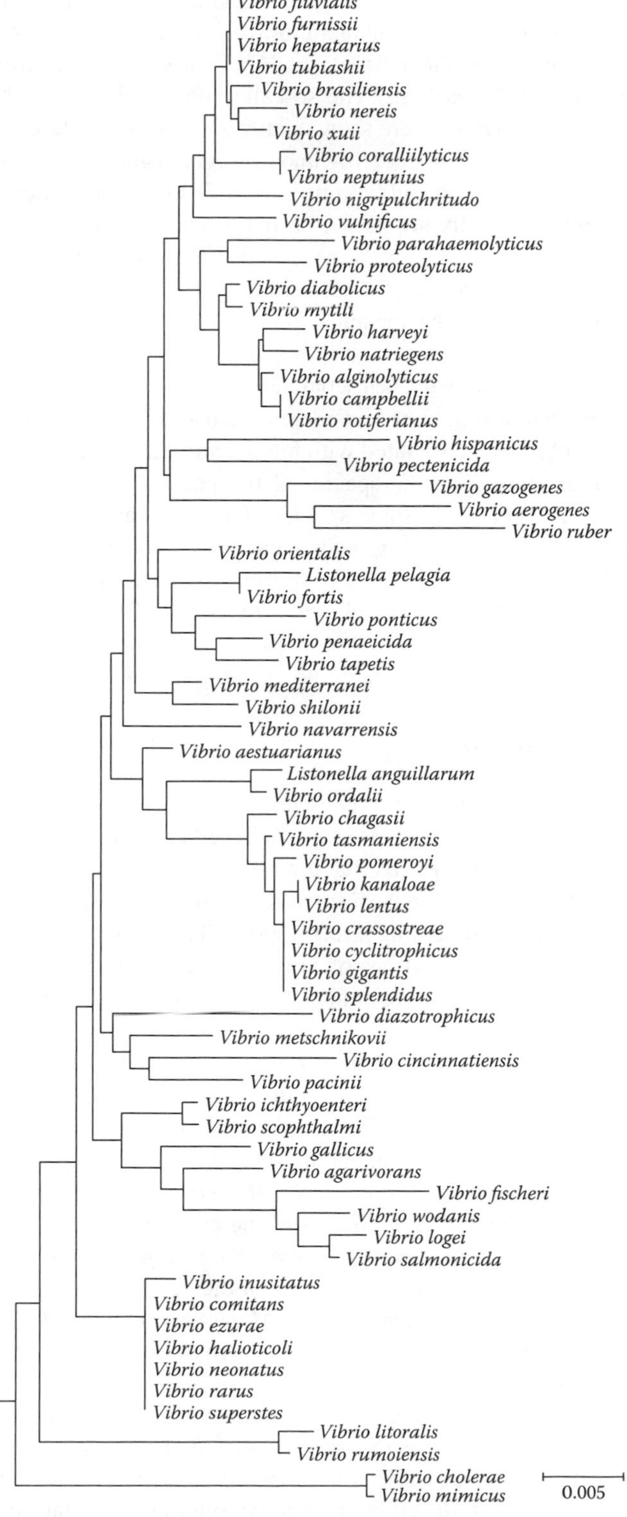

FIGURE 21.1 Phylogenetic tree based on 16S rRNA gene sequences showing the relationship among *Vibrio* spp. The tree was generated using the neighbor-joining method. The bar represents 0.005 nucleotide substitution per site.

In the past, many cholerae-like vibrios were given separate species status because they were isolated from patients suffering diarrhea, not cholera, or from water and foods. *Vibrio proteus*, *Vibrio metschnikovii*, *Vibrio berolinensis*, *Vibrio albensis*, *Vibrio mimicus*, and *Vibrio paracholerae* can be considered biotypes of *Vibrio cholerae*. The so-called non-agglutinable (NAG) or non-cholera (NCV) vibrios, including *Vibrio eltor*, were shown many years ago by Citarella and Colwell,[18] using DNA/DNA hybridization, to be *Vibrio cholerae* and this conclusion has been substantiated.

Since its initial isolation by Fujino and Fukumi[19] in 1953, *Vibrio parahaemolyticus* has received considerable attention and rightfully so, as it is a major cause of seafood-borne illness. *Vibrio parahaemolyticus* (syn. *Pasteurella parahaemolytica*, *Pseudomonas enteritis*, *Oceanomonas parahaemolytica*) is the causative agent of food poisoning arising from ingestion of contaminated seafood and is autochthonous to the marine and estuarine environment.

The species *Vibrio alginolyticus*, although previously considered by some investigators as synonymous with *Vibrio parahaemolyticus*, is a separate species.

Other species of *Vibrio* that should be considered are listed in Figure 21.1 and Table 21.1. Vibrios are found in seawater and are associated with human and marine animals.[20, 21]

Distinctions can be made among the species of the genus *Vibrio*, and useful differentiating characteristics are given in Table 21.1. Some species of *Vibrio* may demonstrate sheathed flagella (i.e., a flagellum with a central core and an outer sheath) and, under certain conditions of growth, peritrichous flagella.[22] "Round bodies" or sphaeroplasts are commonly present during various stages of growth,[23, 24] and fimbriae (pili) have been observed in strains of *Vibrio cholerae*.[25]

Separation of *Vibrio* from *Aeromonas*, *Plesiomonas*, and *Photobacterium* can, on occasion, prove difficult. Features useful for such differentiation are given in Table 21.2.

GENUS *SPIRILLUM* EHRENBERG 1832

The genus *Spirillum* has been defined more restrictedly in recent years. *Spirillum* Ehrenberg 1832, as described in earlier editions of *Bergey's Manual*, included crescent shaped to spiral cells that are frequently united into spiral chains of cells that are not embedded in zoogloeal masses. In the latest edition of *Bergey's Manual*,[1] the genus is described as including spiral cells, usually motile by means of polar flagellation (i.e., possessing a tuft of polar flagella at one or both ends of the cells). The cells form either long screws or portions of a turn. Intracellular granules of polyhydroxybutyrate are present in most species. Spirilla are either aerobic, growing well on ordinary culture media, or microaerophilic.

Nine species were listed in the 7th editon of *Bergey's Manual*[4] but only a single species (i.e., *Spirillum volutans*) remains. The genus *Aquaspirillum* was proposed by Hylemon et al.[2] to harbor strains that had been known as spirilla. The members of the genus were found to be polyphyletic and to have undergone substantial taxonomic changes. Several species were transferred to the genera *Comamonas*, *Curvibacter*, *Giesbergeria*, *Herbaspirillum*, *Hylemonella*, *Microvirgula*, *Prolinoborus*, *Magnetospirillum*, and *Simplicispira*. The current members — namely, *Aquaspirillum arcticum*, *A. itersonii*, *A. peregrinum*, *A. polymorphum*, *A. putridiconchylium*, and *A. serpens* — are still polyphyletic.[26] Similarly, several members of the genus *Oceanospirillum* Hylemon et al. 1973 were reclassified to the genera *Marinomonas*, *Marinobacterium*, *Marinospirillum*, *Pseudospirillum*, *Oceanobacter*, and *Terasakiella*. The genus forms a phylogenetically coherent group and includes *O. beijerinckii*, *O. linum*, *O. maris*, and *O. multiglobuliferum*.

In general terms, the genus can be described as comprising rigid, helical, Gram-negative cells, motile with bipolar polytrichous flagella. Spirilla possess a strictly respiratory metabolism, with oxygen as the terminal electron acceptor; they are oxidase- and catalase-positive, and usually phosphatase-positive. H_2S is usually produced from cysteine, but indole, sulfatase, and amylase are not produced. Other reactions usually negative are hydrolysis of casein, hippurate, and gelatin, and production of urease and acid from sugars. Although the species cannot utilize sugars, various organic acids, alcohols, or amino acids can suffice as sole carbon sources.

TABLE 21.1
Differential Characteristics among Important *Vibrio* spp.

	V. alginolyticus	*V. carchariae*	*V. cholerae*	*V. cincinnatiensis*	*V. damsela*	*V. fluvialis*	*V. furnissii*	*V. hollisae*	*V. metschnikovii*	*V. mimicus*	*V. parahaemolyticus*	*V. vulnificus*
Swarming (marine agar, 25°C)	+	+	-	+	-	-	-	-	-	-	+	-
Yellow pigment at 25°C	–	–	–	–	–	–	–	–	–	–	–	–
Indole production	v	+	+	–	–	v	v	+	v	+	+	+
Voges-Proskauer reaction	+	v	v	–	+	–	–	–	+	–	–	–
Arginine dehydrogenase	–	–	–	–	+	+	+	–	v	–	–	–
Lysine dehydrogenase	+	+	+	v	v	–	–	–	v	+	+	+
Ornithine dehydrogenase	v	–	+	–	–	–	–	–	–	+	+	v
D-Glucose, acid production	+	v	+	+	+	+	+	+	+	+	+	+
D-Glucose, gas production	–	–	–	–	–	–	+	–	–	–	–	–
Nitrate nitrite	+	+	+	+	+	+	+	+	–	+	+	+
Oxidase	+	+	+	+	+	+	+	+	–	+	+	+
Lipase (corn oil)	v	–	+	v	–	+	v	–	+	v	+	+
ONPG test	–	–	+	v	–	v	v	–	v	+	–	v
Tyrosine clearing	v	–	v	–	–	v	v	–	–	v	v	v
Acid production from:												
L-Arabinose	–	–	–	+	–	+	+	+	–	–	v	–
D-Arabitol	–	–	–	–	–	v	v	–	–	–	–	–
Cellobiose	–	v	–	+	–	v	v	–	–	–	–	+
Lactose	–	–	–	–	–	–	–	–	v	v	–	v
Maltose	+	+	+	+	+	+	+	–	+	+	+	+
D-Mannitol	+	v	+	+	–	+	+	–	+	+	+	v
Salicin	–	–	–	+	–	–	–	–	–	–	–	+
Sucrose	+	v	+	+	–	+	+	–	+	–	–	v
Growth in nutrient broth:												
0% NaCl	–	–	+	–	–	–	–	–	–	+	–	–
1% NaCl	+	+	+	+	+	+	+	+	+	+	+	+
6% NaCl	–	s	v	+	+	+	+	v	v	v	+	v
8% NaCl	–	–	–	v	–	v	v	–	v	–	v	–
10% NaCl	v	–	–	–	–	–	–	–	–	–	–	–
Sensitivity to:												
O/129	v	+	+	v	+	v	–	v	+	+	v	+
Polymyxin B	v	+	v	+	v	+	v	+	+	v	v	–

Symbols: +, 90 to 100% positive; –, 0 to 10% positive. v, 11 to 89% positive. Data obtained from Farmer and Hickman-Brenner.[27]

TABLE 21.2

Differentation of Related Genera Frequently Isolated From the Same Source in Nature

Characteristic	*Vibrio*	*Aeromonas*[a]	*Plesiomonas*	*Photobacterium*	*Pseudomonas*	*Spirillum*	*Campylobacter*
Morphology	Straight or curved rod	Straight rod	Straight rod	Straight rod	Straight rod	Helical	Spirally curved rod
Diffusible pigment	None	None	None	None	None or green fluorescent	None or green fluorescent	None
Motility	+	+	+	+	+	+	+
Flagella	Polar	Polar	Lophotrichous	Polar	Polar	Lophotrichous	Polar
Carbohydrate metabolism	Fermentative	Fermentative	Fermentative	Fermentative	Respiratory or not metabolized	Respiratory	Not metabolized
Gas production from carbohydrates	–	v	–	+	–	–	–
Luminescence	v	–	–	+	–	–	–
Oxidase	+	+	+	v	+	+	+
0/129 sensitivity	+	–	+	+	–	–	–
"Round bodies" or "cysts" produced	+	–	–	–	–	+	+

[a] Species of *Aeromonas* may produce a brown pigment.

Symbol code: + = positive or present; – = negative or absent; v = variable.

DIFFERENTIATION FROM OTHER BACTERIA

Characteristics useful for distinguishing vibrios from several genera with which they are frequently confused are given in Table 21.2.

Genera of uncertain affiliation include *Desulfovibrio, Butyrivibrio, Succinivibrio, Succinimonas, Bdellovibrio, Ancylobacter* (formerly *Microcyclus*), '*Pelosigma*,' and '*Brachyarcus*.' Of these, the *Bdellovibrio, Ancylobacter, Pelosigma*, and *Brachyarcus* spp. are included with the spiral and curved.

Bdellovibrio spp. are, in general, small, curved, Gram-negative motile rods, parasitic for other bacteria; that is, they attach to and penetrate into the host cell — the outstanding characteristic of the genus. Various Gram-negative and Gram-positive bacteria can serve as hosts, with a specific host range often characteristic of given species of *Bdellovibrio*. All strains of *Bdellovibrio* isolated from nature are parasitic, but host-independent strains frequently are developed in the laboratory. *Bdellovibrio* spp. possess a polar sheathed flagellum, usually monotrichous.

The genera *Ancylobacter*, '*Pelosigma*,' and '*Brachyarcus*' are described elsewhere and are found in aquatic environments, with *Ancylobacter* being noted as curved aerobic rods, on occasion forming rings.

REFERENCES

1. Buchanan, R.E. and Gibbons, N.E., *Bergey's Manual of Determinative Bacteriology*, 8th ed., Williams & Wilkins, Baltimore, 1974.
2. Hylemon, P.B., Wells, J.S., Krieg, N.R., and Jannasch, H.W., *Int. J. Syst. Bacteriol.*, 23, 340, 1973.
3. Carney, J.F, Wan, L., Lovelace, T.E., and Colwell, R.R., *Int. J. Syst. Bacteriol.*, 25, 38, 1975.
4. Breed, R.S, Murray, E.G.D., and Smith, N.R., *Bergey's Manual of Determinative Bacteriology*, 7th ed., Williams & Wilkins, Baltimore, 1957.
5. Davis, G.H.G. and Park, R.W.A., *J. Gen. Microbiol.*, 27, 101, 1962.
6. Hugh, R. and Leifson, E., *J. Bacteriol.*, 66, 24, 1953.
7. Sebald, M. and Véron, M., *Ann. Inst. Pasteur Lille*, 105, 897, 1963.
8. Colwell, R.R. and Liston, J., *J. Bacteriol.*, 82, 1, 1961.
9. Colwell, R.R. and Mandel, M., *J. Bacteriol.*, 87, 1412, 1964.
10. Ferguson, W.W. and Henderson, N.D., *J. Bacteriol.*, 54, 178, 1947.
11. Ewing, W.H., Hugh, R., and Johnson, J.G., *Studies on the Aeromonas Group,* Communicable Disease Center, Atlanta, 1961.
12. Habs, H. and Schubert, R.H.W., *Zentralbl. Bakteriol. Parasitenkd. Infektionskr. Hyg. Abt. 1 Orig.*, 186, 316, 1962.
13. Eddy, B.R. and Carpenter, K.P., *J. Appl. Bacteriol.*, 27, 96, 1964.
14. Hendrie, M.S., Shewan, J.M., and Véron, M., *Int. J. Syst. Bacteriol.*, 21, 25, 1971.
15. MacDonell, M.T. and Colwell, R.R., *Syst. Appl. Microbiol.*, 6, 171, 1985.
16. Thompson, F.L., Thompson, C.C., Naser, S., Hoste, B., Vandemeulebroecke, K., Munn, C., Bourne, D., and Swings, J., *Int. J. Syst. Evol. Microbiol.* 55, 913, 2005.
17. Hugh, R. and Feeley, J.C., Report (1966-1970) of the Subcommittee on Taxonomy of Vibrios to the International Committee on Nomenclature of Bacteria, *Int. J. Syst. Bacteriol.*, 22, 123, 1972.
18. Citareila, R.V. and Colwell, R.R., *J. Bacteriol.*, 104, 434, 1970.
19. Fujino, T. and Fukumi, H. (Eds.), *Vibrio parahaemolyticus*, 2nd ed., Naya Shoten, Tokyo, 1967 (in Japanese).
20. Colwell, R.R. and Morita, R.Y., *J. Bacteriol.*, 88, 831, 1964.
21. Bianchi, M.A.G., *Arch. Mikrobiol.*, 77, 127, 1971.
22. Baumann, P., Baumann, L., and Mandel, M., *J. Bacteriol.*, 107, 268, 1971.
23. Felter, R.A., Kennedy, S.F., Colwell, R.R., and Chapman, G.B., *J. Bacteriol.*, 102, 552, 1969.
24. Kennedy, S.F., Colwell, R.R., and Chapman, G.B., *Can. J. Microbiol.* 16, 1027, 1970.
25. Tweedy, J.M., Park, R.W.A., and Hodgkiss, W., *J. Gen. Microbiol.*, 51, 235, 1968.
26. Ding, L. and Yokota, A., *FEMS Microbiol. Lett.* 212, 165, 2002.
27. Farmer, J.J. and Hickman-Brenner, F.W., *Prokaryotes,* 2nd ed., 6, 508, 2006.

22 *Staphylococcus aureus* and Related Staphylococci

Olaf Schneewind and Dominique Missiakas

CONTENTS

INTRODUCTION

Staphylococcus aureus was first characterized as a human isolate in the late 19th century. Sir Alexander Ogston, in his lecture at the Ninth Surgical Congress in Berlin in 1880, reported the presence of "Micrococci" associated with pus in surgical wound infections. Later, Sir Ogston would use the word "staphylococci" to refer to these particular organisms.[86] He also used eggs to isolate pure cultures of staphylococci and showed that rabbits inoculated with these cultures developed abscesses, thereby fulfilling Koch's postulates for the identification of the etiological agent of suppurative abscesses.[86] *S. aureus* is a physiological commensalism of the human skin, nares, and mucosal surfaces, and bacteriological culture of the nose and skin of healthy humans invariably reveals staphylococci. In 1884, Rosenbach isolated two colony types of staphylococci found on humans and based on their pigmentation proposed the nomenclature *S. aureus* and *S. albus* for the yellow and white isolates, respectively.[98] The latter species is now named *S. epidermidis;* and until the early 1970s, *S. aureus, S. epidermidis,* and *S. saprophyticus* were the only three species in the genus *Staphylococcus*. Genotypic properties and refined taxonomy have led to the distinction of more than 30 species during the past four decades.[35] This chapter briefly describes various *Staphylococcus* species and also describes in more detail methods for use of *S. aureus* and *S. epidermidis* in the laboratory.

CLASSIFICATION, NATURAL HABITAT, AND EPIDEMIOLOGY

Classification and Natural Habitat

The genus *Staphylococcus* is in the bacterial family Staphylococcaceae, which includes the lesser known genera *Gemella, Jeotgalicoccus, Macrococcus,* and *Salinicoccus.*[31] Nearby phylogenetic relatives include members of the genus *Bacillus* and *Listeria* in the family Bacillaceae and Listeriaceae, respectively.[31] Thirty-six species and eighteen subspecies of the genus *Staphylococcus* are distinguished based on biochemical analysis and DNA-DNA hybridization studies.[33,55] Various species are often distinguished as coagulase-positive and -negative, and novobiocin susceptible or resistant.[35]

Members of the genus *Staphylococcus* inhabit the skin, skin glands, and mucous membranes of humans, animals, and birds. Staphylococci found on the skin are either resident or transient. Transient bacteria are encountered upon exposure with exogenous source and upon contact between different host species. Such transient organisms are eliminated unless the normal defense barriers of the host are compromised. *S. epidermidis* is the most prevalent and persistent *Staphylococcus* species on human skin. Although it is present all over the body surface, it prefers moistured sites such as the anterior nares and apocrine glands. Similarly, *S. aureus, S. hominis,* and *S. haemolyticus,* and to a lesser extent *S. warneri,* share the same habitat as a result of its adaptability. *S. hominis* colonizes drier regions on the human skin. *S. aureus* is found in the nares and other anatomical locales of humans and can exist as a resident or transient member of the natural flora. *S. aureus* can also colonize domestic animals and birds. *S. capitis* colonizes the human scalp following puberty and large populations can be found in areas with numerous sebaceous glands. *S. auricularis* demonstrates a strong preference for the adult human external auditory meatus. Some subspecies of *S. auricularis* preferentially colonize the ears of non-human primates. *S. intermedius* and *S. felis* are major species of the domestic dog and cat, respectively. The host range of these staphylococci is, however, not restricted to the skin of dogs and cats and may encompass other animals or even humans. Other staphylococci, including *S. xylosus, S. saprophyticus,* and *S. cohnii,* are more prevalent on lower primates and mammals while *S. delphini, S. caprae,* and *S. lutrae* have been specifically isolated from dolphins and domestic goats with suppurative skin abscesses, and dead otters, respectively (source: www.bacterio.cict.fr). Interestingly, pigmented *S. xylosus* strains are used as fermenting agents in production facilities and the activity of nitrate reductase contributes to the development of the red color characteristic of dried sausages or the orange color that coats some cheeses.

Epidemiology of *Staphylococcus* Infections

Staphylococcus aureus and *S. epidermidis* are always found associated with humans and bear the greatest pathogenic potential. Likewise, they are the best and most studied species. Among the other human commensals, *S. haemolyticus*, *S. simulans*, *S. cohnii*, *S. warneri*, and *S. lugdunensis* are occasional causative agents of infections in humans, and *S. saprophyticus* is often associated with urinary tract infection. Similarly, commensals of animals such as *S. intermedius, S. hyicus,* or *S. sciuri* display virulence traits in infected animals. Because they colonize the human skin continuously, staphylococcal strains are exposed to all antibiotic therapies.[75] *S. aureus* is a mavjor cause of nosocomial and community-acquired infections. Whenever drug-resistant microbes emerge, these strains can spread by direct contact very rapidly among human populations, as exemplified by the threat of methicillin-resistant *S. aureus* (MRSA) and vancomycin-resistant *S. aureus* (VRSA) worldwide.[13, 17, 49] *S. aureus* isolates collected during outbreaks are traditionally characterized by a phage-typing method based on phenotypic markers in the envelope. This typing method requires maintenance of a large number of phage stocks and is technically challenging. Various molecular typing methods are used for the epidemiological analysis of methicillin-resistant strains (MRSA) because such strains encode defined genetic markers. Molecular typing relying on plasmid analysis is often used but often lacks accuracy because plasmids are easily lost especially when strains are grown in laboratory media. The most reliable method is pulsed field gel electrophoresis, where genomic DNA is cut with a restriction enzyme that generates large fragments of 50 to 700 kb.

Staphylococci should always be considered potential pathogens because the difference between commensal and pathogen is a tenuous one, or at least one that is difficult to define at the molecular or genetic level. The spectrum of human diseases caused by staphylococci ranges from soft tissue infections, abscesses in organ tissues, osteomyelitis, endocarditis, toxic shock syndrome, to necrotizing pneumonia.[1, 65] *Staphylococcus aureus* is a versatile pathogen and specific strains have been isolated from many different pathological-anatomical sites or disease entities.[58] Such versatility is enabled by a long list of virulence factors, including cell wall-anchored proteins, secreted toxins including hemolysins, leukocidins, and enterotoxins, capsular- and exo-polysaccharides, iron-transport systems, and modulators of host immune functions in addition to antibiotic-resistance genes.[1, 22, 67, 78] For example, some strains of the fermenting *S. xylosus* isolated from milk produce enterotoxins D, C, or E. Other *S. xylosus* strains have been isolated in nosocomial infections and display resistance to diverse antibiotics. Our understanding of staphylococci-host interactions remains limited. For example, epidemiological studies suggested that a single toxin, Panton-Valentine leukocidin, could account for the hypervirulence of community-associated MRSA such as strain USA300. Recent conflicting reports have failed to establish the epidemiological link between PVL and hypervirulence[61, 123] and rather suggest that such strains may simply produce more of the well-characterized and canonical exotoxins and virulence factors.[16]

BIOCHEMICAL PROPERTIES

General Properties

Members of the genus *Staphylococcus* are Gram-positive cocci with sizes ranging between 0.5 and 1.5 μm in diameter. Staphylococci divide in two perpendicular planes and new daughter cells may remain associated in irregular grape-like clusters, a feature that distinguishes them from the slightly oblong streptococci that divide in one plane and usually grow in chains. Cocci also occur singly, in pairs, tetrads, and short chains. Staphylococci are nonmotile, non-sporeforming, and although genes encoding for capsule formation are found in the genus, usually staphylococci grow non-encapsulated. Most species are facultative anaerobes that grow by aerobic respiration or by fermentation that yields principally lactic acid. Staphylococci grow more rapidly under aerobic conditions with the exception of *S. saccharolyticus* and *S. aureus* ssp. *anaerobius*. Aerobic respiration is achieved by the presence of a- and b-type cytochromes, and electron transfer in the membrane is mediated by menaquinones. Thus far, c-type cytochromes whereby heme is covalently attached to cysteine residues have only been identified in *S. lentus, S. sciuri,* and *S. vitulus*. Most species are oxidase-negative (exceptions include *S. lentus, S. sciuri,* and *S. vitulus*) and catalase-positive (exceptions include the anaerobic *S. saccharolyticus* and *S. aureus* ssp. *anaerobius*). The catalase test helps distinguish the catalase-negative streptococci from staphylococci. The test is performed by adding 3% hydrogen peroxide to a colony on an agar plate or slant. Catalase-positive cultures produce O_2 and bubble. Most staphylococci are susceptible to lysostaphin (200 μg mL^{-1} and above). Lysostaphin is a glycyl-glycine endopeptidase secreted by *S. simulans* that digests the peptidoglycan of target cells, an activity that results in lysis of growing cells. *S. capitis* produces the enzyme ALE-1 a functional homologue of lysostaphin.[112] Both *S. simulans* and *S. capitis* protect themselves from their secreted endopeptidases by modifying the pentaglycine cross-bridges of the peptidoglycan.[117] Most staphylococci are also susceptible to furazolidone (100 μg on disk), display low resistance to erythromycin (0.04 μg mL^{-1}), and produce acid from glycerol oxidation.[101]

Major Differences between *Staphylococcus aureus* and *S. epidermidis*

Staphylococcus aureus forms a fairly large yellow colony on rich medium caused by the production of the orange carotenoid staphyloxanthin.[68, 88] Production of this membrane-bound pigment helps scavenge reactive oxygen species and protects *S. aureus* from neutrophil killing.[19] All strains of *S. aureus* produce the enzyme coagulase. Nearly all strains of *S. epidermidis* lack this enzyme. *S. aureus* should always be considered a potential pathogen; most strains of *S. epidermidis* are nonpathogenic. *S. aureus* is hemolytic on blood agar, whereas *S. epidermidis* is not. All *S. aureus* sequenced genomes examined thus far encode *hla,* the structural gene for hemolysin, on their chromosome.[84, 85] Yet, some strains of *S. aureus* appear non-hemolytic on blood agar. This can likely be attributed to mutations in the *agr* locus (see section entitled "Virulence Factors and Regulation") or in other regulatory loci that control globally the expression of secreted proteins. The *hla* gene encodes a 293 amino acid long protein secreted as a water-soluble monomer.[10] Staphylococcal α-hemolysin (Hla or α-toxin) is the founding member of bacterial pore-forming β-barrel toxins.[109] Hla is thought to engage surface receptors of sensitive host cells, thereby promoting its oligomerization into a heptameric prepore and insertion of a β-barrel with 2-nm pore diameter into the plasma membrane.[36] Hla pores form in lymphocytes, macrophages, type I alveolar cells, pulmonary endothelium, and erythrocytes, whereas granulocytes and fibroblasts appear resistant to lysis.[10, 71] Instillation of purified Hla into rabbit or rat lung tissue triggers vascular leakage and pulmonary hypertension,[71, 105, 106] which has been attributed to the release of different signaling molecules (e.g., phosphatidyl inositol, nitric oxide, prostanoids, and thromboxane A_2).[97, 105, 106, 114]

Cell Wall Envelope

The Gram stain broadly differentiates bacteria into Gram-positive and Gram-negative groups. Staphylococci have a thick (60 to 80 nm) continuous cell wall also called the murein sacculus. The cell wall is composed largely of peptidoglycan, also known as mucopeptide or murein. Other cell wall polymers include wall teichoic acid and surface proteins, both of which are covalently attached to the peptidoglycan. In contrast, the lipoteichoic acid polymer is modified by a diglucosyl diacylglycerol moiety and much like lipoproteins is tethered non-covalently to the plasma membrane and protrudes in the cell wall. The cell wall presents a unique problem in cellular architecture as new incoming units must be inserted into a peptidoglycan sheet that is held together by covalent bonds. Random openings in the wall would otherwise result in cell lysis. Further, the cell wall must grow into a specific shape. This process is achieved by a regulated pattern of enzymatic activities that control opening and closure of covalent bonds, in particular during cell division. Distinct morphological features of the cell wall have been revealed by electron microscopy of the newly emerging cross walls of dividing cells. In particular, a ring of periodically arranged tubules has been observed in the center of the cross wall,[32] suggesting that the tubules may contain the autolytic centers of the cell splitting system. After completion of the cross wall, electron micrographs of dividing *Staphylococcus aureus* cells reveals tiny holes named murosomes.[33] The murosomes are surmised to represent puncture sites along the division plane of future daughter cells.[32]

Peptidoglycan

Composition: Most bacterial peptidoglycans, with the exception of archebacteria, have a very similar structure. The peptidoglycan consists of a backbone of glycan chains of the repeating disaccharides N-acetylglucosamine (NAG) and its lactic ether N-acetylmuramic acid (NAM). Each disaccharide carries a tetrapeptide substituent of alternating L- and D-amino acids. A peptide bridge links the terminal COOH of D-alanine of one tetrapeptide to an NH_2 group of a tetrapeptide on a neighboring glycan chain (Figure 22.1). In *S. aureus*, each interbridge peptide consists of a pentaglycine (Figure 22.1). *Staphylococcus aureus* contains the most extensively crosslinked peptidoglycan of all bacteria.[46] In rapidly growing cells, more than 95% of the subunits are crosslinked and bear almost no free carboxyl groups, because the D-glutamic acid α-carboxyl group is amidated.[119] *S. aureus* peptidoglycan is extensively modified on the *N*-acetylmuramic acid residues with *O*-acetyl substituents such as the 4-*N*, 6-*O*-diacetyl derivative.[121] This modification renders the peptidoglycan resistant to egg-white lysozyme and other muramidases.[128]

Biosynthesis: Four main stages can be distinguished for the synthesis of peptidoglycan (Figure 22.1):

1. Synthesis of a water-soluble nucleotide-linked precursor in the cytoplasm.
2. Transfer from the nucleotide to a membrane lipid and addition of substituents.
3. Addition of this block to the linear glycan chain, or transglycosylation.

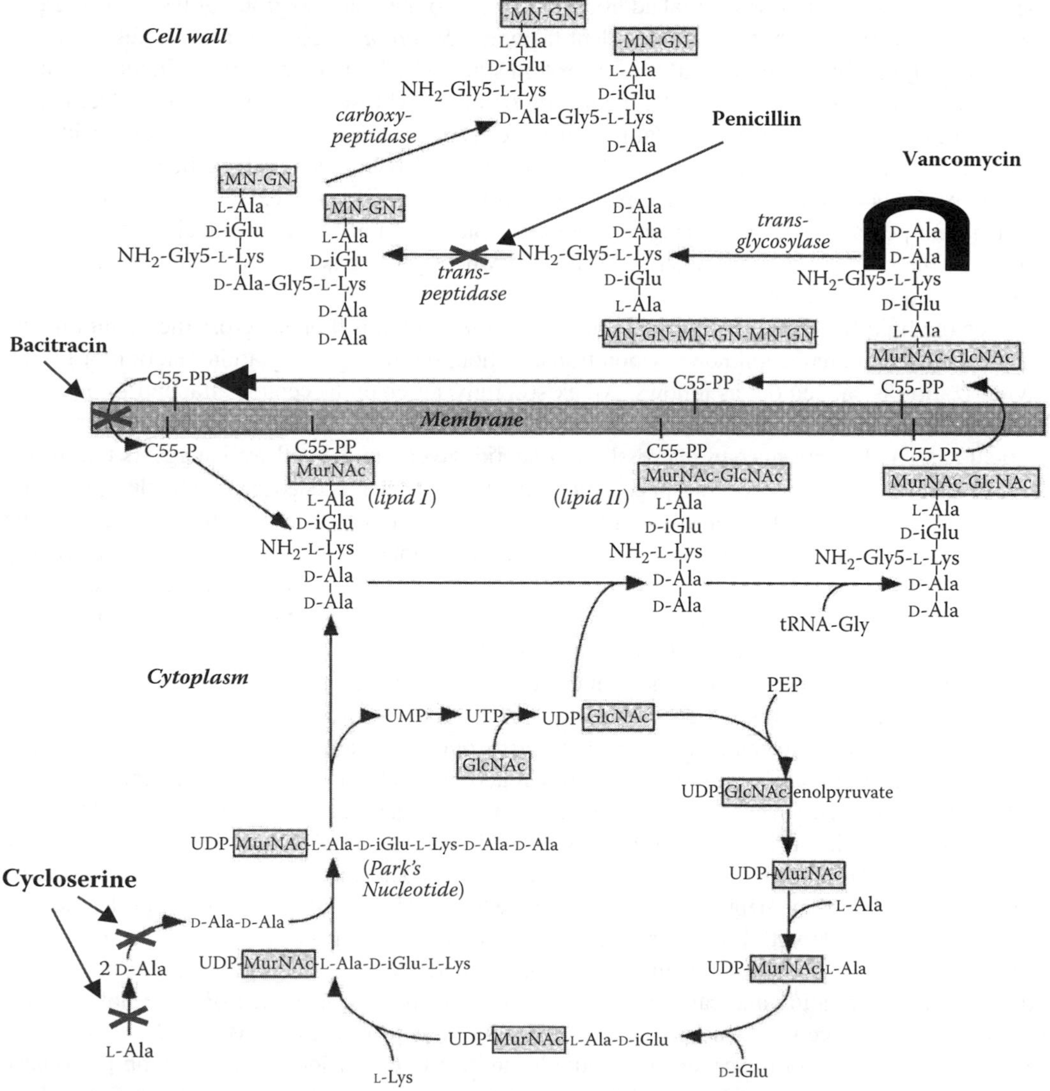

FIGURE 22.1 Model of peptidoglycan biogenesis. (*Source:* Adapted from Ref. 74.)

4. Cross-linking to an adjacent chain, or transpeptidation.

Stage 1 of peptidoglycan synthesis is initiated in the cytoplasm by modifying the uridine nucleotide of NAM (UDP-NAM) with the pentapeptide at its lactic carboxyl end. This product is also called UDP-NAM-pentapeptide or Park's nucleotide and accumulates in *Staphylococcus aureus* inhibited by penicillin.[50, 87] It is formed by the sequential addition of L-alanine, D-glutamic acid, L-lysine and D-alanyl-D-alanine to the D-lactyl group of UDP-NAM, and each step is catalyzed by a specific synthetase.[50, 51] Further, as for *Escherichia coli,* the enzyme MurI converts the L-Glu to D-Glu,[23] while L-Ala is converted to D-Ala by the alanine-racemase Alr in *S. aureus*.[59] A dedicated ligase generates a D-alanyl-D-alanine dimer.[51] The final soluble intermediate is UDP-NAM-L-Ala-D-*iso*-Glu-L-Lys-D-Ala-D-Ala. In stage 2, this intermediate is transferred from the nucleotide to the phosphate of a membrane carrier lipid, undecaprenol-P (also called bactoprenol-P) to yield C55–PP-NAM-L-Ala-D-*i*Gln-L-Lys-D-Ala- D-Ala, or lipid I.[47, 91] Lipid I faces the cytoplasm and UDP-NAG is linked to the muramoyl moiety at the expense of UPD to generate the disaccharide lipid II precursor [C55- PP-NAM-(-L-Ala–D-*i*Gln–L-Lys(Gly5)-DAla-D-Ala)-α1-4-NAG]. Lipid II precursors are modified by sequential addition of glycine to the ε-amino group of lysine, using glycyl-tRNA as donor, in a ribosome independent fashion.[53] *S. aureus* encodes three species of glycyl-tRNA for cell wall biosynthesis and one for protein synthesis.[37] The Fem factors (*f*actors *e*ssential for *m*ethicillin resistance[9]) catalyze the incorporation of glycyl residues in the cross bridge. FemX links the first glycine residue to the ε-amino group of lysine, FemA incorporates glycyl residues 2 and 3 and FemB, glycyl residues 4 and 5.[46, 95, 96] Other FemAB-like factors have been identified in staphylococci, such as those conferring resistance to lysostaphin cleavage.[113, 117] In *S. epidermidis,* four different pentapeptide cross-bridges have been isolated.[118] Each contains glycine and L-serine residues in the ratio 3:2 in a characteristic sequence where glycine is always the initial substituent of the ε-amino group of L-lysine.

Once the lipid II-pentaglycine intermediate is synthesized, it is flipped across the membrane by a mechanism that remains unknown. Upon translocation, the transglycosylation reaction of glycan polymerization is carried out as follows. An existing glycan chain (acceptor) attached to one molecule of lipid carrier is transferred from it to the disaccharide of the incoming subunit (donor) on an identical carrier. The presence of uncrosslinked peptidoglycan in the cell wall suggests that transglycosylation reaction precedes transpeptidation reaction. The elements governing the length of glycans remain elusive.[126, 127] The released carrier undecaprenol-PP then loses a P, thereby regenerating the form that can accept a precursor. The whole reaction appears to be promoted by the energy released by this P-P hydrolysis. The final step of the reaction is the formation of the peptide cross-links between adjacent glycan chains. This reaction is called transpeptidation and is sensitive to β-lactams.[120] In the presence of penicillin, uncrosslinked peptide monomer units accumulate with intact D-alanyl-D-alanine termini. The transpeptidases are also known as penicillin binding proteins (PBPs), and most organisms encode multiple PBPs that are involved in peptidoglycan growth, cell shape, and septum formation. Some PBPs carry out both transglycosylation and transpeptidation, the β-lactam inactivating only the latter reaction. Such enzymes are essential for viability. The effect of β-lactam inactivation explains why cells accumulate amorphous peptidoglycan. Antibiotics that block specifically transglycosylation also prevent transpeptidation, suggesting once more that the incoming subunit must be transglycosylated before it can be cross-linked. Smaller PBPs encode D,D-carboxypeptidases that release the terminal D-Ala from a pentapeptide and thus generate a tetrapeptide with a free terminus (that cannot enter a cross-link). These enzymes also have endopeptidase activity on existing cross-links. They function only in guiding morphogenesis and not in polymerization and can be eliminated without loss of viability. PBPs are not the only enzymes involved in cell wall morphogenesis. Lysis by penicillin was long assumed to result from a mechanistic bursting of a weakened peptidoglycan during expansion of the growing protoplast. However, lysis is the result of two autolytic enzymatic activities: (1) amidase and (2) glycosidase. Amidase splits the bond between tetrapeptides and the glycan, whereas glycosidase cleaves the glycan. Autolysins and other murein hydrolases that attack other peptidoglycan bonds are important for morphogenesis rather than self-destruction. Autolysins play an important role during cell division when new cell wall synthesis must take place and be added to the preexisting wall. Small openings in the wall are created by autolysins possibly at sites of murosome assembly. This could explain why peptidoglycan synthesis must occur for penicillin to induce cell lysis. In non-growing cells, the action of autolysins does not occur and breakdown of the cell wall peptidoglycan is prevented.

Protoplasts: The cytoplasmic membrane is pressed against the wall by an internal osmotic pressure (turgor pressure) that exceeds that of the external environment. Digestion of the wall or inhibition of its synthesis in growing cells leads to cell lysis. The cell wall of Gram-positive bacteria can be destroyed by digestion of the peptidoglycan using lysosyme, a protein that breaks the 1,4-glycosidic bonds between N-acetylglucosamine and N-acetylmuramic acid. Because of extensive peptidoglycan modification, staphylococci are insensitive to lysozyme and, instead, lysostaphin is used to digest the cell wall. This procedure is necessary for the extraction of cytoplasmic components such as DNA, RNA, and proteins. Cell lysis will not occur in the presence of lysostaphin when the medium is sufficiently hypertonic. Typically, a solute such as sucrose (0.5 M) in the buffer solution will balance the concentration of solutes inside the cell. This leads to the formation of an osmotically sensitive sphere called a protoplast. Such sucrose-stabilized bodies can grow briefly without cell division. However, if placed in hypotonic solution, lysis occurs immediately.

Peptidoglycan biosynthesis steps inhibited by antibiotics: Penicillin interferes with wall formation by inhibiting the transpeptidation reaction. Electron microscopy reveals that penicillin-treated cells accumulate amorphous material between the membrane and the wall.[32] There is a close structural analogy between the D-Ala-D-Ala and the β-lactam ring of penicillin or other β-lactam antibiotics. This ring reacts with the transpeptidase to form a stable covalent bond that inactivates the enzyme. Cycloserine, a structural analog of D-Ala, is a competitive inhibitor of both racemization of L-Ala to D-Ala and conversion of the latter to the dipeptide D-Ala-D-Ala. Fosfomycin, an analog of P-enolpyruvate, irreversibly inhibits the conversion of P-enolpyruvate to the ether-linked lactyl group of UDP-NAM. Neomycin blocks transfer of muramyl peptide to the growing chain. Vancomycin blocks transpeptidation by binding to D-Ala-D-Ala in several intermediates and in the peptidoglycan form. Bacitracin also binds to a substrate instead of an enzyme. It binds undecaprenol-PP, blocking its dephosphorylation and thereby preventing its recycling to undecaprenol-P.

In response to vancomycin selective pressure, some staphylococcal strains acquired mutations that trigger altered cell wall envelope structures and low-level vancomycin resistance.[48, 130] Other strains such as VRSA acquired the *vanA* resistance gene from enterococci.[17, 116] VanA ligates β-hydroxy carboxylic acids (e.g., D-lactate) to D-Ala and the product, D-Ala-D-Lactate is then incorporated into wall peptides and lipid II: L-Ala-D-iGln-(NH_2-Gly_5)-L-Lys-D-Ala-D-Lactate-COOH as opposed to the wild-type wall peptide L-Ala-D-iGln-(NH_2-Gly_5)-L-Lys-D-Ala-D-Ala-COOH.[15] The net result of staphylococcal *vanA* gene acquisition, vancomycin resistance, stems from the reduced affinity of the glycopeptide antibiotic to the depsipeptide bond between D-Ala-D-Lactate.[125]

Resistance to methicillin by MRSA results from the acquisition of the oxacillin/methicillin-resistance gene *mecA*. *mecA* was possibly acquired from *Staphylococcus sciuri* by an unknown mechanism.[131] MecA encodes a penicillin-binding protein (PBP), PBP2a, that displays unusual low affinity for all β-lactam antibiotics.[2, 42] The genetic determinant of methicillin resistance is located on a mobile genetic element called the staphylococcal cassette chromosome (SCC) in *S. aureus (*21 to 67-kb fragment).[54] Because MRSA and VRSA strains have also acquired macrolide, tetracycline, aminoglyocoside, chloramphenicol, and fluoroquinolone resistance mechanisms,[60] the pharmaceutical, scientific, and medical communities have suddenly begun to realize that the therapeutic arsenals are close to depletion, causing a threat for infectious disease therapy.[92]

Teichoic Acids

Cell wall teichoic acid (WTA) and lipoteichoic acid (LTA) are secondary wall polymers that traverse the envelope of Gram-positive bacteria.[3] In *Staphylococcus aureus,* about 8% of peptidoglycan is covalently modified with WTA by phosphodiester linkages on the C6 carbon of muramic acid residues.[43, 44] TA consists of chains of as many as 30 glycerol or ribitol (a 5-C polyhydric alcohol) residues with phosphodiester links and with various substituents (sugars, choline, D-alanine).[77] D-alanyl esterification of TA is promoted by the Dlt locus and is dispensable for cell viability; however, *dlt* mutants of *S. aureus* are more susceptible to neutrophil killing during infection.[20] WTA mutants of *S. aureus* are viable *in vitro* but are resistant to bacteriophage that use WTA for receptors and have been shown to be unable to colonize cotton rat nares.[129] Unlike WTA, LTA is a 1-3 linked glycerol-phosphate polymer linked to a glycolipid in the membrane. *S. aureus* LTA is retained by a glycolipid anchor [diglucosyl diacylglycerol (Glc_2-DAG)] in bacterial membranes.[24, 26] Fischer and co-workers proposed a model for LTA synthesis whereby polyglycerol-phosphate is polymerized on the outer surface of bacterial membranes from the phosphatidyl glycerol substrate.[56] Staphylococci can produce polyglycerol-phosphate polymers even in the absence of glycolipids, and instead anchor LTA via diacylglycerol.[38] This less-preferred anchoring is a survival strategy used by *S. aureus* as it appears that LTA is required for bacterial growth and cell division. Indeed, *ltaS,* the gene responsible for synthesis of glycerol-phosphate LTA, cannot be deleted without loss of viability.[39]

Cell Wall-Anchored Surface Proteins

Staphylococci contain different proteins that function as adhesins and receptors that mediate attachment to specific tissues or surfaces and molecule uptake, respectively. For example, in *Staphylococcus aureus*, a collagen adhesin mediates attachment to connective tissues and bone, thereby allowing staphylococci to attach to the bone and cause osteomyelitis. Other surface proteins of *S. aureus* have the ability to bind eukaryotic serum proteins. Protein A, for example, binds the constant region of immunoglobulins, Clumping factor binds plasminogen and stimulates blood clotting on the cell surface.[67] In this manner, the staphylococci can camouflage themselves within the host tissues to avoid phagocytic killing.

The mechanism of anchoring was first solved for *Staphylococcus aureus* protein A. Protein A could be released as a homogeneous population of *S. aureus* cell wall digested with lysostaphin.[107] When protein A was released mechanically from the cell wall fraction, it displayed an increase in molecular mass that could be attributed to the amino sugars *N*-acetylglucosamine and *N*-acetylmuramic acid, suggesting that protein A must be linked to the staphylococcal cell wall.[108] The amino acid sequence of protein A reveals two hydrophobic domains, one at the N-terminus and one at the C-terminus. Sequence comparison with other cell wall-anchored proteins revealed that the C-terminus carries an almost invariant LPXTG sequence motif, where X is any amino acid, followed by a C-terminal hydrophobic domain and a tail of mostly positively charged residues.[27] The LPXTG motif and C-terminal tail are necessary for cell wall anchoring.[103] Following translocation, the LPXTG motif sorting signal is cleaved between the threonine (T) and the glycine (G) residues[73] by the enzyme sortase A.[69] Sortase A also catalyzes the transpeptidation reaction between the carboxyl of threonine and the free amino group of pentaglycine cross-bridges in the staphylococcal cell wall,[102] using lipid II peptidoglycan intermediates as a substrate.[90] A second sortase, SrtB, is responsible for the anchoring of Isd (iron-regulated surface determinant) factors of *S. aureus*. These factors are responsible for hemoglobin binding and passage of heme-iron to the cytoplasm and also act as an import apparatus that uses cell wall-anchored proteins to relay heme-iron across the bacterial envelope.[70]

Bioinformatic searches identify 18 to 22 genes encoding putative sortase A-anchored surface proteins in the genomes of *Staphylococcus aureus*.[67] The function of cell wall-anchored proteins is not exhaustively known; nonetheless, many of them appear to function as microbial surface components recognizing adhesive matrix molecules (MSCRAMMs) and thus represent bacterial elements of tissue adhesion and immune evasion.[30]

Physiology

Carbohydrate uptake and metabolism have been best characterized for *Staphylococcus aureus* and *S. xylosus,* or may be deduced from genome analysis of *S. aureus* and *S. epidermidis* strains. Two types of carbohydrate transport systems have been investigated, the phosphoenol pyruvate (PEP): sugar phosphotransferase system (PTS) and the PTS-independent carbohydrate transport whereby sugars taken up by a dedicated permease are subsequently phosphorylated by an ATP-dependent kinase (instead of PEP-dependent phosphorylation). Both systems have evolved to take up glucose as demonstrated in *S. aureus* and *S. xylosus.*[93] Glucose oxidation is mediated by the fructose 1,6-bisphosphate and the oxidative pentose phosphate pathways. There is no evidence for the existence of the Entner-Doudoroff pathway and, in the absence of oxygen, staphylococci will behave like facultative anaerobes. Enzymes of the entire tricarboxylic acid (TCA) cycle are encoded in the genome of *S. aureus* and reducing equivalent generated by the TCA cycle is reoxidized by an electron transport chain containing a- and b-type cytochromes. A typical F0F1-ATPase is encoded in the genome of staphylococci. In addition to glucose, *S. aureus* can use D-galactose in a pathway that converts it to D-galactose 6-P and next tagatose. Tagatose-6-P kinase, the key enzyme of the tagatose pathway, is inducible by growth on galactose or lactose and is found in *S. aureus, S. epidermidis,* and *S. hominis* but absent in *S. intermedius, S. saprophyticus,* and *S. xylosus.* In these organisms, galactose is oxidized via the Leloir pathway.[100] Other sugars can be metabolized by staphylococci, and a summary of these other metabolic pathways can be found in Ref. 35. Upon glucose depletion, *S. aureus* cells growing in aerobic conditions will oxidize acetate, succinate, and malate.[111] Most strains examined thus far encode a lactate dehydrogenase that is activated by fructose-1,6-bisphosphate, itself up-regulated during anaerobic conditions in the presence of glucose. Staphylococci can use nitrate as an electron acceptor in the absence of oxygen, a metabolic activity that has broad application in food technology; however, little is known about the biochemistry and genetic requirement for nitrate reduction with the exception of *S. carnosus.*[76]

VIRULENCE FACTORS AND REGULATION

Staphylococcus aureus is the leading cause of hospital-acquired infections.[65] The spectrum of human diseases caused by staphylococci ranges from soft tissue infections, abscesses in organ tissues, osteomyelitis, endocarditis, and toxic shock syndrome, to necrotizing pneumonia, all of which can be tested using the small animal model of infection.[1,65] *S. aureus* is a versatile pathogen, and specific strains have been isolated from many different pathological-anatomical sites or disease entities.[58] Research over the past several decades identified *S. aureus* cell wall-anchored proteins, secreted toxins, capsular- and exo-polysaccharides, iron-transport systems, and modulators of host immune functions in addition to antibiotic-resistance genes as important virulence factors.[1,67,78] Furthermore, such versatility is enabled by a complex regulation of virulence determinants. For example, staphylococci perform a bacterial census via the secretion of auto-inducing peptides (AgrD-derived substances processed and secreted via AgrB) that bind to a cognate histidine kinase sensor (AgrC) at threshold concentration, thereby activating phospho-relay reactions and transcriptional activation (AgrA mediated) of a plethora of exotoxin genes.[78] During infection, this bacterial census ensures massive secretion of virulence factors when staphylococcal counts are high, presumably increasing the likelihood of bacterial spread in infected tissues and/or systemic dissemination. Additional transcription factors of the Sar family as well as two component regulatory systems play important roles in regulating expression of *agr* or production of other toxin genes, albeit that the mechanisms of signal transduction or environmental signaling are not yet known in molecular detail.[18] The unifying theme of all these observations is the multi-factorial nature of *S. aureus* pathogenesis, where many virulence factors contribute overlapping and often redundant roles to the establishment of a wide variety of infectious syndromes and intoxications in humans.

GENOME AND GENOMIC VARIABILITY

Several different *Staphylococcus aureus* and *S. epidermidis* strains have been sequenced and encompass between 2550 and 2870 genes[5, 21, 28, 60] and display up to 22% of DNA sequence variability.[28] The vast majority of staphylococcal virulence functions (including surface proteins, sortase, α-hemolysin, exotoxins, and *agr* regulon) are encoded on the bacterial chromosome.[60] These genes are not associated with mobile DNA elements and are found in all staphylococcal strains. Most of the genome variability is brought about by insertions of transposons and mobile elements, as well as prophages and plasmids.[80] For example, MRSA strains carry a known hotspot for the insertion of transposons and insertion sequences of the methicillin resistance cassette.[40, 54] Genomic analysis of strain USA300 shows that the same hotspot acquired the ACME (Arginine Catabolic Mobile Element) gene cluster.[21] ACME may enable the strain to evade host immune responses. USA300 also encodes a pathogenicity island that carries two enterotoxins not found in other sequenced genomes except *S. aureus* strain COL and three plasmids with multiple antibiotic resistance determinants.[21] Pathogenicity islands are common in *S. aureus* and encode toxins associated with specific clinical conditions (toxinoses), some of which are superantigens.[82] Secretion of superantigens stimulates large T-cell populations and causes massive cytokine release that leads to toxic shock. Pathogenicity islands are 15- to 20-kbp DNA elements that occupy constant positions in the chromosomes of toxigenic strains. Phage-related features and the presence of flanking direct repeats characterize these mobile elements.[82]

Recent work identified *Staphylococcus aureus* bacteriophages as important contributors to pathogenesis and the evolution of staphylococcal genomes.[78, 122] All sequenced genomes of clinical *S. aureus* harbor at least one or up to three prophages.[14] Phage excision or transduction has been observed during the course of chronic lung infections in cystic fibrosis patients and are cause of genomic variation.[34] Phages have been shown to transfer pathogenicity islands between strains, which is best exemplified by φ80α-mediated excision and transduction of SaPI1.[99] Innate immune modulators such as chemotaxis inhibitory proteins (CHIPs) and staphylococcal complement inhibitor (SCINs) have been shown to be encoded by β-hemolysin converting phages.[122] Prophages are likely to encode additional factors that favor their replication in infected hosts. Recently, a variant of the clinical isolate *S. aureus* Newman lacking all prophages was constructed and shown to be unable to cause disease in animals.[6]

PRACTICAL GROWTH AND MANIPULATIONS

Growth in Laboratory Media

Staphylococci can be grown in many non-selective rich media in the laboratory, such as trypticase soy broth (TSB), brain heart infusion (BHI), or Luria Bertani broth (LB). Most staphylococci can grow in a temperature range of 15 to 45°C and at NaCl concentrations as high as 15 to 20%. Because most staphylococci are coagulase-positive, several coagulase tests have been developed. Isolated colonies should be transferred in 0.2 mL brain heart infusion (BHI) broth, and the cultures should be incubated for 18 to 24 hr at 37°C. After incubation, the culture should be mixed with 0.5 mL reconstituted coagulase plasma with ethylenediaminetetraacetic acid (EDTA) (plasma is commercially available and reconstituted according to manufacturer's directions). Incubation should be continued at 37°C and examined at 6-hr intervals for clot formation. Direct enumeration of coagulase-positive staphylococci can be made on Baird-Parker agar containing rabbit plasma-fibrinogen tellurite.[11] Other media derived or not from Baird-Parker medium can be used for the non-selective enrichment of coagulase-positive staphylococci from food sources and the like. Schleifer-Krämer

(SK) agar is used for the selective isolation and enumeration of staphylococci from foods. Recipes and methods for using such media can be found in Ref. 35.

Biosafety Measures

Staphylococcus aureus is an opportunistic and adaptable pathogen with the ability to infect, invade, persist, and replicate in any human tissue, including skin, bone, visceral organs, or vasculature. Manipulation with most staphylococci should be performed following biosafety level 2 (BSL2) measures. Guidelines for BSL2 practice can be obtained from the following CDC web link: www.cdc.gov/od/ohs/safety/S2.pdf.

Molecular Biology and Genetic Manipulation

Cloning Plasmids

Several staphylococcal plasmids can be used for genetic manipulations and cloning purposes in *Staphylococcus aureus*. These can be grouped in four families with the following representative plasmids: pC194, pT181, pSN2, and pE194.[79] Derivatives of these plasmids carrying mutations with conditional replication phenotypes also have been isolated. For example, pTS1 carries a chloramphenicol resistance determinant for plasmid selection and the pE194ts *ori* for replication in staphylococci, in addition to the Gram-negative bacterial plasmid pMB1 replication functions and aTEM *bla* resistance marker.[83]

Lee and colleagues[62] also developed single-copy integration vectors suitable for cloning in *Staphylococcus aureus*. This technology takes advantage of the site-specific recombination system of staphylococcal phage L54a.[63] The vectors contain a replicon for the *Escherichia coli* host but not for *S. aureus*. As a consequence, expression of determinants carried by these vectors in *S. aureus* can only be achieved by the integration system of the phage. Integration is very stable, requires no selection, and allows single-copy gene dosage of the cloned sequences.

Due to the host restriction-modification system, plasmid DNA cannot be introduced into *Staphylococcus aureus* clinical isolates directly, but needs to be passaged through the laboratory strain RN4220.[57] RN4220 is a vital intermediate for laboratory *S. aureus* manipulations, as it can accept *Escherichia Coli*-propagated plasmid DNA due to nitrosoguanidine-induced mutation(s) in its restriction-modification system,[89] recently mapped to thc *sau1hsdR* gene.[124] Plasmids extracted from strain RN4220 can be electroporated in most staphylococcal isolates (see Tables 22.1 and 22.2 for plasmid electroporation and transformation).

Allelic Replacement

Allelic replacement has been used extensively to generate mutations in chromosomal genes. In bacterial cells, mutated alleles of a gene of interest, commonly deletions of the gene, or insertions of an antibiotic resistance gene into it, are cloned into plasmids carrying replication-defective conditional mutations. Under non-permissive conditions, plasmid integration into the chromosome occurs via homologous recombination.[29] The products of single cross-over recombination, merodiploid cells harboring both wild-type and mutant alleles, are resolved under permissive conditions in which recombination and plasmid excision are thought to be stimulated by rolling circle replication of the plasmid. As a result of double cross-over recombination, wild-type or mutant alleles reside on the chromosome while the reciprocal alleles replicate with the excised plasmid.

pTS1 is frequently used for allelic replacement in staphylococci. Staphylococcal target DNA sequences are cloned in the shuttle plasmid pTS1 and amplified in *Escherichia coli*. Typically, 1 kbp of DNA sequences upstream and downstream of the region target for deletion are ligated and cloned together in plasmid pTS1. Often, a marker, typically *ermC*, is cloned in place of the missing target sequence. Recombinant pTS1 plasmids are transferred to staphylococci via electroporation

TABLE 22.1
Transformation of Plasmid DNA in Electrocompetent Cells of *Staphylococcus aureus*

1. Streak *S. aureus* strain to be transformed from a frozen stock on a tryptic soy agar (TSA) plate and incubate overnight at 37°C.
2. Pick an isolated colony with a sterile loop and inoculate 2 mL tryptic soy broth (TSB) in a 100-mL flask.
3. Incubate overnight at 37°C with shaking.
4. Transfer the overnight culture into a 2-L flask containing 200 mL TSB.
5. Grow cells to mid-log phage (OD600 = 0.5) with vigorous shaking (approximate incubation time 2.5–3 hr).
6. Transfer the culture into sterile spin bottles and collect cells by centrifugation at 5000 *x g* for 15 min.
7. Discard the supernatant and suspend the cell pellet in 40 mL of ice-cold sterile 0.5 M sucrose in deionized water.
8. Transfer the cell suspension to a prechilled 50-mL sterile centrifuge tube and keep on ice.
9. Collect cells by centrifugation at 8000 *x g*, 10 min, 4°C.
10. Discard supernatant and suspend the cell pellet in 20 mL of the ice-cold 0.5 M sucrose solution as above.
11. Collect cells by centrifugation at 8,000 *x g*, 10 min, 4°C.
12. Repeat Steps 10 and 11 once more.
13. Resuspend the pellet in 2 mL ice-cold 0.5 M sucrose solution.
14. Transfer 100 μL aliquots of the prepared electrocompetent cells into microcentrifuge tubes chilled on ice.
15. Freeze tubes by plunging them in a dry ice-ethanol bath and store cells at –80°C until use (this protocol can be adapted to prepare larger volumes of competent cells).
16. For electroporation with plasmid DNA, retrieve a tube of competent cells from the freezer and place tube on ice.
17. When cells are thawed, add 100–500 ng of purified plasmid.
18. Transfer the cell and DNA mix into a 0.1-cm prechilled electroporation cuvette (Bio-Rad, Hercules, CA).
19. Use the following settings for electroporation: voltage = 2.5 kV, resistance = 100 ohms capacity = 25F.
20. Immediately following the pulse, add 1 mL TSB kept at room temperature and transfer entire contents to sterile eppendorf tube.
21. Incubate for 1 hr at 30°C (no shaking required).
22. Pellet cells in a microcentrifuge (8,000 *x g*, 3 min, RT) and remove most of the supernatant by flipping the tube upside-down.
23. Suspend cell pellet in remaining medium (50–100 μL) and spread cells on a TSA plate containing appropriate antibiotic for plasmid selection.
24. Incubate plate at 30° or 37°C for at least 16 hr (or until colonies are visible).

TABLE 22.2
Plasmid Extraction

1. Pick isolated colonies and grow cells in TSB with appropriate antibiotic overnight at 37°C.
2. Transfer 1.5 mL of the overnight culture and collect cells by centrifugation (5000 *x g*, 3 min, RT).
3. Suspend cell pellet in 50 μL TSM buffer.
4. Add 2.5 μL lysostaphin solution (2 mg/mL stock) and incubate for 15 min at 37°C (this will yield protoplasts).
5. Collect protoplasts by centrifugation (8000 *x g*, 5 min, RT) and discard supernatant.
6. Extract plasmid DNA from protoplasts using a QIAprep Spin Miniprep Kit (Qiagen) following the manufacturer's recommendations.
7. Analyze extracted plasmid by agarose gel electrophoresis.

(Table 22.1) and colonies arise on agar medium containing chloramphenicol at 30°C, a condition that is permissive for plasmid replication. Following temperature shift to the non-permissive condition at 43°C, bacterial colonies arise on agar medium containing erythromycin and chloramphenicol. Cultures inoculated with merodiploid variants, i.e., chromosomal plasmid cointegrates, are then grown at 30°C and plated on agar medium with erythromycin. Colonies are picked from the erythromycin-containing plates and transferred to plates containing medium with or without inhibiting chloramphenicol. Those able to grow in the absence of chloramphenicol, but not in its presence, should be examined for the presence of mutant alleles and loss of plasmid.[29]

Recently, pMAD was developed to facilitate isolation of allelic replacement mutants in Gram-positive bacteria, including *Staphylococcus aureus*.[4] pMAD carries the pE194ts *ori* and *bgaB* genes, encoding thermostable β-galactosidase; *bgaB* expression allows staphylococci to cleave the chromogenic substrate X-gal (5-bromo-4-chloro-3-indoyl-β-d-galactopyranoside), thereby generating colonies with blue color. This technology affects neither frequency nor selection of mutations, but provides a screening tool for excision and loss of plasmid. The new pKOR1 plasmid (pTS1 derived) employs antisense *secY* RNA expression for counter-selection during allelic replacement procedures.[8] *secY* expression is essential for bacterial growth and survival, and expression of *secY* antisense RNA inhibits colony formation on agar plates.[52] In plasmid pKOR1, expression of the antisense *secY* is controlled by the Pxyl/tetO promoter and can be induced with 1 μg mL^{-1} anhydrotetracycline (ATc). When this plasmid is used for allelic replacement, the target DNA sequence is cloned; and after electroporation at 30°C, the temperature is shifted to the non-permissive condition at 43°C. The plasmid integrates on the chromosome by homologous recombination. Plasmid eviction is induced by growing cells in liquid culture at the permissive temperature. The culture is diluted 10^4-fold with sterile broth and 100-μL aliquots are spread on TSA containing 0, 1, or 2 μg mL^{-1} ATc. Expression of antisense *secY* RNA suppresses the growth of cells containing pKOR1 plasmid DNA and hence selects against co-integrants that are not resolved. The complete procedure for pKOR1-mediated allelic replacement can be found in the supplementary material published in Ref. 8.

Bacteriophage

Staphylococcus aureus strains are lysogenic and carry temperate phages that are UV inducible. Such phages integrate at unique chromosomal sites by the Campbell mechanism.[64] *S. aureus* phages fall into three main serological groups — A, B, and F — of which group B contains most of the known transducing phages, including φll[81] and typing phages 53, 79, 80, and 83.[79]. φll is a prototypical group B transducing phage that has a burst size of approximately 250 plaque-forming units and exhibits low lysogenization frequency (1 to 10%).[81] Use of φll is ideal for transducing alleles marked with a selectable marker between *S. aureus* isolates. Strains should be grown in TSB in the presence of $CaCl_2$ (5 mM) to favor phage adsorption. Not all strains are susceptible to phage φll. Transduction in strain Newman is best achieved with phage φ85. Typically, phage lysates on a donor strain are generated by infecting bacteria grown in TSB containing $CaCl_2$ (5 mM) at a multiplicity of infection of 1:1. Lysis can be performed in liquid broth or in soft agar poured over TSA-containing plates. Lysis is observed after 16-hr incubation at 37°C. The phage can be recovered in TSB and used for transduction of a recipient strain at a multiplicity of infection of one phage particle to ten bacteria. Bacteria and phage stock should be incubated for 30 min at 37°C and before plating, the phage should be eliminated by centrifugation, whereby bacteria are pelleted and culture supernatants containing the phage discarded. Bacteria in the pellet should be washed twice with TSB containing the calcium chelating agent, sodium citrate (40 mM). For transductant selection, bacteria should be plated on TSA with appropriate selectable antibiotic in the presence of sodium citrate (40 mM). Isolated colonies arise within 48 hr of incubation and should be streaked on citrate containing plate to eliminate residual lytic phage.

Phage typing is practiced for epidemiological monitoring of staphylococcal strains. All phage receptors are located in the peptidoglycan-teichoic acid complex. Differences in phage susceptibility can be accounted for by post-adsorption events that include restriction-modification systems and other intracellular processes that prevent phage development.

Transposon Mutagenesis

Staphylococcus aureus can be mutagenesized using various transposons. Most transposons are delivered on plasmids carrying a temperature replicon. The transposon and transposase genes can be carried on one or two plasmids. Plasmid pTV1 carries Tn917,[41] a transposon that has been used for insertional mutagenesis in Gram-positive microbes.[72, 104] Recently, we developed bursa aurealis, a mariner-based transposon that can be used for random mutagenesis of Gram-positive bacteria.[7, 115] After transformation with plasmid-bearing transposons and temperature-induced selection of transposon insertions, mutants can be generated randomly and analyzed as described.[7]

THE CHALLENGE IN STAPHYLOCOCCAL RESEARCH

Gram-positive bacteria present a serious therapeutic challenge to human infections due to the emergence of antibiotic-resistant strains.[75] Of concern are infections with *Staphylococcus aureus*, *S. epidermidis*, and *Enterococcus faecalis*, microorganisms that are the most common cause of bacterial disease in American hospitals.[12, 13] Thus, a large body of studies has been done to understand the composition and biogenesis of the staphylococcal cell wall envelope and to identify the targets of antibiotics used until now. Some of these elements have been emphasized in this chapter. The inability of infected hosts to elicit immunity against staphylococci poses a puzzling problem. Whole-cell live or killed vaccines largely fail to generate protective immune responses.[94] Purified capsular polysaccharide, types 5 and 8 (which represent about 80% of all capsular types found in clinical isolates), showed promise when used as a conjugate vaccine in experimental animals or in patients with end-stage renal disease.[25] Immunization with poly-*N*-acetylglucosamine, an *S. aureus* surface carbohydrate synthesized by *icaABC* products,[45] has been shown to protect mice against staphylococcal disease.[66] Recent data suggest that immunization with four surface proteins (IsdA, IsdB, SdrD, and SdrE) generated the highest level of protection in mice compared with 15 other antigens.[110] Possibly, for *S. aureus* vaccines to be successful, genetic determinants for specific antigens must be essential for staphylococcal virulence. Yet, our ability to generate a staphylococcal vaccine remains an open question.

REFERENCES

1. Archer, G.L. 1998. *Staphylococcus aureus*: a well-armed pathogen. *Clin. Infect. Dis.*, 26:1179–1181.
2. Archer, G.L., and D.M. Niemeyer. 1994. Origin and evolution of DNA associated with resistance to methicillin in staphylococci. *Trends Microbiol.*, 2:343–247.
3. Archibald, A.R., J. Baddiley, and J.E. Heckels. 1973. Molecular arrangment of teichoic acid in the cell wall of *Staphylococcus aureus*. *Nature New Biol.*, 241:29-31.
4. Arnaud, M., A. Chastanet, and M. Debarbouille. 2004. New vector for efficient allelic replacement in naturally nontransformable, low-GC-content, Gram-positive bacteria. *Appl. Environ. Microbiol.*, 70:6887–6891.
5. Baba, T., F. Takeuchi, M. Kuroda, H. Yuzawa, K. Aoki, A. Oguchi, Y. Nagai, N. Iwama, K. Asano, T. Naimi, H. Kuroda, L. Cui, K. Yamamoto, and K. Hiramatsu. 2002. Genome and virulence determinants of high virulence community-acquired MRSA. *Lancet*, 359:1819–1827.
6. Bae, T., T. Baba, K. Hiramatsu, and O. Schneewind. 2006. Prophages of *Staphylococcus aureus* Newman and their contribution to virulence. *Mol. Microbiol.*, 62:1035–1047.
7. Bae, T., A.K. Banger, A. Wallace, E.M. Glass, F. Aslund, O. Schneewind, and D.M. Missiakas. 2004. *Staphylococcus aureus* virulence genes identified by *bursa aurealis mutagenesis* and nematode killing. *Proc.. Natl. Acad. Sci. U.S.A.*, 101:12312–12317.
8. Bae, T. and O. Schneewind. 2006. Allelic replacement in *Staphylococcus aureus* with inducible counter-selection. *Plasmid*, 55:58–63.
9. Berger-Bächi, B. 1994. Expression of resistance to methicillin. *Trends Microbiol.*, 2:389–309.
10. Bhakdi, S. and J. Tranum-Jensen. 1991. Alpha-toxin of *Staphylococcus aureus*. *Microbiol. Rev.*, 55:733–751.
11. Boothby, D., L. Daneo-Moore, and G.D. Shockman. 1971. A rapid, quantitative, and selective estimation of radioactively labeled peptidoglycan in Gram-positive bacteria. *Anal. Biochem.*, 44:645–653.
12. Boyce, J.M. 1990. Increasing prevalence of methicillin-resistant *Staphylococcus aureus* in the United States. *Infect. Control Hosp. Epidemiol.*, 11:639–642.

13. Brumfitt, W. and J. Hamilton-Miller. 1989. Methicillin-resistant *Staphylococcus aureus. N. Engl. J. Med.*, 320:1188–1199.
14. Brussow, H., C. Canchaya, and W.D. Hardt. 2004. Phages and the evolution of bacterial pathogens: from genomic rearrangements to lysogenic conversion. *Microbiol. Mol. Biol. Rev.*, 68:560–602.
15. Bugg, T.D.H., G.D. Wright, S. Dutka-Malen, M. Arthur, P. Courvalin, and C.T. Walsh. 1991. Molecular basis for vancomycin resistance in *Enterococcus faecium* BM4147: biosynthesis of a depsipeptide peptidoglycan precursor by vancomycin resistance proteins VanH and VanA. *Biochemistry*, 30:10408–10415.
16. Burlak, C., C.H. Hammer, M.A. Robinson, A.R. Whitney, M.J. McGavin, B.N. Kreiswirth, and F.R. Deleo. 2007. Global analysis of community-associated methicillin-resistant *Staphylococcus aureus* exoproteins reveals molecules produced *in vitro* and during infection. *Cell Microbiol.*, 9:1172–1190.
17. Chang, S., D.M. Sievert, J.C. Hageman, M.L. Boulton, F.C. Tenover, F.P. Downes, S. Shah, J.T. Rudrik, G.R. Pupp, W.J. Brown, D. Cardo, S.K. Fridkin, and V.-R. S.a.I. Team. 2003. Infection with vancomycin-resistant *Staphylococcus aureus* containing the vanA resistance gene. *N. Engl. J. Med.*, 348:1342–1347.
18. Cheung, A.L., A.S. Bayer, G. Zhang, H. Gresham, and Y.Q. Xiong. 2004. Regulation of virulence determinants *in vitro* and *in vivo* in *Staphylococcus aureus. FEMS Immunol. Med. Microbiol.*, 40:1–9.
19. Clauditz, A., A. Resch, K.P. Wieland, A. Peschel, and F. Gotz. 2006. Staphyloxanthin plays a role in the fitness of *Staphylococcus aureus* and its ability to cope with oxidative stress. *Infect. Immun.*, 74:4950–4953.
20. Collins, L.V., S.A. Kristian, C. Weidenmaier, M. Faigle, K.P. Van Kessel, J.A. Van Strijp, F. Gotz, B. Neumeister, and A. Peschel. 2002. *Staphylococcus aureus* strains lacking D-alanine modifications of teichoic acids are highly susceptible to human neutrophil killing and are virulence attenuated in mice. *J. Infect. Dis.*, 186:214–219.
21. Diep, B.A., S.R. Gill, R.F. Chang, T.H. Phan, J.H. Chen, M.G. Davidson, F. Lin, J. Lin, H.A. Carleton, E.F. Mongodin, G.F. Sensabaugh, and F. Perdreau-Remington. 2006. Complete genome sequence of USA300, an epidemic clone of community-acquired methicillin-resistant *Staphylococcus aureus. Lancet*, 367:731–739.
22. Dinges, M.M., P.M. Orwin, and P.M. Schlievert. 2000. Exotoxins of *Staphylococcus aureus. Clin. Microbiol. Rev.*, 13:16–34, table of contents.
23. Doublet, P., J. van Heijenoort, and D. Mengin-Lecreulx. 1992. Identification of the *Escherichia coli murl* gene, which is required for the biosynthesis of D-glutamic acid, a specific component of bacterial peptidoglycan. *J. Bacteriol.*, 174:5772–5779.
24. Duckworth, M., A.R. Archibald, and J. Baddiley. 1975. Lipoteichoic acid and lipoteichoic acid carrier in *Staphylococcus aureus H. FEBS Lett.*, 53:176–179.
25. Fattom, A.I., G. Horwith, S. Fuller, M. Propst, and R. Naso. 2004. Development of StaphVAX, a polysaccharide conjugate vaccine against *S. aureus* infection: from the lab bench to phase III clinical trials. *Vaccine*, 22:880–887.
26. Fischer, W. 1990. In K. Morris (Ed.), *Glycolipids, Phosphoglycolipids and Sulfoglycopids*. Plenum Press, New York, Vol. 6, p. 123–234.
27. Fischetti, V.A., V. Pancholi, and O. Schneewind. 1990. Conservation of a hexapeptide sequence in the anchor region of surface proteins from Gram-positive cocci. *Mol. Microbiol.*, 4:1603–1605.
28. Fitzgerald, J.R., D.E. Sturdevant, S.M. Mackie, S.R. Gill, and J.M. Musser. 2001. Evolutionary genomics of *Staphylococcus aureus*: insights into the origin of methicillin-resistant strains and the toxic shock syndrome epidemic. *Proc. Natl. Acad. Sci. U.S.A.*, 98:8821–8826.
29. Foster, T.J. 1998. Molecular genetic analysis of staphylococcal virulence. *Methods Microbiol.*, 27:432–454.
30. Foster, T.J. and M. Höök. 1998. Surface protein adhesins of *Staphylococcus aureus. Trends Microbiol.*, 6:484–488.
31. Garrity, G.M. and J.G. Holt. 2001. The road map to the manual. In R.D. Boone, R.W. Castenholz, and G.M. Garrity (Eds.), *Bergey's Manual of Systematic Bacteriology*, 2nd ed. Springer-Verlag, New York, p. 119–166.
32. Giesbrecht, P., T. Kersten, H. Maidhof, and J. Wecke. 1998. Staphylococcal cell wall: morphogenesis and fatal variations in the presence of penicillin. *Microbiol. Mol. Biol. Rev.*, 62:1371–1414.
33. Giesbrecht, P., T. Kersten, and J. Wecke. 1992. Fan-shaped ejections of regularly arranged murosomes involved in penicillin-induced death of staphylococci. *J. Bacteriol.*, 174:2241–2252.

34. Goerke, C., C. Wirtz, U. Fluckiger, and C. Wolz. 2006. Extensive phage dynamics in *Staphylococcus aureus* contributes to adaptation to the human host during infection. *Mol. Microbiol.*, 61:1673–1685.
35. Gotz, F., T. Bannerman, and K.-H. Schleifer. 2006. The Genera Staphylococcus and Macrococcus. In M. Dworkin, S. Falkow, E. Rosenberg, K.-H. Schleifer, and E. Stackebrandt (Eds.), *The Prokaryotes*. Springer New York, Vol. 4, p. 5–75.
36. Gouaux, E., M. Hobaugh, and L. Song. 1997. alpha-Hemolysin, gamma-hemolysin, and leukocidin from *Staphylococcus aureus*: distant in sequence but similar in structure. *Protein Sci.*, 6:2631–2635.
37. Green, C.I. and B.S. Vold. 1993. *Staphylococcus aureus* has clustered tRNA genes. *J. Bacteriol.*, 175:5091–5096.
38. Grundling, A. and O. Schneewind. 2007. Genes required for glycolipid synthesis and lipoteichoic acid anchoring in *Staphylococcus aureus*. *J. Bacteriol.*, 189:2521–2530.
39. Grundling, A. and O. Schneewind. 2007. Synthesis of glycerol-phosphate lipoteichoic acid in *Staphylococcus aureus*. *Proc. Natl. Acad. Sci. U.S.A.*, 104:8478–8483..
40. Hanssen, A.M. and J.U. Ericson Sollid. 2006. SCCmec in staphylococci: genes on the move. *FEMS Immunol. Med. Microbiol.*, 46:8–20.
41. Hartley, R.W. and C J. Paddon. 1986. Use of plasmid pTV1 in transposon mutagenesis and gene cloning in *Bacillus amyloliquefaciens*. *Plasmid*, 16:45–51.
42. Hartman, B.J. and A. Tomasz. 1984. Low affinity penicillin binding protein associated with β-lactam resistance in *Staphylococcus aureus*. *J. Bacteriol.*, 158:513–516.
43. Hay, J.B., A.R. Archibald, and J. Baddiley. 1965. The molecular structure of bacterial walls. The size of ribitol teichoic acids and the nature of their linkage to glycosaminopeptides. *Biochem. J.*, 97:723–730.
44. Hay, J.B., N.B. Davey, A.R. Archibald, and J. Baddiley. 1965. The chain length of ribitol teichoic acids and the nature of their association with bacterial cell walls. *Biochem. J.*, 94:7C–9C.
45. Heilmann, C., O. Schweitzer, C. Gerke, N. Vanittanakom, D. Mack, and F. Gotz. 1996. Molecular basis of intercellular adhesion in the biofilm-forming *Staphylococcus epidermidis*. *Mol. Microbiol.*, 20:1083–1091.
46. Henze, U., T. Sidow, J. Wecke, H. Labischinski, and B. Berger-Bachi. 1993. Influence of femB on methicillin resistance and peptidoglycan metabolism in *Staphylococcus aureus*. *J. Bacteriol.*, 175:1612–1620.
47. Higashi, Y., J.L. Strominger, and C.C. Sweeley. 1967. Structure of a lipid intermediate in cell wall peptidoglycan synthesis: a derivative of a C55 isoprenoid alcohol. *Proc. Natl. Acad. Sci. U.S.A.*, 55:1878–1884.
48. Hiramatsu, K., N. Aritaka, H. Hanaki, S. Kawasaki, Y. Hosoda, S. Hori, Y. Fukuchi, and I. Kobayashi. 1997. Dissemination in Japanese hospitals of strains of *Staphylococcus aureus* heterogeneously resistant to vancomycin. *Lancet*, 350:1670–1673.
49. Hiramatsu, K., H. Hanaki, T. Ino, K. Yabuta, T. Oguri, and F.C. Tenover. 1997. Methicillin-resistant *Staphylococcus aureus* clinical strain with reduced vancomycin susceptibility. *J. Antimicrob. Chemother.*, 40:135–136.
50. Ito, E. and J.L. Strominger. 1960. Enzymatic synthesis of the peptide in a uridine nucleotide from *Staphylococcus aureus*. *J. Biol. Chem.*, 235:PC5–PC6.
51. Ito, E. and J.L. Strominger. 1973. Enzymatic synthesis of the peptide in bacterial uridine nucleotides. VII. Comparative biochemistry. *J. Biol. Chem.*, 248:3131–3136.
52. Ji, Y., B. Zhang, S.F. Van, Horn, P. Warren, G. Woodnutt, M.K. Burnham, and M. Rosenberg. 2001. Identification of critical staphylococcal genes using conditional phenotypes generated by antisense RNA. *Science*, 293:2266–2269.
53. Kamiryo, T. and M. Matsuhashi. 1972. The biosynthesis of the cross-linking peptides in the cell wall peptidoglycan of *Staphylococcus aureus*. *J. Bacteriol.*, 247:6306–6311.
54. Katayama, Y., T. Ito, and K. Hiramatsu. 2000. A new class of genetic element, staphylococcus cassette chromosome mec, encodes methicillin resistance in *Staphylococcus aureus*. *Antimicrob. Agents Chemother.*, 44:1549–1555.
55. Kloos, W., K.-H. Schleifer, and F. Gotz. 1992. The genus *Staphylococcus*. In A. Balows, H.G. Truper, M. Dworkin, W. Harder, and K.-H. Schleifer (Eds.), *The Prokaryotes*. Springer-Verlag, New-York, p. 1369–1420.
56. Koch, H.U., R. Haas, and W. Fischer. 1984. The role of lipoteichoic acid biosynthesis in membrane lipid metabolism of growing *Staphylococcus aureus*. *Eur. J. Biochem.*, 138:357–363.
57. Kreiswirth, B.N., S. Lofdahl, M.J. Betley, M. O'Reilly, P.M. Schlievert, M.S. Bergdoll, and R.P. Novick. 1983. The toxic shock syndrome exotoxin structural gene is not detectably transmitted by a prophage. *Nature*, 305:709–712.

58. Kuehnert, M.J., D. Kruszon-Moran, H.A. Hill, G. McQuillan, S.K. McAllister, G. Fosheim, L.K. McDougal, J. Chaitram, B. Jensen, S.K. Fridkin, G. Killgore, and F.C. Tenover. 2006. Prevalence of *Staphylococcus aureus* nasal colonization in the United States, 2001–2002. *J. Infect. Dis.*, 193:172–179.
59. Kullik, I., R. Jenni, and B. Berger-Bachi. 1998. Sequence of the putative alanine racemase operon in *Staphylococcus aureus*: insertional interruption of this operon reduces D-alanine substitution of lipoteichoic acid and autolysis. *Gene*, 219:9–17.
60. Kuroda, M., T. Ohta, I. Uchiyama, T. Baba, H. Yuzawa, L. Kobayashi, L. Cui, A. Oguchi, K. Aoki, Y. Nagai, J. Lian, T. Ito, M. Kanamori, H. Matsumaru, A. Maruyama, H. Murakami, A. Hosoyama, Y. Mizutani-Ui, N. Takahashi, T. Sawano, R. Inoue, C. Kaito, K. Sekimizu, H. Hirakawa, S. Kuhara, S. Goto, J. Yabuzaki, M. Kanehisa, A. Yamashita, K. Oshima, K. Furuya, C. Yoshino, T. Shiba, M. Hattori, N. Ogasawara, H. Hayashi, and K. Hiramatsu. 2001. Whole genome sequencing of methicillin-resistant *Staphylococcus aureus*. *Lancet*, 357:1225–1240.
61. Labandeira-Rey, M., F. Couzon, S. Boisset, E.L. Brown, M. Bes, Y. Benito, E. M. Barbu, V. Vazquez, M. Hook, J. Etienne, F. Vandenesch, and M.G. Bowden. 2007. *Staphylococcus aureus* Panton-Valentine leukocidin causes necrotizing pneumonia. *Science*, 315:1130–1133.
62. Lee, C.Y., S.L. Buranen, and Z.-H. Ye. 1991. Construction of single-copy integration vectors for *Staphylococcus aureus*. *Gene*, 103:101–105.
63. Lee, C.Y. and J.J. Iandolo. 1986. Lysogenic conversion of staphylococcal lipase is caused by insertion of the bacteriophage L54a genome into the lipase structural gene. *J. Bacteriol.*, 166:385–391.
64. Lee, C.Y. and J.J. Iandolo. 1988. Structural analysis of staphylococcal bacteriophage phi 11 attachment sites. *J. Bacteriol.*, 170:2409–2411.
65. Lowy, F.D. 1998. *Staphylococcus aureus* infections. *N. Engl. J. Med.*, 339:520–532.
66. Maira-Litran, T., A. Kropec, D.A. Goldmann, and G.B. Pier. 2005. Comparative opsonic and protective activities of *Staphylococcus aureus* conjugate vaccines containing native or deacetylated staphylococcal Poly-N-acetyl-beta-(1-6)-glucosamine. *Infect. Immun.*, 73:6752–6762.
67. Marraffini, L.A., A.C. Dedent, and O. Schneewind. 2006. Sortases and the art of anchoring proteins to the envelopes of gram-positive bacteria. *Microbiol. Mol. Biol. Rev.*, 70:192–221.
68. Marshall, J.H., and G.J. Wilmoth. 1981. Proposed pathway of triterpenoid carotenoid biosynthesis in *Staphylococcus aureus*: evidence from a study of mutants. *J. Bacteriol.*, 147:914–919.
69. Mazmanian, S.K., G. Liu, H. Ton-That, and O. Schneewind. 1999. *Staphylococcus aureus* sortase, an enzyme that anchors surface proteins to the cell wall. *Science*, 285:760–763.
70. Mazmanian, S.K., E.P. Skaar, A.H. Gaspar, M. Humayun, P. Gornicki, J. Jelenska, A. Joachmiak, D.M. Missiakas, and O. Schneewind. 2003. Passage of heme-iron across the envelope of *Staphylococcus aureus*. *Science*, 299:906-909.
71. McElroy, M.C., H.R. Harty, G.E. Hosford, G.M. Boylan, J.F. Pittet, and T.J. Foster. 1999. Alpha-toxin damages the air-blood barrier of the lung in a rat model of *Staphylococcus aureus*-induced pneumonia. *Infect. Immun.*, 67:5541–5544.
72. Mei, J.M., F. Nourbakhsh, C.W. Ford, and D.W. Holden. 1997. Identification of *Staphylococcus aureus* virulence genes in a murine model of bacteraemia using signature-tagged mutagenesis. *Mol. Microbiol.*, 26:399–407.
73. Navarre, W.W. and O. Schneewind. 1994. Proteolytic cleavage and cell wall anchoring at the LPXTG motif of surface proteins in gram-positive bacteria. *Mol. Microbiol.*, 14:115-121.
74. Navarre, W.W. and O. Schneewind. 1999. Surface proteins of Gram-positive bacteria and the mechanisms of their targeting to the cell wall envelope. *Microbiol. Mol. Biol. Rev.*, 63:174–229.
75. Neu, H C. 1992. The crisis in antibiotic resistance. *Science*, 257:1064–1073.
76. Neubauer, H. and F. Gotz. 1996. Physiology and interaction of nitrate and nitrite reduction in *Staphylococcus carnosus*. *J. Bacteriol.*, 178:2005-2009.
77. Neuhaus, F C. and J. Baddiley. 2003. A continuum of anionic charge: structures and functions of D-alanyl-teichoic acids in Gram-positive bacteria. *Microbiol. Mol. Biol. Rev.*, 67:686–723.
78. Novick, R.P. 2003. Autoinduction and signal transduction in the regulation of staphylococcal virulence. *Mol. Microbiol.*, 48:1429–1449.
79. Novick, R.P. 1991. Genetic systems in staphylococci. *Methods Enzymol.*, 204:587–636.
80. Novick, R.P. 2003. Mobile genetic elements and bacterial toxinoses: the superantigen-encoding pathogenicity islands of *Staphylococcus aureus*. *Plasmid*, 49:93–105.
81. Novick, R.P. 1967. Properties of a cryptic high-frequency transducing phage in *Staphylococcus aureus*. *Virology*, 33:155–166.

82. Novick, R.P. and A. Subedi. 2007. The SaPIs: mobile pathogenicity islands of *Staphylococcus. Chem. Immunol. Allergy*, 93:42–57.
83. O'Connell, D., P.A. Pattee, and T.J. Foster. 1993. Sequence and mapping of the aroA gene of *Staphylococcus aureus* 8325-4. *J. Gen. Microbiol.,* 139:1449–1460.
84. O'Reilly, M., J.C. de Azavedo, S. Kennedy, and T.J. Foster. 1986. Inactivation of the alpha-haemolysin gene of *Staphylococcus aureus* 8325-4 by site-directed mutagenesis and studies on the expression of its haemolysins. *Microb. Pathog.*, 1:125–138.
85. O'Reilly, M., B.N. Kreiswirth, and T.J. Foster. 1990. Cryptic alpha-toxin gene in toxic shock syndrome and septicemia strains of *Staphylococcus aureus. Mol. Microbiol.*, 4:1947–1955.
86. Ogston, A. 1883. Micrococcus poisoning. J. *Anat. Physiol.*, 17:24–58.
87. Park, J.T. 1952. Uridine-5′-pyrophosphate derivatives. II. Isolation from *Staphylococcus aureus. J. Biol. Chem.*, 194:877–884.
88. Pelz, A., K.P. Wieland, K. Putzbach, P. Hentschel, K. Albert, and F. Gotz. 2005. Structure and biosynthesis of staphyloxanthin from *Staphylococcus aureus. J. Biol. Chem.*, 280:32493–32498.
89. Peng, H.L., R.P. Novick, B. Kreiswirth, J. Kornblum, and P. Schlievert. 1988. Cloning, characterization, and sequencing of an accessory gene regulator (agr) in *Staphylococcus aureus. J. Bacteriol.*, 170:4365–4372.
90. Perry, A.M., H. Ton-That, S.K. Mazmanian, and O. Schneewind. 2002. Anchoring of surface proteins to the cell wall of *Staphylococcus aureus.* III. Lipid II is an *in vivo* peptidoglycan substrate for sortase-catalyzed surface protein anchoring. *J. Biol. Chem.,* 277:16241–16248.
91. Pless, D.D. and F.C. Neuhaus. 1973. Initial membrane reaction in peptidoglycan synthesis. Lipid dependence of phospho-n-acetylmuramyl-pentapeptide translocase (exchange reaction). *J. Biol. Chem.*, 248:1568–1576.
92. Projan, S.J. 2003. Why is big Pharma getting out of antibacterial drug discovery?. *Curr. Opin. Microbiol.*, 6:427–430.
93. Reizer, J., C. Hoischen, F. Titgemeyer, C. Rivolta, R. Rabus, J. Stulke, D. Karamata, M.H. Saier, Jr., and W. Hillen. 1998. A novel protein kinase that controls carbon catabolite repression in bacteria. *Mol. Microbiol.*, 27:1157–1169.
94. Rogers, D. E. and M.A. Melly. 1965. Speculations on the immunology of staphylococcal infections. *Ann. N.Y. Acad. Sci.*, 128:274–284.
95. Rohrer, S. and B. Berger-Bachi. 2003. FemABX peptidyl transferases: a link between branched-chain cell wall peptide formation and beta-lactam resistance in Gram-positive cocci. *Antimicrob. Agents Chemother.*, 47:837–846.
96. Rohrer, S., K. Ehlert, M. Tschierske, H. Labischinski, and B. Berger-Bächi. 1999. The essential *Staphylococcus aureus* gene fmhB is involved in the first step of peptidoglycan pentaglycine interpeptide formation. *Proc. Natl. Acad. Sci. U.S.A.*, 96:9351–9356.
97. Rose, F., G. Dahlem, B. Guthmann, F. Grimminger, U. Maus, J. Hanze, N. Duemmer, U. Grandel, W. Seeger, and H.A. Ghofrani. 2002. Mediator generation and signaling events in alveolar epithelial cells attacked by *S. aureus* alpha-toxin. *Am. J. Physiol. Lung Cell Mol. Physiol.*, 282:L207–L214.
98. Rosenbach, F.J. 1884. Mikroorganismen bei den Wundinfections-Krankheiten des Menschen.
99. Ruzin, A., J. Lindsay, and R.P. Novick. 2001. Molecular genetics of SaPI1 — a mobile pathogenicity island in *Staphylococcus aureus. Mol. Microbiol.,* 41:365–377.
100. Schleifer, K.-H., A. Hartinger, and F. Götz. 1978. Occurrence of D-tagatose-6-phosphate pathway of D-galactose metabolism among staphylococci. *FEMS Microbiol. Lett.*, 3:9–11.
101. Schleifer, K.H. and W.E. Kloos. 1975. A simple test system for the separation of staphylococci from micrococci. *J. Clin. Microbiol.*, 1:337–8.
102. Schneewind, O., A. Fowler, and K.F. Faull. 1995. Structure of the cell wall anchor of surface proteins in *Staphylococcus aureus. Science*, 268:103–106.
103. Schneewind, O., P. Model, and V.A. Fischetti. 1992. Sorting of protein A to the staphylococcal cell wall. *Cell*, 70:267–281.
104. Schwan, W.R., S.N. Coulter, E.Y. Ng, M.H. Lnghorne, H.D. Ritchie, L.L. Brody, S. Westbrock-Wadman, A.S. Bayer, K.R. Folger, and C.K. Stover. 1998. Identification and characterization of the PutP proline permease that contributes to *in vivo* survival of *Staphylococcus aureus* in animal models. *Infect. Immun.,* 66:567–572.
105. Seeger, W., M. Bauer, and S. Bhakdi. 1984. Staphylococcal alpha-toxin elicits hypertension in isolated rabbit lungs. Evidence for thromboxane formation and the role of extracellular calcium. *J. Clin. Invest.*, 74:849–858.

106. Seeger, W., R.G. Birkemeyer, L. Ermert, N. Suttorp, S. Bhakdi, and H.R. Duncker. 1990. Staphylococcal alpha-toxin-induced vascular leakage in isolated perfused rabbit lungs. *Lab. Invest.*, 63:341–349.
107. Sjöquist, J., B. Meloun, and H. Hjelm. 1972. Protein A isolated from *Staphylococcus aureus* after digestion with lysostaphin. *Eur. J. Biochem.*, 29:572–578.
108. Sjöquist, J., J. Movitz, I.-B. Johansson, and H. Hjelm. 1972. Localization of protein A in the bacteria. *Eur. J. Biochem.*, 30:190–194.
109. Song, L., M.R. Hobaugh, C. Shustak, S. Cheley, H. Bayley, and J.E. Gouaux. 1996. Structure of staphylococcal alpha-hemolysin, a heptameric transmembrane pore. *Science*, 274:1859–1866.
110. Stranger-Jones, Y.K., T. Bae, and O. Schneewind. 2006. Vaccine assembly from surface proteins of *Staphylococcus aureus*. *Proc. Natl. Acad. Sci. U.S.A.*, 103:16942–16947.
111. Strasters, K.C. and K C. Winkler. 1963. Carbohydrate metabolism of *Staphylococcus aureus*. *J. Gen. Microbiol.*, 33:213–229.
112. Sugai, M., T. Fujiwara, T. Akiyama, M. Ohara, H. Komatsuzawa, S. Inoue, and H. Suginaka. 1997. Purification and molecular characterization of glycylglycine endopeptidase produced by *Staphylococcus capitis EPK1*. *J. Bacteriol.*, 179:1193–1202.
113. Sugai, M., T. Fujiwara, K. Ohta, H. Komatsuzawa, M. Ohara, and H. Suginaka. 1997. epr, which encodes glycylglycine endopeptidase resistance, is homologous to femAB and affects serine content of peptidoglycan cross bridges in *Staphylococcus capitis* and *Staphylococcus aureus*. *J. Bacteriol.*, 179:4311–4318.
114. Suttorp, N., W. Seeger, E. Dewein, S. Bhakdi, and L. Roka. 1985. Staphylococcal alpha-toxin-induced PGI2 production in endothelial cells: role of calcium. *Am. J. Physiol.*, 248:C127–C134.
115. Tam, C., E.M. Glass, D.M. Anderson, and D. Missiakas. 2006. Transposon mutagenesis of *Bacillus anthracis* strain Sterne using *Bursa aurealis*. *Plasmid*, 56:74–77.
116. Tenover, F.C., J.W. Biddle, and M.V. Lancaster. 2001. Increasing resistance to vancomycin and other glycopeptides in *Staphylococcus aureus*. *Emerg. Infect. Dis.*, 7:327–332.
117. Thumm, G. and F. Gotz. 1997. Studies on prolysostaphin processing and characterization of the lysostaphin immunity factor (Lif) of *Staphylococcus simulans* biovar *staphylolyticus*. *Mol. Microbiol.*, 23:1251–1265.
118. Tipper, D.J. and M.F. Berman. 1969. Structures of the cell wall peptidoglycans of Staphylococcus epidermidis Texas 26 and *Staphylococcus aureus* Copenhagen. I. Chain length and average sequence of cross-bridge peptides. *Biochemistry*, 8:2183–2191.
119. Tipper, D.J., W. Katz, J.L. Strominger, and J.M. Ghuysen. 1967. Substituents on the alpha-carboxyl group of D-glutamic acid in the peptidoglycan of several bacterial cell walls. *Biochemistry*, 6:921–929.
120. Tipper, D.J. and J.L. Strominger. 1968. Biosynthesis of the peptidoglycan of bacterial cell walls. XII. Inhibition of cross-linking by penicillins and cephalosporins: studies in *Staphylococcus aureus in vivo*. *J. Biol. Chem.*, 243:3169–3179.
121. Tipper, D.J., M. Tomoeda, and J.L. Strominger. 1971. Isolation and characterization of 1,4-N-acetylmuramyl-N-acetylglucosamine and its O-acetyl derivative. *Biochemistry*, 10:4683–4690.
122. van Wamel, W.J., S.H. Rooijakkers, M. Ruyken, K.P. van Kessel, and J.A. van Strijp. 2006. The innate immune modulators staphylococcal complement inhibitor and chemotaxis inhibitory protein of *Staphylococcus aureus* are located on beta-hemolysin-converting bacteriophages. *J. Bacteriol.*, 188:1310–1315.
123. Voyich, J.M., M. Otto, B. Mathema, K.R. Braughton, A.R. Whitney, D. Welty, R.D. Long, D.W. Dorward, D.J. Gardner, G. Lina, B.N. Kreiswirth, and F.R. DeLeo. 2006. Is Panton-Valentine leukocidin the major virulence determinant in community-associated methicillin-resistant *Staphylococcus aureus* disease?, *J. Infect. Dis.*, 194:1761–1770.
124. Waldron, D.E., and J.A. Lindsay. 2006. Sau1: a novel lineage-specific type I restriction-modification system that blocks horizontal gene transfer into *Staphylococcus aureus* and between *S. aureus* isolates of different lineages. *J. Bacteriol.*, 188:5578–5585.
125. Walsh, C.T. 1993. Vancomycin resistance: decoding the molecular logic. *Science*, 261:308–309.
126. Ward, J.B. 1973. The chain length of the glycans in bacterial cell walls. *Biochem. J.*, 133:395–398.
127. Ward, J.B. and H.R. Perkins. 1973. The direction of glycan synthesis in a bacterial peptidoglycan. *Biochem. J.*, 135:721–728.
128. Warren, G.H. and J. Gray. 1965. Effect of sublethal concentrations of penicillins on the lysis of bacteria by lysozyme and trypsin. *Proc. Soc. Exp. Biol. Med.*, 120:504–511.
129. Weidenmaier, C., J. F. Kokai-Kun, S.A. Kristian, T. Chanturiya, H. Kalbacher, M. Gross, G. Nicholson, B. Neumeister, J.J. Mond, and A. Peschel. 2004. Role of teichoic acids in *Staphylococcus aureus* nasal colonization, a major risk factor in nosocomial infections. *Nat. Med.*, 10:243–245.

130. Weigel, L.M., D.B. Clewell, S.R. Gill, N.C. Clark, L.K. McDougal, S.E. Flannagan, J.F. Kolonay, J. Shetty, G.E. Killgore, and F.C. Tenover. 2003. Genetic analysis of a high-level vancomycin-resistant isolate of *Staphylococcus aureus*. *Science*, 302:1569–1571.
131. Wu, S.W., H. de Lencastre, and A. Tomasz. 2001. Recruitment of the mecA gene homologue of *Staphylococcus sciuri* into a resistance determinant and expression of the resistant phenotype in *Staphylococcus aureus*. *J. Bacteriol.*, 183:2417–2424.

23 The Genus *Streptococcus*

Vincent A. Fischetti and Patricia Ryan

CONTENTS

INTRODUCTION

Streptococci will appear under the microscope as round bacteria arranged in pairs or in chains. By the Gram-staining technique, they will be Gram-positive. The chaining characteristic is best observed when organisms are grown in liquid media or isolated from infected body fluids such as blood. Streptococci are differentiated as α, β, and γ types, based on their activity on the surface of blood agar, which could differ somewhat based on the species and age of the red blood cells. α-Hemolytic streptococcal colonies are surrounded by a narrow zone of hemolysis that shows green discoloration based on the hemolysin's action on the hemoglobin; β-hemolytic streptococci show a well-defined clear zone of hemolysis around the colony; while γ-hemolytic streptococci have no effect on the red blood cells. *Streptococcus pyogenes* (or group A) are nearly always β-hemolytic, whereas closely related groups B and C streptococci usually appear as β-hemolytic colonies; however, different strains can vary in their hemolytic activity. Nearly all strains of *S. pneumoniae* are α-hemolytic but have been shown to exhibit β-hemolysis during anaerobic incubation. Most oral streptococci and enterococci are non-hemolytic, and thus considered γ types. The property of hemolysis is used in rapid screens for identification of *S. pyogenes* and *S. pneumoniae,* but may be unreliable for general differentiation of other streptococci.

HEMOLYTIC STREPTOCOCCI

Hemolysins are considered virulence determinants, and as such, hemolytic streptococci are found to be human and animal pathogens. Most clinical isolates appear as β-hemolytic streptococci when grown on blood agar. *Streptococcus pyogenes,* groups B and C streptococci, and *Streptococcus pneumoniae* (described below in more detail) are the most prevalent streptococcal pathogens isolated from humans. The human oropharynx is the sole known natural reservoir for *S. pyogenes* in the environment. These organisms are maintained in the throats of humans (usually children) in a "carrier state," with up to a 30% frequency.[56] Group B streptococci are predominantly found colonizing the human vagina, but are also associated with animal infections. Although resistance to penicillin-based antibiotics has yet to emerge in *S. pyogenes*, infections caused by this organism continue to occur with occasional widespread outbreaks.

CLASSIFICATION

The classification of hemolytic streptococci was considerably simplified when Lancefield showed that surface antigens (carbohydrates) extracted from the cell wall of streptococci could react with carbohydrate-specific antisera prepared in rabbits. Lancefield's group A, for example, are the *Streptococcus pyogenes* strains responsible for human diseases such as scarlet fever, streptococcal pharyngitis, erysipelas, puerperal sepsis, and wound infections. N-acetylglucosamine is the group A-specific carbohydrate, whereas N-acetylgalactosamine is the group C determinant. Extended studies of the cell wall carbohydrates of streptococci from several sources have shown at least 13 different serologic groups encompassing human and animal pathogens and commensals.

The group A streptococci can be subdivided into more than 120 different types based on a variable surface protein, called the M protein. M protein was used as a serologic determinant for early classification of group A strains, but recently the sequence of the variable region of the M protein gene has been used instead with identical results. M typing is used for epidemiologic purposes in studying the spread of streptococcal disease.

Capsular serotyping has been used to differentiate different strains of group B streptococci. To date, nine capsular serotypes have been described (Ia, Ib, II, III, IV, V, VI, VII, and VIII). Worldwide, serotype III strains are of particular importance because they are responsible for the majority of infections, including neonatal meningitis.

STREPTOCOCCUS PYOGENES EXTRACELLULAR PRODUCTS

Streptococcus pyogenes produce a wide array of extracellular products, many of which are considered virulence factors (reviewed in Ref. 2).

Streptococcal pyrogenic exotoxins (Spes) are produced by most *Streptococcus pyogenes* strains. The most potent, SpeA, is responsible for the rash in cases of scarlet fever, and its gene is carried by a prophage.[65] Currently, nearly ten different Spes have been reported. In all cases, Spes are superantigens, or mitogenic proteins with the capacity to crosslink the V_b domain of the T-cell receptor and the major histocompatibility complex of class II molecules on the surface of an antigen-presenting cell. This T-cell stimulation results in high systemic levels of proinflammatory cytokines and T-cell mediators, causing hypotension, fever, and shock. As a result, Spes have been suggested to be the prime mediators in streptococcal toxic shock syndrome. Interestingly, except for SpeB, Spes are not part of the bacterial genome but are carried on prophage widely found in streptococci.

Streptolysin O (SLO) is an oxygen-labile protein with the capacity to lyse red blood cells under anaerobic conditions. It is highly antigenic, such that the antibodies produced during infection can be quantified in relation to their capacity to neutralize the hemolytic activity of SLO. It has been shown that the titer of antibodies to streptolysin O (ASO) is related to a recent *Streptococcus pyogenes* infection. Streptolysin S (SLS), on the other hand, is stable in air and is the molecule responsible for the hemolytic zone around streptococcal colonies on blood plates. SLS is a non-antigenic

2.8-kDa peptide that is tightly associated with the bacterial cell surface with potent toxic effects on experimental animals *in vivo* and on leukocytes *in vitro*.[1]

Streptokinase is a secreted molecule with no intrinsic enzymatic activity; however, when bound to plasminogen, it initiates its conversion to plasmin, the active protease with fibrinolytic activity. Because of this activity, streptokinase has had some use in human medicine as an agent to lyse fibrin clots in coronary arterial thrombosis, acute pulmonary embolism, and deep venous thrombosis. However, because of its antigenicity, it has experienced limited use.

Hyaluronidase is an enzyme that acts on hyaluronic acid, the ground substance in connective tissue. Because of this activity, hyaluronidase was called the "spreading factor" for these streptococci. Surprisingly, the identical hyaluronic acid is also the composition of the capsule that is found on many strains of *Streptococcus pyogenes,* a capsule that enables the organism to resist phagocytic attack by human leukocytes.[2]

Streptococcus pyogenes also produces four enzymes that are able to cleave nucleic acids. These DNases, sometimes referred to as streptodornase A through D, all possess deoxyribonuclease activity while streptodornases B and D also possess ribonuclease activity. It has been suggested that by digesting the DNA released from dead mammalian cells at an infected site, the enzyme reduces the local viscosity, allowing the organism greater mobility.

Surface Proteins on *Streptococcus pyogenes*

In addition to secreted molecules, *Streptococcus pyogenes* express a variety of molecules that are displayed on the cell surface, which are critical for colonization and infectivity. Among these are M protein, IgA binding protein, IgG binding protein, C5a peptidase, serum opacity factor, fibronectin binding protein, plasmin binding protein, and five glycolytic enzymes (glyceraldehyde-3-phosphate dehydrogenase, α-enolase, phosphoglycerate mutase, phosphoglycerate kinase, and triose phosphate isomerase). All these proteins have been implicated in some way with streptococcal pathogenesis; however, we discuss the major virulence determinant, M protein, in detail.

M Protein

The streptococcal M protein is probably one of the best-defined molecules among the known bacterial virulence determinants. It was discovered more than 70 years ago by Rebecca Lancefield.[39] It is clear that protective immunity to group A streptococcal infection is achieved through antibodies directed to the M protein.[18] The A- and B-repeats located within the N-terminal half are antigenically variable among the more than 120 known streptococcal types with the N-terminal non-repetitive region and A-repeats exhibiting hypervariability. The C-terminal C-repeats, the majority of which are surface exposed, contain epitopes that are highly conserved among the identified M proteins.[35] Due to its antigenically variable N-terminal region, the M protein provides the basis for the Lancefield serological typing scheme for group A streptococci.[18]

The M protein is considered the major virulence determinant because of its ability to prevent phagocytosis when present on the streptococcal surface and thus, by this definition, all clinical isolates express M protein. This function may be attributed in part to the specific binding of complement factor H to both the conserved C-repeat domain[19] and the fibrinogen bound to the B-repeats,[27] preventing the deposition of C3b on the streptococcal surface. It is proposed that when the streptococcus contacts serum, the factor H bound to the M molecule inhibits or reverses the formation of C3b,Bb complexes and helps convert C3b to its inactive form (iC3b) on the bacterial surface, preventing C3b-dependent phagocytosis. This is supported by studies showing that antibodies directed to the B- and C-repeat regions of the M protein are unable to promote phagocytosis.[34] This may be the result of the ability of factor H to also control the binding of C3b to the Fc receptors on these antibodies, resulting in inefficient phagocytosis.[14] Antibodies directed to the hypervariable N-terminal region are opsonic, perhaps because they cannot be controlled by the factor H bound to the

B- and C-repeat regions. Thus, it appears that the streptococcus has devised a method to protect its conserved region from being used against itself by binding factor H to regulate the potentially opsonic antibodies that bind to these regions. Both the N-terminal hypervariable region and the conserved region are targets for vaccine development.

Streptococcal Diseases

Streptococcus pyogenes is responsible for a number of diseases, including pharyngitis, otitis media, impetigo, meningitis, necrotizing fasciitis, scarlet fever, erysipelas, rheumatic fever, and acute glomerulonephritis, to name a few.[7] The latter two, rheumatic fever and acute glomerulonephritis, are sequelae of an *S. pyogenes* infection. Although a great deal of research has focused on these diseases, little is known regarding their etiology except that an *S. pyogenes* infection preceded the disease, suggesting that an immune component may be responsible for the observed symptoms.

The latent period for rheumatic fever is about 3 weeks after a streptococcal pharyngitis.[6] Curiously, none of the other *Streptococcus pyogenes* infections culminates in this disease, which ultimately results in damage to the mitral heart valve. It has been proven that aggressive treatment of the streptococcal pharyngitis with penicillin within 10 days of its onset will prevent the cardiac damage. If untreated, recurrent rheumatic attacks following streptococcal infections may occur, resulting in further cardiac damage and ultimately necessitating valvular surgery.

Acute glomerulonephritis follows a *Streptococcus pyogenes* infection by about 1 week, resulting in the malfunction of the glomeruli in the kidneys. However, in this disease, the patients usually undergo spontaneous recovery. Treatment is usually managing the reduced kidney function until the condition resolves itself.

In recent years, streptococcal toxic shock syndrome (STSS) has been a concern worldwide. It is a severe illness associated with invasive or noninvasive *Streptococcus pyogenes* infection. The disease may result from an infection at any site but most often in association with infection of a cutaneous lesion. In some cases, a traumatic injury without any external breaks has been the focus, suggesting that organisms circulating in the blood of carriers could be the cause. Toxicity and a rapidly progressive clinical course (24 to 48 hr) are characteristic, having a case-fatality rate of greater than 50%. In most instances, patients present with shock-like symptoms and a necrotic lesion or painful abscess. Treatment is usually management of the shock, debridement of the lesion, and antibiotic. In some severe cases, amputation may be necessary.

About 100 years ago, scarlet fever was considered a deadly disease, and a severe complication of a streptococcal infection. Today, perhaps due to changes in the bacteria and treatment options, it is simply a streptococcal pharyngitis accompanied by a rash. Erysipelas (cellulitis with fever and toxicity) is also less common today, again perhaps due to changes in the bacteria, the host, and the availability of antibiotics.

Neonatal meningitis is a disease that has been increasing in the past few decades. This disease, caused by group B streptococci (GBS), results from the contamination of newborns of vaginally colonized mothers during natural childbirth. The contaminated newborn becomes infected (through the nasal and oral route), resulting in neurological damage from bacterial localization in the brain. Identification of colonized mothers prior to delivery and antibiotic treatment to remove the GBS and/or antibiotic treatment of the newborn immediately after delivery has helped control the disease.

Laboratory Diagnosis

The hemolytic reaction and colonial morphology exhibited on blood agar has been used as the standard to classify streptococci. *Streptococcus pyogenes* is associated with β-hemolysis, or complete lysis of blood cells around the colony. *S. pyogenes* (or group A streptococci) are nearly always β-hemolytic as opposed to group B streptococci, which can exhibit α-, β-, or γ-hemolysis. Serological identification is the most accurate method of streptococcal identification; however, these reagents are for research purposes and are unavailable for routine use. Because *S. pyogenes* are sensitive to bacitracin, and most other β-hemolytic streptococci are not, using a bacitracin disk on a blood agar plate can be used for identification.

Rapid tests: There are a number of rapid tests for *Streptococcus pyogenes* in use today that rely on the group-specific carbohydrate found in the cell wall of this organism.[40] These tests, which are routinely employed in doctors' offices and hospitals, use acid to extract the carbohydrate determinant (N-acteylglucosamine) from the streptococci on a swab and antibodies to that sugar for positive identification. Because *S. pyogenes* is the only streptococcus that has this determinant, most of these rapid tests are greater than 95% specific and about 90% sensitive for these bacteria.

Other β-Hemolytic Streptococci

Groups B, C, and G Streptococci: Group C and G streptococci are predominantly isolated from animals, with the exception of *Streptococcus dygalactiae*, which is found in humans. These organisms (both human and animal species) have been isolated from severe systemic infections in humans that are usually associated with food-borne contamination (primarily from milk and cheese). More commonly, however, groups C and G streptococci may be found colonizing the human nasopharynx and have been associated with outbreaks of acute pharyngitis. However, because of the specificity of antigen-specific tests for the group A streptococci, group C and G streptococci colonizing and infecting the nasopharynx are usually missed. Recently, group G streptococci have been identified as an important opportunistic and nosocomial pathogen. In certain tropical climates (e.g., the Caribbean, and northern territory of Australia), group G as well as group C streptococci have been associated with cases of rheumatic fever, a disease only related to group A streptococcal pharyngitis.[43] Because both groups C and G streptococci have surface M protein, they are able to be subtyped based on the differences in the molecule.

OTHER STREPTOCOCCI

Viridans Streptococci

The viridans streptococci are a heterogeneous group that includes both α-hemolytic and non-hemolytic streptococci. Many species are components of the normal flora of humans and animals; however, the members of this group are capable of causing human disease if given access to normally sterile sites in the body. These opportunistic pathogens are primarily associated with the formation of dental caries (*Streptococcus mutans)*, infective endocarditis (*S. sanguis*), and bone, joint, and liver abcesses (*S. anginosis*).[10]

Viridans streptococci include at least 26 species that belong to one of five general groups: *Streptococcus mutans, S. salivarius, S. sanguis, S. mitis*, or *S. anginosus*.[17] The members of the viridans group are historically considered difficult to differentiate on the species level; and because they do not contain C-carbohydrate antigens in their cell walls, they cannot be serologically identified using the traditional methods described for β-hemolytic streptococci. It is generally accepted that the viridans group is first differentiated from other streptococci by exclusion: all members are bile insoluble, non-β-hemolytic, pyrrolidonylarylamidase (PYR) negative, cannot grow in 6.5% NaCl, and with the exception of the salivarius group, are bile-esculin negative.

The differentiation of the viridans streptococci into groups or species requires a combination of molecular and biochemical tools. A number of species-specific biochemical characteristics may prove useful. For example, members of the *Streptococcus mutans* species/group produce acid in nearly all carbohydrate broths, and the strains belonging to the *S. salivarius* group are the only viridans to hydrolyze urea. The *S. mitis* group members are the only viridans that are negative for all of the following biochemical tests: they do not ferment mannitol or sorbitol, they do not hydrolyze arginine or urea, and they are negative in the Voges-Proskauer and bile-esculin tests. The species within each of the major groupings can be further differentiated by additional biochemical characteristics.[15, 17]

The clinical significance of viridans group members is generally species dependent and, as such, accurate species differentiation is the first step in properly diagnosing the etiologic agent of an infection in order to prescribe the proper treatment course. Commercially available kits for biochemical identification of streptococci are available, such as the API 20 Strep kit and Rapid ID 32 Strep kit (both from bioMérieux, Hazelwood, MO). These tests, however, are generally more reliable for the non-viridans streptococci, and do not always provide accurate results for the identification of individual viridans group species.[23] In addition to biochemical analysis, several molecular assays have been developed for viridans identification, including PCR-based techniques, ribotyping, and sequence comparisons of various target genes, including 16s rRNA,[3, 25] manganese-dependent superoxide dismutase, and *groESL* genes.[57]

Various automated systems are now used to identify clinical isolates of the viridans group, including VITEK 2 (bioMérieux) and Phoenix systems (Becton Dickinson Diagnostic Systems, Sparks, MD). The VITEK system however, has recently been reported[22] to often be unreliable for the viridans group (particularly *Streptococcus mitis* and *S. sanguis*),[25] with only about 40% of clinical isolates correctly identified at the species level. The performance of the Phoenix system for identification of *Streptococci* and *Enterococci* was recently evaluated for reproducibility and reliability in comparison to the API system.[4] Phoenix correctly identified 67% of clinical isolates of viridans streptococci at the species level, with most accurate results obtained for *S. anginosus* (100%) and *S. sanguis* groups (75%), and the least correct results for the *S. mitis* group (53%). The discrepancies resulting from all automated systems, however, can generally be resolved with further testing using molecular techniques or other biochemical tests.

Streptococcus pneumoniae

Streptococcus pneumoniae are Gram-positive cocci often arranged in pairs (diplococci) or in short chains that commonly asymptomatically colonize the human nasopharynx. Like other streptococci, the pneumococci are non-motile, catalase-negative, facultative anaerobes that ferment glucose to lactic acid. *S. pneumoniae* produce a large polysaccharide capsule, which is considered one of the major virulence factors (reviewed in Ref. 41), as it confers antiphagocytic properties on the organism, and was recently shown to limit mucus-mediated clearance of the organism from human mucosal surfaces.[50] The composition of the polysaccharide capsule forms the basis for pneumococcal serotyping by the Neufeld Quelleng reaction, which involves agglutination of the organism with type-specific capsular antibodies.[55] To date, at least 90 serotypes have been identified.[41]

Pneumococci (and other streptococci in the mitis group[38]) contain phosphorylcholine that is covalently linked to teichoic acid in the cell wall and to the lipoteichoic acid component of the cell membrane;[60] thus, choline is an absolute growth requirement of the organism. A group of surface-exposed choline-binding proteins (as many as 15 in some strains[58]) interacts non-covalently with the choline residues in the cell wall, and exhibits a number of different functions essential to the overall biology of the organism, including host tissue adherence and autolysis.[24]

Pneumococcal Diseases

Streptococcus pneumoniae cause a wide variety of illnesses, including lobar pneumonia, meningitis, otitis media, bacteremia, and sinusitis. The elderly and children under the age of 2 years carry the major disease burden, and World Health Organization (http://www.who.int/vaccine_research/diseases/ari/en/index5.html) estimates from 2003 indicate that nearly a million children (most of whom are from underdeveloped nations) die each year of pneumococcal infections. In the years prior to 2000, the Centers for Disease Control and Prevention (www.cdc.gov) estimated that *S.*

pneumoniae caused over 100,000 cases of pneumonia requiring hospitalization, 6 to 7 million cases of otitis media, and 60,000 cases of invasive disease in the United States alone.

Pneumococci produce a number of virulence determinants that contribute to the overall pathogenicity of the organism, and the anti-phagocytic polysaccharide capsule has long been considered one of the most important. There are, however, numerous cell-surface associated and secreted proteins involved in various disease processes, including (but not limited to) hyaluronate lyase, pneumolysin, choline binding proteins, two neuraminidases, autolysin LytA, pneumococcal surface antigen PsaA, and pneumococcal surface proteins PspA and PspC (reviewed in Ref. 20 and Ref. 32). Also contributing to virulence is the ability of pneumococci to undergo spontaneous phase variation between an opaque and a transparent colony phenotype. Although the mechanisms of this process are still not fully understood, opaque variants have more capsular polysaccharide than the transparent phenotype and are more commonly associated with invasive infections. The transparent phenotype is generally better able to colonize the nasopharynx than the opaque variants.[62, 63]

The pathogenic program of pneumococci is a complicated, multifaceted process that depends on the interaction between host and bacterium, as well as the genetic and physiological characteristics of both. Recent findings[11, 53] indicate that the capsular types, as well as the overall genetic background of the pneumococci located at specific infection sites, contribute to the degree to which particular isolates are virulent. Furthermore, a relationship has been shown to exist between capsular type and risk for, and outcome of, invasive pneumococcal disease.

Identification

Streptococcus pneumoniae are fragile organisms, lysing within 18 to 24 hr after the start of growth due to the production of autolysin, a cell-wall degrading lytic enzyme.[21, 46] Pneumococci are fastidious, growing best in 5% CO_2 on agar containing blood as a source of catalase. On blood agar, colonies are round and about 1 mm in diameter. Colonies initially appear raised but, as incubation continues, become flattened and shiny (or mucoid) in appearance with depressed centers (resulting from the action of autolysin).[36] In 5% CO_2, colonies are surrounded by large zones of α-hemolysis (resulting from the production of pneumolysin), a characteristic that helps to distinguish it from the β-hemolytic group A streptococci.

Further tests are necessary to distinguish *Streptococcus pneumoniae* from the α-hemolytic viridans streptococci and classically include differentiation based on colony morphology, bile (deoxycholate) solubility,[28] and optochin (cuprein hydrochloride) sensitivity.[61]

The latter two tests are still considered generally reliable for initial identification, with reported sensitivities of greater than 98% and 90–100%, respectively.[36] The bile solubility test is based on the observation that broth cultures of *Streptococcus pneumoniae* lyse (solubilize) when treated with a 2% solution of the bile acid salt, sodium deoxycholate. This assay remains one of the most sensitive and specific tests for the identification of pneumococci.[36] Optochin (ethylhydrocupreine hydrochloride) sensitivity is determined using disks impregnated with the compound that are placed on a freshly streaked agar plate of bacteria. After incubation, a zone of inhibition (greater than 14 mm) surrounding the disk indicates that the organism is a pneumococcus.[17] Growth of other α-hemolytic streptococci in the area surrounding the disk will not be inhibited. For cases in which discrepancies in either test occur, other identification methods should be included, as there have been reports of optochin-susceptible viridans group streptococci as well as bile-insoluble *S. pneumoniae*.[8] The traditional Quelleng (capsular swelling) reaction for identification is cumbersome and labor-intensive, and has been generally restricted to specialized laboratories.

Numerous diagnostic tests, identification, and typing methods are now available for *Streptococcus pneumoniae* (reviewed in Ref. 20) that facilitate rapid identification of pneumococci. An overview of a number of such tests is provided here.

Detection of the polysaccharide capsule by latex agglutination forms the basis of a number of the currently available serological tests, including Pneumoslide (Becton-Dickinson Diagnostic Systems, Sparks, MD), Directigen (Becton-Dickinson), and Pneumotest-Latex (Statens Serum Institut, Copenhagen, Denmark). However, many false positives involving α- and non-hemolytic streptococci have been reported with both Pneumoslide and Directigen.[36] In contrast, the Pneumo test-Latex method has recently been shown to detect 90 different pneumococcal serotypes with 95% accuracy).[54] Serological tests that detect capsular antigens by co-agglutination techniques are also available, including Phadebact (Bactus AB, Huddinge, Sweden), which, in independent tests, correctly identified 98% of pneumococcal isolates.[36] Obviously, the major limiting factor for all these serological tests is that they can only be used on encapsulated isolates.

A number of diagnostic methods have been developed that are based on specific virulence factors of *Streptococcus pneumoniae*. Currently, pneumolysin-derived peptides can be detected immunologically (ELISA and agglutination) and genetically (nested-PCR and real-time PCR). PCR-based assays targeting autolysin are also available. These methods (reviewed in Ref. 20) have facilitated the identification of the organism from cultured specimens and clinical samples such as urine, sputum, and pleural fluid.

For clinical cases in which it is difficult to obtain isolates of *Streptococcus pneumoniae* (low recovery after antibiotic treatment, negative subcultures, sampling procedures that are too invasive), an immunochromatographic membrane assay (Binax NOW, Binax, Portland, ME) is available that, with great rapidity, sensitivity, and specificity (80 to 95%), detects the presence of the C polysaccharide cell wall antigen (common to all serotypes) in urine,[5, 48] blood culture bottles,[51] bronchoalveolar lavage fluid,[31] and cerebrospinal fluid.[42]

Treatment and Vaccines

For most hospital and outpatient settings, penicillin, erythromycin, and tetracycline have been recommended for the treatment of pneumococcal infections; however, the CDC currently estimates that nearly a third of pneumococcal strains from patients with pneumonia are resistant to common antibiotics (www.cdc.gov/drpsurveillancetoolkit/docs/PneumococcalDisease.pdf and www.cdc.gov/ncidod/dbmd/abcs/survreports/spneu05). In 2005, the data from the Active Bacterial Core Surveillance project at the CDC revealed that 25% of *S. pneumoniae* strains are resistant to penicillin and 20% are resistant to erythromycin. The prevalence of resistance to tetracycline is reportedly 9% and to trimethoprim-sulfamethoxazole is as great as 15%.

There are currently two pneumococcal vaccines, both based on capsular polysaccharides, that are licensed for use in the United States: (1) a 23-valent polysaccharide vaccine recommended for adults age 65 or older, and (2) a heptavalent protein-polysaccharide conjugate vaccine for the prevention of systemic infections in children. Unfortunately, the 23-valent vaccine is not immunogenic in children under 2 years of age and is underused in the target adult population; however, since the introduction of the heptavalent vaccine in 2001, the CDC reports a decline in invasive disease caused by the serotypes represented in the vaccine.

Despite the introduction of the conjugate vaccine and the decrease in invasive illness in the developed world, the WHO warns that the heptavalent vaccine is not cost-effective for use in poorer nations and is designed to protect against invasive serotypes that are more prevalent in industrialized nations than in less-developed countries. Furthermore, there is new concern about the long-term efficacy of the conjugate vaccine, as increased carriage of non-vaccine serotypes is already being reported.[26]

Streptococcus pneumoniae is a naturally transformable organism, and as such, during a physiological state known as competence,[9, 45, 59] the bacterium can readily acquire exogenous DNA, thereby increasing its genetic plasticity. This trait is one of the underlying factors that contributes to capsular switching, increased antibiotic resistance, and virulence factor exchange.[33] Furthermore, pneumococci harbor bacteriophages, and recent reports indicate that at least 70%, and as many as 90%, of the pneumococcal genomes from clinical isolates contain prophages or phage remnants. Therefore, through both transformation and transduction, pneumococci are able to readily exchange and acquire genetic material, including novel capsular genes, antibiotic resistance markers, and encoded virulence determinants — factors that may potentially compound our ability to treat and to prevent infections caused by this organism.

Enterococci

Members of the genus *Enterococcus* are catalase-negative, Gram-positive cocci that appear in short chains, in pairs, or as single cocci. Enterococci were included in genus *Streptococcus* until 1984, when Schleifer and Kilpper-Balz[57] used DNA-DNA hybridization and DNA-rRNA hybridization techniques to determine that these two groups were too distinct and distant to be classified in the same genus. Most strains react with Lancefield group D typing serum, grow in 6.5% NaCl, and produce pyrrolidonylarylamidase. Colonies on agar plates are generally larger than streptococci, growing to 2 to 3 mm in diameter within 2 days, and appear raised and gray/white in color.

Enterococci share a number of characteristics with other Gram-positive cocci and, as such, a battery of biochemical tests (reviewed in Ref. 49) can be performed to differentiate enterococcal isolates from less commonly encountered Gram-positive cocci. Examples of such tests include growth on bile-esculin media (most enterococci are positive), production of gas from glucose (most enterococci are negative), and the ability to grow at both 45°C and 10°C (enterococci generally grow at both temperatures).

Enterococci are normal components of the flora of the intestinal tract, oral cavity, and vaginal canal of humans and animals, and have historically been considered commensals with low pathogenic potential. The past two decades, however, have seen an increase in enterococcal infections, including those of the urinary tract, burn wounds, surgical incisions, and heart valves. Enterococci colonize catheters and implanted medical devices, which often results in endocarditis and septicemia. At least 12 different species have been associated with various illnesses; however, two species — *Enterococcus faecalis* and *E. faecium* — have emerged in recent decades as a major cause of nosocomial infections (reviewed in Ref. 47), with *E. faecalis* responsible for the vast majority (80 to 90%) of the infections caused by this genus.

As most human infections are caused by either *Enterococcus faecalis* or *E. faecium*, it is often necessary to specifically differentiate these two species. Clinical laboratories use automated and rapid identification systems such as the VITEK II system (bioMérieux), BBL Crystal kits (Becton Dickinson, BD), and more recently, the Phoenix system (Becton Dickinson, BD). Given reported concerns about the reliability of such systems,[30] molecular techniques and species-specific biochemical differences can facilitate the identification process. PCR-based techniques for the amplification and downstream sequencing of the 16S rRNA genes have been used for a number of years to distinguish enterococci from lactococci and to identify specific enterococcal species.[13] Similar techniques involving the sequencing of species-specific regions of other genes have been described (e.g., heat shock protein 60 and manganese-dependent superoxide dismutase, *sodA,* genes).[30] Recently, a novel multiplex PCR technique, incorporating both 16S rRNA and *sodA*-specific primers, has been developed that allows for the simultaneous identification of both genus and species from a variety of sample types (feces, retail foods, animal carcasses). In terms of simple biochemical differences, *E. faecalis*, but not *E. faecium,* will grow on medium containing 0.4% telluride, reduce tetrazolium to formazan, and most isolates will produce acid from sorbitol, glycerol, and D-tagatose. On the other hand, most strains of *E. faecium*, but not *E. faecalis,* will produce acid from L-arabinose and melibiose.[49]

Enterococcus faecalis and *E. faecium* elaborate a number of virulence factors that contribute to the infection process. Although both species have acquired antibiotic resistance determinants (discussed below), the best-studied virulence factors are predominantly associated only with *E. faecalis*. For example, this species elaborates a cytolytic toxin (hemolysin), gelatinase, aggregation substance, and an enterococcal surface protein (Esp), each of which enhance virulence but are either not produced (or very rarely produced) by *E. faecium*.[47] The enterococcal cytolysin is a proinflammatory, acutely lethal hemolytic toxin, which also functions as a bacteriocin against other Gram-positive bacteria. The aggregation substance is a surface protein that is multifunctional: it not only promotes aggregate formation during bacterial conjugation to facilitate efficient contact between donor and recipient strains for plasmid transfer, but it is also associated with binding to, and subsequent intracellular survival within, human neutrophils. The role of this virulence factor as well as the numerous others elaborated by *E. faecalis* are reviewed extensively elsewhere.[47]

The clinical relevance of enterococci has increased due to the emergence of vancomycin-resistant strains (VRE). Recent CDC estimates from 2004 indicate that one out of every three infections in American intensive care units is caused by vancomycin-resistant enterococci, whereas in the years prior to 1990, VRE accounted for less than 0.5% of hospital-acquired infections (www.cdc.gov). The organism is spread through direct contact with stool, contaminated urine and blood, and indirectly via healthcare workers and contaminated hospital surfaces.

The explanation for the rapid emergence of VRE is markedly different in different parts of the world. In Europe, for example, the emergence of vancomycin resistance has been associated with avoparcin, a vancomycin-related antibiotic, used as a growth promoter in various agricultural processes and added to animal feed until it was banned in 1997.[64] Colonization of humans with vancomycin-resistant strains likely occurred after ingestion of meat products resulting from such practices. In the United States, where avoparcin was never approved for use in animal feed, the major reservoirs for VRE are hospital staff and patients.[47] A root cause of vancomycin resistance in the United States appears, at least in part, to be the improper use of antimicrobial agents, and VRE spread seems to be directly linked to clinical settings where the organism is most prevalent.

The recent emergence of multidrug resistance in enterococci presents an even bigger challenge to the health care profession. Multidrug-resistant isolates generally emerge from intrinsic baseline resistance to particular antibiotics or by acquiring resistance through mobile genetic element transfers (transposons and plasmids).[47] A recent report (published in 2007) from the SENTRY Antimicrobial Surveillance Program details multidrug antimicrobial resistance for VRE isolated in Europe and North America in 2003.[12] Approximately 13% of VRE isolates from North America were multidrug resistant; 28% of *Enterococcus faecalis* strains were also resistant to chloramphenicol, whereas 99% were also resistant to ciprofloxacin. VRE isolates from Europe exhibited a different but equally dangerous pattern: 21% of vancomycin-resistant isolates of *E. faecalis* exhibited rifampin resistance, compared to those from North America, of which 5.4% were resistant. Of the *E. faecium* VRE strains from Europe, 15% were chloramphenicol resistant, compared to 0.5% of the North American *E. faecium* isolates. Studies have found a link between ampicillin and vancomycin resistance, particularly in *E. faecium;* and it has been suggested that β-lactam exposure is a predisposing factor for multidrug resistance. Never before has the need to control the spread of VRE in nosocomial settings been more apparent than after a recent report detailing the horizontal transfer of one such vancomycin-resistant gene (*vanA*) from *E. faecalis* to methicillin-resistant strains of *Staphylococcus aureus* (MRSA).[64] Not only are the reports of multidrug-resistant VRE increasing, but it is clear that the danger of such strains is no longer restricted to other enterococci, as the passage of resistance markers to other genera is now a documented reality.

Lactococci

Lactococci are Gram-positive, catalase-negative cocci that grow in chains and produce lactic acid from the fermentation of lactose. The two major species, *Lactococcus lactis* ssp. *lactis* and *L. lactis* ssp. *cremoris* were previously categorized as lactic acid streptococci until they were moved to their own genus in 1985.[52] They can be distinguished from other similar cocci by specific phenotypic characteristics: they do not produce gas from glucose fermentation, most strains can grow in 6.5% NaCl, grow poorly at 45°C (distinguishing them from enterococci) but well at 10°C, and exhibit positive reactions on bile-esculin media. Further differentiation from the enterococci relies on additional biochemical tests (reviewed in Ref. 16): lactococcal strains do not produce acid in arabinose broth (as do *Enterococcus faecalis*), but will produce acid in mannitol broth (unlike *E. durans* and *E. hirae*).

Lactococci are common environmental bacteria that are most notable for their use in the food industry for food preservation and flavor. The biochemical process that produces lactic acid during carbohydrate fermentation has a number of specific applications: the byproducts of the fermentative pathway can enhance food flavor; the decrease in pH during lactic acid production can precipitate proteins that help to change or improve food texture. The increased acidity (which can be as low as 4.0) inhibits the growth of many microorganisms, thereby increasing the shelf life of such fermented foods. *Lactococcus lactis* is used in the production of fermented milk products (such as buttermilk) and is the major starter bacterium in the cheese industry.

Beyond the food industry, *Lactococcus lactis* has an important place in the biotechnology industry due to the production of nisin, a 34 amino acid potent bacteriocin with broad-spectrum activity against a number of bacterial species (reviewed in Ref. 29 and Ref. 37). The lactococci are also involved in other applications, including the expression of antigens for the development of mucosal vaccines, gene expression systems (NICE), and the production of human proteins such as cytokines for *in situ* applications.[44]

REFERENCES

1. Alouf, J.E. 1980. Streptococcal toxins (streptolysin O, streptolysin S, erythrogenic toxin). *Pharmacol. Ther.*, 11:617–717.
2. Bisno, A.L., M.O. Brito, and C.M. Collins. 2003. Molecular basis of group A streptococcal virulence. *Lancet Infect..Dis.*, 3:191–200.
3. Bosshard, P.P., S. Abels, M. Altwegg, and E.C.Z.R. Bottger. 2004. Comparison of conventional and molecular methods for identification of aerobic catalase-negative Gram-positive cocci in the clinical laboratory. *J. Clin. Microbiol.*, 42:2065–2073.
4. Brigante, G., F. Luzzaro, A. Bettacini, G. Lombardi, F. Meacci, B. Pini, S. Stefani, and A. Toniolo. 2006. Use of the phoenix automated system for identification of *Streptococcus* and *Enterococcus* spp. *J. Clin. Microbiol.*, 44:3263–3267.
5. Briones, M., J. Blanquer, D. Ferrando, M.L. Blasco, C. Gimeno, and J. Marin. 2006. Assessment of analysis of urinary pneumococcal antigen by immunochromatography for etiologic diagnosis of community-acquired pneumonia in adults. *Clin. Vacc. Immunol.*, 13:1092–1097.
6. Carapetis, J.R., M. McDonald, and N.J. Wilson. 2005. Acute rheumatic fever. *Lancet,* 366:155–168.
7. Carapetis, J.R., A.C. Steer, E.K. Mulholland, and M. Weber. 2005. The global burden of group A streptococcal diseases. *Lancet Infect. Dis.*, 11:685–694.
8. Carvalho, M., A.G. Steigerwalt, T. Thompson, D. Jackson, and R.R. Facklam. 2003. Confirmation of nontypeable *Streptococcus pneumoniae*-like organisms isolated from outbreaks of epidemic conjunctivitis as *Streptococcus pneumoniae. J. Clin. Microbiol.*, 41:4415–4417.
9. Claverys, J.-P., M. Prudhomme, and B. Martin. 2006. Induction of competence regulons as a general response to stress in Gram-positive bacteria. *Annu. Rev. Microbiol.*, 60:451–475.
10. Coykendall, A.L. 1989. Classification and identification of the viridans. *Clin. Microbiol.*, 2:315–328.
11. Crook, D.W. 2006. Capsular type and the pneumococcal human host-parasite relationship. *Clin. Infect. Dis.*, 42:460–462.
12. Deshpande, L. M., T. R. Fritsche, G. J. Moet, D. J. Biedenbach, and R. N. Jones. 2007. Antimicrobial resistance and molecular epidemiology of vancomysin-resistant enterococci from North America and Europe: A report from teh SENTRY antimivrobial surveillance program. *Diagn. Microbiol. Infect. Dis.*, 58:163–170.
13. Deasy, B.M., M.C. Rea, G.F. Fitzgerald, T.M. Cogan, and T.P. Beresford. 2000. A rapid PCR based method to distinguish between *Lactococcus* and *Enterococcus. Syst. Appl. Microbiol.*, 23:510–522.
14. Ehlenberger, A. G. and V. Nussenzweig. 1977. Role of C3b and C3d receptors in phagocytosis. *J. Exp. Med.,* 145:357–371.
15. Facklam, R. 2002. What happened to the streptococci: overview of taxonomic and nomenclature changes. *Clin. Microbiol.,* 15:613–630.
16. Facklam, R. and J.A. Elliot. 1995. Identification, classification, and clinical relevance of catalase negative, gram positive cocci, excluding the streptococci and enterococci. *Clin. Microbiol. Rev.*, 8:479–495.

17. Facklam, R.R. and J.A. Washington, II. 1991. *Streptococcus* and related catalase-negative gram-positive cocci, in A. Balows, W.J. Hausler Jr., K.L. Hermann, H.D. Isenberg, and H.J. Shadomy (Eds.), *Manual of Clinical Microbiology, 5th ed.* American Society for Microbiology, Washington, DC, 1991, p. 238–257.
18. Fischetti, V.A. 1989. Streptococcal M protein: molecular design and biological behavior. *Clin. Microbiol. Rev.*, 2:285–314.
19. Fischetti, V.A., R.D. Horstmann, and V. Pancholi. 1995. Location of the complement factor H binding site on streptococcal M6 protein. *Infect. Immun.*, 63:149–153.
20. Garcia-Suarez, M., F. Vaquez, and F. Mendez. 2006. *Streptococcus pneumonia* virulence factors and their clinical impact: an update. *Enferm. Infecc. Microbiol. Clin.*, 24:512–517.
21. Garcia, E., J.L. Garcia, C. Ronda, P. Garcia, and R. Lopez. 1985. Cloning and expression of the pneumococcal autoysin gene in *Escherichia coli. Mol. Gen. Genet.*, 201:225–230.
22. Gavin, P.J., J.R. Warren, A.A. Obias, S.M. Collins, and L.R. Peterson. 2002. Evaluation of the VITEK 2 for rapid identification of clinical isolates of gram-negative bacilli and members of the family Streptococcaceae. *Eur. J. Clin. Microbiol. Infect. Dis.*, 21:869–874.
23. Gorm Jensen, T., H. Bossen Konradsen, and B. Brunn. 1999. Evaluation of the Rapid ID 32 Strep system. *Clin. Microbiol. Infec.*, 5:417–423.
24. Gosink, K.K., F.R. Mann, C. Guglielmo, E.I. Tuomanen, and H.R. Masure. 2000. Role of novel choline binding proteins in virulence of *Streptococcus pneumoniae. Infect. Immun.*, 68:5690–5695.
25. Haanpera, M., J. Jalava, P.M.O. Huovinen, and K. Rantakokko-Jalava. 2007. Identification of alpha-hemolytic streptococci by pyrosequencing the 16S rRNA gene and by use of VITEK 2. *J. Clin. Microbiol.*, 45:762–770.
26. Hanage, W.P., S.S. Huang, M. Lipsitch, C.J.G D. Bishop, S.I. Pelton, R. Goldstein, H. Huot, and J.A. Finkelstein. 2007. Diversity and antibiotic resistance among nonvaccine serotypes of *Streptococcus pneumoniae* carriage isolates in the post-heptavalent conjugate vaccine era. *J. Infect. Dis.*, 195:347–352.
27. Horstmann, R.K., H.J. Sievertsen, M. Leippe, and V.A. Fischetti. 1992. Role of fibrinogen in complement inhibition by streptococcal M protein. *Infect. Immun.*, 60:5036–5041.
28. Howden, R. 1979. A rapid bile solubility test for pneumococci. *J. Clin. Pathol.*, 12:1293–1294.
29. Jack, R.W., J.R. Tagg, and B. Ray. 1995. Bacteriocins of Gram-positive bacteria. *Microbiol. Rev.*, 59:171–200.
30. Jackson, C.R., P.J. Fedorka-Cray, and J.B. Barrett. 2004. Use of genus- and species-specific multiplex PCR for identification of enterococci. *J. Clin. Microbiol.*, 42:3558–3565.
31. Jacobs, J.A., E.E. Stobberingh, E.I.M. Cornelissen, and M. Drent. 2005. Detection of *Streptococcus pneumoniae* antigen in bronchoalveolar lavage fluid samples by a rapid immunochromatographic membrane assay. *J. Clin. Microbiol.*, 43:4037–4040.
32. Jedrzejas, M. J. 2001. Pneumococcal virulence factors: structure and function. *Microbiol. Mol. Biol. Rev.*, 65:187–207.
33. Jeffries, J.M., A. Smith, S.C. Clarke, C. Dowson, and T.J. Mitchell. 2004. Genetic analysis of diverse disease-causing pneumococci indicates high levels of diversity within serotypes and capsule switching. *J. Clin. Microbiol.*, 42:5681–5688.
34. Jones, K.F. and V.A. Fischetti. 1988. The importance of the location of antibody binding on the M6 protein for opsonization and phagocytosis of group A M6 streptococci. *J. Exp. Med.*, 167:1114–1123.
35. Jones, K.F., B.N. Manjula, K.H. Johnston, S.K. Hollingshead, J.R. Scott, and V.A. Fischetti. 1985. Location of variable and conserved epitopes among the multiple serotypes of streptococcal M protein. *J. Exp. Med.*, 161:623–628.
36. Kellogg, J.A., D. Bankert, C. Elder, J. Gibbs, and MC. Smith. 2001. Identification of *Streptococcus pneumoniae* revisited. *J. Clin. Microbiol.*, 39:3373–3375.
37. Klaenhammer, T.R. 1993. Genetics of bacteriocins produced by lactic acid bacteria. *FEMS Microbiol. Rev.*, 12:39–85.
38. Kolberg, J.H. E.A. Holby, and E. Jantzen. 1997. Detection of phosphorylcholine eptiope in streptococci, Haemophilus and pathogenic Neisseriae by immunoblotting. *Microb. Pathog.*, 22:321–329.
39. Lancefield, R.C. 1928. The antigenic complex of *Streptococcus hemolyticus*. I. Demonstration of a type-specific substance in extracts of *Streptococcus hemolyticus. J. Exp. Med.*, 47:91–103.
40. Leung, A.K., R. Newman, A. Kumar, and H.D. Davies. 2006. Rapid antigen detection testing in diagnosing group A beta-hemolytic streptococcal pharyngitis. *Expert Rev. Mol. Diagn.*, 6:761–766.
41. Lopez, R. 2006. Pneumococcus: the sugar coated-bacteria. *Int. Microbiol.*, 9:176–190.

42. Marcos, M.A., E. Martinez, M. Almela, J. Mensa, and M.T. Jimenez de Anta. 2001. New rapid antigen test for diagnosis of pneumococcal meningitis. *Lancet,* 357:1499–1500.
43. McDonald, M.I., R.J. Towers, R. Andrews, N. Benger, B.J. Currie, and J.R. Carapetis. 2006. Low rates of streptococcal pharyngitis and high rates of pyoderma in Australian aboriginal communities where acute rheumatic fever is hyperendemic. *Clin. Inf. Dis.,* 43:683–689.
44. Mierau, I. and M. Kleerebezem. 2005. 10 years of the nisin-controlled gene expression system (NICE) in *Lactococcus lactis. Appl. Microbiol. Biotechnol.*, 68:705–717.
45. Morrison, D.A. 1997. Streptococcal competence for genetic transformation: regulation by peptide pheromones. *Microb. Drug Resist.*, 3:27–37.
46. Mosser, J.L. and A. Tomasz. 1970. Choline-containing teichoic acid as a structural component of pneumococcal cell wall and its role in sensitivity to lysis by an autolytic enzyme. *J. Biol. Chem.,* 245:287–298.
47. Mundy, L.M., D.F. Sahm, and M. Gilmore. 2006. Relationships between enterococcal virulence and antimicrobial susceptibility. *Clin. Microbiol. Rev.*, 13:513–522.
48. Murdoch, D.R., R. Laing, G.D. Mills, N.C. Karalus, G.I. Town, M.S. Reller, and L.B. Reller. 2001. Evaluation of a rapid immunochromatogenic test for detection of *Streptococcus pneumoniae* antigen in urine samples from adults with community-acquired pneumonia. *J. Clin. Microbiol.*, 39:3495–3498.
49. Murray, B.E. 1990. The life and times of the Enterococcus. *Clin. Microbiol. Rev.,* 3:46–65.
50. Nelson, A.L., A.M. Roche, J.M. Gould, K. Chim, A.J. Ratner, and J. N. Weiser. 2007. Capsule enhances pneumococcal colonization by limiting mucus-mediated clearance. *Infect. Immun.,* 75:83–90.
51. Petti, C.A., C.W. Woods, and L.B. Reller. 2005. *Streptococcus pneumoniae* antigen test using positive blood culture bottles as an alternative method to diagnose pneumococcal bacteremia. *J. Clin. Microbiol.,* 43:2510–2512.
52. Schleifer, K.H., J. Kraus, C. Dvorak, R. Kilpper-Balz, M.D. Collins, and W. Fischer. 1985. Transfer of *Streptococcus lactis* and related streptococci to the genus *Lactococcus* gen. nov. *Syst. Appl. Microbiol.*, 6:183–195.
53. Sjostrom, K., C. Spindler, A. Ortqvist, M. Kalin, A. Sandgren, and S.H.-N. Kuhlmann-Berenzen. 2006. Clonal and capsular types decide whether pneumococci will act as a primary or opportunistic pathogen. *Clin. Infect. Dis.*, 42:451–456.
54. Slotved, H.C., N. Kaltoft, I. Skovsted, M.B. Kerrn, and F. Espersen. 2004. Simple, rapid latex agglutination test for serotyping of pneumococci (Pneumotest-Latex). *Clin. Microbiol.*, 42:2518–2522.
55. Sorensen, U.B. 1993. Typing of pneumococci by using 12 pooled antisera. *J. Clin. Microbiol.*, 8:2097–2100.
56. Stromberg, A., A. Schwan, and O. Cars. 1988. Throat carrier rates of beta-hemolytic streptococci among healthy adults and children. *Scand. J. Infect. Dis.*, 20:411–417.
57. Teng, L.J., P.R. Hsueh, J.C. Tsai, P.W.H J.C. Chen, H.C. Lai, C.N. Chun-Nan Lee, and S.W. Ho. 2002. *groESL* Sequence determination, phylogenetic analysis, and species differentiation for viridans group streptococci. *J. Clin. Microbiol.*, 40:3172–3178.
58. Tettelin, H. 2001. Complete genome sequence of a virulent isolate of *Streptococcus pneumoniae. Science,* 293:498–506.
59. Tomasz, A. 1965. Control of the competent state in *Pneumococcus* by a hormone-like cell product: an example for a new type of regulatory mechanism in bacteria. *Nature*, 208:155–159.
60. Tomasz, A. 1967. Choline in the cell wall of a bacterium: novel type of polymer-linked choline in pneumococcus. *Science*, 157:694–697.
61. Wasilauskas, B.L. and K.D. Hampton. 1984. An analysis of *Streptococcus pneumoniae* identification using biochemical and serological procedures. *Diagn. Microbiol. Infect. Dis.*, 2:301–307.
62. Weiser, J.N., R. Austrian, P.K. Sreenivasan, and H.R. Masure. 1994. Phase variation in pneumococcal opacity: relationship between colonial morphology and nasopharyngeal colonization. *Infect.Immun.,* 62:2582–2589.
63. Weiser, J.N., Z. Markiewicz, E. Tuomanen, J.H. Wanji, and . 1996. Relationship between phase variation in colony morphology, intrastrain variation in cell wall physiology, and nasopharyngeal colonization by *Streptococcus pneumoniae. Infect. Immun.,* 64:2240–2245.
64. Willems, R.J.L., J. Top, M. van Santen, D.A. Robinson, T.M. Coque, F. Baquero, H. Grundmann, and M.J.M. Bontem. 2005. Global spread of vancomysin-resistant *Enterococcus faecium* from distinct nosocomial genetic complex. *Emerg. Infect. Dis.,* 11:821–828.
65. Zabriskie, J.B. 1964. The role of temperate bacteriophage in the production of erythrogenic toxin by group A streptococci. *J. Exp. Med.*, 119:761–779.

24 The Genus *Bacillus*

Daniel R. Zeigler and John B. Perkins

CONTENTS

INTRODUCTION

The genus *Bacillus* represents a very large, diverse set of bacteria that have one common yet distinct feature: the ability to make dormant endospores aerobically when challenged with unfavorable growth conditions. A photograph of sporulating cells of *Bacillus subtilis* is shown in Figure 24.1A. In 1989, Ruth Gordon authored a chapter for the *Practical Handbook of Microbiology* (CRC Press) that reviewed the systematics of the genus *Bacillus* using the most extensive, up-to-date information at that time (Gordon, 1989). Then, the various species of *Bacillus* were divided into three groups based on the phenotype of the sporangia (e.g., swollen vs. non-swollen) and the mature spore (e.g., spherical, cylindrical, or ellipsoidal), and on biochemical tests (e.g., catalase, starch hydrolysis) and growth properties (e.g., 7% NaCl). The species within these groups were highly heterogeneous. Since that time, however, there have been several major reclassifications of species within *Bacillus* caused in large part by the introduction of more sophisticated testing methods that rely on comparison at the genome level and a constant influx of new species isolated from extreme environments. *Bacillus* now represents just a small part of a larger taxonomic consortium of endospore-producing bacteria referred to as *Bacillus* sensu lato. Consequently, it is not possible in a review such as this one to provide an in-depth description of each *Bacillus* species. On the contrary, the goal of this work is to provide the reader with a higher level (over-arching) summary on *Bacillus* sensu lato, describing in large brush-strokes the common characteristics of its current members. However, the review will also attempt to focus on those key species, specifically *B. subtilis*, that play an increasingly important role in industry, medicine, and basic science. This two-tier approach should provide the investigator with a greater appreciation of this highly interesting and adaptable group of bacteria.

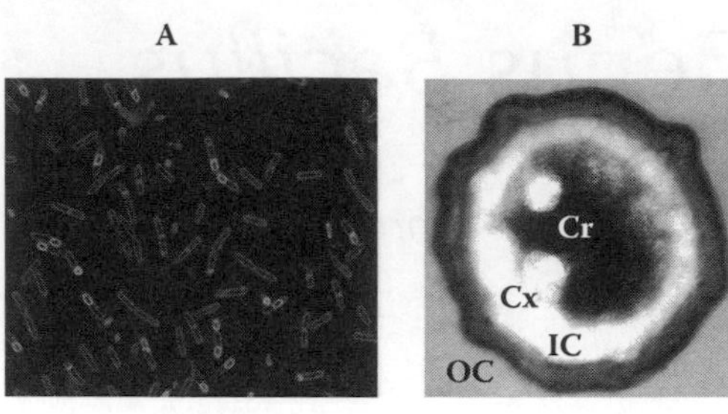

FIGURE 24.1 Formation of endospores in *Bacillus subtilis*. (A) Fluorescence micrograph of *Bacillus subtilis* sporangia. The engulfment membrane is visualized with the fluorescent protein fused to a sporulation protein that surrounds the growing spore (forespore). (The cytoplasmic membrane is stained with the vital stain FM 4-64. The micrograph was taken by E. Angert and K. Price (Harvard University) and reprinted with permission from Prof. Richard Losick (Harvard University) and the American Society for Microbiology. (B) Cross-section showing the structure of a mature endospore of a wild-type derivative of *B. subtilis* 168. The endospore is approximately 1.2 μm in diameter and is composed of a thick spore coat that lies just underneath a small exosporium (not labeled). The spore coat contains up to 60 proteins, which are cross-linked together to form two sets of layers, a thick outer coat (OC) and a less dense layered inner coat (IC). The number and thickness of layers vary from spore to spore (Driks, 2004). The spore coat encases the spore cortex (Cx) that consists of a large peptidoglycan layer with dipicolinic acid (DPA); this structure is thought to keep the interior of the endospore dehydrated, protecting the nucleoid from heat and radiation. The interior of the endospore, called the core (Cr), contains the ribosomes and the spore nucleoid. The highly condensed nucleoid is wrapped with small, acid-soluble spore proteins (SASP) into a strongly protective structure not observed in growing cells. (The electron micrograph is reproduced from Silvaggi et al. (2004). With permission from the American Society for Microbiology.)

THE SYSTEMATICS OF *BACILLUS* SENSU LATO

Bacillus Systematics: The Classical Paradigm

The bacterial genus *Bacillus* has a long and rich history in the annals of microbiology. During the 1870s, Ferdinand Cohn, working at the University of Breslau, isolated a small, motile, aerobic bacterium from boiled hay infusions. He named it *Bacillus subtilis,* meaning "thin rod." Cohn provided the first accurate description of the *Bacillus* life cycle, demonstrating the formation, heat resistance, germination, and outgrowth of endospores (Cohn, 1872). When Robert Koch established that a phenotypically similar organism, *B. anthracis,* was the causative agent of anthrax (Koch, 1876), the scientific importance of this novel group of organisms became obvious. For the next 50 years, many new endospore-forming bacteria were identified; however, due to limited tools, the naming and classification of these bacterial species were often a confusing and disorderly process.

The formidable challenge of bringing order to *Bacillus* taxonomy was met with admirable dedication by Nathan R. Smith, Francis E. Clark, and Ruth E. Gordon in the 1930s and 1940s. They adopted a clear working definition of the genus as comprising "rod-shaped bacteria capable of aerobically forming refractile endospores that are more resistant than vegetative cells to heat, drying, and other destructive agencies" (Gordon et al., 1973b). Smith assembled a collection of 1134 mesophilic *Bacillus* strains bearing 158 different species names. With his colleagues, he began to gather a meticulous and rigorous set of morphological and physiological data on each strain. After analyzing the data — with a deliberate bias toward "lumping" rather than "splitting" — they assigned each isolate to one of 19 rigorously defined species (Smith et al., 1946). Soon afterward, Gordon and Smith repeated the process with a collection of 206 thermophilic *Bacillus* isolates, assigning them to only two additional species (Gordon and Smith, 1949).

In 1973, in the now-classic monograph *The Genus Bacillus* (Gordon et al., 1973b), Ruth Gordon and colleagues provided precise, detailed descriptions for 18 *Bacillus* species, together with a battery of 33 morphological and physiological tests and an identification key that could unambiguously assign most known isolates to one of those species. *Bacillus* taxonomy had grown into a logical, orderly, and lucid science. With this landmark publication, what one could call the classical period of *Bacillus* systematics, which largely succeeded in bringing order out of chaos, was drawing to a close.

Bacillus Systematics: The Molecular Paradigm

In the 1970s and early 1980s, Woese and colleagues ushered in a bold new age of systematics that led to the reordering of all known organisms into three domains of life, two of them prokaryotic (Fox et al., 1977; Woese and Fox, 1977). The core technology that underpinned this revolution was the sequencing of 16S rRNA. When the new tools of molecular taxonomy were applied to the genus *Bacillus,* the resulting data challenged the existing paradigm that the genus was comprised of a small number of rigorously defined species identifiable by certain phenotypic characteristics. Ash and co-workers (Ash et al., 1991) utilized this new technology to scrutinize most of the known species of *Bacillus* and concluded that they fell into five distinct 16S rRNA sequence similarity groups, which they suggested could logically correspond to five novel genera. *B. subtilis*, *B. licheniformis*, *B. amyloliquefaciens*, *B. megaterium*, *B. cereus*, *B. anthracis*, *B. thuringiensis*, and many of the other best-known and earliest studied species were placed in Group 1. This group can be termed *Bacillus* sensu stricto, because members of this group are considered to have the classic characteristics known for aerobic Gram-positive sporeforming bacilli. It is important to note that additional species that were phenotypically similar to members of these five groups, yet were nonetheless sequence outliers, were thought to "represent the nuclei of other hitherto unrecognized genera."

An Overview of *Bacillus* sensu lato

This prescience has been realized by a new generation of taxonomists who have been mining the rich biodiversity of the Earth's ecosystem in search of novel microorganisms. It is no exaggeration to say that virtually everywhere they have looked, microbiologists have uncovered Gram-positive endospore-formers closely related to the classical genus *Bacillus.* Such novel species have been recovered from ocean sediments thousands of meters below sea level (Bae et al., 2005; Lu et al., 2001; Rüger et al., 2000) and from stratospheric air samples tens of kilometers above it (Shivaji et al., 2006). Other *Bacillus* species have been isolated from acidic geothermal pools and peat bogs (Albert et al., 2005a; Simbahan et al., 2004) and from highly alkaline groundwater (Tiago et al., 2004). Several *Bacillus* species have been found in hypersaline terminal lakes, some contaminated with heavy metals (Arahal et al., 1999; Lim et al., 2006; Switzer Blum et al., 1998; Vargas et al., 2005). Others have been discovered in human-created niches both ancient and modern, from Mexican shaft tombs and deteriorating Roman wall paintings (Gatson et al., 2006; Heyrman et al., 2003), to ultra-clean rooms in spacecraft assembly facilities (La Duc et al., 2004; Satomi et al., 2006; Venkateswaran et al., 2003). Plants continue to be a rich source of novel *Bacillus* species, some endophytic and others rhizosphere-associated (Olivera et al., 2005; Reva et al., 2002).

As a result of these efforts, the systematics of the old genus *Bacillus* has been radically transformed. With the exponential influx of validly described endospore-forming species since the 1990s (see Figure 24.2), the five *Bacillus* groups developed by Ash et al. (1991) and later refined by Priest (1993) have not only been assigned novel genus names, but many have been subdivided into several additional genera, and many outliers from the original similarity groups were assigned to novel genera as well. This enlarged category of endospore formers is now referred to as *Bacillus* sensu lato, with many of the original *Bacillus* species still defined as *Bacillus* sensu stricto. Moreover, during the period from 2004 to 2006, novel species of *Bacillus* sensu lato have been described at an average rate of about one per week! At the time of this writing, the trend shows no signs of abating.

In a review of this size, it is no longer practical even to *list* the species, let alone catalog and describe them. The reader is directed, however, to several excellent books and reviews that describe these new genera comprehensively (Fritze, 2004; Goto et al., 2000; Logan and Rodriguez-Diaz, 2006; Logan and Turnbull, 2003).

The oldest and nomenclatural type species for the genus *Bacillus* remains *B. subtilis* (Conn, 1930) (Figure 24.1A). Descriptive studies (Gordon et al., 1973a) have characterized *B. subtilis* as a mesophile, although a hardy one, growing at temperatures as low as 5 to 20°C and as high as 45 to 55°C and in salinities as high as 7% NaCl (wt/vol). Cells are moderately sized rods, about 0.7 to 0.8 μm by 2.0 to 3.0 μm, typically with many long peritrichous flagella and strong motility. Endospores are usually rounded but moderately elongated and are located at or near the end of the mother cell. The mature endospore, shown in Figure 24.1B, is released upon mother cell lysis. *B. subtilis* is one of the most thoroughly studied of all bacteria, both as an industrial powerhouse (see section entitled "Industrial Applications of *Bacillus*") and a model system for cell development (see section entitled "Cell Physiology and Development"). A cluster of closely related species, identified by an arrow in Figure 24.3, is often referred to as the "*Bacillus subtilis* group." One of these species, *B. licheniformis,* is also an important industrial microorganism. The species is named for its lichen-like appearance when grown on solid media, where it typically forms rough, dry, irregular colonies that adhere stubbornly to the agar surface (Gibson, 1944).

Another very commonly studied cluster of species is known as the "*Bacillus cereus* group" (Figure 24.3). Descriptive studies of these species (Gordon et al., 1973a) show that their cells are somewhat larger than those of *B. subtilis,* with widths of 1.0 to 1.2 μm and lengths of 3.0 to 5.0 μm. Most members of the group form colonies with a distinctive "ice crystal" surface texture, although strains of *B. mycoides* usually form complex rhizoid colonies. The *B. cereus* group of species is perhaps best known for its pathogenic members. *B. thuringiensis* forms large parasporal crystals

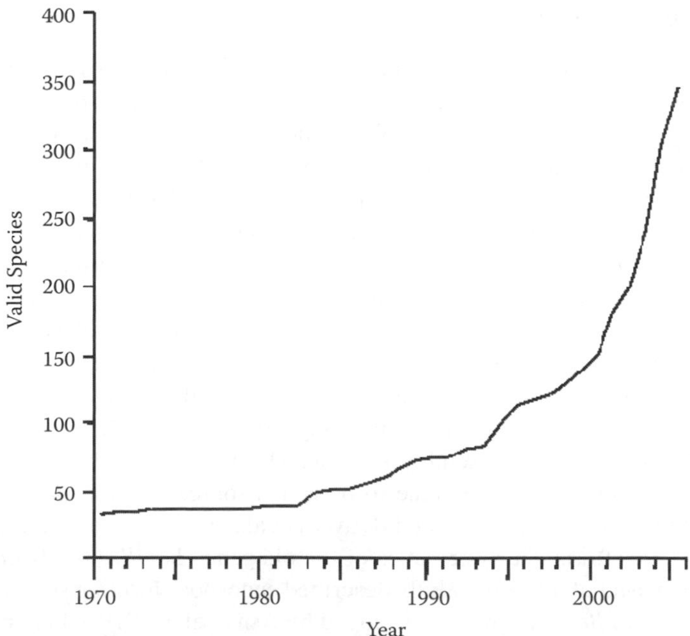

FIGURE 24.2 Number of validly described species belonging to the genus *Bacillus* sensu lato, cumulative by year. Current members of the families *"Alicyclobacillaceae," Bacillaceae,* and *"Paenibacillaceae"* are considered to belong to *Bacillus* sensu lato. Dates were tallied from the effective publication dates listed in the "List of Prokaryotic Names with Standing in Nomenclature" (http://www.bacterio.cict.fr/) (Euzéby, 1997). Revived and amended taxa were counted as being validly published in their original publication.

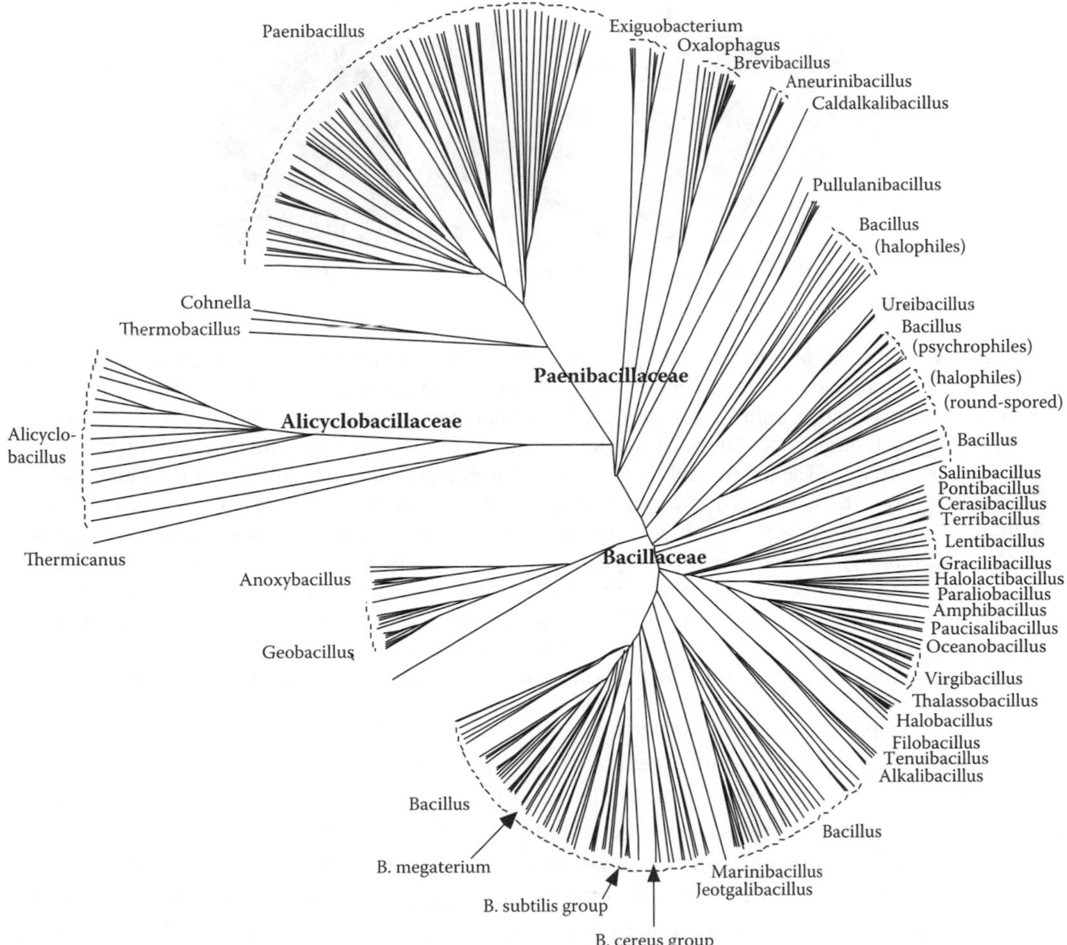

FIGURE 24.3 Phylogeny of *Bacillus* sensu lato from 16S rRNA gene sequences. GenBank DNA sequences for species type strains were aligned with ClustalW (Higgins et al., 1994). An unrooted phylogenetic tree was constructed from the ClustalW distance matrix with the PHYLIP Neighbour application (Felsenstein, 1989) and visualized with PhyloDraw. Dashed lines group sequences that have been assigned to the same taxon. Species mentioned frequently in the text are indicated. The *"B. subtilis* group" consists of 14 species, including *B. licheniformis,* that are closely related to *B. subtilis.* The *"B. cereus* group" consists of six species, including *B. anthracis* and *B. thuringiensis,* that are closely related to *B. cereus.*

composed of protoxins targeting specific invertebrate organisms, notably insect larvae (see subsection entitled "Enzymes"). *B. anthracis* and certain strains of *B. cereus* are causative agents for human diseases (see section entitled "Medical Applications of *Bacillus*"). Some strains from the *B. cereus* group are known to produce spores with caps, spikes, filaments, and other bizarre appendages (Rampersad et al., 2003) (see Figures 24.4A and 24.4B). Unusual spore morphologies and complex developmental structures can also be found in isolates throughout the *Bacillus* sensu lato (Ajithkumar et al., 2001) (see Figure 24.4C).

Although these commonly studied species are illustrative of *Bacillus*, they are by no means fully representative of its diversity. Currently there are no fewer than 37 genera that make up *Bacillus* sensu lato (see Table 24.1), and these genera have been grouped into three taxonomic families, joining seven other non-*Bacillus* families in the order Bacillales. Many of these novel genera of *Bacillus* sensu lato contain organisms adapted to extreme conditions, bacteria largely unknown to the classical taxonomists of the 1930s and 1940s. Thermophiles, psychrophiles, acidophiles, alka-

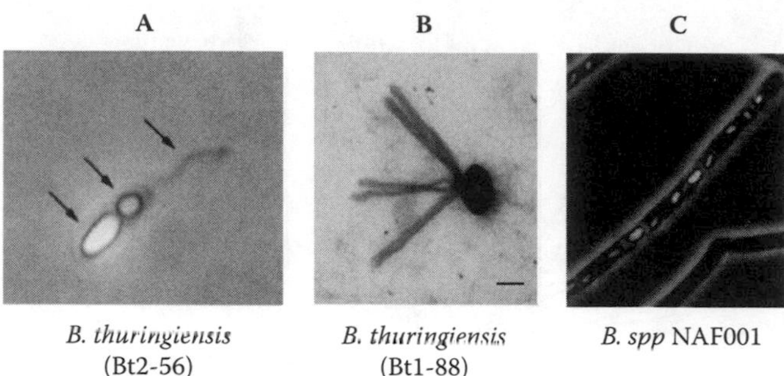

FIGURE 24.4 Diversity of *Bacillus* endospores. (A) Phase contrast micrograph of a *Bacillus thuringiensis* Bt2-56 cell, which produces a complex endospore consisting of spore, a spherical parasporal body, and a long filament (arrows from left to right respectively), which upon germination remains attached to the cell wall of the outgrowing bacterium. This filament structure is unique and is not observed with other commonly known *B. thuringiensis* strains. (The photograph is reprinted from Rampersad et al. (2003). With permission from Springer-Verlag New York, Inc.) (B) Electron micrograph of *Bacillus thuringiensis* Bt1-88 spores with multiple filaments. This species is related to *B. thuringiensis* Bt2-56, but produces multiple filaments as opposed to singlets. Bar represents 500 nm. (The micrograph is reproduced with permission from Dra. Luz Irene Rojas Avelizapa, Departamento de Microbiología, Escuela National de Ciencias Biológicas, México.) (C) Phase contrast micrograph of *Bacillus* spp. NAF001 filaments with endospores. Cells are also able to produce spore-like resting cells (SLRCs), which are heat resistant and can bud to form short filaments. (The photograph is reprinted from Ajithkumar et al. (2001). With permission from the Society for General Microbiology.)

lophiles, and halophiles, as well as isolates that are neither rod-shaped nor spore-forming, all have a place within these newly described genera. If we repeat the analysis of Ash et al. and construct a phylogram from the 16S rRNA gene sequences of every available type strain within *Bacillus* sensu lato, some patterns begin to emerge (Figure 24.3). First, there is a reassuring — although by no means totally consistent — congruity between certain broad phenotypic traits and position on the phylogenetic tree. For example, two thermophilic genera, the aerobic *Geobacillus* and the anaerobic *Anoxybacillus,* cluster together on the phylogram. Several moderately alkalophilic and halophilic genera also tend to cluster together in another branch of the tree. There is some degree of convergence between classical phenotypic approaches and modern genotypic approaches to bacterial systematics after all. Second, species still classified as *Bacillus* sensu stricto are nevertheless scattered in distantly related clusters all over the tree. Despite being subdivided multiple times, the genus is still too phylogenetically deep. There remains much work to be done in reorganizing the taxonomy of these bacteria, even without taking into consideration the novel species that continue to be discovered. Careful systematic studies of *Bacillus* are still absolutely necessary if their taxonomy is to retain any significance and usefulness.

Characterization and Classification of Novel *Bacillus* Isolates

In today's era, what do such studies consist of? What should a researcher do to characterize a novel *Bacillus* isolate adequately and to perhaps assign it to a novel species or even genus? The answer depends on the exact goals he or she has for the study.

Often the aim is to characterize the isolate on a strain-specific level. It may be imperative, for example, for an epidemiologist to determine whether a certain *Bacillus cereus* isolate from a patient with emetic food poisoning is identical with a previous isolate or represents a novel pathogenic strain. In that case, or in any number of applications where closely related members of the same species must be compared, a molecular fingerprinting technique using nucleic acid, peptide, or fatty acid profiles is perhaps the most efficient approach. Several of these methods, together with references that illustrate their practical application to *Bacillus* sensu lato, are given in Table 24.2.

TABLE 24.1
Genera of Bacillus sensu lato.

Genus	Year	Ref.	Meaning of name	Brief description
Bacillus	1872	(Skerman et al. 1980)	"small rod"	aerobic or facultatively anaerobic, endospore-forming, rod-shaped bacteria
Exiguobacterium	1984	(Collins et al. 1983)	"small" bacterium	non-sporeforming; some alkaliphilic or psychrophilic species
Saccharococcus	1984	(Nystrand 1984)	"sweet" coccus	obligate thermophilic
Amphibacillus	1990	(Niimura et al. 1990)	"both ways" bacillus	facultatively anaerobic
Sulfobacillus	1991	(Golovacheva and Karavaiko 1978)	"sulfur" bacillus	thermophilic
Alicyclobacillus	1992	(Wisotzkey et al. 1992)	"cyclic fatty acid" bacillus	acidophilic, moderately thermophilic
Paenibacillus	1994	(Ash et al. 1994; Shida et al. 1997)	"almost" a bacillus	facultatively anaerobic, mesophilic
Oxalophagus	1994	(Collins et al. 1994)	"oxalate eater"	strictly anaerobic
Aneurinibacillus	1996	(Shida et al. 1996)	"thiamine" bacillus	aerobic, mesophilic
Brevibacillus	1996	(Shida et al. 1996)	"short" bacillus	mostly aerobic, mesophilic
Halobacillus	1996	(Spring et al. 1996)	"salt" bacillus	moderately halophilic
Virgibacillus	1998	(Heyndrickx et al. 1998)	"branch" of bacillus	mesophilic, moderately halotolerant
Gracilibacillus	1999	(Wainø et al. 1999)	"slender" bacillus	mesophilic; some extremely halotolerant
Thermicanus	2000	(Gößner et al. 1999)	"hot capable"	fermentative microaerophilic
Anoxybacillus	2000	(Pikuta et al. 2000)	"no oxygen" bacillus	strictly or facultatively anaerobic, alkaliphilic, moderately thermophilic
Thermobacillus	2000	(Touzel et al. 2000)	"hot" bacillus	aerobic, thermophilic
Ureibacillus	2001	(Fortina et al. 2001)	"urea" bacillus	aerobic, thermophilic
Geobacillus	2001	(Nazina et al. 2001)	"earth" bacillus	aerobic or facultatively anaerobic, thermophilic
Filobacillus	2001	(Schlesner et al. 2001)	"thread" bacillus	aerobic, alkali-tolerant and halophilic
Jeotgalibacillus	2001	(Yoon et al. 2001)	"jeotgal" bacillus	mesophilic, round endospores in swollen sporangia
Marinibacillus	2001	(Yoon et al. 2001)	"marine" bacillus	moderately halophilic
Oceanobacillus	2002	(Lu et al. 2001)	"ocean" bacillus	extremely halotolerant and alkaliphilic
Lentibacillus	2002	(Yoon et al. 2002)	"slow" bacillus	aerobic, moderately halophilic
Paraliobacillus	2003	(Ishikawa et al. 2002)	"shoreline" bacillus	slightly halophilic, extremely halotolerant, facultative anaerobe
Cerasibacillus	2004	(Nakamura et al. 2004)	"cherry" bacillus	moderately thermophilic and alkaliphilic
Thalassobacillus	2005	(García et al. 2005)	"sea" bacillus	moderately halophilic
Halolactibacillus	2005	(Ishikawa et al. 2006)	"salt lactic acid" bacillus	halophilic, alkaliphilic marine lactic acid bacteria
Alkalibacillus	2005	(Jeon et al. 2005)	"alkaline" bacillus	halophilic, alkaliphilic
Vulcanibacillus	2005	(L'Haridon et al. 2006)	"volcanic" bacillus	strictly anaerobic, thermophilic
Pontibacillus	2005	(Lim et al. 2005)	"sea" bacillus	moderately halophilic
Salinibacillus	2005	(Ren and Zhou 2005a)	"salted" bacillus	moderately halophilic
Tenuibacillus	2005	(Ren and Zhou 2005b)	"slender" bacillus	moderately halophilic
Cohnella	2006	(Kämpfer et al. 2006)	"Cohn's" bacillus	thermophilic
Ornithinibacillus	2006	(Mayr et al. 2006)	"ornithine" bacillus	mesophilic
Caldalkalibacillus	2006	(Xue et al. 2006)	"hot, alkaline" bacillus	thermophilic, alkaliphilic
Pullulanibacillus	2006	(Hatayama et al. 2006)	"pullalan" bacillus	thermophilic
Paucisalibacillus	2006	(Nunes et al. 2005)	"not much salt" bacillus	moderately halophilic
Terribacillus	2007	(An et al. 2007)	"soil" bacillus	strictly aerobic, some moderately halophilic

When the aim is to address basic questions in environmental, evolutionary, or systematic microbiology, broader analyses are generally required. Often the very first information that a researcher will acquire is the 16S rRNA gene sequence of the isolate. While those data may be sufficient to place an isolate in a given genus — or perhaps to demonstrate that it is rather unlike any known genus — they may not be adequate for a reliable assignment at the species level or below. The 16S sequence is a powerful tool for tracing deeper phylogenies, but may show too little variability to differentiate among closely related species (Zeigler, 2005). In fact, there are known examples when two *Bacillus* isolates may be clearly distinguishable by whole genome comparison techniques yet share essentially identical 16S rRNA genes (Fox et al., 1992). For this reason, a 16S sequence alone is not considered sufficient grounds to propose that a bacterial isolate belongs to a novel species; it should be supplemented with experimental measures of whole-genome relatedness (Stackebrandt

TABLE 24.2
Common Methods for Identifying and Characterizing Novel *Bacillus* Isolates

Method, with selected examples	Useful range[a]	Application to *Bacillus* sensu lato
Physiological and morphological tests		
Microscopy and 26-test battery	species	(Gordon et al. 1973b)
Microscopy and 26-test battery	species	(Priest and Alexander 1988)
Microscopy and 20-test battery	species	(Reva et al. 2001)
Carbohydrate utilization (commercial kit)	species	(Logan and Berkeley 1984)
DNA sequencing		
Single locus sequence analysis		
16S rRNA	domain-genus	(Ash et al. 1991; Bavykin et al. 2004; Zeigler 2005)
23S rRNA	domain-genus	(Bavykin et al. 2004)
groEL	species-subspecies	(Chang et al. 2003)
gyrB	species-subspecies	(Bavykin et al. 2004; Goto et al. 2003)
recN	species-subspecies	(Zeigler 2005)
rpoB	species-subspecies	(Palmisano et al. 2001)
spoIIA	species-subspecies	(Sung and Yudkin 1997)
hag	subspecies	(Xu and Cote 2006)
Multilocus Sequence Typing (MLST)		
adk, ccpA, ftsA, glpT, pyrE, recF, sucC	genus-subspecies	(Helgason et al. 2004)
glpF, gmk, ilvD, pta, pur, pycA, tpi	genus-subspecies	(Priest et al. 2004)
rpoB, gyrB, pycA, mdh, mbl, mutS, plcR	genus-subspecies	(Ko et al. 2004)
DNA fingerprinting		
Restriction Fragment Length Polymorphism (RFLP)		
rRNA operons	subspecies-strain	(Coelho et al. 2003; Joung and Cote 2001; Shaver et al. 2002)
amplified secY genes	genus-species	(Palmisano et al. 2001)
amplified *gyrB* genes	genus-species	(Manzano et al. 2003)
Pulsed Field Gel Electrophoresis (PFGE)	subspecies-strain	(Gaviria Rivera and Priest 2003a)
Repetitive DNA PCR fingerprinting	subspecies-strain	(Satomi et al. 2006)
Random Amplified Polymorphic DNA (RAPD)	subspecies-strain	(Gaviria Rivera and Priest 2003b)
Oligonucleotide microarray fingerprinting	subspecies-strain	(Chandler et al. 2006)
Other molecular methods		
Fatty acid profiling	species-subspecies	(Kämpfer 1994; Palmisano et al. 2001; Schraft et al. 1996)
MALDI-TOF mass spectroscopy of spore proteins	species-subspecies	(Dickinson et al. 2004)
Multilocus Enzyme Electrophoresis (MLEE)	subspecies-strain	(Coelho et al. 2003; Helgason et al. 2000; Zahner et al. 1989; Zahner et al. 1994)

[a] Approximate range of taxonomic levels over which the method has maximum utility for both distinguishing and grouping related bacterial isolates.

and Goebel, 1994). A preliminary comparative genomics approach offers the encouraging suggestion that other highly conserved genes may offer greater resolution when comparing bacteria at the species or subspecies level (Zeigler, 2003). Still, DNA-DNA cross-hybridization, together with %G+C genome content comparison, remains the "gold standard" in defining new species (Stackebrandt et al., 2002; Wayne et al., 1987), although a variety of other methods are also useful (Gürtler and Mayall, 2001). However, with the advent of low-cost DNA sequencing, whole-genome sequencing may someday be the new gold standard (see below). In any event, the ideal approach for systematics includes a polyphasic analysis, in which data gathered from complementary techniques are compared in order to draw a consensus conclusion (Brenner et al., 2001). Physiological and morphological phenotypic testing, together with 16S rRNA gene sequence comparison, still provide a solid foundation for bacterial taxonomy (Ludwig and Klenk, 2001). Table 24.2 also lists a variety of methods that have found application in *Bacillus* systematics.

Although these methods have increased our understanding of aerobic endospore formers in *Bacillus* sensu lato, still only a handful of these species have been intensely studied for academic and commercial use. Nevertheless, as described in the following section, this small group of bacteria has led the way in various fields of science and industry, and thus has made a profound impact on our society.

INDUSTRIAL APPLICATIONS OF *BACILLUS*

Developing *Bacillus* Strains for Industry

The *Bacillus* species have played a crucial role in establishing a wide range of sustainable industrial fermentation processes, which in some cases (e.g., riboflavin) have supplanted older, less-efficient chemical processes. Commercial products produced by *Bacillus* species range from specialty chemicals and antibiotics to vitamins and food enzymes (Schallmey et al., 2004). Table 24.3 lists the key industrial fermentation processes that use *Bacillus* species.

In large part, many of these processes were established or improved using a combination of classical genetic methods and more modern recombinant DNA technology. Implementation of these latter techniques has been facilitated by the excellent natural competence and DNA recombination mechanisms exhibited by many of these species (Dubnau, 1991; Hamoen et al., 2003). However, the most extensive array of genetic tools has been developed for *Bacillus subtilis*. These tools include numerous plasmid cloning vectors (Bruckner, 1992; Choi et al., 2002; Errington, 1988; Janniere et al., 1990; Kobayashi and Kawamura, 1992; Vary, 1992; Wang, 1992); transposon mutagenesis systems (Le Breton et al., 2006; Perkins and Youngman, 1986; Steinmetz and Richter, 1994; Youngman et al., 1983); antibiotic resistance markers to select for transformed cells (Dale et al., 1995; Guérout-Fleury et al., 1995; Horinouchi and Weisblum, 1982; Itaya et al., 1989; Itaya et al., 1990; Perkins and Youngman, 1983); constitutive and regulated promoters (Goldstein and Doi, 1995; Lee and Pero, 1981; Nguyen et al., 2005; Terpe, 2006; Udaka and Yamagata, 1993–1994); and single and multiply copy gene expression cassettes, including those with promoter and signal sequence combinations to secrete high-valued proteins and enzymes (see below). Introduction of genetic material into the cell can occur by natural competence, protoplast fusion, and electroporation transformation methods, and in some instances by conjugation (Aquino de Muro and Priest, 2000; Jensen et al., 1996; Poluektova et al., 2004). Although most of these tools were first developed in *B. subtilis*, many have been adapted for use in related *Bacillus* species, facilitating the rapid development of new processes. Moreover, the utility of these tools has been amplified by the availability of the complete genome sequence of *B. subtilis* and related species. The importance of genomics is addressed at the end of this review. So in many ways the development and ease of use of modern recombinant genetic tools in *Bacillus* rivals that of the other well-known production hosts, such as *E. coli* and *S. cerevisiae*. Extensive description of the key genetic modification and

TABLE 24.3
Industrial Fermentation Product Produced by *Bacillus* Species

Product	*Bacillus* Species	Industrial Application
Fine chemicals		
Riboflavin	*B. subtilis*	Food, Pharma
Purine nucleosides	*B. subtilis*	Food
Poly-γ-glutamic acid	*B. subtilis*	Food, Feed, Pharma
D-ribose	*B. subtilis, B. pumilus*	Food, Feed, Cosmetics and Pharma
Thaumatin	*B. subtilis*	Food and Pharma
Polyhydroxybutyrate	*B. megaterium*	Plastics
Streptavidin	*B. subtilis*	Microarrays
2-actyl-1-pyrroline	*B. cereus*	Food
Enzymes		
α-Acetolactate decarboxylase	*Bacillus* sp.	Beverage
α-Amylase	*B. licheniformis, B. amyloliquefaciens, B. circulans, B. subtilis, Geobacillus stearothermophilus*[a]	Food, Paper, Starch, Textile, Brewing
β-Amylase	*Paenibacillus polymyxa*[b], *B. cereus, B. megaterium*	Textile
Alkaline phosphatase	*B. licheniformis*	Detergent
Cyclodextrin glucanotransferase	*Paenibacillus macerans*[c], *B. megaterium, Bacillus sp.*	Food, Pharma, Cosmetics
β-Galactosidase	*G. stearothermophilus*[a]	Food, Beverage
β-Glucanase	*B. subtilis, B. circulans, B. amyloliquefaciens*	Beverage
β-Glucosidase	*Bacillus* sp.	Brewing
Glucose isomerase	*B. coagulans, B. acidipullulyticus, B. deramificans*	Starch
Glucosyl transferase	*B. megaterium*	
Glutaminase	*B. subtilis*	Food, Flavor
Galactomannase	*B. subtilis*	Feed, Beverage
β-Lactamase	*B. licheniformis*	Pharma
Lipases	*Bacillus* sp.	Detergent
Neutral (metallo-) protease	*B. lentus, P. polymyxa, B. subtilis, B. thermoproteolyticus, B. amyloliquefaciens*	Detergent, Food
Penicillin acylase	*Bacillus* sp.	Pharma
Pullulanase	*Bacillus* sp., *B. acidopullulans*	Starch, Food, Beverage
Alkaline (serine-) protease	*B. amyloliquefaciens, B. amylosaccharicus, B. licheniformis, B. subtilis*	Detergent, Textile
Urease	*Bacillus* sp.	Analysis, Beverage
Xylanases	*Bacillus* sp.	Baking, Feed, Beverage, Brewing, Food
Insect pathogen	*B. thuringiensis, B. sphaericus, Paenibacillus popilliae*[d] *Paenibacillus lentimorbuse*[e]	Agriculture
Fermentation	*B. natto*	Food
Animal probiotic	*B. licheniformis, B. subtilis, B. cereus var toyoi, B. clausii*[f], *B. cereus*[f]	Feed
Aqua probiotic	*B. subtilis, B. megaterium, B. licheniformis, P. polymyxa, Bacillus* sp.	Feed

This table presents a condensed description of industrial products produced by Bacillus species. The data were compiled from previously published lists: enzymes from Meima et al. 2004 with permission from Horizon Scientific Press/Horizon Bioscience, UK; fine chemicals, insect pathogens, and fermentation from Schallmey et al. 2004, with permission from NRC Research Press, Canada; and probiotics from Hong et al. 2005 with permission from Blackwell Publishing, UK (Hong et al. 2005; Meima et al. 2004; Schallmey et al. 2004); enzyme applications were also obtained from Association of Manufacturers and Formulators of Enzyme Products (AMFEP) at http://www.amfep.org/main.html and Quax 2003. This table is by no means comprehensive since new products are continually being introduced into the market.

[a] formerly *B. stearothermophilus*; [b] formerly *B. polymyxa*; [c] formerly *B. macerans*; [d] formerly *B. popilliae*; [e] formerly *B. lentimorbus*; [f] not approved for use in the EU.

recombinant cloning techniques used for *Bacillus* can be found in recent reviews (Harwood and Cutting, 1990; Meima et al., 2004; Perkins et al., 2008).

Enzymes

The largest product group produced by *Bacillus* species by far is the bulk and specialty enzymes, making up about 50% of the total market size. Applications include household detergents, starch hydrolysis, textile, baking, and beverages. Many of these enzymes have unique properties, such as tolerance to high temperatures (thermostability), alkaline conditions, broad pH ranges, or inhibitory levels of byproducts, which make them the enzymes of choice in many industrial processes and household functions (Meima et al., 2004). Moreover, with the availability of classical and recombinant methods, new protein-engineered variants with new or improved catalytic properties have been developed in order to expand their use into new commercial applications (Nielsen and Borchert, 2000; Olempska-Beer et al., 2006; Schmidt-Dannert et al., 1988; Takagi 1993). Due to their propensity to secrete proteins into the extracellular medium, *Bacillus* species are also good hosts for the production of heterologous proteins; this product class is addressed in the next section. Numerous plasmid-borne or chromosomal integration expression cassettes have been reported that combine transcriptional and translation-controlling signals with secretion signal sequences that allow transport of the candidate proteins into the culture medium for subsequent isolation. In addition, a variety of protease-deficient mutants have been constructed to prevent degradation of the heterologous protein; strains with up to eight major and minor protease-encoding gene deletions have been constructed (Pero and Sloma, 1993; Wu et al., 2002).

Vitamins and Fine Chemicals

Another commercial area in which several *Bacillus* species play an important role as a fermentation host is the production of vitamins and fine chemicals. Commercialized processes include those for purine nucleosides, riboflavin, poly-γ-glutamic acid (Ashiuchi et al., 2006), D-ribose, and an array of minor products such as thaumatin, a low-calorie sugar substitute, polyhydroxybutyrate (PHB), streptavidin, and 2-acetyl-1-pyrroline. All but PHB (from *B. cereus*) utilize *B. subtilis* as the production host. *B. subtilis*-based processes for other vitamins, including biotin, pantothenate, thiamin, and folic acid have been reported, but none have been commercialized (Perkins et al., 2008).

Products produced by these *Bacillus* species have a long history of safe use and as such have gained GRAS (Generally Regarded As Safe) status in the United States. A status similar to GRAS, called QPS (Qualified Presumption of Safety), has been recently proposed by the EU regulatory authorities (Barlow et al., 2005). GRAS status is based either on a recognized history of safe use in foods prior to 1958 or on scientific procedures similar to those required for a food additive regulation. An important criterion in determining GRAS status is the lack of pathogenicity of the production microorganism. In this regard, these *Bacillus* species are found not to be harmful to humans or animals. Another regulatory issue that impacts upon the use of *Bacillus* industrial processes that utilize GMO-derived strains is the physical containment of the engineered bacteria (EU Council Directive 90/219/EEC of 23 April 1990 on the contained use of genetically modified microorganisms). Because spores can persist in the environment for years due to their unique resistance to heat, chemicals, and UV radiation, commercial strains are required to contain one or more mutations that arrest this terminal developmental stage. Consequently, most commercialized production strains contain a mutation that prevents the formation of a mature, resistant spore (Perkins and Pragai, 2004; Perkins et al., 1999; Priest et al., 1997).

Agriculture

Bacillus is also an important commercial product in the agricultural business. *B. thuringiensis*, *B. sphaericus*, *B. popilliae*, and *B. lentimorbus* have long been recognized as important sources of natural biopesticides effective against a wide range of insect pests (El-Bendary, 2006; Schallmey et al., 2004; Schnepf et al., 1998). By far the largest and most widely used is *B. thuringiensis*, which produces the δ-endotoxin pesticide, or Cry protein, as a highly visible crystal during the process of sporulation. The species is equipped with a potent arsenal of highly specific pesticides that vary widely in target specificity from strain to strain. Cry proteins are classified according to their amino acid sequence, but the classifications frequently correspond to the insect order against which the toxin is effective. For example, the Cryl proteins of *B. thuringiensis* ssp. *kurstaki* are toxic mainly to Lepidoptera species, the Cry3 protein of *B. thuringiensis* biovar *tenebrionis* is toxic to certain Coleoptera, and the Cry4 protein of *B. thuringiensis* ssp. *israelensis* is highly toxic against mosquitoes and black flies (Schnepf et al., 1998). As of 2004, there were at least 26 *B. thuringiensis*-based products registered by the United States Environmental Protection Agency (EPA). Moreover, with the advent of recombinant genetic techniques, the toxicity of these insecticides has been improved and the killing spectrum widened (Ghribj et al., 2004; Liu and Dean, 2006; Sanchis et al., 1996). Ultimately, the genes that encode these highly diverse toxins are now expressed directly in a variety of agricultural plants, such as corn, tobacco, and soy bean, which further extends the usefulness of these *Bacillus* species in agriculture (Christou et al., 2006; Mehlo et al., 2005).

The use of *Bacillus* as a promoter of plant growth is an upcoming area of agricultural importance. Application of *B. subtilis* on plant seeds has been shown to greatly enhance the survival and robustness of the seedlings (Cavaglieri et al., 2005; Ugoji et al., 2005). In addition, *Bacillus* species, such as *B. polymyxa* (now called *Paenibacillus polymyxa*), *B. cereus*, and *B. subtilis*, are also used to facilitate the establishment and durability of sport turf grass. The mechanisms by which *Bacillus* promotes plant growth have yet to be firmly established, but it appears that production of plant growth hormones, antimicrobial agents, and enzymes, such as phytase (Tye et al., 2002), that increase the nutrient availability to the grass, play a critical role (Gange and Hagley, 2004).

MEDICAL APPLICATIONS OF *BACILLUS*

The importance of *Bacillus* in the field of medicine was first established in the early history of microbiology by the identification of *Bacillus anthracis* by Pasteur and Koch as the causative agent of anthrax in the late 1800s. However, the genus as a whole is considered relatively benign in terms of human and animal disease when compared to other Gram-positive genera, such as *Staphylococcus* and *Streptococcus*. *Bacillus cereus* and the highly related species *B. weihenstephanensis* can be responsible for gastrointestinal illnesses caused by contaminated food or dairy products. Pathogenesis requires a common set of enterotoxin genes (Phelps and McKillip, 2002). Moreover, on rare occasions, *B. cereus* can be an opportunistic human pathogen in immunocompromised patients (Kotiranta et al., 2000; Schoeni and Wong, 20005). Recently, the general public's awareness of *Bacillus* was heightened by the use of *B. anthracis* in acts and threats of bioterrorism (Schneider et al., 2004).

Despite this heightened awareness of *Bacillus*, what is not nearly so well known are the numerous beneficial medical-related products produced by this genus (Table 24.4). By far the most predominant product class is antibiotics. Most of these are peptide antibiotics that are synthesized either by non-ribosomal enzymatic processes or by ribosomal synthesis of linear peptides that then undergo posttranslational modification and proteolytic processing (Stein, 2005). Another important product class consists of human proteins and vaccines, production processes of which take advantage of *Bacillus'* natural ability to produce and secrete enzymes as previously described for industrial and food enzymes. Almost all these processes have been developed using *B. subtilis*, *B. megaterium,* and *Brevibacillus brevis* (Ferreira et al., 2005; Meima et al., 2004; Terpe, 2006; Udaka and Yamagata, 1993–1994; Vary, 1994; Westers et al., 2004). The advantage of *B. brevis* is that extracellular protease production is low and heterologous protein expression is high. In sum, although the number of implemented processes using bacilli is not as extensive as for other, better-known microbial protein production host such as *Escherichia coli* or yeast, there are nevertheless

TABLE 24.4
Medical and Human Health Products Produced by *Bacillus* Species

Product	*Bacillus* Species
Antibiotics	
Bacitracin	*B. licheniformis*
Gramicidin	*Brevibacillus brevis*[a]
Tyrocidines	*B. brevis*
Subtilin	*B. subtilis*
Polymixins	*P. polymyxa*
Edeines	*B. brevis*
Butirosins	*B. circulans*
Diagnostic Test	
AIDS coat proteins	*B. megaterium*
Antigens for Vaccines	
PA (B. anthracis)	*B. subtilis*
Pneumolysin (Streptococcus pneumoniae)	*B. subtilis*
P1 (Neisseria meningitides)	*B. subtilis*
OmpP2 (Haemophilus influenzae)	*B. subtilis*
PT subunits (Bordetella pertussis)	*B. subtilis*
OmpS and Hsp60 (Chlamydia pneumonis)	*B. subtilis*
Human Proteins	
Growth hormone	*B. subtilis*
Interferon	*B. subtilis*
Proinsulin	*B. subtilis*
Tissue plasminogen activator	*B. subtilis*
scFv	*B. subtilis*
Human probiotic	*B. subtilis, B. licheniformis, B. cereus, B. pumilus, B. polyfermenticus SCD, B. clausii, Brevibacillus laterosporus,*[b] *B. coagulans, P. polymyxa, Bifidobacterium bifidum*[c]

Note: This table presents a condensed description of products produced by *Bacillus* species. The data were compiled from previously published lists: antigens for vaccines from Ferreira et al. (2005) with permission from Anais da Academia Brasileira de Ciências, Brazil; antibiotic and human proteins from Schallmey et al. (2004), with permission from NRC Research Press, Canada; probiotics from Hong et al. (2005) with permission from Blackwell Publishing, U.K. This table is by no means comprehensive as new products are continually being introduced into the market.

[a] Formerly *Bacillus brevis*; [b] formerly *Bacillus laterosporus*; [c] formerly *Bacillus bifidum*.

several successful examples, such as the use of *B. megaterium* to produce AIDS proteins for commercial diagnostic tests (Vary, 1994).

As mentioned in the preceding section, compounds produced by *Bacillus subtilis* and related species have GRAS status. However, the microorganisms themselves can also be of medical use as a probiotic. Two major human applications are for prophylactic use and for use as health food supplements. Key species found in these preparations are *B. subtilis, B. licheniformis, B. coagulans, B. pumilus, B. clausii*, nonpathogenic *B. cereus* strains, and other spore-formers from the *Bacillus* sensu lato group. In Europe (excluding the United Kingdom), *Bacillus* probiotics are used prophylactically against gastrointestinal ailments. In Southeast Asia, they are used as an adjunct to antibiotic treatment (Hong et al., 2005; Sanders et al., 2003). Perhaps use in Western societies, especially in the United States, could increase with the realization that *B. subtilis* natto, a strain closely related to *B. subtilis* 168, is used to produce "natto," a traditional Japanese fermented soy bean delicacy (Hosoi and Kiuchi, 2004; Ueda, 1989). *B. cereus*, *B. subtilis*, and *B. licheniformis* have also been shown to be very potent animal probiotics, especially in the pig and poultry sectors and in aquaculture (Alexopoulos et al., 2004; Barbosa et al., 2004; Cartman and La Ragione, 2004; Farzanfar, 2006; Tam et al., 2006). The effectiveness of these *Bacillus* species is thought to lie in their ability to be retained in the gut of the animal by undergoing repeated rounds of vegetative growth, sporulation, and germination, during which time the production of bacteriocins, antibiotics, or other antimicrobial agents effectively prevents the establishment of pathogenic bacteria in the gut.

Bacillus subtilis is also used to identify and validate novel antibacterial targets. *In silico* genome analysis and functional testing have been used to identify essential pathways as potential targets for new antimicrobial agents, including amino acid and vitamin biosynthetic pathways, proteases (Gerdes et al., 2002; Kobayashi et al., 2003; Perkins et al., 2004; Yocum and Patterson, 2005), as well as regulatory elements such as riboswitches (Blount and Breaker, 2007). Again, due to its well-developed genetics and recombinant engineering, *B. subtilis* has also been used as a screening host to identify new antimicrobial agents (Bunyapaiboonsri et al., 2003).

CELL PHYSIOLOGY AND DEVELOPMENT

By virtue of its ability to make endospores, a primitive yet intricate developmental cycle, the genus *Bacillus* has been intensely studied in the academic arena. Most of the early research was dominated by systematic characterization of the genus (Fritze, 2004), anthrax pathogenicity (Scorpio et al., 2006; Turnbull, 1991), and sporulation in terms of food safety (Russell, 1982). One highlight of this period of *Bacillus* research was the development of the anthrax vaccine in the 1930s, which is still the main course of prevention today. Beginning in the second half of the 20th century, the research focus shifted to deciphering the process of sporulation at the molecular level as a means to understand the basic mechanisms that underpin cell differentiation. In particular, *B. subtilis* has received most of this attention (and funding) due to the demonstration of a DNA transformation system by Spizizen in 1958 (Spizizen, 1958). Because of this intense academic research interest in *B. subtilis*, it is considered today as a paradigm for Gram-positive molecular and developmental biology, second only to *Escherichia coli* as a bacterial model organism. Because this work is voluminous, this section focuses only on highlighting the key research areas pertaining to *B. subtilis*. The reader is directed to many excellent reviews that cover the study of other *Bacillus* species in the field of enzyme and antibiotic production, insect pathogenicity, and cell physiology and development (Sonenshein et al., 1993, 2001).

Practically all *Bacillus subtilis* strains used in academic research are derived from strain 168, a *trpC2* mutant (defective in tryptophan biosynthesis) of wild-type *B. subtilis* Marburg, which was isolated by x-ray mutagenesis (Burkholder and Giles, 1947). It was with this strain that DNA transformation was first demonstrated. The ability to genetically map sporulation mutations in *B. subtilis* 168 (Hoch, 1971) was the first step in a long process (that is still not completed), not only to decipher the protein products encoded by these genes and how they function to create a spore, but also to understand how the cell regulates expression of these genes and in response to environmental conditions that trigger the process. Figure 24.5 shows the well-defined developmental stages that lead from a vegetative cell to a mature spore. At least 239 genes are involved in this intricate process that converts the growing cells into a two-cell compartment into which are assembled the macromolecular structures that constitute the spore (Piggot and Losick, 2001). This number is likely an underestimate because sporulation overlaps with many other areas of bacterial physiology. To help explain the choreography of this process, concepts such as sigma-factor cascades (Losick and Pero, 1981; Stragier and Losick, 1996), multicomponent phosphorelays (Burbulys et al., 1991; Hoch, 1993), intercompartmental cross-talk (Bassler and Losick, 2006), and cell cannibalism and

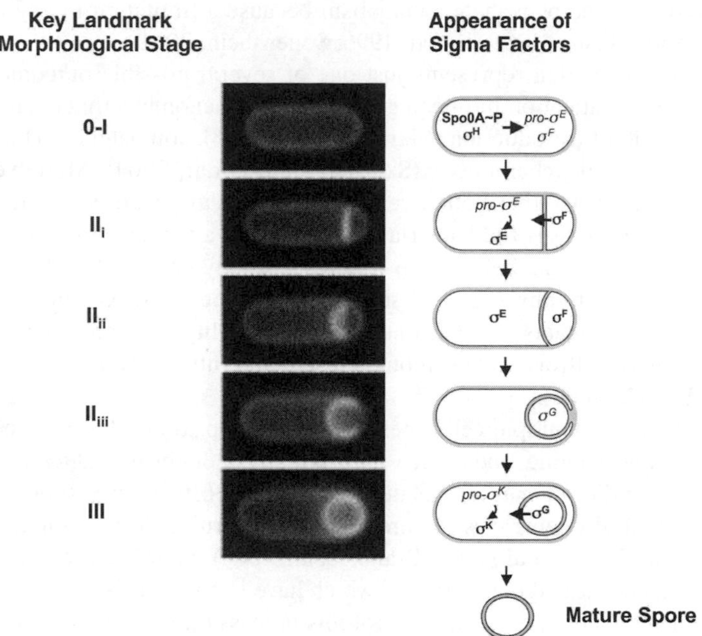

FIGURE 24.5 Key landmark morphological stages in sporulation. Sporulating cells of *B. subtilis* stained with the membrane dye TMA-DPH are depicted in the middle panel. The sequence from top to bottom shows a cell at the predivisional stage (0-I) progressing through stages of polar division (II) to engulfment (III); stages (IV–VI) from spore maturation to final release of the mature spore from the mother cell are not shown. The right panel shows a schematic representation of the landmark morphological events and cascade-like appearance of sporulation-specific sigma factors. Activation and inactivation factors are highlighted in bold and italics, respectively. Spo0A~P and sigma H (σ^H) are first activated in the predivisional cell (stage 0) by external environmental signals (Piggot and Losick, 2001). During this stage, terminal DNA replication of the genome takes place. Spo0A~P and σ^H direct the transcription of early sporulation genes, including those involved in formation of the asymmetric septum and the synthesis of inactive sigma F (σ^F) and pro-σ^E (stage I). Formation of the sporulation septum triggers the activation of σ^F (stage II). This regulatory factor directs the transcription of forespore-specific genes, including those involved in eliciting an intercompartmental signal that emanates from the forespore to the mother cell and activates sigma E (σ^E) through processing of inactive proprotein precursor (pro-σ^E). Translocation of the genome into the pre-forespore compartment occurs during this time. Inactive sigma G (σ^G), whose transcription is initiated by σ^F, is then synthesized during engulfment (stage II$_{i\text{-}iii}$) or shortly thereafter, but requires engulfment completion (stage III) for activation. Again, intercompartmental signal transduction, controlled by σ^G, is required to activate sigma K (σ^K) via processing of the inactive proprotein form (pro-σ^K). Final stages involve maturation of the spore cortex, including depositing of the spore coat proteins, and eventual release of the mature spore from the mother cell. (The illustration is reprinted from the journal *Developmental Cell*, 1:733–742, 2001, Rudner, D.Z. and Losick, R. With permission from Elsevier.)

fratricide (Ellermeier et al., 2006; Gonzalez-Pastor et al., 2003) have been introduced. More importantly, these concepts can be applied to understand other, more complex developmental processes such as fruiting body formation in *Myxococcus* and swarmer-stalked cell differentiation in *Caulobacter* (Ryan and Shapiro, 2003; Shapiro and Losick, 2000).

Because sporulation is an energy-intensive process and requires a variety of molecular building blocks, substantial work has been undertaken to elucidate central carbon metabolism, pentose phosphate pathway, nitrogen and phosphate assimilation, and the biosynthesis of amino acids, vitamins, nucleotides, and macromolecules in *Bacillus*. Special emphasis has been placed on understanding carbon, nitrogen, and phosphate metabolism because a limitation in any of these compounds induces sporulation (Fisher, 1999; Hulett, 1996; Sonenshein, 2001).

Importantly, sporulation represents just one of several possible outcomes of late stationary growth. Considerable attention has been given to other phenomena that occur during this growth period, such as antibiotic production (Marahiel et al., 1993), competence (Dubnau, 1991; Hamoen et al., 2003), and motility/chemotaxis (Szurmant and Ordal, 2004). Moreover, current thinking now places sporulation and these other cellular behaviors as part of an overarching developmental scheme that encompasses the bacterial population as a whole. This concept is referred to as bistability and is defined as a behavior in which seemingly genetically identical cells can reversibly switch into different physiological states or cell types, for example, one that is geared to sporulate and the other that is not. The outcome of bistability is to maximize the cell population's freedom of action to confront and overcome present and future adverse environmental conditions (Dubnau and Losick, 2006).

Differentiation into multiple cell types is the first step to develop multicellular communities such as biofilms and fruiting bodies. It was originally thought that *Bacillus subtilis* lacked this ability. However, studies of natural isolates (e.g., NCBI 3610) clearly show that this species can form such architecturally complex structures, in which sporulation, competence, and swimming are tightly interwoven (Branda et al., 2001; Branda et al., 2004). Moreover, comparison of these natural isolates to their domesticated counterparts, which have lost multicellularity, has led to the identification of master regulatory genes that control this process (Kearns et al., 2005). Other controlling elements include extracellular peptides ("pheromones") that maintain cell-to-cell communication via quorum sensing and facilitate recruitment of cells for specific developmental functions. So far, peptide pheromones have been identified for sporulation, competence, and cell cannibalism/fratricide (Cormella and Grossman, 2005; Gonzalez-Pastor et al., 2003).

To regulate the expression of genes necessary to carry out these developmental processes, three key regulatory mechanisms are used in *Bacillus subtilis*. One mechanism involves the deployment of alternate sigma subunits to temporally and spatially control the expression of specific groups of genes by altering the specificity of the RNA polymerase to recognize particular sets of promoters. In all, *B. subtilis* contains 17 sigma factors (Goldstein and Doi, 1995; Haldenwang, 1995; Helman and Moran, 2001; Helmann, 2002; Missiakas and Raina, 1998). In most cases, the genes encoding these factors are themselves under control of a sigma subunit that is expressed at an earlier time in cell growth or in a preceding step in a developmental process. Such is the case in sporulation in which five sigma factors are expressed in a cascade-like manner to link the expression of specific gene packets to the landmark stages (see Figure 24.4). As an added layer of regulation, the activity of the sigma factors themselves is modulated through protein processing or by inhibition of anti-sigma factors (Hughes and Mathee, 1998). The two-component regulatory mechanism found in *Escherichia coli* and other bacteria also plays a major regulatory role in *B. subtilis* by permitting it to respond to environmental signals and convert these stimuli into an actionable response at the cellular level. In all, 34 two-component systems have been identified in *B. subtilis* (Perego and Hoch, 2001) and control all key developmental activities, including competence, sporulation, and protein secretion, and are also found to be essential for growth. A third regulatory mechanism used in *B. subtilis* is called the riboswitch. This mechanism involves the direct use of metabolites and not proteins to regulate gene expression by binding directly to the nascent mRNA molecules. Depending on the secondary structure of the 5′ leader, binding directly leads to termination of the nascent RNA transcript (or in Gram-negative bacteria to the prevention of translation by blocking access of the ribosome binding site to the 16S RNA). In the case of regulation of the riboflavin biosynthetic genes, the riboswitch repressor has been identified as FMN. Riboswitch elements have also been detected to regulate gene expression involved in SAM, lysine, thiamine, purine, pyrimidine, and other metabolic pathways (Mandal and Breaker, 2004; Sudarsan et al., 2006; Vitreschak et al., 2004) . Other riboswitches involving uncharged tRNA molecules have also been identified with genes involved in amino acid metabolism (Henkin, 1994; Henkin and Yanofsky, 2002). Interestingly, it has been proposed that this mechanism is a vestige of an ancient time predating DNA, when living organisms used RNA as their main heredity material.

All the metabolic and developmental processes described thus far occur during aerobic growth. Like most members of its genus, *Bacillus subtilis* forms catalase, a hallmark of aerobic growth, which traditionally has distinguished this bacterial group from clostridia. Although originally thought to be a strict aerobe, *B. subtilis* has been recently found to be a facultative anaerobe, capable of either anaerobic growth using nitrate or fumarate as the final electron acceptor, or fermentative growth from glucose and pyruvate resulting in various end products such as ethanol, acetate, lactate, acetoin, and 2,3-butanediol (Nakano et al., 1997; Nakano and Zuber, 1998). Other species able to grow or at least respire under anaerobic conditions include *B. licheniformis* (Bulthuis et al., 1991; Clements et al., 2002); *B. megaterium* (Warren, 2007); *B. cereus*; and *B. mojavensis*. Extensive studies of *B. subtilis* have identified and characterized the two-component regulatory mechanism (*resED*) controlling the expression of a select number of genes that permit growth under anaerobic conditions (Nakano et al., 1995; Nakano et al., 1996; Sun et al., 1996). This regulatory mechanism is conserved among *Bacillus* species that are facultative anaerobes. It will be interesting to see how this anaerobic behavior fits with the extensive repertoire of aerobic cell development displayed by *B. subtilis* in the overall ecology of this bacterium.

FUTURE TECHNOLOGY AND APPLICATIONS

It has been 18 years since the last chapter on the genus *Bacillus* was written for the *Practical Handbook of Microbiology*. During this time span, there has been an explosion of knowledge pertaining to the genetics, physiology, and developmental biology of these species that has increased not only our knowledge of their biology, but also has facilitated the introduction of new industrial and medical applications. So it is fair to speculate what the future holds for the genus *Bacillus* as a whole. Clearly, genomics and other "'omics" technologies will have a vital impact. As expected, *Bacillus subtilis* has led the way in being the first Gram-positive bacterium to have its genome sequenced and annotated (Kunst et al., 1997). With the genome sequence in hand, this has led to the establishment of gene knock-out libraries (Kobayashi et al., 2003; Schumann et al., 2001), whole genome microarrays (Caldwell et al., 2001; Jürgen et al., 2005; Lee et al., 2001), regulatory element databases (Makita et al., 2004), proteomic libraries (Bernhardt et al., 1999; Hecker and Volker, 2004), global mRNA half-life determinations (Hambraeus et al., 2003), minimal genome determinations (Westers et al., 2003), and extensive metabolic flux determination (Fischer and Sauer, 2005; Sauer and Elkmanns, 2005), all of which have provided invaluable insights into *B. subtilis* as a developmental system and as an industrially important platform strain. Future developments will see further improvements in the existing 'omics platforms, such as comprehensive metabolic flux measurements (i.e., fluxome) and mathematical modeling (Dasika et al., 2004; Dettwiler et al., 1993; Lee et al., 1997; Sauer, 2006) and the development of new platforms. Emerging platforms under consideration include those to identify new global regulatory elements, such as sRNA (Silvaggi et al., 2005; Silvaggi et al., 2006), to catalog global protein-protein interactions (Noirot and Noirot-Gros, 2004), and to establish and comprehensively analyze metabolite compound libraries (i.e., metabolomics; Koek et al., 2006; Soga et al., 2003). Another emerging improvement, especially of industrial value, is lower-cost genomic (re)sequencing technologies (Albert et al., 2005b; Bennett et al., 2005; Margulies et al., 2005). Application of this sequencing technology to improved strains recovered from classical mutagenesis programs can rapidly identify those critical mutations that result in higher metabolic or enzyme overproduction. Moreover, this technology is being applied to compare natural isolates of *B. subtilis* as a way to better understand the genomic diversity (Earl et al., 2007; Perkins and Zeigler, 2008).

Since the release of the *Bacillus subtilis* 168 genome sequence, seven additional genomes of other *Bacillus* species have been determined and released publicly (*B. thuringiensis, B. clausii, B. licheniformis, B. cereus, B. megaterium, B. anthracis*, and *B. halodurans)* with an additional 45 genomes in progress (Liolios et al., 2006). Of course, this opens the door to use 'omics and resequencing technologies to study the physiology and molecular biology of these species as has already been demonstrated for *B. subtilis*, perhaps finding new developmental behaviors or industrial applications along the way. But in a broader sense, such studies also have a significant impact on how we perceive evolutionary changes at the genomic level. For example, based on the high percentage of conserved genes among *Bacillus* species, one can consider genomes from highly related species to be actually composed of two components: (1) a "core genome" shared by all members of the group, and (2) a much larger "dispensable genome" consisting of partially shared or strain-specific genes (Tettelin et al., 2005). An important implication of this concept, referred to as "pan-genome," is that bacteria are not completely clonal; that is, they do not inherit their genetic material (and therefore phenotypic traits) solely from a parental line, with a gradual accumulation of random mutations fixed by chance or natural selection. Instead, much of their genetic content is obtained through horizontal gene transfer from non-related bacterial sources. One reasonable estimate, based on variations in G+C content in the DNA sequence, is that 552 of the roughly 4112 protein-encoding genes in the *B. subtilis* chromosome entered the genome via horizontal gene transfer from non-*Bacillus* sources (Garcia-Vallve et al., 2003). In addition, *B. cereus*, *B. thuringiensis*, and *B. anthracis* are now known to share a common core genome but differ in their approach to pathogenicity due to the presence of different plasmid-borne toxin genes (Rasko et al., 2005).

Because the science of comparative genomics is still in its infancy, the implications of these insights for bacterial systematics are only now beginning to be understood. Moreover, as the cost of DNA sequencing decreases through the introduction of new technologies (e.g., high-throughput pyrosequencing; Nyren, 2006) genomes of new and existing endospore-forming species will be determined, which future studies in *Bacillus* systematics will have to take into account. Thus, it is likely that the phylogenetic tree of *Bacillus* will undergo yet further changes as we learn more about these intriguing, aerobic endospore-forming bacteria.

ACKNOWLEDGMENTS

We thank Wim Quax, Adam Driks, and Adriano Henriques for helpful discussions; and Markus Wyss and Richard Losick for critical reading of the manuscript. The *Bacillus* Genetic Stock Center is supported by the National Science Foundation under Grant No. 0234214.

REFERENCES

Ajithkumar, V.P., B. Ajithkumar, K. Mori, K. Takamizawa, R. Iriye, and S. Tabata. 2001. A novel filamentous *Bacillus* sp., strain NAF001, forming endospores and budding cells. *Microbiology*, 147:1415–1423.

Albert, R.A., J. Archambault, R. Rosselló-Mora, B.J. Tindall, and M. Matheny. 2005a. *Bacillus acidicola* sp. nov., a novel mesophilic, acidophilic species isolated from acidic Sphagnum peat bogs in Wisconsin. *Int. J. Syst. Evol. Microbiol.*, 55:2125–2130.

Albert, T.J., D. Dailidiene, G. Dailide, et al. 2005b. Whole genome scanning for point mutations by hybridization-based "comparative genome resequencing": change associated with high-level metronidazole resistance in *Helicobacter pylori*. *Nat. Meth.*, 2:951–953.

Alexopoulos, C., I.E. Georgoulakis, A. Tzivara, S.K. Kritas, A. Siochu, and S.C. Kyriakis. 2004. Field evaluation of the efficacy of a probiotic containing *Bacillus licheniformis* and *Bacillus subtilis* spores, on the health status and performance of sows and their litter. *J. Anim. Physiol. Anim. Nutr.* (Berl.), 88:381–392.

An, S.Y., M. Asahara, K. Goto, H. Kasai, and A. Yokota. 2007. *Terribacillus saccharophilus* gen. nov., sp. nov. and *Terribacillus halophilus* sp. nov., spore-forming bacteria isolated from field soil in Japan. *Int. J. Syst. Evol. Microbiol.*, 57:51–55.

Aquino de Muro, M. and F.G. Priest. 2000. Construction of chromosomal integrants of *Bacillus sphaericus* 2362 by conjugation with *Escherichia coli*. *Res. Microbiol.*, 151:547–555.

Arahal, D.R., M.C. Márquez, B.E. Volcani, K.H. Schleifer, and A. Ventosa. 1999. *Bacillus marismortui* sp. nov., a new moderately halophilic species from the Dead Sea. *Int. J. Syst. Evol. Microbiol.*, 49:521–530.

Ash, C., J.A.E. Farrow, S. Wallbanks, and M.D. Collins. 1991. Phylogenetic heterogeneity of the genus *Bacillus* revealed by comparative analysis of small-subunit-ribosomal RNA sequences. *Lett. Appl. Microbiol.*, 13:202–206.

Ash, C., F.G. Priest, and M.D. Collins. 1994. Molecular identification of rRNA group 3 bacilli (Ash, Farrow, Wallbanks and Collins) using a PCR probe test. Proposal for the creation of a new genus *Paenibacillus*. *Antonie Van Leeuwenhoek*, 64:253–260.

Ashiuchi, M., K. Shimanouchi, T. Horiuchi, T. Kamei, and H. Misono. 2006. Genetically engineered poly-gamma-glutamate producer from *Bacillus subtilis* ISW1214. *Biosci. Biotechnol. Biochem.*, 70:1794–1797.

Bae, S.S., J.H. Lee, and S.J. Kim. 2005. *Bacillus alveayuensis* sp. nov., a thermophilic bacterium isolated from deep-sea sediments of the Ayu Trough. *Int. J. Syst. Evol. Microbiol.*, 55:1211–1215.

Barbosa, T.M., R. Serra, and A.O. Henriques. 2004. Gut sporeformers. In *Bacterial Spore Formers: Probiotics and Emerging Application*, Eds. E. Ricca, A.O. Henriques, and S.M. Cutting, pp. 183–191. Norfolk, U.K.: Horizon Bioscience.

Barlow, S., A. Chesson, J. Collins, et al. 2005. Opinion of the Scientific Committee on a request from EFSA related to a generic approach to the safety assessment by EFSA of microorganisms used in food/feed and the production of food/feed additives. *EFSA J.*, 226:1–12.

Bassler, B.A. and R. Losick. 2006. Bacterially speaking. *Cell*, 125:237–246.

Bavykin, S.G., Y.P. Lysov, V. Zakhariev, et al. 2004. Use of 16S rRNA, 23S rRNA, and gyrB gene sequence analysis to determine phylogenetic relationships of *Bacillus cereus* group microorganisms. *J. Clin. Microbiol.*, 42:3711–3730.

Bennett, S.T., C. Barnes, A. Cox, L. Davies, and C. Brown. 2005. Toward the $1,000 human genome. *Pharmacogenomics*, 6:373–382.

Bernhardt, J., K. Buttner, C. Scharf, and M. Hecker. 1999. Dual channel imaging of two-dimensional electropherograms in *Bacillus subtilis. Electrophoresis*, 20:2225–2240.

Blount, K.F. and R.R. Breaker. 2007. Riboswitches as antibacterial drug targets. *Nature Biotechnol.*, 24:1558–1564.

Branda, S.S., J.E. Gonzalez-Pastor, S. Ben-Yehuda, R. Losick, and R. Kolter. 2001. Fruiting body formation by *Bacillus subtilis. Proc. Natl. Acad. Sci. U.S.A.*, 98:11621–11626.

Branda, S.S., J.E. Gonzalez-Pastor, E. Dervyn, S.D. Ehrlich, R. Losick, and R. Kolter. 2004. Genes involved in formation of structured multicellular communities by *Bacillus subtilis. J. Bacteriol.*, 186:3970–3979.

Brenner, D.J., J.T. Staley, and N.R. Krieg. 2001. Classification of procaryotic organisms and the concept of bacterial speciation. In *Bergey's Manual of Systematic Bacteriology*, 2nd ed., Eds. G.M. Garrity, pp. 27–31. New York: Springer.

Bruckner, R. 1992. A series of shuttle vectors for *Bacillus subtilis* and *Escherichia coli*. *Gene*. 122:187–192.

Bulthuis, B.A., C. Rommens, G.M. Koningstein, A.H. Stouthamer, and H.W. van Verseveld. 1991. Formation of fermentation products and extracellular protease during anaerobic growth of *Bacillus licheniformis* in chemostat and batch-culture. *Antonie Van Leeuwenhoek*, 62:355–371.

Bunyapaiboonsri, T., H. Ramstroem, O. Ramstroem, J. Haiech, and J.-M. Lehn. 2003. Generation of bis-cationic heterocyclic inhibitors of *Bacillus subtilis* HPr kinase/phosphatase from a ditopic dynamic combinatorial library. *J. Med. Chem.*, 46:5803–5811.

Burbulys, D., K.A. Trach, and J.A. Hoch. 1991. Initiation of sporulation in *B. subtilis* is controlled by a multicomponent phosphorelay. *Cell*, 64:545–552.

Burkholder, P.R. and N.H. Giles. 1947. Induced biochemical mutations in *Bacillus subtilis Am. J. Bot.*, 34:345–348.

Caldwell, R., R. Sapolsky, W. Weyler, R.R. Maile, S.C. Causey, and E. Ferrari. 2001. Correlation between *Bacillus subtilis* ScoC phenotype and gene expression determined using microarrays for transcriptome analysis. *J. Bacteriol.*, 183:7329–7340.

Cartman, S.T. and R.M. La Ragione. 2004. Spore probiotics as animal feed supplements. In *Bacterial Spore Formers: Probiotics and Emerging Application,* Eds. E. Ricca, H. O. Henriques, and S.M. Cutting, pp. 155–161. Norfolk, U.K.: Horizon Bioscience.

Cavaglieri, L., J. Orlando, M.I. Rodríguez, S. Chulze, and M. Etcheverry. 2005. Biocontrol of *Bacillus subtilis* against *Fusarium verticillioides in vitro* and at the maize root level. *Res. Microbiol.*, 156:748–754.

Chandler, D.P., O. Alferov, B. Chernov, et al. 2006. Diagnostic oligonucleotide microarray fingerprinting of *Bacillus* isolates. *J. Clin. Microbiol.*, 44:244–250.

Chang, Y.-H., Y.-H. Shangkuan, H.-C. Lin, and H.-W. Liu. 2003. PCR assay of the groEL gene for detection and differentiation of *Bacillus cereus* group cells. *Appl. Environ. Microbiol.*, 69:4502–4510.

Choi, Y.-J., T.-T. Wang, and B.H. Lee. 2002. Positive selection vectors. *Crit. Rev. Biotechnol.*, 22:225–244.

Christou, P., T. Capell, A. Kohli, J.A. Gatehouse, and A.M. Gatehouse. 2006. Recent developments and future prospects in insect pest control in transgenic crops. *Trends Plant Sci.*, 11:302–308.

Clements, L.D., B.S. Miller, and U.N. Streips. 2002. Comparative growth analysis of the facultative anaerobes *Bacillus subtilis*, *Bacillus licheniformis*, and *Escherichia coli*. *Syst. Appl. Microbiol.*, 25:284–286.

Coelho, M.R.R., I. von der Weid, V. Zahner, and L. Seldin. 2003. Characterization of nitrogen-fixing *Paenibacillus* species by polymerase chain reaction-restriction fragment length polymorphism analysis of part of genes encoding 16S rRNA and 23S rRNA and by multilocus enzyme electrophoresis. *FEMS Microbiol. Letts.*, 222:243–250.

Cohn, F. 1872. Untersuchen über Bacterien. *Beitr. z. Biol. Der Pflanz.*, 1 (Heft 2):127–224.

Collins, M.D., P.A. Lawson, A. Willems, et al. 1994. The phylogeny of the genus *Clostridium*: proposal of five new genera and eleven new species combinations. *Int. J. Syst. Bacteriol.*, 44:812–826.

Collins, M.D., B.M. Lund, J.A.E. Farrow, and K.H. Schleifer. 1983. Chemotaxonomic study of an alkalophilic bacterium, *Exiguobacterium aurantiacum* gen. nov., sp. nov. *J. Gen. Microbiol.*, 129:91–92.

Conn, H.J. 1930. The identity of *Bacillus subtilis*. *J. Infect. Dis.*, 46:341–350.

Cormella, N. and A.D. Grossman. 2005. Conservation of genes and processes controlled by the quorum response in bacteria: characterization of genes controlled by the quorum-sensing transcription factor ComA in *Bacillus subtilis*. *Mol. Microbiol.*, 57:1159–1174.

Dale, G.E., H. Langen, M.G. Page, R.L. Then, and D. Stuber. 1995. Cloning and characterization of a novel, plasmid-encoded trimethoprim-resistant dihydrofolate reductase from *Staphylococcus haemolyticus* MUR313. *Antimicrob. Agents Chemother.*, 39:1920–1924.

Dasika, M.S., A. Gupta, and C.D. Maranas. 2004. A mixed integer linear programming (MLP) MILP framework for inferring time delay in gene regulatory networks. *Pac. Symp. Biocomput.*, http://helix-web.stanford.edu/psb04/dasika2.pdf.

Dettwiler, B.I., E. Heinzl, and J.E. Prenosil. 1993. A simulation model for the continuous production of acetoin and butanediol using *Bacillus subtilis* with integrated pervaporation separation. *Biotechnol. Bioeng.* 41:791–800.

Dickinson, D.N., M.T. La Duc, M. Satomi, J.D. Winefordner, D.H. Powell, and K. Venkateswaran. 2004. MALDI-TOFMS compared with other polyphasic taxonomy approaches for the identification and classification of *Bacillus pumilus* spores. *J. Microbiol. Meth.*, 58:1–12.

Driks, A. 2004. The *Bacillus* spore coat. *Phytopathology*, 94:1249–1251.

Dubnau, D. 1991. Genetic competence in *Bacillus subtilis*. *Microbiol. Rev.*, 55:395–424.

Dubnau, D. and R. Losick. 2006. Bistability in bacteria. *Mol. Microbiol.*, 61:564–572.

Earl, A.M., R. Losick, and R. Kolter. 2007. *Bacillus subtilis* genome diversity. *J. Bacteriol.*, 189:1163–1170.

El-Bendary, M.A. 2006. *Bacillus thuringiensis* and *Bacillus sphaericus* biopesticides production. *J. Basic Microbiol.*, 46:158–170.

Ellermeier, C.D., E.C. Hobbs, J.E. Gonzalez-Pastor, and R. Losick. 2006. A three-protein signaling pathway governing immunity to a bacterial cannibalism toxin. *Cell*, 124:549–559.

Errington, J. 1988. Generalized cloning vectors for *Bacillus subtilis*. *Biotechnology*, 10:345–362.

Euzéby, J.P. 1997. List of bacterial names with standing in nomenclature: a folder available on the Internet. *Int. J. Syst. Bacteriol.*, 47:590–592.

Farzanfar, A. 2006. The use of probiotic in shrimp aquaculture. *FEMS Immunol. Med. Microbiol.*, 48:149–158.

Felsenstein, J. 1989. PHYLIP — Phylogeny Inference Package (Version 3.2). *Cladistics*, 5:164–166.

Ferreira, L.C.S., R.C.C. Ferreira, and W. Schumann. 2005. *Bacillus subtilis* as a tool for vaccine development: from antigen factories to delivery vectors. *An. Acad. Bras. Ciênc.*, 77:113–124.

Fischer, E. and U. Sauer. 2005. Large-scale in vivo flux analysis shows rigidity and suboptimal performance of *Bacillus subtilis* metabolism. *Nat. Genet.*, 37:636–640.

Fisher, S.H. 1999. Regulation of nitrogen metabolism in *Bacillus subtilis*: vive la difference. *Mol. Microbiol.*, 32:223–232.

Fortina, M.G., R. Pukall, P. Schumann, et al. 2001. *Ureibacillus* gen. nov., a new genus to accommodate *Bacillus thermosphaericus* (Andersson et al. 1995), emendation of *Ureibacillus thermosphaericus* and description of *Ureibacillus terrenus* sp. nov. *Int. J. Syst. Evol. Microbiol.*, 51:447–455.

Fox, G.E., K.R. Pechman, and C.R. Woese. 1977. Comparative cataloging of 16S ribosomal ribonucleic acid: molecular approach to procaryotic systematics. *Int. J. System. Bacteriol.*, 27:44–57.

Fox, G.E., J.D. Wisotzkey, and P. Gurtshuk Jr. 1992. How close is close: 16S rRNA sequence identity may not be sufficient to guarantee species identity. *Int. J. Syst. Bacteriol.*, 41:166–170.

Fritze, D. 2004. Taxonomy and systematics of the aerobic endospore forming bacteria: *Bacillus* and related genera. In *Bacterial Spore Formers: Probiotics and Emerging Application*, Eds. E. Ricca, A.O. Henriques, and S.M. Cutting, pp. 17–34. Norfolk, U.K.: Horizon Bioscience.

Gange, A.C. and K.J. Hagley. 2004. The potential for use of *Bacillus* spp. in sports turf management. In *Bacterial Spore Formers: Probiotics and Emerging Application,* Eds. E. Ricca, A.O. Henriques, and S.M. Cutting, pp. 163–170. Norfolk, UK: Horizon Bioscience.

Garcia-Vallve, S., E. Guzman, M.A. Montero, and A. Romeu. 2003. HGT-DB: a database of putative horizontally transferred genes in prokaryotic complete genomes. *Nucleic Acids Res*., 31:187–189.

García, M.T., V. Gallego, A. Ventosa, and E. Mellado. 2005. *Thalassobacillus devorans* gen. nov., sp. nov., a moderately halophilic, phenol-degrading, Gram-positive bacterium. *Int. J. Syst. Evol. Microbiol.*, 55:1789–1795.

Gatson, J.W., B.F. Benz, C. Chandrasekaran, M. Satomi, K. Venkateswaran, and M.E. Hart. 2006. *Bacillus tequilensis* sp. nov., isolated from a 2000-year-old Mexican shaft-tomb, is closely related to *Bacillus subtilis. Int. J. Syst. Evol. Microbiol*., 56:1475–1484.

Gaviria Rivera, A.M. and F.G. Priest. 2003a. Pulsed field gel electrophoresis of chromosomal DNA reveals a clonal population structure to *Bacillus thuringiensis* that relates in general to crystal protein gene content *FEMS Microbiol. Lett*., 223:61–66.

Gaviria Rivera, A.M. and F.G. Priest. 2003b. Molecular typing of *Bacillus thuringiensis* Serovars by RAPD-PCR. *Appl. Environ. Microbiol*., 26:254–261.

Gerdes, S.Y., M.D. Scholle, M. D'Souza, et al. 2002. From genetic footprinting to antimicrobial drug targets: examples in cofactor biosynthetic pathways. *J. Bacteriol*., 184:4555–4572.

Ghribj, D., N. Zouari, and S. Jaoua. 2004. Improvement of bioinsecticides production through mutagenesis of *Bacillus thuringiensis* by UV and nitrous acid affecting metabolic pathways and/or delta-endotoxin synthesis. *J. Appl. Microbiol*., 97:338–346.

Gibson, T. 1944. A study of *Bacillus subtilis* and related organisms. *J. Dairy Res*., 13:248–260.

Goldstein, M.A. and R.H. Doi. 1995. Prokaryotic promoters in biotechnology. *Biotechnol. Annu. Rev*., 1:105–128.

Golovacheva, R.S. and G.I. Karavaiko. 1978. *Sulfobacillus*, a new genus of thermophilic sporeforming bacteria. *Mikrobiologiya, 47:815–822.*

Gonzalez-Pastor, J.E., E.C. Hobbs, and R. Losick. 2003. Cannibalism by sporulating bacteria. *Science*, 301:510–513.

Gordon, R.E. 1989. The Genus *Bacillus*. In *Practical Handbook of Microbiology*, Ed. W.M. O'Leary, pp. 109–126. Boca Raton, FL: CRC Press.

Gordon, R.E., W.C. Haynes, and C.H.-N. Pang. 1973a. The genus *Bacillus*. *Agricultural Handbook* No. 427. Washington, DC: U.S. Department of Agriculture.

Gordon, R.E., W.C. Haynes, and C.H.-N. Pang. 1973b. The genus *Bacillus*. Washington, DC: U.S. Government Printing Office.

Gordon, R.E. and N.R. Smith. 1949. Aerobic sporeforming bacteria capable of growth at high temperatures. *J. Bacteriol*., 58:327–341.

Gößner, A.S., R. Devereux, N. Ohnemüller, G. Acker, E. Stackebrandt, and H.L. Drake. 1999. *Thermicanus aegyptius* gen. nov., sp. nov., isolated from oxic soil, a fermentative microaerophile that grows commensally with the thermophilic acetogen *Moorella thermoacetica*. *Appl. Environ. Microbiol*., 65:5124–5133.

Goto, K., K. Mochida, M. Asahara, M. Suzuki, H. Kasai, and A. Yokota. 2003. *Alicyclobacillus pomorum* sp. nov., a novel thermo-acidophilic, endospore-forming bacterium that does not possess omega-alicyclic fatty acids, and emended description of the genus *Alicyclobacillus*. *Int. J. Syst. Evol. Microbiol*., 53:1537–1544.

Goto, K., T. Omura, Y. Hara, and Y. Sadaie. 2000. Application of the partial 16S rDNA sequence as an index for rapid identification of species in the genus *Bacillus*. *J. Gen. Appl. Microbiol*., 46:1–8.

Guérout-Fleury, A. M., K. Shazand, N. Frandsen, and P. Stragier. 1995. Antibiotic-resistance cassettes for *Bacillus subtilis*. *Gene*, 167:335–336.

Gürtler, V. and B.C. Mayall. 2001. Genomic approaches to typing, taxonomy and evolution of bacterial isolates. *Int. J. Syst. Evol. Microbiol*., 51:3–16.

Haldenwang, W.G. 1995. The sigma factors of *Bacillus subtilis*. *Microbiol. Rev*., 59:1–30.

Hambraeus, G., C. von Wachenfeldt, and L. Hederstedt. 2003. Genome-wide survey of mRNA half-lives in *Bacillus subtilis* identifies extremely stable mRNAs. *Mol. Genet. Genomics*, 269:706–714.

Hamoen, L.W., G. Venema, and O.P. Kuipers. 2003. Controlling competence in *Bacillus subtilis*: shared use of regulators. *Microbiology*, 149:9–17.

Harwood, C. and S.M. Cutting, Eds. 1990. *Molecular Biological Methods for Bacillus*. Chichester, U.K.: John Wiley & Sons, Ltd.

Hatayama, K., H. Shoun, Y. Ueda, and A. Nakamura. 2006. *Tuberibacillus calidus* gen. nov., sp. nov., isolated from a compost pile and reclassification of *Bacillus naganoensis* Tomimura et al. 1990 as *Pullulanibacillus naganoensis* gen. nov., comb. nov. and *Bacillus laevolacticus* Andersch et al. 1994 as *Sporolactobacillus laevolacticus* comb. nov. *Int. J. Syst. Evol. Microbiol.*, 56:2545–2551.

Hecker, M. and U. Volker. 2004. Towards a comprehensive understanding of *Bacillus subtilis* cell physiology by physiological proteomics. *Proteomics*, 4.3727–3750.

Helgason, E., O.A. Økstad, D.A. Caugant et al. 2000. *Bacillus anthracis*, *Bacillus cereus*, and *Bacillus thuringiensis*—one species on the basis of genetic evidence. *Appl. Environ. Microbiol.*, 66:2627–2630.

Helgason, E., N. Tourasse, R. Meisal, D. Caugant, and A.-B. Kolstø. 2004. Multilocus sequence typing scheme for bacteria of the *Bacillus cereus* group. *Appl. Environ. Microbiol.*, 70:191–201.

Helman, J.D. and C.P. Moran. 2001. RNA polymerase and sigma factors. In *Bacillus subtilis and Its Closest Relatives: From Genes to Cells*, Eds. A.L. Sonenshein, J.A. Hoch, and R. Losick, pp. 289–312. Washington, DC: ASM Press.

Helmann, J.D. 2002. The extracytoplasmic function (ECF) sigma factors. *Adv. Microb. Physio.*, 46:47–110.

Henkin, T.M. 1994. tRNA-directed transcription antitermination. *Mol. Microbiol.*, 131:381–387.

Henkin, T.M. and C. Yanofsky. 2002. Regulation by transcription attenuation in bacteria: how RNA provides instructions for transcription termination/antitermination decisions. *Bioessays*, 24:700–707.

Heyndrickx, M., L. Lebbe, K. Kersters, P. De Vos, G. Forsyth, and N.A. Logan. 1998. *Virgibacillus*: a new genus to accommodate *Bacillus pantothenticus* (Proom and Knight, 1950). Emended description of *Virgibacillus pantothenticus*. *Int. J. Syst. Bacteriol.*, 48:99–106.

Heyrman, J., A. Balcaen, M. Rodriguez-Diaz, N.A. Logan, J. Swings, and P. De Vos. 2003. *Bacillus decolorationis* sp. nov., isolated from biodeteriorated parts of the mural paintings at the Servilia tomb, Roman necropolis of Carmona, Spain. *Int. J. Syst. Evol. Microbiol.*, 53:459–463.

Higgins, D., J. Thompson, T. Gibson, J.D. Thompson, D.G. Higgins, and T.J. Gibson. 1994. CLUSTAL W: improving the sensitivity of progressive multiple sequence alignment through sequence weighting, position-specific gap penalties and weight matrix choice. *Nucl. Acids Res.*, 22:4673–4680.

Hoch, J.A. 1971. Genetic analysis of pleiotropic negative sporulation mutants in *Bacillus subtilis*. *J. Bacteriol.*, 105:896–901.

Hoch, J.A. 1993. Regulation of the phosphorelay and the initiation of sporulation in *Bacillus subtilis*. *Annu. Rev. Microbiol.*, 47:441–465.

Hong, H.A., L.H. Duc, and S.M. Cutting. 2005. The use of bacterial spore formers as probiotics. *FEMS Microbiol. Rev.*, 29:813–835.

Horinouchi, S. and B. Weisblum. 1982. Nucleotide sequence and functional map of pC194, a plasmid that specifies inducible chloramphenicol resistance. *J. Bacteriol.*, 150:815–825.

Hosoi, T. and K. Kiuchi. 2004. Production and probiotic effects of natto. In *Bacterial Spore Formers: Probiotics and Emerging Applications*, Eds. E. Ricca, A.O. Henriques, and S.M. Cutting, pp. 143–154. Norfolk, U.K.: Horizon Bioscience.

Hughes, K.T. and K. Mathee. 1998. The anti-sigma factors. *Annu. Rev. Microbiol*,. 52:231–286.

Hulett, F.M. 1996. The signal-transduction network for Pho regulation in *Bacillus subtilis*. *Mol. Microbiol.*, 19:933–939.

Ishikawa, M., S. Ishizaki, Y. Yamamoto, and K. Yamasato. 2002. *Paraliobacillus ryukyuensis* gen. nov., sp. nov., a new Gram-positive, slightly halophilic, extremely halotolerant, facultative anaerobe isolated from a decomposing marine alga. *J. Gen. Appl. Microbiol.*, 48:269–279.

Ishikawa, M., K. Nakajima, Y. Itamiya, S. Furukawa, Y. Yamamoto, and K. Yamasato. 2006. *Halolactibacillus halophilus* gen. nov., sp. nov. and *Halolactibacillus miurensis* sp. nov., halophilic and alkaliphilic marine lactic acid bacteria constituting a phylogenetic lineage in Bacillus rRNA group 1. *Int. J. Syst. Evol. Microbiol.*, 55:2427–2439.

Itaya, M., K. Kondo, and T. Tanaka. 1989. A neomycin resistance gene cassette selectable in a single copy state in the *Bacillus subtilis* chromosome. *Nucl. Acids Res.*, 17:4410.

Itaya, M., I. Yamaguchi, K. Kobayashi, T. Endo, and T. Tanaka. 1990. The blasticidin S resistance gene (bsr) selectable in a single copy state in the *Bacillus subtilis* chromosome. *J. Biochem.*, 107:799–801.

Janniere, L., C. Bruand, and S. D. Ehrlich. 1990. Structurally stable *Bacillus subtilis* cloning vectors. *Gene*, 87:53–61.

Jensen, G.B., L. Andrup, A. Wilcks, L. Smidt, and O.M. Poulsen. 1996. The aggregation-mediated conjugation system of *Bacillus thuringiensis* subsp. *israelensis*: host range and kinetics of transfer. *Curr. Microbiol.*, 33:228–236.

Jeon, C.O., J.M. Lim, J.M. Lee, L.H. Xu, C.L. Jiang, and C.J. Kim. 2005. Reclassification of *Bacillus haloalkaliphilus* Fritze 1996 as *Alkalibacillus haloalkaliphilus* gen. nov., comb. nov. and the description of *Alkalibacillus salilacus* sp. nov., a novel halophilic bacterium isolated from a salt lake in China. *Int. J. Syst. Evol. Microbiol.*, 55:1891–1896.

Joung, K.B. and J.C. Cote. 2001. A phylogenetic analysis of *Bacillus thuringiensis* serovars by RFLP-based ribotyping. *J. Appl. Microbiol.*, 91:279–289.

Jürgen, B., S. Tobisch, M. Wumpelmann, et al. 2005. Global expression profiling of *Bacillus subtilis* cells during industrial-close fed-batch fermentations with different nitrogen sources. *Biotechnol. Bioeng.*, 92:277–298.

Kämpfer, P. 1994. Limits and possibilities of total fatty acid analysis for classification and identification of *Bacillus* species. *Syst. Appl. Microbiol.*, 17:86–98.

Kämpfer, P., R. Rosselló-Mora, E. Falsen, H.J. Busse, and B.J. Tindall. 2006. *Cohnella thermotolerans* gen. nov., sp. nov., and classification of '*Paenibacillus hongkongensis*' as *Cohnella hongkongensis* sp. nov. *Int. J. Syst. Evol. Microbiol.*, 56:781–786.

Kearns, D.B., F. Chu, S.S. Branda, R. Kolter, and R. Losick. 2005. A master regulator for biofilm formation by *Bacillus subtilis*. *Mol. Microbiol.*, 55:739–749.

Ko, K.S., J.-W. Kim, J.-M. Kim, et al. 2004. Population structure of the *Bacillus cereus* group as determined by sequence analysis of six housekeeping genes and the plcR gene. *Infect. Immun.*, 72:5253–5261.

Kobayashi, K., S.D. Ehrlich, A. Albertini, et al. 2003. Essential *Bacillus subtilis* genes. *Proc. Natl. Acad. Sci.* U.S.A., 100:4678–4683.

Kobayashi, Y. and F. Kawamura. 1992. Molecular cloning. *Biotechnology*, 22:123–141.

Koch, R. 1876. Die Ätiologie der Milbrandkrankheit, begrüdet die Entwicklunsgesicht des Bacillus Anthracis. *Beitrag Biol. Pflanzer*, 2:277–310.

Koek, M.M., B. Muilwijk, M.J. van der Werf, and T. Hankemeier. 2006. Microbial metabolomics with gas chromatography/mass spectrometry. *Anal. Chem.*, 78:1272–1281.

Kotiranta, A., K. Lounatmaa, and M. Haapasalo. 2000. Epidemiology and pathogenesis of *Bacillus cereus* infections. *Microbes Infect.*, 2:189–198.

Kunst, F., N. Ogasawara, I. Moszer, et al. 1997. The complete genome sequence of the Gram-positive bacterium *Bacillus subtilis*. *Nature*, 390:249–256.

L'Haridon, S., M.L. Miroshnichenko, N.A. Kostrikina, et al. 2006. *Vulcanibacillus modesticaldus* gen. nov., sp. nov., a strictly anaerobic, nitrate-reducing bacterium from deep-sea hydrothermal vents. *Int. J. Syst. Evol. Microbiol.*, 56:1047–1053.

La Duc, M.T., M. Satomi, and K. Venkateswaran. 2004. *Bacillus odysseyi* sp. nov., a round-spore-forming bacillus isolated from the Mars Odyssey spacecraft. *Int. J. Syst. Evol. Microbiol.*, 54:195–201.

Le Breton, Y., N.P. Mohapatra, and W.G. Haldenwang. 2006. *In vivo* random mutagenesis of *Bacillus subtilis* by use of TnYLB-1, a mariner-based transposon. *Appl. Envir. Microbiol.*, 72:327–333.

Lee, G. and J. Pero. 1981. Conserved nucleotide sequences in temporally controlled bacteriophage promoters. *J. Mol. Biol.*, 152:247–265.

Lee, J., AA. Goel, M.M. Ataai, and M.M. Domach. 1997. Supply-side analysis of growth of *Bacillus subtilis* on glucose-citrate medium: feasible network alternatives and yield optimality. *Appl. Environ. Microbiol.*, 63:710–718.

Lee, J.M., S. Zhang, S. Saha, et al. 2001. RNA expression analysis using an antisense *Bacillus subtilis* genome array. *J. Bacteriol.*, 183:7371–7380.

Lim, J.M., C.O. Jeon, S.M. Lee et al. 2006. *Bacillus salarius* sp. nov., a halophilic, spore-forming bacterium isolated from a salt lake in China. *Int. J. Syst. Evol. Microbiol.*, 56:373–377.

Lim, J.M., C.O. Jeon, S.M. Song, and C.J. Kim. 2005. *Pontibacillus chungwhensis* gen. nov., sp. nov., a moderately halophilic Gram-positive bacterium from a solar saltern in Korea. *Int. J. Syst. Evol. Microbiol.*, 55:165–170.

Liolios, K., N. Tavernarakis, P. Hugenholtz, and N.C. Kyrpides. 2006. The Genomes On Line Database (GOLD) v.2: a monitor of genome projects worldwide. *Nucl. Acids Res.*, 34:D332–334

Liu, X.S. and D.H. Dean. 2006. Redesigning *Bacillus thuringiensis* Cry1Aa toxin into a mosquito toxin. *Protein Eng. Des. Sel.*, 19:107–111.

Logan, N.A. and R.C. Berkeley. 1984. Identification of *Bacillus* strains using the API system. *J. Gen. Microbiol*,. 130:1871–1882.

Logan, N.A. and M. Rodriguez-Diaz. 2006. *Bacillus, Alicyclobacillus* and *Paenibacillus*. In *Principles and Practice of Clinical Bacteriology*, 2nd edition, Eds. S.H. Gillespie and P.M. Hawkey, pp. 139–158. Chichester, U.K.: John Wiley.

Logan, N.A. and P.C.B. Turnbull. 2003. *Bacillus* and related genera. In *Manual of Clinical Microbiology*, 8th edition, Ed. P.R. Murray, E.J. Baron, J.H. Jorgensen, M.A. Pfaller, and R.H. Yolken, pp. 445–460. Washington, DC: American Society for Microbiology.

Losick, R. and J. Pero. 1981. Cascades of sigma factors. *Cell*, 25:582–584.

Lu, J., Y. Nogi, and H. Takami. 2001. *Oceanobacillus iheyensis* gen. nov., sp. nov., a deep-sea extremely halotolerant and alkaliphilic species isolated from a depth of 1050 m on the Iheya Ridge. *FEMS Microbiol. Lett.*, 205:291–297.

Ludwig, W. and H.-P. Klenk. 2001. Overview: a phylogenetic backbone and taxonomic framework for prokaryotic systematics. In *Bergey's Manual of Systematic Bacteriology*, 2nd ed., Ed. G. M. Garrity, pp. 49–65. New York: Springer.

Makita, Y., M. Nakao, N. Ogasawara, and K. Nakai. 2004. DBTBS: database of transcriptional regulation in *Bacillus subtilis* and its contribution to comparative genomics. *Nucl. Acids Res*., 32:D75–77.

Mandal, M. and R.R. Breaker. 2004. Gene regulation by riboswitches. *Nat. Rev. Mol. Cell. Biol.*, 5:451–463.

Manzano, M., L. Cocolin, C. Cantoni, and G. Comi. 2003. *Bacillus cereus, Bacillus thuringiensis* and *Bacillus mycoides* differentiation using a PCR-RE technique. *Int. J. Food Microbiol.*, 81:249–254.

Marahiel, M.A., M.M. Nakano, and P. Zuber. 1993. Regulation of peptide antibiotic production in *Bacillus. Mol. Microbio*l., 7:631–636.

Margulies, M., M. Egholm, W.E. Altman, et al. 2005. Genome sequencing in microfabricated high-density picolitre reactors. *Nature*, 437:376–380.

Mayr, R., H.J. Busse, H.L. Worliczek, M. Ehling-Schulz, and S. Scherer. 2006. *Ornithinibacillus* gen. nov., with the species *Ornithinibacillus bavariensis* sp. nov. and *Ornithinibacillus californiensis* sp. nov. *Int. J. Syst. Evol. Microbiol.*, 56:1383–1389.

Mehlo, L., D. Gahakwa, P.T. Nghia, et al. 2005. An alternative strategy for sustainable pest resistance in genetically enhanced crops. *Proc. Natl. Acad. Sci. U.S.A.*, 102:7812–7816.

Meima, R.B., J.M. van Dijl, S. Holsappel, and S. Bron. 2004. Expression systems in *Bacillus*. In *Protein Expression Technologies: Current Status and Future Trends*, Ed. F. Baneyx, pp. 199–252. Wymondham, U.K.: Horizon Bioscience.

Missiakas, D. and S. Raina. 1998. The extracytoplasmic function sigma factors: role and regulation. *Mol. Microbiol*, 28:1059–1066.

Nakamura, K., S. Haruta, S. Ueno, M. Ishii, A. Yokota, and Y. Igarashi. 2004. *Cerasibacillus quisquiliarum* gen. nov., sp. nov., isolated from a semi-continuous decomposing system of kitchen refuse. *Int. J. Syst. Evol. Microbiol.*, 54:1063–1069.

Nakano, M.M., Y.P. Dailly, P. Zuber, and D.P. Clark. 1997. Characterization of anaerobic fermentative growth of Bacillus subtilis: identification of fermentation end products and genes required for growth. *J. Bacteriol.*, 179:6749–6755.

Nakano, M.M., F. Yang, P. Hardin, and P. Zuber. 1995. Nitrogen regulation of *nasA* and the *nasB* operon, which encode genes required for nitrate assimilation in *Bacillus subtilis. J. Bacteriol.*, 177:573–579.

Nakano, M.M. and P. Zuber. 1998. Anaerobic growth of a "strict aerobe." *Annu. Rev. Microbiol.*, 52:165–190.

Nakano, M.M., P. Zuber, P. Glaser, A. Danchin, and F. M. Hulett. 1996. Two-component regulatory proteins ResD-ResE are required for transcriptional activation of fnr upon oxygen limitation in *Bacillus subtilis. J. Bacteriol.*, 178:3796–3802.

Nazina, T.N., T.P. Tourova, A.B. Poltaraus, et al. 2001. Taxonomic study of aerobic thermophilic bacilli: descriptions of *Geobacillus subterraneus* gen. nov., sp. nov. and *Geobacillus uzenensis* sp. nov. from petroleum reservoirs and transfer of *Bacillus stearothermophilus, Bacillus thermocatenulatus, Bacillus thermoleovorans, Bacillus kaustophilus, Bacillus thermoglucosidasius and Bacillus thermodenitrificans* to *Geobacillus* as the new combinations *G. stearothermophilus, G. thermocatenulatus, G. thermoleovorans, G. kaustophilus, G. thermoglucosidasius* and *G. thermodenitrificans. Int. J. System. Evol. Microbiol.*, 51:433–446.

Nguyen, H.D., Q.A. Nguyen, R.C. Ferreira, L.C.S. Ferreira, L.T. Tran, and W. Schumann. 2005. Construction of plasmid-based expression vectors for *Bacillus subtilis* exhibiting full structural stability. *Plasmid*. 54:241–248.

Nielsen, J.E. and T.V. Borchert. 2000. Protein engineering of bacterial alpha-amylases. *Biochim. Biophys. Acta*, 1543:253–274.

Niimura, Y., E. Koh, F. Yanagida, K.-I. Suzuki, K. Komagata, and M. Kozaki. 1990. *Amphibacillus xylanus* gen. nov., sp. nov., a facultatively anaerobic sporeforming xylan-digesting bacterium which lacks cytochrome, quinone, and catalase. *Int. J. Syst. Bacteriol.*, 40:297–301.

Noirot, P. and M.F. Noirot-Gros. 2004. Protein interaction networks in bacteria. *Curr. Opin. Microbiol,*. 7:505–512.

Nunes, I., I. Tiago, A.L. Pires, M.S. Da Costa, and A. Veríssimo. 2005. *Paucisalibacillus globulus* gen. nov., sp. nov., a Gram-positive bacterium isolated from potting soil. *Int. J. Syst. Evol. Microbiol.*, 56:1841–1845.

Nyren, P. 2006. The history of pyrosequencing. *Methods Mol. Biol.*, 373:1–14.

Nystrand, R. 1984. *Saccharococcus thermophilus* gen. nov., sp. nov., isolated from beet sugar extraction. *Syst. Appl. Microbiol.,* 5:204–219.

Olempska-Beer, Z.S., R.I. Merker, M.D. Ditto, and M.J. DiNovi. 2006. Food-processing enzymes from recombinant microorganisms — A review. *Regul. Toxicol. Pharmacol.*, 45:144–158.

Olivera, N., F. Siñeriz, and J.D. Breccia. 2005. *Bacillus patagoniensis* sp. nov., a novel alkalitolerant bacterium from the rhizosphere of Atriplex lampa in Patagonia, Argentina. *Int. J. Syst. Evol. Microbiol.*, 55:443–447.

Palmisano, M.M., L.K. Nakamura, K.E. Duncan, C.A. Istock, and F.M. Cohan. 2001. *Bacillus sonorensis* sp. nov., a close relative of *Bacillus licheniformis*, isolated from soil in the Sonoran Desert, Arizona. *Int. J. Syst. Evol. Microbiol.*, 51:1671–1679.

Perego, M. and J. A. Hoch. 2001. Two-component systems, phosphorelays, and regulation of their activities. In *Bacillus subtilis and Its Closest Relatives: From Genes to Cells*, Ed. A. L. Sonenshein, J. A. Hoch and R. Losick, pp. 473-482. Washington, DC: ASM Press

Perkins, J.B., M. Goese, and G. Schyns. 2004. Thiamin production by fermentation. World Patent Application, WO 2004/106557.

Perkins, J.B. and Z. Pragai. 2004. Production of pantothenate using microorganisms incapable of sporulation. World Patent Application, WO 2004/113510.

Perkins, J.B., A. Sloma, T. Hermann, et al. 1999. Genetic engineering of *Bacillus subtilis* for the commercial production of riboflavin. *J. Ind. Microbiol. Biotechnol.*, 22:8–18.

Perkins, J.B., M. Wyss, W. Sauer, and H.-P. Hohmann. 2008. Metabolic engineering in *Bacillus subtilis.* In *The Metabolic Pathway Engineering Handbook*, Ed. C. Smolke, sect. 5, 3, in press. Boca Raton, FL: CRC Press.

Perkins, J.B. and P.J. Youngman. 1983. *Streptococcus* plasmid pAMβ1 is a composite of two separable replicons, one of which is closely related to *Bacillus* plasmid pBC16. *J. Bacteriol.*, 155:607–615.

Perkins, J.B. and P.J. Youngman. 1986. Construction and properties of Tn917–lac, a transposon derivative that mediates transcriptional gene fusions in *Bacillus subtilis. Proc. Natl. Acad. Sci. U.S.A.*, 83:140–144.

Perkins, J.B. and D.R. Zeigler. 2008. Unpublished data.

Pero, J. and A. Sloma. 1993. Proteases. In *Bacillus subtilis and Other Gram-Positive Bacteria: Biochemistry, Physiology, and Molecular Genetics*, Eds. A.L. Sonenshein, J.A. Hoch, and R. Losick, pp. 939–952. Washington, D.C.: American Society for Microbiology.

Phelps, R.J. and J.L. McKillip. 2002. Enterotoxin production in natural isolates of Bacillaceae outside the *Bacillus cereus* group. *Appl. Environ. Microbiol.*, 68:3147–3151.

Piggot, P. and R. Losick. 2001. Sporulation genes and intercompartmental regulation. In *Bacillus subtilis and Its Closest Relatives: From Genes to Cells*, Eds. A.L. Sonenshein, J.A. Hoch, and R. Losick, pp. 483–517. Washington, DC: ASM Press.

Pikuta, E., A. Lysenko, N. Chuvilskaya, et al. 2000. *Anoxybacillus pushchinensis* gen. nov., sp. nov., a novel anaerobic, alkaliphilic, moderately thermophilic bacterium from manure, and description of *Anoxybacillus flavithermus* comb. nov. *Int. J. Syst. Evol. Microbiol.*, 50:2109–2117.

Poluektova, E.U., E.A. Fedorina, O.V. Lotareva, and A.A. Prozorov. 2004. Plasmid transfer in bacilli by a self-transmissible plasmid p19 from a *Bacillus subtilis* soil strain. *Plasmid*, 52:212–217.

Priest, F.G. 1993. Systematics and ecology of *Bacillus.* In *Bacillus subtilis and Other Gram-Positive Bacteria: Biochemistry, Physiology, and Molecular Genetics*, Eds. A. L. Sonenshein, J.A. Hoch, and R. Losick, p. 1–16. Washington, DC: American Society for Microbiology.

Priest, F.G. and B. Alexander. 1988. A frequency matrix for probabilistic identification of some Bacilli. *J. Gen. Microbiol.*, 134:3001–3018.

Priest, F.G., M. Barker, L.W.J. Baillie, E.C. Holmes, and M.C.J. Maiden. 2004. Population structure and evolution of the *Bacillus cereus* group. *J. Bacteriol.*, 186:7959–7970.

Priest, F.G., A.B. Fleming, M. Tangney, P.L. Jorgensen, and B. Diderichsen. 1997. Production of proteins using *Bacillus* incapable of sporulation. World Patent Application. WO/1997/003185.

Quax, W.J. 2003. Bacterial Enzymes. *In The Prokaryotes: An Evolving Electronic Source for the Microbiological Community*, 3rd edition, Eds. M. Dworkin. Epub. New York: Springer-Verlag.

Rampersad, J., A. Khan, and D. Ammons. 2003. A *Bacillus thuringiensis* isolate possessing a spore-associated filament. *Curr. Microbiol.*, 47:355–357.

Rasko, D.A., M.R. Altherr, C.S. Han, and J. Ravel. 2005. Genomics of the *Bacillus cereus* group of organisms. *FEMS Microbiol. Rev.*, 29:303–329.

Ren, P.G. and P.J. Zhou. 2005a. *Salinibacillus aidingensis* gen. nov., sp. nov. and *Salinibacillus kushneri* sp. nov., moderately halophilic bacteria isolated from a neutral saline lake in Xin-Jiang, China. *Int. J. Syst. Evol. Microbiol.*, 55:949–953.

Ren, P.G. and P.J. Zhou. 2005b. *Tenuibacillus multivorans* gen. nov., sp. nov., a moderately halophilic bacterium isolated from saline soil in Xin-Jiang, China. *Int. J. Syst. Evol. Microbiol.*, 55:95–99.

Reva, O.N., V.V. Smirnov, B. Pettersson, and F.G. Priest. 2002. *Bacillus endophyticus* sp. nov., isolated from the inner tissues of cotton plants, *Gossypium* sp. *Int. J. Syst. Evol. Microbiol.*, 52:101–107.

Reva, O.N., I.B. Sorokulova, and V.V. Smirnov. 2001. Simplified technique for identification of the aerobic spore-forming bacteria by phenotype. *Int. J. Syst. Evol. Microbiol.*, 51:1361–1371.

Rüger, H.J., D. Fritze, and C. Spröer. 2000. New psychrophilic and psychrotolerant *Bacillus marinus* strains from tropical and polar deep-sea sediments and emended description of the species. *Int. J. Syst. Evol. Microbiol.*, 50:1305–1313.

Russell, A.D. 1982. *The Destruction of Bacterial Spores*. London: Academic Press.

Ryan, K.R. and L. Shapiro. 2003. Temperal and spatial regulation in prokaryotic cell cycle progression and development. *Annu. Rev. Biochem.*, 72:367–394.

Sanchis, V., H. Agaisse, J. Chaufaux, and D. Lereclus. 1996. Construction of new insecticidal *Bacillus thuringiensis* recombinant strains by using the sporulation non-dependent expression system of cryIIIA and a site specific recombination vector. *J. Biotechnol.*, 48:81–96.

Sanders, M.E., L. Morelli, and T.A. Tompkins. 2003. Sporeformers as human probiotics: *Bacillus, Sporolactobacillus,* and *Brevibacillus*. Comprehens. *Rev. Food Sci. Food Safety*, 2:101–110.

Satomi, M., M.T. La Duc, and K. Venkateswaran. 2006. *Bacillus safensis* sp. nov., isolated from spacecraft and assembly-facility surfaces. *Int. J. Syst. Evol. Microbiol.* 56:1735–1740.

Sauer, U. 2006. Metabolic networks in motion: 13 C-based flux analysis. *Mol. Syst. Biol.*, 2:1–10.

Sauer, U. and B.J. Elkmanns. 2005. The PEP-pyruvate-oxaloacetate node as the switch point for carbon flux distribution in bacteria. *FEMS Microbiol.* Rev., 29:765–794.

Schallmey, M., A. Singh, and O.P. Ward. 2004. Developments in the use of *Bacillus* species for industrial production. *Can. J. Microbiol.*, 50:1–17.

Schlesner, H., P.A. Lawson, M.D. Collins, et al. 2001. *Filobacillus milensis* gen. nov., sp. nov., a new halophilic spore-forming bacterium with Orn-D-Glu-type peptidoglycan. *Int. J. Syst. Evol. Microbiol.*, 51:425–431.

Schmidt-Dannert, C., J. Pleiss, and R.D. Schmid. 1988. A toolbox of recombinant lipases for industrial applications. *Ann. N.Y. Acad. Sci.*, 864:14–22.

Schneider, K.R., M.E. Parish, R.M. Goodrich, and T. Cookingham. 2004. Preventing foodborne illness: *Bacillus cereus* and *Bacillus anthracis,* University of Florida (FSHN04-05 http://edis.ifas.ufl.edu/FS103).

Schnepf, E., N. Crickmore, J. Van Rie, et al. 1998. *Bacillus thuringiensis* and its pesticidal crystal proteins. *Microbiol. Mol. Biol. Rev.*, 62:775–806.

Schoeni, J.L. and A. Wong. 20005. *Bacillus cereus* food poisoning and its toxins. *J. Food Prot.*, 68:636–648.

Schraft, H., M. Steele, B. McNab, J. Odumeru, and M.W. Griffiths. 1996. Epidemiological typing of *Bacillus* spp. isolated from food. *Appl. Environ. Microbiol.*, 62:4229–4232.

Schumann, W., S.D. Ehrlich, and N. Ogasawara, Eds. 2001. *Functional Analysis of Bacterial Genes: A Practical Manual*. West Sussex, U.K.: John Wiley & Sons, Ltd.

Scorpio, A., T.E. Blank, W.A. Day, and D.J. Chabot. 2006. Anthrax vaccines: Pasteur to the present. *Cell. Mol. Life Sci.*, 63:2237–2248.

Shapiro, L. and R. Losick. 2000. Dynamic spatial regulation in the bacterial cell. *Cell*, 100:89–98.

Shaver, Y., M.L. Nagpal, R. Rudner, L.K. Nakamura, K.F. Fox, and A. Fox. 2002. Restriction fragment length polymorphism of rRNA operons for discrimination and intergenic spacer sequences for cataloging of *Bacillus subtilis* sub-groups. *J. Microbiol. Meth.*, 50:215–223.

Shida, O., H. Takagi, K. Kadowaki, and K. Komagata. 1996. Proposal for two new genera, *Brevibacillus* gen. nov. and *Aneurinibacillus* gen. nov. *Int. J. Syst. Bacteriol.*, 46:939–946.

Shida, O., H. Takagi, K. Kadowaki, L.K. Nakamura, and K. Komagata. 1997. Transfer of *Bacillus alginolyticus, Bacillus chondroitinus, Bacillus curdlanolyticus, Bacillus glucanolyticus, Bacillus kobensis, and Bacillus thiaminolyticus* to the genus *Paenibacillus* and emended description of the genus *Paenibacillus. Int. J. Syst. Bacteriol.*, 47:289–298.

Shivaji, S., P. Chaturvedi, K. Suresh, et al. 2006. *Bacillus aerius* sp. nov., *Bacillus aerophilus* sp. nov., *Bacillus stratosphericus* sp. nov. and *Bacillus altitudinis* sp. nov., isolated from cryogenic tubes used for collecting air samples from high altitudes. *Int. J. Syst. Evol. Microbiol.*, 56:1465–1473.

Silvaggi, J.M., D.L. Popham, A. Driks, P. Eichenberger, and R. Losick. 2004. Unmasking novel sporulation genes in *Bacillus subtilis. J. Bacteriol.*, 186:8089–8095.

Silvaggi, J.M., J.B. Perkins, and R. Losick. 2005. Small untranslated RNA antitoxin in *Bacillus subtilis. J. Bacteriol.*, 187:6641–6650.

Silvaggi, J.M., J.B. Perkins, and R. Losick. 2006. Genes for small, noncoding RNAs under sporulation control in *Bacillus subtilis. J. Bacteriol.*, 188:532–541.

Simbahan, J., R. Drijber, and P. Blum. 2004. *Alicyclobacillus vulcanalis* sp. nov., a thermophilic, acidophilic bacterium isolated from Coso Hot Springs, California, USA. *Int. J. Syst. Evol. Microbiol.*, 54:1703–1707.

Skerman, V.B.D., V. McGowan, and P.H.A. Sneath. 1980. Approved lists of bacterial names. *Int. J. Syst. Bacteriol.*, 30:225–420.

Smith, N.R., R.E. Gordon, and F.E. Clark. 1946. Aerobic Mesophilic Sporeforming Bacteria. Washington, D.C.: United States Department of Agriculture.

Soga, T., Y. Ohashi, Y. Ueno, H. Naraoka, M. Tomita, and T. Nishioka. 2003. Quantitative metabolome analysis using capillary electrophoresis mass spectrometry. *J. Proteome Res.*, 2:488–494.

Sonenshein, A.L. 2001. The Kreb citric acid cycle. In *Bacillus subtilis and Its Closest Relatives: From Genes to Cells*, Eds. A.L. Sonenshein, J.A. Hoch, and R. Losick, pp. 151–162. Washington, DC: ASM Press.

Sonenshein, A.L., J.A. Hoch, and R. Losick, Eds. 1993. *Bacillus subtilis and Other Gram-Positive Bacteria: Biochemistry, Physiology, and Molecular Genetics.* Washington DC: American Society for Microbiology.

Sonenshein, A.L., J.A. Hoch, and R. Losick, Eds. 2001. *Bacillus subtilis and Its Closest Relatives: From Genes to Cells.* Washington, DC, ASM Press.

Spizizen, J. 1958. Transformation of biochemically deficient strains of *Bacillus subtilis* by deoxyribonucleate. *Proc. Natl. Acad. Sci. U.S.A.*, 44:1072–1078.

Spring, S., W. Ludwig, M.C. Marquez, A. Ventosa, and K.H. Schleifer. 1996. *Halobacillus* gen. nov., with descriptions of *Halobacillus litoralis* sp. nov. and *Halobacillus trueperi* sp. nov., and transfer of Sporosarcina halophila to *Halobacillus halophilus* comb. nov. *Int. J. Syst. Bacteriol.*, 46:492–496.

Stackebrandt, E., W. Frederiksen, G.M. Garrity, et al. 2002. Report of the ad hoc committee for the re-evaluation of the species definition in bacteriology. *Int. J. Syst. Evol. Microbiol.*, 52:1043–1047.

Stackebrandt, E. and B.M. Goebel. 1994. Taxonomic note: a place for DNA-DNA reassociation and 16S rRNA sequence analysis in the present species definition in bacteriology. *Int. J. Syst. Bacteriol.*, 44:846–849.

Stein, T. 2005. *Bacillus subtilis* antibiotics: structures, syntheses and specific functions. *Mol. Microbiol.*, 56:845–857.

Steinmetz, M. and R. Richter. 1994. Easy cloning of mini-Tn10 insertions from *Bacillus subtilis* chromosome. *J. Bacteriol.*, 176:1761–1763.

Stragier, P. and R. Losick. 1996. Molecular genetics of sporulation in *Bacillus subtilis. Annu. Rev. Genet.*, 30:297–341.

Sudarsan, N., M.C. Hammond, K.F. Block, et al. 2006. Tandem riboswitch architectures exhibit complex gene control functions. *Science*, 314:300–304.

Sun, G., E. Sharkova, R. Chesnut, et al. 1996. Regulators of aerobic and anaerobic respiration in *Bacillus subtilis*. J. *Bacteriol.*, 178:1374–1385.

Sung, G.P. and M.D. Yudkin. 1997. Sequencing and phylogenetic analysis of the spoIIA operon from diverse *Bacillus* and *Paenibacillus* species. *Gene*, 194:25–33.

Switzer Blum, J., A. Burns Bindi, J. Buzzelli, J.F. Stolz, and R.S. Oremland. 1998. *Bacillus arsenicoselenatis*, sp. nov., and *Bacillus selenitireducens*, sp. nov.: two haloalkaliphiles from Mono Lake, California that respire oxyanions of selenium and arsenic. *Arch. Microbiol.*, 171:19–30.

Szurmant, H. and G.W. Ordal. 2004. Diversity in chemotaxis mechanisms among the bacteria and archaea. *Microbiol. Mol. Rev.*, 68:301–319.

Takagi, H. 1993. Protein engineering on subtilisin. *Int. J. Biochem.*, 25:307–312.

Tam, N.K.M., N. Q. Uyen, H.A. Hong, et al. 2006. The intestinal life cycle of *Bacillus subtilis* and close relatives. *J. Bacteriol.*, 188:2692–2700.

Terpe, K. 2006. Overview of bacterial expression systems for heterologous protein production: from molecular and biochemical fundamentals to commercial systems. *Appl. Microbiol. Biotechnol.*, 72:211–222.

Tettelin, H., V. Masignani, M.J. Cieslewicz, et al. 2005. Genome analysis of multiple pathogenic isolates of *Streptococcus agalactiae*: implications for the microbial "pan-genome." *Proc. Natl. Acad. Sci. U.S.A.*, 102:13950–13955.

Tiago, I., A.P. Chung, and A. Veríssimo. 2004. Bacterial diversity in a nonsaline alkaline environment: heterotrophic aerobic populations. *Appl. Environ. Microbiol.*, 70:7378–7387.

Touzel, J.P., M.O'Donohue, P. Debeire, E. Samain, and C. Breton. 2000. *Thermobacillus xylanilyticus* gen. nov., sp. nov., a new aerobic thermophilic xylan-degrading bacterium isolated from farm soil. *Int. J. Syst. Evol. Microbiol.*, 50:315–320.

Turnbull, P.C.B. 1991. Anthrax vaccines: past, present and future. *Vaccine*, 9:533–539.

Tye, A., F. Siu, T. Leung, and B. Lim. 2002. Molecular cloning and the biochemical characterization of two novel phytases from *B. subtilis 168* and *B. licheniformis*. *Appl. Microbiol. Biotechnol.*, 59:190–197.

Udaka, S. and H. Yamagata. 1993-1994. Protein secretion in *Bacillus brevis*. *Antonie van Leeuwenhoek*, 64:137–143.

Ueda, S. 1989. Industrial applications of *B. subtilis*: utilization of soybean as natto, a traditional Japanese food. In *Bacillus subtilis: Molecular Biology and Industrial Applications*, Eds. B. Maruo and H. Yoshikawa, p. 143–161. Amsterdam: Elsevier Science B.V.

Ugoji, E.O., M.D. Laing, and C.H. Hunter. 2005. Colonization of *Bacillus* spp. on seeds and in the plant rhizoplane. *J. Environ. Biol.*, 26:459–466.

Vargas, V.A., O.D. Delgado, R. Hatti-Kaul, and B. Mattiasson. 2005. *Bacillus bogoriensis* sp. nov., a novel alkaliphilic, halotolerant bacterium isolated from a Kenyan soda lake. *Int. J. Syst. Evol. Microbiol.*, 55:899–902.

Vary, P. 1992. Development of genetic engineering in *Bacillus megaterium*. *Biotechnology*, 22:251–310.

Vary, P. 1994. Prime time for *Bacillus megaterium*. *Microbiology*, 140:1001–1013.

Venkateswaran, K., M. Kempf, F. Chen, M. Satomi, W. Nicholson, and R. Kern. 2003. *Bacillus nealsonii* sp. nov., isolated from a spacecraft-assembly facility, whose spores are γ-radiation resistant. *Int. J. Syst. Evol. Microbiol.*, 53:165–172.

Vitreschak, A.G., D.A. Rodionov, A.A. Mironov, and M.S. Gelfand. 2004. Riboswitches: the oldest mechanism for the regulation of gene expression. *Trends Genet.*, 20:44–50.

Wainø, M., B.J. Tindall, P. Schumann, and K. Ingvorsen. 1999. *Gracilibacillus* gen. nov., with description of *Gracilibacillus halotolerans* gen. nov., sp. nov.; transfer of *Bacillus dipsosauri* to *Gracilibacillus dipsosauri* comb. nov., and *Bacillus salexigens* to the genus *Salibacillus* gen. nov., as *Salibacillus salexigens* comb. nov. *Int. J. Syst. Evol. Microbiol.*, 49:821–831.

Wang, L.F. 1992. Useful *Bacillus* strains and plasmids. *Biotechnology*, 22:339–347.

Warren, M. 2007. personal communication.

Wayne, L.G., D.J. Brenner, R.R. Colwell, et al. 1987. Report of the Ad-Hoc-Committee on Reconciliation of Approaches to Bacterial Systematics. *Int. J. Syst. Bacteriol.*, 37:463–464.

Westers, H., R. Dorenbos, J.M. van Dijl et al. 2003. Genome engineering reveals large dispensable regions in *Bacillus subtilis*. *Mol. Biol. Evol.*, 20:2076–2090.

Westers, L., W. Westers, and W.J. Quax. 2004. *Bacillus subtilis* as a cell factory for pharmaceutical proteins: a biotechnological approach to optimize the host organism. *Biochim. Biophys. Acta,* 1694:299–310.

Wisotzkey, J.D., P. Jurtshuk, Jr., G.E. Fox, G. Deinhard. and K. Poralla. 1992. Comparative sequences analyses on the 16S rRNA (rDNA) of *Bacillus acidocaldarius, Bacillus acidoterrestris*, and *Bacillus cycloheptanicus* and proposal for creation of a new genus, *Alicyclobacillus* gen. nov. *Int. J. Syst. Bacteriol.*, 42:263–269.

Woese, C.R. and G.E. Fox. 1977. Phylogenetic structure of the prokaryotic domain: the primary kingdoms. *Proc. Natl. Acad. Sci. U.S.A.*, 74:5088–5090.

Wu, S.C., J.C. Yeung, Y. Duan et al. 2002. Functional production and characterization of a fibrin-specific single-chain antibody fragment from *Bacillus subtilis*: effects of molecular chaperones and a wall-bound protease on antibody fragment production. *Appl. Environ. Microbiol.*, 68:3261–3269.

Xu, D. and J. Cote. 2006. Sequence diversity of the *Bacillus thuringiensis* and *B. cereus* sensu lato flagellin (H antigen) protein: comparison with H serotype diversity. *Appl. Environ. Microbiol.,* 72:4653–4662.

Xue, Y., X. Zhang, C. Zhou, et al. 2006. *Caldalkalibacillus thermarum* gen. nov., sp. nov., a novel alkalithermophilic bacterium from a hot spring in China. *Int. J. Syst. Evol. Microbiol.*, 56:1217–1221.

Yocum, R.R. and T.A. Patterson. 2005. Microorganisms and Assays for the Identification of Antibiotics. U.S. Patent Application US2005158842.

Yoon, J.H., K.H. Kang, and Y. -H. Park. 2002. *Lentibacillus salicampi* gen. nov., sp. nov., a moderately halophilic bacterium isolated from a salt field in Korea. *Int. J. Syst. Evol. Microbiol.*, 52:2043–2048.

Yoon, J.H., N. Weiss, K.C. Lee, I.S. Lee, K.H. Kang, and Y.H. Park. 2001. *Jeotgalibacillus alimentarius* gen. nov., sp. nov., a novel bacterium isolated from jeotgal with L-lysine in the cell wall, and reclassification of *Bacillus marinus* Rüger 1983 as *Marinibacillus marinus* gen. nov., comb. nov. *Int. J. Syst. Evol. Microbiol.*, 51:2087–2093.

Youngman, P.J., J.B. Perkins, and R. Losick. 1983. Genetic transposition and insertional mutagenesis in *Bacillus subtilis* with *Streptococcus faecalis* transposon Tn917. *Proc. Natl. Acad. Sci. U.S.A.*, 80:2305–2309.

Zahner, V., H. Momen, C.A. Salles, and L. Rabinovitch. 1989. A comparative study of enzyme variation in *Bacillus cereus* and *Bacillus thuringiensis*. *J. Appl. Bacteriol.*, 67:275–282.

Zahner, V., L. Rabinovitch, C.F. Cavados, and H. Momen. 1994. Multilocus enzyme electrophoresis on agarose gel as an aid to the identification of entomopathogenic *Bacillus sphaericus* strains. *J. Appl. Bacteriol.*, 76:327–335.

Zeigler, D.R. 2003. Gene sequences useful for predicting relatedness of whole genomes in bacteria. *Int. J. Syst. Evol. Microbiol.*, 53:1893–1900.

Zeigler, D.R. 2005. Application of a *recN* sequence similarity analysis to the identification of species within the bacterial genus *Geobacillus*. *Int. J. Syst. Evol. Microbiol.*, 55:1171–1179.

25 The Genus *Clostridium*

Peter Dürre

CONTENTS

INTRODUCTION

Clostridium is one of the largest bacterial genera, including more than 150 validly described species. Among these are several with enormous biotechnological potential (e.g., for production of biofuels, bulk chemicals, and important enzymes, as well as for usage in cancer treatment) and also a few well-known pathogens. However, even some of their toxins proved meanwhile to be valuable in medical and cosmetic applications. Clostridia thus belong to the avant-garde of industrially useful microbes. Members of this genus, in general, stain Gram-positive, are more or less strictly anaerobic bacteria, employ an impressive number of varying fermentation routes, and are able to degrade numerous natural and artificial substances. Due to required precautions for excluding oxygen during handling, clostridia were for a long time virtually inaccessible at the genetic level. This situation has completely changed. Gene cloning, DNA transfer, gene expression modulation, and gene knockout systems have been successfully established. Thus, the road is paved for further elucidation and commercial exploitation of the enormous metabolic potential of the clostridia.

In the past 18 years, five books have been published, completely devoted to clostridia (Minton and Clarke, 1989; Woods, 1993; Rood et al., 1997; Bahl and Dürre, 2001; Dürre, 2005). Due to the limited space of this chapter, the interested reader is referred to these references for additional and more detailed information.

HISTORICAL BACKGROUND

Hippocrates (460–377 BC), a physician living on the Greek island of Kos, was probably the first to report clostridia-effected diseases. His documentation describes gas gangrene (Sussman, 1958), caused by *Clostridium histolyticum*, and lockjaw (also called opisthotonus or tetanus) (Kiple, 1993), caused by *C. tetani*. Until the Middle Ages, indigo dyeing in Europe was based on the woad plant. It contains two indigo precursors, isatan B (indoxyl-5-ketogluconate) and indican (indoxyl-β-D-glucoside), that are converted in a two-step process to indigo. Scientific research during the past decade revealed that the responsible bacteria are *Enterobacter agglomerans* for the initial aerobic process and *C. isatidis* for the successive fermentation (Ewerdwalbesloh and Meyer, 1995; Padden et al., 1998; Padden et al., 1999). Similarly, it is now clear that flax and hemp retting, in use for thousands of years by mankind for clothing fabrication, are essentially dependent on clostridia (Bahl and Dürre, 1993; Tamburini et al., 2003). Microbiological investigations of these bacteria only started in 1861 when the famous French microbiologist Louis Pasteur discovered that microbes exist that can grow without oxygen (Pasteur, 1861). This represented a revolutionary finding at that time. It is usually concluded that the term "*Clostridium*" stems from Prazmowski (1880). However, before him, Trécul had already used this expression (compare Dürre, 2001). It is based on a Greek root, later became Latinized, and means "small spindle." Originally, it was used exclusively to describe the bacterial morphology, not any metabolic properties (Dürre, 2001). The first pure culture obtained was *C. butyricum* (Prazmowski, 1880), which now represents the type species of the genus *Clostridium*. Without a doubt, one of the world's most famous bacteria is *C. acetobutylicum*. It had both enormous biotechnological as well as political impact. The acetone-butanol fermentation ranks second in size to ethanol production and was thus one of the largest bioprocesses ever performed. Until about 1950 two-thirds of the world's butanol production was obtained by fermentation, and the solvent was an important bulk chemical for industry (Jones and Woods, 1986; Dürre and Bahl, 1996). Originally, however, acetone was the desired product. Weizmann isolated the organism during World War I in the United Kingdom (Weizmann, 1915), when this chemical was in demand for production of ammunition. Despite his important contribution, he refused any honors but made clear that he was in favor of a Jewish homeland in Palestine. There is no doubt that the Balfour Declaration of 1917 on this very subject was affected by Weizmann's achievements, who later became the first president of the State of Israel.

METABOLISM

Clostridia in the public opinion are associated with production of a bad smell. In most cases, this is caused by butyric acid, one of the major fermentation products. Butyrate is produced by saccharolytic as well as proteolytic pathways (Table 25.1). For the former, the uptake of glucose (and other sugars) is performed by a phosphoenolpyruvate:phosphotransferase system. Glycolytic reactions then lead to pyruvate, which is split into acetyl-CoA, CO_2, and reduced ferredoxin by pyruvate: ferredoxin-oxidoreductase. Ferredoxin is a small iron-sulfur protein with a very negative redox potential (approx. −400 mV). This is of considerable advantage for the anaerobic clostridia, as they can "blow off" reducing equivalents as hydrogen (employing an additional hydrogenase) and do not need to sacrifice carbon compounds as acceptor molecules. Acetyl-CoA is the starting point for the production of acetate (via phosphotransacetylase and acetate kinase, thereby also yielding ATP), ethanol (via acetaldehyde and alcohol dehydrogenases, thereby consuming reducing equivalents), and C_4 compounds such as butyrate and butanol. The latter step is accomplished by fusion of two acetyl-CoA to acetoacetyl-CoA and free coenzyme A. Acetoacetyl-CoA is reduced to butyryl-CoA, which is transformed into butyrate or butanol by reactions analogous to those of the C_2 compounds. The ATP-yielding reaction might either be catalyzed by a butyrate kinase or by transfer of the CoA moiety to acetate and then conversion of acetyl-CoA via acetyl phosphate to acetate by acetate kinase. For caproate formation in the case of *Clostridium kluyveri*, a third molecule of acetyl-CoA is fused to butyryl-CoA. Except for the substrate combination ethanol/acetate, this organism can also grow on ethanol/succinate, thereby forming acetate and butyrate (in this case, directly from a C_4 compound). Acetone is made from acetoacetate after transfer of the CoA moiety to either acetate or butyrate, and isopropanol (e.g., by *C. beijerinckii*) by further reduction of acetone.

Proteolytic production of butyrate follows either the pathways described above starting with acetyl-CoA, or it is made directly from the C_4 compound threonine by a 2-,3-elimination reaction of the amino group (Table 25.1) (Buckel, 1990). For utilization of amino acids, fermentation in parallel oxidative and reductive branches is very common. Examples are the so-called Stickland reaction, in which pairs of amino acids are degraded, or the Se-dependent glycine fermentation, in which one

TABLE 25.1
Clostridial Fermentations

Fermentation Type	Typical Fermentation Balance	Example for Respective Species
saccharolytic:		
butyrate	glucose → 0.8 butyrate + 0.4 acetate + 2 CO_2 + 2.1 H_2	*C. butyricum*
butanol	glucose → 0.6 butanol + 0.2 acetone + 2.2 CO_2 + 1.4 H_2 + 0.04 butyrate + 0.14 acetate + 0.07 ethanol	*C. acetobutylicum*
homoacetate	fructose → 3 acetate	*C. aceticum*
acidotrophic:		
propionate	lactate → 0.66 propionate + 0.33 acetate + 0.33 CO_2	*C. propionicum*
propionate, Na^+-dependent	succinate → propionate + CO_2	probably *C. mayombei*
alcoholotrophic	2 ethanol + acetate → butyrate + 0.33 caproate + 0.66 H_2	*C. kluyveri*
autotrophic	2 CO_2 + 4 H_2 → acetate	*C. aceticum*
proteolytic		
pairs of amino acids		
Stickland reaction	alanine + 2 glycine → 3 acetate + CO_2 + 3 NH_4^+	*C. sporogenes*
single amino acids:		
2-,3-elimination	threonine → 0.66 propionate + 0.33 butyrate + 0.66 CO_2 + NH_4^+	*C. propionicum*
B_{12}-dependent	glutamate → 0.5 acetate + 0.25 butyrate + 0.5 CO_2 + NH_4^+	*C. tetanomorphum*
SAM-dependent	lysine → butyrate + acetate + 2 NH_4^+	*C. subterminale*
Se-dependent	glycine → 0.75 acetate + 0.5 CO_2 + NH_4^+	*C. purinilyticum*
heteroaromaticotrophic		
purine	adenine → acetate + formate + 2 CO_2 + 5 NH_4^+	*C. purinilyticum*
pyrimidine	orotic acid → aspartate + CO_2 + NH_4^+	*C. oroticum*

Note: SAM, S-adenosylmethionine.

glycine is completely oxidized to CO_2 and the reducing equivalents are used to reduce three glycines to acetate via the selenoenzyme glycine reductase and producing ATP from acetyl-phosphate via acetate kinase (Dürre and Andreesen, 1982). In many cases, amino acids are transformed into the respective 2-oxo compounds, removing the amino group by 2-,3-elimination as already mentioned, oxidation, or transamination (Buckel, 1990). The 2-oxo acid can then be converted into a saturated fatty acid by a number of reduction, CoA activation, and dehydration steps. Other pathways employ B_{12}-dependent C-C or S-adenosylmethionine-dependent C-N rearrangements.

Another important route in some saccharolytic clostridia is homoacetate fermentation. By reutilization of the CO_2 formed in pyruvate degradation, three acetates are formed per hexose. Species such as *Clostridium aceticum* are also able to grow autotrophically by converting CO_2 and H_2 into acetate. It is not yet definitely clear how energy conservation is achieved under such conditions. However, the presence of cytochromes indicates the generation of a proton gradient that might serve for ATP production.

A variety of organic acids, alcohols, polymers, aromatics, and halogenated substances can also be used as substrates by clostridia. For a detailed description, see Bahl and Dürre (1993). Several typical examples are listed in Table 25.1. Also, some clostridia such as *Clostridium pasteurianum* are able to fix molecular nitrogen, thereby producing NH_4^+ and H_2.

Lactate is metabolized by a number of clostridia. *Clostridium propionicum* converts this organic acid to propionate and acetate by branched pathways. Two lactates are activated with coenzyme A and dehydrated to yield acrylyl-CoA, which then is reduced to propionyl-CoA. Transfer of the CoA moiety to lactate leads to propionate formation. The reducing equivalents stem from oxidation of a third lactate to acetate and CO_2. The acetate kinase reaction of this branch generates the only ATP of this fermentation. It is thus clearly less efficient than the methylmalonyl-CoA pathway employed by *Propionibacterium.* Succinate decarboylation is another possibility to produce propionate from an organic acid. This way is employed by *Propionigenium modestum* and based on a membrane-located decarboxylation of methylmalonyl-CoA, which generates a Na^+-gradient. This gradient can be directly used for ATP generation by a Na^+-dependent F_1F_O-ATPase (as in *P. modestum*) or indirectly after conversion into a H^+-gradient by action of a Na^+/H^+ antiporter. The other decarboxylation product propionyl-CoA is used to activate succinate, and succinyl-CoA is converted into methylmalonyl-CoA. Thus, only three enzymes are required for this pathway. Within the clostridia, probably *Clostridium mayombei* degrades succinate by Na^+-dependent decarboxylation (Bahl and Dürre, 1993). A dangerous organism for the cheese industry is *C. tyrobutyricum*, which forms butyrate from lactate. The so-called late blowing ruins the structure and taste of contaminated cheese.

Heteroaromates stemming from the degradation of nucleic acids can also be used for growth by clostridia. *Clostridium acidurici*, *C. cylindrosporum*, and *C. purinilyticum* are almost completely specialized on fermenting purines (*C. purinilyticum* also grows well on glycine). The pyrimidine moiety is split first, degraded to CO_2 and NH_4^+, and then the imidazole moiety is hydrolytically cleaved, yielding finally acetate (via glycine reductase and acetate kinase), formate, CO_2, and NH_4^+ (Dürre and Andreesen, 1983). Pyrimidine fermentation is accomplished by *C. glycolicum* and *C. oroticum,* and is based on similar reactions (Vogels and van der Drift, 1976).

Clostridia belong to those bacteria that are able to form endospores. These are the most resistant biological survival forms known and protect the genome against environmental dangers such as desiccation, heat, radiation, and hazardous compounds. Retrieval of *Clostridium aceticum*, a species considered lost from culture collections during World War II, provided evidence for successful germination even after four decades (Braun et al., 1981). Clostridia thus belong to the few prokaryotes able to perform cell differentiation. The sporulation process very much resembles that of *Bacillus* (Dürre and Hollergschwandner, 2004), which is well understood. Initiation is achieved via a number of phosphate transfers from sensor kinases to the response regulator Spo0A, the so-called phosphorelay (Hoch, 1993). Such a chain of transfer components is obviously not present in clostridia, as shown by the numerous genome sequencing projects, but the master regulator Spo0A is conserved and fulfills the same functions. In cooperation with the sporulation-specific sigma factor σ^H, Spo0A~P induces transcription of further forespore- and mother cell-specific sporulation sigma factors. In both compartments, pairs of these proteins become active successively, to guarantee a sequential number of steps, required for mature spore synthesis. In the forespore, σ^F and σ^G are responsible for this task and are regulated via anti- and anti-anti-sigma factors. In the mother cell, their companions are σ^E and σ^K, which are proteolytically activated. While most endospore-forming bacteria produce just one such compartment per cell, there are a few notable exceptions. *Anaerobacter polyendosporus* and *Metabacterium polyspora*, phylogenetically close relatives of *Clostridium*, produce up to five endospores in a single mother cell (Siunov et al., 1999; Angert et al., 1996). An evolutionary link to production of already-living offspring might be represented by *Epulopiscium fishelsonii.* This is not only one of world's largest bacteria, surpassing many eukaryotic microorganisms in size, but also belongs phylogenetically to the clostridia and gives birth to several internal daughter cells (Angert et al., 1993).

An important task in food preservation is to prevent contamination with spores of pathogenic clostridia. For this purpose, preservatives can be added such as nitrite, which converts important redox-active iron-sulfur clusters (e.g., in ferredoxin) into inactive iron-nitric oxide complexes (Reddy et al., 1983; Carpenter et al., 1987), benzoic acid, sodium chloride, and sorbate. Acidification, desiccation, heating, ionizing radiation, and refrigeration are other methods to achieve extended shelf life (Lund and Peck, 2000). Hygienic conditions during preparation are of utmost importance, as failure in this respect might even affect vacuum-packed materials (production of gas and/or sulfides). In the future, high-pressure treatment might become an important additional means to prevent clostridial spoilage of foods.

PHYLOGENY AND TAXONOMY

Phylogenetically, the clostridia belong to the low G+C Gram-positive phylum. This classification is based on 16 S rRNA gene sequence analysis. It reflects ordering of the species according to common ancestry, disregarding physiological and morphological features. Relatedness is usually displayed as a dendrogram. Historically, however, members of the genus *Clostridium* had been defined as performing an anaerobic lifestyle, having a Gram-positive-type cell wall, producing endospores, and being unable to perform dissimilatory sulfate reduction. Thus, the genus represents a problem for taxonomists, and many bacteria, originally described as a *Clostridium*, have been reclassified. Taxonomy follows the scheme: order *Firmicutes*, class *Clostridia*, order *Clostridiales*, family *Clostridiaceae*, genus *Clostridium*. Compilations of validly described clostridia (currently more than 150 species, making *Clostridium* one of the largest bacterial genera) can be found on the Internet (http://www.bacterio.cict.fr/c/clostridium.html by J.P. Enzéby: List of Prokaryotic Names with Standing in Nomenclature — Genus *Clostridium*; http://dx.doi.org/10.1007/bergeysoutline200310 by G.M. Garrity, J.A. Bell, and T.G. Lilburn: Taxonomic Outline of the Prokaryotes, *Bergey's Manual of Systematic Bacteriology, 2nd ed.*, release 5 May 2004). There, and also in catalogs of culture collections such as ATCC (American Type Culture Collection) and DSMZ (Deutsche Sammlung von Mikroorganismen und Zellkulturen), are numerous indications of new genera with which former clostridia are now affiliated. These genera are *Caloramator, Dendrosporobacter, Eubacterium, Filifactor, Fusobacterium, Moorella, Oxalophagus, Oxobacter, Paenibacillus, Thermoanaerobacter, Thermoanaerobacterium, Sedimentibacter, Sporohalobacter,* and *Syntrophospora* (Stackebrandt, 2004; Wiegel et al., 2006). A further complication is that not all taxonomically validly described *Clostridium* species form phylogenetically coherent clusters. 16 S rDNA analyses of the *Bacillus/Clostridium* subphylum of the Gram-positives allowed identification of 20 clusters, designated I through XX. Clostridia are found in Clusters I (with a number of subdivisions and regarded as genus *Clostridium* sensu stricto (Stackebrandt and Hippe, 2001; Wiegel et al., 2006)), III, IV, XIa, XIb, XII, XIVa, XIVb, XVI, XVIII, and XIX. Cluster I contains the type species of the genus, *Clostridium butyricum*. Different classifications of single species can be found (e.g., *C. limosum* in Cluster II (Stackebrandt and Hippe, 2001) or Cluster I (Wiegel et al., 2006)), and the phylum *Firmicutes* will probably still remain the "greatest challenge for taxonomists" (Stackebrandt, 2004) for quite some time.

PATHOGENICITY

Despite the bad public reputation of clostridia, caused by well-known species such as *Clostridium botulinum* and *C. tetani*, only a few members of this genus (less than 10%) form dangerous toxins. Some species contain more than one (outstanding is *C. perfringens* with 14 toxins); altogether, 58 clostridial toxins have been identified thus far (Popoff and Stiles, 2005). This number is slightly less than a fifth of all bacterial toxins. The clostridial toxins are proteins and can be classified as either pore-forming, zinc-dependent metalloproteases, glycosyltransferases, ADP-ribosyltransferases, or phospholipases (Table 25.2).

One of the most intensively studied representatives is the *Clostridium perfringens* enterotoxin (CPE) (McClane, 2005). It is a common cause of food-borne disease. However, it should be kept in mind that the vast majority of strains of this species do not carry a *cpe* gene. While *in vitro* studies revealed that CPE can form pores in artificial lipid membranes, the *in vivo* effect relies on recruiting a number of host proteins to trigger massive changes in membrane permeability. Members of the claudin family are most probably the primary target and serve as receptors of CPE. With the help of additional membrane proteins, first a small complex of approx. 90 kDa and then large complexes of approx. 135, 155, and 200 kDa are formed. Interaction of a large complex with tight junction proteins leads to massive damage of the latter. CPE is one of the clostridial toxins whose synthesis

TABLE 25.2
Clostridial Toxins

Class of Toxins	Type	Typical Representative	Pathogenic Species
pore-forming	enterotoxin	CPE	*C. perfringens*
	thiol-activated	perfringolysin O (θ-toxin)	*C. perfringens*
	aerolysin-like	α-toxin	*C. septicum*
	hemolysin-like	β-toxin	*C. perfringens*
zinc-dependent metalloprotease	neurotoxin	botulinum toxin	*C. botulinum A-F*
			C. baratii
			C. argentinense
		tetanus toxin	*C. tetani*
	collagenase	ColG, ColH	*C. histolyticum*
		ColA	*C. perfringens*
glycosyltransferase	UDP-glucose-dependent	ToxA, ToxB	*C. difficile*
	(large clostridial toxins)	lethal toxin (LT)	*C. sordellii*
		hemorrhagic toxin (HT)	*C. sordellii*
	UDP-N-acetyl-glucosamine-dependent	α-Toxin	*C. novyi*
ADP-ribosyltransferase	actin ADP-ribosylating	C2 toxin	*C. botulinum C, D*
	(binary toxins)	iota toxin	*C. perfringens*
		spiroforme toxin	*C. spiroforme*
		CDT	*C. difficile*
	Rho ADP-ribosylating	C3	*C. botulinum C, D*
		C3	*C. limosum*
phospholipase	phosphalidylcholine-, sphingomyelin-degrading	α-Toxin	*C. perfringens*
		PLC	*C. absonum*
		PLC	*C. bifermentans*
		γ-Toxin	*C. novyi*
		PLC	*C. haemolyticum*
	additionally phosphatidylserine-degrading	α-Toxin	*C. perfringens*
		γ-Toxin	*C. novyi*
	additionally phosphatidylethanol-amine-, phosphatidylinositol-degrading	γ-Toxin	*C. novyi*

is tightly linked to sporulation. The protein is not secreted, but only set free upon lysis of the mother cell, together with the mature endospore.

The prototype of thiol-activated pore-forming toxins is perfringolysin O (Popoff and Stiles, 2005), which shows highest activity in the presence of reducing compounds. Elucidation of the structure revealed an elongated, rod-shaped molecule. Upon binding to its membrane target cholesterol, oligomerization of 40 to 50 molecules results in large pore formation. *Clostridium septicum* α-toxin is secreted as a nontoxic precursor and becomes active by proteolysis. *C. perfringens* β-toxin does not destroy erythrocyte membranes, but rather forms cation-specific membrane pores. It is a major cause of necrotic enteritis in domesticated livestock and also humans.

Most public attention is directed to the clostridial neurotoxins (Johnson, 2005). Botulinum toxin is the most poisonous substance of biological origin known to date (for human beings, a concentration of approx. 0.1 ng kg^{-1} has been calculated to be lethal) (Arnon, 1997). There are three forms of botulism: (1) infant (probably the most common with up to 100 cases per year in the United States, honey feeding up to approx. 6 months is a risk factor); (2) food-borne (classical, ingestion of toxin); and (3) wound (very rare, bacterial spores infect a tissue with damaged blood and thus oxygen supply, germinate, and subsequently produce toxin). Botulinum toxin (BoNT) is on the list of bioterrorist agents because of its extreme toxicity. In 1995, the Japanese Aum Shinrikyo sect wanted to use this compound to attack passengers on the Tokyo subway. Due to problems with aerolization, they finally decided to use sarin gas. Also, following the first Gulf War in 1990 and during United Nations weapons inspections, Iraq conceded the production of botulinum toxin. About half the total amount of 19,000 L had already been used for the preparation of SCUD missile warheads and artillery shells (Arnon et al., 2001). However, even this extremely dangerous compound is now extensively used in medical therapy and cosmetic applications. Blepharospasm (involuntary eyelid contractions), cervical dystonia (involuntary contraction of neck and shoulder muscles), and strabismus (crossed eyes) are uncontrolled muscle spasms that can be treated by injecting sublethal doses of toxin. Economically even more important are cosmetic applications. Commercial preparations of botulinum toxin ("Botox," "Dysport") are used in the treatment of face aging (relaxation of muscle contractions leading to frown lines and wrinkles) as well as hyperhidrosis and hypersalivation (Keller and Vann, 2004). The injected toxin (in very dilute concentration) is active for only a few months; then injection must be repeated. Seven serotypes of botulinum toxin are known. The proteolytic group I of *Clostridium botulinum* produces serotypes A, B, and F (whose genes are all located on the chromosome); the nonproteolytic group II produces B, E, and F (also from chromosomally located genes); and group III produces C and D (encoded by bacteriophages). Group IV is actually *C. argentinense* and produces serotype G (from a plasmid-located gene). In addition, strains of *C. barati* can produce F and strains of *C. butyricum* can produce E (both from chromosomally located genes). The molecular action starts by lysis of bacterial cells, release of the progenitor toxin, and cleavage of the large precursor protein into a heavy and a light chain (approx. 100 and 50 kDa, respectively). The light chain carries the metalloprotease domain. Linkage of the chains is achieved by a disulfide bond. The heavy chain then binds to receptors at the nerve terminal and the complex of heavy and light chain is internalized as an endosomal vesicle. Acidification of the endosome leads to a conformational change of the heavy chain and subsequent translocation of the light chain into the cytosol. There, several SNARE (soluble N-ethylmaleimide sensitive factor attachment protein receptor) protein family members are proteolytically inactivated. Substrate specificity depends on the serotype of botulinum toxin. Serotypes A, C, and E cleave SNAP-25; C cleaves syntaxin; and B, D, F, and G cleave VAMP (vesicle-associated membrane protein). Due to the large number of serotypes, vaccination is a problem. Only a pentavalent toxoid (A–E) is available. Current treatment is injection of equine antiserum for absorbing circulating toxin. Medical countermeasure for infant botulism is an antitoxin-based therapy with a novel immune globulin.

With tetanus toxin, the situation is different. The genome of the causative agent, *Clostridium tetani*, has been sequenced (see below), and the gene encoding the toxin (*tetX*) is located on the 74-kb plasmid pE88. Thus, only one serotype is known and vaccination is the preferred medical countermeasure. It is performed with a toxoid originating from formaldehyde-inactivated 150-kDa toxin. Tetanus toxin (TeTx) is less toxic than BoNT. The human lethal dose is estimated at about 1 ng kg^{-1}. The disease is a spastic paralysis and almost exclusively caused by infections of deep wounds. Thereby, blood vessels are disrupted and the tissue becomes hypoxic, allowing the endospores to germinate and the arising vegetative cells to produce toxin. The mode of action is similar to that of BoNT, except that TeTx does not act on the peripheral, but rather on the central nervous system. TeTx also is cleaved into a light and a heavy chain, which are then linked by a disulfide bond. The complex binds to the neuromuscular junction of motorneurons. After internalization, it is transported to the spinal cord and migrates into inhibitory interneuron terminals. There, the catalytic activity of the light chain blocks release of inhibitory neurotransmitters by cleaving VAMP (also designated synaptobrevin).

Collagenases also belong to the class of zinc-dependent proteases. The causative agent of gas gangrene and myonecrosis, *Clostridium histolyticum,* produces six such enzymes, which are now used in medical therapies (see following paragraph). They have molecular masses of 68 to 125 kDa, are also called α-clostripain, and specifically cleave collagen and gelatin at the glycine residue of a PXGP motif. *Clostridium perfringens* also produces a collagenase (ColA), which resembles ColG of *C. histolyticum.*

The paradigm of a glycosyltransferase-producer is *Clostridium difficile*, a major contributor to healthcare-associated illnesses. Due to their molecular mass of up to 308 kDa, these proteins are also designated as large clostridial cytotoxins (Barth and Aktories, 2005). *C. difficile* produces ToxA and ToxB, which use UDP-glucose as a cofactor and glycosylate G-proteins such as Rho, Rac, Cdc42, Rap, and Ral. Similar enzymes are synthesized by *C. novyi* (α-toxin uses UDP-*N*-acetylglucosamine as a cofactor) and *C. sordellii* (Hemorrhagic and Lethal toxins, the latter ones act additionally on Ras, but not on Rho). The toxins are bound by a receptor of the host cell and taken up by endocytosis. After processing (the location of this process still awaits elucidation), acidification of the endosome leads to secretion of the enzyme domain into the cytosol of the host cell, where it modifies its target proteins. Glycosylated G-proteins become inactive, and signal chains important for proliferation, cell differentiation, organization of actin cytoskeleton, enzyme activation, and transcription are blocked.

ADP-ribosyltransferases target either actin or Rho (Barth and Aktories, 2005). The best-studied example of the first group is the C2 toxin of *Clostridium botulinum*. It is comprised of two proteins, and the group is therefore referred to as binary toxins. In addition to type C and D *C. botulinum*, such proteins are produced by *C. diffile* (CDT), *C. perfringens* (iota toxin), and *C. spiroforme* (spiroforme toxin). To become cytotoxic, the binding component of C2 (C2II) is cleaved at its N-terminus (removal of an approx. 20-kDa peptide) to create C2IIa. This component oligomerizes into heptamers, binds to a receptor, and attaches to the host cell membrane. The other component, C2I, docks to this complex, which is then translocated into the host cell by endocytosis. Acidification of the endosome leads to a conformational change of C2IIa, its insertion into the membrane, and finally the release of C2I into the cytosol of the host cell, where it ADP-ribosylates G-actin monomers and thus prevents their polymerization.

The paradigm of the second group is C3 toxin of *Clostridium botulinum*. It lacks a cell binding and transport domain. It ADP-ribosylates Rho GTPases, thereby inactivating them. A similar toxin is found in *C. limosum*. Because free toxin is poorly taken up by eukaryotic cells, invasion of pathogen into the host and subsequent release of toxin is a prerequisite for cytotoxic action.

Phospholipases produced by clostridia all belong to class C (Titball and Tweten, 2005). Such enzymes have been detected in *Clostridium absonum*, *C. barati*, *C. bifermentans*, *C. haemolyticum*, *C. novyi*, *C. perfringens*, and *C. sordellii*. As yet characterized, all cleave phosphatidylcholine and sphingomyelin. Diseases caused by these toxins are gas gangrene and necrotic enteritis. The latter occurs in farmed fowl and increased significantly over the past years.

BIOTECHNOLOGY

An outstanding historical application of biotechnology employing clostridia was the acetone-butanol fermentation by *Clostridium acetobutylicum* (see above). Plant sizes were enormous, the largest facilities in the United States were located in Terre Haute, Indiana (52 fermenters) and Peoria, Illinois (96 fermenters with a capacity of 189,250 L each) (Gabriel, 1928). South Africa operated a plant in Germiston until 1982, consisting of 12 production fermenters with a working volume of 90,000 L each (Jones, 2001). Facilities were also located in Australia, Brazil, Canada, China, Egypt, India, Japan, the former Soviet Union, and the United Kingdom (Jones and Woods, 1986). Production in China was only stopped in 2004 (Chiao and Sun, 2007). In addition to acetone and 1-butanol, 2-propanol (or isopropanol) is another solvent produced by clostridia, namely *C. aurantibutyricum*, *C. beijerinckii*, and a variant of *C. butyricum* (Dürre and Bahl, 1996). Due to increasing crude oil prices, the fermentative production of butanol is again able to compete economically with the petrochemical process. As a consequence, BP and DuPont will start in 2007 to operate a "Biobutanol" plant in the United Kingdom (http://www2.dupont.com/Biofuels/en_US/). In addition to its being an important bulk chemical for industrial purposes (Dürre, 2005), butanol can also serve as a biofuel. Mixtures with gasoline are superior to those with ethanol, as butanol has a lower vapor pressure and a higher energy content, can be blended at higher concentrations than ethanol, and does not require specific adjustments of vehicle and engine technologies.

The clostridial metabolic pathways leading to solvent production are tightly coupled to regulation of endospore formation (Paredes et al., 2005; Dürre, 2005). This makes sense, as formation of acids during exponential growth leads to acidification of the surrounding environment. Anaerobic bacteria are unable to keep their internal pH constant, so it will drop in parallel with the external one, but being more alkaline by about 1 pH unit (Dürre et al., 1988). Reaching an external pH of about 4, undissociated acid from the outside will diffuse into the cytoplasm and dissociate there. As a consequence, the proton gradient over the membrane will collapse, and the cell will die. Converting acids into neutral products will avoid this lethal effect, and thus respective species have an ecological advantage. However, because especially butanol in higher concentrations is also toxic, sporulation should be initiated at the same time as solvent formation, in order to guarantee long-time survival. Overall, the cells thus manage to stay metabolically active for a longer period and to postpone entering a dormant stage.

Enzymes required for solventogenesis are coenzyme A transferase, butyraldehyde and butanol dehydrogenases for butanol formation, as well as coenzyme A transferase and acetoacetate decarboxylase for acetone production (an additional alcohol dehydrogenase allows 2-propanol synthesis from acetone in *Clostridium beijerinckii*). In *C. acetobutylicum*, three dehydrogenases are known. Butanol dehydrogenases A and B are encoded by chromosomal genes, forming a monocistronic operon each (Walter et al., 1992). BdhA seems to function as an electron sink, removing excess reducing equivalents by alcohol formation, while BdhB plays a major role in massive butanol production (Sauer and Dürre, 1995). Initiation of solventogenesis is mediated by AdhE, a bifunctional butyraldehyde/butanol dehydrogenase, whose gene is located on a large plasmid (pSOL1, 192 kbp) and forms a common operon (*sol*) with two genes encoding the subunits of coenzyme A transferase. Directly downstream, but with convergent direction of transcription, lies a monocistronic operon encoding the gene for acetoacetate decarboxylase (*adc*). This arrangement is unique, as in other solventogenic species such as *C. beijerinckii*, *C. saccharobutylicum*, and *C. saccharoperbutylacetonicum,* the *sol* operon consists of genes encoding butyraldehyde dehydrogenase, coenzyme A transferase, and acetoacetate decarboxylase (Dürre, 2005). The signal(s) responsible for inducing solventogenesis (and also sporulation) have not been unequivocally identified, although pH, concentration of undissociated acids, metabolic intermediates, cofactors, and salts, as well as DNA supercoiling, all play a role (Dürre, 1998, Girbal and Soucaille, 1998). The master regulator of sporulation, phosphorylated Spo0A, also acts as a transcription factor during onset of solventogenesis. In addition to other regulatory proteins, RNA processing and protein modification represent further levels of regulation (Dürre, 2005). The use of DNA microarrays allows a new approach for studying this complex regulatory network (Tummala et al., 2005) and, consequently, construction of production strains with superior properties (Dürre, 2005; Tomas et al., 2005).

Because some clostridia are able to degrade halogenated and hazardous compounds partially or completely, they may also be used in bioremediation processes. Examples are lindane (hexachlorocyclohexane, used as an insecticide), which is completely dechlorinated by *Clostridium butyricum*; PCBs (polychlorinated biphenyls, used in insulating fluids), which is metabolized by *C. hydroxybenzoicum*; and TNT (trinitrotoluene, used as an explosive), which is reduced to triaminotoluene (Bahl and Dürre, 1993; Ahmad et al., 2005).

A medical application of a clostridial enzyme is wound debridement (Brett, 2005). Chronic, nonhealing wounds require removal of necrotic tissue, which is anchored by collagen to the wound surface, to achieve successful curing. Such treatment might become necessary in case of severe burns or dermal ulcer. One of the agents used for this goal is collagenase from *Clostridium histolyticum*, of which six different such proteins are known. They cleave interstitial collagens into small peptides, mostly tripeptides. Collagen accounts for approx. 75% of the skin tissue (dry weight), and the enzymes act at physiological pH and temperature. Commercial preparations of collagenase (e.g., "Santyl," "Iruxol," or "Novuxol") are therefore particularly useful in the removal of detritus and contribute toward the formation of granulation tissue and subsequent epithelization. Collagen in healthy tissue or in newly formed granulation tissue is not attacked.

Oncolysis caused by clostridia can be used for treatment of solid cancers (Minton, 2003). It has been successfully tested at the laboratory stage with animals and is about to enter clinical trials (Brown and Liu, 2004). As clostridia perform an anaerobic lifestyle, their spores will only germinate in an anaerobic or at least hypoxic environment. This is a situation normally not occurring in mammals. Consequently, injected clostridial spores are cleared from blood within a few days (Brown and Liu, 2004). However, the environment of solid tumors in a mammal is hypoxic, thus allowing spores reaching such an area to germinate (only in this place, which is a very specific targeting) and to produce vegetative cells, which are able to multiply in that region. The bacteria proliferate at the expense of only necrotic tissue, as healthy, oxygenated tissue does not allow colonization. Recombinant spores will give rise to recombinant clostridia, which are able to produce specific proteins at the tumor. Examples are cytosine deaminase, carboxypeptidase G2, and nitroreductase, which are all able to convert a harmless prodrug, injected into and delivered by the bloodstream, into an effective anti-cancer agent, which is generated only at the tumor and causes oncolysis (Minton, 2003). Another possibility is the transformation and expression of genes that encode cytotoxic proteins. After secretion, the gene products will then be able to exert anti-tumor activity. Currently investigated are tumor necrosis factor α (TNFα) (van Mellaert et al., 2005) and, especially for treatment of pancreatic cancer, *Clostridium perfringens* enterotoxin (CPE) (Dürre, 2007).

HANDLING AND MOLECULAR BIOLOGY TECHNIQUES

Due to the large number of species in the genus *Clostridium*, almost the entire range of microbial metabolic properties is represented (Bahl and Dürre, 1993; Dürre, 2007). Clostridia are found within acidophiles, neutrophiles, and alkaliphiles (thus covering the pH range from 4 to 10.5) as well as in psychrophiles, mesophiles, and thermophiles (thus covering most of the temperature range). Although in general being obligate anaerobes, some species are able to detoxify oxygen (to varying extent) by employing superoxide reductase, peroxidases, and/or superoxide dismutase (Dürre, 2001). In addition to using specific substrates already mentioned, clostridia can be grown in complex media commercially available (e.g., CMC, chopped meat-carbohydrate; PYG, peptone-yeast extract-glucose). Detailed media compositions are provided by culture collections distributing the species and in species-specific original publications.

The technique originally invented by Robert Hungate allows easy handling of clostridia using anaerobic and prereduced media (with the help of titanium(III) citrate, a mixture of sodium sulfide and cysteine-HCl, or sodium thioglycolate) (Breznak and Costilow, 1994). Media are boiled to remove oxygen and cooled under a constant flow of nitrogen. After addition of the aforementioned reducing agents and a redox indicator (such as resorufin) for visual detection of potential oxygen contamination, the liquid is poured into tubes or bottles, and then a butyl rubber stopper is secured in place. After sterilization, inoculation or sampling can easily be performed with sterile syringes and needles. Solid media can be poured and streaked in an anaerobic chamber.

A number of clostridial genomes have meanwhile been sequenced or are in various stages of completion (Table 25.3). Mining of this information is in progress; one of the first general conclusions that can be drawn is that initiation of endospore formation is not triggered by a phosphorelay as in bacilli (Dürre and Hollergschwandner, 2004). Plasmids are quite common in clostridia, with about a third of the investigated species containing extrachromosomal elements (Lee et al., 1987). A number of shuttle vectors have been constructed, in general making use of erythromycin and chloramphenicol resistance genes (Davis et al., 2005; Tummala et al., 2001). However, in practice, these two antibiotics can be inactivated by clostridia so that clarithromycin (a pH-stable derivative of erythromycin) and thiamphenicol (a chloramphenicol derivative without the reducible nitro group) are employed. Transformation of free DNA into clostridia can be achieved by electroporation, and conjugation (direct cell contact with a donor cell) is also possible (Davis et al., 2005). Shelter against

TABLE 25.3
***Clostridium* Genome Sequencing Projects**

Sequencing Status	Species	Strain	Metabolic Feature	Chromosome (Mbp)	Number of Plasmids (designation) [size in kbp]	Sequencing Institution
Complete	*C. acetobutylicum*	ATCC 824	Acetone-butanol producer	3.94	1 (pSOL1) [192]	Oscient Pharmaceuticals Company
Complete	*C. perfringens*	ATCC 13124	Causative agent of gas gangrene	3.26	-	TIGR
Complete	*C. perfringens*	SM 101	Causative agent of food poisoning	2.90	2 [12.4; 12.2]	TIGR
Complete	*C. perfringens*	13	Causative agent of gas gangrene	3.03	1 (pCP13) [5.4]	Kyushu University
Complete	*C. tetani*	E 88	Causative agent of tetanus	2.80	1 (pE88) [74.1]	Göttingen Genomics Laboratory
Draft assembly	*C. beijerinckii*	NCIMB 8052	Solvent producer	5.95	-	DOE Joint Genome Institute
Draft assembly	*C. difficile*	QCD-32g58	Important pathogen	3.83	-	Washington University, St. Louis
Draft assembly	*C. phytofermentans*	ISDg (ATCC 700394)	Cellulose degrader	4.53	-	DOE Joint Genome Institute
Draft assembly	*Clostridium* sp.	OhILAs	Arsenate reducer	3.0	-	DOE Joint Genome Institute
Draft assembly	*C. thermocellum*	ATCC 27405	Thermophilic cellulose degrader	3.78	-	DOE Joint Genome Institute
In progress	*C. bolteae*	ATCC BAA-613	Human gut bacterium	n.d.a.	n.d.a.	Washington University, St. Louis
In progress	*C. botulinum*	A (ATCC 3502)	Botulinum toxin producer	3	-	Sanger Institute
In progress	*C. cellulolyticum*	H10 (ATCC 35319)	Cellulose degrader	n.d.a.	n.d.a.	JGI/JGI-PGF
In progress	*C. difficile*	630	Important pathogen	n.d.a	1(pCD630) [7.9]	Sanger Institute
In progress	*C. ljungdahlii*	ATCC 49587	Acetogen	n.d.a.	n.d.a.	Göttingen Genomics Laboratory
In progress	*C. scindens*	ATCC 35704	Human gut bacterium	n.d.a.	n.d.a.	Washington University, St. Louis
In progress	*Clostridium* sp.	A2-232	Human fecal butyrate producer	n.d.a.	n.d.a.	Washington University, St. Louis
In progress	*Clostridium* sp.	BC1	N_2 fixation, radionuclide bioremediation	4	-	Brookhaven National Laboratory
In progress	*Clostridium* sp.	L2-50	Human fecal butyrate producer	n.d.a.	n.d.a.	Washington University, St. Louis
In progress	*Clostridium* sp.	M62/1	Human fecal butyrate producer	n.d.a.	n.d.a.	Washington University, St. Louis
In progress	*Clostridium* sp.	SS2/1	Human fecal butyrate producer	n.d.a.	n.d.a.	Washington University, St. Louis
In progress	*C. symbiosum*	ATCC 14940	Human gut bacterium	n.d.a.	n.d.a.	Washington University, St. Louis

Note: Data from http://www.ncbi.nlm.nih.gov/entrez/, November 2006; n.d.a., no data available.

host restriction endonucleases can be artificially provided by a *Bacillus* phage methylase expressed in the intermediate host *Escherichia coli* (Mermelstein and Papoutsakis, 1993).

Random mutagenesis has been tried with *Clostridium acetobutylicum* and *C. tetani* using conjugative transposons from streptococci such as Tn*916*, Tn*925*, and Tn*1545* (for a review, see Dürre, 1993). Several mutants have been obtained; but because these elements prefer AT-rich regions found especially in promoters, structural gene inactivations are not frequent. Natural transposons are also found in clostridia. A conjugative element of 20.7 kb (Tn*5397*) is present in *C. difficile*, integrative mobilizable elements of 6.3 kb (Tn*4451* and Tn*4453*) have been detected in *C. perfringens* and *C. difficile*, a trans-mobilizable element of 9.6 kb (Tn*5398*) was found in *C. difficile*, and a compound transposon of 6.3 kb (Tn*5565*) has been described for *C. perfringens* (Lyras and Rood, 2005). Chemical mutagenesis by nitrite is not possible, as this compound destroys essential iron-sulfur centers, as mentioned before. However, in 2006, a targeted gene knock-out system with a high success rate was developed, which is based on the use of a mobile group II intron from *Lactococcus lactis* ("ClosTron," https://hcai.nottingham.ac.uk/minton.pdf). Modulation of gene expression is possible by means of antisense RNA constructs (Tummala et al., 2005). In *C. saccharobutylicum*, a naturally expressed asRNA participates in regulation of nitrogen assimilation. The 43-bp transcript is complementary to a region at the start of the *glnA* gene (encoding glutamine synthetase) and thus inhibits translation during nitrogen-rich conditions (Fierro-Monti et al., 1992). The toolbox for genetic manipulation of clostridia also contains a number of well-suited reporter gene systems. Successful use of *catP* (encoding chloramphenicol acetyltransferase), *eglA* (encoding β-1,4-endoglucanase), *gusA* (encoding β-glucuronidase), *lacZ* (encoding β-galactosidase), and *luxAB* or *lucB* (encoding luciferase) has been reported (Tummala et al., 2001; Feustel et al., 2004). The latter one is especially well suited, as no interfering metabolic equivalent is known from clostridia. In summary, all technologies required for genetic analysis and manipulation of members of this large bacterial genus are now at hand.

ACKNOWLEDGMENTS

Work in this laboratory was supported by grants from the BMBF GenoMik and GenoMikPlus projects (Competence Network Göttingen), the transnational BMBF SysMO project ((PtJ-BIO/SysMO/P-D-01-06-13), and the European Community (contract No. QLK3-CT-2001-01737).

REFERENCES

Ahmad, F., J.B. Hughes, and G.N. Bennett. 2005. Biodegradation of hazardous materials by clostridia. In *Handbook on Clostridia*, Ed. P. Dürre, 831–854, Boca Raton, FL: CRC Press, Taylor & Francis Group.

Angert, E.R., K.D. Clements, and N.R. Pace. 1993. The largest bacterium. *Nature* 362: 239–241.

Angert, E.R., A.E. Brooks, and N.R. Pace. 1996. Phylogenetic analysis of *Metabacterium polyspora*: clues to the evolutionary origin of daughter cell production in *Epulopiscium* species, the largest bacteria. *Journal of Bacteriology* 178: 1451–1456.

Arnon, S.S. 1997. Human tetanus and human botulism. In *The Clostridia: Molecular Biology and Pathogenesis*, Ed. J.I. Rood, B.A. McClane, J.G. Songer, and R.W. Titball, pp. 95–115, San Diego: Academic Press.

Arnon, S.S., R. Schechter, T.V. Inglesby, et al. 2001. Botulinum toxin as a biological weapon: medical and public health management. *The Journal of the American Medical Association* 285: 1059–1070.

Bahl, H. and P. Dürre. 1993. *Clostridia*. In *Biotechnology*, 2nd ed., Vol. 1, Ed. H. Sahm, pp. 285–323, Weinheim: VCH Verlagsgesellschaft mbH.

Bahl, H. and P. Dürre. 2001. *Clostridia. Biotechnology and Medical Applications.* Weinheim: Wiley-VCH.

Barth, H. and K. Aktories. 2005. Clostridial cytotoxins. In *Handbook on Clostridia*, Ed. P. Dürre, pp. 407–449, Boca Raton, FL: CRC Press, Taylor & Francis Group.

Braun, M., F. Mayer, and G. Gottschalk. 1981. *Clostridium aceticum* (Wieringa), a microorganism producing acetic acid from molecular hydrogen and carbon dioxide. *Archives of Microbiology* 128: 288–293.

Brett, D.W. 2005. Clostridial collagenase in wound repair. In *Handbook on Clostridia*, Ed. P. Dürre, pp. 855–876, Boca Raton, FL: CRC Press, Taylor & Francis Group.

Breznak, J.A. and R.N. Costilow. 1994. Physicochemical factors in growth. In *Methods for General and Molecular Bacteriology*, Ed. P. Gerhardt, R.G.E. Murray, W.A. Wood, and N.R. Krieg, pp. 137–154, Washington, DC: American Society for Microbiology.

Brown, J.M. and S.-C. Liu. 2004. Use of anaerobic bacteria for cancer therapy. In *Strict and Facultative Anaerobes. Medical and Environmental Aspects*, Ed. M.M. Nakano and P. Zuber, pp. 211–219, Wymondham: Horizon Bioscience.

Buckel, W. 1990. Amino acid fermentations: coenzyme B_{12}-dependent and –independent pathways. In *The Molecular Basis of Bacterial Metabolism*, Ed. G. Hauska and R. Thauer, pp. 21–30, Heidelberg: Springer-Verlag.

Carpenter, C.E., D.S.A. Reddy, and D.P. Cornforth. 1987. Inactivation of clostridial ferredoxin and pyruvate-ferredoxin oxidoreductase by sodium nitrite. *Applied and Environmental Microbiology* 53: 549–552.

Chiao, J.-S, and Z.-H. Sun. 2007. History of the acetone-butanol-ethanol fermentation industry in China: development of continuous production technology. *Journal of Molecular Microbiology and Biotechnology* 13: 12–14.

Davis, I.J., G. Carter, M. Young, and N.P. Minton. 2005. Gene cloning in clostridia. In *Handbook on Clostridia*, Ed. P. Dürre, pp. 37–52, Boca Raton, FL: CRC Press, Taylor & Francis Group.

Dürre, P. and J.R. Andreesen. 1982. Selenium-dependent growth and glycine fermentation by *Clostridium purinolyticum*. *Journal of General Microbiology* 128: 1457–1466.

Dürre, P. and J.R. Andreesen. 1983. Purine and glycine metabolism by purinolytic clostridia. *Journal of Bacteriology* 154: 192–199.

Dürre, P., H. Bahl, and G. Gottschalk. 1988. Membrane processes and product formation in anaerobes. In *Handbook on Anaerobic Fermentations*, Ed. L.E. Erickson and D.Y.-C. Fung, pp. 187–206, New York: Marcel Dekker Inc.

Dürre, P. 1993. Transposons in clostridia. In *The Clostridia and Biotechnology*, Ed. D.R. Woods, pp. 227–246, Stoneham: Butterworth-Heinemann.

Dürre, P. and H. Bahl. 1996. Microbial production of acetone/butanol/isopropanol. In *Biotechnology*, 2nd ed., Vol. 6, Ed. M. Roehr, pp. 229–268, Weinheim: VCH Verlagsgesellschaft mbH.

Dürre, P. 1998. New insights and novel developments in clostridial acetone/butanol/isopropanol fermentation. *Applied Microbiology and Biotechnology* 49: 639–648.

Dürre, P. 2001. From Pandora's box to cornucopia: clostridia—a historical perspective. In *Clostridia. Biotechnology and Medical Applications*, Ed. H. Bahl and P. Dürre, pp. 1–17, Weinheim: Wiley-VCH.

Dürre, P. and C. Hollergschwandner. 2004. Initiation of endospore formation in *Clostridium acetobutylicum*. *Anaerobe* 10: 69–74.

Dürre, P. 2005. *Handbook on Clostridia*. Boca Raton, FL: CRC Press, Taylor & Francis Group.

Dürre, P. 2005. Formation of solvents in clostridia. In *Handbook on Clostridia*, Ed. P. Dürre, pp. 671–693, Boca Raton, FL: CRC Press, Taylor & Francis Group.

Dürre, P. 2007. Clostridia. *Encyclopedia of Life Sciences*, http://www.els.net/ [doi:10.1002/9780470015902.a0020370].

Ewerdwalbesloh, I. and O. Meyer. 1995. Bacteriology and enzymology of the woad fermentation. Proceedings of the 2nd International Symposium on Woad, Indigo and other Natural Dyes: Past, Present and Future, Toulouse.

Feustel, L., S. Nakotte, and P. Dürre. 2004. Characterization and development of two reporter gene systems for *Clostridium acetobutylicum*. *Applied and Environmental Microbiology* 70: 798–803.

Fierro-Monti, I.P., S.J. Reid, and D.R. Woods. 1992. Differential expression of a *Clostridium acetobutylicum* antisense RNA: implications for regulation of glutamine synthetase. *Journal of Bacteriology* 174: 7642–7647.

Gabriel, C.L. 1928. Butanol fermentation process. *Industrial and Engineering Chemistry* 28: 1063–1067.

Girbal, L. and P. Soucaille. 1998. Regulation of solvent production in *Clostridium acetobutylicum*. *Trends in Biotechnology* 16: 11–16.

Hoch, J.A. 1993. Regulation of the onset of the stationary phase and sporulation in *Bacillus subtilis*. *Advances in Microbial Physiology* 35: 111–133.

Johnson, E.A. 2005. Clostridial neurotoxins. In *Handbook on Clostridia*, Ed. P. Dürre, pp. 491–525, Boca Raton, FL: CRC Press, Taylor & Francis Group.

Jones, D.T. and D.R. Woods. 1986. Acetone-butanol fermentation revisited. *Microbiological Reviews* 50: 484–524.

Jones, D.T. 2001. Applied acetone-butanol fermentation. In *Clostridia. Biotechnology and Medical Applications*, Ed. H. Bahl and P. Dürre, pp. 125–168, Weinheim: Wiley-VCH.

Keller, J.E. and W.F. Vann. 2004. Botulinum toxin. *Encyclopedia of Life Sciences*, http://www.els.net/ [doi:10.1038/npg.els.0004218].

Kiple, K.F. 1993. *The Cambridge World History of Human Disease*. Cambridge: Cambridge University Press.

Lee, C.-K., P. Dürre, H. Hippe, and G. Gottschalk. 1987. Screening for plasmids in the genus *Clostridium*. *Archives of Microbiology* 148: 107–114.

Lund, B.M. and M.W. Peck. 2000. *Clostridium botulinum*. In *The Microbiological Safety and Quality of Foods*, Ed. B.M. Lund, A.C. Baird-Parker, and G.W. Gould, 1057–1109, Gaithersburg, MD: Aspen Publ. Inc.

Lyras, D. and J.I. Rood. 2005. Transposable genetic elements of clostridia. In *Handbook on Clostridia*, Ed. P. Dürre, pp. 631–643, Boca Raton, FL: CRC Press, Taylor & Francis Group.

McClane, B.A. 2005. Clostridial enterotoxins. In *Handbook on Clostridia*, Ed. P. Dürre, pp. 385–406, Boca Raton, FL: CRC Press, Taylor & Francis Group.

Mermelstein, L.D. and E.T. Papoutsakis. 1993. *In vivo* methylation in *Escherichia coli* by the *Bacillus subtilis* phage ϕ3T I methyltransferase to protect plasmids from restriction upon transformation of *Clostridium acetobutylicum* ATCC 824. *Applied and Environmental Microbiology* 59: 1077–1081.

Minton, N.P. and D.J. Clarke. 1989. *Clostridia*. New York: Plenum Press.

Minton, N.P. 2003. Clostridia in cancer therapy. *Nature Reviews in Microbiology* 1: 237–242.

Padden, A.N., V.M. Dillon, P. John, J. Edmonds, M.D. Collins, and N. Alvarez. 1998. *Clostridium* used in medieval dyeing. *Nature* 396: 225.

Padden, A.N., V.M. Dillon, J. Edmonds, M.D. Collins, N. Alvarez, and P. John. 1999. An indigo-reducing moderate thermophile from a woad vat, *Clostridium isatidis* sp. nov. *International Journal of Systematic Bacteriology* 49: 1025–1031.

Paredes, C.J., K.V. Alsaker, and E.T. Papoutsakis. 2005. A comparative genomic view of clostridial sporulation and physiology. *Nature Reviews in Microbiology* 3: 969–978.

Pasteur, L. 1861. Animacules infusoires vivant sans gaz oxygène libre et déterminant des fermentations. *Comptes Rendus Hebdomadaires des Séances de l'Académie des Sciences* 52: 344–347.

Popoff, M.R. and B.G. Stiles. 2005. Clostridial toxins vs. other bacterial toxins. In *Handbook on Clostridia*, Ed. P. Dürre, 323–383, Boca Raton, FL: CRC Press, Taylor & Francis Group.

Prazmowski, A. 1880. Untersuchungen über die Entwickelungsgeschichte und Fermentwirkung einiger Bacterien-Arten. Ph.D. thesis, University of Leipzig.

Reddy, D., L.R. Lancaster, Jr., and D.P. Cornforth. 1983. Nitrite inhibition of *Clostridium botulinum*: electron spin resonance detection of iron-nitric oxide complexes. *Science* 221: 769–770.

Rood, Julian I., B.A. McClane, J.G. Songer, and R.W. Titball. 1997. *The Clostridia: Molecular Biology and Pathogenesis*. San Diego: Academic Press.

Sauer, U. and P. Dürre. 1995. Differential induction of genes related to solvent formation during the shift from acidogenesis to solventogenesis in continuous culture of *Clostridium acetobutylicum*. *FEMS Microbiology Letters* 125: 115–120.

Siunov, A.V., D.V. Nikitin, N.E. Suzina, V.V. Dimitriev, N.P. Kuzmin, and V.I. Duda. 1999. Phylogenetic status of *Anaerobacter polyendosporus*, an anaerobic, polysporic bacterium. *International Journal of Systematic Bacteriology* 49: 1119–1124.

Stackebrandt, E. and H. Hippe. 2001. Taxonomy and systematics. In *Clostridia. Biotechnology and Medical Applications*, Ed. H. Bahl and P. Dürre, pp. 19–48, Weinheim: Wiley-VCH.

Stackebrandt, E. 2004. The phylogeny and classification of anaerobic bacteria. In *Strict and Facultative Anaerobes. Medical and Environmental Aspects*, Ed. M. M. Nakano and P. Zuber, pp. 1–25, Wymondham: Horizon Bioscience.

Sussman, M. 1958. A description of *Clostridium histolyticum* gas-gangrene in The Epidemics of Hippocrates. *Medical History* 2: 226.

Tamburini, E., A.G. Leon, B. Perito, and G. Mastromei. 2003. Characterization of bacterial pectinolytic strains involved in the water retting process. *Environmental Microbiology* 5: 730–736.

Titball, R.W. and R.K. Tweten. 2005. Membrane active toxins. In *Handbook on Clostridia*, Ed. P. Dürre, pp. 451–489, Boca Raton, FL: CRC Press, Taylor & Francis Group.

Tomas, C.A., S.B. Tummala, and E.T. Papoutsakis. 2005. Metabolic engineering of solventogenic clostridia. In *Handbook on Clostridia*, Ed. P. Dürre, pp. 813–830, Boca Raton, FL: CRC Press, Taylor & Francis Group.

Tummala, S.B., C. Tomas, L.M. Harris, et al. 2001. Genetic tools for solventogenic clostridia. In *Clostridia. Biotechnology and Medical Applications*, Ed. H. Bahl and P. Dürre, pp. 105–123, Weinheim: Wiley-VCH.

Tummala, S.B., C.A. Tomas, and E.T. Papoutsakis. 2005. Gene analysis of clostridia. In *Handbook on Clostridia*, Ed. P. Dürre, pp. 53–70, Boca Raton, FL: CRC Press, Taylor & Francis Group.

van Mellaert, L., J. Theys, O. Pennington, et al. 2005. Clostridia as production systems for prokaryotic and eukaryotic proteins of therapeutic value in tumor treatment. In *Handbook on Clostridia*, Ed. P. Dürre, pp. 877–893, Boca Raton, FL: CRC Press, Taylor & Francis Group.

Vogels, G.D., and C. van der Drift. 1976. Degradation of purines and pyrimidines by microorganisms. *Bacteriological Reviews* 40: 403–468.

Walter, K.A., G N. Bennett, and E.T. Papoutsakis. 1992. Molecular characterization of two *Clostridium acetobutylicum* ATCC 824 butanol dehydrogenase isozyme genes. *Journal of Bacteriology* 174: 7149–7158.

Weizmann, C. 1915. Improvements in the bacterial fermentation of carbohydrates and in bacterial cultures for the same. British Patent 4945.

Wiegel, J., R. Tanner, and F.A. Rainey. 2006. An introduction to the family *Clostridiaceae*. In *The Prokaryotes—A Handbook on the Biology of Bacteria*, Vol. 4, Ed. M. Dworkin, S. Falkow, E. Rosenberg, K.-H. Schleifer, E. Stackebrandt, pp. 654–678, New York: Springer Science+Business Media, LLC.

Woods, D.R. 1993. *The Clostridia and Biotechnology*. Stoneham: Butterworth-Heinemann.

26 The Genus *Corynebacterium*

Lothar Eggeling and Michael Bott

CONTENTS

GENERAL FEATURES

Corynebacterium was originally defined in 1896 by Lehmann and Neumann to accommodate non-motile parasitic and pathogenic bacteria, including diphtheroid bacilli [1]. *Diphther* is the Greek word for "membrane" and describes the fact that these latter bacteria can typically be isolated from a false membrane in the pharynx developed due to the necrotic action of the toxin made by *Corynebacterium diphtheriae*. Before the establishment of modern systematics, the taxon *Corynebacterium* accommodated a number of heterogeneous bacteria. However, based on the extended use of chemotaxonomic markers, mainly in the last third of the past century, like cell wall chemistry, lipid composition and DNA base composition, the taxon *Corynebacterium* was brought into sharper focus and it also was recognized that it is a member of the so-called CMN-group of bacteria including, in addition to *Corynebacterium,* also *Mycobacterium* and *Nocardia*. It is now clear that based on 16S rRNA/rDNA sequence patterns, *Corynebacterium* forms a robust and well-defined monophyletic group that together with *Dietzia, Gordonia, Mycobacterium, Nocardia, Rhodococcus, Skermania, Tsukamurella,* and *Williamsia*, forms the suborder Corynebacterineae belonging to the order Actinomycetales [2]. Recent reviews covering several aspects of pathogenic and non-pathogenic *Corynebacterium* species are available [3–6].

The entire family Corynebacteriaceae currently (early 2007) contains 67 validated species [7]. A phylogenetic tree using a maximum likelihood analysis, based on approx. 2500 sequences and evaluated among others by maximum parsimony and distance procedures, is given in Figure 26.1. In addition, two type strains for the closely related families Dietziaceae and Tsukamurellaceae are included as a reference. It can be seen, for example, that the amino acid producing *Corynebacterium* species *C. glutamicum, C. efficiens,* and *C. callunae* cluster closely together.

Corynebacterium species can be isolated from a number of sources. Nonpathogenic species are found in a broad variety of habitats such as dairy products, animal fodder, rotting plant material, and also in soil as is the case with the biotechnologically important *C. glutamicum.* Pathogenic species are isolated from human or animal specimens, where they occur either as true pathogens or as cutaneous or mucocutaneous contaminants. They may be part of the normal flora of the skin as

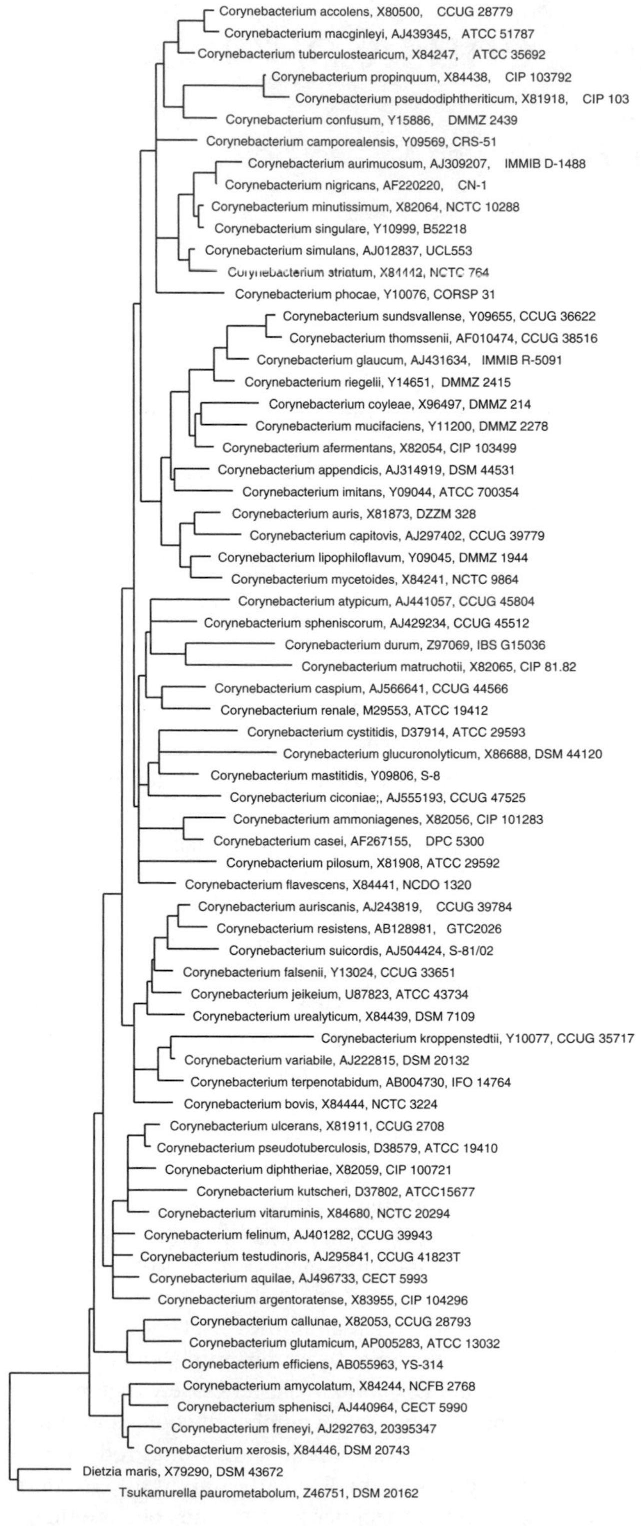

FIGURE 26.1 Clustering of *Corynebacterium* species. The scale bar indicates 10% estimated sequence divergence. (This tree was kindly provided by Dr. W. Ludwig, Technical University Munich.)

is the case with *C. xerosis,* which, however, in immunocompromised hosts, may cause infections, such as endocarditis, pneumonia, empyema, wound infection, and peritonitis. Although there is no selective medium or enrichment procedure known specifically suited for *Corynebacterium* species, a number of rapid identification and differentiation methods are commercially available, which are based on physiological traits. These include the API CORYNE (bioMerieux), API ZYM (bio Merieux), BIOLOG (Biolog), MINITEK (Becton Dickinson), and RapID CB Plus (Remel) identification systems. They are possibly biased toward the identification of clinically relevant species.

For the in-depth identification of *Corynebacterium* species and the study of their phylogenetic relationship, DNA and 16S rRNA analyses are used. In addition to total DNA characterization by either thermal denaturation or hybridization, most informative and convenient is 16S rDNA sequence analysis. For this purpose, a large fragment of the 16S rRNA gene is amplified via PCR (polymerase chain reaction), using the universal primers pA 5′-AGAGTTTGATCCTGGCTCAG (positions 8 to 27 of the *Escherichia coli* sequence) and pH* 5′-AAGGAGGTGATCCAGCCGCA (positions 1541 to 1522). These bind near the 5′ and 3′ ends of the rRNA gene, and the PCR product is subsequently sequenced and analyzed by comparison with known 16S rDNA sequences [8], as has been done for the *Corynebacterium* species compiled in Figure 26.1.

Corynebacterium cells appear rod-shaped with a somewhat irregular ("coryneform") morphology. Two cells can be arranged in V-formation as a consequence of cell division ("snapping mode"). Also, packages of several cells in parallel arrangement ("pallisades") occur. Cells are nonmotile; and mycelia, capsules, and spores are absent. Some species show piliation, such as *C. diphtheriae.* Corynebacteria are catalase-positive, aerobic, or facultatively anaerobic bacteria. They grow in the range of 15 to 42°C.

CELL WALL

Interestingly, in some properties the architecture and function of the cell wall of the Gram-positive *Corynebacterium* resembles those of the Gram-negative cell envelope, although the molecular details are strikingly different. The major macromolecular cell wall components of *Corynebacterium* species are peptidoglycan, arabinogalactan (AG), lipomannan, and lipoarabinomannan. Whereas the peptidoglycan is similar to that found in other bacteria, like *Escherichia coli* for example [9], the sugarpolymers and liposugarpolymers deserve closer attention because they are common to *Corynebacterium* and other taxons within the *Actinomycetales,* giving them their unique and well-defined position within the bacteria [2, 7].

The sugarpolymers are of relevance because they might get in contact with the environment, respectively the host, and therefore decide on immunological properties and persistence of the bacterium. Much progress has been made in recent years on the enzymology and structure of these polymers, such as AG. AG is covalently linked via its galactan part to the C-6 position of MurNGlyc residues of the peptidoglycan via the specific linker unit L-Rha*p*-(1→3)-α-D-GlcNAc. To the arabinose part of AG mycolic acids are bound, which are 2-alkyl-branched-3-hydroxy-acids. Thus, AG is part of a large mycolyl-AG-peptidoglycan (mAGP) complex. It can be considered the major scaffold within the cell wall, whose absence or disorder causes severe growth defects [10–12], and existing antibiotics, such as ethambutol, have AG as target as might have future antibiotics hoped for. For isolation of the mAGP complex, cells are suspended in phosphate-buffered saline containing 2% Triton X-100 (pH 7.2), disrupted by sonication and centrifuged [13]. The pelleted material is then extracted with 2% SDS in phosphate-buffered saline at 95°C to remove associated proteins, successively washed with water, 80% (vol/vol) acetone in water, and acetone, and finally lyophilized to yield a highly purified cell wall preparation. Recently, a modified protocol for mAGP isolation of the related mycobacteria has been reported [14].

Access to the glycosyl composition of AG is by the further analysis of the isolated cell wall preparation. This is hydrolyzed with 2 M trifluoroacetic acid (TFA) and reduced with NaB_2H_4 to result in alditols that are per-*O*-acetylated and examined by gas chromatography [10, 13, 15]. An

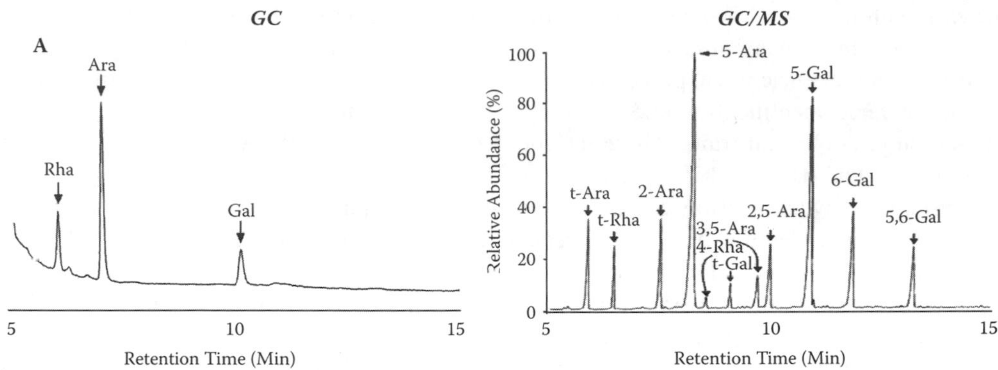

FIGURE 26.2 On the left is shown a GC analysis of arabinogalactan of *C. glutamicum* showing the basic building blocks arabinose, galactose and rhamnose of this heterosugarpolymer, whereas on the right the individually linked types of the sugars are shown as analyzed by GC/MS.

example for such a cell wall analysis for *Corynebacterium glutamicum* is shown in Figure 26.2 (left). A recent comparison of *C. glutamicum, C. diphtheriae, C. xerosis,* and *C. amycolatum* by such methodology confirmed that all species have primarily arabinose (Ara) and galactose (Gal) in their arabinogalactan. In addition, there are small but significant amounts of glucose linked to AG of *C. xerosis* and *C. amycolatum*, whereas AG of *C. diphtheriae* additionally contains mannose [16] and AG of *C. glutamicum* contains rhamnose.

In addition to the total composition of AG, further information concerning the structure of this sugar polymer comes from a linkage analysis of the sugars. This is done by per-*O*-methylation of cell wall preparations using dimethylsulfinyl carbanion [10, 13, 15], followed by hydrolysis using 2 M TFA and reduction with NaB_2H_4. Subsequently, the free hydroxyl groups of per-*O*-methylated sugar derivatives are per-*O*-acetylated and examined by gas chromatography/mass spectrometry. Such an analysis for *C. glutamicum* is shown in Figure 26.2, right. It reveals that 5-Gal*f*, 6-Gal*f*, and 5,6-Gal*f* residues are present. They form a linear chain of alternating β(1→6) linked β(1→5) Gal*f* residues [17, 18]. To this backbone, arabinose chains are attached at the 8th, 10th, and 12th β(1→6) Gal*f* residue [13]. Thus, the latter Gal*f* residue with arabinose attached has three linkages and results in the 5,6-Gal*f* residue (for convention, the 1-position is not given). The arabinan domain consists of α(1→5), α(1→3), and β(1→2) arabinofuranosyl linkages to form a branched polymer of approximately 25 residues in size [13].

To the *t*-Ara*f* and 2-Ara*f* residues mycolic acids are bound, representing the unique lipophilic components of the cell envelopes of all genera of the suborder Corynebacterineae. In *Corynebacterium* species, these 2-alkyl-branched-3-hydroxy-acids are of rather simple structure and consist of two condensed linear fatty acid chains. Interestingly, there are a few atypical species not possessing mycolic acids. These are *C. amycolatum* [19], *C. kroppenstedtii* [20], and *C. atypicum* [21]. The mycolic acids in *Corynebacterium* species are collectively called corynomycolic acids and consist of up to 38 carbon atoms. The most prominent mycolic acid in *C. glutamicum* has the formula 34:1 followed by 32:0 [22, 23], the latter apparently the result of a condensation of two palmitic acid molecules, and the former the result of one palmitic plus one oleic acid molecule. The mycolic acids formed may depend on the culture conditions. Thus, addition of oleic acid to *C. glutamicum* results in a dominant content of more than 90% of the mycolic acid of the type 36:2.

In addition to the mycolic acids bound to arabinogalactan, mycolic acids exist bound to trehalose to form trehalose mono- and dimycolates, respectively. These latter mycolic acid conjugates are solvent extractable and can be separated by thin layer chromatography. Together with the mycolic acids bound to AG, they form an outer hydrophobic layer around *Corynebacterium*, and visualization of a second fracture plane in freeze-etch electron microscopy studies suggests that these adopt an outer lipid bilayer membrane in addition to the cytoplasmic membrane [24]. The existence of an outer lipid layer is entirely consistent with the presence of a range of pore-forming proteins (porins) within the cell walls of these Gram-positive bacteria [25].

Mycolic acid synthesis requires the activity of a specific enzyme complex, as evident from a recent incentive genetic analysis of *Corynebacterium glutamicum.* Whereas the carboxylase β-subunit D1 is a component of the ubiquitous acetyl CoA carboxylase required for fatty acid synthesis, the paralogous carboxylase β-subunits D2 and D3 present in all *Corynebacterium* species sequenced are essential for mycolic acid synthesis [26]. The acetyl CoA carboxylase consists of the carboxylase β-subunit D1, the biotin-carrying α-subunit, and an additional ε-subunit. It carboxylates acetyl CoA in *Corynebacterium* to provide malonyl CoA necessary for fatty acid synthesis, as is similarly the case in Gram-negative bacteria and other organisms. However, the carboxylase involved in mycolic acid synthesis is apparently a specific evolutionary incidence, involving gene duplications, to derive an enzyme complex structurally and mechanistically related to the ubiquitous acetyl CoA carboxylase. The acyl carboxylase consists of the carboxylase β-subunits D2 and D3 forming a complex together with the same α- and ε-subunit as required for acetyl CoA complex formation. Although the details are not yet fully understood, it is clear that one linear fatty acid is carboxylated, forming an intermediate during condensation of the two linear fatty acid chains by the single polyketide synthase present in *Corynebacterium* [22, 23, 26]. It is clear that due to the importance of pathogenic *Corynebacterium* species such as *C. diphtheriae* or *C. jeikeium* and the related *Mycobacterium* species, in particular *M. tuberculosis*, which also possess mycolic acids, considerable efforts are currently being made to understand the biochemical details of mycolic acid synthesis. Structural information on one of the carboxylase β-subunits involved is already available [27–30]. In this respect, the ease of handling *C. glutamicum* and its rather small genome structure just containing the genes required for core reactions of cell wall synthesis and not overloaded with paralogous genes, as is the case in *Mycobacterium,* offers a great advantage in understanding the basic principles of mycolic acid synthesis.

GENOMES

The *Corynebacterium* genomes currently published are those of *C. glutamicum*, *C. efficiens*, *C. diphtheriae,* and *C. jeikeium* [31–35, 37]. As a fifth genome, that of *C. ureolyticum* is ready but not yet published [36], and further genome sequences are expected soon. From *C. glutamicum,* two sequences were released at almost the same time, reflecting the commercial interest in this organism [31, 37]. They are not perfectly identical but differ roughly by 27 kb, which is mainly due to additional copies of insertion elements and a putative prophage inserted in the larger genome. The basic features of the published genomes are listed in Table 26.1. With 63.4%, the *C. efficiens* chromosome has the highest G+C-content, which is attributed to its high growth temperature of up to 45°C [38]. This trait is of interest for the large-scale production of amino acids by *C. glutamicum* or *C. efficiens* because cooling costs in fermentations are reduced. *C. efficiens* inherits two large plasmids, and *C. jeikeium* inherits one plasmid, displaying a G+C content significantly greater than that of the chromosome, which is indicative of their recent acquisition. The genomes of the two pathogenic species *C. diphtheriae* and *C. jeikeium* are much smaller than that of the nonpathogenic species, due to a more specialized lifestyle and a restricted metabolic versatility. Within the four species, as much as 52% of the genes are considered to be orthologous, apparently representing the basic gene equipment that is conserved within the investigated *Corynebacterium* species, and these genes also can be anticipated to be present in most other members of the family.

Not unexpectedly, the overall genome sequence organization of the *Corynebacterium* species also shows that genes are largely syntenic. The comparison of *C. glutamicum* vs. *C. diphtheriae* is shown in Figure 26.3 (left). It reveals one insertion in *C. glutamicum* compared to *C. diphtheriae,* in the range from 1,775,332 to 1,973,456 bp. Nothing in that approximately 200-kb range of the *C. glutamicum* genome is found in *C. diphtheriae.* In contrast, *C. diphtheriae* vs. *C. jeikeium,* or *C.*

TABLE 26.1
Genome Features and Transcriptional Regulatory Repertoire of Sequenced *Corynebacterium* Species

Feature	*C. glutamicum* **ATCC 13032**	*C. efficiens* **YS-314**	*C. diphtheriae* **NCTC13129**	*C. jeikeium* **K411**	*C. urealyticum* **DSM7109**	*C. urealyticum* **DSM7111**
Reason of Interest	**Industrial Amino Acid Producer**	**Potential Amino Acid Producer**	**Human Pathogen, Causative Agent of Diphtheria**	**Multiresistant Nosocomial Pathogen**	**Nosocomial Pathogen Causing Urinary Tract Infections**	**Nosocomial Pathogen Causing Urinary Tract Infections**
Genome size (bp)	3,282,708	3,147,090	2,488,635	2,462,499	2,294,755	2,268,483
Number of protein-coding regions	3002	2950	2320	2104	2027	1896
Number of DNA-binding transcriptional regulators	128	103	62	55	50	46
Number of response regulators	13	11	11	9	9	10
Number of other regulators	10	10	9	9	9	9
Number of sigma factors	7	7	9	9	6	6
Sum total of potential regulators	158	131	91	82	74	71
Percentage of potential regulators (%)	5.3	4.4	3.9	3.9	3.7	3.7

Source: From Brinkrolf, K., Brune, I., and Tauch, A. *J. Biotechnol.*, 129, 191–211, 2007. With permission.

glutamicum vs. *C. jeikeium* (Figure 26.3, middle and right), reveals one major large inversion in *C. jeikeium*, which is interestingly symmetric about the origin of replication, creating an "X" pattern (or the beginning of one). Such features of genome structures are known for moderate related bacteria and were also shown to occur in *Escherichia coli* and *Salmonella typhimurium* under laboratory conditions [39]. In *C. jeikeium,* there are also some smaller breakpoints of synteny present, indicating that this species is the most distant of the four analyzed thus far in detail. *C. jeikeium* is characterized by a restricted carbohydrate utilization pattern as compared to the other *Corynebacterium* species, and it lacks phosphoenolpyruvate:carbohydrate phosphotransferase systems (PTSs), which might be one reason for the smaller genome and might also reflect an adaptation of *C. jeikeium* to the availability of nutrients in the predominantly colonized areas of the human skin. Also, regarding the reactions around the pyruvate/phosphoenolpyruvate node, *C. jeikeium* seems to be particularly limited because it possesses neither the pyruvate carboxylase gene *pyc* nor *mez*, encoding malic enzyme, whereas these are present in *C. glutamicum* and *C. efficiens*, with *pyc* being also present in *C. diphtheriae.*

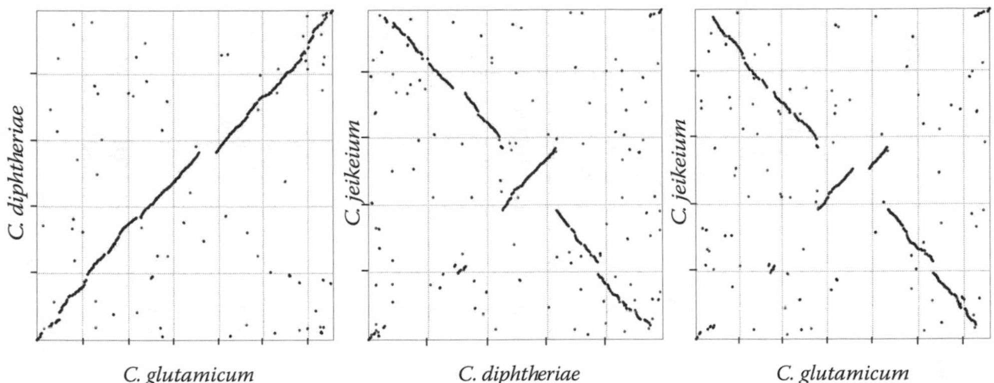

FIGURE 26.3 Synteny between the genomes of *C. glutamicum* ATCC 13032, *C. jeikeium* K411, and *C. diphtheriae* NCTC 13129. The diagram shows scatterplots of the conserved protein sequences between each pair of species forming syntenic regions between the corynebacterial genomes. The x/y axes are scaled in 0.5 Mbp. (Alignments were kindly provided by S. Salzberg, Bioinformatics and Computational Biology, University of Maryland.)

TRANSCRIPTIONAL REGULATORS

As is evident from Table 26.1, the number of sigma factors (between 6 and 9) and two-component signal transduction systems (between 9 and 13) are quite similar in the sequenced *Corynebacterium* strains; however, there is a large difference in the number of "one-component systems" [40], meaning classical transcriptional regulators. The nonpathogenic species *C. glutamicum* and *C. efficiens* harbor 128 and 103 of these proteins, respectively, whereas the pathogenic species *C. diphtheriae*, *C. jeikeium*, and *C. urealyticum* possess only between 46 and 62. This difference is probably due to the higher versatility in catabolism and lifestyle of the former species. The common set of 25 transcriptional regulators present in all currently sequenced *Corynebacterium* species is shown in Table 26.2.

The first regulator identified in a *Corynebacterium* species was DtxR, the repressor of the diphtheria toxin gene *tox*, which is located on the corynebacteriophage β [41]. After the availability of the *C. glutamicum* genome sequence, DNA microarray analysis [42, 43] and proteome analysis [44, 45] were established for this species. In particular, the former method is extremely useful to search for target genes of a specific regulator by comparing a mutant lacking the regulatory gene of interest and the wild-type. A further approach that proved efficient in identifying the transcriptional regulators of certain target genes is DNA affinity chromatography, which led, for example, to the identification of ClgR, a regulator controlling genes involved in proteolysis and DNA repair [46], or of RamB [47] and RamA [48], both of which control genes involved in acetate metabolism, but also in additional functions. Currently, 33 transcriptional regulators have been analyzed experimentally in *C. glutamicum*, which are listed in Table 26.3 together with a short description of their function. Several of these act as global regulators, such as RamA, RamB, and GlxR in carbon metabolism; AmtR in nitrogen metabolism; PhoR in phosphorus metabolism; McbR in sulfur metabolism; or DtxR in iron metabolism. As described above, DtxR was originally identified as the repressor of the diphtheria toxin gene but it is now clear that its function is much broader [49, 50]. In *C. glutamicum*, more than 50 genes are repressed by DtxR under iron excess, including those for three other transcriptional regulators (RipA, CgtR11, and the ArsR-type regulator Cg0527). Whereas the function of the latter two has not yet been determined, the AraC-type regulator RipA was shown to repress a set of prominent iron proteins under iron limitation: for example, aconitase, succinate dehydrogenase, nitrate reductase, isopropylmalate dehydratase, and others. The anticipated function of RipA is therefore to reduce the cell's iron demand under iron-limiting conditions.

TABLE 26.2
The Common Set of DNA-Binding Transcriptional Regulators in Corynebacteria (Excluding Response Regulators)

Functional Module	Orthologous Genes of the Common Set in:						Gene Name	Protein Family
	C. glutamicum **ATCC 13032**	*C. efficiens* **YS-314**	*C. diphtheriae* **NCTC 13129**	*C. jeikeium* **K411**	*C. urealyticum* **DSM 7109**	*C. urealyticum* **DSM 7109**		
Cell division and septation	*cg0337*	*ce0283*	*dip0299*	*jk1976*	*cu09 1262*	*cu11 1587*	*whiB4*	WhiB
	cg0850	*ce0758*	*dip0684*	*jk1644*	*cu09 1543*	*cu11 0455*	*whiB2*	WhiB
Carbohydrate metabolism	*cg0350*	*ce0287*	*dip0303*	*jk1972*	*cu09 1258*	*cu11 1583*	*glxR*	Crp
	cg1738	*ce1663*	*dip1284*	*jk0970*	*cu09 0837*	*cu11 1050*	*acnR*	TetR
	cg2115	*ce1824*	*dip1427*	*jk1107*	*cu09 0278*	*cu11 0918*	*sugR*	DeoR
	cg2831	*ce2445*	*dip1889*	*jk0397*	*cu09 1430*	*cu11 0694*	*ramA*	LuxR
	cg2910	*ce2511*	*dip1969*	*jk0329*	*cu09 0673*	*cu11 1476*	—	LacI
Specific biosynthesis pathways and transport systems	*cg0454*	*ce0397*	*dip0937*	*jk1455*	*cu09 0753*	*cu11 0591*	—	TetR
	cg1486	*ce1426*	*dip1126*	*jk1222*	*cu09 0562*	*cu11 1897*	—	IclR
	cg1585	*ce1531*	*dip1172*	*jk0846*	*cu09 1119*	*cu11 0328*	*argR*	ArgR
	cg2112	*ce1820*	*dip1424*	*jk1105*	*cu09 0281*	*cu11 0916*	*nrdR*	YbaD/NrdR
	cg3261	*ce2809*	*dip2280*	*jk0088*	*cu09 0174*	*cu11 0711*	—	GntR
Macroelement and metal homeostasis	*cg1631*	*ce1574*	*dip1205*	*jk0904*	*cu09 0785*	*cu11 1000*	—	MerR
	cg1633	*ce1576*	*dip1207*	*jk0906*	*cu09 0786*	*cu11 1001*	—	MerR
	cg2103	*ce1812*	*dip1414*	*jk1097*	*cu09 0289*	*cu11 0909*	*dtxR*	DtxR
	cg2502	*ce2180*	*dip1710*	*jk0612*	*cu09 0386*	*cu11 1836*	*furB*	Fur
	cg3253	*ce2788*	*dip2274*	*jk0101*	*cu09 1211*	*cu11 0215*	*mcbR*	TetR
SOS and stress response	*cg0878*	*ce0783*	*dip0712*	*jk1618*	*cu09 1518*	*cu11 0480*	*whcE*	WhiB
	cg1765	*ce1687*	*dip1296*	*jk0985*	*cu09 0851*	*cu11 1063*	*sufR*	ArsR
	cg2109	*ce1817*	*dip1421*	*jk1102*	*cu09 0283*	*cu11 0914*	*oxyR*	LysR
	cg2114	*ce1823*	*dip1426*	*jk1106*	*cu09 0279*	*cu11 0917*	*lexA*	LexA
	cg2152	*ce1855*	*dip1456*	*jk1122*	*cu09 0263*	*cu11 0933*	*clgR*	HTH 3
	cg2516	*ce2190*	*dip1721*	*jk0600*	*cu09 0374*	*cu11 1824*	*hrcA*	HrcA
	cg3097	*ce2626*	*dip2117*	*jk0184*	*cu09 1976*	*cu11 1727*	*hspR*	MerR
	cg3315	*ce2826*	*dip2296*	*jk2072*	*cu09 0118*	*cu11 1644*	—	MarR

Source: From Brinkrolf, K., Brune, I., and Tauch, A. *J. Biotechnol.* 129, 191–211, 2007.

TABLE 26.3
Experimentally Analyzed Transcriptional Regulators (Including Response Regulators of Two-Component Signal Transduction Systems) of *Corynebacterium glutamicum*

Gene Number	Gene Name	Functional Module[a]	Regulatory Protein Family	Regulatory Role and Physiological Function of Transcriptional Regulator	Ref.
cg0196	*iolR*	CM	GntR	Repressor of inositol metabolism	82
cg0350	*glxR*	CM	Crp	Repressor of acetate/gluconate metabolism and *dtsR1 gene*	83–85
cg0444	*ramB*	CM	HTH 3	Regulator of acetate metabolism	47
cg1308	—	CM	TetR	Repressor of hydroxyquinol pathway	86–87
cg1738	*acnR*	CM	TetR	Repressor of aconitase gene *acn*	88
cg2115	*sugR*	CM	DeoR	Repressor of *ptsG* gene for EII of glucose PTS	89
cg2615	*vanR*	CM	PadR	Repressor of vanillate metabolism	86, 89
cg2624	*pcaR*	CM	IclR	Repressor of protocatechuate degradation	86
cg2831	*ramA*	CM	LuxR	Regulator of acetate metabolism and *cspB* gene	41, 48
cg3352	*nagR*	CM	IclR	Activator of gentisate uptake and degradation	92
cg0313	*lrp*	SBPTS	AsnC	Activator of amino acid export system BrnEF	93
cg0897	*pdxR*	SBPTS	GntR	Activator of pyridoxine biosynthesis genes	44
cg1053	—	SBPTS	TetR	Repressor of MmpL-type transporter Cg1054	82
cg1425	*lysG*	SBPTS	LysR	Activator of amino acid export system LysE	95, 96
cg1482	*ltbR*	SBPTS	IclR	Regulator of leucine and tryptophan biosynthesis genes	97
cg1585	*argR*	SBPTS	ArgR	Repressor of arginine biosynthesis	98
cg1817	*pyrR*	SBPTS	PyrR	Repressor of *pyrH* and *pyrBC* genes	99
cg1846	—	SBPTS	TetR	Repressor of *cg1844–cg1847* genes	82
cg2894	—	SBPTS	TetR	Repressor involved in antibiotic resistance	100
cg0012	*ssuR*	MMH	ROK	Activator of sulfonate and sulfonate ester utilization	101
cg0156	*cysR*	MMH	ROK	Activator of assimilatory sulfate reduction	102
cg0317	*arsR2*	MMH	ArsR	Repressor of arsenite permease and arsenate reductase	103
cg0986	*amtR*	MMH	TetR	Global regulator of nitrogen metabolism	104
cg1120	*ripA*	MMH	AraC	Repressor of iron protein expression	106
cg1704	*arsR1*	MMH	ArsR	Repressor of arsenite permease and arsenate reductase	103
cg2103	*dtxR*	MMH	DtxR	Global regulator of iron metabolism	49, 50
cg2888	*phoR*	MMH	RR	Regulator of phosphate starvation-inducible genes	107, 108
cg3253	*mcbR*	MMH	TetR	Global regulator of sulfur metabolism	109
cg0862	*mtrA*	SSR	RR	Regulator of genes involved in cell wall metabolism and osmoregulation	110, 111
cg0878	*whcE*	SSR	WhiB	Activator of thioredoxin genes	112
cg2114	*lexA*	SSR	LexA	Repressor of SOS response	82
cg2152	*clgR*	SSR	HTH 3	Activator of proteolysis and DNA repair genes	46, 113
cg3097	*hspR*	SSR	MerR	Repressor of heat shock response at HAIR	46

[a] Abbreviation of functional modules: CM, carbohydrate metabolism; SBPTS, specific biosynthesis pathways and transport systems; MMH, macroelement and metal homeostasis; SSR, SOS and stress response.

Source: From Brinkrolf, K., Brune, I., and Tauch, A. *J. Biotechnol.* 129, 191–211, 2007.

DtxR is not only a repressor, but presumably also an activator of about a dozen genes that encode the iron storage proteins ferritin and Dps and proteins required for the assembly of iron-sulfur clusters in proteins.

In view of the pace currently made in the characterization of new transcriptional regulators from *Corynebacterium glutamicum*, it can be envisaged that the majority of them will be analyzed in the near future. Based on these data, a detailed view of the regulatory network in this organism will be possible and educated guesses can be made regarding the function of the orthologs in other *Corynebacterium* species.

POSTTRANSCRIPTIONAL REGULATION

The genomes of *Corynebacterium glutamicum*, *C. efficiens*, *C. diphtheriae,* and *C. jeikeium* contain four or five (*C. jeikeium*) genes for serine/threonine protein kinases. Those common to all are designated PknA, PknB, PknL, and PknG; the one specific for *C. jeikeium* has been named PknJ (*jk1732*). In addition, a gene for a phosphoserine/phosphothreonine-specific phosphoprotein phosphatase (*ppp*) is present in these four species. The genes for PknA, PknB, and Ppp are part of a gene cluster that is highly conserved in the suborder Corynebacterineae and might be implicated in cell division. Whereas PknA, PknB, PknL, and PknJ are predicted to contain a single transmembrane helix and thus are membrane-integral proteins, PknG appears to be a cytosolic protein. Recent proteome studies with *C. glutamicum* led to the identification of the first target of PknG, a 15-kDa protein with a forkhead-associated domain that was designated OdhI [51]. In its unphosphorylated state, OdhI binds to the E1 subunit of the 2-oxoglutarate dehydrogenase complex and inhibits the activity of this enzyme complex, which catalyzes the oxidative decarboxylation of 2-oxoglutarate to succinyl-CoA in the citric acid cycle. Phosphorylation of OdhI at threonine residue 14 by PknG relieves the inhibitory function of OdhI and thus allows for a high activity of 2-oxoglutarate dehydrogenase. This happens when glutamine is used as carbon source, which is catabolized via glutamate and 2-oxoglutarate. However, there might be also many other stimuli that effect the phosphorylation status of OdhI. Most interestingly, not only PknG, but also one or several of the other serine/threonine protein kinases can phosphorylate OdhI, as even in a *pknG* mutant part of OdhI is present in a phosphorylated state. Thus, OdhI might be able to integrate diverse stimuli to optimally control the flux at the 2-oxoglutarate node of metabolism. This novel control mechanism for 2-oxoglutarate dehydrogenase via PknG and OdhI, probably operating in a number of Corynebacterianeae, is of prime importance for glutamate production [52], as a reduced 2-oxoglutarate dehydrogenase activity was shown to be the key factor in the metabolic network leading to glutamate. A *C. glutamicum* mutant lacking *odhI* was severely inhibited (90 to 100%) in its ability to secrete glutamate under different conditions. On the other hand, a mutant lacking *pknG* showed significantly improved glutamate production under certain conditions. Thus, first insights into the molecular mechanisms responsible for glutamate production are available now.

PLASMIDS AND TRANSPOSON USE

Based on the strong demand to improve the amino acid production properties of *Corynebacterium glutamicum,* a broad set of vectors has been developed that are anticipated to be useful for other *Corynebacterium* species. They include shuttle vectors, shuttle expression vectors, vectors for integration in the chromosome, promoter- and terminator-probe vectors, vectors for site-specific integration, and vectors for self-cloning. They are compiled in the *Handbook of Corynebacterium glutamicum* published in 2005, which is also a comprehensive source for additional information on *Corynebacterium* [4]. The plasmids are mostly based on the two different replicons pBL1 and pCG1 and their close relatives occurring in *Corynebacterium*, each of these plasmids replicating by a rolling circle mechanism. For more details on the plasmids isolated from amino acid producing coryneform bacteria, excellent reviews are available [53, 54, 64]. It has already been demonstrated that

TABLE 26.4
Vectors of Different Origin Shown to Replicate in an Other *Corynebacterium* Species

Original Isolation	Plasmid Name	Plasmid Family	Demonstrated Replication	Ref.
C. glutamicum	pBL1	pBL1	*C. callunae*	55
			C. diphtheriae	56
C. glutamicum	pCG1	pCG1	*C. ammoniagenes*	57
			C. callunae	57
			C. diphtheriae	56
			C. pilosum	57
C. glutamicum	pGA1	pCG1	*C. callunae*	58
			C. diphtheriae	56
B. stationins	pBY503	pCG1	*C. glutamicum*	59
C. callunae	pCC1	pBL1	*C. glutamicum*	60
C. diphtheriae	pNG2	pCG1	*C. ulcerans*	61
			C. glutamicum	61
C. renale	pCR1		*C. glutamicum*	62
			C. lilium	62
			C. acetoacidophilum	62

derivatives of pBL1 and pCG1 also replicate in other *Corynebacterium* species, including pathogenic ones, thus presuming that they might be useful for molecular work in probably all *Corynebacterium* strains. There are also a few examples of vectors originally developed for use in species others than *C. glutamicum* and shown to replicate in *C. glutamicum* (Table 26.4). To these vectors belongs pNG2, the archetype of vectors like pCG1, and originally isolated from *C. diphtheriae*. The smallest plasmid isolated to date from *Corynebacterium* is the cryptic plasmid pCR1 isolated from *C. renale*, shown to replicate in *C. glutamicum*, *C. lilium*, and *C. acetoacidophilum*. These data corroborate the belief that the molecular tools developed for *C. glutamicum* are of broader use.

It is important to realize that the developed vectors might be present in different copy numbers in the host. This is of particular relevance in case they encode proteins that due to structure or localization are deleterious to the cells, as can be the case for membrane proteins, or whose function is deleterious, as can be the case for deregulated biosynthesis enzymes [63]. For representative examples, their relative copy numbers can be compared in Figure 26.4. The lowest copy numbers in *Corynebacterium glutamicum* have pWK0 [63] and pBHK18 [64], both using the replicon of pNG2 [61], and pKW0 using the replicon of pGA2 [58]. Based on homoserine dehydrogenase and threonine dehydratase genes cloned in pWK0 and enzyme activity determinations, pWK0 is present in three to four copies in *C. glutamicum* [63, 65]. pEKEx2 might be of intermediate copy number, whereas the highest copy numbers are present for pJC1 and pECT18mob2. An assessment of the band intensities from the gels, and transformations with aliquots of plasmids isolated from *C. glutamicum*, suggests that the copy numbers of pJC1 and pECT18mob2 are at least tenfold increased relative to pKW0. As can be seen from Figure 26.4, the copy numbers in *Escherichia coli* are very much different. As a control, pUC18, isolated in an identical manner to other vectors from *E. coli*, is given on the far right. pKW0 is also present in *E. coli* at a low copy number; but because replication of this vector relies in *E. coli* on *repA*, which is under control of the temperature-sensitive repressor cI^{857} [66, 67], a shift to higher cultivation temperatures allows "run-away replication" and therefore easy preparation of sufficient vector for further use in recombinant work. Further details on the vec-

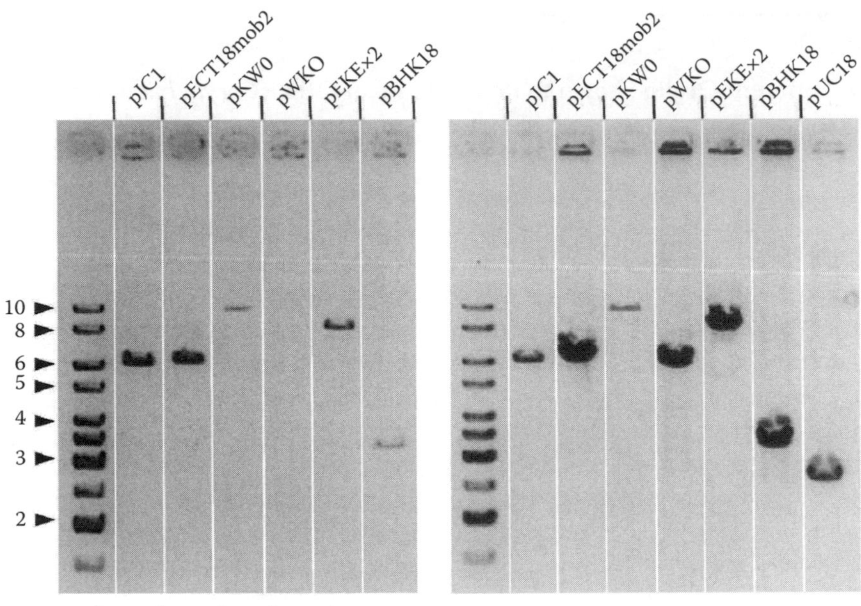

FIGURE 26.4 Comparison of relative copy numbers. Plasmids were isolated from recombinant strains of 100-mL LB cultures and isolated via the QIAfilter Plasmid Midi Kit according to the protocol of the supplier (QIAGEN, Hilden, Germany). 2.5 μL preparation and digested with EcoRI was applied in each lane for plasmids isolated from *Corynebacterium Glutamicum,* and 0.5 μL preparation and digested with EcoRI applied for plasmids isolated from *Escherichia coli.* With plasmid pWKO isolated from *C. glutamicum,* the band is present but not visible.

tors can be taken from the cited literature or comprehensive overviews available on vectors of use for *C. glutamicum* [53, 54, 64].

For transformation of *Corynebacterium,* a number of methods have been used over time. Again, the moving force to derive efficient protocols was the wish of molecular engineering *C. glutamicum.* The most efficient protocol only recently established is also characterized by its simplicity. It uses as medium brain-heart infusion containing 91 g sorbitol/L (BHIS), upon which cells are grown to only a low optical density (O.D. 578 nm) of 1.75 [59]. Cells are carefully washed and transformed via electroporation. Applying the same protocol to *C. diphtheriae,* a very high transformation efficiency of up to 5.6 10^5 colony forming units per microgram DNA is obtained when DNA directly prepared from *Escherichia coli* is used. Such high transformation efficiency is of particular relevance when vectors will be integrated in the chromosome of the recipient because additional host-specific recombination processes reduce the chance of plasmid once present in the cytosol being integrated in the chromosome. Also, for *C. pseudotuberculosis,* similar transformation protocols have been described [68, 69].

For genome-scale analyses, application of transposon mutagenesis is a valuable tool. It is established and used in a number of applications for *Corynebacterium glutamicum* with Tn*5531* [70–72]. This transposon is composed of *aph3* conferring kanamycin resistance, flanked by two *IS1207* sequences encoding the transposase. Because Tn*5531* relies on the IS*1207* sequences, only strains without *IS1207* present in the chromosome are appropriate hosts to generate mutants. Whereas the type strain *C. glutamicum* ATCC13032 contains seven copies of highly similar sequences related to *IS1207,* hosts such as *C. glutamicum* ATCC14752, *C. efficiens* YS-314, and *Corynebacterium* sp. 2262 are appropriate. Another transposon operating in *C. glutamicum* is based on *IS6100,* which is present on plasmid pTET3 isolated from strain *C. glutamicum* LP-6. The transposon is useful in isolating mutants from the type strain ATCC 13032 [73]. Also, other transposons were isolated and constructed and shown to operate in *C. glutamicum* [74, 75]. There is little information on the application of these transposon delivery vectors in other *Corynebacterium* species. However, a transposon derivative based on Tn*4001* and originally developed for *Streptococcus pyogenes* was profitably used in *C. pseudotuberculosis* to identify genes encoding extracytosolic genes, such as fimbriae or transporter subunits [76].

Whereas the *in vivo* systems described depend on nonreplicative delivery vectors, a recent *in vitro* system has been developed [77] that might represent an attractive alternative, in particular due to its ease of handling, its general applicability, and its commercial availability (Epicentre Biotechnologies, Madison, WI, USA). It involves the *in vitro* formation of released Tn5 transposition complexes, called transposomes, followed by introduction of the complexes into the target cell of choice by electroporation. This simple technology has been used for construction and characterization of transposon insertion mutations in *Corynebacterium diphtheriae* and the isolation of mutations that affect expression of the diphtheria toxin repressor (DtxR) [78]. The transposome technology was also successfully applied to *C. matruchotii,* leading to the identification of an integral membrane transporter probably linked to corynomycolic acid synthesis [79], and leading to mutants that synthesize unnatural hybrid fatty acids that functionally replace corynomycolic acid. The transposome-mediated insertions in selected *Corynebacterium* species, and their established function in a wide range of organisms (including *Acinetobacter*, *Mycobacterium*, *Proteus*, and even *Saccharomyces* and *Trypanosoma*) reveals that this molecular device also will be instrumental for the analysis of other *Corynebacterium* species.

CONCLUSION

Nonpathogenic *Corynebacterium* species find many applications, such as amino acid production with *C. glutamicum*, *C. efficiens,* and *C. callunae* [4], or nucleotide production with *C. ammoniagenes* [81]. This strong industrial interest explains the interest of companies and associated research groups to understand *Corynebacterium* at the physiological, molecular, and genomic levels. A recent vision is to use the ability of *C. glutamicum* to make, in addition to the "bulk" products already mentioned, additional metabolites that are cell inherent or whose synthesis capacity is artificially expanded, summarized under the term "white biotechnology." Fortunately, the fast development of standard techniques such as genome sequence establishments has also generated the basic blueprint information for a number of pathogenic *Corynebacterium* species. Together with the fact that the *Corynebacterium* species share basic molecular features, this assists in deepening the knowledge of pathogenic *Corynebacterium* species, with *C. diphtheriae* probably representing the most important species with respect to its epidemiological properties. However, *C. xerosis*, *C. ureolyticum*, *C. ulcerans*, *C. pseudotuberculosis*, *C. striatum,* and further species also pose severe problems as nosocomial pathogens — for example, where a molecular understanding of physiology and antibiotic resistance mechanisms might assist in controlling these bacteria.

REFERENCES

1. Lehmann, K.B. and Neumann, R. *Atlas und Grundriss der Bakteriologie und Lehrbuch der speciellen bakteriologischen Diagnostik*. 1st ed. Munich, 1896.
2. Stackebrandt, E., Rainey, F.A., and Ward-Rainey, N.L. Proposal for a new hierarchal classification system, *Actinobacteria* classis nov. *Int. J. Syst. Bacteriol.*, 47, 479–449, 1997.
3. Eggeling, L. and Sahm, H. The cell wall barrier of *Corynebacterium glutamicum* and amino acid efflux. *J. Biosci. Bioeng.*, 92, 201–213, 2001
4. Eggeling, L. and Bott, M. *Handbook of Corynebacterium glutamicum*. CRC Press, Taylor and Francis Group, Boca Raton, FL, 2005.
5. Dover, L.G., Cerdeno-Tarraga, A.M., Pallen, M.J., Parkhill, J., and Besra, G.S. Comparative cell wall core biosynthesis in the mycolated pathogens, *Mycobacterium tuberculosis* and *Corynebacterium diphtheriae*. *FEMS Microbiol. Rev.*, 28, 225–250, 2004.

6. Izurieta, H.S., Strebel, P.M., Youngblood, T., Hollis, D.G., and Popovic, T. Exudative pharyngitis possibly due to *Corynebacterium pseudodiphtheriticum*, a new challenge in the differential diagnosis of diphtheria. *Emerg. Infect. Dis.*, 3, 65–68, 1997.
7. http://www.bacterio.cict.fr/c/corynebacterium.html
8. Pascual, C., Lawson, P.A., Farrow, J.A., Gimenez, M.N., and Collins, M.D. Phylogenetic analysis of the genus *Corynebacterium* based on 16S rRNA gene sequences. *Int. J. Syst. Bacteriol.*, 45, 724–728,.1995.
9. Janczura, E., Leyh-Bouille, M., Cocito, C., and Ghuysen, J.M. Primary structure of the wall peptidoglycan of leprosy-derived corynebacteria. *J. Bacteriol.*, 145, 775–779, 1981.
10. Alderwick, L.J., Radmacher, E., Seidel, M., Gande, R., Hitchen, P.G., Dell, A., Sahm, H., Eggeling, L., and Besra, G.S. Deletion of Cg-emb in Corynebacterianeae leads to a novel truncated cell wall arabinogalactan, whereas inactivation of Cg-ubiA results in an arabinan-deficient mutant with a cell wall galactan core. *J. Biol. Chem.*, 280, 32362–32371, 2006.
11. Alderwick, L.J., Seidel, M., Sahm, H., Besra, G.S., and Eggeling, L. Identification of a novel arabinofuranosyl transferase (AftA) involved in cell wall arabinan biosynthesis in *Mycobacterium tuberculosis. J. Biol. Chem.*, 281, 15653–15661, 2006.
12. Seidel, M., Alderwick, L.J., Sahm, H., Besra, G.S., and Eggeling, L. Topology and mutational analysis of the single Emb arabinofuranosyltransferase of *Corynebacterium glutamicum* as a model of Emb proteins of *Mycobacterium tuberculosis. Glycobiology,* 17, 210–219, 2007.
13. Besra, G.S., Khoo, K.H., McNeil, M.R., Dell, A., Morris, H.R., and Brennan, P.J. A new interpretation of the structure of the mycolyl-arabinogalactan complex of *Mycobacterium tuberculosis* as revealed through characterization of oligoglycosylalditol fragments by fast-atom bombardment mass spectrometry and 1H nuclear magnetic resonance spectroscopy. *Biochemistry*, 34, 4257–4266, 1995.
14. Rezwan, M., Laneelle, M.A., Sander, P., and Daffe, M. Breaking down the wall: ractionation of mycobacteria. *J. Microbiol. Meth.*, 68, 32–39, 2007.
15. Daffe, M,, Brennan, P.J., and McNeil, M. Predominant structural features of the cell wall arabinogalactan of *Mycobacterium tuberculosis* as revealed through characterization of oligoglycosyl alditol fragments by gas chromatography/mass spectrometry and by ^{1}H and ^{13}C NMR analyses. *J. Biol. Chem.,* 265, 6734–6743, 1990.
16. Puech, V., Chami, M., Lemassu, A., Laneelle, M.A., Schiffler, B., Gounon, P., Bayan, N., Benz, R., and Daffe, M. Structure of the cell envelope of corynebacteria: importance of the noncovalently bound lipids in the formation of the cell wall permeability barrier and fracture plane. *Microbiology,* 147, 1365–1382, 2001.
17. Kremer, L., Dover, L.G., Morehouse, C., Hitchin, P., Everett, M., Morris, H.R., Dell, A., Brennan, P.J., McNeil, M.R., Flaherty. C., Duncan. K., and Besra. G.S. Galactan biosynthesis in *Mycobacterium tuberculosis.* Identification of a bifunctional UDP-galactofuranosyltransferase. *J. Biol. Chem.*, 276, 26430–26440, 2001.
18. Mikusova, K., Yagi, T., Stern, R., McNeil, M.R., Besra, G.S., Crick, D.C., and Brennan, P.J. Biosynthesis of the galactan component of the mycobacterial cell wall. *J. Biol. Chem.*, 275, 33890–33897, 2000.
19. Collins, M.D., Burton, R.A., and Jones, D. *Corynebacterium amycolatum*, sp. nov. a new mycolic acid-less *Corynebacterium* species from human skin. *FEMS Microbiol. Lett.*, 49, 349–352, 1988.
20. Collins, M.D., Falsen, E., Akervall, E., Sjöden, B., and Alvarez, A. *Corynebacterium kroppenstedtii* sp. nov., a novel corynebacterium that does not contain mycolic acids. *Int. J. Syst. Bacteriol.,* 48, 1449–1454, 1998.
21. Hall, V., Collins, M.D., Hutson, R.A., Lawson, P.A., Falsen, E., and Duerden, B.I. *Corynebacterium atypicum* sp. nov., from a human clinical source, does not contain corynomycolic acid. *Int. J. Syst. Evol. Microbiol.,* 53, 1065–1068, 2003.
22. Radmacher, E., Alderwick, L.J., Besra, G.S., Brown, A.K., Gibson, K.J., Sahm, H., and Eggeling, L. Two functional FAS-I type fatty acid synthases in *Corynebacterium glutamicum. Microbiology,* 151, 2421–2427, 2005.
23. Portevin, D., De Sousa-D'Auria, C., Houssin, C., Grimaldi, C., Chami, M., Daffe, M., and Guilhot, C. A polyketide synthase catalyzes the last condensation step of mycolic acid biosynthesis in mycobacteria and related organisms. *Proc. Natl. Acad. Sci. U.S.A.*, 101, 314–319, 2004.
24. Puech, V., Chami, M., Lemassu, A., Laneelle, M.A., Schiffler, B., Gounon, P., Bayan, N., Benz, R., and Daffe, M. Structure of the cell envelope of corynebacteria: importance of the non-covalently bound lipids in the formation of the cell wall permeability barrier and fracture plane. *Microbiology,* 147, 1365–1382, 2001.

25. Hunten, P., Schiffler, B., Lottspeich, F., and Benz, R. PorH, a new channel-forming protein present in the cell wall of *Corynebacterium efficiens* and *Corynebacterium callunae*. *Microbiology*, 151, 2429–2438, 2005.
26. Gande, R., Gibson, K.J., Brown, A.K., Krumbach, K., Dover, L.G., Sahm, H., Shioyama, S., Oikawa, T., Besra, G.S., and Eggeling, L. Acyl-CoA carboxylases (*accD2* and *accD3*), together with a unique polyketide synthase (Cg-*pks*), are key to mycolic acid biosynthesis in *Corynebacterianeae* such as *Corynebacterium glutamicum* and *Mycobacterium tuberculosis*. *J. Biol. Chem.*, 279, 44847–44857, 2004.
27. Oh, T.J., Daniel, J., Kim, H.J., Sirakova, T.D., and Kolattukudy, P.E.. Identification and characterization of Rv3281 as a novel subunit of a biotin-dependent acyl-CoA carboxylase in *Mycobacterium tuberculosis* H37Rv. *J. Biol. Chem.*, 281, 3899–3908, 2006.
28. Gago, G., Kurth, D., Diacovich, L., Tsai, S.C., and Gramajo, H. Biochemical and structural characterization of an essential acyl coenzyme A carboxylase from *Mycobacterium tuberculosis*. *J. Bacteriol.*, 188, 477–486, 2006.
29. Lin, T.W., Melgar, M.M., Kurth, D., Swamidass, S.J., Purdon, J., Tseng, T., Gago, G., Baldi, P., Gramajo, H., and Tsai, S.C. Structure-based inhibitor design of AccD5, an essential acyl-CoA carboxylase carboxyltransferase domain of *Mycobacterium tuberculosis*. *Proc. Natl. Acad. Sci. U.S.A.*, 103, 3072–3077, 2006.
30. Holton, S.J, King-Scott, S., Nasser Eddine, A., Kaufmann, S.H., and M. Wilmanns. Structural diversity in the six-fold redundant set of acyl-CoA carboxyltransferases in *Mycobacterium tuberculosis*. *FEBS Lett.*, 580, 6898–6902, 2006.
31. Kalinowski, J., Bathe, B., Bartels, D., Bischoff, N., Bott, M., Burkovski, A., Dusch, N., Eggeling, L., Eikmanns, B.J., Gaigalat, L., Goesmann, A., Hartmann, M., Huthmacher, K., Kramer, R., Linke, B., McHardy, A. C., Meyer, F., Mockel, B., Pfefferle, W., Puhler, A., Rey, D.A., Ruckert, C., Rupp, O., Sahm, H., Wendisch, V. F., Wiegrabe, I., and Tauch, A. The complete *Corynebacterium glutamicum* ATCC 13032 genome sequence and its impact on the production of L-aspartate-derived amino acids and vitamins. *J. Biotechnol.*, 104, 5–25, 2003.
32. Nishio, Y., Nakamura, Y., Kawarabayasi, Y., Usuda, Y., Kimura, E., Sugimoto, S., Matsui, K., Yamagishi, A., Kikuchi, H., Ikeo, K., and Gojobori, T. Comparative complete genome sequence analysis of the amino acid replacements responsible for the thermostability of *Corynebacterium efficiens*. *Genome Res.*, 13, 1572–1579, 2003.
33. Cerdeno-Tarraga, A.M., Efstratiou, A., Dover, L.G., Holden, M.T., Pallen, M., Bentley, S.D., Besra, G.S., Churcher, C., James, K.D., De Zoysa, A., Chillingworth, T., Cronin, A., Dowd, L., Feltwell, T., Hamlin, N., Holroyd, S., Jagels, K., Moule, S., Quail, M.A., Rabbinowitsch, E., Rutherford, K.M., Thomson, N.R., Unwin, L., Whitehead, S., Barrell, B.G., and Parkhill, J. The complete genome sequence and analysis of *Corynebacterium diphtheriae* NCTC13129. *Nucleic Acids Res.*, 31, 6516–6523, 2003.
34. Tauch, A., Kaiser, O., Hain, T., Goesmann, A., Weisshaar, B., Albersmeier, A., Bekel, T., Bischoff, N., Brune, I., Chakraborty, T., Kalinowski, J., Meyer, F., Rupp, O., Schneiker, S., Viehoever, P., and Puhler, A. Complete genome sequence and analysis of the multiresistant nosocomial pathogen *Corynebacterium jeikeium* K411, a lipid-requiring bacterium of the human skin flora. *J. Bacteriol.*, 187, 4671–4682, 2005.
35. Tauch, A., Trost, E., Bekel, T., Goesmann, A., Ludewig, U., and Pühler, A. Ultrafast *de novo* sequencing of *Corynebacterium urealyticum* using the Genome Sequencer 20 System. *Biochemica*, 4, 4–6, 2006.
36. Brinkrolf, K., Brune, I., and Tauch, A. The transcriptional regulatory network of the amino acid producer *Corynebacterium glutamicum*. *J. Biotechnol.*, 129, 191–211, 2007.
37. Ikeda, M. and Nakagawa, S. The *Corynebacterium glutamicum* genome: features and impacts on biotechnological processes. *Appl. Microbiol. Biotechnol.*, 62, 99–109, 2003.
38. Fudou, R., Jojima, Y., Seto, A., Yamada, K., Kimura, E., Nakamatsu, T., Hiraishi, A., and Yamanaka, S. *Corynebacterium efficiens* sp. nov., a glutamic-acid-producing species from soil and vegetables. *Int. J. Syst. Evol. Microbiol.*, 52, 1127–1131, 2002.
39. Eisen, J.A., Heidelberg, J.F., White, O., and Salzberg, S.L. Related evidence f*or symmetric chromosomal inversions around the replication origin in bacteria. Genome Biol.*, 1, RESEARCH0011. Epub 2000 Dec 4. (http://genomebilogy.com/2000/1/6/research/0011).
40. Ulrich, L.E., Koonin, E.V., and Zhulin, I.B. One-component systems dominate signal transduction in prokaryotes. *Trends Microbiol.*, 13, 52–56, 2005.
41. Boyd, J., Oza, M.N., and Murphy, J.R. Molecular cloning and DNA sequence analysis of a diphtheria tox iron-dependent regulatory element (*dtxR*) from *Corynebacterium diphtheriae*. *Proc. Natl. Acad. Sci. U.S.A.*, 87, 5968–5972, 1990.
42. Wendisch, V.F. Genome-wide expression analysis in *Corynebacterium glutamicum* using DNA microarrays. *J. Biotechnol.*, 104, 273–285, 2003.

43. Hüser, A.T., Becker, A., Brune, I., Dondrup, M., Kalinowski, J., Plassmeier, J., Puhler, A., Wiegrabe, I., and Tauch, A. Development of a *Corynebacterium glutamicum* DNA microarray and validation by genome-wide expression profiling during growth with propionate as carbon source. *J. Biotechnol.*, 106, 269–286, 2003.
44. Hermann, T., Pfefferle, W., Baumann, C., Busker, E., Schaffer, S., Bott, M., Sahm, H., Dusch, N., Kalinowski, J., Puhler, A., Bendt, A.K., Kramer, R., and Burkovski, A. Proteome analysis of *Corynebacterium glutamicum*. *Electrophoresis*, 22, 1712–1723, 2001.
45. Bendt, A.K., Burkovski, A., Schaffer, S., Bott, M., Farwick, M., and Hermann, T. Towards a phosphoproteome map of *Corynebacterium glutamicum*. *Proteomics*, 3, 1637–1646, 2003.
46. Engels, S., Schweitzer, J.E., Ludwig, C., Bott, M., and Schaffer, S. *clpC* and *clpP1P2* gene expression in *Corynebacterium glutamicum* is controlled by a regulatory network involving the transcriptional regulators ClgR and HspR as well as the ECF sigma factor sigma H. *Mol. Microbiol.*, 52, 285–302, 2004.
47. Gerstmeir, R., Cramer, A., Dangel, P., Schaffer, S., and Eikmanns, B.J. RamB, a novel transcriptional regulator of genes involved in acetate metabolism of *Corynebacterium glutamicum*. *J. Bacteriol.*, 186, 2798–2809, 2004.
48. Cramer, A., Gerstmeir, R., Schaffer, S., Bott, M., and Eikmanns, B.J. Identification of RamA, a novel LuxR-type transcriptional regulator of genes involved in acetate metabolism of *Corynebacterium glutamicum*. *J. Bacteriol.*, 188, 2554–2567, 2006.
49. Brune, I., Werner, H., Hüser, A.T., Kalinowski, J., Pühler, A., and Tauch, A. The DtxR protein acting as dual transcriptional regulator directs a global regulatory network involved in iron metabolism of *Corynebacterium glutamicum*. *BMC Genomics*, 7, 21, 2006.
50. Wennerhold, J. and Bott, M. The DtxR regulon of *Corynebacterium glutamicum*. *J. Bacteriol.*, 188, 2907–2918, 2006.
51. Niebisch, A., Kabus, A., Schultz, C., Weil, B., and Bott, M. Corynebacterial protein kinase G controls 2-oxoglutarate dehydrogenase activity via the phosphorylation status of the OdhI protein. *J. Biol. Chem.*, 281, 12300–12307, 2006.
52. Schultz, C., Niebisch, A. Gebel, L., and Bott, M. Glutamate production by *Corynebacterium glutamicum*: dependence on the oxoglutarate dehydrogenase inhibitor protein Odhl and protein kinase PknG. *Appl. Microbiol. Biotechnol.*, 76, 691–700, 2007.
53. Tauch, A., Puhler, A., Kalinowski, J., and Thierbach, G. Plasmids in *Corynebacterium glutamicum* and their molecular classification by comparative genomics. *J. Biotechnol.*, 104, 27–40, 2003.
54. Eggeling, L. and Reyes, O. Experiments. In Eggeling, L. and M. Bott, Eds., *Handbook of Corynebacterium glutamicum*. CRC Press, Taylor and Francis Group, Boca Raton, FL, 2005, p. 535–566.
55. Sandoval, H., del Real, G., Mateos, L.M., Aguilar, A., and Martín, J.F. Screening of plasmids in non-pathogenic *corynebacteria*. *FEMS Microbiol. Lett.*, 27, 93–98, 1985.
56. Tauch, A., Kirchner, O., Löffler, B., Götker, S., Pühler, A., and Kalinowski, J. Efficient electrotransformation of *Corynebacterium diphtheriae* with a mini-replicon derived from the *Corynebacterium glutamicum* plasmid pGA1. *Curr. Microbiol.*, 45, 362–367, 2002.
57. Schäfer, A., Kalinowski, J., Simon, R., Seep-Feldhaus, A.-H., and Pühler, A. High-frequency conjugal plasmid transfer from gram-negative *Escherichia coli* to various Gram-positive coryneform bacteria. *J. Bacteriol.*, 172:1663–1666, 1990.
58. Sonnen, H., Thierbach, G., Kautz, S., Kalinowski, J., Schneider, J., Pühler, A., and Kutzner. H.J. Characterization of pGA1, a new plasmid from *Corynebacterium glutamicum* LP-6. *Gene*, 107, 69–74, 1991.
59. Satoh, Y., Hatakeyama, K., Kohama, K., Kobayashi, M., Kurusu, Y., and Yukawa, H. Electrotransformation of intact cells of *Brevibacterium flavum* MJ-233. *J. Indust. Microbiol.*, 5, 159–165, 1990.
60. Sandoval, H., Aguilar, A., Paniagua, C., and Martín, J.F. Isolation and physical characterization of plasmid pCC1 from *Corynebacterium callunae* and construction of hybrid derivatives. *Appl. Microbiol. Biotechnol.*, 19, 409–413, 1984.
61. Serwold-Davis, T., Groman, N., and Rabin, M. Transformation of *Corynebacterium diphtheriae, Corynebacterium ulcerans, Corynebacterium glutamicum,* and *Escherichia coli* with the *C. diphtheriae* plasmid pNG2. *Proc. Natl. Acad. Sci. U.S.A.*, 84, 4964–4968, 1987.
62. Srivastava, P., Nath, N., and Deb, J.K. Characterization of broad host range cryptic plasmid pCR1 from *Corynebacterium renale*. *Plasmid*, 56, 24–34, 2006.
63. Reinscheid, D.J., Kronemeyer, W., Eggeling, L., Eikmanns, B.J., and Sahm, H. Stable expression of hom-1-thrB in *Corynebacterium glutamicum* and its effect on the carbon flux to threonine and related amino acids. *Appl. Environ. Microbiol.*, 60, 126–132, 1994.

64. Kirchner, O. and Tauch, A. Tools for genetic engineering in the amino acid-producing bacterium *Corynebacterium glutamicum*. *J. Biotechnol.*, 104, 287–299, 2003.
65. Morbach, S., Eggeling, L., and Sahm, H. Use of feedback-resistant threonine dehydratases of *Corynebacterium glutamicum* to increase carbon flux towards L-isoleucine. *Appl. Environ. Microbiol.*, 61, 4315–4320, 1995.
66. Light, J. and Molin, S. Replication control functions of plasmid R1 act as inhibitors of expression of a gene required for replication. *Mol. Gen. Genet.*, 184, 56–61, 1981.
67. Larsen, J.E., Gerdes, K., Light, J., and Molin, S. Low-copy-number plasmid-cloning vectors amplifiable by derepression of an inserted foreign promoter. *Gene*, 28, 45–54, 1984.
68. Songer, J.G., Hilwig, R.W., Leeming, M.N., Iandolo, J.J., and Libby, S.J. Transformation of *Corynebacterium pseudotuberculosis* by electroporation. *Am. J. Vet. Res.*, 52, 1258–1261, 1991.
69. Dorella, F.A., Estevam, E.M., Cardoso, P.G., Savassi, B.M., Oliveira, S.C., Azevedo, V., and Miyoshi, A. An improved protocol for electrotransformation of *Corynebacterium pseudotuberculosis*. *Vet. Microbiol.*, 114, 298–303, 2006.
70. Ankri, S., Serebrijski, I., Reyes, O., and Leblon, G. Mutations in the *Corynebacterium glutamicum* proline biosynthetic pathway: a natural bypass of the *proA* step. *J. Bacteriol.*, 178, 4412–4419, 1996.
71. Bonamy, C., Labarre, J. Cazaubon, L., Jacob, C., Le Bohec, F., Reyes O., and Leblon, G. The mobile element IS1207 of *Brevibacterium lactofermentum* ATCC21086: isolation and use in the construction of Tn5531, a versatile transposon for insertional mutagenesis of *Corynebacterium glutamicum*. *J. Biotechnol.*, 104, 301–309, 2003.
72. Simic, P., Sahm, H., and Eggeling L. L-threonine export: use of peptides to identify a new translocator from *Corynebacterium glutamicum*. *J. Bacteriol.*, 183, 5317–5324, 2001.
73. Mormann, S., Lomker, A., Ruckert, C., Gaigalat, L., Tauch, A., Puhler, A., and Kalinowski, J. Random mutagenesis in *Corynebacterium glutamicum* ATCC 13032 using an IS6100-based transposon vector identified the last unknown gene in the histidine biosynthesis pathway. *BMC Genomics,* 10, 205, 2006.
74. Inui, M., Tsuge, Y., Suzuki, N., Vertes, A.A., and Yukawa, H. Isolation and characterization of a native composite transposon, Tn14751, carrying 17.4 kilobases of *Corynebacterium glutamicum* chromosomal DNA. *Appl. Environ. Microbiol.*, 71, 407–416, 2005.
75. Tsuge, Y., Ninomiya, K., Suzuki, N., Inui, M., and Yukawa, H. A new insertion sequence, IS14999, from *Corynebacterium glutamicum*. *Microbiology*, 151, 501–508, 2005.
76. Dorella, F.A., Estevam, E.M., Pacheco, L.G., Guimaraes, C.T., Lana, U.G., Gomes, E.A., Barsante, M.M., Oliveira, S.C., Meyer, R., Miyoshi, A., and Azevedo, V. *In vivo* insertional mutagenesis in *Corynebacterium pseudotuberculosis*: an efficient means to identify DNA sequences encoding exported proteins. *Appl. Environ. Microbiol.*, 72, 7368–7372, 2006.
77. Goryshin, I.Y., Jendrisak, J., Hoffman, L.M., Meis, R., and Reznikoff, W.S. Insertional transposon mutagenesis by electroporation of released Tn5 transposition complexes. *Nat. Biotechnol.*, 18, 97–100, 2000.
78. Oram, D.M., Avdalovic, A., and Holmes, R.K. Construction and characterization of transposon insertion mutations in *Corynebacterium diphtheriae* that affect expression of the diphtheria toxin repressor (DtxR). *J. Bacteriol.*, 184, 5723–5732, 2002.
79. Wang, C., Hayes, B., Vestling, M.M., and Takayama, K. Transposome mutagenesis of an integral membrane transporter in *Corynebacterium matruchotii*. *Biochem. Biophys. Res. Commun.*, 340, 953–960, 2006.
80. Takayama, K., Hayes, B., Vestling, M.M., and Massey, R.J. Transposon-5 mutagenesis transforms *Corynebacterium matruchotii* to synthesize novel hybrid fatty acids that functionally replace corynomycolic acid. *Biochem. J.*, 373, 465–474, 2003.
81. Abbouni, B., Elhariry, H.M., and Auling, G. Overproduction of NAD+ and 5′-inosine monophosphate in the presence of 10 microM Mn^{2+} by a mutant of *Corynebacterium ammoniagenes* with thermosensitive nucleotide reduction (nrd(ts)) after temperature shift. *Arch. Microbiol.*, 182, 119–125, 2004.
82. Baumbach, J., Brinkrolf, K., Czaja, L.F., Rahmann, S., and Tauch, A. CoryneRegNet: An ontology-based data warehouse of corynebacterial transcription factors and regulatory networks. *BMC Genomics*, 7, 24, 2006.
83. Kim, H.J., Kim, T.H., Kim, Y., and Lee, H.S. Identification and characterization of glxR, a gene involved in regulation of glyoxylate bypass in *Corynebacterium glutamicum*. *J. Bacteriol.*, 186, 3453-3460, 2004.

84. Letek, M., Valbuena, N., Ramos, A., Ordonez, E., Gil, J.A., and Mateos, L.M. Characterization and use of catabolite-repressed promoters from gluconate genes in *Corynebacterium glutamicum. J. Bacteriol.*, 188, 409-423, 2006.
85. Kimura, E. Triggering mechanism of L-glutamate overproduction by DtsR1 in coryneform bacteria. *J. Biosci. Bioeng.*, 94, 545-551, 2002.
86. Brinkrolf, K., Brune, I., and Tauch, A. Transcriptional regulation of catabolic pathways for aromatic compounds in *Corynebacterium glutamicum. Genet. Mol. Res.*, 5, 773-789, 2006.
87. Huang, Y., Zhao, K.X., Shen, X.H., Chaudhry, M.T., Jiang, C.Y., and Liu, S.J. Genetic characterization of the resorcinol catabolic pathway in *Corynebacterium glutamicum. Appl. Environm. Microbiol.*, 72, 7238-7245, 2006.
88. Krug, A., Wendisch, V.F., and Bott, M. Identification of AcnR, a TetR-type repressor of the aconitase gene acn in *Corynebacterium glutamicum. J. Biol. Chem.*, 280, 585-595, 2005.
89. Engels, V., and Wendisch, V.F. The DeoR-type regulator SugR represses expression of ptsG in *Corynebacterium glutamicum. J. Bacteriol.,* 189, 2955-2966, 2007.
90. Merkens, H., Beckers, G., Wirtz, A., and Burkovski, A. Vanillate metabolism in *Corynebacterium glutamicum. Curr. Microbiol.,* 51, 59-65, 2005.
91. Hansmeier, N., Albersmeier, A., Tauch, A., Damberg, T., Ros, R., Anselmetti, D., Pühler, A., and Kalinowski, J. The surface (S)-layer gene cspB of *Corynebacterium glutamicum* is transcriptionally activated by a LuxR-type regulator and located on a 6 kb genomic island absent from the type strain ATCC 13032. *Microbiology*, 152, 923-935, 2006.
92. Shen, X.H., Jiang, C.Y., Huang, Y., Liu, Z.P., and Liu, S.J. Functional identification of novel genes involved in the glutathione-independent gentisate pathway in *Corynebacterium glutamicum. Appl. Environ. Microbiol.*, 71, 3442-3452, 2005.
93. Kennerknecht, N., Sahm, H., Yen, M.R., Patek, M., Saier, M.H., and Eggeling, L. Export of L-isoleucine from *Corynebacterium glutamicum*: a two-gene-encoded member of a new translocator family. *J. Bacteriol.*, 184, 3947-3956., 2002.
94. McHardy, A.C., Tauch, A., Rückert, C., Pühler, A., and Kalinowski, J. Genome-based analysis of biosynthetic aminotransferase genes of *Corynebacterium glutamicum.* J. Biotechnol., 104, 229-240, 2003.
95. Vrljic, M., Sahm, H., and Eggeling, L. A new type of transporter with a new type of cellular function: L-lysine export from *Corynebacterium glutamicum. Mol. Microbiol.*, 22, 815-826, 1996.
96. Bellmann, A., Vrljic, M., Patek, M., Sahm, H., Krämer, R., and Eggeling, L. Expression control and specificity of the basic amino acid exporter LysE of *Corynebacterium glutamicum. Microbiology,* 147, 1765-1774, 2001.
97. Brune, I., Jochmann, N., Brinkrolf, K., Hüser, A.T., Gerstmeir, R., Eikmanns, B.J., Kalinowski, J., Pühler, A., and Tauch, A. The IclR-type transcriptional repressor LtbR regulates the expression of leucine and tryptophan biosynthesis genes in the amino acid producer *Corynebacterium glutamicum. J. Bacteriol.*, 189, 2720-2733, 2007.
98. Kim, S.H., Yim, S.H., Lee, M.S. Organization and regulation of arginine biosynthesis genes in *Corynebacterium glutamicum.* In: International Meeting of the Microbiological Society of Korea, The Microbiological Society of Korea, p. 187, 2006.
99. Chung, S.S. Molecular analysis of *pyrH* gene encoding UMP kinase in *Corynebacterium glutamicum* In: *International Meeting of the Microbiological Society of Korea*, The Microbiological Society of Korea, p. 138, 2006.
100. Itou, H., Okada, U., Suzuki, H., Yao, M., Wachi, M., Watanabe, N., and Tanaka, I. The CGL2612 protein from *Corynebacterium glutamicum* is a drug resistance-related transcriptional repressor - Structural and functional analysis of a newly identified transcription factor from genomic DNA analysis. *J. Biol. Chem.,* 280, 38711-38719, 2005.
101. Koch, D.J., Rückert, C., Albersmeier, A., Hüser, A.T., Tauch, A., Pühler, A., and Kalinowski, J. The transcriptional regulator SsuR activates expression of the *Corynebacterium glutamicum* sulphonate utilization genes in the absence of sulphate. *Mol. Microbiol.,* 58, 480-494, 2005.
102. Rückert, C., Koch, D., Rey, A., Pühler, A., and Kalinowski, J. The transcriptional activator CysR of *Corynebacterium glutamicum* and its regulon. 2nd European Conference on Prokaryotic Genomes. Dechema e. V., Frankfurt am Main, p. 80, 2006.
103. Ordónez, E., Letek, M., Valbuena, N., Gil, J.A., Mateos, L.M. Analysis of genes involved in arsenic resistance in *Corynebacterium glutamicum* ATCC 13032. *Appl. Environ. Microbiol.,* 71, 6206-6215, 2005.

104. Burkovski, A. Ammonium assimilation and nitrogen control in *Corynebacterium glutamicum* and its relatives: an example for new regulatory mechanisms in actinomycetes. *FEMS Microbiol. Rev.*, 27, 617-628, 2003.
105. Silberbach, M., and Burkovski, A. Application of global analysis techniques to *Corynebacterium glutamicum*: new insights into nitrogen regulation. *J. Biotechnol.*, 126, 101-110, 2006.
106. Wennerhold, J., Krug, A., and Bott, M. The AraC-type regulator RipA represses aconitase and other iron proteins from *Corynebacterium* under iron limitation and is itself repressed by DtxR. *J. Biol. Chem.*, 280, 40500-40508, 2005.
107. Kocan, M., Schaffer, S., Ishige, T., Sorger-Herrmann, U., Wendisch, V.F., and Bott, M. Two-component systems of *Corynebacterium glutamicum*: Deletion analysis and involvement of the PhoS-PhoR system in the phosphate starvation response. *J. Bacteriol.*, 188, 724-732. 2006.
108. Schaaf, S., and Bott, M. (2007) Target genes and DNA-binding sites of the response regulator PhoR from *Corynebacterium glutamicum*. *J. Bacteriol.*, 189, 5002-5011.
109. Rey, D.A., Nentwich, S.S., Koch, D.J., Rückert, C., Pühler, A., Tauch, A., and Kalinowski, J. The McbR repressor modulated by the effector substance S-adenosylhomocysteine controls directly the transcription of a regulon involved in sulphur metabolism of *Corynebacterium glutamicum* ATCC 13032. *Mol. Microbiol.*, 56, 871-887, 2005.
110. Möker, N., Brocker, M., Schaffer, S., Krämer, R., Morbach, S., and Bott, M. Deletion of the genes encoding the MtrA-MtrB two-component system of *Corynebacterium glutamicum* has a strong influence on cell morphology, antibiotics susceptibility and expression of genes involved in osmoprotection. *Mol. Microbiol.*, 54, 420-438. 2004.
111. Brocker, M., and Bott, M. Evidence for activator and repressor functions of the response regulator MtrA from *Corynebacterium glutamicum*. *FEMS Microbiol. Lett.*, 264, 205-212, 2006.
112. Kim, T.H., Park, J.S., Kim, H.J., Kim, Y., Kim, P., and Lee, H.S. The whcE gene of *Corynebacterium glutamicum* is important for survival following heat and oxidative stress. *Biochem. Biophys. Res. Commun.*, 337, 757-764, 2005.
113. Engels, S., Ludwig, C., Schweitzer, J.-E., Mack, C., Bott, M., and Schaffer, S. The transcriptional activator ClgR controls transcription of genes involved in proteolysis and DNA repair in *Corynebacterium glutamicum*. *Mol. Microbiol.*, 57, 576-591, 2005.

27 The Actinobacteria

Alan C. Ward and Nagamani M. Bora

CONTENTS

SUMMARY

The Actinobacteria are a major evolutionary line in the 16S phylogenetic tree, characterized as Gram-positive, high G+C bacteria. However, the members of this major taxon, defined phylogenetically, are morphologically and physiologically diverse with few characters identified, phenotypically or metabolically, in common. Whole genome sequencing of many Actinobacteria is beginning to provide sequence data, signature nucleotides and indels, and signature proteins that are characteristic of the whole phylum, or taxa within it. And whole genome sequences are being analyzed to reconstruct evolutionary history. Whether it is the origin of the eukaryotes[1] or the invasion of the land,[2] these broad timescale analyses provide the big picture of what has made Actinobacteria, although still leaving conclusions, over more than 4 billion years, speculative.

BACKGROUND

Members of the high G+C, Gram-positive bacteria were among the earliest bacteria isolated and/or described ("*Schinzia alni*"* = *Frankia alni* Woronin 1866,[3] "*Bacteridium luteum*" = *Micrococcus luteus* Schroeter 1872,[4] "*Microsporon diphthericum*" = *C. diphtheriae* Klebs 1875,[5] *Actinomyces* Harz 1877,[6] *Mycobacterium tuberculosis* Zopf 1883,[7] *Micrococcus agilis* = *Arthrobacter agilis* Ali-Cohen 1889,[8] "*Cladothrix asteroides*" = *Nocardia asteroides* Eppinger 1891,[9] *Rhodococcus rhodochrous* Zopf 1891,[10] "*Streptothrix albidoflava*" = *Streptomyces albidoflavus* Rossi Doria 1891,[11] *Actinomadura madurae* Vincent 1894,[12] *Corynebacterium* Lehmann and Neumann 1896,[13] *Bifidobacteria* Tissier 1900,[14] "*Bacillus acnes*" = *Propionibacterium acnes* Gilchrist 1900,[15] *Brevibacterium linens* Wolff 1910[16]). Many were recognized as belonging to a major taxon in the domain Bacteria, the Actinomycetales[17] from the earliest taxonomic descriptions.

However, many of these early higher bacterial taxa, which still often influence our concept of the major groups of bacteria, were defined, and redefined, before the advent of molecular phylogeny, and have proved phylogenetically diverse. Classical bacterial taxonomists recognized their inability to define a phylogenetic taxonomy, and deliberately eschewed it, but the advent of 16S rRNA/DNA sequence analysis, enabling a phylogenetic framework, has been seized upon by taxonomists and non-taxonomists alike — leaving the field with a hint of underlying schizophrenia. The evolutionary breadth in these higher bacterial taxa is captured by the term "phylum," although it has been argued that these lines represent the equivalent of kingdoms (see eukaryotes in Figure 27.1). However, the Actinobacteria phylum[18] is not a taxonomic rank covered by the Bacteriological Code, and the idea

* Although 90 years before its cultivation.

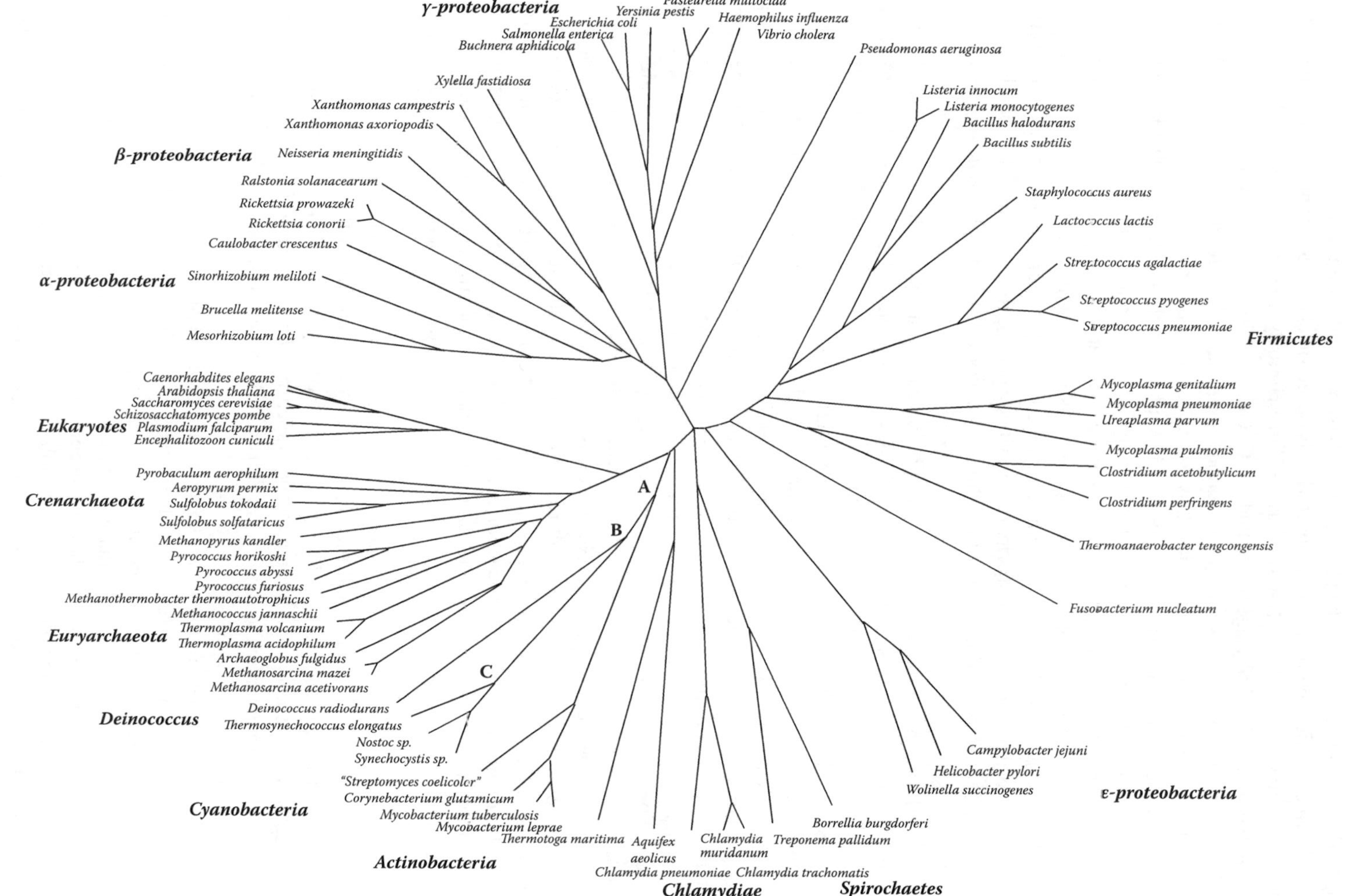

FIGURE 27.1 Schematic representation of the main lines of evolution based on whole genome analyses based on Refs. 2, 50, and 51 and a timescale of actinobacterial evolution based on Ref. 2 A = 2784 Ma; B = 2558 Ma, and C = 1039 Ma.

of the unification of the high G+C and low G+C Gram-positives, the Actinobacteria and Firmicutes, as a kingdom[19] may not, in the end, be phylogenetically supportable (Figure 27.1).

The modern classification of the Actinobacteria[19] is based on chemotaxonomy[20–23] and the ability to define species using DNA:DNA similarity,[24] but primarily 16S gene sequence determination.[25–28] On the basis of 16S phylogeny, the high G+C, Gram-positive bacteria were recognized, in more or less their current form, as a major evolutionary lineage.[25] The foundation of this current, validly described,* taxonomic group is the class *Actinobacteria* as proposed by Stackebrandt et al. (1997),[19] and the validly described composition is kept updated by Euzeby (1997; http://www.bacterio.cict./fr).[29] The phylogenetic depth of this clade is not as great as some of the other major bacterial lineages but the phenotypic diversity is large. Notwithstanding whether organisms are Gram-positive or whether their DNA base composition is high, membership of this main line of descent, and the class *Actinobacteria*, is defined phylogenetically.[19] It encompasses the order Actinomycetales; this order contains the majority of described species, the actinomycetes, but does not capture the largely undescribed diversity of the "unseen majority"[30] of deeply branching members of the clade.

Members of the *Actinobacteria* prove morphologically and physiologically diverse and include deep branching orders: the Acidimicrobiales, Rubrobacterales, Coriobacteriales, and Bifidobacteriales. The Acidimicrobiales includes *Acidimicrobium ferrooxidans* the only validly described species in the order, a ferrous ion oxidizing, acidophilic, chemolithotroph, while Rubrobacterales, also a major environmental clade, on the basis of molecular clones detected, has only a few mesophilic, heterotrophic isolates and described species — (five families, each with a single genus and eight species in total). Coriobacteriales and Bifidobacteriales are deep branching orders containing genera and species associated with the guts of, mostly, warm-blooded animals — although a few isolates from insect guts and the environment hint at a wider distribution. The Coriobacteriales are a deep-branching order, like the Acidimicrobiales and Rubrobacterales, in its own subclass, while the order Bifidobacteriales is in the subclass Actinobacteridae with the Actinomycetales. Nevertheless, the bifidobacteria as anaerobic, fermentative lactic acid producing bacteria were not recognized as actinobacteria but were traditionally linked with the other lactic acid fermenters in the Lactic Acid Bacteria (LAB) with *Lactobacillus*. Similarly, members of the Micrococcineae, the Micrococcus suborder, although phylogenetically within the order Actinomycetales, and only a deep-branching clade within that context, were traditionally grouped with the morphologically similar staphylococci and only recognized as actinomycetes by phylogenetic analysis.

Morphologically, members of the *Actinobacteria* span the range from coccoid (e.g., *Micrococcus* or members of the G-bacteria) and rods (*Conexibacter*, *Demequina*) through more complex forms such as: rod-coccus cycle (*Arthrobacter*); diphtheroid/coryneform (*Corynebacterium*); bifid, bifurcating cells (*Bifidobacterium*); bifurcating or branching (*Propionibacterium*); nocardioform, branching mycelium fragmenting to rods (*Nocardia*); to fully complex mycelium (*Frankia*) and differentiating growth (*Streptomyces*).

Physiologically, members of the *Actinobacteria* are often mesophilic, neutrotolerant, aerobic heterotrophs isolated from the terrestrial environment. They often exhibit the ability to degrade complex and recalcitrant organic compounds (*Arthrobacter*, *Rhodococcus*), secreting extensive arrays of extracellular enzymes (e.g., cellulases, xylanases, and chitinases[31–33]) and synthesize many taxon-specific secondary metabolites,[34] showing sweeping metabolic versatility.

However, psychrophilic and thermophilic actinobacteria are not uncommon, and when sought, acidophilic and alkalophilic are readily found. Examples of different major nutritional modes, anaerobic and fermentative (*Actinomyces*, bifidobacteria, propionibacteria), chemolithotrophic

* Strictly even the class Actinobacteria is illegitimate as no nomenclatural type was described.[29]

(*Acidimicrobium ferrooxidans*), nitrogen-fixing (*Frankia*), as well as aerobic heterotrophs, are found in the taxon.

Although historically associated with the terrestrial soil environment, marine actinobacteria and commensals, symbionts and pathogens of marine invertebrates, as well as plants, insects, and animals generally, are increasingly recognized, isolated, and characterized.

HISTORY

Buchanan (1918)[17] described the order Actinomycetales with one family Actinomycetaceae containing four genera: *Actinobacillus*, Brumpt, 1900;[35] *Leptotrichia*, Trevisan, 1879;[36] *Actinomyces*, Harz, 1877;[6] and *Nocardia*, Trevisan, 1889.[37]

Ørskov (1923)[38] classified actinomycetes into three large groups on the basis of morphology. Many of the organisms in his second group would now contain strains belonging to the genera *Actinomyces, Nocardia,* and *Rhodococcus*, with morphological similarity to corynebacteria and mycobacteria.

In 1931, Jensen proposed the genus *Proactinomyces*[39] for Ørskov's group II actinomycetes.

In the classification by Waksman and Henrici[40] in 1943, three families were proposed: the family Mycobacteriaceae containing the genus *Mycobacterium*; while in the family Actinomycetaceae, the genus names proposed as having priority were *Actinomyces* and *Nocardia* for the anaerobic and aerobic actinomycetes, respectively; and the family Streptomycetaceae contained the genera *Streptomyces* and *Micromonospora*. Waksman and Henrici agreed with Stanier and van Niel[40] in considering *Corynebacterium* as part of the Lactobacteriaceae.

On the basis of morphology and chemotaxonomy, six major groups were recognized and eighteen genera of aerobic actinomycetes[41] by the Lechavaliers, who introduced many chemotaxonomic methods.

By 1997,[19] there were six orders (Acidimicrobiales, Rubrobacterales, Coriobacteriales, Sphaerobacterales, Actinomycetales, Bifidobacteriales) and within the order Actinomycetales, ten suborders Actinomycineae (1)*, Micrococcineae (9), Corynebacterineae (6), Micromonosporineae (1), Propionibacterineae (2), Pseudonocardineae (1), Streptomycineae (1), Streptosporangineae (3), Frankineae (5), and Glycomycineae (1). The taxonomy was defined phylogenetically, based on 16S sequences and taxa characterized with signature nucleotides.

DEFINING THE ACTINOBACTERIA

Members of the class Actinobacteria have 16S sequence similarity greater than 80% and the signature nucleotide A at position 906,† and, with the exception of the members of the subclass Rubrobacteridae, have either A or C at position 955, members of the Rubrobacteridae have a U at position 955. The patterns of signature nucleotides for the lower taxa, subclasses, orders, families, and genera are more comprehensive.[19]

The presence of a characteristic sequence insertion in the 23S rRNA gene[42] has been diagnostic of the high G+C, Gram-positives and appears to be almost universally present in the actinomycetes. The advent of whole genome sequencing (Table 27.1) has enabled a more comprehensive search for characteristic signature sequences in other genes.[43] The near-universal occurrence of the large insert (typically 90 to 100 nt) in the 23S rRNA gene was confirmed for all the actinobacteria tested except members of the genus *Rubrobacter*. However, of the other deep-branching orders, only the Bifidobacteriales was tested, and it contained the insert.

Gao and Gupta[43] also identified a two amino acid deletion in the cytochrome c oxidase gene *cox1*, at positions 332 and 333 (in the cox1 protein in *Paracoccus dentrificans*), as unique in actinobacteria. In the gene for CTP synthetase, a three or four amino acid insert, unique to actinobac-

---

* Number of families in each suborder shown in brackets.

† *Escherichia coli* numbering

TABLE 27.1
Completed Actinobacterial Whole Genome Sequences

Organism	Phenotype	Entrez PID	Database	Genome Attributes
Acidothermus cellulolyticus	Acidophile, thermophile, cellulolytic, aerobe, sporulating	16097	NCBI, Joint Genome Institute	2.44 Mb, 66.9% GC, 2157 genes
Arthrobacter aurescens	Atrazine metabolism, motile, aerobe, rod-shaped, nonsporulating	12512	NCBI	5.23 Mb, 62.3% GC, 4587 genes
Arthrobacter sp.	Radiation resistant, hydrocarbon degrading, aerobe, rod-shaped, motile	12640	NCBI, Joint Genome Institute	5.07 Mb, 65.5% GC, 4516 genes
Bifidobacterium adolescentis	Probiotic, anaerobe, rod-shaped, nonsporulating, nonmotile, non-pathogen	16321	NCBI	2.08 Mb, 59.2% GC, 1631 genes
Bifidobacterium longum	Probiotic, rod-shaped, anaerobe, nonmotile, nonsporulating, non-pathogen	328	http://www.uga.edu/mib/mschell.htm	2.25 Mb, 60% GC, 1813 genes
Corynebacterium diphtheriae gravis	Pathogen, facultative, nonmotile, rod-shaped, nonsporulating	87	http://www.sanger.ac.uk/Users/parkhill/	2.48 Mb, 53.5% GC, 2320 genes
Corynebacterium efficiens	Rod-shaped, nonmotile, nonsporulating, facultative	305	NCBI, NITE	3.13 Mb, 63.1% GC, 3031 genes
Corynebacterium glutamicum	Rod-shaped, nonmotile, facultative, nonsporulating	13760	NCBI	3.28 Mb, 53.8% GC, 3002 genes
Corynebacterium glutamicum	Rod-shaped, nonmotile, facultative, nonsporulating	307	NCBI, TIGR	3.30 Mb, 53.8% GC, 3040 genes
Corynebacterium jeikeium	Pathogen, rod-shaped, motile, facultative, lipophilic	13967	NCBI	2.46 Mb, 61% GC, 2137 genes
Frankia alni	Aerobe, mesophilic, chemoorganotroph, nitrogen fixation, non-pathogen	17403	Genoscope, NCBI, FrankiaGenomes	7.49 Mb, 72% GC, 6711 genes
Frankia sp.	Aerobe, mesophilic, chemoorganotroph, nitrogen fixation, non-pathogen, plant symbiont, nonmotile, sporulating	13963	NCBI, FrankiaGenomes, Joint Genome Institute	5.43 Mb, 70% GC, 4499 genes
Leifsonia xyli xyli	Pathogen, rod-shaped, nonmotile, aerobe	212	NCBI, UNICAMP	2.58 Mb, 67.7% GC, 2003 genes
Mycobacterium avium	Pathogen, aerobe, chemoorganotroph, rod-shaped, nonmotile, nonsporulating	88	NCBI	5.47 Mb, 69% GC, 5120 genes
Mycobacterium avium paratuberculosis	Pathogen, aerobe, chemoorganotroph, rod-shaped, nonmotile, nonsporulating	91	http://www.cvm.umn.edu/departments/vpb/faculty/kapur.htm	4.32 Mb, 69.3% GC, 4350 genes
Mycobacterium bovis	Pathogen, rod-shaped, nonmotile, nonsporulating, aerobe	89	NCBI, Sanger Institute, Pasteur	4.34 Mb, 65.6% GC, 3955 genes

Mycobacterium bovis BCG	Pathogen, aerobe, chemoorganotroph, rod-shaped, nonmotile, nonsporulating	18059	NCBI, Institut Pasteur	4.37 Mb, 65.6% GC, 3952 genes
Mycobacterium leprae	Pathogen, aerobe, chemoorganotroph, rod-shaped, nonmotile	90	NCBI, Sanger Institute, Leproma, PEDANT, GIB, KEGG, TIGR	3.26 Mb, 57.8% GC, 1604 genes
Mycobacterium smegmatis	Pathogen, aerobe, chemoorganotroph, rod-shaped, nonmotile, nonsporulating	92	NCBI	6.98 Mb, 67.4% GC, 6716 genes
Mycobacterium sp.	Rod-shaped, nonmotile, nonsporulating, aerobe, PAH-degrading	16079	NCBI	6.04 Mb, 68% GC, 5739 genes
Mycobacterium sp.	Rod-shaped, nonmotile, nonsporulating, aerobe, PAH-degrading	16081	NCBI, Joint Genome Institute	6.22 Mb, 68% GC, 5975 genes
Mycobacterium sp.	Rod-shaped, nonmotile, nonsporulating, aerobe, PAH-degrading	15762	NCBI, Joint Genome Institute	5.70 Mb, 68% GC, 5615 genes
Mycobacterium tuberculosis	Pathogen, aerobe, chemoorganotroph, rod-shaped, nonmotile	223	NCBI, TIGR, GIB, KEGG	4.40 Mb, 65.6% GC, 4187 genes
Mycobacterium tuberculosis	Pathogen, aerobe, chemoorganotroph, rod-shaped, nonmotile, nonsporulating	5	NCBI, Sanger Institute, TIGR, PEDANT, KEGG, TubercuList, GCat, IBM-Annotation, MBGD, GIB	4.41 Mb, 65.6% GC, 4402 genes
Mycobacterium ulcerans	Pathogen, rod-shaped, nonmotile, nonsporulating, aerobe	16230	NCBI, Institut Pasteur, BuruList	5.8Mb, 65.5% GC, 4160 genes
Mycobacterium vanbaalenii	Rod-shaped, nonmotile, nonsporulating, PAH-degrading	15761	NCBI, Joint Genome Institute	6.49 Mb, 67.8% GC, 5979 genes
Nocardia farcinica	Pathogen, filament-shaped, sporulating, nonmotile, aerobe	13117	NCBI, NIID	6.29 Mb, 70.8% GC, 5933 genes
Nocardioides sp.	Aerobe, rod-shaped, nonmotile	12738	NCBI, Joint Genome Institute	4.98 Mb, 71.7% GC, 4909 genes
Propionibacterium acnes	Anaerobe, nonsporulating, pathogen, rod-shaped, nonmotile	12460	NCBI, Entrez	2.56 Mb, 60% GC, 2294 genes
Rhodococcus sp.	Degrades PCBs, aerobe, coccus-shaped	13693	NCBI, Genome BC, Genome British Columbia	7.80 Mb, 67% GC, 7211 genes
Rubrobacter xylanophilus	Thermophile, radiation resistant, rod-shaped, nonmotile, aerobe, nonsporulating	10670	NCBI, Joint Genome Institute	3.22 Mb, 70.4% GC, 3140 genes

—*continued*

TABLE 27.1 (continued)
Completed Actinobacterial Whole Genome Sequences

Organism	Phenotype	Entrez PID	Database	Genome Attributes
Streptomyces avermitilis	Nonmotile, sporulating, aerobe	189	http://www.kuleuven.ac.be/gih/satoshi.htm	9.02 Mb, 70.7% GC, 7573 genes
Streptomyces coelicolor	Nonmotile, sporulating, aerobe	242	NCBI, Sanger Institute, TIGR	8.66 Mb, 72.1% GC, 7825 genes
Symbiobacterium thermophilum	Thermophile, symbiont, rod-shaped	12994	NCBI	3.56 Mb, 68.7% GC, 3310 genes
Thermobifida fusca	Pathogen, degrades organic material, rod-shaped, nonmotile, sporulating, aerobe, thermophile	94	NCBI, Joint Genome Institute	3.64 Mb, 67% GC, 3110 genes
Tropheryma whipplei	Pathogen, rod-shaped, nonmotile, aerobe	95	NCBI, Genoscope	927 kb, 46% GC, 808 genes
Tropheryma whipplei	Pathogen, rod-shaped, nonmotile, aerobe	354	http://www.sanger.ac.uk/Users/parkhill/	925 kb, 46.3% GC, 817 genes

Source: The Genomes On Line Database (GOLD) v. 2 .

teria, has been identified. And, in the gene for glutamyl-tRNA synthetase (*GluRS*), a five amino acid insert in GluRS is only present in actinobacteria. Five putative actinobacteria show variability in these sequence signatures (indels): *Symbiobacterium thermophilum*, *Rubrobacter xylanophilus*, *Trophymera whipplei*, *Thermobifida fusca*, and *Propionibacterium acnes*. *Symbiobacterium thermophilum*, although formally still in the actinobacteria, is clearly more closely related to the Firmicutes.[44] *Rubrobacter xylanophilus* is the only representative of the deepest branching orders in the Actinobacteria included in the study[43] and consistently differs, in not containing the 23S rRNA insert or the insert in glutamyl-tRNA synthetase; the presence or absence of the other two indels was not determined. *Trophymera whipplei* is an intracellular pathogen with a reduced genome; it contains the cox1 2 aa deletion and GluRS 5 aa insertion but has a reduced insert in the 23S rRNA gene (79 nt) and CTP synthetase (3 aa). *Propionibacterium acnes* and *Thermobifida fusca* are missing the GluRS insertion.

Whole genome sequences enable a comprehensive search within whole genomes for diagnostic signatures, but are limited in the representation of the organisms, even in a well-represented group like the Actinobacteria (Table 27.1). This limitation will remain, even with the explosion in whole genome sequencing. Nevertheless, these comprehensive snapshots offer tremendous insight into the biology and evolutionary relationships of organisms. An unexpected feature has been the extent to which even closely related organisms contain unique genes, ORFans.[45]

Whole genome sequences thus offer another route to understanding "what is an actinobacterium?" — the identification of signature proteins,[46] proteins uniquely present in the Actinobacteria. Of the 233 actinobacterial-specific proteins identified (selected using candidate proteins present in *Mycobacterium leprae*, *Leifsonia xyli*, *Bifidobacterium longum,* and *Thermobifida fusca*, perhaps chosen as **not** representative of actinobacterial genomes*), only 29 were uniquely present in most actinobacteria. Five proteins are present in all actinobacteria, including *Rubrobacter xylanophilus,*† supporting the relationship of this deep-branching order to the other Actinobacteria seen in the 16S rDNA phylogenetic tree. A further 10 proteins are present in all the other sequenced actinobacterial species, including *Bifidobacterium longum* and *Tropheryma whipplei,* while 14 more proteins are present in all these species except *Tropheryma whipplei,* which, as an intracellular pathogen, has a reduced genome of only 0.93 Mb. Six further proteins are found in all the other actinobacterial sequences except *Bifidobacterium longum*; although the Bifidobacterales is a deep-branching order in the 16S phylogenetic tree, it is much more clearly related to the actinomycetes than the other deep-branching orders, largely represented in environmental clones. None of these diagnostic signature proteins are present in *Symbiobacterium thermophilum.*

Almost all these signature proteins are proteins of unknown function and annotated as hypothetical. Two signature proteins identified are from annotated genes: whiB and merR. WhiB was identified as a transcription factor-like protein[47] essential for sporulation in "*Streptomyces coelicolor*"‡ and part of an extensive family of whiB-like proteins present in actinobacteria.[48] They seem to be involved in regulating processes from sporulation in *Streptomyces* to septum formation in *Mycobacterium* and response to heat and oxidative stress in *Corynebacterim.* MerR is a transcriptional regulator of mercury resistance; although merR family proteins[49] are widely distributed across other bacteria, they do not possess much sequence similarity to the actinobacterial *merR* gene.

Battistuzzi et al.[2] identified 32 proteins present, in common, across 72 genomes of prokaryotes and eukaryotes and have performed phylogenetic analysis of the concatenated and aligned dataset (Figure 27.1), using a local clock method, and dated the evolutionary nodes. Their analysis groups three phylogenetic lines — *Actinobacteria*, *Deinococcus-Thermus,* and *Cyanobacteria* — into a major lineage, not dissimilar to that constructed from gene composition data,[50] although the

* So may not optimally lead to the identification of a complete set of interesting actinobacterial-specific proteins.

† The *Rubrobacter xyalnophilus* genome was not complete at the time of the analysis.

‡ The first streptomycete whole genome sequenced, the model organism for the *Streptomyces* but famously not validly described, and indeed with the same name as a different streptomycete that **is** validly named (*S. coelicolor* Müller).

relationships between major clades remain difficult to determine and different whole genome methods derive differing trees.[49, 51–53] Battistuzzi et al.[2] identify common features of resistance to desiccation[54] and the synthesis of photoprotective compounds,[55] and correlate this with the colonization of the land and the evolution of oxygenic photosynthesis. This analysis puts the common ancestor of actinobacteria as one of the "*Terrabacter*,"[2] and the evolutionary forces that have shaped the actinobacterial clade as the invasion of the land perhaps 3 billion years ago — although the evolutionary radiation we perceive as the actinomycetes probably only dates back to half that. This exposed environment and the evolution of oxygen in plant-like photosynthesis by the cyanobacteria must have imposed new stresses on these proto-actinobacteria, most famously illustrated in the radiation- and desiccation-resistance of *Deinococcus*[56] but equally found in some actinobacteria, such as *Rubrobacter radiotolerans.*[57]

Glutathione plays a key role in protecting against oxygen toxicity in eukaryotes but in the prokaryotes seems to be mainly restricted to the cyanobacteria and purple bacteria.[58] Most bacteria produce other low-molecular-weight thiols[59] and actinobacteria contain a unique thiol, mycothiol, as a replacement for glutathione.[60, 61] Another unique Actinobacterial character, the possession of multiple paralogous, *whiB*-like genes (*wbl*) now seems to be linked to oxidative stress by the demonstration that at least some whiB proteins are disulfide reductases.[62,63]

Actinobacteria have been proposed as progenitors of eukaryotes,[1] a controversial proposal based primarily on the possession of a 20S proteosome, a unique characteristic among the eubacteria. Core proteins of the proteosome are ubiquitous among all three kingdoms of life;[64] however, Archaea and actinomycetes[65,66] contain a 20S proteosome whereas other eubacteria possess a proteosome-related system, HslV. Although, in contrast to eukaryotes, the 20S proteosome of archaea and actinomycetes is a much more simply organized core particle (only one type of a and one type of b subunit), *Actinobacteria* also possess serine-threonine kinase regulatory proteins, once thought to be characteristic of eukaryotes. However, although this class of regulatory protein is turning up in bacteria across all phylogenetic diversity, their extensive occurrence in a few major clades (actinobacteria, cyanobacteria, myxobacteria) is a major characteristic.

DISTRIBUTION

Actinobacteria, or, more strictly, actinomycetes, the more readily isolated component, have been found widely distributed in the terrestrial environment[67] and were believed to be predominantly terrestrial, and washed in to aquatic environments.[68] However, again once sought, not only are actinomycetes found in the marine environment, but found as true marine organisms[69–72] and in freshwater.[73, 74]

As well as free-living organisms, plant, animal, and human pathogens are found throughout the actinobacterial genera, including some of the most devastating human pathogens, *Mycobacterium tuberculosis* and *M. leprae* (see Chapter 31) causing tuberculosis and leprosy, and *Corynebacterium diphtheriae* (see Chapter 26) causing diphtheria. But many other disease-causing organisms are found, for example, *Propionibacterium acne* causing acne,[75, 76] *Trophymera whipplei*[77–79] causing Whipple's disease,[80] *Streptomyces somaliensis* and *Actinomadura madurae* causing mycetoma,[81] *Nocardia farcinica* and other nocardia causing nocardiosis,[82] or the CDC collection of coryneform bacteria isolated from clinical specimens.[83 ,84]

Many actinobacteria are part of the normal skin flora (Brevibacteria, Corynebacteria, Micrococcus, and Propionibacteria)[85] and present in the gut flora (Bifidobacteria and Coriobacteri).[86] Coryneform bacteria are widely distributed in food products such as fermented meats and dairy products.[87]

Animal pathogens include a wide array of mycobacteria with differing degrees of host specificity, such as *Mycobacterium bovis* (cows), *M. microti* (voles), or *M. pinnepedia* (seals) but, in the mycobacteria as a whole, hosts as diverse as fish[88] and elephants.[89] *Rhodococcus equi* is a major pathogen for foals, and an emerging human pathogen,[90] while many more esoteric and unexpected pathogens of race horses are emerging.[91, 92] The pathogenic *Actinomyces bovis* was first described by Harz in 1877 and given the designation, after the Greek for "ray," based on the disk-like rays of mycelial growth observed.[6]

Plant pathogens include potato scab (*Streptomyces scabies*),[93] bean wilt (*Curtobacterium flaccumfaciens*),[94] or the sugar cane pathogen causing ratoon stunting disease (*Leifsonia xyli*).[95]

Many actinobacteria are also plant associated, commensal, endophytes or symbionts, most famously *Frankia,* an obligate N-fixing symbiont of actinorhizal plants,[96] diverse woody plants that contribute significantly to fixed nitrogen input to temperate forests. *Micromonospora*, normally isolated from soils, also seem to be associated with plant roots,[97] and *Streptomyces* are found as plant endophytes.[98]

In the soil environment, actinobacteria play a key role in decomposition and humus formation, and many of the genera isolated over the previous century are closely associated with the soil environment, for example, *Arthrobacter* and *Streptomyces.*

However, molecular ecology has revolutionized our view of microbial diversity in natural environments, and this cultured fraction is not an accurate reflection of the true diversity or its quantitative representation. In 16S clone libraries, the *Actinobacteria* are a significant component making up, on average, 13%, with a typical range of 0 to 34% of the 16S clones analyzed.[99] Of the clones assigned to major taxa, only the Proteobacteria (a deep, phylogenetically heterogeneous taxonomic group) and the Acidobacteria (a major uncultivated phylogenetic lineage) account for a larger fraction of the clones.

However, although just under half of the actinobacterial sequences are assigned to the Actinobacteridae, the subclass that contains the majority of well-described and readily cultivated genera, in each study, less than half, and often only a quarter, of these cloned sequences can be assigned to these well-described genera. The other, more than half, cloned sequences are assigned to the subclasses Acidimicrobidae and Rubrobacteridae with very few described species. Only one species is validly described in the Acidimicrobidae, *Acidimicrobium ferrooxidans,*[100] but other isolates include "*Ferrimicrobium acidophilus,*" another Fe^{2+} oxidizing acidophile[101] and 'candidatus *Microthrix parvicella*' (a well-recognized but originally uncultured filamentous bacteria from activated sludge),[102] now joined by 'candidatus *Microthrix calida.*[103] Similarly, only eight species are described in genera in the *Rubrobacteridae*, three species in the genus *Rubrobacter,* two species in *Thermoleophilum* (*T. minutum* and *T. album*[104]), and one species in each of the other genera: *Patulibacter minatonensis,*[105] *Solirubrobacter pauli,*[106] and *Conexibacter woesi.*[107] Necessarily, most clones in these deep-rooted actinobacterial clades represent novel, uncultivated actinobacteria.

EXPLOITABLE BIOLOGY

The Actinobacteria are also important industrially and in providing environmental services. They are a major source of exploitable biology for biotechnology.[108] Most notably, the genus *Streptomyces* has been the source of more than 60% of the commercially important bioactive metabolites used as antibiotics, anticancer drugs, and immunosuppressants. Currently, the marine environment is providing a new source of novel actinomycetes for search and discovery of new bioactive metabolites.[109]

Corynebacterium glutamicum is the basis for the production of massive quantities of amino acids for feed supplements, 1.2×10^6 tonnes of L-glutamate and 6×10^5 tonnes of L-lysine, as well as L-cysteine.

Many actinobacteria are the source for enzymes; *Thermonospoa fusca* produces a wide range of thermostable, broad pH range, secreted cellulases and xylanases that have been cloned and expressed with a view to degrading agricultural waste to sugars and subsequently ethanol.[110] Rhodococcal strains are the source for many nitrile hydratases and nitrilases to produce amides and carboxylic acids for chemical syntheses;[111] the production of acrylamide from acrylonitrile is a major biotechnological application of enzymes.[112]

The bifidobacteria are used extensively, as live bacteria, as probiotics[113] mainly in yogurt-based foods, with claimed benefits in maintaining a healthy balance in gastro-intestinal tract commensals; there has been an enormous increase in commercial interest over the past 20 years.

Within the environment, *Thermobifida fusca*, as well as being an exploitable source of extracellular enzymes for biotechnology, is also a major component of the flora that degrade plant cell material in compost.[31] It is representative of many of the actinobacterial taxa, each producing its own complement of extracellular enzymes to degrade complex plant- and insect-derived polysaccharides and recycle biomass.

Rhodococci and mycobacteria are well-known examples of soil organisms that are important in the biodegradation of complex organic pollutants and xenobiotics.[114]

DESCRIPTIONS OF MAJOR TAXA

The subclasses, orders, and suborders described in the class Actinobacteria are listed in Table 27.2 with the numbers of families, genera, and species.

Since 1997 there have been few changes in the higher phyla; there are four subclasses, five orders, and fifteen suborders (not all valid), with fifty families. There were 195 validly described genera and 1936 validly described species as of mid-2007; many families contain one genus and many genera contain one species; descriptions of some of the larger or more significant taxa are given below.

THE DEEP ACTINOBACTERIAL PHYLETIC LINES

SUBCLASSES ACIDIMICROBIDAE, SPHAEROBACTERIDAE, AND RUBROBACTERIDAE

The deeply branching phylogenetic lines in the Actinobacteria are problematic, as illustrated by *Symbiobacterium thermophilum* and *Sphaerobacter thermophilus*. In the case of *S. thermophilum* a whole genome sequence has made clear that it is more closely related to the low G+C Gram-positives,[44] and, in the comparison of signature proteins and nucleotides, made possible by the

TABLE 27.2
Actinobacterial Orders

Phylum	**Class**				
"Actinobacteria"	Actinobacteria	**Number of Members in Each Taxonomic Rank**			
Subclass	**Order**	**Suborder**	**Family**	**Genus**	**Species**
Acidimicrobidae	*Acidimicrobiales*	*"Acidimicrobineae"*	1	1	1
Rubrobacteridae	*Rubrobacterales*	*"Rubrobacterineae"*	5	5	8
Coriobacteridae	*Coriobacteriales*	*"Coriobacterineae"*	1	8	19
Actinobacteridae	*Bifidobacteriales*	—	1	5	37
	Actinomycetales	*Actinomycineae*	1	5	49
		Catenulisporineae	2	2	3
		Corynebacterineae	7	13	318
		Frankineae	7	12	24
		Glycomycineae	1	2	7
		Micrococcineae	15	72	311
		Micromonosporineae	1	15	73
		Propionibacterineae	2	18	96
		Pseudonocardineae	2	19	99
		Streptomycineae	1	3	749
		Streptosporangineae	3	17	142

whole genome sequence, it possesses none of the characteristic signatures.[43, 46] *Sphaerobacter thermophilus* was the most deeply branching phyletic line in the Actinobacteria but a reevaluation of its phylogeny, with a new full-size 16S sequence and inclusion of other new sequence data, shows it most closely related to *Thermomicrobium roseum*; *T. roseum* was classified in its own phylum Thermomicrobia but both strains, which are phenotypically similar, fall within the phylum Chloroflexi. Thermomicrobia is proposed as a class within the phylum Chloroflexi with two orders (Thermomicrobiales and Sphaerobacterales).[115]

The Acidimicrobidae and Rubrobacteridae are now the two most deep-rooted taxa still included in the Actinobacteria. Both are represented by just a few species, one in the Acidimicrobidae and eight (in five families) in the Rubrobacteridae (Table 27.3). They reflect a relatively recent but significant theme in current biology, the discrepancy between environmental diversity observed by culture and by using molecular ecological techniques.[30, 116]

The Acidimicrobidae

The combination of a deeply branched phyletic line with a single validly described representative occurs in the Acidimicrobidae, with the only validly described species, *Acidimicrobium ferrooxidans,*[100] representing the whole class. However, a number of other isolates (e.g., "*Ferrimicrobium acidophilum*"[101]) and sequenced clones[117] are available. *A. ferrooxidans* is a moderately thermophilic, acidophilic, mixotrophic Fe^{2+} oxidizer,[100] while "*Ferrimicrobium acidophilum*" is a mesophilic, acidophilic, heterotrophic Fe^{2+} oxidizer.[101] These chemolithotrophic actinobacteria are associated with acid mine waste and metal leaching operations. The microbiology of bio-mining has been reviewed recently.[118] However, these extreme environments usually have limited diversity, and these actinobacteria may or may not be represented in clone libraries as significant players.

However, organisms that fall within the phylogenetic compass of this clade are not only found in these extreme environments; "candidatus *Microthrix parvicella*" has long been recognized morphologically[119] in wastewater treatment plants, associated with operational problems of bulking and foaming.[120] It was first isolated by van Veen (1973)[121] but it was 20 years before a strain was isolated again[102] and characterized as an actinobacteria. Since then, a number of isolates have been studied and a new candidatus species described, "candidatus *Microthrix calida,*"[103] and compared with other available strains, phylogenetically and physiologically. "*Microthrix parvicella*" and "*M. calida*" are aerobic, perhaps optimally microaerophilic, chemoorganotrophs capable of reducing nitrate to nitrite. They are very slow-growing, and generate poor biomass yields on solid or liquid media (often not detectable in liquid media) with and without a carbon source. Microautoradiography indicates that growth occurs on long-chain fatty acids[103] but the carbon sources detected as utilized by such studies differed between isolates (see Ref. 103 for references).[103] Related 16S clones have been found in clone libraries from soils,[122] lakes,[123] marine sponges,[124, 125] and in bacterial communities associated with black band diseased coral.[126]

The Rubrobacteridae

Rubrobacter radiotolerans was isolated from a Japanese hot spring sample after irradiation with gamma rays[127] as *Arthrobacter radiotolerans*; it was reclassified as *Rubrobacter radiotolerans* in 1988.[57] Two further species have been isolated from a hot water factory effluent in the United Kingdom (*Rubrobacter xylanophilus*[128]) and a hot spring in Taiwan (*Rubrobacter taiwanensis*[129]) — without pretreatment by irradiation, but which are also radiation resistant.

Most Actinobacteria are a well-defined group; many are readily cultured and as a result there are extensive 16S sequences available representing well-defined taxa; this makes recognizing 16S clone sequences in environmental libraries straightforward. However, the subclass Rubrobacteridae is highly divergent, deeply branching, and poorly represented in the databases, and their presence in clone libraries easily overlooked; Holmes et al.[130] demonstrated from literature data[131–135] the

TABLE 27.3
Deep Branching Orders in the Actinobacteria

Class Actinobacteria					
Subclass	**Order**	**Suborder**	**Family**	**Genus**	**Species**
Acidimicrobidae	Acidimicrobiales	"Acidimicrobineae"	Acidimicrobiaceae	*Acidimicrobium*	*A. ferrooxidans*^{T}
Rubrobacteridae	Rubrobacterales	"Rubrobacterineae"	Conexibacteraceae	*Conexibacter*	*C. woesei*^{T}
			Patulibacteraceae	*Patulibacter*	*P. minatonensis*^{T}
			Rubrobacteraceae	*Rubrobacter*	*R. radiotolerans*^{T}
					R. taiwanensis
					R. xylanophilus
			Solirubrobacteraceae	*Solirubrobacter*	*S. pauli*^{T}
			Thermoleophilaceae	*Thermoleophilum*	*T. album*^{T}
					T. minutum
Coriobacteridae	Coriobacteriales	"Coriobacterineae"	Coriobacteriaceae	*Atopobium*	*A. minutum*^{T}
					A. fossor
					A. parvulum
					A. rimae
					A. vaginae
				Collinsella	*C. aerofaciens*^{T}
					C. intestinalis
					C. stercoris
				Coriobacterium	*C. glomerans*^{T}
				Cryptobacterium	*C. curtum*^{T}
				Denitrobacterium	*D. detoxificans*^{T}
				Eggerthella	*E. lenta*^{T}
					E. hongkongensis
					E. sinensis
				Olsenella	*O. uli*^{T}
					O. profusa
				Slackia	*S. exigua*^{T}
					S. faecicanis
					S. heliotrinireducens

presence of diverse *Rubrobacter*-related clones in samples from agricultural soil to peat bog samples, and the presence of *Rubrobacter* as a dominant group in four arid desert soils in Australia.[130] They[130] define three subgroups of *Rubrobacter*-related clones, of which group 1, which was dominant in the Australian arid soil samples, contained the two isolates then in culture (*R. radiotolerans* and *R. xylanophilus*) — only these group 1 clones matched the signature nucleotides described by Stackebrandt et al.[19]

Since the description of these subgroups, isolates previously described as green non-sulfur bacteria,[104] *Thermoleophilum album* and *Thermoleophilum minutum,* have been described as cultured representatives of group 2 of the Rubrobacteridae.[136] Representatives of three other new genera from group 2 have been isolated and described: *Solirubrobacter pauli,*[106] *Conexibacter woesei,*[107] and *Patulibacter minatonensis.*[105] Representatives of group 3 remain uncultured.

The diversity encompassed by these isolates is reflected in the assignment, explicitly or implicitly, of these few isolates to different families, based on the phylogenetic depth and taxonomic space encompassed (Figure 27.2) and the phenotypic diversity. *Thermoleophilum album* and *T. minutum* are obligately aerobic, thermophilic bacteria that grow on a few *n*-alkanes, and are isolated from hotsprings (Yellowstone) and soil. *Solirubrobacter pauli* is an aerobic, mesophilic heterotroph that is desiccation sensitive and isolated from the burrow of the earthworm *Lumbricus rubellus* in agricultural soil. *Conexibacter woesei* is a mesophilic, heterotroph isolated from forest soil with long peritrichous flagella. *Patulibacter minatonensis* is a mesophilic, aerobic, motile, heterotroph isolated from soil on agar medium supplemented with superoxide dismutase. *Rubrobacter taiwanensis*[129] is

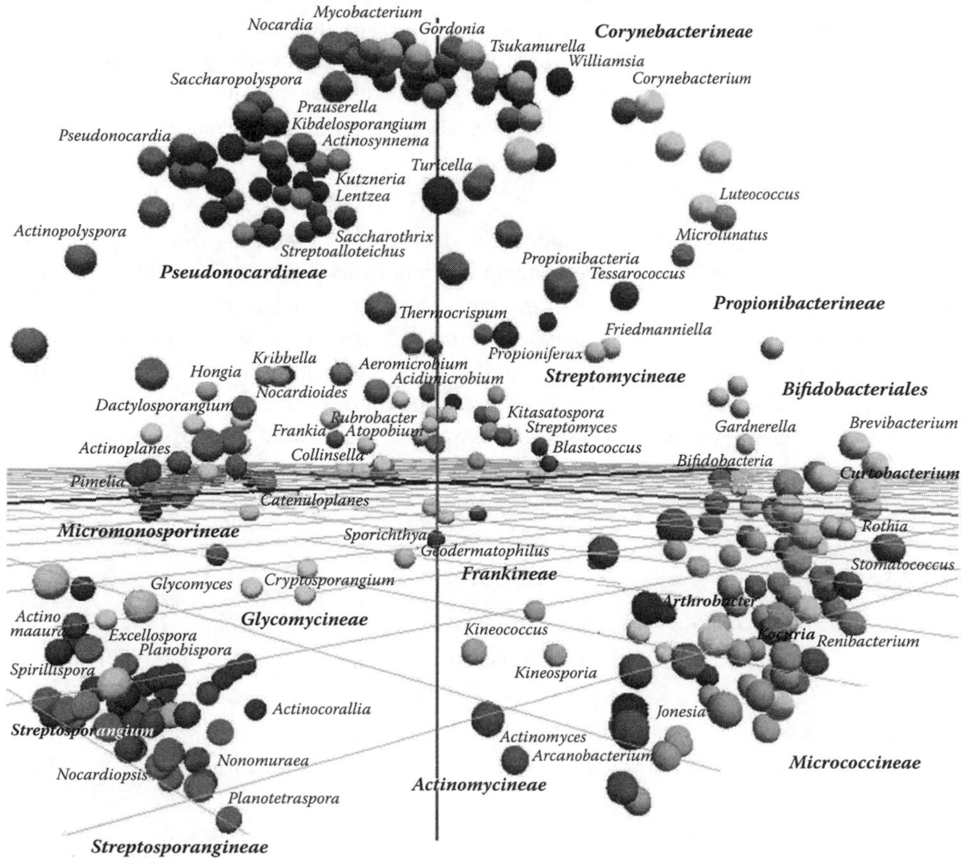

FIGURE 27.2 Three D ordination of 16S sequence data for representatives of the Actinobacterial genera showing the major groups. Principal coordinates of Jukes and Cantor distance matrix.

the most recent isolate in the genus *Rubrobacter*, a radiation-resistant, thermophilic heterotroph isolated from hot springs in Taiwan. The chemotaxonomy of these isolates is also diverse.[105–107, 129, 136]

The support for these major, environmentally significant but largely uncultured clades as members of the Actinobacteria is almost exclusively phylogenetic, and the examples of *Symbiobacterium thermophilum* and *Sphaerobacter thermophilus* illustrate the potential problems. Similarly, *Thermoleophilum* were identified as green non-sulfur bacteria[104] pre-16S phylogeny, and then classified in the Pseudomonadaceae[137] before being identified as group 2 rubrobacter.

The Subclass Coriobacteridae and the Order Bifidobacteriales in the Subclass Actinobacteridae

The gastrointestinal tract (GIT) contains a complex bacterial flora[86, 138] that constitutes a major fraction of the biomass of animals, including humans,[139] is believed to have co-evolved with the host, and is important in nutrition, health, disease,[140] and interaction with pathogens.[141] Two major Actinobacterial taxa that fall outside the order Actinomycetales (which includes nearly all the other major cultivated representatives in the Actinobacteria) are the orders Coriobacteriales and Bifidobacteriales, which include the major actinobacterial representatives in the GIT of animals. Coriobacteriales is the only order in the subclass Coriobacteridae with one family, the *Coriobacteriaceae,* containing 8 genera with 19 species (Table 27.3). *Bifidobacteriales* (in the subclass Actinobacteridae, with the Actinomycetales), with one family, Bifidobacteriaceae, contains five genera, with 33 species in the genus *Bifidobacterium* and four other species (Table 27.4). These two groups include *Collinsella aerofaciens* (previously *Eubacterium aerofaciens*), the most predominant and ubiquitous microorganism isolated from the human gut,[142, 143] and the bifidobacteria, a major component of the flora of especially infants but also adults, and the center of biotechnological and commercial interest as probiotic commensals (see genome sequencing projects).

The Subclass Coriobacteridae

The genus *Coriobacterium* was created to accommodate *Coriobacterium glomerans* isolated from the gut of the red soldier bug,[144] while the genus *Atopobium* was proposed[145] to accommodate phylogenetically misplaced lactobacilli, *L. minutus* and *L. rimae*, and *Streptococcus parvulus* as *A. minutum*, *A. rimae,* and *A. parvulum*. *Coriobacterium* and *Atopobium* were identified as phylogenetic neighbors within the actinobacteria[146, 147] and placed in the class *Actinobacteria* and subclass *Coriobacteridae.*[19] The group is largely isolated from human oral cavities and the GIT, it has been expanded by the transfer of existing isolates to new genera and new isolates. *Eubacterium aerofaciens* and *Eubacterium lentum*, two of the dominant isolates from the human gut, were transferred to *Collinsella aerofaciens*[142] and to *Eggerthella lenta,*[148] respectively. New isolates (e.g., *Slackia*

TABLE 27.4
The Order *Bifidobacteriales*

Class Actinobacteria					
Subclass	**Order**	**Suborder**	**Family**	**Genus**	**Species**
Actinobacteridae	Bifidobacteriales		Bifidobacteriaceae	*Aeriscardovia*	*A. aeriphila*^{T}
				Bifidobacterium	*B. bifidum*^{T} and 32 species
				Gardnerella	*G. vaginalis*^{T}
				Parascardovia	*P. denticolens*^{T}
				Scardovia	*S. inopinata*^{T}

faecicanis from canine gut[149]) were added to *S. exiguum* and *S. heliotrinireducens*[148] or have been assigned to new genera (e.g., *Denitrobacterium detoxificans*[150] isolated from the rumen or *Olsenella profuse*[151] and *Cryptobacterium curtum,*[152] both from the human mouth).

Collinsella aerofaciens and *Eggerthella lenta* are readily isolated but almost never detected in molecular studies,[135] using specific probes and fluorescent *in situ* hybridization. Harmsen et al.[153] confirmed their presence as a numerically significant group. Suau et al.[154] ascribe this discrepancy to PCR conditions and high G+C DNA, a bias in molecular studies relevant in many environmental studies.

The members of this deep-branching actinobacterial lineage are unremarkable, anaerobic, Gram-positive rods or cocci, easily confused phenotypically with *Eubacterium*, a group defined on negative fermentation characters, or *Lactobacilli.* Unlike the bifidobacteria (see next), they are associated with the gut flora of older humans and formula-fed rather than breast-fed infants. They are part of the oral flora and isolated from periodontitis samples (*Atopobium*, *Cryptobacterium*, *Slackia*) and can cause bacteremia (*Eggerthella*).[155]

The Order Bifidobacteriales

Although the order Bifidobacteriales is a deep-branching clade, it is clearly identified within the Actinobacteria both phylogenetically, by signature nucleotides,[19] and by comparative genomics.[43, 46] As well as the genus *Bifidobacterium,* there are four other genera, each with a single species: *Aeriscardovia aeriphila*, *Gardnerella vaginalis*, *Parascardovia denticolens,* and *Scardovia inopinata.*

"*Bacillus bifidus*" was described by Tissier (1900)[14] as a Gram-positive anaerobe with bifid morphology, abundant in the feces of breast-fed babies but absent in bottle-fed babies, in his studies of infant diarrhea. The recognition of a relationship with actinomycetes is evident from the earliest synonyms.[29] Orla-Jensen (1924)[156] proposed it as a new genus *Bifidobacterium* — but their fermentation products, lactic and acetic acids, meant it has been linked with *Lactobacillus.*

The link with health benefits dates from Tissier (1906)[157] and Metchnikoff (1907);[158] and, with other lactic acid bacteria, they are linked with the concepts of probiotics, prebiotics, and functional foods.[159] Recent application in food with health-related and therapeutic claims has drawn extensive scientific attention and many whole genome sequencing projects,[160, 161] giving significant insight into the genus.[162]

The identification of species[163] is based on morphology, phenotypic properties, chemotaxonomy, and DNA:DNA reassociation.[164] Bifidobacteria are Gram-positive, with irregular Y-shaped, bifid morphology, anaerobic, and saccharolytic with sugars and oligosaccharides fermented to acetic and lactic acids with no gas produced; sugar fermentation profiles are variable and used for species identification; and wall chemotaxonomy is also highly variable and can vary within species.

Bifidobacteria are found in six ecological niches:[163] the human, animal, and insect GIT; the human oral cavity; food; and sewage; but animal and insect gut seems to be the primary habitat. On the basis of multigene sequencing, Ventura et al.[165] concluded that bifidobacteria are derived from an ancestor of the *Bifidobacterium asteroides* insect group. The first complete genomes of bifidobacteria show how bifidobacteria are adapted to their ecological niche in the distal GIT,[163] an anaerobic environment, with a complex microflora, supplied with complex polysaccharides recalcitrant to digestion. About 10% of the genome of *B. longum* biotype *longum* NCC2705[161] is predicted to encode genes involved in carbohydrate transport and metabolism, one-third more than *Escherichia coli* or *Lactobacillus lactis*, but similar to another colon commensal *Bacteroides.* The human genome contains just one predicted glycosyl hydrolase and dietary fiber-derived oligosaccharides, as well as complex, host-derived mucin oligosaccharides and glycosphingolipids, are predicted to have selected the genetic repertoire of the gut flora. As such, fructo-oligosaccharides, galacto-oligosaccharides, xylo-oligosaccharides, lactulose, inulin, and raffinose are examples of candidate prebiotics predicted to enhance beneficial gut flora such as bifidobacteria,[140, 166] whereas *Lactobacillus*, more prevalent in the upper GIT, ferment simpler mono-, di-, and trisaccharides.[86] The genomes of bifidobacteria encode many gene clusters containing a LacI sugar-responsive regulator, an ABC MalEFG oligosaccharide-type transporter, and one to six genes encoding glycosyl hydrolases.[163] These extensive complements of enzymes not only vary between species, but are included in regions of DNA identified as probably acquired by horizontal gene transfer, indicating that acquisition of new metabolic potential is involved in speciation and opens up new ecological niches (e.g., as shown for *E. coli*)[167] and are variable between strains.[168]

Breast-fed newborn babies rapidly acquire a gut flora dominated (95%) by bifidobacteria.[169] Human milk contains complex, soluble oligosaccharides[170] that can be degraded by bifidobacterial genes[171] and bifidogenic human milk peptides, and proteins that digest to bifidogenic peptides.[172] *Bifidobacterium breve* is the dominant strain found in infants; all *B. breve* strains tested possess amylopullulanase activity; it may be important for weaning, when new complex carbohydrates are introduced to the diet.[173] The infant GIT microflora changes dramatically at weaning to become more complex; and bifidobacteria, typically *B. longum*, constitute 3 to 6% of the adult GIT microflora until they decrease in old age.[174]

THE SUBCLASS ACTINOBACTERIDAE

The Suborder Actinomycineae

For much of the early history of the actinomycetes there was a division between the anaerobic actinomycetes and the aerobic actinomycetes. The anaerobes still represent a relatively small subset of the class (Table 27.5); the genus *Actinomyces* was among the earliest recognized and is still probably the best known.

Genus *Actinomyces*

The genus *Actinomyces* is in the suborder *Actinomycineae* with one family Actinomycetaceae and five genera (Table 27.5). *Actinomyces bovis* was isolated by Harz (1877),[6] and these diverse largely facultatively anaerobic (from strict anaerobes to aerotolerant) Gram-positive rods are isolated from the mucous membranes of mammals. Isolates from the oral cavity and dental plaque[175] of humans, including clinical material (actinomycoses, dental caries, and periodontal disease) are prevalent, but isolates

TABLE 27.5
The Actinomycetaceae
Class *Actinobacteria*

Subclass	Order	Suborder	Family	Genus	Species
Actinobacteridae	Actinomycetales	Actinomycineae	Actinomycetaceae	*Actinobaculum*	*A. suis*[T] *A. massiliense* *A. schaalii* *A. urinale*
				Actinomyces	*A. bovis*[T] and 33 species
				Arcanobacterium	*A. haemolyticum*[T] and 7 species
				Falcivibrio	→ *Mobiluncus*
				Mobiluncus	*M. curtisii*[T] and *M. mulieris*
				Varibaculum	*V. cambriense*[T]

from the intestinal and urogenital tracts of man and other mammals (cats, dogs, cows, horses, pigs, and marine mammals — seals and porpoises) are described. The diversity described has increased with improved taxonomy[84] but probably under-represents the true diversity.[176] There are 34 species validly described (as of 2007) and another 16 species in the other four genera in the family.

The best-known species is probably *Actinomyces naeslundii*[177] found in dental plaque, although *A. bovis* is the type species and *A. israeli* often associated with pathogenity. Oral streptococci (e.g., *Streptococcus sanguis*) are pioneer colonizers of the salivary pellicle on the tooth surface (up to 80% of initial plaque), but followed in the succession by increasing numbers of *Actinomyces* that also attach directly to the salivary pellicle via type I fimbriae associated adhesins. They then contribute to interbacterial co-adhesion, which builds up the complex plaque biofilm succession, with more than 500 species identified in human plaque. Specific mechanisms of adhesion have been identified between *A. naeslundii* and *Streptococcus gordonii*, *S. oralis,* and *Porphyromonas gingivalis.*[178] Although many species are associated with clinical samples, *Actinomyces* are probably opportunistic pathogens.

The genus is described as Gram-positive, non-spore-forming, non-acid-fast facultative or obligate anaerobes with cells usually irregular rods. They are high G+C (51 to 70% but typically greater than 65%) and fall into a phylogenetically coherent clade within the Actinobacteria in the 16S tree. They are identified largely phenotypically, and commercial test systems for carbohydrate utilization and acid production can be used, and by 16S sequence or restriction analysis. The genus is described in Schaal (1992),[179] and the taxonomy is updated in Funke et al. (1997)[84] but an up-to-date overview is best found in a recent species description.[180]

The Suborder Corynebacterineae

The *Corynebacterium-Mycobacterium-Nocardia* (CMN) group contains many important actinobacterial groups (Table 27.6), including groups covered in other chapters. Chemotaxonomy has played a major role in identifying coherent major taxa in the aerobic actinomycetes (Tables 27.7, 27.8, 27.9, and 27.10). The primacy of 16S sequencing makes identification by 16S sequencing and phylogenetic analysis a key technique to identify major groups — this is not so easy for identification in complex communities, and signature nucleotides can be specified and specific primers designed to detect specific taxa. These have been largely developed to support search and discovery strategies for aerobic actinomycetes (Table 27.11).

Genus *Corynebacterium*

(See Chapter 26.)

Genus *Gordonia*

The genus *Gordonia* (previously *Gordona*) is part of the suborder Corynebacterineae — the CMN group (Corynebacterium, Mycobacterium, and Nocardia), or the mycolata, the mycolic acid-containing actinomycetes. The original members of this taxon were isolated from clinical samples[189] and considered opportunistic pathogens. However, most species have been isolated as metabolically versatile species capable of degrading toxic environmental pollutants and xenobiotics, from diverse terrestrial and aquatic samples including soils, mangrove swamps, contaminated soils, oil well water, rubber, and wastewater treatment and waste gas biofilters. Examples of their potential application include (1) degradation of persistent triazines, such as atrazine, probably the most widely used herbicide in the world. The Environmental Protection Agency concluded, from its cumulative risk assessment on traizine herbicides, that "no harm that would result to the general U.S. population, infants, children or other ... consumers." but they have been linked to endocrine disruption in amphibians,[182] and few microbial isolates are capable of transforming substituted *s*-triazines. *Gordonia rubripertincta* (= *Rhodococcus corallinus*) produces an inducible *s*-triazine hydrolase;[183]

TABLE 27.6
The Corynebacterineae — The CMN Group
Class Actinobacteria

Subclass	Order	Suborder	Family	Genus	Species
Actinobacteridae	Actinomycetales	Corynebacterineae	Corynebacteriaceae	*Bacterionema*	*B. matruchotii*^{T} → *Corynebacterium matruchotii*
				Caseobacter	*C. polymorphus*^{T} → *Corynebacterium variabile*
				Corynebacterium (see Chapter 26)	*C. diphtheriae*^{T} and 69 species
				Dietzia	*D. maris*^{T}
					D. cinnamea
					D. kunjamensis
					D. natronolimnaea
					D. psychralcaliphila
			Gordoniaceae	*Gordonia*	*G. bronchialis*^{T} and 22 species
				Millisia	*M. brevis*^{T}
				Skermania	*S. piniformis*^{T}
			Mycobacteriaceae	*Mycobacterium*	*M. tuberculosis*^{T} and 127 species
			Nocardiaceae	*Micropolyspora*	→ *Nocardia*
				Nocardia	*N. asteroides*^{T} and 47 species
				Rhodococcus	*R. rhodochrous*^{T} and 26 species
				Smaragdicoccus	*S. niigatensis*^{T}
			Segniliparaceae	*Segniliparus*	*S. rotundus*^{T}
					S. rugosus
			Tsukamurellaceae	*Tsukamurella*	*T. inchonensis*^{T}
					T. paurometabola
					T. pseudospumae
					T. pulmonis
					T. spumae
					T. strandjordae
					T. tyrosinosolvens
			"Williamsiaceae"	*Williamsia*	*W. muralis*^{T}
					W. deligens
					W. marianensis
					W. maris
					W. serinedens

TABLE 27.7
Classification of Actinomycetes Based on Wall Chemotype

Wall Chemotype	Diagnostic Amino Acids and Sugars	Representative Taxa
I	*LL*-A_2pm, glycine	*Streptomycineae*
II	*meso*-A_2pm and/or hydroxyl-A_2pm, glycine	*Micromonosporineae*
IIIA	*meso*-A_2pm, madurose	*Dermatophilaceae, Frankiaceae, Streptosporangiaceae*
IIIB	*meso*-A_2pm	*Actinosynnemataceae, Brevibacteriaceae, Thermomonosporaceae*
IVA	*meso*-A_2pm, arabinose, galactose and mycolic acid	*Corynebacterineae*
IVB	*meso*-A_2pm, arabinose, galactose	*Pseudonocardiaceae*
V	Lysine and ornithine	*Actinomyces israelii*
VI	Lysine with aspartic acid and galactose variable	*Microbacterium, Oerskovia*
VII	Diaminobutyric acid, glycine with lysine variable	*Agromyces Clavibacter*
VIII	Ornithine	*Bifidobacterium*
IX†	*LL* and *meso*-A_2pm with galactose and glycine	*Kitasatosporia*

† Wall chemotype proposed for *Kitasatosporae* (Wellington et al., 1992).

Wall sugar subtypes with meso-DAP

A = arabinose and galactose

B = madurose (3-O-methyl-D-galactose)

C = no diagnostic sugars

D = arabinose and xylose (Lechevalier et al., 1971)

Source: Data modified from Lechevalier and Lechevalier (1970, 1980)[206] and Goodfellow and O'Donnell (1989)

and (2) desulfurization of hydrocarbon fuels in which sulfur-containing components are desulfurized by a sulfur-scavenging microorganism. *Gordonia desulfuricans*[184] was isolated by its ability to grow on benzothiophene as a sole source of sulfur and could be used in the desulfurization of diesel, which contains high levels of sulfur in benzothiophenes.

Their role as opportunistic pathogens may limit their application in bioremediation, and several new species isolated from clinical material[185, 186] emphasize this aspect of their biology. *Gordonia araii*, *G. bronchialis*, *G. effuse*, *G. otitidis*, *G. polyisoprenivorans*, *G. rubripertincta*, *G. sputa,* and *G. terrae* have all been implicated in human infections.

Members of *Gordonia* are aerobic and catalase-positive; they are Gram-positive or Gram-variable, slightly acid-fast, are non-motile, and exhibit a nocardioform growth pattern (mycelia growth with fragmentation into rods or cocci) although *Gordonia* are characterized by a lack of mycelium.

They have a cell wall chemotype IV with *meso*-diaminopimelic acid (*meso*-DAP) and wall sugars arabinose and galactose, and peptidoglycan type A1 γ with peptidoglycan directly cross-linked with *meso*-DAP, and muramic acid has an N-glycolyl substituent. The mycolic acids are typically 44-66 C atoms long.

Polar lipids are typically diphosphatidylglycerol, phosphatidylethanolamine, and phosphatidylinositol and phosphatidylinositol mannosides containing straight-chain saturated and monounsaturated fatty acids and tuberculostearic acid. The major menaquinone is MK-9[H_2], although small amounts of MK-8[H_2] can occur.

TABLE 27.8
Peptidoglycan Types

A	Cross-linked between positions 3 and 4	
A1	Direct – no interpeptide bridge	
	A1α	L-Lysine
	A1β	L-Ornithine
	A1γ	*meso*-Diaminopimelic acid (A_2pm)
A2	Polymerized peptide bridge	
	A2α	L-Lysine
A3	Monocarboxylic acid and/or glycine	
	A3α	L-Lysine
	A3β	L-Ornithine
	A3γ	LL-Diaminopimelic acid
A4	Dicarboxylic acid	
	A4α	L-Lysine
	A4β	L-Ornithine
	A4γ	*meso*-Diaminopimelic acid
	A4δ	L-Diaminobutyric acid
A5	Dicarboxlic acid and lysine interpeptide bridge	
B	Cross-linked between positions 2 and 4	
B1	L-Diamino acid	
	B1α	L-Lysine
	B1β	L-Homoserine
	B1γ	L-Glutamic acid
	B1δ	L-Alanine
B2	D-diamino acid	
	B2α	L-Ornithine
	B2β	L-Homoserine
	B2γ	L-Diaminobutyric acid

Greek letters = diversity of amino acid at position 3 indicates replacement of alanine by glycine

From Schleifer and Kandler, 1972; Schleifer and Stackebrandt, 1983; Schleifer and Seidl, 1985.

Muramic acid type:

A = N-acetylmuramic acid
B = N-glycolylmuramic acid

From Uchida and Aida, 1977, 1984.

TABLE 27.9
Fatty Acid Types

Type 1	Pathway to straight-chain fatty acids
1a	Saturated and unsaturated straight chain
1b	10-Methyl branched fatty acids
1c	Cyclopropane fatty acids
Type 2	Pathway to terminally branched fatty acids
2	*iso*- and *anteiso*-Fatty acids
Type 3	Complex branched fatty acids
3	*iso*- and *anteiso*-fatty acids and 10-methyl branched

Source: Kroppenstedt, 1985.[292]

TABLE 27.10
Polar Lipid Types

PI	Nitrogenous phospolipid absent (phosphatidylglycerol variable)
PII	Phosphatidyl ethanolamine only
PIII	Phosphatidyl choline (phosphatidyl ethanolamine, phosphatidyl methylethanolamine, phosphatidyl glycerol variable)
PIV	Phospholipids containing glucosamine (with phosphatidyl ethanolamine variable)
PV	Phospholipids containing glucosamine with phosphatidyl glycerol (with phosphatidyl ethanolamine variable)

Note: All contain phosphatidyl inositol.
Source: From Lechevalier et al., 1977, 1981.[23]

The DNA G+C ratio ranges from 63 to 69%. They form a monophyletic clade within the radiation of the Corynebacterineae on the basis of phylogenetic analysis of 16S sequence data.

A recent review of the biology of the genus has been published.[187]

Genus *Mycobacterium*

(See Chapter 31.)

Genus *Nocardia*

Nocardia farcinica was isolated from a case of bovine farcy in 1888 by Nocard;[188] it was named a year later and made the type species in 1954,[189] but confusion about the identity of the deposited strains led to a new type strain, *N. asteroides,* being designated. Many clinical isolates from human nocardiosis, identified as *N. asteroides*, were heterogeneous, different from the designated type strain, leaving the *N. asteroides* complex as a problem taxonomic group. Many members of the complex have since been described as new species.[189] There are 65 valid species as of 2007, most isolated from clinical specimens causing pulmonary nocardiosis; the former *N. asteroides* complex (*N. abscessus*, *N. brevicatena* complex, *N. nova* complex, *N. transvalensis* complex, *N. farcinica*, *N. asteroides* complex, *N. cyriacigeorgica*) is often associated with infection in arid,

TABLE 27.11
Specific Primers and Probes Developed for Identification of Some Members of the Order Actinomycetales

Target Taxa	Primer	Sequence 5′ 3′	Position*	Ref.
Actinobacteria	ACF254	CGCGGCCTATCAGCTTGTTG	235–254	Stach et al. (2002)
	ACR894	CCGTACTCCCCAGGCGGGG	894–875	
Actinobacteria	F243	GGATGAGCCCGCGGCCTA	243–261	Heuer et al. (1997)
	R513	CGGCCGCGGCTGCTGGCACGTA	513–491	
Amycolatopsis spp.	AS1	CACGCAGTCGAGTTGCAGACTG	1295–1317	McVeigh et al.
	ATOP	GTATCGCAGCCCTCTGTACCAGC	1227–1250	(1994)
Dactylosporangium spp.	DF3	GCGGCTTGTTGCGTCAG	585–601	Monciardini et al.
	D2R	CCGCTGGCAACATCGAACA	1134–1116	(2002)
Micromonosporaceae	M2F	SAGAAGAAGCGCCGGCC	492–508	Monciardini et al.
	A3R	CCAGCCCCACCTTCGAC	1430–1414	(2002)
Pseudonocardia spp.	AMP2	GTGGAAAGTTTTTTCGGCTGGGG	197–219[a]	Moron et al. (1999)
	AMP3	GCGGCACAGAGACCGTGGAAT	833–813[a]	
Saccharomonospora spp.	SM1	ACGGCACGGGACACGTGMACAGC	833–811[a]	Salazar et al. (2000)
	SM2	CGTCTGCCGTGAAAACCTGCGGC	566–588[a]	
S. azurea	1fA	CCCAACCCGCTTGC		
S. cyanea	2fG	ATGGCCGGTACAATGGGCT	1424–1436	
S. glauca	3fC	CGGATAGGACGCCTCACC	1220–1238[b]	
S. viridis	4Fv	AGCCGTCTTCGGGC	169–185[b]	Yoon et al. (1996)
			75–86[b]	
Saccharopolyspora spp.	SP1 SP3	GTGGAACCCATCCCCACACC	819–801[a]	Moron et al. (1999)
		GGTGACGGTAGGTGTAGAAG	450–469[a]	
Saccharothrix spp.	STX1	TCGACCGCAGGCTCCACG	611–594[a]	Salazar et al. (2002)
	STX2	AAGGCCCTTCGGGGTACACGAG	80–101[a]	
Streptomycetaceae	Sm6F	GGTGGCGAAGGCGGAGA	721–735	Monciardini et al.
	Sm5R	ACTGAGACCGGCTTTTTGA	1303–1283	(2002)
Streptomyces spp.	StrepB	ACAAGCCCTGGAAACGGGGT	139–158[a]	
	StrepE	CACCAGGAATTCCGATCT	657–640[a]	
	StrepF	ACGTGTGCAGCCCAAGACA	1212–1194[a]	Rintala et al. (2001)
Streptosporangiaceae	21F	GACGAARNTGACGTGTA	407–424	Monciardini et al.
	959R	CGTTGCGTCTAATTAAGCAA	971–952	(2002)
Thermomonosporaceae	T3F	GGGAGAATGGAATTCCC	665–681	Monciardini et al.
	T8R	CCCCACCTTCGACC	1426–1413	(2002)

* *Escherichia coli* numbering position (Brosius et al., 1978) unless otherwise stated.

[a] 16S rRNA numbering of *Streptomyces ambofaciens* (Pernodet et al., 1989).

[b] 16S rRNA numbering of *Streptomyces griseus* ssp. *griseus* (Kim et al., 1993).

warm climates such as the southwestern United States[190] and in immunocompromised hosts. *N. brasiliensis* and *N. pseudobrasiliensis* are associated with cutaneous nocardiosis and usually infect immunocompetent hosts; infections may result in mycetomas similar to those caused by other actinomycetes and fungi.[191] Diagnosis and antimicrobial susceptibility patterns are critical for treatment. The other major source of isolates is from soil, where most clinically associated species also can be isolated, but described species have also been isolated from oysters and fish[192, 193] and can be isolated from fresh and salt water, soil, and decaying vegetative and animal detritus.

The nocardia form a homogenous cluster among the CMN group in the Corynebacteriaceae in the 16S phylogenetic tree. They possess mycolic acids and tuberculostearic acid but can be distinguished from members of the genus *Mycobacterium* by the chain length of mycolic acids (40 to 60 carbon atoms) and their quite extensive filamentous branching with fragmentation (nocardioform growth). The cell wall contains *meso*-diaminopimelic acid (*meso*-DAP) and wall sugars arabinose and galactose. Sequence data from 16S and sequence or restriction pattern analysis of *hsp65* is essential,[194] but not sufficient even with phenotypic characteristics, which may be negative. For example, *Nocardia kruczakiae*[195] showed very high sequence similarity to other members of the *N. nova* complex (99.8% 16S similarity to *N. veterana*) but DNA:DNA hybridization showed it to be a unique species. The genus has been reviewed recently.[189]

Genus *Rhodococcus*

The name *Rhodococcus* was introduced by Zopf (1891)[10] for two species of red bacteria that had been described by Overbeck (1891)[196] as *Micrococcus erythromyxa* and *Micrococcus rhodochrous*. The genus *Rhodococcus* was recognized in early editions of *Bergey's Manual* (1923, 1925, 1930, 1934).[197–200] However, in the fifth edition (1939),[201] *R. agilis, R. cinnebareus, R. corallinus, R. rhodochrous, R. rosaceus,* and *R. roseus* were transferred to the genus *Micrococcus* and remained there in the next edition (1948).[202]

In the 1950s, Gordon applied a polythetic approach to the study of corynebacteria, mycobacteria, and nocardiae. Species received as *Bacillus, Bacterium, Erythrobacillus, Micrococcus, Mycobacterium, Nocardia, Proactinomyces, Rhodococcus,*[203, 204] and later *Jensenia*[205] were recognized as a distinct group and assigned as a species, *Mycobacterium rhodochrous*, a name derived from *Micrococcus rhodochrous,*[196] the oldest named strain in the group, and placed in the genus *Mycobacterium*, but described as a species in search of a genus.

The introduction of chemotaxonomy with the determination of wall chemotypes[206] and peptidoglycan types[22] showed corynebacteria, mycobacteria, and nocardiae contained *meso*-diaminopimelic acid (*meso*-A2*pm*), arabinose and galactose, wall chemotype IV *sensu* Lechevalier & Lechevalier (1970),[206] and an A1 peptidoglycan,[22] and all contained mycolic acids.[207] Chemotaxonomic and numerical taxonomic studies[208] established the rhodochrous group as a distinct, if heterogeneous taxon, but equivalent in rank to other aggregate clusters in the mycobacteria and nocardia. Some strains that would later be classified in this group were assigned to the proposed genus *Gordona*[181] before the genus *Rhodococcus* was resurrected by Tsukamura (1974)[209] and the core of the current genus described in 1977.[210] The genus was reviewed in 1992[211] and 1998.[212]

The confused and tortuous taxonomy is typical of many older actinomycete genera and is well described for *Rhodococcus* in Goodfellow et al. (1998),[208] although the genus is not yet completely described by 16S-based polyphasic taxonomy but contains members included prior to reliable 16S sequencing and many isolates assigned by 16S with no other taxonomic description.[213] Whole genome sequences are likely to refine the structure of what is still a diverse and probably heterogeneous genus.[211, 214]

Rhodococci have been isolated from contaminated and uncontaminated soil, groundwater and wastewater treatment plants, healthy and diseased animals, humans and plants, insect guts, and marine samples. From 12 species in 1998[212] there are now 27 validly described species, including, for example, *Rhodococcus coprophilus* from animal feces and a fecal indicator in water, *R. equi* a foal pathogen and emerging human pathogen, *R. fascians,* a plant pathogen, *R. rhodnii* a gut symbiont from *Rhodnius prolixus* the insect vector involved in Chagas' disease, *R. percolatus* isolated from a chlorophenol-fed bioreactor, *R. erythropolis,* a biosurfactant producer, and *R. rhodochrous* JI used to produce acrylamide.[111, 112]

Genus *Tsukamurella*

Corynebacterium paurometabolum isolated from the bed bug (*Cimex lectularius*) by Steinhaus (1941)[215] and "*Gordona aurantiacus*"[216] were unusual in possessing very long, highly unsaturated mycolic acids and assigned to a single species, *T. paurometabola,* in a new genus, *Tsukamurella*.[217]

Most *Tsukamurella* are isolated, like "*G. aurantiacus*," from patients compromised with a primary debilitating disease (e.g., *T. strandjordae*)[218] but *T. spumae* and *T. pseudospumae*[219] are just two species described from *Tsukamurella* isolates from an activated sludge plant, which included at least four other, well-defined centers of variation to be described as new species.

The genus *Tsukamurella* contains aerobic, Gram-positive, weakly acid-fast, nonmotile, straight to curved rods — cells may aggregate to give a pseudomycelial appearance and form some cocci. The cell wall peptidoglycan contains *meso*-diaminopimelate, N-glycolated muramic acid, and sugars arabinose and galactose. The mycolic acids have 62 to 78 C atoms that are highly unsaturated, and α- and α'-mycolates. The fatty acids are straight-chain saturated and unsaturated fatty acids with tuberculostearic acid, with major phospholipids: phosphatidylglycerol, phosphatidylethanolamine, phosphatidylinositol, and phosphatidylinositol mannosides. The major menaquinone is fully unsaturated with nine isoprene units, MK-9.

The Suborder Frankineae

The suborder Frankineae contains 6 families and 12 genera but only 24 species (Table 27.12).

Genus *Frankia*

The family Frankiaceae is proposed with one genus *Frankia*[220] containing one valid species, *Frankia alni*, first described in 1866[3] but not isolated until 1956,[221] an isolate that was lost. They were regarded as obligate symbionts[222] until an infective strain was isolated in 1978.[223] Since then, many isolates from actinorhizal plants, more than 200 species in eight families[224] have been isolated, but some symbionts remain resistant to cultivation efforts. Three whole genome sequences[225] should provide the basis for evaluating lumping these host associated strains into a single species.

This nitrogen-fixing symbiosis is of worldwide significance with well-known plants in the genera *Alnus, Casuarina, Elaeagnus,* and *Myrica* nodulated by nitrogen-fixing *Frankia* endosymbionts.[224]

They form a distinct phyletic taxon almost equally distant from *Geodermatophilus* and *Blastococcus*, within the suborder Frankineae, and *Actinoplanes* in the Micromonosporineae.[220] With *Acidothermus cellulolyticus* and *Sporichthya polymorpha,* they form a coherent phyletic line in the order Actinomycetales. They have a distinct morphology with complex differentiation and sporulation, hyphae with a specialized thick-walled diazovesicle and a multi-locular sporangia with nonmotile spores (unlike *Actinoplanes*). The cell wall contains *meso*-diaminopimelic acid, wall amino acids alanine and glutamate, and wall sugar 2-*O*-methyl-D-mannose; phospholipids include phosphatidyl-inositol, phosphatidyl-inositol mannosides, and diphosphatidylglycerol.

The Suborder Micrococcineae

The genus *Micrococcus* was described by Cohn (1872)[226] and the distinctive coccal cell morphology and classic Gram-staining of the readily cultured aerobic, catalase-positive cocci make them instantly recognizable and familiar to any microbiology student. However, it was not until the mid-1960s that strains fermenting glucose were assigned to group 1, staphylococci, and those not oxidizing glucose to group 2, micrococci[227] and the low G+C of staphylococci and high G+C of micrococci recognized.[228] The clear assignment of the high G+C micrococci to the actinomycete line of descent depended on 16S phylogenetic analysis,[229] also revealing the difficulty of delineating micrococci, phylogenetically, from *Arthrobacter*[230] with classic rod-coccus transition morphology, part of the evidence that Gram-positive cocci, and bacterial morphology, can fail to map to phylogenetically consistent taxa.[231]

The extensive diversity, especially in wall and lipid chemotaxonomy, has led to the extensive taxonomic dissection of the micrococci,[232] leaving the genus *Micrococcus* with, currently, only four species.

TABLE 27.12
The Suborder Frankineae
Class *Actinobacteria*

Subclass	Order	Suborder	Family	Genus	Species
Actinobacteridae	Actinomycetales	Frankineae	Acidothermaceae	*Acidothermus*	*A. cellulolyticus*^{T}
			Frankiaceae	*Frankia*	*F. alni*^{T}
			Geodermatophilaceae	*Blastococcus*	*B. aggregatus*^{T}
					B. jejuensis
					B. Saxobsidens
				Geodermatophilus	*G. obscurus*^{T}
				Modestobacter	*M. multiseptatus*^{T}
			"Kineosporiaceae"	*Cryptosporangium*	*C. arvum*^{T}
					C. aurantiacum
					C. japonicum
					C. minutisporangium
				Kineococcus	*K. aurantiacus*^{T}
					K. marinus
					K. radiotolerans
				Kineosporia	*K. aurantiaca*^{T}
					K. mikuniensis
					K. rhamnosa
					K. rhizophila
					K. succinea
			Nakamurellaceae	*Humicoccus*	*H. flavidus*^{T}
				Nakamurella	*N. multipartita*^{T}
				Quadrisphaera	*Q. granulorum*^{T}
			Sporichthyaceae	*Sporichthya*	*S. polymorpha*^{T}
					S. brevicatena
		Glycomycineae	Glycomycineae	*Glycomyces*	*G. harbinensis*^{T}
					G. algeriensis
					G. arizonensis
					G. lechevalierae
					G. rutgersensis
					G. tenuis
				Stackebrandtia	*S. nassauensis*^{T}

The family Micrococcaceae has 9 genera and 82 species (51 in the genus *Arthrobacter*). The suborder Micrococcineae has 14 families, 65 genera, and 289 species, but 42 genera have only one or two species, only seven genera have ten or more species, and only four have more than ten species (Table 27.13).

Genus *Micrococcus*

The genus *Micrococcus* currently has four species. *M. luteus* is the best known, with strain ATCC 9341 designated as a quality-control strain and a standard culture in official methods and manuals. However, this strain has been reclassified as *Kocuria rhizophila*.[233] At one time the only other species in the genus was *M. lylae*; it was so different, on the basis of chemotaxonomic characters used to separate other genera, it was only the relative genomic similarity that persuaded Stackebrandt et al.[232] not to assign *M. lylae* to a novel genus. Since then, two new species — *M. antarcticus*, a psychrophile from Antarctica, and *M. flavus*, from an activated sludge plant — have been described. However, a set of strains isolated from indoor air at a museum, a wall painting, and an activated sludge plant were all identified, by 16S sequence analysis, as *M. luteus*. This relationship was confirmed by DNA:DNA similarity. Nevertheless, these *M. luteus* strains differed substantially in cell wall type and menaquinone composition, differences that had suggested *M. lylae* represented a novel genus. But Wieser et al.[234] assigned these variants to three biovars of *M. luteus*.

In many actinomycete taxa, chemotaxonomic characters can be used to group higher taxa — for example, the taxa in the mycolata can only be separated completely on the basis of their menaquinones; however, chemotaxonomy is highly variable within the Micrococcaceae, even to the extent of varying within a species.

Genus *Brevibacterium*

The genus *Brevibacterium*, Breed 1953,[235] includes pleomorphic, strictly aerobic, Gram-positive bacteria with a high mol% G+C content (63%). *Brevibacterium* comes from Latin *brevis,* short and Greek *rod*. This genus belongs to the family Brevibacteriaceae.[19, 137, 236] At present there are 38 validly described species in this genus. Intensive taxonomic investigation by Collins et al. (1980)[237] significantly improved the taxonomic status of the genus, which currently has *Brevibacterium linens* as type species. *B. linens* is extensively used in the ripening of cheese.[238] Some strains of *B. linens* have bacteriocidal properties,[239] whereas others are considered inhibited by probiotics in blue veined cheeses.[240] Bacteria of this genus are isolated from soil, food, and human skin. Members of the genus have biotechnological relevance and are used in dairy milk products, for vitamin production (but vitamin-producing strains have recently been transferred to *Corynebacterium glutamicum*), and in wastewater treatment. They are also encountered in humans as opportunistic pathogens[241] and/or found to be common residents of various human environments (e.g., poultry[242]). Most of them are nonsporulating, nonmotile, halo-tolerant, and exhibit rod-to-coccus morphology. Species in the genus exhibit a high degree of heterogeneity and thus cannot be identified by classical methods.[238] The discriminatory power of DNA-based methods is utilized to differentiate species of this genus.[241, 243] These organisms are also noted for their capacity to metabolize heterocyclic and polycyclic ring structures, a trait that is not associated with other bacteria but is common in fungi. Of particular note is the degradation of insecticides (including DTT and DDE). These organisms also metabolize amino acids to produce plant growth hormones, especially from aromatic amino acids.[238]

Genus *Tetrasphaera*

Phosphate removal in activated sludge processes is an important activity that has been associated with microbes with characteristic coccal morphology, often in tetrads. Gram-negative isolates, G-bacteria, have been identified as α-proteobacteria (*Amaricoccus*), but Gram-positive, polyphosphate accumulating cocci have also been isolated, including *Microlunatus*[244] and *Tessaracoccus*[245] in the family Propionibacteriaceae, *Nakamurella*[246, 247] and *Quadrisphaera*[248] in the Frankineae, and *Tetrasphaera*[249] in the family Intrasporangiaceae of the suborder Micrococcineae.

TABLE 27.13
The Suborder Micrococcineae
Class *Actinobacteria*

Subclass	Order	Suborder	Family	Genus	Species
Actinobacteridae	Actinomycetales	Micrococcineae	"Beutenbergiaceae"	*Beutenbergia*	*B. cavernae*^{T}
				Georgenia	*G. muralis*^{T}
				Salana	*S. multivorans*^{T}
			Bogoriellaceae	*Bogoriella*	*B. caseilytica*^{T}
			Brevibacteriaceae	*Brevibacterium*	*B. linens*^{T} *and 18 species*
			Cellulomonadaceae	*Cellulomonas*	*C. flavigena*^{T} *and 14 species*
				Oerskovia	*Oerskovia turbata*^{T}
					O. enterophila
					O. jenensis
					O. paurometabola
				Tropheryma	*T. whipplei*^{T}
			Dermabacteraceae	*Brachybacterium*	*B. faecium*^{T} *and 10 species*
				Dermabacter	*D. hominis*^{T}
			Dermacoccaceae	*Demetria*	*D. terragena*^{T}
				Dermacoccus	*D. nishinomiyaensis*^{T}
					D. abyssi
					D. barathri
					D. profundi
				Kytococcus	*K. sedentarius*^{T}
					K. schroeteri
			Dermatophilaceae	*Dermatophilus*	*D. congolensis*^{T}
					D. chelonae
				Kineosphaera	*K. limosa*^{T}
			Intrasporangiaceae	*Arsenicicoccus*	*A. bolidensis*^{T}
				Intrasporangium	*I. calvum*^{T}
				Janibacter	*J. limosus*^{T}
					J. anophelis
					J. melonis
					J. terrae
				Knoellia	*K. sinensis*^{T}
					K. subterranean
				Kribbia	*K. dieselivorans*^{T}
				Ornithinicoccus	*O. hortensis*^{T}
				Ornithinimicrobium	*O. humiphilum*^{T}
					O. kibberense
				Oryzihumus	*O. leptocrescens*^{T}
				Serinicoccus	*S. marinus*^{T}

—continued

TABLE 27.13 (continued)
The Suborder Micrococcineae
Class *Actinobacteria*

Subclass	Order	Suborder	Family	Genus	Species
					T. terrae
				Terracoccus	*T. luteus*^{T}
				Tetrasphaera	*T. japonica*^{T}
					T. australiensis
					T. duodecadis
					T. elongate
					T. jenkinsii
					T. vanveenii
					T. veronensis
			Jonesiaceae	*Jonesia*	*J. denitrificans*^{T}
					J. quinghaiensis
			Microbacteriaceae	*Agreia*	*A. bicolorata*^{T}
					A. pratensis
				Agrococcus	*A. jenensis*^{T}
					A. baldri
					A. casei
					A. citreus
					A. lahaulensis
				Agromyces	*A. ramosus*^{T} *and 21 species*
				Aureobacterium	*Aureobacterium* → *Microbacterium*
				Clavibacter	*C. michiganensis*^{T}
					C. michiganensis ssp. insidiosus, ssp. michiganensis, ssp. nebraskensis, ssp. sepedonicus, ssp. tessellarius
				Cryobacterium	*C. psychrophilum*^{T}
				Curtobacterium	*C. citreum*^{T} *6 out of 9 species, including type, are synonyms of other genera*
				Frigoribacterium	*F. faeni*^{T}
				Gulosibacter	*G. molinativorax*^{T}
				Leifsonia	*L. aquatica*^{T}
					L. aurea

		L. ginsengi
		L. naganoensis
		L. poae
		L. rubra
		L. shinshuensis
		L. xyli ssp. xyli
		L. xyli ssp. cynodontis
	Leucobacter	*L. komagatae*^{T}
		L. albus
		L. alluvii
		L. aridicollis
		L. chromiireducens
		L. luti
	Microbacterium	*M. lacticum*^{T} *and 42 species*
	Microcella	*M. putealis*^{T}
		M. alkaliphila
	Mycetocola	*M. saprophilus*^{T}
		M. lacteus
		M. tolaasinivorans
	Okibacterium	*O. fritillariae*^{T}
	Plantibacter	*P. flavus*^{T}
		P. auratus
	Pseudoclavibacter	*P. helvolus*^{T}
	Rathayibacter	*R. rathayi*^{T}
		R. caricis
		R. festucae
		R. iranicus
		R. toxicus
		R. tritici
	Rhodoglobus	*R. vestalii*^{T}
	Salinibacterium	*S. amurskyense*^{T}
	Subtercola	*S. boreus*^{T}
		S. frigoramans
	Yonghaparkia	*Y. alkaliphila*^{T}
	Zimmermannella	*Z. helvola*^{T} → *Pseudoclavibacter helvolus*
		Z. alba
		Z. bifida
		Z. faecalis
Micrococceae	*Acaricomes*	*A. phytoseiuli*^{T}
	Arthrobacter	*A. globiformis*^{T} *and 47 species*

—continued

TABLE 27.13 (continued)
The Suborder Micrococcineae
Class *Actinobacteria*

Subclass	Order	Suborder	Family	Genus	Species
				Citricoccus	*C. muralis*^{T}
					C. alkalitolerans
				Kocuria	*K. rosea*^{T}
					K. aegyptia
					K. carniphila
					K. himachalensis
					K. kristinae
					K. marina
					K. palustris
					K. polaris
					K. rhizophila
					K. varians
				Micrococcus	*M. luteus*^{T}
					M. antarcticus
					M. flavus
					M. lylae
				Nesterenkonia	*N. halobia*^{T}
					N. aethiopica
					N. halotolerans
					N. jeotgali
					N. lacusekhoensis
					N. lutea
					N. sandarakina
					N. xinjiangensis
				Renibacterium	*R. salmoninarum*^{T}
				Rothia	*R. dentocariosa*^{T}
					R. aeria
					R. amarae
					R. mucilaginosa
					R. nasimurium
				Stomatococcus	*Stomatococcus* → *Rothia*
			Promicromonosporaceae	*Cellulosimicrobium*	*C. cellulans*^{T}
					C. funkei
				Isoptericola	*I. variabilis*^{T}
					I. dokdonensis
					I. halotolerans

		I. hypogeus
	Myceligenerans	*M. xiligouense*^{T}
		M. crystallogenes
	Promicromonospora	*P. citrea*^{T}
		P. aerolata
		P. sukumoe
		P. vindobonensis
	Xylanibacterium	*X. ulmi*^{T}
	Xylanimonas	*X. cellulosilytica*^{T}
Rarobacteraceae	*Rarobacter*	*R. faecitabidus*^{T}
		R. incanus
Sanguibacteraceae	*Sanguibacter*	*S. keddieii*^{T}
		S. inulinus
		S. marinus
		S. suarezii
Yaniaceae	*Yania*	*Y. halotolerans*^{T}
		Y. flava
Unclassified	*Actinotalea*	*A. fermentans*^{T}
	Demequina	*D. aestuarii*^{T}
	Phycicoccus	*P. jejuensis*^{T}

The genus *Tetrasphaera* is described as Gram-positive, aerobic, slow-growing, nonmotile, nonsporeforming cocci mostly as tetrads and clusters. The cell wall contains *meso*-diaminopimelic acid and peptidoglycan type A1γ; the major menaquinone is MK-8[H_4], with *iso*- and *anteiso*-branched-chain fatty acids and no mycolic acids. It is distinguished from other genera in the Intrasporangiaceae by a combination of 16S phylogeny and chemotaxonomy.

Genus *Leifsonia*

Leifsonia was created[250,251] to accommodate 2,4-diaminobutyric acid-containing actinomycetes "*Corynebacterium aquaticum*"[252] and *Clavibacter xyli,*[253] the plant pathogen, as *Leifsonia xyli*. The genus *Leifsonia* comprises eight species isolated from plants, soil, distilled water, and an Antarctic pond.[250, 252, 254,255] *Leifsonia xyli* has two subspecies: (1) *L. xyli* ssp. *cynodontis*, a pathogen causing stunting in Bermuda grass (*Cynodon dactylon*), and (2) *L. xyli* ssp. *xyli*, causal agent of ratoon stunting disease, affecting sugarcane (*Saccharum* spp.).

Leifsonia are bacteria with coryneform morphology, filaments that fragment into rods and coccoid elements; a few species are motile. The peptidoglycan contains 2,4-diaminobutyric acid; the major cell wall sugar is rhamnose with minor amounts of glucose, galactose, and mannose, and sometimes fucose. The major menaquinone is MK-11, and phosphatidylglycerol and diphosphatidylglycerol are the major phospholipids with *anteiso*- and *iso*-branched saturated fatty acids. The DNA G-C base composition is 66 to 73 mol%.

Genus *Arthrobacter*

Conn (1928)[256] described an organism, under the name of *Bacterium globiforme*, which appears as Gram-negative, short rods in 24-hr agar slant cultures but as Gram-positive cocci after the cultures are 3 to 4 days old. Later, Conn and Dimmick (1947)[257] proposed a new genus, *Arthrobacter,* with *Arthrobacter globiformis* as the type species.

The genus *Arthrobacter* was limited to those species that, like the type species *A. globiformis*, contained lysine as the cell wall diamino acid. The wall peptidoglycan contains N-acetylated muramic acid, lysine in position-3 or -4 of the peptide subunit, and an interpeptide bridge linking the peptide subunits of adjacent glycan strands between the L-lysine and the terminal D-alanine.[22] Considerable variation is found in the number and nature of the amino acids in the interpeptide bridges. Within these numerous different peptidoglycan types, two main groups occur, which are referred to as A3α and A4α variations (Table 27.8). In the A3α variation, found in *A. globiformis, A. citreus,* and most other species (the "*A. globiformis/A. citreus*" group[258]), the interpeptide bridge contains only monocarboxylic acids and/or glycine. However, in the A4α variation found in *A. nicotianae* and three other species (the "*A. nicotianae*" group[258]), the interpeptide bridge always contains a dicarboxylic acid and usually in most strains also contains alanine.[22] There are also another four peptidoglycan types in *Arthrobacter.*

In those strains containing peptidoglycans of variation A3α (*Arthrobacter simplex* and *A. tumescens*), LL-diaminopimelic acid occurs instead of L-lysine in position-3 of the peptide subunit, and the interpeptide bridges consist of a single or several glycine residues. The A4α variation, containing *meso*-diaminopimelic acid in position-3, has only been detected in *A. duodecadis*. However, the directly cross-linked *meso*-diaminopimelic acid containing peptidoglycan type A1α has been found in *Arthrobacter roseoparaffinus*, *A. viscosus,* and *A. variabilis.*

The predominant wall sugar is galactose but glucose, mannose, and rhamnose are found in some strains.[259] Duxbury[260] found only glucose, galactose, and rhamnose in the hydrolysates of *Arthrobacter globiformis* NCIMB10683.

The Suborder Micromonosporineae

The suborder has only one family but 14 genera (Table 27.14).

TABLE 27.14
The Suborder Micromonosporineae
Class *Actinobacteria*

Subclass	Order	Suborder	Family	Genus	Species
Actinobacteridae	Actinomycetales	Micromonosporineae	Micromonosporaceae	*Actinocatenispora*	***A. thailandica***^{T}
				Actinoplanes	***A. philippinensis***^{T} and 23 species
				Amorphosporangium	*Amorphosporangium* → *Actinoplanes*
				Ampullariella	*Ampullariella* → *Actinoplanes*
				Asanoa	***A. ferruginea***^{T}
					A. iriomotensis
					A. ishikariensis
				Catellatospora	***C. citrea***^{T}
					C. bangladeshensis
					C. chokoriensis
					C. coxensis
					C. koreensis
					C. methionotrophica
					C. sunoense
				Catenuloplanes	***C. japonicus***^{T}
					C. atrovinosus
					C. castaneus
					C. crispus
					C. indicus
					C. nepalensis
					C. niger
				Couchioplanes	***C. caeruleus***^{T} ssp. *caeruleus*
					C. caeruleus ssp. *Azureus*
				Dactylosporangium	***D. aurantiacum***^{T}
					D. fulvum
					D. matsuzakiense
					D. roseum
					D. vinaceum
				Longispora	***L. albida***^{T}
				Micromonospora	***M. chalcea***^{T} and 14 species
				Pilimelia	***P. terevasa***^{T}
					P. anulata
					P. columellifera ssp. columellifera
					P. columellifera ssp. pallida
				Planopolyspora	Planopolyspora → Actinoplanes
				Salinispora	***S. arenicola***^{T}
					S. tropica
				Spirilliplanes	***S. yamanashiensis***^{T}
				Verrucosispora	***V. gifhornensis***^{T}
				Virgisporangium	*V. ochraceum*^{T}, *V. aurantiacum*

Micromonosporaceae

The family Micromonosporaceae was described by Krasil'nikov (1938)[261] and emended by Stackebrandt et al.,[19] the only family in the suborder Micromonosporineae. It is phylogenetically distinct but its members are a chemotaxonomically and morphologically diverse group of filamentous organisms. The genera *Actinoplanes*, *Catellatospora*, *Catenuloplanes*, *Couchioplanes*, *Dactylosporangium*, *Micromonospora,* and *Pilimelia* were classified in the family Micromonosporaceae by Koch et al. (1996)[262] and the genera *Spirilliplanes,*[263] *Verrucosispora,*[264] *Virgisporangium,*[265] *Asanoa,*[266] *Longispora,*[267] *Salinispora,*[268] *Actinocatenispora,*[269] and *Polymorphospora*[270] added subsequently.

The different genera can be differentiated on the basis of phenotypic and chemotaxonomic characters. The wall peptidoglycan contains *meso-* and/or 3-hydroxydiaminopimelic acid and is of the A1γ type. Except for *Pilimelia* species, which contain acetate as the first amino acid of the peptide chain attached to muramic acid, all other members have glycine; whole-organism hydrolysates are rich in arabinose and xylose, with variable amounts of other sugars. The organisms produce complex mixtures of saturated, *iso-*, and *anteiso*-fatty acids and have diphosphatidylglycerol, phosphatidylethanolamine, phosphatidylglycerol, and phosphatidylinositol as major lipids; phosphatidylcholine occurs in addition in *Catenuloplanes* species. Mycolic acids are absent. Menaquinone profiles are heterogeneous. Tetrahydrogenate menaquinones of the MK-9[H_4] type predominate in members of the genera *Actinoplanes*, *Dactylosporangium*, *Pilimelia*, and *Couchioplanes*, but major amounts of tetra-, hexa-, and/or octahydrogenated menaquinones with nine, ten, and/or twelve isoprene units are found in *Micromonospora* strains. *Asano ferruginea* contains MK-10[H_8] and MK-10[H_6], while *Catenuloplanes* species contain predominantly MK-9[H_8] and MK-10[H_8]. The guanine (G)-plus-cytosine (C) content of the DNA is within the range of 71 to 73 mol%. The family can be distinguished from other suprageneric groups of actinomycetes by 16S rDNA similarity.

Genus *Actinoplanes*

The genus *Actinoplanes* Couch 1950 emend. Stackebrandt and Kroppenstedt 1988[271] is a member of the family Micromonosporaceae Krassil'nikov 1938 emend. Stackebrandt et al. (1997).[19] The genus *Actinoplanes* produces characteristic spherical, cylindrical, or very irregular sporangia with little aerial hyphae. The sporangiospores are motile with multiple polar flagella. The peptidoglycan contains *meso*-diaminopimelic acid, which may be replaced by hydroxyl-diaminopimelic acid, as the major diamino acid. Phosphatidylethanolamine is the diagnostic phospholipid. *Iso-* and *anteiso*-branched and monounsaturated fatty acids, including *cis*-9-octadecenoic acid (oleic acid), are the predominant fatty acids. The major isoprenoid quinone is MK-9[H_4]. The type species is *Actinoplanes philippinensis* Couch 1950,[272] and 28 species with validly published names have been described to date. A phenotypic analysis, including the chemotaxonomy, has been reported;[273] the authors determined the chemotaxonomic and phenotypic characteristics and performed a numerical taxonomic analysis that supported the integrity of the genus. A comprehensive phylogenetic analysis of the genus has been given.[274]

Genus *Catellatospora*

The genus *Catellatospora* Asano & Kawamoto (1986)[275] contains actinomycete strains with short chains of nonmotile spores borne directly on the substrate mycelium without the formation of aerial mycelium. The genus contained *C. citrea*; subsequently *C. citrea* ssp. *methionotrophica,*[276] *C. ferruginea*, *C. matsumotoense*, *C. tsunoense,*[277] and *C. koreensis* [278] were described. However, *C. matsumotoense* has been transferred to the genus *Micromonospora* as *Micromonospora matsumotoense*, based on 16S rRNA gene sequence analysis and phenotypic characteristics,[279] and *C. ferruginea* [275] has been transferred to the genus *Asanoa* as *Asanoa ferruginea.*[266] Molecular systematics, numerical taxonomy, and chemotaxonomy have led to the assignment of three species — *C. citrea* (with *C. citrea* ssp. *citrea* and *C. citrea* ssp. *methionotrophica*), *C. tsunoense*, and *C. koreensis* to the genus.

Genus *Catenuloplanes*

Planopolyspora crispa was isolated from decomposing leaf litter from *Prunus persica* and described[280] as producing tubular, curly, twisted, and sometimes branched sporangia in clusters with numerous spores arranged in a single row that became motile with subpolar flagella. The genus *Catenuloplanes* was proposed[281] as a new genus and revived name. Members of this genus bore arthrospores in chains arising either from the substrate mycelium or aerial hyphae; they did not produce sporangia and the spores were motile by peritrichous flagella.

According to chemotaxonomic analyses in the original description of *Planopolyspora*, *meso*-2,6-diaminopimelic acid and madurose (3-O-methyl-D-galactose) were detected in the whole-cell hydrolysate. The cell wall of *Catenuloplanes* contained lysine but not *meso*-diaminopimelic acid as the diamino acid in the peptidoglycan with a cell wall sugar of xylose, but madurose was not detected. The major menaquinones were reported to be MK-9(Hx)) and MK-10(Hx)) and characteristic phospholipids were phosphatidylethanolamine and phosphatidylcholine (type PIII[23]). In addition to the type species *Catenuloplanes japonicus*, five species (*Catenuloplanes niger*, *Catenuloplanes indicus*, *Catenuloplanes atrovinosus*, *Catenuloplanes castaneus*, and *Catenuloplanes nepalensis*) were proposed.[282] Subsequently, the genus *Catenuloplanes* was classified in the family Micromonosporaceae Krasil'nikov 1938 emend. Koch et al. 1996 on the basis of 16S rDNA sequences.[262] The genera *Planopolyspora* and *Catenuloplanes* were described differently but scanning electron micrographs with the original descriptions of both genera showed similar morphological properties. The type strains of *P. crispa* and *Catenuloplanes* species were studied, and the incorporation of the genus *Planopolyspora* into the genus *Catenuloplanes* as *Catenuloplanes crispus* comb. nov. was proposed.[283]

Genus *Micromonospora*

Ørskov (1923)[38] described actinomycete strains with single spores on sporophores branching from the substrate hyphae. Members of the genus are well-known producers of antibiotics such as gentamicin,[284–288] with nearly 150 aminoglycoside antibiotics isolated. Representatives of many chemical families of antibiotics have been isolated from members of the genus, including novel therapeutic compounds such as the enediyne calicheamicin[289, 290] and the antiviral megalomicin.[291]

The genus has been well circumscribed phylogenetically, chemotaxonomically, and morphologically.[262, 271, 292, 293] But Kasai et al. (2000)[294] observed that 16S rDNA phylogeny did not always agree with other taxonomic characteristics within the genus and they reclassified *Micromonospora* using *gyrB* gene sequence data and DNA–DNA hybridization into 14 species. However, the number of species has increased significantly since then,[295–298] with 28 validly described species.[97]

Micromonospora are isolated from soil, freshwater and marine habitats, including association with plant roots,[97] rivers and streams,[299] lakes,[300] coastal regions,[301] marine sediments,[302] and deep-sea sediments.[71, 303–305]

Genus *Salinispora*

Actinomycetes in marine environments were attributed to terrestrial wash-in;[68] this belief probably limited efforts to isolate marine actinomycetes and was not challenged by the analysis of, at least initial, molecular studies, perhaps with some bias against high G+C DNA.[155] The widespread and persistent presence of indigenous actinomycetes in the marine environment was first reported by Mincer et al. (2002).[302] Large numbers of strains designated MAR 1 were isolated from geographically diverse tropical and/or subtropical locations showing an obligate requirement for seawater.

The strains form a monophyletic 16S rRNA clade within the family Micromonosporaceae, with an obligate requirement for seawater for growth. The MAR 1 isolates were provisionally assigned to a taxon informally designated '*Salinospora*'[302, 306] and validly described as *Salinispora arenicola* and *Salinispora tropica*[71] including their chemotaxonomic characterization and its comparison with other taxa in the Micromonosporaceae family (Table 27.15).

They produce the cytotoxic proteosome inhibitor salinosporamide,[306] which is in clinical trials as an anticancer drug,[109] and marked the beginning of a new phase of recognition of marine actinomycetes[69] and microbial marine drug discovery.[307]

Salinispora are aerobic, Gram-positive, non-acid-fast actinomycetes forming a highly branched substrate hyphae with smooth-surfaced spores singly or in clusters. Seawater, or a sodium-supplemented medium, is required for growth. Strains contain *meso*-diaminopimelic acid in peptidoglycan that is glycolated; arabinose, galactose, and xylose in whole-organism hydrolysates; diphosphatidylglycerol, phosphatidylethanolamine, phosphatidylglycerol, and phosphatidylinositol as diagnostic polar lipids; tetrahydrogenated menaquinones with nine isoprene units as the major isoprenolog; and complex mixtures of saturated *iso*- and *anteiso*-fatty acids; but they lack mycolic acids. The G+C content is 70 to 73 mol%, and unique nucleotide signatures are present at positions 207 (A), 366 (C), 467 (U), and 1456 (G) of the 16S rRNA gene.[71]

The Suborder Propionibacterineae

The Propionibacterineae contains two families — the Propionibacteriaceae and the Nocardioidaceae — with 18 genera (Table 27.16).

Genus *Propionibacterium*

Genus *Nocardioides*

(See Table 27.16.)

The Suborder Pseudonocardineae

The suborder Pseudonocardineae contains two families — the Pseudonocardiaceae[308] and the Actinosynnemataceae[309] — plus some genera in search of a family (Table 27.17).

The taxonomic integrity of the family Pseudonocardiaceae is supported by 16S sequence data and amino acid sequence analysis of the ribosomal protein AT-L30.[310] However, it exhibits different cell wall chemotypes, phospholipid types, and principal menaquinone components.

Labeda and Kroppenstedt (2000)[309] proposed the family Actinosynnemataceae with the genera *Actinokineospora,*[311] *Actinosynnema,*[312] and *Saccharothrix*[313] as members; the genera *Lechevalieria*[314] and *Lentzea*[315] were subsequently added to the family.[314] The family forms a homogeneous group as well as a distinct clade in the 16S phylogenetic tree.

The genera *Actinoalloteichus,*[316] *Crossiella,*[317] *Kutzneria,*[318] and *Streptoalloteichus*[319] fall into the suborder Pseudonocardineae but do not easily fit as members of these two families (Table 27.17).

Pseudonocardiaceae

The family Pseudonocardiaceae was proposed[308] on the basis of 16S rRNA phylogeny and other data.[320] It contained the genera *Actinopolyspora,*[321] *Amycolata,*[322] *Amycolatopsis,*[322] *Faenia,*[323] *Pseudonocardia,*[324] the type genus *Saccharomonospora*[328] and *Saccharopolyspora,*[326] but the genera *Amycolata* and *Pseudonocardia* were combined[327] in an emended genus *Pseudonocardia*, and species in *Faenia* were placed into *Saccharopolyspora*·[328] New genera have been added to the family: *Actinoalloteichus;*[316] *Actinobispora,*[329] but it is proposed to combine this genus with *Pseudono-*

TABLE 27.15
Characteristics that Differentiate the Genera of the Family Micromonosporaceae

Characteristic	1	2	3	4	5	6	7	8	9	10	11	12	13	14
Spore Motility	–	+	–	–	+	+	+	–	–	+	–	+	–	+
Spore Vesicles	–	+	–	–	–	–	+	–	–	+	–	–	–	+
Cell wall Chemotype	II	II	II	II	VI	VI	II	II	II	II	II	II	II	II
Major Whole Organism Sugars	Xyl	Ara, Xyl	Ara, Gal, Xyl	Ara, Gal and Xyl or only Xyl	Xyl	Ara, Gal, Xyl	Ara, Xyl	Ara, Gal, Xyl	Ara, Xyl	Ara, Xyl	Ara, Gal, Xyl	Gal, Xyl	Man, Xyl	Ara, Gal, Man, Rha, Xyl
Fatty Acid Type	2a	2c	2d	3b	2c	2c	2d	2d	3b	2b	3a	2d	2b	2d
Major Menaquinones (MK-)	10(H_6, H_4), 9(H_6, H_4)	9(H_4), 10(H_4)	10(H_6, H_8)	9(H_4, H_6) or 10(H_4)	10(H_4), 11(H_4)	9(H_4)	9(H_4, H_6, H_8)	10(H_4, H_6)	9(H_4, H_6), 10(H_4, H_6)	9(H_2, H_4)	9(H_4)	10(H_4)	9(H_4)	10(H_4, H_6, H_8)
Phospholipid Type	II	II	II	II	III	II	II	II	II	II	II	II	II	II
DNA G+C Content (mol%)	71	72–73	71–72	70–72	70–73	70–72	72–73	70	71–73	ND	70–73	69	70	71

Note: Taxa: 1, Polymorphospora; 2, Actinoplanes; 3, Asanoa; 4, Catellatospora; 5, Catenuloplanes; 6, Couchioplanes; 7, Dactylosporangium; 8, Longispora; 9, Micromonospora; 10, Pilimelia; 11, Salinispora; 12, Spirilliplanes; 13, Verrucosispora; 14, Virgisporangium. Data are taken from Kroppenstedt (1985), Asano and Kawamoto (1986, 1988), Asano et al. (1989), Goodfellow (1989), Yokota et al. (1993), Tamura et al. (1994, 1995, 1997, 2001), Rheims et al. (1998), Kudo et al. (1999), Lee et al. (2000), Lee and Hah (2002), Matsumoto et al. (2003), and Maldonado et al. (2005). Cell wall chemotype is based on the classification of Lechevalier and Lechevalier (1970). Fatty acid type is according to the classification of Kroppenstedt (1985), and phospholipid type is assigned according to Lechevalier et al. (1977). +, Positive; –, negative; ND, not determined; Ara, arabinose; Gal, galactose; Man, mannose; Rha, rhamnose; Xyl, xylose.

TABLE 27.16
The Suborder Propionibacterineae
Class *Actinobacteria*

Subclass	Order	Suborder	Family	Genus	Species
Actinobacteridae	Actinomycetales	Propionibacterineae	Nocardioidaceae	*Actinopolymorpha*	*A. singaporensis*^{T}
				Aeromicrobium	*A. erythreum*^{T}
					A. alkaliterrae
					A. fastidiosum
					A. marinum
					A. Tamlense
				Friedmanniella	*F. antarctica*^{T}
					F. capsulate
					F. lacustris
					F. Spumicola
				Hongia	*Hongia* → *Kribbella*
				Kribbella	*K. flavida*^{T}
					K. alba
					K. antibiotica
					K. jejuensis
					K. karoonensis
					K. koreensis
					K. lupine
					K. solani
					K. swartbergensis
					K. Yunnanensis
				Marmoricola	*M. aurantiacus*^{T}
				Micropruina	*M. glycogenica*^{T}
				Nocardioides	*N. albus*^{T} and 19 species
				Pimelobacter	*P. simplex*^{T}
				Propionicicella	*P. superfundia*^{T}
				Propionicimonas	*P. paludicola*^{T}
			Propionibacteriaceae	*Arachnia*	*Arachnia* → *Propionibacterium*
				Brooklawnia	*B. cerclae*^{T}
				Jiangella	*J. gansuensis*^{T}
				Luteococcus	*L. japonicus*^{T}
					L. peritonei
					L. sanguinis
				Microlunatus	*M. phosphovorus*^{T}
				Propionibacterium	*P. freudenreichii*^{T} and 11 species
				Propioniferax	*P. innocua*^{T}
				Propionimicrobium	*P. lymphophilum*^{T}
				Tessaracoccus	*T. bendigoensis*^{T}

cardia,[330] *Kibdelosporangium,*[331]*Prauserella,*[332] *Thermobispora,*[333] and *Thermocrispum.*[334] These genera share common chemotaxonomic and morphological features (Table 27.18). They are aerobic, Gram-positive, non-acid-fast, catalase-positive actinomycetes and form extensively branched vegetative and aerial hyphae. Smooth, spiny, or hairy spores, singly, in pairs, or in chains of variable length, or in sporangia-like structures on aerial mycelia, are produced. Some species produce spore-like structures on vegetative hyphae. The mycelium may fragment but is generally not significant. Most organisms are chemoorganotrophic, using diverse organic compounds as the sole source of carbon for assimilation and energy production, but there are also some facultative autotrophs. They contain *meso*-diaminopimelic acid (A_2pm), have the peptidoglycan type A1γ (direct cross-linkage between positions-3 and -4 of two peptide subunits with *meso*-A_2pm in position-3), and N-acetylmuramic acid. The G+C content of the DNA ranges between 64 and 79%.

Genus *Amycolatopsis*

Like the *Streptomyces*, described below, members of the genus *Amycolatopsis* have been prolific producers of bioactive metabolites, including medically and commercially significant antibiotics (Table 27.19). They are found in soils and on plant materials,[335] although members of the taxon are beginning to be found from marine sources.[336] The genus *Amycolatopsis*[322] contains nocardioform actinomycetes with *meso*-diaminopimelic acid and sugars arabinose and galactose (wall chemotype IV *sensu* Lechevalier & Lechevalier, 1970),[206] an A1γ type peptidoglycan, and muramic acid in the *N*-acetylated form. They lack mycolic acids. Phosphatidylethanolamine is the diagnostic phospholipid; the polar lipid pattern consists of phosphatidylethanolamine and perhaps phosphatidylmethylethanolamine, and phosphatidylglycerol with variable occurrence of diphosphatidylglycerol, phosphatidylinositol, and phosphatidylinositol mannosides (phospholipid type II *sensu* Lechevalier et al., 1977);[23] phosphatidylcholine and glucosamine-containing phospholipids are absent. The fatty acids are mostly straight-chain, monounsaturated, *iso*-, *anteiso*-, and 10-methyl branched saturated fatty acids.[337] Predominant isoprenologs are di-, tetra-, and hexahydrogenated menaquinones with nine isoprene units.[338]

They are aerobic, Gram-positive, catalase positive, non-acid-fast, nonmotile actinomycetes that form branched substrate hyphae that may fragment into squarish subunits. Aerial hyphae may be formed on some media but may be either sterile or break up into squarish to oval fragments or spore-like structures; spore chains are also produced on vegetative hyphae. The guanine-plus-cytosine (G+C) content of the DNA is 66 to 69 mol%. The type species of the genus is *Amycolatopsis orientalis.*[322, 339]

Members of the genus *Amycolatopsis* form a distinct phyletic line within the evolutionary radiation of the family Pseudonocardiaceae in the suborder Pseudonocardineae. Chemotaxonomically, they are defined as strains that contain arabinose and galactose in whole-organism extracts, tetrahydrogenated manaquinones with nine isoprene units as the predominant isoprenologue, and major amounts of phosphatidylethanolamine but which lack phosphatidylcholine (Table 27.18).

The Suborder Streptomycineae

The suborder Streptomycineae contains one family (the Streptomycetaceae) and three genera; with 724 species in the genus *Streptomyces,* this group is the antithesis of much of the class, many families containing a single genus with one species (Table 27.20). Even this large number of species is the result of extensive efforts to remove synonyms and discard poorly described species from the more than 3000 species named following the discovery of antibiotics in members of the genus by Waksman.

TABLE 27.17
The Suborder Pseudonocardineae

Class Actinobacteria

Subclass	Order	Suborder	Family	Genus	Species
Actinobacteridae	Actinomycetales	Pseudonocardineae	Actinosynnemataceae	*Actinokineospora*	*A. riparia*^{T}
					A. auranticolor
					A. diospyrosa
					A. enzanensis
					A. globicatena
					A. inagensis
					A. terrae
				Actinosynnema	*A. mirum*^{T}
					A. pretiosum ssp. *pretiosum*
					A. pretiosum ssp. *auranticum*
				Lechevalieria	*L. aerocolonigenes*^{T}
					L. flava
				Lentzea	*L. albidocapillata*^{T}
					L. albida
					L. californiensis
					L. flaviverrucosa
					L. violacea
					L. waywayandensis
				Saccharothrix	*S. australiensis*^{T} and 12 species
			Pseudonocardiaceae	*Actinoalloteichus*	*A. cyanogriseus*^{T}
					A. hymeniacidonis
					A. spitiensis
				Actinobispora	*Actinobispora* → *Pseudonocardia*
				Actinopolyspora	*A. halophila*^{T}
					A. iraqiensis
					A. Mortivallis
				Amycolata	*Amycolata* → *Pseudonocardia*
				Amycolatopsis	*A. orientalis*^{T} and 33 species
				Crossiella	*C. cryophila*^{T}
					C. equi
				Faenia	*Faenia* → *Saccharopolyspora*
				Goodfellowia	*G. coeruleoviolacea*^{T}
				Kibdelosporangium	*K. aridum*^{T} and ssp. *largum*

	K. albatum
	K. philippinense
Kutzneria	*K. viridogrisea*^{T}
	K. albida
	K. kofuensis
Prauserella	*P. rugosa*^{T}
	P. alba
	P. halophila
Pseudoamycolata	*Pseudoamycolata* → *Pseudonocardia*
Pseudonocardia	*P. thermophila*^{T} and 25 species
Saccharomonospora	*S. viridis*^{T}
	S. azurea
	S. cyanea
	S. glauca
	S. halophila
	S. paurometabolica
	S. xinjiangensis
Saccharopolyspora	*S. hirsuta*^{T} and ssp. *kobensis*
	S. erythraea
	S. flava
	S. gregorii
	S. hordei
	S. rectivirgula
	S. spinosa
	S. spinosporotrichia
	S. taberi
	S. thermophila
Streptoalloteichus	*S. hindustanus*^{T}
Thermobispora	*T. bispora*^{T}
Thermocrispum	*T. municipale*^{T}
	T. agreste

TABLE 27.18
Chemical and Morphological Markers Found in Members of the Genera Classified in the Suborder Pseudonocardineae

Genus	Wall Chemotype[a]	Whole	Cell Sugar Pattern	Fatty Acids[b]	Predominant Menaquinone(s)[c]	Phospholipid Type[d]	DNA G+C Content (mol%)	Acropetal Budding	Aerial Hyphae	Spores on Aerial Hyphae	Fragmentation of Substrate Mycelium	Sporangia like Structures
Family Actinosynnemataceae												
Actinokineospora	III	Galactose, mannose, rhamnose	A, I, S, U	MK	$9(H_4)$	II	73	–	+	In chains	–	–
Actinosynnema	III	Galactose, mannose	A, S, U	MK	$9(H_4,H_6)$	II	71–73	–	+	In chains	+	–
Lechevalieria	III	Galactose, mannose, rhamnose	A, I, S, U	MK	$9(H_4)$	II	ND	–	Sparse	Sparse	–	–
Lentzea	III	Galactose, mannose, ribose	A, I, S, U	MK	$9(H_4)$	II	ND	–	+	In chains	+	–
Saccharothrix	III	Galactose, mannose, rhamnose	A, I, S, U	MK	$9(H_4)$, $10(H_4)$	II, IV	70–76	–	+	In chains	+	–
Family Pseudonocardiaceae												
Actinopolyspora	IV	Arabinose, galactose	A, I, M, S	MK	$9(H_4)$, $10(H_4)$	III	64–68	–	+	In chains	v	–
Amycolatopsis	IV	Arabinose, galactose	A, I, M, S	MK	$9(H_4, H_6)$	II	66–69	–	+	In chains	+	–
Kibdelosporangium	IV	Arabinose, galactose	A, I	MK	$9(H_4)$	II	66	–	+	In chains	+	+
Prauserella	IV	Arabinose, galactose	A, I, M, S	MK	$9(H_2, H_4)$	II	67–68.9	–				
Pseudonocardia	IV	Arabinose, galactose	A, I, M, S	MK	$8(H_4)$	III/II	68–79	+	+	In chains	v	–
Saccharomonospora	IV	Arabinose, galactose	A, I, S	MK	$9(H_4)$	II	69–74	–	+	Singly	–	–
Saccharopolyspora	IV	Arabinose, galactose	A, I, M, S	MK	$9(H_4)$	III	70–77	–	+	In chains	+	–

Thermobispora	III	None	ND	MK	$9(H_0, H_2)$	IV	71	–	+	In pairs	–	–
Thermocrispum	III	None	A, I, H, M	MK	$9(H_4)$	II	69–73	–	+	In chains	–	+
Genera in Search of a Family												
Actinoalloteichus	III	Galactose, mannose, ribose	A, I	MK	$9(H_4)$	II	73	–	+	In chains	+	–
Crossiella	III	Galactose, mannose, rhamnose, ribose	A, I, S	MK	$9(H_4)$	IV	71.4	–	+			+
Kutzneria	III	Galactose, trace rhamnose	A, I, S	MK	$9(H_4)$	II	70–71	–	+			–
Streptoalloteichus	III	Galactose, mannose, ribose	ND	MK	$10(H_4, H_6)$	II	ND	–	+	In chains	+	+

Note: Data taken from Gochnauer et al. (1975), Lechevalier et al. (1986), Tomita et al. (1987, 1993), Korn-Wendisch et al. (1989, 1995), Stackebrandt et al. (1994), Wang et al. (1996), Kim and Goodfellow (1999), Reichert et al. (1999), Tamura et al. (2000), Labeda et al. (2000, 2001), and Labeda (2001). ND, not determined.

[a] Major constituents of wall chemotypes: I, LL-A_2pm and glycine; II, meso-A_2pm and glycine; III, meso-A_2pm; IV, meso-A_2pm, arabinose, and galactose; V, lysine and ornithine; VI, lysine (with variable presence of aspartic acid and galactose); VII, diaminobutyric acid and glycine (lysine variable); VIII, ornithine; IX, LL- and meso-A_2pm (Lechevalier and Lechevalier, 1970, 1980).

[b] A, anteiso-branched fatty acids; I, iso-branched fatty acids; H, hydroxy fatty acids; M, 10-methyl-branched fatty acids; S, straight-chain fatty acids; U, monounsaturated and unsaturated fatty acids.

[c] Abbreviations exemplified by MK $9(H_4)$, menaquinone with two of the nine isoprene units hydrogenated.

[d] Characteristic polar lipids: I, nitrogenous phospholipids absent (with phosphatidylglycerol variable); II, only phosphatidylethanolamine; III, phosphatidylcholine (with phosphatidylethanolamine, phosphatidylmethylethanolamine and phosphatidylglycerol variable, phospholipids containing glucosamine absent); IV, phospholipids containing glucosamine (with phosphatidylethanolamine variable); V, phospholipids containing glucosamine and phosphatidylglycerol (with phosphatidylethanolamine variable); all preparations contain phosphatidylinositol (Lechevalier et al., 1977).

TABLE 27.19

Bioactive Compounds Produced by Members of the Genus *Amycolatopsis*

Strain/Taxon	Product(s)	Feature(s)	Ref.
A. alba	Glycopeptide antibiotic	Inhibits peptidoglycan synthesis	Mertz & Yao (1993)
A. azurea	Azureomycins A and B	Glycopeptide antibiotic	Omura et al. (1979)
A. coloradensis	Avoparcin LL-AV290	Glycopeptide antibiotic	Labeda (1995)
A. fastidiosa	Antibiotic 41,034 and 41,494	Macrobicyclic peptide	Celmer et al. (1976)
A. japonica	(S,S)-N,N′ ethylene-diaminedisuccinic acid	Inhibits phospholipase C	Nishikori et al. (1984)
'*A. lactamdurans*'	Cephamycin C	β-Lactam antibiotic	Stapley et al. (1972)
	Efrotomycin	β-Isomer	Nielsen & Arison (1989)
	3-Methylpseudouridine	Polyether	Wax et al. (1976)
A. mediterranei	Rifamycin A, B, C, D, E, G, O, SV, Y	Clinically useful ansamycins, active against *Mycobacterium* spp.	Margalith & Beretta (1960); Lancini et al. (1967); Birner et al. (1972); Lancini & Sartori (1976)
	Balhimycin	Glycopeptide antibiotic	Pelzer et al. (2001)
	31-Homorifamycin W	Ansamycin antibiotic	Wang et al. (1994)
A. mediterranei MI710 51F6	Dethymicin	Immunosupressant	Ueno et al. (1992)
A. mediterranei R-21	3-Hydroxyrifamycin S	Ansamycin antibiotic	Traxler et al. (1981)
'*A. mediterranei* var. *kanglensis*' 1747-64	Kanglemycin A	Ansamycin antibiotic	Wang et al. (1988)
A. mediterranei N813	Protorifamycin I, Proansamycin B-M1, Protorifamycin I-M1,	Ansamycin antibiotic	Ghisalba et al. (1978, 1979, 1980)
A.methanolica	Aromatic amino acids	Suitable for strain improvement	de Boer et al. (1990)
A. orientalis ssp. *Orientalis*	Vancomycin	Clinically useful glycopeptide antibiotic, active against severe bacterial infections	Pittenger & Brigham (1956); U.S. Patent No. 3,067,099 (Eli Lilly);
	Glycopeptide Compounds	Glycosyltransferase gene, gtfA	Baltz (2000); U.S. Patent No. 6,087,143 (Eli Lilly)
A. orientalis ssp. *Orientalis*	Muraceins	Angiotensin converting enzyme inhibitors	Bush et al. (1984)
A.orientalis sp. NRRL 15232	N-Demethylvancomycin	Vancomycin analog	Boeck et al. (1984)
A. orientalis NCIB 12608 and NCIB 40089	MM47761, MM49721 and MM55266, MM55268	Glycopeptide antibiotics	Box et al. (1990, 1991)
A. orientalis Q427-8 (ATCC 53884)	Quartromicin A1, A2, A3, D1, D2, D3	Antiviral antibiotics	Tsunakawa et al. (1992)
A. orientalis ATCC 53550	UK-69,753	Related to efrotomycin and factumycin	Pacey et al. (1989)

—continued

TABLE 27.19 (continued)

Bioactive Compounds Produced by Members of the Genus *Amycolatopsis*

Strain/Taxon	Product(s)	Feature(s)	Ref.
A. orientalis ssp. *Lurida*	Ristocetin Benzathrins	Glycopeptide antibiotic Quinone antibiotic with antitumor activity	Grundy et al. (1957) Theriault et al. (1986)
A. sulphurea	Chelocardin	Tetracycline derivative	Oliver & Sinclair (1964) U.S. patent No. 3,155,582
'*A. trehalostatica*'	Trehalostatin	Insecticide	Murao & Shin (1995) U.S. Patent No. 5,372,816
Amycolatopsis sp. NNO 21702	Tigloside	Tigloylated tetrasaccharide	Breinholt et al. (1998)
Amycolatopsis sp. K104-1	Polylactic acid depolymerise	Polymer degrading enzymes	Nakamura et al. (2001)
Amycolatopsis sp. MK299-95F4	Epoxyquinomicins A, B, C, D	Improves collagen induced arthritis *in vivo*	Tsuchida et al. (1996) ;Matsumoto et al. (1997)
Amycolatopsis MJ347-81F4	MJ347-81F4 A, MJ347-81F4 B	Cyclic thiazolyl peptide antibiotics	Sasaki et al. (1998)
Amycolatopsis sp. Y-86,21022	Balhimycin 4-oxovancosamine containing glycopeptides	Glycopeptide antibiotics	Nadkarni et al. (1994) Vertesy et al. (1996)
Amycolatopsis sp. MJ126-NF4	Azicemicins A and B	Inhibits Gram-positive bacteria, including mycobacteria	Tsuchida et al. (1993, 1995)
Amycolatopsis sp. MI481-42F4	Amythiamicins A, B, C, D	Inhibits Gram-positive bacteria	Shimanaka et al. (1994)
Amycolatopsis sp. PA-45052	Chloroorienticins	Glycopeptide antibiotics	Tsuji et al. (1988a, b)
Amycolatopsis sp. MG398-hF9	Octacosamicins A, B	Antifungal antibiotics	Dobashi et al. (1988)

Genus *Streptomyces*

The genus *Streptomyces* was introduced by Waksman and Henrici (1943)[40] based upon morphology and pigmentation, characteristics that made members of the genus distinctive [e.g., *Streptothrix foersteri*[340] (Cohn, 1875); *Streptothrix alba*[11] (Rossi Doria, 1891); *Streptomyces coelicolor*[341] (Müller, 1908); and *Streptomyces violaceoruber*[342] (Waksman and Curtis, 1916)], if also confused. The type strain of the genus is *Streptomyces albus* Rossi Doria, although the original strain from Rossi Doria (1891) was no longer extant in 1943 when the genus was introduced and the type strain designated[40] (Waksman and Henrici, 1943), and many related strains were confused with *S. griseus.*[343,344] Similarly *Streptomyces coelicolor* Müller, *Streptomyces violaceoruber,* and "*Streptomyces coelicolor*" A3(2) have been confused for most of the past century.[345,346]

Following the discovery of actinomycin,[347] streptothicin,[348] and streptomycin[349] in 1941–1944, extensive isolation and screening led to the description of more than 3000 species, often only in patents, largely based on non-standard procedures, with type strains that were largely unavailable, and the potential for confusion illustrated by the inability to distinguish *S. albus*/*S. griseus*[40] and *S. violaceoruber*/*S. coelicolor.*[345]

TABLE 27.20
The Suborder Streptomycineae

Class Actinobacteria

Subclass	Order	SubOrder	Family	Genus	Species
Actinobacteridae	Actinomycetales	Streptomycineae	Streptomycetaceae	*Actinopycnidium*	*Actinopycnidium* → *Streptomyces*
				Actinosporangium	*Actinosporangium* → *Streptomyces*
				Chainia	*Chainia* → *Streptomyces*
				Elytrosporangium	*Elytrosporangium* → *Streptomyces*
				Kitasatoa	*Kitasatoa* → *Streptomyces*
				Kitasatospora	***K. setae***^{T} and 19 species
				Microellobosporia	*Microellobosporia* → *Streptomyces*
				Streptacidiphilus	***S. albus***^{T}
					S. carbonis
					S. jiangxiensis
					S. neutrinimicus
					S. oryzae
				Streptomyces	***S. albus***^{T} and 700 species
				Streptoverticillium	*Streptoverticillium* → *Streptomyces*

The International Streptomyces Project (ISP)[350–353] aimed to establish reliable standard procedures but introduced no new methodology. The most systematic attempt at a reliable classification was based on application of the principles of numerical taxonomy[354, 355] to *Streptomyces*.[356–358] The underlying expectation behind the ISP study and these numerical taxonomic studies was the reduction in the numbers of species. These studies preceded extensive 16S rDNA sequencing[359] and the recognition of the enormous diversity of prokaryotes.[30] The phenotypic classifications provided by the numerical taxonomic studies still provide the basis for the sub-generic taxonomy of the *Streptomyces*.[360]

Streptomyces have a distinctive morphology, producing extensive branching mycelial growth, which does not fragment, followed by the formation of vertical aerial hyphae and then septation of the multinucleate hyphae into chains of three or more uninucleate arthrospores. The chain morphology may be straight, flexous, hooked, looped, or spiral; and the spore surfaces may be hairy, knobby, ridged, rugose, smooth, spiny, or warty. The detailed morphological characters give initially smooth colonies that develop floccose, granular, powdery, or velvety surfaces as the vegetative hyphae differentiate to aerial hyphae and into spore chains. The prolific production of secondary metabolites, many of which are pigmented, results in variable colors of substrate and aerial mycelium, and diffusible pigments. Although the resulting morphology is distinctive and recognizable, it is also variable but it is not definitive; and simplistic identification based on morphology can lead to overspeciation,[361] misidentification within the genus,[345] and misidentification of other genera as *Streptomyces*.[362] Nevertheless, the genus is well circumscribed by its chemotaxonomy; the cell wall contains major amounts of LL-diaminopimelic acid (cell wall chemotype I) with a glycine cross-link (A3γ). The phospholipids typically include phosphatidylethanolamine, phosphatidylglycerol, phosphatidylinositol, and phosphatidylinositol-mannosides with saturated, *iso*-, and *anteiso*- fatty acids and menaquinones with nine isoprene units, hexa- and octa-hydrogenated, MK-9[H_6] and MK-9[H_8], as the major menaquinones.

Nevertheless, the genus is defined phylogenetically by its 16S rDNA sequence,[19] a process that has led to the accretion of morphologically defined genera *Actinopycnidium*,[363] *Actinosporangium*,[361] *Chainia*,[363] *Elytrosporangium*,[365] *Kitasatosporia*,[366] *Microellobosporia*,[365] *Microstreptospora*,[367] *Streptoverticillium*[368] into the genus, and the continued description of new species — now more than 700 validly described species. With a deeper understanding of prokaryotic diversity, this process is likely to reverse with the reinstatement of *Kitasatosporia* as *Kitasatospora*,[369] the description of *Streptacidiphilus*[370] and distinctive clades within this large genus such as the streptoverticillia,[371] and the rugose spored *Streptomyces violaceusniger*/*S. hygroscopicus* clade[372] — a process that is driven by the continued importance of *Streptomyces* as a source of novel, bioactive natural products,[373, 374] as well as enzymes.[32]

The Suborder Streptosporangineae

The suborder Streptosporangineae has three families and fourteen genera (Table 27.21). Stackebrandt et al. (1997)[19] created the suborder Streptosporangineae with the families Nocardiopsaceae,[375] Streptosporangiaceae,[376] and Thermomonosporaceae.[377]

Genus *Thermobifida*

Thermobifida fusca (formerly *Thermomonospora fusca*) and in the family Nocardiopsaceae was isolated from rotting wood[378] and is a major degrader of plant cell walls in compost and rotting, heated organic matter. It produces an extensive array of cellulases and xylanases, many studied for their thermostability and broad pH range but the whole genome sequence[110] has revealed 29 putative glycoside hydrolases in addition to the previously studied cellulases and xylanases, as well as many carbohydrate transport systems and transcriptional regulators controlling the expression of the transporters and glycosylhydrolases.

TABLE 27.21
The Suborder Streptosporangineae

Class Actinobacteria

Subclass	Order	Suborder	Family	Genus	Species
Actinobacteridae	Actinomycetales	Streptosporangineae	Nocardiopsaceae	*Nocardiopsis*	***N. dassonvillei***^{T} and 25 species
				Streptomonospora	*S. salina*^{T}
					S. alba
				Thermobifida	*T. alba*^{T}
					T. cellulosilytica
					T. fusca
			Streptosporangiaceae	*Acrocarpospora*	*A. pleiomorpha*^{T}
					A. corrugate
					A. macrocephala
				Herbidospora	*H. cretacea*^{T}
				Microbispora	*M. rosea*^{T} and 11 species
				Microtetraspora	*M. glauca*^{T}
					M. fusca
					M. malaysiensis
					M. niveoalba
					M. tyrrhenii
				Nonomuraea	*N. pusilla*^{T} and 17 species
				Planobispora	*P. longispora*^{T}
					P. Rosea
				Planomonospora	*P. parontospora*^{T} and ssp. *antibiotica*
					P. alba
					P. sphaerica
					P. venezuelensis

	Planotetraspora	*P. mira*^{T}
		P. silvatica
	Streptosporangium	*S. roseum*^{T} and 16 species
	Thermopolyspora	*T. flexuosa*^{T}
Thermomonosporaceae	*Actinocorallia*	*A. herbida*^{T}
		A. aurantiaca
		A. cavernae
		A. glomerata
		A. libanotica
		A. longicatena
	Actinomadura	*A. madurae*^{T} and 35 species
	Excellospora	*Excellospora* → *Actinomadura*
	Spirillospora	*S. albida*^{T}
		S. rubra
	Thermomonospora	*T. curvata*^{T}
		T. chromogena

Genus *Nonomuraea*

The genus *Nonomuraea*[379, 380] forms a coherent clade in the 16S rRNA gene tree within the Streptosporangiaceae family. They form a substrate mycelium, aerial mycelium chains of spores that can vary in morphology (spiral, hooked, or straight) and spore ornamentation (smooth to irregular or warty). The cell wall is chemotype IIIB, phospholipid type IV, and menaquinones MK-9[H_2], MK-9[H_4], and MK-9[H_6]. The phenotypic properties that can be used to separate species in the genus are also tabulated in Quintana et al. (2003).[381]

This genus is an example of the "rare" actinomycetes targeted for natural product search and discovery.[382] The type strain *Nonomuraea pusilla* makes an antitumor compound myxochelin A,[383] *N. longicatena* makes indolocarbazole K-252a,[384] and a *Nonomuraea* species produces natural glycopeptide A40926,[385] which is chemically modified to produce dalbavancin.[386]

Genus *Streptosporangium*

The genus *Streptosporangium* is the type genus of the family Streptosporangiaceae[376] in the suborder Streptosporangineae.[19] They produce globose/spherical sporangia on aerial mycelium; septation of the single hypha coiled up in the sporangium forms nonmotile spores that are spherical, oval, or rods. The cell wall contains *N*-acetylmuramic acid and *meso*-DAP and whole-cell hydrolysates contain madurose. Major phospholipids include unknown glucosamine-containing compounds, but no phosphatidylcholine or phosphatidylglycerol (PG).

As for *Nonomuraea,* the genus has been targeted for search and discovery.[382]

Genus *Actinomadura*

Streptothrix madurae was described as the causative agent of Madura foot. It was in the genus *Nocardia* and then transferred to the *Streptomyces.* Cell-wall analysis demonstrated that they differed from *Streptomyces* and set up the genus *Actinomadura.*[206] However, this characterization by cell wall containing *meso*-DAP and no characteristic sugars led to a heterogenous taxon with diverse actinomycete strains, clarified by 16S rRNA sequence analysis.[377]

The type species is *Actinomadura madurae* and contains 35 species (Table 27.15). The genus belongs to the family Thermomonosporaceae, also including the genera *Actinocorallia*, *Spirillospora,* and *Thermomonospora*. The genus *Actinomadura* forms well-developed, nonfragmenting substrate mycelia and differentiates to form aerial hyphae with spiral, hooked, or straight spore chains and irregular, smooth, spiny, or warty spores. The cell wall is chemotype IIIB and contains *N*-acetyl muramic acid, major phospholipids diphosphatidylglycerol and phosphatidylinositol, and MK-9[H_6] as major menaquinones.

REFERENCES

1. Cavalier-Smith, T. (2006) Rooting the tree of life by transition analyses. *Biology Direct*, 1, 19.
2. Battastuzzi, F.U., Feijao, A., and Hedges, S.B. (2004) A genomic timescale of prokaryote evolution: insights into the origin of methanogenesis, phototrophy, and the colonization of the land. *BMC Evol. Biol.*, 4, 44.
3. Woronin, M. (1866) Über die bei der Schwarzerle (*Alnus glutinosa*) und der gewöhnlichen Garten-Lupine (*Lupinus mutabilis*) auftretenden, Wurzelanschwellungen. *Mémoires de l'Académie des Sciences, St Petersburg*, 10, 1.
4. Schroeter, J. (1872) Über einige durch Bacterien gebildete Pigmente, 1872. In: F. Cohn (Ed.), *Beitrage zur Biologie der Pflanzen*, J.U. Kern's Verlag, Breslau, 1875, pp. 109–126.
5. Klebs, 1875 cited in http://www.bacterio.cict.fr/c/corynebacterium.html.
6. Harz, C.O. (1877) *Actinomyces bovis* ein neuer schimmel in den geweben des rindes. *Deutsche Zeitschrift für Thiermedizin,* 5, 125.
7. Zopf, W. (1883) *Die Spaltpilze*, Edward Trewendt, Breslau, pp. 1–100.

8. Ali-Cohen, C.H. (1889). Eigenbewegung bei Mikrokokken. Zentralbl. Bakteriol Parasitenkd. *Infektionskr. Hyg. Abt.* 1 Orig., 6, 33–36.
9. Eppinger, H. (1891) Über eine neue pathogene *Cladothrix* und eine durch sie hervorgerufene Pseudotuberculosis (Cladothrichica). *Beitrage zur Pathologischen Anatomie*, 9, 287–328.
10. Zopf, W. (1891) Über Ausscheidung von Fettfarbstoffen (Lipochromen) seitens gewisser Spaltpilze. *Berichte der Deutschen Botanischen Gesellschaft*, 9, 22–28.
11. Rossi Doria, T. (1891) Su di alcune specie di "*Streptothrix*" trovate nell'aria studate in rapporto a quelle giá note a specialmente all' "*Actinomyces*". *Annali dell'Istituto d'Igiene Sperimentale*, Universita Roma, 1, 399–438.
12. Vincent, H. (1894) Étude sur le parasite du pied le madura. *Annales de l'Institut Pasteur* (Paris), 8, 129–151.
13. Lehmann, K.B., and Neumann, R. (1896) *Atlas und Grundriss der Bakteriologie und Lehrbuch der speciellen bakteriologischen Diagnostik*, 1st ed., J.F. Lehmann, München.
14. Tissier, H. (1900) Recherchers sur la flora intestinale normale et pathologique du nourisson. Thesis. University of Paris, Paris, France.
15. Gilchrist, T.C. (1900) A bacteriological and microscopical study of over three hundred vesicular and pustular lesions of the skin, with a research upon the etiology of *Acne vulgaris*. *Johns Hopkins Hospital Report*, 9, 409–430.
16. Wolff, M. (1910) Über eine neve Krankheit der Raupe von *Bupalus piniarius* L. *Mitteilungen des Kaiser-Wilhelm-Instituts fur Landwirtschaft in Bromberg*, 3, 69–92.
17. Buchanan, R.E. (1918) Studies in the nomenclature and classification of the bacteria. VII. The subgroups and genera of the Actinomycetales. *J. Bact.*, 3, 403–406.
18. Boone, D.R., Castenholtz, R.W., and Garrity, G.M. (Editors). (2001) *Bergey's Manual of Systematic Bacteriology*. 2nd edition, Vol. 1. New York, Springer.
19. Stackebrandt, E., Rainey, F.A., and Ward-Rainey, N.L. (1997) Proposal for a new hierarchic classification system, *Actinobacteria classis* nov. *Int. J. Syst. Bacteriol.*, 47, 479–491.
20. Lechevalier, H.A., and Lechevalier, M.P. (1989) CRC *Practical Handbook of Prokaryotes*. CRC Press, Boca Raton, FL.
21. Collins, M.D., and Jones, D. (1983) Distribution of isoprenoid quinone structural types in bacteria and their taxonomic implications. *Microbiol. Rev.* 45, 316-354
22. Schleifer, K.H., and Kandler, O. (1972) Peptidoglycan types of bacterial cell walls and their taxonomic implications. *Bacteriol. Rev.* 36, 407-477
23. Lechevalier, M.P., Biève, C., and Lechevalier, H. (1977) Phospholipids in the taxonomy of actinomycetes. Zentrabl. Bakteriol. Parasitenkd. *Infektionskr. Hyg. I Abr. Suppl.*, 11, 111–116.
24. Wayne, L.G., Brenner, D.J., Colwell, R.R., Grimont, P.A.D., Kandler, P., Krichevsky, M.I., Moore, L.H., Moore, W.E.C., Murray, R.G.E., Stackebrandt, E., Starr, M.P., and Truper, H.G. (1987) Report of the ad hoc committee on reconciliation of approaches to bacterial systematics. *Int. J. Syst. Bacteriol.*, 37, 463–464.
25. Woese, C.R., Kandler, O., and Wheelis, M.L. (1990) Towards a natural system of organisms. Proposal for the domains Archaea, Bacteria and Eucarya. *Proc. Natl. Acad. Sci. U.S.A.*, 18, 399–416.
26. Ludwig, W., and Schleifer, K.-H. (1994) Bacterial phylogeny based on 16S and 23S rRNA sequence analysis. *FEMS Microbiol. Rev.* 15, 155-173
27. Stackebrandt, E., and Goebel, B.M. (1994) A place for DNA:DNA reassociation and 16S rRNA sequence analysis in the present species definition in bacteriology. *Int. J. Syst. Bacteriol.*, 44, 846–849.
28. Embley, T.M., and Stackebrandt, E. (1994) The molecular phylogeny and systematics of the actinomycetes. *Annu. Rev. Microbiol.*, 48, 257–289.
29. Euzéby, J.P. (1997) List of bacterial names with standing in nomenclature: a folder available on the Internet. *Int. J. Syst. Bacteriol.*, 47, 590–592 (URL: http://www.bacterio.net).
30. Whitman, W.B., Coleman, D.C., and Wiebe, W.J. (1998) Prokaryotes: the unseen majority *Proc. Natl. Acad. Sci. U.S.A.*, 95, 6578–6583.
31. Bachmann, S.L., and McCarthy, A.J. (1991). Purification and cooperative activity of enzymes constituting the xylan-degrading system of Thermomonospora fusca. *Appl. Environ. Microbiol.*, 57, 2121–2130.
32. Bhattacharya, D., Nagpure, A., and Gupta, R.K. (2007) Bacterial chitinases: Properties and potential. *Crit. Rev. Biotech.*, 27, 21-28
33. Schrempf, H. (2001) Recognition and degradation of chitin by streptomycetes. *Antonie van Leeuwenhoek* 79, 285–289
34. Berdy, J. (2005) Bioactive microbial metabolites – A personal view. *J. Antibiot.*, 58, 1–26.

35. Brumpt, E. (1900) Cited in Brumpt, E. (1910) *Precis de Parasitologie*, Paris.
36. Trevisan, V. (1879) Prime linee d'introduzione allo studio dei Batterj italiani. *Rend. reale ist. Lombardo sci.,* Ser. II, 12, 133–151.
37. Trevisan, V. (1889) *I. Generie le Specie delle Batteriacee,* Zanaboni and Gabuzzi, Milano.
38. Ørskov, J. (1923). *Investigation into the Morphology of the Ray Fungi.* Copenhagen: Levin and Munksgaard.
39. Jensen, H.L. (1931) The definition and subdivision of the genus Actinomyces. *Proc. Linnean Soc. N.S. Wales*, 57, 364–376.
40. Waksman, S., and Henrici, A. (1943) The nomenclature and classification of the actinomycetes. *J. Bact.*, 46, 337–341.
41. Lechevalier, H.A., and Lechevalier, M.P. (1970). A critical evaluation of the genera of aerobic actinomycetes, pp. 393–405. In H. Prauser (Ed.), The *Actinomycetales.* Jena, German Democratic Repulbic: VEB Gustav Fischer Verlag.
42. Roller, C., Ludwig, W., and Schleifer, K.H. (1992) Gram positive bacteria with a high DNA GC content are characterised by a common insertion within their 23S rRNA genes. *J. Gen. Microbiol.*, 138, 167–175.
43. Gao, B., and Gupta, R.S. (2005) Conserved indels in protein sequences that are characteristic of the phylum Actinobacteria. Int. *J. Syst. Evol Microbiol.*, 55, 2401–2412.
44. Ueda, K., Yamasha, A., Ishikawa, J., Shimada, M., Watsuji, T., Morimura, K., Ikeda, H., Hatori, M., and Beppu, T. (2004) Genome sequence of *Symbiobacterium thermophilum*, an uncultivable bacterium that depends upon microbial commensalisms. *Nucleic Acid Res.*, 32, 4937–4944.
45. Gollery, M., Harper, J., Cushman, J., Mittler, T., Girke, T., Zhu, J.K., Bailey-Serres, J., and Mittler, R. (2006) What makes species unique? The contribution of proteins with obscure features. *Genome Biol.*, 7, Art. No. R57.
46. Gao, B., Paramanathan, R., and Gupta, R.S. (2006) Signature proteins that are distinctive characteristics of Actinobacteria and their subgroups. *Antonie van Leeuwenhoek*, 90, 69–91.
47. Davis, N.K., and Chater, K.F. (1992) The Streptomyces coelicolor whiB gene encodes a small transcription factor-like protein dispensable for growth but essential for sporulation. *Mol. Gen. Genet.*, 232, 351–358.
48. Soliveri, J.A., Gomez, J., Bishai, W.R., and Chater, K.F. (2000) Multiple paralogous genes related to the *Streptomyces coelicolor* developmental regulatory gene whiB are present in *Streptomyces* and other *actinomycetes. Microbiol-UK*, 146, 333–343.
49. Ravel, J., DiRuggiero, J., Robb, F.T., and Hill, R.T. (2000) Cloning and sequence analysis of the mercury resistance operon of *Streptomyces* sp. strain CHR28 reveals a novel putative second regulatory gene. *J. Bact.*, 182, 2345–2349.
50. Yu, Z.G., Zhou, L.Q., Anh, V.V., Chu, K.H., Long, S.C., and Deng, J.Q. (2005) Phylogeny of prokaryotes and chloroplasts revealed by a simple composition approach on all protein sequences from complete genomes without sequence alignment. *J. Mol. Evol.*, 60, 538–545.
51. Henz, S.R., Huson, D.H., Auch, A.F., Nieselt-Struwe, K., and Schuster, S.C. (2005) Whole-genome prokaryotic phylogeny. *Bioinformatics*, 21, 2329–2335.
53. Wang, M.L., and Caetano-Anolles, G. (2006) Global phylogeny determined by the combination of protein domains in proteomes. *Mol. Biol. Evol.*, 23, 2444–2454.
54. Potts, M. (1994) Desiccation tolerance of prokaryotes. *Microbiol. Rev.*, 58, 755–805.
55. Wynn-Williams, D.D., Edwards, H.G., Newton, E.M., and Holder, J.M. (2002) Pigmentation is a survival strategy for ancient and modern photosynthetic microbes under high ultraviolet stress on planetary surfaces. *Int. J. Astrobiol.*, 1, 39–49.
56. Battista, J.R. (1997). Against all odds: the survival strategies of *Deinococcus radiodurans. Annu. Rev. Microbiol.*, 51, 203–224.
57. Suzuki, K., Collins, M.D., Iijiama, E., and Komagata, K. (1988) Chemotaxonomic characterization of a radiotolerant bacterium, *Arthrobacter radiotolerans*: description of *Rubrobacter radiotolerans* gen. nov., comb. nov. *FEMS Microbiol. Lett.*, 52, 33–40.
58. Newton, G.L., and Fahey, R.C. (1999) Glutathione in prokaryotes. pp. 69–77 in J. Vina (Ed.), *Glutathione, Metabolism and Physiological Function.* Boca Raton: CRC Press.
59. Fahey, R.C. (2001) Novel thiols of prokaryotes. *Ann. Rev. Microbiol.*, 55, 333–356.
60. Newton, G.L., Arnold, K., Price, M.S., Sherrill, C., Delcardayre, S.B., Aharonowitz, Y., Cohen, G., Davies, J., Fahey, R.C., and Davis, C. (1996) Distribution of thiols in microorganisms: mycothiol is a major thiol in most actinomycetes. *J. Bact.*, 178, 1990–1995.

61. Rawat, M., and Av-Gay, Y. (2007) Mycothiol-dependent proteins in actinomycetes. *FEMS Microbiol. Rev.*, 31, 278–292.
62. Alam, M.S., Garg, S.K., and Agrawal, P. (2007) Molecular function of WhiB4/Rv3681c of *Mycobacterium tuberculosis* H37Rv: a [4Fe-4S] cluster co-ordinating protein disulphide reductase. *Mol. Microbiol.*, 63, 1414–1431.
63. Garg, S.K., Alam, M.S., Soni, V., Kishan, K.V.R., and Agrawal, P. (2007) Characterisation of Mycobacterium tuberculosis WhiB1/Rv3219 as a protein disulfide reductase. *Prot. Exp. Purif.*, 52, 422–432.
64. Groll, M., Bochtler, M., Brandstetter, H., Clausen, T., and Huber, R. (2005) Molecular machines for protein degradation. *ChemBiochem.*, 6, 222–256.
65. Hu, G., Lin, G., Wang, M., Dick, L., Xu, R.-M., Nathan, C., and Li, H. (2006) Structure of the Mycobacterium tuberculosis proteasome and mechanism of inhibition by a peptidyl boronate. *Mol. Microbiol.*, 59, 1417–1428.
66. DeMot, R., Nagy, I., and Baumeiste, W. (1998) A self-compartmentalizing protease in Rhodococcus: the 20S proteasome. *Antonie van Leeuwenhoek*, 74, 83–87.
67. Williams, S.T., Goodfellow, M., Alderson, G., Wellington, E.M.H., Sneath, P.H.A., and Sackin, M.J. (1983) Numerical classification of *Streptomyces* and related genera. *J. Gen. Microbiol.*, 129, 1743–1813.
68. Goodfellow, M., and Haynes, J.A. (1984) Actinomycetes in marine sediments. In L. Ortiz-Ortiz (Ed.), *Biological, Biochemical and Biomedical Aspects of Actinomycetes.* Academic Press Inc., p. 453–472.
69. Ward, A.C., and Bora, N. (2006) Diversity and biogeography of marine actinobacteria . *Curr. Opin. Microbiol.* 9, 279-286
70. Fenical, W., and Jensen, P.R. (2006) Developing a new resource for drug discovery: marine actinomycete bacteria. *Nat. Chem. Biol.* 2, 666-673
71. Maldonado, L.A,, Fenical, W., Jensen, P.R., Kauffman, C.A., Mincer, T.J., Ward, A.C., Bull, A.T., and Goodfellow, M. (2005) *Salinispora arenicola* gen. nov., sp nov and *Salinispora tropica* sp nov., obligate marine actinomycetes belonging to the family Micromonosporaceae. *Int. J. Syst. Evol. Microbiol.* 55, 1759-1766
72. Pathom-aree, W., Stach, J.E.M., Ward, A.C., Horikoshi, K., Bull, A.T., Goodfellow, M. (2006) Diversity of actinomycetes isolated from Challenger Deep sediment (10,898 m) from the Mariana Trench. *Extremophiles*, 10, 181-189
73. Allgaier, M., and Grossart, H.P. (2006) Diversity and seasonal dynamics of Actinobacteria populations in four lakes in northeastern Germany. *Appl. Env. Microbiol.*, 72, 3489–3497.
74. Warnecke, F., Amann, R., and Pernthaler, J. (2004) Actinobacterial 16S rRNA genes from freshwater habitats cluster in four distinct lineages. *Env. Microbiol.*, 6, 242–253.
75. Leyden, J.J. (2001) The evolving role of *Propionibacterium acnes* in acne. *Sem. Cutan. Med. Surg.*, 20, 139–143.
76. Eady, E.A., and Ingham, E. (1994) *Propionibacterium acne* — friend or foe? *Rev. Med. Microbiol.*, 5, 163–175.
77. La Scola, B., Fenollar, F., Fournier, P.E., Altwegg, M., Mallet, M.N., and Raoult, D. (2001) Description of Tropheryma whipplei gen. nov., sp. nov., the Whipple's disease bacillus. *Int. J. Syst. Evol. Microbiol.*, 51, 1471-1479
78. Bentley, S.D., Maiwald, M., Murphy, L.D., Pallen, M.J., Yeats, C.A., Dover, L.G., Norbertczak, H.T., Besra, G.S., Quail, M.A., Harris, D.E., von Herbay, A., Goble, A., Rutter, S., Squares, R., Squares, S., Barrell, B.G., Parkhill, J., and Relman, D.A. (2003) Sequencing and analysis of the genome of the Whipple's disease bacterium *Tropheryma whipplei. Lancet* 361, 637-644.
79. Raoult, D., Ogata, H., Audic, S., Robert, C., Krsten, S., Drancourt, M., and Claverie, M. (2003) Tropheryma whipplei Twist : a human pathogenic Actinobacteria with a reduced genome. *Genome Res.* 13, 1800-1809
80. Fenolla, F., Puechal, X., and Raoult, D. (2007) Medical Progress: Whipple's disease. *New Eng. J. Med.*, 356, 55–66.
81. Fahal, A.H. (2004) Mycetoma: a thorn in the flesh. Trans. *Roy. Soc. Trop. Med. Hyg.*, 98, 3–11.
82. Munoz, J., Mirelis, B., Aragon, L.M., Gutierrez, N., Sanchez, F. Espanol, M., Esparcia, O., Gurgui, M., Domingo, P., and Coll, P. (2007) Clinical and microbiological features of nocardiosis 1997–2003 J. *Med. Microbiol.*, 56, 545–550.
83. Rollis, D.C., and Weaver, R.E. (1981) *Gram Positive Organisms: A Guide to Identification.* Special Bacteriology Section, Centers for Disease Control, Atlanta, GA.

84. Funke, G., von Graevenitz, G.R., Clarridge, J.E. III, and Bernard, K.A. (1997) Clinical microbiology of coryneform bacteria. *Clin. Microbiol. Rev.*, 10, 125–159.
85. Bojar, R.A., and Holland, K.T. (2002) Review: the human cutaneous microflora and factors controlling colonisation. *World J. Microbiol. Biotech.*, 18, 889–903.
86. Vaughan, E.E., Schut, F., Heilig, H.G., Zoetendal, E.G., de Vos, W.M., Akkermans, A.D. (2000) A molecular view of the intestinal ecosystem. *Curr. Issues Intest. Microbiol.*, 1, 1–12.
87. Brennan, N.M., Ward, A.C., Beresford, T.P., Fox, P.F., and Cogan, T.M. (2002) Biodiversity of the bacterial flora on the surface of a smear cheese. *Appl. Environ. Microbiol.*, 68, 820–830.
88. Yip, M.J., Porter, J.L., Fyfe, J.A.M., Lavender, C.J., Portaels, F., Rhodes, M., Kator, H., Colorni, A., Jenkin, G.A., and Stinear, T. (2007) Evolution of *Mycobacterium ulcerans* and other mycolactone-producing mycobacteria from a common *Mycobacterium marinum* progenitor. *J. Bact.*, 189, 2021–2029.
89. Shojaei, H., Magee, J.G., Freeman, R., Yates, M., Horadagoda, N.U., and Goodfellow, M. (2000) *Mycobacterium elephantis* sp nov., a rapidly growing non-chromogenic Mycobacterium isolated from an elephant. *Int. J. Syst. Evol. Microbiol.*, 50, 1817–1819.
90. Muscatello, G., Gilkerson, J.R., and Browning, G.F. (2007) Comparison of two selective media for the recovery, isolation, enumeration and differentiation of *Rhodococcus equi*. *Vet. Microbiol.*, 119, 324–329.
91. Donahue, J.M., Williams, N.M., Sells, S.F., and Labeda, D.P. (2002) *Crossiella equi* sp nov., isolated from equine placentas. *Int. J. Syst. Evol. Microbiol.*, 52, 2169–2173.
92. Labeda, D.P., Donahue, J.M., Williams, N.M., Sells, S.F., and Henton, M.M. (2003) Amycolatopsis kentuckyensis sp. nov., *Amycolatopsis lexingtonensis* sp. nov. and *Amycolatopsis pretoriensis* sp. nov., isolated from equine placentas. *Int. J. Syst. Evol. Microbiol.*, 53, 1601–1605.
93. Loria, R., Kers, J., and Joshi, M. (2006) Evolution of plant pathogenicity in *Streptomyces*. *Annu. Rev. Phytopath.*, 44, 469–487.
94. Guimaraes, P.M., Smith, J.J., Palmano, S., and Saddler, G.S. (2003) Characterisation of *Curtobacterium flaccumfaciens* pathovars by AFLP, rep-PCR and pulsed-field gel electrophoresis. *Eur. J. Plant Path.*, 109, 817–825.
95. Brumbley, S.M., Petrasovits, L.A., Hermann, S.R. Young, A.J., and Croft, B.J. (2006) Recent advances in the molecular biology of *Leifsonia xyli* subsp xyli, causal organism of ratoon stunting disease. *Austral. Plant Path.*, 35, 681–689.
96. Normand, P., Lapierre, P., Tisa, L.S., Gogarten, J.P., Alloisio, N., Bagnarol, E., Bassi, C.A., Berry, A.M., Bickhart, D.M., Choisne, N., Couloux, A., Cournoyer, B., Cruveiller, S., Daubin, V., Demange, N., Francino, M.P., Goltsman, E., Huang, Y., Kopp, O.R., Labarre, L., Lapidus, A., Lavire, C., Marechal, J., Martinez, M., Mastronunzio, J.E., Mullin, B.C., Niemann, J., Pujic, P., Rawnsley, T., Rouy, Z., Schenowitz, C., Sellstedt, A., Tavares, F., Tomkins, J.P., Vallenet, D.,Valverde, C., Wall, L.G., Wang, Y., Medigue, C., and Benson, D.R. (2007) Genome characteristics of facultatively symbiotic Frankia sp. strains reflect host range and host plant biogeography. *Genome Res.*, 17, 7–15.
97. Trujillo, M.E., Kroppenstedt, R.M., Schumann, P., Carro, L., and Martinez-Molina, E. (2006) *Micromonospora coriariae* sp. nov., isolated from root nodules of *Coriaria myrtifolia*. *Int. J. Syst. Evol. Microbiol.*, 56, 2381–2385.
98. Castillo, U.F., Browne, L., Strobel, G., Hess, W.M., Ezra, S., Pacheco, G., Ezra, D. (2007) Biologically active endophytic streptomycetes from *Nothofagus* spp. and other plants in Patagonia. *Microbial Ecol.*, 53, 12–19.
99. Janssen, P.H. (2006) Identifying the dominant soil bacterial taxa in libraries of 16S rRNA and 16S rRNA genes. *Appl. Environ. Microbiol.*, 72, 1719–1728.
100. Clark, D.A., and Norris, P.R. (1996) *Acidimicrobium ferrooxidans* gen. nov., sp. nov. mixed culture ferrous iron oxidation with *Sulfobacillus* species. *Microbiology,* 142, 785–790.
101. Johnson, D.B., Bacelar-Nicolau, P., Okiba, N., Yahya, A., and Haberg, K.B. (2001) Role of pure and mixed cultures of Gram-positive eubacteria in mineral leaching pp. 461–470. In (Eds. V.S.T. Cimimelli and O. Garcia) *Biohydrometallurgy: Fundamentals, Technology and Sustainable Development*. Process Metallurgy IIA. Elsevier, Amsterdam.
102. Blackall, L.L., Seviour, E.M., Cunnigham, M.A., Seviour, R.J., and Hugenholtz, P. (1995) "*Microthrix parvicella*" is a deep branching member of the actinomycete subphylum. *Syst. Appl. Microbiol.*,17, 513–518.

103. Levantesi, C., Rossetti, S., Thelan, K., Kragelund, C., Krooneman, J., Eikelboom, D., Nielsen, P.H., and Tandoi, V. (2006) Phylogeny, physiology and distribution of "Candidatus *Microthrix calida*", a new *Microthrix* species isolated from industrial activated sludge waste water treatment plants. *Environ Microbiol.*, 8, 1552–1563.
104. Zarilla, K.A., and Perry, J.J. (1984) *Thermoleophilum album*, gen. nov. and sp. nov., a bacterium obligate for thermophily and n-alkane substrates. *Arch. Microbiol.*, 137, 286–290.
105. Takahashi, Y., Matsumoto, A., Morisaki, K., and Ömura, S. (2006) *Patulibacter minatonensis* gen. nov., sp. nov., a novel actinobacterium isolated using an agar medium supplemented with superoxide dismutase and proposal of *Patulibacteraceae fam.* nov. *Int. J. Syst. Evol. Microbiol.*, 56, 401–406.
106. Singleton, D.R., Furlong, A., Peacock, M.A., White, D.C., Coleman, D.C., and Whitman, W.B. (2003) *Solirubrobacter pauli* gen. nov., sp. nov. a mesophilic bacterium within the *Rubrobacteridae* related to common soil clones. *Int. J. Syst. Evol. Microbiol.*, 53, 485–490.
107. Monciardini, P., Covaletti, L., Shumann, P., Rohde, M., and Donadio, S. (2003) *Conexibacter woesi* gen. nov., sp. nov. a novel representative of a deep evolutionary line of descent within the class Actinobacteria. *Int. J. Syst. Evol. Microbiol.*, 53, 569–576.
108. Bull, A.T., Ward, A.C., and Goodfellow, M. (2000) Search and discovery strategies for biotechnology: the paradigm shift. *Microbiol. Mol. Biol. Rev.*, 64, 573–606.
109. Lam, K.S. (2006) Discovery of novel metabolites from marine actinomycetes. *Curr. Opin. Microbiol.*, 9, 245–251.
110. Lykidis, A., Mavromatis, K., Ivanova, N., Anderson, I., Land, M., DiBartolo, G., Martinez, M., Lapidus, A., Lucas, S., Copeland, A., Richardson, P., Wilson, D.B., and Kyrpides, N. (2007) Genome sequence and analysis of the soil cellulolytic actinomycete *Thermobifida fusca*. *J. Bact.*, 189, 2477–2486.
111. Bunch, A.W. (1998) Biotransformation of nitriles by rhodococci. *Antonie van Leeuwenhoek*, 74, 89–97.
112. Yamada, H., and Kobayashi, M. (1996) Nitrile hydratase and its application to the industrial production of acrylamide. *Biosci. Biotech. Biochem.*, 60, 1391–1400.
113. Ouwehand, A.C., Salminen, S., and Isolauri, E. (2002) Probiotics: an overview of beneficial effects. *Antonie van Leeuwenhoek,* 82, 279–289.
114. Larkin, M.J., Kulakov, L.A., and Allen, C.C.R. (2005) Biodegradation and Rhodococcus – masters of catabolic versatility. *Curr. Opin. Biotech.*, 16, 282–290.
115. Hugenholtz, P., and Stackebrandt, E. (2004) Reclassification of *Sphaerobacter thermophilus* from the subclass *Sphaerobacteridae* in the phylum Actinobacteria to the class *Thermomicrobia* (emended description) in the phylum *Chloroflexi* (emended description). *Int. J. Syst. Evol. Microbiol.*, 54, 2049–2051.
116. Rappe, M.S., and Giovannoni, S.J. (2003) The uncultured microbial majority. *Annu. Rev. Microbio*l., 57, 369–394.
117. Bond, P.L., Smriga, S.P., and Banfield, J.F. (2000) Phylogeny of microorganisms populating a thick, subaerial, predominantly lithotrophic biofilm at an extreme acid mine drainage site. *Appl. Environ. Microbiol.*, 66, 3842–3849.
118. Rawlings, D.E., and Johnson, D.B. (2007) The microbiology of biomining: development and optimization of mineral-oxidising microbial consortia. *Microbiology–UK*, 153, 315–324.
119. Eikelboom, D.H. (1975) Filamentous organisms observed in activated sludge. *Water Res.*, 9, 365–385.
120. Rossetti, S., Tomei, M.C., Nielsen, P.H., and Tandoi, V. (2005) "*Microthrix parvicella*," a filamentous bacterium causing bulking and foaming in activated sludge systems: a review of current knowledge. *FEMS Microbiol. Rev.*, 29, 49–64.
121. van Veen, W.L. (1973) Bacteriology of activated sludge, in particular, the filamentous bacteria. *Antonie van Leeuwenhoek*, 39, 189–205.
122. Joseph, S.J., Hugenholtz, P., Sangwan, P., Osborne, C.A., and Janssen, P. (2003) Laboratory cultivation of widespread and previously uncultured soil bacteria. *Appl. Environ. Microbiol.*, 69, 7210–7215.
123. Glöckner, F.O., Zaichikov, E., Belkova, N., Denissova, L., Pernthaler, J., Pernthaler, A., and Amann, R. (2000) Comparative 16S rRNA analysis of *lake bacterioplankton* reveals globally distributed phylogenetic clusters including an abundant group of Actinobacteria. *Appl. Environ. Microbiol.*, 66, 5053–5065.
124. Montalvo, N.F., Mohamed, N.M., Enticknap, J.J., and Hill, R.T. (2005) Novel actinobacteria from marine sponges. *Antonie van Leeuwenhoek*, 87, 29–36.

125. Hentschel, U., Hopke, J., Horn, M., Friedrich, A.B., Wagner, M., Hacker, J., and Moore, B.S. (2002) Molecular evidence for a uniform microbial community in sponges from different oceans. *Appl. Environ. Microbiol.*, 68, 4411–4444.
126. Frias-Lopez, J., Zerkle, A.L., Bonheyo, G.T., and Fouke, B.W. (2002) Partitioning of bacterial communities between seawater and healthy, black band diseased, and dead coral surfaces. *Appl. Environ. Microbiol.*, 68, 2214–2228.
127. Yoshinaka, T., Yano, K., and Yamaguchi, H. (1973) Isolation of a highly radioresistant bacterium *Arthrobacter radotolerans* nov. sp. *Agric. Biol. Chem.*, 37, 2269–2275.
128. Carreto, L., Moore, E., Nobre, M.F., Wait, R., Riley, P.W., Sharp, R.J., and Costa, M.S. (1996). *Rubrobacter xylanophilus* sp. nov., a new thermophilic species isolated from a thermally polluted effluent. *Int. J. Syst. Bacteriol.*, 46, 460–465.
129. Chen, M.-Y., Wu, S.-H., Lin, G.-H., Lu, C.-P., Lin, Y.-T., Chang, W.-C., and Tsay, S.-S. (2004) *Rubrobacter taiwanensis* sp. nov., a novel thermophilic, radiation-resistant species isolated from hot springs. *Int. J. Syst. Evol. Microbiol.,* 54, 1849–1855.
130. Holmes, A.J., Bowyer, J., Holley, M.P., O'Donoghue, M., Montgomery, M., and Gillings, M.R. (2000) Diverse, yet-to-be-cultured members of the *Rubrobacter* subdivision of the Actinobacteria are widespread in Australian arid soils. *FEMS Microbiol. Ecol.*, 33, 111–120.
131. Bond, P.L., Hugenholtz, P., Keller, J., and Blackall, L.L. (1995) Bacterial community structure of phosphate-removing and non-phosphate-removing activated sludges from sequencing batch reactors. *Appl. Environ. Microbiol.* 61, 1910-1916
132. Ueda, T., Suga, Y., and Matsuguchi, T. (1995) Molecular phylogenetic analysis of a soil microbial community in a soybean field. *Eur. J. Soil. Sci.* 46, 415–421
133. Rheims, H., Felske, A., Siefert, S., and Stackebrandt, E. (1996) Molecular biological evidence for the occurrence of uncultured members of the actinomycete line of descent in different environments and geographical locations. *Microbiology-UK* 142, 2863-2870
134. McCaig, A.E., Glover, L.A., and Prosser, J.L. (1999) Molecular analysis of bacterial community structure and diversity in unimproved and improved upland grass pastures. *Appl. Environ. Microbiol.* 65, 1721-1730
135. Leisack, W., and Stackebrandt, E. (1992) Occurrence of novel groups of the domain bacteria as revealed by analysis of genetic material isolated from an Australian terrestrial environment. *J. Bact.* 174, 5072-5078
136. Yakimov, M.M., Lünsdorf, H., and Golyshin, P.N. (2003) *Thermoleophilum album* and *Thermoleophilum minutum* are culturable representatives of group 2 of the *Rubrobacteridae* (Actinobacteria). *Int. J. Syst. Evol. Microbiol.,* 53, 377–380.
137. Garrity, G.M., and Holt, J.G. (2001) The road map to the manual. pp. 119–166. In *Bergey's Manual of Systematic Bacteriology* (Eds. Boone, D.R., Castenholtz, R.W., and Garrity, G.M.) Springer-Verlag, New York.
138. Falk, P.G., Hooper, L.V., Midtvedt, T., and Gordon, J.L. (1998) Creating and maintaining the gastrointestinal ecosystem: what we know and need to know from gnotobiology. *Microbiol. Mol. Biol. Rev.*, 62, 1157–1170.
139. Tannock, G.W. (1997) Probiotic properties of lactic acid bacteria: plenty of scope for fundamental R&D. *Trends Biotechnol.*, 15, 270–274.
140. Guarner, F., and Malagelada, J.R. (2003) Gut flora in health and disease. *Lancet*, 361, 512–519.
141. Servin, A.L. (2004) Antagonistic activities of lactobacilli and bifidobacteria against microbial pathogens. *FEMS Microbiol. Rev.*, 28, 405–440.
142. Kageyama, A., Benno, Y., and Nakase, T. (1999) Phylogenetic and phenotypic evidence for the transfer of *Eubacterium aerofaciens* to the genus *Collinsella* as *Collinsella aerofaciens* gen. nov., comb. nov. *Int. J. Syst. Bact.*, 49, 557–565.
143. Benno, Y., Suzuki, K., Suzuki, K., Narisawa, K., Bruce, W., Mitsuoka, T. (1986) Comparison of the fecal microflora in rural Japanese and urban Canadians. *Microbiol. Immunol.*, 30, 521–532.
144. Haas, F., and Konig, H. (1988) *Coriobacterium glomerans* gen. nov., sp. nov. from the intestinal tract of the red soldier bug. *Int. J. Syst. Bact.*, 38, 382–384.
145. Collins, M.D., and Wallbanks, S. (1992) Comparative sequence analysis of the 16S rRNA of *Lactobacillus minutus, Lactobacillus rimae* and *Streptococcus parvalus*: proposal for the creation of a new genus *Atopobium. FEMS Microbiol. Lett.*, 95, 235–240.
146. Rainey, F.A., Weiss, N., and Stackebrandt, E. (1994) *Coriobacterium* and *Atopobium* are phylogenetic neighbours within the actnomycetes line of descent. *Syst. Appl. Microbiol.*, 17, 202–205.

147. Stackebrandt, E., and Ludwig, W. (1994) The importance of using outgroup reference organisms in phylogenetic studies: the *Atopobium* case. *Syst. Appl. Microbiol.*, 17, 39–43.
148. Wade, W.G., Downes, J., Dymock, D., Hiom, S.J., Weightman, A.J., Dewhirst, F.E., Paster, B.J., Tzellas, N., and Coleman, B. (1999) The family *Coriobacteriaceae*: reclassification of *Eubacterium exiguum* (Poco et al. 1996) and *Peptostreptococcus helioreducens* (Lanigan 1976) as *Slackia helioreducens* gen. nov., comb. nov., and *Eubacterium lentum* (Prevot 1938) as *Eggerthella lenta* gen. nov., comb. nov. *Int. J. Syst. Bact.,* 49, 595–600.
149. Lawson, P.A., Greetham, H.L., Gibson, G.R., Giffard, C., Falsen, E., Collins, M.D. (2005) *Slackia faecicanis* sp. nov., isolated from canine faeces. *Int. J. Syst. Evol. Microbiol.*, 55, 1243–1246.
150. Anderson, R.C., Rasmussen, M.A., Jensen, N.S., and Milton, J.A. (2000) *Denitrobacterium detoxificans* gen. nov., sp. nov., a ruminal bacterium that respires on nitrocompounds. *Int. J. Syst. Evol. Microbiol.,* 50, 633–638.
151. Dewhirst, F.E., Paster, B.J., Tzellas, N., Cleman, B., Downes, J., Spratt, D.A., and Wade, W.G. (2001) Characterisation of novel human oral isolates and cloned 16S rDNA sequences that fall in the family Coriobacteriaceae: description of *Olsenella gen.* nov., reclassification of *Lactobacillus uli as Olsenella uli* comb. nov. and description of *Olsenella profuse* sp. nov. *Int. J. Syst. Evol. Microbiol.*, 51, 1797–1804.
152. Nakazawa, F., Poco, S.E., Ikeda, T., Sato, M., Kalfas, S., Sundqvist, G., and Hoshino, E. (1999) *Cryptobacterium curtum* gen. nov., sp. nov., a new genus of Gram-positive anaerobic rod isolated from human oral cavities. *Int. J. Syst. Bact.*, 49, 1193–1200.
153. Harmsen, H.J.M., Wildeboer-Veloo, A.C.M., Grupstra, J., Knol, J., Degener, J.E., and Welling, G.W. (2000) Development of 16S rRNA-based probes for the *Coriobacterium* group and the *Atopobium* cluster and their application for enumeration of *Coriobacteriaceae* in human faeces from volunteers of different age groups. *Appl. Environ. Microbiol.* 66, 4523–4527.
154. Suau, A., Bonnet, R., Sutren, M., Gordon, J.J., Gibson, G.R., Collins, M.D., and Dore, J. (1999) Direct analysis of gene encoding 16S rRNA from complex communities reveals many novel molecular species within the human gut. *Appl. Environ. Microbiol.*, 65, 4799–4807.
155. Lau, S.K.P., Woo, P.C.Y., Woo, G.K.S., Fung, A.M.Y., Wong, M.K.M., Chan, K., Tam, D.M.W., and Yuen, K. (2004) *Eggerthella hongkongensis* sp. nov. and *Eggerthella sinensis* sp. nov., two novel Eggerthella species, account for half of the cases of *Eggerthella bacteremia. Diag. Microbiol. Infect. Dis.*, 49, 253–263.
156. Orla-Jensen, S. (1924) La classification des bactéries lactiques. *Lait,* 4, 468–474.
157. Tissier, H. (1906) Traitement des infections intestinales par la méthode de la flore bactérienne de l'intestin. *Crit. Rev. Soc. Biol.*, 60, 359–361.
158. Metchnikoff, E. (1907) *The Prolongation of Life.* G.P. Putman's Sons, New York.
159. Leahy, S.C., Higgins, D.G., Fitzgerald, G.F., and van Sinderen, D. (2005) Getting better with bifidobacteria. *J. Appl. Microbiol.*, 98, 1303–1315.
160. Liu, M., van Enckevort, F.H., and Siezen, R.J. (2005) Genome update: lacic acid bacteria genome sequencing is booming. *Microbiolgy–UK*, 151, 3811–3814.
161. Schell, M.A., Karmirantzou, M., Snel, B.,Vilanova, D., Berger, B., Pessi, G., Zwahlen, M.C., Desiere, F., Bork, P., Delley, M., Pridmore, R.D., and Arigoni, F. (2002) The genome sequence of *Bifidobacterium longum* reflects its adaptation to the human gasrointestinal tract. *Proc. Natl. Acad. Sci. U.S.A.* 99, 14422–14427.
162. Ventura, M., Canchaya, C., Fitzgerald, G.F., Gupta, R.S., and van Sinderen, D. (2007) Genomics as a means to understand bacterial phylogeny and ecological adaptation: the case of bifidobacteria. *Antonie van Leeuwenhoek*, 91, 351–372.
163. Ventura, M., van Sinderen, D., Fitzgerald, G.F., and Zink, R. (2004) Insights into the taxonomy, genetics and physiology of bifidobacteria. *Antonie van Leeuwenhoek,* 86, 205–223.
164. Klein, G., Pack, A., Bonaparte, C., and Reuter, G. (1998) Taxonomy and physiology of probiotic lactic acid bacteria. *Int. J. Food Microbiol.*, 41, 103–125.
165. Ventura, M., Canchaya, C., del Casale, A., Dellaglio, F., Neviani, E., Fitzgerald, G.F., and van Sinderen, D. (2006) Analysis of bifidobacterial evolution using a multilocus approach. *Int. J. Syst. Evol. Microbiol.*, 56, 2783–2792.
166. Hooper, L.V., Midtvedt, T., and Gordon, J.I. (2002) How host-microbial interactions shape the nutrient environment of the mammalian intestine. *Annu. Rev. Nutr.,* 22, 283–307.

167. Chang, D.E., Smalley, D.J., Tucker, D.L., Leatham, M.P., Norris, W.E., Stevenson, S.J., Anderson, A.B., Grissom, J.E., Laux, D.C., Cohen, P.S., Conway, T. (2004) Carbon nutrition of *Escherichia coli* in the mouse intestine. *Proc. Natl. Acad. Sci. U.S.A.*, 101, 7427–7432.
168. Klijn, A., Mercenier, A., and Arigoni, F. (2005) Lessons from the genomes of bifidobacteria. *FEMS Microbiol. Rev.*, 29, 491–509.
169. Favier, C.F., Vaughan, E.E., de Vos, W.M., and Akkermans, A.D. (2002) Molecular monitoring of succession of bacterial communities in human neonates. *Appl. Environ. Microbiol.*, 68, 219–226.
170. Kunz, C., Rudloff, S., Baier, W., Klein, N., Strobel, S. (2000) Oligosaccharides in human milk: structural, functional and metabolic aspects. *Annu. Rev. Nutr.*, 20, 699–722.
171. Hinz, S.W., Pastink, M.I., van den Broek, L.A., Vincken, J.P., and Voragen, A.G. (2005) *Bifidobacterium longum endogalactanase* liberates galactotriose from type I galactans. *Appl. Environ. Microbiol.*, 71, 5501–5510.
172. Liepke, C., Adermann, K., Raida, M., Magert, H.J., Forssmann, W.G., Zucht H.D. (2002) Human milk provides peptides highly stimulating the growth of bifidobacteria. *Eur. J. Biochem.*, 269, 712–718.
173. Ryan, S.M., Fitzgerald, G.F., and van Sinderen, D. (2006) Screening and identification of starch, amylopectin and pullulan-degrading activities in bifidobacterial strains. *Appl. Environ. Microbiol.*, 72, 5289–5296.
174. Satakori, R.M., Vaughan, E.E., Smidt, H., Saarela, M., Matto, J., and de Vos, W.M. (2003) Molecular approaches for the detection and identification of bifidobacteria and lactobacilli in the human gastrointestinal tract. *Syst. Appl. Microbiol.*, 26, 572–584.
175. Rosan, B., and Lamont, R.J. (2000) Dental plaque formation. *Microbes and Infect.*, 2, 1599–1607.
176. Hall, V., Talbot, P.R., Stubbs, S.L., and Duerden, B.I. (2001) Identification of clinical isolates of Actinomyces species by amplified 16S ribosomal DNA restriction analysis (ARDRA). *J. Clin. Microbiol.*, 39, 3555–3562.
177. Thompson, L., and Lovestedt, S.A. (1951) An actinomyces-like organism obtained from the human mouth. *Proc. Staff Meetings Mayo Clinic*, 26, 169–175.
178. Yeung, M.K. (1999) Molecular and genetic analysis of *Actinomyces* spp. *Crit. Rev. Oral Biol.*, 10, 120–138.
179. Schaal, K.P. (1992) The genera *Actinomyces*, *Arcanobacterium* and *Rothia*. pp. 850–905. In *The Prokaryotes,* Vol. 1. Eds. Balows, A., Trüper, H.G., Dworkin, M., Harder, W., and Schleifer, K.H. New York: Springer.
180. An, D., Cai, S., and Dong, X. (2006) *Actinomyces ruminicola* sp. nov., isolated from cattle rumen. *Int. J. Syst. Evol. Microbiol.*, 56, 2043–2048.
181. Tsukamura, M. (1971) Proposal of a new genus, *Gordona*, for slightly acid fast organisms occurring in sputa of patients with pulmonary disease and in soil. *J. Gen. Microbiol.*, 68, 15–26.
182. Hayes, T., Haston, K., Tsui, M., Hoang, A., Haeffele, C., and Vonk, A. (2002) Herbicides: feminization of male frogs in the wild. *Nature*, 419, 895–896.
183. Mulbry, W.W. (1994) Purification and characterisation of an inducible s-triazine hydrolase from *Rhodococcus corallines* NRRL B-15444R. *Appl. Environ. Microbiol.*, 60, 613–618.
184. Kim, S.B., Brown, R., Oldfield, C., Gilbert, S.C., and Goodfellow, M. (1999) *Gordonia desulfuricans* sp. nov., a benzothiophene-desulphurizing actinomycete *Int. J. Syst. Bact.*, 49, 1845–1851.
185. Iida, S., Taniguchi, H., Kageyama, A., Yazawa, K., Chibana, H., Murata, S., Nomura, F., Kroppenstedt, R.M., and Mikami, Y. (2005) *Gordonia otitidis* sp. nov., isolated from a patient with external otitis. *Int. J. Syst. Evol. Microbiol.*, 55, 1871–1876.
186. Kageyama, A., Iida, S., Yazawa, K., Kudo, T., Suzuki, S.I., Koga, T., Saito, H., Inagawa, H., Wada, A., Kroppenstedt, R.M., and Mikami, Y. (2006) *Gordonia araii* sp. nov. and *Gordonia effusa* sp. nov., isolated from patients in Japan. *Int. J. Syst. Evol. Microbiol.*, 56, 1817–1821.
187. Arenskötter, M., Broker, D., Steinbuchel, A. (2004) Biology of the metabolically diverse genus *Gordonia. Appl. Environ. Microbiol.*, 70, 3195–3204.
188. Nocard, E. (1888) Note sur la maladie des boeufs de la Gouadeloupe connue sous le nom farcin. *Ann. Inst. Pasteur,* 2, 293–302.
189. Brown-Elliott, B.A., Brown, J.M., Conville, P.S., and Wallace, R.J. (2006) Clinical and laboratory features of the *Nocardia* spp. based on current molecular taxonomy. *Clin. Microbiol. Rev.,* 19, 259-282
190. Fergie, J.E., and Purcell, K. (2001) Nocardiosis in south Texas children. *Pediatr. Infect. Dis. J.*, 20, 711–714.
191. Salinas-Carmona, M.C. (2000) *Nocardia brasiliensis*: from microbe to human and experimental infections. *Microbes Infect.*, 2, 1373–1381.

192. Friedman, C.S., Beaman, B.L., Chun, J., Goodfellow, M., Gee, A., and Hedrick, R.P. (1998) *Nocardia crassostreae* sp. nov., the causal agent of nocardiosis in Pacific oysters. *Int. J. Syst. Bacteriol.*, 48, 237–246.
193. Maldonado, L., Hookey, J.V., Ward, A.C., and Goodfellow, M. (2000) The *Nocardia salmonicida* clade, including descriptions of *Nocardia cummidelens* sp. nov., *Nocardia fluminea* sp. nov. and *Nocardia soli* sp. nov. *Antonie van Leeuwenhoek,* 78, 367–377.
194. Steingrube, V.A., Brown, B.A., Gibson, J.L., Wilson, R.W., Brown, J., Blacklock, Z., Jost, K., Locke, S., Ulrich, R.F., and Wallace, R.J. (1995) DNA amplification and restriction endonuclease analysis for differentiation of 12 species and taxa of *Nocardia* including recognition of four new taxa within the *Nocardia asteroides* complex. *J. Clin. Microbiol.,* 33, 3096–3101.
195. Conville, P.S., Brown, J.M. Steigerwalt, A.G., Lee, J.W., Anderson, V.L., Fishbain, J.T., Holland, S.M., and Witebsky, F.G. (2004) Nocardia kruczakia sp. nov., a pathogen in immunocompromised patients and a member of the "N. *nova complex.*" *J. Clin. Microbiol.*, 42, 5130–5145.
196. Overbeck A (1891) Zur Kenntnis der Fettfarbstoff – Production bei Spaltpilzen. *Nov. Acta Leopold.* 55, 399–416.
197. Bergey, D.H. (1939) *Manual of Determinative Bacteriology*, 5th ed. Baltimore, MD: Williams & Wilkins.
198. Bergey, D.H. (1925) *Manual of Determinative Bacteriology*, 2nd ed. The Williams & Wilkins Co.
199. Bergey, D.H. (1930) *Manual of Determinative Bacteriology,* 3rd ed. The Williams & Wilkins Co.
200. Bergey, D.H. (1934) *Manual of Determinative Bacteriology*, 4th ed. The Williams & Wilkins Co.
201. Bergey, D.H. (1939) *Manual of Determinative Bacteriology*, 5th ed. The Williams & Wilkins Co.
202. Breed, R.S., Murray, E.G.D., and Hitchens, A.P., Eds. (1948) Bergey's *Manual of Determinative Bacteriology.* Baltimore, MD: Williams & Wilkins.
203. Gordon, R.E., and Mihm, J.M. (1957) A comparative study of some strains received as nocardiae. *J. Bacteriol.,* 73, 15–27.
204. Gordon, R.E., and Mihm, J.M. (1959) A comparison of four species of mycobacteria. *J. Gen. Microbiol.*, 21, 736–748.
205. Gordon, R.E., and Mihm, J.M. (1961) The specific identity of *Jensenia canicruria. Can. J. Microbiol.*, 7, 108–110.
206. Lechevalier, M.P., and Lechevalier, H.A. (1970) Chemical composition as a criterion in the classification of aerobic actinomycetes. *Int. J. Syst. Bacteriol.*, 20, 435–443.
207. Cummins, C.S., and Harris, H. (1958) Studies on the cell wall composition and taxonomy of *Actinomycetales* and related groups. *J. Gen. Microbiol.*, 18, 173–189.
208. Goodfellow, M., Alderson, G., and Chun, J.S. (1998) Rhodococcal systematics: problems and developments. *Antonie van Leeuwenhoek* 74, 3-20
209. Tsukamura, M. (1974) A further numerical taxonomic study of the rhodochrous group. *Jap. J. Microbiol.*, 18, 37–44.
210. Goodfellow, M., and Alderson, G. (1977) The actinomycete-genus *Rhodococcus*: a home for the 'rhodochrous' complex. *J. Gen. Microbiol.*, 100, 99–122.
211. Finnerty, W.M. (1992) The biology and genetics of the genus R*hodococcus. Annu. Rev. Microbiol.*, 46, 193–218.
212. Bell, K.S., Philip, J.C., Aw, D.W.J., and Christofi, N., (1998) The genus *Rhodococcus. J. Appl. Microbiol.*, 85, 195–210.
213. Gürtler, V., Mayall, B.C., and Seviour, R. (2004) Can whole genome analysis refine the taxonomy of the genus *Rhodococcus*? *FEMS Microbiol. Rev.*, 28, 377–403.
214. McMinn, E.J., Alderson, G., and Ward, A.C. (2000) Genomic and phenomic differentiation of *Rhodococcus equi* and related strains. *Antonie van Leeuwenhoek*, 78, 331–340.
215. Steinhaus, E.A. (1941). A study of bacteria associated with thirty species of insects. *J. Bacteriol.*, 42, 757–790.
216. Tsukamura, M., and Mizuno, S. (1971) A new species *Gordona aurantiacus* occurring in sputa of patients with pulmonary disease. *Kekkaku*, 46, 93–98 (in Japanese).
217. Collins, M.D., Smida, J., Dorsch, M., and Stackebrandt, E. (1988) *Tsukamurella* gen. nov., harbouring *Corynebacterium paurometabolum* and *Rhodococcus aurantiacus. Int. J. Syst. Bacteriol.*, 38, 385–391.
218. Kattar, M.M., Cookson, B.T., Carlson, L.C., Stiglich, S.K., Schwartz, M.A., Nguyen, T.T., Daza, R., Wallis, C.K., Yarfutz, S.L., and Coyle, M.B. (2001) *Tsukamurella strandjordae* sp. nov., a proposed new species causing sepsis. *J. Clin. Microbiol.,* 39, 1467–1476.

219. Nam, S.W., WonYong, K., Chun, J., and Goodfellow, M. (2004) *Tsukamurella pseudospumae* sp. nov., a novel actinomycete isolated from activated sludge foam. *Int. J. Syst. Evol. Microbiol.*, 54, 1209–1212.
220. Normand, P., Orso, S., Cournoyer, B., Jeannin, P., Chapelon, C., Dawson, J., Evtushenko, L., and Misra, A.K. (1996) Molecular phylogeny of the genus *Frankia* and related genera and emendation of the family *Frankiaceae*. *Int. J. Syst. Bact.*, 46, 1–9.
221. Pommer, E.H. (1959) Über die isolierung des endophytes aus den wurzel-knöllchen *Alnus glutinosa* Gaertn. und über erfolgreiche re-infektionsversuche. *Ber. Dtsch. Bot. Ges.*, 72, 136–150.
222. Becking, J.H. (1970) *Frankiaceae* fam. nov. (*Actinomycetales*) with one new combination and six new species of the genus *Frankia Brunchorst* 1886. *Int. J. Syst. Bact.*, 20, 201–220.
223. Callaham, D.D., del Tridici, P., and Torrey, J.G. (1978) Isolation and cultivation in vitro of the actinomycete causing root nodulation in *Comptonia*. *Science*, 199, 899–902
224. Benson, D.R., and Silvester, W.R. (1993) Biology of *Frankia* strains, actinomycete symbionts of actinorhizal plants. *Microbiol. Rev.*, 57, 293–319.
225. Normand. P., Queiroux, C., Tisa, L.S., Benson, D.R., Rouy, Z., Cruveiller, S., and Medigue, C. (2007) Exploring the genomes of Frankia. *Physiologia Plantarum*, 130, 331–343.
226. Cohn, F. (1872) Untersuchungen über bakterien. *Beitr. Biol. Pflanz,* 1, 127–244.
227. Baird-Parker, A.C. (1965) The classification of staphylococci and micrococci from world-wide sources. *J. Gen. Microbiol.*, 38, 363–387.
228. Rosypal, S., Rosylpalová, A., and Horejs, J. (1966) The classification of micrococci and staphylococci based on their DNA base composition and Adansonian analysis. *J. Gen. Microbiol.,* 44, 281–292.
229. Stackebrandt, E., Lewis, R.J., and Woese, C.R. (1980) The phylogenetic structure of the coryneform group of bacteria. Zentralbl. Bakteriol. Parasiterkd. *Infektionskr. Hyg. Abt. 1 Orig. Reihe* C, 2, 137–149.
230. Stackebrandt, E., and Woese, C.R. (1979) A phylogenetic dissection of the family *Micrococcaceae*. *Curr. Microbiol.*, 2, 317–322.
231. Siefert, J.L., and Fox, G.E. (1998) Phylogenetic mapping of bacterial morphology. *Microbiology,* 144, 2803–2808.
232. Stackebrandt, E., Koch, C., Gvozdiak, O., and Schumann P. (1995) Taxonomic dissection of the genus *Micrococcus*: *Kocuria* gen. nov., *Nesterenkonia* gen. nov., *Kytococcus* gen. nov., *Dermacoccus* gen. nov., and *Micrococcus* Cohn 1872 gen. emend. *Int. J. Syst. Bacteriol.*, 45, 682–692.
233. Tang, J.S., and Gillevet, P.M. (2003) Reclassification of ATCC 9341 from *Micrococcus luteus* to *Kocuria rhizophila*. *Int. J. Syst. Evol. Microbiol.,* 53, 995–997.
234. Wieser, M., Denner, E.B.M., Kämpfer, P., Schumann, P., Tindall, B., Steiner, U., Vybiral, D, Lubitz, W., Maszenan, A.M., Patel, B.K.C., Seviour, R.J., Radax, C., and Busse, H.-J, (2002) Emended descriptions of the genus *Micrococcus, Micrococcus luteus* (Cohn, 1872) and *Micrococcus lylae* (Kloos et al. 1974). *Int. J. Syst. Evol. Microbiol.*, 52, 629–637.
235. Breed, R.S. (1953) The families developed from *Bacteriaceae* Cohn with a description of the family *Brevibacteriaceae*. *Rias Commun VI Int Cong Microbiol Roma*, 1, 10–15.
236. Cai, J., and Collins, M.D. (1994) Phylogenetic analysis of species of the meso-diaminopimelic acid-containing genera *Brevibacterium* and *Dermabacter*. *Int. J. Syst. Bacteriol.*, 44, 583–585.
237. Collins, M.D., Jones, D., Keddie, R.M., and Sneath, P.H.A. (1980) Reclassification of *Chromobacterium iodinum* (Davis) in a redefined genus B*revibacterium* (Breed) as B*revibacterium iodinum* nom. rev.; comb. nov. *J. Gen. Microbiol.*, 120, 1–10.
238. Onraedt, A., Soetaert, W., and Vandamme, E. (2005) Industrial importance of the genus *Brevibacterium*. *Biotechnol. Lett.*, 27, 527–533.
239. Valdes-Stauber, N., and Scherer, S. (1996) Nucleotide sequence and taxonomical distribution of the bacteriocin gene lin cloned from *Brevibacterium linens* M18. *Appl Environ Microbiol.*, 62, 1283–1286.
240. Knox, A.M., Viljoen, B.C., and Lourens-Hattingh, A. (2005). Inhibition of *Brevibacterium linens* by probiotics from dairy products. *Food Technol. Biotechnol.*, 43, 393–396.
241. Hoppe-Seyler, T.S., Jaeger, B., Bockelmann, W., Geis, A., and Heller, K.J. (2007) Molecular identification and differentiation of *Brevibacterium* species and strains. *Syst. Appl. Microbiol.,* 30, 50–57.
242. Pascual, C., and Collins, M.D. (1999) *Brevibacterium avium* sp. nov., isolated from poultry. *Int. J. Syst. Bacteriol.,* 49, 1527–1530.
243. Gelsomino, R., Vancanneyt, M., Vandekerckhove, T.M., and Swings, J. (2004) Development of a 16S rRNA primer for the detection of *Brevibacterium* spp. *Lett. Appl. Microbiol.*, 38, 532–535.

244. Nakamura, K., Hiraishi, A., Yoshimi, Y., Kawaharasaki, M., Masuda, K., and Kamagata, Y., (1995) *Microlunatus phosphovorus* gen. nov., sp. nov., a new Gram-positive polyphosphate-accumulating bacterium isolated from activated sludge. *Int. J. Syst. Bacteriol.*, 45, 17–22.
245. Maszenan, A.M., Seviour, R.J., Patel, B.K.C., Schumann, P., and Rees, G.N. (1999) *Tessaracoccus bendigoensis* gen. nov., sp. nov., a new Gram-positive coccus occurring in regular packets or tetrads isolated from activated sludge. *Int. J. Syst. Bacteriol.*, 49, 459–468.
246. Yoshimi, Y., Hiraishi, A., and Nakamura, K. (1996) Isolation and characterization of *Microsphaera multipartite* gen. nov., sp. nov., a polysaccharide-accumulating Gram-positive bacterium from activated sludge. *Int. J. Syst. Bacteriol.*, 46, 519–525.
247. Tao, T.S., Yue, Y.Y., Chen, W.X., and Chen, W.F. (2004) Proposal of *Nakamurella* gen. nov. as a substitute for the bacterial genus *Microsphaera* Yoshimi et al. 1996 and N*akamurellaceae* fam. nov. as a substitute for the illegitimate bacterial family *Microsphaeraceae* Rainey et al. 1997. *Int. J. Syst. Evol. Microbiol.*, 54, 999–1000.
248. Maszenan, A.M., Tay, J.H., Schumann, P., Jiang, H.L., and Tay, S.T.L. (2005) *Quadrisphaera granulorum* gen. nov., sp. nov., a Gram-positive polyphosphate-accumulating coccus in tetrads or aggregates isolated from aerobic granules. Int. *J. Syst. Evol. Microbiol.*, 55, 1771–1777.
249. Maszenan, A.M., Seviour, R.J., Patel, B.K.C., Schumann, P., Burghardt, J., Tokiwa, Y., and Stratton, H.M. (2000) Three isolates of novel polyphosphate-accumulating Gram-positive cocci, obtained from activated sludge, belong to a new genus, *Tetrasphaera gen.* nov., and description of two new species, *Tetrasphaera japonica* sp. nov., and *Tetrasphaera australiensis* sp. nov. *Int. J. Syst. Evol. Microbiol.*, 50, 593–603.
250. Suzuki, K., Suzuki, M., Sasaki, J., Park, Y.H., and Komagata, K. (1999) *Leifsonia* gen. nov., a genus for 2,4-diaminobutyric acid-containing actinomycetes to accommodate "C*orynebacterium aquaticum*" Leifson 1962 and *Clavibacter xyli* subsp. cynodontis Davis et al. 1984. *J. Gen. Appl. Microbiol.*, 45, 253–262.
251. Evtushenko, L.I., Dorofeeva, L.V., Subbotin, S.A., Cole, J.R., and Tiedje, J.M., (2000) *Leifsonia poae gen.* nov., sp. nov., isolated from nematode galls on Poa annua, and reclassification of '*Corynebacterium aquaticum'* Leifson 1962 as *Leifsonia aquatica* (ex Leifson 1962) gen. nov., nom. rev., comb. nov. and *Clavibacter xyli* Davis et al. 1984 with two subspecies as *Leifsonia xyli* (Davis et al. 1984) gen. nov., comb. nov. *Int. J. Syst. Evol. Microbiol.,* 50, 371–380.
252. Leifson, E. (1962). The bacterial flora of distilled and stored water. III. New species of the genera *Corynebacterium, Flavobacterium, Spirillum* and *Pseudomonas. Int. Bull. Bacteriol. Nomencl. Taxon,*, 12, 161–170.
253. Davis, M.J., Gillaspie, A.G., Vidaver, A.K., and Harris, R.W. (1984). *Clavibacter*: a new genus containing some phytopathogenic coryneform bacteria, including *Clavibacter xyli* subsp. *xyli* sp. nov., subsp. nov. and *Clavibacter xyli* subsp. *cynodontis* subsp. nov., pathogens that cause ratoon stunting disease of sugarcane and bermudagrass stunting disease. *Int. J. Syst. Bacteriol.*, 34, 107–117.
254. Janssen, P.H., Yates, P.S., Grinton, B.E., Taylor, P.M., and Sait, M. (2002) Improved culturability of soil bacteria and isolation in pure culture of novel members of the divisions *Acidobacteria, Actinobacteria, Proteobacteria*, and *Verrucomicrobia*. *Appl. Environ. Microbiol.*, 68, 2391–2396.
255. Reddy, G.S., Prakash, J.S.S., Srinivas, R., Matsumoto, G.I., and Shivaji, S. (2003). *Leifsonia rubra* sp. nov. and *Leifsonia aurea* sp. nov., psychrophiles from a pond in Antarctica. *Int. J. Syst. Evol. Microbiol.*, 53, 977–984.
256. Conn, H.J. (1928) A type of bacteria abundant in productive soils, but apparently lacking in certain soils of low productivity. *New York State Agricultural Experimental Station Technical Bulletin* 138, 3–26.
257. Conn, H.J., and Dimmick, I. (1947) Soil bacteria similar in morphology to *Mycobacterium* and *Corynebacterium. J. Bacteriol.*, 54, 291–303.
258. Keddie, R.M., Collins, M.D., and Jones, D. (1986) Genus *Arthrobacter* Conn and Dimmick 1947. In *Bergey's Manual of Systematic Bacteriology,* Vol. 2. Sneath, P.H.A., Mair, N.S., Sharpe, N.E., and Holt, J.G., Eds. Williams and Wilkins, Baltimore, MD, pp. 1288–1301.
259. Keddie, R.M., and Cure, G.L. (1978). Cell wall composition of coryneform bacteria. In I.J. Bousfield and A.G. Callely (Eds.), *Coryneform Bacteria*, pp. 47–83. London: Academic Press.
260. Duxbury, T., Gray, T.R.G., and Sharples, G.P. (1977) Structure and chemistry of walls of rods, cocci and cystites of *Arthrobacter-globiformis. J. Gen. Microbiol.*, 103, 91–99.
261. Krasil'nikov, N.A. (1938) *Ray Fungi and Related Organisms – Actinomycetales.* Akademii Nauk SSSR, Moscow, p. 272.

262. Koch, C., Kroppenstedt, R.M., and Stackebrandt, E. (1996). Intrageneric relationships of the actinomycete genus *Micromonospora. Int. J. Syst. Bacteriol.*, 46, 383–387.
263. Tamura, T., Hayakawa, M., and Hatano, K. (1997) A new genus of the order *Actinomycetales, Spirilliplanes* gen. nov., with description of *Spirilliplanes yamanashiensis* sp. nov. *Int. J. Syst. Bacteriol.*, 47, 97–102.
264. Rheims, H., Schumann, P., Rohde, M., and Stackebrandt, E. (1998) *Verrucosispora gifhornensis* gen. nov., sp. nov., a new member of the actinobacterial family *Micromonosporaceae. Int. J. Syst. Bacteriol.*, 48, 1119–1127.
265. Tamura, T., Hayakawa, M., and Hatano, K. (2001) A new genus of the order *Actinomycetales, Virgosporangium* gen. nov., with descriptions of *Virgosporangium ochraceum* sp. nov. and *Virgosporangium aurantiacum* sp. nov. *Int. J. Syst. Evol. Microbiol.*, 51, 1809–1816.
266. Lee, S.D., and Hah, Y.C. (2002) Proposal to transfer *Catellatospora ferruginea* and '*Catellatospora ishikariense*' to *Asanoa* gen. nov. as *Asanoa ferruginea* comb. nov. and *Asanoa ishikariensis* sp. nov., with emended description of the genus *Catellatospora. Int. J. Syst. Evol. Microbiol.*, 52, 967–972.
267. Matsumoto, A., Takahashi, Y., Shinose, M., Seino, A., Iwai, Y., and Omura, S. (2003) *Longispora albida* gen. nov., sp. nov., a novel genus of the family *Micromonosporaceae. Int. J. Syst. Evol. Microbiol.*, 53, 1553–1559.
268. Maldonado, L.A., Fenical, W., Jensen, P.R., Kauffman, C.A., Mincer, T.J., Ward, A.C., Bull, A.T., and Goodfellow, M. (2005) *Salinispora arenicola* gen. nov., sp. nov. and *Salinispora tropica* sp. nov., obligate marine actinomycetes belonging to the family *Micromonosporaceae. Int. J. Syst. Evol. Microbiol.*, 55, 1759–1766.
269. Thawai, C., Tanasupawat, S., Itoh, T., and Kudo, T. (2006) *Actinocatenispora thailandica* gen. nov., sp. nov., a new member of the family *Micromonosporaceae. Int. J. Syst. Evol. Microbiol.*, 56, 1789–1794.
270. Tamura, T., Hatano, K., and Suzuki, K. (2006) A new genus of the family *Micromonosporaceae, Polymorphospora* gen. nov., with description of *Polymorphospora rubra* sp. nov. *Int. J. Syst. Evol. Microbiol.*, 56, 1959–1964.
271. Stackebrandt, E., and Kroppenstedt, R.M. (1987) Union of the genera *Actinoplanes* Couch, *Ampullariella* Couch, and *Amorphosporangium* Couch in a redefined genus *Actinoplanes. Syst. Appl. Microbiol.*, 9, 110–114.
272. Couch, J.N. (1950) *Actinoplanes.* A new genus of the *Actinomycetales. J. Elisha Mitchell Scientific Society*, 66, 87–92.
273. Goodfellow, M., Stanton, L.J., Simpson, K.E., and Minnikin, D.E. (1990) Numerical and chemical classification of *Actinoplanes* and some related actinomycetes. *J. Gen. Microbiol.*, 136:,19–36.
274. Tamura, T., and Hatano, K. (2001). Phylogenetic analysis of the genus *Actinoplanes* and transfer of *Actinoplanes minutisporangius* Ruan et al. 1986 and '*Actinoplanes aurantiacus*' to *Cryptosporangium minutisporangium comb.* nov. and *Cryptosporangium aurantiacum* sp. nov. *Int. J. Syst. Evol. Microbiol.*, 51, 2119–2125.
275. Asano, K., and Kawamoto, I. (1986) *Catellatospora*, a new genus of the *Actinomycetales. Int. J. Syst. Bacteriol.*, 36, 512–517.
276. Asano, K., and Kawamoto, I. (1988) *Catellatospora citrea* subsp *methionotrophica* subsp. nov, a methionine- deficient auxotroph of the *Actinomycetales. Int. J. Syst. Bacteriol.*, 38, 326–327.
277. Asano, K., Masunaga, I., and Kawamoto, I. (1989) *Catellatospora matsumotoense* sp. nov. and *Catellatospora tsunoense* sp. nov., actinomycetes found in woodland soils. *Int. J. Syst. Bacteriol.*, 39, 309–313.
278. Lee, S.D., Kang, S.O., and Hah, Y.C. (2000) *Catellatospora koreensis* sp. nov., a novel actinomycete isolated from a gold-mine cave. *Int. J. Syst. Evol. Microbiol.*, 50, 1103–1111.
279. Lee, S.D., Goodfellow, M., and Hah, Y.C. (1999). A phylogenetic analysis of the genus *Catellatospora* based on 16S ribosomal DNA sequence, including transfer of *Catellatospora matsumotoense* to the genus *Micromonospora* as *Micromonospora matsumotoense* comb. nov. *FEMS Microbiol. Lett.*, 178, 349–354.
280. Petrolini, B., Quaroni, S., Saracchi, M., and Sardi, P. (1993) A new genus of the maduromycetes: *Planopolyspora* gen. nov. *Actinomycetes*, 4, 8–16.
281. Yokota, A., Tamura, T., Hasegawa, T., and Huang, L.H. (1993) *Catenuloplanes japonicus* gen. nov., sp. nov., nom. rev., a new genus of the order *Actinomycetales. Int. J. Syst. Bacteriol.*, 43, 805–812.

282. Tamura, T., Yokota, A., Huang, L.H., Hasegawa, T., and Hatano, K. (1995) Five new species of the genus *Catenuloplanes*: *Catenuloplanes niger* sp. nov., *Catenuloplanes indicus* sp. nov., *Catenuloplanes atrovinosus* sp. nov., *Catenuloplanes castaneus* sp. nov., and *Catenuloplanes nepalensis* sp. nov. *Int. J. Syst. Bacteriol.*, 45, 858–860.
283. Kudo, T., Nakajima, Y., and Suzuki, K. (1999) *Catenuloplanes crispus* (Petrolini et al. 1993) comb. nov.: incorporation of the genus *Planopolyspora* Petrolini 1993 into the genus *Catenuloplanes* Yokota et al. 1993 with an amended description of the genus *Catenuloplanes*. *Int. J. Syst. Bacteriol.*, 49, 1853–1860.
284. Luedemann, G.M., and Brodsky, B.C. (1964). Taxonomy of gentamicin -producing *Micromonospora*. *Antimicrob. Agents Chemother.*, 1, 116–124.
285. Wagman, G.H., Testa, R.T., Marquez, J.A.,.and Weinstein, M.J. (1974) Antibiotic G-418, a new *Micromonospora*-produced aminoglycoside with activity against protozoa and helmints: fermentation, isolation, and preliminary characterization. *Antimicrob Agents Chemother.*, 6, 144-149
286. Wagman, G.H. (1980) Antibiotics from *Micromonospora*. *Ann Rev Microbiol.*, 34, 537-557
287. Horan, A.C., and Brodsky, B.C. (1986) *Micromonospora rosaria* sp. nov. nom. rev., the rosaramicin producer. *Int. J. Syst. Bacteriol.*, 36, 478-480
288. He, H., Ding, W.D., Bernan, T. (2001) Lomaiviticins A and B, potent antitumor antibiotics from *Micromonospora lomaivitiensis*. *J. Am. Chem. Soc.*, 123, 5362–5363.
289. Thorson, J.S., Sievers, E.L., Ahlert, J., Shepard, E., Whitwam, R.E., Onwueme, K.C., and Ruppen, M. (2000) Understanding and exploiting nature's chemical arsenal: the past, present and future of calicheamicin research. *Curr. Pharm. Design*, 6, 1841–1879.
290. Ahlert, J., Shepard, E., Lomovskaya, N., Zazopoulos, E., Staffa, A., Bachmann, B.O., Huang, K.X., Fonstein, L., Czisny, A., Whitwam, R.E., Farnet, C.M., Thorson, J.S. (2002) The calicheamicin gene cluster and its iterative type I enediyne PKS. *Science*, 297, 1173–1176.
291. Alarcon, B., Gonzalez, M.E., and Carrasco, L. (1988) Megalomycin-C, a macrolide antibiotic that blocks protein glycosylation and shows antiviral activity. *FEBS Lett.*, 231, 207–211.
292. Kroppenstedt, R.M. (1985). Fatty acid and menaquinone analysis of actinomycetes and related organisms. In *Chemical Methods in Bacterial Systematics*, p. 173–199. M. Goodfellow and D.E. Minnikin, Eds. London: Academic Press.
293. Kawamoto, I. (1989). Genus *Micromonospora*. In *Bergey's Manual of Systematic Bacteriology*, Vol. 4, p. 2442–2450. S. T. Williams, M.E. Sharpe, and J.G. Holt. Baltimore, MD: Williams & Wilkins.
294. Kasai, H., Tamura, T., and Harayama, S. (2000) Intrageneric relationships among *Micromonospora* species deduced from gyrB-based phylogeny and DNA relatedness. *Int. J. Syst. Evol. Microbiol.*, 50, 127–134
295. Kroppenstedt, R.M., Mayilraj, S., Wink, J.M., Kallow, W., Schumann, P., Secondini, C., and Stackebrandt, E. (2005). Eight new species of the genus *Micromonospora, Micromonospora citrea* sp. nov., *Micromonospora echinaurantiaca* sp. nov., *Micromonospora echinofusca* sp. nov., *Micromonospora fulviviridis* sp. nov., *Micromonospora inyonensis* sp. nov., *Micromonospora peucetia* sp. nov., *Micromonospora sagamiensis* sp. nov., and *Micromonospora viridifaciens* sp. nov. *Syst. Appl. Microbiol.*, 28, 328–339.
296. Thawai, C., Tanasupawat, S., Itoh, T., Suwanborirux, K., Suzuki, K., and Kudo, T. (2005). *Micromonospora eburnea* sp. nov., isolated from a Thai peat swamp forest. *Int. J. Syst. Evol. Microbiol.*, 55, 417-422
297. Thawai, C., Tanasupawat, S., Itoh, T., Suwanborirux, K., and Kudo, T. (2005). *Micromonospora siamensis* sp. nov., isolated from Thai peat swamp forest. *J. Gen. Appl. Microbiol.*, 51, 229-234
298. Trujillo, M.E., Fernandez-Molinero, C., Velazquez, E., Kroppenstedt, R.M., Schumann, P., Mateos, P.F., and Martinez-Molina, E. (2005). *Micromonospora mirobrigensis* sp. nov. *Int. J. Syst. Evol. Microbiol.*, 55, 877–880.
299. Rowbotham, T.J., and Cross, T. (1977) Ecology of *Rhodococcus coprophilus* and associated actinomycetes in fresh water and agricultural habitats. *J. Gen. Microbiol.*, 100, 231-240.
300. Potter, L.F., and Baker, G.E. (1956) The microbiology of Flathead and Rogers lakes. I. Preliminary survey of microbial populations. *Ecology,* 37, 351–355.
301. Watson, E.T., and Williams, S.T. (1974) Studies on the ecology of actinomycetes in soil. VII. *Soil Biol. Biochem.*, 6, 43–52.
302. Mincer, T.J., Jensen, P.R., Kauffman, C.A., and Fenical, W. (2002) Widespread and persistent populations of a major new marine actinomycete taxon in ocean sediments. *Appl. Env. Microbiol.,* 68, 5005–5011.
303. Weyland, H. (1969) Actinomycetes in North Sea and Atlantic Ocean sediment. *Nature*, 223, 858.

304. Weyland, H. (1981) Distribution of actinomycetes on the sea floor. *Zentral Bakteriol., Suppl.,* 11, 185–193.
305. Maldonado, L.A., Stach, J.E.M., Pathom-Aree, W., Ward, A.C., Bull, A.T., and Goodfellow, M. (2005) Diversity of cultivable actinobacteria in geographically widespread marine sediments. *Antonie van Leeuwenhoek, 87, 11–18.*
306. *Feling, R*.H., Buchanan, G.O., Mincer, T.J., Kauffman, C.A., Jensen, P.R., Fenical, W. (2003) Salinosporamide A: a highly cytotoxic proteasome inhibitor from a novel microbial source, a marine bacterium of the new genus *Salinospora. Angew. Chem. Int. Ed.*, 42, 355–357.
307. Newman, D.J., and Hill, R.T. (2006) New drugs from marine microbes: the tide is turning. *J. Ind. Microbiol. Biotech.*, 33, 539–544
308. Embley, T.M., Smida, J., and Stackebrandt, E. (1988) Reverse-transcriptase sequencing of 16S ribosomal-RNA from *Faenia rectivirgula, Pseudonocardia thermophila* and *Saccharopolyspora hirsuta*, 3 wall type-IV actinomycetes which lack mycolic acids. *J. Gen. Microbiol.*, 134, 961–966.
309. Labeda, D.P., and Kroppenstedt, R.M. (2000) Phylogenetic analysis of Saccharothrix and related taxa: proposal for *Actinosynnemataceae* fam. nov. *Int. J. Syst. Evol. Microbiol.*, 50, 331–336.
310. Ochi, K. (1995) Amino-acid-sequence analysis of ribosomal-protein AT-L30 from members of the family *Pseudonocardiaceae. Int. J. Syst. Bacteriol.*, 45, 110–115.
311. Hasegawa, T. (1988) *Actinokineospora*: a new genus of the *Actinomycetales. Actinomycetologica*, 2, 31–45.
312. Hasegawa, T., Lechevalier, M.P., and Lechevalier, H.A. (1978) A new genus of *Actinomycetales*: *Actinosynnema* gen. nov. *Int. J. Syst. Bacteriol.*, 28, 304–310.
313. Labeda, D.P., Testa, R.T., Lechevalier, M.P., and Lechevalier, H.A. (1984) *Saccharothrix*: a new genus of the *Actinomycetales* related to *Nocardiopsis. Int. J. Syst. Bacteriol.*, 34, 426–431.
314. Labeda, D.P., Hatano, K., Kroppenstedt, R.M., and Tamura, T. (2001) Revival of the genus *Lentzea* and proposal for *Lechevalieria* gen. nov. *Int. J. Syst. Evol. Microbiol.*, 51, 1045–1050.
315. Yassin, A.F., Rainey, F.A., Brzezinka, H., Jahnke, K.D., Weissbrodt, H., Budzikiewicz, H., Stackebrandt, E., and Schaal, K.P. (1995) *Lentzea* gen. nov., a new genus of the order *Actinomycetales. Int. J. Syst. Bacteriol.*, 45, 357–363.
316. Tamura, T., Zhiheng, L., Yamei, Z., and Hatano, K. (2000) *Actinoalloteichus cyanogriseus* gen. nov., sp. nov. *Int. J. Syst. Evol. Microbiol.*, 50, 1435–1440.
317. Labeda, D.P. (2001) *Crossiella* gen. nov., a new genus related to *Streptoalloteichus. Int. J. Syst. Evol. Microbiol.*, 51, 1575–1579.
318. Stackebrandt, E., Kroppenstedt, R.M., Jahnke, K.D., Kemmerling, C., and Gürtler, H., (1994) Transfer of *Streptosporangium viridogriseum* (Okuda et al. 1966), *Streptosporangium viridogriseum* subsp. *kofuense* (Nonomura and Ohara 1969), and *Streptosporangium albidum* (Furumai et al. 1968) to *Kutzneria* gen. nov. as *Kutzneria viridogrisea* comb. nov., *Kutzneria kofuensis* comb. nov., and *Kutzneria albida* comb. nov., respectively, and emendation of the genus *Streptosporangium. Int. J. Syst. Bacteriol.*, 44, 265–269.
319. Tomita, K., Nakakita, Y., Hoshino, Y., Numata, K., and Kawaguchi, H. (1987) New genus of the *Actinomycetales: Streptoalloteichus hindustanus* gen. nov., nom. rev.; sp. nov., nom. rev. *Int. J. Syst. Bacteriol.,* 37, 211–213.
320. Embley, T.M., O'Donnell, A.G., Rostron, J., and Goodfellow, M. (1988) Chemotaxonomy of wall type-IV actinomycetes which lack mycolic acids. *J. Gen. Microbiol.*, 134: 953–960.
321. Gochnauer, M.B., Leppard, G.G., Komaratat, P., Kates, M., Novitsky, T., and Kushner, D.J. (1975) Isolation and characterization of *Actinopolyspora halophila*, gen. et sp. nov., an extremely halophilic actinomycete. *Can. J. Microbiol.*, 21, 1500–1511.
322. Lechevalier, M.P., Prauser, H., Labeda, D.P., and Ruan, J.S. (1986) Two new genera of nocardioform actinomycetes: *Amycolata* gen. nov. and *Amycolatopsis* gen. nov. *Int. J. Syst. Bacteriol.*, 36, 29–37.
323. Kurup, V.P., and Agre, N.S. (1983) Transfer of *Micropolyspora rectivirgula* (Krasil'nikov and Agre 1964) Lechevalier, Lechevalier, and Becker 1966 to *Faenia* gen. nov. *Int. J. Syst. Bacteriol.*, 33, 663–665.
324. Henssen, A. (1957) Beiträge zur Morphologie und Systematik der thermophilen Actinomyceten. *Archiv fur Mikrobiologie*, 26, 373–414.
325. Nonomura, H., and O'Hara, Y. (1971) Distribution of actinomycetes in soil. X. New genus and species of monosporic actinomycetes in soil. *J. Ferment. Technol.*, 49, 895–903.
326. Lacey, J., and Goodfellow, M. (1975) A novel actinomycete from sugar-cane bagasse, *Saccharopolyspora hirsuta* gen. et sp. nov. *J. Gen. Microbiol.*, 88, 75–85.

327. Warwick, S., Bowen, T., Mcveigh, H., and Embley, T.M. (1994) A phylogenetic analysis of the family *Pseudonocardiaceae* and the genera *Actinokineospora* and *Saccharothrix* with 16S rRNA sequences and a proposal to combine the genera Amycolata and *Pseudonocardia* in an emended genus *Pseudonocardia. Int. J. Syst. Bacteriol.*, 44, 293-299.
328. Korn-Wendisch, F., Kempf, A., Grund, E., Kroppenstedt, R.M., and Kutzner, H.J. (1989) Transfer of *Faenia rectivirgula* Kurup and Agre 1983 to the genus *Saccharopolyspora* Lacey and Goodfellow 1975, elevation of *Saccharopolyspora hirsuta* subsp. *taberi* Labeda 1987 to species level, and emended description of the genus *Saccharopolyspora. Int. J. Syst. Bacteriol.*, 39, 430–441.
329. Jiang, C., Xu, L., Yang, Y.R., Guo, G.Y., Ma, J., and Liu, Y. (1991) *Actinobispora*, a new genus of the order *Actinomycetales. Int. J. Syst. Bacteriol.,* 41, 526–528.
330. Huang, Y., Wang, L., Lu, Z., Hong, L., Liu, Z., Tan, G.Y.A., and Goodfellow, M. (2002) Proposal to combine the genera *Actinobispora* and *Pseudonocardia* in an emended genus *Pseudonocardia*, and description of *Pseudonocardia zijingensis* sp. nov. *Int. J. Syst. Evol. Microbiol.*, 52, 977–982.
331. Shearer, M.C., Colman, P.M., Ferrin, R.M., Nisbet, L.J., and Nash, III, C.H. (1986) New genus of the *Actinomycetales: Kibdelosporangium aridum* gen. nov., sp. nov. *Int. J. Syst. Bacteriol.*, 36, 47–54.
332. Kim, S.B., and Goodfellow, M. (1999) Reclassification of *Amycolatopsis rugosa* Lechevalier et al. 1986 as *Prauserella rugosa* gen. nov., comb. nov. *Int. J. Syst. Bacteriol.*, 49, 507–512.
333. Wang, Y., Zhang, Z., and Ruan, J. (1996) A proposal to transfer *Microbispora bispora* (Lechevalier 1965) to a new genus, *Thermobispora* gen. nov., as *Thermobispora bispora* comb. nov. *Int. J. Syst. Bacteriol.*, 46, 933–938.
334. Korn-Wendisch, F., Rainey, F., Kroppenstedt, R.M., Kempf, A., Majazza, A., Kutzner, H.J., and Stackebrandt, E. (1995) *Thermocrispum* gen. nov., a new genus of the order *Actinomycetales*, and description of *Thermocrispum municipale* sp. nov. and *Thermocrispum agreste* sp. nov. *Int. J. Syst. Bacteriol.*, 45, 67–77.
335. Tan, G.Y.A., Ward, A.C., and Goodfellow, M. (2006) Exploration of *Amycolatopsis* diversity in soil using genus-specific primers and novel selective media. *Syst. Appl. Microbiol.*, 29, 557–569.
336. Stach, J.E.M., Maldonado, L.A., Masson, D.G., Ward, A.C., Goodfellow, M., and Bull, A.T. (2003) Statistical approaches to estimating bacterial diversity in marine sediments. *Appl. Environ. Microbiol.*, 69, 6189–6200.
337. Yassin, A.F., Haggenei, B., Budzikiewicz, H., and Schaal, K.P. (1993) Fatty-acid and polar lipid-composition of the genus *Amycolatopsis* – application of fast-atom-bombardment mass-spectrometry to structure-analysis of underivatized phospholipids. *Int. J. Syst. Bacteriol.*, 43, 414–420.
338. Yassin, A.F., Schaal, K.P., Brzezinka, H., Goodfellow, M., Pulverer, G., (1991) Menaquinone patterns of *Amycolatopsis* species. *Zentralblatt fur Bakteriologie-Int. J. Med. Microbiol. Virol. Parasitol. Infect. Dis.*, 274, 465–470.
339. Pittenger, R.C., and Brigham, R.B. (1956). *Streptomyces orientalis*, n. sp., the source of vancomycin. *Antibiot. Agents Chemother.*, 6, 642–647.
340. Cohn, F. (1875) Untersuchungen uber Bacterien. II. In *Beitrge zur Biologie der Pflanzen*, Vol. 1, Dritte Heft, p. 185–188; 202–207; Plate V. J.U. Kern's Verlag (Max Muller), Breslau, Germany.
341. Müller, R. (1908) Eine Diphtheridee und eine Streptothrix mit gleichem Farbstoff sowie Untersuchungen über Streptothrixarten im allgemeinen. *Zentr. Bakteriol. Parasiten*k., Abst. I. 6, 195–212
342. Waksman, S.A., and Curtis, R.E. (1916) The Actinomyces of the soil. *Soil Sci.* 1, 99–134.
343. Pridham, T.G., and Lyons, A.J. Jr. (1961) Streptomyces albus (Rossi-Doria) Waksman et Henrici: taxonomic study of strains labeled *Streptomyces albus. J. Bact.*, 81, 431–441.
344. Lyons, A.J. Jr., and Pridham, T.G. (1962) Proposal to designate strain ATCC 3004 (IMRU 3004) as the neotype strain of *Streptomyces albus* (Rossi Doria) Waksman and Henrici. *J. Bact.*, 83, 370–380.
345. Kutzner, H.J., and Waksman, S.A. (1959) *Streptomyces coelicolor* Müller and *Streptomyces violaceoruber* Waksman and Curtis, two distinctly different organisms. *J. Bact.*, 78, 528—538.
346. Hopwood, D.A. (1999) Forty years of genetics with *Streptomyces*: from *in vivo* through *in vitro* to *in silico. Microbiology-UK*, 145, 2183–2202.
347. Waksman, S.A., and Woodruff, H.B. (1941) *Actinomyces antibioticus*, a new soil organism antagonistic to pathogenic and non-pathogenic bacteria. *J. Bact.*, 42, 231–249.
348. Waksman, S.A., and Woodruff, H.B. (1942) Streptothricin, a new selective bacteriostatic and bactericidal agent, particularly active against Gram-negative bacteria. *Proc. Soc. Exptl. Biol. Med.*, 49, 207–210.
349. Schatz, A., Bugie, E., and Waksman, S.A. (1944) Streptomycin, a substance exhibiting antibiotic activity against Gram-positive and Gram-negative bacteria. *Proc. Soc. Exp. Biol. Med.*, 55, 66–69.

350. Shirling, E.B., and Gottlieb, D. (1966) Methods for characterization of *Streptomyces* species. *Int. J. Syst. Bact.*, 16, 313–340.
351. Shirling, E.B., and Gottlieb, D. (1966) Cooperative description of type cultures of *Streptomyces*. II. Species descriptions from first study. *Int. J. Syst. Bact.*, 18, 69-189
352. Shirling, E.B., and Gottlieb, D. (1966) Cooperative description of type cultures of *Streptomyces*. III. Additional species descriptions from first and second studies. *Int. J. Syst. Bact.*, 18, 279-392
353. Shirling, E.B., and Gottlieb, D. (1969) Cooperative description of type cultures of *Streptomyces*. IV. Species description from the second, third and fourth studies. *Int. J. Syst. Bact.*, 19, 391–512.
354. Sneath, P.H.A., and Sokal, R.R. (1973) *Numerical Taxonomy: The Principles and Practice of Numerical Classification*. San Francisco, CA: Freeman.
355. Sneath, P.H.A. (1995) 30 Years of numerical taxonomy. *Syst. Biol.*, 44, 281–298.
356. Locci, R., Rogers, J., Sardi, G., and Schofield, G.M. (1981). A preliminary numerical study on named species of the genus *Streptoverticillium*. *Ann. Microbiol.*, 31, 115–121.
357. Williams, S.T., Goodfellow, M., Alderson, G., Wellington, E.M.H., Sneath, P.H.A., and Sackin, M.J. (1983) Numerical classification of *Streptomyces* and related genera. *J. Gen. Microbiol.* 129, 1743-1813
358. Kämpfer, P., Kroppenstedt, R.M., and Dott, W. (1991) A numerical classification of the genera *Streptomyces* and *Streptoverticillium* using miniaturized physiological tests. *J. Gen. Microbiol.*, 137, 1831–1891.
359. Fox, G.E., Stackebrandt, E., Hespell, R.B., Gibson, J., Maniloff, J., Dyer, T.A., Wolfe, R.S., Balch, W.E., Tanner, R.S., Magrum, L.J., Zablen, L.B., Blakemore, R., Gupta, R., Bonen, L., Lewis, B.J., Stahl, D.A., Luehrsen, K.R., Chen K.N., and Woese, C.R. (1980) The phylogeny of prokaryotes. *Science*, 209, 457–463.
360. Anderson, A.S., and Wellington, E.M.H. (2001) The taxonomy of *Streptomyces* and related genera. *Int. J. Syst. Evol. Microbiol.*, 51, 797–814.
361. Lanoot, B., Vancanneyt, M., Schoor, A.V., Liu, Z., and Swings, J. (2005) Reclassification of Streptomyces nigrifaciens as a later synonym of Streptomyces flavovirens; Streptomyces citreofluorescens, *Streptomyces chrysomallus* subsp. *chrysomallus* and *Streptomyces fluorescens* as later synonyms of *Streptomyces anulatus*; *Streptomyces chibaensis* as a later synonym of *Streptomyces corchorusii*; *Streptomyces flavisclerotίcus* as a later synonym of *Streptomyces minutisclerotίcus*; and *Streptomyces lipmanii*, *Streptomyces griseus* subsp. alpha, *Streptomyces griseus* subsp. cretosus and *Streptomyces willmorei* as later synonyms of *Streptomyces microflavus*. *Int. J. Syst. Evol. Microbiol.*, 55, 729–731.
362. Currie, C.R., Scott, J.A., Summerbell, R.C., and Malloch, D. (2003) Corrigendum: fungus-growing ants use antibiotic-producing bacteria to control garden parasites, *Nature*, 423, 461.
363. Goodfellow, M., Williams, S.T., and Alderson, G. (1986) Transfer of *Actinosporangium violaceum* Krasil'nikov and Yuan, *Actinosporangium vitaminophilum* Shomura et al. and *Actinopycnidium caeruleum* Krasil'nikov to the genus Streptomyces with amended descriptions of the species. *Syst. Appl. Microbiol.*, 8, 61–64.
364. Goodfellow, M., Williams, S.T., and Alderson, G. (1986) Transfer of *Chainia* species to the genus *Streptomyces* with emended description of species. *Syst. Appl. Microbiol.*, 8, 55–60.
365. Goodfellow, M., Williams, S.T., and Alderson, G. (1986) Transfer of *Elytrosporangium brasiliense* Falcão de Morais et al., *Elytrosporangium carpinense* Falcão de Morais et al., *Elytrosporangium spirale* Falcão de Morais, *Microellobosporia cinerea* Cross et al., *Microellobosporia flavea* Cross et al., *Microellobosporia grisea* (Konev et al.) Pridham and *Microellobosporia violacea* (Tsyganov et al.) Pridham to the genus Streptomyces, with emended descriptions of the species. *Syst. Appl. Microbiol.*, 8, 48–54.
366. Wellington, E.M.H., Stackebrandt, E., Sanders, D., Wolstrup, J., and Jorgensen, N.O.G. (1992) Taxonomic status of *Kitasatosporia*, and proposal unification with *Streptomyces* on the basis of phenotypic and 16S rRNA analysis and emendation of *Streptomyces* Waksman and Henrici 1943, 339AL. *Int. J. Syst. Bacteriol.*, 42, 156–160.
367. Liu, Z., Shi, Y., Zhang, Y., Zhou, Z., Lu, Z., Li, W., Huang, Y., Rodríguez, C., and Goodfellow, M. (2005) Classification of S*treptomyces griseus* (Krainsky 1914) Waksman and Henrici 1948 and related species and the transfer of "*Microstreptospora cinerea*" to the genus *Streptomyces* as *Streptomyces* yanii sp. nov. *Int. J. Syst. Evol. Microbiol.*, 55, 1605–1610.
368. Witt, D., and Stackebrandt, E. (1990) Unification of the genera *Streptoverticillium* and Streptomyces, and amendation of *Streptomyces* Waksman and Henrici 1943, 339AL. *Syst. Appl. Microbiol.*, 13, 361–371.
369. Zhang, Z., Wang, Y., and Ruan, J. (1997) A proposal to revive the genus *Kitastospora* (Omura, Takahashi, Iwai, and Tanaka 1982). *Int. J. Syst. Bacteriol.*, 47, 1048–1054.

370. Kim, S.B., Lonsdale, J., Seong, C.-N., and Goodfellow, M. (2003). *Streptacidiphilus* gen. nov., acidophilic actinomycetes with wall chemotype I and emendation of the family Streptomycetaceae (Waksman and Henrici 1943AL) emend Rainey et al. 1997. *Antonie van Leeuwenhoek*, 83, 107–116.
371. Labeda, D.P. (1996) DNA relatedness among verticil–forming *Streptomyces* species (formerly *Streptoverticillium* species). *Int. J. Syst. Bacteriol.*, 46, 699–703.
372. Sembiring, L., Goodfellow, M., and Ward, A.C. (2000). The selective isolation and characterization of the *Streptomyces violaceusniger* clade associated with the roots of *Paraserianthes falcataria*. *Antonie van Leeuwenhoek*, 78, 353–366.
373. Chater, K.F. (2006) *Streptomyces* inside-out: a new perspective on the bacteria that provide us with antibiotics. *Phil. Trans. Roy. Soc. B, Biol. Sci.*, 361, 761–768.
374. Horinouchi, S. (2007) Mining and polishing of the treasure trove in the bacterial genus Streptomyces. *BioSci. Biotech. Biochem.*, 71, 283–299.
375. Rainey, F.A., WardRainey, N., Kroppenstedt, R.M., and Stackebrandt, E. (1996) The genus *Nocardiopsis* represents a phylogenetically coherent taxon and a distinct actinomycete lineage: proposal of *Nocardiopsaceae* fam. nov. *Int. J. Syst. Bacteriol.*, 46, 1088–1092.
376. Goodfellow, M., Stanton, L.J., Simpson, K.E., and Minnikin, D.E. (1990) Numerical and chemical classification of *Actinoplanes* and some related actinomycetes *J. Gen. Microbiol.*, 136, 19–36.
377. Zhang, Z., Kudo, T., Nakajima, Y., and Wang, Y. (2001) Clarification of the relationship between the members of the family Thermomonosporaceae on the basis of 16S rDNA, 16S-23S rRNA internal transcribed spacer and 23S rDNA sequences and chemotaxonomic analyses. *Int. J. Syst. Evol. Microbiol.*, 51, 373–383.
378. Bellamy, W.D. (1977) Cellulose and lignocellulose digestion by thermophilic actinomycetes for single cell protein production. *Dev. Ind. Microbiol.*, 18, 249–254.
379. Zhang, Z.S., Wang, Y., and Ruan, J.S. (1998) Reclassification of Thermomonospora and Microtetraspora. Int. J. Syst. Bacteriol., 48, 411–422.
380. Chiba, S., Suzuki, M., and Ando, K. (1999) Taxonomic re-evaluation of '*Nocardiopsis*' sp K-252(T) (=NRRL 15532(T)): a proposal to transfer this strain to the genus *Nonomuraea* as *Nonomuraea longicatena* sp nov. *Int. J. Syst. Bacteriol.*, 49, 1623–1630.
381. Quintana, E., Maldonado, L., and Goodfellow, M. (2003) *Nonomuraea terrinata* sp nov., a novel soil actinomycete. *Antonie van Leeuwenhoek*, 84, 1–6.
382. Lazzarini, A., Cavaletti, L., Toppo, G., and Marinelli, F. (2000) Rare genera of actinomycetes as potential producers of new antibiotics. *Antonie van Leeuwenhoek*, 78, 399–405.
383. Miyanaga, S., Obata, T., Onaka, H., Fujita, T., Saito, N., Sakurai, H., Saiki, I., Furumai, T., and Igarashi, Y. (2006) Absolute configuration and antitumor activity of myxochelin A produced by *Nonomuraea pusilla* TP-A0861. *J. Antibiot.*, 59, 698–703.
384. Kim, S.Y., Park, J.S., Chae, C.S., Hyun, C.G., Choi, B.W., Shin, J., and Oh, K.B. (2007) Genetic organization of the biosynthetic gene cluster for the indolocarbazole K-252a in *Nonomuraea longicatena* JCM 11136. *Appl. Microbiol. Biotech.*, 75, 1119–1126.
385. Sosio, M., and Donadio, S. (2006) Understanding and manipulating glycopeptide pathways: the example of the Dalbavancin precursor A40926. *J. Ind. Microbiol. Biotech.*, 33, 569–576.
386. Scheinfeld, N. (2007) Dalbavancin: A review. *Drugs Today*, 43, 305–316.

28 The Family Rickettsiaceae

Abdu F. Azad, Magda Beier-Sexton, and Joseph J. Gillespie

CONTENTS

INTRODUCTION

The order Rickettsiales is comprised of Gram-negative, small, rod-shaped coccoid obligate intracellular bacterial parasites of vertebrates and invertebrates, as well as protists. Prior to the advent of DNA sequencing, Rickettsiales was distinguished from other bacteria primarily based on chemical composition and morphology. Additionally, intraordinal taxonomy (i.e., generic characteristics) was founded on five major biological properties: (1) human disease and geographic distribution, (2) natural vertebrate hosts and other animal vectors, (3) experimental infections and serology reactions and cross-reactions, (4) strain cultivation and stability, and (5) energy production and biosynthesis. Traditionally, the order Rickettsiales consisted of three families containing nine obligate and facultative intracellular pathogenic genera: (1) Rickettsiaceae (genera *Rickettsia*, *Coxiella*, *Rochalima*, and *Ehrlichia*); (2) Bartonellaceae (genera *Bartonella*, *Haemobartonella*, *Eperythrozoon*, and *Grahamella*); and (3) Anaplasmataceae (genus *Anaplasma*). However, rickettsial taxonomy has undergone substantial revisionary reorganization in the past decade as several new rickettsial species were described and molecular approaches, such as the comparison of DNA sequences and other genomic characteristics, were applied. Our modern taxonomy of this group is radically different from the traditional classification, with such extreme differences as the removal of the well-studied *Coxiella burnetii*, the agent of Q-fever, and *Bartonella* spp., the causative agents of several human diseases such as endocarditis, Trench fever, and Cat-scratch disease, from the entire class of α-proteobacteria.

This chapter focuses exclusively on the family Rickettsiaceae as it is currently defined to include the genera *Rickettsia* and *Orientia* (Table 28.1). However, it also touches upon three other families that are either currently included in the order Rickettsiales (Anaplasmataceae and Holosporaceae) or were once a member (e.g., Bartonellaceae). The work presented here is a tribute to the earlier pioneers of the rickettsial field and in particular to the memory of Drs. Robert Traub and Charles L. Wisseman, Jr.

Rickettsiaceae

Within Rickettsiaceae, only *Rickettsia*, an extraordinarily unique bacteria with highly reduced genomes (Andersson and Andersson, 1999) and a close evolutionary with the ancestor to all mitochondria (Emelyanov, 2003), remain from the original four genera. *Coxiella*, a monotypic genus originally considered a sister to *Rickettsia*, has since been shown to belong to γ- and *not* α-proteobacteria (Weisburg et al., 1989). Similarly, *Rochalima* was unified with *Bartonella* in the Bartonellaceae, with its species renamed as *Bartonella* spp. (Brenner et al., 1993). Finally, *Ehlrichia* has been moved to the Anaplasmataceae, as it shares greater affinity with *Anaplasma* than *Rickettsia* (Dumler et al., 2001). Of note, one former species of *Rickettsia*, the scrub typhus agent *R. tsutsugamushi*, has since been split from *Rickettsia* and assigned to the novel genus *Orientia* (Tamura et al., 1995). Species of the genus *Rickettsia* have traditionally been grouped into either the typhus group (TG) or the spotted fever group (SFG) rickettsiae. Some *Rickettsia* spp., particularly the nonpathogenic strains (Perlman et al., 2006), are ancestral to the more commonly known pathogenic strains and have been classified as ancestral group (AG) rickettsiae. Several molecular phylogenetic

TABLE 28.1
Some Epidemiological Features of Rickettsial Diseases

	Agent: Rickettsia	Disease/Dominant Symptoms	Vector/Reservoir Host	Geographic Distribution
Typhus fevers	*R. prowazekii*	Epidemic typhus,	Human body louse/humans	Africa, Asia, America
		Sylvatic typhus	Squirrel flea and louse/ flying squirrels	U.S. (only)
	R. typhi	Murine (endemic) typhus	Rodent and cat fleas / rats, mice Opossums (U.S.)	Worldwide
Transitional Group	*R. akari*	Rickettsial pox	Mite/house mice	Worldwide
	R. felis	Cat flea rickettsiosis	Fleas/domestic cats & opossums (U.S.)	Worldwide
	R. australis	Queensland tick typhus	Tick/rodents	Australia, Tasmania
Tick-transmitted spotted fevers	*R. rickettsii*	Rocky Mountain spotted fever	Tick/rodents, rabbits	North, Central and South America
	R. parkeri	Mild spotted fever	Tick/rodents	U.S., Brazil, Uruguay
	R. conorii	Mediterranean spotted fever	Tick rodents, hedgehogs	Europe, Asia, Africa, India, Israel, Sicily
	R. sibirica	North Asian tick typhus, Siberian tick typhus	Tick rodents	Russia, China, Mongolia, Europe
	R. africae	African tick-bite fever	Tick/rodents	Sub-Saharan Africa, Caribbean
	R. japonica	Oriental spotted fever	Tick	Japan
	R. slovaca	Necrosis, erythema,	Tick/rodents & lagomorphs	Europe
	R. helvetica	Aneruptive fever	Tick/rodents	Old World
	R. honei	Finders Island spotted fever / Thai tick typhus	Ticks	Australia, Thailand
	O. tsutsugamushi	Scrub typhus	Larval mites / rodents	Indian subcontinent, Asia, Australia

studies have hinted at a distinct clade within the SFG rickettsiae comprised of *R. felis*, *R. akari*, and *R. australis* (e.g., Sekeyova et al., 2001), as well as other parasites of wasps and booklice (Perlman et al., 2006), and a recent investigation classified this lineage as transitional group (TRG) rickettsiae based on its monophyly, presence of plasmids and associated genes, and affinity with AG rickettsiae (Gillespie et al., 2007). Relationships among 12 *Rickettsia* spp. are shown in a phylogeny estimation of 21 members of Rickettsiales and one marine α-proteobacterium outgroup (Figure 28.1B).

Bartonellaceae

Bartonellaceae was removed from Rickettsiales because DNA hybridization studies and early DNA and RNA sequence comparisons showed that its members were more closely related to *Brucella* spp. in the order Rhizobiales (Brenner et al., 1993) (Figure 28.1A, Table 28.2). Like *Rochalima*, species of *Grahamella* became part of the genus *Bartonella* (Birtles et al., 1995). *Haemobartonella* and *Eperythrozoon* have since been demonstrated to belong in the mollicutes family Mycoplasmaceae in distantly related Gram-positive Firmicutes (Rikihisa et al., 1997). Thus, Bartonellaceae, now of

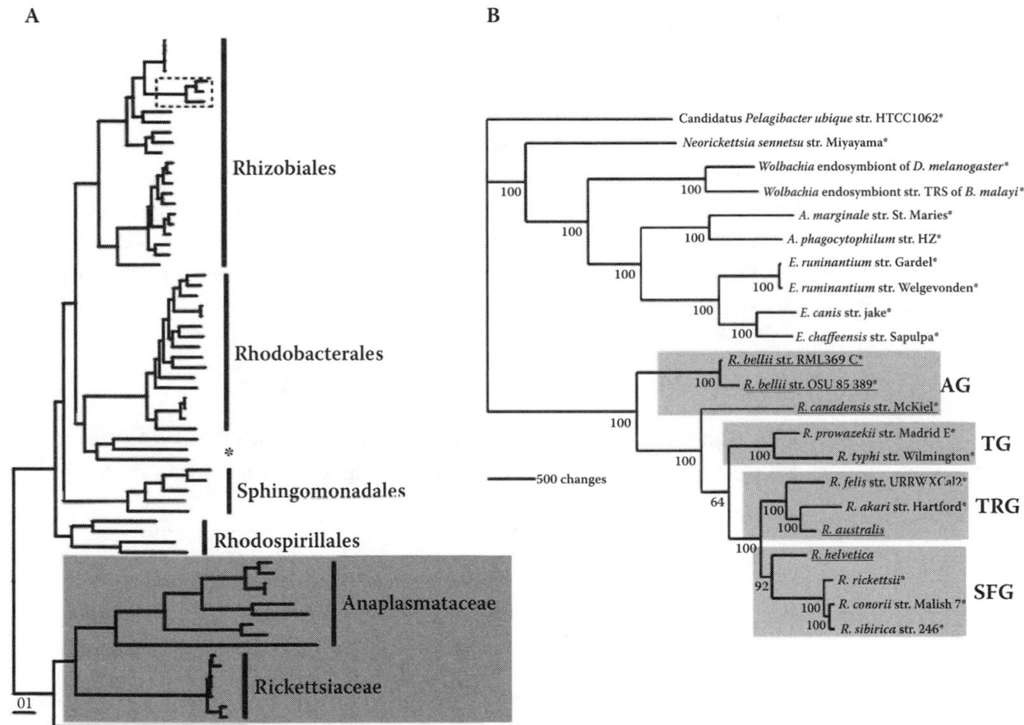

FIGURE 28.1 Phylogeny of Rickettsiales and its placement within the α-proteobacterial tree. (A) Schematic of estimated phylogeny of α-proteobacterial from a Bayesian analysis of the concatenation of masked alignments for 104 selected protein families (33730 characters). Dashed box shows the position of three *Bartonella* species. Asterisk depicts a clade comprised of Caulobacterales, Parvularculales, and Rhodobacterales. Shaded box comprises the Rickettsiales. (*Source:* Redrawn with permission from Williams et al., unpublished, Virginia Bioinformatics Institute at Virginia Tech). (B) Estimated phylogeny of 21 species of Rickettsiales and one marine α-proteobacterium, Candidatus *Pelagibacter ubique* str. HTCC1062. Tree estimated from a heuristic search under parsimony of 16 concatenated protein datasets (5765 parsimony informative characters) that resulted in one parsimonious tree of length 21394. Branch support is from 1 million bootstrap replications (Felsenstein, 1985). Analyses performed in PAUP* version 4.10 (Altivec) (Swofford, 1999). Taxa with an asterisk depict species for which a genome sequence is currently available. Underscored taxa were not included in the larger phylogeny estimated in A.

the order Rhizobiales, consist of only the genus *Bartonella*, some unclassified and environmental specimens, and the incorrectly named *Wolbachia melophagi.*

Anaplasmataceae

Based on sequence analysis, the once monotypic Anaplasmataceae seemed to contain four genetic groups comprising five genera: *Anaplasma*, *Cowdria*, *Ehrlichia*, *Neorickettsia,* and *Wolbachia.* Dumler et al. (2001) meticulously placed the majority of the species in these five genera into *Anaplasma*, *Ehrlichia*, *Neorickettsia,* and *Wolbachia*; and our phylogeny estimation from nine Anaplasmataceae species for which a complete genome exists supports this classification (Figure 28.1B; Table 28.2).

One other previously described genus of Rickettsiales, *Rickettsiella*, had long been hypothesized not to belong in the order because it differs in cell morphology (Weiser and Zizka, 1968), mode of reproduction (Götz, 1971, 1972), and an altogether unusual morphogenic cycle most similar to *Chlamydia* spp. (Federici, 1980). Sequence analysis confirmed the misplacement of *Rickettsiella* and, like *Coxiella*, it was subsequently moved to γ-proteobacteria, as it is more similar to *Legionella* spp. than to any Rickettsiales (Roux et al., 1997).

Holosporaceae

A novel third family, Holosporaceae, was recently added to the Rickettsiales (Boone et al., 2001). This group is comprised primarily of endosymbionts of protists, particularly amoebae, and forms a distinct rickettsial lineage in the α-proteobacterial tree (Baker et al., 2003). The family contains one described species, *Holospora obtusa* (Gromov and Ossipov, 1981), and several unclassified genera and environmental samples. We do not include any species of Holosporaceae in our phylogeny estimation because few gene sequences exist for the family.

RICKETTSIAL GENOTYPIC AND PHENOTYPIC CHARACTERISTICS

As discussed above, the long-standing two-group system (TG and SFG) for classifying rickettsiae has come to an end, as new insight from genomic comparisons suggests at least two more distinct

TABLE 28.2

Other Rickettsial Pathogens No Longer Associated with Rickettsiaceae

	Agent	Disease/ Dominant Symptoms	Vector/Reservoir Host	Geographic Distribution
Bartonella	*B. henselae*	Cat-scratch disease	Cat flea/domestic cat	Worldwide
	B. quintana	Trench fever	Human body louse / humans	Worldwide
	B. bacilliformis	Oroya fever	Sand fly/?	Peru, Ecuador, Colombia
Ehrlichia	*E. chaffeensis*	Monocytic Ehrlichiosis	Tick/mammals (deer, rodents)	Worldwide*
	E. ewingii	Ehrlichiosis	Tick/deer?	U.S.
Anaplasma	*A. phagocytphilum*	Anaplasmosis	Tick/rodents and other small mammals	U.S., Europe, Asia, Africa
Neorickettsia	*N. sennetsu*	Sennetsu fever	Snail/fish	Japan, Malaysia
Coxiella	*C. burnetii*	Q-fever	Inhalation and tick/goats, sheep, cattle and domestic cats	Worldwide

groups (AG and TRG). Among ten sequenced rickettsial genomes, *Rickettsia typhi* str. Wilmington has the smallest genome and *Rickettsia felis* str. URRWXCal2 the largest (Figure 28.2). In addition, the two known members of TG rickettsiae, *R. prowazekii* and *R. typhi*, comprise the most degraded and rapidly evolving genomes of rickettsiae (Blanc et al., 2007). Other genomic characteristics, such as the number of coding genes/genome and nucleotide composition, distinguish TG from SFG rickettsiae (Figure 28.2). However, some rickettsiae do not (and have not historically) fit well within the TG and SFG rickettsiae. Based on a phylogeny estimation of over 30 proteins, we previously grouped *R. canadensis* with both *R. bellii* strains in the AG rickettsiae (Gillespie et al., 2007), a result consistent with several other studies (Stothard and Fuerst, 1995; Vitorino et al., 2007). Interestingly, Vitorino et al. (2007) recently demonstrated an affinity between *R. canadensis* and *R. helvetica* based on phylogeny estimation from eight genes, although they concluded that the phylogenetic position of *R. canadensis* was unstable, which is consistent with previous studies. For example, similar to SFG rickettsiae, *R. canadensis* infects ticks and is maintained transstadially and transovarially (Burgdorfer, 1968; Brinton and Burgdorfer, 1971); grows within the nuclei of its host (Burgdorfer, 1968); and contains both rOmpA and rOmpB genes (Dasch and Bourgeois, 1981; Ching et al., 1990). However, similar to TG rickettsiae, *R. canadensis* grows abundantly in yolk sac, hemolyzes red blood cells, is susceptible to erythromycin, and forms smaller plaques as compared

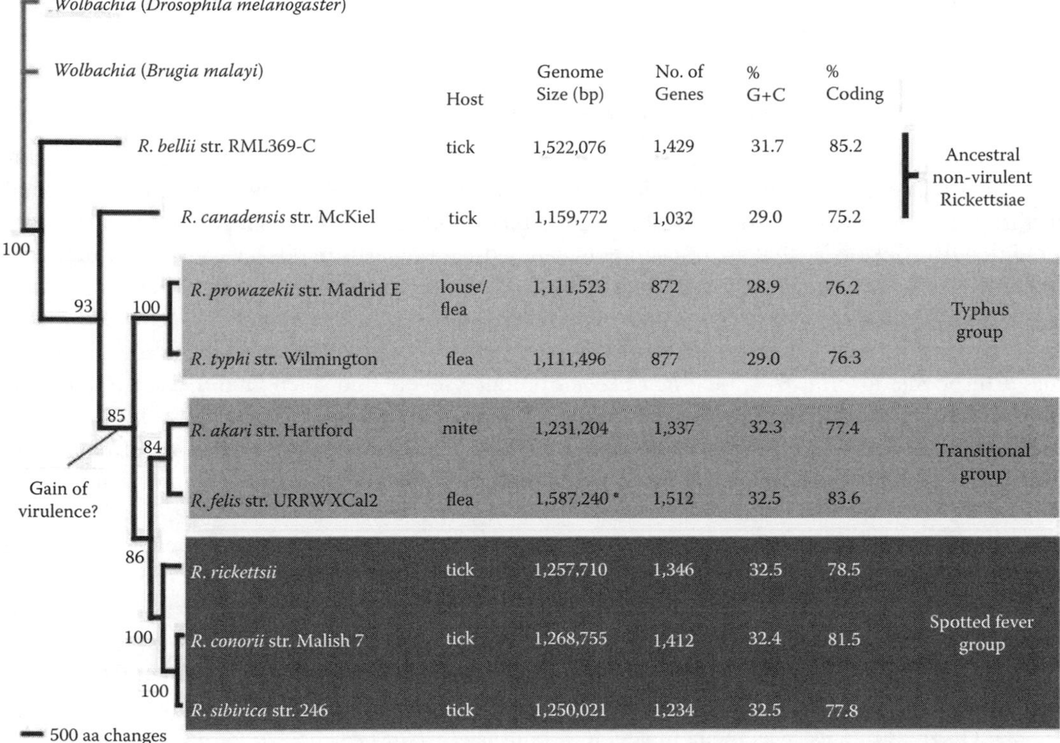

FIGURE 28.2 Phylogeny estimation from analysis of fifteen *Rickettsia felis* proteins (hypothetical protein RF_0005, threonyl-tRNA synthetase, preprotein translocase SecA subunit, uncharacterized low-complexity protein RF_0864, pyruvate phosphate dikinase precursor, leucyl-tRNA synthetase, hypothetical protein RF_0556, NAD-specific glutamate dehydrogenase, DNA polymerase III alpha chain, O-antigen export system permease protein RfbA, thioredoxin, NADPH-dependent glutamate synthase beta chain and related oxidoreductases, putative TIM-barrel protein in nifR3 family, and UDP-3-O-[3-hydroxymyristoyl] glucosamine) from nine rickettsial species (*Rickettsia bellii, R. canadensis, R. prowazekii, R. typhi, R. akari, R. felis, R. conorii, R. rickettsii,* and *R. sibirica*) and two strains of *Wolbachia*. See Gillespie et al. (2007) for methods for tree estimation.

to SFG rickettsiae (Myers and Wisseman Jr., 1981). Genomic characteristics are just as anomalous as, despite sharing the same G+C% as TG rickettsiae (Myers and Wisseman Jr., 1981; Eremeeva et al., 2005) and being more similar to TG rickettsiae in the percentage of coding sequence per genome and number of predicted genes (Figure 28.2), *R. canadensis* shares more small repetitive elements with SFG rickettsiae than TG rickettsiae (Eremeeva et al., 2005). However, until further evidence presents an alternative position for *R. canadensis* in the rickettsial tree other than that shown in Figures 28.1 and 28.2, we retain *R. canadensis* within AG rickettsiae.

We previously reported on the genomic similarities shared between AG rickettsiae (particularly *Rickettsia bellii*) and *R. felis*, the causative agent of murine-like typhus in cats and their associated fleas (Gillespie et al., 2007). This, coupled with the strong support for a unique clade distinct from the SFG rickettsiae comprised of *R. felis*, *R. akari*, *R. australis* (Figure 28.2), and several other insect-associated rickettsiae seemingly not pathogenic in vertebrates, is leading to the characterization of a larger group of rickettsiae with species that either harbor plasmids or have a history of containing plasmids throughout their evolution. The genomic comparisons of at least *R. felis*, *R. akari* and two strains of *R. bellii* hint at the exclusive presence of plasmid-associated genes, such as conjugative machinery and toxin-antitoxin genes (unpublished data), that illustrate an interesting and unique phenomenon within the rickettsial tree. Indeed, plasmids are now known from members of the SFG rickettsiae (Baldrige et al., 2007a, 2007b). The little-understood contributions that plasmids have had in shaping the diversity of rickettsiae and the potential role plasmids have in virulence are exciting aspects of future work on TRG rickettsiae.

RICKETTSIAL VIRULENCE AND PATHOGENESIS

All known and extant members of the genus *Rickettsia* are likely associated with arthropods during some phase of their life cycle. Rickettsial maintenance, however, involves either horizontal (arthropod-vertebrate cycle) or vertical transmission or both as in all known pathogenic members. Traditionally, rickettsiae that are transmitted via blood-feeding arthropods have been classified into either the SFG rickettsiae, which include a diverse array of more than 20 tick-transmitted species, or the TG rickettsiae, which include only *R. prowazekii* and *R. typhi*. Thus, genus *Rickettsia* includes species with various degrees of pathogenicity for eukaryotic hosts, including mammals and arthropods. *R. prowazekii* and *R. rickettsii* produce severe and often fatal disease in humans, and their maintenance in their arthropod vectors (human body lice and tick, respectively) is lethal to the host (Azad and Beard, 1998). Yet, infection with *R. typhi*, while symptomatic and even fatal in humans, has no effect on the vector fleas. While both *R. prowazekii* and *R. typhi* infection is lethal for human body lice, infection with two pathogenic SFG (*R. rickettsii* and *R. conorii*) had no effect on louse fitness (Azad and Beard, 1998). In contrast to TG, not all SFG rickettsiae cause disease in vertebrate hosts; for example, *R. montanensis, R. peacockii,* and *R. rhipicephali* are proposed to be nonpathogenic for mammals and lack any apparent adverse effects on the survival and fitness of their tick hosts (Azad and Beard, 1998).

In the past 10 years, the application of tools of molecular biology and recombinant DNA technology for rickettsial diagnosis resulted in the discovery of several new species of pathogenic rickettsiae. Of great interest was the completion of both *Rickettsia prowazekii* and *R. conorii* genome sequences (Andersson et al., 1998; Ogata et al., 2000, 2001). Today, GenBank contains more than a dozen genomes of the members of the genus *Rickettsia*. While whole genome sequencing revealed some intriguing information (e.g., phylogenetically, rickettsiae are closest to mitochondria; extensive numbers of pseudogenes in varying states of degradation; genome reduction and loss of many genes needed for free living status), it also confirmed many features of rickettsial pathogens that are already known (e.g., ATP/ADP novel transport system, absence of rickettsial bacteriophages, lack of evidence for conjugal acquisition of pathogenicity islands, etc.). Although rickettsial invasion and destruction of the eukaryotic host cells is considered the basis for rickettsial pathogenicity, the precise role of presumed "virulence" genes requires further elucidation. The question that remains to be answered is: what are the underlying mechanisms that make some strains of *Rickettsia* so virulent and others avirulent? While the relevance of these findings to virulence is still an open question, a microarray approach may provide insights into the underlying molecular basis of rickettsial virulence. It is just a matter of time to define the genetic basis of rickettsial pathogenicity.

CHANGING ECOLOGY OF RICKETTSIAL PATHOGENS

Historically, and in the present, rickettsial importance in terms of morbidity and mortality has been underestimated worldwide due to misdiagnosis. The advent of molecular diagnosis and the importance of selected rickettsial pathogens as biothreat agents have rekindled interest in rickettsioses. Recent serosurveys have demonstrated a high prevalence of rickettsial diseases worldwide, particularly in warm and humid climates. Additionally, new molecular diagnostic technology helped to narrow the gap of the continental divide in rickettsial distributions. For example, several rickettsial species that were known only in the United States are now reported in Central and South America. Because of the diversity of rickettsial pathogens throughout the world and variable clinical manifestations, we limit our coverage to four important rickettsial pathogens: (1) the louse-borne *Rickettsia prowazekii*, (2) the flea-borne *R. typhi*, (3) the tick-borne *R. rickettsii*, and (4) the scrub typhus agent, *Orientia tsutsugamushi*.

Classic epidemic typhus is one of the most virulent diseases known to humanity. Symptoms appear about 10 days after an infected body louse has bitten a person, and include a high fever of about 42°C, extreme pain in the muscle and joints, stiffness, and cerebral impairment. During the second week, the patient may become delirious with neurological symptoms and may experience stupor. Gangrene and necrosis may occur due to thrombosis of the small vessels in the extremities. Mortality rate in untreated patients is approximately 20%. In severe epidemics, the mortality rate is often as high as 40%. The human body louse, *Pediculus humanus corporis*, is the principal vector for *Rickettsia prowazekii*. Although the head louse, *P. h. capitis*, is capable of maintaining *R. prowazekii* experimentally, its role in the transmission of this rickettsiosis is not well established. Body lice feed only on humans although the laboratory-adapted colony can be maintained on rabbit, and all three stages of the louse life cycle (i.e., eggs, nymphs, and adults) may occur on the same host. The lice prefer a lower temperature (20°C) and are normally found in the folds of clothing. The body louse will abandon a patient with a temperature greater than 40°C to seek another host. This attribute is a major factor in the transmission of typhus within the susceptible population. Humans serve as host to *R. prowazekii* and lice, and are reservoirs of the rickettsiae. In addition, humans serve as a mobile component of the louse-borne typhus cycle, such that behavior influences the pattern of typhus transmission. The conditions that allow for the coexistence of body lice and a susceptible population could be the starting point for the epidemic of *R. prowazekii* to flare in refugee camps.

The louse-borne epidemic typhus is still endemic in highlands and cold areas of Africa, Asia, Central and South America, and in part of Eastern Europe. Another face of typhus, hardly studied in depth, is recrudescent typhus (Brill-Zinsser disease), in which the symptoms are less pronounced and the mortality rate is less than 1%. However, these patients could serve as a long-range source of *Rickettsia prowazekii*, permitting transmission of rickettsiae to occur months to years after the primary infection. In the United Sates, *R. prowazekii* is maintained in the sylvatic form involving the flying squirrel and its fleas and lice. Fortunately, immunity to typhus rickettsiae develops after recovery from infection. Furthermore, successful treatment with tetracycline and doxycycline is the approach of choice to reduce morbidity and eliminate mortality in susceptible populations. Although vaccination against *R. prowazekii* has been partially achieved with inoculation of inactivated rickettsiae or attenuated strains (e.g., str. Madrid E), unfortunately these vaccine approaches have been accompanied with undesirable toxic reactions and difficulties in standardization.

Murine typhus. In the United States, a major concern is the changing ecology of murine typhus in both south Texas and southern California where the classic cycle of *Rickettsia typhi,* which involves commensal rats and primarily the rat flea (*Xensopsylla cheopis*), has been replaced by the Virginia opossum (*Didelphis virginiana*)/cat flea (*Ctenocephalides felis*) cycle. Curiously, however, infected rats and their fleas are difficult to document within Texas and California's murine typhus foci (Azad et al., 1997; Azad and Beard, 1998). Similarly, the association of 33 cases of locally acquired murine typhus in Los Angeles County with seropositive domestic cats and opossums was also confirmed. Additionally, based on serological surveys, *R. typhi* infections also occur in inland cities in Oklahoma, Kansas, and Kentucky where urban and rural dwelling-opossums thrive. Thus, the maintenance of *R. typhi* in the cat flea/opossum cycle is of potential public health importance and a major health risk considering the distribution of opossum, which spans across the United States, Mexico, Central America, and Canada. Our search for the rural reservoir of murine typhus also resulted in the discovery of *R. felis*, the second flea-borne rickettsiosis in opossums (Azad et al., 1997; Azad and Beard, 1998). Urban rat/flea populations are still the main reservoir of *R. typhi* worldwide and particularly in many cities where urban settings provide a constellation of factors for the perpetuation of the *R. typhi* cycle, including declining infrastructures, increased immunocompromised populations, homelessness, and high population density of rats and fleas.

Rocky Mountain spotted fever (RMSF) is one of the most virulent human infections in the United States. *Rickettsia rickettsii*, the causative agent of RMSF, is a true zoonotic bacterium that cycles between ixodid ticks and wildlife populations, not only in the United States, but also in Mexico and in Central and South America. RMSF, like all rickettsial infections, is classified as a zoonosis requiring a biological vector such as tick to be transmitted between animal hosts (and accidentally to humans). Human infection occurs via the bite of an infected tick. Initial signs and symptoms of the disease include fever, headache, and muscle pain. This is followed by rash and organ-specific symptoms such as nausea, vomiting, and abdominal pain. Delayed treatment results in hospitalization and sequelae, such as amputation, deafness, and permanent learning impairment. The disease can be difficult to diagnose in the early stages due to nonspecific presentations and, unfortunately, in the absence of prompt and appropriate treatment, it kills the infected patients. Mortality of up to 75% had been reported before antibiotic discovery, and even today 5 to 10% of children and adults die from RMSF and many more require intensive care. RMSF is a reportable disease in the United States although the number of reported cases annually ranges between 250 and 1200.

All SFG rickettsiae are transmitted by ixodid ticks (Azad and Beard, 1998; Parola et al., 2005). In addition to *R. rickettsii*, the etiologic agent of RMSF, several other tick-borne rickettsial species are also human pathogens (Table 28.1). In the United States, at least four other SFG rickettsiae — *R. amblyommii, R. montanensis, R. peackockii,* and *R. rhipicephali* — are isolated only from ticks and cause limited or no known pathogenicity to humans or certain laboratory animals. However, *R. parkeri*, which was originally considered nonpathogenic, was recently identified as a human pathogen; thus, the other four rickettsial species could also be etiologic agents of as-yet-undiscovered, less severe rickettsioses.

Distribution of SFG rickettsiae is limited to that of their tick vectors. In the United States, a high prevalence of SFG species in ticks cannot be explained without the extensive contributions of transovarial transmission. The transovarial and transstadial passage of SFG rickettsiae within tick vectors in nature ensures rickettsial survival without requiring the complexity inherent in an obligate multihost reservoir system (Azad and Beard, 1998). Although many genera and species of ixodid ticks are naturally infected with rickettsiae, *Dermacentor andersoni* and *D. variabilis* are the major vectors of *Rickettsia rickettsii* in the United States.

Scrub typhus is an acute, febrile, infectious illness caused by *Orientia* (formerly *Rickettsia*) *tsutsugamushi*. The genus *Orientia* contains only a single species, *O. tsutsugamushi*, which is an obligate intracellular Gram-negative bacterium with a different cell wall structure, lack of peptidoglycan and lipopolysaccharide, and a different genetic composition as compared to *Rickettsia*. Humans acquire the disease via the bite of infected larval stage of numerous species of trombiculid mites (*Leptotrombidium* spp.). Scrub typhus is endemic in regions of eastern Asia and the southwestern Pacific (Korea to Australia) and from Japan to India and Pakistan. Although no significant morbidity or mortality occurs in patients who receive appropriate treatment, complications such as pneumonia, myocarditis, and disseminated intravascular coagulation that results in fatality could range from 0 to 30% in untreated patients or those infected with antibiotic resistant *O. tsutsugamushi* strains.

The current urban and rural cycles of rickettsial pathogens have a potential for future outbreaks amid the drastic increase in housing construction and related services that provide mammalian reservoirs and their blood-sucking ectoparasites ample harborage and proximity to human habitations. Considering the existing trends, rickettsial agents will continue to be introduced into human populations, and there are compelling reasons why this trend will be continued, as susceptible populations will be likely targets for emerging pathogens. In addition, the completed annotation of over a dozen sequenced rickettsial genomes has advanced molecular diagnosis to the extent that it has now changed the eco-dynamics and compositional pattern of rickettsial species throughout the world as well as the United States.

RICKETTSIAL PATHOGENS AS BIOTHREAT AGENTS

Throughout history the epidemics of louse-borne typhus have been important in the molding of human destiny, and are credited with causing more deaths than all the wars in history (reviewed in Azad and Radulovic, 2003). For example, more than 30 million cases of louse-borne typhus occurred during and immediately after World War I, causing an estimated 3 million deaths. The explosive spread of the brutal epidemic of louse-borne typhus within crowded human populations in the wake of war, famine, flood, and other disasters made a deep impression on the commanders of the Russian Red Army, and by 1928, *Rickettsia prowazekii* was transformed into a battlefield weapon. Many years later, the Japanese Army successfully tested biobombs, containing *R. prowazekii,* causing outbreaks of typhus. Thus, the precedent exists in utilizing pathogenic rickettsiae in warfare, and recent increased risk of misuse of bacterial pathogens as weapons of terror is no longer fiction (reviewed in Azad and Radulovic, 2003).

The rickettsial diseases vary from mild to very severe clinical presentations, with case fatality ranging from 2% to over 30% (Table 28.4). The severity of rickettsial disease has been associated with age, delayed diagnosis, hepatic and renal dysfunction, CNS abnormalities, and pulmonary compromise. Despite the variability in clinical presentations, pathogenic rickettsiae cause debilitating diseases, and several rickettsial pathogens could be used as a potential biological weapon. Realistically, only *Rickettsia prowazekii* and *R. rickettsii* pose serious problems. Table 28.4 also compares several salient features of pathogenic rickettsial species with selected category A agents. Despite the fact that effective chemotherapy is available and effective control measures are known, rickettsial diseases continue to be a problem in the United States and many other parts of the world.

ACKNOWLEDGMENTS

The work presented in this chapter is supported in part by NIH/NIAID grants to AFA. JJG acknowledges support from NIAID Contract HHSN266200400035C awarded to Bruno Sobral (Virginia Bioinformatics Institute at Virginia Tech).

TABLE 28.3

Phenotypic and Genotypic Differences between Typhus (TG), Spotted Fever (SFG), Transitional (TRG), and Ancestral (AG) Group Rickettsiae

Characteristics	TG	SFG	TRG	AG
Vector(s)	Flea/louse	Tick	Flea/mite/tick	Tick
In vitro growth[a]	Y	Y	Y	Y
Plaque formation	Y	Y/N	Y/N	N
Intranuclear growth	N	Y	Y	Y/N
Hemolytic activity	Y	N	Y/N	Y/N
rOmpA[b]	N	Y	Y*	Y
RickA[c]	N	Y	Y	Y
Plasmid	N	Y/N	Y/N	Y/N

Y/N: consistently yes or no. [a]Vertebrate/invertebrate established cell lines; [b]rOmpA: *sca0* gene and/or protein expression; [c]RickA: host actin polymerization and cell-to-cell motility. * rOmpA is truncated.

TABLE 28.4

Comparison of Selected Rickettsial Pathogens to Examples of Biological Warfare Bacteria

Bacteria	Disease/ Incubation[a]	Natural Hosts	Treatment/ Vaccine	Rapid Diagnosis	Transmission[b]	Mortality[c] Rate
Bacillus anthracis	Anthrax/ variable	+	+/+	+	Inh/Ing	5–80%
Clostridium botulinum	Botulism/12–36 hr	-	+/–	+	Ing	70%
Yersinia pestis	Plague 2–7 d	+	+/±	+	V/Inh	30–90%
Burkholderia mallei	Glanders/3–6 d	+	+/–	+	Inh/C	95%
Francisella tularensis	Tularemia/2–10 d	+	+/–	+	C/V	5%
Rickettsia prowazekii	Epidemic typhus/ 6–14 d	+	+/–	+	V/inh	30%
Rickettsia rickettsii	Rocky Mountain/ 3–15 d spotted fever	+	+/–	+	V/inh	20–25%
Rickettsia typhi	Endemic typhus/ 6–14 d	+	+/–	+	V/Inh	4%
Coxiella burnetti	Q-fever/7–14 d	+	+/–	+	Inh/V	2-3%

[a] Incubation period: dependent upon bacterial dose and mode of exposure; [b]Mode of exposure: C: cutaneous; Inh: inhalation (aerosol); Ing: ingestion; V: vector-borne; [c]Fatality rate (dependent upon the mode of exposure).

REFERENCES

Andersson, S.G., Zomorodipour, A., Andersson, J.O., Sicheritz-Ponten, T., Alsmark, U.C., Podowski, R.M., Naslund, A.K., Eriksson, A.S., Winkler, H.H., Kurland, C.G. (1998). The genome sequence of *Rickettsia prowazekii* and the origin of mitochondria. *Nature*, 396:133–140.

Andersson, J.O. and Andersson, S.G.E. (1999). Insights into the evolutionary process of genome degradation. *Curr. Opin. Genet. Dev.*, 9:664–671.

Azad, A.F., Radulovic, S., Higgins, J.A., Noden, B.H., and Troyer, M.J. (1997). Flea-borne rickettsioses: some ecological considerations. *Emerg, Infect. Dis.*, 3:319–328.

Azad, A.F. and Beard, C.B. (1998). Interactions of rickettsial pathogens with arthropod vectors. *Emerg. Infect. Dis.*, 4:179–186.

Azad, A.F. and Radulovic, S. (2003). Pathogenic rickettsiae as bioterrorism agents. *Ann. NY Acad. Sci.*, 990:1–5.

Baker, B.J., Hugenholtz, P., Dawson, S.C., and Banfield, J.F. (2003). Extremely acidophilic protists from acid mine drainage host Rickettsiales-lineage endosymbionts that have intervening sequences in their 16S rRNA genes. *Appl. Environ. Microbiol.,* 69:5512–5518.

Baldridge, G.D., Burkhardt, N.Y., Felsheim, R.F., Kurtti, T.J., and Munderloh, U.G. (2007a). Transposon insertion reveals pRM, a plasmid of *Rickettsia monacensis. Appl. Environ. Microbiol.*, 73:4984–4995.

Baldridge, G.D., Burkhardt, N.Y., Felsheim, R.F., Kurtti, T.J., and Munderloh, U.G. (2007b). Plasmids of the pRM/pRF family occur in dierse *Rickettsia* species. *Appl. Environ. Microbiol.*, 74:645–652.

Birtles, R.J., Harrison, T.G., Saunders, N.A., and Molyneux, D.H. (1995). Proposals to unify the genera *Grahamella* and *Bartonella*, with descriptions of *Bartonella talpae* comb. nov., *Bartonella peromysci* comb. nov., and three new species, *Bartonella grahamii* sp. nov., *Bartonella taylorii* sp. nov., and *Bartonella doshiae* sp. nov. *Int. J. Syst. Bacteriol.*, 45:1–8.

Blanc, G., Ogata, H., Robert, C., Audic, S., Suhre, K., et al. (2007). Reductive genome evolution from the mother of *Rickettsia*. PLoS Genetics 3: e14.

Boone, D.R., Castenholz, R.W., and Garrity, G.M. (2001). *Bergey's Manual of Systematic Bacteriology.* Springer, New York.

Brenner, D.J., O'Connor, S.P., Winkler, H.H., and Steigerwalt, A.G. (1993). Proposals to unify the genera *Bartonella* and *Rochalimaea*, with descriptions of *Bartonella quintana* comb. nov., *Bartonella vinsonii* comb. nov., *Bartonella henselae* comb. nov., and *Bartonella elizabethae* comb. nov., and to remove the family Bartonellaceae from the order Rickettsiales. *Int. J. Syst. Bacteriol.*, 43:777–786.

Brinton, L.P. and Burgdorfer, W. (1971). Fine structure of *Rickettsia canada* in tissues of *Dermacentor andersoni* Stiles. *J. Bacteriol.,* 105:1149–1159.

Burgdorfer, W. (1968). Observations on *Rickettsia canada*, a recently described member of the typhus group rickettsiae. *J. Hyg. Epid. Microbiol. Immunol.*, 12:26–31.

Ching, W.M., Dasch, G.A., Carl, M., and Dobson, M.E. (1990). Structural analyses of the 120-kDa serotype protein antigens of typhus group rickettsiae: comparison with other S-layer proteins. *Ann. NY Acad. Sci.*, 590:334–351.

Dasch, G.A. and Bourgeois, A.L. (1981). Antigens of the typhus group of rickettsiae: importance of the species-specific surface protein antigens in eliciting immunity. In *Rickettsiae and Rickettsial Diseases*, W. Burgdorfer and R.L. Anacker, Eds., Academic Press, New York, p. 61–70.

Dumler, J.S., Barbet, A.F., Bekker, C.P., Dasch, G.A., Palmer, G.H., Ray, S.C., Rikihisa, Y., and Rurangirwa, F.R. (2001). Reorganization of genera in the families Rickettsiaceae and Anaplasmataceae in the order Rickettsiales: unification of some species of *Ehrlichia* with *Anaplasma*, *Cowdria* with *Ehrlichia* and *Ehrlichia* with *Neorickettsia*, descriptions of six new species combinations and designation of *Ehrlichia equi* and 'HGE agent' as subjective synonyms of *Ehrlichia phagocytophila. Int. J. Syst. Evol. Microbiol.*, 51:2145–2165.

Emelyanov, V.V. (2003). Mitochondrial connection to the origin of the eukaryotic cell. *Eur. J. Biochem.*, 270:1599–1618.

Eremeeva, M.E., Madan, A., Shaw, C.D., Tang, K., and Dasch, G.A. (2005). New perspectives on rickettsial evolution from new genome sequences of rickettsia, particularly *R. canadensis,* and *Orientia tsutsugamushi. Ann. NY Acad. Sci.*, 1063:47–63.

Federici, B.A. (1980). Reproduction and morphogenesis of *Rickettsiella chironomi*, an unusual intracellular procaryotic parasite of midge larvae. *J. Bacteriol.*, 143:995–1002.

Felsenstein, J., (1985). Confidence limits on phylogenies: An approach using the bootstrap. *Evolution* 39: 783–791.

Gillespie, J.J., Beier, M.S., Rahman, M.S., Ammerman, N.C., Shallom, J.M., Purkayastha, A, Sobral, B.S., and Azad, A.F. (2007). Plasmids and Rickettsial Evolution: Insight from *Rickettsia felis*. PLoS ONE 2: e266.

Götz, P. (1971). "Multiple cell division" as a mode of reproduction of a cell-parasitic bacterium. *Naturwissenschaften*, 58: 569–570.

Götz, P. (1972). "*Rickettsiella chironomi*:" an unusual bacterial pathogen which reproduces by multiple cell division. *J. Invertebr. Pathol.*, 20:22–30.

Gromov, B.V. and Ossipov, D.V. (1981). "*Holospora* (ex Hafkine 1890) nom. rev., a genus of bacteria inhabiting the nuclei of paramecia." *Int. J. Syst. Bacteriol.*, 31:348–352.

Hackstadt, T. (1996). The biology of rickettsiae. *Infect. Agents Dis.*, 5:127–143.

Higgins, J.A., Radulovic, S., Schriefer, M.E., and Azad, A.F. 1996. *Rickettsia felis*: a new species of pathogenic rickettsia isolated from cat fleas. *J. Clin. Microbiol.*, 34:671–674.

Myers, W.F. and Wisseman, C.L. Jr. (1981). The taxonc relationship of *Rickettsia canada* to the typhus and spotted fever groups of the genus *Rickettsia*. In *Rickettsiae and Rickettsial Diseases*, W. Burgdorfer and R.L. Anacker, Eds., Academic Press, New York, p. 313–325.

Ogata, H., Audic, S., Renesto-Audiffren, P., Fournier, P.E., Barbe, V., Samson, D., Roux, V., Cossart, P., Weissenbach, J., Claverie, J.M., and Raoult, D. (2001). Mechanisms of evolution in *Rickettsia conorii* and *R. prowazekii*. *Science*, 293:2093–2098.

Ogata, H., Audic, S., Barbe, V., Artiguenave, F., Fournier, P.E., Raoult, D., and Claverie, J.M. (2000). Selfish DNA in protein-coding genes of *Rickettsia*. *Science*, 290:347–350.

Perlman, S.J., Hunter, M.S., and Zchori-Fein, E. (2006). The emerging diversity of *Rickettsia*. *Proc. Biol. Sci.*, 273:2097–2106.

Rikihisa, Y., Kawahara, M., Wen, B., Kociba, G., Fuerst, P., Kawamori, F., Suto, C., Shibata, M., and Futohashi, M. (1997). Western immunoblot analysis of *Haemobartonella muris* and comparison of 16S rRNA gene sequences of *H. muris*, *H. felis*, and *Eperythrozoon suis*. *J. Clin. Microbiol.*, 35:823–829.

Roux, V., Bergoin, M., Lamaze, N., and Raoult, D. (1997). Reassessment of the taxonomic position of *Rickettsiella grylli*. *Int. J. Syst. Bacteriol.*, 47:1255–1257.

Sekeyova, Z., Roux, V., and Raoult, D. (2001). Phylogeny of *Rickettsia* spp. inferred by comparing sequences of 'gene D' which encodes an intracytoplasmic protein. *Int. J. Syst. Evol. Microbiol.*, 51:1353–1360.

Stothard, D.R. and Fuerst, P.A. (1995). Evolutionary analysis of the spotted fever and typhus groups of *Rickettsia* using 16S rRNA gene sequences. *Syst. Appl. Microbiol.*, 18:52–61.

Swofford, D. (1999). PAUP*: Phylogenetic Analysis Using Parsimony (*and Other Methods), version 4 ed. Sinauer, Sunderland, MA.

Tamura, A., Ohashi, N., Urakami, H., and Miyamura, S. (1995). Classification of *Rickettsia tsutsugamushi* in a new genus, *Orientia* gen. nov., as *Orientia tsutsugamushi* comb. nov. *Int. J. Syst. Bacteriol.*, 45:589–591.

Vitorino, L., Chelo, I.M., Bacellar, F., and Zé-Zé, L. (2007). Rickettsiae phylogeny: a multigenic approach. *Microbiology,* 153:160–168.

Weisburg, WG, Dobson, ME, Samuel, JE, Dasch, GA, Mallavia, LP, Baca, O, Mandelco, L., Sechrest, J.E., Weiss, E., and Woese, C.R. (1989). Phylogenetic diversity of the Rickettsiae. *J. Bacteriol.*, 171:4202 4206.

Weiser, J. and Zizka, Z. (1968). Electron microscope studies of *Rickettsiella chironomi* in the midge *Camptochironomous tentans*. *J. Invertebr. Pathol.*, 12:222–230.

29 Chlamydia

Lourdes G. Bahamonde

CONTENTS

INTRODUCTION

The last CRC series chapter on Chlamydiae was written by Leslie A. Page and published in 1989. Since then, significant advances in Chlamydia diagnostics, genomics, pharmacotherapeutics, classification of new species, and bioinformatics have come to light. This chapter provides a review of our current understanding of chlamydial identification, while addressing how its classification has evolved to date.

In 1966, Page was the first to publish evidence for the order Chlamydiales, the family Chlamydiaceae, and the genus *Chlamydia*.[1] The classification system described at the time was based on a chlamydia-specific developmental cycle that ultimately separated the *Chlamydiae* from the *Rickettsia* species[2] — both obligate intracellular parasites. The identification of *C. trachomatis* and *C. psittaci* species were subsequently possible after the American Society of Microbiology adopted stricter numerical — plus (+) and minus (–) — taxonomic criteria for bacterial classification.[2, 3] These numerical criteria were organized into matrices and only included phenotypic characteristics of bacteriological "morphology, biochemistry, culture composition, physiology, nutritional requirements, antigenic composition, and phage sensitivity."[2] However, relying on phenotypic characteristics and numerical matrices is problematic because divergent organisms sharing the same environment typically have similar phenotypes, without evidence of evolutionary relatedness. Thus, phenotypic criteria may not necessarily imply relatedness, but rather evolutionary convergence.

In 1989, DNA-DNA hybridization studies of *Chlamydia* revealed far more diversity in the order than previously thought. Most *C. psittaci* strains for example, had less than 70% hybridization similarity with each other, a common cutoff for the establishment of new species.[4] Based on this criterion, two new species were proposed: *C. pecorum* and *C. pneumoniae*. Within the *C. trachomatis* species, similar difficulties were being observed, leading to a need for more consistent species criteria.

To definitively establish the number and type of chlamydia species, researchers at the U.S. Department of Agriculture (USDA) National Animal Disease Center set about developing a DNA-based classification system based on 16S and 23S ribosomal RNA genes and the major outer membrane protein (MOMP) gene. Based on the gene analysis findings, the order *Chlamydiales* was redefined. Instead of one family, Chlamydiaceae, there are now four families: Chlamydiaceae, Parachlamydiaceae, Simkaniaceae, and Waddliaceae. Strains with greater than 90% 16S rRNA sequence identity remained in the Chlamydiaceae family, while those with 80 to 90% identity were separated into the three new families.[4]

Within the Chlamydiaceae family, the genus *Chlamydia* was expanded to include not just *C. trachomatis*, but also *C. suis*, and *C. muridarum*. The species *C. psittaci* and its associated species *C. pecorum* and *C. pneumoniae* were moved to a new genus (*Chlamydophila*) and joined by three new species (*Chlamydophila felis*, *Chlamydophila caviae*, and *Chlamydophila abortus*) (See Figure 29.1 for an overview of taxonomy). *Chlamydophila* is distinguishable as a separate genus based on 16S and 23S rRNA sequences, which among the *Chlamydophila* are over 95% identical. The *Chlamydophila* genome is slightly larger than the *Chlamydia* genome and contains a single ribosomal operon, while the *Chlamydia* genome contains two ribosomal operons.[4]

Life Cycle

Chlamydia are obligate intracellular bacteria that exist in two morphologically distinct forms and are characterized by a biphasic life cycle under cell culture conditions.[5, 6] The bacteria alternate between a small (300 to 400 nm) infectious extracellular form — the elementary body (EB) and a noninfectious, larger (800 to 1000 nm) intracellular replicating form — the reticulate body (RB).[5, 7] Resistant to environmental factors, EB are spore-like and infectious, albeit unable to replicate outside of a host. EB are comprised of DNA wrapped around histone-like proteins and enveloped by an outer membrane of cross-linked proteins with disulfide bonds between cysteine residues for osmotic stability.[5]

The EB life cycle begins when they attach to and stimulate phagocytosis by host cells (Figure 29.2). Within the phagosome, named an inclusion, the EB differentiate into metabolically active RB, which later undergo multiple cycles of binary fission leading to their secondary differentiation back to an EB form.[6, 8] Lysis of the host cell releases a new generation of infectious EB that propagate among neighboring cells.[8] This life cycle often lasts 2 to 3 days, with some chlamydia-like strains taking up to 14 days to propagate. The conditions under which this propagation occurs can lead to aberrant RB and can determine whether the host will continue to express infective EB forms with each bacterial cell division.[8] Propagation is linked to host immune responses *in vivo*; therefore, proinflammatory chemokines, cytokines, nutrient availability, and antibiotics are among

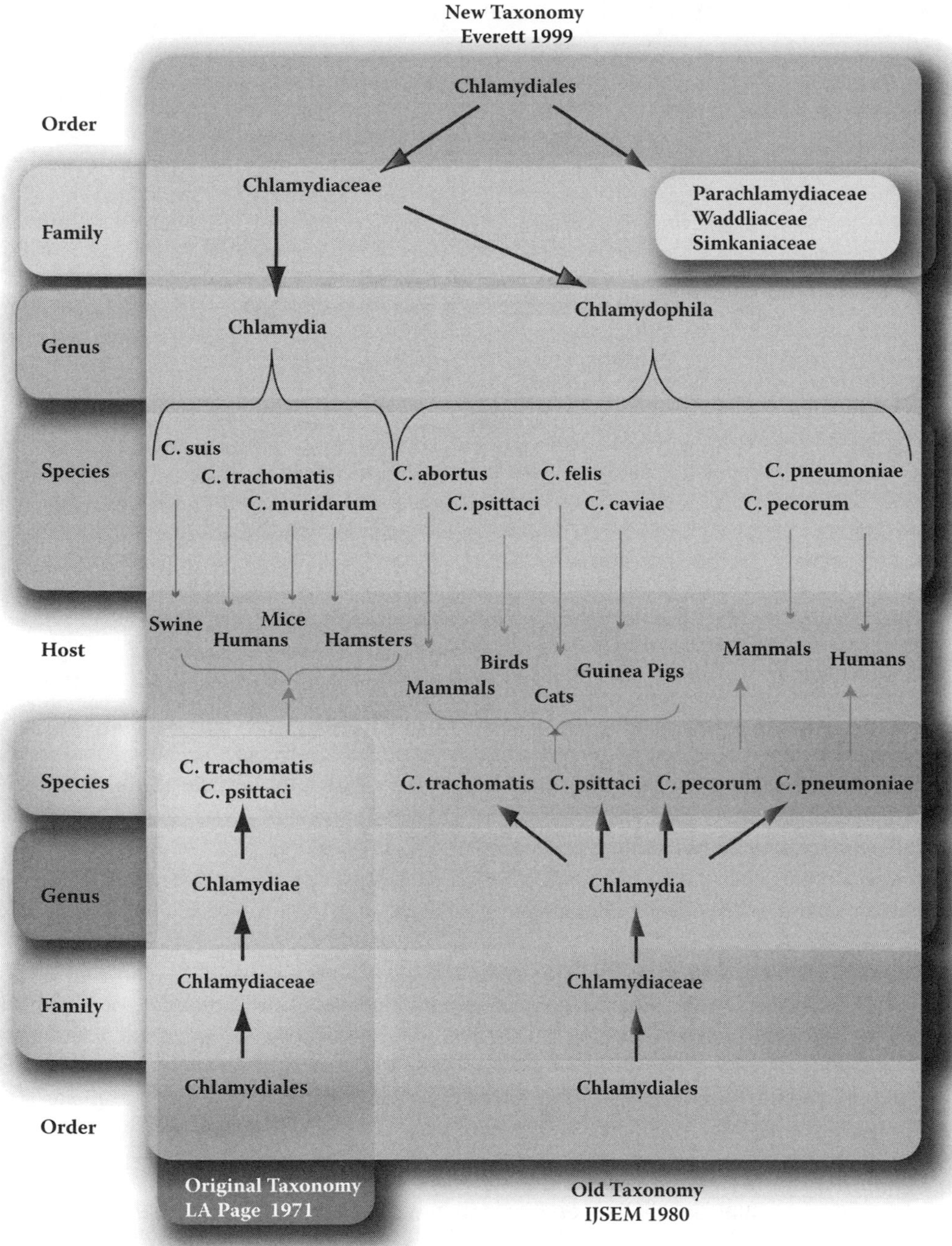

FIGURE 29.1 Overview showing the evolution of *Chlamydia* taxonomy.

some of the environmental factors that can disrupt the EB life cycle.[9] Characterization of the *in vitro* persistent phase of infection and *in vivo* models suggests that chlamydia can persist in an altered form.[6] Recurrent chlamydial infection may result from either re-infections or persistent, unresolved infections.[6, 9]

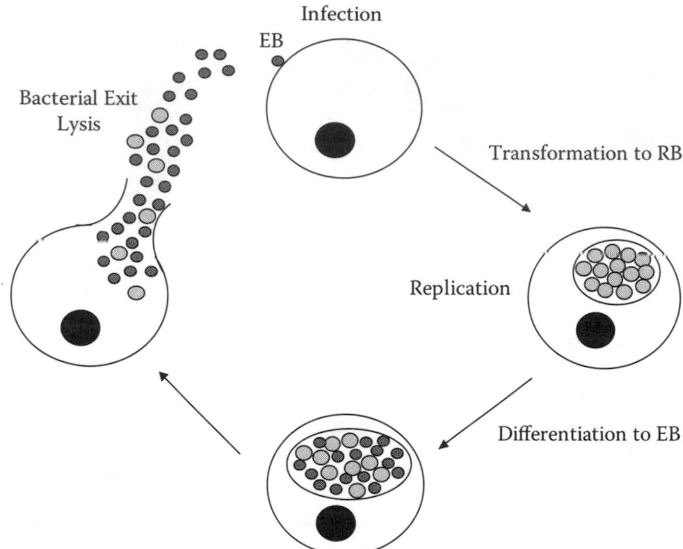

FIGURE 29.2 The *Chlamydia* life cycle.

IDENTIFICATION

Phenotypic Markers

The genus *Chlamydia* is marked by the presence of intracellular glycogen; unfortunately, the genus *Chlamydophila* is marked by neither the presence nor absence of glycogen because some species show transient levels of glycogen during their life cycle. Originally, resistance to sulfadiazine distinguished chlamydial species, but this too has turned out to be highly variable, with resistance being the exclusive domain of neither *Chlamydia* nor *Chlamydophila*.

It is worth noting that the classification of antibiotic resistance among species prior to 1999 is unreliable because a DNA-typing, taxonomic classification system of species had not yet been adopted; for example, data on antibiotic sensitivity and resistance for *C. trachomatis* prior to 1999 may have inadvertently included *C. trachomatis* and *C. suis* strains.

The Chlamydiaceae family is marked by detection with monoclonal antibodies (mAbs) to the group-specific lipopolysaccharide αKdo-(2→8)-αKdo-(2→4)-αKdo (originally called the genus-specific epitope). *Chlamydia* can be differentiated from *Chlamydophila* by antibodies that recognize epitopes (NPTI, TLNPTI, LNPTIA, LNPTI) in the variable region 4 of the major outer membrane protein (MOMP) of *Chlamydia,* but not *Chlamydophila*. Because these mAbs are routinely used to serotype human *C. trachomatis* isolates, cross-reaction may lead to errors in species identification.[10]

DNA Markers

Phenotypic markers cannot reliably differentiate between the nine chlamydial species, thus DNA typing holds the best chance of identifying them. This makes sense because the nine species were established via DNA sequence analysis in the first place. Polymerase chain reaction (PCR) primers are available and able to amplify the MOMP, 16S rRNA, and 23S rRNA genes.[10] The MOMP primers are specific for all Chlamydiaceae, while the 16S and 23S primers are specific for all Chlamydiales.[10] To identify species-specific DNA, sequencing would need to be performed after amplification. An established polymerase chain reaction-restriction fragment length polymorphism (PCR-RFLP) system can detect and differentiate between all nine Chlamydiaceae species.[11] In addition, established real-time PCR protocols can detect and quantitate the four species: *Chlamydia trachomatis*, *Chlamydophila psittaci*, *Chlamydophila pneumoniae*, and *Chlamydophila pecorum*. As little as 250 fg of Chlamydial DNA is detectable, showing that real-time PCR has the potential to be both a rapid and economical means of clinical detection.[12]

CHLAMYDIACEAE FAMILY

Chlamydiaceae, the original family, now includes two genera — *Chlamydia* and *Chlamydophila*, encompassing nine species. All are Gram-negative and express the family-specific lipopolysaccharide epitope αKdo-(2→8)-αKdo-(2→4)-αKdo. The Chlamydiaceae EB's extracellular osmotic stability is maintained by a complex of disulfide-cross-linked envelope proteins that include the 40-kDa major outer membrane protein, a hydrophilic cysteine-rich 60-kDa protein, and a low-molecular-mass, cysteine-rich lipoprotein.[2]

CHLAMYDIA TRACHOMATIS

Chlamydia trachomatis has only been isolated from humans and is comprised of two biovars that are transmitted by sexual or other contact. The complete gene sequences of two strains (A/HAR-13 and D/UW-3/CX) have been determined.[13, 14] *C. trachomatis* strains are generally sensitive to sulfadiazine and tetracyclines, and most have an extrachromosomal plasmid.

CHLAMYDIA MURIDARUM

Isolated from mice and hamsters, the two strains of *Chlamydia muridarum* (MoPn and SFPD) had originally belonged to *C. trachomatis*.[4] SFPD is not known to cause disease, while MoPn may produce pneumonia in mice or cause an asymptomatic infection. The MoPn genome has been sequenced, and the isolate used contained an extrachromosomal plasmid.[15] MoPn is sensitive to sulfadiazine.

CHLAMYDIA SUIS

Chlamydia suis strains were originally referred to as *C. trachomatis* due to DNA sequence homology in the MOMP gene (*ompA*).[10] They have only been isolated from swine, where they cause conjunctivitis, enteritis, pneumonia, and asymptomatic infections. Several strains of *C. suis* carry an extrachromosomal plasmid. Some strains are resistant to sulfadiazine and/or tetracycline.

CHLAMYDOPHILA PSITTACI

Eight *Chlamydophila psittaci* serovars (A through F, M56, WC) and, likely, nine genotypes have been identified among specific bird groups.[16] *C. psittaci* previously contained strains that not only infected birds, but also were associated with abortion and disease in house cats and guinea pigs. After reclassification, the species primarily became associated with infections in birds.[10] The infection is often systemic and can be unapparent, severe, acute, or chronic associated with intermittent shedding and passed through eggs. Stress often triggers the onset of severe symptoms, resulting in rapid deterioration and death. One serovar is found among psittacine birds, while another is endemic to pigeons and has been found in turkeys. Several *C. psittaci* strains have an extrachromosomal plasmid and are sensitive to tetracycline and doxycycline.[10]

CHLAMYDOPHILA PNEUMONIAE

Chlamydophila pneumoniae was previously synonymous with TWAR (a human biovar). However, *C. pneumoniae* can cause respiratory disease not only in humans, but also in koalas and horses, each with a distinct biovar — TWAR (humans), Koala, and Equine.[4] Of the three biovars, only the horse (N16 strain) is known to contain an extrachromosomal plasmid.[4] *C pneumoniae* are sensitive to tetracycline and macrolides and the complete gene sequence of four strains (AR39, CWL029, J138, and TW183) is now known.[15, 17, 18]

Chlamydophila pecorum

Originally designated *Chlamydia pecorum*, *Chlamydophila pecorum* has been isolated only from mammals and is serologically and pathogenically diverse. In many animals, it is associated with a range of diseases such as abortion, conjunctivitis, encephalomyelitis, enteritis, pneumonia, and polyarthritis. In koalas, *C. pecorum* causes reproductive disease, infertility, and urinary tract disease.[10, 2]

Chlamydophila felis

Endemic among domestic cats, *Chlamydophila felis* primarily causes conjunctivitis, rhinitis, and respiratory disease, with zoonotic infection reported in humans.[2] Strains differ in their pathogenesis although the causes of these pathogenic differences are not known.[10] The genome of strain Fe/C-56 has been sequenced.[19] Some strains, including the one sequenced, have an extrachromosomal plasmid. *C felis* is sensitive to tetracycline and macrolides.

Chlamydophila caviae

C. caviae specifically infects guinea pigs, causing conjunctivitis.[10] The genome of strain GPIC has been sequenced and it contains an extrachromosomal plasmid.[20]

Chlamydophila abortus

Chlamydophila abortus has been removed from under *C. psittaci* and made its own species.[4] Associated primarily with cases of abortion and poor prognosis in neonates, *C. abortus* infections attack the placenta. *C. abortus* is endemic among a variety of ruminants and has been associated with cases of abortion in animals ranging from horses, to rabbits, guinea pigs, mice, and pigs. In humans, documented cases of zoonotic abortion due to *C. abortus* have been seen in women working with sheep.[10] The complete genome has be sequenced, and no strain of *C. abortus* has been identified to contain an extrachromosomal plasmid.[21]

CHLAMYDIA-LIKE FAMILIES

Unlike with the Chlamydiaceae family, antibodies directed against the group-specific lipopolysaccharide αKdo-(2→8)-αKdo-(2→4)-αKdo do not recognize members of the Parachlamydiaceae, Simkaniaceae, or Waddliaceae families. These families belong to the order Chlamydiales due to the 80 to 90% identity within their 16S rRNA sequences.[4, 22]

Parachlamydiaceae

Originally described as *Candidatus Parachlamydia acanthamoebae*, the Parachlamydiaceae has been declared a new family. Isolated from amoebae, the Parachlamydiaceae can also infect *Dictyostelium* and can be grown in cultures of African Green Monkey Kidney cells (Vero).[4] Two genera have been isolated, *Parachlamydia* and *Neochlamydia,* with other isolates being considered.[10] Among case reports, at least one of these isolates has been found to increase the cytotoxicity of the amoeba they infect and was isolated from humans during an outbreak of humidifier fever in Vermont (United States) ("Hall's coccus").[4] Other members of this family have been detected in cats, Australian marsupials, reptiles, and fish.[22]

Simkaniaceae

The Simkaniaceae family originally contained only one strain: *Simkania negevensis.* First identified as a bacterial contaminant in cell cultures, the strain is associated with pneumonia in humans. The strain lacks any extrachromosomal plasmids and has a longer than normal developmental cycle in Vero cells (2 weeks vs. the normal 2 to 3 days).[10] Newly discovered members of this family, *Fritschea bemisiae* and *Fritschea eriococci*, have been found to infect invertebrates.[22]

Waddliaceae

The defining strain of this family, *Waddlia chondrophila,* was identified in 1986 as an agent of bovine abortion. Serological studies have shown that titers of anti-*Waddlia* antibodies are statistically associated with cows that have aborted.[10] A second family member, *Waddlia malaysiensis,* was identified in urine samples from Malaysian fruit bats (*Eonycteris spelaea*).[22]

HUMAN INFECTION AND DISEASE

Chlamydiae were believed to exclusively cause infection and disease in humans. However, as new and old species are genetically identified and classified, the epidemiology of *Chlamydia* in a vast number of animal and human hosts has come to light. There is also evidence that chlamydia-like strains can be found in soil and that chlamydia-infected amoeba can be transmitted to animals and humans alike.[10, 23] This demonstrates that chlamydia may have both direct and an indirect roles in disease.[10]

Chlamydophila felis

Chlamydial infection from a domestic cat with signs of conjunctivitis and rhinitis to a human owner who developed acute follicular conjunctivitis was first demonstrated by Schachter et al. in 1969.[24] Zoonotic chlamydial infection is rare and it may be underestimated secondary to faulty isolation and detection practices.[24] Seroepidemiological data on *C. felis* reported sera tested for anti-*C. felis* antibodies by microimmunofluorescence test (MIF), using purified elementary bodies and fluorescein-conjugated goat anti-cat immunoglobulin G sera (Euroclone).[24] However, sera positive for *C. felis* was also positive for *C. pneumoniae* and *C. psittaci*, demonstrating significant cross-reactivity between the species.[24] Given the higher seroprevalence of *C. pneumoniae* among humans, cross-reactivity is expected,[24] and thus molecular classification techniques should be used to improve the accuracy of evidence for zoonotic infection of *C. felis.*[25]

Recently, Parachlamydiaceae have been identified in cats with neutrophilic and eosinophilic conjunctivitis,[26] and the treatment of choice was found to be doxycycline, with azithromycin affording less efficacy.[26]

Chlamydophila abortus

Although *C. abortus* is mainly associated with late-gestation abortions in ruminants and pigs,[21] it has been implicated in rare but severe cases of septicemia in pregnant women, often resulting in abortions with a fetal mortality of 94% and maternal mortality of 6.3%.[27] After aborting, most mammals acquire immunity and rebreed successfully.[28] Early treatment with erythromycin should be considered the treatment of choice in suspected zoonotic exposure to *C. abortus*, especially because fetal abortions may be preventable with early administration of macrolides or tetracyclines.[27] Serologic diagnosis of *C. abortus* has focused on surface antigens, with MOMP and POMPS having the most diagnostic relevance.[28] MOMP VS1, and VS2 sequences have the greatest diversity between *C. abortus* and *C. pecorum*, making them sufficiently distinguishable for the development of specific competitive ELISA.[25, 28] POMPS has improved specificity to *C. abortus* serovars.[29]

Given that *C. abortus* zoonotic infections are rare, *C. abortus*-specific PCR has been considered impractical.[27] Instead, the use of broad-spectrum primers to all Chlamydiaceae has been found to be more realistic, because *C. trachomatis* infection and/or co-infection may also be detected in routine screening tests.[27]

Chlamydia trachomatis

Immunoreactivity of antibodies against epitopes on the *C. trachomatis* MOMP has enabled the identification of 18 distinct serovars, divided into three classes: B class (serovars B, D, E, L1, L2, L2a); C class (A, C, H, I, Ia, J, Ja, K, L3); and Intermediate class (F, G).[4, 30] Specifically, A, B, Ba, C are implicated in chronic eye infections (or trachoma) and serovars Ba, D–K, Da, Ia in urogenital infections.[30] The lymphogranuloma venerum (LGV) serovars L1–L3, L2a can cause more invasive urogenital disease, including painful buboes and proctatitis.[30, 31] *C. trachomatis* has also been implicated in some forms of arthritis, neonatal conjunctivitis, and pneumonia.[4]

Although clinicians consider chlamydia culture methods the gold standard for identification of chlamydial infection, it has low specificity and it is laboratory expertise-dependent.[31] Serotyping using direct fluorescent staining with monoclonal antibodies (DFA) is highly specific, but time and labor intensive.[31] *C. trachomatis* routinely cross-react with *C. suis* epitopes with DFA serotyping leading to errors in species identification.[10] Conversely, PCR has a sensitivity of 90% and specificity of 99%.[31] Treatment of choice against *C. trachomatis* includes macrolides (azithromycin, erythromycin); tetracyclines; quinolones (ciprofloxacin, ofloxacin); and penicillins (amoxicillin).[31]

Chlamydophila pneumoniae

The human *C. pneumoniae* strain TWAR is primarily a pathogen of the respiratory tract capable of causing acute or chronic bronchitis and pneumonia.[4] *Chlamydophila pneumoniae* has been identified in asymptomatic and co-infected individuals.[32] It has been implicated in obstructive pulmonary disease, cystic fibrosis, Alzheimer's disease, asthma, erythema nodosum, reactive airway disease, Reiter's syndrome, and sarcoidosis.[4] There is evidence that *C. pneumoniae* may be involved in atherosclerosis and the acceleration of atherosclerotic-lesion development in humans and mice.[33] Although *C. pneumoniae* primarily infects humans, mice have been found to be susceptible via intranasal inoculations, making them a useful animal model for the study of pneumonitis.[34] The atherogenic properties of *C. pneumoniae* were not influenced by antibiotic therapy, unless the antibiotics were administered in the early, acute state of infection.[33] Immunization against *C. pneumoniae* is the focus of current research.[25, 33] At present, the most reliable assay for the detection of *C. pneumoniae* is real-time PCR; however, no well-standardized assay has been approved by the U.S. Food and Drug Administration (FDA).[35]

Chlamydophila psittaci

Infection by *C. psittaci* causes psittacosis.[16] It has been particularly prevalent among bird slaughterhouse workers and owners of pet birds — with samples mainly of serovar A in psittacine birds and patients with psittacosis.[4, 16, 29] Zoonotic infection is thought to occur horizontally by diseased or subclinically infected birds through contaminated aerosols and droppings.[36] *C. psittaci* is implicated in respiratory disease; however, there are case reports of other infected organs leading to myocardiditis, endocarditis, hepatitis, encephalitis, and meningitis.[25] Currently, no vaccines have been developed against *C. psittaci*.[25] Treatment options for *C. psittaci* include bacteriostatic antibiotics effective against other Chlamydiaceae.

CONCLUSIONS

Due to being obligate pathogens, coupled with their inability to be genetically manipulated like other bacteria, chlamydia are notoriously difficult to study.[37] The Chlamydiales order has nonetheless undergone tremendous reorganization in the past few years due to the application of DNA sequence comparison and genome sequencing. These techniques will also allow for greater understanding of the chlamydia life cycle and its pathogenicity. The new taxonomy as outlined here is the subject of debate among Chlamydiologists, but they are in line with a great number of taxonomic studies.[3, 38] Continued genomic and proteomic studies will expand our understanding of the Chlamydiales order,[5] leading to better treatment options that can benefit both human and animal health.

ACKNOWLEDGEMENT

Much appreciation to my husband whose brilliance, endless support, and clear direction motivates me to challenge myself in ways unimaginable…

REFERENCES

1. Page, L.A., Revision of the family Chlamydiaceae Rake (Rickettsiales): unification for the psittacosis-lymphogranuloma venerum-trachoma group of organisms in the genus Chlamydia, Jones, Rake and Stearns, 1945. *Int. J. Syst. Bacteriol.*, 16, 223–252, 1966.
2. Ward, M.E., www.chlamydiae.com, 2004.
3. Everett, K.D.E. and Andersen, A.A., Radical changes to chlamydial taxonomy are not necessary just yet — reply (Letter), *Int. J. Syst. Evol. Microbiol.*, 51(1), 251–253, 2001.
4. Everett, K.D., Bush, R.M., and Andersen, A.A., Emended description of the order Chlamydiales, proposal of Parachlamydiaceae fam. nov. and Simkaniaceae fam. nov., each containing one monotypic genus, revised taxonomy of the family Chlamydiaceae, including a new genus and five new species, and standards for the identification of organisms, *Int. J. Syst. Bacteriol.*, 49(Pt. 2), 415–440, 1999.
5. Vandahl, B.B., Birkelund, S., and Christiansen, G., Genome and proteome analysis of Chlamydia, *Proteomics*, 4(10), 2831–2842, 2004.
6. Hogan, R.J., Mathews, S.A., Mukhopadhyay, S., Summersgill, J.T., and Timms, P., Chlamydial persistence: beyond the biphasic paradigm, *Infect. Immun.*, 72(4), 1843–1855, 2004.
7. Murray, P.R., *Medical Microbiology*, 3rd ed. Mosby, St. Louis, MO, 1998.
8. Abdelrahman, Y.M. and Belland, R.J., The chlamydial developmental cycle, *FEMS Microbiol. Rev.*, 29(5), 949–959, 2005.
9. Stephens, R.S., The cellular paradigm of chlamydial pathogenesis, *Trends Microbiol.*, 11(1), 44–51, 2003.
10. Everett, K.D., Chlamydia and Chlamydiales: more than meets the eye, *Vet. Microbiol.*, 75(2), 109–126, 2000.
11. Everett, K.D. and Andersen, A.A., Identification of nine species of the Chlamydiaceae using PCR-RFLP, *Int. J. Syst. Bacteriol.*, 49(Pt. 2), 803–813, 1999.
12. Yang, J.M., Liu, H.X., Hao, Y.X., He, C., and Zhao, D.M., Development of a rapid real-time PCR assay for detection and quantification of four familiar species of Chlamydiaceae, *J. Clin. Virol.*, 36(1), 79–81, 2006.
13. Stephens, R.S., Kalman, S., Lammel, C., Fan, J., Marathe, R., Aravind, L., Mitchell, W., Olinger, L., Tatusov, R.L., Zhao, Q., Koonin, E.V., and Davis, R.W., Genome sequence of an obligate intracellular pathogen of humans: Chlamydia trachomatis, *Science*, 282(5389), 754–759, 1998.
14. Carlson, J.H., Porcella, S.F., McClarty, G., and Caldwell, H.D., Comparative genomic analysis of Chlamydia trachomatis oculotropic and genitotropic strains, *Infect. Immun.*, 73(10), 6407–6418, 2005.
15. Read, T.D., Brunham, R.C., Shen, C., Gill, S.R., Heidelberg, J.F., White, O., Hickey, E.K., Peterson, J., Utterback, T., Berry, K., Bass, S., Linher, K., Weidman, J., Khouri, H., Craven, B., Bowman, C., Dodson, R., Gwinn, M., Nelson, W., DeBoy, R., Kolonay, J., McClarty, G., Salzberg, S.L., Eisen, J., and Fraser, C.M., Genome sequences of Chlamydia trachomatis MoPn and Chlamydia pneumoniae AR39, *Nucleic Acids Res.*, 28(6), 1397–1406, 2000.
16. Heddema, E.R., van Hannen, E.J., Duim, B., Vandenbroucke-Grauls, C.M., and Pannekoek, Y., Genotyping of Chlamydophila psittaci in human samples, *Emerg. Infect. Dis.*, 12(12), 1989–1990, 2006.
17. Kalman, S., Mitchell, W., Marathe, R., Lammel, C., Fan, J., Hyman, R.W., Olinger, L., Grimwood, J., Davis, R.W., and Stephens, R.S., Comparative genomes of Chlamydia pneumoniae and C. trachomatis, *Nat. Genet.*, 21(4), 385–9, 1999.
18. Shirai, M., Hirakawa, H., Kimoto, M., Tabuchi, M., Kishi, F., Ouchi, K., Shiba, T., Ishii, K., Hattori, M., Kuhara, S., and Nakazawa, T., Comparison of whole genome sequences of Chlamydia pneumoniae J138 from Japan and CWL029 from USA, *Nucleic Acids Res.*, 28(12), 2311–2314, 2000.

19. Azuma, Y., Hirakawa, H., Yamashita, A., Cai, Y., Rahman, M. A., Suzuki, H., Mitaku, S., Toh, H., Goto, S., Murakami, T., Sugi, K., Hayashi, H., Fukushi, H., Hattori, M., Kuhara, S., and Shirai, M., Genome sequence of the cat pathogen, Chlamydophila felis, *DNA Res.*, 13(1), 15–23, 2006.
20. Read, T.D., Myers, G.S., Brunham, R.C., Nelson, W.C., Paulsen, I.T., Heidelberg, J., Holtzapple, E., Khouri, H., Federova, N.B., Carty, H.A., Umayam, L.A., Haft, D.H., Peterson, J., Beanan, M.J., White, O., Salzberg, S.L., Hsia, R.C., McClarty, G., Rank, R.G., Bavoil, P.M., and Fraser, C.M., Genome sequence of Chlamydophila caviae (Chlamydia psittaci GPIC): examining the role of niche-specific genes in the evolution of the Chlamydiaceae, *Nucleic Acids Res.*, 31(8), 2134–2147, 2003.
21. Thomson, N.R., Yeats, C., Bell, K., Holden, M.T., Bentley, S.D., Livingstone, M., Cerdeno-Tarraga, A.M., Harris, B., Doggett, J., Ormond, D., Mungall, K., Clarke, K., Feltwell, T., Hance, Z., Sanders, M., Quail, M.A., Price, C., Barrell, B.G., Parkhill, J., and Longbottom, D., The Chlamydophila abortus genome sequence reveals an array of variable proteins that contribute to interspecies variation, *Genome Res.*, 15(5), 629–640, 2005.
22. Corsaro, D. and Greub, G., Pathogenic potential of novel Chlamydiae and diagnostic approaches to infections due to these obligate intracellular bacteria, *Clin. Microbiol. Rev.*, 19(2), 283–297, 2006.
23. Fritsche, T.R., Horn, M., Wagner, M., Herwig, R.P., Schleifer, K.H., and Gautom, R.K., Phylogenetic diversity among geographically dispersed Chlamydiales endosymbionts recovered from clinical and environmental isolates of Acanthamoeba spp., *Appl. Environ. Microbiol.*, 66(6), 2613–2619, 2000.
24. Di Francesco, A., Donati, M., Mazzeo, C., Battelli, G., Piva, S., Cevenini, R., and Baldelli, R., Feline chlamydiosis: a seroepidemiological investigation of human beings with and without contact with cats, *Vet. Rec.*, 159(23), 778–779, 2006.
25. Longbottom, D. and Livingstone, M., Vaccination against chlamydial infections of man and animals, *Vet. J.* 171(2), 263–275, 2006.
26. Sykes, J.E., Feline chlamydiosis, *Clin. Tech. Small Anim. Pract.*, 20(2),129–134, 2005.
27. Walder, G., Hotzel, H., Brezinka, C., Gritsch, W., Tauber, R., Wurzner, R., and Ploner, F., An unusual cause of sepsis during pregnancy: recognizing infection with Chlamydophila abortus, *Obstet. Gynecol.*, 106(5 Pt. 2), 1215–1217, 2005.
28. Vretou, E., Radouani, F., Psarrou, E., Kritikos, I., Xylouri, E., and Mangana, O., Evaluation of two commercial assays for the detection of Chlamydophila abortus antibodies, *Vet. Microbiol.*, 123(1–3), 153–161, 2007.
29. Longbottom, D., Psarrou, E., Livingstone, M., and Vretou, E., Diagnosis of ovine enzootic abortion using an indirect ELISA (rOMP91B iELISA) based on a recombinant protein fragment of the polymorphic outer membrane protein POMP91B of Chlamydophila abortus, *FEMS Microbiol. Lett.*, 195(2), 157–161, 2001.
30. Gomes, J.P., Nunes, A., Bruno, W.J., Borrego, M.J., Florindo, C., and Dean, D., Polymorphisms in the nine polymorphic membrane proteins of Chlamydia trachomatis across all serovars: evidence for serovar Da recombination and correlation with tissue tropism, *J. Bacteriol.*, 188(1), 275–286, 2006.
31. Manavi, K., A review on infection with Chlamydia trachomatis, *Best Pract. Res. Clin. Obstet. Gynaecol.*, 20(6), 941–951, 2006.
32. Cevenini, R., Donati, M., and Sambri, V., Chlamydia trachomatis — the agent, *Best Pract. Res. Clin. Obstet. Gynaecol.*, 16(6), 761–773, 2002.
33. de Kruif, M.D., van Gorp, E.C., Keller, T.T., Ossewaarde, J.M., and ten Cate, H., Chlamydia pneumoniae infections in mouse models: relevance for atherosclerosis research, *Cardiovasc. Res.*, 65(2), 317–327, 2005.
34. Yang, Z.P., Kuo, C.C., and Grayston, J.T., A mouse model of Chlamydia pneumoniae strain TWAR pneumonitis, *Infect. Immun.*, 61(5), 2037–2040, 1993.
35. Kumar, S. and Hammerschlag, M.R., Acute respiratory infection due to Chlamydia pneumoniae: current status of diagnostic methods, *Clin. Infect. Dis.*, 44(4), 568–576, 2007.
36. Chahota, R., Ogawa, H., Mitsuhashi, Y., Ohya, K., Yamaguchi, T., and Fukushi, H., Genetic diversity and epizootiology of Chlamydophila psittaci prevalent among the captive and feral avian species based on VD2 region of ompA gene, *Microbiol. Immunol.*, 50(9), 663–678, 2006.
37. Subtil, A. and Dautry-Varsat, A., Chlamydia: five years A.G. (after genome), *Curr. Opin. Microbiol.*, 7(1), 85–92, 2004.
38. Olsen, G.J., Woese, C.R., and Overbeek, R., The winds of (evolutionary) change: breathing new life into microbiology, *J. Bacteriol.*, 176(1), 1–6, 1994.

30 *Mycoplasma* and Related Organisms

Meghan May, Robert F. Whitcomb, and Daniel R. Brown*

CONTENTS

INTRODUCTION

Mycoplasma spp. are members of the class Mollicutes, have the smallest cells among free-living eubacteria, and have genomes presumed to approach the minimal essential information for independent cellular life. *Mycoplasma* and the other mollicutes probably evolved from Gram-positive ancestors (Maniloff, 1992, 2002) by reductive processes that resulted in essentially obligate commensalism or parasitism of eukaryotic host cells. The distinguishing characteristics of mollicutes include small cell size (200 to 500 nm), small genome size (580 to 2200 kbp), low G+C content (typically in the range of 23 to 34 mol%, but 40 mol% in *Mycoplasma pneumoniae*), 16S rDNA sequences clearly affiliated with the class (Weisburg et al., 1989), unique codon usage (e.g., UGA as a tryptophan codon) in some lineages, minimal metabolic capabilities, lack of any cell wall, and species-specific serology. Although their cells are bounded only by a unit membrane, resulting in a general cellular pleomorphism, cytoskeletal elements confer helicity or polarity in

* Deceased

some species. Some species exhibit rotatory, flexional, or gliding motility. The best-studied mollicutes are significant pathogens of humans, domesticated animals, or plants, and several species are common contaminants of *in vitro* eukaryotic cell cultures (Baseman and Tully, 1997). Division G of the American Society for Microbiology (www.asm.org) encompasses the genetic, pathogenic, immunogenic, taxonomic, biochemical, and clinical aspects of the animal, human, plant, and insect mollicutes. The International Organization for Mycoplasmology (www.the-iom.org) and its International Research Programme on Comparative Mycoplasmology promote the cooperative international study of mycoplasmas and mollicute diseases. The purpose of this chapter is to introduce scientists who have been trained in other disciplines to *Mycoplasma* and related organisms, and to provide an entry to the literature of practical mycoplasmology with emphasis on vaccinology and methods of genetic manipulation for microbiologists who specialize in other species.

THE SPECIES CONCEPT FOR *MYCOPLASMA* AND RELATED ORGANISMS

The species concept for *Mycoplasma* and related organisms is similar to the general species concept for other bacteria (Rosselló-Mora and Aman, 2001; Rosselló-Mora, 2003, 2005, 2006). Its modern foundation is justified by three sources of evidence: DNA–DNA hybridization (DDH), serology, and 16S rDNA sequence analyses. DDH, the gold standard for bacterial species circumscription (Stackebrandt et al., 2002), has been used to determine interspecies relationships in the mollicute genera *Acholeplasma*, *Mycoplasma*, and *Spiroplasma* (Aulakh et al., 1983; Bové et al., 1983; Bonnet et al., 1993). Rosselló-Mora and Amann (2001) recommended that, as long as a coherent cluster was maintained, as little as 50% DDH between strain pairs should be acceptable to circumscribe a species frontier. That opinion was endorsed by an ad hoc committee for the reevaluation of the species definition in bacteriology (Stackebrandt et al., 2002). As a result of differences in the biology and ecology among *Mycoplasma* and related species, which manifest a general pattern of "lumpy" genomic diversity (Stackebrandt et al., 2002), a frontier as high as 80% DDH might embrace some coherent species, whereas in cases like *Acholeplasma* spp., there are clusters in which some strain pairs have as little as 40% DDH (Aulakh et al., 1983; Stephens et al., 1983a, 1983b).

In the studies of *Acholeplasma*, *Mycoplasma*, and *Spiroplasma* cited, species relationships suggested by the DDH technique were generally in good accord with serological data (e.g., reciprocal growth or metabolism inhibition by specific polyclonal antisera). Because of the general phenotypic simplicity of mollicutes, serology had been established from the earliest days of mycoplasmology as the most important basis for defining species of the class. Serological heterogeneity and DDH heterogeneity in the range of 20 to 80% tend to be closely correlated in mollicutes. With serology thus justified as a proteomic surrogate for the gold standard DDH to define mollicute species in most cases, the cumbersome DDH technique (Sachse and Hotzel, 1998) can be reserved for special cases in which the affinities of closely related or serologically cross-reactive strains are questioned (Abalain-Colloc et al., 1993; Gasparich et al., 1993b). Serology does not work well to discriminate among infraspecific groups where DDH is relatively high (>80%).

As with other bacteria, in *Mycoplasma* the correlation between 16S rDNA sequence similarity and gold standard DDH values is practically useful although imprecise (Stackebrandt and Gobel, 1994; Keswani and Whitman, 2001; Rosselló-Mora and Amann, 2001). For example, *Mycoplasma gallisepticum* and *Mycoplasma imitans* have >99% 16S rDNA similarity, yet show only 40% DDH (Bradbury et al., 1993). Some *Mycoplasma hominis* strains show as little as 50% intraspecific DDH but also have >99% 16S rDNA similarity (Blanchard et al., 1993). *Spiroplasma insolitum* and *Spiroplasma* sp. 277F have about 95% 16S rDNA sequence similarity with *Spiroplasma citri* but show only about 20% DDH with *S. citri*. Nevertheless, the 16S rDNA sequence now provides the primary basis for assignment to hierarchical rank and can guide directed serological screening. Stackebrandt and Goebel (1994) proposed an upper limit of 97% 16S rDNA similarity as a threshold, which if exceeded, would indicate a need for additional tests to determine whether strains should be regarded as a single or separate species. For *Mycoplasma* and related organisms, experience suggests that a frontier of about 94% 16S rDNA sequence similarity indicates either a spectrum of related species from which an isolate would have to be distinguished by serology or other means, or an upper limit for similarity values that constitute persuasive evidence of novelty of a proposed new species. In summary, mycoplasmology employing a combination of 16S rDNA sequence analyses, directed serology, and supplementary phenotypic data to identify species is consistent with the most recently revisited standards of the microbial systematics community for a species concept (Stackebrandt et al., 2002).

DESCRIPTION OF THE CLASS MOLLICUTES

Mollicutes have been observed in close association with plants, invertebrates, and all classes of vertebrate hosts except amphibians. The general morphological and metabolic simplicity of mycoplasmas and related organisms has led to a current reliance principally on the combination of 16S rDNA analyses and serology for species identification. Consistent with their limited genome size, mycoplasmas have very limited capacity for intermediary metabolism, which severely restricts the utility of conventional biochemical tests for purposes of identification (Poveda, 1998). The lack of anabolic pathways in mycoplasmas is compensated for by an overrepresentation of transport-associated proteins in their proteomes. Pathways for energy generation are also minimal, as mycoplasmas lack the tricarboxylic acid cycle and other mediators of oxidative phosphorylation (Razin et al., 1998). Sugar-fermenting mycoplasmas generate much of their energy from glycolysis and the pyruvate dehydrogenase pathway. The acidic byproducts of these pathways cause a characteristic downward pH shift of the culture medium following growth of those species. Some mycoplasmas hydrolyze arginine as a means of energy generation through the relatively simple arginine dihydrolase pathway, which requires only three enzymes. Hydrolysis of arginine results in the accumulation of ammonia, resulting in the characteristic upward pH shift of the culture medium seen following growth of, for example, *Mycoplasma arginini* and *Mycoplasma salivarium*. Others, such as *Mycoplasma bovis*, do not cause any shift in pH of the culture medium following growth. Some mycoplasmas catabolize organic acids and alcohols such as lactic acid, pyruvic acid, oxobutyric acid, ethanol, glycerol, or isopropanol (Taylor et al., 1994; Razin et al., 1998; Khan et al., 2005). Catabolism of glycerol leads to the excretion of hydrogen peroxide, which is suspected to be a virulence factor of some species. The hydrolysis of urea for energy is a characteristic of species of *Ureaplasma*. It remains up to the community of mycoplasmologists to discover which traits are most meaningful to discriminate among species, and additional metabolic tests may be shown to be useful in the future.

THE ORDER MYCOPLASMATALES AND FAMILY MYCOPLASMATACEAE

The order Mycoplasmatales presently contains the single family Mycoplasmataceae. Approximately 130 species have been named. Its polyphyletic genus *Mycoplasma* is divided into hominis and pneumoniae phylogenetic groups (Johansson and Pettersson, 2002) plus a mycoides cluster. The mycoides cluster contains six species whose 16S rDNA sequences are affiliated unambiguously with those of the family Entomoplasmataceae rather than with the Mycoplasmataceae. For that reason, the nomenclature of those species is a matter of continuing controversy. Neimark et al. (2005) explained why those species should not simply be renamed. Confusion and peril would result, especially with regard to those pathogenic members of the mycoides cluster that are currently U.S. Departments of Health and Human Services and Agriculture Select Agents subject to numerous strict international regulations.

Genus *Mycoplasma*

Mycoplasma spp. are aerobic or facultatively anaerobic mollicutes usually isolated from vertebrate hosts, fluids of vertebrate origin, or on rare occasions from plants or insects. Their genome sizes range from 580 to 1360 kbp, and the G+C content of their chromosome varies from 23 to 40 mol%. They require cholesterol and/or other sterols for growth, a need that is usually met by adding serum to the culture medium. A variety of culture media have been described (Tully, 1995). Nonmotile mycoplasmas, or species with only gliding motility, tend to form umbonate ("fried egg-type") colonies on solid media. Satellite colonies are often observed for motile species on agar. Most *Mycoplasma* spp. grow best at temperatures reflecting their usual habitat. Optimum growth at 37°C is a common feature of species isolated from homeothermic vertebrates, and the lower part of the permissive temperature range even of *Mycoplasma* spp. isolated from poikilothermic fish and reptiles is only 20 to 25°C. A flask-shaped cellular morphology, terminal adhesion-related structures, and a complex cytoskeleton are common in the lineages of the pneumoniae and *Mycoplasma sualvi* clusters. The differential utilization of glucose and arginine is an important diagnostic feature of *Mycoplasma* spp. (Table 30.1).

Genus *Ureaplasma*

Ureaplasma spp. are, in general, phenotypically similar to *Mycoplasma* spp. but all possess urease activity, which is an important diagnostic character. A variety of culture media have been described (Shepard, 1983). Fewer than a dozen species have been named. The vast majority of ureaplasmas have been isolated from the urogenital tract of vertebrates. Their 16S rDNA sequences, genome sizes of 760 to 1170 kbp, and G+C values of 27 to 30 mol% support phylogenetic affiliation with the pneumoniae group of *Mycoplasma*.

The Order Entomoplasmatales and Families Spiroplasmataceae and Entomoplasmataceae

The order Entomoplasmatales currently encompasses two families and three genera that will soon include close to 70 species and putative species. All members of the Entomoplasmatales utilize glucose as an energy source. They usually grow best at 30 to 32°C, but may grow from 10 to 41°C. The family Spiroplasmataceae is monotypic, so its characteristics are essentially those of the genus *Spiroplasma*. A serogroup classification was established for spiroplasmas to accommodate the large number of emerging isolates (Junca et al., 1980; Whitcomb et al., 1987). In the most recent revision (Williamson et al., 1998), that classification listed 34 groups. Three of those groups were divided into a total of 11 subgroups. All of those groups and subgroups represent species or putative species. Members of the Entomoplasmataceae are nonhelical mollicutes that are associated with invertebrates and/or plant surfaces. About 20 species have been named. The Entomoplasmataceae are widely accepted, on the basis of 16S rDNA sequence analysis, to have evolved from spiroplasmal ancestors by processes resulting in loss of phenotypic complexity.

Genus *Spiroplasma*

Spiroplasmas are helical, aerobic or facultatively anaerobic, usually motile mollicutes associated with invertebrates or plants. Their genome sizes are about 780 to 2,200 kbp, and the G+C content of their chromosomes varies from 25 to 32 mol% (Williamson et al., 1998). A significant amount of the total DNA in *Spiroplasma* spp. occurs as free plasmids (Renaudin, 2002), which may be lost during strain passages. Spiroplasmas have well-defined cytoskeletal features (Trachtenberg, 2004; Kürner et al., 2005). A variety of culture media have been described (Whitcomb, 1983; Hackett and Whitcomb, 1995). All spiroplasmas utilize glucose as an energy source. Arginine hydrolysis can be demonstrated in some cases, but only if glucose or another energy source is supplied at the same time (Townsend, 1976). A few spiroplasmas grow well at 37°C but most have optima below

TABLE 30.1
Taxa of the *Mycoplasmatales-Entomoplasmatales* Clade: Major Characters

Taxon	Cytoskeleton	Glucose Fermentation	Arginine Hydrolysis	Genome Size (MB)[b]	%G + C
Order Mycoplasmatales:					
Mycoplasma pneumoniae group[a]					
Mycoplasma fastidiosum cluster[a]	+	2/2	0/2	ND[c]	31 (30–32)
Mycoplasma pneumoniae cluster[a]	+	6/6	1/6	8.04 (0.58–1.01)	32 (26–40)
Ureaplasma spp. cluster[a]	–	6/6	0/6	0.72–1.14	27–30
Mycoplasma muris cluster[a]	+	3/3	3/3	1.33 (1.30–1.36)	28 (25–31)
Mycoplasma hominis group[a]					
Mycoplasma hominis cluster[a]	–	0/21	21/21	0.73 (0.67–0.91)	28 (25–31)
Mycoplasma pulmonis cluster[a]	–	3/3	0/3	0.96	26
Mycoplasma neurolyticum cluster[a]	–	10/12	1/12	0.90 (0.78–1.10)	27 (23–33)
Mycoplasma sualvi cluster[a]	+	3/3	2/3	0.78	25 (24–26)
Mycoplasma equigenitalium cluster[a]	–	2/2	0/2	ND[c]	28 (24–32)
Mycoplasma synoviae cluster[a]	–	0/22	–	0.96 (0.82–1.05)	28 (26–34)
Mycoplasma lipophilum cluster[a]	–	0/2	2/2	ND[c]	27 (24–30)
Mycoplasma bovis cluster[a]	–	3/21	15/21	0.82 (0.61–1.08)	29 (25–32)
Order Entomoplasmatales:					
Spiroplasma spp.					
ixodetis cluster[a]	+	1/1	0/1	2.22	25
citri-mirum cluster[a]	+	9/9	6/9	1.78 (1.46–2.04)	27 (26–28)
chrysopicola group	+	3/3	3/3	1.22 (1.17–1.23)	30 (29–30)
apis cluster[a]	+	29/29	1/29	1.28 (.84–1.72)	27 (24–31)
Entomoplasma	–	6/6	0/6	0.93 (.83–1.20)	29 (27–34)
Mesoplasma	–	12/12	0/12	0.92 (.83–1.60)	29 (28–32)
Mycoplasma mycoides cluster[a]	–	6/6	0/6	1.10 (1.00–1.21)	27 (23–29)

[a] Terminology of Johansson and Pettersson (2002).
[b] Mean (range).
[c] Not determined.

this temperature. There is no "typical" colony morphology in *Spiroplasma*, but satellite colonies are commonly observed on agar. Because of their motility, which can be readily discerned by dark-field microscopy, strain cloning is performed in liquid media by the limiting dilution method.

As with most mollicutes, the *Spiroplasma* species concept is based on serological data. Spiroplasmas exhibit substantial serologic heterogeneity, which is extremely valuable for diagnostic and serogroup-based systematics. Species differentiation can be accomplished serologically by the spiroplasma deformation test (Williamson et al., 1979) but, unfortunately, serological data are insufficient to establish stable infraspecific units in some cases. The genus is divided into the Ixodetis, Citri-Chrysopicola-Mirum, and Apis clades on the basis of 16S rDNA sequence analysis (Gasparich et al., 2004). However, relationships among strains of genomically variable species, such as the group VIII spiroplasma assemblage, cannot be determined by sequence analysis of the 16S rDNA or the 16S/23S rDNA intergenic spacer region (Regassa et al., 2004).

The Genera *Entomoplasma* and *Mesoplasma*

Species are affiliated with the family Entomoplasmataceae on the basis of 16S rDNA similarities, nonhelical morphology, lack of motility, and carriage in insects or on plant surfaces. Species of this family that are unable to grow in the presence of 0.04% polyoxyethylene sorbitan are currently placed in the genus *Entomoplasma*, while species able to do so are placed in *Mesoplasma* (Rose et al., 1993; Tully et al., 1993). However, the genome sizes of known *Entomoplasma* and *Mesoplasma* species overlap at about 790 to 1140 kbp; the G+C content of their chromosomes overlaps at 27 to 30 mol%; and the 16S rDNA sequences of species of the two genera are very similar (Johansson and Pettersson, 2002; Gasparich et al., 2004). Phylogenetic studies indicate that they are polyphyletic. The genera *Entomoplasma* and *Mesoplasma* thus probably represent only a single spectrum of related species.

The Order Acholeplasmatales, Family Acholeplasmataceae, and Genus *Acholeplasma*

The properties of this order and family are essentially that of the genus *Acholeplasma*. About 20 species have been named. Acholeplasmas are aerobic or facultative anaerobic mollicutes that do not require a cholesterol or serum supplement for growth in artificial media. They invariably utilize glucose as an energy source, and some are able to hydrolyze esculin. Known species grow well at 25 to 37°C. Acholeplasmas seem to have a comparatively more flexible ecological association with eukaryotic hosts than the Mycoplasmatales or Entomoplasmatales, because *Acholeplasma* spp. have been isolated from soils and sewage. Some species such as *Acholeplasma axanthum* contain carotenoid pigments. They have G+C values of 27 to 38 mol% and comparatively larger genome sizes of about 1500 to 2100 kbp.

The Order Anaeroplasmatales, Family Anaeroplasmataceae, and Genera *Anaeroplasma* and *Asteroleplasma*

This order encompasses the strictly anaerobic mollicutes. The genus *Anaeroplasma* currently contains four species of sterol-requiring mollicutes that were isolated from bovine rumen fluid (Robinson et al., 1975; Robinson, 1983; Robinson and Freundt, 1987). A variety of culture media have been described (Robinson, 1983); but because cultivation requires obligate anaerobiosis, very little is known about this group. The single described asteroleplasma, the type species *Asteroleplasma anaerobium*, does not require sterol (Robinson et al., 1975). The genome sizes of the two genera are similar (Robinson and Freundt, 1987), but the G+C content of chromosomal DNA in *Anaeroplasma* ranges from 29 to 34 mol%, in contrast to about 40 mol% in *A. anaerobium*. Recent work by Relman and colleagues (Eckburg et al., 2005; Gill et al., 2006) has suggested that members of this order also occur in the human gut and thus may occur in the digestive tract of many vertebrates.

Uncultivated Mollicutes

The indefinite rank Candidatus is used in the class Mollicutes for putative taxa that have not been described in sufficient detail to warrant establishment of a novel taxon (Murray and Stackebrandt, 1995). For example, the plant-pathogenic mollicutes previously referred to as "mycoplasma-like organisms" but now properly called phytoplasmas (IRPCM Phytoplasma/Spiroplasma Working Team, 2004), have not yet been cultivated *in vitro,* in part because, among other biochemical lesions, they lack the capacity for ATP synthesis (Oshima et al., 2004). They are maintained in biological cycles involving the phloem of host plants and homopterous insect vectors. Dozens of species candidates have been proposed. Two complete '*Candidatus* Phytoplasma' genomes have been sequenced and annotated to date (Oshima et al., 2004; Bai et al., 2006), and several more are expected to be completed in the near future. Hemotrophic mycoplasmas like '*Candidatus* Mycoplasma haemofelis' and '*Candidatus* Mycoplasma wenyonii,' which were originally affiliated to the rickettsial genera *Hemobartonella* or *Eperythrozoon* based on phenotypic traits, also remain uncultivated in artificial media. About a dozen different hemotrophic mycoplasmas are distinguishable by host origin and 16S rDNA sequences. Others not yet cultivated are phylotypes of insect and vertebrate commensals that have been discovered through global 16S rDNA PCR. For example, Eckburg et al. (2005) recently determined the 16S rRNA gene sequences of at least ten novel phylotypes among the human intestinal microbiota that cluster distinctly enough to suggest the existence of a previously undiscovered order within the class Mollicutes (Figure 30.1).

Standards for Descriptions of New Species

Increasing emphasis is being placed on objective standards for descriptions of new species of bacteria (Stackebrandt et al., 2002; Kämpfer et al., 2003). Adherence to the spirit of standards for establishing species of *Mycoplasma* and related organisms, which have been refined over the past 50 years, has been generally excellent. The minimum standards for new species of the class Mollicutes were recently revised by the International Committee on Systematics of Prokaryotes, Subcommittee on the Taxonomy of Mollicutes (Brown et al., 2007). The mandatory requirements will be (1) public deposition of a type strain and antiserum to it; (2) comparison of the 16S rDNA sequences of the new species and its neighbors; (3) demonstration through a combination of 16S rDNA, serological, and supplementary phenotypic analyses that the type strain differs from all named species; and (4) assignment to higher taxa in Mollicutes with a specific epithet. The 16S rDNA sequence usually establishes an accurate basis for taxonomic assignment, and may also indicate species novelty, but serological and supplementary phenotypic data must be presented to substantiate it. If the similarity of the 16S rDNA to one or more existing species is greater than 97%, then species status may require close scrutiny, and in certain cases may require DDH studies (Stackebrandt and Goebel, 2004). The investigator is given responsibility and freedom to justify which characters are most meaningful for assignment to the part of the mollicute taxonomy in which a novel species is to be located. With certain strain assemblages (e.g., group VIII spiroplasmas), it has thus far proved impossible to identify stable infraspecific units. In such cases, the assemblage may be regarded as a single variable species (*sensu* Rosselló-Mora and Amann, 2001). In the future, bacterial taxonomy will no doubt encompass multiple genetic loci, and the complete genome sequences especially of economically significant species seem certain to become known, but it currently seems doubtful that such sequences will ever be available for more than a minority of known mollicute species.

CULTURE MEDIA

The nutritional requirements of mollicutes were reviewed extensively by Miles (1992a, 1992b). Tully (1995) described in detail the most commonly used complex media formulations for cultivation of *Mycoplasma* spp. and related organisms. American Type Culture Collection medium 988 (SP-4; Table 30.2) is a rich, undefined medium that has proved superior for meeting their comparatively fastidious nutritional requirements during primary isolation, and for the maintenance of many different mollicutes. It is prepared as an aqueous base, cooled to 56°C after autoclaving, to which is then added aseptically a mixture of serum, yeast extract, and other supplements that has been sterilized by sequential passage through 0.80-, 0.45-, and 0.22-μm filters. Penicillin G or other cell wall-targeting antibiotics are included to discourage growth of other bacteria. The final pH (7.6 to 7.8) of the complete medium facilitates phenol red-based detection of a pH shift caused by growth of mollicutes that excrete acidic end-products of fermentation, although not all species do so. Isolation and growth of arginine-hydrolyzing species can be enhanced by including 5 mL/L^{-1} of a 42% (w/v) arginine solution in the supplements, in which case the final pH of the complete medium should be between 7.0 and 7.2 to facilitate detection of an alkaline shift. Some species, such as *Mycoplasma fermentans*, are capable of utilizing both sugars and arginine as carbon sources, so an initial decrease in pH may reverse later during their course of growth in a complex medium (Razin et al., 1998). Quality control tests for the serum and yeast extract components and for the complete SP-4 media are particularly important because serum quality is susceptible to variation and yeast extract contains labile components (Tully and Rose, 1983; Tully, 1995; Windsor and Windsor, 1998).

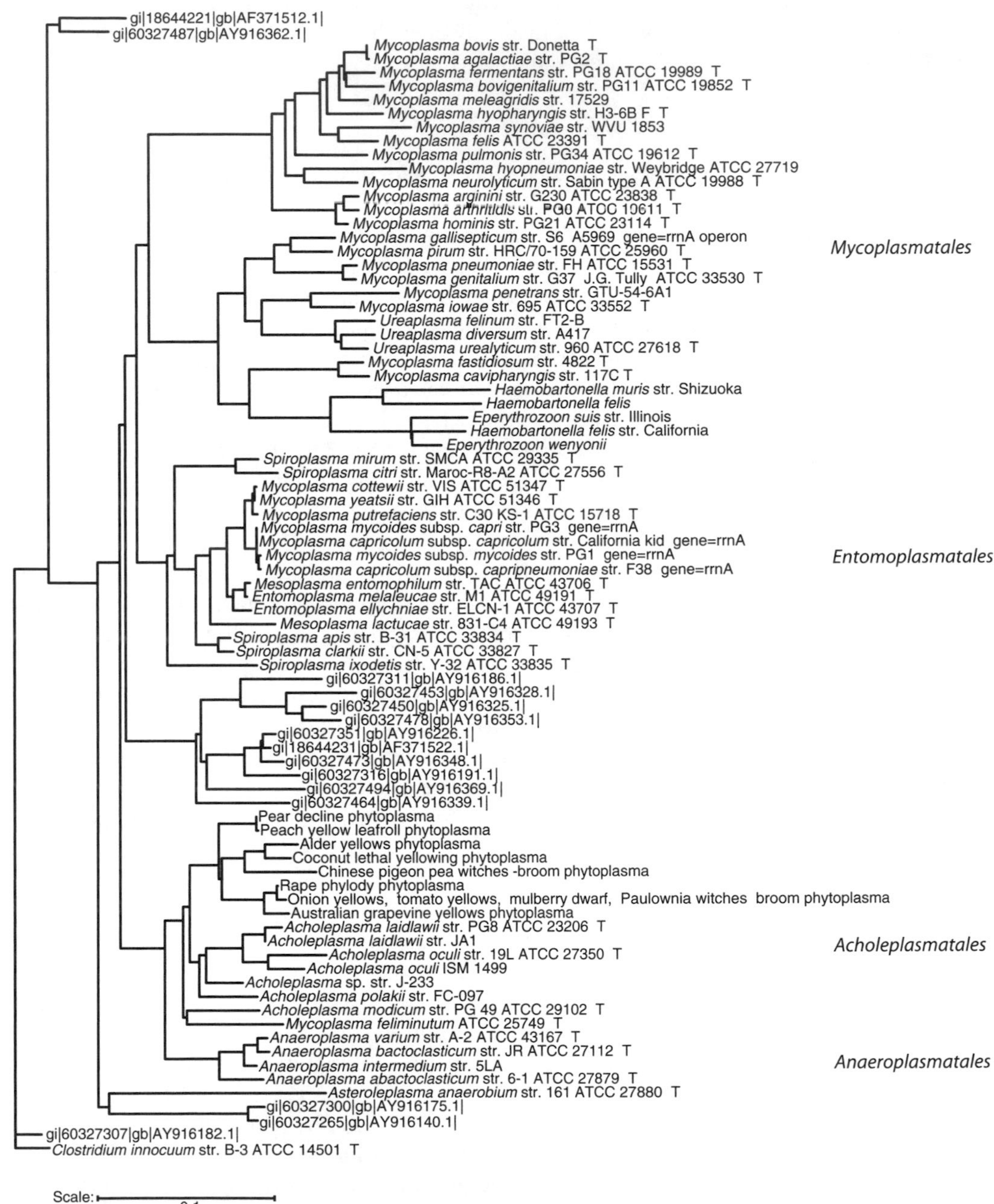

FIGURE 30.1 Unrooted neighbor-joining phylogram of 16S rDNA sequences of representative *Mycoplasma* spp. and related organisms from the Ribosomal Database Project release 8.1 (Cole et al., 2003) and GenBank. The phylogram was generated using the PHYLIP program Neighbor with default parameters (Felsenstein, 1989) and *Clostridium innocuum* as an outgroup. Scale = substitutions site^{-1}. Sequences represented by GenBank GenInfo Identifier (gi) numbers were described by Eckburg et al. (2005).

TABLE 30.2
American Type Culture Collection Medium 988 (SP-4)

Base Medium[a]	
BBL Mycoplasma broth base (Becton, Dickinson)	3.5 g/L
Tryptone (Difco)	10 g/L
Peptone (Difco)	5.3 g/L
0.1% (wt/vol) phenol-red aqueous solution	20 mL/L
Ultrapure water	660 mL/L
Sterile Supplements	
10X CMRL 1066 tissue culture medium (Gibco)	50 mL/L
50% (w/v) glucose aqueous solution	10 mL/L
Yeast extract (Gibco)	35 mL/L
4% (wt/vol) Bacto TC yeastolate aqueous solution (Becton, Dickinson)	50 mL/L
Heat-inactivated fetal bovine serum (Gibco)	170 mL/L
100,000 U/mL aqueous penicillin G	5 mL/L

[a] For solid medium, add 8 g/L noble agar (Difco) to the base medium before autoclaving.

Commonly used alternatives to SP-4, such as Frey's, Hayflick's, and Friis' media differ mainly in the proportions of inorganic salts, amino acids, serum sources, and types of antibiotic supplements included. Defined mycoplasma culture media have been described in detail (Rodwell, 1983), but provision of lipids and amino acids in favorable ratios is technically difficult (Miles, 1992b).

Urea must be supplied for primary isolation and cultivation of ureaplasmas, and they require a medium with pH 5.5 to 6.5. A variety of liquid and solid media formulations for *Ureaplasma* spp. culture from human and animal sources were described in detail by Shepard (1983). Phenol red and bromothymol blue indicator broths have been developed to detect urease activity. Incorporation of calcium chloride in the differential agar medium A8 distinguishes *Ureaplasma* spp. colonies, which appear deep brown or brown-black in color (as a result of ammonia production through urease activity) from *Mycoplasma* spp., which remain colorless on A8 agar.

Liquid and solid culture media for *Spiroplasma* spp. were described in detail by Whitcomb (1983) and Hackett and Whitcomb (1995). SP-4 or M1D media, which substitutes *Drosophila* tissue culture medium for CMRL 1066, and also includes fructose, sucrose, and sorbitol, are suggested for routine culture of a wide array of spiroplasmas from various serogroups. Sorbitol is used to adjust the osmolality to approximately 500 mOsm. Spiroplasmas have also been cultured in defined media or co-cultured with mammalian or insect cells (Hackett and Lynn, 1995).

DETECTION, SURVEILLANCE, TREATMENT, AND CONTROL

In addition to detection by culture, since the early 1990s PCR employing universal primers or primers complementary to 16S rDNA or 16S/23S rDNA spacer sequences conserved in the class has been used widely to detect mollicutes in clinical material and in cell cultures (Razin, 2002). PCR avoids the limitations of culturing, which include sometimes very slow growth, false-negative results due to the complex nutritional requirements of many species, or uninformative results due to overgrowth with fungal or other bacterial contaminants during primary isolation. An additional advantage is that a large number of mollicute 16S rDNA and 16S/23S rDNA spacer sequences are available in GenBank, which in many cases enable preliminary species identification through technically simple, inexpensive, standardized, and fast AFLP, RFLP, or partial DNA sequencing approaches. Examples of the numerous PCR primer combinations that have been used successfully are shown in Table 30.3, and certain combinations are currently available commercially along with

TABLE 30.3

Examples of PCR Primers for Detection of *Mycoplasma* and Related Organisms

Forward (5′ - 3′)	Reverse (5′ - 3′)	Target
Un-nested		
GAGTTTGATCCTGGCTCAG	TCTCGGCCCGGCTAAACATCAT	16S rDNA[a]
ACTCCTACGGGAGGCAGCAGTA	TGTATTACCGCGGCTGCTG	
CTTAAAGGAATTGACGGGAACCCG	TGCACCATCTGTCACTCTGTTAACCTC	
CATCTATTGCGACGCTA		
AGAGTTTGATCCTGGCTCAGGA	GGTAGGGATACCTTGTTACGACT	16S rDNA[b]
CGCCTGAGTAGTACGT(T/A)CGC	GCGGTGTGTACAAGACCCGA	16S rDNA[c]
TGCCTG(A/G)GTAGTACATTCGC	GCGGTGTGTACAAAACCCGA	
CGCCTGGGTAGTACATTCGC	GCGGTGTGTACAAACCCCGA	
CGCCTGAGTAGTATGCTCGC		
AGAGTTTGATCCTGGCTCAG	GACGGGCGGTGTGTRCA	Universal 16S rDNA[d]
Nested		
First stage		
ACACCATGGGAG(C/T)TGGTAAT	CTTC(A/T)TCGACTT(C/T)CAGACCCAAGGCAT TCACGCTTAGATGCTTTCAGCG[e]	16S/23S rDNA spacer[e]
AAAGTGGGCAATACCCAACGC[e]		
Second stage		
GTG(C/G)GG(A/C)TGGATCACCTCCT	GCATCCACCA(A/T)A(A/T)AC(C/T)CTT CCACTGTGTGCCCTTTGTTCCT[e]	

[a] van Kuppeveld et al., 1992; [b]Robertson et al., 1993; [c]Uphoff and Drexler, 2002; [d]Eckburg et al., 2005; [e]Tang et al., 2000 (patented sequences).

standardized PCR protocols. Species-specific PCR-based detection methods also have been developed that target sites such as unique 16S rDNA motifs, the P1 adhesin of *M. pneumoniae*, or the urease of *Ureaplasma* spp. The PCR-based approaches achieve sensitivity and specificity far superior to staining with Hoechst 33258, bisbenzimide, 4′,6-diamidino-2-phenylindole, or similar nucleic acid-binding dyes described by Masover and Becker (1996). A qualitative method of detection, based on mycoplasma-specific ATP synthesis activity present in tissue culture supernatant medium, is also currently available commercially with a standardized protocol for research purposes (Myco-Alert®, Lonza Group Ltd.). In part because of the technical challenges to culturing certain species, the severely restricted usefulness of conventional metabolic diagnostic tests of isolates, and because sample quality can affect the sensitivity of PCR-based analysis of clinical specimens, ELISA is an important method of routine clinical surveillance for mycoplasmosis (Kenny, 1992). ELISA techniques have been described in detail for the major human (Cassell et al., 1995; Wang and Lo, 1995) and veterinary (Brown et al., 1995; Nicolet and Martel, 1995) pathogens.

Standardized *in vitro* methods to determine antibiotic susceptibility of isolates of *Mycoplasma* and related organisms have been described in detail (Davis, 1981; Bébéar and Robertson, 1996; Hannan, 2000). Special challenges in comparative susceptibility testing for mycoplasmas include standardization of the inoculum, and their variable growth requirements and rates (Kenny, 1996).

Mycoplasmas are intrinsically resistant to many antibiotics. Lacking a cell wall, mycoplasmas are not affected by β-lactams, vancomycin, or fosfomycin. In addition, mycoplasmas as a whole are resistant to sulfonamides and trimethoprim because they do not synthesize their own nucleotides; they are resistant to rifampicin because they express a form of RNA polymerase unrecognized by the antibiotic; and they are resistant to polymyxins (Bébéar and Kempf, 2005). Individual species exhibit even a broader spectrum of antibiotic resistance. Examples include the resistance to erythromycin and azithromycin exhibited by *M. hominis*, *M. fermentans*, *Mycoplasma hyopneumoniae*, *Mycoplasma flocculare*, *Mycoplasma pulmonis*, *Mycoplasma hyorhinis*, *Mycoplasma hyosynoviae*, *Mycoplasma mealagridis*, and *M. bovis*, which is apparently mediated by a change in sequence of the 23S RNA (Pereyre et al., 2002).

Antibiotic treatment options for mycoplasmosis are limited due to their intrinsic resistance and the toxicity of those antibiotics that are effective. Treatment of mycoplasmosis often involves the use of antibiotics that inhibit DNA replication or protein synthesis, such as tetracyclines, fluoroquinolones, aminoglycosides, and certain macrolides or ketolides. Tetracyclines, most often doxycycline, are used commonly to treat human and animal mycoplasmoses, based on their low cost and the fact that resistance is not intrinsic (Bébéar and Kempf, 2005). Macrolides are used for treatment when tetracyclines or fluoroquinolones are not appropriate, such as infections of infants, children, or pregnant women. Macrolides less readily penetrate the blood-brain barrier, making them suboptimal treatments for disseminated mycoplasmosis. Fluoroquinolones are less widely used to treat human mycoplasmosis at present but they do exhibit a bacteriocidal effect on *Mycoplasma* spp. (Bébéar et al., 1998, 2000; Duffy et al., 2003; Waites et al., 2003). The use of fluoroquinolones in veterinary medicine is more common, although the threat of generating resistant zoonotic agents may dictate against their use (Teuber, 1999). Aminoglycosides, pleuromutilins, and phenicols are not currently used against human mycoplasmosis, with the exception of chloramphenicol for neonates with mycoplasmosis of the central nervous system unresponsive to other antibiotics (Waites et al., 1993). Typically, those treatments for mycoplasmosis are employed only in veterinary medicine (Bébéar and Kempf, 2005). Mycoplasmal infections require long-term antimicrobial therapy, as short-term treatment often results in recurrence of the infection. This is speculated as due to sequestration of a small number of bacteria in privileged sites, potentially including inside host cells. Aminoglycosides and fluoroquinolones have been used successfully for the eradication of *Mycoplasma* or *Acholeplasma* spp. from contaminated cell or tissue cultures (Del Giudice and Gardella, 1996). Strategies for prevention and control of mycoplasmal contamination of cell cultures were described in detail by Smith and Mowles (1996).

Vaccination is the most promising approach to the control of mycoplasma infections in humans and animals (Ellison et al., 1992). Although vaccination against human mycoplasmosis is not available at present, several vaccines are commercially available for the protection of livestock and poultry against infection by *Mycoplasma* spp. Bacterins are the simplest vaccines to produce; and although many *Mycoplasma* spp. are capable of antigenic variation, potentially lowering the efficacy of a bacterin, a commercially available bacterin appears to protect against contagious caprine pleuropneumonia (CCPP). Sonicated *Mycoplasma capricolum* ssp. *capripneumoniae* adjuvanted with saponin dramatically reduced endemic CCPP in goat herds (King et al., 1992). A similar bacterin composed of lyophilized *M. capricolum* ssp. *capripneumoniae* adjuvanted with saponin was both completely protective and stable at ambient temperature, an important factor in the design of vaccines for diseases prevalent in developing countries (Rurangirwa et al., 1987). Far more commonly, however, bacterins were described as providing only partial protection from the avian pathogen *M. gallisepticum* (Hildebrand et al., 1983); *Mycoplasma crocodyli* from Nile crocodiles (Mohan et al., 1997); the human respiratory tract pathogen *M. pneumoniae* (Mogabgab, 1968, 1973); *M. bovis* from ruminants (Nicholas et al., 2002); and *M. pulmonis* from rodents (Atobe and Ogata, 1977). The only mycoplasmal bacterin used in the United States targets porcine enzoonotic pneumonia caused by *M. hyopneumoniae*. Although multiple bacterins exist (e.g., Stellamune-ONCE, Ingelvac Mhyo), their efficacy is equivocal. Vaccinated pigs exhibit only a modest weight gain as compared to unvaccinated pigs (Dawson et al., 2002). Despite the availability and use of these vaccines, it is estimated that upward of 80% of commercial pigs are infected with *M. hyopneumoniae* (Thacker, 2006).

Multiple live attenuated vaccines protective against mycoplasmosis are currently available. The superior protection generated by live attenuated vaccines is, in some cases, offset by residual virulence or the threat of reversion to pathogenic capabilities. The vaccine strain T1/44 of *Mycoplasma mycoides* ssp. *mycoides* "small colony" (SC), the agent of contagious bovine pleuropneumonia (CBPP), was generated by serial passage *in ovo*. The attenuating mutations of this strain are not known. T1/44 appears efficacious, and is commonly used in *M. mycoides* ssp. *mycoides* (SC)-endemic regions (Browning et al., 2005). Although T1/44 is widely considered avirulent, inoculation site reactions are not uncommon (Thiaucourt et al., 2003). Outbreaks of CBPP occurring in vaccinated herds as well as T1/44 challenge studies suggest that residual pathogenicity exists (Hubschle et al., 2002). A vaccine strain of the ruminant pathogen *Mycoplasma agalactiae*, AIK40, was generated by serial passage *in vitro*. Although not widely used, AIK40 appears to protect vaccinated sheep and goats against arthritis and keratoconjunctivitis (Browning et al., 2005). It was not completely protective against agalactia, however, as mild mastitis was observed in some challenged dams (Foggie et al., 1971). The attenuated LKR strain of *M. hyopneumoniae* was protective against experimental challenge but exhibited moderate pathogenic effects itself (Etheridge, 1982; Lloyd et al., 1989).

Strain MS-H, the vaccine strain of the avian pathogen *Mycoplasma synoviae*, is a temperature-sensitive mutant generated by chemical mutagenesis (Morrow et al., 1998). Eyedrop administration of MS-H appears both efficacious and safe in chickens and turkeys (Markham et al., 1998a, b). MS-H did appear to colonize the upper respiratory tract of inoculated turkeys but did not cause any gross or microscopic lesions (Noormohammadi et al., 2003). Three live attenuated vaccine strains of *M. gallisepticum* are currently commercially available. The F strain is a naturally occurring isolate that exhibits reduced virulence in chickens. Although the protection generated by the F strain is the greatest of the three strains (Abd-el-Motelib and Kleven, 1993), it is not commonly used because it is mildly pathogenic in chickens and highly pathogenic in turkeys (Lin and Kleven, 1982). Strains ts-11 and 6/85 were generated by chemical mutangenesis and serial passage, respectively, and are of greater safety, but lower efficacy, than F (Evans and Hafez, 1992; Noormohammadi et al., 2002; Browning et al., 2005). Two additional candidate vaccine strains of *M. gallisepticum* (GT-5 and K5054) have been described. GT-5 was generated by the complementation of an attenuated, high-passage variant of strain R with the primary cytadhesin GapA. This strain did not produce respiratory lesions when administered intratracheally, and protected chickens from subsequent *M. gallisepticum* challenge (Papazisi et al., 2002b). The protection appeared to be mediated primarily by a humoral response, as evidenced by high levels of serum IgG and greater numbers of IgG- and IgA-secreting plasma cells localized in the trachea (Javed et al., 2005). Although it appears to be a safe and efficacious vaccine candidate, it is unlikely that GT-5 can be used commercially because it carries a gene conferring resistance to gentamicin on a transposable element. K5054 is a naturally occurring strain of *M. gallisepticum* known to be pathogenic to house finches. Aerosol challenge produced minimal lesions in chickens and turkeys, and protected both from subsequent challenge (Ferguson et al., 2004).

Attempts to generate live attenuated vaccines against *M. pneumoniae* have been unsuccessful. The strain MAC, derived by serial passage *in vitro*, appeared both safe and protective in the hamster model of human primary atypical pneumonia (Fernald and Clyde, 1970; Barile et al., 1988). When examined in humans, however, strain MAC exhibited a virulent phenotype. Temperature-sensitive mutants either retained virulence in humans despite safety and efficacy in hamsters, or were so severely attenuated that they failed to invoke an immune response in humans (Steinberg et al., 1969, 1971; Brunner et al., 1973; Barile et al., 1988).

Although often elegantly designed, "next-generation" vaccines — including protein subunit, DNA, capsular polysaccharide, and bacterial or viral vector vaccines — have been hampered by problems, including low immunogenicity, highly specific responses that are not protective, and lack of efficacy against heterologous challenge. At present, none have been developed for commercial use against mycoplasmosis. Immunization with purified proteins or purified capsular polysaccharides had been attempted in early models with the primary cytadhesins P1 of *M. pneumoniae* and P97 of *M. hyopneumoniae*, and the capsular polysaccharide (CPS) of *M. mycoides* ssp. *mycoides*. Adjuvanted CPS and P97 both elicited immune responses in mouse and pig models, respectively, but neither conferred any protection against challenge (King et al., 1997; Waite and March, 2002). Hamsters were not protected by the P1 subunit vaccine, and its administration appeared to exacerbate their clinical signs after experimental challenge (Jacobs et al., 1988). DNA vaccines designed to protect against *M. hyopneumoniae* have been described. Mice immunized with a construct encoding the heat shock protein p42 displayed an IgG response, and the antibodies neutralized *M. hyopneumoniae in vitro* (Chen et al., 2003). In another study, mice were immunized with a construct encoding a fusion protein of the *M. hyopneumoniae* surface antigen P71 and the *Mycobacterium tuberculosis* mitogen ESAT-6. Although ESAT-6 had the adjuvant effect intended (the mice seroconverted at a higher rate, and exhibited altered cytokine profiles), no evaluation was made of the effect of the antibodies generated on *M. hyopneumoniae* (Minion et al., 2003). These proof-of-concept experiments in the mouse model are encouraging but demonstration of the protective effect of these vaccines in pigs remains to be undertaken.

Two bacterial vector candidate vaccines aimed at protecting pigs from *M. hyopneumoniae* have been created. *Erysipelothrix rhusiopathiae* expressing the P97 adhesin was administered to pigs. Following challenge with *M. hyopneumoniae*, vaccinated pigs seemed partially protected, despite a lack of serum antibody (Shimoji et al., 2003). An attenuated strain of *Salmonella typhimurium* expressing a fragment of the *M. hyopneumoniae* ribonucleotide reductase subunit NrdF appeared to invoke a local IgA response in mice following oral inoculation (Fagan et al., 1996). Its protectiveness remains to be evaluated in a challenge study.

MOLECULAR GENETIC MANIPULATION

Because of their small genome sizes and the range of hosts they colonize as commensals or pathogens, mollicutes constitute an ideal class of bacteria for comparative genomic studies (Dandekar et al., 2002). More than two dozen complete genomes have been partially or completely sequenced and annotated (Table 30.4) and many more are currently underway. A Web-accessible database named MolliGen (http://cbi.labri.fr/outils/molligen/) was designed to integrate all the complete genomes of mollicutes for comparative analyses as they become available (Barré et al., 2004). However, a major hurdle that has yet to be overcome is the limited ability at present to manipulate mycoplasmas genetically. With few exceptions, mycoplasmas do not appear naturally to harbor self-replicating plasmids, and attempts to transform them with commercial plasmid cloning vectors have failed. The plasmids pKMK1 and pADB201, first isolated from *M. mycoides* ssp. *mycoides*, are also able to replicate in *M. capricolum* (Bergemann and Finch, 1988; Dybvig and Khaled, 1990). Replicons from *Escherichia coli* plasmids that were cloned into pKMK1 were stable and could be shuttled from *E. coli* to *M. mycoides* and back without genetic variation, but did not function as gene delivery systems as desired (King and Dybvig, 1994). Descriptions of multiple extrachromosomal DNA elements of *Spiroplasma* spp. have been published (Barber et al., 1983; Mouches et al., 1984; Gasparich et al., 1993a). The most detailed information is for two cryptic plasmids isolated from *S. citri* and *Spiroplasma kunkelii*. The plasmid of *S. citri* (pBJS-O) appears to encode and express an adhesion protein (SARP1), making this plasmid a factor in the virulence of the organism. The plasmid of *S. kunkelii* (pSKU146) appears to encode a homologue of SARP1, although its role in adherence has not been determined (Joshi et al., 2005). Elements necessary for transfer also appear to be encoded by pSKU146, indicating that this plasmid may be amenable to conjugation (Davis et al., 2005).

Success with targeted mutagenesis has been extremely rare. Whether this stems from an inherent inability of mycoplasmas to produce the necessary proteins to mediate homologous recombination (Labarère, 1992; Zou and Dybvig, 2002), or from the instability of constructs delivered on

TABLE 30.4
Characteristics of Sequenced Mollicute Genomes

Mycoplasma or Related Species	Strain	Size (MB)	%G+C	Open Reading Frames	Hypothetica	Coding Density	Ref.
M. capricolum ssp. *capricolum*	27343	1.00	23	812	ND[a]	88	Glass et al., 2005
M. gallisepticum	R	0.99	31	742	37	91	Papazisi et al., 2003
M. genitalium	G37	0.58	31	484	21	90	Fraser et al., 1995
M. hyopneumoniae[b]	232	0.89	29	691	56	91	Minion et al., 2004
M. mobile	163K	0.78	25	635	27	ND[a]	Jaffe et al., 2004
M. mycoides . ssp. *mycoides* (SC)	PG1	1.21	24	985	41	80	Westberg et al., 2004
M. penetrans	HF-2	1.36	26	1037	41	88	Sasaki et al., 2002
M. pneumoniae	M129	0.82	40	689	38	87	Himmelreich et al., 1996
M. pulmonis	UAB CTIP	0.96	26	782	38	90	Chambaud et al., 2001
M. synoviae	53	0.80	28	672	33	91	Vasconcelos et al., 2005
'*Candidatus* Phytoplasma asteris'[b]	AYWB	0.71	26	671	ND[a]	73	Bai et al., 2006
Mesoplasma florum	L1	0.79	27	687	ND[a]	92	Knight and Fournier, 2004
Ureaplasma urealyticum[c]	700970	0.75	26	614	33	93	Glass et al., 2000

[a] Not determined.

[b] *M. hyopneumoniae* strains 7448 and J were also sequenced by Vasconcelos et al. (2005). '*Candidatus* Phytoplasma asteris' AYWB refers to the aster yellows witches broom strain; the onion yellows strain was also sequenced by Oshima et al. (2004).

[c] Revised *Ureaplasma* nomenclature was subsequently proposed by Robertson et al. (2002).

short-lived plasmids, experience has shown that homologous recombination is not a reliable method of targeted mutagenesis in mycoplasmas. There have only been four reported successes in gene displacement via homologous recombination, indicating that while it does occur, it is very rare (Dhandayuthapani et al., 1999, 2001; Dybvig and Woodard, 1992; Markham et al., 2003; Burgos et al., 2006). In *M. gallisepticum*, recombination involving a construct cloned into the commercial vector pGEMT occurred before the loss of the plasmid (Markham et al., 2003). The failure of mycoplasmas to undergo homologous recombination may be a result of delivering constructs on suicide vectors, leading to their loss prior to integration. A factor complicating the cloning and expression of mycoplasmal genes for analysis and manipulation in other bacterial hosts is the usage by many species of the UGA codon to encode tryptophan, rendering expression without site-directed mutagenesis impossible if this codon is present.

Some advances have been made in recent years toward expanding the repertoire of molecular tools available for use in mycoplasmas. At present, the preferred method of obtaining isogenic mutants is transposon-mediated insertional mutagenesis. The first successful transposition of a mollicute occurred in *Acholeplasma laidlawii* using the streptococcal transposon Tn*916* (Dybvig and Cassell, 1987), which confers resistance to tetracycline. Shortly thereafter, the staphylococcal transposon Tn*4001* was also found to integrate into mycoplasmal genomes (Mahairas et al., 1989). Tn*4001* quickly became the preferred transposon due to its smaller size, natural gentamicin resistance gene, and much higher efficiency. Several modifications have since been made to Tn*4001*, rendering it the most widely used molecular tool in mycoplasmology. The most important derivative is Tn*4001*mod (Knudtson and Minion, 1993). This modified transposon contains a *Bam*H1/*Sma*I cloning site in one of the insertion sequences that does not seem to impair its mobility. This modification allows Tn*4001*mod to function as a cloning vector, delivering gene constructs for expression in cis. Complementation of genes inactivated by spontaneous point mutations allowed the confirmation of their function, as was seen, for example, with *gapA* and *crmA* of *M. gallisepticum* (Papazisi et al., 2002a). A second valuable trait of Tn*4001*mod is its usefulness for oligonucleotide primer-directed DNA sequencing from only one of its terminal insertion sequences. The multiple cloning site represents a unique sequence that allows design of outward-reading primers that anneal to only one end of the integrated transposon, which permits precise identification of the insertion point of Tn*4001*mod (Hudson et al., 2006).

Another derivative of Tn*4001* is Tn*4001*lac, which carries a promoterless *lacZ* gene. This transposon is designed to examine relative promoter strengths in mycoplasmas by the amount of β-galactosidase they produce (Knudtson and Minion, 1993). A third derivative generated for use in *M. pneumoniae*, Tn*4001*cat, offered chloramphenicol resistance as an alternative selectable marker (Hahn et al., 1999). Similarly, a tetracycline resistance derivative was generated for use in mollicute strains found to be innately resistant to gentamicin. In that derivative, mini-Tn*4001*tet, the transposase gene is located outside the insertion sequences that flank the *tet* gene, rendering secondary transposition after integration into the bacterial chromosome extremely rare (Pour-El et al., 2002). Additionally, Tn*4001*cat and mini-Tn*4001*tet can be used to deliver genes into strains already carrying Tn*4001* or Tn*4001*mod. Major characteristics of the transposons described are shown in Table 30.5.

An entirely different class of plasmids for molecular genetic manipulation of mollicutes, the artificial OriC plasmids, were first developed in 1994. The principle is to duplicate exactly the origin of replication from the chromosome of a particular species, and use that to generate a self-replicating extrachromosomal element. Perhaps not surprisingly, these vectors tend to exhibit specificity for the species from which the OriC was cloned, but notable exceptions have been reported. The first OriC plasmid developed for a mollicute, pOT1, consisted of the *S. citri* OriC ligated to the *tetM* tetracycline resistance marker of Tn*916* (Ye et al., 1994). Once pOT1 was shown to replicate autonomously in *S. citri*, an *E. coli* replicon was cloned in pOT1 to generate pBOT1, a plasmid that could replicate in either *E. coli* or *S. citri*. This enabled propagation and/or manipulation of pBOT1 to occur in *E. coli* while maintaining the ability to replicate in *S. citri* (Renaudin et al., 1995). The use of pBOT1 as a successful cloning vector was first described in 1997. A nonmotile *S. citri* mutant (G540) carrying a transposon insertion in the putative motility-associated gene *scm1* was used in the first cloning and expression studies. The plasmid pCJ6 was generated by cloning *scm1* into pBOT1, and subsequently used to transform mutant G540. A motile phenotype was restored, indicating not only that *scm1* was responsible for motility, but also that OriC plasmids could deliver and drive gene expression in a spiroplasma (Jacob et al., 1997).

An OriC plasmid designed for use in *M. pulmonis* was used to deliver a construct intended to displace the gene *hlyA*, which encodes a hemolysin. The development of plasmids pMPO1 and pMPO5 was accomplished in a similar fashion to pBOT1. The OriC was amplified and cloned into pSRT2, an *E. coli* replicon containing *tetM* from Tn*916* driven by a homologous promoter. The pMPO1 was found to integrate into the *M. pulmonis* chromosome after very few generations (Cordova et al., 2002). Reducing the size of the cloned OriC and orienting *tetM* between DnaA boxes rendered pMPO5 a stably maintained extrachromosomal plasmid that was used to deliver

TABLE 30.5

Transposons Commonly Used for Molecular Genetic Manipulation of Mollicutes

Transposon	Origin	Selective Marker	Reporter Gene	Distinctions
Tn*916*	*Streptococcus*	Tetracycline	None	Large; inefficient; first mycoplasma transpositions
Tn*4001*	*Staphylococcus*	Gentamicin	None	Efficient; widely used; replaced Tn*916*
Tn*4001*mod	*Staphylococcus* Tn*4001*	Gentamicin	None	Multiple cloning site; gene delivery and expression
Tn*4001*cat	*Staphylococcus* Tn*4001*	Chloramphenicol	None	Multiple cloning site; alternative resistance
Mini-Tn*4001*tet	*Staphylococcus* Tn*4001*	Tetracycline	None	Multiple cloning site; alternative resistance; limited 2° mobility
Tn*4001*lac	*Staphylococcus* Tn*4001*	Gentamicin	β-Galactosidase	Promoterless *lacZ*; assessment of promoter strengths

a *hlyA* recombination cassette. However, the construct did not accomplish targeted mutangenesis of *hlyA*, despite being carried on a stable OriC plasmid rather than a suicide vector. Similar studies with *S. citri*, however, did result in targeted mutagenesis of *scm1*, *fruA*, and *fruK* (Duret et al., 1999; Gaurivaud et al., 2000). Although the results of the use of OriC plasmids for use in delivering constructs for homologous recombination have been mixed, it seems apparent that this approach has tremendous potential to expand the field of targeted mutagenesis in mycoplasmas.

The plasmids generated for use in *M. myoides* SC, *M. myoides* LC, and *M. capricolum* are pMYSO1, pMYCO1, and pMCO3, respectively. Each was generated in an identical manner to pMPO1 (i.e., the OriC region from each was cloned into pSRT2). Although the OriC regions of these very closely related species are nearly identical, their plasmids did not behave in the same manner. While pMCO3 and pMYSO1 were stably maintained in an extrachromosomal form, pMYCO1 rapidly integrated into the chromosome of *M. myoides* (LC) at the OriC (Lartigue et al., 2003). Of these three plasmids, successful expression of genes *in trans* has only been reported for pMCO3. Spiralin (*S. citri*) and β-galactosidase (*E. coli*) have been expressed from pMCO3 by *M. capricolum* (Renaudin and Lartigue, 2005). Interestingly, the plasmids of these closely related species have been shown to replicate in one another. Perhaps more strikingly, *M. capricolum* was shown to support replication of an *S. citri* OriC plasmid (Lartigue et al., 2003).

A different approach was taken for generating the OriC plasmids of *M. agalactiae* and *M. gallisepticum*. The OriC from *M. agalactiae* was cloned with *tetM* into pBluescript, yielding pMM20-1. Transformation of *M. agalactiae* yielded tetracycline-resistant clones, but pMM20-1 integrated into the chromosome at the OriC (indicating that the integration was a result of homologous recombination) after 16 passages. A derivative plasmid containing a much smaller fragment of the OriC, pMM21-7, generated tetracycline-resistant clones and was stably extrachromosomal (Chopra-Dewasthaly et al., 2005). Cloning of the OriC from *M. gallisepticum* strain S6 into pGEM-T created pGTLoriC. This plasmid stably replicated outside the chromosome for up to ten passages, but no pGTLoriC could be detected extrachromosomally after 15 passages (Lee et al., 2004).

Mycoplasma bacteriophage MAV1 of *Mycoplasma arthritidis*, Br1 of *Mycoplasma bovirhinis*, Hr1 of *M. hyorhinis*, and P1 of *M. pulmonis* have not been demonstrated to have any transducing capabilities (Dybvig et al., 1987; Voelker et al., 1995; Roske et al., 2004). The virus SpV1 infects *S. citri*, and its use for gene cloning in *S. citri* was described in detail by Renaudin and Bové (1995). Transduction by SpV1 was first demonstrated by cloning the *cat* gene onto the replicative form (SpV1-RF) and infecting *S. citri*. Expression of *cat* was demonstrated in membrane extracts of infected cells (Stamburski et al., 1991). Also, a portion of the gene encoding the primary cytadhesin P1 of *M. pneumoniae* was transduced after cloning in SpV1 into *S. citri*. Expression of the P1 epitope by *S. citri* was confirmed by Northern and Western blotting (Marais et al., 1993). Further work with SpV1 has been hampered by the instability of its replicative form, however, and it is thus not often used to deliver genes (Marais et al., 1996). In summary, molecular manipulation of mycoplasmal genes and targeted mutagenesis are still major barriers in mycoplasmology. However, progress may be imminent based on the potential of OriC plasmids to deliver recombination cassettes.

ACKNOWLEDGMENTS

Supported by Public Health Service grant 1R01GM076584-01A1 from the NIH National Institute of General Medical Sciences (DRB).

REFERENCES

Abalain-Colloc, M.L., D.L. Williamson, P. Carle, et al. 1993. Division of group XVI spiroplasmas into subgroups. *Int. J. Syst. Bacteriol.*, 43:342–346.

Abd-el-Motelib, T.Y. and S.H. Kleven. 1993. A comparative study of *Mycoplasma gallisepticum* vaccines in young chickens. *Avian Dis.*, 37:981–987.

Atobe, H. and M. Ogata. 1977. Protective effect of killed *Mycoplasma pulmonis* vaccine against experimental infection in mice. *Nippon Juigaku Zasshi*, 39:39–46.

Aulakh, G.S., E.B. Stephens, J.G. Tully, D.L. Rose, and M.F. Barile. 1983. Nucleic acid relationships among *Acholeplasma* species. *J. Bacteriol.*, 153:1338–1341.

Bai, X., J. Zhang, A. Ewing, et al. 2006. Living with genome instability: the adaptation of phytoplasmas to diverse environments of their insect and plant hosts. *J. Bacteriol.*, 188:3682–3696.

Barber, C.E., D.B. Archer, and M.J. Daniels. 1983. Molecular biology of spiroplasma plasmids. *Yale J. Biol. Med.*, 56:777–781.

Barile, M.F. 1983. Arginine hydrolysis. In *Methods in Mycoplasmology* Vol. I, S. Razin and J.G. Tully, Eds. New York: Academic Press, p. 345–349.

Barile, M.F., D.K. Chandler, H. Yoshida, M.W. Grabowski, and S. Razin. 1988. Hamster challenge potency assay for evaluation of *Mycoplasma pneumoniae* vaccines. *Infect. Immun.*, 56:2450–2457.

Barré, A., A. de Daruvar, and A. Blanchard. 2004. MolliGen, a database dedicated to the comparative genomics of Mollicutes. *Nucleic Acids Res.*, 32:D307–D310.

Baseman, J.B. and J.G. Tully. 1997. Mycoplasmas: sophisticated, reemerging, and burdened by their notoriety. *Emerg. Infect. Dis.*, 3:21–32.

Bébéar, C.M. and I. Kempf. 2005. Antimicrobial therapy and antimicrobial resistance. In *Mycoplasmas: Pathogenicity, Molecular Biology, and Strategies for Control*, A. Blanchard and G. Browning, Eds. Norfolk, U.K.: Horizon Bioscience, p. 535–568.

Bébéar, C.M., H. Renaudin, A. Boudjadja, and C. Bébéar. 1998. *In vitro* activity of BAY 12–8039, a new fluoroquinolone against mycoplasmas. *Antimicrob. Agents Chemother.*, 42:703–704.

Bébéar, C.M., H. Renaudin, A. Bryskier, and C. Bébéar. 2000. Comparative activities of telithromycin (HMR 3647), levofloxacin, and other antimicrobial agents against human mycoplasmas. *Antimicrob. Agents Chemother.*, 44:1980–1982.

Bébéar, C. and J.A. Robertson. 1996. Determination of minimal inhibitory concentration. In *Molecular and Diagnostic Procedures in Mycoplasmology,* Vol. II, J.G. Tully and S. Razin, Eds. San Diego, CA: Academic Press, p. 189–197.

Bergemann, A.D. and L.R. Finch. 1988. Isolation and restriction endonuclease analysis of a mycoplasma plasmid. *Plasmid,* 19:68–70.

Blanchard, A., A. Yáñez, K. Dybvig, H.L. Watson, G. Griffiths, and G.H. Cassell. 1993. Evaluation of intraspecies genetic variation within the 16S rRNA gene of *Mycoplasma hominis* and detection by polymerase chain reaction. *J. Clin. Microbiol.,* 31:1358–1361.

Bonnet, F., J.M. Bové, R.H. Leach, G.S. Cottew, D.L. Rose, and J.G. Tully. 1993. Deoxyribonucleic acid relatedness between field isolates of mycoplasma F38 group, the agent of contagious caprine pleuropneumonia and strains *of Mycoplasma capricolum. Int. J. Syst. Bacteriol*., 43:597–602.

Bové, J.M., C. Mouches, P. Carle-Junca, F.R. Degorce-Dumas, J.G. Tully, and R.F. Whitcomb. 1983. Spiroplasmas of group I — The *Spiroplasma citri* cluster. *Yale J. Biol. Med.,* 56:573–582.

Bradbury, J.M., O. Abdul-Wahab, M. Saed, C.A. Yavari, J.-P. Dupiellet, and J.M. Bové. 1993. *Mycoplasma imitans* sp. nov. is related to *Mycoplasma gallisepticum* and found in birds. *Int. J. Syst. Bacteriol.,* 43:721–728.

Brown, D.R., R.F. Whitcomb, and J.M. Bradbury. 2007. Revised minimal standards for description of new species of the class *Mollicutes* (division *Tenericutes*). *Int. J. Syst. Evol. Microbiol.,* 57:2703–2719.

Brown, M.B., J.M. Bradbury, and J.K. Davis. 1996. ELISA in small animal hosts, rodents, and birds. In *Molecular and Diagnostic Procedures in Mycoplasmology,* Vol. II. J.G. Tully and S. Razin, Eds. San Diego, CA: Academic Press, p. 93–104.

Browning, G.F., K.G. Whithear, and S.J. Geary. 2005. Vaccines to Control Mycoplasmosis. In *Mycoplasmas: Pathogenicity, Molecular Biology, and Strategies for Control*, A. Blanchard and G. Browning, Eds. Norfolk, U.K.: Horizon Bioscience, p. 569–598.

Brunner, H., H.B. Greenberg, W.D. James, R.L. Horswood, R.B. Couch, and R.M. Chanock. 1973. Antibody to *Mycoplasma pneumoniae* in nasal secretions and sputa of experimentally infected human volunteers. *Infect. Immun.,* 8:612–620.

Burgos, R., O.Q. Pich, M. Ferrer-Navarro, J.B. Baseman, E. Querol, and J. Piñol. 2006. *Mycoplasma genitalium* P140 and P110 cytadhesins are reciprocally stabilized and required for cell adhesion and terminal-organelle development. *J. Bacteriol.,* 188:8627–8637.

Cassell, G.H., G. Gambill, and L. Duffy. 1996. ELISA in respiratory infections of humans. In *Molecular and Diagnostic Procedures in Mycoplasmology,* Vol. II, J.G. Tully and S. Razin, Eds. San Diego, CA: Academic Press, p. 123–136.

Chambaud, I., R. Heilig, S. Ferris, et al. 2001. The complete genome sequence of the murine respiratory pathogen *Mycoplasma pulmonis. Nucleic Acids Res*., 29:2145–2153.

Chen, Y.L., S.N. Wang, W.J. Yang, Y.J. Chen, H.H. Lin, and D. Shiuan. 2003. Expression and immunogenicity of *Mycoplasma hyopneumoniae* heat shock protein antigen P42 by DNA vaccination. *Infect. Immun.,* 71:1155–1160.

Chopra-Dewasthaly, R., M. Marenda, R. Rosengarten, W. Jechlinger, and C. Citti. 2005. Construction of the first shuttle vectors for gene cloning and homologous recombination in *Mycoplasma agalactiae. FEMS Microbiol. Lett.,* 253:89–94.

Cordova, C.M., C. Lartigue, P. Sirand-Pugnet, J. Renaudin, R.A. Cunha, and A. Blanchard. 2002. Identification of the origin of replication of the *Mycoplasma pulmonis* chromosome and its use in oriC replicative plasmids. *J. Bacteriol.,* 184:5426–5435.

Dandekar, T., B. Snel, S. Schmidt, et al. 2002. Comparative genome analysis of the mollicutes. In *Molecular Biology and Pathogenicity of Mycoplasmas,* S. Razin and R. Herrmann, Eds. London: Kluwer, p. 255–278.

Davis, R.E. 1981. Antibiotic sensitivities *in vitro* of diverse spiroplasma strains associated with plants and insects. *Appl. Environ. Microbiol*., 41:329–333.

Davis, R.E., E.L. Dally, R. Jomantiene, Y. Zhao, B. Roe, S. Lin, and J. Shao. 2005. Cryptic plasmid pSKU146 from the wall-less plant pathogen *Spiroplasma kunkelii* encodes an adhesin and components of a type IV translocation-related conjugation system. *Plasmid,* 53:179–190.

Dawson, A., R.E. Harvey, S.J. Thevasagayam, J. Sherington, and A.R. Peters. 2002. Studies of the field efficacy and safety of a single-dose *Mycoplasma hyopneumoniae* vaccine for pigs. *Vet. Rec.,* 151:535–538.

Del Giudice, R.A. and R.S. Gardella. 1996. Antibiotic treatment of mycoplasma-infected cell cultures. In *Molecular and Diagnostic Procedures in Mycoplasmology* Vol. II, J.G. Tully and S. Razin, Eds. San Diego, CA: Academic Press, p. 439–443.

Dhandayuthapani, S., M.W. Blaylock, C.M. Bébéar, W.G. Rasmussen, and J.B. Baseman. 2001. Peptide methionine sulfoxide reductase (MsrA) is a virulence determinant in *Mycoplasma genitalium. J. Bacteriol.,* 183:5645–5650.

Dhandayuthapani, S., W.G. Rasmussen, and J.B. Baseman. 1999. Disruption of gene mg218 of *Mycoplasma genitalium* through homologous recombination leads to an adherence-deficient phenotype. *Proc. Nat. Acad. Sci. U.S.A.,* 96:5227–5232.

Duffy, L.B., D.M. Crabb, X. Bing, and K.B. Waites. 2003. Bactericidal activity of levofloxacin against *Mycoplasma pneumoniae. J. Antimicrob. Chemother.,* 52:527–528.

Duret, S., J.L. Danet, M. Garnier, and J. Renaudin. 1999. Gene disruption through homologous recombination in *Spiroplasma citri*: an scm1-disrupted motility mutant is pathogenic. *J. Bacteriol.,* 181:7449–7456.

Dybvig, K. and G.H. Cassell. 1987. Transposition of gram-positive transposon Tn*916* in *Acholeplasma laidlawii* and *Mycoplasma pulmonis. Science,* 235:1392–1394.

Dybvig, K. and M. Khaled. 1990. Isolation of a second cryptic plasmid from *Mycoplasma mycoides* subsp. *mycoides. Plasmid,* 24:153–155.

Dybvig, K., A. Liss, J. Alderete, R.M. Cole, and G.H. Cassell. 1987. Isolation of a virus from *Mycoplasma pulmonis. Israel J. Med. Sci.,* 23:418–422.

Dybvig, K. and A. Woodard. 1992. Construction of recA mutants of *Acholeplasma laidlawii* by insertional inactivation with a homologous DNA fragment. *Plasmid,* 28:262–266.

Eckburg, P.B., E.M. Bik, C.N. Bernstein, et al. 2005. Diversity of the human intestinal microbial flora. *Science,* 308:1635–1638.

Ellison, J.S., L.D. Olson, and M.F. Barile. 1992. Immunity and vaccine development. In *Mycoplasmas Molecular Biology and Pathogenesis*, J. Maniloff, R.N. McElhaney, L.R. Finch, and J.B. Baseman, Eds., Washington, DC: American Society for Microbiology, p. 491–504.

Etheridge, J.R. and L.C. Lloyd. 1982. A method for assessing induced resistance to enzootic pneumonia of pigs. *Res. Vet. Sci.,* 33:188–191.

Evans, R.D. and Y.S. Hafez. 1992. Evaluation of a *Mycoplasma gallisepticum* strain exhibiting reduced virulence for prevention and control of poultry mycoplasmosis. *Avian Dis.,* 36:197–201.

Fagan, P.K., S.P. Djordjevic, G.J. Eamens, J. Chin, and M.J. Walker. 1996. Molecular characterization of a ribonucleotide reductase (nrdF) gene fragment of *Mycoplasma hyopneumoniae* and assessment of the recombinant product as an experimental vaccine for enzootic pneumonia. *Infect. Immun.,* 64:1060–1064.

Felsenstein, J. 1993. Phylogeny Inference Package (PHYLIP). Version 3.5. University of Washington, Seattle.

Ferguson, N.M., V.A. Leiting, and S.H. Kleven. 2004. Safety and efficacy of the avirulent *Mycoplasma gallisepticum* strain K5054 as a live vaccine in poultry. *Avian Dis.,* 48:91–99.

Fernald, G.W. and W.A. Clyde. 1970. Protective effect of vaccines in experimental *Mycoplasma pneumoniae* disease. *Infect. Immun.,* 1:559–565.

Foggie, A., J.R. Etheridge, O. Erdag, and F. Arisoy. 1971. Contagious agalactia of sheep and goats studies on live and dead vaccines in lactating sheep. *J. Comp. Pathol.,* 81:165–172.

Fraser, C.M., J.D. Gocayne, O. White, et al. 1995. The minimal gene complement of *Mycoplasma genitalium. Science,* 270:397–403.

Gasparich, G.E., K.J. Hackett, E.A. Clark, J. Renaudin, and R.F. Whitcomb. 1993a. Occurrence of extrachromosomal deoxyribonucleic acids in spiroplasmas associated with plants, insects, and ticks. *Plasmid,* 29:81–93.

Gasparich, G.E., C. Saillard, E.A. Clark, et al. 1993b. Serologic and genomic relatedness of group VIII and group XVII and subdivision of spiroplasma group VIII into subgroups. *Int. J. Syst. Bacteriol.,* 43:338–341.

Gasparich, G.E., R.F. Whitcomb, D. Dodge, F.E. French, J.I. Glass, and D.L. Williamson. 2004. The genus *Spiroplasma* and its nonhelical descendants, phylogenetic classification, correlation with phenotype and roots of the *Mycoplasma mycoides* clade. *Int. J. Syst. Evol. Microbiol.,* 54:893–918.

Gaurivaud, P., F. Laigret, E. Verdin, M. Garnier, and J. M. Bové. 2000. Fructose operon mutants of *Spiroplasma citri. Microbiology (Reading),* 146:2229–2236.

Gill, S.R., M. Pop, R.T. DeBoy, et al. 2006. Metagenomic analysis of the human distal gut microbiome. *Science,* 312:1355–1359.

Glass, J.I., E.J. Lefkowitz, J.S. Glass, C.D.R. Heiner, E.Y. Chen, and G.H. Cassell. 2000. The complete sequence of the mucosal pathogen *Ureaplasma urealyticum. Nature,* 407:757–762.

Glass, J.I., C. Lartigue, C. Pfannkoch, et al. 2005. The complete genome of *Mycoplasma capricolum* subsp. *capricolum*. http://www.ncbi.nlm.nih.gov.

Gundersen, D.E., I.-M. Lee, S.A. Rehner, R.E. Davis, and D.T. Kingsbury. 1994. Phylogeny of mycoplasma-like orgasnisms (phytoplasmas), a basis for their classification. *J. Bacteriol.,* 176:5244–5254.

Hackett, K.J. and D.E. Lynn. 1985. Cell-assisted growth of a fastidious spiroplasma. *Science,* 230:825–827.

Hackett, K.J. and R.F. Whitcomb. 1995. Cultivation of spiroplasmas in undefined and defined media. In *Molecular and Diagnostic Procedures in Mycoplasmology,* Vol. I, S. Razin and J. G. Tully, Eds. p. 41–54. San Diego, CA: Academic Press.

Hahn, T.W., E.A. Mothershed, R.H. Waldo, 3rd, and D.C. Krause. 1999. Construction and analysis of a modified Tn4001 conferring chloramphenicol resistance in *Mycoplasma pneumoniae. Plasmid,* 41:120–124.

Hannan, P.C.T. 2000. Guidelines and recommendations for antimicrobial minimum inhibitory concentration (MIC) testing against veterinary mycoplasma species. *Vet. Res.,* 31:373–395.

Hildebrand, D.G., D.E. Page, and J.R. Berg. 1983. *Mycoplasma gallisepticum* (MG) — laboratory and field studies evaluating the safety and efficacy of an inactivated MG bacterin. *Avian Dis.,* 27:792–802.

Himmelreich, R., H. Hilbert, H. Plagens, E. Pirkl, B.C. Li, and R. Herrmann. 1996. Complete sequence analy sis of the genome of the bacterium *Mycoplasma pneumoniae. Nucleic Acids Res.*, 24:4420–4449.

Hubschle, O., R. Lelli, J. Frey, and R. Nicholas. 2002. Contagious bovine pleuropneumonia and vaccine strain T1/44. *Vet. Rec.,* 150:615.

Hudson, P., T.S. Gorton, L. Papazisi, K. Cecchini, S. Frasca, Jr., and S.J. Geary. 2006. Identification of a virulence-associated determinant, dihydrolipoamide dehydrogenase (lpd), in *Mycoplasma gallisepticum* through *in vivo* screening of transposon mutants. *Infect. Immun.* 74:931–939.

Inamine, J.M., K.-C. Ho, S. Loechel, and P.C. Hu. 1990. Evidence that UGA is read as a tryptophane codon rather than as a stop codon by *Mycoplasma pneumoniae*, *Mycoplasma genitalium*, and *Mycoplasma gallisepticum. J. Bacteriol.,* 172:504–506.

International Committee on Systematic Bacteriology, Subcommittee on the Taxonomy of *Mollicutes.* 1995. Revised minimum standards for descriptions of new species of the class *Mollicutes* (Division *Tenericutes*). *Int. J. Syst. Bacteriol.,* 45:605–612.

IRPCM Phytoplasma/Spiroplasma Working Team — Phytoplasma taxonomy group. 2004. '*Candidatus* Phytoplasma', a taxon for the wall-less, non-helical prokaryotes that colonize plant phloem and insects. *Int. J. Syst. Evol. Microbiol.,* 54:1243–1255.

Jacob, C., F. Nouzieres, S. Duret, J.M. Bové, and J. Renaudin. 1997. Isolation, characterization, and complementation of a motility mutant of *Spiroplasma citri. J. Bacteriol.,* 179:4802–4810.

Jacobs, E., M. Drews, A. Stuhlert, et al. 1988. Immunological reaction of guinea-pigs following intranasal *Mycoplasma pneumoniae* infection and immunization with the 168 kDa adherence protein. *J. Gen. Microbiol.,* 134:473–479.

Jaffe, J.D., N. Stange-Thomann, C. Smith, et al. 2004. The complete genome and proteome of *Mycoplasma mobile. Genome Res.*, 14:1447–1461.

Javed, M.A., S. Frasca, Jr., D. Rood, et al. 2005. Correlates of immune protection in chickens vaccinated with *Mycoplasma gallisepticum* strain GT5 following challenge with pathogenic *M. gallisepticum* strain R(low). *Infect. Immun.,* 73:5410–5419.

Johansson, K.-E. and B. Pettersson. 2002. Taxonomy of *Mollicutes.* In *Molecular Biology and Pathogenicity of Mycoplasmas*, S. Razin and R. Herrmann, Eds. London: Kluwer, p. 1–29.

Joshi, B.D., M. Berg, J. Rogers, J. Fletcher, and U. Melcher. 2005. Sequence comparisons of plasmids pBJS-O of *Spiroplasma citri* and pSKU146 of *S. kunkelii*: implications for plasmid evolution. *BMC Genomics,* 6:175. http://www.biomedcentral.com/1471-2164/6/175.

Junca, P., C. Saillard, J.G. Tully, et al. 1980. Charactérisation de spiroplasmes isolés d'insectes et de fleurs de France continentale, de Corse et du Maroc. Proposition pour une classification des spiroplasmas. *C. R. Acad. Sci. Ser. D* , 290:1209–1212.

Kämpfer, P., S. Buczolits, A. Albrecht, H.-J. Busse, and E. Stackebrandt. 2003. Towards a standardized format for the description of a novel species (of an established genus): *Ochrobactrum gallinifaecis* sp. nov. *Int. J. Syst. Evol. Microbiol.,* 53:893–896.

Kenny, G.E. 1992. Serodiagnosis. In *Mycoplasmas Molecular Biology and Pathogenesis*, J. Maniloff, R.N. McElhaney, L.R. Finch, and J.B. Baseman, Eds. Washington, DC: American Society for Microbiology, p. 505–512.

Kenny, G.E. 1996. Problems and opportunities in susceptibility testing of mollicutes. In *Molecular and Diagnostic Procedures in Mycoplasmology* Vol. II, J.G. Tully and S. Razin, Eds. San Diego, CA: Academic Press, p. 185–188.

Keswani, J. and W.B. Whitman. 2001. Relationship of 16S rRNA sequence similarity to DNA hybridization in prokaryotes. *Int. J. Syst. Evol. Microbiol.*, 51:667–678.

Khan, L.A., G.R. Loria, A.S. Ramirez, R.A. Nicholas, R.J. Miles, and M.D. Fielder. 2005. Biochemical characterisation of some non fermenting, non arginine hydrolysing mycoplasmas of ruminants. *Vet. Microbiol.,* 109:129–134.

King, G.J., M. Kagumba, and D.P. Kariuki. 1992. Trial of the efficacy and immunological response to an inactivated mycoplasma F38 vaccine. *Vet. Rec.,* 131:461–464.

King, K.W., and K. Dybvig. 1994. Mycoplasmal cloning vectors derived from plasmid pKMK1. *Plasmid,* 31:49–59.

King, K.W., D.H. Faulds, E.L. Rosey, and R.J. Yancey, Jr. 1997. Characterization of the gene encoding Mhp1 from *Mycoplasma hyopneumoniae* and examination of Mhp1's vaccine potential. *Vaccine,* 15:25–35.

Knight,T., Jr. and G. Fournier. 2004. The complete genome sequence of *Mesoplasma florum.* http://www.ncbi.nlm.nih.gov.

Knudtson, K.L., and F.C. Minion. 1993. Construction of Tn4001lac derivatives to be used as promoter probe vectors in mycoplasmas. *Gene,* 137:217–222.

Kürner, J., A.S. Frangakis and W. Baumeister. 2005. Cryo-electron tomography reveals the cytoskeletal structure of *Spiroplasma melliferum. Science,* 307:436–438.

Labarère, J. 1992. DNA replication and repair. In *Mycoplasmas Molecular Biology and Pathogenesis*, J. Maniloff, R.N. McElhaney, L.R. Finch, and J.B. Baseman, Eds. Washington, DC: American Society for Microbiology, p. 309–324.

Lartigue, C., A. Blanchard, J. Renaudin, F. Thiaucourt, and P. Sirand-Pugnet. 2003. Host specificity of mollicutes oriC plasmids: functional analysis of replication origin. *Nucleic Acids Res.*, 31:6610–6618.

Lee, S.-W., P.E. Markham, and G.F. Browning. 2004. Development of a replicapable OriC plasmid for *Mycoplasma gallisepticum.* Presented at the *15th International Congress of the International Organization for Mycoplasmology* (abstr. 57).

Lin, M.Y. and S.H. Kleven. 1982. Cross-immunity and antigenic relationships among five strains of *Mycoplasma gallisepticum* in young Leghorn chickens. *Avian Dis.,* 26:496–507.

Lloyd, L.C., G.S. Cottew, and D.A. Anderson. 1989. Protection against enzootic pneumonia of pigs: intraperitoneal inoculation with live LKR strain of *Mycoplasma hyopneumoniae. Australian Vet. J.,* 66:9–12.

Mahairas, G.G. and F.C. Minion. 1989. Random insertion of the gentamicin resistance transposon Tn4001 in *Mycoplasma pulmonis. Plasmid,* 21:43–47.

Maniloff, J. 1992. Phylogeny of mycoplasmas. In *Mycoplasmas Molecular Biology and Pathogenesis*, J. Maniloff, R.N. McElhaney, L.R. Finch, and J.B. Baseman, Eds. 549–559. Washington, DC: American Society for Microbiology, p. 549–559.

Maniloff, J. Phylogeny and evolution. 2002. In *Molecular Biology and Pathogenicity of Mycoplasmas*, S. Razin and R. Herrmann, Eds. London: Kluwer, p. 31–43.

Marais, A., J.M. Bové, S.F. Dallo, J.B. Baseman, and J. Renaudin. 1993. Expression in *Spiroplasma citri* of an epitope carried on the G fragment of the cytadhesin P1 gene from *Mycoplasma pneumoniae. J. Bacteriol.*, 175:2783–2787.

Marais, A., J.M. Bové, and J. Renaudin. 1996. *Spiroplasma citri* virus SpV1-derived cloning vector: deletion formation by illegitimate and homologous recombination in a spiroplasmal host strain which probably lacks a functional recA gene. *J. Bacteriol.*, 178:862–870.

Markham, P.F., M.D. Glew, G.F. Browning, K.G. Whithear, and I.D. Walker. 1998a. Expression of two members of the pMGA gene family of *Mycoplasma gallisepticum* oscillates and is influenced by pMGA-specific antibodies. *Infect. Immun.,* 66:2845–2853.

Markham, J.F., C.J. Morrow, and K.G. Whithear. 1998b. Efficacy of a temperature-sensitive *Mycoplasma synoviae* live vaccine. *Avian Dis.,* 42:671–676.

Markham, P.F., A. Kanci, G. Czifra, B. Sundquist, P. Hains, and G.F. Browning. 2003. Homologue of macrophage-activating lipoprotein in *Mycoplasma gallisepticum* is not essential for growth and pathogenicity in tracheal organ cultures. *J. Bacteriol.,* 185:2538–2547.

Masover, G.K., and F.A. Becker. 1996. Detection of mycoplasmas by DNA staining and fluorescent antibody methodology. In *Molecular and Diagnostic Procedures in Mycoplasmology,* Vol. II. J. G. Tully and S. Razin, Eds. San Diego, CA: Academic Press, p. 419–429.

McCoy, R.E., A. Caudwell, C.J. Chang, et al. 1989. Plant diseases associated with mycoplasmalike organisms. In *The Mycoplasmas,* Vol. 5. R.F. Whitcomb and J.G. Tully, Eds. San Diego, CA: Academic Press, p. 545–640.

Miles, R.J. 1992a. Catabolism in mollicutes. *J. Gen. Microbiol.,* 138:1773–1783.

Miles, R.J. 1992b. Cell nutrition and growth. In *Mycoplasmas Molecular Biology and Pathogenesis*, J. Maniloff, R.N. McElhaney, L.R. Finch, and J.B. Baseman, Eds. Washington, DC: American Society for Microbiology, p. 23–40.

Minion, F.C., E.J. Lefkowitz, M.L. Madsen, B.J. Cleary, S.M. Swartzell, and G.G. Mahairas. 2004. The genome sequence of *Mycoplasma hyopneumoniae* strain 232, the agent of swine mycoplasmosis. *J. Bacteriol.,* 186:7123–7133.

Minion, F.C., S.A. Menon, G.G. Mahairas, and M.J. Wannemuehler. 2003. Enhanced murine antigen-specific gamma interferon and immunoglobulin G2a responses by using mycobacterial ESAT-6 sequences in DNA vaccines. *Infect. Immun.,* 71:2239–2243.

Mogabgab, W.J. 1968. Protective effects of inactive *Mycoplasma pneumoniae* vaccine in military personnel, 1964–1966. *Am. Rev. Resp. Dis.,* 97:359–365.

Mogabgab, W.J. 1973. Protective efficacy of killed *Mycoplasma pneumoniae* vaccine measured in large-scale studies in a military population. *Am. Rev. Resp. Dis.,* 108:899–908.

Mohan, K., C.M. Foggin, P. Muvavarirwa, and J. Honywill. 1997. Vaccination of farmed crocodiles (*Crocodylus niloticus*) against *Mycoplasma crocodyli* infection. *Vet. Rec.,* 141:476.

Morrow, C.J., J.F. Markham, and K.G. Whithear. 1998. Production of temperature-sensitive clones of *Mycoplasma synoviae* for evaluation as live vaccines. *Avian Dis.,* 42:667–670.

Mouches, C., G. Barroso, A. Gadeau, and J.M. Bové. 1984. Characterization of two cryptic plasmids from *Spiroplasma citri* and occurrence of their DNA sequences among various spiroplasmas. *Annales de Microbiologie,* 135A:17–24.

Murray, R.G.E. and E. Stackebrandt. 1995. Taxonomic note: implementation of the provisional status Candidatus for incompletely described procaryotes. *Int. J. Syst. Bacteriol.,* 45:186–187.

Neimark, H., W. Peters, B.L. Robinson, and L.B. Stewart. 2005. Phylogenetic analysis and description of *Eperythrozoon coccoides*, proposal to transfer to the genus *Mycoplasma* as *Mycoplasma coccoides* comb. nov. and request for an opinion. *Int. J. Syst. Evol. Microbiol.,* 55:1385–1391.

Nicholas, R.A., R.D. Ayling, and L.P. Stipkovits. 2002. An experimental vaccine for calf pneumonia caused by *Mycoplasma bovis*: clinical, cultural, serological and pathological findings. *Vaccine,* 20:3569–3575.

Nicolet, J. and J.L. Martel. 1996. ELISA in large animals. In *Molecular and Diagnostic Procedures in Mycoplasmology,* Vol. II, J.G. Tully and S. Razin, Eds. San Diego, CA: Academic Press, p. 105–114.

Noormohammadi, A.H., J.E. Jones, G. Underwood, and K.G. Whithear. 2002. Poor systemic antibody response after vaccination of commercial broiler breeders with *Mycoplasma gallisepticum* vaccine ts-11 not associated with susceptibility to challenge. *Avian Dis.,* 46:623–628.

Noormohammadi, A.H., J.F. Jones, K.E. Harrigan, and K.G. Whithear. 2003. Evaluation of the non-temperature-sensitive field clonal isolates of the *Mycoplasma synoviae* vaccine strain MS-H. *Avian Dis.,* 47:355–360.

Oshima, K., S. Kazikawa, H. Nishigawa, et al. 2004. Reductive evolution suggested from the complete genome sequence of a plant-pathogenic phytoplasma. *Nature Genetics,* 36:27–29.

Papazisi, L., S. Frasca, Jr., M. Gladd, X. Liao, D. Yogev, and S.J. Geary. 2002a. GapA and CrmA coexpression is essential for *Mycoplasma gallisepticum* cytadherence and virulence. *Infect. Immun.,* 70:6839–6845.

Papazisi, L., L.K. Silbart, S. Frasca, et al. 2002b. A modified live *Mycoplasma gallisepticum* vaccine to protect chickens from respiratory disease. *Vaccine,* 20:3709–3719.

Papazisi, L., T.S. Gorton, G. Kutish, et al. 2003. The complete genome sequence of the avian pathogen *Mycoplasma gallisepticum* strain R(low). *Microbiology,* 149:2307–2316.

Pereyre, S., P. Gonzalez, B. De Barbeyrac, et al. 2002. Mutations in 23S rRNA account for intrinsic resistance to macrolides in *Mycoplasma hominis* and *Mycoplasma fermentans* and for acquired resistance to macrolides in *M. hominis. Antimicrob. Agents Chemother.,* 46:3142–3150.

Pour-El, I., C. Adams, and F.C. Minion. 2002. Construction of mini-Tn4001tet and its use in *Mycoplasma gallisepticum. Plasmid,* 47:129–137.

Poveda, J.B. 1998. Biochemical characteristics in mycoplasma identification. In *Mycoplasma Protocols. Methods in Molecular Biology,* Vol. 104, R. Miles and R. Nicholas, Eds. Totowa, NJ: Humana Press, p. 69–78.

Razin, S. 2002. Diagnosis of mycoplasmal infections. In *Molecular Biology and Pathogenicity of Mycoplasmas*, S. Razin and R. Herrmann, Eds. London: Kluwer, p. 531–544.

Razin, S., D. Yogev, and Y. Naot. 1998. Molecular biology and pathogenicity of mycoplasmas. *Microbiol. Mol. Biol. Rev.,* 62:1094–1156.

Regassa, L.B., K.M. Stewart, A.C. Murphy, F.E. French, T. Lin, and R.F. Whitcomb. 2004. Differentiation of group VIII *spiroplasma* strains with sequences of the 16S-23S rDNA intergenic spacer region. *Can. J. Microbiol.,* 50:1061–1067.

Renaudin, J. 2002. Extrachromosomal elements and gene transfer. In *Molecular Biology and Pathogenicity of Mycoplasmas*, S. Razin and R. Herrmann, Eds. London: Kluwer, p. 347–370.

Renaudin, J. and J.M. Bové. 1995. Plasmid and viral vectors for gene cloning and expression in *Spiroplasma citri*. In *Molecular and Diagnostic Procedures in Mycoplasmology,* Vol. I, S. Razin and J.G. Tully, Eds. San Diego, CA: Academic Press, p. 167–178.

Renaudin, J. and C. Lartigue. 2005. OriC plasmids as gene vectors for mollicutes. In *Mycoplasmas: Pathogenicity, Molecular Biology, and Strategies for Control,* A. Blanchard and G. Browning, Eds. Norfolk, U.K.: Horizon Bioscience, p. 3–30.

Renaudin, J., A. Marais, E. Verdin, et al. 1995. Integrative and free *Spiroplasma citri* oriC plasmids: expression of the *Spiroplasma phoeniceum* spiralin in *Spiroplasma citri*. *J. Bacteriol.,* 177:2870–2877.

Robertson, J.A., A. Vekris, C. Bébéar, and G.W. Stemke. 1993. Polymerase chain reaction using 16S rRNA gene sequences distinguishes the two biovars of *Ureaplasma urealyticum*. *J. Clin. Microbiol.,* 31:824–830.

Robertson, J.A., G.W. Stemke, J.W. Davis, Jr., et al. 2002. Proposal of *Ureaplasma parvum* sp. nov. and emended description of *Ureaplasma urealyticum* (Shepard et al. 1974) Robertson et al. 2001. *Int. J. Syst. Evol. Microbiol.,* 52:587–97.

Robinson, I.M. 1983. Culture media for anaeroplasmas. In *Methods in Mycoplasmology,* Vol. I, S. Razin and J.G. Tully, Eds. New York: Academic Press, p. 159–162.

Robinson, I.M. and E.A. Freundt. 1987. Proposal for an amended classification of anaerobic mollicutes. *Int. J. Syst. Bacteriol.,* 37:78–81.

Robinson, I.M., M.J. Allison, and P.A. Hartman. 1975. *Anaeroplasma abactoclasticum* gen. nov., sp. nov., an obligately anaerobic mycoplasma from the rumen. *Int. J. Syst. Bacteriol.,* 25:173–181.

Rose, D.L., J.G. Tully, J.M. Bové, and R.F. Whitcomb. 1993. A test for measuring growth responses of mollicutes to serum and polyoxyethylene sorbitan. *Int. J. Syst. Bacteriol.,* 43:527–532.

Roske, K., M.J. Calcutt, and K.S. Wise. 2004. The *Mycoplasma fermentans* prophage phiMFV1: genome organization, mobility and variable expression of an encoded surface protein. *Mol. Microbiol.,* 52:1703–1720.

Rosselló-Mora, R. 2003. The species problem, can we achieve a universal concept? *Syst. Appl. Microbiol.,* 26:323–326.

Rosselló-Mora, R. 2005. Updating prokaryotic taxonomy. *J. Bacteriol.,* 187:6255–6257.

Rosselló-Mora, R. 2006. DNA–DNA reassociation methods applied to microbial taxonomy and their critical evaluation. In *Molecular Identification, Systematics, and Population Structure of Prokaryotes,* E. Stackebrandt, Ed. Berlin: Springer-Verlag, p. 23–50.

Rosselló-Mora, R. and R. Amann. 2001. The species concept for prokaryotes. *FEMS Microbiol. Rev.,* 25:39–67.

Rurangirwa, F.R., T.C. McGuire, A. Kibor, and S. Chema. 1987. An inactivated vaccine for contagious caprine pleuropneumonia. *Vet. Rec.,* 121:397–400.

Sachse, K. and H. Hotzel. 1998. Classification of isolates by DNA-DNA hybridization. In *Mycoplasma Protocols. Methods in Molecular Biology* Vol. 104, R. Miles and R. Nicholas, Eds. Totowa, NJ: Humana Press, p. 189–196.

Sasaki, Y., J. Ishikawa, A. Yamashita, et al. 2002. The complete genomic sequence of *Mycoplasma penetrans*, an intracellular bacterial pathogen in humans. *Nucleic Acids Res.,* 30:5293–5300.

Shepard, M.C. 1983. Culture media for ureaplasmas. In *Methods in Mycoplasmology,* Vol. I, S. Razin and J.G. Tully, Eds. New York: Academic Press, p. 137–146.

Shimoji, Y., E. Oishi, Y. Muneta, H. Nosaka, and Y. Mori. 2003. Vaccine efficacy of the attenuated *Erysipelothrix rhusiopathiae* YS-19 expressing a recombinant protein of *Mycoplasma hyopneumoniae* P97 adhesin against mycoplasmal pneumonia of swine. *Vaccine,* 21:532–537.

Smith, A. and J. Mowles. 1996. Prevention and control of mycoplasma infection of cell cultures. In *Molecular and Diagnostic Procedures in Mycoplasmology,* Vol. II, J.G. Tully and S. Razin, Eds. San Diego, CA: Academic Press, p. 445–451.

Stackebrandt, E., W. Frederiksen, G.M. Garrity, et al. 2002. Report of the ad hoc committee for the re-evaluation of the species definition in bacteriology. *Int. J. Syst. Evol. Microbiol.,* 52:1043–1047.

Stackebrandt, E. and B.M. Goebel. 1994. Taxonomic note: a place for DNA-DNA reassociation and 16S rRNA sequence analysis in the present species definition in bacteriology. *Int. J. Syst. Bacteriol.* 44:846–849.

Stamburski, C., J. Renaudin, and J.M. Bové. 1991. First step toward a virus-derived vector for gene cloning and expression in spiroplasmas, organisms which read UGA as a tryptophan codon: synthesis of chloramphenicol acetyltransferase in *Spiroplasma citri*. *J. Bacteriol.*, 173:2225–2230.

Steinberg, P., R.L. Horswood, and R.M. Chanock. 1969. Temperature-sensitive mutants of *Mycoplasma pneumoniae*. I. *In vitro* biologic properties. *J. Infect. Dis.,* 120:217–224.

Steinberg, P., R.L. Horswood, H. Brunner, and R.M. Chanock. 1971. Temperature-sensitive mutants of *Mycoplasma pneumoniae*. II. Response of hamsters. *J. Infect. Dis.,* 124:179–187.

Stephens, E.B., G.S. Aulakh, D.L. Rose, J.G. Tully, and M.F. Barile. 1983a. Intraspecies genetic relatedness among strains of *Acholeplasma laidlawii* and of *Acholeplasma axanthum* by nucleic acid hybridization. *J. Gen. Microbiol.,* 129:1929–1934.

Stephens, E.B., G.S. Aulakh, D.L. Rose, J.G. Tully, and M. F. Barile. 1983b. Interspecies and intraspecies DNA homology among established species of *Acholeplasma*, a review. *Yale J. Biol. Med.,* 56:729–735.

Tang, J. , M. Hu, S. Lee, and R. Roblin. 2000. A polymerase chain reaction based method for detecting *Mycoplasma/Acholeplasma* contaminants in cell culture. *J. Microbiol. Meth.,* 39:121–126.

Taylor, R.R., H. Varsani, and R.J. Miles. 1994. Alternatives to arginine as energy sources for the non-fermentative *Mycoplasma gallinarum*. *FEMS Microbiol. Lett.,* 115:163–167.

Teuber, M. 1999. Spread of antibiotic resistance with food-borne pathogens. *Cell Mol. Life Sci.,* 56:755–763.

Thacker, E.L. 2006. Mycoplasmal diseases. In *Diseases of Swine*, B.E. Straw, J.J. Zimmerman, S. D'allaire, W. Mengeline, and D.J. Taylor, Eds. Ames, IA: Blackwell Publishing, p. 701–718.

Thiaucourt, F., L. Dedieu, J.C. Maillard, et al. 2003. Contagious bovine pleuropneumonia vaccines, historic highlights, present situation and hopes. *Dev. Biol.,* 114:147–160.

Townsend, R. 1976. Arginine metabolism by *Spiroplasma citri*. *J. Gen. Microbiol.,* 94:417–420.

Trachtenberg, S. 2004. Shaping and moving a *Spiroplasma*. *J. Mol. Microbiol. Biotechnol.,* 7:78–87.

Tully, J.G. 1995. Culture medium formulation for primary isolation and maintenance of mollicutes. In *Molecular and Diagnostic Procedures in Mycoplasmology*, Vol. I, S. Razin and J.G. Tully, Eds., San Diego, CA: Academic Press, p. 33–39.

Tully, J.G., J.M. Bové, F. Laigret, and R.F. Whitcomb. 1993. Revised taxonomy of the class *Mollicutes*, proposed elevation of a major cluster of arthropod-associated mollicutes to ordinal rank (*Entomoplasmatales* ord. nov.), with provision for familial rank to separate species with nonhelical morphology (*Entomoplasmataceae*) from helical species (*Spiroplasmataceae*) and emended description of the order *Mycoplasmatales*, family *Mycoplasmasmataceae*. *Int. J. Syst. Bacteriol.,* 43:378–385.

Tully, J.G. and D.L. Rose. 1983. Sterility and quality control of mycoplasma culture media. In *Methods in Mycoplasmology,* Vol. I, S. Razin and J.G. Tully, Eds. New York: Academic Press, p. 121–126.

Uphoff, C.C. and H.G. Drexler. 2002. Detection of mycoplasma in leukemia-lymphoma cell lines using polymerase chain reaction. *Leukemia,* 16:289–293.

van Kuppeveld, F.J.M., J.T.M. van der Logt, A.F. Angulo, et al. 1992. Genus- and species-specific identification of mycoplasmas by 16S rRNA amplification. *Appl. Environ. Microbiol.,* 58:2606–2615.

Vasconcelos, A.T., H.B. Ferreira, C.V. Bizarro, et al. 2005. Swine and poultry pathogens: the complete genome sequences of two strains of *Mycoplasma hyopneumoniae* and a strain of *Mycoplasma synoviae*. *J. Bacteriol.,* 187:5568–5577.

Voelker, L.L., K.E. Weaver, L.J. Ehle, and L.R. Washburn. 1995. Association of lysogenic bacteriophage MAV1 with virulence of *Mycoplasma arthritidis*. *Infect. Immun.,* 63:4016–4023.

Waite, E.R. and J.B. March. 2002. Capsular polysaccharide conjugate vaccines against contagious bovine pleuropneumonia: immune responses and protection in mice. *J. Comp. Pathol.,* 126:171–82.

Waites, K.B., D.T. Crouse, and G.H. Cassell. 1993. Therapeutic considerations for *Ureaplasma urealyticum* infections in neonates. *Clin. Infect. Dis.,* 17(Suppl. 1):S208–S214.

Waites, K.B., D.M. Crabb, and L.B. Duffy. 2003. Comparative *in vitro* susceptibilities and bactericidal activities of investigational fluoroquinolone ABT-492 and other antimicrobial agents against human mycoplasmas and ureaplasmas. *Antimicrob. Agents Chemother.,* 47:3973–3975.

Wang, R.Y.H. and S.-C. Lo. 1996. ELISA in human urogenital infections and AIDS. In *Molecular and Diagnostic Procedures in Mycoplasmology* Vol. II, J.G. Tully and S. Razin, Eds. San Diego, CA: Academic Press, p. 115–122.

Weisburg, W.G., J.G. Tully, D.L. Rose, et al. 1989. A phylogenetic analysis of the mycoplasmas: basis for their classification. *J. Bacteriol.,* 171:6455–6467.

Westberg, J., A. Persson, A. Holmberg, et al. 2004. The genome sequence of *Mycoplasma mycoides* subsp. *mycoides* SC type strain PG1T, the causative agent of contagious bovine pleuropneumonia (CBPP). *Genome Res.,* 14:221–227.

Whitcomb, R.F. 1983. Culture media for spiroplasmas. In *Methods in Mycoplasmology,* Vol. I, S. Razin and J.G. Tully, Eds. New York: Academic Press, p. 147–158.

Whitcomb, R.F., J.M. Bové, T.A. Chen, J.G. Tully, and D.L. Williamson. 1987. Proposed criteria for an interim serogroup classification for members of the genus *Spiroplasma* (class *Mollicutes*). *Int. J. Syst. Bacteriol.,* 37:8284.

Williamson, D.L., J.G. Tully, and R.F. Whitcomb. 1979. Serological relationships of spiroplasmas as shown by combined deformation and metabolism inhibition tests. *Int. J. Syst. Bacteriol.,* 29:345–351.

Williamson, D.L., J.G. Tully, G.E. Gasparich, et al. 1998. Revised group classification of the genus *Spiroplasma*. *Int. J. Syst. Bacteriol.*, 48:1–12.

Windsor, D. and H. Windsor. 1998. Quality control testing of mycoplasma media. In *Mycoplasma Protocols. Methods in Molecular Biology* Vol. 104, R. Miles and R. Nicholas, Eds. Totowa, NJ: Humana Press, p. 61–68.

Ye, F., J. Renaudin, J.M. Bové, and F. Laigret. 1994. Cloning and sequencing of the replication origin (oriC) of the *Spiroplasma citri* chromosome and construction of autonomously replicating artificial plasmids. *Curr. Microbiol.,* 29:23–29.

Zou, N. and K. Dybvig. 2002. DNA replication, repair and stress response. In *Molecular Biology and Pathogenicity of Mycoplasmas*, S. Razin and R. Herrmann, Eds. London: Kluwer, p. 303–323.

31 The Genus *Mycobacteria*

Vincent J. LaBombardi

CONTENTS

INTRODUCTION

Mycobacteria belong to the family *Mycobacteriaceae*. These organisms are characterized by possessing an elevated lipid content, most notably a high level of waxes called mycolic acids [1]. These mycolic acids are responsible for the organism being resistant to decolorization by acid alcohols. These cells are referred to as being acid-fast. The cells often appear as beaded rods when stained with an acid-fast stain such as the Ziehl-Neelsen or modified Kinyoun stains. Mycobacteria should not be confused with members of the genus *Nocardia* or *Rhodococcus*, which can be partially acid-fast.

The genus *Mycobacteria* consists of organisms that are true pathogens, opportunistic pathogens, and saprophytes. *Mycobacterium tuberculosis* is a worldwide cause of morbidity and mortality. Other members of the genus are ubiquitous and can be found in water and other environmental sources. These organisms vary in their growth requirements and generation times. Many of the pathogenic species have a generation time of 20 to 24 hr, which is reflected in the weeks sometimes required before these organisms will exhibit visible growth. Due to the nature of the diseases these organisms cause, it is essential that the cultural requirements of this genus be understood and conditions employed that will maximize recovery. Therapy for infections caused by these organisms will vary depending on the isolate identification. For the slow-growing species, anti-tubercular agents are required; whereas for the rapidly growing species, routine antibiotics may suffice. The importance of the *Mycobacteria* cannot be overemphasized and continued study is required to further delineate the role of these organisms in disease.

CLINICAL IMPORTANCE OF *MYCOBACTERIA* SPP.

MYCOBACTERIUM TUBERCULOSIS

According to the World Health Organization, between the year 2000 and 2020 nearly 1 billion people will become newly infected with *Mycobacterium tuberculosis*, 200 million of which will progress to active disease with an estimated 35 million deaths [2]. Statistics for 2002 show that the incidence of tuberculosis in the Americas is 43 cases per 100,000 in population, while in Africa the rate is 350 per 100,000. In many areas, especially in sub-Sahara Africa, a large number of these patients are co-infected with HIV [2].

The *Mycobacterium tuberculosis* complex consists of four species: *M. tuberculosis, M. bovis, M. africanum,* and *M. microti. M. bovis* is found in cattle but can cause a tuberculosis-like illness in humans. An attenuated strain of *M. bovis*, *M. bovis* BCG, is used as a vaccine to protect against acquiring tuberculosis. *M. africanum* also causes human tuberculosis while *M. microti* is found in rodents and, except in rare instances, has not been associated with human disease [3].

Anti-Tubercular Drug Resistance

Of added importance is the number of strains of *Mycobacterium tuberculosis* that are resistant to one or more of the anti-tubercular agents. Newly infected, treatment-naïve patients are now presenting with resistant strains. In certain areas of the world, more than one-third of these newly acquired infections are with isolates exhibiting resistance to at least one of the anti-tubercular agents. In a joint CDC-WHO study conducted in 2002–2004 with an international network of tuberculosis laboratories, 20% of 17,600 isolates were multi-drug resistant and 2% were extensively drug resistant (resistant to >3 classes of anti-tubercular agents) [4]. This makes treating tuberculosis, especially in a resource-poor environment, very difficult if not impossible.

NON-TUBERCULOUS MYCOBACTERIA

This group of organisms includes true pathogens, opportunistic pathogens, colonizers, and contaminates. These organisms can be placed into two broad categories based on their rate of growth. The slow-growing mycobacteria will require anywhere from 1 to 12 weeks before exhibiting growth on solid media, whereas the rapid-growing mycobacteria require 72 hr to 1 to 2 weeks to grow on solid media.

The following are some of the slow-growing mycobacteria that have been shown to cause clinical disease.

Mycobacterium leprae is the causative agent of leprosy or Hansen's disease. The disease can range from mild and self-limiting (tuberculoid type) to severely debilitating (lepromatous type). This is an ancient disease with cases depicted around 600 BC. There has always been a large social stigma associated with this disease, which resulted in victims being segregated to leper colonies. This is a very slow-growing organism with a long incubation period. Infected individuals might not show disease for 5 to 20 years after exposure.

The elimination of leprosy has been a goal of the WHO. The WHO reports that in 2005 there were 296,499 new cases of leprosy worldwide [5]. The majority of these cases were found in nine countries in Asia, Africa, and Latin America. This organism can be found in large quantities in the nasal passages of infected patients, who transmit the disease via droplet nuclei. This disease is now totally treatable with a combination of three drugs: dapsone, rifampicin, and clofazimine. These drugs are made available free of charge through WHO.

Mycobacterium leprae does not grow on laboratory media. Diagnosis can be made using a Fite stain on a nasal biopsy and/or by clinical presentation.

Mycobacterium ulcerans is the causative agent of Buruli ulcer. This disease, which occurs in tropical and sub-tropical climates, begins as a painless nodule, but if left untreated can progress to a debilitating ulcer. Most of the lesions appear on the limbs, especially on the lower extremities. The exact means of transmission of this organism is unknown but most infections appear to be found adjacent to bodies of water. There is also some evidence that mosquitoes may be responsible for transmission of the disease. This disease is treatable with combination therapy using rifampicin and streptomycin or amikacin. Surgical debridement is often required as well [6].

Although the disease can often be diagnosed on clinical presentation, this organism can grow on standard mycobacteriological media. The growth temperature optimum for this organism is 30°C, which is below normal body temperature. As a rule, media inoculated with specimens obtained from skin lesions sent for mycobacterial culture should always be incubated at this lower temperature.

Mycobacterium avium complex (MAC) is composed of the closely related species of *M. avium* and *M. intracellulare*. These two species are biochemically indistinguishable and cause similar diseases. They are rather ubiquitous organisms and are commonly found in the water supply. These organisms are not considered pathogenic in normal hosts but cause a variety of illnesses in immunosuppressed patients. MAC has been observed to cause a cervical lymphadenitis in immunocompetent patients, as well as pulmonary infections and disseminated disease [7]. Infections can be acquired by inhalation or ingestion of the organism. Because this organism is so ubiquitous, the isolation of MAC from respiratory specimens does not necessarily correlate with clinical disease. Isolation from blood or from a normally sterile body site is a clear indication of an infection.

MAC pulmonary disease can be manifested in a condition called "hot tub lung." This is seen in normal hosts who may use a hot tub indoors when there is poor ventilation. The organism is present in the water of the hot tub with inadequate levels of disinfectant. The organisms are aerosolized, inhaled, and subsequently establish a pulmonary infection.

MAC infections are treated with combination therapy, usually with clarithromycin, a fluoroquinolone such as moxifloxacin, ethambutol, and sometimes amikacin in cases of disseminated disease. Susceptibility tests can be performed against any of these agents but only tests with clarithromycin are recommended because data on clinical correlation with *in vitro* tests exist only with this drug.

Mycobacterium haemophilum was first isolated in 1978 from a patient with Hodgkin's disease. This organism was then shown to be a cause of skin lesions and disseminated infections in patients with AIDS in the early to mid 1990s. In addition, this organism has been described in immunosuppressed patients, including those with lymphoma, those who underwent bone marrow or cardiac transplantation, and patients receiving steroid therapy [8, 9]. It also has been described as a cause of lymphadenitis in immunocompetent children [10]. There is no standard therapy for infections with *M. haemophilum* but regimens including clarithromycin, doxycycline, ciprofloxacin, amikacin, isoniazid, and rifampin have led to resolution of the disease.

This organism is rather unique among the mycobacteria in that it requires hemin for growth. This organism also has an optimal growth temperature of 30°C. This probably accounts for the relative infrequency of its isolation prior to the 1990s. It is advisable that all patients presenting with a skin lesion have their specimen cultured onto medium containing a source of hemin and be incubated at 30°C.

Mycobacterium kansasii causes a disease that resembles tuberculosis. Typically, the patient is an older male with a prior history of tuberculosis, chronic obstructive pulmonary disease, or other chronic lung diseases. Cavitary disease may be seen on x-ray. The patient presents with cough, fever, and night sweats as is seen in cases of tuberculosis [11]. This disease has been described in HIV-positive individuals but is not the most common mycobacterial infection associated with HIV disease [12]. Extra-pulmonary disease caused by *M. kansasii* has also been described [13]. Treatment involves the use of first-line anti-tubercular drugs.

Mycobacterium malmoense is rarely seen in the United States but is recognized to cause pulmonary disease most often in northern Europe. In one study of 106 HIV-negative patients, 74% of patients had a cavity on chest x-ray and 58% of sputum specimens were smear-positive for acid-fast bacilli. The typical patient was a male in his 50s with prior or underlying lung disease [14]. There have also been reports of this organism isolated from children with lymphadenitis [15]. Therapy has included the use of some of the first line anti-tubercular agents such as rifampin, ethambutol, isoniazid, as well as clarithromycin.

Mycobacterium xenopi, like many of the other non-tuberculous mycobacteria, is a cause of pulmonary disease. This disease has become prominent in the United Kingdom, Canada, and the United States [16]. This species of mycobacteria is fairly ubiquitous and can be found in water, including the water supplies of hospitals [17]. The typical patient is an older male with a prior history of tuberculosis or other lung disease. The patient often presents with cavitary lung disease. This organism is easily grown in culture but prefers an elevated growth temperature (42°C) and forms typical "bird's nest" colonies when grown on solid media. This organism is not easily treatable, and limited success has been reported using rifampin, ethambutol, isoniazid, and clarithromycin.

Mycobacterium celatum was first described in the early 1990s as a cause of disseminated disease in AIDS patients [18]. These patients were similar in that they had very low CD4 counts, which is associated with profound immunosuppression. Subsequently, this organism has been identified as a cause of pulmonary disease in immunocompetent individuals [19]. In the latter cases, the disease resembles that caused by tuberculosis and other species of NTM. This isolate can be misidentified as either *M. tuberculosis* or *M. xenopi*. *M. celatum* isolates can be falsely positive in the AccuProbe test for *M. tuberculosis*, which detects species-specific nucleic acid [20]. These isolates are also very close biochemically and in mycolic acid profile to *M. xenopi*. Isolates of *M. celatum* are generally resistant to rifampin but susceptible to other anti-tubercular agents and clarithromycin.

Mycobacterium marinum was first observed in 1926 causing infections in saltwater fish in the Philadelphia aquarium. The first reported cases in humans were from an epidemic in Sweden in 1954. These patients, who all swam in the same swimming pool, developed lesions, mainly on the elbow. These lesions began as papules at the site of infection, which often became ulcerated and/or exudative [21]. This infection has been referred to as "swimming pool granuloma." This disease has also been associated with fish tank ownership. The infection usually is on the upper extremities, including the hands. The organism gains entrance through scrapes and abrasions on the hands and arms. Infection can spread from the initial site of inoculum but usually remains localized. Therapy with regimens containing ethambutol, rifampin, clarithromycin, amikacin, or doxycycline are usually effective. In some instances, surgical intervention is required [22]. This organism can be cultured from the infected lesion but the plated media should be placed at 30°C, which is the optimum growth temperature for this species.

Other slowly growing *Mycobacteria* spp. can be found associated with clinical disease. Additional clinically significant species can be found in Table 31.1.

CLINICALLY SIGNIFICANT SPECIES OF RAPIDLY GROWING MYCOBACTERIA

The rapidly growing species of mycobacteria are so named due to their shorter generation time compared to the slowly growing mycobacterial species. Many of these species can exhibit visible growth within 72 hr and can be found growing on routine bacteriological culture media. There are two major groups of rapidly growing mycobacteria: (1) the *Mycobacterium fortuitum* group, which consists of *M. fortuitum*, *M. peregrinum*, and *M. fortuitum*, 3rd biovar, and (2) the *Mycobacterium chelonae* group, consisting *of M. chelonae*, *M. abscessus*, and *M. mucogenicum*. The *M. fortuitum* group can be found in the water supply and has been associated with a number of pseudo-outbreaks in hospitals [23]. *M. fortuitum* has also been associated with surgical site infections, skin lesions, otitis media, and catheter-related bacteremia [24, 25].

TABLE 31.1
Other Clinically Significant Species of Slowly Growing Mycobacteria

Species	Clinical Disease
M. bohemicum	Lymphadenitis
M. branderi	Pulmonary
M. conspicuum	Pulmonary, sepsis
M. genavense	Sepsis, disseminated disease
M. heidelbergense	Pulmonary, disseminated disease
M. interjectum	Lymphadenitis
M. lentiflavum	Pulmonary, lymphadenitis
M. scrofulaceum	Lymphadenitis, sepsis
M. shimoidei	Pulmonary
M. simiae	Pulmonary, sepsis, osteomyelitis
M. szulgai	Pulmonary
M. triplex	Pulmonary

Mycobacterium abscessus has been associated with surgical site infections. There have been recent outbreaks due to contaminated materials used in plastic surgery performed by unlicensed practitioners. These infections have been manifested by necrotic cutaneous wounds that have required antibiotic therapy and surgical debridement. This organism is also being isolated with increasing frequency from the respiratory tract of cystic fibrosis patients. In these cases it can either be causing an infection or colonization [26].

Therapy for infections with rapidly growing species of mycobacteria includes the use of antibacterial agents and not anti-tubercular agents. *Mycobacterium abscessus* can be very resistant to antibiotics while *M. fortuitum* is usually more susceptible. As mentioned above, surgical intervention is sometimes required to effect a cure.

Table 31.2 lists other species of potentially clinically significant rapidly growing species of mycobacteria. A listing of mycobacterial species that are normally considered contaminates and not clinically significant is found in Table 31.3.

THE CULTURE AND ISOLATION OF MYCOBACTERIA FROM CLINICAL SPECIMENS

Specimen Selection

Respiratory specimens constitute the most common type of specimen submitted for the culture and isolation of mycobacteria. Sputum should be obtained early in the morning and prior to the patient expectorating. Three early-morning specimens are required to rule out mycobacterial disease. The patient should be instructed to rinse his/her mouth and expectorate into a sputum collection device specifically designed for mycobacterial culture. If the patient cannot produce sputum, a sputum induction can be performed. In this case, the patient is placed inside a special chamber or a hood is placed over the patient's head and nebulized saline is introduced, causing the patient to cough up sputum.

Specimens can also be obtained by the use of a bronchoscope. In this case, a bronchial wash and/or a bronchial alveolar lavage (BAL) can be obtained. Using this method, the area of the lung that appeared to be involved on x-ray can be sampled. If warranted, specimens can be obtained by performing an open lung biopsy.

TABLE 31.2
Other Clinically Significant Species of Rapidly Growing Mycobacteria

Species	Clinical Disease
M. goodii	Pulmonary, osteomyelitis, surgical site infection
M. immunogenum	Pulmonary, sepsis, disseminated disease
M. mageritense	Wound, sinusitis
M. mucogenicum	Sepsis
M. thermoresistibile[a]	Skin, surgical site infection
M. wolinskyi	Wound, surgical site infection

[a] Rapidly growing species when incubated at 42°C.

TABLE 31.3
Species of Unknown Clinical Significance

M. flavecsens
M. gastri
M. gordonae
M. nonchromogenicum
M. phlei
M. smegmatis
M. terrae complex

Biopsy specimens can be obtained from any site suspected of being infected with mycobacteria. Biopsies of an organ can be performed either by ultrasound guidance, a fine needle aspirate, or through general surgery. Skin lesions can be sampled by punch biopsy or by taking scrapings from the margin of the lesion. It is rarely productive to collect samples from the necrotic center of the lesions. To rule out leprosy, a nasal biopsy should be performed and stained using the Fite method.

Blood specimens should be obtained either using a specialized tube called an isolator tube or collecting a specimen in a tube containing an anti-coagulant. The isolator tube can be centrifuged to concentrate the specimen, or the buffy coat can be collected from the anticoagulated tube. The concentrated specimen can then be inoculated onto the appropriate media. Blood can be inoculated directly into blood culture bottles of some of the automated mycobacterial culture systems.

Specimen Processing

Sputa and other specimens collected from the respiratory tract are contaminated with normal flora. In addition, sputum is a viscous nonhomogenous specimen. Both of these facts makes it impossible to culture sputum without first processing the specimen. Sputum specimens are treated with a mucolytic agent to break down the mucus and a decontaminating agent to suppress the growth of the normal flora present in the sample. If this were not done, the normal flora would overgrow the more slowly growing species of mycobacteria, making their isolation from clinical samples impossible.

In the author's laboratory, a solution of sodium citrate plus N-acetyl L-cysteine (NALC) is added to an equal volume of a 4% sodium hydroxide solution. The NALC acts to digest the mucus while the sodium hydroxide acts to decontaminate the sample. An equal volume of this solution is added to the specimen, vortexed, and allowed to stand for 15 min at room temperature. After this time, an equal volume of 0.067 M phosphate buffer (pH 6.8) is added to the specimen to neutralize the alkali. The specimen is then centrifuged at 3500 × g to pellet the specimen [27].

The amount of sodium hydroxide used and the time allowed for decontamination are critical because both contaminating bacteria and mycobacteria can be affected by the sodium hydroxide treatment. Too high a concentration or too long before neutralization can result in the killing of the mycobacteria in the sample, thus resulting in a false-negative report.

Specimens obtained for normally sterile body sites need not be processed before inoculation into the appropriate media.

Specimen Staining

The composition of the mycobacterial cell wall renders the organism resistant to decolorization of the primary dye when subjected to an acid-alcohol wash. This is termed "acid-fastness," hence the name acid-fast bacteria (AFB).

This is often the first test performed on a specimen. In the case of specimens obtained from the respiratory tract, this test is used to determine if the patient is potentially infectious. In tuberculosis-endemic areas, this smear result is considered diagnostic. In areas of low prevalence, a positive smear may indicate the presence of tuberculosis or another mycobacterial disease. Any patient with an acid-fast positive smear would be considered potentially infectious and would require isolation. There are several methods that are used to perform an acid-fast stain:

1. The classic method, the Ziehl-Neelsen stain, uses carbolfuchsin as the primary dye. In this method, the stain must be steamed into the cell.
2. The modified or cold Kinyoun stain also uses carbolfuchsin as the primary stain but does not require heating. In both methods, acid-fast organisms will appear red.

There are also fluorescent stains available that can increase the sensitivity of the stain. These stains consist of the fluorochrome stain auramine, which can be combined with another fluorochrome, rhodamine. These stained smears must be read using a fluorescent microscope where mycobacteria will fluoresce a bright yellow. Using this method, smears can be screened at a lower magnification than the traditional stains. However, care must be taken to distinguish fluorescing artifacts from true organisms.

DIRECT TESTING OF SPECIMENS FOR *MYCOBACTERIUM TUBERCULOSIS* USING MOLECULAR METHODS

Molecular tests have been developed that can be used to test specimens directly for *Mycobacterium tuberculosis*. There are two assays available commercially: (1) the MTD test by Gen-Probe (San Diego, CA) and (2) a polymerase chain reaction- (PCR-) based assay from Roche Molecular Systems (Indianapolis, IN). These assays were initially approved for use on AFB smear-positive respiratory specimens. These assays both exhibit good sensitivity and specificity when testing smear-positive specimens. The sensitivity of the assays is observed to decrease when smear-negative specimens are tested. PCR-based assays have a disadvantage in that specimens can occasionally contain inhibitors of amplification, yielding a false-negative result. All molecular tests performed directly from the specimen must be followed up by a culture. The MTD test has also been used to test CSF specimens. The sensitivity of the MTD assay with this specimen type will be lower due to the paucity of bacilli usually found in this specimen type. Nonetheless, these methods can be very useful in diagnosing patients with tuberculosis.

Routine Culture Media and Automated Mycobacterial Culture Systems

For the optimal isolation of mycobacteria from clinical specimens, both a liquid medium and one or more pieces of solid media should be used. Solid media used for mycobacterial culture can be divided into two main categories: (1) those that are egg-based and (2) those that are agar-based. The egg-based media consist of eggs, salts, potato flour, amino acids, glycerol, and malachite green. The media are solidified by heating the mixture to a temperature of 85° to 90°C for approximately 45 min in a process called inspissation. The most common form of egg-based media used in the United States is Lowenstein-Jensen media. This media can also be obtained with antibiotics (Gruft formulation) or with other additives (pyruvic acid or RNA).

Agar-based media commonly employed include Middlebrook 7H10 and 7H11 media. They both contain defined salt concentrations; growth factors such as oleic acid, albumin, catalase glycerol; and a low concentration of malachite green as an inhibitor of bacterial growth. This media also is available with antibiotics.

Middlebrook 7H9 broth is also available but is rarely used manually for culture. It is incorporated into some of the automated systems.

In resource-poor settings, specimens can be cultured using only Lowenstein-Jensen media. However, for the optimal isolation of mycobacteria from clinical specimens, both liquid and solid media should be employed. To decrease the time required to isolate mycobacteria from clinical specimens, several automated systems have been developed and are now used routinely in many laboratories.

BACTEC 460 System

The original semi-automated system for mycobacterial culture was the BACTEC 460 (BD Diagnostic Systems, Sparks, MD). This utilized media containing a radiolabeled substrate that would be hydrolyzed by the mycobacterial isolate and release radioactivity. The radioactivity would be measured by the instrument and translated into a Growth Index (GI). A GI greater than 10 was indicative of a positive culture. This system is no longer available and will not be discussed further.

The VersaTREK (formerly ESP) Myco System

This system consists of a Myco bottle containing 7H9 broth and cellulose sponges to which growth supplement and an antibiotic mixture are added prior to inoculation with a specimen. This system has no limitations as to specimen type. The cellulose sponges act as a growth platform for the mycobacteria. After inoculation, the bottle is fitted with a connector and placed into the VersaTREK instrument. This instrument can detect changes in gas pressure within the headspace of the Myco bottle. Mycobacteria will consume oxygen as they grow, which results in a decrease in gas pressure. After three successive readings indicate a change in the slope of the barometric pressure line, the bottle will give a positive signal. As with any liquid culture, the positive bottle is smeared to confirm the presence of acid-fast bacilli and subcultured to solid media. Growth from the positive bottle can be identified directly using several methodologies (see below).

The MGIT 960 System

This system is composed of a tube containing 7H9 broth with a fluorescent indicator embedded in silicone in the bottom the tube. Prior to specimen inoculation, a growth supplement including an antibiotic mixture must be added. This system has not been validated with all specimen types, and blood must be inoculated into a different type of bottle, the MYCO/F Lytic F, and incubated in a different instrument, the BACTEC 9240.

This system also works on the principle of oxygen consumption. Oxygen present in the media quenches the fluorescent indicator in the tube. When the oxygen is consumed, the tubes will begin to fluoresce in the presence of UV light. This instrument will then flag the tube as positive. Because bacterial contamination can lead to a false-positive result, any positive must be confirmed by acid-fast staining.

MB Myco System

This system is composed of two types of media bottles. The MB Process System consists of broth, growth factors, and an antibiotic supplement and is utilized for respiratory specimens. The MB Blood System contains an enrichment fluid, a lytic agent, and growth factors. This can be inoculated with 3 to 5 mL whole blood. Both are incubated in the MB instrument where they are monitored for a change in pH associated with carbon dioxide production. As with any of the automated systems, contamination with bacteria must be ruled out by the performance of an acid-fast stain on an aliquot from a bottle that has been called positive by the instrument.

IDENTIFICATION OF MYCOBACTERIA

Mycobacteria are initially characterized using several criteria. The first, as mentioned above, is by growth rate. The next is by pigmentation. They are described as either buff, meaning a white- to cream-colored colony or a chromogen, which refers to their yellow to orange pigmentation. The chromogens are classified as either a scotochromogen or photochromogen. Scotochromogens will be pigmented regardless if growing in the light or dark, whereas a photochromogen will only develop pigment if exposed to light.

Many mycobacteria can be speciated by their biochemical profile. This can be very tedious, requiring, in some cases, several weeks before a reaction can be determined as positive or negative. It is also not always possible to distinguish species biochemically.

Because it is imperative that *Mycobacterium tuberculosis* be rapidly distinguished from other species of mycobacteria, tests have been developed that can accomplish this in a short period of time. The AccuProbe test produced by Gen-Probe, (San Diego, CA) is a nucleic acid probe that hybridizes with ribosomal RNA from several species of Mycobacteria. There are probes specific for *M. tuberculosis* complex, *M. avium* complex, *M. kansasii,* and *M. gordonae.* These probes can be used directly from positive cultures, which will allow the identification of isolates as *M. tuberculosis* within clinically relevant time frames [28].

While the AccuProbe test can reliably differentiate *Mycobacterium tuberculosis* complex from other mycobacterial species, it can identify only a limited number of species. Another method utilizes the mycolic acid profiles in the mycobacterial cell wall to identify the species. This method, high performance (or pressure) liquid chromatography (HPLC), has been shown to reliably identify many of the species of mycobacteria encountered in the clinical mycobacteriology laboratory [29]. The mycolic acids elute from the column at different rates, yielding either one, two, or three peak clusters. These species can be ascertained by calculating the relative peak retention times and comparing those to profiles previously run. One commercial system, the Sherlock Midi (Midi Labs, Newark, DE), contains a database to which the profile of the unknown organism can be matched to speciate the isolate.

While the use of HPLC increases the number of species of mycobacteria that can be identified, there still will be some species that cannot be distinguished from one another using this method. Also, new species of mycobacteria could not be identified by HPLC alone. The use of 16S ribosomal RNA gene sequencing has been used to routinely speciate mycobacteria [30]. However, as with HPLC, databases must contain the sequence of the isolate in question to be identified. Databases can vary as to the sequences contained in that listing. However, novel species can be recognized using this method.

SUSCEPTIBILITY TESTING OF MYCOBACTERIA SPECIES

Susceptibility testing of isolates of *Mycobacterium tuberculosis* is advisable even in newly diagnosed infections, as an individual may have been infected with a resistant strain. Resistance rates also vary widely, depending on the region of the world. Classically, anti-tubercular susceptibility testing was performed by an agar dilution proportion method. The first line drugs, isoniazid (INH), rifampin, streptomycin, and ethambutol, but not pyrazinamide (PZA), can be tested by this method. In this assay, anti-tubercular agents were added to agar medium and each drug-containing quadrant inoculated with a suspension of the isolate to be tested. A drug-free control quadrant was inoculated with the organism at the same time. Resistance was defined as the growth on a drug-containing quadrant of greater than 1% of the growth observed on the drug-free control. This method, while being reliable, required 3 weeks of incubation, which is not in keeping with current recommendations.

The first semi-automated system for performing anti-tubercular susceptibility testing was the BACTEC 460. This system reduced the time to final susceptibility results to about 1 week. This system also was able to perform susceptibility tests against PZA, which requires that the test be run at a more acidic pH to maintain activity of the drug. Again, this system is no longer commercially available.

The ESP, now VersaTREK, was the first continuously monitored system able to perform susceptibility tests against isolates of *Mycobacterium tuberculosis*. In this system, drug-containing and control bottles are inoculated with the isolate to be tested and placed in the instrument. The bottles are monitored for evidence of growth. Once the control bottle has signaled positive, the drug-containing bottles are incubated for an additional 3 days. In the case of a susceptible strain, no growth will be observed within those 3 days. If a drug-containing bottle signals positive within 3 days of the control bottle, that isolate is considered resistant to that drug at that concentration. This system can also determine the isolate's susceptibility to PZA [31].

The MGIT system has also been used to perform anti-tubercular susceptibility tests. This system can test the four first-line drugs as well as PZA. As above, the tubes containing drugs and the drug-free control are inoculated with the isolate to be tested. They are placed in the instrument in a fixed order and monitored for growth. The MGIT 960 instrument compares the growth in the control and drug-containing tubes, and automatically determines the susceptibility profile of the isolate. To be considered susceptible, the isolate in a drug-containing tube cannot turn positive within 2 days of the control tube.

Some investigators perform genetic-based assays to look for genes that code for resistance to some of the drugs. This is a rapid method that, if positive, predicts resistance. However, because the isolate can be resistant to a drug by other means, negative results must be viewed with caution.

The reliability of susceptibility testing of species of NTM is controversial. There are data to support testing of MAC isolates against clarithromycin. Other testing lacks a clear correlation with susceptibility test results and clinical outcomes. However, testing of NTM isolates has been performed using different methods, such as the VersaTREK Myco system. There is some data as to the clinical usefulness of this testing [32].

Susceptibility testing of the rapid growing mycobacteria can be accomplished using microtiter plates containing antibiotics. These tests are performed in either Mueller-Hinton or 7H9 broth and incubated for 72 hr. There are special panels available from Sensititre (TREK Diagnostics, Cleveland, OH). Interpretative criteria exist for a number of antibiotics when testing rapidly growing species of mycobacteria [33]. For those antibiotics where interpretations do not exist, minimum inhibitory concentration (MIC) values should be reported.

SUMMARY

Mycobacteria are the cause of significant morbidity and mortality in both developed and emerging nations. Tuberculosis affects millions of people worldwide. There is also a growing list of mycobacteria species that can cause clinical disease in both immunocompetent and immunosuppressed individuals. Although there have been advances in technology, culturing specimens for mycobacteria still remains an important tool for clinicians and microbiologists. There have been improvements made in the methodologies for culturing, isolating, and identifying mycobacteria. These improvements have greatly reduced the time required to identify an isolate and provide susceptibility data. All these factors will aid in the control of diseases caused by mycobacteria.

REFERENCES

1. Tuboly, S., The lipid composition of pathogenic and saprophytic mycobacteria, *Acta Microbiol. Hung.*, 15, 207, 1968.
2. World Health Organization, Tuberculosis, Fact Sheet No. 104. March 2004, www.who.int/mediacentre/factsheets/fs104/en/print.html.
3. Jenkins, N.E. et al., Immune restoration disease in HIV patient, *Emerg. Infect. Dis.*, 12, 689, 2006.
4. Emergence of *Mycobacterium tuberculosis* with extensive resistance to second-line drugs-worldwide, 2000–2004, *MMWR*, 55, 301, 2006.
5. World Health Organization, Leprosy, Fact Sheet No. 101. October 2005, www.who.int/mediacentre/factsheets/fs101/en/print.html.
6. World Health Organization, Buruli ulcer disease, Fact Sheet No. 199. September 2005, www.who.int/mediacentre/factsheets/fs199/en/print.html.
7. Horsburgh, C.R., Epidemiology of *Mycobacterium avium* complex disease, *Am. J. Med.*, 102, 11, 1997.
8. LaBombardi, V.J. and Nord, J.A., Clinical and laboratory aspects of *Mycobacterium haemophilum* infections, *Rev. Med. Microbiol.*, 9, 49, 1998.
9. Saubolle, M.A. et al., *Mycobacterium haemophilum*: microbiology and expanding clinical and geographic spectra of disease in humans, *Clin. Microbiol. Rev.*, 9, 435, 1996.
10. Armstrong, K. et al., *Mycobacterium haemophilum* causing perihilar and cervical lymphadenitis in healthy children, *J. Pediatr.*, 121, 202, 1992.
11. Maliwan, N. and Zvetina, J.R., Clinical features and follow up of 302 patients with *Mycobacterium kansasii* pulmonary infection: a 50 year experience, *Postgrad. Med. J.*, 81, 530, 2005.
12. Marras, T.K. and Daley, C.L., A systematic review of the clinical significance of pulmonary *Mycobacterium kansasii* isolates in HIV infection, *J. Acquir. Immune Defic. Syndr.*, 36, 883, 2004.
13. Nakamura, T. et al., *Mycobacterium kansasii* arthritis of the foot in a patient with systemic lupus erythematosis, *Intern. Med.*, 40, 1045, 2001.
14. The Research Committee of the British Thoracic Society, Pulmonary disease caused by *M. malmoense* in HIV negative patients: 5-yr follow-up of patients receiving standardized treatment, *Eur. Respir. J.*, 21, 478, 2003.
15. Lopez-Calleja, A.I. et al., *Mycobacterium malmoense* lymphadenitis in Spain: first two cases in immunocompetent patients, *Eur. J. Clin. Microbiol. Infect. Dis.*, 23, 567, 2004.
16. Jenkins, P.A. and Campbell, I.A., Pulmonary disease caused by *Mycobacterium xenopi* in HIV-negative patients: five year follow-up of patients receiving standardised treatment, *Respir. Med.*, 97, 439, 2003.
17. Faress, J.A. et al., *Mycobacterium xenopi* pneumonia in the southeastern United States, *South. Med. J.*, 96, 596, 2003.
18. Tortoli, E. et al., Isolation of the newly described species *Mycobacterium celatum* from AIDS patients, *J. Clin. Microbiol.*, 33, 137, 1995.
19. Piersimoni, C. et al., *Mycobacterium celatum* pulmonary infection in the immunocompetent: case report and review, *Emerg. Infect. Dis.*, 9, 399, 2003.
20. Butler, W.R. et al., Cross-reactivity of genetic probe for detection of *Mycobacterium tuberculosis* with the newly described species *Mycobacterium celatum, J. Clin. Microbiol.*, 32, 536, 1994.
21. Zeligman, I., *Mycobacterium marinum* granuloma. A disease acquired in the tributaries of Chesapeake Bay, *Arch. Derm.*, 106, 26, 1972.
22. Edelstein, H., *Mycobacterium marinum* skin infections. Report of 31 cases and review of the literature, *Arch. Intern. Med.*, 154, 1359, 1994.
23. LaBombardi, V.J., O'Brien, A.M., and Kislak, J. W., Pseudo-outbreak of *Mycobacterium fortuitum* due to contaminated ice machines, *Am. J. Infect. Control*, 30, 184, 2002.
24. Moreno, A. et al., *Mycobacterium fortuitum* bacteremia in an immunocompromised patient with a long-term venous catheter, *Eur. J. Clin. Microbiol. Infect. Dis.*, 15, 423, 1996.

25. Plemmons, R.M. et al., Otitis media and mastoiditis due to *Mycobacterium fortuitum*: case report, review of four cases and a cautionary note, *Clin. Infect. Dis.*, 22, 1105, 1996.
26. Sermet-Gaudelus, I. et al., *Mycobacterium abscessus* and children with cystic fibrosis, *Emerg. Infect. Dis.*, 9, 2003, at: www.cdc.gov/ncidod/EID/vol9no12/02-0774.htm.
27. Kent, P.T. and Kubica, G.S., *Public Health Mycobacteriology. A Guide to the Level III Laboratory*, United States Dept. of Health and Human Services, CDC, Atlanta, GA, 1985, p. 31–70.
28. LaBombardi, V.J., Carter, L., and Massarella, S., Use of nucleic acid probes to identify mycobacteria directly from Difco ESP bottles, *J. Clin. Microbiol.*, 35, 1002, 1997.
29. LaBombardi, V.J., Katariwala, R., and Pipia, G., The identification of mycobacteria from solid media and directly from VersaTREK Myco bottles using the Sherlock Mycobacteria Identification HPLC system, *Clin. Microbiol. Infect.*, 12, 478, 2006.
30. Hall, L. et al., Evaluation of the MicroSeq system for the identification of mycobacteria by 16S ribosomal DNA sequencing and its integration into a routine clinical mycobacteriology laboratory, *J. Clin. Microbiol.*, 41, 1447, 2003.
31. LaBombardi, V.J., Comparison of the ESP and BACTEC systems for testing susceptibilities of *Mycobacterium tuberculosis* complex isolates to Pyrazinamide, *J. Clin. Microbiol.*, 40, 2238, 2002.
32. Lui, A.Y. et al., The ESP Culture System for drug susceptibilities of *Mycobacterium avium* complex, *Clin. Microbiol. Infect.*, 6, 649, 2000.
33. National Committee for Clinical Laboratory Standards, Susceptibility Testing of Mycobacteria, Nocardiae and Other Aerobic Actinomyces, Approved Standard, NCCLS (now CLSI) Document M24-A, Clinical and Laboratory Standards Institute, Wayne, PA, 2005.

32 The Genus *Legionella*

Eva M. Campodonico and Craig R. Roy

CONTENTS

THE GENUS *LEGIONELLA*

Members of the genus *Legionella* are Gram-negative, aerobic bacilli that belong to the γ-subgroup of proteobacteria. *Legionella* are facultative intracellular bacteria that are ubiquitous in freshwater ponds and streams where they colonize phagocytic protozoa. The relevance of these organisms to human health was first recognized following the eponymous 1976 outbreak at a meeting of the American Legion in Philadelphia. During the weeks following the convention, 221 attendees fell ill with unexplained, pneumonia-like symptoms; of these, 34 later died (Garrett, 1994). Legionnaires' disease was ultimately linked to infection by a bacterium called *Legionella pneumophila*. An examination of the air conditioning system at the convention site revealed high levels of *Legionella pneumophila*-laden protozoa, indicating that exposure to these bacteria via inhalation of contaminated aerosols could result in severe respiratory disease.

Subsequent outbreaks of *Legionella* infection have been traced to several types of man-made water reservoirs, including cooling towers, industrial air conditioners, hot water systems, and fountains. If improperly maintained, biofilms can develop on the accumulated sediment in these structures, creating an ecosystem that supports both protozoa and *Legionella* growth. Members of the genus *Legionella* are found in numerous natural ecosystems proximal to human populations, but their ability to colonize human hosts depends on inhalation of aerosolized water-borne bacteria. Due to this somewhat artificial means of transmission, *Legionella* is a relatively recent addition to the list of human pathogens.

The diseases resulting from exposure to *Legionella*, collectively termed legionellosis, include Legionnaires' disease and Pontiac fever. Legionnaires' disease is a severe infection of the respiratory tract that is fatal in 5 to 30% of cases (CDC, 2005). Pontiac fever is a milder illness, with flu-like symptoms that may resolve within a few hours to a few days of onset. Whereas Legionnaires' disease is thought to result from an infection in which bacteria are actively replicating in the lung,

Pontiac fever is thought to result from an acute exposure to high levels of bacteria without bacterial multiplication. All forms of legionellosis can be successfully treated with a range of antibiotics, including macrolides (erythromycin, azithromycin, clarithromycin), tetracyclins (doxycyclin), or broad-spectrum fluoroquinolones (Amsden, 2005).

Legionella infection is one of the most common causes of community-acquired pneumonia (CAP) and is second only to bacteremic pneumococcal pneumonia in causing severe CAP that requires treatment in intensive care. *Legionella* has also emerged as a common nosocomial-acquired pneumonia, a phenomenon of particular concern because mortality resulting from legionellosis is highest in elderly and immunocompromised patients (Amsden, 2005). In the United States, between 8000 and 18,000 people are diagnosed with legionellosis each year, although only a small number of these cases are reported (CDC, 2005).

To date, 48 *Legionella* species, consisting of 65 serogroups, have been identified (Benson and Fields, 1998; Fields et al., 2002). While many of these species can occasionally cause disease in humans, analysis of culture-confirmed cases of legionellosis indicates that *Legionella pneumophila* is responsible for greater than 90% of disease, with *L. longbeacheae*, *L. bozemanii*, *L. micdadei*, *L. feeleii*, *L. dumoffii*, *L. wadsworthii*, and *L. anisa* also periodically isolated from immunocompromised patients (Yu et al., 2002). Of the 15 serogroups identified for *L. pneumophila*, the vast majority of disease results from infection by serogroup 1, with serogroups 4 and 6 periodically found in nosocomial outbreaks (Amsden, 2005). Due to the marked clinical prevalence of *L. pneumophila* serogroup 1, the majority of scientific research has focused on this single *Legionella* strain.

LEGIONELLA PNEUMOPHILA GROWTH IN NATURE AND IN THE LABORATORY

In freshwater ecosystems, *Legionella pneumophila* colonizes free-living protozoa and has been found to be capable of replicating within various protozoan species, including *Acanthamoeba*, *Hartmonella,* and the model amoeba *Dictyostelium* (Fields et al., 1993; Moffat and Tompkins, 1992; Solomon et al., 2000). To study the basis of human infection in the laboratory, *L. pneumophila* can be cultured within a range of mammalian cell lines. Primary macrophages derived from the bone marrow of A/J mice have been used in numerous studies (Yamamoto et al., 1991), in addition to macrophage-like cells lines such as U937, RAW, and J774 (Diez et al., 2000; Husmann and Johnson, 1992; King et al., 1991). *L. pneumophila* can also replicate within cells that are not inherently phagocytic, provided they have a means of internalization. For example, *L. pneumophila* that have been opsonized with specific antibody can be internalized by Chinese hamster ovary (CHO) cells expressing the FcγRII receptor. Cell biological assays indicate that the intracellular life-cycle events that unfold in this cell line parallel those seen in protozoan and phagocyte hosts (Kagan et al., 2004).

The nutrients required for *Legionella pneumophila* to establish and maintain its intracellular niche are yet to be fully described. Microarray analysis of differential gene expression in *L. pneumophila* growing extracellularly in broth culture or intracellularly in *Acanthamoeba castellanii* indicate that intracellular growth involves scavenging for amino acids by up-regulation of factors governing amino acid uptake and catabolism. In addition, *L. pneumophila* may also acquire carbohydrate from its host, as intracellular growth also leads to an up-regulation of the gene cluster encoding the Entner-Douderoff pathway, a putative glucokinase and a sugar transporter (Bruggemann et al., 2006b).

The growth medium commonly used for extracellular culture of *Legionella* spp. consists of yeast extract and ACES buffer, with potassium hydroxide added to adjust the pH to 6.9 (Feeley et al., 1979). This medium is then supplemented with cysteine and iron; for solid medium, activated charcoal is added to chelate an unknown factor in agar that is toxic to *Legionella*. Bacterial growth is optimal at 37°C, but lower temperatures (20 to 25°C) are often required for studies involving protozoan host species. Streptomycin, chloramphenicol, and kanamycin are frequently used for antibiotic selection; however, *L. pneumophila* is resistant to even high doses of ampicillin.

BIPHASIC LIFE CYCLE OF *LEGIONELLA PNEUMOPHILA*

Studies of *Legionella pneumophila* grown in broth or in protozoa have shown that this bacterium has (at least) two different stages in its life cycle. While growing exponentially in broth culture, and within the first several hours following host cell entry, *L. pneumophila* adopts a non-motile, non-flagellated state that is competent to replicate to high levels but shows a low level of infectivity of new host cells. When cultured in broth to high density in post-exponential phase, *L. pneumophila* become motile due to the production of flagella and these bacteria are highly infectious. This phase roughly correlates with late time points of *in vivo* infection of host cells when *L. pneumophila* have completed several rounds of replication and exist within densely populated replicative compartments. With nutrient concentrations and host cell viability on the decline, these virulent, motile bacteria are primed to carry out new rounds of infection in neighboring host cells. In addition, the transition of *L. pneumophila* from its replicative form to infectious form is associated with increased resistance to osmotic and heat stress, sensitivity to sodium chloride, increased pigmentation, and a reduction in bacterial cell size (Byrne and Swanson, 1998).

These broadly defined stages of *Legionella pneumophila*'s life cycle have been termed the replicative and transmissive phases, and the regulatory mechanisms underlying this phase switch are beginning to be described. A screen for mutants that expressed low levels of flagellin during post-exponential growth conditions revealed that a two-component regulatory system (LetA/LetS), the sigma factor FliA, and the stationary-phase sigma factor RpoS, function in concert to regulate *L. pneumophila*'s transition from its replicative to transmissive form (Hammer et al, 2002; Lynch et al., 2003). The two component LetA/LetS system is thought to respond to ppGpp, an alarmone produced as part of a stringent response under conditions of nutrient starvation, such as low amino acid levels (Hammer and Swanson, 1999; Jain et al., 2006). In turn, LetA/LetS induces transition to the transmissive phase by relieving repression by CsrA (Molofsky and Swanson, 2003). Csr regulatory systems are found in numerous bacterial species and are responsible for post-transcriptional regulation of gene expression on a global level (Majdalani et al, 2005; Romeo, 1998). However, the requirement for LetA/LetS signaling in *L. pneumophila* virulence appears to be host cell type dependent. *LetA* mutants that showed a replication defect in the protozoan host *Acanthamoeba castellanii* were able to grow efficiently in A/J mouse bone marrow-derived macrophages (Hammer et al., 2002; Lynch et al., 2003).

The role of the sigma factor RpoS in phase switching is not yet fully understood. Due to the requirement for this transcription factor for intracellular survival and replication, RpoS function has been placed amidst a cascade of replicative phase regulatory events (Abu-Zant et al., 2006; McNealy et al., 2005). However, seemingly contradictory data reveal an up-regulation of RpoS expression during the stationary phase as *L. pneumophila* prepares to transform into the transmissive phase (Bachman and Swanson, 2001, 2004; Gal-Mor and Segal, 2003; Hales and Shuman, 1999; Molofsky and Swanson, 2003).

Finally, the successful transformation of replicating *Legionella pneumophila* to virulent, motile bacteria requires the signaling cascade of the flagella regulon. Upon entering the transmissive phase, the alternative sigma factor RpoN and the transcriptional activator FleQ positively regulate expression of *fleN* (FleN may be an anti-activator of FleQ), *fliM*, *fleSR*, and numerous other genes whose products are required for flagellum biosynthesis (Bruggemann et al., 2006b; Dasgupta and Ramphal, 2001; Jacobi et al., 2004). RpoN and the FleS/FleR two-component system act in concert to trigger expression of the alternative sigma factor FliA, which controls the last stages of flagellum biosynthesis, including production of the FlaA flagellin subunit (Bruggemann et al., 2006b; Heuner and Steinert, 2003).

INTRACELLULAR LIFE CYCLE OF *LEGIONELLA PNEUMOPHILA*

BACTERIAL ENTRY

Since *Legionella pneumophila* was identified as the causative agent of the fatal pneumonia outbreak in 1976, the intracellular life-cycle events of this bacterium have been extensively dissected. Early work by Horwitz and colleagues provided an essential descriptive model of *L. pneumophila*'s intracellular trafficking within both protozoan and macrophage hosts (Horwitz, 1983a, b). Subsequent studies revealed that *L. pneumophila* is able to evade the degradative endocytic compartments of the host cell (Berger et al., 1994) and establish a replicative niche by remodeling its phagosome into a ribosome-studded vacuole that resembles the host cell endoplasmic reticulum (ER) (Kagan and Roy, 2002; Robinson and Roy, 2006; Tilney et al., 2001) (Figure 32.1).

Legionella pneumophila enter host cells by phagocytosis, engaging host cell surface receptors that appear to vary with host cell type. In mammalian monocytes, bacterial entry can be partially inhibited by the presence of antibodies specific for the C1 and C3 complement receptors on the host cell surface (Payne and Horwitz, 1987). It has also been shown that opsonization of *L. pneumophila* with a bacteria-specific antibody allows for Fc-receptor-mediated phagocytosis (Horwitz and Silverstein, 1981), although the absence of *L. pneumophila*-specific IgG in the lungs of naïve human hosts renders this pathway of uncertain clinical relevance. In addition, levels of complement in the lung are normally low (Reynolds and Newball, 1974), suggesting that *L. pneumophila* also may engage as yet unidentified surface receptors to induce uptake into mammalian host cells.

Legionella pneumophila's natural hosts — freshwater protozoa — lack both Fc and complement receptors. Studies of *L. pneumophila* uptake by the social protozoan *Hartmannella vermiformis* suggest that the Gal/GalNAc lectin receptor can be used by the bacterium to mediate entry (Venkatara-

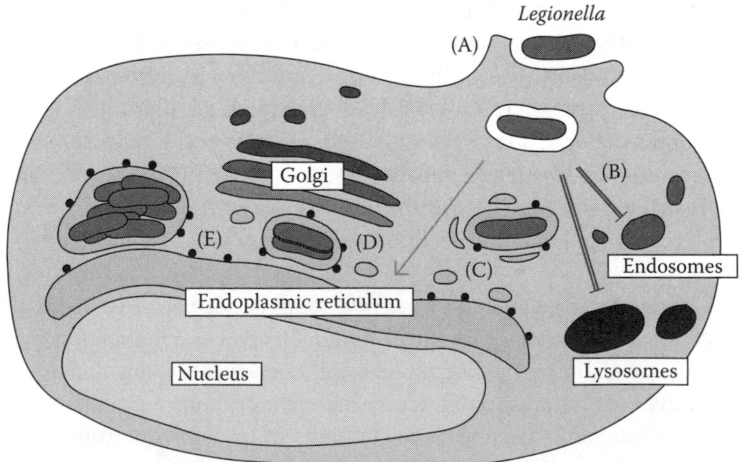

FIGURE 32.1 Intracellular life cycle of *Legionella pneumophila*. (A) *L. pneumophila* engage protozoan or mammalian host cell surface receptors and are internalized by phagocytosis. (B) The bacterium inhibits transport of its phagosome down the endocytic pathway to degradative lysosomes and (C) associates with membrane vesicles and mitochondria within minutes of internalization. (D) *L. pneumophila* completes remodeling of its vacuole and begins to replicate in a compartment that is morphologically indistinguishable from the ER. (E) Bacterial replication proceeds for several hours until the *L. pneumophila*-containing vacuole is lysed and *L. pneumophila* exit host cells to initiate new rounds of infection. (Figure adapted from Roy and Tilney, 2002.)

man et al., 1997). However, similar experiments in another protozoan, *Acanthamoeba polyphaga*, revealed little effect on bacterial entry when access to this receptor is limited (Harb et al., 1998). Therefore, as is likely the case with mammalian phagocytes, *L. pneumophila* appears to be capable of interacting with a variety of host cell surface components to facilitate entry into protozoa.

Following adherence to host cells, internalization of *Legionella pneumophila* proceeds via conventional zippering phagocytosis that can be efficiently blocked by depolymerization of the actin cytoskeleton with Cytochalasin D (Elliott and Winn, 1986). There is also evidence that *L. pneumophila* can induce a novel form of entry termed "coiling phagocytosis," in which the bacterium is enveloped in multiple layers of host cell membrane during internalization (Horwitz, 1984). This phenomenon has been observed when professional phagocytes are infected with relatively high doses of live or heat-killed *L. pneumophila* and can be disrupted if the bacteria are pretreated with anti-*L. pneumophila* antibody. This suggests that high concentrations of a stable bacterial surface component may be required to perturb phagocyte membrane and induce coiling, a model supported by the observation that only certain *L. pneumophila* serotypes induce coiling phagocytosis (Rechnitzer and Blom, 1989).

The role of coiling phagocytosis during *Legionella pneumophila*'s life cycle is unclear. Bacterial viability has been shown to be required for all *L. pneumophila* virulence factors identified to date, including Type IV secretion (discussed below) and changes in morphology, motility, and gene expression associated with *L. pneumophila*'s transition from a replicating, noninfectious state to a stationary and transmissive one. Because heat-killed bacteria can induce coiling phagocytosis but are quickly delivered to lysosomes and degraded (Horwitz, 1984), it is unlikely that this novel mode of entry plays a significant role in the later stages of *L. pneumophila*'s unique pattern of intracellular trafficking.

Intracellular Trafficking and Replicative Vacuole Formation

Once inside its host cell, *Legionella pneumophila* must quickly execute two essential tasks to successfully establish its intracellular niche. The bacterium must avoid the fate of most cargo internalized by phagocytic cells by preventing its rapid delivery to lysosomes, the degradative terminal compartment of the endocytic pathway. Concurrently, *L. pneumophila* must establish a subcellular niche that supports bacterial replication.

Early experiments by Horwitz and colleagues showed that phagosomes containing live *Legionella pneumophila* acquire lysosomal markers at greatly reduced levels compared to phagosomes containing fixed *L. pneumophila* or *Escherichia coli,* and this separation from the endocytic pathway is maintained for several hours following infection (Horwitz, 1983a, b). Subsequently, however, various lines of experimentation have generated conflicting data regarding late-stage fusion events of *L. pneumophila*'s compartment with acidic lysosomes and the importance of acidic pH for efficient bacterial replication.

Evidence supporting a need for acidification of the *Legionella pneumophila*-containing vacuole (LCV) is provided by the reduced numbers of bacteria recovered from host cells treated with the vacuolar ATPase inhibitor Bafilomycin-A (BFA), which prevents acidification of endocytic compartments (Sturgill-Koszycki and Swanson, 2000). However, BFA treatment has been shown to disregulate a number of cellular pathways (Clague et al., 1994; Reaves and Banting, 1994; van Weert et al., 1997; Yamamoto et al., 1998), calling into question the ability to directly interpret these data as proof of *L. pneumophila*'s requirement for an acidified compartment. In addition, bacterial growth assays conducted in monocyte cell lines with differing capacities to acidify their endocytic compartments yielded indistinguishable levels of bacterial growth (Wieland et al., 2004). Therefore, it is likely that the only prudent action that *L. pneumophila* can take relative to the endocytic pathway is to avoid it.

With the degradative effects of the lysosome held at bay, *Legionella pneumophila* conducts an extensive remodeling of its plasma membrane-derived phagosome. Within 15 minutes of internalization, the LCV is surrounded by numerous smooth vesicles and mitochondria (Horwitz, 1983b).

As the infection proceeds, these vesicles flatten and fuse with the limiting membrane of the LCV, which also becomes studded with ribosomes (Horwitz, 1983b; Robinson and Roy, 2006; Tilney et al., 2001). Localization of various ER proteins to this compartment shortly after infection, including calnexin, BiP, and KDEL epitope-containing proteins, indicate that these vesicles are derived from the ER (Derre and Isberg, 2004; Kagan and Roy, 2002; Swanson and Isberg, 1995). Conversely, LCVs do not accumulate the Golgi protein Giantin (Kagan and Roy, 2002), suggesting that *L. pneumophila* gains access to secretory membrane traffic directly, rather than circuitously through the Golgi as is seen with the trafficking of the Cholera and Shiga bacterial toxins (Sandvig and van Deurs, 2002).

Disruption of early secretory traffic using chemical inhibitors has allowed for a further narrowing of the window in which *Legionella pneumophila* intercepts the membrane traffic used to remodel its phagosome. Brefeldin A inhibits ADP ribosylation factor 1 (ARF1)-dependent vesicle budding from the ER. The microtubule-based transport of these vesicles from the ER through the intermediate compartment (ERGIC) to the Golgi can be blocked with the microtubule disrupting agent nocodozol. The LCV is able to recruit the ERGIC protein p58 in the presence of nocodozol, but not when cells are treated with Brefeldin A (Kagan and Roy, 2002). Thus, following internalization, *L. pneumophila* remodels its phagosome into a ribosome-studded, ER-like vacuole by intercepting and fusing with ER vesicles before they are transported through the ERGIC to the Golgi.

Another model proposed to describe the generation of *Legionella pneumophila*'s unique vacuole invokes the involvement of host autophagic machinery to convert the LCV into an autophagosomal compartment. Common features shared by autophagosomes and LCVs include a double membrane and the presence of ER proteins such as calnexin and BiP (Horwitz, 1983b; Swanson and Isberg, 1995; Tilney et al., 2001). In addition, the induction of autophagy in macrophages by either amino acid starvation or rapamycin treatment results in both the recruitment of autophagy proteins Atg7 and Atg8 to LCVs and an increase in colony-forming units recovered from autophagy-induced cells compared to untreated conditions (Amer and Swanson, 2005).

However, a genetic assessment of the role of autophagy in LCV remodeling using the protozoan model host *Dictyostelium discoideum* suggests that this pathway may not play a critical role in the creation of *Legionella pneumophila*'s membrane-bound replicative niche. Various mutants of *D. discoideum* lacking essential components of the autophagy machinery were tested for their ability to support bacterial replication. *L. pneumophila* were able to form normal vacuoles and replicate to similar levels in both wild-type and autophagy-deficient *D. discoideum* strains (Otto et al., 2004). Unlike macrophages, amino acid-starved *D. discoideum* infected with bacteria did not redistribute autophagy proteins to the LCV, further suggesting that *L. pneumophila*'s ability to remodel its phagosome and replicate intracellularly does not absolutely require interaction with the autophagic pathway (Otto et al., 2004).

Egress

Little is known about the final phase of *Legionella pneumophila*'s intracellular existence. At 12 to 24 hr after infection, vacuoles containing multiplying *L. pneumophila* have grown substantially, housing dozens of bacteria. At this point, it is unclear whether the bacteria actively lyse their compartment and subsequently their host cell, or if the integrity of these membrane-bound vacuoles simply fails due to physical stress and the destructive effects of being converted into a bacterial nursery. Electron microscopy studies have revealed what appear to be cytosolic *L. pneumophila* at late stages of infection, indicating that there may be a delay between rupture of the LCV and egress of bacteria from the host cell to go on to further rounds of infection (Molmeret et al., 2004).

IDENTIFICATION OF THE DOT/ICM SECRETION SYSTEM AS A KEY COMPONENT OF *LEGIONELLA PNEUMOPHILA'S* INTRACELLULAR LIFE CYCLE

To determine the mechanisms underlying *Legionella pneumophila*'s unique intracellular life cycle, numerous screens have been conducted to identify the bacterial genes required for intracellular growth. Various mutagenesis strategies were used to generate *L. pneumophila* strains unable to replicate within and lyse host cells, to prevent fusion of bacteria-containing phagosomes with lysosomes, or to recruit host cell vesicles and organelles.

One such set of mutants was generated by repeated passage of *Legionella pneumophila* on suboptimal laboratory media to allow for the loss of genes that are not essential for extracellular replication (Marra et al., 1992). One mutant strain isolated by this method, named 25D, was unable to grow inside a macrophage-like cell line, to avoid delivery to lysosomes, to recruit host vesicle traffic, and to kill host cells — a read-out for host cell lysis. By transforming this mutant strain with a cosmid library of *L. pneumophila* genomic DNA, a locus of genes capable of complementing all these defects was defined and subsequently named the *intracellular multiplication* (*icm*) locus.

Another strategy seeking a similar category of intracellular growth-deficient mutants was devised using *Legionella pneumophila* thymine auxotrophs that are only viable if they are not replicating when cultured in the absence of thymine (Berger and Isberg, 1993). Mutagenesis of this *thy* strain followed by several rounds of infection and bacterial recovery produced a pool of mutants enriched for strains that were deficient when tested in intracellular growth and phagosome trafficking assays. Complementation by transformation of these mutants with a *L. pneumophila* genomic library also identified a locus of relevant genes, which was subsequently called the *defect in organelle trafficking* (*dot*) locus.

The *icm* and *dot* loci defined by these screening techniques encompass similar although not identical sets of *Legionella pneumophila* genes (Figure 32.2). Sequencing of nearby open reading frames and further analysis of salt-sensitive mutants produced a list of 26 *dot/icm* genes that are essential for *L. pneumophila*'s intracellular growth and trafficking (Andrews et al., 1998; Segal and Shuman, 1997). Homology found within *dotB*, *dotG*, *dotL*, and *dotM* to known components of *Agrobacterium tumefaciens* conjugation machinery led to the discovery that the *dot/icm* genes collectively encode the components of a type IVb secretion system (Vogel et al., 1998). Members of this family of specialized secretion systems are ancestrally related to bacterial conjugation systems and are found across a wide range of pathogenic bacterial species (Christie, 2001). *Helicobactor pylori*, the causative agent of stomach ulcers, uses a type IV secretion system to deliver the effector protein CagA into mammalian cells (Odenbreit et al., 2000). Via its type IV secretion system, the

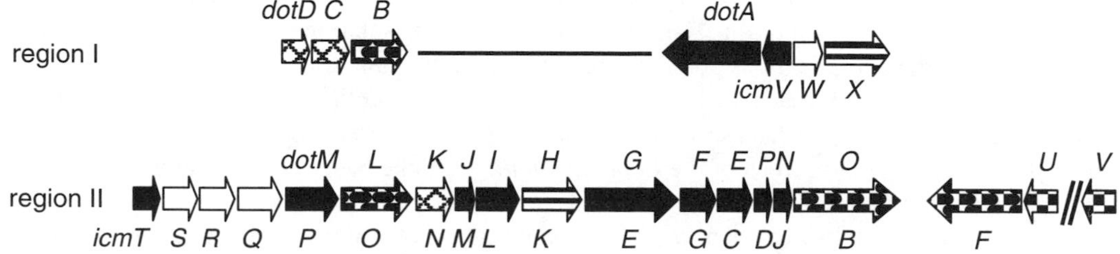

FIGURE 32.2 The *Legionella pneumophila dot/icm* locus encodes a type IVb secretion system. The 24 dot/icm genes are located in two regions of the *L. pneumophila* chromosome. Subcellular localization predictions for each gene product are indicated as follows: inner membrane (black), outer membrane (hatched), periplasmic (striped), or soluble (white). Proteins with predicted ATP/GTP binding sites are checkered. *dot/icm* genes bearing homology to components of the *Agrobacterium tumefaciens* Ti plasmid Tra/Trb conjugation machinery include: *dotB* (*trbB*), *dotG* (*trbI*), *dotL* (*trbC*), and *dotM* (*trbA*). (Figure adapted from Segal and Shuman, 1998.)

plant pathogen *Agrobacterium tumefaciens* injects both protein effectors as well as oncogenic T-DNA to produce the tumorous growth characteristic of infected plant tissues (Christie, 2001).

Further studies have found that in addition to the Dot/Icm apparatus, *Legionella pneumophila* encodes a veritable laundry list of other secretion systems, including a second type IV system called *lvh*, a type II system called *lsp*, a type I system called *lss*, and numerous secretion systems with homology to the *tra* conjugation machinery. Analysis of the Paris strain of *L. pneumophila* has also revealed a type V secretion that serves as an autotransporter (Bruggemann et al., 2006a; Cazalet et al., 2004). The *lsp* type II system has been shown to be important for efficient growth of *L. pneumophila* in protozoan and mammalian cells, and is thought to control secretion of factors with protease, RnaseA, and lipase activities, among others (Rossier et al., 2004). The role of the other secretion systems in *L. pneumophila*'s intracellular growth has yet to be studied in depth.

IDENTIFICATION OF DOT/ICM EFFECTOR SUBSTRATES

Proper functioning of the Dot/Icm secretion apparatus has since been shown to be absolutely required for *Legionella pneumophila*'s intracellular replication. Therefore, the identification of the substrates of this transporter is essential for the understanding of *L. pneumophila* pathogenesis. However, the screens that successfully identified the Dot/Icm transporter as the predominant tool required for *L. pneumophila*'s intracellular growth failed to isolate mutants lacking individual Dot/Icm substrates. Therefore, numerous novel strategies to hunt for Dot/Icm secreted proteins have been developed.

As described above, these original mutagenesis screens employed both gross intracellular replication defects and phagosome mistrafficking as phenotypic read-outs. Although the mutants isolated in this manner only contained lesions within individual *dot/icm* genes, it is probable that the absence of any one component of the apparatus results in a Dot/Icm complex that is misassembled, improperly functioning, or both. Therefore, the strong intracellular growth and trafficking defects displayed by these mutants most likely reflect their inability to secrete most if not all of *Legionella pneumophila*'s effector proteins into the host cell cytosol.

Given the varied and complex aspects of its intracellular life cycle, it is probable that *Legionella pneumophila* requires the coordinate function of numerous effector proteins to establish its replicative vacuole. Within this repertoire there may be proteins of redundant or overlapping function and proteins that mimic host cell activities. A mutation in any one of these genes may not cripple the bacterium sufficiently for it to be selected using such severe defects as overall intracellular growth or trafficking as a marker. Therefore, novel strategies to identify Dot/Icm substrates have allowed for the isolation of *L. pneumophila* genes whose products are responsible for more subtle phenotypes such as acquisition of a single cellular marker or minor perturbations of host cell pathways. In addition, strategies that permit the analysis of candidate genes in isolation have provided a means by which to sidestep the complications posed by functionally redundant effector proteins.

This more targeted approach was used to identify the *Legionella pneumophila* protein RalF as a secreted substrate of the Dot/Icm secretion apparatus (Nagai et al., 2002). This protein was identified using an *in silico* screen of the *L. pneumophila* genome based on the presence of a Sec7 domain, a catalytic motif conserved throughout the family of guanine nucleotide exchange factors for eukaryotic ARF proteins. Dot/Icm-mediated translocation of RalF into host cells was found to be required for the recruitment of detectable levels of host ARF1 to the *L. pneumophila* phagosome. However, when tested for growth in multiple host cell types, *L. pneumophila* lacking RalF showed no defect in their ability to grow compared to wild type bacteria, supporting the hypothesis that single effector mutants are unlikely to elicit gross defects in such assays.

In recent years, numerous strategies have been used to identify a growing list of Dot/Icm substrates. Bacterial two-hybrid screening using IcmG/DotF (predicted to be an inner membrane protein of the Dot/Icm complex that interacts with the carboxy terminus of RalF) was conducted and candidates were tested for translocation using the Cre/loxP system. These experiments yielded a large number of substrates of Dot/Icm transporter (Sid) proteins (Luo and Isberg, 2004). *Saccharomyces cerevisiae* has been used in two different studies to identify *Legionella pneumophila* effectors based on their presumed ability to affect (and interfere with) host membrane transport when present in high levels. The VPS inhibitor proteins (Vip) were isolated using a well-established assay that measures perturbation of membrane transport via missorting of vacuolar cargo to the cell surface (Shohdy et al., 2005). Yeast lethal factors (Ylf) A and B are two Dot/Icm substrates that were identified based upon their ability to inhibit yeast growth when conditionally overexpressed (Campodonico et al., 2005). In addition, yeast two-hybrid analysis using IcmW (a soluble protein predicted to mediate association of translocated effectors with the Dot/Icm apparatus) as bait isolated the IcmW interacting proteins (Wip) that were subsequently shown to be translocated into host cells in a Dot/Icm-dependent manner (Ninio et al., 2005). The list of Dot/Icm apparatus substrates has grown considerably, now presenting researchers with the challenge of assigning function to each protein or family of proteins in the context of *L. pneumophila*'s intracellular life cycle.

EVOLUTION OF *LEGIONELLA PNEUMOPHILA* AS AN INTRACELLULAR PATHOGEN

The origin of *Legionella pneumophila*'s translocated effectors is unknown but it is likely that both convergent evolution and horizontal gene transfer (HGT) have provided the bacterium with its impressive repertoire of Dot/Icm substrates. Analysis of the sequenced genomes of three serogroup 1 species, Philadelphia-1, Paris, and Lens has revealed that *L. pneumophila* displays a high degree of genome plasticity between species (Cazalet et al., 2004; de Felipe et al., 2005). Paris and Lens alone differ by three plasmids and approximately 13% genomic sequence (Cazalet et al., 2004). G-C content calculations for all three species indicate that *L. pneumophila* has likely acquired a large number of its predicted effector proteins from other organisms (Cazalet et al., 2004; de Felipe et al., 2005). The identification of a large group of *L. pneumophila* G-C biased open reading frames that encode proteins containing eukaryotic-like motifs strongly suggests that HGT from eukaryotic host cells themselves have supplied *L. pneumophila* with a number of these open reading frames.

De Felipe and colleagues have described a group of 46 genes termed *Legionella pneumophila eukaryotic-like genes* (*leg*) by screening the Philadelphia-1 genome for open reading frames bearing eukaryotic sequences or motifs with a low prevalence in other prokaryotes. This strategy yielded a diverse collection of proteins containing common protein-protein interaction domains, including Ankyrin repeats, coiled coils, and Leucine-rich repeats. Several other genes encode proteins with motif or active site homology, suggesting they may serve as serine-threonine kinases, lipid signaling factors, GDP-GTP exchange proteins, and factors involved in ubiquitination. Preliminary experiments have shown that a number of these proteins are translocated into host cells in a Dot/Icm-dependent manner, providing evidence that *L. pneumophila*'s ability to manipulate its host cell is partially dependent upon effector proteins built using components of the host cell itself (de Felipe et al., 2005).

CONCLUSIONS

Legionella pneumophila is one of several facultative intracellular bacterial species that survive by subverting host defenses while engaging and co-opting other host pathways to create a replicative niche. With few exceptions, studies of *L. pneumophila* in both protozoan and mammalian host cell systems have revealed remarkable parallels in the intracellular life cycle, indicating that the host proteins and pathways targeted by *L. pneumophila* to accomplish this feat are well conserved across eukarya. In recent years, several groups have generated a long list of proteins that can be translocated by the Dot/Icm system into host cells. Characterization of these proteins will provide the mechanistic details underlying the unique phenomena observed during *L. pneumophila* infection, trafficking, replication, and egress.

REFERENCES

Abu-Zant, A., Asare, R., Graham, J.E., and Abu Kwaik, Y. (2006). Role for RpoS but not RelA of *Legionella pneumophila* in modulation of phagosome biogenesis and adaptation to the phagosomal microenvironment. *Infect. Immun.*, 74: 3021–3026.

Amer, A.O. and Swanson, M.S. (2005). Autophagy is an immediate macrophage response to *Legionella pneumophila*. *Cell Microbiol.*, 7: 765–778.

Amsden, G.W. (2005). Treatment of Legionnaires' disease. *Drugs*, 65: 605–614.

Andrews, H.L., Vogel, J.P., and Isberg, R.R. (1998). Identification of linked *Legionella pneumophila* genes essential for intracellular growth and evasion of the endocytic pathway. *Infect. Immun.*, 66: 950–958.

Bachman, M.A., and Swanson, M.S. (2001). RpoS co-operates with other factors to induce Legionella pneumophila virulence in the stationary phase. *Mol Microbiol* 40: 1201–1214.

Bachman, M.A., and Swanson, M.S. (2004). Genetic evidence that *Legionella pneumophila* RpoS modulates expression of the transmission phenotype in both the exponential phase and the stationary phase. *Infect. Immun.*, 72: 2468–2476.

Benson, R.F. and Fields, B.S. (1998). Classification of the genus *Legionella*. *Semin Respir. Infect.*, 13: 90–99.

Berger, K.H., and Isberg, R.R. (1993). Two distinct defects in intracellular growth complemented by a single genetic locus in *Legionella pneumophila*. *Mol. Microbiol.*, 7: 7–19.

Berger, K.H., Merriam, J.J., and Isberg, R.R. (1994). Altered intracellular targeting properties associated with mutations in the *Legionella pneumophila dotA* gene. *Mol. Microbiol.*, 14: 809–822.

Bruggemann, H., Cazalet, C., and Buchrieser, C. (2006a). Adaptation of *Legionella pneumophila* to the host environment: role of protein secretion, effectors and eukaryotic-like proteins. *Curr. Opin. Microbiol.*, 9: 86–94.

Bruggemann, H., Hagman, A., Jules, M., Sismeiro, O., Dillies, M.A., Gouyette, C., Kunst, F., Steinert, M., Heuner, K., Coppee, J.Y., and Buchrieser, C. (2006b). Virulence strategies for infecting phagocytes deduced from the *in vivo* transcriptional program of *Legionella pneumophila*. *Cell Microbiol.*, 8: 1228–1240.

Byrne, B. and Swanson, M.S. (1998). Expression of *Legionella pneumophila* virulence traits in response to growth conditions. *Infect. Immun.*, 66: 3029–3034.

Campodonico, E.M., Chesnel, L., and Roy, C.R. (2005). A yeast genetic system for the identification and characterization of substrate proteins transferred into host cells by the *Legionella pneumophila* Dot/Icm system. *Mol. Microbiol.*, 56: 918–933.

Cazalet, C., Rusniok, C., Bruggemann, H., Zidane, N., Magnier, A., Ma, L., Tichit, M., Jarraud, S., Bouchier, C., Vandenesch, F., Kunst, F., Etienne, J., Glaser, P., and Buchrieser, C. (2004). Evidence in the *Legionella pneumophila* genome for exploitation of host cell functions and high genome plasticity. *Nat. Genet.*, 36: 1165–1173.

CDC. (2005). Legionellosis: Legionnaire's Disease (LD) and Pontiac Fever. Vol. 2007. Disease, C.C.f.I.D.D.o.B.a.M. (Ed.): Centers for Disease Control and Prevention.

Christie, P.J. (2001). Type IV secretion: intercellular transfer of macromolecules by systems ancestrally related to conjugation machines. *Mol. Microbiol.*, 40: 294–305.

Clague, M.J., Urbe, S., Aniento, F., and Gruenberg, J. (1994). Vacuolar ATPase activity is required for endosomal carrier vesicle formation. *J. Biol. Chem.*, 269: 21–24.

Dasgupta, N. and Ramphal, R. (2001). Interaction of the antiactivator FleN with the transcriptional activator FleQ regulates flagellar number in *Pseudomonas aeruginosa*. *J. Bacteriol.*, 183: 6636–6644.

de Felipe, K.S., Pampou, S., Jovanovic, O.S., Pericone, C.D., Ye, S.F., Kalachikov, S., and Shuman, H.A. (2005). Evidence for acquisition of Legionella type IV secretion substrates via interdomain horizontal gene transfer. *J. Bacteriol.*, 187: 7716–7726.

Derre, I. and Isberg, R.R. (2004). *Legionella pneumophila* replication vacuole formation involves rapid recruitment of proteins of the early secretory system. *Infect. Immun.*, 72: 3048–3053.

Diez, E., Yaraghi, Z., MacKenzie, A., and Gros, P. (2000). The neuronal apoptosis inhibitory protein (Naip) is expressed in macrophages and is modulated after phagocytosis and during intracellular infection with *Legionella pneumophila*. *J. Immunol.*, 164: 1470–1477.

Elliott, J.A. and Winn, W.C., Jr. (1986). Treatment of alveolar macrophages with cytochalasin D inhibits uptake and subsequent growth of *Legionella pneumophila*. *Infect. Immun.*, 51: 31–36.

Feeley, J.C., Gibson, R.J., Gorman, G.W., Langford, N.C., Rasheed, J.K., Mackel, D.C., and Blaine, W.B. (1979). Charcoal-yeast extract agar: primary isolation medium for *Legionella pneumophila. J. Clin. Microbiol.,* 10: 437–441.

Fields, B.S., Fields, S.R., Loy, J.N., White, E.H., Steffens, W.L., and Shotts, E.B. (1993). Attachment and entry of *Legionella pneumophila* in *Hartmannella vermiformis. J. Infect. Dis.,* 167: 1146–1150.

Fields, B.S., Benson, R.F., and Besser, R.E. (2002). Legionella and Legionnaires' disease: 25 years of investigation. *Clin. Microbiol. Rev.,* 15: 506–526.

Gal-Mor, O. and Segal, G. (2003). The *Legionella pneumophila* GacA homolog (LetA) is involved in the regulation of icm virulence genes and is required for intracellular multiplication in *Acanthamoeba castellanii. Microb. Pathog.,* 34: 187–194.

Garrett, L. (1994). *The Coming Plague: Newly Emerging Diseases in a World Out of Balance.* New York: Farrar, Straus, Giroux.

Hales, L.M. and Shuman, H.A. (1999). The *Legionella pneumophila* rpoS gene is required for growth within *Acanthamoeba castellanii. J. Bacteriol.,* 181: 4879–4889.

Hammer, B.K. and Swanson, M.S. (1999). Co-ordination of *Legionella pneumophila* virulence with entry into stationary phase by ppGpp. *Mol. Microbiol.,* 33: 721–731.

Hammer, B.K., Tateda, E.S., and Swanson, M.S. (2002). A two-component regulator induces the transmission phenotype of stationary-phase *Legionella pneumophila. Mol. Microbiol.,* 44: 107–118.

Harb, O.S., Venkataraman, C., Haack, B.J., Gao, L.Y., and Kwaik, Y.A. (1998). Heterogeneity in the attachment and uptake mechanisms of the Legionnaires' disease bacterium, *Legionella pneumophila*, by protozoan hosts. *Appl. Environ. Microbiol.,* 64: 126–132.

Heuner, K. and Steinert, M. (2003). The flagellum of *Legionella pneumophila* and its link to the expression of the virulent phenotype. *Int. J. Med. Microbiol.,* 293: 133–143.

Horwitz, M.A. and Silverstein, S.C. (1981). Interaction of the Legionnaires' disease bacterium (*Legionella pneumophila*) with human phagocytes. II. Antibody promotes binding of *L. pneumophila* to monocytes but does not inhibit intracellular multiplication. *J. Exp. Med.,* 153: 398–406.

Horwitz, M.A. (1983a). The Legionnaires' disease bacterium (*Legionella pneumophila*) inhibits phagosome lysosome fusion in human monocytes. *J. Exp. Med.,* 158: 2108–2126.

Horwitz, M.A. (1983b). Formation of a novel phagosome by the Legionnaires' disease bacterium (*Legionella pneumophila)* in human monocytes. *J. Exp. Med.,* 158: 1319–1331.

Horwitz, M.A. (1984). Phagocytosis of the Legionnaires' disease bacterium (*Legionella pneumophila)* occurs by a novel mechanism: engulfment within a pseudopod coil. *Cell,* 36: 27–33.

Husmann, L.K. and Johnson, W. (1992). Adherence of *Legionella pneumophila* to guinea pig peritoneal macrophages, J774 mouse macrophages, and undifferentiated U937 human monocytes: role of Fc and complement receptors. *Infect. Immun.,* 60: 5212–5218.

Jacobi, S., Schade, R., and Heuner, K. (2004). Characterization of the alternative sigma factor sigma54 and the transcriptional regulator FleQ of *Legionella pneumophila*, which are both involved in the regulation cascade of flagellar gene expression. *J. Bacteriol.,* 186: 2540–2547.

Jain, V., Kumar, M., and Chatterji, D. (2006). ppGpp: stringent response and survival. *J. Microbiol.,* 44: 1–10.

Kagan, J.C. and Roy, C.R. (2002). Legionella phagosomes intercept vesicular traffic from endoplasmic reticulum exit sites. *Nat. Cell. Biol.,* 4: 945–954.

Kagan, J.C., Stein, M.P., Pypaert, M., and Roy, C.R. (2004). Legionella subvert the functions of rab1 and sec22b to create a replicative organelle. *J. Exp. Med.,* 199: 1201–1211.

King, C.H., Fields, B.S., Shotts, E.B., Jr., and White, E.H. (1991). Effects of cytochalasin D and methylamine on intracellular growth of *Legionella pneumophila* in amoebae and human monocyte-like cells. *Infect., Immun.,* 59: 758–763.

Luo, Z.Q. and Isberg, R.R. (2004). Multiple substrates of the *Legionella pneumophila* Dot/Icm system identified by interbacterial protein transfer. *Proc. Natl. Acad. Sci. U.S.A.,* 101: 841–846.

Lynch, D., Fieser, N., Gloggler, K., Forsbach-Birk, V., and Marre, R. (2003). The response regulator LetA regulates the stationary-phase stress response in *Legionella pneumophila* and is required for efficient infection of *Acanthamoeba castellanii. FEMS Microbiol. Lett.,* 219: 241–248.

Majdalani, N., Vanderpool, C.K., and Gottesman, S. (2005). Bacterial small RNA regulators. *Crit. Rev. Biochem. Mol. Biol.,* 40: 93–113.

Marra, A., Blander, S.J., Horwitz, M.A., and Shuman, H.A. (1992). Identification of a *Legionella pneumophila* locus required for intracellular multiplication in human macrophages. *Proc. Natl. Acad. Sci. U.S.A.,* 89: 9607–9611.

McNealy, T.L., Forsbach-Birk, V., Shi, C., and Marre, R. (2005). The Hfq homolog in *Legionella pneumophila* demonstrates regulation by LetA and RpoS and interacts with the global regulator CsrA. *J. Bacteriol.*, 187: 1527–1532.

Moffat, J.F. and Tompkins, L.S. (1992). A quantitative model of intracellular growth of *Legionella pneumophila* in *Acanthamoeba castellanii*. *Infect. Immun.*, 60: 296–301.

Molmeret, M., Bitar, D.M., Han, L., and Kwaik, Y.A. (2004). Disruption of the phagosomal membrane and egress of *Legionella pneumophila* into the cytoplasm during the last stages of intracellular infection of macrophages and Acanthamoeba polyphaga. *Infect. Immun.*, 72: 4040–4051.

Molofsky, A.B. and Swanson, M.S. (2003). *Legionella pneumophila* CsrA is a pivotal repressor of transmission traits and activator of replication. *Mol. Microbiol.*, 50: 445–461.

Nagai, H., Kagan, J.C., Zhu, X., Kahn, R.A., and Roy, C.R. (2002). A bacterial guanine nucleotide exchange factor activates ARF on *Legionella* phagosomes. *Science,* 295: 679–682.

Ninio, S., Zuckman-Cholon, D.M., Cambronne, E.D., and Roy, C.R. (2005). The Legionella IcmS-IcmW protein complex is important for Dot/Icm-mediated protein translocation. *Mol. Microbiol.,* 55: 912–926.

Odenbreit, S., Puls, J., Sedlmaier, B., Gerland, E., Fischer, W., and Haas, R. (2000). Translocation of *Helicobacter pylori* CagA into gastric epithelial cells by type IV secretion. *Science,* 287: 1497–1500.

Otto, G.P., Wu, M.Y., Clarke, M., Lu, H., Anderson, O.R., Hilbi, H., Shuman, H.A., and Kessin, R.H. (2004). Macroautophagy is dispensable for intracellular replication of *Legionella pneumophila* in *Dictyostelium discoideum*. *Mol. Microbiol.,* 51: 63–72.

Payne, N.R. and Horwitz, M.A. (1987). Phagocytosis of *Legionella pneumophila* is mediated by human monocyte complement receptors. *J. Exp. Med.,* 166: 1377–1389.

Reaves, B., and Banting, G. (1994). Vacuolar ATPase inactivation blocks recycling to the trans-Golgi network from the plasma membrane. *FEBS Lett.,* 345: 61–66.

Rechnitzer, C. and Blom, J. (1989). Engulfment of the Philadelphia strain of *Legionella pneumophila* within pseudopod coils in human phagocytes. Comparison with other *Legionella* strains and species. *Apmis,* 97: 105–114.

Reynolds, H.Y. and Newball, H.H. (1974). Analysis of proteins and respiratory cells obtained from human lungs by bronchial lavage. *J. Lab. Clin. Med.,* 84: 559–573.

Robinson, C.G. and Roy, C.R. (2006). Attachment and fusion of endoplasmic reticulum with vacuoles containing *Legionella pneumophila*. *Cell Microbiol.,* 8: 793–805.

Romeo, T. (1998). Global regulation by the small RNA-binding protein CsrA and the non-coding RNA molecule CsrB. *Mol. Microbiol.,* 29: 1321–1330.

Rossier, O., Starkenburg, S.R., and Cianciotto, N.P. (2004). *Legionella pneumophila* type II protein secretion promotes virulence in the A/J mouse model of Legionnaires' disease pneumonia. *Infect. Immun.,* 72: 310–321.

Roy, C.R. and Tilney, L.G. (2002). The road less traveled: transport of *Legionella* to the endoplasmic reticulum. *J. Cell. Biol.,* 158: 415–419.

Sandvig, K. and van Deurs, B. (2002). Membrane traffic exploited by protein toxins. *Annu. Rev. Cell. Dev. Biol.,* 18: 1–24.

Segal, G. and Shuman, H.A. (1997). Characterization of a new region required for macrophage killing by *Legionella pneumophila*. *Infect. Immun.,* 65: 5057–5066.

Segal, G. and Shuman, H.A. (1998). How is the intracellular fate of the *Legionella pneumophila* phagosome determined? *Trends Microbiol.,* 6: 253–255.

Shohdy, N., Efe, J.A., Emr, S.D., and Shuman, H.A. (2005). Pathogen effector protein screening in yeast identifies *Legionella* factors that interfere with membrane trafficking. *Proc. Natl. Acad. Sci. U.S.A.,* 102: 4866–4871.

Solomon, J.M., Rupper, A., Cardelli, J.A., and Isberg, R.R. (2000). Intracellular growth of *Legionella pneumophila* in *Dictyostelium discoideum*, a system for genetic analysis of host-pathogen interactions. *Infect. Immun.,* 68: 2939–2947.

Sturgill-Koszycki, S. and Swanson, M.S. (2000). Legionella pneumophila replication vacuoles mature into acidic, endocytic organelles. *J. Exp. Med.,* 192: 1261–1272.

Swanson, M.S. and Isberg, R.R. (1995) Association of *Legionella pneumophila* with the macrophage endoplasmic reticulum. *Infect. Immun.,* 63: 3609–3620.

Tilney, L.G., Harb, O.S., Connelly, P.S., Robinson, C.G., and Roy, C.R. (2001). How the parasitic bacterium *Legionella pneumophila* modifies its phagosome and transforms it into rough ER: implications for conversion of plasma membrane to the ER membrane. *J. Cell. Sci.,* 114: 4637–4650.

van Weert, A.W., Geuze, H.J., and Stoorvogel, W. (1997). Heterogeneous behavior of cells with respect to induction of retrograde transport from the trans-Golgi network to the Golgi upon inhibition of the vacuolar proton pump. *Eur. J. Cell. Biol.,* 74: 417–423.

Venkataraman, C., Haack, B.J., Bondada, S., and Abu Kwaik, Y. (1997). Identification of a Gal/GalNAc lectin in the protozoan *Hartmannella vermiformis* as a potential receptor for attachment and invasion by the Legionnaires' disease bacterium. *J. Exp. Med.,* 186: 537–547.

Vogel, J.P., Andrews, H.L., Wong, S.K., and Isberg, R.R. (1998). Conjugative transfer by the virulence system of *Legionella pneumophila. Science,* 279: 873–876.

Wieland, H., Goetz, F., and Neumeister, B. (2004) Phagosomal acidification is not a prerequisite for intracellular multiplication of Legionella pneumophila in human monocytes. *J. Infect. Dis.,* 189: 1610–1614.

Yamamoto, A., Tagawa, Y., Yoshimori, T., Moriyama, Y., Masaki, R., and Tashiro, Y. (1998). Bafilomycin A1 prevents maturation of autophagic vacuoles by inhibiting fusion between autophagosomes and lysosomes in rat hepatoma cell line, H-4-II-E cells. *Cell Struct. Funct.,* 23: 33–42.

Yamamoto, Y., Klein, T.W., and Friedman, H. (1991). *Legionella pneumophila* growth in macrophages from susceptible mice is genetically controlled. *P.S.E.B.M.,* 196: 405–409.

Yu, V.L., Plouffe, J.F., Pastoris, M.C., Stout, J.E., Schousboe, M., Widmer, A., Summersgill, J., File, T., Heath, C.M., Paterson, D.L., and Chereshsky, A. (2002). Distribution of *Legionella* species and serogroups isolated by culture in patients with sporadic community-acquired legionellosis: an international collaborative survey. *J. Infect. Dis.,* 186: 127–128.

33 The Genus *Haemophilus*

Elisabeth E. Adderson

CONTENTS

HAEMOPHILUS SPECIES

The genus *Haemophilus* currently includes 12 species: *Haemophilus aegyptius*, *H. ducreyi*, *H. felis*, *H. haemoglobinophilus*, *H. haemolyticus*, *H. influenzae*, *H. paracuniculus*, *H. parahaemolyticus*, *H. parainfluenzae*, *H. paraphrohaemolyticus*, *H. parasuis*, and *H. pittmaniae* (International Committee on Systematics of Prokaryotes, 2006). Pfeiffer first described the type species, *H. influenzae*, during the 1889–1892 influenza pandemic and, until 1918, it was incorrectly presumed to be the etiologic agent of these infections. The taxonomy of the genus has evolved rapidly since sensitive genotyping methods have become available. The most recent changes include the recognition that *H. aphrophilus* and *H. paraphrophilus* are synonymous, and that these strains, along with *H. segnis*, are more appropriately classified within a distinct genus *Aggregatibacter* gen. nov., as *Aggregatibacter aphrophilus* comb. nov., and *Aggregatibacter segnis* comb. nov., respectively (Norskov-Lauritsen, 2006). Genotypic and phenotypic studies have also prompted the reassignment of *H. paragallinarum* and *H.* (*Pasteurella*) *avium* to a novel genus, *Avibacterium* gen. nov., as *A. paragallinarum* comb. nov., and *A. avium* comb. nov., respectively (Blackall, 2005). The significant phenotypic and genotypic differences between *H. ducreyi* and other *Haemophilus* species indicate that, while currently still included in the genus, *H. ducreyi* would be more properly recognized as a distinct member of the Pasteurellaceae (Sturm, 1981). The taxonomy of *H. parainfluenzae*, *H. haemoglobinophilus*, *H. parahaemolyticus*, *H. parasuis*, *H. paraphrohaemolyticus*, *H. paracuniculus*, and *H. felis* has also been questioned. *H. aegyptius* and nonencapsulated *H. influenzae* may be indistinguishable by standard phenotypic tests but are genetically distinct and continue to be regarded as separate species. *Haemophilus* species causing human disease include *H. influenzae*,

H. aegyptius, *H. ducreyi*, *H. haemolyticus*, *H. parahaemolyticus*, *H. parainfluenzae*, *H. paraphrohaemolyticus*, and *H. pittmaniae*.

Haemophilus species constitute part of the normal flora of the upper respiratory tract of mammalian hosts. All are facultatively anaerobic, non-sporeforming, and fastidious, requiring rich growth media supplemented with V factor [nicotinamide adenine dinucleotide (NAD), nicotinamide adenine dinucleotide phosphate, or nicotinamide mononucleotide] and/or X factor (a source of protoporphyrin IX). The genus name "*Haemophilus,*" from the Greek "blood-loving," reflects the fact that these growth factors are traditionally provided by erythrocytes — X factor by hemin or hemoglobin and Y factor by NAD. Most *Haemophilus* species grow best when incubated at 35 to 37°C in a moist environment supplemented with 5 to 10% CO_2, and have an optimal pH of 7.6. *H. ducreyi* grows better at 33°C and a pH of 6.5 to 7.0, and *H. felis*, *H. paracuniculus*, and *H. paraphrohaemolyticus* have a strict requirement for CO_2 (Kilian, 2003). All species except *H. ducreyi* ferment carbohydrates, producing acid and, in some cases, gas. Some *H. influenzae* strains elaborate a polysaccharide capsule.

COLLECTION, TRANSPORT, AND IDENTIFICATION

Specimens should be processed promptly because most species, especially *Haemophilus ducreyi*, are highly susceptible to drying and other environmental changes. The highest recovery rate is obtained if clinical material is inoculated directly onto appropriate media. If this cannot be done expediently, specimens should be transferred to the laboratory in a suitable transport system. Cerebrospinal fluid (CSF) cultures should be maintained at room temperature. Optimal recovery of *H. ducreyi* has been reported with the use of thioglycolate-hemin-based transport medium supplemented with selenium dioxide, L-glutamine, and albumin at 4°C.

Growth of *Haemophilus* requires a source of X and V factors. X factor is immediately available to bacteria from erythrocytes but V factor must be released from cells, typically by gentle heat lysis ("chocolating") or peptic digestion (Fildes enrichment). GC agar with 5% chocolated sheep blood and 1% yeast extract is recommended for the recovery of all *Haemophilus* species except *H. ducreyi* from clinical specimens (Rennie, 1992). Supplementing conventional blood agar with NAD results in somewhat lower rates of recovery. Inoculation of two media is recommended for isolation of *H. ducreyi* from genital specimens — Mueller-Hinton agar with 5% chocolated horse blood, 1% IsoVitale X, and 3 mg/L vancomycin, and either GC agar supplemented with 1% hemoglobin, 5% fetal bovine serum, 1% IsoVitale X, and 3 mg/L vancomycin or Columbia agar supplemented with 1% hemoglobin, 1% activated charcoal, 5% fetal bovine serum, 1% IsoVitale X, and 3 mg/L vancomycin (Pillay, 1998; Trees, 1995). Selective media, such as chocolate agar incorporating vancomycin; vancomycin, bacitracin, and clindamycin; or cefsulodin are useful in optimizing recovery of bacteria from respiratory specimens. Automated blood culture systems readily isolate *Haemophilus* species from blood but are less successful in recovering bacteria from other normally sterile body fluids inoculated into blood culture bottles if this media is not enriched with blood or X and V factors (Pennekamp, 1996).

IDENTIFICATION

Most *Haemophilus* species grow to 1- to 2-mm-diameter colonies after 24 to 48 hr of incubation. *Haemophilus influenzae* form grayish, semi-opaque, smooth, convex colonies. Encapsulated strains may become confluent and demonstrate iridescence, particularly on clear agar. Colonies of *H. parainfluenzae* may be rough and wrinkled. Those of *H. aegyptius* rarely grow larger than 1 mm in diameter and typically require more prolonged incubation than less fastidious species. *H. ducreyi* colonies are smooth, cohesive, and grow to a diameter of 0.1 to 0.5 mm after incubation for 3 to 4 days. Indole-producing *Haemophilus* strains have a characteristic pungent smell, like that of *Escherichia coli*. The odor of other strains is described as resembling that of a mouse nest. Isolates can be stored at ambient temperature after lyophilization in skim milk and by freezing broth cultures or bacterial suspensions in 10% glycerol at –70°C, or heavily inoculated cotton swabs at –135°C.

In Gram-stained smears, *Haemophilus* species are small (0.2–0.3 × 0.5–1.0 μm) pleomorphic Gram-negative rods. *H. ducreyi* often form characteristic parallel chains described as "railroad tracks," "schools of fish," or "fingerprints," particularly in direct smears of clinical specimens and in broth culture.

Identification of *Haemophilus* species is confirmed by demonstrating dependence on X and V factors. X factor-independent species, but not those requiring both V and X factors, will grow on sheep blood agar. β-Hemolytic *Staphylococcus aureus* secretes NAD and releases hemin from erythrocytes. Some laboratories, therefore, determine X and Y factor dependence by observing satellite colonies around a streak of *S. aureus* on blood agar. The satellite phenomenon, however, is not exclusive to *Haemophilus* species, and it is sometimes difficult to detect small satellite colonies. The requirement for X and V factors also can be tested by growth around factor-impregnated filter paper disks or strips. "Quad" and "tri" plates, in which unsupplemented blood agar and media incorporating one or more factors are separated into individual compartments, are also commercially available. Care must be taken with each of these methods to avoid carry-over of X factor from the primary isolation media.

The porphyrin test is the most reliable method to detect X factor dependence and has the additional advantage of being independent of bacterial growth. δ-Aminolevulinic acid hydrochloride is mixed with bacteria from an agar plate culture. X factor-independent strains are able to convert this substrate to porphyrin and porphobilinogen, which are detected after 4 to 24 hr of incubation by the demonstration of red fluorescence in ultraviolet light or a red color change when the suspension is mixed with Kovacs' reagent. The observation of hemolysis on blood agar is chiefly useful to distinguish *Haemophilus influenzae* from *H. haemolyticus.* Carbohydrate fermentation and the expression of catalase and β-galactosidase are key reactions for further differentiation of *Haemophilus* species (Table 33.1).

Both *Haemophilus influenzae* and *H. parainfluenzae* can be subdivided into eight biotypes on the basis of indole production and the expression of urease and ornithine decarboxylase (Table 33.2) (Kilian, 1976). Each of these is readily determined by simple colorimetric assays. Several commercial systems for identifying and biotyping *Haemophilus* species are available but may require additional tests to speciate isolates other than *H. influenzae* and *H. parainfluenzae*.

Serotyping of encapsulated *Haemophilus influenzae* is most commonly performed by slide agglutination, counter-immunoelectrophoresis, or enzyme-linked immunoassay, using antisera specific for individual capsular polysaccharides. Typing systems based on PCR (polymerase chain reaction) amplification of serotype-specific segments of the capsulation locus and serotype-specific gene probes are more sensitive and specific than serological assays and are available through reference laboratories.

Nucleic acid amplification techniques (NAATs) have been developed for the identification of cultured bacteria and to directly detect bacteria in blood, other normally sterile body fluids, and respiratory secretions. These tests are rapid and, because bacterial nucleic acids may persist at sites of infection or in clinical specimens longer than viable bacteria, they may be particularly valuable when the yield of conventional cultures may be suboptimal. Currently, however, clinical experience with NAATs for *Haemophilus* species other than *H. ducreyi* is limited and these tests are not generally available. A chemiluminescent nucleic acid probe (AccuProbe® Haemophilus influenzae Culture Identification Test, Gen-Probe Incorporated San Diego, CA) is useful for the rapid identification of *H. influenzae* after primary isolation.

The genetic diversity and population structure of *Haemophilus*, most notably *H. influenzae*, has been studied by a variety of techniques, including multilocus enzyme electrophoresis, ribotyping, pulsed-field gel electrophoresis, and 16S rRNA and multilocus sequence typing.

TABLE 33. 1

Major Differential Characteristics of *Haemophilus* Species

Species	Factor Requirement		Hemolysis	Fermentation of				Catalase	ONPG[a]	CO_2[b]
	X	V		Glucose	Sucrose	Lactose	Xylose			
H. influenzae	+	+	–	+	–	–	+	+	–	–
H. aegyptius	+	+	–	+	–	–	–	+	–	–
H. ducreyi	+	–	–	–	–	–	–	–	–	–
H. felis	+	+	V[c]	–	+	+	-	+	+	+
H. haemoglobinophilus	+	–	–	–	+	–	+	+	V	–
H. haemolyticus	+	+	+	+	–	–	+	+	–	–
H. paracuniculus	+	+	–	–	+	–	–	+	+	+
H. parahaemolyticus	–	+	+	+	+	–	–	+	–	–
H. parainfluenzae	–	+	–	+	+	–	–	V	V	–
H. paraphrohaemolyticus	–	+	+	+	V	V	V	+	+	+
H. parasuis	–	+	–	+	+	–	–	+	+	–
H. pittmaniae	+	+	+	+	+	–	–	–	+	–

[a] 0-nitrophenyl-β-D-galactopyranoside, production of β-galactosidase.

[b] Growth enhanced by CO_2.

[c] Variable.

Source: Adapted from Kilian, 2003.

TABLE 33.2
Biotypical Classification of *H. influenzae* and *H. parainfluenzae*

Species	Biotype	Indole	Urease	Ornithine Decarboxylase	Associated Capsular Type	Clinical Association
H. influenzae	I	+	+	+	a, b, f	Respiratory, invasive infection
	II	+	+	–	c	Respiratory
	III	–	+	–		Respiratory
	IV	–	+	+	d, e	Genitourinary, obstetrical, neonatal infection
	V	+	–	+		Invasive infection
	VI	–	–	+		
	VII	+	–	–		
	VIII	–	–	–		
H. influenzae Biogroup *aegyptius*		–	+	–		Brazilian purpuric fever
H. parainfluenzae	I	–	–	–		
	II	–	+	+		
	III	–	+	–		
	IV	+	+	+		
	V	–	–	–		
	VI	+	–	–		
	VII	+	+	–		
	VIII	+	–	–		

Source: Adapted from Kilian, 2003.

CLINICAL ASSOCIATIONS

HAEMOPHILUS INFLUENZAE

Haemophilus influenzae are divided into encapsulated and nonencapsulated or nontypeable (NTHi) strains. Encapsulated strains express one of six antigenically distinct polysaccharide capsules (types a–f). *H. influenzae* colonizes the nasopharyngeal tract and, occasionally, the female genital tract. Whereas less than 5% of healthy persons are colonized by encapsulated *H. influenzae* strains, NTHi may be found in the respiratory tract of up to 80% of adults and children. The vast majority of these persons are asymptomatic. Bacteria are transmitted by inhalation of airborne droplets or by direct contact with infected respiratory tract secretions. Detailed epidemiological studies have revealed that *H. influenzae* colonization is a dynamic process. Individual strains are carried for variable periods of time before loss or replacement by new strains, and several distinct strains may co-colonize individuals (Murphy, 1999; Smith-Vaughan, 1996).

Serotype b *Haemophilus influenzae* (Hib) is a major cause of serious invasive infections, including bacteremia, meningitis, epiglottitis, and soft tissue and skeletal infections (Wilfert, 1990). Serum antibody concentrations of anti-type b capsular antibody correlate inversely with the risk of infection. Children under 4 years of age are most susceptible to infection because of an age-dependent impairment in the acquisition of specific immunity against Hib and other bacterial polysaccharides (Anderson, 1977). Hib infection is also more frequent in certain ethnic populations, including Australian and some North American indigenous peoples. Highly immunogenic polysaccharide-protein conjugate vaccines introduced in the 1980s have dramatically reduced both the incidence of Hib infection and the prevalence of nasopharyngeal carriage of serotype b strains in immunized populations, but globally these bacteria remain a serious child health problem (Chandran, 2005). Non-type b encapsulated *H. influenzae*, particularly serotypes a and f, are occasional causes of bacteremia and other invasive disease (Bisgard, 1998; Millar, 2005).

NTHi are very common causes of upper respiratory tract infections and important causes of lower respiratory tract infections, particularly in adults with chronic obstructive pulmonary disease and patients with cystic fibrosis. Rare cases of invasive disease have also been reported (Murphy, 2003). A cryptic biotype IV genospecies has been implicated in serious infections of parturient women and neonates. Nonencapsulated *Haemophilus influenzae* biogroup *aegyptius* causes Brazilian purpuric fever, a syndrome characterized by purulent conjunctivitis, shock, and hemorrhagic skin lesions in young children.

Diagnostic Tests for *Haemophilus influenzae*

Several rapid tests have been developed for the diagnosis of Hib infection. Encapsulated *Haemophilus* species shed capsular polysaccharide during growth. Detection of this antigen by latex agglutination or enzyme-linked immunoassay is possible in up to 90% of cases of Hib meningitis. Because antigen may persist for longer than a week after bacteria cease to be cultivatable, this assay is particularly useful in situations where the yield of conventional culture may be poor, such as in patients who have received antibiotics. Antigen detection can be performed on other normally sterile body fluids and concentrated urine; however, the sensitivity and specificity of the assay in these settings are not known.

Single and multiplex NAAT have been developed for the direct identification of *Haemophilus influenzae* and the simultaneous detection of *H. influenzae* and other common bacterial pathogens from blood, cerebrospinal fluid, and respiratory secretions. Although promising, these assays have not been validated in large clinical trials and are currently available only on a research basis.

Pathogenesis of *Haemophilus influenzae* Infection

Encapsulated *Haemophilus influenzae* colonize the nasopharynx of up to 5% of healthy persons, particularly children. It is not clear why Hib are more likely to cause serious disease than other serotypes of encapsulated strains. The pathogenesis of Hib infections can be divided into four stages: (1) colonization of respiratory epithelium, (2) invasion of epithelial and endothelial cells, (3) persistence in the bloodstream, and (4) central nervous system invasion. Bacterial factors contributing to pharyngeal colonization include hemagglutinating pili, which adhere to mucin; extracellular matrix proteins and epithelial cells; the major non-pilus adhesin, Hsf; and phosphocholine (Farley, 1990; St. Geme, 1995; Weiser, 1998). The persistence of bacteria at mucosal surfaces is promoted by bacterial products that interfere with the host's physical and immune defenses, including protein D and peptidoglycan, which damage ciliated epithelial cells and impair ciliary function, and IgA1 protease, which specifically inactivates secretory and serum IgA1 (Janson, 1999; Male, 1979; Wilson, 1988). Adherence of *H. influenzae* to epithelial cells results in the loss of integrity of tight junctions and sloughing of cells. This epithelial damage, which is attributable in part to the bacterial lipooligosaccharide (LOS), exposes nonluminal epithelial cells and basement membrane, to which bacteria adhere in large numbers. Endothelial invasion is an active process in which bacteria are taken up within vacuoles and translocated to the basal cell surface and bloodstream. The type b capsule plays a critical role in the persistence of bacteria in the bloodstream. The hydrophilic capsule may provide a physical barrier to phagocytosis, preventing clearance of Hib from the bloodstream by complement-mediated phagocytosis. Supporting this hypothesis is the observation that the opsonophagocytic killing of Hib is inversely proportional to the amount of capsule expressed by the bacteria. The α-D-galactose[1-4]-β-D-galactose and sialic acid components of lipopolysaccharide (LPS), the major component of the Gram-negative outer membrane, also contribute to resistance to bacterial clearance (Hood, 1999; Weiser, 1998). After gaining access to the bloodstream, Hib adhere to and damage the blood-brain barrier, allowing bacteria to translocate across intercellular tight junctions to the cerebrospinal fluid. LPS increases blood-brain barrier permeability and also contributes to the systemic symptoms of Hib infection.

NTHi colonize the upper respiratory tract of up to 80% of healthy adults and children. The ability of the bacteria to adhere to respiratory epithelium and resist clearance by host immune defenses is critical to their survival in this niche, and multiple and redundant mechanisms facilitate NTHi persistence in the upper respiratory tract. Binding to mucin is mediated by pili and the P2 and P5 outer membrane proteins (Gilsdorf, 1996; Kubiet, 2000; Reddy, 1996). Other adhesions involved in adherence to epithelial cells include the HMW1 and HMW2 high-molecular-weight surface proteins, Hia, and opacity-associated protein A (OapA) (Barenkamp, 1996; St. Geme, 1993; Weiser, 1995).

NTHi LPS and peptidoglycan impair host clearance of bacteria, and most strains possess one or more IgA proteases (Fernaays, 2006). Some NTHi strains persist at sites of chronic infection or colonization by growth as a sialic acid-containing biofilm (Greiner, 2004; Murphy, 2002). Although considered an extracellular pathogen, NTHi, like Hib, may invade and multiply in epithelial cells. Bacteria may be protected from killing by both host immune defenses and antimicrobial agents in this intracellular compartment (Forsgren, 1994).

Variation of surface antigenicity is another strategy by which NTHi avoids immune recognition by the host. Structural diversity may arise through genetic exchange between NTHi or closely related bacteria, or as a result of point mutations in surface-exposed structures (Duim, 1994; Shen, 2005). Some NTHi strains with defects in components of the methyl-directed mismatch DNA repair system have exceptionally high rates of mutation that may provide these bacteria with a particular survival advantage (Watson, 2004). *Haemophilus influenzae* has few two-component and global regulatory systems. An alternative strategy for gene regulation is provided by what have been described as "simple contingency loci," which contain di- or tetranucleotide tandem sequence repeats within or 5′ to coding regions (Bayliss, 2001). These repeats undergo rapid, reversible, recombinase-independent mutation, resulting in alterations in promoter activity or shifts in the translational reading frame. *H. influenzae* have at least 18 such loci, including genes involved in LPS biosynthesis, adhesion, Fe acquisition, and restriction-modification enzyme systems (Bayliss, 2001). This phase variation increases fitness by permitting bacteria to very rapidly alter immunogenic surface structures and their nutritional requirements. The structure of the lipooligosaccharide (LOS) can also be modified independently of phase variation in response to environmental cues and growth rate. Finally, the structure of some forms of LOS is shared by host glycolipids and glycosphingolipids, and may fail to be recognized as foreign by the host (Harvey, 2001).

Haemophilus parainfluenzae

Haemophilus parainfluenzae are the most prevalent *Haemophilus* species in normal flora of the human oropharynx and are a common commensal of the genitourinary tract (Kuklinska, 1984). These bacteria occasionally cause respiratory and invasive disease, most notably endocarditis and biliary tract infections (Darras-Joly, 1997). They are frequently isolated from the sputum of adults with chronic obstructive pulmonary disease but their contribution to exacerbations of this disease remains controversial (Sethi, 2001). Antigen and antibody directed against *H. parainfluenzae* outer membrane antigen preparations have also been detected in the glomeruli of patients with IgA nephropathy, suggesting that colonization or infection may trigger immune-complex mediated disease (Suzuki, 1994).

Like *Haemophilus influenzae*, *H. parainfluenzae* can be subdivided into eight biotypes, although no clear associations with specific forms of infection have been described.

Haemophilus ducreyi

First described by Auguste Ducrey in 1889, *Haemophilus ducreyi* is the etiologic agent of the sexually transmitted genital ulcer disease chancroid (soft chancre). Chancroid is endemic in Asia, Africa, the Caribbean, and some regions of North America, and has gained increasing importance because, like other genital ulcer diseases, it facilitates the transmission of human immunodeficiency virus (HIV) (Greenblatt, 1988). The infectious inoculum is likely to be less than 100 organisms, which enter the skin or mucosa through minor abrasions. In males, a painful papule develops at the site of inoculation following an incubation period of 3 to 10 days. This erodes, forming a superficial ulcer with a friable base, which is often covered by a gray or yellow necrotic exudate. Painful unilateral inguinal lymph nodes (buboes) are common. Most infected women are asymptomatic. Bacteria remain localized to the superficial tissues and invasive infection has not been described.

Diagnostic Tests for *Haemophilus ducreyi*

Although not routinely available, NAATs are preferred for the diagnosis of chancroid. PCR amplification of a single target, such as 16S rRNA or ribosomal intergenic spacer regions, is highly sensitive but more specific when used for the confirmation of identification than for detection of organisms in clinical specimens. Multiplex assays (M-PCR) for the simultaneous detection of *Haemophilus ducryei* and other etiologic agents of genital ulcer disease have a sensitivity and specificity greater than 96%, but are available only on a research basis. Definitive diagnosis of *H. ducreyi* requires isolation from a genital lesion. Diagnosing chancroid by culture, however, is difficult because appropriate media may not be readily available and the sensitivity is estimated to be less than 75% when M-PCR is considered the gold standard (Lewis, 2003). Antigen detection using a monoclonal antibody against a 29-kDa outer membrane protein or LOS is more sensitive than culture, but cross-reactivity with other bacteria may limit this test's usefulness in populations with a low prevalence of disease. Because serum antibody may persist for many months after infection, serology is not useful for diagnostic purposes.

Pathogenesis of *Haemophilus ducreyi* Infection

The pathogenesis of *Haemophilus ducreyi* infection has been studied using human and animal models of cutaneous inoculation. The ability of *Haemophilus ducreyi* to cause disease depends on its ability to adhere to genital epithelium, damage host cells, and to avoid phagocytosis. Four adhesins have been described: *H. ducreyi* serum resistance protein A (DsrA), a major serum resistance factor that also mediates binding to keratinocytes; LOS; the *H. ducreyi* homolog of the GroEL heat shock protein, which binds glycosphingolipids exposed on the surface of epithelial cells; and NcaA, a type I collagen-binding outer membrane protein (Alfa, 1997; Cole, 2002; Fulcher, 2006; Pantzar, 2006). *H. ducreyi* expresses two extracellular toxins, which are likely to contribute to genital ulceration. Hemolysin (HhdA) is cytotoxic to keratinocytes, fibroblasts, macrophages, and lymphocytes. The cytolethal distending toxin (CTD) causes cell cycle arrest and apoptosis in fibroblasts and lymphocytes (Cope, 1997).

Haemophilus ducreyi is highly resistant to complement-mediated killing by human serum. Four outer membrane proteins contribute to serum resistance, including DsrA, DltA (*H. ducreyi* lectin A), LspA1, and LspA2 (Leduc, 2004; Mock, 2005). LspA1 and LspA2 are highly homologous secreted proteins that interfere with protein tyrosine kinase signaling in the early phases of Fc receptor-mediated phagocytosis. Lysis of immune cells by HdtA and CTD also may contribute to evasion of host immune responses.

Other *Haemophilus* Species

Haemophilus aegyptius (Koch-Weeks bacillus) is a frequent cause of acute bacterial conjunctivitis in humans and a rare agent of invasive human and veterinary infections (Pittman, 1950). *H. haemolyticus* and H. *paraphrohaemolyticus* are human respiratory tract commensals and anecdotal causes of invasive infection (Murphy, 2007). *H. pittmaniae* is the most recently described *Haemophilus* species. It colonizes the oropharynx of humans and is a rare opportunistic pathogen (Norskov-Lauritsen, 2005). *H. felis* is part of the normal flora of the respiratory tract of cats and may cause respiratory tract infections and conjunctivitis (Olsson, 1994). *H. haemoglobinophilus* is a commensal of the canine genital tract (Rivers, 1922). *H. paracuniculus* has been associated with enteric disease in rabbits (Targowski, 1979). *H. parasuis* commonly colonizes the respiratory tract of swine and is an important cause of respiratory and invasive disease in stressed animals (MacInnes, 1999).

TREATMENT

Serious invasive *Haemophilus influenzae* and *H. parainfluenzae* infections are typically treated with parenteral third-generation cephalosporins, carbapenems, or ampicillin. Less serious infections can be treated with oral ampicillin or cephalosporins. Approximately one-third of strains elaborate β-lactamase, and are treated with β-lactam-resistant cephalosporins, macrolides, or antimicrobial formulations combining penicillins with a β-lactamase inhibitor. Drugs recommended for the treatment of chancroid include erythromycin or azithromycin, ceftriaxone, and ciprofloxacin. Mucosal lesions typically improve rapidly, but suppurative inguinal lymph nodes may require needle or open drainage to hasten resolution. *H. parasuis* is typically susceptible to penicillin, ceftiofur, and sulphonamides.

GENOMES AND GENETIC DIVERSITY

The first bacterial genome completely sequenced was that of *Haemophilus influenzae* Rd KW20 (Rd), a non-encapsulated variant of a serotype d strain (Fleischmann, 1995). The complete genomes of 86-028NP (an NTHi strain recovered from a child with chronic otitis media) and *H. ducreyi* strain 35000HP have since been reported, and additional genome projects are in progress (Challacombe, 2006; Harrison, 2005). The size of the *H. influenzae* genome varies between 1.7 and 1.92 Mb. A total of 180 open reading frames (ORFs) present in the 86-028NP genome are absent in that of the non-pathogenic Rd, and 169 ORFs are exclusive to the Rd genome (Harrison, 2005). Major differences between these genomes include the presence of both a 471-kb inversion in 86-028NP relative to RD and nine unique insertions of larger than 5 kb. Comparison of the genome of *H. ducreyi* to that of Rd demonstrates these species have 1113 coding sequences in common, 411 that are unique to *H. ducreyi* and 228 that are unique to Rd (Challacombe, 2006). Not surprisingly, many genes known to be present in other NTHi strains and absent in most encapsulated *H. influenzae* strains are found in the 86-028NP genome, including those encoding the P5 outer membrane protein, Hap, a type IV pilus, HMW1/2, and the ICEhin1056 gene cluster, a type IV secretion system involved in propagation of genetic islands (Juhas, 2007). Consistent with the limited genetic similarity previously observed between *H. ducreyi* and other *Haemophilus* species, the *H. ducreyi* chromosome has no lengthy regions of synteny with those of Rd or 86-028NP. Heterogeneity in genes involved in lipooligosaccharide syntheses, carbohydrate and amino acid utilization, nucleotide biosynthesis, iron acquisition, and oxidative stress can account for the phenotypic differences between these *Haemophilus* strains and their adaptation to particular environments. Further analysis is likely to identify additional virulence factors that contribute to the wide range of infections caused by *Haemophilus* species.

The population structure of encapsulated serotypes of *Haemophilus influenzae* is relatively clonal, whereas NTHI isolates are genetically distinct from encapsulated isolates and more genetically diverse (Meats, 2003). It is hypothesized that this heterogeneity results from the acquisition of different members of a complete complement of "contingency genes" available from the NTHi population supergenome (Shen, 2005). This genetic exchange is made possible by the ability of *H. influenzae,* which is naturally transformable, to take up DNA from highly related organisms by recognition of an uptake signal sequence (USS) (Danner, 1980). USS are numerous and widely distributed throughout the genome (Smith, 1995). Examples of horizontal gene transfer between NTHi strains and between *H. influenzae* and other respiratory pathogens have been described (Hiltke, 2003; Kroll, 1998). The rate of recombination in *H. influenzae* is not known but appears to be higher in NTHi than among encapsulated isolates and multiple genetically distinct strains of *H. influenzae* can be present simultaneously in the respiratory tract. The ability to undergo continuous reassortment may be another strategy that allows these bacteria to adapt to a changing environment.

REFERENCES

Alfa, M.J. and Degagne, P. (1997). Attachment of *Haemophilus ducreyi* to human foreskin fibroblasts involves LOS and fibronectin, *Microbial Pathogenesis,* 22: 39–46.

Anderson, P., Smith, D.H., Ingram, D.L., Wilkins, J., Wehrle, P.F., and Howie, V.M. (1977). Antibody of polyribophate of *Haemophilus influenzae* type b in infants and children: effect of immunization with polyribophosphate, *J. Infect. Dis.,* 136: S57–S62.

Barenkamp, S.J. and St Geme III, J.W. (1996). Identification of a second family of high-molecular-weight adhesion proteins expressed by non-typable *Haemophilus influenzae*, *Molec. Microbiol.,* 19: 1215–23.

Bayliss, C.D., Field, D., and Moxon, E.R. (2001). The simple sequence contingency loci of *Haemophilus influenzae* and *Neisseria meningitidis*, *J. Clin. Invest.,* 107: 657–662.

Bisgard, K.M., Kao, A., Leake, J., Strebel, P.M., Perkins, B.A., and Wharton, M. (1998). *Haemophilus influenzae* invasive disease in the United States, 1994–1995: near disappearance of a vaccine-preventable childhood disease, *Emerg. Infect. Dis,* 4: 229–237.

Blackall, P.J., Christensen, H., Beckenham, T., Blackall, L.L., and Bisgaard, M. (2005). Reclassification of *Pasteurella gallinarum*, [*Haemophilus*] *paragallinarum*, *Pasteurella avium* and *Pasteurella volantium* as *Avibacterium gallinarum* gen. nov., comb. nov., *Avibacterium paragallinarum* comb. nov., *Avibacterium avium* comb. nov. and *Avibacterium volantium* comb. nov, *Int. J. Syst. Evol. Microbiol.,* 55: 353–362.

Challacombe, J.F., Duncan, A.J., Brettin, T.S., Bruce, D., Chertkov, O., Detter, J.C., Han, C.S., Misra, M., Richardson, P., Tapia, R., Thayer, N., Xie, G., and Inzana, T.J. (2006). Complete genome sequence of *Haemophilus somnus* (*Histophilus somni*) strain 129Pt and comparison to *H. ducreyi* 35000HP and *H. influenzae* Rd, *J. Bacteriol.,* Published online ahead of print on 15 December 2006, <http://jb.asm.org/cgi/content/abstract/JB.01422-06v1> (accessed 16 January 2007).

Chandran, A., Watt, J.P., and Santosham, M. (2005). Prevention of *Haemophilus influenzae* type b disease: past success and future challenges, *Expert Rev. Vaccines,* 4: 819–827.

Cole, L.E., Kawula, T.H., Toffer, K.L., and Elkins, C. (2002). The *Haemophilus ducreyi* serum resistance antigen DsrA confers attachment to human keratinocytes, *Infect. Immun.,* 70: 6158–6165.

Cope, L.D., Lumbley, S., Latimer, J.L., Klesney-Tait, J., Stevens, M.K., Johnson, L.S., Purven, M., Munson Jr., R.S., Lagergard, T., Radolf, J.D., and Hansen, E.J. (1997). A diffusible cytotoxin of Haemophilus ducreyi, *Proc. Natl. Acad. Sci. U.S.A.,* 94: 4056–4061.

Danner, D.B., Deich, R.A., Sisco, K.L., and Smith, H.O. (1980). An eleven-base-pair sequence determines the specificity of DNA uptake in *Haemophilus* transformation, *Gene,* 11: 311–318.

Darras-Joly, C., Lortholary, O., Mainardi, J.L., Etienne, J., Guillevin, L., and Acar, J. (1997). Haemophilus endocarditis: report of 42 cases in adults and review. Haemophilus Endocarditis Study Group, *Clin. Infect. Dis.,* 24: 1087–1094.

Duim, B., Van Alphen, L., Eijk, P., Jansen, H.M. and Dankert, J. (1994). Antigenic drift of non-encapsulated Haemophilus influenzae major outer membrane protein P2 in patients with chronic bronchitis is caused by point mutations, *Molec. Microbiol.,* 11: 1181–1189.

Farley, M.M., Stephens, D.S., Kaplan, S.L., and Mason, E.O., Jr. (1990). Pilus- and non-pilus-mediated interactions of *Haemophilus influenzae* type b with human erythrocytes and human nasopharyngeal mucosa, *J. Infect. Dis.,* 161: 274–280.

Fernaays, M.M., Lesse, A.J., Cai, X., and Murphy, T.F. (2006). Characterization of igaB, a second immunoglobulin A1 protease gene in nontypeable *Haemophilus influenzae, Infect. Immun.,* 74: 5860–5870.

Fleischmann, R.D., Adams, M.D., White, O., Clayton, R.A., Kirkness, E.F., Kerlavage, A.R., Bult, C.J., Tomb, J.F., Dougherty, B.A., Merrick, J.M., et al. (1995). Whole-genome random sequencing and assembly of Haemophilus influenzae Rd, *Science,* 269: 496–512.

Forsgren, J., Samuelson, A., Ahlin, A., Jonasson, J., Rynnel-Dagoo, B., and Lindberg, A. (1994). *Haemophilus influenzae* resides and multiplies intracellularly in human adenoid tissue as demonstrated by in situ hybridization and bacterial viability assay, *Infect. Immun.,* 62: 673–679.

Fulcher, R.A., Cole, L.E., Janowicz, D.M., Toffer, K.L., Fortney, K.R., Katz, B.P., Orndorff, P.E., Spinola, S.M., and Kawula, T.H. (2006). Expression of *Haemophilus ducreyi* collagen binding outer membrane protein NcaA is required for virulence in swine and human challenge models of chancroid, *Infect. Immun.,* 74: 2651–2658.

Gilsdorf, J.R., Tucci, M., and Marrs, C.F. (1996). Role of pili in *Haemophilus influenzae* adherence to, and internalization by, respiratory cells, *Pediatr. Res.,* 39: 343–348.

Greenblatt, R.M., Lukehart, S.A., Plummer, F.A., Quinn, T.C., Critchlow, C.W., Ashley, R.L., D'costa, L.J., Ndinya-Achola, J.O., Corey, L., Ronald, A.R., et al. (1988). Genital ulceration as a risk factor for human immunodeficiency virus infection, *AIDS,* 2: 47–50.

Greiner, L.L., Watanabe, H., Phillips, N.J., Shao, J., Morgan, A., Zaleski, A., Gibson, B.W., and Apicella, M.A. (2004). Nontypeable *Haemophilus influenzae* strain 2019 produces a biofilm containing N-acetylneuraminic acid that may mimic sialylated O-linked glycans, *Infect. Immun.,* 72: 4249–4260.

Harrison, A., Dyer, D.W., Gillaspy, A., Ray, W.C., Mungur, R., Carson, M.B., Zhong, H., Gipson, J., Gipson, M., Johnson, L.S., Lewis, L., Bakaletz, L.O., and Munson Jr., R.S. (2005). Genomic sequence of an otitis media isolate of nontypeable *Haemophilus influenzae*: comparative study with *H. influenzae* serotype d, strain KW20, *J. Bacteriol.,* 187: 4627–4636.

Harvey, H.A., Swords, W.E., and Apicella, M.A. (2001). The mimicry of human glycolipids and glycosphingolipids by the lipooligosaccharides of pathogenic neisseria and haemophilus, *J. Autoimmunity,* 16: 257–262.

Hiltke, T.J., Schiffmacher, A.T., Dagonese, A.J., Sethi, S., and Murphy, T.F. (2003). Horizontal transfer of the gene encoding outer membrane protein P2 of nontypeable *Haemophilus influenzae*, in a patient with chronic obstructive pulmonary disease, *J. Infect. Dis.,* 188: 114–117.

Hood, D.W., Makepeace, K., Deadman, M.E., Rest, R.F., Thibault, P., Martin, A., Richards, J.C., and Moxon, E.R. (1999). Sialic acid in the lipopolysaccharide of *Haemophilus influenzae*: strain distribution, influence on serum resistance and structural characterization, *Mol. Microbiol.,* 33: 679–92.

International Committee on Systematics of Prokaryotes Subcommittee on the Taxonomy of *Pasteurellaceae.* (2006). Taxa covered by the ICSP Subcommittee on the Taxonomy of *Pasteurellaceae* — November 2006, Online. Available HTTP: http://www.the-icsp.org/taxa/Pasteurellaceaelist.htm> (accessed 25 January 2007).

Janson, H., Carl N.B., Cervin, A., Forsgren, A., Magnusdottir, A.B., Lindberg, S., and Runer, T. (1999). Effects on the ciliated epithelium of protein D-producing and -nonproducing nontypeable *Haemophilus influenzae* in nasopharyngeal tissue cultures, *J. Infect. Dis.,* 180: 737–746.

Juhas, M., Crook, D.W., Dimopoulou, I.D., Lunter, G., Harding, R.M., Ferguson, D.J., and Hood, D.W. (2007). Novel type IV secretion system involved in propagation of genomic islands, *J. Bacteriol.,* 189: 761–771.

Kilian, M. (1976). A taxonomic study of the genus *Haemophilus*, with the proposal of a new species, *J. Gen. Microbiol.,* 93: 9–62.

Kilian, M. (2003). Haemophilus, in Murray, P.R., Baron, E.J., Jorgensen, J.H., Pfaller, M.A., and Yolken, R.H. (Eds.). *Manual of Clinical Microbiology*, 8th ed, Washington, DC: ASM Press, 623–635.

Kroll, J.S., Wilks, K.E., Farrant, J.L., and Langford, P.R. (1998). Natural genetic exchange between *Haemophilus* and *Neisseria*: intergeneric transfer of chromosomal genes between major human pathogens, *Proc. Natl. Acad. Sci. U.S.A.,* 95: 12381–12385.

Kubiet, M., Ramphal, R., Weber, A., and Smith, A. (2000). Pilus-mediated adherence of *Haemophilus influenzae* to human respiratory mucins, *Infect. Immun.,* 68: 3362–3367.

Kuklinska, D. and Kilian, M. (1984). Relative proportions of *Haemophilus* species in the throat of healthy children and adults, *Eur. J. Clin. Microbiol.,* 3: 249–252.

Leduc, I., Richards, P., Davis, C., Schilling, B., and Elkins, C. (2004). A novel lectin, DltA, is required for expression of a full serum resistance phenotype in *Haemophilus ducreyi*, *Infect. Immun.,* 72: 3418–3428.

Lewis, D.A. (2003). Chancroid: clinical manifestations, diagnosis, and management, *Sexual. Transmit. Infect.*, 79: 68–71.

MacInnes, J.I. and Desrosiers, R. (1999). Agents of the "suis-ide diseases" of swine: *Actinobacillus suis, Haemophilus parasuis*, and *Streptococcus suis*, *Cdn. J. Vet. Res.* 63: 83–89.

Male, C.J. (1979). Immunoglobulin A1 protease production by *Haemophilus influenzae* and *Streptococcus pneumoniae*, *Infect. Immun.*, 26: 254–261.

Meats, E., Feil, E.J., Stringer, S., Cody, A.J., Goldstein, R., Kroll, J.S., Popovic, T., and Spratt, B.G. (2003). Characterization of encapsulated and noncapsulated Haemophilus influenzae and determination of phylogenetic relationships by multilocus sequence typing, *J. Clin. Microbiol.*, 41: 1623–1636.

Millar, E.V., O'Brien, K.L., Watt, J.P., Lingappa, J., Pallipamu, R., Rosenstein, N., Hu, D., Reid, R., and Santosham, M. (2005). Epidemiology of invasive Haemophilus influenzae type a disease among Navajo and White Mountain Apache children, 1988–2003, *Clin. Infect. Dis.*, 40: 823–830.

Mock, J.R., Vakevainen, M., Deng, K., Latimer, J.L., Young, J.A., Van Oers, N.S., Greenberg, S., and Hansen, E.J. (2005). *Haemophilus ducreyi* targets Src family protein tyrosine kinases to inhibit phagocytic signaling, *Infect. Immun.*, 73: 7808–7816.

Murphy, T.F. (2003). Respiratory infections caused by non-typeable *Haemophilus influenzae*, *Curr. Opin. Infect. Dis.*, 16: 129–134.

Murphy, T.F., Brauer, A.L., Sethi, S., Kilian, M., Cai, X., and Lesse, A.J. (2007). Haemophilus haemolyticus: A human respiratory tract commensal to be distinguished from Haemophilus influenzae, *J. Infect Dis.*, 195: 81–89.

Murphy, T.F. and Kirkham, C. (2002). Biofilm formation by nontypeable *Haemophilus influenzae*: strain variability, outer membrane antigen expression and role of pili, *BMC Microbiol.*, 2: 7.

Murphy, T.F., Sethi, S., Klingman, K.L., Brueggemann, A.B., and Doern, G.V. (1999). Simultaneous respiratory tract colonization by multiple strains of nontypeable Haemophilus influenzae in chronic obstructive pulmonary disease: implications for antibiotic therapy, *J. Infect. Dis.*, 180: 404–409.

Norskov-Lauritsen, N., Bruun, B., and Kilian, M. (2005). Multilocus sequence phylogenetic study of the genus *Haemophilus* with description of Haemophilus pittmaniae sp. nov, *Int. J. Syst. Evolut. Microbiol.*, 55: 449–456.

Norskov-Lauritsen, N. and Kilian, M. (2006). Reclassification of Actinobacillus actinomycetemcomitans, Haemophilus aphrophilus, Haemophilus paraphrophilus and Haemophilus segnis as Aggregatibacter actinomycetemcomitans gen. nov., comb. nov., Aggregatibacter aphrophilus comb. nov. and Aggregatibacter segnis comb. nov., and emended description of Aggregatibacter aphrophilus to include V factor-dependent and V factor-independent isolates, *Int. J. Syst. Evolut. Microbiol.*, 56: 2135–2146.

Olsson, E. and Falsen, E. (1994). "*Haemophilus felis*": a potential pathogen for cats? *J. Clin. Microbiol.*, 32: 858–859.

Pantzar, M., Teneberg, S., and Lagergard, T. (2006). Binding of *Haemophilus ducreyi* to carbohydrate receptors is mediated by the 58.5-kDa GroEL heat shock protein, *Microbes Infect.*, 8: 2452–2458.

Pennekamp, A., Zbinden, R., and Von Graevenitz, A. (1996). Detection of *Haemophilus influenzae* and *Haemophilus parainfluenzae* from body fluids in blood culture bottles. *J. Microbiol. Meth.*, 25: 303–307.

Pillay, A., Hoosen, A.A., Loykissoonlal, D., Glock, C., Odhav, B., and Sturm, A.W. (1998). Comparison of culture media for the laboratory diagnosis of chancroid, *J. Mol. Microbiol.*, 47: 1023–1026.

Pittman, M. and Davis, D.J. (1950). Identification of the Koch-Weeks bacillus (*Hemophilus aegyptius*), *J. Bacteriol.*, 59: 413–426.

Reddy, M.S., Bernstein, J.M., Murphy, T.F., and Faden, H.S. (1996). Binding between outer membrane proteins of nontypeable *Haemophilus influenzae* and human nasopharyngeal mucin, *Infect. Immun.*, 64: 1477–1479.

Rennie, R., Gordon, T., Yaschuk, Y., Tomlin, P., Kibsey, P., and Albritton, W. (1992). Laboratory and clinical evaluations of media for the primary isolation of *Haemophilus* species, *J. Clin. Microbiol.*, 30: 1917–1921.

Rivers, T.M. (1922). Bacillus hemoglobinophilus canis (Friedberger): Hemophilus canis Emend, *J. Bacteriol.*, 7: 579–581.

Sethi, S. and Murphy, T.F. (2001). Bacterial infection in chronic obstructive pulmonary disease in 2000: a state-of-the-art review, *Clin. Microbiol., Rev.*, 14: 336-363.

Shen, K., Antalis, P., Gladitz, J., Sayeed, S., Ahmed, A., Yu, S., Hayes, J., Johnson, S., Dice, B., Dopico, R., Keefe, R., Janto, B., Chong, W., Goodwin, J., Wadowsky, R.M., Erdos, G., Post, J.C., Ehrlich, G.D., and Hu, F.Z. (2005). Identification, distribution, and expression of novel genes in 10 clinical isolates of nontypeable *Haemophilus influenzae*, *Infect. Immun.*, 73: 3479–3491.

Smith, H.O., Tomb, J.F., Dougherty, B.A., Fleischmann, R.D., and Venter, J.C. (1995). Frequency and distribution of DNA uptake signal sequences in the *Haemophilus influenzae* Rd genome, *Science,* 269: 538–540.

Smith-Vaughan, H.C., Leach, A.J., Shelby-James, T.M., Kemp, K., Kemp, D.J., and Mathews, J.D. (1996). Carriage of multiple ribotypes of non-encapsulated *Haemophilus influenzae* in aboriginal infants with otitis media, *Epidemiol., Infect.,* 116: 177–183.

St. Geme III, J.W., Falkow, S., and Barenkamp, S.J. (1993). High-molecular-weight proteins of nontypable *Haemophilus influenzae* mediate attachment to human epithelial cells, *Proc. Natl. Acad. Sci. U.S.A.,* 90: 2875–2879.

St. Geme III, J.W. and Cutter, D. (1995). Evidence that surface fibrils expressed by Haemophilus influenzae type b promote attachment to human cpithelial cells, *Mol. Microbiol.,* 15: 77–85.

Sturm, A.W. (1981). Identification of *Haemophilus ducreyi, Antonie Van Leeuwenhoek,* 47: 89–90.

Suzuki, S., Nakatomi, Y., Sato, H., Tsukada, H., and Arakawa, M. (1994). Haemophilus parainfluenzae antigen and antibody in renal biopsy samples and serum of patients with IgA nephropathy, *Lancet,* 343: 12–16.

Targowski, S. and Targowski, H. (1979). Characterization of a Haemophilus paracuniculus isolated from gastrointestinal tracts of rabbits with mucoid enteritis, *J. Clin. Microbiol.,* 9: 33–37.

Trees, D.L. and Morse, S.A. (1995). Chancroid and *Haemophilus ducreyi*: An update, *Clin. Microbiol. Rev.,* 8: 357–375.

Watson, M.E., Burns, J.L., and Smith, A.L. (2004). Hypermutable *Haemophilus influenzae* with mutations in mutS are found in cystic fibrosis sputum. *Microbiology,* 150: 2947–2958.

Weiser, J.N., Chong, S.T., Greenberg, D., and Fong, W. (1995). Identification and characterization of a cell envelope protein of Haemophilus influenzae contributing to phase variation in colony opacity and nasopharyngeal colonization, *Mol. Microbiol.,* 17: 555–564.

Weiser, J.N. and Pan, N. (1998). Adaptation of Haemophilus influenzae to acquired and innate humoral immunity based on phase variation of lipopolysaccharide, *Mol. Microbiol.,* 30: 767–775.

Weiser, J.N., Pan, N., McGowan, K.L., Musher, D., Martin, A., and Richards, J. (1998). Phosphorylcholine on the lipopolysaccharide of Haemophilus influenzae contributes to persistence in the respiratory tract and sensitivity to serum killing mediated by C-reactive protein, *J. Exp. Med.,* 187: 631–640.

Wilfert, C.M. (1990). Epidemiology of Haemophilus influenzae type b infections, *Pediatrics,* 85: 631–635.

Wilson, R. and Cole, P.J. (1988). The effect of bacterial products on ciliary function, *Am. Rev. Resp. Dis.,* 138: S49–S53.

34 The Genus *Listeria*

Sukhadeo Barbuddhe, Torsten Hain, and Trinad Chakraborty

CONTENTS

INTRODUCTION

The genus *Listeria* consists of a group of Gram-positive bacteria of low G+C content closely related to *Bacillus, Clostridium, Enterococcus, Streptococcus*, and *Staphylococcus. Listeria* spp. are isolated from a diversity of environmental sources, including soil, water, effluents, a large variety of foods, and the feces of humans and animals. The natural habitat of these bacteria is thought to be decomposing plant matter, in which they live as saprophytes.

Historically, *Listeria* was first isolated in 1926 from a natural disease of rabbits characterized by mononuclear leucocytosis and therefore named as *Bacterium monocytogenes* by Murray and coworkers at Cambridge. Later, Pierie, who isolated similar bacilli from the liver of gerbils in 1927, named it *Listerella hepatolytica* in honor of surgeon Joseph Lister, and finally renamed it *Listeria* in 1940 due to taxonomic reasons.

Listeria monocytogenes contamination is one of the leading microbiological causes of food recalls, mainly of meat, poultry, seafood, and dairy products. Prevention and control measures are based on hazard analysis and critical control point programs throughout the food industry, and on specific recommendations for high-risk groups.

Listeria monocytogenes is morphologically indistinguishable from other *Listeria* species, and often causes non-specific clinical symptoms; therefore, laboratory testing is required to differentiate *L. monocytogenes* from other *Listeria* species and to diagnose listeriosis. The earlier diagnostic methods for *L. monocytogenes* are largely phenotype based, and characterize the bacteria by their fermentation, antigenic, and bacteriophage properties. Following recent advances in molecular genetics techniques, methods targeting unique genes in *Listeria* have been designed for the specific differentiation of *L. monocytogenes* from other *Listeria* species. These methods are intrinsically more precise and less affected by natural variation than the phenotypic methods. Extensive research in recent decades has led to significant insights regarding *Listeria* species and listeriosis [1]. The establishment of animal models and *in vitro* cell culture systems for listeriosis has helped in the delineation of key stages in *L. monocytogenes* infection and pathogenesis [2].

In-depth reviews on the taxonomy, isolation, identification, epidemiology, pathogenesis, virulence determinants, and genomics of *Listeria monocytogenes* are available [1–7].

THE GENUS *LISTERIA*

The genus *Listeria* (Group 19, *Bergey's Manual*, 9th ed.) includes six species, namely, *Listeria monocytogenes, L. ivanovii, L. innocua, L. welshimeri, L. seeligeri,* and *L. grayi.* Of these, *L. monocytogenes* is an opportunistic pathogen in human beings and various animal species, whereas

L. ivanovii mainly affects ruminants, causing abortion, only occasionally occurring in man. Phylogenetic analyses based on the *16S* and *23S rRNA* genes and the *iap, prs, vclA, vclB*, and *ldh* genes indicate that *L. innocua* is highly related to *L. monocytogenes*, while *L. welshimeri* is more distant, exhibiting the deepest branching of this group. The second group has *L. ivanovii*, together with *L. seeligeri*; *L. grayi* seems very distant from these two groups [8].

Characteristics

Listeria species are small, Gram-positive, nonsporulating, facultatively anaerobic rods that measure 1 to 2 by 0.5 μ and show characteristic tumbling motility at or around 25°C. *Listeria* are able to multiply at high salt concentrations (10% NaCl) and in a broad pH range (pH 4.5 to 9) and temperature (0 to 45°C, optimum 30 to 37°C) [9]. The morphology of *Listeria* as seen by Gram staining and electron microscopy is depicted in Figures 34.1A and 1B, respectively.

Listeria species are closely related bacteria that share many morphological and biochemical characteristics. *Listeria* species are catalase and Voges-Proskauer reaction-positive, and indole- and oxidase-negative. *Listeria* species can hydrolyze esculin, but not urea or reduce nitrates. A rapid test strip for identification has been developed [10]. *Listeria* species show variations in their ability to hemolyze horse or sheep red blood cells, and in their ability to produce acid from L-rhamnose, D-xylose, and α-methyl-D-mannoside [11].

Methods for Confirmation/Differentiation of *Listeria*

As *Listeria monocytogenes* is morphologically indistinguishable from other *Listeria* species, additional laboratory testing is required to differentiate *L. monocytogenes* from other *Listeria* species. Table 34.1 shows a scheme for identification of *Listeria* spp. *L. ivanovii* is differentiated biochemi-

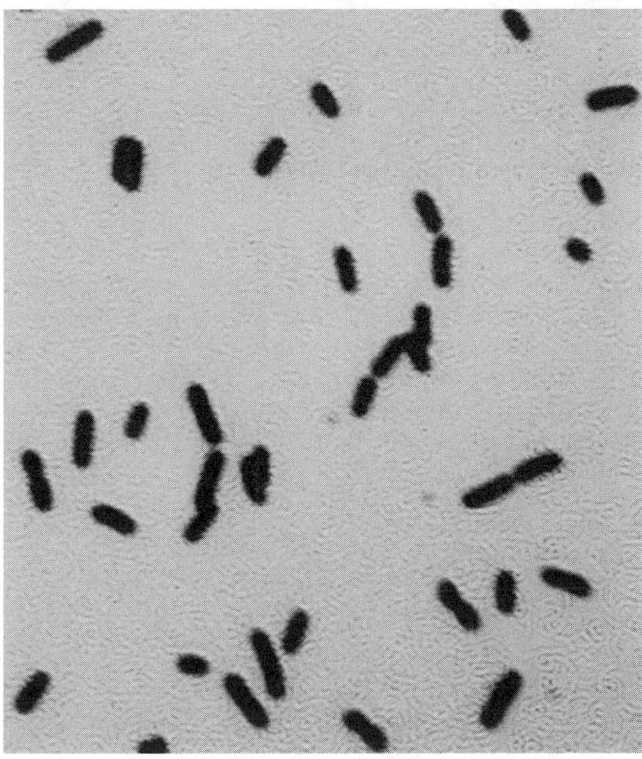

FIGURE 34.1A Gram-stained *Listeria monocytogenes* (magnification 100X).

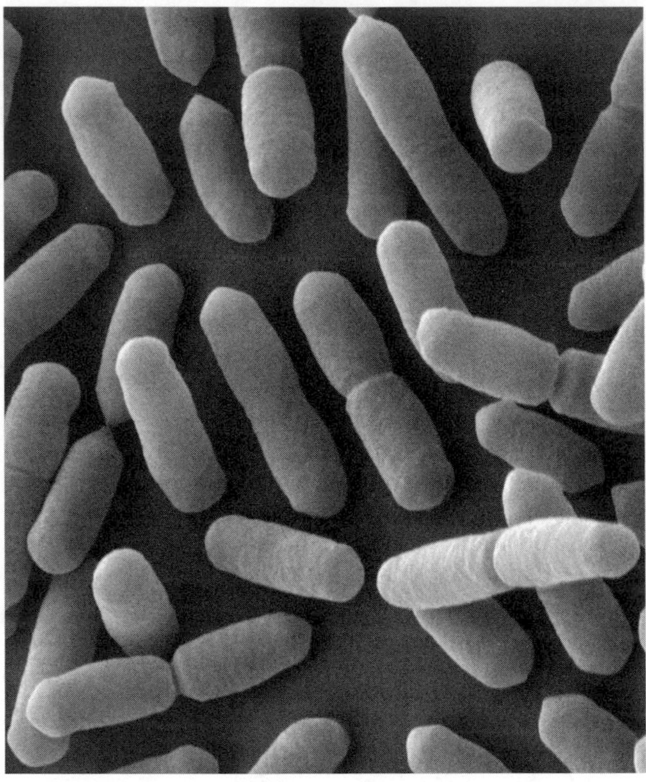

FIGURE 34.1B Scanning electron microscopy of *Listeria monocytogenes* (magnification 16000X).

TABLE 34.1
A Scheme for Phenotypic Identification of *Listeria*

	L. monocytogenes	*L. innocua*	*L. ivanovii*	*L. seeligeri*	*L. welshimeri*	*L. grayi*
Hemolysin	+	–	+	+	–	–
Catalase	+	+	+	+	+	+
Oxidase	–	–	–	–	–	–
CAMP with						
S. aureus	+	–	–	–	–	–
R. equi	+	–	+	–	–	–
PI-PLC	+	–	+	–	–	–
Fermentation of:						
L-Rhamnose	+	+/–	–	–	+/–	+/–
D-Mannitol	–	–	–	–	–	+
D-Xylose	–	–	+	+	+	–
α-D-Methyl-Mannoside	+	+	–	–	+	+
Mice virulence	+	–	+	–	–	–

cally from *L. monocytogenes* and other *Listeria* species by its production of a wide, clear or double zone of hemolysis on sheep or horse blood agar, strong lecithinase reaction with or without charcoal in the medium, a positive Christie–Atkins–Munch-Petersen (CAMP) reaction with *Rhodococcus equi* but not with hemolytic *Staphylococcus aureus*, and fermentation of D-xylose but not L-rhamnose [12]. *L. monocytogenes* requires charcoal for its lecithinase reaction [13]. *L. innocua* is distinguished from *L. monocytogenes* on the basis of its negative CAMP reaction and its failure to cause α-hemolysis or to show PI-PLC activity on chromogenic media [11].

An assay based on phosphatidylinositol-specific phospholipase C (PI-PLC) activity has been described for the discrimination of pathogenic and nonpathogenic *Listeria* species based on which hemolytic but non-pathogenic species (i.e., *L. seeligeri*) can be separated from the hemolytic and pathogenic species (i.e., *L. monocytogenes* and *L. ivanovii*) [14]. The application of a multiplex PCR assay that selectively amplifies a commonly shared region of the *iap* gene that facilitates the differentiation of all six *Listeria* species in a single test has been developed [15].

Biology of *Listeria*

Listeria monocytogenes is a remarkable bacterium that has evolved over a long time period to acquire a diverse collection of virulence factors, each with unique properties and functions. Its life cycle reflects its remarkable adaptation to intracellular survival and multiplication in professional phagocytic and nonphagocytic cells of vertebrates and invertebrates.

Pathogenic *Listeria* species are able to breach endothelial and epithelial barriers of the infected hosts, including the intestinal, blood-brain, and fetoplacental barrier, and are able to invade and replicate in phagocytic and nonphagocytic cells [1]. Entry of the pathogen is mediated by internalins A and B, which are expressed on the surface of the bacterium. Two virulence-associated molecules are responsible for lysis of the primary single-membraned vacuoles and subsequent escape by *Listeria monocytogenes*: (1) listeriolysin (LLO) and (2) PI-PLC. After lysis of the primary single-membraned vacuoles, *L. monocytogenes* is released to the cytosol, where it undergoes intracellular growth and multiplication. The bacterial surface protein ActA leads to the intracellular mobility and cell-to-cell spread, which is co-transcribed with PC-PLC and mediates the formation of polarized actin tails that propel the bacteria toward the cytoplasmic membrane, which enables the bacterium to infect the second cell [1]. Immunofluorescence staining of both bacteria and the host cell actin in infected cells reveals the unique ability of these bacteria to move intracellularly (Figure 34.1C).

The genes encoding the virulence-associated proteins PI-PLC, LLO, Mpl, ActA, and PC-PLC are located in a 9.6-kb virulence gene cluster [16], which is principally regulated by a pleiotropic virulence regulator, *prfA* (a 27-kDa protein encoded by *prfA*). In addition to these virulence-associated genes and proteins, several other genes (such as *iap, bsh, vip, inlJ, auto, ami,* and *bilA*) are also involved in *Listeria monocytogenes* virulence and pathogenicity. The resistance of *L. monocytogenes* to acidic conditions and to bile salts makes it particularly adept at infecting the gastrointestinal tract [17].

Intracellular gene expression profiling of *Listeria monocytogenes* indicated that many of the genes are up-regulated in the cytosol and allowed the identification of activation of several *L. monocytogenes* genes, including the *plcA* and *prfA* after infection of the host cells. Indeed, *L. monocytogenes* modulates the expression of approximately 500 genes for its survival in this cellular compartment [18, 19].

DISEASE IN HUMANS

The majority of cases in adults and juveniles occur among the immunosuppressed, that is, patients receiving steroid or cytotoxic therapy or with malignant neoplasms. Other "at-risk" groups include AIDS patients, diabetics, elderly people, kidney dialysis patients, individuals with prosthetic heart valves or replacement joints, and individuals with alcoholism or alcoholic liver disease. Approxi-

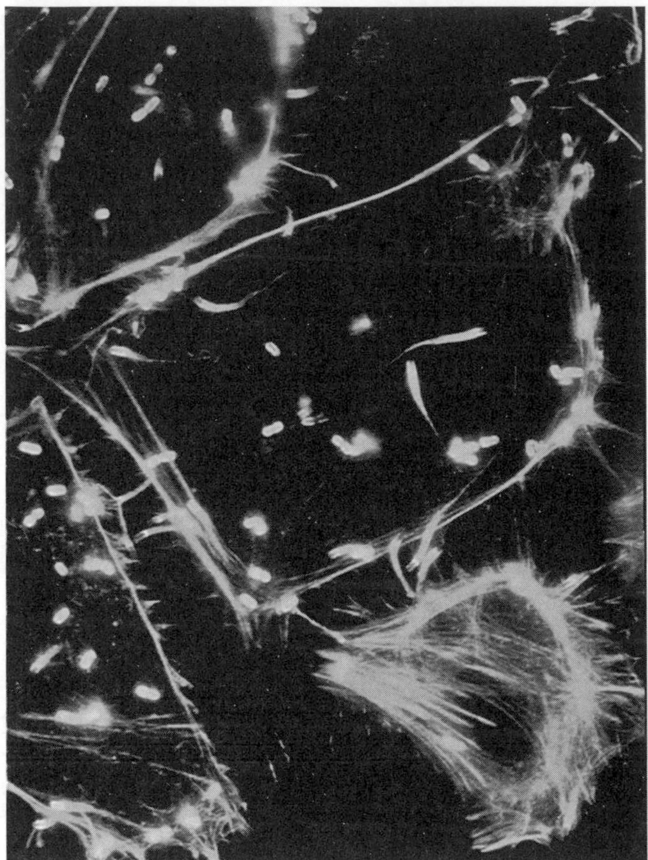

FIGURE 34.1C Immunofluorescence staining of intracellular motile *Listeria*. (Bacteria are stained red while the actin tail of *Listeria monocytogenes* is stained green with FITC-labeled phalloidin.)

mately one-third of patients with listerial meningitis and around 10% with primary bacteremia are immunocompetent [20]. Rarer presentations in this patient group include meningoencephalitis and encephalitis, together with infections with identifiable foci, that is, endocarditis, pneumonia, peritonitis, and deep-seated abscess formation. During the early stages of infection, human listeriosis often displays nonspecific flu-like symptoms (e.g., chills, fatigue, headache, and muscular and joint pain) and gastroenteritis. However, without appropriate antibiotic treatment, it can develop into septicemia, meningitis, encephalitis, abortion and, in some cases, death [1]. Minor skin infections, particularly in farmers or veterinarians after handling bovine calvings or abortions, have been recorded. Symptoms such as vomiting and conjunctivitis also have been observed.

Because of its high case fatality rate, listeriosis ranks among the most frequent causes of death due to food-borne illness. *Listeria monocytogenes* infections are responsible for the highest hospitalization rates (91%) among known food-borne pathogens, and have been linked to sporadic episodes and large outbreaks of human illness worldwide. Listeriosis has been identified as third to *Campylobacter* and *Salmonella* infections as an indigenous food-borne infectious agent contributing to the numbers of hospital bed days lost, as well as the fourth most common cause of death [21]. Figures 34.2A and B show a stillborn with granulomatosis infantiseptica characterized by the presence of a military-disseminated pyogranulomatous lesion on the body surface and a cut from the liver showing two subcapsular pyogranulomes, respectively.

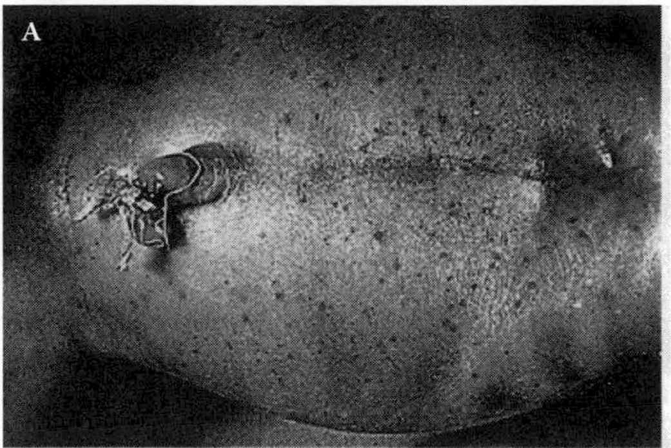

FIGURE 34.2 (A) Stillborn with granulomatosis infantiseptica characterized by the presence of miliary-disseminated pyogranulomatous lesions on the body surface. (B) Hematoxilin-eosin-stained section from the liver showing two subcapsular pyogranulomes.

DISEASE IN ANIMALS

The spectrum of the disease in animals is broad ranging, from asymptomatic infection and carriage to uncommon cutaneous lesions or various focal infections such as conjunctivitis, urethritis, endocarditis, and severe disturbance of the gait, followed by death. Abortions and perinatal deaths are common in cattle and sheep. *Listeria monocytogenes* is a well-recognized cause of mastitis, abortion, repeat breeding, infertility, encephalitis, and septicemia in cattle [22, 23]. *L. ivanovii* has also been implicated as a cause of abortion in cattle and sheep but occurs less frequently than *L. monocytogenes.* Listeric infections and abortions usually develop in the late winter or early spring. Abortions are most commonly recognized in the last trimester of pregnancy, frequently in the absence of other clinical signs.

IMMUNITY TO *LISTERIA*

Listeria infection in mice has been exploited extensively as a model to study molecular mechanisms of early innate as well as adaptive immunity. Restriction of parenchymal and systemic spreading of *Listeria monocytogenes* is achieved by containment in granulomatous structures or abscesses containing cellular and acellular components. The major role in this process has been attributed to an intricate cooperation between macrophages, neutrophils, and T cells. During systemic *L. monocytogenes* infection in mice, macrophages, neutrophils, natural killer (NK) cells and dendritic cells are central to the early innate immune response, acting as both the host for and forming a major line of defense against this pathogen [24].

The development of a protective immune response to *Listeria monocytogenes* has been ascribed to the involvement of $CD4^+$ and $CD8^+$ T-cell populations. In this regard, class I MHC and class II MHC-restricted epitopes of LLO for $CD8^+$ and $CD4^+$ T-cells, respectively, have been identified [25]. $CD8^+$ cells have been shown to lyse macrophages infected with *L. monocytogenes* or pulsed with *L. monocytogenes* peptides, the latter requiring additional *in vitro* stimulation for expressing significant cytolytic activity [26]. It also now is known that CD4 T-cell help is required during priming of CD8 T-cells to generate a life-long protective memory response to *L. monocytogenes* infection [27]. Increased levels of non-MHC-restricted cytotoxic activity, mediated by NK cells, have also been observed following *in vivo* challenge with *L. monocytogenes* in a mouse model [28].

Activation of TH_1 cells, a subset of $CD4^+$ T helper lymphocytes, by various antigens has been found to produce γ-interferon (IFN-γ), which stimulates macrophages [26]. Tumor necrosis factor (TNF) and IFN-γ are essential for *in vivo* immune defense against primary infection with *Listeria monocytogenes* [29]. Full induction of protective immunity requires infection with live, cytosol-invasive *L. monocytogenes. L. monocytogenes* possesses many attributes that would make it a good antigen delivery system for vaccine strategies aimed at the induction of CD8 T-cell responses [27].

ANTIGENS OF *LISTERIA*

Conventional Antigens

Somatic (O) and flagellar (H) antigens of *Listeria monocytogenes* have been described and used for serological groupings. The conventional antigens employed in various serological assays include heat-killed suspensions to detect antibodies to somatic antigens and formalin-treated organisms for flagellar antigens [30, 31]. These antigens exhibit cross-reactions with *Staphylococcus aureus, Corynebacterium pyogenes, Escherichia coli, Streptococcus faecalis,* and *Bacillus* species [32] when employed in serological tests such as agglutination, complement fixation, and immunoprecipitation. Trypsinization of heat-killed antigen of *L. monocytogenes* has been reported to increase its specificity and sensitivity in addition to avoiding cross-reactions [30].

Listeriolysin (LLO)

Of the various factors incriminated for the virulence of *Listeria monocytogenes*, LLO, an extracellular hemolysin produced by all pathogenic strains of *L. monocytogenes,* has been identified as a candidate antigen for a serological assay [33] and DNA probes encoding part of listeriolysin gene [34].

LLO has been characterized as a 58-kDa protein belonging to the group of sulfhydryl-activated toxins and antigenically related to streptolysin-O (SLO), pneumolysin, and perfringolysin [35]. It contained 529 amino acids and showed strong regional homologies to SLO and pneumolysin-O [36].

The purification of LLO by column chromatography with cross-linked polydextrans and carboxymethyl cellulose indicated a molecular size of 58 kDa [37]. When expressed in *Listeria innocua,* expression of LLO was found to increase 500-fold, and purification of LLO by ion exchange chromatography yielded a homogeneous protein free of p60 [38].

The demonstration that LLO could be detected in serum samples after absorption with SLO [39] was suggestive of specific antigenic epitopes in LLO; therefore, an immunoassay based on purified LLO would be a reliable indicator of listeric infection [40, 41]. LLO was shown to be a reliable indicator for serodiagnosis of listeriosis in sheep by ELISA [40] and in human beings by dot-blot assay, especially when bacteria could not be isolated [39, 42].

Protein p60 (cwhA)

Antibodies directed against the major secreted protein of *Listeria monocytogenes*, termed p60, were found to be more frequent than anti-listeriolysin antibodies in sera of listeriosis patients [43], which also were found to recognize the native and denatured secreted p60 protein of all serotypes of this species [44]. Antibodies raised against synthetic peptides derived from p60 could be useful for the development of an immunological assay specific for the food-borne pathogen *L. monocytogenes* [44].

Surface Proteins/Outer Membrane Proteins

An approach in immunological detection of pathogenic *Listeria monocytogenes* has been the use of antibodies directed against specific cell surface proteins/outer membrane proteins of microorganisms. Certain bacterial outer membrane proteins/antigens have been reported to be specific [45]. Characteristic banding patterns of the cell surface proteins of various *Listeria* spp. by SDS-PAGE and immunoblotting have been reported [46]. The cell surface proteins are specific for species and serovars with molecular weights of 64 and 68 kDa. Monoclonal antibodies associated with 66-kDa cell surface antigen specific for *L. monocytogenes* have been produced [47]; they could be useful in detecting the organism in clinical samples and foods.

Phosphatidylinositol-specific phospholipase C (PI-PLC) is an essential determinant of *Listeria monocytogenes* pathogenesis and is secreted by only pathogenic species of *Listeria* [14]. It has been purified to homogeneity [48]. It is a 33-kDa protein with an isoelectric point of 9.4. Its secretion can be enhanced by the addition of divalent cations to the culture medium. Expression of PI-PLC in the host tissues can be studied by detection of specific antibodies against it, which in turn may be useful in the diagnosis of *L. monocytogenes* infection.

DIAGNOSIS OF LISTERIOSIS

Listeriosis can be tentatively diagnosed on the basis of clinical symptoms and demonstration of the organisms in smear by Gram's staining, or by immunofluorescence. The organism can be isolated from clinical specimens such as blood, cerebrospinal fluid (CSF), and the meconium of newborns (or the fetus in abortion cases), and feces, vomitus, food stuffs/animal feed, and vaginal secretions of infected individuals or animals.

Detection of soluble antigen in CSF, especially in human meningitis cases, may be useful but it is not reliable. Serodiagnostic methods such as serum agglutination, complement fixation test, hemagglutination test, hemagglutination inhibition test, antibody precipitation test, growth inhibition test, ELISA exist but crude antigens employed in the tests show cross-reactivity. Listeriosis should be differentiated from influenza, tuberculous meningitis (in humans), and rabies, brucellosis, pasteurellosis, toxoplasmosis (in animals), especially in abortion cases. In view of the high cost, increased time and skills required by isolation procedures, along with their impracticability to screen large numbers of samples, the serological, novel, and pathogenicity marker based methods play an important role in its rapid and reliable diagnosis.

ISOLATION OF *LISTERIA*

Attempts to isolate and identify *Listeria monocytogenes* and to limit its proliferation in foods have been the focus of a significant international effort [49]. Historically, it has been challenging to isolate *Listeria* from food or other samples, and this explains why it remained unnoticed as a major food pathogen until recently. The earliest method available was the cold enrichment technique [50], which remained the only available method for many years. This required inoculation of the sample into a nutrient broth lacking selective agents, followed by incubation at 4°C for long periods, primarily to isolate the pathogen from infected animal or human tissue.

The key issues — enrichment/isolation time and the recovery of stressed *Listeria* cells — were addressed when further methods of enrichment and isolation were developed. Owing to many foodborne outbreaks of listeriosis, a zero tolerance level (absence in 25 g of food) has been implemented. Therefore, the tests must be able to detect one *Listeria* organism per 25 g food if it is to be approved by the regulatory agencies. *Listeria* cells are fastidious and can be rapidly outgrown by competitors. The recovery of *Listeria monocytogenes* from food, animals/human beings, and environmental samples requires the use of enrichment cultures followed by selective plating; and where injured organisms are likely to be present, a pre-enrichment step is required [51]. These methods are sensitive but often time consuming, laborious, and may take 5 or 6 days before the result is available. Two of the most widely used culture reference methods for detection of *Listeria* in all foods are the FDA bacteriological and analytical method (BAM) and the International Organization of Standards (ISO) 11290 method.

In the FDA BAM method the sample (25 g) is enriched for 48 hr at 30°C in *Listeria* enrichment broth (LEB) [52] containing the selective agents acriflavin and nalidixic acid, and the antifungal agent cycloheximide, followed by plating onto selective agar (Oxford, PALCAM, MOX, or LPM) [6]. The ISO 11290 method employs a two-stage enrichment process: the first enrichment in half

Fraser broth [53] for 24 hr, then an aliquot is transferred to full-strength Fraser broth for further enrichment, followed by plating on Oxford and PALCAM agars. Fraser broth also contains the selective agents acriflavin and nalidixic acid as well as esculin, which allows detection of β-D-glucosidase activity by *Listeria*, observed as blackening of the growth medium [6]. Primary or pre-enrichment broths usually contain lower amounts of the selective agents to aid the resuscitation of possibly injured cells. Both primary and secondary broths contain a phosphate buffering system.

The USDA and the Association of Analytical Chemists (AOAC/IDF) methods use a modification of University of Vermont Medium (UVM) [54] containing acriflavin and nalidixic acid for primary enrichment, followed by secondary enrichment in Fraser broth and plating onto Modified Oxford (MOX) agar containing the selective agents moxalactam and colistin sulfate [6].

The FDA method was designed for processing dairy products, whereas the USDA method was designed and has been officially recommended primarily for meat and poultry products [55], the latter being slightly superior for detection of *Listeria monocytogenes* in food and environmental samples. The detection rate of *L. monocytogenes* in milk using the FDA method by changing the final Oxford and PALCAM plating agar to *L. monocytogenes* blood agar (LMBA) has been improved [56].

Selective or Differential Plating Media

Use of potassium tellurite to inhibit Gram-negative bacteria was the first significant step in producing a *Listeria*-selective agar (LSA] [57]. Later, another LSA, referred to as McBride *Listeria* agar (MLA), was formulated in 1960 by substituting phenyl ethanol agar containing lithium chloride, glycine, and blood [58].

Later, modified McBride *Listeria* agar (MMLA) [59] and LiCl-phenylethanol-moxalactam agar (LPMA) [51] were developed. The FDA method employed MMLA while the USDA method used LiCl-phenylethanol-moxalactam agar (LPMA) as isolation agars [55].

A range of media have been developed, including Oxford agar [60], LiCl-ceftazidime agar, modified (LCAM) [61], polymyxin-acriflavine-lithium chloride-ceftazidime-esculin-mannitol (PALCAM), polymixin-acriflavine-lithium chloride-ceftazidime-esculin-mannitol egg yolk [62], Dominguez-Rodriguez isolation agar [63], Dominguez- Rodriguez *Listeria*-selective agar medium, modified [64], enhanced hemolysis agar (EHA) [60], modified Vogel Johnson agar (MVJ) [65], and MVJ modified further [66].

Chromogenic Media

Phosphotidylinositol-specific phospholipase C is an enzyme produced only by *Listeria monocytogenes* and *L. ivanovii*, which hydrolyzes a specific substrate added to the medium, producing an opaque halo around the colonies. Substrates for the detection of β-glucosidase, common to all *Listeria* spp., can be used as an elective feature for the detection of all *Listeria* colonies. Lithium chloride, nalidixic acid, and/or cyclohexamide are added to the media to obtain sufficient selectivity [67]. The chromogenic media commercially available include Agar *Listeria* according to Ottoviani and Agosti, the BCM *L. monocytogenes* detection system, CHROM agar, and Rapid' L.mono. Chromogenic media are simple, cost effective, and easy to interpret [6].

Virulence Determination

Many *Listeria monocytogenes* strains are highly pathogenic, while others are relatively avirulent and cause little harm in the host. A variety of methods have been developed to gauge the virulence of *L. monocytogenes* strains. In addition, hemolysin, mouse pathogenicity, tissue culture systems, and the detection of virulence-associated proteins and genes have been used to identify virulent *Listeria*.

Animal Inoculation

Mouse virulence assay has been frequently used for assessing *Listeria monocytogenes* virulence. The mouse virulence assay is done by inoculating groups of mice with varying doses of *L. monocytogenes* bacteria via the oral, nasal, intraperitoneal, intravenous, or subcutaneous routes [68]. Tests using oral inoculation in mice must be interpreted with caution because the main bacterial invasion protein internalin A does not bind to the mouse E-cadherin gene, resulting in inefficient translocation of these bacteria from the gastrointestinal tract to deeper tissues and organs [69]. The virulence of *L. monocytogenes* for man has been correlated with pathogenicity in mice [70], particularly in those made immunocompromised by treatment with carrageenan [71]. However, a large difference in the 50% of lethal doses (LD_{50}) between virulent and less virulent *L. monocytogenes* strains have been observed in the immunocompromised model, unlike the normal mice, which allows a clear and rapid means of distinguishing between the strains with a single-dose inoculum and with the sole criterion being death of mice [72]. In some instances, the relatively high numbers of animals required have rendered LD_{50} determination impracticable; therefore, it has been emphasized that the criterion for assessing virulence of *Listeria* strains on the basis of death is wrong and should be replaced more appropriately with persistence of microorganisms in liver or spleen following inoculation [73]. Recently, relative virulence (%) has been described as an alternative to LD_{50} measurement of the mouse virulence assay for *L. monocytogenes* [74]. The relative virulence (%) is obtained by dividing the number of dead mice by the total number of mice tested for a particular strain, using a known virulent strain (e.g., *L. monocytogenes* EGD) as reference. All the nonhemolytic species of *Listeria* (*L. innocua*, *L. welshimeri*, and *L. grayi*) and the weakly hemolytic *L. seeligeri* are avirulent in mouse pathogenicity tests [70].

In addition to mice, Sprague-Dawley rats pretreated with cimetidine have also been developed as an experimental model for gastric intubation [75]. The chick embryo test has been reported to agree with the mouse bioassay for assessment of the pathogenicity of *Listeria* species [76] as embryos inoculated with pathogenic strains through the chorioallontoic membrane (CAM) route died within 72 hr, whereas those inoculated with nonpathogenic strains survived [77]. The LD_{50} has been reported to be less than 6×10^2 cells for virulent strains [78]. Yolk sac challenge of 7-day-old chick embryos has been found to be less suitable than the CAM challenge for assessing virulence because of nonspecific deaths encountered in the former route [76].

Esophageal inoculation of juvenile rats with 10^6 CFU (colony forming unit) *Listeria monocytogenes* showed about a 50% infection rate in the liver or spleen [75]. An experimental keratoconjunctivitis test (Anton's eye test) can be performed in either guinea pigs or rabbits by inoculating a live bacterial suspension onto the eye [79].

Approximation of the infectious dose of *Listeria monocytogenes* in foods using cynomolgus monkey as a non-human primate model indicated that animals receiving 10^9 cells of *L. monocytogenes* became noticeably ill with symptoms of septicemia, irritability, loss of appetite, and occasional diarrhea [80].

In vitro Cell Assays

Listeria monocytogenes is able to infect and grow intracellularly in a range of mammalian cell types growing *in vitro*, including enterocytes, macrophages, hepatocytes, neuronal cells, and fibroblasts [1, 81] as well as invertebrate *Drosophila melanogaster*-derived cells [82]. These methods measure the ability of *L. monocytogenes* to cause cytopathogenic effects in the enterocyte-like cell line Caco-2 [83], to form plaques in the human adenocarcinoma cell line HT-29 [84] and L929 mouse fibroblast [85], and also help to study the heterogeneity of virulence factors. *L. monocytogenes, L. ivanovii,* and *L. seeligeri* show properties of invasion and spreading but other *Listeria* species do not [83, 86].

The main advantages of *in vitro* cell assays include their relatively low cost and ease of use. However, these tests are time-consuming, and occasionally variable, which have prevented them from being adopted in clinical laboratories for determining *Listeria monocytogenes* virulence and pathogenic potential [2].

Expression of Virulence-Associated Factors

In vitro demonstration of LLO, PC-PLC, and PI-PLC activities often provides general guidance on the pathogenic potential of *Listeria monocytogenes* strains. However, its reliability as a virulence indicator is not satisfactory. Detection of *L. monocytogenes* virulence associated genes by PCR and genetic lineage analysis has not resulted in a clear correlation between these genes and the underlying virulence of *L. monocytogenes* [22, 87, 88]. However, mutations in virulence-associated genes, resulting in the expression of truncated or nonfunctional proteins [89, 90], targeting these gene mutations as a means of determining *L. monocytogenes* virulence, often experience difficulties. An optimal strategy for *L. monocytogenes* virulence testing should be the detection of virulence-specific gene(s) that are present only in virulent strains, but absent in avirulent strains [2].

Immunological Tests

Conventional Tests

Serological methods have been reported to be nonspecific because of antigenic cross-reactivity between *Listeria monocytogenes* and other Gram-positive bacteria, such as *Staphylococci, Enterococci,* and *Bacillus* spp. [32]. Such methods also lack sensitivity and therefore cannot be used for reliable diagnosis of listeriosis [39].

Enzyme-Linked Immunosorbent Assay (ELISA)

ELISAs for *Listeria monocytogenes* detection are either based on polyclonal antibody [91] or on monoclonal antibodies (mAbs) [92]. There are a number of ELISA formats, including direct ELISAs, sandwich ELISAs, and competitive ELISAs. MAbs-based ELISA and dot-blot assays have been reported to identify *Listeria* species in food [93, 94] and clinical samples [95]. However, some of these mAbs have been shown to react with all species of *Listeria* [93, 94].

The *Listeria*-Tek assay, an ELISA-based method, has been reported to be a rapid and simple procedure for determination of *Listeria* spp. in foods [96]. The ELISA and the USDA procedure have been proved equally sensitive for processing the food samples having a *Listeria monocytogenes* count greater than 3 CFU g^{-1} and had detection limits of approximately 10^6 CFU mL^{-1} and 10^4 CFU mL^{-1} in pure cultures, respectively [97]. The ELISA for detection of *L. monocytogenes* in dairy, seafood, and meat products has been adopted by AOAC International [98].

The most commonly used immunoassays for the detection of the pathogen are based on the use of whole cells. Often for detection of *Listeria monocytogenes* from foods, the sample is enriched, heat-killed [99, 100], or formalin-fixed [101]. However, many of the cell-surface antigens are genus specific rather than *L. monocytogenes* specific [102, 103]. Detection of the flagella of the bacterium has also been attempted [104, 105].

ELISAs using the O and H antigens [106] as well as whole cell protein extracts [107] have also been used to detect the pathogen. Monoclonal antibodies are used in the Vitek Immuno Diagnostic Assay System (VIDAS)-LMO (bioMerieux Vitek, Hazeltown, MO) and Lister test (Vicam, Watertown, MA) [99, 108].

Immunofluorescence Test

The fluorescent antibody technique for the detection of *Listeria monocytogenes* in smears, impression smears from dead tissues, meat, and milk was described by Khan et al. [109]. However, they reported cross-reaction with micrococci and streptococci. Later, immunofluorescence tests based on monoclonal antibodies for the detection of *L. monocytogenes* from clinical specimens such as necropsy tissue [95, 110] and specific protein p60 were developed [111].

Cellular Assays

Cell-mediated immune (CMI) effector functions have been shown to play an important role in the host defense against different facultative intracellular bacterial pathogens through cytokine production and perhaps by recognition and lysis of macrophages and other cells infected with these pathogens, including *Listeria monocytogenes* [112]. The measurement of CMI can be accomplished by several methods, such as delayed type hypersensitivity, lymphocyte proliferation assay, cytotoxicity assay, and cytokine assay, which are either generalized or specialized in nature [113].

Delayed-type hypersensitivity (DTH) is characterized by a localized swelling that occurs 48 to 72 hr after intradermal injection with the microbial antigen under test. DTH carried out with soluble crude antigen fractions of nonpathogenic *Listeria innocua* (serotype 6a) or *L. monocytogenes* (serotype 4 b) has been shown to be useful in detecting listeriosis at an early stage [114] using an experimental mouse model.

Cytotoxicity assays are based on the measurement of the ability of lymphocytes to kill target cells. Cytotoxic T-lymphocytes have been found to play an important role in the immunity of hosts to facultative intracellular bacteria [115].

Cytokine released by activated lymphocytes have been assayed to serve as a measure of CMI. Cytokine assays might be in the form of bioassays or a direct concentration measurement as in ELISA, radioimmunoassays, or precipitation assays or detection of specific mRNA in lymphocytes [113]. IFN-γ in the bloodstreams and spleens of mice has been detected from day 1 to day 4 post infection with *Listeria monocytogenes* by double sandwich ELISA as well as immunohistochemical techniques [116]. IFN-γ appeared as early as day 7 of an oral infection in bovines [117].

MOLECULAR METHODS

DNA Probes

The presence of a target sequence is detected using a single-stranded nucleic acid that is enzyme- or radiolabeled. Datta et al. [118] reported the first DNA probe wherein a Hind III-Hind II fragment of a presumptive hemolysin gene was used in a trial for specific detection of *Listeria monocytogenes.* Nonisotopic colorimetric detection has been applied in a rapid nucleic acid dipstick hybridization assay for detection of *Listeria* sp. in food and environmental samples [119]. Different DNA probes, such as digoxigenin-labeled synthetic oligonucleotide probe [120], encoding part of listeriolysin gene [34], on a *16S rRNA* sequence [121], have been developed. DNA probes targeting the *inlA* and *plcA* [122, 123], and the *prfA* [124] genes have also been developed. As this procedure exploits differences among *Listeria* species at the genetic level, it is more specific than biochemical and serological methods that are phenotype based.

Fluorescence *in situ* hybridization (FISH) is used to study the presence and distribution of specific strains in microbial communities. FISH can be used in phylogenetic studies, and in assessing the spatial distribution of target microbes in communities such as biofilms [125]. FISH using oligonucleotide probes specific for the virulence gene transcript *iap*-mRNA has been carried out; this allowed for analysis of virulence gene expression of *Listeria monocytogenes* within a mixed microbial community [125].

Nucleic Acid Amplification

In the first reported PCR for identification of *Listeria monocytogenes,* the *hly* sequence published by Mengaud et al. [126] was used. This PCR was used to detect *L. monocytogenes* in water, whole milk, and human cerebrospinal fluid. Standard PCR followed by dot-blot hybridization has come out as an encouraging approach for the diagnosis of *Listeria* meningitis [127].

PCRs have been developed for the identification of *Listeria monocytogenes* in food samples [128, 129] and in vegetables [130]. The minimum detection limits of *L. monocytogenes* in a 25-g sample after 48 hr incubation have been reported to be 10 cells [131], and 4 to 10 cells after overnight incubation [132]. Using a two-step PCR with nested primers, 1 colony forming unit (CFU) *L. monocytogenes* could be directly detected in 25 mL raw milk [133]. However, false-negative results have also been reported with PCR while analyzing foods containing high populations of *L. monocytogenes* [132].

As PCR has the ability to selectively amplify specific targets present in low concentrations, it offers exquisite specificity, unsurpassed sensitivity, rapid turnover, and ease of automation for laboratory detection of *Listeria monocytogenes* from clinical specimens [2]. The molecular differences within *16S* and *23S rRNA* genes, intergenic spacer regions, *hly, inlA, inlB, iap,* and other genes can be used to differentiate *L. monocytogenes* from other *Listeria* species and common bacteria [134]. Table 34.2 lists the genes employed for PCR-based identification of *Listeria.*

Reverse Transcription (RT)-PCR

Testing of food or environmental samples for pathogenic *Listeria* should only target living organisms because only live *Listeria* cells can cause disease. The choice of RNA or mRNA as a target for food pathogen testing has gathered increasing favor because the presence of mRNA is an indication of the living state of the cell [135, 136].

TABLE 34.2
Identification of *Listeria* Species by PCR-Based Procedures

Target Gene	Ref
16S rRNA gene	197, 198
23S rRNA gene	172, 199
16S/23S rRNA intergenic regions	200, 201
hly	22, 202–206
plcA, plcB	14, 22, 88, 122
actA	22, 88, 207
prfA	14, 208, 209
inlA, inlB, inl AB	88, 123, 203, 210, 211
iap	197, 212, 213
lma/dth18	214–216
fbp	217
flaA	218
pepC	219
clpE	122
sigB	220, 221
vip	222
lse24-315	223
lin2483	224
liv22-228	225

Listeria monocytogenes has been detected using RT-PCR in artificially inoculated meat samples (Klein and Juneja, 1997) by targeting mRNA transcripts of the *hly, prfA,* and *iap* genes in waste samples [137] by targeting the transcripts for *rRNA* genes, and also for the detection of heat-injured *L. monocytogenes* by targeting the *hly* transcript [138].

Real-Time PCR

Real-time PCR has been used to identify and quantify *Listeria monocytogenes* in foods and clinical samples in several studies [139–142]. Real-time PCR is quantitative, which is a significant advantage over other molecular methods, and so this technology is extremely attractive for food testing and epidemiological investigations.

Other Methods

Other rapid methods for detecting *Listeria* spp. include the use of flow cytometry [54], nucleic acid sequence-based amplification [143], and electrical impedence [144].

Immunomagnetic separation used primarily to isolate strains of *Listeria monocytogenes* from pure cultures as well as from heterogenous suspensions has been viewed as a new approach for extraction and isolation of pathogenic bacteria directly from foods [145].

The combined use of micro- and nano-fabrication techniques in the area of biosensors holds great promise in the area of detection of food-borne pathogens [146].

METHODS FOR SUBTYPING OF *LISTERIA*

The availability of subtyping procedures to track individual strains involved in listeriosis outbreaks, and to examine the epidemiology and population genetics of *Listeria monocytogenes* bacteria, is integral to control and prevention programs aimed at limiting listeriosis. Application of subtyping methods also provides insight into the population genetics, epidemiology, ecology, and evolution of *Listeria monocytogenes.* A variety of conventional, phenotypic, and DNA-based subtyping methods have been described for differentiation of *L. monocytogenes* beyond the species and subspecies levels [147]. While phenotype-based methods have been used for many years to subtype *L. monocytogenes* and other food-borne pathogens, DNA-based subtyping methods are generally more discriminatory and amenable to inter-laboratory standardization and are thus increasingly replacing phenotype-based subtyping methods [148]. Commonly used phenotype-based subtyping methods for *L. monocytogenes* and other food-borne pathogens include serotyping, phage typing, and multilocus enzyme electrophoresis (MLEE) [147, 149]. The genetic subtyping approach encompasses PCR-based approaches (e.g., random amplified polymorphic DNA and amplified fragment length polymorphism), PCR-restriction fragment length polymorphism (PCR-RFLP), ribotyping, pulsed-field gel electrophoresis, and DNA sequencing-based subtyping techniques (e.g., multilocus sequence typing (MLST)) [150–153]. The phenotypic subtyping approach occasionally suffers from low discrimination and reproducibility; the genetic subtyping approach is highly sensitive, discriminatory, and reproducible. For improved subtyping discrimination, a combination of two or more subtyping techniques, be they gene or phenotype based, is often used in practice for epidemiologic investigation of *L. monocytogenes* outbreaks. The methods for subtyping of *Listeria* are listed in Table 34.3.

PHENOTYPING TYPING METHODS

Serotyping

Serotyping is a universally accepted subtyping method for *Listeria monocytogenes.* Identification of the strain serotype permits differentiation between important food-borne strains. *Listeria*

TABLE 34.3
Methods for Subtyping of *Listeria* Species

Methods	Refs.
Phenotypic Methods	
Serotyping	106, 155
Phage typing	160, 226, 227
MLEE	160
Esterase typing	228
Genotypic Methods	
PFGE	152, 168, 229
Ribotyping	164, 165, 230
RFLP	164
RAPD	166, 167
AFLP	169, 171
PCR-RFLP	172
REP-PCR	173
DNA sequencing based	151, 164, 231
DNA microarray	178–180

species possess group-specific surface proteins, such as somatic (O) and flagellar (H) antigens, that are useful targets for serological detection with corresponding monoclonal and polyclonal antibodies. While there are 15 *Listeria* somatic (O) antigen subtypes (I–XV), flagellar (H) antigens comprise four subtypes (A–D) [154, 155] with the serotypes of individual *Listeria* strains being determined by their unique combinations of O and H antigens. An ELISA has recently been developed to improve the efficiency of serotyping [106].

Serotyping may potentially be useful for tracking *Listeria monocytogenes* strains involved in disease outbreaks. Indeed, it has been observed that *L. monocytogenes* serotypes 1/2a, 1/2b, and 4b are responsible for 98% of documented human listeriosis cases, whereas serotypes 4a and 4c are rarely associated with outbreaks of the disease [156, 157]. However, due mainly to the high cost of acquiring sub-type-specific antisera, serotyping methods are not routinely performed in clinical laboratories. Serotyping does not correlate with the species distinctions. Serotyping methods have largely been superseded by molecular procedures that are intrinsically more specific and sensitive for the identification and differentiation of *Listeria* species. The development of PCR-based serotyping procedures has provided additional tools for the identification and grouping of *L. monocytogenes* [158, 159].

Phage Typing

Phage typing of *Listeria monocytogenes* has been used as a discriminatory typing system. Bacteriophages have the capacity to lyse closely related *Listeria* because of their host specificity, independently of the species and serovar identities. *Listeria* strains can be separated into distinct phage groups and phagovars, which are useful for tracking the origin and course of listeriosis outbreaks [160]. With close to 10% of *Listeria* strains being untypable (especially serovar 3 and *L. grayi* strains), the usefulness of phage typing as an independent tool for epidemiological investigations is severely constrained.

MLEE (Multilocus Enzyme Electrophoresis)

MLEE is a protein-based, isoenzyme typing method that correlates specific protein band patterns with genotypes. MLEE differentiates bacterial strains by detecting variations in the patterns of the electrophoretic mobilities of various constitutive enzymes. Based on the similar electrophoretic types detected in MLEE, *Listeria monocytogenes* serovars 1/2b, 3b, and 4b are classified into one distinct division, and serovars 1/2a, 1/2c, and 3a in another division [161, 162]. The detection of a large number of electrophoretic types in *L. monocytogenes* strains by MLEE necessitates careful optimization and standardization of the test procedure so that run-to-run variations are minimized.

GENETIC TYPING METHODS

PFGE (Pulsed-Field Gel Electrophoresis)

PFGE is a highly reproducible, discriminatory, and effective molecular typing method based on restriction fragment length polymorphisms (RFLPs) of bacterial DNA [2] and provides a "gold standard" for comparing isolates analyzed in different labs and with different techniques. PFGE uses selected restriction enzymes to yield fewer and larger fragments of DNA, resulting in a higher level of fragment resolution. PFGE provides sensitive subtype discrimination and is often considered the standard subtyping method for *Listeria monocytogenes* [163]. However, this method is not automated and is labor intensive.

Pulsenet, a national network of public health and food regulatory laboratories, is established in United States to detect clusters of food-borne disease and respond quickly to food-borne outbreak investigations [151]. The laboratories use a highly standardized 1-day PFGE to subtype the bacteria and exchange normalized DNA fingerprint patterns via the Internet.

Ribotyping

Ribotyping is similar to RFLP in that it uses restriction endonuclease digestion of DNA to create a pattern that can be analyzed. It involves the restriction enzyme digestion of chromosomal DNA, followed by DNA hybridization using an *rRNA* gene probe. Weidmann et al. [164] used ribotyping on different isolates of *Listeria monocytogenes* to group different lineages and relate the pathogenicity of the organism. *EcoRI* ribotyping was used to demonstrate the level of genetic diversity among *L. monocytogenes* contamination of Gorgonzola cheeses [165].

RAPD and Arbitrarily Primed PCR (AP-PCR)

Randomly amplified polymorphic DNA (RAPD) and AP-PCR analysis make use of a short arbitrary primer (usually 10 bases long for AP-PCR, and 10 to 15 bases long for RAPD) that anneals randomly along genomic DNA to amplify a number of fragments within the genome at a relatively low temperature (around 36°C). RAPD was used to follow the incidence and typing of *L. monocytogenes* in both poultry [166], and vegetable [167] processing plants. RAPD is more economical and faster than other typing methods, and is particularly suitable for testing fewer than 50 strains.

AFLP (Amplified Fragment Length Polymorphism)

AFLP is a modification of RFLP through the addition of adaptors to restriction enzyme-digested DNA, followed by PCR amplification and electrophoretic separation of PCR products [2]. AFLP can be used to differentiate strains of *Listeria monocytogenes* on a more discriminating basis than serotyping [168, 169]. AFLP is sensitive and reproducible, thus representing a valuable tool in the characterization of *L. monocytogenes* strains, and also in the identification of *Listeria* species [170, 171].

PCR-RFLP

PCR-RFLP involves the PCR amplification of one or more *Listeria monocytogenes* housekeeping or virulence-associated genes (e.g., *hly, actA,* and *inlA*), digestion with selected restriction enzymes, and separation by agarose gel electrophoresis. Subsequent examination of the distinct band patterns permits differentiation of *L. monocytogenes* subtypes [164]. Paillard et al. [172] used PCR–RFLP on the *23S rRNA* gene to determine the species of *Listeria* in sludge samples. PCR-RFLP provides a sensitive, discriminatory, and reproducible method for tracking and epidemiological investigation of *L. monocytogenes* bacteria, if used in combination with other subtyping procedures.

REP-PCR

Repetitive extragenic palindromic (REP) and enterobacterial repetitive intergenic consensus (ERIC) sequences represent useful primer binding sites for PCR amplification of the *Listeria monocytogenes* genome to achieve species and strain discrimination and divide *L. monocytogenes* strains into four clusters that match the origin of isolation, each consisting of multiple subtypes [173]. REP-PCR showed a higher discriminative potential than ERIC-PCR and a comparable discriminative potential as RAPD combining 3 or 4 primers [173].

DNA Sequencing-Based Subtyping Techniques

DNA sequencing-based methods are being developed and increasingly used for subtyping and characterizing bacterial isolates. In these methods, complete or partial nucleotide sequences are determined for one or more bacterial genes or chromosomal regions, thus providing unambiguous and discrete data. The advantages of sequencing methods over DNA fragment size-based typing methods include their ability to generate unambiguous data that are portable through Web-based databases and that can be used for phylogenetic analyses [174, 175]. While a variety of DNA sequence-based subtyping strategies targeting virulence genes, housekeeping genes, or other chromosomal genes and regions are feasible, multilocus sequence typing (MLST), which is an extension of MLEE, represents a widely used strategy [176]. The implementation of DNA sequence-based subtyping approaches for routine characterization of human, animal, and food *Listeria monocytogenes* isolates will not only allow for sensitive and standardized subtyping for outbreak detection, but will also provide an opportunity for using subtyping data to probe the evolution of this food-borne pathogen and to track the spread of clonal groups [148]. With the cost of DNA sequencing decreasing rapidly, MLST is poised to play a more important role in *L. monocytogenes* subtyping and phylogenetic studies.

Using various subtyping procedures, *Listeria monocytogenes* strains have been grouped into three genetic lineages (or divisions), with lineage I consisting of serotypes 1/2b, 3b, 4b, 4d, and 4e; lineage II of serotypes 1/2a, 1/2c, 3a, and 3c; and lineage III of serotypes 4a and 4c [177]. Interestingly, *L. monocytogenes* isolates from sporadic and endemic human listeriosis mostly belong to lineages I and II, whereas those from animal and environmental specimens are of lineage III.

DNA Microarrays

Microarrays consist of a number of discretely located DNA probes corresponding to an oligonucleotide specific to a target DNA sequence fixed on a solid substrate such as glass. DNA microarrays are used to investigate microbial evolution and epidemiology and can serve as a diagnostic tool for clinical, environmental, or food testing. Two approaches are used; one is based on sequence specific oligonucleotides and the other employs specific PCR products.

Several microarray-based strategies were developed to differentiate between the six listerial species [122] and for discrimination among *Listeria monocytogenes* serovars [178, 179] and phylogenetic lineages [180, 181]. Microarray-based assays were also applied to investigate the genome evolution within the genus *Listeria* [177] and for identification of natural atypical *L. innocua* strains harboring genes of the LIPI-1 [182].

Oligonucleotide microarrays based on the *iap* and *hly* genes have been used simultaneously to detect and discriminate between *Listeria* species [122]. Borucki et al. [178] developed a mixed genome microarray to identify gene sequences that differentiate different serotypes of *Listeria monocytogenes*. The most attractive feature of these microarrays is the capability of simultaneous identification and typing of *Listeria* strains in one test.

GENOMICS: THE BEGINNING OF A NEW ERA

The availability of genome sequences of *Listeria monocytogenes* serovars, *L. innocua,* and *L. welshimeri* [183–185] has provided insight into the molecular basis of the pathogenesis determinants of *Listeria* species. Further, sequencing of the strains comprising other species of this genus will provide a rich resource for understanding the sources of variation and evolutionary history. Comparative genomics could reveal genetic loci that confer specific pathogenic traits to epidemic strains. The information provides access to niche-specific candidate genes and proteins that will need to be evaluated by traditional physiological, biochemical, and genetic approaches, in turn enabling the development of novel ways to interrupt transmission and prevent food-borne transmission.

PREVENTION AND CONTROL

Listeriosis can be prevented in humans by taking care during handling of abortion cases in both humans and in animals, avoiding consumption of contaminated foodstuffs, and avoiding cross-infections, especially among infants in hospitals. Culling infected animals should be advocated as they secrete the organisms in secretions and excretions, especially in the cases of mastitis. Care in the use and preparation of silage is important because the pathogen grows luxuriantly at a pH greater than 5, particularly when fermentation is ineffective and molds grow. Thus far, no vaccine is available against listeriosis.

The government, industry, and consumers are the most important parties in ensuring food safety. The government should establish standards, develop codes of practice, and ensure that these will be followed. Food producers and food preparers should produce safe food, which can be achieved by adequate hygiene standards, good manufacturing practices (GMP), and implementation of Hazard Analysis and Critical Control Points (HACCPs). Consumers, in particular those who buy food and prepare meals, should have a basic knowledge of safe food preparation.

TREATMENT

The organism is, thus far, usually sensitive to a wide range of antibiotics [186]. Ampicillin, amoxicillin, tetracyclines, chloramphenicol, β-lactam antibiotics, together with an aminoglycoside, trimethoprim, and sulfamethoxazole, are recommended. However, ampicillin is the drug of choice in cases of encephalitis. Ampicillin along with gentamicin are recommended for prolonged treatment regimes. Recently, it was found that fosfomycin is extremely effective in controlling listerial growth in a mouse model of infection. The basis of this exquisite sensitivity has been elucidated by the finding that following entry of the bacteria into the cytoplasm, intracellularly growing *Listeria* express a permase for the uptake of hexose phosphates (UhpT), and that fosfomycin is preferentially translocated into growing bacteria via this transporter [187].

Listeria monocytogenes infections are usually treated with a single antimicrobial agent, and drugs are only combined for treatment of immunocompromised patients [188]. Recently, reports have described clinical strains resistant to chloramphenicol, erythromycin, streptomycin, tetracycline, vancomycin, and trimethoprim [189–191]. Resistant *L. monocytogenes* strains have also been found in food samples [192].

BIOTECHNOLOGICAL APPLICATIONS

Listeria monocytogenes has become an important paradigm for immunological investigation and also an important model system for analysis of the molecular mechanisms of intracellular parasitism and pathophysiology [17].

As a facultative intracellular bacterium, *Listeria monocytogenes* survives within cells after phagocytosis, and it is an ideal vector for the delivery of antigens to be processed and presented through both the class I and II antigen-processing pathways. It is possible that the virulence factors released in cytosol can enhance the immunogenicity of tumor-associated antigens, which are poorly immunogenic [193]. In several mouse models, *Listeria*-based vaccines have been demonstrated as an effective method of influencing tumor growth and eliciting potent anti-tumor immune responses [194].

LLO incorporated into liposomes can be used as an efficient vaccine delivery system carrying a viral antigenic protein to generate protective antiviral immunity [195].

Listeria has been engineered to express a number of HIV/SIV antigens and has been tried as a live bacterial vaccine vector for the delivery of HIV antigens [196].

OUTLOOK AND PERSPECTIVE

Listeria remains among the deadliest known food-borne pathogens worldwide. Great progress has been made in eliciting the mechanisms leading to pathogenesis of this bacterium, and a large battery of tests is now available to detect and diagnostically classify *Listeria,* both at the species level and even permit discrimination between strains associated with human and animal infections or even sporadic from epidemic strains. Much more, however, remains to be learned about the ability of this bacterial species to persist in the environment and in food processing plants and their ability to grow at refrigeration temperatures. The advent of functional listerial genomics is providing new possibilities to comprehensively catalog genes and their products involved in the transition from life in the environment to life within infected cells. The availability of genome sequences will provide us with an opportunity to define the evolutionary paths taken to pathogenesis and help us understand the emergence of *Listeria* spp. as major food-borne pathogens.

ACKNOWLEDGMENTS

We thank M. Rohde (Hemholtz Institute for Infectious Disease Research, Braunschweig) for electron microscopy. The work reported in this manuscript was made possible by grants from the Bundesministerium fuer Bildung und Forschung (BMBF) and the DeutscheForschungsGemeinschaft (DFG) to T.H. and T.C. S.B. is a scholar of the Department of Biotechnology, Government of India (BT/IN/BTOA/17/2006).

REFERENCES

1. Vazquez-Boland, J.A. et al. *Listeria* pathogenesis and molecular virulence determinants, *Clin. Microbiol. Rev.*, 14, 584, 2001.
2. Liu, D. Identification, subtyping and virulence determination of *Listeria monocytogenes*, an important foodborne pathogen, *J. Med. Microbiol.*, 55, 645, 2006.
3. Low, J.C. and Donachie, W. A review of *Listeria monocytogenes* and listeriosis, *Vet. J.*, 153, 9, 1997.
4. Kathariou, S. *Listeria monocytogenes* virulence and pathogenicity, a food safety perspective, *J. Food Prot.*, 65, 1811, 2002.
5. Churchill, R.L., Lee, H., and Hall, J.C. Detection of *Listeria monocytogenes* and the toxin listeriolysin O in food, *J. Microbiol. Methods*, 64, 141, 2006.
6. Gasanov, U., Hughes, D., and Hansbro, P.M. Methods for the isolation and identification of *Listeria* spp. and *Listeria monocytogenes*: a review, *FEMS Microbiol. Rev.*, 29, 851, 2005.

7. Hain, T., Steinweg, C., and Chakraborty, T. Comparative and functional genomics of *Listeria* spp, *J. Biotechnol.*, 126, 37, 2006.
8. Schmid, M.W. et al. Evolutionary history of the genus *Listeria* and its virulence genes, *Syst. Appl. Microbiol.*, 28, 1, 2005.
9. Grau F.H. and Vanderlinde P.B. Growth of *Listeria monocytogenes* on vacuum packaged beef, *J. Food Prot.*, 53, 739, 1990.
10. Bille, J. et al. API *Listeria*, a new and promising one-day system to identify *Listeria* isolates, *Appl. Environ. Microbiol.*, 58, 1857, 1992.
11. Robinson, R.K., Batt, C.A., and Patel, P.D. Eds. *Encyclopedia of Food Microbiology*. San Diego, CA: Academic Press, 2000.
12. Rocourt, J. and Catimel, B. Biochemical characterization of species in the genus, *Listeria*, *Zentralbl. Bakteriol. Mikrobiol. Hyg.[A]*, 260, 221, 1985.
13. Ermolaeva, S. et al. A simple method for the differentiation of *Listeria monocytogenes* based on induction of lecithinase activity by charcoal, *Int. J. Food Microbiol.*, 82, 87, 2003.
14. Notermans, S.H. et al. Phosphatidylinositol-specific phospholipase C activity as a marker to distinguish between pathogenic and nonpathogenic *Listeria* species, *Appl. Environ. Microbiol.*, 57, 2666, 1991.
15. Bubert, A. et al. Differential expression of *Listeria monocytogenes* virulence genes in mammalian host cells, *Mol. Gen. Genet.*, 261, 323, 1999.
16. Gouin, E., Mengaud, J., and Cossart, P. The virulence gene cluster of *Listeria monocytogenes* is also present in *Listeria ivanovii*, an animal pathogen, and *Listeria seeligeri*, a nonpathogenic species, *Infect. Immun.*, 62, 3550, 1994.
17. Hamon, M., Bierne, H., and Cossart, P. *Listeria monocytogenes*: a multifaceted model, *Nat. Rev. Microbiol.*, 4, 423, 2006.
18. Chatterjee, S.S. et al. Intracellular gene expression profile of *Listeria monocytogenes, Infect. Immun.*, 74, 1323, 2006.
19. Joseph, B. et al. Identification of *Listeria monocytogenes* genes contributing to intracellular replication by expression profiling and mutant screening, *J. Bacteriol.*, 188, 556, 2006.
20. Smerdon, W.J. et al. Surveillance of listeriosis in England and Wales, 1995–1999, *Commun. Dis. Public Health.*, 4, 188, 2001.
21. Adak, G.K., Long, S.M., and O'Brien, S.J. Trends in indigenous foodborne disease and deaths, England and Wales: 1992 to 2000, *Gut*, 51, 832, 2002.
22. Rawool, D.B. et al. Detection of multiple virulence-associated genes in *Listeria monocytogenes* isolated from bovine mastitis cases, *Int. J. Food Microbiol.*, 113, 201, 2007.
23. Shakuntala, I. et al. Isolation of *Listeria monocytogenes* from buffaloes with reproductive disorders and its confirmation by polymerase chain reaction, *Vet. Microbiol.*, 117, 229, 2006.
24. Popov, A. et al. Indoleamine 2,3-dioxygenase-expressing dendritic cells form suppurative granulomas following *Listeria monocytogenes* infection, *J. Clin. Invest.*, 116, 3160, 2006.
25. Safley, S.A. et al. Role of listeriolysin-O (LLO) in the T lymphocyte response to infection with *Listeria monocytogenes*. Identification of T cell epitopes of LLO, *J. Immunol.*, 146, 3604, 1991.
26. Kaufmann, S.H. Immunity to intracellular bacteria, *Annu. Rev. Immunol.*, 11, 129, 1993.
27. Lara-Tejero, M. and Pamer, E.G.T cell responses to *Listeria monocytogenes*, *Curr. Opin. Microbiol.*, 7, 45, 2004.
28. Kearns, R.J. and Leu, R.W. Modulation of natural killer activity in mice following infection with *Listeria monocytogenes, Cell Immunol.*, 84, 361, 1984.
29. Pamer, E.G. Immune responses to *Listeria monocytogenes. Nat. Rev. Immunol.*, 4, 812, 2004.
30. Osebold, J.W., Aalund, O., and Crisp, C.E., Chemical and immunological composition of surface structures of *Listeria monocytogenes, J. Bacteriol.*, 89, 84, 1965.
31. Larsen, S.A., Wiggins, G.A., and Albritton, W.L. Immune response to *Listeria,* in *Manual of Clinical Immunology,* Rose, N.R. and Friedman, H., Eds, 2nd edition, American Society for Microbiology, 1980, 506.
32. Gray, M.L. and Killinger, A.H. *Listeria monocytogenes* and listeric infections, *Bacteriol. Rev.*, 30, 309, 1966.
33. Low, J.C. and Donachie, W. Clinical and serum antibody responses of lambs to infection by *Listeria monocytogenes, Res. Vet. Sci.*, 51, 185, 1991.
34. Chenevert, J. et al. A DNA probe specific for *L. monocytogenes* in the genus *Listeria, Int. J. Food Microbiol.*, 8, 317, 1989.

35. Geoffroy, C. et al. Purification, characterization, and toxicity of the sulfhydryl-activated hemolysin listeriolysin O from *Listeria monocytogenes, Infect. Immun.*, 55, 1641, 1987.
36. Leimeister-Wachter, M. and Chakraborty, T. Detection of listeriolysin, the thiol-dependent hemolysin in *Listeria monocytogenes, Listeria ivanovii*, and *Listeria seeligeri., Infect. Immun.*, 57, 2350, 1989.
37. Mengaud, J. et al. A genetic approach to demonstrate the role of listeriolysin O in the virulence of *Listeria monocytogenes, Acta Microbiol. Hung.*, 36, 177, 1989.
38. Darji, A. et al. Hyperexpression of listeriolysin in the nonpathogenic species *Listeria innocua* and high yield purification, *J. Biotechnol.,* 43, 205, 1995.
39. Berche, P. et al. Detection of anti-listeriolysin O for serodiagnosis of human listeriosis, *Lancet*, 335, 624, 1990.
40. Low, J.C., Davies, R.C., and Donachie, W. Purification of listeriolysin O and development of an immunoassay for diagnosis of listeric infections in sheep, *J. Clin. Microbiol.*, 30, 2705, 1992.
41. Barbuddhe, S.B., Malik, S.V.S., and Gupta, L.K. Kinetics of antibody production and clinical profiles of calves experimentally infected with *Listeria monocytogenes, J. Vet. Med. B.,* 47, 497, 2000.
42. Barbuddhe, S.B., Malik, S.V.S., and Kumar, P. High seropositivity against listeriolysin O in humans, *Ann. Trop. Med. Parasitol.*, 93, 537, 1999.
43. Gentschev, I. et al. Identification of p60 antibodies in human sera and presentation of this listerial antigen on the surface of attenuated salmonellae by the HlyB-HlyD secretion system, *Infect. Immun.*, 60, 5091, 1992.
44. Bubert, A. et al. Synthetic peptides derived from the *Listeria monocytogenes* p60 protein as antigens for the generation of polyclonal antibodies specific for secreted cell-free *L. monocytogenes* p60 proteins, *Appl. Environ. Microbiol.*, 60, 3120, 1994.
45. Chang, T.C., Chen, C.H., and Chen, H.C., Development of a latex agglutination test for the rapid identification of *Vibrio parahaemolyticus, J. Food Prot.*, 57, 31, 1994.
46. Tabouret, M., de Rycke, J., and Dubray, G. Analysis of surface proteins of *Listeria* in relation to species, serovar and pathogenicity, *J. Gen. Microbiol.*, 138, 743, 1992.
47. Bhunia, A.K. and Johnson, M.G. Monoclonal antibody specific for *Listeria monocytogenes* associated with a 66-kilodalton cell surface antigen, *Appl. Environ. Microbiol.*, 58, 1924, 1992.
48. Goldfine, H. and Knob, C. Purification and characterization of *Listeria monocytogenes* phosphatidylinositol-specific phospholipase C, *Infect. Immun.*, 60, 4059, 1992.
49. Farber, J.M. and Peterkin, P.I. *Listeria monocytogenes*, a food-borne pathogen, *Microbiol. Rev.*, 55, 476, 1991.
50. Gray, M.L. et al. A new technique for isolating listerelle from the bovine brain, *J. Bacteriol*, 55, 471, 1948.
51. Curtis, G.D. and Lee, W.H. Culture media and methods for the isolation of *Listeria monocytogenes, Int. J. Food Microbiol.*, 26, 1, 1995.
52. Lovett, J., Francis, D.W., and Hunt, J.M. *Listeria monocytogenes* in raw milk: detection, incidence and pathogenicity, *J. Food Prot.*, 50, 185, 1987.
53. Fraser, J.A. and Sperber, W.H. Rapid detection of *Listeria* spp. in food and environmental samples by esculin hydrolysis, *J. Food Prot.*, 51, 762, 1988.
54. Donnelly, C.W. and Baigent, G.J. Method for flow cytometric detection of *Listeria monocytogenes* in milk, *Appl. Environ. Microbiol.*, 52, 689, 1986.
55. Brackett, R.E. and Beuchat, L.R. Methods and media for the isolation and cultivation of *Listeria monocytogenes* from various foods, *Int. J. Food Microbiol.*, 8, 219, 1989.
56. Kells, J. and Gilmour, A. Incidence of *Listeria monocytogenes* in two milk processing environments, and assessment of *Listeria monocytogenes* blood agar for isolation, *Int. J. Food Microbiol.*, 91, 167, 2004.
57. Gray, M.L., Stafseth, H.J., and Thorp, F. Jr. The use of potassium tellurite, sodium azide and acetic acid in a selective medium for the isolation of *Listeria monocytogenes, J. Bacteriol.*, 59, 443, 1950.
58. McBride, M.E. and Girard, K.F. A selective method for the isolation of *Listeria monocytogenes* from mixed bacterial populations, *J. Lab. Clin. Med.*, 55, 153, 1960.
59. Lee, W.H. and McClain, D. Improved *Listeria monocytogenes* selective agar, *Appl. Environ. Microbiol.*, 52, 1215, 1986.
60. Curtis, G.D.W., Nicholas, W.W., and Falla, T.J., Selective agents for *Listeria* can inhibit their growth, *Lett. Appl. Microbiol.*, 8, 169, 1989.
61. Lachica, R.V. Selective plating medium for quantitative recovery of food-borne *Listeria monocytogenes, Appl. Environ. Microbiol.*, 56, 167, 1990.

62. van Netten, P. et al. Liquid and solid selective differential media for the detection and enumeration of *L. monocytogenes* and other *Listeria* spp, *Int. J. Food Microbiol.*, 8, 299, 1989.
63. Dominguez-Rodriguez, L. et al. New methodology for the isolation of *Listeria* microorganisms from heavily contaminated environments, *Appl. Environ. Microbiol.*, 47, 1188, 1984.
64. Blanco, M.B. et al. A technique for the direct identification of haemolytic pathogenic *Listeria* on selective plating media, *Lett. Appl. Microbiol.*, 9, 125, 1989.
65. Buchanan, R.L. et al. Comparison of lithium chloride-phenylethanol-moxalactam and modified Vogel Johnson agars for detection of *Listeria* spp. in retail-level meats, poultry, and seafood, *Appl Environ. Microbiol.*, 55, 599, 1989.
66. Smith, J.L. and Buchanan, R.L. Identification of supplements that enhance the recovery of *Listeria monocytogenes* on modified Vogel Johnson agar, *J. Food Safety*, 10, 155, 2007.
67. Vlaemynck, G., Lafarge, V., and Scotter, S. Improvement of the detection of *Listeria monocytogenes* by the application of ALOA, a diagnostic, chromogenic isolation medium, *J. Appl. Microbiol.*, 88, 430, 2000.
68. Menudier, A., Bosiraud, C., and Nicolas, J.A. Virulence of *Listeria monocytogenes* serovars and *Listeria* spp. in experimental infection in mice, *J. Food Prot.*, 54, 917, 1991.
69. Lecuit, M. et al. A single amino acid in E-cadherin responsible for host specificity towards the human pathogen *Listeria monocytogenes, EMBO J.*, 18, 3956, 1999.
70. Mainou-Fowler, T., MacGowan, A.P., and Postlethwaite, R. Virulence of *Listeria* spp.: course of infection in resistant and susceptible mice, *J. Med. Microbiol.*, 27, 131, 1988.
71. Stelma, G.N., Jr. et al. Pathogenicity test for *Listeria monocytogenes* using immunocompromised mice, *J. Clin. Microbiol.*, 25, 2085, 1987.
72. Tabouret, M. et al. Pathogenicity of *Listeria monocytogenes* isolates in immunocompromised mice in relation to listeriolysin production, *J. Med. Microbiol.*, 34, 13, 1991.
73. Bracegirdle, P. et al. A comparison of aerosol and intragastric routes of infection with *Listeria* spp., *Epidemiol. Infect.*, 112, 69, 1994.
74. Liu, D. *Listeria monocytogenes*: comparative interpretation of mouse virulence assay, *FEMS Microbiol. Lett.*, 233, 159, 2004.
75. Schlech, W.F., III, Chase, D.P., and Badley, A. A model of food-borne *Listeria monocytogenes* infection in the Sprague-Dawley rat using gastric inoculation: development and effect of gastric acidity on infective dose, *Int. J. Food Microbiol.*, 18, 15, 1993.
76. Notermans, S., et al. The chick embryo test agrees with the mouse bioassay for assessment of the pathogenicity of *Listeria* species, *Lett. Appl. Microbiol.*, 161, 1991.
77. Terplan, G. and Steinmeyer, S. Investigations on the pathogenicity of *Listeria* spp. by experimental infection of the chick embryo, *Int. J. Food Microbiol.*, 8, 277, 1989.
78. Schonberg, A. Method to determine virulence of *Listeria* strains, *Int. J. Food Microbiol.*, 8, 281, 1989.
79. Anton, W. Kritisch-experimenteller beitrag zur biologie des *Bacterium monocytogenes*, *Zentralbl. Bakteriol. Parasitenkd. Infektionskr.*, 131, 89, 1934.
80. Farber, J.M. et al. Feeding trials of *Listeria monocytogenes* with a nonhuman primate model, *J. Clin. Microbiol.*, 29, 2606, 1991.
81. Dons, L. et al. Rat dorsal root ganglia neurons as a model for *Listeria monocytogenes* infections in culture, *Med. Microbiol. Immunol. (Berlin),* 188, 15, 1999.
82. Mansfield, B.E. et al. Exploration of host-pathogen interactions using *Listeria monocytogenes* and *Drosophila melanogaster, Cell Microbiol.*, 5, 901, 2003.
83. Pine, L. et al. Cytopathogenic effects in enterocytelike Caco-2 cells differentiate virulent from avirulent *Listeria* strains, *J. Clin. Microbiol.*, 29, 990, 1991.
84. Roche, S.M. et al. Assessment of the virulence of *Listeria monocytogenes*: agreement between a plaque-forming assay with HT-29 cells and infection of immunocompetent mice, *Int. J. Food Microbiol.*, 68, 33, 2001.
85. Chatterjee, S.S. et al. Invasiveness is a variable and heterogeneous phenotype in *Listeria monocytogenes* serotype strains, *Int. J. Med. Microbiol.*, 296, 277, 2006.
86. Van Langendonck, N. et al. Tissue culture assays using Caco-2 cell line differentiate virulent from non-virulent *Listeria monocytogenes* strains, *J. Appl. Microbiol.*, 85, 337, 1998.
87. Nishibori, T. et al. Correlation between the presence of virulence-associated genes as determined by PCR and actual virulence to mice in various strains of *Listeria* spp., *Microbiol. Immunol.*, 39, 343, 1995.

88. Jaradat, Z.W., Schutze, G.E., and Bhunia, A.K. Genetic homogeneity among *Listeria monocytogenes* strains from infected patients and meat products from two geographic locations determined by phenotyping, ribotyping and PCR analysis of virulence genes, *Int. J. Food Microbiol.*, 76, 1, 2002.
89. Roberts, A., Chan, Y., and Wiedmann, M. Definition of genetically distinct attenuation mechanisms in naturally virulence-attenuated *Listeria monocytogenes* by comparative cell culture and molecular characterization, *Appl. Environ. Microbiol.*, 71, 3900, 2005.
90. Roche, S.M. et al. Investigation of specific substitutions in virulence genes characterizing phenotypic groups of low-virulence field strains of *Listeria monocytogenes, Appl. Environ. Microbiol.*, 71, 6039, 2005.
91. Olapedo, D.K. Detection of *Listeria monocytogenes* using polyclonal antibody, *Lett. Appl. Microbiol.*, 14, 26, 1992.
92. Beumer, R.R. Detection of *Listeria* spp. with a monoclonal antibody based ELISA, *Food Microbiol.*, 6, 171, 1989.
93. Farber, J.M. and Speirs, J.I. Monoclonal antibodies directed against the flagellar antigens of *Listeria* spp. and their potential in EIA based methods, *J. Food Prot.*, 50, 479, 1987.
94. Mattingly, J.A. et al. Rapid monoclonal antibody-based enzyme-linked immunosorbent assay for detection of *Listeria* in food products, *J. Assoc. Off. Anal. Chem.*, 71, 679, 1988.
95. McLauchlin, J. et al. Monoclonal antibodies show *Listeria monocytogenes* in necropsy tissue samples, *J. Clin. Pathol.*, 41, 983, 1988.
96. Walker, S.J., Archer, P., and Aleyard, J. Comparison of the ListeriaTek ELISA kit with cultural procedures for the detection of *Listeria* species in foods, *Food Microbiol.*, 7, 335, 1990.
97. Norrung, B. et al. Evaluation of an ELISA test for the detection of *Listeria* spp., *J. Food Prot.*, 54, 752, 1991.
98. Curiale, M.S., Lepper, W., and Robison, B. Enzyme-linked immunoassay for detection of *Listeria monocytogenes* in dairy products, seafoods, and meats: collaborative study, *J. AOAC Int.*, 77, 1472, 1994.
99. Sewell, A.M. et al. The development of an efficient and rapid enzyme linked fluorescent assay method for the detection of *Listeria* spp. from foods, *Int. J. Food Microbiol.*, 81, 123, 2003.
100. Silbernagel, K.M. et al. Evaluation of the VIDAS *Listeria* (LIS) immunoassay for the detection of *Listeria* in foods using demi-Fraser and Fraser enrichment broths, as modification of AOAC Official Method 999.06 (AOAC Official Method 2004.06), *J. AOAC Int.*, 88, 750, 2005.
101. Solve, M., Boel, J., and Norrung, B. Evaluation of a monoclonal antibody able to detect live *Listeria monocytogenes* and *Listeria innocua, Int. J. Food Microbiol.*, 57, 219, 2000.
102. Durham, R.J. et al. A monoclonal antibody enzyme immunoassay (ELISA) for the detection of *Listeria* in foods and environmental samples, in *Foodborne Listeriosis,* Miller, A.J., Smith, J.L., and Somkuti, G.A. Eds., Elsevier Science Publishers, New York, 1990, 105.
103. Knight, M.T. et al. TECRA *Listeria* Visual Immunoassay (TLVIA) for detection of *Listeria* in foods: collaborative study, *J. AOAC Int.*, 79, 1083, 1996.
104. Kim, S.H. et al. Development of a sandwich ELISA for the detection of *Listeria* spp. using specific flagella antibodies, *J. Vet. Sci.*, 6, 41, 2005.
105. Skjerve, E., Bos, W., and van der, Guag, B. Evaluation of monoclonal antibodies to *Listeria monocytogenes* flagella by checkerboard ELISA and cluster analysis, *J. Immunol. Methods*, 144, 11, 1991.
106. Palumbo, J.D. et al. Serotyping of *Listeria monocytogenes* by enzyme-linked immunosorbent assay and identification of mixed-serotype cultures by colony immunoblotting, *J. Clin. Microbiol.*, 41, 564, 2003.
107. Bourry, A., Cochard, T., and Poutrel, B. Serological diagnosis of bovine, caprine, and ovine mastitis caused by *Listeria monocytogenes* by using an enzyme-linked immunosorbent assay, *J. Clin. Microbiol.*, 35, 1606, 1997.
108. Allerberger, F. *Listeria*: growth, phenotypic differentiation and molecular microbiology, *FEMS, Immunol. Med. Microbiol.*, 35, 183, 2003.
109. Khan, M.A., Seaman, A., and Woodbine, M. Immunofluorescent identification of *Listeria monocytogenes, Zentralbl. Bakteriol. [Orig.A]*, 239, 62, 1977.
110. McLauchlin, J., Ridley, A.M., and Taylor, A.G. The use of monoclonal antibodies against *Listeria monocytogenes* in a direct immunofluorescence technique for the rapid presumptive identification and direct demonstration of *Listeria* in food, *Acta Microbiol. Hung.*, 36, 467, 1989.
111. Ruhland, G.J. et al. Cell-surface location of *Listeria*-specific protein p60 — detection of *Listeria* cells by indirect immunofluorescence, *J. Gen. Microbiol.*, 139, 609, 1993.

112. Czuprynski, C.J. Host defence against *Listeria monocytogenes*: implications for food safety, *Food Microbiol.*, 11, 131, 1994.
113. Clough, N.E. and Roth, J.A. Methods for assessing cell-mediated immunity in infectious disease resistance and in the development of vaccines, *J. Am. Vet. Med. Assoc.*, 206, 1208, 1995.
114. Klink, M. et al. Specific cellular and humoral reactions as markers of *Listeria monocytogenes* infections, *Acta Microbiol. Pol.*, 43, 335, 1994.
115. Kaufmann, S.H.E. CD8+ lymphocytes in intracellular microbial infections, *Immunol. Today*, 9, 168, 1988.
116. Nakane, A. et al. Evidence that endogenous gamma interferon is produced early in *Listeria monocytogenes* infection, *Infect. Immun.*, 58, 2386, 1990.
117. Barbuddhe, S.B. et al. Kinetics of interferon-gamma production and its comparison with anti-listeriolysin O detection in experimental bovine listeriosis, *Vet. Res. Commun.*, 22, 505, 1998.
118. Datta, A.R., Wentz, B.A., and Hill, W.E. Identification and enumeration of beta-hemolytic *Listeria monocytogenes* in naturally contaminated dairy products, *J. Assoc. Off. Anal. Chem.*, 71, 673, 1988.
119. King, W. et al. A new colorimetric nucleic acid hybridization assay for *Listeria* in foods *Int. J. Food Microbiol.*, 8, 225, 1989.
120. Kim, C. et al. Rapid confirmation of *Listeria monocytogenes* isolated from foods by a colony blot assay using a digoxigenin-labeled synthetic oligonucleotide probe, *Appl. Environ. Microbiol.*, 57, 1609, 1991.
121. Bobbit, J.A. and Betts, R.P. Confirmation of *Listeria monocytogenes* using a commercially available nucleic acid probe, *Food Microbiol.*, 9, 311, 1992.
122. Volokhov, D. et al. Identification of *Listeria* species by microarray-based assay, *J. Clin. Microbiol.*, 40, 4720, 2002.
123. Ingianni, A. et al. Rapid detection of *Listeria monocytogenes* in foods, by a combination of PCR and DNA probe, *Mol. Cell Probes*, 15, 275, 2001.
124. Wernars, K. et al. Suitability of the *prfA* gene, which encodes a regulator of virulence genes in *Listeria monocytogenes*, in the identification of pathogenic *Listeria* spp, *Appl. Environ. Microbiol.*, 58, 765, 1992.
125. Wagner, M. et al. *In situ* detection of a virulence factor mRNA and *16S rRNA* in *Listeria monocytogenes, FEMS Microbiol. Lett.*, 160, 159, 1998.
126. Mengaud, J. et al. Expression in *Escherichia coli* and sequence analysis of the listeriolysin O determinant of *Listeria monocytogenes, Infect. Immun.*, 56, 766, 1988.
127. Jaton, K., Sahli, R., and Bille, J. Development of polymerase chain reaction assays for detection of *Listeria monocytogenes* in clinical cerebrospinal fluid samples, *J. Clin. Microbiol.*, 30, 1931, 1992.
128. Rossen, L. et al. A rapid polymerase chain reaction (PCR)-based assay for the identification of *Listeria monocytogenes* in food samples, *Int. J. Food Microbiol.*, 14, 145, 1991.
129. Niederhauser, C. et al. Use of polymerase chain reaction for detection of *Listeria monocytogenes* in food, *Appl. Environ. Microbiol.*, 58, 1564, 1992.
130. Torriani, S. and Pallotta, M.L. Use of polymerase chain reaction to detect *Listeria monocytogenes, Biotechnol. Tech.*, 8, 157, 1994.
131. Bohnert, M. et al. Use of specific oligonucleotides for direct enumeration of *Listeria monocytogenes* in food samples by colony hybridization and rapid detection by PCR, *Res. Microbiol.*, 143, 271, 1992.
132. Wang, R.F., Cao, W.W., and Johnson, M.G. *16S rRNA*-based probes and polymerase chain reaction method to detect *Listeria monocytogenes* cells added to foods, *Appl. Environ. Microbiol.*, 58, 2827, 1992.
133. Herman, L.M., De Block, J.H., and Moermans, R.J. Direct detection of *Listeria monocytogenes* in 25 milliliters of raw milk by a two-step PCR with nested primers, *Appl. Environ. Microbiol.*, 61, 817, 1995.
134. Aznar, R. and Alarcon, B. On the specificity of PCR detection of *Listeria monocytogenes* in food: a comparison of published primers, *Syst. Appl. Microbiol.*, 25, 109, 2002.
135. Novak, J.S. and Juneja, V.K. Detection of heat injury in *Listeria monocytogenes* Scott A., *J. Food Prot.*, 64, 1739, 2001.
136. Keer, J,T. and Birch, L. Molecular methods for the assessment of bacterial viability, *J. Microbiol. Methods*, 53, 175, 2003.
137. Burtscher, C. and Wuertz, S. Evaluation of the use of PCR and reverse transcriptase PCR for detection of pathogenic bacteria in biosolids from anaerobic digestors and aerobic composters, *Appl. Environ. Microbiol.*, 69, 4618, 2003.

138. Koo, K. and Jaykus, L.A. Selective amplification of bacterial RNA: use of a DNA primer containing mismatched bases near its 30 terminus to reduce false-positive signals, *Lett. Appl. Microbiol.,* 31, 187, 2000.
139. Hein, I. et al. Detection and quantification of the *iap* gene of *Listeria monocytogenes* and *Listeria innocua* by a new real-time quantitative PCR assay, *Res. Microbiol.*, 152, 37, 2001.
140. Hough, A.J. et al. Rapid enumeration of *Listeria monocytogenes* in artificially contaminated cabbage using real-time polymerase chain reaction, *J. Food Prot.*, 65, 1329, 2002.
141. Bhagwat, A.A. Simultaneous detection of *Escherichia coli O157:H7, Listeria monocytogenes* and *Salmonella* strains by real-time PCR, *Int. J. Food Microbiol.*, 84, 217, 2003.
142. Guilbaud, M. et al. Quantitative detection of *Listeria monocytogenes* in biofilms by real-time PCR, *Appl. Environ. Microbiol.*, 71, 2190, 2005.
143. Uyttendaele, M. et al. Development of NASBA, a nucleic acid amplification system, for identification of *Listeria monocytogenes* and comparison to ELISA and a modified FDA method, *Int. J. Food Microbiol.*, 27, 77, 1995.
144. Hancock, I.B., Bointon, M, and McAthey, P. Rapid detection of *Listeria* spp. by selective impedimetric assay, *Lett. Appl. Microbiol.*, 16, 311, 1993.
145. Skjerve, E., Rorvik, L.M., and Olsvik, O. Detection of *Listeria monocytogenes* in foods by immunomagnetic separation, *Appl. Environ. Microbiol.*, 56, 3478, 1990.
146. Lazcka, O., Campo, F.J., Munoz, F.X. Pathogen detection: a perspective of traditional methods and biosensors, *Biosens Bioelectron.*, 22, 1205, 2007.
147. Graves, L.M., Swaminathan, B., and Hunter, S.B. Subtyping *Listeria monocytogenes*, in *Listeria, Listeriosis, and Food Safety*, Ryser, E. and Marth, E., Eds., Marcel Dekker, New York, 1999, 279.
148. Wiedmann, M. Molecular subtyping methods for *Listeria monocytogenes, J. AOAC Int.*, 85, 524, 2002.
149. Schonberg, A. et al. Serotyping of 80 strains from the WHO multicentre international typing study of *Listeria monocytogenes, Int. J. Food Microbiol.*, 32, 279, 1996.
150. Bruce, J.L. et al. Sets of *EcoRI* fragments containing ribosomal RNA sequences are conserved among different strains of *Listeria monocytogenes, Proc. Natl. Acad. Sci. U.S.A.*, 92, 5229, 1995.
151. Graves, L.M. and Swaminathan, B. PulseNet standardized protocol for subtyping *Listeria monocytogenes* by macrorestriction and pulsed-field gel electrophoresis, *Int. J. Food Microbiol.*, 65, 55, 2001.
152. Jeffers, G.T. et al. Comparative genetic characterization of *Listeria monocytogenes* isolates from human and animal listeriosis cases, *Microbiology*, 147, 1095, 2001.
153. Katzav, M. et al. Pulsed-field gel electrophoresis typing of *Listeria monocytogenes* isolated in two Finnish fish farms, *J. Food Prot.*, 69, 1443, 2006.
154. Seeliger, H.P.R. and Hohne, K. Serotyping of *Listeria monocytogenes* and related species, *Methods Microbiol.*, 13, 31, 1979.
155. Seeliger, H.P.R. and Jones, D. *Listeria*. in *Bergey's Manual of Systematic Bacteriology*, Vol. 2, Sneath, P.H.A. et al., Eds., Baltimore: Williams and Wilkins, 1986, 1235.
156. Jacquet, C. et al. Expression of *ActA, Ami, InlB*, and listeriolysin O in *Listeria monocytogenes* of human and food origin, *Appl. Environ. Microbiol.*, 68, 616, 2002.
157. Wiedmann, M. et al. Ribotype diversity of *Listeria monocytogenes* strains associated with outbreaks of listeriosis in ruminants, *J. Clin. Microbiol.,* 34, 1086, 1996.
158. Borucki, M.K. and Call, D.R. *Listeria monocytogenes serotype* identification by PCR, *J. Clin. Microbiol.*, 41, 5537, 2003.
159. Doumith, M. et al. Differentiation of the major *Listeria monocytogenes* serovars by multiplex PCR, *J. Clin. Microbiol.*, 42, 3819, 2004.
160. Audurier, A. et al. A phage typing system for *Listeria monocytogenes* and its use in epidemiological studies, *Clin. Invest. Med.*, 7, 229, 1984.
161. Bibb, W.F. et al. Analysis of clinical and food-borne isolates of *Listeria monocytogenes* in the United States by multilocus enzyme electrophoresis and application of the method to epidemiologic investigations, *Appl. Environ. Microbiol.*, 56, 2133, 1990.
162. Piffaretti, J.C. et al. Genetic characterization of clones of the bacterium *Listeria monocytogenes* causing epidemic disease, *Proc. Natl. Acad. Sci. U.S.A.*, 86, 3818, 1989.
163. Graves, L.M. et al. Comparison of ribotyping and multilocus enzyme electrophoresis for subtyping of *Listeria monocytogenes* isolates, *J. Clin. Microbiol.*, 32, 2936, 1994.
164. Wiedmann, M. et al. Ribotypes and virulence gene polymorphisms suggest three distinct *Listeria monocytogenes* lineages with differences in pathogenic potential, *Infect. Immun.*, 65, 2707, 1997.

165. Manfreda, G. et al. Occurrence and ribotypes of *Listeria monocytogenes* in Gorgonzola cheeses, *Int. J. Food Microbiol.*, 102, 287, 2005.
166. Lawrence, L.M. and Gilmour, A. Characterization of *Listeria monocytogenes* isolated from poultry products and from the poultry-processing environment by random amplification of polymorphic DNA and multilocus enzyme electrophoresis, *Appl. Environ. Microbiol.*, 61, 2139, 1995.
167. Aguado, V., Vitas, A.I., and Garcia-Jalon, I. Characterization of *Listeria monocytogenes* and *Listeria innocua* from a vegetable processing plant by RAPD and REA, *Int. J. Food Microbiol.*, 90, 341, 2004.
168. Aarts, H.J., Hakemulder, L.E., and Van Hoef, A.M. Genomic typing of *Listeria monocytogenes* strains by automated laser fluorescence analysis of amplified fragment length polymorphism fingerprint patterns, *Int. J. Food Microbiol.*, 49, 95, 1999.
169. Vogel, B.F. et al. 2004. High-resolution genotyping of *Listeria monocytogenes* by fluorescent amplified fragment length polymorphism analysis compared to pulsed-field gel electrophoresis, random amplified polymorphic DNA analysis, ribotyping, and PCR-restriction fragment length polymorphism analysis, *J. Food Prot.,* 67, 1656, 2004.
170. Guerra, M.M., Bernardo, F., and McLauchlin, J. Amplified fragment length polymorphism (AFLP) analysis of *Listeria monocytogenes, Syst. Appl. Microbiol.*, 25, 456, 2002.
171. Keto-Timonen, R.O., Autio, T.J., and Korkeala, H.J. An improved amplified fragment length polymorphism (AFLP) protocol for discrimination of *Listeria* isolates, *Syst. Appl. Microbiol.*, 26, 236, 2003.
172. Paillard, D. et al. Rapid identification of *Listeria* species by using restriction fragment length polymorphism of PCR-amplified *23S rRNA* gene fragments, *Appl. Environ. Microbiol.*, 69, 6386, 2003.
173. Jersek, B. et al. Typing of *Listeria monocytogenes* strains by repetitive element sequence-based PCR, *J. Clin. Microbiol.*, 37, 103, 1999.
174. Chan, M.S., Maiden, M.C., and Spratt, B.G. Database-driven multi locus sequence typing (MLST) of bacterial pathogens, *Bioinformatics*, 17, 1077, 2001.
175. Enright, M.C. et al. Multilocus sequence typing of *Streptococcus pyogenes* and the relationships between emm type and clone, *Infect. Immun.*, 69, 2416, 2001.
176. Spratt, B.G. Multilocus sequence typing: molecular typing of bacterial pathogens in an era of rapid DNA sequencing and the internet, *Curr. Opin. Microbiol.*, 2, 312, 1999.
177. Doumith, M. et al. New aspects regarding evolution and virulence of *Listeria monocytogenes* revealed by comparative genomics and DNA arrays, *Infect. Immun.*, 72, 1072, 2004.
178. Borucki, M.K. et al. Discrimination among *Listeria monocytogenes* isolates using a mixed genome DNA microarray, *Vet. Microbiol.*, 92, 351, 2003.
179. Rudi, K., Katla, T., and Naterstad, K. Multi locus fingerprinting of *Listeria monocytogenes* by sequence-specific labeling of DNA probes combined with array hybridisation, *FEMS Microbiol. Lett.*, 220, 9, 2003.
180. Call, D.R., Borucki, M.K., and Besser, T.E. Mixed-genome microarrays reveal multiple serotype and lineage-specific differences among strains of *Listeria monocytogenes, J. Clin. Microbiol.*, 41, 632, 2003.
181. Zhang, C. et al. Genome diversification in phylogenetic lineages I and II of *Listeria monocytogenes*: identification of segments unique to lineage II populations, *J. Bacteriol.*, 185, 5573, 2003.
182. Johnson, J. et al. Natural atypical *Listeria innocua* strains with *Listeria monocytogenes* pathogenicity island 1 genes, *Appl. Environ. Microbiol.*, 70, 4256, 2004.
183. Glaser, P. et al. Comparative genomics of *Listeria* species, *Science*, 294, 849, 2001.
184. Nelson, K.E. et al. Whole genome comparisons of serotype 4b and 1/2a strains of the food-borne pathogen *Listeria monocytogenes* reveal new insights into the core genome components of this species, *Nucleic Acids Res.*, 32, 2386, 2004.
185. Hain, T. et al. Whole-genome sequence of *Listeria welshimeri* reveals common steps in genome reduction with *Listeria innocua* as compared to *Listeria monocytogenes, J. Bacteriol.*, 188, 7405, 2006.
186. Jones, E.M. and MacGowan, A.P. Antimicrobial chemotherapy of human infection due to *Listeria monocytogenes, Eur. J. Clin. Microbiol. Infect. Dis.*, 14, 165, 1995.
187. Scortti, M. et al. Coexpression of virulence and fosfomycin susceptibility in *Listeria*: molecular basis of an antimicrobial *in vitro-in vivo* paradox, *Nat. Med.*, 12, 515, 2006.
188. Moellering, Jr., R.C., et al. Synergy of penicillin and gentamicin against *Enterococci, J. Infect. Dis.*, 124, S207, 1971.
189. Biavasco, F. et al. *In vitro* conjugative transfer of *VanA* vancomycin resistance between *Enterococci* and Listeriae of different species, *Eur. J. Clin. Microbiol. Infect. Dis.*, 15, 50, 1996.

190. Charpentier, E. and Courvalin, P. Emergence of the trimethoprim resistance gene *dfrD* in *Listeria monocytogenes* BM4293, *Antimicrob. Agents Chemother.*, 41, 1134, 1997.
191. Poyart-Salmeron, C. et al. Transferable plasmid-mediated antibiotic resistance in *Listeria monocytogenes, Lancet*, 335, 1422, 1990.
192. Roberts, M.C. et al. Transferable erythromycin resistance in *Listeria* spp. isolated from food, *Appl. Environ. Microbiol.*, 62, 269, 1996.
193. Singh, R. and Paterson, Y. *Listeria monocytogenes* as a vector for tumor-associated antigens for cancer immunotherapy, *Expert. Rev. Vaccines*, 5, 541, 2006.
194. Singh, R. et al. Fusion to listeriolysin O and delivery by *Listeria monocytogenes* enhances the immunogenicity of HER-2/neu and reveals subdominant epitopes in the FVB/N mouse, *J. Immunol.*, 175, 3663, 2005.
195. Mandal, M. et al. Cytosolic delivery of viral nucleoprotein by listeriolysin O-liposome induces enhanced specific cytotoxic T lymphocyte response and protective immunity, *Mol. Pharm.*, 1, 2, 2004.
196. Paterson, Y. and Johnson, R.S. Progress towards the use of *Listeria monocytogenes* as a live bacterial vaccine vector for the delivery of HIV antigens, *Expert. Rev. Vaccines*, 3, S119, 2004.
197. Wesley, I.V. et al. Application of a multiplex polymerase chain reaction assay for the simultaneous confirmation of *Listeria monocytogenes* and other *Listeria* species in turkey sample surveillance, *J. Food Prot.,* 65, 780, 2002.
198. Somer, L. and Kashi, Y. A PCR method based on *16S rRNA* sequence for simultaneous detection of the genus *Listeria* and the species *Listeria monocytogenes* in food products, *J. Food Prot.*, 66, 1658, 2003.
199. Sallen, B. et al. Comparative analysis of *16S* and *23S rRNA* sequences of *Listeria* species, *Int. J. Syst. Bacteriol.*, 46, 669, 1996.
200. Graham, T.A. et al. Inter- and intraspecies comparison of the *16S-23S rRNA* operon intergenic spacer regions of six *Listeria* spp., *Int. J. Syst. Bacteriol.*, 47, 863, 1997.
201. O'Connor, L, et al. Rapid polymerase chain reaction/DNA probe membrane-based assay for the detection of *Listeria* and *Listeria monocytogenes* in food, *J. Food Prot.,* 63, 337, 2000.
202. Furrer, B. et al. Detection and identification of *Listeria monocytogenes* in cooked sausage products and in milk by *in vitro* amplification of haemolysin gene fragments, *J. Appl. Bacteriol.*, 70, 372, 1991.
203. Lunge, V.R. et al. Factors affecting the performance of 5′ nuclease PCR assays for *Listeria monocytogenes* detection, *J. Microbiol. Methods*, 51, 361, 2002.
204. Thimothe, J. et al. Tracking of *Listeria monocytogenes* in smoked fish processing plants, *J.Food Prot.*, 67, 328, 2004.
205. Paziak-Domanska, B. Evaluation of the API test, phosphatidylinositol-specific phospholipase C activity and PCR method in identification of *Listeria monocytogenes* in meat foods, *FEMS Microbiol. Lett.,* 171, 209, 1999.
206. Greer, G.G. Bacterial contamination of recirculating brine used in the commercial production of moisture-enhanced pork, *J. Food Prot.*, 67, 185, 2004.
207. Longhi, C. et al. Detection of *Listeria monocytogenes* in Italian-style soft cheeses, *J. Appl. Microbiol.*, 94, 879, 2003.
208. D'Agostino, M. A validated PCR-based method to detect *Listeria monocytogenes* using raw milk as a food model—towards an international standard, *J. Food Prot.,* 67, 1646, 2004.
209. Liu, D. et al. Characterization of virulent and avirulent *Listeria monocytogenes* strains by PCR amplification of putative transcriptional regulator and internalin genes, *J. Med. Microbiol.*, 52, 1065, 2003.
210. Jung, Y.S. et al. Polymerase chain reaction detection of *Listeria monocytogenes* on frankfurters using oligonucleotide primers targeting the genes encoding internalin AB, *J. Food Prot.*, 66, 237, 2003.
211. Pangallo, D. et al. Detection of *Listeria monocytogenes* by polymerase chain reaction oriented to *inlB* gene, *New Microbiol.*, 24, 333, 2001.
212. Cocolin, L. et al. Direct identification in food samples of *Listeria* spp. and *Listeria monocytogenes* by molecular methods, *Appl. Environ. Microbiol.*, 68, 6273, 2002.
213. Schmid, M. et al. Nucleic acid-based, cultivation-independent detection of *Listeria* spp and genotypes of *L monocytogenes, FEMS Immunol. Med. Microbiol.*, 35, 215, 2003.
214. Johnson, W.M. et al. Detection of genes coding for listeriolysin and *Listeria monocytogenes* antigen A (*ImaA*) in *Listeria* spp. by the polymerase chain reaction, *Microb.Pathog.*, 12, 79, 1992.
215. Wernars, K. Use of the polymerase chain reaction for direct detection of *Listeria monocytogenes* in soft cheese, *J. Appl. Bacteriol.,* 70, 121, 1991.
216. Fluit, A.C. et al. Detection of *Listeria monocytogenes* in cheese with the magnetic immuno-polymerase chain reaction assay, *Appl. Environ. Microbiol.,* 59, 1289, 1993.

217. Gilot, P. and Content, J. Specific identification of *Listeria welshimeri* and *Listeria monocytogenes* by PCR assays targeting a gene encoding a fibronectin-binding protein, *J. Clin. Microbiol.*, 40, 698, 2002.
218. Gray, D.I. and Kroll, R.G. Polymerase chain reaction amplification of the *flaA* gene for the rapid identification of *Listeria* spp, *Lett. Appl. Microbiol.*, 20, 65, 1995.
219. Winters, D.K.. Maloney, T.P., and Johnson, M.G. Rapid detection of *Listeria monocytogenes* by a PCR assay specific for an aminopeptidase, *Mol. Cell Probes,* 13, 127, 1999.
220. Moorhead, S.M., Dykes, G.A., and Cursons, R.T. An SNP-based PCR assay to differentiate between *Listeria monocytogenes* lineages derived from phylogenetic analysis of the sigB gene, *J. Microbiol. Methods*, 55, 425, 2003.
221. Nightingale, K. et al. Combined *sigB* allelic typing and multiplex PCR provide improved discriminatory power and reliability for *Listeria monocytogenes* molecular serotyping, *J. Microbiol. Methods*, 68, 52, 2007.
222. Cabanes, D. et al. Gp96 is a receptor for a novel *Listeria monocytogenes* virulence factor, Vip, a surface protein, *EMBO J.,* 24, 2827, 2005.
223. Liu, D. et al. Species-specific PCR determination of *Listeria seeligeri, Res. Microbiol.*, 155, 741, 2004.
224. Rodriguez-Lazaro, D. et al. Quantitative detection of *Listeria monocytogenes* and *Listeria innocua* by real-time PCR: assessment of *hly, iap*, and *lin02483* targets and AmpliFluor technology, *Appl. Environ. Microbiol.,* 70, 1366, 2004.
225. Liu, D. et al. PCR detection of a putative N-acetylmuramidase gene from *Listeria ivanovii* facilitates its rapid identification, *Vet. Microbiol.*, 101, 83, 2004.
226. Loessner, M.J. and Busse, M. Bacteriophage typing of *Listeria* species, *Appl. Environ. Microbiol.*, 56, 1912, 1990.
227. Capita, R. et al. Evaluation of the international phage typing set and some experimental phages for typing of *Listeria monocytogenes* from poultry in Spain. *J. Appl. Microbiol.*, 92, 90, 2002.
228. Harvey, J. and Gilmour, A. Characterization of *Listeria monocytogenes* isolates by esterase electrophoresis, *Appl. Environ. Microbiol.*, 62, 1461, 1996.
229. Aarnisalo, K. et al. Typing of *Listeria monocytogenes* isolates originating from the food processing industry with automated ribotyping and pulsed-field gel electrophoresis, *J. Food Protect.,* 66, 249, 2003.
230. Suihko, M.L. et al. Characterization of *Listeria monocytogenes* isolates from the meat, poultry and seafood industries by automated ribotyping, *Int. J. Food Microbiol.,* 30, 137, 2002.
231. Nadon, C. A. et al. Correlations between molecular subtyping and serotyping of *Listeria monocytogenes, J. Clin. Microbiol.*, 39, 2704, 2001.

35 The Genus *Campylobacter*

*Collette Fitzgerald, Jean Whichard, and Patricia I. Fields**

CONTENTS

TAXONOMIC HISTORY OF CAMPYLOBACTERS

Spiral campylobacter-like organisms were first observed microscopically from the stool of children by Escherich in 1886. Between 1909 and 1944, there was a growing number of reports of similar "vibrio-like" organisms isolated from bovine and ovine sources, but they were not isolated from humans until 1938, in association with a milk-borne outbreak of gastroenteritis where blood cultures were positive for organisms resembling "*Vibrio jejuni.*"[14] The microaerobic vibrios were assigned to the new genus *Campylobacter* in 1963,[79] and included just two species: *Campylobacter fetus* and *Campylobacter bubulus* (now *Campylobacter sputorum*). Campylobacters were first successfully isolated from stool in the late 1960s using a filtration technique.[14] Later, the development of selective media brought the routine isolation of *Campylobacter* into the clinical microbiology setting and *Campylobacter* spp. rapidly became recognized as a common cause of bacterial gastroenteritis. The taxonomic structure of *Campylobacter* has changed substantially since its inception in 1963, first with a comprehensive study of the taxonomy of the genus in 1973.[86] By the late 1980s there was a rapid increase in classification at the species level, and 14 species had been described (see review by Penner[68]).

Advances in DNA technology, particularly 16S rRNA sequence analysis and DNA-DNA hybridization,[91] clarified the systematics of campylobacters and resulted in the extensive restructuring of the genus and assignment of campylobacters to the phylogenetic group rRNA superfamily VI.[90] The proposal that the genera *Campylobacter* and *Arcobacter* should be included in a separate family, the Campylobacteraceae, was also made in 1991 by Vandamme and Levy.[89] As of 2006, 17 *Campylobacter* species and 6 subspecies have been described (Table 35.1). The disease spectrum of this genus ranges from well established for *C. jejuni,* to emerging for species such as *C. fetus* ssp. *fetus,* to speculative for species such as *C. concisus.*

* Disclaimer: Use of trade names is for identification only and does not imply endorsement by the Public Health Service or by the U.S. Department of Health and Human Services.

TABLE 35.1
Currently Described *Campylobacter* Species

Campylobacter Species	Disease Association	
	Known Sources	Human
C. jejuni ssp. *jejuni*	Poultry, cattle Sheep, birds, pigs	Gastroenteritis, Meningitis Septicemia, GBS
C. jejuni ssp. *doylei*	Humans	Gastroenteritis, Septicemia
C. coli	Pigs, poultry, sheep Birds	Gastroenteritis Septicemia
C. lari	Birds, poultry, dogs Cats, water	Gastroenteritis Septicemia
C. fetus ssp. *fetus*	Cattle, sheep	Gastroenteritis, Septicemia Abortion, meningitis
C. fetus ssp. *venerealis*	Cattle	Septicemia
C. upsaliensis	Dogs, cats	Gastroenteritis Septicemia
C. hyointestinalis ssp. *hyointestinalis*	Pigs, cattle	Gastroenteritis
C. hyointestinalis ssp. *lawsonii*	Pigs	None at present
C. lanienae	Humans, pigs	None at present
C. sputorum bv. *sputorum*	Cattle, pigs	Abscesses Gastroenteritis
C. sputorum bv. *faecalis*	Sheep, bulls	None at present
C. sputorum bv. *paraureolyticus*	Cattle	Gastroenteritis
C. helveticus	Cats, dogs	None at present
C. hominis	Humans	None at present
C. mucosalis	Pigs	None at present
C. concisus	Humans	Periodontal disease Gastroenteritis
C curvus	Humans	Periodontal disease Gastroenteritis
C. rectus	Humans	Periodontal disease
C. showae	Humans	Periodontal disease
C. gracilis	Humans	Periodontal disease Abscesses
C. insulaenigrae	Marine mammals	None at present

The family Campylobacteraceae also contains the genetically misclassified species *Bacteroides ureolyticus* and strains originally described as *Sulfurospirillum* species.[78, 87] Species belonging to the genus *Sulfurospirillum* have no known pathogenicity for humans or animals, and are considered environmental organisms.

The most commonly recognized *Campylobacter* spp., *C. jejuni*, is the leading cause of *Campylobacter* enteritis cases in humans; a number of other *Campylobacter* species, including *C. coli, C. upsaliensis, C. lari,* and *C. fetus,* have also been linked with enteritis (see Table 35.1). The true proportion of illness attributable to each of these species is not clear because campylobacters are difficult to differentiate and many clinical laboratories do not identify *Campylobacter* isolates to the species level. *C. jejuni* is divided into two subspecies: *C. jejuni* ssp. *jejuni* (referred to as *C. jejuni* and the main focus of the chapter) and *C. jejuni* ssp. *doylei* (seldom found as the cause of human disease). *C. jejuni, C. coli* and *C. lari, C. upsaliensis,* and *C. helveticus* form a genetically close group as shown by 16S rRNA analysis[62] and are known as the thermotolerant or thermophilic campylobacters because they all grow optimally at 42°C.

Another group that is closely related genetically is the hydrogen-requiring species *Campylobacter sputorum, C. concisus, C. curvus, C. showae, C. rectus, C. gracilis,* and the more recently proposed *C. hominis*. Most of these species have been associated with periodontal disease, although *C. hominis* has been isolated only from healthy human feces[52] and *C. sputorum* is also found in the reproductive and gastrointestinal tracts of various production animals. The taxonomic history of *C. sputorum* is complex and was revised in 1998 to include three biovars[64]: bv. *sputorum*, bv. *faecalis,* and bv. *paraureolyticus* based on the ability of a given strain to produce catalase or urease. Strains previously described as C. *bubulus* were reclassified as bv. *sputorum* as the tests that had been used to differentiate these two taxa were not reproducible.[63]

Campylobacter mucosalis and *C. sputorum* are closely related based on DNA-DNA hybridization studies, and they share a common source (porcine).[62] Because *C. mucosalis* strains require a hydrogen-enriched microaerobic atmosphere in order to grow, they will not be detected using culture methods used in most clinical laboratories, so their clinical significance is unknown. There are two *C. fetus* subspecies; C. *fetus* ssp. *fetus* is associated with both ovine/bovine reproductive disorders as well as human infections, and *C. fetus* ssp. *venerealis* causes bovine-associated abortion and infertility. This species is most similar phenotypically and genetically to the two *C. hyointestinalis* subspecies: *C. hyointestinalis* ssp. *hyointestinalis* and *C. hyointestinalis* ssp. *lawsonni.* Extensive 16S rRNA diversity has been reported for *C. hyointestinalis* strains.[40] The more recently described *C. lanienae*, isolated first from healthy human feces during a hygiene survey of abattoir workers[54] and then from beef cattle[44] and pigs,[77] was suggested to be closely related to *C. hyointestinalis.*[54] Additional reports suggest that the taxonomic status of this species remains unclear and that it may in fact represent a novel subspecies of *C. hyointestinalis.*[62] *C. gracilis* was originally described as a bacteriodes species associated with peridontal disease but formally assigned to the genus *Campylobacter* in 1995.[88] *C. gracilis* is often not recognized as being a *Campylobacter* because of its unusual morphology and metabolic characteristics; *C. gracilis* has an anaerobic growth requirement, a straight rod-shaped morphology, and is oxidase-negative.

GENERAL CHARACTERISTICS

Campylobacter species are Gram-negative, non-sporeforming rods. *Campylobacter* cells are typically curved, S-shaped, or spiral rods that are 0.2 to 0.9 μm wide and 0.5 to 5 μm long. Some species, such as *C. hominis* and *C. gracilis,* form straight rods. Organisms are usually motile by means of a single, polar, unsheathed flagellum at one or both ends. Cells of some species are nonmotile (*C. gracilis*) or have multiple flagella (*C. showae*). Most species require a microaerobic atmosphere for optimal growth; however, some species grow aerobically or anaerobically. An atmosphere containing increased hydrogen appears to be a growth requirement for other species such as *C. sputorum, C. concisus, C. mucosalis, C. curvus, C. showae, C. rectus, C. gracilis,* and *C. hominis.*[57]

EPIDEMIOLOGY

Campylobacters are a leading cause of bacterial enteritis worldwide and can be transmitted directly from animal to person, through ingestion of fecally contaminated water, food, or by direct contact with animal feces or contaminated environmental surfaces. *Campylobacter* species are primarily zoonotic, with a variety of animals implicated as reservoirs of infection, including a diverse range of domestic and wild animals and birds.[4, 13] In addition, water and the environment play a significant but poorly understood role in the epidemiology of campylobacteriosis.[47]

In the United States, requirements for reporting incidence of culture-confirmed infections vary by state. The active food-borne disease surveillance program FoodNet (www.FoodNet.gov) provides uniform reporting from a panel of sentinel sites, giving an accurate incidence of diagnosed infections. The reported incidence of *Campylobacter* infections in the United States has been declining for several years, from 24.7 cases per 100,000 persons in 1997 to 12.7 cases per 100,000 persons in 2005, a rate lower than reported for salmonellosis.[15] This decline occurred at the same time as several important food safety initiatives and increased efforts to educate the consumer and food handlers in basic food safety.[3]

While sporadic infections with *Campylobacter* are extremely common, *Campylobacter* outbreaks, defined as two or more persons epidemiologically linked by time and exposure, are relatively rare compared with other enteric pathogens. Outbreaks in the United States have been linked to a variety of foods, including poultry and other meats, and fruit. However, the most common single food implicated in *Campylobacter* outbreaks is unpasteurized milk.[33] Outbreaks have also been caused by contamination of untreated or poorly disinfected water supplies, and by drinking water from unprotected sources such as lakes or streams.[60, 73] In contrast to other enteric pathogens, person-to-person transmission is unusual for outbreaks of campylobacteriosis despite the low infectious dose of *C. jejuni*. The seasonal distribution of sporadic cases, both in the United States and in England, shows a well-defined peak in the summer. However, *Campylobacter* outbreaks typically peak in late spring and fall.[33, 34] Handling or consumption of contaminated poultry meat is a major risk factor for human infection with *C. jejuni* and is thought to be responsible for making an important contribution to the number of sporadic *Campylobacter* infections seen annually.[32, 76] While poultry is a principal source of campylobacteriosis, the contribution of other reservoirs to the overall burden of sporadic human disease has not been established.

In a typical case of diagnosed *Campylobacter* infection, illness lasts 5 to 7 days;[80] the infection resolves without antimicrobial treatment in the great majority of cases. Symptoms of infection typically begin 2 to 4 days after ingesting *C. jejuni* bacteria and include diarrhea with cramping abdominal pain, often with fever and some nausea. Extraintestinal *Campylobacter* infections are rare for people in good health, but are more likely to occur if the person is immunocompromised, elderly, or pregnant. Other complications of *Campylobacter* illness include bacteremia, Guillain Barré syndrome (GBS), reactive arthritis, and irritable bowel syndrome.[23]

ISOLATION AND IDENTIFICATION OF *CAMPYLOBACTER* SPP.

Diagnosis of *Campylobacter* infection is usually made by isolating the organism from stool, blood, or other specimens. Fastidious growth requirements can make isolation of *Campylobacter* from clinical specimens difficult. However, once isolated, presumptive identification of the genus *Campylobacter* is easily made based on morphology and darting motility. Further differentiation to species level is difficult because of the complex and rapidly evolving taxonomy and biochemical inertness of these bacteria. These problems have resulted in a proliferation of phenotypic and genotypic methods for identifying members of this group.

Direct Examination of Specimens

Microscopy: Presumptive diagnosis of a *Campylobacter* infection can be made directly by microscopic detection of the organism in a fresh stool specimen from patients with acute enteritis. Dark-field microscopy reveals the characteristic morphology and darting motility of campylobacters. Gram-stain analysis of stool specimens looking for organisms with typical *Campylobacter* morphology is also a highly sensitive and specific test that can be performed for rapid preliminary diagnosis of *Campylobacter* infection[94] and is currently underutilized.

Antigen Detection: Nonculture-based methods for the direct detection of campylobacters in clinical specimens allow the identification of this fastidious organism without the specialized media and equipment that is needed for *Campylobacter* culture. The ProSpecT *Campylobacter* Microplate assay for stool specimens (Remel Inc., Lenexa, KS) is an enzyme immunoassay (EIA) for the detection of *Campylobacter*-specific antigen found in *C. jejuni* and *C. coli*. There is no differentiation between the two species. The test cross-reacts with *C. upsaliensis*, but not other species of *Campy-*

lobacter.[41] Performance analysis has shown the assay to have sensitivity ranging from 66 to 94%[30] and specificity above 95% with a cost in excess of culture.

Polymerase Chain Reaction (PCR): Many PCR-based assays have been described to directly detect campylobacters in clinical and environmental specimens, although thus far they have been used only for research applications.[51, 93] Advantages of using a PCR approach include same-day detection and identification of *Campylobacter* to the species level and identification of the less-common *Campylobacter* species that are often missed by conventional culture.[48] However, this approach is more expensive than culture and does not provide an isolate for further characterization.

Transport Media

For the clinical microbiology laboratory whose main goal is to detect *Campylobacter* in the stool of patients with diarrhea, fresh stool (less than 2 hr old) is the ideal specimen. If a delay is anticipated, the specimen should be put in transport media such as modified Cary-Blair and kept at 4°C, not frozen. Transport media should also be used for rectal swab specimens. *Campylobacter* can be isolated from routine blood culture systems or wound specimens, so blood cultures should be done if the clinical features of the patient suggest campylobacteriosis.

Isolation from Human Specimens

Selective Media for Isolation: Diagnosis of *Campylobacter* gastroenteritis is commonly based on direct plating of stool onto selective media. No single medium will isolate all *Campylobacter* species. *C. jejuni* and *C. coli* can be isolated from stool samples by plating the stool on selective media and incubation in a microaerobic atmosphere at 42°C for at least 48 hr. A variety of media with different formulations, including both blood-based and charcoal-based selective media, are currently used for the isolation of these two species, most of which are commercially available. See Table 35.2 for the formulation of some of the commonly used selective media for isolation of *Campylobacter jejuni* and *C. coli*. The history of the development of selective media for isolation of campylobacters was reviewed by Corry et al.[18] All the selective media contain a basal media, either blood or other

TABLE 35.2
Formulation of Commonly Used Selective Media for Isolation of *C. jejuni* and *C. coli*

	Campy BAP	CVA	mCCDA	Karmali (CSM)
Basal agar	Brucella	Brucella	Nutrient	Columbia
Sheep blood	10%	5%	Charcoal/Na pyruvate + ferrous sulfate	Charcoal/Na pyruvate + hematin
Trimethoprim (5 mg)	+			
Vancomycin (10 mg)	+	+		+ (20 mg)
Poymyxin B (2500 IU)	+			
Amphotericin B (2 mg)	+	+	+ (10 mg)	+ (10 mg)*
Cephalothin (15 mg)	+			
Cefoperazone (20 mg)		+	+ (32 mg)	+ (32 mg)
Cyclohexamide (100 mg)				(+)

* A modified version of the Karmali Selective Supplement uses amphotericin B as the antifungal agent instead of cycloheximide.

agents such as charcoal, to quench oxygen toxicity and a cocktail of antimicrobial agents to inhibit the normal flora associated with fecal specimens.

The use of antimicrobial-containing selective media and 42°C as the incubation temperature can inhibit the growth of some *Campylobacter* spp. For example, the antimicrobial agent cephalothin that is present in some selective media formulations such as Campy-blood agar plates (BAPs) is inhibitory to *C. fetus* ssp. *fetus*. *C. jejuni* ssp. *doylei*, *C. upsaliensis*, and *C. concisus*. Therefore, these species generally do not grow on cephalothin-containing media. For primary isolation of *Campylobacter* from fecal samples, the use of cefoperazone-containing media instead of cephalothin-containing formulations is recommended.[30]

Enrichment Culture: A number of enrichment broths have been formulated to enhance the recovery of *Campylobacter* from stool samples.[18] Inclusion of an enrichment step may be beneficial in instances where low numbers of organisms are expected due to delayed transport to the laboratory, or after the acute stage of disease when the concentration of organisms may be low, such as in the investigation of GBS following acute *Campylobacter* infection.[30] Enrichment culture also is used to enhance the culture sensitivity of potentially environmentally stressed organisms. After

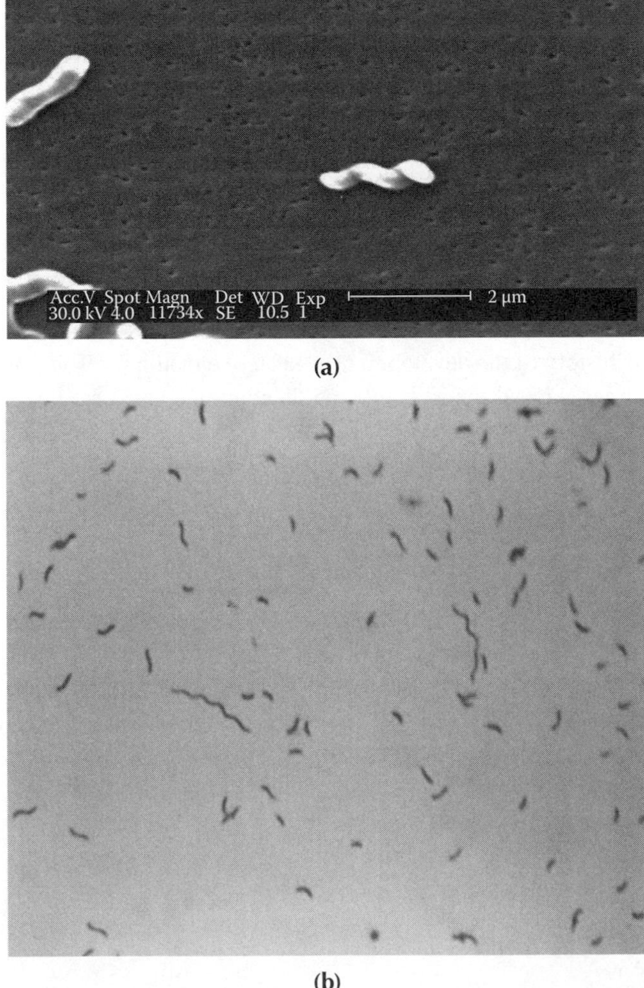

FIGURE 35.1 (a) Electron micrograph of *Campylobacter jejuni*. (b) Gram-stain image showing spiral morphology of *Campylobacter fetus* ssp. *fetus*. (Courtesy of Public Health Image Library, www.http://phil.cdc.gov.)

selective enrichment, suspect *Campylobacter* colonies are subcultured onto *Campylobacter*-selective media such as mCCDA (modified charcoal cefoperazone deoxycholate agar).

Filtration Method for Isolation: A passive filtration technique that uses a nonselective medium developed by Steele and McDerrmott[82] is an effective approach to isolate campylobacters, particularly *Campylobacter* species that are susceptible to the antibiotics in selective agar isolation procedures.[50, 92] The method is based on the principle that campylobacters can pass through membrane filters (0.45 to 0.65 µm) with relative ease (because of their small size and motility), while other stool flora are retained during the short processing time. Complete details of the procedure can be found in the *Manual of Clinical Microbiology*.[30] Incubation of nonselective media at 37°C under microaerobic conditions, preferably with an atmosphere containing increased hydrogen (for the hydrogen-requiring species) will allow isolation of the non-*C. jejuni*/*C. coli* species. When this method is used, a variety of other *Campylobacter* taxa can be identified in diarrheic stool, although their pathogenic potential remains to be clarified.[27] Stool samples containing ~10^5 CFU mL^{-1} *Campylobacter* will be detected with this method, and thus the filtration method is not as sensitive as primary culture with selective media.

Incubation Variables

Atmosphere: Most *Campylobacter* species require a microaerobic atmosphere containing approximately 5% O_2, 10% CO_2, and 85% N_2 for optimal recovery. There are a number of ways to generate microaerobic conditions; several manufacturers produce commercial microaerobic gas packs that are suitable for routine use in conjunction with an anaerobic jar, the evacuation-replacement method in conjunction with the approximate gas mixture, a tri-gas incubator, or the anoxomat system (www.anoxomat.com) can also be used for routine cultures.[30] Some species of *Campylobacter*, such as *C. sputorum*, *C. concisus*, *C. mucosalis*, *C. curvus*, *C. rectus,* and *C. hyointestinalis* require an increased hydrogen concentration for isolation and growth. Typically, these species will not be recovered using commercial gas packs because the amount of hydrogen generated in properly used, commercial gas packs is less than 2%. A gas mixture containing at least 6% H_2 is sufficient for isolating hydrogen-requiring species.

Temperature and Length of Incubation: Species within the genus *Campylobacter* have different optimal temperatures for growth. Therefore, the choice of incubation temperature for routine stool cultures is critical in determining the spectrum of species that will be recovered.[30] It is common practice for most laboratories to use 42°C as the primary incubation temperature for *Campylobacter*, as this temperature allows growth of *C. jejuni* and *C. coli* on selective media. The non-thermophilic species will not grow at 42°C. In contrast, most *Campylobacter* species grow well at 37°C. Many of the selective media containing blood, such as Skirrow medium, were devised for use at 42°C and have poor selective properties at 37°C, whereas CCDA and Karmali show good selective properties at 37°C.[18] Growth of *C. jejuni* and *C. coli* on selective media is typically visible within 24 to 48 hr. However, several studies report increased isolation rates when the incubation time is extended to 72 hr. Plates should be incubated a minimum of 72 hr before reporting as negative for *Campylobacter*.

Isolation of Campylobacters from Non-Human Sources

Protocols for the isolation of *Campylobacter* species from food and water are available in the U.S. Food and Drug Administration's (FDA) *Bacteriological Analytical Manual* (Ref. 43; http://www.cfsan.fda.gov). An international standard procedure for the detection of thermotolerant *Campylobacter* in food and animal foodstuff was first made by the ISO (International Organization for Standardization) in 1995.[7] This method was recently updated (ISO10272-1:2006).[6] Standard methods for the isolation of campylobacters from live animals are available from the World Organization of Animal Health (OIE) (www.oie.int).

Isolation of campylobacters from water can be difficult because campylobacters can be present in low numbers, and may be stressed as a consequence of low temperature, osmotic stress, nutrient depletion, and/or competition from other organisms. For this reason, recovery of *Campylobacter* from water samples from contaminated surface and ground water sources implicated in *Campylobacter* outbreaks worldwide has not always been successful using conventional culture techniques.[11, 17] ISO 17995 describes a standardized procedure for the isolation of *Campylobacter* from water.[8]

IDENTIFICATION

Identification of *Campylobacter* to the species level can be difficult because campylobacters are biochemically inert; few phenotypic tests are available to identify them to the species level. A presumptive identification of *Campylobacter* spp. can be made if oxidase-positive colonies exhibiting a characteristic Gram-stain appearance (curved gram-negative rods) grow on selective media incubated at 42°C under microaerobic conditions. All *Campylobacter* species except *C. gracilis* are typically oxidase-positive. Campylobacters neither oxidize nor ferment glucose. Isolates that are hippurate-positive can be further identified as *C. jejuni* because it is the only *Campylobacter* sp. that produces the enzyme hippuricase. However, isolates of *C. jejuni* that are hippuricase-negative have been reported.[70] Hippuricase-negative *C. jejuni* strains cannot be differentiated from *C. coli* by phenotypic testing (see Figure 35.1). Detection of species-specific sequences via PCR can be helpful in these instances to differentiate between these two closely related species.[12, 65] The number of available PCR assays for species-specific identification of *C. jejuni* and *C. coli* continues to grow. Recently published evaluations of some of these assays highlight the importance of validating their sensitivity and specificity.[12, 65] Figure 35.2 shows a flowchart for identification of *C. jejuni* and *C. coli*. Additional tests are needed to determine the species of hippurate-negative, indoxyl acetate-negative isolates.[30]

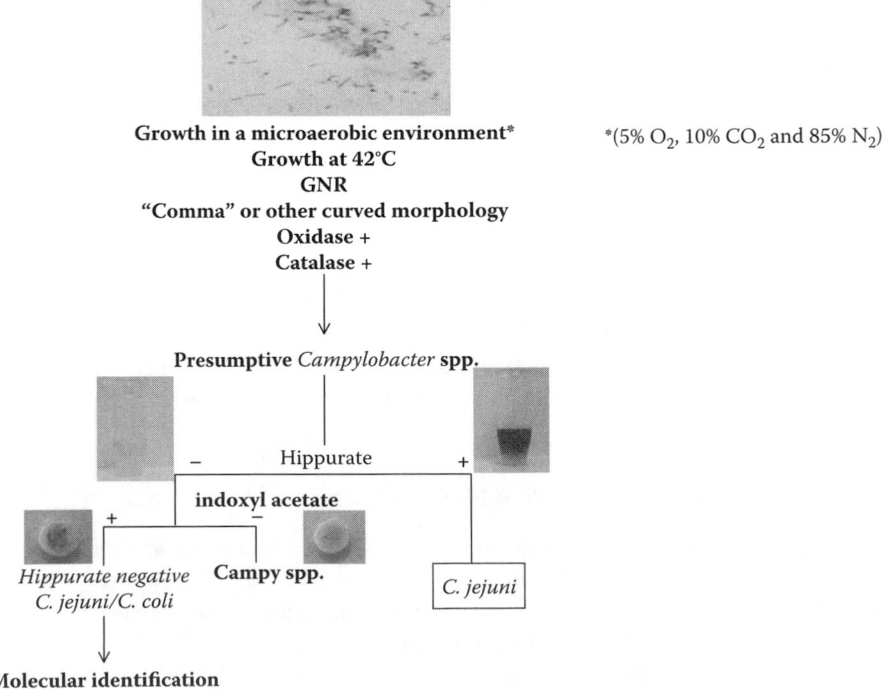

FIGURE 35.2 Identification of hippurate negative *C. jejuni/C. coli*.

Nalidixic acid and cephalothin susceptibility testing have been used in species identification in the past.[10] A presumptive *Campylobacter* strain that was hippurate-positive, sensitive to nalidixic acid, and resistant to cephalothin was identified as a *C. jejuni*. However, fluoroquinolone resistant and cross-resistant to nalidixic acid *Campylobacter* species have become increasingly common, with rates reported to be as high as 80%.[25] Therefore, antimicrobial susceptibility tests can no longer be relied upon for the phenotypic identification of *Campylobacter* isolates. Cellular fatty acid (CFA) analysis can be helpful in identifying *Campylobacter* and in differentiating some species; however, it cannot differentiate between *C. jejuni* and *C. coli*, the two most common species.[49]

Several commercial kits are available for identification of *Campylobacter* in stool cultures. These assays are typically used by clinical laboratories, where rapid diagnosis of *Campylobacter*-associated diarrhea is the primary objective. DNA-based kits included the Accuprobe assay (Gen-Probe, Inc.) and *Campylobacter jejuni* BioProbe (Enzo Life Sciences, Inc.). The Accuprobe assay detects *C. jejuni*, *C. coli*, and *C. lari* colonies without differentiating between the three species; the BioProbe assay detects only *C. jejuni*. Immunologic assays based on detection of cell wall antigens include Campyslide (B-D Diagnostics Systems), Dryspot *Campylobacter* test (Oxoid, Inc.), and Meritec Campy (Meridian Diagnostics). The Dryspot test detects *C. jejuni, C. coli, C. lari*, and *C. upsaliensis,* as well as some isolates of other species. The Meritec Campy test detects *C. jejuni, C. coli, and C. lari.* While these immunological methods are relatively inexpensive and can be easily performed in a routine laboratory, only a limited number of species can be definitively identified by antibody-based approaches; further, cross-reactivity of these tests with closely related taxa and/or more newly described species needs to be determined.

ANTIMICROBIAL SUSCEPTIBILITY TESTING

The unique growth requirements of *Campylobacter* present challenges for antimicrobial susceptibility determination. The Clinical and Laboratory Standards Institute (CLSI, formerly NCCLS) and the European Committee for Antimicrobial Susceptibility Testing (EUCAST) are two widely used resources for standardized susceptibility testing methods,[5, 58] and the CLSI provides specific guidance for *Campylobacter* testing. Before 2005, the CLSI described *Campylobacter* agar dilution testing.[59] Given that agar dilution is cumbersome and can be costly to perform, broth microdilution and diffusion methods have also been used.[16, 20, 83] In 2005, broth microdilution quality control minimum inhibitory concentration (MIC) ranges for *C. jejuni* ATCC 33560 were developed.[56] CLSI published eight quality control ranges as well as clinical breakpoints for four antimicrobial agents: ciprofloxacin, erythromycin, doxycycline, and tetracycline. Methods for performing broth microdilution as well as disk diffusion testing on *Campylobacter* species were also provided (CLSI, M45AE). Agar dilution and disk diffusion methods require media supplemented with 5% sheep blood, whereas broth microdilution tests require supplementation with 2.5 to 5% lysed horse blood. The tests can either be incubated at 36°C for 48 hr or at 42°C for 24 hr under microaerobic conditions.

Antimicrobial Resistance among *Campylobacter*: Antimicrobial resistance among *Campylobacter* varies by species, perhaps because *Campylobacter* species can be associated with different animal reservoirs. Resistance rates can vary between *C. jejuni* and *C. coli* because *C. jejuni* is more common in poultry and cattle while *C. coli* predominates in pigs[2] and antimicrobial use patterns vary by food animal species. The differences between *C. jejuni* and *C. coli* resistance are most notable in the case of the macrolides.

Macrolide Resistance: Macrolide resistance among *Campylobacter jejuni* and *C. coli* was reviewed by Gibreel and Taylor.[38] This antimicrobial class includes erythromycin, azithromycin, and tylosin (a veterinary pharmaceutical), and resistance is attributed to chromosomal point mutations in the 23S rDNA that result in an altered ribosome binding site. These alterations most likely affect binding of the macrolide to the ribosome and are nontransmissible. The CmeABC multidrug resistance efflux pump plays some role in resistance to several drug classes and, based on reduction of tylosin and azithromycin MIC for strains in which the pump is inactivated, is thought to be involved in low-level macrolide resistance.[38] Almost all studies report higher macrolide resistance among *C. coli* than among *C. jejuni*. Because *C. coli* are associated with a swine reservoir, this raises questions about macrolide-selective pressure in those production settings.[1, 24] The use of macrolides for growth promotion in animals was banned by the European Union in 1999 but they may still be used therapeutically.[19]

Fluoroquinolone Resistance: Fluoroquinolones interfere with DNA gyrase and topoisomerase IV function, which prevents DNA replication. DNA gyrase is a tetramer of *gyrA* and *gyrB* gene products, and topoisomerase IV is a tetramer of *parC* and *parE* products.[55] As is the case with the macrolides, resistance to fluoroquinolones among *Campylobacter* is most often due to a chromosomal mutation and is nontransmissible. Although mutation of *parC* was proposed as a mechanism of resistance,[37] subsequent attempts to amplify this gene both from the original strain and other strains have been unsuccessful. Functional mutations in *gyr*A are sufficient to confer resistance to fluoroquinolones and nalidixic acid.[72] CmeABC-mediated efflux can contribute to fluoroquinolone resistance when other mechanisms are present.[67]

Campylobacter jejuni- and *C. coli*-specific mutations and MIC results were reviewed by Engberg et al.[26] Resistance to fluoroquinolones was rarely reported among human *Campylobacter* isolates in most of the world prior to 1989. Since that time, many countries have reported a dramatic increase in fluoroquinolone-resistant *Campylobacter* infections. Extremely high rates of resistance have been reported from Thailand (84%)[42] and Spain (75%).[74] Resistance rates are much lower in most of Europe and the United States, with many resistant infections being associated with travel to areas where resistance is more common.[16, 20] The emergence of fluoroquinolone-resistant *Campylobacter* infections has been temporally related to the approval of fluoroquinolones (primarily enrofloxacin) for use in food animal production (particularly poultry).[81] It is widely accepted that poultry is a common source of *C. jejuni* infections in humans, and fluoroquinolone-resistant *C. jejuni* has been isolated from retail poultry products.[35, 81]

Tetracycline Resistance: Tetracyclines affect protein synthesis primarily by binding to the 30S ribosomal subunit. Resistance to tetracycline is common in *C. jejuni* and *C. coli*.[16, 35, 84] Tetracycline resistance in *Campylobacter* is due to the presence of the *tet*O gene, usually on a plasmid.[26] Transfer of *tet*O genes between *C. jejuni* strains in the digestive tract of chickens has been demonstrated, suggesting that other resistance genes carried on the same plasmid could be similarly spread.[9]

Antimicrobial Resistance in Other *Campylobacter* Species: There is very little information available on the epidemiology of antimicrobial resistance among *Campylobacter* species other than *C. jejuni* and *C. coli*. Thwaites and Frost[84] reported susceptibility results for 25 isolates of *C. lari*. Resistance was observed for ampicillin (36%), kanamycin (60%), and tetracycline (12%); none were erythromycin-resistant. All the isolates were resistant to nalidixic acid, due to the long-recognized intrinsic resistance. They were also resistant to ciprofloxacin. In a study of 104 human isolates of *C. fetus* ssp. *fetus*, nonsusceptibility (intermediate or resistant) was found for cefotaxime (13%), erythromycin (71%), ciprofloxacin (5%), and tetracycline (34%).[85] In a 2006 study of 994 *C. fetus* isolated from cattle feces in Alberta, Canada, 39% were resistant to doxycycline and tetracycline, and 97% were resistant to nalidixic acid.[45] However, ciprofloxacin resistance was not found among the nalidixic acid-resistant *C. fetus* isolates, a phentoype typical for *C. fetus*. Ciprofloxacin resistance has been seen in *C. fetus* due to a *gyrA* mutation that has also been found in a ciprofloxacin-resistant *C. coli* strain.[26] Vandenberg et al.[92] found 100% nalidixic acid resistance among the eleven *C. fetus* strains isolated from humans in Belgium, but no resistance to the other five drugs tested, including ciprofloxacin. Of the 85 *C. upsaliensis* included in this study, 12% were nalidixic acid-resistant, 6% were ciprofloxacin-resistant, and 13% were erythromycin-resistant. Of 20 *C. concisus* isolates tested, 80% were nalidixic acid-resistant but only one isolate was ciprofloxacin-resistant.

EPIDEMIOLOGIC SUBTYPING OF *C. JEJUNI*

The ability to discriminate or subtype campylobacters below the level of species has been successfully applied to aid the epidemiological investigation of outbreaks of campylobacteriosis.[29] Subtyping provides information to recognize outbreaks of infection, match cases with potential vehicles of infection, and discriminate these from unrelated strains. In addition, these methods are important in epidemiological research projects to identify potential reservoirs of strains that cause disease in humans, identify routes of transmission, and improve our understanding of *Campylobacter* epidemiology.[31]

A number of criteria are used to evaluate subtyping methods, including typability, reproducibility, discriminatory power, ease of use, ease of interpretation, and cost. The subtyping methods available to subtype *Campylobacter jejuni* vary considerably using these criteria, depending on the method used. At present there is no definitive gold-standard method for subtyping *C. jejuni*. The subtyping method of choice is ultimately determined by consideration of the basic microbiology of the organism in question, the nature of the microbiological question being asked, and, importantly, the ability of the typing method to detect significant epidemiological differences. Phenotypic subtyping methods that have been used to differentiate *C. jejuni* include biotyping, phage typing, and serotyping. There are two generally accepted, well-evaluated serotyping schemes that were developed in the 1980s for epidemiological characterization of *Campylobacter* isolates; the Penner scheme is based on heat-stable (HS) antigens using a passive hemagglutination technique,[69] and the Lior scheme is based on heat-labile (HL) antigens and a bacterial agglutination method.[53] The application of these techniques to *C. jejuni* isolates from human, farm animal, and environmental specimens has proven useful in outbreak investigations and for identification of potential reservoirs for human infection. However, the maintenance of a large panel of antisera that is not commercially available has prevented its routine application for strain characterization and not all isolates are typable.[28] Biotyping and phage typing schemes have also been described but these methods are not as discriminatory as serotyping.

The limitations associated with phenotypic subtyping methods and the rapid growth of molecular biology techniques led to the development of a wide range of molecular subtyping methods. These molecular methods have become widely applied to subtype *Campylobacter jejuni* and are often more universally applicable, reduce problems encountered with typability, and provide a much higher level of discriminatory power. Many different DNA-based subtyping schemes such as pulsed field gel electrophoresis (PFGE) of macrorestricted chromosomal DNA and amplified fragment length polymorphism (AFLP) analysis that allow us to investigate strain variation throughout the genome of *C. jejuni* have been described. A review of molecular subtyping methods used for characterization of *Campylobacter* isolates is described in Ref. 31 and 95. These methods demonstrate large diversity among human *C. jejuni* isolates, indicative of the sporadic nature of campylobacteriosis and the variety of different sources for *Campylobacter* infection. Many of the advantages offered by highly sensitive and discriminatory molecular methods for subtyping *C. jejuni* are, however, negated by the lack of a standardized approach to both procedures used and adoption of universal nomenclature schemes for the resulting subtyping data generated, thus making it impossible to compare and exchange data and determine if the same strains are circulating in different places. For some commonly used methods such as PFGE, this has already been addressed with the PusleNet network (Ref. 36 and 71; www.cdc.gov/pulseNet). The application of higher-resolution molecular subtyping methods such as PFGE has resulted in questions regarding the (in)stability of the campylobacter genome and implications of this for molecular epidemiological investigations.[95] Current evidence suggests that genomic rearrangements do play a role in strain diversity, which can make interpretation of PFGE data difficult.

Advances in DNA sequencing technology have provided a means to investigate strain variation at the nucleotide level. DNA sequence analysis is highly reproducible and does not rely on the interpretation of gel patterns; DNA sequencing provides more precise information on strain relatedness, which is only suggested by restriction-based typing methods such as PFGE. One such method, multilocus sequencing typing (MLST), which involves the sequencing of short (400 to 500) nucleotide sequences within seven housekeeping genes, has become increasingly popular over the past few years; MLST is not only highly discriminatory, but also enables investigation of the population structure of the organism.[21] In addition, the subtyping data generated is electronically portable and amenable to storage in Internet-accessible databases. *Campylobacter* MLST databases are available for several *Campylobacter* species, including *C. jejuni,* at www.pubmlst.org/campylobacter. MLST studies have confirmed that *C. jejuni* strains are genetically diverse, exhibit a high degree of plasticity, and have a weakly clonal population structure. However, the utility of MLST as an epidemiological tool for outbreak investigations of campylobacteriosis has yet to be established.[75]

Sequencing of the complete genome sequence of a *Campylobacter jejuni* strain in 2000[66] and the subsequent sequencing of several more *Campylobacter* genomes since (information on these genomes can be found at a online database for *Campylobacter* genome analysis: http://campy.bham.ac.uk/genome), and additional *Campylobacter* genomes in progress (http://msc.tigr.org/campy) has resulted in a new range of approaches to investigate strain variation. Genomotyping is a method that uses DNA microarrays to compare interstrain- and intrastrain-specific variation in *C. jejuni* strains at the genome level.[22, 46] In this post-genomic era, the application of these powerful new molecular tools for comparative genomic studies of *C. jejuni* will facilitate the identification of strains with phenotypic and molecular markers that may contribute to virulence, different disease outcomes, and host and ecological niche specificities. This, in turn, will greatly enhance much-needed attribution studies to determine the fraction of *Campylobacter* infections attributable to different food sources, which will aid greatly the design of targeted control strategies to eliminate *C. jejuni* from the food chain so that food safety and public health risks are reduced.

REFERENCES

1. Aarestrup, F.M. 2000. Occurrence, selection and spread of resistance to antimicrobial agents used for growth promotion in Denmark. APMIS, 108:1–48.
2. Aarestrup, F.M., E.M. Nielsen, M. Madsen, and J. Engberg. 1997. Antimicrobial susceptibility patterns of thermophilic *Campylobacter* spp. from humans, pigs, cattle, and broilers in Denmark. *Antimicrob. Agents Chemother.*, 41:2244–2250.
3. Allos, B.M., M.R. Moore, P. M. Griffin, and R.V. Tauxe. 2004. Surveillance for sporadic foodborne disease in the 21st century: the FoodNet perspective. *Clin. Infect. Dis.*, 38(Suppl. 3):S115–S120.
4. Altekruse, S.F., N.J. Stern, P.I. Fields, and D.L. Swerdlow. 1999. *Campylobacter jejuni* — an emerging foodborne pathogen. *Emerg. Infect. Dis.*, 5:28–35.
5. Anon. 2000. EUCAST Definitive Document E.DEF 3.1, June 2000: Determination of minimum inhibitory concentrations (MICs) of antibacterial agents by agar dilution. *Clin. Microbiol. Infect.*, 6:509–515.
6. Anon. 2006. Microbiology of food and animal feeding stuffs — Horizontal method for detection of *Campylobacter* spp. Geneva: International Organization for Standardization.
7. Anon. 1995. Microbiology of food and animal feeding stuffs — Horizontal method for detection of thermotolerant *Campylobacter.* Geneva: International Organization for Standardization.
8. Anon. 2005. Water Quality — Detection and enumeration of thermotolerant *Campylobacter* species. Geneva: International Organization for Standardization.
9. Avrain, L., C. Vernozy-Rozand, and I. Kempf. 2004. Evidence for natural horizontal transfer of tetO gene between *Campylobacter* jejuni strains in chickens. *J. Appl. Microbiol.*, 97:134–140.
10. Barrett, T.J., C.M. Patton, and G.K. Morris. 1988. Differentiation of *Campylobacter* species using phenotypic characterization. *Lab. Med.*, 19:96–102.
11. Bopp, D.J., B.D. Sauders, A.L. Waring, J. Ackelsberg, N. Dumas, E. Braun-Howland, D. Dziewulski, B.J. Wallace, M. Kelly, T. Halse, K.A. Musser, P.F. Smith, D.L. Morse, and R.J. Limberger. 2003. Detection, isolation, and molecular subtyping of *Escherichia coli* O157:H7 and *Campylobacter jejuni* associated with a large waterborne outbreak. *J. Clin. Microbiol.*, 41:174–180.
12. Burnett, T.A., A.H.M., P. Kuhnert, and S.P. Djordjevic. 2002. Speciating *Campylobacter jejuni* and *Campylobacter coli* isolates from poultry and humans using six PCR-based assays. *FEMS Microbiol. Lett.*, 216:201–209.

13. Butzler, J.P. 1984. *Campylobacter Infections in Man and Animals.* CRC Press, Boca Raton, FL.
14. Butzler, J.P. 2004. *Campylobacter*, from obscurity to celebrity. *Clin. Microbiol. Infect.*, 10:868–876.
15. CDC. 2005. Preliminary Foodnet data on the incidence of infection with pathogens transmitted commonly through food — 10 sites, 2005. *Morbidity Mortality Weekly Report* (MMWR), 55:392–395.
16. Centers for Disease Control and Prevention. 2006 posting date. NARMS Human Isolates Final Report, 2003. Division of Bacterial and Mycotic Diseases. [Online.]
17. Clark, C., L. Price, R. Ahmed, D.L. Woodward, P.L. Melito, F.G. Rodgers, F. Jamisson, B. Clebin, A. Li, and A. Ellis. 2003. Characterization of waterborne outbreak-associated *Campylobacter jejuni*, Walkerton, Ontario. *Emerg. Infect. Dis.*, 9:1232–1241.
18. Corry, J.E., D.E. Post, P. Colin, and M.J. Laisney. 1995. Culture media for the isolation of campylobacters. *Int. J. Food Microbiol.*, 26:43–76.
19. Council of the European Union. 1998. Council Regulation (EC) No. 2821/98 of 17 December 1998 amending, as regards withdrawal of the authorisation of certain antibiotics, Directive 70/524/EED concerning additives in feedstuffs. Official Journal of the European Communities L351:4–8.
20. Danish Integrated Antimicrobial Resistance Monitoring and Research Programme. 2004. DANMAP 2003 — Use of antimicrobial agents and occurrence of antimicrobial resistance in bacteria from food animals, foods and humans in Denmark.
21. Dingle, K.E. and M.C.J. Maiden. 2005. Population genetics of *Campylobacter jejuni*, p. 43–57, *Campylobacter: Molecular and Cellular Biology*. Horizon Bioscience, Norwich, U.K.
22. Dorrell, N., O.L. Champion, and B.W. Wren. 2005. Advances in *Campylobacter jejuni* comparative genomics through whole genome DNA microarrays, p. 79–99, *Campylobacter Molecular and Cellular Biology*. Horizon Bioscience, Norwich, U.K.
23. Dunlop, S., D. Jenkins, K. Neal, and R. Spiller. 2003. Relative importance of enterochromaffin cell hyperplasia, anxiety, and depression in post infectious IBS. *Gastroenterology*, 125:1651–1659.
24. Engberg, J., F.M. Aarestrup, D.E. Taylor, P. Gerner-Smidt, and I. Nachamkin. 2001. Quinolone and macrolide resistance in *Campylobacter jejuni* and *C. coli:* resistance mechanisms and trends in human isolates. *Emerg. Infect. Dis.*, 7:24–34.
25. Engberg, J., F.M. Aarestrup, D.E. Taylor, P. Gerner-Smidt, and I. Nachamkin. 2001. Quinolone and macrolide resistance in *Campylobacter jejuni* and *C. coli*: resistance mechanisms and trends in human isolates. *Emerg. Infect. Dis.*, 7:24–34.
26. Engberg, J., M. Keelan, P. Gerner-Smidt, and D.E. Taylor. 2006. Antimicrobial Resistance in Campylobacter, p. 269–292. In F.M. Aarestrup (Ed.), *Antimicrobial Resistance in Bacteria of Animal Origin.* ASM Press, Washington, DC.
27. Engberg, J., S.L.W. On, C.S. Harrington, and P. Gerner-Smidt. 2000. Prevalence of *Campylobacter, Arcobacter, Helicobacter*, and *Sutterella* spp. in human fecal samples as estimated by reevaluation of isolation methods for *Campylobacters. J. Clin. Microbiol.*, 38:286–291.
28. Fitzgerald, C., A. Sails, P.I. Fields. 2005. *Campylobacter jejuni* strain variation, p. 59–77, *Campylobacter Molecular and Cellular Biology*. Horizon Biosciences, Norwich, U.K.
29. Fitzgerald, C., L.O. Helsel, M.A. Nicholson, S.J. Olsen, D.L. Swerdlow, R. Flahart, J. Sexton, and P.I. Fields. 2001. Evaluation of methods for subtyping *Campylobacter jejuni* during an outbreak involving a food handler. *J. Clin. Microbiol.*, 39:2386–2390.
30. Fitzgerald, C. and Nachamkin, I. 2006. *Campylobacter* and *Arcobacter*, p. 902–914, *Manual of Clinical Microbiology*, 9th ed. ASM Press, Washington, DC.
31. Fitzgerald, C., Swaminathan, B., and Sails, A. 2003. Genetic techniques: molecular subtyping methods, p. 271–293. *Detecting Pathogens in Food.* CRC Press, Boca Raton, FL.
32. Friedman, C.R., R.M. Hoekstra, M. Samuel, R. Marcus, J. Bender, B. Shiferaw, S. Reddy, S.D. Ahuja, D.L. Helfrick, F. Hardnett, M. Carter, B. Anderson, and R.V. Tauxe. 2004. Risk factors for sporadic *Campylobacter* infection in the United States: a case-control study in FoodNet sites. *Clin. Infect. Dis.*, 38(Suppl. 3):S285–296.
33. Friedman, C.R., J. Neimann, H.C. Wegener, and R.V. Tauxe. 2000. Epidemiology of *Campylobacter jejuni* infections in the United States and other industrialized nations, p. 121–138. In M. Blaser and I. Nachamkin (Eds.), *Campylobacter*, 2nd ed. American Society for Microbiology, Washington, DC.
34. Frost, J., I. Gillespie, and S. O'Brien. 2002. Public health implications of *campylobacter* outbreaks in England and Wales, 1995–9: epidemiological and microbiological investigations. *Epidemiol. Infect.*, 128:111–118.
35. Ge, B., D.G. White, P.F. McDermott, W. Girard, S. Zhao, S. Hubert, and J. Meng. 2003. Antimicrobial-resistant *Campylobacter* species from retail raw meats. *Appl. Environ. Microbiol.*, 69:3005–3007.

36. Gerner-Smidt, P., K. Hise, J. Kincaid, S. Hunter, S. Rolando, E. Hyytia-Trees, E.M. Ribot, and B. Swaminathan. 2006. PulseNet USA: a five-year update. *Foodborne Pathog. Dis.*, 3:9–19.
37. Gibreel, A., E. Sjogren, B. Kaijser, B. Wretlind, and O. Skold. 1998. Rapid emergence of high-level resistance to quinolones in *Campylobacter jejuni* associated with mutational changes in gyrA and parC. *Antimicrob. Agents Chemother.*, 42:3276–3278.
38. Gibreel, A. and D.E. Taylor. 2006. Macrolide resistance in *Campylobacter jejuni* and *Campylobacter coli. J. Antimicrob. Chemother.*, 58:243–255.
39. Goossens, H., B. Pot, L. Vlaes, C. Van den Borre, R. Van den Abbeele, C. Van Naelten, J. Levy, H. Cogniau, P. Marbehant, J. Verhoef, et al. 1990. Characterization and description of "*Campylobacter upsaliensis*" isolated from human feces. *J. Clin. Microbiol.*, 28:1039–1046.
40. Harrington, C.S. and S.L. On. 1999. Extensive 16S rRNA gene sequence diversity in *Campylobacter hyointestinalis* strains: taxonomic and applied implications. *Int. J. Syst. Bacteriol.*, 49(Pt. 3):1171–1175.
41. Hindiyeh, M., S. Jense, S. Hohmann, H. Benett, C. Edwards, W. Aldeen, A. Croft, J. Daly, S. Mottice, and K.C. Carroll. 2000. Rapid detection of *Campylobacter jejuni* in stool specimens by an enzyme immunoassay and surveillance for *Campylobacter upsaliensis* in the greater Salt Lake City area. *J. Clin. Microbiol.,* 38:3076–3079.
42. Hoge, C.W., J.M. Gambel, A. Srijan, C. Pitarangsi, and P. Echeverria. 1998. Trends in antibiotic resistance among diarrheal pathogens isolated in Thailand over 15 years. *Clin. Infect. Dis.*, 26:341–345.
43. Hunt, J.M. et al. 1998. *Campylobacter, Food and Drug Administration Bacteriological Analytical Manual*, 8th edition. AOAC, Arlington, VA.
44. Inglis, G.D. and L.D. Kalischuk. 2004. Direct quantification of *Campylobacter jejuni* and *Campylobacter lanienae* in feces of cattle by real-time quantitative PCR. *Appl. Environ. Microbiol.*, 70:2296–2306.
45. Inglis, G.D., D.W. Morck, T.A. McAllister, T. Entz, M.E. Olson, L.J. Yanke, and R.R. Read. 2006. Temporal prevalence of antimicrobial resistance in *Campylobacter* spp. from beef cattle in Alberta feedlots. *Appl. Environ. Microbiol.*, 72:4088–4095.
46. Klena, J.D. and Konkel, M.E. 2005. Methods for epidemiological analysis of *Campylobacter jejuni*, p. 165–179. *Campylobacter Molecular and Cellular Biology*, Horizon Bioscience, Norwich, U.K.
47. Koenraad, P.M., F.M. Rombouts, and S.H.W. Notermans. 1997. Epidemiological aspects of thermophilic *Campylobacter* in water-related environments: a review. *Water Environ. Res.*, 69:52.
48. Kulkarni, S.P., S. Lever, J.M. Logan, A.J. Lawson, J. Stanley, and M.S. Shafi. 2002. Detection of *campylobacter* species: a comparison of culture and polymerase chain reaction based methods. *J. Clin. Pathol.*, 55:749–753.
49. Lambert, M.A., C.M. Patton, T.J. Barrett, and C.W. Moss. 1987. Differentiation of *Campylobacter* and *Campylobacter*-like organisms by cellular fatty acid composition. *J. Clin. Microbiol.*, 25:706–713.
50. Lastovica, A.J. and Skirrow, M.B. 2000. Clinical significance of *Campylobacter* and related species other than *Campylobacter jejuni* and *C. coli,* p. 89–120. *Campylobacter*, second ed. ASM Press, Washington, DC.
51. Lawson, A.J., J.M. Logan, L. O'Neill, M. Desai, and J. Stanley. 1999. Large-scale survey of *Campylobacter* species in human gastroenteritis by PCR and PCR-enzyme-linked immunosorbent assay. *J. Clin. Microbiol.*, 37:3860–3864.
52. Lawson, A.J., S.L. On, J.M. Logan, and J. Stanley. 2001. *Campylobacter hominis* sp. nov., from the human gastrointestinal tract. *Int. J. Syst. Evol. Microbiol.*, 51:651–660.
53. Lior, H., D.L. Woodward, J.A. Edgar, L.J. Laroche, and P. Gill. 1982. Serotyping of *Campylobacter jejuni* by slide agglutination based on heat-labile antigenic factors. *J. Clin. Microbiol.*, 15:761–768.
54. Logan, J.M., A. Burnens, D. Linton, A.J. Lawson, and J. Stanley. 2000. *Campylobacter lanienae* sp. nov., a new species isolated from workers in an abattoir. *Int. J. Syst. Evol. Microbiol.,* 50(Pt. 2):865–872.
55. Mascaretti, O.A. (Ed.). 2003. *Bacteria versus Antibacterial Agents, An Integrated Approach.* ASM Press, Washington, DC.
56. McDermott, P.F., S.M. Bodeis-Jones, T.R. Fritsche, R.N. Jones, and R.D. Walker. 2005. Broth microdilution susceptibility testing of *Campylobacter jejuni* and the determination of quality control ranges for fourteen antimicrobial agents. *J. Clin. Microbiol.*, 43:6136–6138.
57. Nachamkin, I. 2003. *Campylobacter* and *Arcobacter*, p. 902–914. *Manual of Clinical Microbiology*, 8th ed. ASM Press, Washington, DC.
58. NCCLS. 2003. *Methods for Dilution Antimicrobial Susceptibility Tests for Bacteria that Grow Aerobically*; Approved Standard — sixth edition. NCCLS, Wayne, PA.
59. NCCLS. 2004. *Performance Standards for Antimicrobial Susceptibility Testing;* Fourteenth Informational Supplement. NCCLS, Wayne, PA.

60. O'Reilly, C.E., A.B. Bowen, N.E. Perez, J.P. Sarisky, C.A. Shepherd, M.D. Miller, B.C. Hubbard, M. Herring, S.D. Buchanan, C.C. Fitzgerald, V. Hill, M.J. Arrowood, L.X. Xiao, R.M. Hoekstra, E.D. Mintz, and M.F. Lynch. 2007. A waterborne outbreak of gastroenteritis with multiple etiologies among resort island visitors and residents: Ohio, 2004. *Clin. Infect. Dis.*, 44:506–512.
61. On, S.L. 1996. Identification methods for *campylobacters, helicobacters*, and related organisms. *Clin. Microbiol. Rev.*, 9:405–422.
62. On, S.L. 2001. Taxonomy of *Campylobacter, Arcobacter, Helicobacter* and related bacteria: current status, future prospects and immediate concerns. *Symp. Ser. Soc. Appl. Microbiol.*, p. 1S–15S.
63. On, S.L. 2005. Taxonomy, phylogeny and methods for the identification of *Campylobacter* species, p. 14–42. *Campylobacter Molecular and Cellular Microbiology*. Horizon Bioscience, Norwich, U.K.
64. On, S.L., H.I. Atabay, J.E. Corry, C.S. Harrington, and P. Vandamme. 1998. Emended description of *Campylobacter sputorum* and revision of its infrasubspecific (biovar) divisions, including *C. sputorum* biovar *paraureolyticus*, a urease-producing variant from cattle and humans. *Int. J. Syst. Bacteriol.*, 48(Pt. 1):195–206.
65. On, S.L. and P.J. Jordan. 2003. Evaluation of 11 PCR assays for species-level identification of *Campylobacter jejuni* and *Campylobacter coli*. *J. Clin. Microbiol.*, 41:330–336.
66. Parkhill, J., B.W. Wren, K. Mungall, J.M. Ketley, C. Churcher, D. Basham, T. Chillingworth, R.M. Davies, T. Feltwell, S. Holroyd, K. Jagels, A.V. Karlyshev, S. Moule, M.J. Pallen, C.W. Penn, M.A. Quail, M.A. Rajandream, K.M. Rutherford, A.H. van Vliet, S. Whitehead, and B.G. Barrell. 2000. The genome sequence of the food-borne pathogen *Campylobacter jejuni* reveals hypervariable sequences. *Nature*, 403:665–668.
67. Payot, S., J.M. Bolla, D. Corcoran, S. Fanning, F. Megraud, and Q. Zhang. 2006. Mechanisms of fluoroquinolone and macrolide resistance in *Campylobacter* spp. *Microbes Infect.*, 8:1967–1971.
68. Penner, J.L. 1988. The genus *Campylobacter*: a decade of progress. *Clin. Microbiol. Rev.*, 1:157–172.
69. Penner, J.L. and J.N. Hennessy. 1980. Passive hemagglutination technique for serotyping *Campylobacter fetus* subsp. *jejuni* on the basis of soluble heat-stable antigens. *J. Clin. Microbiol.*, 12:732–737.
70. Rautelin, H., J. Jusufovic, and M.L. Hanninen. 1999. Identification of hippurate-negative thermophilic campylobacters. *Diagn. Microbiol. Infect. Dis.*, 35:9–12.
71. Ribot, E.M., C. Fitzgerald, K. Kubota, B. Swaminathan, and T.J. Barrett. 2001. Rapid pulsed-field gel electrophoresis protocol for subtyping of *Campylobacter jejuni*. *J. Clin. Microbiol.*, 39:1889–1894.
72. Ruiz, J., P. Goni, F. Marco, F. Gallardo, B. Mirelis, T. Jimenez De Anta, and J. Vila. 1998. Increased resistance to quinolones in *Campylobacter jejuni:* a genetic analysis of gyrA gene mutations in quinolone-resistant clinical isolates. *Microbiol. Immunol.*, 42:223–226.
73. Sacks, J.J., S. Lieb, L.M. Baldy, S. Berta, C.M. Patton, M.C. White, W.J. Bigler, and J.J. Witte. 1986. Epidemic campylobacteriosis associated with a community water supply. *Am. J. Pub. Health*, 76:424–428.
74. Saenz, Y., M. Zarazaga, M. Lantero, M.J. Gastanares, F. Baquero, and C. Torres. 2000. Antibiotic resistance in *Campylobacter* strains isolated from animals, foods, and humans in Spain in 1997–1998. *Antimicrob. Agents Chemother.*, 44:267–271.
75. Sails, A.D., B. Swaminathan, and P.I. Fields. 2003. Utility of multilocus sequence typing as an epidemiological tool for investigation of outbreaks of gastroenteritis caused by *Campylobacter jejuni*. *J. Clin. Microbiol.*, 41:4733–4739.
76. Samuel, M.C., D.J. Vugia, S. Shallow, R. Marcus, S. Segler, T. McGivern, H. Kassenborg, K. Reilly, M. Kennedy, F. Angulo, and R.V. Tauxe. 2004. Epidemiology of sporadic *Campylobacte*r infection in the United States and declining trend in incidence, FoodNet 1996–1999. *Clin. Infect. Dis.*, 38(Suppl. 3): S165–S174.
77. Sasaki, Y., T. Fujisawa, K. Ogikubo, T. Ohzono, K. Ishihara, and T. Takahashi. 2003. Characterization of *Campylobacter lanienae* from pig feces. *J. Vet. Med. Sci.,* 65:129–131.
78. Schumacher, W., P.M.H. Kroneck, and N. Pfennig. 1992. Comparative systematic study on Spirillum 5175, *Campylobacter* and *Wolinella* species. Description of *Spirillum* 5175 as *Sulfurospirillum deleyanum* gen. Nov., sp. nov. *Arch. Microbiol.*, 158:287–293.
79. Sebald, M. and M. V'Eron. 1963. Base DNA Content and Classification of Vibrios. *Ann. Inst. Pasteur* (Paris), 105:897–910.
80. Skirrow, M. and M. Blaser. 2000. Clinical aspects of *Campylobacter* infection, Chapter 4, p. 69–88. In M. Blaser and I. Nachamkin (Eds.), *Campylobacter*, 2nd edition. American Society for Microbiology, Washington, DC.

81. Smith, K.E., J.M. Besser, C.W. Hedberg, F.T. Leano, J.B. Bender, J.H. Wicklund, B.P. Johnson, K.A. Moore, and M.T. Osterholm. 1999. Quinolone-resistant *Campylobacter jejuni* infections in Minnesota, 1992–1998. Investigation Team. *N. Engl. J. Med.*, 340:1525–1532.
82. Steele, T.W. and S.N. McDermott. 1984. The use of membrane filters applied directly to the surface of agar plates for the isolation of *Campylobacter jejuni* from feces. *Pathology*, 16:263–265.
83. Tenover, F.C., C.N. Baker, C.L. Fennell, and C.A. Ryan. 1992. Antimicrobial resistance in *Campylobacter* species, p. 66–73. In I. Nachamkin, M.J. Blaser, and L.S. Tompkins (Eds.), *Campylobacter jejuni Current Status and Future Trends*. American Society for Microbiology, Washington, DC.
84. Thwaites, R.T. and J.A. Frost. 1999. Drug resistance in *Campylobacter jejuni, C. coli*, and *C. lari* isolated from humans in north west England and Wales, 1997. *J. Clin. Pathol.*, 52:812–814.
85. Tremblay, C., C. Gaudreau, and M. Lorange. 2003. Epidemiology and antimicrobial susceptibilities of 111 *Campylobacter fetus* subsp. *fetus* strains isolated in Quebec, Canada, from 1983 to 2000. *J. Clin. Microbiol.,* 41:463–466.
86. V'Eron, M. and R. Chatelain. 1973. Taxonomic study of the genus *Campylobacter* Sebald and Veron, and designation of the neotype strain for the type species *C. fetus* (Smith and Taylor) Sebald and Veron. *Int. J. Sys. Bacteriol.*, 23:122.
87. Vandamme, P. 2000. Taxonomy of the family Campylobacteraceae, p. 3–26. In M. Blaser and I. Nachamkin (Eds.), *Campylobacter*, 2nd ed. ASM Press, Washington, DC.
88. Vandamme, P., M.I. Daneshvar, F.E. Dewhirst, B.J. Paster, K. Kersters, H. Goossens, and C.W. Moss. 1995. Chemotaxonomic analyses of *Bacteroides gracilis* and *Bacteroides ureolyticus* and reclassification of *B. gracilis* as *Campylobacter gracilis* comb. nov. *Int. J. Syst. Bacteriol.*, 45:145–152.
89. Vandamme, P., and J. de Ley. 1991. Proposal for a new family *Campylobacteraceae. Int. J. Syst. Bacteriol.*, 41:451–455.
90. Vandamme, P., E. Falsen, R. Rossau, B. Hoste, P. Segers, R. Tytgat, and J. dte Ley. 1991. Revision of *Campylobacter*, *Helicobacter*, and *Wolinella* taxonomy: emendation of generic descriptions and proposal of Arcobacter gen. nov. *Int. J. Syst. Bacteriol.*, 41:88–103.
91. Vandamme, P., B. Pot, M. Gillis, P. de Vos, K. Kersters, and J. Swings. 1996. Polyphasic taxonomy, a consensus approach to bacterial systematics. *Microbiol. Rev.,* 60:407–438.
92. Vandenberg, O., K. Houf, N. Douat, L. Vlaes, P. Retore, J.P. Butzler, and A. Dediste. 2006. Antimicrobial susceptibility of clinical isolates of non-*jejuni/coli* campylobacters and arcobacters from Belgium. *J. Antimicrob. Chemother.*, 57:908–913.
93. Waage, A.S., T. Vardund, V. Lund, and G. Kapperud. 1999. Detection of small numbers of *Campylobacter* jejuni and *Campylobacter coli* cells in environmental water, sewage, and food samples by a seminested PCR assay. *Appl. Environ. Microbiol.*, 65:1636–1643.
94. Wang, H. and D.R. Murdoch. 2004. Detection of *Campylobacter* species in faecal samples by direct Gram stain microscopy. *Pathology*, 36:343–344.
95. Wassenaar, T.M. and D.G. Newell. 2000. Genotyping of *Campylobacter* spp. *Appl. Environ. Microbiol.*, 66:1–9.

36 The Genus *Helicobacter*

Ernestine M. Vellozzi and Edmund R. Giugliano

CONTENTS

INTRODUCTION

Many observations of spiral bacteria colonizing the gastric mucosa of animals and humans have been noted throughout the 20th century, some of which suggested their role in gastric pathology.[1–3] Despite the presence of these microorganisms, the stomach was still considered a sterile environment. It was the discovery of *Helicobacter pylori* that forever changed our perception of the stomach, and made us view this organ as a habitat for specialized bacteria. Current knowledge suggests that *H. pylori* has most probably colonized the human stomach for thousands of years.[4]

Isolated from the human stomach in 1982, this organism was first categorized as a spiral-shaped bacterium resembling *Campylobacter*.[5] The organism bore many similarities to the campylobacters, including morphology, growth under microaerophilic conditions, G+C content, among others. As a result, the organism first referred to as a *Campylobacter*-like organism was later named *Campylobacter pylordis* in 1985.[6] The species name was subsequently revised to "*pylori*" in 1987.[7] Electron microscopy and fatty acid analysis of *C. pylori* soon demonstrated that it showed marked differences from the other *Campylobacter* species.[8] *C. pylori* was later reclassified into the new genus *Helicobacter*. This new classification was supported by sequencing studies of 16S rRNA.[9] Early studies described this bacterium as a probable cause of gastritis and peptic ulcer disease, and this etiology would later be validated.[10] Now, there is much evidence to support the role of *H. pylori* in gastric cancers such as adenocarcinoma and development of non-Hodgkin's lymphoma.[11, 12] The clinical significance of *H. pylori* has been emphasized by the World Health Organization, via the classification of this organism as a Class I definite carcinogen.[13] The isolation, characterization, and recognition of *H. pylori* have generated a renewed interest in the possible association of other spiral mucus-associated bacteria with disease. Organisms such as *H. mustelae* and *H. acinonychis* were discovered serendipitously while studying animal pathology, while others were uncovered when studying some of the gastric micro-ecosystems of various animals.[14, 17]

Virtually the stomachs of all animals are now known to be colonized by a wide range of highly adapted bacteria that belong to the genus *Helicobacter.*[18] The *Helicobacter* species as a group can be broadly categorized according to their site of colonization in the host. Gastric *Helicobacter* species are widely distributed in mammalian hosts and are nearly universally present in the gastric mucosa. This group of organisms is capable of causing an inflammatory response resembling that seen with *H. pylori* in humans.[19] Although not usually pathogenic to their natural host, the gastric helicobacters often serve as good models of human disease. Enterohepatic *Helicobacter* species are a diverse group of organisms that have been identified in the intestinal tract and liver of humans, other mammals, and birds. These organisms have been linked with inflammation or malignant transformation in immunocompetent hosts, whereas in immunocompromised humans and animals more severe clinical disease is observed.[20]

This chapter describes the genus *Helicobacter* and characterizes the role of these organisms in the pathogenesis of gastric and enterohepatic disease. Epidemiology, transmission, pathogenesis, virulence factors, along with methods in culture, identification, detection, and treatment, are reviewed. Subsequent discussion of animal models in this chapter illustrates how these models have helped investigators gather pertinent information to further our understanding of this fascinating group of organisms comprising the genus *Helicobacter.*

GENUS DESCRIPTION

The genus *Helicobacter* is comprised of more than 20 species, with additional provisionally named species waiting to be validated. Others have been isolated but await classification. New and more sophisticated techniques, especially in the field of molecular biology, have helped in the further delineation of the genus.[21–23] These Gram-negative, non-sporeforming rods typically have spiral, curved, or fusiform morphology ranging from 0.3 to 1.0 µm in width to 1.5–10.0 µm in length. The morphology of the spiral-shaped body varies greatly among the species. For example, *H. acinonychis* is an "S"-shaped organism similar to *H. pylori,* while *H. felis* and *H. heilmannii* have tightly spiraled bodies that resemble a corkscrew. Also, it is not uncommon for *Helicobacter* cells to develop a coccoid morphology with age.[24] Motility in the genus *Helicobacter* is due to the presence of flagella, which give them a characteristic rapid corkscrew-like or slower wave-like motion. Most species have bundles of multiple sheathed flagella, having either a polar or bipolar distribution, while others have a single polar or bipolar flagellum. *H. pylori* isolates, in contrast to other *Helicobacter* species, have multiple monopolar sheathed flagella. See Table 36.1. Flagellar distribution can also be peritrichous (*H. mustelae*) or non-sheathed (*H. pullorum, H. rodentium,* and *H. mesocricetorum*).[21] These basic morphology and motility characteristics are thought to play an important role in the ability of the organism to maneuver successfully in the mucous layer of their host gastrointestinal tracts.

The host range of colonization varies among the gastric *Helicobacter* species, with some being more restrictive than others. For example, *H. pylori* is associated with humans and other primates, *H. felis* with cats and dogs, *H. mustelae* with ferrets, *H. salomonis* with dogs, and *H. acinonychis,* which specifically colonizes the cheetah. Other gastric species, for example, *H. bizzozeronii, H. heilmannii,* and related bacteria, exhibit a much broader host specificity, naturally colonizing, cats, dogs, and primates, including humans.[25, 26] The presence of gastric *Helicobacter* species is associated with inflammation and/or ulceration, and these bacteria are recognized as important pathogens in gastroduodenal disease. In contrast to the gastric helicobacters, the enterohepatic helicobacters inhabit the intestinal and heptatobiliary tracts of mammals and birds. In humans, enterohepatic helicobacters, such as *H. canadensis, H. canis, H. cinaedi, H. fennelliae, H. pullorum* and *H. winghamensis,* and *Helicobacter* sp. *strain flexispira taxon 8,* have been isolated from rectal swabs and feces. Enterohepatic helicobacters have been linked to severe inflammatory lesions in the lower bowel, gall bladders, and livers of infected mammalian hosts including humans.[27–30]

In the laboratory, most species of *Helicobacter* will grow at 37°C under microaerophilic conditions in an atmosphere of reduced oxygen concentration (5 to 10% O_2), and increased humidity and carbon dioxide levels (5 to 12% CO_2). The role of atmospheric hydrogen in the microaerophilic incubation of *Helicobacter* is not fully understood. The requirement appears to be strain dependent, being required by some strains and having a growth stimulating effect in others.[21] Routine aerobic atmospheres do not support the growth of most *Helicobacter* species. Some species grow poorly, at best, in these conditions whereas several *Helicobacter* species, such as *H. rodentium*, can grow both microaerobically and anaerobically.[31]

Biochemically, helicobacters display a respiratory type of metabolism. Neither oxidation nor fermentation is observed when examining standard methods of carbohydrate utilization. However, glucose oxidation appears to occur in *H. pylori,* and oxidase activity is found in all species.[32] The principal means of energy production and biosynthesis is believed to occur via glycolysis and gluconeogenesis, respectively. Other catabolic pathways such as the Entner-Douderoff pathway, the pentose phosphate shunt, and the tricarboxcylic acid cycle are believed to be at least partially present while the glyoxylate shunt is absent in these organisms.[32, 33]

EPIDEMIOLOGY AND TRANSMISSION OF *HELICOBACTER PYLORI* AND OTHER HELICOBACTERS

Preventing the spread of infectious pathogenic agents has historically relied on public health measures that were based on sound epidemiological data. In the case of *Helicobacter*, in particular *H. pylori*, the epidemiology and especially the transmission of helicobacters are incompletely understood. A thorough understanding of these processes must be first elucidated prior to the possible implementation of control measures for this group of organisms. There is evidence to suggest direct person-to-person transmission of *H. pylori.* This mode of transmission is supported by the clustering of cases within families, the similarity of genotypes found among isolates of related persons, and the failure to find definitive evidence of an environmental reservoir.[34–36] However, the mechanism by which *H. pylori* is transmitted from one host to another remains speculative. In developing countries, fecal/oral transmission of *H. pylori* by a contaminated food or water supply appears to be a likely means of spread of *Helicobacter.*[37] Detection of *H. pylori* in sewage in Lima, Peru, by polymerase chain reaction and recovery of viable *H. pylori* in the stool of children lend support to this means of transmission.[37, 38] One of the major difficulties in attempting to culture *H. pylori* from feces or the oral cavity is the presence of normal flora organisms. These tend to grow more rapidly than *H. pylori* and may mask the presence of the organism.[36] Person-to-person transmission by way of saliva may also be possible as *H. pylori* has been isolated from dental plaque, saliva, and vomitus.[39, 40] Although *Helicobacter* organisms have never been cultured from environmental sources, molecular and immunological methods have been used to detect *Helicobacter* DNA in contaminated municipal water sources. Failure to consistently isolate *H. pylori* from reservoirs other than humans suggests that direct person-to-person contact is the most likely mode of transmission.[41, 42] In humans, *H. pylori* is believed to be acquired early in life where children are the primary source of infection in families.[43, 44]

In a similar fashion, ferret colonization with *H. mustelae* also demonstrates a relationship of prevalence to age that mimics that of *H. pylori* in humans. Studies have shown that *H. mustelae*

TABLE 36.1
Habitats and Phenotypic Characteristics of *Helicobacter* Species[a]

Helicobacter Taxon	Source(s)	Primary Site	Catalase Production	Nitrate Reduction	Alkaline Phosphatase	Urease	Indoxyl-Acetate Hydrolysis	γ Glutamyl tranferase	Growth at 42°C	Growth with 1% Glycine	Resistance to Naladixic Acid[b]	Resistance to Cephalothin[b]	Flagella
Human													
H. bizzozeronii[c]	Human, cat, dog, primate	Stomach	+	+	+	+	+	+	+	–	R	S	Bipolar
H. canis	Human, cat, dog	Intestine	–	–	+	–	+	+	+	–	S	I	Bipolar
H. canadensis	Human	Intestine	+	+/–	–	–	+	–	+	+	R	R	Mono/ Bipolar
H. cinaedi	Human, hamster, macaque	Intestine	+	+	–	–	–	–	–	+	S	I	Bipolar
H. fennelliae	Human	Intestine	+	–	+	–	+	–	–	+	S	S	Bipolar
H. pullorum	Human, chicken	Intestine	+	+	–	–	–	ND[d]	+	–	R	S	Monopolar
H. pylori	Human, macaque, cat	Stomach	+	–	+	+	–	+	–	–	R	S	Monopolar
Helicobacter sp. strain flexispira taxon 8[e]	Human, dog, sheep, mouse	Intestine	+/–	–	–	+	–	+	+	–	R	R	Bipolar
H. winghamensis	Human	Intestine	–	–	–	–	+	ND	–	+	R	R	Bipolar
Non-human													
H. acinonychis	Cheetah	Stomach	+	–	+	+	–	+	–	–	R	S	Bipolar
H. bilis	Mouse, dog, rat	Intestine	+	+	–	+	–	+	+	+	R	R	Bipolar
H. cholecystus	Hamster	Liver	+	+	+	–	–	–	+	+	I	R	Monopolar
H. felis	Cat, dog	Stomach	+	+	+	+	–	+	+	–	R	S	Bipolar
H. hepaticus	Mouse	Intestine	+	+	–	+	+	–	–	+	R	R	Bipolar
H. mesocricetorum	Hamster	Intestine	+	+	+	–	ND	–	+	–	S	R	Bipolar
H. muridarum	Mouse, rat	Intestine	+	–	+	+	+	+	–	–	R	R	Bipolar

H. mustelae	Ferret, mink	Stomach	+	+	+	+	+	+	+	–	S	R	Peritrichous
H. pametensis	Birds, swine	Intestine	+	+	+	–	–	–	+	+	S	S	Bipolar
H. rodentium	Mouse	Intestine	+	+	–	–	–	–	+	+	R	R	Bipolar
H. salomonis	Dog	Stomach	+	+	+	+	+	+	–	ND	R	S	Bipolar
H. trogontum	Rat	Intestine	+	+	–	+	–	+	+	ND	R	R	Bipolar

[a] All *Helicobacter* species are oxidase positive and lack the ability to oxidize or ferment carbohydrates in routine reactions.

[b] Resistance is determined by disk diffusion. Isolates are incubated for several days at 37°C on blood-containing medium containing 30-μg antibiotic disks. Microaerobic conditions are typically used, and exact incubation times vary among organisms. Resistance (R) is defined as the complete absence of an inhibition zone, whereas intermediate (I; zones usually < 15 mm) and susceptible (S; zones usually >20 mm in diameter) isolates have visible inhibition zones of various sizes.

[c] Probably the same as "*H. heilmannii*." "*H. heilmannii*" (formerly *Gastrospirillum hominis*) has the same phenotype as listed here for *H. bizzozeronii*. Only a single "*H. heilmannii*" strain has been isolated by culture, and so is not included in this table.

[d] ND, not determined.

[e] Formerly regarded as "*Flexispira rappini*," now has been subgrouped into ten taxa.

Source: Adapted from Versalovic, J. and Fox, J.G., *Helicobacter*, in *Manual of Clinical Microbiology*, 8th ed, Vol. I, Murray, P.R., Ed., ASM Press, Washington, DC, 2003. With permission.

infection ranges from 0% in ferrets less than 1 month of age to 100% in adult animals over 1 year of age.[45, 46] Fecal-oral spread of *H. mustelae* in the ferret population has likewise been proposed, as well as the possible zoonotic transmission of these gastric helicobacters to humans.[47] Naturally infected cats and sheep have been found to have *H. pylori* in their secretions and feces, also raising the possibility of transmission to humans.[48, 49] Likewise, the zoonotic food-borne transmission of enterohepatic helicobacters, such as *H. pullorum* and *H. canadensis,* has been speculated.[29, 50] The significance of enterohepatic helicobacters in transmission of disease and the true prevalence of these organisms in human and animal populations are yet to be determined. What has become apparent over the years is that *Helicobacter* species are commonly present as part of the gastric, enteric, and hepatobiliary microbiota in humans and other animals.[51] See Table 36.1.

PATHOGENESIS

The gastric helicobacters have evolved and acquired a number of traits that enable them to survive in a harsh environment, and the complex mechanisms used by these organisms to establish infection and maintain colonization is just beginning to be understood. Knowledge of the prevalence and natural history of *Helicobacter* infections as they occur in nature in non-humans is likewise limited.[52] Gastric *Helicobacter* species exhibit different degrees of host specificity; hence, manifestations of disease are host specific in most cases. High levels of host specificity in gastric helicobacters are most evident in those species that adhere tightly to host tissues and are able to agglutinate red blood cells.[53, 54] The wide range of *Helicobacter* infections seems to indicate that these bacteria may not always have a true pathogenic relationship with their host.[55] Although some *Helicobacter* strains clearly cause more damage to cells than others, much of the pathogenesis seen in *Helicobacter* infections such as *H. pylori* may be due to the host response to the bacterium, rather than any significant toxicity mediated by the bacterium itself. Some studies suggest that colonization of the stomach by this organism serves as a protective factor against gastroesophageal reflux disease and adenocarcinoma of the esophagus.[55] These types of pathology have increased at an alarming rate over the years and coincide with the dramatic decrease in gastric cancer and increased eradication of *H. pylori*.[56] This view of the role of *Helicobacter* is one of a commensal displaying a symbiotic host-parasite relationship.[57] Most of the pathology seen with gastric helicobacters such as *H. mustelae, H. acinonychis,* and *H. felis* occurs when the organism is placed in an unnatural host or when infection has occurred for a long period of time. Of course there are exceptions; for example, *H. cinaedi* and *H. fennelliae* clearly produce disease in the primate model and have also been found to be associated with human disease.[58] Unlike *H. pylori*, *H. cinaedi* has been cultured from human blood and joint fluid, as has *H. fennelliae* from human blood.[59] It is believed that helicobacters that invade the bloodstream do so via colonization of the lower gastrointestinal tract.[60] *H. heilmannii* appears to cause minimal inflammation in non-human primates and other natural hosts, as do the multiple species that infect dogs and cats. However, *H. heilmannii,* which has been found in human gastric biopsy specimens, has also been cultured from human stomach tissue.[21, 61, 62] This organism has also been associated with mild to moderate gastritis in cats and dogs; peptic ulcer disease in swine; and gastritis, peptic ulcer disease, and gastric mucosa-associated lymphoid tissue (MALT) lymphomas in humans.[63–65] In the naturally infected ferret, the pathology resulting from *H. mustelae* is usually mild, whereas most of the severe pathology observed usually occurs through experimental infection by this species. However, infection with *H. mustelae* in ferrets can also lead to precancerous lesions and has also been linked to gastric adenocarcinoma and MALT lymphoma in these animals.[66]

In addition, severe pathologic changes can occur without apparent clinical manifestations of disease. A good example of this observation occurs in immunocompetent mice where *H. hepaticus* can often cause chronic active hepatitis and formation of hepatocellular tumors. Although the animals are infected and have developed severe pathology, the animals nevertheless appear to be clinically fit.[67,68] In humans, natural infection with *H. pylori* is likely acquired quite differently from animal infection models, which require large and often repeated inocula.[68,69] It is widely believed that in the absence of treatment, *H. pylor*i infection, once established, persists for life. In the elderly, however, it is likely that infection can disappear. This is most likely due to the stomach mucosa becoming increasingly atrophic with age, rendering it inhospitable to colonization.[70] The proportion of acute infection that persists is not known. However, in studies that have followed the natural history of disease progression in populations, spontaneous elimination has been reported.[70,71]

Helicobacter can cause direct damage to epithelial cells in the gastric mucosa as well as induce an inflammatory response in the host. Both host factors and organism factors determine the phenotypic expression of thc infection over time. Essentially, all persons colonized with *H. pylori* develop gastric inflammation, known as chronic superficial gastritis. The condition can develop into chronic inflammation of the gastric mucosa if the bacteria are not eradicated. However, this chronic inflammation is clinically silent in most infected persons.[43] This type of gastritis usually predisposes the host to hyperacidity and duodenal ulcer disease. In cases where the gastritis results from long-standing infection, it becomes more generalized and affects the corpus, leading to glandular atrophy and intestinal metaplasia. This is a recognized precursor state for gastric ulcer disease, gastric carcinoma, and MALT lymphoma.[72–75]

Additional species of *Helicobacter* have been implicated as causes of human gastroenteritis, especially in immunocompromised individuals.[76,77] Human cases of gastroenteritis have been associated with the following novel *Helicobacter* species including *H. canadensis*, *H. canis*, *H. pullorum,* and *H. winghamensis.*[28,29,36,50,78,79] *Helicobacter* species, such as *H. bilis*, *H. pullorum,* and *H. hepaticus* have also been detected in the human hepatobiliary tract.[80,81] Similar helicobacters have also been found to colonize and cause inflammation in other mammalian species. This includes, for example, *H. cinaedi* in the rhesus monkey and *H. hepaticus* in mice.[82–84]

The role of *Helicobacter* is less clear in non-ulcer dyspepsia, nonsteroidal antiinflammatory drug-induced (NSAID) ulcer, gastroesophageal reflux disease (GERD), chronic inflammation in coronary disease, and pancreatic cancer.[85–89] Conditions in which there is a suggested role for *Helicobacter* include iron-deficiency anemia and idiopathic thrombocytopenic purpura.[90–92]

VIRULENCE FACTORS

Many virulence factors have been described in *Helicobacter* over the years, as summarized in Table 36.2. These factors enable the organism to survive the extreme acidic environment of the gastric tract, to reach the more neutral environment of the mucus layer, to colonize the gastric epithelium, and to evade the human immune response. This progression of events ultimately results in persistence of the organism, leading to possible infection and subsequent disease. Flagella, urease, and adhesins are essential colonizing factors. Among the other unique adaptations of *H. pylori* that enable survival in the hostile environment of the stomach is the presence of vacuolating cytotoxin (VacA) and oncogenic cytotoxin associated antigen A (cagA) protein. These virulence factors are discussed in greater detail below.[21,56,93–141]

Flagella and Motility

Movement is an important function of the gastric colonizers. First, it prevents the organisms from being washed out of the stomach and, second, allows the organisms to find suitable nutrients, while evading the local hostile environment. A fully functioning flagellum structure is essential for optimal survival. Studies involving the genes responsible for the flagella filament structure reveal that successful colonization and organism survival became compromised when these genes were disrupted.[142] Most of the gastric colonizers possess a spiral-shaped body with one exception, *H. mustelae,* which is a short rod. The morphology of the bacterium and the type of flagella dictate how

TABLE 36.2
Virulence Factors of *Helicobacter*

Virulence Factor	Type	Proposed Function	Ref.
Flagella	Organelle	Colonization	21, 93
Urease	Enzyme, immunogen	Colonization, neutralization, tissue damage (tight junctions)	93–98
Adhesins	Cell ultrastructure	Anchor mechanism	99–101
Superoxide dismutase	Enzyme	Prevents phagocytosis and bacterial cell death	102
Catalase	Enzyme	Prevents phagocytosis and bacterial cell death	94, 103
Glycosulfatase	Proteolytic enzyme	Mucin degradation	104
Phospholipase A	Enzyme	Tissue damage (cell membrane disruption)	105
Lipopolysaccharide (LPS)	Endotoxin	Epithelial cell apoptosis	106–108
Lewis blood group antigens	ABO binding adhesins	Molecular mimicry and autoimmunity	109–113
Vacuolating cytotoxin (VacA)	Cytotoxin	Tissue damage	114–124
Cytolethal distending toxin (CDT)	Cytotoxin	Tissue damage (found in enterohepatic species of *Helicobacter*)	125–129
Cytotoxin associated antigen (cagA)	Antigen	Oncogenic activity	56, 130–133
Coccoid forms	Morphological adaptation	Possible bacterial cell survival	134–141

the organism moves and ultimately where in the gastric mucosa the organism will be located. For example, the movement of *H. felis* and *H. heilmannii* differs completely from that of *H. mustelae,* and therefore these bacteria are found in different regions of the stomach. These spiral organisms do not adhere to the cell surface, but instead are found free tracking backward and forward along the mucus strands with their corkscrew motion.[143] The curved morphology of *Helicobacter* and the polar flagella of species such as *H. pylori* create screw-like motion, which may enable the organism to penetrate the mucin layer. Motility in helicobacters has been found to be pH dependent and becomes impaired at a pH below 4.0.[93]

Urease

All gastric helicobacters possess the urease enzyme. Among the proposed functions for the presence of urease in *Helicobacter* is the ability of the enzyme to protect the bacterium from the hostile acid environment of the stomach. Urease is also strongly immunogenic and chemotaxic for phagocytes.[94] The enzyme catalyzes the breakdown of urea to alkaline ammonia and bicarbonate. The byproduct ammonia has the potential to cause tissue injury; but more importantly, the local alkaline microenvironment created offers a protective mechanism against the acid environment. This protective function enables the organism to reach its final location in the gastric mucus and therefore urease is believed to be a critical virulence factor necessary for colonization. Support for this theory has been demonstrated in studies with urease-negative mutants, which were shown to lack the ability to colonize the gastric mucosa at normal physiological pH.[95, 96, 144] *H. pylori* produces approximately 5 to 10% of its protein as urease enzyme.[97] The urease produced by *H. pylori* is of special interest as it is 10 to 100 times more active than other bacterial ureases and is located in the membrane, rather than the cytoplasm, of the organism. In addition, the Michaelis-Menten constant (K_m) of the urease is much lower than that of other bacterial ureases, a finding consistent with the very low urea levels in the stomach.[98] The ureases of *H. pylori*, *H. felis*, and *H. heilmannii* are more closely related to one another than they are to the urease of *H. mustelae*.[145] Studies have shown that the enzyme of *H. heilmannii* might be different from those of other gastric *Helicobacter* species. However, there is also evidence to suggest that the *H. heilmannii* enzyme may be more similar to the urease of *H. felis*.[146] Unlike helicobacters that colonize the stomach, expression of urease in enterohepatic *Helicobacter* species is variable. See Table 36.1.

Adhesins

Adhesins are bacterial proteins, glycoconjugates, or lipids that are involved in the initial stages of colonization by mediating the interaction between the bacterium and the host cell surface.[100] Host cell receptors are composed of lipids, proteins, glycolipids, or glycoproteins. Adherence of bacteria to host cell receptors triggers cellular changes that include signal transduction cascades, leading to infiltration of inflammatory cells and to the local persistence of the organism. *H. pylori* has evolved adherence mechanisms to efficiently maintain itself in the gastric mucosa despite cellular turnover and has been demonstrated experimentally both *in vivo* and *in vitro*.[100, 101] *In vitro* studies of adherence have demonstrated nonspecific attachment, although it has been shown that attachment mechanisms *in vivo* appear to have a greater host tissue specificity.[147, 148] The attachment mechanisms of *H. pylori* to gastric epithelial cells are associated with cytoskeletal rearrangements and modification of host proteins, which are known to occur in enteropathogenic *Escherichia coli* adherence processes.[149] Similar structures have also been observed in ultrastructural studies of *H. mustelae* in its adherence to ferret gastric mucosa.[150] The outer membrane proteins required for *in vitro* attachment of *H. pylori* to gastric epithelium have not been demonstrated in other gastric helicobacters such as *H. mustelae* and *H. felis*.[151] However, *H. mustelae* and *H. pylori* appear to bind common lipid receptors such as phophatidylethanolamine (PE). The adhesion to intact eukaryotic cells has been found to correlate with the amount of PE present in the membrane.[152, 153] Other cell surface properties, such as hydrophobicity, may also contribute to the gastric epithelial binding of helicobacters, as in the case of *H. mustelae*.[154] *H. heilmannii,* which is typically found in the mucus layer above surface epithelial cells, does not show the closely associated adherence structures seen in *H. pylori* adherence.[155]

Other Enzymes

Additional enzymes have been identified in the genus *Helicobacter* that enhance survival in the host environment. The action of superoxide dismutase, for example, breaks down the superoxide produced in polymorphonuclear leukocytes and macrophages, preventing their destruction of organisms. This enzyme has been isolated from *H. pylori*, for example, and is believed to play a role in its survival.[102] In a similar fashion, catalase is believed to protect *H. pylori* and other helicobacters from the damaging effects of hydrogen peroxide released from phagocytes.[103] Also, both urease and catalase, when excreted into the surrounding environment, are believed to have a somewhat protective role against the humoral immune response.[94] Proteolytic enzymes such as glycosulfatase may cause degredation of gastric mucin, and the production of phospholipase A may contribute to the degradation of cell membranes by helicobacters.[104, 105]

Lipopolysaccharide

The cell wall of most Gram-negative bacteria contains lipopolysaccharide (LPS). This endotoxin is known to play a key role in the induction of host immune responses. The LPS component of the *H. pylori* cell wall has a lower biological activity when compared to other helicobacters and to other Gram-negative bacteria.[156] For example, *H. felis*-associated LPS appears to exert an inflammatory effect on host cells, whereas *H. pylori* endotoxin is a poor inducer of cytokine production by inflammatory cells *in vitro*.[157] This observation is explained in part by the unusual structure of lipid A in *Helicobacter*.[106, 107] The LPS from *H. pylori* is of special interest because of evidence that it expresses human Lewis (Le) antigens that are also present on the gastric epithelium.[113] The lower biological activity of the LPS, combined with host antigens on its surface, may be a mechanism for *H. pylori* to evade the host inflammatory response. This property could enable long-term colonization by the organism.[158] Auto-antibodies targeted against the bacterial Le antigens may play a role in the pathogenesis of gastric disease.[159] This role is supported by studies that show that *Helicobacter* strains isolated from patients with ulcer disease express an increased number of Lewis antigens as compared to those isolates from patients with dyspepsia.[111, 112] In contrast, the LPS of the gastric helicobacter *H. mustelae* does not express Le antigens, nor are the antigens expressed on ferret gastric epithelial cells. However, both ferret gastric epithelium and *H. mustelae* LPS express blood group antigen A. This antigenic expression may be considered a mechanism of molecular mimicry similar to the expression of Le antigens by *H. pylori*.[160, 161]

Vacuolating Cytotoxin A

The toxins of *Helicobacter* species have been implicated in lesion formation associated with gastric and enterohepatic infections. The *H. pylori* VacA protein was the first of these toxins to be identified and characterized. This protein was initially described as inducing a vacuolating cytopathic effect in HeLa cell tissue cultures.[115, 116] The *H. pylori* strains most frequently associated with duodenal ulcer disease were found to produce toxogenic forms of VacA. Both purified VacA protein and sonicated cell extracts of VacA-producing *H. pylori* were found to induce ulcer-like lesions when administered to mice.[162] This finding supports the possible role of VacA in mediation of ulcer formation in the host. Peptic ulcers have also been observed in gnotobiotic piglets and Mongolian gerbils that were colonized with *H. pylori*. Interestingly, these effects were also observed in mice, ferrets, and swine that were infected with non-*H. pylori* species.[53, 163]

Some evidence exists that supports a role for VacA in the colonization process. The cellular morphologic changes induced by the protein as well as the gastric epithelial cell necrosis appear to create a nutrient-rich environment that optimizes organism survival.[117–119] Immunization of mice with VacA prevented infection when mice were subsequently challenged with *H. pylori*. However, the relationship of this cytotoxin to precancerous lesions, if any, is yet to be elucidated.[122–124] The *vacA* gene appears to be absent from *H. felis,* one of the gastric helicobacters.[164] Similarly, *H. mustelae* does not appear to produce vacuolating cytotoxin activity *in vitro*.[165] This observation suggests that there may be other factors involved in ulcer disease production. Hence, the definitive role of VacA in disease processes is not yet known.

Cytolethal Distending Toxin

A new toxin activity has been identified among the enterohepatic helicobacters that, like VacA, causes vacuole formation in cell lines. This was first demonstrated in *H. hepaticus,* which produced a toxic effect in a murine liver cell line resulting in a granular appearance of the affected cells.[125] This granulating cytotoxin is a heat-labile, secreted protein that is distinct from the vacuolating cytotoxin of *H. pylori*. This cytolethal distending toxin (CDT) activity identified in *H. hepaticus* was found to closely resemble the CDT of *Campylobacter* species.[126, 127] The role of CDT in *H. hepaticus* pathogenesis needs to be delineated. The expression of CDT activity has also been observed in other enterohepatic *Helicobacter* species, such as *H. canis, H. pullorum,* and *H. bilis*.[128, 129]

Cytotoxin-Associated Gene Antigen

One of the most intensely studied proteins of *H. pylori* is Cytotoxin-Associated Gene Antigen (CagA); however, the complete function of this protein is not fully understood. The *cagA* gene of *H. pylori* is believed to be partially responsible for initiating signaling mechanisms that lead to the development of gastric carcinoma. It has been shown that the risk of gastric carcinoma development appears to be much greater in *H. pylori*-infected populations than in those that are non-infected.[56] Both *cagA*-positive and *cagA*-negative *H. pylori* strains exist in nature. *cagA*-containing strains are associated with higher grades of gastric or duodenal ulceration and are more virulent than *cagA*-negative counterparts.[130] Epidemiological studies have demonstrated the roles of *cagA*-positive *H. pylori* strains in the development of atrophic gastritis, peptic ulcer disease, and gastric carcinoma.[131, 132] It is believed that gene *cagA* is translocated to the gastric epithelial cells and undergoes kinase-mediated tyrosine phosphorylation. Upon phosphorylation, *cagA* binds and activates a specific phosphatase believed to act as a human oncoprotein. This protein can then trigger oncogenic activities.[133, 166]

In animal hosts infected with non-*H. pylori* gastric helicobacters, such as *H. felis*, development of severe, chronic gastritis is known to occur. However, the presence of a *cagA* equivalent in these bacteria has not yet been definitively demonstrated. The induction of chronic gastritis in *Helicobacter*-infected animal hosts may therefore be independent of this cytotoxin.[164]

Coccoid Forms

Like other spiral-shaped bacteria, members of the *Helicobacter* genus convert to a coccoid morphology as cells begin to age. *In vitro* cell culture studies as well as *in vivo* animal models have been used to help divulge the significance and viability of coccoid forms.[134–136] However, controversy still exists as to the role of coccoid forms in transmission, colonization, infection, and pathogenesis.[137–141]

CULTURE OF *HELICOBACTER*

The culture and isolation of *Helicobacter* in the laboratory is technically demanding and requires proper specimen transport, and specific media and growth conditions.[21] Given the effectiveness of currently available *Helicobacter* detection methods, it could be argued that culture is not an essential routine procedure in the diagnosis of *H. pylori* infections. However, there are many instances in which culture remains a valuable tool. For example, culture is indicated when evaluating antimicrobial resistance to prevent antibiotic treatment failures. There is also an increasing need to fingerprint isolates of *H. pylori* and other *Helicobacter* species for epidemiological purposes. Many of these methods of molecular typing are relatively simple, well validated, and readily available, requiring pure cultures of newly isolated organisms.[34, 35] Culture is also a basic research tool in the study of virulence factors such as urease and cytotoxin, among others. It is important to note that continuous passage on artificial media may cause isolates to lose properties essential for growth in the host and induction of disease. For this reason, those undertaking research with laboratory strains of *Helicobacter* isolates should repeat their studies using newly isolated organisms.[167]

Specimens for Culture of *Helicobacter*

Blood, feces, gastric, and tissue biopsies are sources for *Helicobacter* isolation. Although, *H. pylori* and other gastric helicobacters are rarely isolated from human blood, the enterohepatic helicobacters can often cause invasive infections. If bacteremia is suspected, peripheral venous blood should be collected as recommended by the manufacturer, in commercially available aerobic and anaerobic blood culture bottles. The gastric helicobacters *H. pylori* and *H. heilmannii* are rarely isolated from human fecal specimens. However, enterohepatic helicobacters, like campylobacters, can be cultured routinely by modified stool culture, preferably within 4 hr of collection. Glycerol-containing media can serve as both transport and storage media for biopsy specimens. Brucella broth with 20% glycerol is recommended as it is readily available commercially. Biopsy material should be minced or homogenized in 0.9% saline or 20% glucose prior to plating.[167]

Culture Techniques

In the laboratory, *Helicobacter* strains will typically grow under microaerophilic conditions at 37°C with increased humidity. Growth does not occur in aerobic conditions, with the exception of species like *H. rodentium,* which grows both microaerophilically and anaerobically. Although some species may take up to 7 days to grow, many others produce colonies in 3 to 4 days of incubation.[167] Microaerophilic environments can be created using a variable-atmosphere incubator, partially evacuated anaerobic jars with defined gas mixtures, or commercial gas-generating packets.[21] *Helicobacter* can be grown on a variety of rich agar bases supplemented with 5% whole blood or serum. These media include chocolate agar, Skirrow *Campylobacter* medium, Meuller Hinton agar with 5% sheep blood, and Marshall brain heart infusion medium with 7% horse blood. Vancomycin, nalidixic acid, and amphotericin are common additives when a selective medium is required. When isolating *H. pylori*, the antibiotic cephalothin should be omitted, due to its susceptibility to this agent.[167, 168] Most media that will support the growth of *H. pylori* can be used to grow *H. felis* as well. Fastidious species such as *H. bizzozeronii* and *H. salomonis* prefer very moist environments, and a thin broth layer can be poured on the top of the agar surface to stimulate growth.[169] Successful isolation of most enterohepatic helicobacters from feces requires a defined microaerobic atmosphere (5 to 10% each of CO_2, H_2, O_2) using partially evacuated anaerobic jars. Commercially available gas-generating packets may be inadequate in supplying the increased atmospheric hydrogen required by this group of helicobacters.[21]

IDENTIFICATION

Helicobacters produce varying colonial phenotypes on blood agar. *H. pylori* and gastric helicobacters tend to produce entire, gray, and transluscent colonies, whereas the intestinal helicobacters, such as *H. cinaedi, H. fennelliae, H. pullorum,* and *H. canadensis*, produce various swarming phenotypes as most isolates are motile. *Helicobacter* morphology resembles that of other Gram-negative curved and spiral bacteria. See Figure 36.1. Enterohepatic helicobacters cultured from blood may require special staining techniques to visualize the organisms. Such is the case when culturing *H. cinaedi,* which is generally not seen in a Gram stain preparation, but instead requires dark-field microscopy, Giemsa or acridine orange staining for the best visualization of the bacterial cells.[59] A modified Gram stain using Carbol-fuchsin (0.5% wt/vol) counterstain has been found to augment detection of most helicobacters.[21] When microscopically identifying helicobacters from biopsy material, Gram stains of histologic sections or smears of gastric mucus often can reveal the presence of the organisms. The typical histologic stain, hematoxylin and eosin (H&E), is excellent for identifying the presence of gastritis and other gastric pathologies, but is not considered reliable for identifying the actual bacterium. The Warthin-Starry and modified Giemsa stains make identification of the bacterial cells easier but are still not considered ideal staining techniques. Triple stains, which combine H&E, Alcian blue, and a third stain such as the Genta or the El-Zimaity triple stain, have been found to be superior for identification of *H. pylori* and allow for easier delineation of gastric morophology.[170, 171] Helicobacters are routinely tested for cytochrome oxidase, catalase, and urease activity. All are oxidase-positive and most species, including *H. pylori,* are catalase-positive and urease-positive.[168] Susceptibility to naladixic acid and cephalothin are also used as phenotypic markers of identification. See Table 36.1.

ANTIBIOTIC SUSCEPTIBILITY AND ERADICATION THERAPY

Various susceptibility testing methods have been used to assess antimicrobial resistance in *Helicobacter.* These methods include broth microdilution, disk diffusion, the Epsilometer test (E-test®), and agar dilution.[172, 173] Multiple-drug regimens have been found to have the highest eradication rates for *H. pylori.* Triple therapy using a nitroimidazole such as metronidazole, or a macrolide

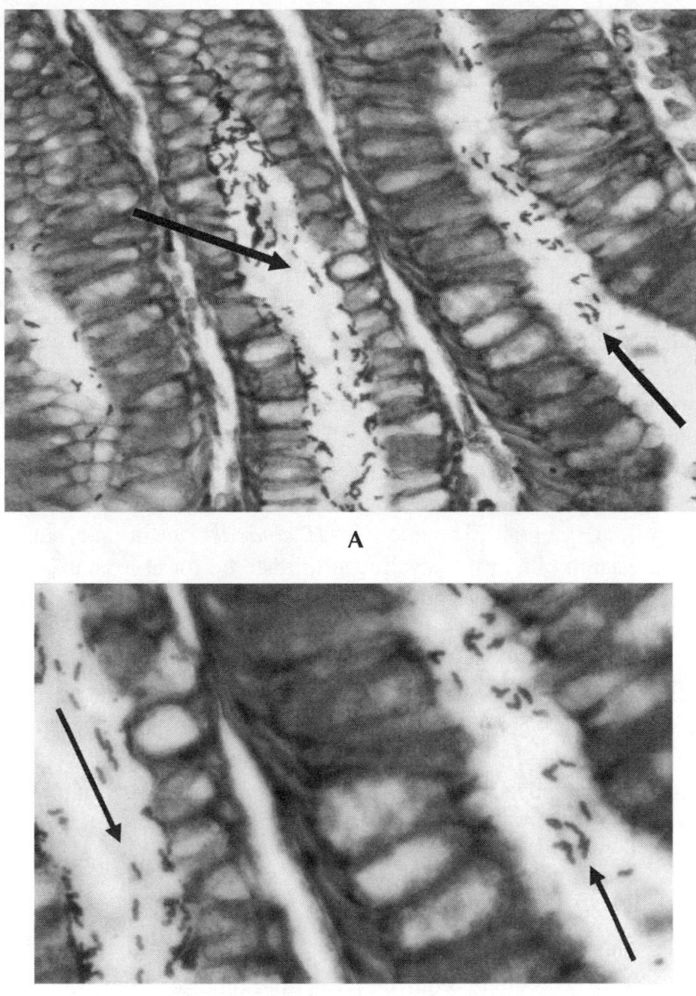

FIGURE 36.1 Warthin Starry stain of *Helicobacter pylori* in the gastric mucosa demonstrating typical curved cell morphology. (Photomicrograph courtesy of Long Island Jewish Medical Center, North Shore Long Island Jewish Health System, Department of Pathology, New Hyde Park, New York.)

such as clarithromycin, are among the antibiotics that can be included as part of the multiple-drug regimen to achieve high cure rates. Antacid medications such as proton pump inhibitors or H_2 antagonists are usually added to the regimen to reduce acid output and to encourage ulcer healing. Proton pump inhibitors suppress acid production by halting the mechanism that pumps acid into the stomach. H_2 blockers work by blocking histamine, which stimulates acid secretion. Bismuth-containing compounds, such as bismuth subsalicylate, can be added to the regimen to protect the stomach lining from acid damage. The compound also has some inhibitory effect on the organism. Triple therapy has been reported to eradicate *H. pylori* in approximately 90% of individuals.[174]

The Clinical and Laboratory Standards Institute (CLSI) recommendations for antimicrobial susceptibility testing are currently available for *H. pylori* only. The CLSI recommends agar dilution susceptibility testing of clarithromycin when testing *H. pylori*. Susceptibility testing for metronidazole has not yet been standardized.[175] The prevalence of *H. pylori* strains resistant to either metronidazole or clarithromycin has been increasing over the years and is associated with a greater incidence of treatment failure.[176] Antibiotic susceptibility data for *H. heilmannii* are somewhat limited. However, *H. heilmannii* isolates have been reported as susceptible to amoxicillin, ciprofloxacin, erythromycin, and tetracycline, as well as to naladixic acid and metronidazole.[26] Although, *H. heilmannii* appears to be sensitive to triple therapy in humans, long-term eradication from animal hosts such as cats, dogs, and primates has not been as effective.[173, 177] *H. felis*, which has been isolated from cats, dogs, and occasionally, the human stomach, is sensitive to a variety of antimicrobial agents. These compounds include metronidazole, cephalothin, erythromycin, ampicillin, and the antimicrobial agents used in triple therapy. Clearance of *H. felis* from mice using the triple therapy combination has led to the development of an animal model that can be used for the study of new antimicrobial regimens against *H. pylori*.[178] The antimicrobial therapy that is required to clear *H. mustelae* infection from ferrets is similar to that used for *H. pylori* in humans.[179] *H. acinonychis* is resistant to naladixic acid and sensitive to ampicillin, penicillin, erythromycin, gentamicin, and chloramphenicol. *In vivo* triple antibiotic therapy has been found to be unsuccessful in totally clearing this organism from the cheetah host. This observation is also true for *H. salomonis* and *H. bizzozeronii,* which can be suppressed but not be totally eradicated from the stomach of dogs.[17]

Intestinal helicobacters, which can be causative agents of bacteremia, require combination antibiotic therapies that include aminoglycosides such as amikacin or gentamicin. Prolonged therapy is often necessary to eradicate infection. Treatment for *H. cinaedi*, for example, can include antibiotics such as ciprofloxacin, gentamicin, or tetracycline administered for at least a 2- to 3-week period.[59] Although data are limited, gentamicin and ampicillin-sulbactam have been found to successfully treat bacteremia caused by *H. fennelliae*.[60, 180] Likewise, limited information is available for *in vitro* susceptibilities and treatment regimens for gastroenteritis caused by *H. canis* and *H. pullorum*. In addition to antibiotics, there is now evidence indicating that some natural compounds may have an inhibitory effect against *Helicobacter.* In particular, studies have shown that regular consumption of broccoli sprouts may eradicate *H. pylori*. Similarly, green tea extract has been shown to suppress *H. pylori* growth in animal models.[181] Green tea extract has also been found to inhibit *H. pylori* adhesion to human epithelial cells *in vitro*.[182]

DETECTION METHODS

There are a number of both invasive and noninvasive approaches for the detection of *Helicobacter,* including *H. pylori*. Invasive methods require endoscopic biopsies and include rapid urease testing, histological examination, culture, and polymerase chain reaction. Noninvasive testing methods include serologic testing, the urea breath test, and the stool antigen test.[183] Although the details of these test methods are beyond the scope of this chapter, additional information can be found in Table 36.3.[21, 167–171, 184–194]

ANIMAL MODELS FOR THE STUDY OF *HELICOBACTER*

Much of the current understanding of *Helicobacter* pathogenesis, and especially that of *H. pylori*, has derived from *in vitro* studies. However, these methods often fail to demonstrate the true complexities of the host-pathogen relationship. The availability of small animal models has augmented and improved our understanding of the *in vivo* mechanisms of host adaptation by these organisms. Many different animal models, such as mice, ferrets, dogs, and cats, have been used over the years for studying *Helicobacter* organisms. These animal models have generated important information that has also been physiologically relevant to human disease. Mice, rats, and rabbits were some of the first animal models used for the establishment of *H. pylori* infection, but they did not prove successful.[195] Animal models were then sought based on their natural infectivity with closely related *Helicobacter* species, such as the ferret or non-human primates, or by their ability to be experimentally infected with *H. pylori*, such as gnotobiotic piglets and dogs.[196–199] Experimental *H. pylori* infections have also been developed in Mongolian gerbils, guinea pigs, cats, and macaque monkeys.[200–203] Likewise, several animal models have been developed to study other gastric heli-

TABLE 36.3
Methods for Detection of *Helicobacter*

Test Method	Principle	Comments	References
Histological examination	Direct observation of organism in tissue	Requires invasive procedures such as endoscopy	170, 171
Culture	Isolation of *Helicobacter* from specimens	100% sensitivity and specificity; requires invasive procedures; technically demanding	21, 167, 168
Rapid urease	Indirect detection of *Helicobacter* by demonstration of urea breakdown products	Rapid results and easy to perform; false-positive results possible; endoscopy required	184, 185
Radiolabeled C^{13} or C^{14} urea breath tests	Measurement of labeled CO_2 in the breath; derived from urease hydrolysis of ingested C-labeled urea	Non-invasive; well tolerated; high sensitivity and specificity; used in diagnosis of H. pylori infection	186, 187
Serology	Measurement of *Helicobacter*-specific immunoglobulins IgG, IgA, and IgM by ELISA and latex agglutination	Noninvasive; good epidemiologic study tool; may not differentiate active vs. past infection; widely available for *H. pylori* antibodies; not commercially available for *H. cineadii* or other enterohepatic helicobacters	188–190
Antigen detection	Direct detection of *Helicobacter* antigen from feces using monoclonal antibodies by ELISA	Noninvasive; rapid assays available; high sensitivity and specificity; used in diagnosis of *H. pylori* infection	191, 192
DNA amplification	Polymerase chain reaction assays are available for speciation, determination of resistance genes, and 16S rDNA relatedness	Not commercially available; not evaluated against all species	169, 193, 194

cobacters, as well as the enterohepatic helicobacters such as *H. hepaticus*, *H. cinaedi*, *H. bilis,* and *H. fennelliae*.[164, 204–211]

The ferret, having a stomach with anatomic and physiological similarities to that of humans, was also found to develop naturally occurring gastritis and gastric ulcer disease caused by *Helicobacter mustelae*.[196] As a result, the *H. mustelae* ferret model became one of the best-studied animal models of gastric *Helicobacter* infection in the natural host.[212, 213] Although the ferret model is an excellent research tool, it is not always readily accessible.[213] Other animal models, such as the gnotobiotic piglets and primates, are likewise expensive and impractical for widespread use although they have been successfully utilized.[136, 144, 214]

Although there is no one model that is perfect for all aspects of research applications, the mouse model, based on cost and practicality, is the animal model of choice for *Helicobacter* studies. *H.*

felis, which normally colonizes the gastric mucosa of cats and dogs, was found to readily colonize mice, particularly the gastric antrum.[215] This was an important finding because *H. pylori* could not be made to colonize early mouse models.[195] The first demonstration of the potential use of the *H. felis* mouse model was in studies with germ-free mice, where the bacterium was found to induce an active chronic gastritis and inflammatory effects similar to those induced by *H. pylori* in humans.[216] Subsequently, the *H. felis* mouse model provided an opportunity to study and understand an infectious process, which resembled gastric pathological changes in human *H. pylori* infection.[217, 218] Incidentally, *H. felis* is an excellent control bacterium for those research studies designed to investigate the specific properties of *H. pylori* and the broader characteristics of the genus *Helicobacter.*[219]

With the availability of the *H. pylori* genome and subsequently the successful introduction and colonization of *H. pylori* in mice, the *H. felis* mouse model is being replaced with newer study models.[33, 114] These recent developments and the use of sophisticated molecular tools will continue to yield new and pertinent information in our understanding of the helicobacters. This will most especially be seen in the realms of vaccine development, anti-*Helicobacter* agent discovery, and in basic investigations of colonization, inflammation, and pathogenesis.

REFERENCES

1. Blaser, M.J. and Atherton, J.C., *Helicobacter pylori* persistence: biology and disease, *J. Clin. Invest.*, 113, 321, 2004.
2. Doenges, J.L., Spirochetes in the gastric glands of *Macacus rhesus* and of man without related diseases, *Arch. Pathol.*, 27, 469, 1939.
3. Freedberg, A.S. and Baron, L.E., The presence of spirochetes in human gastric mucosa, *Am. J. Dig. Dis.*, 7, 443, 1940.
4. Covacci, A. and Rappuoli, R., *Helicobacter pylori*: molecular evolution of a bacterial quasi-species, *Curr. Opin. Microbiol.*, 1, 96, 1998.
5. Warren, J.R. and Marshall, B., Unidentified curved bacilli on gastric epithelium in active chronic gastritis, *Lancet*, i, 1973, 1983.
6. Anonymous, Validation of the publication of new names and new combinations previously effectively published outside the IJSB: list no. 17, *Int. J. Syst. Bacteriol.*, 35, 223, 1985.
7. Marshall, B.J. and C.S. Goodwin, Revised nomenclature of *C. pyloridis, Int. J. Syst. Bacteriol.*, 37, 68, 1987.
8. Goodwin, C.S. et al., Unusual cellular fatty acids and distinctive ultrastructure in a new spiral bacterium (*Campylobacter pyloridis*) from the human gastric mucosa, *J. Med. Microbiol.*, 19, 257, 1985.
9. Romaniuk, P.J. et al., *Campylobacter pylori*, the spiral bacterium associated with human gastritis is not a true *Campylobacter* sp., *J. Bacteriol.*, 169, 2137, 1987.
10. Marshall, B.J. and Warren, J.R., Unidentified curved bacilli in the stomach of patients with gastritis and peptic ulceration, *Lancet*, i, 1311, 1984.
11. Huang, J.Q. et al., Meta-analysis of the relationship between *Helicobacter pylori* seropositivity and gastric cancer, *Gastroenterology*, 114, 1169, 1998.
12. Zucca, E. et al., Molecular analysis of the progression from *Helicobacter pylori*-associated chronic gastritis to mucosa-associated lymphoid tissue lymphoma of the stomach, *N. Engl. J. Med.*, 338, 804, 1998.
13. Anonymous, Schistosomes, liver flukes and *Helicobacter pylori*. IARC Working Group on the Evaluation of Carcinogenic Risks to Humans, *IARC Monogr. Eval. Carcinog. Risks Hum.*, 61(5), 1, 1994.
14. Eaton, K.A. et al., Gastric spiral bacilli in captive cheetahs, *Scand. J. Gastroenterol.*, 26, 38, 1991.
15. Fox, J.G. et al., *Campylobacter*-like organisms isolated from gastric mucosa of ferrets, *Am. J. Vet. Res.*, 47, 236, 1986.
16. Hanninen, M.L. et al., Culture and characteristics of *Helicobacter bizzozeroni*, a new canine gastric *Helicobacter* species, *Int. J. Syst. Bacteriol.*, 46, 160, 1996.
17. Jalava, K. et al., *Helicobacter salomonis* sp. nov., A canine gastric *Helicobacter* sp. related to *Helicobacter felis* and *Helicobacter bizzozeroni*, *Int. J. Syst. Bacteriol.*, 47, 975, 1997.
18. Fox, J.G., The expanding genus of *Helicobacter*: pathogenic and zoonotic potential, Semin. *Gastrointest. Dis.*, 8, 124, 1997.

19. Fox, J.G. and Lee, A., The role of *Helicobacter* species in newly recognized gastrointestinal diseases of animals, *Lab Anim. Sci.*, 47, 222, 1997.
20. Solnick, J.V. and Schauer, D.B., Emergency of the *Helicobacter* genus in the pathogenesis of gastrointestinal disease, *Clin. Microbiol. Rev.*, 14, 59, 2001.
21. Versalovic, J. and Fox, J.G., *Helicobacter*, in Manual of Clinical Microbiology, 8th edition, Vol. I, Murray, P.R., Ed., ASM Press, Washington, DC, 2003, chap. 58.
22. Dewhirst, F.E. et al., '*Flexispira rappini*' strains represent at least 10 *Helicobacter* taxa, *Int. J. Syst. Evol. Microbiol.*, 50, 1781, 2000.
23. On, S.L., Taxonomy of *Campylobacter, Arcobacter, Helicobacter* and related bacteria: current status, future prospects and immediate concerns, *J. Appl. Microbiol.*, 90, 1S, 2001.
24. Garrity, G.M. et al., *Helicobacteriaceae*, in *Bergey's Manual of Systematic Bacteriology*, Volume Two, The Proteobacteria (Part C), 2nd edition, Garrity, G.M. et al., Eds., Springer-Verlag, New York, 2005, p. 1169.
25. Happonen, I. et al., Occurrence and topological mapping of gastric *Helicobacter*-like organisms and their association with histological changes in apparently healthy dogs and cats, *J. Vet. Med.*, 43, 305, 1996.
26. Andersen, L.P. et al., Characterization of a culturable "*Gastrospirillum hominis*" (*Helicobacter heilmannii*) strain isolated from human gastric mucosa, *J. Clin. Microbiol.*, 37, 1069, 1999.
27. Burnens, A.P. et al., Novel *Campylobacter*-like organism resembling *Helicobacter fennelliae* isolated from a boy with gastroenteritis and from dogs, *J. Clin. Microbio*l., 31, 1916, 1993.
28. Stanley, J. et al., *Helicobacter canis* sp. nov., a new species from dogs: an integrated study of phenotype and genotype, *J. Gen. Microbiol.,* 139, 2495, 1993.
29. Stanley, J. et al., *Helicobacter pullorum* sp. nov. genotype and phenotype of a new species isolated from poultry and from human patients with gastroenteritis, *Microbiology,* 140, 3441, 1994.
30. Melito, P.L. et al., *Helicobacter winghamensis* sp. nov., a novel *Helicobacter* sp. isolated from patients with gastroenteritis, *J. Clin. Microbiol.,* 39, 2412, 2001.
31. Shen, Z. et al,. *Helicobacter rodentium* sp. nov., a urease negative *Helicobacter* species isolated from laboratory mice. *Int. J. Syst. Bacteriol.,* 47, 627, 1997.
32. Chalk, P.A., Roberts, A.D., and Blows, W.M., Metabolism of pyruvate and glucose by intact cells of *Helicobacter pylori* studied by 13C NMR spectroscopy, *Microbiology,* 140, 2085, 1994.
33. Tomb, J.F. et al., The complete genome sequence of the gastric pathogen *Helicobacter pylori*, *Nature*, 388, 539, 1997.
34. Drumm, B. et al., Intrafamilial clustering of *Helicobacter pylori* infection, *N. Engl. J. Med.*, 322, 359, 1990.
35. Wang, J.T. et al., Direct DNA amplification and restriction pattern analysis of *Helicobacter pylori* in patients with duodenal ulcer and their families, *J. Infect. Dis.*, 168, 1544, 1993.
36. Hulten, K. et al., *Helicobacter pylori* in the drinking water in Peru, *Gastroenterology*, 110, 1031, 1996.
37. Thomas, J.E. et al., Isolation of *Helicobacter pylori* from human feces, *Lancet,* 341, 380, 1993.
38. Klein, P.D. et al., Water source as risk factor for *Helicobacter pylori* infection in Peruvian children, *Lancet,* 377, 1503, 1991.
39. Pytko-Polonczyk, J. et al., Oral cavity as permanent reservoir of *Helicobacter pylori* and potential source of reinfection, *J. Physiol. Pharmacol.,* 47, 121, 1996.
40. Parsonnet, J., Shmuely, H., and Haggerty, T., Fecal and oral shedding of *Helicobacter pylori* from healthy infected adults, *JAMA*, 282, 2240, 1999.
41. Hulten, K. et al., *Helicobacter pylori* in the drinking water in Peru, *Gastroenterology*, 110, 1031, 1996.
42. Hegarty, J.P., Dowd, M.T., and Baker, K.H., Occurrence of *Helicobacter pylori* in surface water in the United States, *J. Appl. Microbiol.*, 87, 697, 1999.
43. Vellozzi, E.M., The evolving role of *Helicobacter pylori* in gastric disease, *Clin. Microbiol. Updates*, 6(4), 1, 1995.
44. Raymond, J. et al., A two year study of *Helicobacter pylori* in children, *J. Clin. Microbiol.*, 32, 461, 1994.
45. Fox, J.G. et al., Gastric colonization *by Campylobacter pylori* subsp. *mustelae* in ferrets, *Infect. Immun.*, 56, 2994, 1988.
46. Fox, J.G. et al., Gastric colonization of the ferret with *Helicobacter* species: natural and experimental infections, *Rev. Infect. Dis.*, 13(Suppl. 8), S671, 1991.
47. Fox, J.G. et al., *Helicobacter mustelae* isolation from feces of ferrets: evidence to support fecal-oral transmission of gastric *Helicobacter, Infect. Immun.*, 60, 606, 1992.

48. Dore, M.P. et al., Isolation of *Helicobacter pylori* from sheep—implications for transmission to humans, *Am. J. Gastroenterol.*, 96, 1396, 2001.
49. Fox, J.G. et al., Local immune response in *Helicobacter pylori* infected cats and identification of *H. pylori* in saliva, gastric fluid and feces. *Immunology*, 88, 400, 1996.
50. Fox, J.G. et al., *Helicobacter canadensis* sp. nov. isolated from humans with diarrhea as an example of an emerging pathogen, *J. Clin. Microbiol.,* 38, 2546, 2000.
51. Blaser, M.J. In a world of black and white, *Helilcobacter pylori* is gray, *Ann. Intern. Med.*, 130, 695, 1999.
52. Go, M.F., Natural history and epidemiology of *Helicobacter pylori* infection, *Aliment. Pharacol. Ther.*, 16 (Suppl. 1), 3, 2002.
53. O'Rourke, J.L., Lee, A., and Fox, J.G., *Helicobacter* infection in animals: a clue to the role of adhesion in the pathogenesis of gastroduodenal disease, *Eur. J. Gastroenterol. Hepatol.*, 4, (Suppl. 1), S31, 1992.
54. Taylor, N.S. et al., Haemagglutination profiles of *Helicobacter* species that cause gastritis in man and animals, *J. Med. Microbiol.*, 37, 299, 1992.
55. Blaser, M.J., Hypothesis: the changing relationships of *Helicobacter pylori* and humans: implications for health and disease, *J. Infect. Dis.*, 179, 1523, 1999.
56. Uemura, N. et al., *Helicobacter pylori* infection and the development of gastric cancer, *N. Engl. J. Med.*, 345, 784, 2001.
57. Blaser, M.J., An endangered species in the stomach, *Sci. Am.,* 292(2), 38, 2005.
58. Vandamme, P. et al., Identification of *Campylobacter cinaedi* isolated from blood and feces from children and adult females, *J. Clin. Microbiol.*, 28, 1016, 1990.
59. Kiehlbauch, J.A. et al., *Helicobacter cinaedi*-associated bacteremia and cellulitis in immunocompromised patients, *Ann. Intern. Med.,* 121, 90, 1994.
60. Hsueh, P.R. et al., Septic shock due to *Helicobacter fennelliae* in a non-human immunodeficiency virus-infected heterosexual patient, *J. Clin. Microbiol.,* 37, 2084, 1999.
61. Solnick, J.V. et al., An uncultured gastric spiral organism is a newly identified *Helicobacter* in humans, *J. Infect. Dis.,* 168, 379, 1993.
62. De Groote, D. et al., Detection of non-pylori *Helicobacter* species in "*Helicobacter heilmannii*"-infected humans, *Helicobacter*, 10(5), 398, 2005.
63. Norris, C.R. et al., Healthy cats are commonly colonized with *Helicobacter heilmannii* that is associated with minimal gastritis, *J. Clin. Microbiol.,* 37, 189, 1999.
64. Queiroz, D.M. et al., Association between *Helicobacter* and gastric ulcer disease of the pars esophagea in swine, *Gastroenterology,* 111, 19, 1996.
65. Morgner, A. et al., *Helicobacter heilmannii*-associated primary gastric low-grade MALT lymphoma: complete remission after curing the infection, *Gastroenterology,* 118, 821, 2000.
66. Erdman, S.E. et al., *Helicobacter mustelae*-associated gastric MALT lymphoma in ferrets, *Am. J. Pathol.,* 151, 273, 1997.
67. Ward, J.M. et al., Chronic active hepatitis and associated liver tumors in mice caused by a persistent bacterial infection with a novel *Helicobacter* species, *J. Natl. Cancer Inst.*, 86, 1222, 1994.
68. Whary, M.T. and Fox, J.G., Detection, eradication, and research implications of *Helicobacter* infection in laboratory rodents, *Lab Animal*, 35(7), 25, 2006.
69. Lee, A. et al., A standardized mouse model of *Helicobacter pylori* infection. Introducing the Sydney strain, *Gastroenterology,* 112, 1386, 1997.
70. Goodman, K. and Cockburn, M., The role of epidemiology in understanding the health effect of *Helicobacter pylori, Epidemiology*, 12(2), 266, 2001.
71. Goodman, K. et al., Dymanics of *Helicobacter pylori* infection in a US – Mexico cohort during the first two years of life, *Int. J. Epidemiol.*, 34(6), 1348, 2005.
72. McCarthy, C. et al., Long-term prospective study of *Helicobacter pylori* in nonulcer dyspepsia, *Dig. Dis. Sci.,* 40, 114, 1995.
73. Hansson, L.E. et al., The risk of stomach cancer in patients with gastric or duodenal ulcer disease, *N. Engl. J. Med.,* 335, 242, 1996.
74. Ahmad, A., Govil, Y., and Frank, B.B., Gastric mucosa-associated lymphoid tissue lymphoma, *Am. J. Gastroenterol.,* 98, 975, 2003.
75. Mager, D.L., Bacteria and cancer: cause, coincidence or cure? A review, *J. Transl. Med.*, 4, 14, 2006.
76. Grayson, M.L., Tee, W., and Dwyer, B., Gastroenteritis associated with *Campylobacter cinaedi, Med. J. Aust.,* 150, 214, 1989.

77. Flores, B.M., Fennell, C.L., and Stamm, W.E., Characterization of *Campylobacter cinaedi* and *C. fennelliae* antigens and analysis of the human immune response, *J. Infect. Dis.*, 159, 635, 1989.
78. Ceelen, L. et al., Prevelance of *Helicobacter pullorum* among patients with gastric disease and clinically healthy persons, *J. Clin. Microbiol.*, 43(6), 2984, 2005.
79. Solnick, J.V. et al., Extragastric manifestations of *Helicobacter pylori* infection—other *Helicobacter* species, *Helicobacter*, 11(Suppl. 1), 46, 2006.
80. Fox, J.G. et al., Hepatic *Helicobacter* species identified in bile and gall bladder tissue from Chileans with chronic cholecystitis, *Gastroenterology*, 114, 755, 1998.
81. Nilsson, H.O. et al., Identification of *Helicobacter pylori* and other *Helicobacter* species by PCR, hybridization, and partial DNA sequencing in human liver samples from patients with primary sclerosing cholangitis or primary biliary cirrhosis, *J. Clin. Microbiol.*, 38, 1072, 2000.
82. Fox, J.G. et al., *Helicobacter hepaticus* sp. nov., a microaerophilic bacterium isolated from livers and intestinal mucosal scrapings from mice, *J. Clin. Microbiol.*, 32, 1238, 1994.
83. Fox, J.G. et al., Isolation of *Helicobacter cinaedi* from the colon, liver, and mesenteric lymph node of a rhesus monkey with chronic colitis and hepatitis, *J. Clin. Microbiol.*, 39, 1580, 2001.
84. Fox, J.G. et al., Persistent hepatitis and enterocolitis in germfree mice infected with *Helicobacter hepaticus*, *Infect. Immun.*, 64, 3673, 1996.
85. Laine, I., Schoenfeld, P., and Fennerty, M.B., Therapy for *Helicobacter pylori* for patients with nonulcer dyspepsia. A meta analysis of randomized, control trials, *Ann. Intern. Med.*, 134, 361, 2001.
86. Chan, F.K. et al., Eradication of *Helicobacter pylori* and risk of peptic ulcers in patients starting long-term treatment with nonsteroidal anti-inflammatory drugs: a randomized trial, *Lancet*, 359, 9, 2002.
87. Schwizer, W. et al., *Helicobacter pylori* and symptomatic relapse of gastroesophageal reflux disease: a randomized control trial, *Lancet*, 357, 1738, 2001.
88. Ridker, P.M. et al., A prospective study of *Helicobacter pylori* seropositivity and the risk for future myocardial infarction among socioeconomically similar U.S. men, *Ann. Intern. Med.*, 135, 184, 2001.
89. Stolzenberg-Solomon, R.Z. et al., *Helicobacter pylori* seropositivity as a risk factor for pancreatic cancer, *J. Natl. Cancer Inst.*, 93, 937, 2001.
90. Choe, Y.H. et al., *Helicobacter pylori*-associated iron deficiency anemia in adolescent female athletes, *J. Pediatr.*, 139, 100, 2001.
91. Jaing, T.H. et al., Efficacy of *Helicobacter pylori* eradication on platelet recovery in children with chronic idiopathic thrombocytopenic purpura, *Acta Paediatr.*, 92, 1153, 2003.
92. Franceschi, F., Roccarina, D., and Gasbarrini, A., Extragastric manifestations of *Helicobacter pylori* infection, *Minerva Med.*, 97(1), 39, 2006.
93. Hazell, S.L. et al., *Campylobacter pyloridis* and gastritis: association with intercellular spaces and adaptation to an environment of mucus as important factors in colonization of the gastric epithelium, *J. Infect. Dis.*, 153, 658, 1986.
94. Hawtin, P.R., Stacey, A.R., and Newell, D.G., Investigation of the structure and location of the urease of *Helicobacter pylori* using monoclonal antibodies, *J. Gen. Microbiol.*, 136, 1995, 1990.
95. Takahashi, S. et al., *Helicobacter pylori* urease activity: comparative study between urease positive and urease negative strains, *Jpn. J. Clin. Med.*, 51, 3149, 1993.
96. Tsuda, M. et al., A urease-negative mutant of *Helicobacter pylori* constructed by allelic exchange mutagenesis lacks the ability to colonize the nude mouse stomach, *Infect. Immun.*, 62, 3586, 1994.
97. Bauerfeind, P. et al., Synthesis and activity of *Helicobacter* urease and catalase at low pH, *Gut*, 40, 25, 1997.
98. Mobley, H.L. et al., *Helicobacter pylori* urease: properties and role in pathogenesis, *Scand. J. Gastroenterol.*, Supplement 187, 39, 1991.
99. Simon, P.M. et al., Inhibition of *Helicobacter pylori* binding to gastrointestinal epithelial cells by sialiac acid-containing oligosaccharides, *Infect. Immun.*, 65, 750, 1997.
100. Hessay, S.J. et al., Bacterial adhesion and disease activity in *Helicobacter* associated chronic gastritis, *Gut*, 31, 134, 1990.
101. Tzouvelekis, L.S. et al., *In vitro* binding of *Helicobacter pylori* to human gastric mucin, *Infect. Immun.*, 59, 4252, 1991.
102. Spigelhalder, C. et al., Purification of *Helicobacter pylori* superoxide dismutase and cloning and sequencing of the gene, *Infect. Immune.*, 61, 5315, 1993.
103. Hazell, S.L., Evans, Jr., D.J., and Graham, D.Y., *Helicobacter pylori* catalase, *J. Gen. Microbiol.*, 137, 57, 1991.

104. Slomainy, B.L. et al., Glycosulfatase activity of *Helicobacter pylori* toward gastric mucin, *Biochem. Biophys. Res. Comm.,* 183, 506, 1992.
105. Langton, S.R. and Cesareo, S.D., *Helicobacter pylori* associated phospoholipase A2 activity: a factor in peptic ulcer production? *J. Clin. Pathol.,* 44, 221, 1992.
106. Conrad, R.S. et al., Extraction and biochemical analyses of *Helicobacter pylori* lipopolysaccharide, *Curr. Microbiol.,* 24, 165, 1992.
107. Mattsby-Baitzer, I. et al., Lipid A in *Helicobacter pylori, Infect. Immun.,* 60, 4383, 1992.
108. Piotrowski, J. et al., Induction of gastritis and epithelial apoptosis by *Helicobacter pylori* lipopolysaccharide, *Scand. J.Gastroenterol.,* 32, 203, 1997.
109. Heneghan, M.A., McCarthy, C.F., and Moran, A.P., Relationship of blood group determinants on *Helicobacter pylori* lipopolysaccharide with host Lewis phenotype and inflammatory response, *Infect. Immun.,* 68, 937, 2000.
110. Simoons-Smit, I.M. et al., Typing of *Helicobacter pylori* with monoclonal antibodies against Lewis antigens in lipopolysaccharide, *J. Clin. Microbiol.,* 34, 2196, 1996.
111. Wirth, H.P. et al., Expression of Lewis X and Y blood group antigens by *Helicobacter pylori* strains is related to cagA status, *Gastroenterology,* 110, A296, 1996.
112. Zheng, P.Y. et al., Association of peptic ulcer with increased expression of Lewis antigens but not cagA, iceA and vacA in *Helicobacter pylori* isolates in an Asian population, *Gut,* 47, 18, 2000.
113. Moran, A.P., Appelmelk, B.J., and Aspinall, G.O., Molecular mimicry of host structures by lipopolysaccharides of *Camphylobacter* and *Helicobacter* spp.: implications in pathogenisis, *J. Endotox. Res.,* 3, 521, 1996.
114. Marchetti, M. et al., Development of a mouse model of *Helicobacter pylori* infection that mimics human disease, *Science,* 267, 1655, 1995.
115. Cover, T.L. and Blaser, M.J., Purification and characterization of the vacuolating toxin from *Helicobacter pylori, J. Biol. Chem.,* 267, 10570, 1992.
116. Phadais, S.H. et, al., Pathological significance and molecular characterization of the vacuolating toxin gene of *Helicobacter pylori, Infect. Immun.,* 62, 1557, 1994.
117. Katrenich, C.H. and Chestnut, M.H., Character and origin of vacuoles induced in mammalian cells by the cytotoxin of *Helicobacter pylori, J. Med. Microbiol.,* 37, 389, 1992.
118. Pelicic, V. et al., *Helicobacter pylori* VacA cytotoxin associated with the bacteria increases epithelial permeability independently of its vacuolating activity, *Microbiology,* 145, 2043, 1999.
119. Szabo, L. et al., Formation of anion-selective channels in the cell membrane by the toxin VacA of *Helicobacter pylori* is required for its biological activity, *EMBO J.,* 18, 5517, 1999.
120. van Doorn, L.J. et al., Expanding allelic diversity of *Helicobacter pylori* vacA, *J. Clin. Microbiol.*, 36, 2597, 1998.
121. Molinari, M. et al., The acid activation *of Helicobacter pylori* toxin VacA: structural and membrane binding studies, *Biochem. Biophys. Res. Commun.,* 248, 334, 1998.
122. Shirasaka, D. et al., Analysis of *Helicobacter pylori* VacA gene and serum antibodies to VacA in Japan, *Dig. Dis. Sci.,* 45, 789, 2000.
123. Jang, T.J. and Kim, J.R., Proliferation and apoptosis in gastric antral epithelial cells in patients infected with *Helicobacter pylori, J. Gastroenterol.,* 35, 265, 2000.
124. Houghton, J. et al., Tumor necrosis factor alpha and interleukin 1β up-regulate gastric mucosal fas antigen expression in *Helicobacter pylori* infection, *Infect. Immun.,* 68, 1189, 2000.
125. Taylor, N.S., Fox, J.G. and Yan, L., *In vitro* hepatotoxic factor in *Helicobacter hepaticus, H. pylori* and other *Helicobacter* species, *J. Med. Microbiol.*, 42, 48, 1995.
126. Pickett, C.L. and Whitehouse, C.A., The cytolethal distending toxin family, *Trends Microbiol.*, 7, 292, 1999.
127. Young, V.B., Knox, K.A., and Schauer, D.B., Cytolethal distending toxin sequence and activity in the enterohepatic pathogen *Helicobacter hepaticus, Infect. Immun.*, 68, 184, 2000.
128. Chien, C.C. et al., Identification of cdtB homologues and cytolethal distending toxin activity in enterohepatic *Helicobacter* spp., *J. Med. Microbiol.*, 49, 525, 2000.
129. Young, V.B. et al., Cytolethal distending toxin in avian and human isolates of *Helicobacter pullorum, J. Infect. Dis.*, 152, 620, 2000.
130. Kuipers, E.J. et al., *Helicobacter pylori* and atrophic gastritis: importance of the cagA status, *J. Natl. Cancer Inst.,* 87, 1777, 1995.
131. Blaser, M.J. et al., Infection with *Helicobacter pylori* strain possessing cagA is associated with an increased risk of developing adenocarcinoma of the stomach, *Cancer Res.,* 55, 2111, 1995.

132. Parsonnet, J. et al., Risk for gastric cancer in people with CagA positive and CagA negative *Helicobacter pylori* infection, *Gut*, 40, 297, 1997.
133. Hatakeyama, M., Oncogenic mechanisms of *Helicobacter pylori* cagA protein, *Nat. Rev. Cancer*, 4, 688, 2004.
134. Bode, G., Mauch, F., and Malfertheiner, P., The coccoid forms of *Helicobacter pylori*. Criteria for their viability, *Epidemiol. Infect.*, 111, 483, 1993.
135. Cellini, L. et al., Coccoid *Helicobacter pylori* not culturable *in vitro* reverts in mice, *Microbiol. Immunol.*, 38, 843, 1994.
136. Eaton, K.A. et al., Virulence of coccoid and bacillary forms of *Helicobacter pylori* in gnotobiotic piglets, *J. Infect. Dis.*, 171, 459, 1995.
137. Wang, X. et al., Infection of BALB/cA mice by spiral and coccoid forms of *Helicobacter pylori*, *J. Med. Microbiol.*, 46, 657, 1997.
138. Cellini, L. et al., Recovery of *Helicobacter pylori* ATCC 43504, from a viable but not a culturable state—regrowth or resuscitation, *APMIS*, 106, 571, 1998.
139. Benaissa, M. et al., Changes in *Helicobacter pylori* ultrastructure and antigens during conversion from the bacillary to the coccoid form, *Infect. Immun.*, 64, 2331, 1996.
140. Janas, B. et al., Electron microscopic study of association between coccoid forms of *Helicobacter pylori* and gastric epithelial cells, *Am. J. Gastroenterol.*, 90, 1829, 1995.
141. Khin, M.M. et al., Binding of human plasminogen and lactoferrin by *Helicobacter pylori* coccoid forms, *J. Med. Microbiol.*, 45, 433, 1996.
142. Josenhans, C. et al., Cloning and alleleic exchange mutagensis of two flagellin genes of *Helicobacter felis*, *Mol. Microbiol*, 33, 350, 1999.
143. Lee, A. et al., Isolation of a spiral-shaped bacterium from a the cat stomach, *Infect. Immun.*, 56, 2843, 1988.
144. Eaton, K.A. et al., Essential role of urease in pathogenesis in gastritis induced by *Helicobacter pylori* in gnotobiotic piglets, *Infect. Immun.*, 59, 2470, 1991.
145. Solnick, J.V. et al., Construction and characterization of an isogenic urease-negative mutant of *H. mustelae*, *Infect. Immun.*, 63, 3718, 1995.
146. Solnick, J.V. et al., Molecular analysis of urease genes from a newly identified uncultured species of *Helicobacter*, *Infect. Immun.*, 62, 1631, 1994.
147. Carlsohn, E. et al., HpaA is essential *for Helicobacter pylori* colonization mice, *Infect. Immun.*, 74(2), 920, 2006.
148. Logan, R.P. et al., A novel flow cytometric assay for quantitating adherence of *Helicobacter pylori* to gastric epithelial cells, *J. Immunol. Methods*, 213, 19, 1998.
149. Segal, E.D., Falkow, S., and Thompkins, L.S., *Helicobacter pylori* attachment to gastric cells induces cytoskeletal rearrangements and tyrosine phosphorylation of host proteins, *Proc. Natl. Acad. Sci. U.S.A.*, 93, 1259, 1996.
150. O'Rourke, J., Lee, A., and Fox, J.G., An ultrastructural study of *Helicobacter mustelae* and evidence of a specific association with gastric mucosa, *J. Med. Microbiol.*, 36, 420, 1992.
151. Odenbreit, S. et al., Genetic and functional characterization of the alpAB gene locus for the adhesion of *Helicobacter pylori* to human gastric tissue, *Mol. Microbiol.*, 31, 1537, 1999.
152. Gold, B.D. et al., Comparison of *Helicobacter mustelae* and *Helicobacte pylori* adhesion to eukaryotic cells *in vitro*, *Gastroenterology*, 109, 692, 1995.
153. Gold, B.D., Sherman, P.M., and Lingwood, C.A., *Helicobacter mustelae* and *Helicobacter pylori* bind to common lipid receptors *in vitro*, *Infect. Immun.*, 61, 2632, 1993.
154. Gold, B.D. et al., Surface properties of *Helicobacter mustelae* and ferret gastrointestinal mucosa, *Clin. Investig. Med.*, 19, 92, 1996.
155. Stolte, M. et al., A comparison of *Helicobacter pylori* and *H. heilmannii* gastritis. A matched control study involving 404 patients, *Scand. J. Gastroenterol.*, 32, 28, 1997.
156. Muotiala, A. et al., Low biological activity of *Helicobacter pylori* lipopolysaccharide, *Infect. Immun.*, 60, 1714, 1992.
157. Bliss, C.M.J. et al., *Helicobacter pylori* lipopolysaccharide binds to CD14 and stimulates release of interleukin-8, epithelial neutrophil-activating peptide 78, and monocyte chemotactic protein 1 by human monocytes, *Infect. Immun.*, 66, 5357, 1998.
158. Blaser, M.J. and Parsonnet, J., Parasitism by the "slow" bacterium *Helicobacter pylori* leads to altered homeostasis and neoplasia, *J. Clin. Investi.*, 94, 4, 1994.

159. Appelmelk, B.J. et al., Potential role of molecular mimicry between *Helicobacter pylori* lipopolysaccharide and host Lewis blood group antigens in autoimmunity, *Infect. Immun.*, 64, 2031, 1996.
160. O. Cróinin, T. et al., Molecular mimicry of ferret gastric epithelial blood group antigen A by *Helicobacter mustelae*, *Gastroenterology*, 11, 690, 1998.
161. Monteiro, M.A. et al., The lipopolysaccharide of *Helicobacter mustelae* type strain ATCC 43772 expresses the monofucosyl A type 1 histo-blood group epitope, *FEMS Microbiol. Lett.*, 154, 103, 1997.
162. Telford, J.L. et al., Gene structure of the *Helicobacter pylori* cytotoxin and evidence of its key role in gastric disease, *J. Exp. Med.*, 179, 1653, 1994.
163. Ikeno, T. et al., *Helicobacter pylori* induced chronic active gastritis, intestinal metaplasia and gastric ulcer in Mongolian gerbils, *Am. J. Pathol.*, 154, 951, 1999.
164, Mohammadi, M. et al., Role of the host in pathogenesis in *Helicobacter*-associated gastritis: *H. felis* infection of inbred and congenic mouse strains, *Infect. Immun.*, 64, 238, 1996.
165. Morgan, D. R., Fox, J.G., and Leunk, R.D., Comparison of isolates of *Helicobacter pylori* and *Helicobacter mustelae*, *J. Clin. Microbiol.* 29, 395, 1991.
166. Fox, J.G. and Wang, T.C, Inflammation, atrophy and gastric cancer, *J. Clin. Invest.,* 117(1), 60, 2007.
167. Han, S.W. et al., Transport and storage of *Helicobacter pylori* from gastric mucosal biopsies and clinical isolates, *Eur. J. Clin. Microbiol. Infect. Dis.,* 14, 349, 1995.
168. Schrader, J.A. et al., A role for culture and diagnosis of *Helicobacter pylori*–related gastric disease, *Am. J. Gastroenterol.*, 88, 1729, 1993.
169. Vakil, N. et al., The cost-effectiveness of diagnostic testing strategies for *Helicobacter pylori*, *Am. J. Gastroenterol.*, 95, 1691, 2000.
170. El-Zimaity, H.M., et al., Histologic assessment of *Helicobacter pylori* status after therapy: comparison of Giemsa, Diff-Quik, and Genta stains, *Mod. Pathol.*, 11, 288, 1998.
171. Genta, R.M. and Graham, D.Y., Comparison of biopsy sites for the histopathologic diagnosis of *Helicobacter pylori*: a topographic study of *H. pylori* density and distribution, *Gastrointest. Endosc.*, 40, 342, 1994.
172. Hachem, C.Y. et al., Antimicrobial susceptibility testing of *Helicobacter pylori*. Comparison of E-test, broth microdilution, and disk diffusion for ampicillin, clarithromycin, and metronidazole, *Diagn. Microbiol. Infect. Dis.,* 24, 37, 1996.
173. Danon, S.J. et al., Gastrin release and gastric acid secretion in the rat infected with either *H. felis* or *H. heilmannii*, *J. Gastroenterol. Hepatol.,* 13, 95, 1998.
174. Hunt, R.H., Peptic ulcer disease: defining the treatment strategies in the era of *Helicobacter pylori*, *Am. J. Gastroenterol.* 92, 36S, 1997.
175. Clinical and Laboratory Standards Institute (CLSI), *Performance Standard for Antimicrobial Susceptibility Testing: Seventeenth Informational Supplement*, M100-S17 CLSI, Wayne, PA, 2007.
176. Osato, M.S. et al., Pattern of primary resistance of *Helicobacter pylori* to metronidazole or clarithromycin in the United States, *Arch. Intern. Med.*, 161, 1217, 2001.
177. Neiger, R., Seiler, G., and, Schmassmann, A., Use of a urea breath test to evaluate short term treatments for cats naturally infected with *Helicobacter heilmannii*, *Am. J. Vet. Res.*, 60, 880, 1999.
178. Dick-Hegedus, E. and Lee, A., Use of a mouse model to examine anti-*Helicobacter pylori* agents, *Scand. J. Gastroenterol.*, 26, 909, 1991.
179. Marini, R.P. et al., Ranitidine bismuth citrate and clarithromycin, alone or in combination, for eradication of *Helicobacter mustelae* infection in ferrets, *Am. J. Vet. Res.*, 60, 1280, 1999.
180. Ng, V.L. et al., Successive bacteremias with "*Campylobacter cinaedi*" and "*Campylobacter fennelliae*" in a bisexual male, *J. Clin. Microbiol.*, 25, 2008, 1987.
181. Galan, M.V., Kishan, A.A., and Silverman, A.L., Oral broccoli sprouts for the treatment of *Helicobacter pylori* infection: a preliminary report, *Dig. Dis. Sci.*, 49, 1088, 2004.
182. Lee, J. et al., Inhibition of pathogenic bacterial adhesion by acidic polysaccharide from green tea (*Cameilia sinensis*), *J. Agric. Food Chem.*, 54(23), 871, 2006.
183. Vakil, N. and Fendrick, A.M., How to test for *Helicobacter pylori* in 2005, *Clevland Clinic J. Med.* 72(Suppl. 2), S8, 2005.
184. El-Zimaity, H.M., et al., Confirmation of successful therapy of *Helicobacter pylori* infection: number and site of biopsies or a rapid urease test, *Am. J. Gastroenterol.*, 90, 1962, 1995.
185. Yousfy, M.M. et al., Detection of *Helicobacter pylori* by rapid urease tests: is biopsy size a critical variable? *Gastrointest. Endosc.*, 43, 222, 1996.

186. Cohen, H. et al., Accuracy of four commercially available serologic tests, including two office-based tests and a commercially available ^{13}C urea breath test for diagnosis of *Helicobacter pylori*, *Helicobacter*, 4, 49, 1999.
187. Chara, S. et al., Studies of ^{13}C-urea breath test for diagnosis of *Helicobacter pylori* infection in Japan, *J. Gastroenterol.*, 33, 6, 1998.
188. Faigel, D.O. et al., Evaluation of rapid antibody tests for the diagnosis of *Helicobacter pylori* infection, *Am. J. Gastroenterol.*, 95, 72, 2000.
189. Raymond, J. et al., Immunoblotting and serology for diagnosis of *Helicobacter pylori* infection in children, *Pediatr. Infect. Dis. J.*, 19, 118, 2000.
190. Wilcox, M.H. et al., Accuracy of serology for the diagnosis of *Helicobacter pylori* infection — a comparison of eight kits, *J. Clin. Pathol.*, 49, 373, 1996.
191. Fanti, L. et al., A new simple immunoassay for detection of *Helicobacter pylori* infection: antigen in stool specimens, *Digestion*, 60, 456, 1999.
192. Romero, J.M. et al., Usefulness of *Helicobacter pylori* antigen detection in stools in the diagnosis of infection and confirming eradication after treatment, *Med. Clin.* (Barcelona), 114, 571, 2000.
193. El-Zaatari, F.A., Oweis, S.M., and Graham, D.Y., Uses and cautions for use of polymerase chain reaction for detection of *Helicobacter pylori*, *Dig. Dis. Sci.*, 42, 2116, 1997.
194. Sevin, E. et al., Codetection of *Helicobacter pylori* and of its resistance to clarithromycin by PCR, *FEMS Microbiol. Lett.*, 165, 369, 1998.
195. Cantorna, M.T. and Balishe, E., Inability of human clinical strains of *Helicobacter pylori* to colonize the alimentary tract of germ free rodents, *Can. J. Microbiol.*, 36, 237, 1990.
196. Fox, J.G. et al., *Helicobacter mustelae* associated gastritis in ferrets: an animal model of *Helicobacter pylori* gastritis in humans, *Gastroenterology*, 99, 352, 1990.
197. Dubois, A. et al., Natural gastric infection with *Helicobacter pylori* in monkeys: a model for spiral bacteria infection in humans, *Gastroenterology*, 106, 1405, 1994.
198. Krakowka, S. et al., Establishment of gastric *Campylobacter pylori* infection in the neonatal gnotobiotic piglet, *Infect. Immun.*, 55, 2789, 1987.
199. Radin, J.M. et. al., *Helicobacter pylori* infection in gnotobiotic beagle dogs, *Infect. Immun.*, 58, 2606, 1990.
200. Matsumoto, S. et al., Induction of ulceration and severe gastritis in Mongolian gerbils by *Helicobacter pylori* infection, *J. Med. Microbiol.*, 46, 391, 1997.
201. Shomer, N.H. et al., Experimental *Helicobacter pylori* infection induces antral gastritis and gastric mucosa-associated lymphoid tissue in guinea pigs, *Infect. Immun.*, 65, 4858, 1998.
202. Fox, J.G. et al., *Helicobacter pylori*–induced gastritis in the domestic cat, *Infect. Immun.*, 63, 2674, 1995.
203. Dubois, A. et al., Host specificity of *Helicobacter pylori* strains and host responses in experimentally challenged non-human primates, *Gastroenterology*, 116, 90, 1999.
204. Czinn, S.J., Cai, A., and Nedrud, J.G., Protection of germ-free mice from infection by *Helicobacter felis* after active oral or passive IgA immunization, *Vaccine*, 11, 637, 1993.
205. Foltz, C.J. et al., Spontaneous inflammatory bowel disease in multiple mutant mouse lines: association with colonization by *Helicobacter hepaticus*, *Helicobacter*, 3, 69, 1998.
206. Wang, T. et al., Synergistic interaction between hypergastrinemia and *Helicobacter* infection in a mouse model of gastric carcinoma, *Gastroenterology*, 118, 36, 2000.
207. Chin, E.Y. et al., *Helicobacter hepaticus* infection triggers inflammatory bowel disease in T cell receptor $\alpha\beta$ deficient mice, *Comp. Med.*, 50, 586, 2000.
208. Flores, B.M. et al., Experimental infection of pig-tailed macaques (*Macaca nemestrina*) with *Campylobacter cinaedi* and *Campylobacter fennelliae*, *Infect. Immun.*, 58, 3947, 1990.
209. Fox, J.G. et al., Chronic proliferative hepatitis in A/JCr mice associated with persistent *Helicobacter hepaticus* infection: a model of *Helicobacter*-induced carcinogenesis, *Infect. Immun.*, 64, 1548, 1996.
210. Haines, D.C. et al., Inflammatory large bowel disease in immunodeficient rats naturally and experimentally infected with *Helicobacter bilis*, *Vet. Pathol.*, 35, 202, 1998.
211. Franklin, C.L. et al., Enterohepatic lesions in SCID mice infected with *Helicobacter bilis*, *Lab. Anim. Sci.*, 48, 334, 1998.
212. Fox, J.G., Anatomy of the ferret, in *Biology and Diseases of the Ferret*, Fox, J.G., Ed., The Williams and Wilkins Company, Baltimore, Maryland, 1998, p. 19.

213. Fox, J.G. and Lee, A., Gastric *Helicobacter* infection in animals: natural and experimental infections, in *Helicobacter pylori: Biology and Clinical Practice,* Goodwin, C.S. and Worsley, B.W., Eds., CRC Press, Boca Raton, FL, 1993, p. 407.
214. Euler, A.R. et al., Evaluation of two monkey species (*Macaca mulatta* and *Mucaca fascicularis*) as possible models for human *Helicobacter pylori* disease, *J. Clin. Microbiol.*, 28, 2285, 1990.
215. Dick, E. et al., Use of the mouse for the isolation and investigation of stomach-associated, spiral-helical shaped bacteria from man and other animals, *J. Med. Microbiol.*, 29, 55, 1989.
216. Lee, A. et al., A small animal model of human *Helicobacter pylori* active chronic gastritis, *Gastroenterology*, 99, 1315, 1990.
217. Eaton, K.A., Ringler, S.R., and Danon, S.J., Murine splenocytes induce severe gastritis and delayed-type hypersensitivity and suppress bacterial colonization in *Helicobacter pylori* infected SCID mice, *Infect. Immun.,* 67, 4494, 1999.
218. Roth, K.A. et al., Cellular immune responses are essential for the development of *Helicobacter felis*-associated gastric pathology, *J. Immunol.,* 163, 1490, 1999.
219. Lee, A., *Helicobacter pylori*: microbiological aspects. A plea for a more critical approach to laboratory investigation of this gastroduodenal pathogen, in *Helicobacter pylori, Gastritis and Peptic Ulcer*, Malfertheiner, P. and Ditschuneit, H., Eds., Springer-Verlag, Heidelberg, 1990, p. 9.

37 The Genus *Yersinia*

Susan E. Sharp

CONTENTS

INTRODUCTION

Plague is one of the oldest recorded infectious diseases. It has caused at least three major pandemics at approximately 600-year intervals in history. The first pandemic began in Egypt and spread through the Middle East to Europe in the 6th century. The second plague occurred during the 14th century and was called The Black Death. It began in the area of the Black Sea and spread to Europe. It was responsible for the death of more than a quarter of Europe's population. The third pandemic began in 1855 in China and spread to India (killing more than 10 million people), Egypt, Portugal, Japan, Paraguay, Eastern Africa, Manila, Scotland (Glasgow), Australia (Sydney), and the United States (San Francisco). This infection is believed to have caused more than 150 epidemics of varying degrees until the 1950s.[1–4]

Because the Black Death was, according to historical accounts, characterized by buboes (swellings in the groin), as was the modern third pandemic that started in Asia in the 19th century, scientists and historians have assumed that the earlier pandemic was an outbreak of the same disease. This led to the conclusion that the Black Death was caused by the bacterium *Yersinia pestis* and spread by fleas with the help of rodent reserviors. However, buboes are features of other diseases as well, and this view has been questioned by some.[1]

NOMENCLATURE

These organisms, originally called *Pasteurella* species, were named in 1894 after the French bacteriologist Alexander Yersin. There are three species known to be human pathogens within the genus

Yersinia: *Yersinia pestis, Y. pseudotuberculosis,* and *Y. enterocolitica*. Other species in the genus include *Y. frederiksenii, Y. intermedia, Y. kristensenii, Y. aldovae, Y. bercovieri, Y. molleretti,* and *Y. rohdei*.[3]

EPIDEMIOLOGY

Yersinia pestis organisms are found in animal species. Those of primary importance in humans are found normally in rodents (rats, mice, rabbits, prairie dogs, and ground squirrels), which are the natural reservoirs for this species. Infection is usually spread between the rat reservoirs and to humans by fleas (which are known to harbor the bacteria). Much less routinely, the organisms can be spread to humans through direct contact with animals. The oriental rat flea (*Xenopsylla cheopis*) is considered the classic vector of plague, yet other species of fleas can be major vectors for endemic plague in the western parts of the United States. For example, ground squirrel fleas (*Oropsylla montana*) have been found to be vectors in California; and prairie dog and rock squirrel fleas (*O. hirsuta, O. montana,* and *Hoplopsyllus anomalus*) have been found to be vectors in New Mexico.[2, 3]

Transmission of the organism to the human host depends on masses of the organisms blocking the flea's proventriculus foregut. This blockage causes the flea to be unable to feed. The effect of this is that the fleas will repeatedly bite their human host in an attempt to obtain a blood meal, allowing transmission of the organism to humans. Cats usually become infected with this organism by eating an infected rodent, as opposed to being bitten by and infected by flea exposure. The cat flea is a poor vector for plague, and cats usually infect humans through biting or scratching. In addition, inhaling infected respiratory secretions from the cat can also infect humans. Some 23 countries reported cases of plague to the World Health Organization between 1985 and 1999. These cases have shown an average mortality rate of approximately 11%.[2]

Y. enterocolitica is also found in a wide variety of animals, including humans, but the primary reservoir for this species is swine.[2, 3] In addition, as pathogenic *Yersinia enterocolitica* organisms can survive for a time in the environment, contaminated soil, water, and vegetation can be a part of the transmission of these organisms to animals and humans.[2] *Y. enterocolitica* is an infrequent cause of disease in the United States but it is common in northern Europe.[4]

Yersinia pseudotuberculosis is the most uncommon of these three human pathogens. It is found in a wide variety of domestic and wild animals, including fowl. The main reservoirs for this organism are rodents, rabbits, and wild birds. Most disease with this organism is seen in European children.[4] Although *Y. enterocolitica* and *Y. pseudotuberculosis* species are found worldwide, they are primarily seen in the tropical and subtropical areas of Europe. They can also be found in Asia, South Africa, Australia, and the United States.[2]

Yersinia frederiksenii, Y. intermedia, Y. kristensenii, Y. aldovae, Y. bercovieri, Y. molleretti, and *Y. rohdei* are species that are primarily environmental organisms that can occasionally cause transient colonization in humans and animals.[2]

CHARACTERIZATION

Gram Stain

Yersinia are non-spore-forming, Gram-negative bacilli measuring between 0.5 and 0.8 μ wide by 1 to 3 μ long.[2] They appear in Gram-stained smears as plump Gram-negative bacilli, and on Wright stain or Wayson stain can appear as plump rods showing bipolar staining characteristics (darker on the ends and lighter in the center of the bacterial cell). This is not a defining characteristic for *Yersinia* species, but is more common with this organism than with the other Enterobacteriaceae.[2] *Yersinia pestis* is nonmotile but all other species are motile by peritrichous flagella at 22 to 30°C. All species are nonmotile at 37°C.[2] Although the organisms have an optimal temperature growth at 25 to 28°C, they can grow from 0 to 45°C.[2]

Culture

Yersinia organisms will grow on nonselective media such as tryptic soy agar (TSA), TSA with sheep blood, and Mueller-Hinton agar, as well as on some selective agar, including MacConkey's (MAC) agar and Salmonella-Shigella (SS) agar, where they appear as lactose-negative colonies. Cefsulodin-irgasan-novobiocin (CIN) agar is a medium developed specifically to grow *Yersinia* species while inhibiting other Enterobacteriaceae from stool samples. While this organism also has a propensity to grow better than other bacteria at lower temperatures, it is standard protocol in many laboratories that are looking for the presence of *Yersinia,* especially in stool specimens, to incubate a CIN agar or a MAC agar at room temperature.[3] The organisms originally grow as pinpoint colonies after a 24-hr incubation at 35°C, and as creamy, larger, complete colonies after 48 hr of incubation.[2, 3]

Biochemicals

Yersinia species ferment glucose, are cytochrome oxidase-negative, and reduce nitrate to nitrite; as such, they are considered members of the family Enterobacteriaceae. The biochemical characteristics of the three *Yersinia* species that infect humans are shown below:

- Positive:
  - Catalase production
  - Glucose fermentation with gas production
  - Trehelose fermentation
  - Nitrate reduction
  - Urease (except *Y. pestis*)
  - Motility at 25°C (except *Y. pestis*)
- Negative:
  - Oxidase
  - Arginine dihydrolase
  - Citrate utilization
  - Gelatin hydrolysis
  - Hydrogen sulfide
  - Indole production
  - Lysine decarboxylase
  - Ornithine decarboxylase (except *Y. enterocolitica*)
  - Gelatin
  - Tryptophan deamination
  - Phenyalanine deamination
  - Pigmentation
  - Growth in presence of KCN

DISEASES

Plague

Bubonic Plague

Following a bite from an infected flea, the organism travels to the nearest lymph node (usually the inguinal lymph nodes) where it is ingested by macrophages. Once inside the phagocytic cells, the organism rapidly multiplies to high levels. The host's inflammatory reaction that follows causes the formation of buboes associated with this disease. Buboes are oval swellings that range in size from

1 to 10 cm in length, which elevate the overlying skin and are often hot, red, and painful.[4] These patients also usually present with fever.

Septicemia Plague

The organism also can spread to the bloodstream and cause septicemic plague. This can result in the infection of various organs via hematogenous spread. As a result of this spread, patients can develop disseminated intravascular coagulation, bleeding, organ failure, and irreversible shock.[4] Infected patients often develop necrosis in the peripheral blood vessels, causing the skin to take on a blackish color, which is why this disease was called "The Black Death."[2] The red purpuric skin lesions are normally found on the trunk and extremities, but soon change to a dark purple. Blockage of vessels in the fingers, toes, ears, and nose can lead to gangrene. In addition, with primary septicemia plague, buboes may be absent, leading to diagnostic confusion and delays.[4]

Pneumonic Plague

Primary pneumonic plague is seen when a person inhales the infecting organisms such that the primary infection takes place in the lungs. Persons with either the primary or secondary form of pneumonic plague are very contagious and can easily spread the infection from person to person via aerosols. Pneumonic plague is rapidly fatal, which makes rapid diagnosis critical. These patients normally present with symptoms of cough, hemoptysis, with or without buboes.

Miscellaneous Plague

Plague meningitis is a rare complication of infection that occurs as acute early disease or as a result of inadequate treatment for bubonic plague. This syndrome is characterized by headache, fever, meningismus, and cerebrospinal fluid (CSF) pleocytosis. *Yersinia pestis* can also manifest as pharyngeal plague, which is characterized by a painful, inflamed pharynx and associated local lympadenopathy, which may mimic acute tonsillitis. This form of disease can rarely follow either ingestion or inhalation of the bacteria. Plague can also rarely manifest itself in gastrointestinal symptoms, including nausea, vomiting, diarrhea, and abdominal pain, which may precede the presence of bubo formation.[4]

Enteritis

Infections can be caused by *Yersinia enterocolitica* and *Y. pseudotuberculosis* and can be acquired by drinking or eating contaminated food or water. Symptoms associated with the disease caused by *Y. enterocolitica* include diarrhea, fever and abdominal pain, and nausea and vomiting. Mesenteric adenitis develops during infections, causing symptoms that can mimic acute appendicitis.[2] These symptoms can last 2 to 3 weeks and normally occur in older children. Leukocytes and blood may also be found in stool specimens. A reactive polyarthritis may also develop in patients after 2 to 14 days of infection. This involves two to four joints (knees, ankles, toes, fingers, wrists), and symptoms of polyarthritis can persist for up to 4 months. Less commonly seen complications associated with *Y. enterocolitica* infections include ankylosing spondylitis, reactive arthritis, erythema nodosum, exudative pharyngitis, pneumonia, empyema, lung abscess, septicemia, endocarditis, and mycotic aneurysms.[4]

Adentitis

This is the most common disease caused by *Yersinia pseudotuberculosis*. The infection is usually self-limiting. Rarely, cases of septicemia, erythema nodosum, and polyarthritis have also been described in *Y. pseudotuberculosis* infections.[4]

Diagnosis

Aspirates of buboes and blood should be cultured for *Yersinia pestis* if bubonic plague is suspected. Additional specimens, as indicated, can include sputum, bronchial washes, swabs from skin lesions or the pharyngeal mucosa, and CSF. Aspirates of buboes are obtained using sterile saline and a small gauze needle and syringe. A small amount (up to 2 mL) of sterile saline is infused into the bubo and aspirated several times in and out of the bubo until blood-tinged fluid is recovered.[4] Gram stains or Wayson stains of this material should show polymorphonuclear cells and Gram-negative bacilli or coccobacilli. The "safety-pin" appearance of the organisms is best visualized with the Wayson stain.[4] The remaining specimen should be cultured as previously described.

Serological diagnosis is also available for plague. The passive hemagglutination test using antigens from the organism is performed on acute and/or convalescent serum samples. A fourfold rise in titer from acute to convalescent samples or a single titer of greater than 1:16 is presumptive evidence for disease, and 1:128 is diagnostic.[4] Most patients will have antibodies by the second or third week after infection. Additional diagnostic techniques include polymerase chain reaction (PCR) for detecting the nucleic acids of the organisms, and enzyme immunoassay for the detection of antigens associated with the organisms.[4]

Diagnosis of *Yersinia enterocolitica* and *Y. pseudotuberculosis* infections normally depends on the isolation of the organisms from stool, mesenteric lymph nodes, peritoneal fluid, blood, or abscesses.[4]

Therapy

If left untreated, plague can result in a 50% mortality rate. Streptomycin is the drug of choice for treatment of all types of plague. Alternatively (as streptomycin can be difficult to obtain in some countries), gentamicin or gentamicin in combination with doxycycline can be used. Chloramphenicol can be used when high tissue concentrations are necessary, as with cases of meningitis. Resistant strains of *Yersinia pestis* have rarely been encountered, and treatment does not seem to select for a resistant population of organisms as relapses do not occur.[4]

Infections caused by *Yersinia enterocolitica*, which are β-lactamase-producing organisms, can be treated with aminoglycosides, chloramphenicol, tetracycline, trimethoprim-sulfamethoxazole, piperacillin, ciprofloxacin, and the third-generation cephalosporins. *Y. pseudotuberculosis* is normally susceptible to the aminoglycosides, chloramphenicol, tetracycline, and cephalosporins although therapy is usually not warranted in patients with mesenteric adenitis caused by this organism. Septicemia caused by either of these organisms is usually greater than 50% fatal despite appropriate therapy.[4]

Prevention

A formalin-killed plague vaccine exists but is only available currently from a manufacturer in Australia. It is not protective against the pulmonary form of the disease, and immunity requires several doses. It is rarely used today except for personnel in research laboratories who work with the agent. Research is currently underway to produce a vaccine that would be effective against the pulmonary form of the disease as concerns mount over this organism potentially being used as a bioterror weapon.[4]

Persons living in endemic areas should take precautions against becoming infected with the plague bacteria. Protective measures against exposure to rodents and their fleas is of utmost importance.

Public health measures against acquiring infections from *Yersinia enterocolitica* and *Y. pseudotuberculosis* include appropriate sanitation measures in food preparation, especially involving pork products, and not eating undercooked pork.[4]

REFERENCES

1. Black Death. From Wikipedia; http://en.wikipedia.org/wiki/Black_Death; accessed January 16, 2007.
2. Bockehuhl, J. and J.D. Wong. 2003. Yersinia. In *The Manual of Clinical Microbiology*. P. R. Murray, E.J. Baron, J.H. Jorgensen, M.A. Pfaller, and R.H. Yolken, Eds. 2003, p. 672–683.
3. Winn, Jr., W.. Allen, S., Janda, W., Koneman, E., Procop, G., Schreckenberger, P., and Woods, G., Eds. *Koneman's Color Atlas and Textbook of Diagnostic Microbiology*, 6th edition. The Enterobacteriaceae, 2006, 239–270.
4. Butler, T. and D.T. Dennis. *Yersinia* species including plague. In *Principles and Practice of Infectious Diseases*, 6th edition. G.L. Mandell, J.E. Bennet, and R. Dolin, Eds. 2005, p. 2693–2695, 2699.

38 The Genus *Bordetella*

Trevor H. Stenson and Mark S. Peppler

CONTENTS

INTRODUCTION AND HISTORY

The genus *Bordetella* currently consists of nine members. This includes the more notable and familiar mammalian pathogens *Bordetella pertussis* and *B. parapertussis*$_{hu}$, which cause whooping cough in humans; *B. parapertussis*$_{ov}$, which causes a respiratory disease in sheep; and *B. bronchiseptica,* which causes respiratory diseases in numerous mammals.[1–7] Genetic data suggest that these four are related enough to actually be subspecies of a single ancestor of *B. bronchiseptica* and are now referred to as the *B. bronchiseptica* cluster[2, 3, 8–10] The members of the cluster are all Gram-negative aerobic coccobacilli.

Bordetella avium, more distantly related to the *B. bronchiseptica* cluster, is a respiratory pathogen of chickens, turkeys, and other domesticated fowl.[11] Three other, more recently identified species in the genus — *B. holmesii, B. trematum,* and *B. hinzii* — have also been associated with human infection. Although *B. hinzii* is more commonly isolated from the respiratory tracts of birds, all three species can cause bloodborne infections in immunocompromised individuals.[12–14] *B. holmesii* is more commonly isolated after causing a respiratory disease in humans,[15–20] whereas *B. trematum* more typically causes wound infections.[21, 22] The most recent addition to the *Bordetella* family is *B. petrii,* and is unusual in that it appears to be the only species so far that was initially isolated from environmental sources such as river sediment (Table 38.1).[23–25]

Bordetella pertussis was the first *Bordetella* identified in the early 1900s by Jules Bordet at the Pasteur Institute.[26] It was originally named *Haemophilus pertussis,* based on some phenotypic similarities to members of the *Haemophilus* genus, including growth on special blood agar. Subsequent biochemical and genetic studies showed there was, in fact, little similarity with *Haemophilus*, and in 1952 the distinct genus *Bordetella* was named in honor of Bordet's foundational work.[27, 28]

TABLE 38.1
Major Differences between the Species and Subspecies of *Bordetella*

	Bordetella Species							
Characteristic	*pertussis*	*parapertussis*$_{hu/ova}$	*bronchiseptica*[131]	*avium*	*trematum*	*holmseii*[137, 138]	*hinzii*	*petrii*
Host Range	Human	Human or sheep	Dog, pig, rodents, horse	Fowl	Humans	Humans	Birds, humans	Environment, humans
Diseases	Whooping cough (typical)	Whooping cough (atypical)	Kennel cough, atrophic rhinitis, snuffles	Respiratory tract infection	Wound infection	Bacteremia after respiratory tract infection	Respiratory tract infection in birds, bacteremia in humans	Infectious potential not fully assessed
Visible colonies on agar (37°C)	3–5 days	2–3 days	1–2 days	24 hours	16–24 hours	3–5 days	2 days	2 days 30°C
Agar growth medium	BGA or R-L agar	BGA or R-L agar	BGA or R-L agar	BHI[b] agar, SBA,[a] MacConkey agar	SBA	BGA or SBA	TSA[d]	BGA, MacConkey or SBA
Broth growth medium	SSB	SSB	SSB, Verwey[54]	BHI broth	ND[e]	BHI	BHI	LB broth[f]
Type strain	Tohama 1	12822	RB50	197N	LMG13506	ATCC 51541	LMG13501T	SE-111RT
Genome size(basepairs)	4,086,186	4,773,551	5,338,400	3,732,255	ND	ND	ND	ND
Motility	–	–	+ (Bvg–)	+	+	–	+	–
LPS	No O sidechain	O side chain	O side chain, mostly identical to *B. parapertussis*$_{hu}$	O sidechain	O sidechain	No O sidechain	O sidechain	ND
Pertussis toxin	+ (Bvg+)	Not expressed	Not expressed	Not encoded	ND	ND	ND	ND
bvg operon[154]	*bvgASR*	*bvgASR*	*bvgASR*	*bvgAS*	*bvgAS´*	*bvgAS*	*bvgA´*	ND

[a] *Bordetella parapertussis*$_{hu}$ and *B. parapertussis*$_{ov}$ are listed in one column. However, it should be remembered that they have distinct host specificities and that there is substantial data that result in them being classified as separate subspecies.
[b] Brain heart infusion.
[c] Sheep's blood agar.
[d] Trypticase soy agar.
[e] Not determined.
[f] Luria-Bertani broth.

PHYLOGENETIC RELATIONSHIPS AND PHYSIOLOGY

The major differences among the nine Bordetellae are summarized in Table 38.1. Some of the most easily observed and practical features are the differences in growth, regarding both the medium required and the time needed for visible colonies to appear; the characteristics of growth (colony types, effect of growth on agar medium — see below); and the presence of assayable antigens, including flagella and lipopolysaccharide (LPS).

An important aspect in observing the described features is how the organisms are grown. This is especially true for the three subspecies of *Bordetella pertussis, B. parapertussis,* and *B. bronchiseptica,* for which the most information is known. These organisms are capable of presenting markedly different antigenic profiles, depending on how they are grown. This is particularly important to be aware of if expression of desired proteins, such as those associated with virulence, is being sought for isolation or characterization.[1, 29–36]

The regulation of virulence-associated antigens in the *Bordetella bronchiseptica* cluster is controlled by the *bvg* regulon.[1, 29, 37, 38] "Bvg" is short for *Bordetella vir* (virulence) genes.[1, 39] The Bvg regulon encodes three key gene products (BvgA, BvgS, and BvgR). BvgS and BvgA make up a classic two-component regulatory system involving a sensor and an activator, respectively. However, the BvgS sensor has more complexity than most common two-component sensors, including another phosphorylation site in what is known as the (cytoplasmic) linker domain.[1, 37, 38, 40] The linker region and associated phosphorylation site add complexity to the signaling system that allows it to act more as a "rheostat switch" rather than an "on-off" signaler.[41–43] Evidence indicates that this allows the signaling system to control a more highly defined and complex phenotypic control than some other two-component global gene regulation systems. BvgS is a cytoplasmic membrane protein that senses environment signals such as temperature, certain anions such as sulfate, and special molecules such as nicotinic acid.[40] At body temperatures and at *in vivo* conditions, BvgS is phosphorylated through a cascade relay system and transfers the phosphate group to the cytoplasmic protein BvgA. Phosphorylated BvgA can then bind to gene sequences that allow for transcription ("activation") of the virulence genes downstream from the promoter sites where BvgA binds.[44] The shorthand for genes that are transcribed this way is "*vags,*" which stands for *vir*-activated genes. (Strictly speaking, these should be called "*bvg*"-activated genes" but the old name for the bvg locus was "*vir*" and the term *vags* was coined at that time).[45] Organisms expressing *vags* are said to be in the Bvg+ state. This is the fully virulent state of the organism as it is typically isolated from infected individuals.

One of the genes activated by BvgA is called BvgR, which is a repressor of a second family of genes called "*vrg*s" or *vir*-repressed genes.[36, 46, 47] When *vags* are highly expressed, *vrg* expression is repressed, and vice versa.[45] However, there are a group of genes that can be expressed when environmental conditions result in an intermediate degree of BvgS and BvgA phosphorylation, and these are known as *bvg*-intermediate genes or "*vigs*."[48–52] The amount of phosphorylated BvgA has been shown to directly affect the positive or negative regulation of these *vigs,* independent of the BvgR repressor protein.[49, 50, 53] In sum, depending on the environment in which the organisms find themselves, certain gene products can be expressed or not, and this regulation is referred to as phenotypic modulation (or sometimes antigenic modulation).[30, 31, 42, 50, 54]

Environmental conditions are one way the organisms can change their phenotypes. Another mechanism is through mutation. In particular, a string of cytosines in the *bvgS* gene of *Bordetella pertussis* is vulnerable to modification, causing *bvgS* to go out of frame and to not be expressed.[55] This results in a phenotype in which only *vrgs* are expressed and is therefore phenotypically Bvg–. These mutations are relatively rare (e.g., 1 in 10^6 colonies per subculture), depending on the strain, and are reversible.[34, 55] The process has been called phase variation and importantly, in the pre-antibiotic era has been observed during the course of disease in *B. pertussis* human disease isolates.[56] When growing *Bordetella* organisms, one needs to keep in mind the potential for such alteration of phenotype, both spontaneous phase variation and the influence of media composition on antigenic modulation. Furthermore, despite relatively limited genetic variability, *B. pertussis* is recognized as having the ability to alter its genomic structure through accumulation of various mutations, including point mutations, gene deletions, and chromosomal inversions that are facilitated by insertion elements in its genome.[57–59] A recent gene-array study has demonstrated alteration of gene expression in loci after as few as 12 culture passages.[60] Hence, when cultivating *B. pertussis,* it is of great importance to make a reserve stock culture of isolated strains and take care not to over-passage.

The remainder of this chapter focuses on the most important features of each species, with emphasis on *Bordetella pertussis.*

Bordetella pertussis

Background

As discussed previously, *Bordetella pertussis* was the first member of the *Bordetella* genus to be isolated and characterized. This priority reflects the negative impact the organism has had on human health as the agent of whooping cough in children. Prior to routine immunization in the United States in 1944, the incidence of mortality due to whooping cough averaged around 1 per 100,000 population and nearly 200,000 cases were recorded annually.[61] Most affected were children under the age of 5 years, although whooping cough can occur at any age.[1, 62, 63] Today, mortality is rare in developed countries but an estimated 350,000 deaths still occur annually in developing countries.[64] The disease is preventable by vaccination (see below) but efficient delivery of vaccine to those most in need has been an issue. Needless to say, whooping cough is still a healthcare issue. Even with high vaccination rates with improved vaccines, the overall incidence of pertussis in the United States and Canada is 10 cases per 100,000 population, although deaths are now a rarity.[65, 66]

Disease and Epidemiology

Whooping cough, also known as "pertussis," begins when organisms are transferred from an infected individual to a susceptible person by aerosols generated through coughing or by contact by unwashed hands. *Bordetella pertussis* has no known environmental or animal reservoir, nor does it have a classical carriage state in people where it might be harbored in between outbreaks.[1, 63, 67]

The first symptoms of disease are much like a common cold with runny nose and perhaps a slight fever. This is known as the catarrhal stage and can last 5 to 10 days. In classic whooping cough, the next stage is the most dramatic and its symptoms are the hallmark of the disease. This is called the paroxysmal stage, so named for the bouts of severe coughing fits, or paroxysms, that occur. In a typical case, the patient coughs in rapid succession up to 30 times until they run out of air. In a concentrated attempt to draw in air, the glottis is often narrowed, and the inspiration produces a whoop or crowing sound that gives the disease its name. Episodes of paroxysmal coughing increase during this stage (up to 20 a day) and tend to be most common at night. The lack of oxygen caused by the cough sequence is one of the main dangers of the disease, and can lead to brain damage or even death. Coughing can also result in broken blood vessels in the conjunctivae, hernias, and prolapsed rectums. As the individual's immune defenses gain the upper hand in 1 to 4 weeks, the severity and frequency of the paroxysmal coughs subside slowly, and may take up to 6 months to completely disappear. Even then, subsequent respiratory infections or irritations can trigger a cough reminiscent of the paroxysmal stage. This recovery period is called the convalescent stage; and if the individual has not experienced the more severe consequences of the paroxysmal stage, he/she usually has no other permanent side effects. Individuals who have recovered from pertussis can get the disease again, but generally with milder symptoms.[1, 63]

Treatment

One of the frustrating aspects of whooping cough is the lack of therapeutic support for individuals experiencing the disease. Although there is very little antibiotic resistance reported for *Bordetella pertussis* and it is clinically sensitive to the macrolide antibiotics like erythromycin and azithromycin, once the symptoms start, they do not abate — even when antibiotics clear the bacteria. This is thought to be consistent with the pathogenesis of disease that is mediated through toxin production. Once the protein toxins are bound and internalized into their target human cells, the damage is done. Killing the infecting bacteria will prevent any further toxin production but has no effect on the toxins already doing their jobs.[1, 66–68]

Vaccination

The first vaccine for pertussis was implemented in 1943 in Canada and in 1944 in the United States. Most commonly, this consisted of formalin-treated, whole killed bacteria. The whole cell pertussis vaccine is administered with diphtheria and tetanus toxoids in a single intramuscular injection and is known as DPT or DTP. A form of this vaccine is still used in developing countries and typically follows a three- to five-dose regimen starting at 2 or 3 months of age, with boosters at 2-month intervals for the first 3 doses, and subsequent boosters at different intervals depending on the jurisdiction administering the vaccine.[69]

In developing countries as the number of whooping cough cases declined, in large part due to vaccination, public concern was voiced over the side effects of the whole cell vaccine. This concern peaked in the late 1970s and early 1980s when vaccine-associated deaths in Japan prompted cessation of vaccination in that country. While no scientific study has ever established a clear cause-and-effect relationship between vaccination and brain damage or death, the scientific community responded by pursuing systematic studies of the molecular basis of disease and of protective components in the vaccine.[70–72] In 1981, Japan produced the first "acellular" vaccines, which were made from components released into broth during growth.[73] The Japanese vaccines were used with good effect in Japan; but in North America and Europe, further purification and quantification of individual antigens followed by extensive clinical trials resulted in the first acellular vaccines approved for use in the United States in the 1990s. Currently there are more than ten licensed acellular vaccines for immunization against pertussis in combination with diphtheria and tetanus toxoids. Few side effects are attributed to these vaccines, their effectiveness is comparable to the whole cell vaccines, and public confidence has been regained. Still, the cost of acellular pertussis vaccines is about 10 times the cost per dose of the whole cell vaccine. This difference in cost makes acellular vaccines unaffordable for developing countries, and alternatives such as live, attenuated intranasal or oral vaccines are being tested as affordable, safe, and improved alternatives to whole cell vaccine for developing countries.[74, 75]

Clinical and Laboratory Diagnosis

Isolation and Growth

Diagnosis of whooping cough can be accomplished by defined clinical criteria but is clearly strengthened by laboratory confirmation. Differing clinical definitions of whooping cough have resulted in different interpretations of vaccine efficacy and therefore need to be clearly stated and applied for meaningful results.[73]

The "gold standard" for diagnosis of whooping cough is the isolation of the causative agent by growth on appropriate media. Isolation of living bacteria also provides a resource from which epidemiological studies of changes in genetic and antigenic structure can be monitored over time. However, even in the best laboratories, cultivation of living organisms from active cases is only achieved about 50% of the time. Samples are typically taken by one of two means: (1) a nasopharyngeal (NP) swab or (2) an NP aspirate. *Bordetella pertussis* colonizes the nasopharynx, not the throat, so the long (14 cm) wire and small swab or the thin canula attached to a syringe containing a buffered wash solution is designed to be inserted through the nose to reach the back of the nasopharynx where the organisms reside. To increase the success of isolation, separate samples are often taken from each side of the nose. These are then either directly plated onto appropriate agar medium such as Regan Lowe, containing 40 g mL^{-1} cephalexin to suppress the growth of other intranasal bacteria, or are placed in a suitable transport medium (e.g., Amies) for delivery to a laboratory for further subculture. If positive, white colonies about 1 mm in diameter appear after 4 to 5 days (sometimes longer) of incubation in a humidified environment at 36°C.[67] Once viable organisms are recovered, they are either lyophilized or frozen at –70°C for long-term storage. A Dacron swab is used to remove the pure growth from an entire 9-cm plate, and suspended in a suitable cryo-protective medium, such as 20% (w/v) skim milk.[76] For research purposes, we prefer to grow organisms from stock on Bordet-Gengou medium containing 15% defibrinated sheep blood. This is so the phenotype of the *Bordetella* organisms can be observed and subcultured faithfully. For example, wild-type (Bvg+) clinical isolates of *Bordetella pertussis* grow as small (1 to 2 mm diameter) domed colonies with notable hemolysis on Bordet-Gengou agar (BGA). Bvg– phase variants, in contrast, form larger (3 to 4 mm) flat, non-hemolytic colonies (Figure 38.1). BGA does not have the same shelf life as Regan-Lowe (R-L) agar but the charcoal base of R-L agar prevents the observation of hemolysis. For growth in broth, the preferred medium is Stainer-Sholte Broth (SSB) or modified SSB, which is aerated through shaking or by having a large surface-to-volume ratio for static liquid culture.[77, 78] SSB is a synthetic medium developed for commercial production of pertussis vaccine and has a good track record. A drawback is that it must be made *de novo* in the laboratory because there is no commercial supplier. Other broth media have been used but the defined composition of SSB is a distinct advantage when it comes to reproducibility.

Non-culture Identification of Bordetella pertussis

Polymerase Chain Reaction (PCR)

One of the great advances in the diagnosis of whooping cough has been the introduction of real-time PCR (RT-PCR). Many clinical labs in the United States have adopted this procedure, and it has had a real impact on the number of reported cases.[65] The technique is highly specific and is sensitive enough to detect as few as 10 to 100 organisms, whether they are alive or dead. Care is needed to prevent contaminating DNA from giving false positives. The standardization of techniques (e.g., which gene targets) are still being worked out for inter-laboratory consistency. NP swabs or aspirates[79] are both suitable sources for PCR analysis.[67]

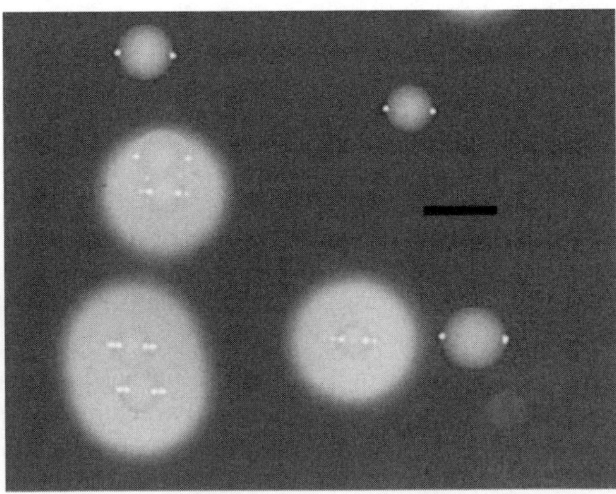

FIGURE 38.1 Colonies of *Bordetella pertussis* phase variants on Bordet-Gengou agar (BGA). Bvg+ (domed, hemolytic) and Bvg– (flat, non-hemolytic). Bar = 3 mm.

Fluorescence Microscopy

Fluorescently labeled monoclonal antibody is commercially available to detect and distinguish between *Bordetella pertussis* and *B. parapertussis.* The specificity is good but it is not as sensitive as PCR. In settings where PCR equipment is not available, the monoclonal direct fluorescent antibody can be an adjunct to culture.[67]

Antibody Titers in Response to Infection

A number of studies have demonstrated the feasibility of using isolated antigens from *Bordetella pertussis* as the foundation of antibody-based diagnosis by enzyme linked immunosorbant assays (ELISAs). Issues for standardization surround the number of serum samples and the number of antigens needed for reliability of the results.[64, 80, 81]

Epidemiology

The age-related profile of pertussis has changed over the past 20 years from a disease of principally non-immunized infants, aged newborn to 6 months, to a disease that includes a large proportion of cases in adolescents. The reasons for this have been debated and appear to have more to do with changing cohorts of susceptible individuals in an age of low disease incidence than it does with the emergence of variant strains of bacteria that can evade the protective immunity provided by vaccination. Regardless, the characterization of different isolates of *Bordetella pertussis* is an important adjunct in monitoring changing trends in pertussis disease.[1, 66, 82]

Serology

An elaborate scheme for characterizing *Bordetella pertussis* isolates by agglutination with adsorbed polyclonal antisera was developed by Eldering in the 1960s.[83, 84] This has since been replaced by other molecular techniques as described below. One of the reasons for this is that *B. pertussis* does not show the same kind of antigenic heterogeneity as observed in other organisms, such as the Enterobacteriaceae, to provide clearly differentiated O (oligosaccharide) antigen types, for example. The other is that monoclonal antibodies or gene-based assays were developed as antigens associated with virulence were characterized genetically and biochemically. These provided much more sensitive and sophisticated tools for observing subtle changes among isolates.[85, 86]

Genomic Polymorphisms

Pulsed-field gel electrophoresis (PFGE) has provided a reliable means by which to fingerprint *Bordetella pertussis* isolates.[87–90] This is a particularly useful tool for confirming the identity of known research strains, akin to karyotyping of tissue culture cells. For epidemiology of disease isolates, PFGE provides an entry point for comparisons but typically does not have the sensitivity to identify changes in individual genes that are associated with virulence. Other more sensitive genomic techniques have been proposed for strain monitoring[91] but none has been uniformly adopted worldwide for comparison of strains.[92]

Allelic Polymorphisms

Changes in gene sequences have been used to determine the genetic relatedness of individual strains of *Bordetella pertussis* and for tracking changes in specific genes. Of special interest are genes associated with virulence that would amount to antigenic drift over time in disease isolates. Changes based on PCR and sequencing or single-strand polymorphisms have demonstrated insertions and deletions in pertussis toxin, pertactin, and fimbria genes that suggest a shift to new allelic types over time.[93–95] To date it does not appear that such changes have compromised the effectiveness of pertussis vaccines, but the recent appearance of non-vaccine capsular type "replacement strains" in invasive *Streptococcus pneumoniae* disease provides fair warning of the potential for routine immunization to select for antigenic types not found in vaccine formulations.[96]

Pathogenesis and Virulence

Overview

Because *Bordetella pertussis* is only found in humans, the lack of a suitable animal model makes it difficult to study the definitive stepwise process of infection and disease. Numerous animal models exist and the one most commonly used is the mouse aerosol challenge model. While mice provide insights into the disease process, it is important to note that the aerosol challenge causes a pneumonia and mice do not experience paroxysmal coughing.[97–99] The following scenario is pieced together from various animal sources and from human clinical data, but there are gaps.

Once *Bordetella pertussis* is inoculated into the human nose by aerosol or by hands contaminated with infected secretions, the organism colonizes the nasopharynx using adhesins, most likely fimbriae or filamentous hemagglutinin (FHA).[1, 100, 101] Organisms also colonize the bronchi where they adhere to the cilia of the respiratory epithelium by means of FHA, and likely other adhesins, and begin to grow. As part of their growth, a number of toxins are secreted that cause the cilia to stop beating and also affect the respiratory epithelium to cause an accumulation of mucus as well as triggering the paroxysmal cough.[1, 67, 102]

Organisms are sequestered in the respiratory tree and are never found in the bloodstream, even in immunocompromised patients. Thus, *Bordetella pertussis* does its damage in the extracellular compartment where in nonlethal cases it is eventually cleared by the host immune response and the respiratory epithelium returns to normal.[1, 63] A potential consequence while the ciliated cells recover is a secondary infection with an organism such as *Streptococcus pneumoniae*. This can be lethal and is more often the cause of death in cases of pertussis than is the hypoxia that can result from the fits of paroxysmal coughing.[63]

An impressive body of work has been devoted to the study of the molecular basis of pathogenesis in whooping cough. Ostensibly, the goal of much of this work was to establish which proteins should be included in acellular vaccines to give the most complete immunity to *Bordetella pertussis*. Not all the molecules known to contribute to virulence are in an acellular vaccine, nor are all acellular vaccines composed of the same number of proteins. The following list will therefore be grouped on the basis of whether the molecules are found in the acellular vaccine or not, and on the basis of how commonly they are found in the vaccine.

Virulence Factors of Bordetella pertussis

Virulence Factors Included in Acellular Vaccines

1. **Pertussis toxin (Ptx)**: The one protein found in all acellular vaccines is pertussis toxin. It has been shown to be necessary to produce disease in most animal models and has been attributed to many biological activities. This is due to its ability to enter cells and modify the inhibitory GTP binding protein of the mammalian cell adenylate cyclase system by covalently adding ADP-ribose derived from intracellular nicotinamide adenine dinucleotide (NAD). The consequence to the target cell is deregulation of normal cellular processes and, depending on the cell affected, can result in altered function that ultimately assists *Bordetella pertussis* in its ability to infect in the lung. The crystal structure of Ptx has been determined; it is a unique A-B type toxin with an A (enzymatically active) subunit called S1 and a heteropentameric B (binding) subunit composed of one S2, one S3, two S4, and one S5 subunit. Despite the clear association of Ptx with virulence, there is no definitive correlation between serum anti-Ptx levels and protection in humans. Ptx, along with cholera toxin, has also been widely used to study the cellular physiology of the 16 mammalian

adenylate cyclase systems and other regulatory systems for which GTP-binding proteins play a key role in cellular homeostasis.[1, 67, 101, 103]

2. **Filamentous hemagglutinin (FHA):** FHA is a unique adhesin with a rod-like structure and an affinity for sulfated glycolipids and cholesterol. It is presumed that antibody to FHA provides protection from colonization with virulent *Bordetella pertussis*.[1, 100, 101]
3. **Pertactin (Prn):** Pertactin is an outer-membrane protein that has adhesive characteristics.[1, 101, 104, 105] Its importance as a vaccine component was strongly supported by results of the early Japanese acellular vaccines that showed the presence of Prn in the most protective vaccines.[105]
4. **Fimbriae (Fim):** Two antigenic types of fimbriae are produced by either *fim2* or *fim3*. Strains of *Bordetella pertussis* typically produce one or the other fimbrial type, but not both; and each is thought to be capable of helping *B. pertussis* attach to respiratory epithelium. Epidemiological evidence supported the importance of having both Fim types in whole cell vaccines, and this observation was applied to the production of some of the licensed acellular vaccines.[1, 101, 105, 106]

Virulence Factors Not Included in Acellular Vaccines

1. **Adenylate cyclase-hemolysin or adenylate cyclase toxin (CyaA):** Early work by Alison Weiss showed that *cya* mutants were as avirulent in mice as mutants in pertussis toxin,[107, 108] but the cloning and characterization of this complex molecule lagged behind that of the other virulence factors and it was never included in any of the acellular vaccines that are currently licensed. CyaA is responsible for the hemolysis produced by Bvg+ organisms on BGA and is also one of the most actively catalytic adenylate cyclases known in the bacterial world.[1, 101, 109, 110] For this reason it has been successfully used as a reporter molecule for bacterial two-hybrid systems.[111] In addition, the adenylate cyclase portion of this bi-functional protein is known to impede phagocytosis of *Bordetella pertussis* by human neutrophils through its enzymatic activity on these host cells.[112–115]
2. **BrkA:** This protein is another *vag* product and gets its name from its ability to confer to Bordetella resistance to killing by serum.[1, 116, 117] The protein is a member of the autotransporter family of proteins and can mediate its own assembly into the outer membrane of *Bordetella pertussis*.[1, 118–120]
3. **Lipopolysaccharide (LPS):** The lipopolysaccharide of *Bordetella pertussis* has the same endotoxic characteristics as other typical Gram-negative bacteria but interestingly does not produce O side chains. On SDS-PAGE and oxidative silver staining, the great majority of clinical isolates produce two distinct bands.[33] The composition of these two species of short LPS is known, as are all of the genes responsible for its production.[121–123] Due to its endotoxicity, LPS was targeted as one of the components that should be absent from any acellular pertussis vaccine.[73]
4. **Tracheal cytotoxin (TCT):** TCT is a breakdown product of peptidoglycan synthesis in most Bordetellae, regardless of phase or phenotypic state. It has a profound effect on causing cilia to stop beating and is thought important in the pathogenesis of Bordetella disease. However, it is poorly immunogenic and has not been included in any acellular vaccine.[1, 101, 106, 124]
5. **Dermonecrotic toxin (DNT)**: DNT, a product of a *vag*, is another true toxin that is sensitive to denaturation by heat and is poorly immunogenic. It causes cell death in low concentrations and is thought to contribute to pertussis pathogenesis in humans although its precise role is still unknown.[1, 101, 106, 125]

Bordetella parapertussis

Bordetella parapertussis exists in two distinct lineages. One, designated *B. parapertussis*$_{hu}$, is a pathogen of humans and causes a milder form of whooping cough. The other lineage is designated *B. parapertussis*$_{ov}$ and is principally a respiratory pathogen of sheep, although human infections have been reported.[8] Each lineage is impressively homogeneous genetically with minor differences among strains, especially *B parapertussis*$_{hu}$.[3] The main differences are noted in Table 38.1.

Of the two, *Bordetella parapertussis*$_{hu}$ is the more consequential for humans. It has caused outbreaks of whooping cough on its own and for a time was responsible for up to 40% of cases in Eastern Europe.[126] A recent PCR study found little evidence for *B. parapertussis*$_{hu}$ disease in Europe but so little is really known about the epidemiology of this organism that it remains to be seen what its impact might be in the future in different parts of the world.[127]

One of the most intriguing aspects of *Bordetella parapertussis*$_{hu}$ is its ability to cause a disease very similar to whooping cough without producing pertussis toxin. In some ways this is akin to shigellosis, where *Shigella sonnei* is perfectly capable of causing disease without producing shiga toxin, whereas the most severe disease is caused by *Shigella dysenteriae*, which does produce shiga toxin. Other interesting differences between *B. parapertussis*$_{hu}$ and *B. pertussis* include the presence of a complete O antigen on *B. parapertussis*$_{hu}$ and the bacterium's ability to grow on blood-free media (such as nutrient agar) where it typically produces a pronounced browning of the medium due to the expression of an enzyme that cleaves tyrosine.[27] *B. parapertussis*$_{hu}$ also grows more rapidly than *B. pertussis*. Because it is a comparatively minor cause of disease in humans, not nearly as much research has been done to understand its mechanisms of pathogenesis. However, in the rat model of whooping cough, *B. pertussis* was able to cause rats to cough paroxysmally; *B. parapertussis*$_{hu}$ was not.[98]

An important question that has been raised in the past is whether *B. parapertussis* might have resurgence with the advent of acellular pertussis vaccines. There is some evidence that whole cell pertussis vaccines provided some cross-protection to *Bordetella parapertussis*$_{hu}$.[128] In contrast, the composition of even the five-component vaccines would not necessarily have enough similarity to antigens in *B. parapertussis*$_{hu}$ to have the same beneficial effect.

Bordetella bronchiseptica

Of all the Bordetellae, *Bordetella bronchiseptica* shows the most genetic diversity. This makes some intuitive sense when one appreciates it has the largest genome (Table 38.1) and the broadest host range. It has even been shown to survive in lake water.[129] It is also the only member of the *B. bronchiseptica* cluster to produce flagella (as the product of a *vrg*). Its growth rate rivals that of *Escherichia coli* and it has full O side chains that are nearly identical to those of *B. parapertussis*$_{hu}$.[121]

The regulation of virulence in *Bordetella bronchiseptica* has been studied in detail; and because the Bvg locus is interchangeable with that of *B. pertussis*, *B. bronchiseptica* has been used as a model for virulence regulation in the Bordetellae. Rodents are part of the natural host range of *B. bronchiseptica*, so infection models are numerous and relevant, and its faster growth rate makes it an attractive organism to study. *B. bronchiseptica* has been shown to invade eukaryotic cells and has even been isolated from human blood.[130–132] For all its capabilities, it is somewhat surprising that it does not cause more than rare zoonotic disease in humans, although these are becoming more common with the increased number of immunocompromised individuals through chemotherapy and AIDS.[130]

Vaccines are available for veterinary use only and mainly for dogs and swine.

Bordetella avium

Of the classically studied Bordetellae, *Bordetella avium* is genetically the least related to *B. pertussis* and yet it shares similar pathogenic mechanisms and some of the same virulence factors[133] (see

Table 38.1). It has an affinity for young turkeys and is therefore of economic consequence to the poultry industry.[11] In terms of environmental durability and in its production of flagella, *B. avium* most resembles *B. bronchiseptica*. However, sequence data show many differences in surface structures that likely contribute to the unique host range for *B. avium*.[133] Both live attenuated and killed bacteria (bacterin) vaccines have been tested, but vaccination of turkeys is not generally practiced and, instead, good hygiene and biosafety practices are usually employed. Fluorescent antibody and PCR tests are available for distinguishing isolates from *B. bronchiseptica* and *B. hinzii*.[11]

OTHER BORDETELLAE

Bordetella holmesii is one of the most intriguing of the less-common bordetellae. In part this is due to its similarity to *B. pertussis* in terms of its more recently appreciated association with human respiratory infection. Whether *B. holmesii* is increasing in prevalence[16] remains to be seen.[134] One of the issues to resolve is the specificity of primers for RT-PCR that can distinguish *B. pertussis* from *B. holmesii*. The pathogenic potential of *B. hinzii* for humans has also been more recently appreciated and its virulence properties have started to be assessed.[135] The same is true for *B. trematum*,[136] and future studies on this, *B. petrii*,[23–25] and other potential new members of the *Bordetella* family will undoubtedly add additional insights into the evolution of pathogenesis in this fascinating genus.

REFERENCES

1. Mattoo, S. and Cherry, J.D., Molecular pathogenesis, epidemiology, and clinical manifestations of respiratory infections due to *Bordetella pertussis* and other *Bordetella* subspecies, *Clin. Microbiol. Rev.,* 18(2), 326–382, 2005.
2. van der Zee, A., Mooi, F., Van Embden, J., and Musser, J., Molecular evolution and host adaptation of *Bordetella* spp.: phylogenetic analysis using multilocus enzyme 500 electrophoresis and typing with three insertion sequences, *J. Bacteriol.,* 179(21), 6609–6617, 1997.
3. Brinig, M.M., Register, K.B., Ackermann, M.R., and Relman, D.A., Genomic features of *Bordetella parapertussis* clades with distinct host species specificity, *Genome Biol.,* 7(9), R81, 2006.
4. Goodnow, R.A., Biology of *Bordetella bronchiseptica*, *Microbiol. Rev.,* 44(4), 722–738, 1980.
5. Heininger, U., Cotter, P.A., Fescemyer, H.W., Martinez de Tejada, G., Yuk, M.H., Miller, J.F., and Harvill, E.T., Comparative phenotypic analysis of the *Bordetella parapertussis* isolate chosen for genomic sequencing, *Infect. Immun.,* 70(7), 3777–3784, 2002.
6. Harvill, E.T., Cotter, P.A., and Miller, J.F., Pregenomic comparative analysis between *Bordetella bronchiseptica* RB50 and *Bordetella pertussis* Tohama I in murine models of respiratory tract infection, *Infect. Immun.,* 67(11), 6109–6118, 1999.
7. Diavatopoulos, D.A., Cummings, C.A., Schouls, L.M., Brinig, M.M., Relman, D.A., and Mooi, F.R., *Bordetella pertussis*, the causative agent of whooping cough, evolved from a distinct, human-associated lineage of *B. bronchiseptica*, *PLoS Pathog* 1(4), e45, 2005.
8. Gerlach, G., von Wintzingerode, F., Middendorf, B., and Gross, R., Evolutionary trends in the genus *Bordetella*, *Microbes Infect.,* 3(1), 61–72, 2001.
9. Aricò, B., Gross, R., Smida, J., and Rappuoli, R., Evolutionary relationships in the genus *Bordetella*, *Mol. Microbiol.,* 1(3), 301–308, 1987.
10. Parkhill, J., Sebaihia, M., Preston, A., Murphy, L.D., Thomson, N., Harris, D.E., Holden, M.T., Churcher, C.M., Bentley, S.D., Mungall, K.L., Cerdeno-Tarraga, A.M., Temple, L., James, K., Harris, B., Quail, M.A., Achtman, M., Atkin, R., Baker, S., Basham, D., Bason, N., Cherevach, I., Chillingworth, T., Collins, M., Cronin, A., Davis, P., Doggett, J., Feltwell, T., Goble, A., Hamlin, N., Hauser, H., Holroyd, S., Jagels, K., Leather, S., Moule, S., Norberczak, H., O'Neil, S., Ormond, D., Price, C., Rabbinowitsch, E., Rutter, S., Sanders, M., Saunders, D., Seeger, K., Sharp, S., Simmonds, M., Skelton, J., Squares, R., Squares, S., Stevens, K., Unwin, L., Whitehead, S., Barrell, B.G., and Maskell, D.J., Comparative analysis of the genome sequences of *Bordetella pertussis*, *Bordetella parapertussis* and *Bordetella bronchiseptica*, *Nat. Genet.,* 35(1), 32–40, 2003.

11. Jackwood, M.A., The Merck Veterinary Manual, 9th ed., in *Bordetellosis*, Kahn, C. M. Merck & Co., Inc., 2006.
12. Gadea, I., Cuenca-Estrella, M., Benito, N., Blanco, A., Fernandez-Guerrero, M. L., Valero-Guillen, P.L., and Soriano, F., *Bordetella hinzii*, a "new" opportunistic pathogen to think about, *J. Infect.*, 40(3), 298–299, 2000.
13. Vandamme, P., Hommez, J., Vancanneyt, M., Monsieurs, M., Hoste, B., Cookson, B., Wirsing von Konig, C.H., Kersters, K., and Blackall, P.J., *Bordetella hinzii* sp. nov., isolated from poultry and humans, *Int. J. Syst. Bacteriol.*, 45(1), 37–45, 1995.
14. Cookson, B.T., Vandamme, P., Carlson, L.C., Larson, A.M., Sheffield, J.V., Kersters, K., and Spach, D.H., Bacteremia caused by a novel *Bordetella* species, "*Bhinzii*", *J. Clin.. Microbiol.*, 32(10), 2569–2571, 1994.
15. Shepard, C.W., Daneshvar, M.I., Kaiser, R.M., Ashford, D.A., Lonsway, D., Patel, J.B., Morey, R.E., Jordan, J.G., Weyant, R.S., and Fischer, M., *Bordetella holmesii* bacteremia: a newly recognized clinical entity among asplenic patients, *Clin. Infect. Dis.*, 38(6), 799–804, 2004.
16. Yih, W.K., Silva, E.A., Ida, J., Harrington, N., Lett, S.M., and George, H., *Bordetella holmesii*-like organisms isolated from Massachusetts patients with pertussis-like symptoms, *Emerg. Infect. Dis.*, 5(3), 441–443, 1999.
17. Tang, Y.W., Hopkins, M.K., Kolbert, C.P., Hartley, P.A., Severance, P.J., and Persing, D.H., *Bordetella holmesii*-like organisms associated with septicemia, endocarditis, and respiratory failure, *Clin. Infect. Dis.*, 26(2), 389–392, 1998.
18. Morris, J.T. and Myers, M., Bacteremia due to *Bordetella holmesii*, *Clin. Infect. Dis.*, (4), 912–913, 1998.
19. Weyant, R.S., Hollis, D.G., Weaver, R.E., Amin, M.F., Steigerwalt, A.G., O'Connor, S.P., Whitney, A.M., Daneshvar, M.I., Moss, C.W., and Brenner, D.J., *Bordetella holmesii* sp. nov., a new Gram-negative species associated with septicemia, *J. Clin. Microbiol.*, 33(1), 1–7, 1995.
20. Lindquist, S.W., Weber, D.J., Mangum, M.E., Hollis, D G., and Jordan, J., *Bordetella holmesii* sepsis in an asplenic adolescent, *Pediatr. Infect. Dis. J.*, 14(9), 813–815, 1995.
21. Daxboeck, F., Goerzer, E., Apfalter, P., Nehr, M., and Krause, R., Isolation of *Bordetella trematum* from a diabetic leg ulcer, *Diabet. Med.*, 21(11), 1247–1248, 2004.
22. Vandamme, P., Heyndrickx, M., Vancanneyt, M., Hoste, B., De Vos, P., Falsen, E., Kersters, K., and Hinz, K.H., *Bordetella trematum* sp. nov., isolated from wounds and ear infections in humans, and reassessment of *Alcaligenes denitrificans* Ruger and Tan 1983, *Int. J. Syst. Bacteriol.*, 46(4), 849–858, 1996.
23. Stark, D., Riley, L.A., Harkness, J., and Marriott, D., *Bordetella petrii* from a clinical sample in Australia: isolation and molecular identification, *J. Med. Microbiol.*, 56(Pt. 3), 435–437, 2007.
24. Fry, N.K., Duncan, J., Malnick, H., Warner, M., Smith, A.J., Jackson, M.S., and Ayoub, A., *Bordetella petrii* clinical isolate, *Emerg. Infect. Dis.*, 11(7), 1131–1133, 2005.
25. von Wintzingerode, F., Schattke, A., Siddiqui, R.A., Rosick, U., Gobel, U.B., and Gross, R., *Bordetella petrii* sp. nov., isolated from an anaerobic bioreactor, and emended description of the genus *Bordetella*, *Int. J. Syst. Evol. Microbiol.*, 51(Pt. 4), 1257–1265, 2001.
26. Bordet, J. and Gengou, U., Le microbe de la coqueluche, *Ann. Inst. Pasteur (Paris),* 20, 731–741, 1906.
27. Pittman, M., *Bordetella*, in *Bergey's Manual of Systemic Bacteriology*, 1st ed., Krieg, N. R. Williams & Wilkins, Baltimore, MD, 1984, p. 338–393.
28. Moreno-López, M., El genero *Bordetella*, *Microbiologia Española,* 5, 177–181, 1952.
29. Stibitz, S. and Miller, J.F., Coordinate regulation of virulence in *Bordetella pertussis* mediated by the *vir* (*bvg*) locus, in *Molecular Genetics of Bacterial Pathogenesis*, 1st ed., ASM Press, Washington, DC, 1994, p. 407–422.
30. McPheat, W.L., Wardlaw, A.C., and Novotny, P., Modulation of *Bordetella pertussis* by nicotinic acid, *Infect. Immun.*, 41(2), 516–522, 1983.
31. Lacey, B.W., Antigenic modulation of *Bordetella pertussis*, *J. Hyg.*, 58, 57–93, 1960.
32. Peppler, M.S. and Schrumpf, M.E., Phenotypic variation and modulation in *Bordetella bronchiseptica*, *Infect. Immun.*, 44(3), 681–687, 1984.
33. Peppler, M.S., Two physically and serologically distinct lipopolysaccharide profiles in strains of *Bordetella pertussis* and their phenotype variants, *Infect. Immun.*, 43(1), 224–232, 1984.
34. Peppler, M.S., Isolation and characterization of isogenic pairs of domed hemolytic and flat nonhemolytic colony types of *Bordetella pertussis*, *Infect. Immun.*, 35(3), 840–851, 1982.

35. Stenson, T.H. and Peppler, M.S., Identification of two *bvg*-repressed surface proteins of *Bordetella pertussis*, *Infect. Immun.*, 63(10), 3780–3789, 1995.
36. Beattie, D.T., Knapp, S., and Mekalanos, J.J., Evidence that modulation requires sequences downstream of the promoters of two *vir*-repressed genes of *Bordetella pertussis*, *J. Bacteriol.*, 172(12), 6997–7004, 1990.
37. Uhl, M.A. and Miller, J.F., *Bordetella pertussis* BvgAS virulence control system, in *Two-Component Signal Transduction*, 1st ed., Hoch, J.A. and Silhavy, T.J., Eds., ASM Press, Washington, DC, 1995, p. 333–349.
38. Cotter, P.A. and DiRita, V.J., Bacterial virulence gene regulation: an evolutionary perspective, *Annu. Rev. Microbiol.*, 54, 519–565, 2000.
39. Aricò, B., Miller, J.F., Roy, C., Stibitz, S., Monack, D., Falkow, S., Gross, R., and Rappuoli, R., Sequences required for expression of *Bordetella pertussis* virulence factors share homology with prokaryotic signal transduction proteins, *Proc. Natl. Acad. Sci. U.S.A.*, 86(17), 6671–6675, 1989.
40. Uhl, M.A. and Miller, J.F., Autophosphorylation and phosphotransfer in the *Bordetella pertussis* BvgAS signal transduction cascade, *Proc. Natl. Acad. Sci. U.S.A.*, 91(3), 1163–1167, 1994.
41. Williams, C.L. and Cotter, P.A., Autoregulation is essential for precise temporal and steady-state regulation by the *Bordetella* BvgAS phosphorelay, *J. Bacteriol.*, 189, 1974–1982, 2007.
42. Veal-Carr, W.L. and Stibitz, S., Demonstration of differential virulence gene promoter activation *in vivo* in *Bordetella pertussis* using RIVET, *Mol. Microbiol.*, 55(3), 788–798, 2005.
43. Cotter, P.A. and Jones, A.M., Phosphorelay control of virulence gene expression in *Bordetella*, *Trends Microbiol.*, 11(8), 367–373, 2003.
44. Roy, C.R., Miller, J.F., and Falkow, S., The *bvgA* gene of *Bordetella pertussis* encodes a transcriptional activator required for coordinate regulation of several virulence genes, *J. Bacteriol.*, 171(11), 6338–6344, 1989.
45. Knapp, S. and Mekalanos, J.J., Two trans-acting regulatory genes (*vir* and *mod*) control antigenic modulation in *Bordetella pertussis*, *J. Bacteriol.*, 170(11), 5059–5066, 1988.
46. Merkel, T.J., Barros, C., and Stibitz, S., Characterization of the *bvgR* locus of *Bordetella pertussis*, *J. Bacteriol.*, 180(7), 1682–1690, 1998.
47. Merkel, T.J. and Stibitz, S., Identification of a locus required for the regulation of *bvg*-repressed genes in *Bordetella pertussis*, *J. Bacteriol.*, 177(10), 2727–2736, 1995.
48. Stockbauer, K.E., Fuchslocher, B., Miller, J.F., and Cotter, P.A., Identification and characterization of BipA, a *Bordetella* Bvg-intermediate phase protein, *Mol. Microbiol.*, 39(1), 65–78, 2001.
49. Williams, C.L., Boucher, P.E., Stibitz, S., and Cotter, P.A., BvgA functions as both an activator and a repressor to control Bvg phase expression of *bipA* in *Bordetella pertussis*, *Mol. Microbiol.*, 56(1), 175–188, 2005.
50. Vergara-Irigaray, N., Chavarri-Martinez, A., Rodriguez-Cuesta, J., Miller, J.F., Cotter, P.A., and Martinez de Tejada, G., Evaluation of the role of the Bvg intermediate phase in *Bordetella pertussis* during experimental respiratory infection, *Infect. Immun.*, 73(2), 748–760, 2005.
51. Cotter, P.A. and Miller, J.F., A mutation in the *Bordetella bronchiseptica bvgS* gene results in reduced virulence and increased resistance to starvation, and identifies a new class of Bvg-regulated antigens, *Mol. Microbiol.*, 24(4), 671–685, 1997.
52. Cummings, C.A., Bootsma, H.J., Relman, D.A., and Miller, J.F., Species- and strain-specific control of a complex, flexible regulon by *Bordetella* BvgAS, *J. Bacteriol.*, 188(5), 1775–1785, 2006.
53. Mishra, M. and Deora, R., Mode of action of the *Bordetella* BvgA protein: transcriptional activation and repression of the *Bordetella bronchiseptica bipA* promoter, *J. Bacteriol.*, 187(18), 6290–6299, 2005.
54. Melton, A.R. and Weiss, A.A., Characterization of environmental regulators of *Bordetella pertussis*, *Infect. Immun.*, 61(3), 807–815, 1993.
55. Stibitz, S., Aaronson, W., Monack, D., and Falkow, S., Phase variation in *Bordetella pertussis* by frameshift mutation in a gene for a novel two-component system, *Nature*, 338(6212), 266–269, 1989.
56. Kasuga, T., Nakase, Y., Ukishima, K., and Takatsu, K., Studies on *Haemophilus pertussis*. V. Relation between the phase of bacilli and the progress of the whooping-cough, *Kitasato Arch. Exp. Med.*, 27(3), 57–62, 1954.
57. Stibitz, S. and Yang, M.S., Genomic plasticity in natural populations of *Bordetella pertussis*, *J. Bacteriol.*, 181(17), 5512–5515, 1999.
58. Caro, V., Hot, D., Guigon, G., Hubans, C., Arrive, M., Soubigou, G., Renauld-Mongenie, G., Antoine, R., Locht, C., Lemoine, Y., and Guiso, N., Temporal analysis of French *Bordetella pertussis* isolates by comparative whole-genome hybridization, *Microbes Infect.*, 8(8), 2228–2235, 2006.

59. Cummings, C.A., Brinig, M.M., Lepp, P.W., van de Pas, S., and Relman, D.A., *Bordetella* species are distinguished by patterns of substantial gene loss and host adaptation, *J. Bacteriol.,* 186(5), 1484–1492, 2004.
60. Brinig, M.M., Cummings, C.A., Sanden, G.N., Stefanelli, P., Lawrence, A., and Relman, D.A., Significant gene order and expression differences in *Bordetella pertussis* despite limited gene content variation, *J. Bacteriol.,* 188(7), 2375–2382, 2006.
61. Linnemann Jr., C.C., Host-parasite interactions in pertussis, in *International Symposium on Pertussis,* National Institutes of Health, U.S.A., Bethesda, MD, 1979.
62. Hewlett, E.L., *Bordetella* species, in *Principles and Practice of Infectious Diseases*, 5th ed., Mandell, G.L., Bennett, J.E., and Dolin, R, Eds. Churchill Livingstone, Philadelphia, PA, 2000, p. 2414–2422.
63. Olson, L.C., Pertussis, *Medicine,* 54(6), 427–468, 1975.
64. Kerr, J.R. and Matthews, R.C., *Bordetella pertussis* infection: pathogenesis, diagnosis, management, and the role of protective immunity, *Eur. J. Clin. Microbiol. Infect. Dis.,* 19(2), 77–88, 2000.
65. Pertussis. Incidence, by year — United States, 1979–2004, Centers for Disease Control, United States Public Health Service, 2004.
66. Raguckas, S.E., VandenBussche, H.L., Jacobs, C., and Klepser, M.E., Pertussis resurgence: diagnosis, treatment, prevention, and beyond, *Pharmacotherapy,* 27(1), 41–52, 2007.
67. Loeffelholz, M.J., *Bordetella*, in *Manual of Clinical Microbiology*, 8th ed., Murray, P.R., Baron, E.J., Jorgensen, J.H., Pfaller, M.A., and Yolken, R.H. ASM Press, Washington, DC, 2003, p. 780–788.
68. Carbonetti, N.H., Immunomodulation in the pathogenesis of *Bordetella pertussis* infection and disease, *Curr. Opin. Pharmacol.*, 7(3), 272–278, 2007.
69. Pertussis vaccines — WHO position paper, *Wkly. Epidemiol. Rec.*, 80(4), 31–39, 2005.
70. Edwards, K.M., Acellular pertussis vaccines — a solution to the pertussis problem? *J. Infect. Dis.,* 168(1), 15–20, 1993.
71. Decker, M.D. and Edwards, K.M., Acellular pertussis vaccines, *Pediatr. Clin. North. Am.,* 47(2), 309–335, 2000.
72. Pichichero, M.E., Edwards, K.M., Anderson, E.L., Rennels, M.B., Englund, J.A., Yerg, D.E., Blackwelder, W.C., Jansen, D.L., and Meade, B.D., Safety and immunogenicity of six acellular pertussis vaccines and one whole-cell pertussis vaccine given as a fifth dose in four- to six-year-old children, *Pediatrics,* 105(1), e11, 2000.
73. Edwards, K.M. and Decker, M.D., Pertussis Vaccine, in *Vaccines, 4th ed.*, Plotkin, S.A. and Orenstein, W.A., Eds. Elsevier, Philadelphia, PA, 2004, p. 471–528.
74. Lee, S.F., Halperin, S.A., Wang, H., and MacArthur, A., Oral colonization and immune responses to *Streptococcus gordonii* expressing a pertussis toxin S1 fragment in mice, *FEMS Microbiol. Lett.,* 208(2), 175–178, 2002.
75. Mielcarek, N., Debrie, A.S., Raze, D., Bertout, J., Rouanet, C., Younes, A.B., Creusy, C., Engle, J., Goldman, W.E., and Locht, C., Live attenuated *B. pertussis* as a single-dose nasal vaccine against whooping cough, *PLoS Pathog* 2(7), e65, 2006.
76. Gherna, R.L., Preservation, in *Manual of Methods for General Bacteriology, 1st ed.*, Gerhardt, P., Murray, R.G.E., Costilow, R.N., Nester, E.W., Wood, W.A., Krieg, N.R., and Phillips, G.B., Eds. American Society for Microbiology, Washington, DC, 1981, p. 208–216.
77. Stainer, D.W. and Scholte, M.J., A simple chemically defined medium for the production of phase I *Bordetella pertussis*, *J. Gen. Microbiol.,* 63(2), 211–220, 1970.
78. Hewlett, E.L., Urban, M.A., Manclark, C.R., and Wolff, J., Extracytoplasmic adenylate cyclase of *Bordetella pertussis*, *Proc. Natl. Acad. Sci. U.S.A.,* 73(6), 1926–1930, 1976.
79. Halperin, S., Kasina, A., and Swift, M., Prolonged survival of *Bordetella pertussis* in a simple buffer after nasopharyngeal secretion aspiration, *Can. J. Microbiol.,* 38(11), 1210–1213, 1992.
80. Hallander, H.O., Microbiological and serological diagnosis of pertussis, *Clin. Infect. Dis.,* 28(Suppl. 2), S99–S106, 1999.
81. Watanabe, M., Connelly, B., and Weiss, A.A., Characterization of serological responses to pertussis, *Clin. Vaccine Immunol.,* 13(3), 341–348, 2006.
82. Halperin, S.A., The control of pertussis — 2007 and beyond, *N. Engl. J. Med.,* 356(2), 110–113, 2007.
83. Eldering, G., Holwerda, J., Davis, A., and Baker, J., *Bordetella pertussis* serotypes in the United States, *Appl. Microbiol.,* 18(4), 618–621, 1969.
84. Robinson, A., Ashworth, L.A., and Irons, L.I., Serotyping *Bordetella pertussis* strains, *Vaccine* 7(6), 491–4, 1989.

85. de Melker, H.E., Schellekens, J.F., Neppelenbroek, S.E., Mooi, F.R., Rumke, H.C., and Conyn-van Spaendonck, M.A., Reemergence of pertussis in the highly vaccinated population of the Netherlands: observations on surveillance data, *Emerg. Infect. Dis.,* 6(4), 348–357, 2000.
86. Van Loo, I.H. and Mooi, F.R., Changes in the Dutch *Bordetella pertussis* population in the first 20 years after the introduction of whole-cell vaccines, *Microbiology,* 148(Pt. 7), 2011–2018, 2002.
87. Bisgard, K.M., Christie, C.D., Reising, S.F., Sanden, G.N., Cassiday, P.K., Gomersall, C., Wattigney, W.A., Roberts, N.E., and Strebel, P.M., Molecular epidemiology of *Bordetella pertussis* by pulsed-field gel electrophoresis profile: Cincinnati, 1989–1996, *J. Infect. Dis.,* 183(9), 1360–1367, 2001.
88. Cassiday, P., Sanden, G., Heuvelman, K., Mooi, F., Bisgard, K.M., and Popovic, T., Polymorphism in *Bordetella pertussis* pertactin and pertussis toxin virulence factors in the United States, 1935–1999, *J. Infect. Dis.,* 182(5), 1402–1408, 2000.
89. Hardwick, T.H., Cassiday, P., Weyant, R.S., Bisgard, K.M., and Sanden, G.N., Changes in predominance and diversity of genomic subtypes of *Bordetella pertussis* isolated in the United States, 1935 to 1999, *Emerg. Infect. Dis.,* 8(1), 44–49, 2002.
90. Hardwick, T.H., Plikaytis, B., Cassiday, P.K., Cage, G., Peppler, M.S., Shea, D., Boxrud, D., and Sanden, G.N., Reproducibility of *Bordetella pertussis* genomic DNA fragments generated by *Xba*I restriction and resolved by pulsed-field gel electrophoresis, *J. Clin. Microbiol.,* 40(3), 811–816, 2002.
91. Schouls, L.M., van der Heide, H.G., Vauterin, L., Vauterin, P., and Mooi, F.R., Multiple-locus variable-number tandem repeat analysis of Dutch *Bordetella pertussis* strains reveals rapid genetic changes with clonal expansion during the late 1990s, *J. Bacteriol.,* 186(16), 5496–5505, 2004.
92. Forsyth, K.D., Wirsing von Konig, C.H., Tan, T., Caro, J., and Plotkin, S., Prevention of pertussis: recommendations derived from the second Global Pertussis Initiative roundtable meeting, *Vaccine,* 25(14), 2634–2642, 2007.
93. Lin, Y.C., Yao, S.M., Yan, J.J., Chen, Y.Y., Hsiao, M.J., Chou, C.Y., Su, H.P., Wu, H.S., and Li, S.Y., Molecular epidemiology of *Bordetella pertussis* in Taiwan, 1993–2004: suggests one possible explanation for the outbreak of pertussis in 1997, *Microbes Infect.,* 8(8), 2082–2087, 2006.
94. Tsang, R.S., Lau, A.K., Sill, M.L., Halperin, S.A., Van Caeseele, P., Jamieson, F., and Martin, I.E., Polymorphisms of the fimbria *fim3* gene of *Bordetella pertussis* strains isolated in Canada, *J. Clin. Microbiol.,* 42(11), 5364–5367, 2004.
95. Tsang, R.S., Sill, M.L., Martin, I.E., and Jamieson, F., Genetic and antigenic analysis of *Bordetella pertussis* isolates recovered from clinical cases in Ontario, Canada, before and after the introduction of the acellular pertussis vaccine, *Can. J. Microbiol.,* 51(10), 887–892, 2005.
96. Singleton, R.J., Hennessy, T.W., Bulkow, L.R., Hammitt, L.L., Zulz, T., Hurlburt, D.A., Butler, J.C., Rudolph, K., and Parkinson, A., Invasive pneumococcal disease caused by nonvaccine serotypes among Alaska native children with high levels of 7-valent pneumococcal conjugate vaccine coverage, *JAMA,* 297(16), 1784–1792, 2007.
97. Halperin, S.A., Heifetz, S.A., and Kasina, A., Experimental respiratory infection with *Bordetella pertussis* in mice: comparison of two methods, *Clin. Invest. Med.,* 11(4), 297–303, 1988.
98. Hall, E., Parton, R., and Wardlaw, A.C., Time-course of infection and responses in a coughing rat model of pertussis, *J. Med. Microbiol.,* 48(1), 95–98, 1999.
99. Woods, D.E., Franklin, R., Cryz, S.J., Jr., Ganss, M., Peppler, M., and Ewanowich, C., Development of a rat model for respiratory infection with *Bordetella pertussis*, *Infect. Immun.,* 57(4), 1018–1024, 1989.
100. Locht, C., Bertin, P., Menozzi, F.D., and Renauld, G., The filamentous haemagglutinin, a multifaceted adhesion produced by virulent *Bordetella* spp., *Mol. Microbiol.,* 9(4), 653–660, 1993.
101. Mooi, F.R., Virulence factors of *Bordetella pertussis*, *Antonie Van Leeuwenhoek,* 54(5), 465–474, 1988.
102. Flak, T.A. and Goldman, W.E., Autotoxicity of nitric oxide in airway disease, *Am. J. Respir. Crit. Care Med.,* 154(4 Pt. 2), S202–S206, 1996.
103. Burnette, W.N., Bacterial ADP-ribosylating toxins: form, function, and recombinant vaccine development, *Behring Inst. Mitt.,* (98), 434–441, 1997.
104. Brennan, M.J. and Shahin, R.D., Pertussis antigens that abrogate bacterial adherence and elicit immunity, *Am. J. Respir. Crit. Care Med.,* 154(4 Pt. 2), S145–S149, 1996.
105. Poolman, J.T. and Hallander, H.O., Acellular pertussis vaccines and the role of pertactin and fimbriae, *Expert Rev. Vaccines,* 6(1), 47–56, 2007.
106. Locht, C., Antoine, R., and Jacob-Dubuisson, F., *Bordetella pertussis*, molecular pathogenesis under multiple aspects, *Curr. Opin. Microbiol.,* 4(1), 82–89, 2001.

107. Weiss, A.A., Hewlett, E.L., Myers, G.A., and Falkow, S., Tn*5*-induced mutations affecting virulence factors of *Bordetella pertussis*, *Infect. Immun.*, 42(1), 33–41, 1983.
108. Weiss, A.A. and Hewlett, E.L., Virulence factors of *Bordetella pertussis*, *Annu. Rev. Microbiol.*, 40, 661–686, 1986.
109. Ladant, D. and Ullmann, A., *Bordatella pertussis* adenylate cyclase: a toxin with multiple talents, *Trends Microbiol.*, 7(4), 172–176, 1999.
110. Ahuja, N., Kumar, P., and Bhatnagar, R., The adenylate cyclase toxins, *Crit. Rev. Microbiol.*, 30(3), 187–196, 2004.
111. Dautin, N., Karimova, G., and Ladant, D., *Bordetella pertussis* adenylate cyclase toxin: a versatile screening tool, *Toxicon*, 40(10), 1383–1387, 2002.
112. Mobberley-Schuman, P.S. and Weiss, A.A., Influence of CR3 (CD11b/CD18) expression on phagocytosis of *Bordetella pertussis* by human neutrophils, *Infect. Immun.*, 73(11), 7317–7323, 2005.
113. Mobberley-Schuman, P.S., Connelly, B., and Weiss, A.A., Phagocytosis of *Bordetella pertussis* incubated with convalescent serum, *J. Infect. Dis.*, 187(10), 1646–1653, 2003.
114. Weingart, C.L., Mobberley-Schuman, P.S., Hewlett, E.L., Gray, M.C., and Weiss, A.A., Neutralizing antibodies to adenylate cyclase toxin promote phagocytosis of *Bordetella pertussis* by human neutrophils, *Infect. Immun.*, 68(12), 7152–7155, 2000.
115. Weingart, C.L. and Weiss, A.A., *Bordetella pertussis* virulence factors affect phagocytosis by human neutrophils, *Infect. Immun.*, 68(3), 1735–1739, 2000.
116. Barnes, M.G. and Weiss, A.A., BrkA protein of *Bordetella pertussis* inhibits the classical pathway of complement after C1 deposition, *Infect. Immun.*, 69(5), 3067–3072, 2001.
117. Fernandez, R.C. and Weiss, A.A., Cloning and sequencing of a *Bordetella pertussis* serum resistance locus, *Infect. Immun.*, 62(11), 4727–4738, 1994.
118. Oliver, D.C., Huang, G., Nodel, E., Pleasance, S., and Fernandez, R.C., A conserved region within the *Bordetella pertussis* autotransporter BrkA is necessary for folding of its passenger domain, *Mol. Microbiol.*, 47(5), 1367–1383, 2003.
119. Oliver, D.C., Huang, G., and Fernandez, R.C., Identification of secretion determinants of the *Bordetella pertussis* BrkA autotransporter, *J. Bacteriol.*, 185(2), 489–495, 2003.
120. Shannon, J.L. and Fernandez, R.C., The C-terminal domain of the *Bordetella pertussis* autotransporter BrkA forms a pore in lipid bilayer membranes, *J. Bacteriol.*, 181(18), 5838–5842, 1999.
121. Caroff, M., Aussel, L., Zarrouk, H., Martin, A., Richards, J.C., Therisod, H., Perry, M.B., and Karibian, D., Structural variability and originality of the *Bordetella* endotoxins, *J. Endotoxin Res.*, 7(1), 63–68, 2001.
122. Caroff, M., Brisson, J., Martin, A., and Karibian, D., Structure of the *Bordetella pertussis* 1414 endotoxin, *FEBS Lett.*, 477(1–2), 8–14, 2000.
123. Preston, A., Thomas, R., and Maskell, D.J., Mutational analysis of the *Bordetella pertussis wlb* LPS biosynthesis locus, *Microb. Pathog.*, 33(3), 91–95, 2002.
124. Luker, K.E., Tyler, A.N., Marshall, G.R., and Goldman, W.E., Tracheal cytotoxin structural requirements for respiratory epithelial damage in pertussis, *Mol. Microbiol.*, 16(4), 733–743, 1995.
125. Matsuzawa, T., Fukui, A., Kashimoto, T., Nagao, K., Oka, K., Miyake, M., and Horiguchi, Y., *Bordetella* dermonecrotic toxin undergoes proteolytic processing to be translocated from a dynamin-related endosome into the cytoplasm in an acidification-independent manner, *J. Biol. Chem.*, 279(4), 2866–2872, 2004.
126. David, S., van Furth, R., and Mooi, F.R., Efficacies of whole cell and acellular pertussis vaccines against *Bordetella parapertussis* in a mouse model, *Vaccine*, 22(15–16), 1892–1898, 2004.
127. Cimolai, N. and Trombley, C., Molecular diagnostics confirm the paucity of *parapertussis* activity, *Eur. J. Pediatr.*, 160(8), 518, 2001.
128. Stehr, K., Cherry, J.D., Heininger, U., Schmitt-Grohe, S., Uberall, M., Laussucq, S., Eckhardt, T., Meyer, M., Engelhardt, R., and Christenson, P., A comparative efficacy trial in Germany in infants who received either the Lederle/Takeda acellular pertussis component DTP (DTaP) vaccine, the Lederle whole-cell component DTP vaccine, or DT vaccine, *Pediatrics*, 101(1 Pt. 1), 1–11, 1998.
129. Porter, J.F. and Wardlaw, A C., Long-term survival of *Bordetella bronchiseptica* in lakewater and in buffered saline without added nutrients, *FEMS Microbiol. Lett.*, 110(1), 33–36, 1993.
130. Dworkin, M.S., Sullivan, P.S., Buskin, S.E., Harrington, R.D., Olliffe, J., MacArthur, R.D., and Lopez, C.E., *Bordetella bronchiseptica* infection in human immunodeficiency virus-infected patients, *Clin. Infect. Dis.*, 28(5), 1095–1099, 1999.

131. Woolfrey, B.F. and Moody, J.A., Human infections associated with *Bordetella bronchiseptica*, *Clin. Microbiol. Rev.,* 4(3), 243–255, 1991.
132. Gueirard, P., Le Blay, K., Le Coustumier, A., Chaby, R., and Guiso, N., Variation in *Bordetella bronchiseptica* lipopolysaccharide during human infection, *FEMS Microbiol. Lett.,* 162(2), 331–337, 1998.
133. Sebaihia, M., Preston, A., Maskell, D.J., Kuzmiak, H., Connell, T.D., King, N.D., Orndorff, P.E., Miyamoto, D.M., Thomson, N.R., Harris, D., Goble, A., Lord, A., Murphy, L., Quail, M.A., Rutter, S., Squares, R., Squares, S., Woodward, J., Parkhill, J., and Temple, L.M., Comparison of the genome sequence of the poultry pathogen *Bordetella avium* with those of *B. bronchiseptica*, *B. pertussis*, and *B. parapertussis* reveals extensive diversity in surface structures associated with host interaction, *J. Bacteriol.,* 188(16), 6002–6015, 2006.
134. Antila, M., He, Q., de Jong, C., Aarts, I., Verbakel, H., Bruisten, S., Keller, S., Haanpera, M., Makinen, J., Eerola, E., Viljanen, M.K., Mertsola, J., and van der Zee, A., *Bordetella holmesii* DNA is not detected in nasopharyngeal swabs from Finnish and Dutch patients with suspected pertussis, *J. Med. Microbiol.,* 55(Pt 8), 1043–1051, 2006.
135. Donato, G.M., Hsia, H.L., Green, C.S., and Hewlett, E.L., Adenylate cyclase toxin (ACT) from *Bordetella hinzii*: characterization and differences from ACT of *Bordetella pertussis*, *J. Bacteriol.,* 187(22), 7579–7588, 2005.
136. Vinogradov, E. and Caroff, M., Structure of the *Bordetella trematum* LPS O-chain subunit, *FEBS Lett.,* 579(1), 18–24, 2005.
137. Vinogradov, E., The structure of the core-O-chain linkage region of the lipopolysaccharide from *Bordetella hinzii*, *Carbohydr. Res.,* 342(3–4), 638–642, 2007.
138. Vinogradov, E., The structure of the carbohydrate backbone of the lipopolysaccharides from *Bordetella hinzii* and *Bordetella bronchiseptica*, *Eur. J. Biochem.,* 267(14), 4577–4582, 2000.
139. Gerlach, G., Janzen, S., Beier, D., and Gross, R., Functional characterization of the BvgAS two-component system of *Bordetella holmesii*, *Microbiology,* 150(Pt. 11), 3715–3729, 2004.

39 Other Zoonotic Bacteria

Sanjay K. Shukla and Steven Foley

CONTENTS

INTRODUCTION

Zoonotic bacteria can be transmitted between humans and other animal species. More than 200 bacterial and viral zoonotic pathogens have been recognized, which include — but are not limited to — many arthropod-, primate-, and rodent-borne agents. Lately, zoonotic bacteria have assumed additional importance because more than two-thirds of all emerging pathogens during the past three decades are zoonotic in origin. These pathogens are represented by diverse taxa and are not restricted to any particular class or group of bacteria. Zoonotic pathogens maintain either an ongoing reservoir life cycle in animals or arthropod vectors without maintaining a permanent life cycle in humans; or in some cases, they have jumped the species barrier and now maintain a permanent life cycle in humans and may not need the animal reservoir. Figure 39.1 shows the interplay of potential zoonotic pathogens with animal reservoirs and human host in the development of zoonotic diseases. Zoonotic pathogens' ability to colonize human organs and their disease-causing capability vary considerably. Because these pathogens have evolved to live in multiple hosts, including both vertebrates and invertebrates, to maintain their enzootic lifestyle, their genomes have evolved to help them sustain life in different host environments. Some of them have a diminished ability to synthesize essential proteins because they can be acquired from the host. However, many of them retain their ability to make all the specific cofactors and vitamins to sustain efficient growth. These pathogens probably exploit receptors that are common in multiple hosts to colonize and/or gain entry into a human host. Several groups of pathogens with representatives that can cause zoonotic disease are described in their own chapters in this book. This chapter briefly discusses the pathogens of the following zoonotic genera: *Anaplasma, Bartonella, Borrelia, Brucella, Coxiella, Francisella,* and *Pasteurella.*

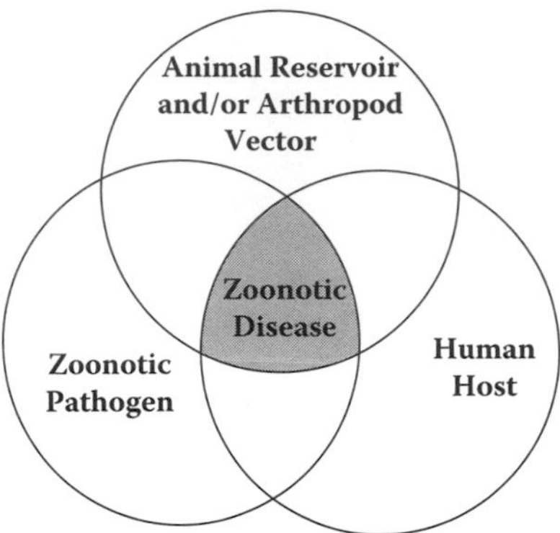

FIGURE 39.1 The interplay among zoonotic pathogen, animal reservoir, and/or arthropod vector, and human host in the development of zoonotic diseases.

GENUS *ANAPLASMA*

Several tick-borne zoonotic pathogens are members of the family Anaplasmataceae. The genus *Anaplasma* is one of four genera placed in the newly organized family Anaplasmataceae. The other three genera in this family are *Ehrlichia, Neorickettsia,* and *Wolbachia*. However, this section primarily focuses on *Anaplasma phagocytophilum*. In the newly proposed taxonomic reclassification of the order Anaplasmatales by Dumler et al. in 2001, the three previously described species (*Ehrlichia equi*, *Ehrlichia phagocytophila,* and human granulocytic agent) were combined and named *Anaplasma phagocytophilum*.[1] This regrouping was justified by Dumler and colleagues following a molecular taxonomic approach that showed at least 99.1% nucleotide similarity in their 16S rRNA genes and identical GroESL amino acid sequences.[1]

Anaplasma phagocytophilum exhibit leukocyte-specific tropisms and infect granulocytes; hence, the disease it causes was named granulocytic anaplasmosis. A wide range of vertebrate animals, including horses, sheep, and humans, can develop granulocytic anaplasmosis. It is a relatively uncommon zoonotic disease in humans and was first recognized in the upper midwestern United States in 1994.[2,3] However, equine granulocytic anaplasmosis was described as early as 1968 in the western United States and Canada[4] and in dogs in 1982.[5] *A. phagocytophilum* infections have been described in California, the upper midwestern and northeastern regions of the United States, British Columbia, and Europe.

This pathogen multiplies in granulocytes and forms membrane-bound, intracytoplasmic colonies known as morulae[6] that can be seen in the peripheral smear of the blood upon staining with Giemsa stain. *Anaplasma phagocytophilum* are pleomorphic, Gram-negative, small obligate intracellular zoonotic bacteria that are 0.5 to 1.5 μm in size. They reside and replicate in membrane-bound vacuoles of neutrophils in eukaryotic cells. The other rickettsia that infects granulocytes is *Ehrlichia ewingii* but it is primarily confined to dogs, although it has been reported to cause infection in humans as well.[7] The symptoms of granulocytic anaplasmosis are not remarkable and overlap with other tick-borne diseases. Symptoms include undifferentiated febrile illness, rigors, headache, myalgia, and malaise, often accompanied by leucopenia and thrombocytopenia. Elevation of liver enzymes such as alanine aminotransferase and aspartate aminotransferase has also been reported. The presence of skin rash is not common.

Because granulocytic anaplasmosis is a tick-borne disease, it is seasonal in nature and often coincides with tick and human outdoor activities (hiking, camping, etc.) in tick-endemic wooded areas. Some of the common natural hosts for *Anaplasma phagocytophilum* include humans, deer, horses, sheep, cattle, bison, and wild rodents. Due to the seasonality of the disease, its transmission and dissemination is intertwined with the 2-year life cycle of the tick vectors, specifically their activities in spring and summer. The disease can be acquired when an *A. phagocytophilum*-infected tick of the *Ixodes ricinus* complex transmits the pathogen to the host during a blood meal. Tick vectors differ in different geographic areas. It is transmitted by *Ixodes pacificus* in the western United States, *Ixodes scapularis* in the midwestern and eastern United States,[8,9] *Ixodes ricinus* in Europe, and *Ixodes persulcatus* in Asia. The tick life cycle can be considered starting with a newly hatched tick larvae parasitizing on white-footed mice in spring. The larvae feed on the mice over the summer and then molt to the nymphal stage. At the nymphal stage, ticks typically acquire the pathogen from the blood meal of their hosts, which can be rodents, deer, or humans. Over winter, nymphs engage in another blood meal before falling off from the host and molting into an adult stage. Adults, both males and females, then search for another host such as white-tailed deer and mate there. Subsequently, females drop off and lay eggs. The eggs hatch and repeat the life cycle next spring. The risk of human infestation is higher when the nymphal tick is searching for a new host, although larvae and adults also are capable of feeding on humans. Humans can get infected upon a blood meal by an infected tick. Studies on mice suggested that infected nymphs may transmit the pathogen within 24 hr of attachment to the host.[10] The main mammalian reservoir for *A. phagocytophilum* is the white-footed mouse (*Peromyscus leucopus*), although white-tailed deer (*Odocoileus virginianus*) also play a role in the maintenance of the pathogen. Other reported vertebrate hosts include humans, cats, sheep, cattle, horses, llamas, bison, etc., whereas ticks are the invertebrate hosts. Granulocytic anaplasmosis is rarely a fatal disease, although it can be fatal in immunocompromised individuals.

A probable diagnosis for granulocytic anaplasmosis includes a febrile illness with a history of tick exposure/bite, the presence of intracytoplasmic morulae in a blood smear of more than 4 times *Anaplasma phagocytophilum* antibodies than the cutoff, or a positive polymerase chain reaction (PCR) result for the pathogen using ehrlichia-specific primers. Because *A. phagocytophilum* is an obligate intracellular pathogen, it is quite difficult to culture this pathogen routinely. However, it can be cultured in HeLa cells[11] and also has been successfully cultured in a tick culture cell line.[12] The availability of the culture-grown cells has advanced our knowledge of the pathogen's physiology and genomics. The genome size of *A. phagocytophilum* has been estimated at 1.47 Mb and include harbor genes that allow them to have a dual existence (as invertebrate symbionts and human and animal pathogens). The genome contains 1369 open reading frames, including genes for major vitamins, cofactors, and nucleotides, but with a somewhat limited ability to make amino acids.[13] Limited genotypic studies on the pathogen suggest that not all *A. phagocytophilum* strains are equal. Disparities exist in disease incidence in different endemic areas, clinical severity, seroprevalences, and disease manifestation in humans, horses, cattle, ruminants, etc. from both North America and Europe, suggesting there may be yet-to-be-understood genotypic differences that exist among strains of this species.[8, 14, 15] Indeed, multiple investigators have identified several genetic variants of *A. phagocytophilum* based on the nucleotide polymorphisms of their 16S RNA genes.[8, 16, 18] Based on their studies of the *A. phagocytophilum* genetic variants of Rhode Island and Connecticut, Massung et al. have hypothesized that some variants are incapable of infecting humans; and in areas where these variants predominate in ticks, a lower incidence of human infection-causing strains would be expected in ticks in these areas and consequently a lower number of human cases.[8, 16, 18] However, a clear picture is yet to emerge relative to their biological significance, specifically with regard to the pathogen's virulence and host specificity. Doxycycline and rifampin are the most effective drugs, based on susceptibility testing.[19] Most people do recover if a timely therapy is instituted.

GENUS *BARTONELLA*

Members of the genus *Bartonella* are part of the α-proteobacteria, along with genera such as *Agrobacterium, Brucella, Rhizobium,* and *Rickettsia*.[20] A number of *Bartonella* species are potential zoonotic pathogens, including *Bartonella henselae, B. elizabethae, B. grahamii, B. vinsonii,* and *B. washoensis*. The bartonellae are small, Gram-negative coccobacilli that are aerobic and fastidious in nature, often requiring weeks to grow in primary culture. In tissues, the organisms are visible with Warthin-Starry silver impregnation stain as tightly compacted clumps. In red blood cells, the organisms can be identified by May-Grünwald Giemsa staining.[21]

The most commonly detected species of the bartonellae is *Bartonella henselae*, the causative agent of cat-scratch disease (CSD). There are 22,000 to 24,000 estimated cases of CSD in the United States each year, with the majority of the cases being relatively mild and having a lesion at the bite/scratch site as well as regional lymphadenopathy. However, approximately 2000 cases require hospitalization to treat the infection.[22] Zoonotic transmission of *B. henselae* to humans generally occurs through a cat bite or scratch. The organisms are likely present in cat flea feces and transmitted to humans during scratching or biting, which disrupts the protective skin barrier. There are two major genotypes of *B. henselae*, Marseille and Houston-1, that are associated with human infection. The distribution of the two genotypes appears to be somewhat segregated. Marseille is the most commonly detected genotype in cats in western Europe, the western United States, and Australia, whereas Houston-1 is most prevalent in Asian countries. Although Marseille is most common in cats in Europe, Houston-1 is often most commonly associated with disease, suggesting that Houston-1 may have a higher level of virulence in humans.[22]

The prevalence of *Bartonella henselae* in cats is generally higher in areas where the climates are warmer and more humid, and lower in colder climates. For example, the prevalence in cats was 68% in the Philippines and 0% in Norway.[22] Cats are generally asymptomatic carriers of the organism. In experimentally infected cats, the organism was able to persist for up to 32 weeks.[23] The organism can persist in the feline bloodstream within erythrocytes and potentially invade endothelial cells. Endothelial invasion has been noted *in vitro* but not yet identified *in vivo*.[24] As mentioned above, the majority of cases of bartonellosis present as typical CSD, with limited regional lymphadenopathy and red to brown papules at the site of infection. In approximately 10 to 20% of cases of typical CSD, multiple lymph nodes may be affected during the disease course. In addition to the skin lesions and lymphadenopathy, a number of patients also report low-grade fever, malaise, headache, and/or sore throat. Atypical CSD occurs in 11 to 12% of patients and involves additional disease manifestations including granulomatous conjunctivitis, hepatitis, splenitis, osteitis, pneumonitis, and neurological pathologies.[20] In addition to CSD, patients infected with *B. henselae* may develop endocarditis or bacillary angiomatosis, a proliferative disease involving vascular tissue that can lead to papular lesions of the skin or affect a number of different internal organs. Bacillary angiomatosis is most commonly seen in immunocompromised patients, such as those with human immunodeficiency virus infections.[24]

Clinically, typical CSD is often diagnosed through the detection of enlarged lymph nodes and granulomatous lesion at the bite/scratch site. Atypical detection of *Bartonella henselae* infections can be done by serological testing, such as by immunofluorescence assay, PCR detection, and direct culture. Direct culture is generally difficult due to the fastidious nature of the organisms and the long incubation periods for growth.[21] Direct culture is beneficial for patients with bacillary angiomatosis, who often do not generate a detectable antibody response required for serological testing. The antimicrobial agents primarily used for the treatment of *B. henselae* infections include the macrolides such as erythromycin, azithromycin or clarithromycin, doxycycline, and rifampin. Antimicrobial agents such as the β-lactams, fluoroquinolones, and trimethoprim/sulfamethoxazole appear to be less effective because lesions can still develop in their presence.[24] Prevention of infection is the most effective way to limit *B. henselae* infections. Therefore, people, especially immunocompromised individuals, should adopt cats that are serologically negative for *B. henselae* and limit flea infestation on cats because fleas likely play a role in the transmission of *B. henselae* to humans.[21]

GENUS *BORRELIA*

The genus *Borrelia*, which presently consists of 11 genospecies, belongs to the family Spirochaetaceae. Not all species are present in all endemic locations. Of the 11 genospecies, *Borrelia andersonii, B. bissettii,* and *B. burgdorferi* sensu stricto have been identified from North America. Seven species, including *B. garinii, B. afzelii, B. valaisiana, B. japonica, B. tanukii, B. turdi,* and *B. sinica* have been identified from Asia. The five European species are *B. burgdorferi* sensu stricto, *B. garinii, B. afzelii, B. valaisiana,* and *B. lustitaniae*.[25] This pathogen is transmitted by tick vectors of the *Ixodes persulcatus* complex (also called the *Ixodes ricinus* complex). *Ixodes scapularis* is the principal vector for this pathogen in the northeastern and north central United States, whereas *Ixodes pacificus* is the primary vector in the western United States. *Ixodes ricinus* and *Ixodes persulcatus* are the principal vectors for Europe and Asia, respectively. Interestingly, not all *Borrelia* species are known to cause infections. Based on a number of genetic and immunologic considerations, all *Borrelia* species can be placed into three phylogenetic distinct genomic clusters: *Borrelia burgdorferi* sensu stricto, *B. garinii,* and *B. afzelii.* In the United States, *B. burgdorferi* sensu stricto is the only species that causes infection in humans.[26, 27]

This spirochetal bacterium was first discovered by Willy Burgdorferi and colleagues in 1981 from the nymph of the tick *Ixodes scapularis.*[28] Subsequently linked to Lyme disease, which was initially recognized by Steere et al. and others,[29, 30] *Borrelia* are highly motile, corkscrew-shaped, endoflagellated bacteria with cells 10 to 30 μm long and 0.2 to 0.5 μm wide. They have a relatively small genome of ~1.5 Mb, which consists of a linear chromosome (~0.91 Mb) and 21 plasmids (9 circular and 12 linear) with a G+C content of 28.6%. Some plasmids are essential and are considered mini chromosomes. Unlike other prokaryotic genomes, the *Borrelia burgdorferi* genome is unique because of its linear chromosome and multiple linear plasmids. Its genomic novelty is also due to the presence of over 150 lipoprotein-encoding genes, including a surface-exposed lipoprotein called VlsE, which undergoes extensive antigenic variation.[31] Analysis of the plasmid sequence suggests evidence of recent genomic rearrangement.[26] *B. burgdorferi* genome encodes for a limited number of biosynthetic proteins and therefore must depend on its host for many of its nutritional needs. Because this pathogen makes no known toxins, it causes infection by attaching to the host cells and migrating through the tissues.

Borrelia burgdorferi sensu lato maintains a complicated enzootic life cycle in nature that involves ticks of the *Ixodes persulactus* complex and a range of animal hosts. Both the transmission vectors and the animal hosts vary, depending on geographic locations. This pathogen causes Lyme disease or Lyme borreliosis, which was first described in the United States in 1975 and has been a notifiable disease since 1990. Its reported incidence in 2002 in the United States was 8.2 cases per 100,000 population. One can acquire Lyme disease after being bitten by a *B. burgdorferi*-infected tick that engages in a blood meal from the host. Small mammals serve as the reservoir for the pathogen for the larva and nymphs, whereas the white-tailed deer acts as the reservoir for the adults. Once infected with the pathogen, larvae can maintain the agent during molting to nymph and adult stages. Lyme disease has multiple phases and typically starts with a spreading skin lesion (erythema migrans) at the tick bite site (phase 1). This typically happens in the spring and/or summer months when tick activity is usually high. In phase 2, which can occur within days to weeks of phase 1, the pathogen may cause additional skin lesions or infect other organs such as the heart, joints, or even the nervous system. Phase 3, which may begin within months to years after the initial bite, is a stage where manifestation of the latent infections takes place.

As expected for vector-borne pathogens, the *Borrelia burgdorferi* sensu lato genome expresses different genes in response to the different microenvironments it lives in as part of its multiple-host associated life cycle. Not all genes expressed in mammalian hosts are expressed in tick vectors, or vice versa.[32] Outer surface proteins (Osp), which are lipoproteins, are among the major antigens of this pathogen. OspA and OspB are expressed in the midgut of the tick vector, whereas OspC is expressed in the salivary gland of the tick once the pathogen migrates there.

The presence of *Borrelia burgdorferi* in a clinical sample can be determined in a laboratory by several approaches, including the direct observation of the pathogen under a microscope. Biopsy material from skin lesions or cerebrospinal fluid or blood samples can be used to isolate the agent in the original Kelly medium[33] or its modified versions such as Barbour-Stoenner-Kelly II (BSK-II) medium, BSK-H, and Kelly medium Preac-Mursic (MKP).[26] *B. burgdorferi* sensu lato has a long generation time of 7 to 20 hr, which necessitates a long incubation period — often up to 12 weeks — to detect a positive culture. However, daily microscopic examination of the culture supernatant can facilitate detection within a week in many cases.[34] The time to infer a positive culture can be further reduced by testing the culture supernatant by PCR even before it is suspected to be positive by microscopy.[35] In general, recovery of the *B. burgdorferi* was achieved best from the skin lesion biopsy followed by plasma, serum, and whole blood, respectively. The latter three also are poor clinical samples for a Lyme PCR assay, probably because of the low number of agents present or the presence of PCR inhibitors in them. A number of qualitative and quantitative PCR-based diagnostic assays have been developed, and these assays can target either chromosomally encoded genes (such as the 16S and 23S rRNA genes, 5S-23S rRNA intergenic region, *recA*, *flab*, *p66*, etc.) or the plasmid-encoded gene *ospA*. Not each target gene has the same level of sensitivity.[36] To confirm Lyme arthritis, PCR-based assays rather than the culture method are recommended from the synovial fluids of the affected joints.[25]

Immunologically, Lyme borreliosis is frequently diagnosed by detecting antibodies in the host's serum against a panel of *Borrelia burgdorferi* antigens that are available as part of the Lyme borreliosis detection kits. Both IgG and IgM antibodies can be detected against *B. burgdorferi* FlaB (41 kDa) and FlaA (37 kDa) antigens within days of exposure to the Lyme agents.[37] Plasmid-encoded OspC, which is heterogeneous in nature (21 to 25 kDa), is another immunodominant antigen during the early stage of the disease.[26] Some of the other commercially available Lyme antibody detection kits use indirect immunofluorescent-antibody assay, enzyme-linked immunosorbent assay, or Western immunoblots or their variations. For the immunofluorescent-antibody assay, diluted serum titers of 124 or 256 that react with fixed Lyme antigen on the glass slide are considered a positive reaction.[38]

Western immunoblots are useful in determining the different stages of Lyme borreliosis as antibody response to different antigens varies. This assay is somewhat variable due to the different strain types and different geographic sources of strains used in the preparation of antigens. Newer enzyme immunoassays that use recombinant antigens of OspC, vlsE, FlaA, P66, etc. have also been developed but their relative sensitivity and reactivity to respective IgM and IgG vary.[26] Borreliacidal antibody assays have also been developed to detect Lyme borreliosis by exploiting the killing properties of OspC in combination with a complex of complements that then act upon the pathogen's outer surface proteins. The borreliacidal assay makes use of flow cytometry and/or staining dyes to distinguish the dead cells from live cells.[39] Overall, all diagnostic tests for Lyme borreliosis should be interpreted in the context of the endemicity of the pathogen in that geographic area and any relevant travel history, clinical samples tested, duration of the infection following tick-bite, and type of assay used.

Lyme disease can be effectively treated with appropriate antibiotics during the early stage of infection. Lyme arthritis and neuroborreliosis can be effectively treated with third-generation cephalosporins. However, prevention of exposure and early treatment should be the strategy for effective management.[40]

GENUS *BRUCELLA*

Members of the genus *Brucella* are known to cause a number of human infections that can manifest themselves in different disease syndromes.[41] A number of different livestock and other animals are the normal hosts for members of the genus *Brucella*. The taxonomic classification of members of this genus is somewhat controversial, with some scientists believing that there is only a single species within the genus, *Brucella melitensis,* which is made up of various biotypes; others believe that the genus is composed of several species. The controversy arises from the high level of genetic similarity among the members of the genus. To aid in the discussion of different hosts and disease manifestations, we utilize the multispecies model in which there are seven different species that are roughly equivalent to the biovars in the monospecific taxonomy.[42] The different species tend to have distinct primary hosts. *B. abortus* is associated with cattle, *B. canis* with dogs, *B. maris* with marine mammals, *B. melitensis* with goats and sheep, *B. neotomae* with rats, *B. ovis* with sheep, and *B. suis* with swine.[43] The species most commonly associated with zoonotic disease are *B. melitensis* and *B. abortus*. However, some of the other species (*B. suis, B. canis*) also have zoonotic potential. Worldwide, the number of human cases of brucellosis is difficult to determine, but it is probable that the annual number of cases is in the hundreds of thousands range. In more developed countries such as the United States, the number of reported cases is relatively small, with the average number of cases being approximately 100 for the past 10 years.[44]

Brucella are Gram-negative, facultative, intracellular bacteria that are members of the α-proteobacteria along with organisms such as *Agrobacterium, Bartenella, Rhizobium,* and *Rickettsia.* The genomes of the organisms are distinctive in that members of the *Brucella abortus, B. melitensis,* and some biotypes of *B. suis* contain two distinct chromosomes, while another *B. suis* biotype has a single larger chromosome.[44] In strains with two chromosomes, there appears to be some duplication in the genetic content of housekeeping genes among the chromosomes. Another interesting feature of the organism is the lack of most of the traditional virulence factors seen in other pathogens, including many toxins and typical secretion systems. The lipopolysaccharide layer of pathogenic *Brucella* is less immunogenic than that of most Gram-negative organisms, thus limiting the activation of the alternative complement pathway and the production of certain cytokines. This mild immune response likely allows for the persistence of the organism in a host.[43]

The transmission of *Brucella* to humans comes primarily through contact with or ingestion of contaminated animal products, or inhalation of aerosolized particles containing brucellae. Occupational exposure and infection of veterinarians, animal workers, and laboratory personnel have also been reported. The most common source of infection in the United States is associated with the consumption of unpasteurized dairy products.[44] To cause human infection, the organisms penetrate the nasal, oral, or pharyngeal mucosa, and are engulfed by lymphocytes and transferred to regional lymph nodes. The organisms that survive phagocytosis are able to persist in the phagosomes and multiply in the endoplasmic reticulum of macrophages.[43] Approximately 2 to 4 weeks post-infection, the organism may be released and disseminated throughout the body and infect localized tissues. This leads to the development of conditions such as endocarditis, meningitis, spondylitis, arthritis, and potential reproductive system problems such as epididymoorchitis and spontaneous abortion if brucellosis is contracted during pregnancy.[44]

The most effective way to limit the effects of brucellosis is to prevent human infections by preventing infection in livestock. Currently there are effective vaccines against *Brucella melitensis* and *B. abortus* that have helped in limiting the number of human infections in the United States. Many current cases in the United States occur in the Southwest and are likely associated with the consumption of unpasteurized dairy products from Mexico, where brucellosis in animals remains a problem.[44] The World Health Organization recommended that treatment of brucellosis involve a multi-drug approach with doxycycline and either streptomycin or rifampin. Other agents with efficacy against *Brucella* include gentamicin, trimethoprim/sulfamethoxazole, and the fluoroquinolones.[44, 45]

GENUS *COXIELLA*

Coxiella burnetii is the causative agent of Q fever, a disease that is found throughout the world. The microorganisms are Gram-negative coccobacilli that are obligate intracellular pathogens. Genetically, the organisms are closely related to *Legionella pneumophila* and are members of the γ-proteobacteria.[46] Some *C. burnetii* cells exist as relatively dormant, small colony variants. These organisms tend to be more resistant to damage by chemical agents and heating. Other cells are transformed into large cell variants that are able to multiply within macrophages and monocytes. As *C. burnetii* are taken up by the phagocytes, they enter the phagosome. Small colony variants are more resistant to the environmental stresses of the external environments and are the primary form of inhaled microorganisms.[47] As the small colony variants enter the phagosome, they appear to delay lysosomal fusion and are transformed into the large cell variant state prior to the formation of the phagolysosome. The low pH environment of the phagolysosome serves as the site of cellular reproduction.[48] *C. burnetii* also can undergo antigenic phase variation following isolation. Organisms recently isolated from humans or other animals display a phase I type antigenic profile consisting of a thick lipopolysaccharide layer capable of blocking antibodies from binding to the surface of the organism. The enhanced lipopolysaccharide layer is one of the major virulence factors employed by

C. burnetii. Upon repeated passage of the microorganisms in cell culture or embryonated eggs, *C. burnetii* can shift to displaying the phase II antigenic structure, which is characterized by having an altered lipopolysaccharide structure and reduced infectivity.[49]

A number of different animal species are capable of being infected with *Coxiella burnetii*, including many mammals, birds, fish, and arthropods such as ticks. Among the mammals, cattle, sheep, goats, cats, and dogs are potential sources of human infections. Spread to humans is generally through direct contact with animals, their byproducts, or through the inhalation of microorganisms that are aerosolized or associated with dust. In animals, such as sheep, the placenta of infected animals has very high levels of *C. burnetii* that can be aerosolized following birth and serve as a key source of infection. Additionally, the hides of infected animals are a potential source of infection, especially for abattoir workers.[46]

Coxiella burnetii infections can present as either an acute illness or as a chronic disease state. Many patients with an acute form of Q fever present with a fever that can be associated with pneumonia and hepatitis; however, in some cases, fever may not be present, while in others, it can last a week or more. In one pre-antibiotic era study, the length of fever ranged from 5 to 57 days with a median of 10 days.[46,50] There is a high level of variation in patient presentation that can confound the diagnosis. The mortality rate of acute infections is around 1 to 2% and is often associated with more severe manifestations of Q fever, such as the development of myocarditis. In addition to fever, many patients with acute Q fever also have severe headaches, night sweats, muscle and joint pain, and anorexia leading to rapid weight loss. The early treatment of patients with an antibiotic such as doxycycline appears to reduce the length of fever and decreases the recovery time in patients who develop pneumonia. Patients who develop pneumonia are characterized by having fever, a nonproductive cough that often expels blood, and typically a severe headache. Radiographic findings are often nonspecific and of limited value for diagnosing Q fever as the cause of the pneumonia.[51]

In patients who develop chronic disease, endocarditis is the major manifestation, accounting for 60 to 70% of the cases. The detection of Q fever-related endocarditis is often delayed up to a year or more from onset to detection of infection. This is associated with significant morbidity in these patients. Patients who do not receive antimicrobial therapy for the endocarditis have a high level of mortality. A primary risk factor for patients to develop endocarditis is having an underlying condition, such as a heart valve abnormality or being immunocompromised.[46] In addition to endocarditis, other presentations can occur, including bone and joint infections; infections of the kidney, liver, or lungs; or a recrudescence of disease during pregnancy. Pregnancy-related infections have been associated with premature birth, miscarriage, and neonatal mortality.[50]

The clinical diagnosis of Q fever can be difficult due to the variable nature of disease presentation. Due to the zoonotic nature, a history of animal exposure to animals known to be associated with Q fever may be helpful. Laboratory testing for *Coxiella burnetii* antibodies is currently the best diagnostic tool. The primary detection method is through indirect fluorescence detection.[49] Other serological tests based on complement fixation and enzyme immunoassays are available but appear to lack the desired specificity and sensitivity for detection. Interestingly, the antibodies against the phase II antigens are generally higher in acute infections and phase I antigens are elevated in chronic infections. Other detection methods are available that utilize PCR detection of the organisms or rely on histological changes in affected tissue. Neither PCR nor histological testing appears to be as valuable as serology for the detection of Q fever at present.[46]

The treatment of Q fever depends on the type of infection. In general, for acute infections, doxycycline treatment for 2 weeks is valuable in reducing the length of fever and/or pneumonia.[49] In children who are infected, treatment with trimethoprim/sulfamethoxazole is preferred. Treatment for patients with chronic endocarditis is somewhat controversial, with some advocating lifelong antibiotic treatment and others recommending shorter courses of 1.5 to 2 years with combination therapy.[46] The combination of doxycycline with hydroxychloroquine appears to be effective in the treatment of chronic endocarditis. Hydroxychloroquine raises the pH of the phagolysosome, allowing the doxycycline to function with greater efficiency in clearing the infection. The use of trimethoprim/sulfamethoxazole to treat *Coxiella burnetii* infections during pregnancy limits the risk for spontaneous abortion and neonatal infections and may prevent recrudescence.[46, 52] In some countries, there are effective vaccines available to prevent Q fever. However, in the United States there is only an investigational vaccine available for laboratory workers who handle live microorganisms. When available, people who are at greatest risk for exposure (abattoir workers, laboratory personnel, animal handlers) should be vaccinated to limit infection.[50, 52]

GENUS *FRANCISELLA*

The genus *Francisella* is part of the γ-proteobacteria and contains the species *Francisella tularensis*, which is the causative agent of the disease tularemia.[53] *F. tularensis* is a small Gram-negative organism that was first isolated in Tulare County, California, in 1911 following an outbreak of disease in small mammals.[54] A large number of animals are known to host *F. tularensis*, including rabbits, ground squirrels, muskrats, and other rodents.[55] *F. tularensis* can be further divided into four subspecies, including *tularensis, holarctica, mediasiatica,* and *novicida*. Each subspecies appears to have varying degrees of virulence for rabbits and humans, with *tularensis* being the most virulent and *novicida* the least.[56]

The incidence of tularemia in the United States has been declining since the 1930s. However, in recent years, *Francisella tularensis* has become a pathogen of high interest as a potential bioterrorism agent. *F. tularensis* was evaluated for use as a potential biological weapon by the United States in the 1950s and was utilized by the biological weapons program of the former Soviet Union.[56, 57] In the natural world, many cases of tularemia likely go undiagnosed because many strains lack significant virulence. In strains that are virulent, an infectious dose of less than 10 colony-forming units (CFU) can cause disease, leading to a mortality rate of up to 5 to 10% without antibiotic treatment.[54] There is a wide range of disease forms of *F. tularensis* infections that vary depending on the route of infection. Ulceroglandular tularemia is the most common, and early disease is characterized by chills, fever, headache, and body pain approximately 3 to 6 days following exposure to the organism. The typical route of infection is through the bite of an insect vector that recently fed on an infected mammal or through non-intact skin following handling of infected game or meat. Following initial flu-like symptoms, an ulcer forms at the site of infection that can persist for a number of months. The bacteria enter the lymphatic system and migrate to the regional lymph nodes, causing swelling of the lymph nodes. The bacteria can also travel and infect other tissues of the body, thereby leading to increased morbidity. Occasionally, inoculation through the skin can occur without the development of an ulcer; this condition is termed glandular tularemia. On some occasions, the conjunctiva of the eye is the primary site of infection and can become ulcerated. This type of infection is called oculargland ular tularemia.[53, 54, 56]

Other routes of infection include the pharynx, gastrointestinal tract, and lungs. Oropharyngeal tularemia is a painful infection of the throat that often involves tonsillitis and enlargement of the cervical lymph nodes. Oropharyngeal tularemia is typically contracted through the consumption of infected food or water.[54, 56] Typhoidal tularemia is a difficult-to-diagnose form of tularemia due to a lack of distinct signs and apparent route of infection. Typically, patients have a high-grade fever but other signs may be more variable. The disease severity is also quite variable, ranging from relatively mild diarrhea to severe bowel ulceration and septicemia that can be fatal. The severity of disease is likely due to differences in infecting strain and the number of infecting organisms to which the patient was exposed.[53, 56] Likely, the most acutely severe form of tularemia is pneumonic tularemia that develops following inhalation of the bacteria. As is the case for typhoidal tularemia, the disease presentation is quite variable, which makes clinical diagnosis quite difficult. People most at risk of developing pneumonic tularemia include farmers, landscapers, and laboratory workers who come in contact with dust containing *Francisella tularensis* or aerosolized microorganisms. Pneumonic tularemia can also develop following hematogenous spread of the organism following ulceroglandular tularemia.[53]

The diagnosis of tularemia depends on the type of disease manifestation. Ulceroglandular tularemia is clinically characterized by a lesion that progresses to develop into an ulcer that heals slowly at the site of infection. Additionally, in ulceroglandular and glandular forms, lymph nodes adjacent to the bite site will be enlarged and tender.[53, 56] Confirmation of infection is typically carried out serologically using agglutination tests. A drawback of the serological test is that there is a delay of approximately 2 to 3 weeks for the antibody titer to increase to significant levels following infection.[55, 56] Other more rapid methods, based on culture or PCR, are available but also have some drawbacks due in large part to the biosafety requirements needed to culture the organism and to the specificity and sensitivity of PCR in certain samples.[54] PCR-based detection has been shown to be highly specific and sensitive in the detection of *Francisella tularensis* from ulceroglandular primary skin lesions.[56] Continued work is being performed to develop more rapid detection methods for *F. tularensis* due to bioterrorism concerns. With respiratory tularemia, radiographic studies can provide insight into the diagnosis of disease, but these too are confirmed serologically or by other means.[54]

In the treatment of severe cases of tularemia, early appropriate antimicrobial therapy is an important complement to the management of potential septic shock and respiratory distress. The primary antibiotics used for the treatment of tularemia are the aminoglycosides streptomycin and gentamicin for 7 to 14 days. Other agents that may be of use for the treatment of less severe infections include doxycycline and the fluoroquinolones. However, relapse of infection is possible in some cases.[53, 56] Work is currently being carried out to develop an effective vaccine against *Francisella tularensis*. The vaccine would likely be beneficial for people in endemic areas who are at higher risk of coming into contact with the microorganism, including hunters, outdoors people, and people working with potentially contaminated soils, such as landscapers.[53]

GENUS *PASTEURELLA*

Microorganisms in the genus *Pasteurella* may be pathogenic to a variety of animals, including humans. The most common species associated with zoonotic disease is *Pasteurella multocida*.[58] *P. multocida* was first identified by Louis Pasteur as the causative agent of the poultry disease fowl cholera.[59] The pasteurellae are Gram-negative facultative anaerobes that are members of the γ-proteobacteria.[60] These nonmotile coccobacilli are closely related to members of the genus *Actinobacillus*.[61] Other species associated with human disease in descending order of prevalence include *P. canis, P. stomatis,* and *P. dagmatis*.[62] Many human infections are associated with animal bites or scratches; however, transmission can occur following contact with infected saliva during licking of the skin or through inhalation of the microorganisms.[63] *P. multocida* can be isolated from the upper respiratory tract of many animal species. Cats and dogs have been found to carry large numbers of the pathogen asymptomatically, which is a potential source for human infections.[64]

It appears that at least 50% of cat bites and nearly 20% of dog bites become infected by pathogens. Of these infections, *Pasteurella* is the most common pathogen associated with cat bites and is one of the more commonly identified pathogens from dog bites. Staphylococcal and streptococcal infections are more common among canine bite wounds. In addition to companion animal bites, other types of animal exposures have also been associated with zoonotic transmission.[62] Many of these non-bite infections lead to severe manifestations, including infections of the bone and joints, skin, soft tissue, meningitis, peritonitis, pneumonia, endocarditis, and septicemia.[58, 62, 63, 65, 66] Often, the route of inoculation for infections not associated with bites and scratches is difficult to ascertain. It is likely that many of the infections occur through initial colonization of the upper respiratory tract, followed by hematogenous spread to the affected organ system. Additionally, between 16 and 31% of *Pasteurella* infections occur without apparent direct animal contact.[67] Some cases are likely associated with human-to-human transmission. The clinical manifestations are similar to those associated with non-bite animal infections.[62]

Initial colonization of non-bite wound associated *Pasteurella* often occurs through fimbriae-associated adherence in the upper respiratory tract to the mucosal epithelial cells. Within the respiratory tract, the tonsils appear to be the primary site of colonization.[62] Other areas of the body also may be colonized. There have been reports of *Pasteurella*-associated urinary tract infections, which are likely due to urinary tract colonization rather than through hematogenous spread.[63] Multiple virulence factors have been discovered in *Pasteurella* strains, including those that allow for the production of fimbriae, toxins, polysaccharide capsules, lipopolysaccharide, and iron acquisition proteins.[59] Toxins detected within virulent strains include the leukotoxins, which may aid in the invasion of lung tissue,[62] and the dermonecrotic toxin (PMT), which is associated with atrophic rhinitis.[59] The production of the capsule aids in resistance to complement and phagocyte-mediated killing, adherence and resistance to drying, and the production of the iron-acquisition compounds, allowing the organisms to capture iron required for growth.[59]

The majority of *Pasteurella* infections are skin and soft tissue infections that most commonly develop following an animal bite and less commonly due to licking of skin or an open wound.[58] Generally following infection, there is swelling and tenderness at the site of infection. Other common (in 20 to 40% of cases) signs and symptoms that occur with infection include regional lymphadenopathy, wound discharge, and fever.[62] Other, more severe manifestations of infection include the development of septic arthritis and/or osteomyelitis. Many bone and joint infections appear to be associated with direct inoculation of the organism from a bite into the periosteum or spread from soft tissue adjacent to the affected bone.[68] Patients with underlying arthritis and receiving corticosteroids may be at increased risk of developing septic arthritis.[62] *Pasteurella multocida* also has been reported as a rare cause of meningitis, with the majority of these patients also being identified as having *Pasteurella* bacteremia.[66, 68] Patients who developed *Pasteurella* septicemia often had underlying health conditions that potentially predisposed them to disease.[65] Following skin and soft tissue, the respiratory tract is the most common site of infection. *P. multocida* is known to cause upper respiratory illnesses such as sinusitis and bronchitis, and lower respiratory tract infections including empyema and pneumonia. Many of the lower respiratory illnesses occur in patients with underlying pathologies. In addition to the respiratory tract, a number of other organ system conditions have been attributable to *Pasteurella* infections, including endocarditis and peritonitis.[68]

The diagnosis of *Pasteurella* infections involves the culture of the organisms from the site of infection or from blood culture.[64] Many of the penicillin drugs appear to be highly effective drugs against most forms of *Pasteurella* infections.[62] The amoxicillin-clavulanic acid combination is a primary recommended therapy. If there is resistance or allergies to the penicillins, macrolides such as clindamycin are recommended for use.[69] The course of treatment of infection depends on the severity of disease. Often, mild soft tissue disease can be treated with antibiotics on an outpatient basis, whereas more severe infections with deeper tissue and/or bone and joint involvement require hospitalization and more aggressive treatment. Intravenous antibiotics and drainage and debridement of infected tissue are often required to clear the infection. Infections involving other portions of the body have relatively high mortality rates. The use of prophylaxis following animal bites remains somewhat controversial. However, many clinicians recommend a short course of antimicrobial therapy to potentially reduce the likelihood of infection with *Pasteurella* or other potential bite wound pathogens.[62]

CONCLUSIONS

This chapter focused on the seven genera of zoonotic pathogens that were not covered in other chapters in this book. The number of other potential zoonotic pathogens is quite substantial; however, our understanding of many of these pathogens is somewhat limited. Over the past decade there has been increasing research interest into understanding the biology of some zoonotic pathogens, such as *Bacillus anthracis, Francisella tularensis,* and *Brucella abortus* because of their potential role as bioterror agents. The taxonomic diversity of zoonotic pathogens is quite substantial as well as with bacterial pathogens, including the mycobacteria, Gram-positive organisms, and Gram-negative organisms. Some pathogens are obligate intracellular bacteria while others are not. Many zoonotic pathogens are spread by direct contact with another vertebrate host through contact or bite wounds, while others are transferred through an arthropod vector, such as a tick or mosquito.

With the high degree of diversity among these pathogens, the control and prevention of zoonotic diseases are more complicated than for pathogens that are transmitted between humans. Our understanding of the ecology of many of the zoonotic pathogens is limited due to the lack of adequate field surveillance systems that include their vectors and animal reservoirs in addition to monitoring the human exposure. Implementation of a comprehensive disease surveillance of wild and domesticated animals and suspected vectors would help to increase our understanding of zoonotic pathogen prevalence and their transmission dynamics. With an increased understanding of the ecological factors that promote the emergence of zoonotic pathogens, scientists will be better prepared to respond to these emerging disease concerns. These zoonotic pathogens and diseases remain a significant cause of morbidity and mortality throughout the world. However, with an increased awareness and understanding of these pathogens, the impact of zoonotic disease can likely be minimized.

REFERENCES

1. Dumler, J.S., Barbet, A.F., Bekker, C.P., Dasch, G.A., Palmer, G.H., Ray, S.C., Rikihisa, Y., and Rurangirwa, F.R., Reorganization of genera in the families *Rickettsiaceae* and *Anaplasmataceae* in the order *Rickettsiales*: unification of some species of *Ehrlichia* with *Anaplasma*, *Cowdria* with *Ehrlichia* and *Ehrlichia* with *Neorickettsia*, descriptions of six new species combinations and designation of *Ehrlichia equi* and "HGE agent" as subjective synonyms of *Ehrlichia phagocytophila*, *Int. J. Syst. Evol. Microbiol.*, 51, 2145–2165, 2001.
2. Bakken, J.S., Dumpler, J.S., Chen, S.M., Eckman, M.R., Van Etta, L.L., and Walker, D.H., Human granulocytic ehrlichiosis in upper Midwest United States. A new species emerging? *JAMA*, 272, 589–595, 1994.
3. Chen, S.M., Dumler, J.S., Bakken, J.S., and Walker, D.H., Identification of a granulocytotropic *Ehrlichia* species as the etiologic agent of human disease, *J. Clin. Microbiol.*, 32, 589–595, 1994.
4. Madigan, J.E. and Gribble, D.H., Equine ehrlichiosis in northern California: 49 cases (1968–1981), *J. Am. Vet. Med. Assoc.*, 190, 445–448, 1987.
5. Madewell, B.R. and Gribble, D.H., Infection in two dogs with an agent resembling *Ehrlichia equi*, *J. Am. Vet. Med. Assoc.*, 180, 512–551, 1982.
6. Rikihisa, Y., The tribe Ehrlichieae and ehrlichial diseases, *Clin. Microbiol. Rev.*, 4, 286–308, 1991.
7. Buller, R.S., Arens, M., Hmiel, S.P., Paddock, C.D., Sumner, J.W., Rikihisa, Y., Unver, A., Gaudreault-Keener, M., Manian, F.A., Liddell, A.M., Schmulewitz, N., and Storch, G.A., *Ehrlichia ewingii,* a newly recognized agent of human ehrlichiosis, *N. Engl. J. Med.*, 341, 148–155, 1999.
8. Massung, R.F., Mauel, M.J, Owens, J.H., Allan, N., Courtney, J.W., Stafford, III, K.C. and Mather, T.N., Genetic variants of *Ehrlichia phagocytophila*, Rhode Island and Connecticut, *Emerg. Infect. Dis.*, 8, 467–472, 2003.
9. Richter, P.J., Kimsey, R.B., Madigan, J.E., Barlough, J.E., Dumler, J.S., and Brooks, D.L., *Ixodes pacificus* (Acari: Ixodidae) as a vector of *Ehrlichia equi* (*Rickettsiales*: *Ehrlichieae*), *J. Med. Entomol.*, 33, 1–5, 1996.
10. des Vignes F., Piesman, J., Heffernan, R., Schulze, T.L., Stafford, K.C., 3rd, Effect of tick removal on transmission of *Borrelia burgdorferi* and *Ehrlichia phagocytophila* by *Ixodes scapularis* nymphs, *J. Infect. Dis.*, 183, 773–778, 2001.
11. Goodman, J.L., Nelson, C., Vitale, B., Madigan, J.E., Dumler, J.S., Kurtti, T.J., and Munderloh, U.G., Direct cultivation of the causative agent of human granulocytic ehrlichiosis, *N. Engl. J. Med.*, 334, 209–215, 1996.
12. Munderloh, U.G., Jauron, S.D., Fingerle, V., Leitritz, L., Hayes, S.F., Hautman, J.M., Nelson, C.M., Huberty, B.W, Kurtti, T.J., Ahlstrand, G.G., Greig, B., Mellencamp, M.A., and Goodman, J.L., Invasion and intracellular development of the human granulocytic ehrlichiosis agent in tick cell culture, *J. Clin. Microbiol.*, 8, 2518–2524, 1999.
13. Hotopp, J.C., Lin, M., Madupu, R., Crabtree, J., Angiuoli, S.V., Eisen, J., Seshadri, R., Ren, Q., Wu, M., Utterback, T.R., Smith, S., Lewis, M., Khouri, H., Zhang, C., Niu, H., Lin, Q., Ohashi, N., Zhi, N., Nelson, W., Brinkac, L.M., Dodson, R.J., Rosovitz, M.J., Sundaram, J., Daugherty, S.C., Davidsen, T.,

Durkin, A.S., Gwinn, M., Haft, D.H., Selengut, J.D., Sullivan, S.A., Zafar, N., Zhou, L., Benahmed, F., Forberger, H., Halpin, R., Mulligan, S., Robinson, J., White, O., Rikihisa, Y., and Tettelin, H., Comparative genomics of emerging human ehrlichiosis agents, *PLoS Genet.*, 2, e21, 2006.

14. Asanovich, K.M., Bakken, J.S., Madigan, J.E., Aguero-Rosenfeld, M., Wormser, G.P., and Dumler, J.S., Antigenic diversity of granulocytic *Ehrlichia* isolates from humans in Wisconsin and New York and a horse in California, *J. Infect. Dis.*, 176, 1029–1034, 1997.
15. Dumler, J.S., Asanovich, K.M., and Bakken, J.S., Analysis of genetic identity of North American *Anaplasma phagocytophilum* strains by pulsed-field gel electrophoresis, *J. Clin. Microbiol.*, 41, 3392–3394, 2003.
16. Belongia, E.A., Reed, K.D., Mitchell, P.D., Kolbert, C.P., Persing, D.H., Gill, J.S., and Kazmierczack, J.J., Prevalence of granulocytic *Ehrlichia* infection among white-tailed deer in Wisconsin, *J. Clin. Microbiol.*, 35, 1465–1468, 1997.
17. Foley, J.E., Crawford-Miksza, L., Dumler, J.S., Glaser, C., Chae, J.S., Yeh, E., Schurr, D., Hood, R., Hunter, W., and Madigan, J.E., Human granulocytic ehrlichiosis in Northern California: two case descriptions with genetic analysis of the *Ehrlichiae*, *Clin. Infect. Dis.*, 29, 388–392, 1999.
18. Poitout, F.M., Shinozaki, J.K., Stockwell, P.J., Holland, C.J., and Shukla, S.K., Genetic variants of *Anaplasma phagocytophilum* infecting dogs in western Washington state, *J. Clin. Microbiol.*, 43, 796–801, 2005.
19. Maurin, M., Bakken, J.S., and Dumler, J.S., Antibiotic susceptibilities of *Anaplasma (Ehrlichia) phagocytophilum* strains from various geographic areas in the United States, *Antimicrob. Agents Chemother.*, 47, 413–415, 2003.
20. Slater, L.N. and Welch, D.F., *Bartonella*, including cat-scratch disease, in *Principles and Practice of Infectious Diseases*, 6th ed., Vol. 2, Mandell, G.L., Bennett, J.E., and Dolin, R., Eds., Elsevier Churchill Livingstone, Philadelphia, PA, 2005, 2733–2748.
21. Boulouis, H.J., Chang, C.C., Henn, J.B., Kasten, R.W., and Chomel, B.B., Factors associated with the rapid emergence of zoonotic *Bartonella* infections, *Vet. Res.*, 36, 383–410, 2005.
22. Chomel, B.B., Boulouis, H.J., Maruyama, S., and Breitschwerdt, E.B., *Bartonella* spp. in pets and effect on human health, *Emerg. Infect. Dis.*, 12, 389–394, 2006.
23. Massei, F., Gori, L., Macchia, P., and Maggiore, G., The expanded spectrum of bartonellosis in children, *Infect. Dis. Clin. North Am.*, 19, 691–711, 2005.
24. Rolain, J.M., Brouqui, P., Koehler, J.E., Maguina, C., Dolan, M.J., and Raoult, D., Recommendations for treatment of human infections caused by *Bartonella* species, *Antimicrob. Agents Chemother.*, 48, 1921–1933, 2004.
25. Aguero-Rosenfield, M.E., Wang, G., Schwartz, I., and Wormser, G.P., Diagnosis of Lyme borreliosis, *Clin. Microbiol. Rev.*, 18, 484–509, 2005.
26. Aguero-Rosenfield, M.E., Nowakowlski, J., McKenna, D.F., Carbonaro, C.A., and Worsmer, G.P., Serodiagnosis of early Lyme diseases, *J. Clin. Microbiol.*, 31, 3090–3095, 1993.
27. Mathiesen, D.A., Oliver, J.H. Jr., Kolbert, C.P., Tullson, E.D., Johnson, B.J., Campbell, G.L., Mitchell, P.D., Reed, K.D., Telford, S.R. 3rd, Anderson, J.F., Lane, R.S., Persing, D.H., Genetic heterogeneity of *Borrelia burgdorferi* in the United States, *J. Infect. Dis.*, 175, 98–107, 1997.
28. Burgdorfer, W., Barbour, A.G., Hayes, S.F., Benach, J.L., Grunwaldt, E., and Davis, J.P., Lyme disease — a tick-borne spirochetosis? *Science*, 216, 1317–1319, 1982.
29. Benach, J.L., Bosler, E.M., Hanrahn, J.P., Coleman, J.L., Habicht, G.S., and Bast, T.F., Spirochetes isolated from two patients with Lyme diseases, *N. Engl. J. Med.*, 308, 740–742, 1983.
30. Steere, A.C., Grodzicki, R.L., Kornblatt, A.N., Craft, J.E., Barbour, A.G., Burgdorfer, W., Schmid, G.P., Johnson, E., and Malawista, S.E., The spirochetal etiology of Lyme disease, *N. Engl. J. Med.*, 308, 733–740, 1983.
31. Zhang, J.R. and Norris, S.J., Genetic variation of the *Borrelia burgdorferi* gene vlsE involves cassette-specific, segmental gene conversion, *Infect. Immun.*, 66, 3698–3704, 1998.
32. Gilmore, R.D. Jr., Mbow, M.L., and Stevenson, B., Analysis of *Borrelia burgdorferi* gene expression during life cycle phases of the tick vector, *Ixodes scapularis*, *Microbes Infect.*, 3, 799–808, 2001.
33. Kelly, R., Cultivation of *Borrelia hermsii*, *Science*, 173, 443–444, 1971.
34. Reed, K.D., Laboratory testing for Lyme disease: possibilities and practicalities, *J. Clin. Microbiol.*, 40, 319–324, 2002.
35. Schwartz, I., Bittker, S., Bowen, S.L., Cooper, D., Pavia, C., and Wormser, G.P., Polymerase chain reaction amplification of culture supernatants for rapid detection of *Borrelia burgdorferi*, *Eur. J. Clin. Microbiol. Infect.*, 12, 879–882, 1993.

36. Schmidt, B.L., PCR in laboratory diagnosis of human *Borrelia burgdorferi* infections, *Clin. Microbiol. Rev.*, 10, 185–201, 1997.
37. Grodzicki, R.L. and Steere, A.L., Comparison of immunoblotting and indirect enzyme-linked immunosorbent assay using different antigen preparations for diagnosing early Lyme disease, *J. Infect. Dis.*, 157, 790–797, 1988.
38. Magnarelli, L.A., Meegan, J.M., Anderson, J.F., and Chappell, W.A., Comparison of an indirect fluorescent-antibody test with an enzyme-linked immunosorbent assay for serological studies of Lyme disease, *J. Clin. Microbiol.*, 20, 181–184, 1984.
39. Callister, S., Jobe, D.A., Agger, W.A., Schell, R.F., Kowalski, T.J., Lovrich, S.D., and Marks, J.A., Ability of the borreliacidal antibody test to confirm Lyme disease in clinical practice, *Clin. Diagn. Lab. Immunol.*, 9, 908–912, 2002.
40. Nathwani, D., Hamlet, N., and Walker, E., Lyme disease: a review, *Br. J. Gen. Pract.*, 40, 72–74, 1990.
41. Ustun, I., Ozcakar, L., Arda, N., Duranay, M., Bayrak, E., Duman, K., Atabay, M., Cakal, B.E., Altundag, K., and Guler, S., *Brucella* glomerulonephritis: case report and review of the literature, *South Med. J.*, 98, 1216–1217, 2005.
42. Cutler, S. J., Whatmore, A.M., and Commander, N.J., Brucellosis — new aspects of an old disease, *J. Appl. Microbiol.*, 98, 1270–1281, 2005.
43. Ko, J. and Splitter, G.A., Molecular host-pathogen interaction in brucellosis: current understanding and future approaches to vaccine development for mice and humans, *Clin. Microbiol. Rev.*, 16, 65–78, 2003.
44. Pappas, G., Akritidis, N., Bosilkovski, M., and Tsianos, E., Brucellosis, *N. Engl. J. Med.*, 352, 2325–2336, 2005.
45. Young, E.J., *Brucella* species, in *Principles and Practice of Infectious Diseases*, 6th ed., Vol. 2, Mandell, G.L., Bennett, J.E., and Dolin, R., Eds., Elsevier Churchill Livingstone, Philadelphia, PA, 2005, 2669–2674.
46. Parker, N.R., Barralet, J.H., and Bell, A.M., Q fever, *Lancet*, 367, 679– 688, 2006.
47. Arricau-Bouvery, N. and Rodolakis, A., Is Q fever an emerging or re-emerging zoonosis? *Vet. Res.*, 36, 327–349, 2005.
48. Howe, D. and Mallavia, L.P., *Coxiella burnetii* exhibits morphological change and delays phagolysosomal fusion after internalization by J774A.1 cells, *Infect. Immun.*, 68, 3815–3821, 2000.
49. Fournier, P.E., Marrie, T.J., and Raoult, D., Diagnosis of Q fever, *J. Clin. Microbiol.*, 36, 1823–1834, 1998.
50. Marrie, T.J. and Raoult, D., *Coxiella burnetii* (Q Fever), in *Principles and Practice of Infectious Diseases*, 6th ed., Vol. 2, Mandell, G.L., Bennett, J.E., and Dolin, R., Eds., Elsevier Churchill Livingstone, Philadelphia, PA, 2005, 2296–2303.
51. Marrie, T.J., *Coxiella burnetii* pneumonia, *Eur. Respir. J.*, 21, 713–719, 2003.
52. Karakousis, P.C., Trucksis, M., and Dumler, J.S., Chronic Q fever in the United States. *J. Clin. Microbiol.*, 44, 2283–2287, 2006.
53. Penn, R.L, *Francisella tularensis* (Tularemia), in *Principles and Practice of Infectious Diseases*, 6th ed., Vol. 2, Mandell, G.L., Bennett, J.E., and Dolin, R., Eds., Elsevier Churchill Livingstone, Philadelphia, PA, 2005, 2674–2685.
54. Tarnvik, A. and Berglund, L., Tularaemia, *Eur. Respir. J.*, 21, 361–373, 2003.
55. Petersen, J.M. and Schriefer, M.E., Tularemia: emergence/re-emergence, *Vet. Res.*, 36, 455–467, 2005.
56. Eliasson, H., Broman, T., Forsman, M., and Back, E., Tularemia: current epidemiology and disease management, *Infect. Dis. Clin. North. Am.*, 20, 289–311, 2006.
57. Farlow, J., Wagner, D.M., Dukerich, M., Stanley, M., Chu, M., Kubota, K., Petersen, J., and Keim, P., *Francisella tularensis* in the United States, *Emerg. Infect. Dis.*, 11, 1835–1841, 2005.
58. Holst, E., Rollof, J., Larsson, L., and Nielsen, J.P., Characterization and distribution of *Pasteurella* species recovered from infected humans, *J. Clin. Microbiol.*, 30, 2984–2987, 1992.
59. Harper, M., Boyce, J.D., and Adler, B., *Pasteurella multocida* pathogenesis: 125 years after Pasteur, *FEMS Microbiol. Lett.*, 265, 1–10, 2006.
60. Prescott, L.M., Harley, J.P., and Klein, D.A., *Microbiology*, 6th ed., McGraw-Hill, New York, 2005.
61. Holmes, B.H., Pickett, M.J., and Hollis, D.G., Unusual Gram-negative bacteria, including *Capnocytophaga, Eikenella, Pasteurella,* and *Streptobacillus*, in *Manual of Clinical Microbiology*, 6th ed., Murray, P.R., Baron, E.J., Pfaller, M.A., Tenover, F.C., and Yolken, R.H., Eds., ASM Press, Washington, DC, 1995, 299–508.

62. Zurlo, J.J., *Pasteurella* species, in *Principles and Practice of Infectious Diseases*, 6th ed., vol. 2, Mandell, G.L., Bennett, J.E., and Dolin, R., Eds., Elsevier Churchill Livingstone, Philadelphia, PA, 2005, 2687–2690.
63. Liu, W., Chemaly, R.F., Tuohy, M.J., LaSalvia, M.M., and Procop, G.W., *Pasteurella multocida* urinary tract infection with molecular evidence of zoonotic transmission, *Clin. Infect. Dis.*, 36, E58– E60, 2003.
64. Tan, J.S., Human zoonotic infections transmitted by dogs and cats, *Arch. Intern. Med.*, 57, 1933–1943, 1997.
65. Ashley, B.D., Noone, M., Dwarakanath, A.D., and Malnick, H., Fatal *Pasteurella dagmatis* peritonitis and septicaemia in a patient with cirrhosis: a case report and review of the literature, *J. Clin. Pathol.*, 57, 210–212, 2004.
66. Boerlin, P., Siegrist, H.H., Burnens, A.P., Kuhnert, P., Mendez, P., Pretat, G., Lienhard, R., and Nicolet, J., Molecular identification and epidemiological tracing of *Pasteurella multocida* meningitis in a baby, *J. Clin. Microbiol.*, 38, 1235–1237, 2000.
67. Hubbert, W.T. and Rosen, M.N., *Pasteurella multocida* infections. II. *Pasteurella multocida* infection in man unrelated to animal bite, *Am. J. Public Health Nations Health,* 60, 1109–1117, 1970.
68. Weber, D.J., Wolfson, J.S., Swartz, M.N., and Hooper, D.C., *Pasteurella multocida* infections. Report of 34 cases and review of the literature, *Medicine*, 63, 133–154, 1984.
69. Gilbert, D.N., Moellering, R.C., and Sande, M.A., *The Sanford Guide to Antimicrobial Therapy*, Antimicrobial Therapy, Inc, Hyde Park, VT, 2003.

40 Other Anaerobic Bacteria
Bacteroides, Porphyromonas, Prevotella, Tannerella, and Fusobacterium

Joseph J. Zambon and Violet I. Haraszthy

CONTENTS

INTRODUCTION

Anaerobic bacteria — those that live in the absence of or in reduced levels of atmospheric oxygen — are both commensals and pathogens. As pathogens, they are typically opportunistic pathogens normally present as part of the endogenous or normal microflora. They can sometimes cause disease if implanted into adjacent anatomical sites as may occur following trauma such as that which occurs during surgical procedures. These endogenous bacteria may also become pathogenic by virtue of gaining access to the bloodstream — bacteremia — and being transported to sites in the body distant from their normal ecologic niche. Recent examples of this include the finding of anaerobic bacteria — members of the genera *Porphyromonas* and *Prevotella* — in atherosclerotic plaques (Haraszthy et al., 2000; Kozarov et al., 2006). A number of species, including *Porphyromonas gingivalis* and *Prevotella intermedia,* have been identified by polymerase chain reaction amplification of bacterial 16S rDNA and DNA hybridization using species-specific DNA probes. Consequently, anaerobic species such as these that have their primary ecologic niche in dental plaque in the human oral cavity are thought to gain access to the vasculature and become embedded into the developing atherosclerotic plaque. The nature of their role in atherogenesis — active or passive — remains to be defined.

As in the case of *Porphyromonas gingivalis* and *Prevotella intermedia* in atherosclerotic plaques, anaerobic infections are typically mixed infections comprised of approximately five or more species almost evenly divided between anaerobes and facultative species. The microbial ecology of these mixed anaerobic infections is mutually sustaining, with the facultative species capable of reducing the local oxygen tension sufficiently to permit the growth of the anaerobic species while the anaerobic species inhibit phagocytosis by host immune cells and provide metabolic end products that are necessary for the growth of the facultative species (Styrt and Gorbach, 1989). The importance of mixed infections was shown in studies by Smith (1930). He found that individual bacterial species isolated from Vincent's infection (now known as necrotizing ulcerative gingivitis) were, alone, unable to cause infection in experimental animals but could cause severe infections when injected in combinations of two or more species into an experimental animal.

Environments with reduced oxygen tension favor infection with anaerobic species. As described above, mixed infections contain facultative bacteria that reduce oxygen levels to facilitate the growth of anaerobic species. Certain disease processes or traumatic injuries can diminish blood flow to a tissue and consequently reduce tissue oxygen levels favoring anaerobic infection. For example, people with diabetes mellitus develop microvascular changes that diminish blood flow, particularly to the extremities, and make them susceptible to anaerobic infections as frequently seen in foot ulcers. In a recent study of 900 clinical isolates collected from patients with intra-abdominal and diabetic foot infections, 61% were anaerobes, of which 56% were *Bacteroides* (Edmiston et al., 2004). Similarly, chronic periodontitis results in the development of epithelial-lined soft tissue pockets around the teeth, sometimes approaching 1 cm in depth. There is a decrease in oxygen tension and an increase in anaerobic bacteria in the dental plaque located in the depths of these periodontal pockets (Tanaka et al., 1998; Mombelli et al., 1996). Finally, there are certain anatomical sites within the human body receiving minimal blood supply that often decreases with age, thus favoring the growth of anaerobic bacteria. The vasculature of dental pulp is reduced over time as the pulp chamber narrows with the deposition of secondary dentin. A carious lesion into the dental pulp can become infected with strict anaerobes such as *Porphyromonas endodontalis,* leading to odontogenic abscess (Siqueira et al., 2001).

Like other bacteria with fastidious growth requirements, studies of anaerobic species have been advanced by the use of techniques that permit maintenance of a continuous anaerobic environment such as roll tubes, anaerobic jars, gas packs, anaerobic chambers, and several commercial systems for the identification of anaerobic bacteria. Even so, routine culture and identification of pathogenic anaerobic bacteria continue to be expensive and consequently limited in the clinical setting (Ortiz and Sande, 2000). The results of clinical microbiological tests for anaerobes often have limited effect on therapy. Less than 1% of anaerobic blood cultures are associated with clinically significant anaerobic bacteremia; and although mortality for these infections is high, patients frequently receive presumptive treatment with appropriate antibiotics based not on the results of anaerobic culture, but on clinical presentation suggestive of an anaerobic infection (Byrd and Roy, 2003). For example, the results of blood cultures of patients with community-acquired pneumonia rarely alter the empiric choice of antibiotics (Ramanujam and Rathlev, 2006).

However, newer DNA-based techniques are rapidly supplanting traditional bacterial culture methods for the identification of bacterial pathogens, including anaerobic bacteria. These methods rely on initial amplification of bacterial DNA in a sample by means of the polymerase chain reaction (PCR), followed by cloning of the amplified DNA, sequencing and, finally, identification of the species by comparison to a database of nucleic acid sequences such as Genbank of the National Center for Biotechnology Information (United States), European Molecular Biology Laboratory (EMBL) Nucleotide Sequence Database, and the DNA Databank of Japan (DDBJ). The direct amplification of microbial nucleic acids, sometimes referred to as broad-range PCR, can identify both cultivable and non-cultivable bacteria (Relman, 2002). This method is based on the highly conserved nature of 16S ribosomal RNA within species and genera. Analysis of the 16S rRNA sequence facilitates the classification of bacteria and defines phylogenetic relationships between species. For example,

Bacteroides splanchnicus, which can be phenotypically distinguished from other members of the genera by its production of n-butyric acid, has recently been shown by analysis of its 16S rRNA to be distinct from the genus *Bacteroides* (Olsen and Shah, 2003). Conversely, bacteria that are phenotypically indistinguishable from other species are frequently being defined as novel species based on 16S rRNA sequence. Thus, nucleic acid sequencing has greatly increased the number of bacterial species, including a large number of previously never-cultured anaerobic bacteria among the *Bacteroides* and *Fusobacteria.* As listed in the Ribosomal Database Project II, of the 6657 entries listed under the genus *Bacteroides*, 6462 (97%) are "uncultured bacteria" identified as *Bacteroides* solely on the basis of 16S sequence (Cole et al., 2007). Consequently, while the identification and characterization of cultivable anaerobic bacteria, including *Bacteroides* and *Fusobacterium,* have previously relied on phenotypic features, biochemical tests, % G+C, and DNA-DNA hybridization, the current "gold standard" for the identification of both cultivable and non-cultivable anaerobes is based on nucleic acid analysis (Clarridge, 2004).

BACTEROIDES, PORPHYROMONAS, AND *PREVOTELLA*

Of the 30 described genera within the family Bacteroidaceae (class: Bacteroidetes; order: Bacteroidales), only four genera have been associated with significant disease in humans. These include *Bacteroides*, *Porphyromonas*, *Prevotella,* and *Fusobacteria*, the latter genus to be considered separately.

Bacterial species within the genera *Bacteroides*, *Porphyromonas*, and *Prevotella* are Gram-negative, obligately anaerobic (sometimes strictly anaerobic, sometimes exhibiting various degrees of aerotolerance), nonmotile (some with twitching motility), and non-sporeforming. The bacterial cells generally range in size from coccobacillus to long rods. The species grow at 37°C, particularly on blood agar. Colonies are generally 1 to 3 mm in diameter, circular, with an entire edge, low convex, smooth, and shiny. Strains are generally not hemolytic but some strains produce small amounts of hemolysin and some are α hemolytic. Strains generally ferment carbohydrates although, as described later, sugar fermentation has been a major feature in distinguishing species and in the taxonomy of this group of microorganisms. They can produce gas and acid, including volatile fatty acids such as acetic and succinic acids. The genera vary in the guanine and cytosine content of the DNA. For the genus *Bacteroides*, the G+C content is 40 to 48 mol%. For the *Porphyromonas,* the G+C content is 40 to 55 mol%; and for the *Prevotella,* the G+C content is 40 to 60 mol%. Both *Bacteroides* and *Prevotella* have a cell wall peptidoglycan based on mesodiaminopimelic acid, similar to other Gram-negative bacteria, and as distinguished from the *Porphyromonas,* which have a cell wall peptidoglycan based on lysine (Shah et al., 1976).

In the human body, species in these three genera colonize a variety of mucosal surfaces — oral, pharyngeal, gastrointestinal, and genitourinary — where they exist as part of the normal (resident) microflora and where they can cause opportunistic infections. As resident species, they are important in stabilizing the microbial ecology on mucosal surfaces and in preventing colonization by exogenous pathogens, a process referred to as colonization resistance (van der Waaij et al., 1972). These species, particularly the *Bacteroides*, are major constituents of the microflora throughout the human gastrointestinal tract where they exist in close contact with the mucosa of the small and large intestines. They outnumber aerobic bacteria in the intestine by 1000:1 and are associated with diseases of the colon. These genera, particularly the *Porphyromonas* and the *Prevotella*, rarely the *Bacteroides*, are also found in the upper gastrointestinal tract. Like *Bacteroides* in feces in the intestine, *Porphyromonas* and *Prevotella* are major components of supragingival and, especially, subgingival dental plaque in the human oral cavity. These genera are associated with oral infectious diseases such as chronic periodontitis, periodontal abscess, endodontic infections, periapical abscess, and sometimes life-threatening soft tissue infections of the head and neck region. In the oropharynx, *Bacteroides*, *Porphyromonas*, and *Prevotella* can be found in tonsillar crypts where they can cause peritonsillar abscess (Kuhn et al., 1995) and in the fissures sometimes found on the dorsal surface of the tongue (Faveri et al., 2006) where they have can cause oral halitosis. Infection with these bacteria in the oropharynx can also be associated with chronic (but not acute) sinusitis and chronic (but not acute) otitis media. These bacteria are present in the upper (but not lower) respiratory tract where they are associated with community-acquired aspiration pneumonia, necrotizing pneumonia, lung abscess, and empyema (Johnson and Hirsch, 2003; Pryor et al., 2002; Brook, 2004a). They are present in the urogenital tract, particularly the vagina, where they can cause bacterial vaginosis, a disease that significantly increases the risk for preterm, low birthweight babies. Bacterial vaginosis is characterized by an alteration in the microbial ecology with decreased numbers of lactobacilli and increased numbers of *Prevotella*. Specific bacteria implicated in bacterial vaginosis include *Bacteroides ureolyticus* and *Prevotella bivia* (Boggess et al., 2005). Finally, *Bacteroides*, *Porphyromonas*, and *Prevotella* are not usually found on the human skin except in the case of human and dog bite wounds (Talan et al., 1999).

TAXONOMY OF *BACTEROIDES, PORPHYROMONAS,* AND *PREVOTELLA*

The taxonomy of the *Bacteroides*, *Porphyromonas*, and *Prevotella* has undergone numerous revisions reflective of changing methodology. Originally, all three genera were classified within a single genera — *Bacteroides* — based on phenotypic and biochemical tests. The bacterial cells were described as Gram-negative, non-sporeforming, anaerobic, often pleomorphic bacilli usually with rounded or pointed ends, sometimes fusiform or filamentous. Growth is stimulated by bile, hemin, and vitamin K. Steroid hormones such as estradiol and progesterone can also stimulate growth. This is important in that people with elevated serum hormone levels such as pregnant women or children during the time of puberty can have oral infections with these species as a result of steroid hormone-stimulated growth (Kornman and Loesche, 1982; Rawlinson et al., 1998).

These bacteria are chemoorganotrophs, fermenting carbohydrates to produce mixed acid fermentation end products including acetate, succinate, lactate, formate, or propionate. Distinct in this regard is *Bacteroides splanchnicus,* which produces butyric acid including isobutyrate and isovalerate. Comparative nucleic acid sequence analysis has shown that *B. splanchnicus* is not a member of the genus (Olsen and Shah, 2003). The absence of butyric acid production differentiates *Bacteroides*, *Porphyromonas*, and *Prevotella* from *Fusobacterium,* which do produce butyric acid.

Prominent among the *Bacteroides* is the species *Bacteroides fragilis,* which at one time consisted of five subspecies: *distasonis, fragilis, ovatus, thetaiotaomicron,* and *vulgatus.* These were subsequently shown by DNA-DNA homology to be poorly related and were then defined as individual species: *Bacteroides distasonis, B. fragilis, B. ovatus, B. thetaiotaomicron,* and *B. vulgatus* (Cato and Johnson, 1976). *Bacteroides merdae, B. splanchnicus,* and *B. stercoris* were identified as closely related species. Recently, 16S rRNA sequence analysis has shown that *B. distasonis*, *B. merdae,* and *B. goldsteinii* should not be classified within *Bacteroides.* They have been proposed to be species within a new genera, *Parabacteroides*, as *Parabacteroides distasonis*, *P. merdae,* and *P. goldsteinii* (Sakamoto and Benno, 2006).

Bacteroides that produce darkly pigmenting colonies constituted an easily distinguishable group that provided the impetus for further taxonomic revisions as well as a relatively easy method for assessing their importance in various diseases such as human chronic periodontitis and, most recently, canine periodontitis (Hardham et al., 2005). Species within the original definition of *Bacteroides* included those within the human oral cavity that produce black or brown pigmented colonies when cultured on blood agar media (sometimes referred to as "black-pigmenting *Bacteroides*"). Colonies from some species exhibit brick-red fluorescence when exposed to UV light, Like the darkly pigmenting colonies, fluorescence has been used as a way of differentiating these species.

In 1921, Oliver and Wherry isolated bacteria producing black or beige pigmented colonies and named them *Bacterium melaninogenicum* ("melanin producing"), mistakenly believing that the pigment in the colonies was melanin. Chemical analysis has since demonstrated that, rather than melanin, the pigments responsible for the dark coloration of the bacterial colonies are protohemin and protoporphyrin, the latter responsible for the brick-red fluorescence of colonies exposed to UV light (Shah et al., 1979; Slots and Reynolds, 1982).

The pigmented *Bacteroides* originally classified as *Bacterium melaninogenicum* were subdivided by Moore and Holdeman into three subspecies based on sugar fermentation: *Bacteroides melaninogenicus* ssp. *melaninogenicus* for strong fermenters, *B. melaninogenicus* ssp. *intermedius* for bacteria with intermediate fermentative ability, and *B. melaninogenicus* ssp. *asaccharolyticus* for non-fermenting species. Subsequently, these subspecies were elevated to species level as *Bacteroides melaninogenicus*, *B. intermedius,* and *B. asaccharolyticus* (Moore and Holdeman, 1973). The sugar-fermenting species *B. melaninogenicus* was divided into *B. loescheii*, *B. denticola,* and *B. melaninogenicus* (Holdeman and Johnson, 1982). The intermediately sugar-fermenting species *Bacteroides intermedius* became two species: *B. intermedius* and *B. corporis* (Johnson and Holdeman, 1983). The sugar non-fermenting strains from the gastrointestinal tract could be distinguished from the sugar non-fermenting strains from the oral cavity both by DNA analysis and by phenotypic traits. Oral strains produced phenylacetic acid while the gastrointestinal strains did not (Kaczmarek and Coykendall, 1980). The asaccharolytic species found in dental plaque around the gingiva were

TABLE 40.1
Classification and Species of *Bacteroides*

Phylum	Bacteroidetes
Class	Bacteroidetes
Order	Bacteroidales
Family	Bacteroidaceae
Genus	*Bacteroides*
Species	*B. acidifaciens*
	B. caccae
	B. capillosus
	B. cellulosolvens
	B. coagulans
	B. (Parabacteroides) distasonis
	B. eggerthii
	B. fragilis
	B. galacturonicus
	B. helcogenes
	B. (Parabacteroides) merdae
	B. ovatus
	B. pectinophilus
	B. polypragmatus
	B. putredinis
	B. pyogenes
	B. salyersiae
	B. splanchnicus
	B. stercoris
	B. suis
	B. tectus
	B. thetaiotaomicron
	B. uniformis
	B. ureolyticus
	B. vulgatus
	B. xylanolyticus

Source: Adapted from Bergey's Taxonomic Outline and from Cole et al. (2007).

first defined as *Bacteroides gingivalis* (Coykendall et al., 1980) and then placed into a new genus *Porphyromonas* as *Porphyromonas gingivalis* (Shah and Collins, 1988). Asaccharolytic species from the oral cavity associated with the root canals of teeth were likewise first defined as *Bacteroides endodontalis* (van Steenbergen et al., 1984) and then placed into the genus *Porphyromonas* as *Porphryromonas endodontalis* (Shah and Collins, 1988). Two weakly fermentative species originally categorized within *Bacteroides intermedius* were later categorized in the genus *Porphyromonas* as *Porphyromonas levii* (Shah et al. 1995) and *Porphyromonas macacae.*

BACTEROIDES

BACTERIODES FRAGILIS

Bacteroides fragilis is the type species and the most important species in the genus *Bacteroides* (Table 40.1) by virtue of its distribution and its pathogenic potential. The cells are 0.5 to 0.8 μm in diameter and 1.5 to 4.5 μm long with rounded ends, and most strains are encapsulated. It is present as part of the normal gastrointestinal microflora (i.e., a commensal) in proportions of about 0.5 to 2% of the gastrointestinal bacteria and it constitutes about 10% of all fecal *Bacteroides.* However, it causes clinical problems in a disproportionately high number of cases relative to its numbers, pointing to the presence of virulence factors in some strains. *B. fragilis* is the most frequently isolated anaerobe from clinical specimens of the gastrointestinal tract of both humans and animals. It is isolated particularly from intra-abdominal abscesses and from cases of peritonitis usually following perforation of the intestines due to pathological processes, trauma, or abdominal surgery. As in other mixed anaerobic infections, it is often isolated together with other anaerobes and facultative species. In contrast to its presence mainly in the gastrointestinal tract, two other members of the genus — *Bacteroides bivius* and *Bacteroides disiens* — are found mainly in the female genital tract.

B. fragilis shares the same cellular and colonial morphology as the *Porphyromonas* and *Prevotella* but it can be distinguished in that its growth is stimulated by the presence of 20% bile. *B. fragilis* is typically nonhemolytic but ten hemolysin genes have been described (Robertson et al., 2006). *B. fragilis* is unusual among *Bacteroides* in possessing a polysaccharide capsule, seen especially in fresh isolates but which is lost on repeated subculture (Kasper, 1976). Like capsules in other bacterial species, the *B. fragilis* capsule can inhibit phagocytosis by host immune cells and can also play a role in abscess formation as demonstrated in animal models (Nakatani et al., 1996). Two types of capsular polysaccharides have been described (Tzianabos et al., 1992). *B. fragilis* also has a lipopolysaccharide notable for a lipid A component chemically distinct from other species; the glucosamine lacks attached phosphate groups and the amino sugars have fewer attached fatty acids. The lipopolysaccharide also lacks β-hydroxymyristic acid. This accounts for its much lower biological activity as compared to other lipopolysaccharide endotoxins (Weintraub et al., 1989; Hofstad, 1992).

Strains of *B. fragilis* can produce an enterotoxin (enterotoxic *B. fragilis*; ETBF) causing a watery diarrhea especially in children between 1 and 5 years of age (Pantosti et al., 1997). While it is present in 6.5% of otherwise healthy people, ETBF accounts for up to 20% of cases of diarrhea. The *B. fragilis* enterotoxin is a 20-kDa zinc metalloprotease. There are three subtypes encoded by the *B. fragilis* toxin (bft) gene (Kato et al., 2000). The enterotoxin binds to intestinal epithelial cells facilitated by pili present on the cell surface of some *B. fragilis* strains. The enterotoxin cleaves the protein E-cadherin that is responsible for intercellular adhesion (Sears, 2002; Wu et al., 2007). *B. fragilis* together with the other species in the *B. fragilis* group — *Bacteroides distasonis*, *B. ovatus*, *B. thetaiotaomicron*, and *B. vulgatus* — frequently produce β-lactamase and are, consequently, resistant to penicillins. There has been a marked increase in antibiotic resistance for *Bacteroides* as well as *Porphyromonas* and *Prevotella* species in recent years.

Bacteroides thetaiotaomicron

Bacteroides thetaiotaomicron is another important member of the genus *Bacteroides*. It is the dominant bacteria in the distal human gastrointestinal tract (Moore and Holdeman, 1974). Next to *B. fragilis*, it is the most important cause of subdiaphragmatic infections in humans. It also has an important symbiotic relationship with its human host. It stimulates both the development of the human gastrointestinal tract and host immune functions (Hooper et al., 2002). Studies of *B. thetaiotaomicron* have been key to understanding the role of the gut microflora in carbohydrate metabolism. *B. thetaiotaomicron* metabolizes the contents of the intestine — otherwise undigestible mainly fibrous dietary carbohydrates as well as carbohydrates produced by humans such as mucins — releasing sugars for its own metabolism and for other gut microorganisms (Gilmore and Ferretti, 2003).

B. thetaiotaomicron is also the first bacterium in this group to be completely sequenced (Xu et al., 2003). The type strain of *B. thetaiotaomicron* (VPI-5482 = ATCC 29148) originally isolated from feces taken from a healthy adult, has a 6.26-Mb genome. The genome encodes 4779 proteins, more than half of which are homologous to other proteins but almost a quarter of which have not been described. Consistent with its role in metabolizing carbohydrates in the gut, a large portion of the *B. thetaiotaomicron* genome encodes enzymes involved in polysaccharide uptake and degradation such as glycosylhydrolases, carbohydrate-binding proteins, and glycosyltransferases involved in capsular polysaccharide formation.

PORPHYROMONAS

Porphyromonas are part of the phylum Bacteroidetes (previously the Cytophaga-Flavobacteria-Bacteroides group (Boone and Castenholtz, 2001)), order Bacteroidales, family Porphyromonadaceae. *Porphyromonas* are named after the porphyrin pigment — protoheme — that results in the colonies becoming darkly pigmented. The *Porphyromonas* include Gram-negative non-sporeforming anaerobic nonmotile rods, sometimes short or coccobacillary rods that produce smooth, shiny, circular, convex colonies with an entire edge. Growth is optimum at 37°C and most species, but not all, produce the protoheme pigment that causes the colonies on blood agar to darken from edge to center over a period of 3 to 6 days. They are asaccharolytic species. The G+C content of the DNA ranges from 40 to 55 mol%. They are primarily found in the oral cavity of humans, non-human primates, and other animals. Species of *Porphyromonas* are shown in Table 40.2.

Porphyromonas asaccharolytica is the type species of the genera and, like *Bacteroides*, is found in the normal human intestinal tract and the human vagina. It is found in mixed infections, including superficial abscesses and ulcers of the genitalia and perineum (Duerden, 1993).

Porphyromonas gingivalis

Porphyromonas gingivalis (formerly *Bacteroides gingivalis*) is an anaerobic, non-fermenting, Gram-negative short rod frequently found in the human oral cavity where it is part of the normal microflora. It can be isolated from dental plaque, particularly the dental plaque that is found underneath the gingival margin (gum line) in adults. An anaerobe, *P. gingivalis* is found in higher numbers in the depths of periodontal pockets where the oxygen tension is lower and which, consequently, favors its growth. *P. gingivalis* is less frequently found in dental plaque in children, and it appears to become established in the human oral cavity following puberty. *P. gingivalis* can also be found, but in fewer numbers, in parts of the oral cavity contacting dental plaque such as in saliva, the buccal mucosa, the lateral borders of the tongue, and the tonsils, particularly if the tonsils are cryptic.

P. gingivalis is considered a periodontal pathogen based on its high prevalence in dental plaque in humans with chronic periodontitis. That is, when *P. gingivalis* is present in the dental plaque, periodontal pockets will often become deeper. There will be greater loss of the connective tissue

TABLE 40.2
Species of *Porphyromonas*

Species	Features	G+C mol%	Growth Characteristics	Type Strain	Refs.
Found Mainly in Humans					
P. asaccharolytica	Type species of the genus Intestinal flora Vaginal flora Diabetic ulcers and gangrene Genital ulcers	52–54	Darkly pigmenting colonies Inhibited by bile and bile salts Fluorescence with UV light	ATCC 25260	Shah and Collins, 1988 Duerden, 1993
P. catoniae	Gingiva	49–51	The only saccharolytic and non-pigmented species in the genus	ATCC 51270	Willems and Collins, 1995
P. endodontalis	Infected dental root canals Periapical abscess	49–51	Darkly pigmenting colonies Fluorescence with UV light	ATCC 35406	van Steenbergen et al., 1981
P. gingivalis	Dental plaque Human periodontal disease Feline periodontal disease	46–48	Darkly pigmenting colonies Produces collagenase Has trypsin-like activity	ATCC 33277	Coykendall et al., 1980 Shah and Collins, 1988 Norris and Love, 1999
Found Mainly in Animals					
P. cangingivalis	Dog subgingival dental plaque Periodontal disease in dogs	49–51	Darkly pigmenting colonies	NCTC 12856	Collins et al., 1994
P. canoris	Gingiva in dogs with periodontal disease Dog bite wounds in humans Periodontal disease in dogs	49–51	Darkly pigmenting colonies Fluorescence with UV light	NCTC 12835	Love et al., 1994 Citron et al., 1996
P. cansulci	Dog subgingival dental plaque Periodontal disease in dogs	49–51	Darkly pigmenting colonies Fluorescence with UV light	NCTC 12858	Collins et al., 1994
P. circumdentaria	Cat gingiva Cat dental plaque Feline periodontal disease	40–42	Darkly pigmenting colonies Fluorescence with UV light	NCTC 12469	Love et al., 1992 Norris and Love, 1999
P. crevioricanis	Dog gingiva Periodontal disease in dogs	44–45	Darkly pigmenting colonies	ATCC 55563	Hirasawa and Takada, 1994

P. denticanis	Periodontal disease in dogs		Darkly pigmenting colonies Most closely related to *B. splanchnicus* by 16S rRNA		Hardham et al., 2005
P. gingivicanis	Dog gingiva Periodontal disease in dogs	41–42	Darkly pigmenting colonies	ATCC 55562	Hirasawa and Takada, 1994
P. gulae	Dental plaque in animals Bear, cat, coyote, dog, wolf, and monkey Canine periodontal disease	51	Darkly pigmenting colonies	ATCC 51700	Fournier et al., 2001 Hardham et al., 2005
P. levii	Intestinal tract of cattle Cow rumen	45–48	Darkly pigmenting colonies Weakly fermentative Fluorescence with UV light	ATCC 29147	Johnson and Holdeman, 1983 Shah et al., 1995
P. macacae	Periodontitis in monkeys	43–44	Differentiable from *P. macacae* based on phenotypic differences Darkly pigmenting colonies Weakly fermentative	ATCC 33141	Slots and Genco, 1980 Love, 1995
P. salivosa	Cat bite wounds in humans Feline periodontal disease Canine periodontal disease	42–44	Differentiable from *P. salivosa* based on phenotypic differences	NCTC 11362	Love et al., 1992 Love, 1995 Citron et al., 1996 Norris and Love, 1999 Hardham et al., 2005

attachment of the gingiva to the teeth and loss of alveolar bone. *P. gingivalis* is considered one of a group of "red complex" bacteria, each of which is found in cases of progressing periodontitis (Socransky et al., 1998). Conversely, healthy gingiva is associated with the absence of *P. gingivalis*. Interestingly, bacteria similar to *P. gingivalis* that infect the oral cavities of other animals such as monkey, dogs, and cats are also associated with periodontitis in those animals.

P. gingivalis is thought to cause periodontitis through a number of virulence factors, including:

1. Cell-surface fimbriae that facilitate bacterial cell adherence to tooth surfaces and to epithelial cells such as those present in the gingival sulcus or periodontal pocket. *P. gingivalis* fimbriae can agglutinate red blood cells. The fimbriae are composed of a 43,000-kDa repeating subunit and can be classified into six genotypes — I to V and Ib — produced by *fim*A genes encoding the major fimbrial subunit FimA. *P. gingivalis* type II fimbria are found more frequently in severe periodontitis (Amano et al., 2004). *P. gingivalis* fimbriae degrade integrin-related signaling molecules, thereby disrupting cellular proliferation, which enables it to persist in the gingival epithelium (Amano, 2007).
2. A bacterial collagenase capable of degrading type I collagen and thought to be responsible for the connective tissue destruction seen in periodontitis (Houle et al., 2003). *P. gingivalis* collagenase is a therapeutic target for low-dose doxycycline therapy used to treat chronic periodontitis (Vernillo et al., 1994) and it is the basis for a diagnostic test for *P. gingivalis* and oral bacteria with similar trypsin-like activities able to hydrolyze N-benzoyl-DL-arginine-2-naphthyl-amide (Lee et al., 2006),
3. A lipopolysaccharide endotoxin associated with alveolar bone loss (Millar et al., 1986) but which, like the LPS from *Bacteroides fragilis*, has relatively weak biological activity based on the chemical structure of the lipid A moiety (Ogawa et al., 2007).
4. A capsular polysaccharide encoded by a glycosyltransferase (Davey and Duncan, 2006), which like the capsule in *B. fragilis*, can inhibit phagocytosis by host immune cells and complement-mediated killing (Slaney et al., 2006) as well as a hemagglutinin, a hemolysin, hyaluronidases, phospholipase, alkaline phosphatase, acid phosphatase (Kantarci and Van Dyke, 2002).

The complete bacterial genome from *Porphyromonas gingivalis* strain W83, originally isolated from a human oral infection by H. Werner (Loos et al., 1993), has been sequenced and found to exhibit similarities to other *Bacteroides* species such as *Bacteroides fragilis* and *B. thetaiotaomicron*, to the Cytophaga-Flavobacteria-Bacteroides phylum, and to the green-sulfur bacteria (Nelson et al., 2003). It exhibits 2,343,476 basepairs and 2053 genes, of which 96.83% are protein encoding. It also exhibits a large number, at least 96, mobile insertion sequences and genes similar to those found in other *Bacteroides,* including immunoreactive surface proteins and proteins for aerotolerance (Duncan, 2003). Whole genome comparison between virulent *P. gingivalis* strain W83 and avirulent *P. gingivalis* strain ATCC 33277 shows 93% similarity, with 7% composed of variant or missing genes, including insertion sequences, the immunoreactive surface protein RagA, enzymes involved in polysaccharide capsule synthesis, and pathogenicity islands possibly acquired by lateral gene transfer (Chen et al., 2004).

Porphyromonas endodontalis

Closely related to and analogous to *P. gingivalis* in the pathogenesis of periodontitis is *Porphyromonas endodontalis*, a bacterium that causes infections of the dental root canal and endodontic abscesses (van Steenbergen and van Winkelhoff, 1984). As opposed to the aerotolerance exhibited by *P. gingivalis*, *P. endodontalis* is a strict anaerobe whose intolerance for atmospheric oxygen has made it difficult to detect on routine culture. For example, *P. endodontalis* can be detected more than twice as frequently in chronically infected dental root canals by the polymerase chain reaction than it can be detected by anaerobic culture (Tomazinho et al., 2007). The lipopolysaccharide from *P. endodontalis* can stimulate cytokine production, and the bacterium can induce dental pulp to produce matrix metalloproteinases (Chang et al., 2002).

PREVOTELLA

The genus *Prevotella* (Shah and Collins, 1990), named after the French anaerobic microbiologist Prevot, includes species previously classified as *Bacteroides*. As described above, these species are obligately anaerobic, non-sporeforming, non-motile, Gram-negative rods. They produce gray, brown, or black colonies that are shiny, circular, and convex with an entire edge although there are also non-pigmenting species within the genus. They "moderately" ferment sugars, require hemin and menadione for growth, and are bile sensitive. The cell walls contain meso-diaminopimelic acid. The G+C content ranges from 40 to 60 mol%.

In humans, *Prevotella* are predominantly oral species (Table 40.3) that can be found in infections of the head and neck regions, including the oral cavity where they can cause periodontal diseases and oral halitosis (Krespi et al., 2006), sinus infections (Brook, 2005a), and infections of the tonsils and middle ear (Brook, 2005) but they can also be found in the gastrointestinal tract (Brook, 2004) and the urogenital tract, particularly *Prevotella bivia* and *Prevotella disiens,* which can cause gynecologic infections including bacterial vaginosis, a risk factor for preterm low birthweight (Marrazzo, 2004). In animals, they comprise the most numerous Gram-negative species in the gastrointestinal tract (Peterka et al., 2003).

Prevotella intermedia

Prior to reclassification in the genus *Prevotella,* this species of intermediately fermentative *Bacteroides* was known as *Bacteroides intermedius.* It can be distinguished on primary culture plates from other darkly pigmenting colonies by the brick-red fluorescence observed in the colonies exposed to UV light. *Prevotella intermedia* can also be distinguished from other darkly pigmenting colonies by its production of β-galactosidase. *P. intermedia* is associated with different forms of human periodontal disease, including moderate to severe gingivitis, chronic periodontitis, and necrotizing ulcerative gingivitis.

TANNERELLA

Most oral species once categorized as *Bacteroides* have been placed into the genus *Porphyromonas* or *Prevotella* as described above. One notable exception is an organism originally isolated by Anne Tanner of The Forsyth Institute from dental plaques in adults with progressing advanced periodontitis. It has subsequently been isolated from gingivitis, periodontitis, endodontic infections, and infections around dental implants. This microorganism was first described as "fusiform" *Bacteroides* based on its cellular appearance — tapered (fusiform) ends with central swellings. The species, originally named *Bacteroides forsythus,* was found to be distinct from *Bacteroides* — it is not resistant to bile — and from *Porphyromonas* based on 16S rRNA phylogenetic analysis. It was placed into a new genus as *Tannerella,* most recently as *Tannerella forsythia* (Maiden et al., 2003). It is a slow-growing, strict anaerobe with unique growth requirements. The colonies are tiny and opaque, and appear on primary culture often as satellites of *Fusobacterium nucleatum* colonies where they appear as speckled pale pink, circular, entire, convex, sometimes with a depressed center. *T. forsythia* requires media containing N-acetylmuramic acid (Wyss, 1989), and it is one of a few oral species from among the up to 800 bacterial species that can inhabit the human oral cavity, such as *Aggregatibacter actinomycetemcomitans* and *Porphyromonas gingivalis* for which there is significant evidence for its

TABLE 40.3
Species of *Prevotella*

Species	Features	G+Cmol%	Growth Characteristics	Type Strain	Refs.
P. albensis	Found in rumen Previously *Prevotella ruminicola* subsp. *ruminicola*	39–43		NCTC 13060	Avgustin et al., 1997
P. baroniae	Isolated from endodontic and periodontal infections, dentoalveolar abscesses, and dental plaque	52		DSM 16972	Downes et al., 2005
P. bivia	Gynecologic and obstetrical infections		Non-pigmented	NCTC 11156	Shah and Collins, 1990
P. bergensis	Infections of the skin and soft tissues	48		DSM 17361	Downes et al., 2005
P. copri	Found in human feces	45	Non-pigmented	DSM 18205	Hayashi et al., 2007
P. brevis	Found in rumen	45–51	Abundant extracellular DNase	NCTC 13061	Avgustin et al., 1997
P. bryantii	Found in rumen	39–43	Abundant extracellular Dnase	NCTC 13062	Avgustin et al., 1997
P. buccae	Found in the human oral cavity		Non-pigmented	ATCC 33574	Shah and Collins, 1990
P. buccalis	Found in the human oral cavity		Non-pigmented	ATCC 35310	Shah and Collins, 1990
P. corporis	Found in the human oral cavity		Pigmenting	ATCC 33547	Shah and Collins, 1990
P, dentalis	Found in the human oral cavity Formerly *Mitsuokella dentalis* and *Hallella seregen*	56–60	Non-pigmenting	ATCC 49559	Willems and Collins, 1995
P. denticola	Found in the human oral cavity, including the gingival crevice	49–51	Pigmenting	ATCC 35308	Shah and Collins, 1990
P. disiens	Found in the human oral cavity		Non-pigmenting	ATCC 29426	Shah and Collins, 1990
P. enoeca	Found in the human gingival crevice		Non-pigmenting	ATCC 51261	Moore et al., 1994
P. intermedia	Found in the oral cavity associated with periodontal disease	40–42	Pigmenting	ATCC 25611	Shah and Collins, 1990
P. loescheii	Found in the gingival crevice	46–48	Pigmenting	ATCC 15930	Shah and Collins, 1990
P. marshii	Isolated from endodontic and periodontal infections and from subgingival dental plaque	51		DSM 16973	Downes et al., 2005
P. melaninogenica	Type species of the genus Found in the human oral cavity and in the vagina	40–42	Pigmenting, moderately saccharolytic, predominantly oral Produces collagenase	ATCC 25845	Wu et al., 1992
P. multisaccharivorax	Human subgingival dental plaque			DSM 17128	Sakamoto et al., 2005

P. nigrescens	Previously a second DNA homology group within *P. intermedia* raised to species status	40–42	Pigmenting; can be distinguished from *P. intermedia* only by oligonucleotide probes	ATCC 33563	Shah and Gharbia, 1992; Matto et al., 1997
P. oralis	Found in the human oral cavity		Non-pigmenting, inhibited by bile	NCTC 11459	Shah and Collins, 1990
P. oris	Found in the human oral cavity		Non-pigmenting	ATCC 33573	Shah and Collins, 1990
P. oulorum	Found in the human oral cavity		Non-pigmenting	ATCC 43324	Shah and Collins, 1990
P. pallens	Found in the human oral cavity		Weak pigmentation	NCTC 13042	Kononen et al., 19998
P. rumincola	Found in rumen	45–51		ATCC 19189	Avgustin et al., 1997
P. salivae	Found in the human oral cavity			DSM 15606	Sakamoto et al., 2004
P. shahii	Found in the human oral cavity			DSM 15611	Sakamoto et al., 2004
P. stercorea	Found in human feces	48	Non-pigmented	DSM 18206	Hayashi et al., 2007
P. tannerae	Found in the human oral cavity		Pigmenting	ATCC 51259	Moore et al., 1994
P. timonensis	Human breast abscess			CCUG 50105	Glazunova et al., 2007
P. veroralis	Found in the human oral cavity		Non-pigmented	ATCC 33779	Wu et al., 1992

role in the etiology of human periodontitis. Although *T. forsythia* is the only member of the genus to have been cultured to date, other members of the genus have been detected by their 16S rRNA sequence. *Tannerella* sequences have been identified in the gut flora from insects such as termites and scarab beetle (Tanner and Izard, 2006). The full sequence of the *T. forsythia* type strain ATCC 43037 has been determined. It contains 3,405,543 basepairs, 3034 predicted open reading frames, 15 pathogenicity islands, and a G+C content of 46.8 mol% (Chen et al., 2005).

FUSOBACTERIUM

Fusobacterium is a genus of Gram-negative non-sporeforming anaerobic bacteria that has a distribution similar to *Bacteroides*. Similar to the species described above, *Fusobacterium* lives on mucous membranes in both humans and other animals. Like *Bacteroides* and related species, some species of *Fusobacterium* are pathogenic and are found in purulent infections and in gangrene. Fusobacterium cells are filamentous or spindle-shaped, varying in size and motility, and its major metabolic end product is butyric acid.

Fusobacterium nucleatum is the type species of the genus. It is indigenous to the human oral cavity where it can be found in dental plaque in association with other Gram-positive and Gram-negative species and where it can cause periodontal disease. The cells are spindle shaped, 5 to 10 μm long, and often paired end to end. *F. nucleatum* can also be found in mixed infections from the head and neck, as well as infections in the chest, lung, liver, and abdomen. The complete genome of *F. nucleatum* strain ATCC 25586 has been sequenced (Kapatral et al., 2002). It has a 27 mol% G+C content and 2.17-Mb encoding 2067 open reading frames.

Fusobacterium necrophorum is also a component of the normal human oropharyngeal, gastrointestinal, and urogenital tract flora. It also can cause occasionally serious diseases, and there is some evidence that *F. necrophorum* disease is increasing (Brazier et al., 2002). The cells are filamentous, curved, with spherical enlargements. *F. necrophorum* is responsible for about 10% of sore throats, second only to Group A streptococci (Aliyu et al., 2004) and can cause a particularly serious disease in healthy young adults known as Lemierre's syndrome — a life-threatening infection that follows an initial sore throat. *F. necrophorum* can also cause meningitis, gastrointestinal and urogenital infections, and a number of clinical syndromes known as necrobacillosis (Hagelskjaer Kristensen and Prag, 2000). *F. necrophorum* causes disease through the production of a number of virulence factors, including a leukotoxin, proteolytic enzymes, lipopolysaccharide, and hemagglutinin.

REFERENCES

Aliyu, S.H., R.K. Marriott, M.D. Curran, S. Parmar, N. Bentley, N.M. Brown, J.S. Brazier, and H. Ludlam. 2004. Real-time PCR investigation into the importance of *Fusobacterium necrophorum* as a cause of acute pharyngitis in general practice. *Journal of Medical Microbiology,* 53: 1029–1035

Amano, A. 2007. Disruption of epithelial barrier and impairment of cellular function by *Porphyromonas gingivalis*. *Frontiers in Bioscience,* 12: 3965–3974.

Amano, A., I. Nakagawa, N. Okahashi, and N. Hamada. 2004. Variations of *Porphyromonas gingivalis* fimbriae in relation to microbial pathogenesis. *Journal of Periodontal Research,* 39:136–142

Avgustin, G., R.J. Wallace, and H.J. Flint. 1997. Phenotypic diversity among ruminal isolates of *Prevotella ruminicola*: proposal of *Prevotella brevis* sp. nov., *Prevotella bryantii* sp. nov., and *Prevotella albensis* sp. nov. and redefinition of *Prevotella ruminicola*. *International Journal of Systematic Bacteriology,* 47: 284–288.

Boggess, K.A., T.N. Trevett, P.N. Madianos, L. Rabe, S.L. Hillier, J. Beck, and S. Offenbacher. 2005. Use of DNA hybridization to detect vaginal pathogens associated with bacterial vaginosis among asymptomatic pregnant women. *American Journal of Obstetrics and Gynecology,* 193: 752–756.

Boone, D.R. and R.W. Castenholtz (Eds.). 2001. *Bergey's Manual of Systematic Bacteriology*, 2nd ed., Vol. 1. Springer-Verlag, New York.

Brazier, J.S., V. Hall, E. Yusuf, and B.I. Duerden. 2002. *Fusobacterium necrophorum* infections in England and Wales 1990–2000. *Journal of Medical Microbiology,* 51: 269–272.

Brook, I. 2004a. Anaerobic pulmonary infections in children. *Pediatric Emergency Care,* 20: 636–640.

Brook, I. 2004b. Urinary tract and genito-urinary suppurative infections due to anaerobic bacteria. *International Journal of Urology,* 11:133–141.

Brook, I. 2005a. Microbiology of intracranial abscesses and their associated sinusitis. *Archives of Otolaryngology - Head and Neck Surgery,* 131: 1017–1019.

Brook, I. 2005b. The role of bacterial interference in otitis, sinusitis and tonsillitis. *Journal of Otolaryngology - Head and Neck Surgery,* 133: 139–146.

Byrd, R.P. Jr. and T.M. Roy. 2003. Anaerobic blood cultures: useful in the ICU? *Chest,* 123: 2158–2159.

Cato, E.P. and J.L. Johnson. 1976. Reinstatement of species rank for *Bacteroides fragilis, B. ovatus, B. distasonis, B. thetaiotaomicron* and *B. vulgatus.* Designation of neotype strains for *B. fragilis* (Veillon and Zuber) Castellani and Chalmers and *B. thetaiotaomicron* (Distaso) Castellani and Chalmers. *International Journal of Systematic Bacteriology,* 26: 230–237.

Chang, Y.C., C.C. Lai, S.F. Yang, Y. Chan, and Y.S. Hsieh. 2002. Stimulation of matrix metalloproteinases by black-pigmented *Bacteroides* in human pulp and periodontal ligament cell cultures. *Journal of Endodontics,* 28: 90–93.

Chen, T., K. Abbey, W.J. Deng, and M.C. Cheng. 2005. The bioinformatics resource for oral pathogens. *Nucleic Acids Research,* 33(Web Server issue):W734–W740.

Chen, T., Y. Hosogi, K. Nishikawa, K. Abbey, R.D. Fleischmann, J. Walling, and M.J. Duncan. 2004. Comparative whole-genome analysis of virulent and avirulent strains of *Porphyromonas gingivalis. Journal of Bacteriology,* 186: 5473–5479.

Citron, D.M., S. Hunt Gerardo, M.C. Claros, F. Abrahamian, D. Talan, and E.J. Goldstein. 1996. Frequency of isolation of *Porphyromonas* species from infected dog and cat bite wounds in humans and their characterization by biochemical tests and arbitrarily primed-polymerase chain reaction fingerprinting. *Clinical Infectious Disease,* 23(Suppl. 1): S78–S82.

Clarridge, J.E. 3rd. 2004. Impact of 16S rRNA gene sequence analysis for identification of bacteria on clinical microbiology and infectious diseases. *Clinical Microbiology Reviews,* 17: 840–862.

Cole, J.R., B. Chai, R.J. Farris, Q. Wang, A.S. Kulam-Syed-Mohideen, D.M. McGarrell, A.M. Bandela, E. Cardenas, G.M. Garrity, and J.M. Tiedje. 2007. The ribosomal database project (RDP-II): introducing myRDP space and quality controlled public data. *Nucleic Acids Research,* 35: D169–D172.

Collins, M.D., D.N. Love, J. Karjalainen, A. Kanervo, B. Forsblom, A. Willems, S. Stubbs, E, Sarkiala, G.D. Bailey, D.I. Wigney, et al. 1994. Phylogenetic analysis of members of the genus *Porphyromonas* and description of *Porphyromonas cangingivalis* sp. nov. and *Porphyromonas cansulci* sp. nov. *International Journal of Systematic Bacteriology,* 44: 674–679.

Coykendall A.L., F.S. Kaczmarek, and J. Slots. 1980. Genetic heterogeneity in *Bacteroides asaccharolyticus* (Holdeman and Moore, 1970) Finegold and Barnes 1977 (Approved List 1980) and proposal of *Bacteroides gingivalis* sp. nov. and *Bacteroides macacae* (Slots and Genco) comb nov. *International Journal of Systematic Bacteriology,* 30: 559–564.

Davey M.E. and M.J. Duncan. 2006. Enhanced biofilm formation and loss of capsule synthesis: deletion of a putative glycosyltransferase in *Porphyromonas gingivalis. Journal of Bacteriology,* 188: 5510–5523.

Downes, J., I. Sutcliffe, A.C.R. Tanner, and W.G. Wade. 2005. *Prevotella marshii* sp. nov. and *Prevotella baroniae* sp. nov., isolated from the human oral cavity. *International Journal of Systematic and Evolutionary Bacteriology,* 55: 1551–1555.

Duerden, B.I. 1993. Black-pigmented Gram-negative anaerobes in genito-urinary tract and pelvic infections. *FEMS Immunology and Medical Microbiology,* 6: 223–227.

Duncan, M.J. 2003. Genomics of oral bacteria. *Critical Reviews in Oral Biology and Medicine,* 14:175–187.

Edmiston, C.E., C. J. Krepel, G.R. Seabrook, L.R. Somberg, A. Nakeeb, R.A. Cambria, and J.B. Towne. 2004. *In vitro* activities of moxifloxacin against 900 aerobic and anaerobic surgical isolates from patients with intra-abdominal and diabetic foot infections. *Antimicrobial Agents and Chemotherapy,* 48: 1012–1016.

Faveri, M., M. Feres, J.A. Shibli, R.F. Hayacibara, M.M. Hayacibara, and L.C. de Figueiredo. 2006. Microbiota of the dorsum of the tongue after plaque accumulation: an experimental study in humans. *Journal of Periodontology,* 77: 1539–1546.

Fournier, D., C. Mouton, P. Lapierre, T. Kato, K. Okuda, and C. Menard. 2001. *Porphyromonas gulae* sp. nov., an anaerobic, Gram-negative coccobacillus from the gingival sulcus of various animal hosts. *International Journal of Systematic and Evolutionary Microbiology,* 51: 1179–1189.

Gilmore, M.S. and J.J. Ferretti. 2003. Microbiology. The thin line between gut commensal and pathogen. *Science,* 299: 1999–2002.

Glazunova, O.O., T. Launay, D. Raoult, and V. Roux. 2007. *Prevotella timonensis* sp. nov., isolated from a human breast abscess. *International Journal of Systematic and Evolutionary Microbiology,* 57: 883–886.

Hagelskjaer Kristensen, L. and J. Prag. 2000. Human necrobacillosis, with emphasis on Lemierre's syndrome. *Clinical Infectious Disease,* 31: 524–532.

Haraszthy, V.I., J.J. Zambon, M. Trevisan, M. Zeid, and R.J. Genco. 2000. Identification of periodontal pathogens in atheromatous plaques. *Journal of Periodontology,* 71: 1554–1560.

Hardham, J., K. Dreier, J. Wong, C. Sfintescu, and R.T. Evans. 2005. Pigmented-anaerobic bacteria associated with canine periodontitis. *Veterinary Microbiology,* 106: 119–128.

Hayashi, H., K. Shibata, M. Sakamoto, S. Tomita, and Y. Benno. 2007. *Prevotella copri* sp. nov. and *Prevotella stercorea* sp. nov., isolated from human faeces. *International Journal of Systematic and Evolutionary Microbiology,* 57: 941–946.

Hirasawa, M. and K. Takada. 1994. *Porphyromonas gingivicanis* sp. nov. and *Porphyromonas crevioricanis* sp. nov., isolated from beagles. *International Journal of Systematic Bacteriology,* 44: 637–640.

Hofstad, T. 1992. Virulence factors in anaerobic bacteria. *European Journal of Clinical Microbiology and Infectious Disease,* 11: 1044–1048.

Holdeman, L.V. and J.L. Johnson. 1982. Description of *Bacteroides loeschii* sp. nov. and emendation of the descriptions of *Bacteroides melaninogenicus* (Oliver and Wherry) Roy and Kelly 1939 and *Bacteroides denticola* Shah and Collins 1981. *International Journal of Systematic Bacteriology,* 32: 399.

Hooper, L.V., T. Midtvedt, and J.I. Gordon. 2002. How host-microbial interactions shape the nutrient environment of the mammalian intestine. *Annual Review of Nutrition*, 22: 283–307.

Houle, M.A., D. Grenier, P. Plamondon, and K. Nakayama. 2003. The collagenase activity of *Porphyromonas gingivalis* is due to Arg-gingipain. *FEMS Microbiological Letters,* 221:181–185.

Johnson, J.L. and C.S. Hirsch. 2003. Aspiration pneumonia. Recognizing and managing a potentially growing disorder. *Postgraduate Medicine,* 113: 99–102.

Johnson J.L. and L.V. Holdeman. 1983. *Bacteroides intermedius* comb. nov., and descriptions of *B. corporis* sp. nov. and *Bacteroides levii* sp. nov. *International Journal of Systematic Bacteriology,* 33: 15–25.

Kaczmarek, F.S. and A.L. Coykendall. 1980. Production of phenylacetic acid by strains of *Bacteroides asaccharolyticus* and *Bacteroides gingivalis* (sp. nov.). *Journal of Clinical Microbiology,* 12: 288–290.

Kantarci, A. and T.E. Van Dyke. 2002. Neutrophil-mediated host response to *Porphyromonas gingivalis. Journal of the International Academy of Periodontology*, 4: 119–125.

Kapatral, V., I. Anderson, N. Ivanova, G. Reznik, T. Los, A. Lykidis, A. Bhattacharyya, A. Bartman, W. Gardner, G. Grechkin, L. Zhu, O. Vasieva, L. Chu, Y. Kogan, O. Chaga, E. Goltsman, A. Bernal, N. Larsen, M. D'Souza, T. Walunas, G. Pusch, R. Haselkorn, M. Fonstein, N. Kyrpides, and R. Overbeek. 2002. Genome sequence and analysis of the oral bacterium Fusobacterium nucleatum strain ATCC 25586. *Journal of Bacteriology,* 184: 2005–2018.

Kasper, D.L. 1976. The polysaccharide capsule of *Bacteroides fragilis* subspecies *fragilis*: immunochemical and morphologic definition. *Journal of Infectious Disease*, 133: 79–87.

Kato, N., C.X. Liu, H. Kato, K. Watanabe, Y. Tanaka, T. Yamamoto, K. Suzuki, and K. Ueno. 2000. A new subtype of the metalloprotease toxin gene and the incidence of the three bft subtypes among *Bacteroides fragilis* isolates in Japan. *FEMS Microbiology Letters*, 182: 171–176.

Kononen, E., E. Eerola, E.V. Frandsen, J. Jalava, J. Matto, S. Salmenlinna, and H. Jousimies-Somer. 1998. Phylogenetic characterization and proposal of a new pigmented species to the genus Prevotella: *Prevotella pallens* sp. nov. *International Journal of Systematic Bacteriology*, 48: 47–51.

Kornman, K.S. and W.J. Loesche. 1982. Effects of estradiol and progesterone on *Bacteroides melaninogenicus* and *Bacteroides gingivalis. Infection and Immunity*, 35: 256–263.

Kozarov, E., D. Sweier, C. Shelburne, A. Progulske-Fox, and D. Lopatin. 2006. Detection of bacterial DNA in atheromatous plaques by quantitative PCR. *Microbes and Infection*, 8: 687–689.

Kuhn, J.J., I. Brook, C.L. Waters, L.W. Church, D.A. Bianchi, and D.H. Thompson. 1995. Quantitative bacteriology of tonsils removed from children with tonsillitis hypertrophy and recurrent tonsillitis with and without hypertrophy. *Annals of Otology, Rhinology, and Laryngology,* 104: 646–652.

Krespi, Y.P., M.G. Shrime, and A. Kacker. 2006. The relationship between oral malodor and volatile sulfur compound-producing bacteria. *Otolaryngology — Head and Neck Surgery,* 135: 671–676.

Lee, Y., W.S. Tchaou, K.B. Welch, and W.J. Loesche. 2006. The transmission of BANA-positive periodontal bacterial species from caregivers to children. *Journal of the American Dental Association*, 137: 1539–1546.

Loos, B.G., D.W. Dyer, T.S. Whittam, and R.K. Selander. 1993. Genetic structure of populations of *Porphyromonas gingivalis* associated with periodontitis and other oral infections. *Infection and Immunity*, 61: 204—212.

Love, D.N., G.D. Bailey, S. Collings, and D.A. Briscoe. 1992. Description of *Porphyromonas circumdentaria* sp. nov. and reassignment of *Bacteroides salivosus* (Love, Johnson, Jones, and Calverley 1987) as *Porphyromonas* (Shah and Collins 1988) *salivosa* comb. nov. *International Journal of Systematic Bacteriology*, 42: 434–438.

Love, D.N. 1995. *Porphyromonas macacae* comb. nov., a consequence of *Bacteroides macacae* being a senior synonym of *Porphyromonas salivosa*. *International Journal of Systematic Bacteriology*, 45: 90–92.

Love, D.N., J. Karjalainen, A. Kanervo, B. Forsblom, E. Sarkiala, G.D. Bailey, D.I. Wigney, and H. Jousimies-Somer. 1994. *Porphyromonas canoris* sp. nov., an asaccharolytic, black-pigmented species from the gingival sulcus of dogs. *International Journal of Systematic Bacteriology*, 44: 204–208.

Maiden, M.F.J., P. Cohee, and A.C. Tanner. 2003. Proposal to conserve the adjectival form of the specific epithet in the reclassification of *Bacteroides forsythus* Tanner et al. 1986 to the genus *Tannerella* Sakamoto et al. 2002 as *Tannerella forsythia* corrig., gen. nov., comb. nov. Request for an Opinion. *International Journal of Systematic and Evolutionary Microbiology*, 53: 2111–2112.

Matto, J., S. Asikainen, M.L. Vaisanen, M. Rautio, M. Doarela, M. Summanen, S. Finegold, and H. Jousimies-Somer. 1997. Role of *Porphyromonas gingivalis*, *Prevotella intermedia*, and *Prevotella nigrescens* in extraoral and some odontogenic infections. *Clinical Infectious Disease,* 25(Suppl. 2): S194–S198.

Marrazzo, J.M. 2004. Evolving issues in understanding and treating bacterial vaginosis. *Expert Review of Anti-Infective Therapy,* 2:913–922.

Millar, S.J., E.G. Goldstein, M.J. Levine, and E. Hausmann. 1986. Modulation of bone metabolism by two chemically distinct lipopolysaccharide fractions from *Bacteroides gingivalis*. *Infection and Immunity,* 51: 302–306.

Mombelli. A., M. Tonetti, B. Lehmann, and N.P. Lang. 1996. Topographic distribution of black-pigmenting anaerobes before and after periodontal treatment by local delivery of tetracycline. *Journal of Clinical Periodontology,* 23 :906–913.

Moore, W.E.C. and L.V. Holdeman. 1973. New names and combinations in the genera *Bacteroides* Castellani and Chalmers, *Fusobacterium* Knorr, *Eubacterium* Prevot, *Propionibacterium* Orla-Jensen, and *Lactobacillus* Beierinck. *International Journal of Systematic Bacteriology,* 23: 69–74.

Moore, W.E.C. and L.V. Holdeman. 1974. Human fecal flora: The normal flora of 20 Japanese-Hawaiians. *Applied Microbiology*, 27: 961–979.

Moore, L.V., J.L. Johnson, and W.E. Moore. 1994. Descriptions of *Prevotella tannerae* sp. nov. and *Prevotella enoeca* sp. nov. from the human gingival crevice and emendation of the description of *Prevotella zoogleoformans*. *International Journal of Systematic Bacteriology,* 44: 599–602.

Nakatani, T., T. Sato, B.F. Trump, J.H. Siegel, and K. Kobayashi. 1996. Manipulation of the size and clone of an intra-abdominal abscess in rats. *Research in Experimental Medicine (Berlin),* 196: 117–126.

Nelson, K.E., R.D. Fleischmann, R.T. DeBoy, I.T. Paulsen, D.E. Fouts, J.A. Eisen, S.C. Daugherty, R.J. Dodson, A.S. Durkin, M. Gwinn, D.H. Haft, J.F. Kolonay, W.C. Nelson, T. Mason, L. Tallon, J. Gray, D. Granger, H. Tettelin, H. Dong, J.L. Galvin, M.J. Duncan, F.E. Dewhirst, and C.M. Fraser. 2003. Complete genome sequence of the oral pathogenic bacterium *Porphyromonas gingivalis* strain W83. *Journal of Bacteriology,* 185: 5591–5601.

Norris, J.M. and D.N. Love. 1999. Associations amongst three feline *Porphyromonas* species from the gingival margin of cats during periodontal health and disease. *Veterinary Microbiology,* 65: 195–207.

Ogawa, T., Y. Asai, Y. Makimura, and R. Tamai. 2007. Chemical structure and immunobiological activity of *Porphyromonas gingivalis* lipid A. *Frontiers in Bioscience,* 12: 3795–3812.

Oliver, W.W. and W.B. Wherry. 1921. Notes on some bacteria parasites of the human mucous membranes. *Journal of Infectious Disease,* 28: 341–344.

Olsen, I. and H.N. Shah. 2003. International committee on systematics of prokaryotes subcommittee on the taxonomy of gram-negative anaerobic rods. *International Journal of Systematic and Evolutionary Microbiology,* 53: 923–924.

Ortiz, E. and M.A. Sande. 2000. Routine use of anaerobic blood cultures: are they still indicated? *American Journal of Medicine,* 108: 445–447.

Pantosti, A., M.G. Menozzi, A. Frate, L. Sanfilippo, F. D'Ambrosio, and M. Malpeli. 1997. Detection of enterotoxigenic *Bacteroides fragilis* and its toxin in stool samples from adults and children in Italy. *Clinical Infectious Diseases,* 24: 12–16.

Peterka, M., K. Tepsic, T. Accetto, R. Kostanjsek, A. Ramsak, L. Lipoglavsek, and G. Avgustin. 2003. Molecular microbiology of gut bacteria: genetic diversity and community structure analysis. *Acta Microbiologica et Immunologica Hungarica,* 50: 395–406.

Pryor, J.P., E. Piotrowski, C.W. Seltzer, and V.H. Gracias. 2001. Early diagnosis of retroperitoneal necrotizing fasciitis. *Critical Care Medicine,* 29: 1071–1073.

Ramanujam, P. and N.K. Rathlev. 2006. Blood cultures do not change management in hospitalized patients with community-acquired pneumonia. *Academic Emergency Medicine,* 13: 740–745.

Rawlinson, A., T.F. Walsh, A. Lee, and S.J. Hodges. 1998. Phylloquinone in gingival crevicular fluid in adult periodontitis. *Journal of Clinical Periodontology,* 25: 662–665.

Relman, D.A. 2002. New technologies, human-microbe interactions, and the search for previously unrecognized pathogens. *Journal of Infectious Disease,* 186(Suppl. 2): S254–S258.

Robertson, K.P., C.J. Smith, A.M. Gough, and E.R. Rocha. 2006. Characterization of *Bacteroides fragilis* hemolysins and regulation and synergistic interactions of HlyA and HlyB. *Infection and Immunity,* 74: 2304–2316.

Sakamoto, M. and Y. Benno. 2006. Reclassification of *Bacteroides distasonis, Bacteroides goldsteinii* and *Bacteroides merdae* as *Parabacteroides distasonis* gen. nov., comb. nov., *Parabacteroides goldsteinii* comb. nov. and *Parabacteroides merdae* comb. nov. *International Journal of Systematic and Evolutionary Microbiology,* 56: 1599–1605.

Sakamoto, M., M. Suzuki, Y. Huang, M. Umeda, I. Ishikawa, and Y. Benno. 2004. *Prevotella shahii* sp. nov. and *Prevotella salivae* sp. nov., isolated from the human oral cavity. *International Journal of Systematic and Evolutionary Microbiology,* 54: 877–883.

Sakamoto, M., M. Umeda, I. Ishikawa, and Y. Benno. 2005. *Prevotella multisaccharivorax* sp. nov., isolated from human subgingival plaque. *International Journal of Systematic and Evolutionary Microbiology,* 55: 1839–1843.

Sears, C.L. 2001. The toxins of *Bacteroides fragilis. Toxicon,* 39: 1737–1746.

Shah, H.N., R. Bonnett, B. Mateen, and R.A. Williams. 1979. The porphyrin pigmentation of subspecies of *Bacteroides melaninogenicus. Biochemical Journal,* 180: 45–50.

Shah, H.N. and M.D. Collins. 1988. Proposal for reclassification of *Bacteroides asaccharolyticus, Bacteroides gingivalis* and *Bacteroides endodontalis* in a new genus, *Porphyromonas. International Journal of Systematic Bacteriology,* 38: 128–131.

Shah, H.N. and D.M. Collins. 1990. *Prevotella,* a new genus to include *Bacteroides melaninogenicus* and related species formerly classified in the genus *Bacteroides. International Journal of Systematic Bacteriology,* 40: 205–208.

Shah, H.N., M.D. Collins, I. Olsen, B.J. Paster, and F.E. Dewhirst. 1995. Reclassification of *Bacteroides levii* (Holdeman, Cato, and Moore) in the genus *Porphyromonas* as *Porphyromonas levii* comb. nov. *International Journal of Systematic Bacteriology,* 45: 586–588.

Shah, H.N. and S.E. Gharbia. 1992. Biochemical and chemical studies on strains designated *Prevotella intermedia* and proposal of a new pigmented species, *Prevotella nigrescens* sp. nov. *International Journal of Systematic Bacteriology,* 42: 542–546.

Shah, H.N., R.A. Williams, G.H. Bowden, and J.M. Hardie. 1976. Comparison of the biochemical properties of *Bacteroides melaninogenicus* from human dental plaque and other sites. *Journal of Applied Bacteriology,* 41: 473–495.

Siqueira, J.F. Jr, I.N. Rocas, J.C. Oliveira, and K.R. Santos. 2001. Molecular detection of black-pigmented bacteria in infections of endodontic origin. *Journal of Endodontology,* 27: 563–566.

Slaney, J.M., A. Gallagher, J. Aduse-Opoku, K. Pell, and M.A. Curtis. 2006. Mechanisms of resistance of *Porphyromonas gingivalis* to killing by serum complement. *Infection and Immunity,* 74: 5352-5361.

Slots, J. and R.J. Genco. 1980. *Bacteroides melaninogenicus* subsp. *macacae,* a new subspecies from monkey periodontopathogenic indigenous microflora. *International Journal of Systematic Bacteriology,* 30: 82–85.

Slots J. and H.S. Reynolds. 1982. Long-wave UV light fluorescence for identification of black-pigmented *Bacteroides* spp. *Journal of Clinical Microbiology,* 16: 1148–1151.

Socransky, S.S., A.D. Haffajee, M.A. Cugini, C. Smith, and R.L. Kent, Jr. 1998. Microbial complexes in subgingival plaque. *Journal of Clinical Periodontology,* 25: 134–144.

Smith, D.T. 1930. Fusospirochetal disease of the lungs produced with cultures from Vincent's angina. *Journal of Infectious Disease,* 46: 303–310.

Styrt, B. and S.L. Gorbach. 1989. Recent developments in the understanding of the pathogenesis and treatment of anaerobic infections (2). *New England Journal of Medicine,* 321: 240–246.

Talan, D.A., D.M. Citron, F.M. Abrahamian, G.J. Moran, and E.J. Goldstein. 1999. Bacteriologic analysis of infected dog and cat bites. Emergency Medicine Animal Bite Infection Study Group. *New England Journal of Medicine,* 340: 85–92.

Tanaka, M., T. Hanioka, K. Takaya, and S. Shizukuishi. 1998. Association of oxygen tension in human periodontal pockets with gingival inflammation. *Journal of Periodontology,* 69: 1127–1130.

Tanner, A.C. and J. Izard. 2006. *Tannerella forsythia*, a periodontal pathogen entering the genomic era. *Periodontology 2000,* 42: 88–113.

Tomazinho, L.F. and M.J. Avila-Campos. 2007. Detection of *Porphyromonas gingivalis*, *Porphyromonas endodontalis*, *Prevotella intermedia*, and *Prevotella nigrescens* in chronic endodontic infection. *Oral Surgery, Oral Medicine, Oral Pathology, Oral Radiology and Endodontology,* 103: 285–288.

Tzianabos, A.O., A. Pantosti, H. Baumann, J.R. Brisson, H.J. Jennings, and D.L. Kasper. 1992. The capsular polysaccharide of *Bacteroides fragilis* comprises two ionically linked polysaccharides. *Journal of Biological Chemistry,* 267: 18230–18235.

van Steenbergen, T.J.M., A.J. van Winkelhoff, D. Mayrand, D. Grenier, and J. de Graaff. 1984. *Bacteroides endodontalis* sp. nov., an asaccharolytic black pigmented *Bacteroides* species from infected dental root canals. *International Journal of Systematic Bacteriology* 34: 118-120.

van der Waaij, D., J.M. Berghuis-de Vries, and J.E.C. Lekkerkerk-van der Wees. 1972. Colonization resistance of the digestive tract and the spread of bacteria to the lymphatic organs in mice. *Journal of Hygiene (London),* 70: 335–342.

Vernillo, A.T., N.S. Ramamurthy, L.M. Golub, and B.R. Rifkin. 1994. The nonantimicrobial properties of tetracycline for the treatment of periodontal disease. *Current Opinion in Periodontology,* 111–118.

Weintraub, A., U. Zähringer, H.W. Wollenweber, U. Seydel, and E.T. Rietschel. 1989. Structural characterization of the lipid A component of *Bacteroides fragilis* strain NCTC 9343 lipopolysaccharide. *European Journal of Biochemistry,* 183: 425–431.

Willems, A. and M.D. Collins. 1995. 16S rRNA gene similarities indicate that *Hallella seregens* (Moore and Moore) and *Mitsuokella dentalis* (Haapasalo et al.) are genealogically highly related and are members of the genus *Prevotella*: emended description of the genus *Prevotella* (Shah and Collins) and description of *Prevotella dentalis* comb. nov *International Journal of Systematic Bacteriology,* 45, 832–836.

Willems, A. and M.D. Collins. 1995. Reclassification of *Oribaculum catoniae* (Moore and Moore 1994) as *Porphyromonas catoniae* comb. nov., and emendation of the genus *Porphyromonas*. *International Journal of Systematic Bacteriology,* 45: 578–581.

Wu, C.C., J.L. Johnson, W.E.C. Moore, and L.V.H. Moore. 1992. Emended descriptions of *Prevotella denticola*, *Prevotella loescheii*, *Prevotella veroralis*, and *Prevotella melaninogenica*. *International Journal of Systematic Bacteriology,* 42: 536–541.

Wu, S., K.J. Rhee, M. Zhang, A. Franco, and C.L. Sears. 2007. *Bacteroides fragilis* toxin stimulates intestinal epithelial cell shedding and {gamma}-secretase-dependent E-cadherin cleavage. *Journal of Cell Science,* 120: 1944–1952.

Wyss, C. 1989. Dependence of proliferation of *Bacteroides forsythus* on exogenous N-acetylmuramic acid. *Infection and Immunity,* 57: 1757–1759.

Xu, J., M.K. Bjursell, J. Himrod, S. Deng, L.K. Carmichael, H.C. Chiang, L.V. Hooper, and J.I. Gordon. 2003. A genomic view of the human-*Bacteroides thetaiotaomicron* symbiosis. *Science*, 299:2074–2076.

41 Introduction to Archaea

Sarah T. Gross

CONTENTS

INTRODUCTION

For years, all prokaryotes were classified under one kingdom: Monera. By the late 1970s, scientists were starting to recognize that this classification system grossly simplified prokaryotic origins. It soon became clear that certain species of prokaryotes known as archaebacteria had characteristics that made them very distinct from common bacteria, or eubacteria. These organisms seemed to inhabit "extreme" environments, and contained unique cell membranes and walls that were not found in any other organisms. Additionally, 16S rRNA analysis clearly indicated that archaebacteria did not share a monophyletic common ancestor with other bacteria, and that archaebacteria and eubacteria diverged before the evolution of the eukaryotic cell.[1]

In 1990, Woese and others proposed a restructuring of the traditional five-kingdom classification system by the addition of a higher taxon known as Domain. In this new system, which includes the domains Bacteria, Archaea, and Eukarya, the phylogenetic origin of the various forms of life are more accurately reflected.[2] Despite opposition from traditional microbiologists who based their classification on morphological and physiological characteristics, the Three-Domain classification system is now widely accepted in the scientific community.

Historically, Archaea have been viewed as organisms that live in "extreme" habitats, such as high temperatures, acidity, or salinity. Indeed, the vast majority of cultured archaeal species to date do inhabit such presumably inhospitable environments. However, 16S rRNA sequences from uncultured environmental samples have demonstrated that archaeal species may be found in a wide variety of habitats, including ocean water, ocean sediment, freshwater lakes, soil, plant roots, petroleum contaminated aquifers, and the human mouth and gut to name but a few.[3]

Archaea share certain characteristics with Bacteria, others with Eukarya, while other characteristics are entirely unique (Table 41.1). It appears that Archaea and Eukarya share many homologous genes involved in information processing (replication, transcription, and translation), whereas Archaea and Bacteria share many morphological structures and metabolic proteins. Various lines of evidence indicate that eukaryotes evolved from Archaea; however, it is clear that a significant amount of lateral gene transfer (LGT) occurred between all the different domains early in their evolution.

ARCHAEAL GENETICS AND MOLECULAR BIOLOGY

Archaeal genetic features display an interesting combination of bacterial and eukaryotic characteristics. Most archaeal species contain a single circular chromosome, similar to bacteria, but the replication initiation machinery appears to be homologous to eukaryotic organisms. Additionally, some species of Archaea contain several origins of replication, as in eukaryotic chromosomes.[4] In fact, the process of replication in many archaeal species appears to be a simplified version of eukaryotic replication, and certain species of Archaea such as the hyperthermophilic *Sulfolobus* are used as model organisms to study eukaryotic replication.[4] Further, chromosomes of several species of Archaea have been found to contain histones and nucleosome-like structures, as in eukaryotic organisms.[4]

Transcription in Archaea shares some features with Bacteria and yet others with Eukarya. Most archaeal genomes consist of operons, as in bacteria, but transcriptional activation in Archaea seems to be more homologous to eukaryotes. For instance, the archaeal RNA polymerase is far more complex than bacterial RNA polymerase. Depending on the species, archaeal polymerases consist of eight to twelve subunits, many of which are homologs to the 12-subunit eukaryotic RNA polymerase II.[5] Transcription initiation in Archaea is governed by binding of transcription factors to

TABLE 41.1

Comparison of Selected Characteristics of the Archaea and the Other Domains of Life

Characteristic	Bacteria	Archaea	Eukarya
Membrane-enclosed nucleus	No	No	Yes
Chromosomal structure	Circular	Circular	Linear
Chromosomal number	1	1	>1
Histone proteins present	No	Yes	Yes
Peptidoglycan in cell wall	Yes	No	No
Membrane lipids	Ester-linked	Ether-linked	Ester-linked
Glycerol	D-glycerol	L-glycerol	D-glycerol
Ribosomes (mass)	70S	70S	80S
Initiator tRNA	Formylmethionine	Methionine	Methionine
Introns	No	Rarely	Yes
Operons	Yes	Yes	No
RNA polymerase	One (4 subunits)	Several (8–12 subunits)	Several (12–14 subunits)
Transcription factors required	No	Yes	Yes
TATA box in promoter	No	Yes	Yes

promoter sequences, thus recruiting the RNA polymerase to bind to the TATA box sequence, as in eukaryotic systems.[5] Additionally, single introns were found in protein-coding regions of genes in at least three species of Archaea — *Aeropyrum pernix, Sulfolobus solfataricus,* and *Sulfolobus tokodaii*[6] — although most archaeal genes are transcribed as polycistronic mRNA and do not undergo RNA processing.

While archaeal ribosomes are of similar size to bacterial ribosomes (70S), their composition and antibiotic sensitivity are more homologous to eukaryotic ribosomes. In addition, translation in Archaea is initiated by the initiator tRNA, which carries methionine and several translation initiation factors, as in eukaryotic organisms, rather than the formyl-methionine used by bacteria. Interestingly, a 22nd amino acid, pyrrolysine, has been identified in certain species of methanogenic Archaea.[7]

ARCHAEAL MEMBRANES AND CELL WALLS

The cell walls and membranes of Archaea are entirely unique and highly diverse. There are several fundamental differences between the membrane architecture of Archaea and those of all other cells. The glycerol used to make archaeal phospholipids is a stereoisomer of the glycerol used to build bacterial and eukaryotic membranes. Archaeal membranes contain L-glycerol, whereas bacteria and eukaryotes have D-glycerol. Also, bacterial and eukaryotic phospholipids contain ester linkages between the glycerol and fatty acid moieties. By contrast, archaeal side chains are bound using an ether linkage, which gives the resulting phospholipids different chemical properties than the membrane lipids of other organisms.

The side chains themselves are also unique in Archaea. Bacterial and eukaryotic phospholipids contain two fatty acid side chains (usually 16 or 18 carbon chains, or even longer). In contrast, the phospholipid side chains of Archaea are 20-carbon branched isoprenes. The branches can link the side chains together and can be intermolecular or intramolecular. The branches can link two phospholipids in the different monolayers of the bilayer, giving rise to transmembrane phospholipids. The isoprenes can also bend around to form a five-carbon ring structure that may function to stabilize the membranes of archaeal species that live in extreme environments. More than 100 different ether-type polar lipids (such as phospholipids and glycolipids) have been identified in Archaea.[3]

The compositions of the cell walls of various species of Archaea are extremely diverse. Because Archaea lack peptidoglycan, they are naturally resistant to antibiotics such as penicillin, which function to destroy or prevent proper synthesis of peptidoglycan. Some species of methanogenic Archaea contain cell walls of pseudo-peptidoglycan, which has a similar structure to bacterial peptidoglycan. Most archaeal species do not contain a peptidoglycan-like molecule in any form, instead covering the outside of the membrane with proteins, glycoproteins, or polysaccharides. The most common cell wall structure found in Archaea is composed of a paracrystalline surface layer termed the S-layer. The S-layer consists of protein or glycoprotein moieties arranged in hexagonal patterns. Although there is a great diversity in archaeal cell walls, the vast majority of archaeal species contain a cell wall of some sort, which is used to define cell shape and prevent osmotic lysis. Certain members of the order Thermoplasmatales are the only known exception; these cells lack a cell wall, but are still able to maintain membrane integrity at low pH.

ARCHAEAL PHYLOGENY

The archaeal phylogenetic tree splits into two prominent phyla, termed Crenarchaeota and Euryarchaeota.[9, 10] At present, Crenarchaeota consists of only one class, Thermoprotei, and five orders (see Table 41.2). Cultured Crenarchaeota typically consist of relatively closely related hyperthermophilic organisms, which have optimal growth temperatures greater than 80°C. These species occupy short branches on the archaeal phylogenetic tree according to 16S rRNA sequence analysis, which may indicate that they are slow evolving. Additionally, several uncultured, cold-dwelling species of Crenarchaea have been identified by community sampling of 16S rRNA in marine environments.

TABLE 41.2
Archaeal Phylogeny and Physiological Characteristics[a]

Phylum	Class	Order	Family	Number of Genera	Type Genus	General Physiological Group
Crenarchaeota	Thermoprotei	Desulfurococcales	Desulfurococcaceae	10	*Desulfurococcus*	Hyperthermophilic neutrophiles
			Pyrodictiaceae	4	*Pyrodictium*	
		Sulfolobales	Sulfolobaceae	5	*Sulfolobus*	Hyperthermophilic acidophiles
		Thermoproteales	Thermoproteaceae	5	*Thermoproteus*	
			Thermofilaceae	1	*Thermofilum*	
		Caldisphaerales	Caldisphaeraceae	1	*Caldisphaera*	Thermoacidophiles
		Cenarchaeales	Cenarchaeaceae	1	*Cenarchaeum*	Mesophiles to psychrophiles
Euryarchaeota	Methanobacteria	Methanobacteriales	Methanobacteriaceae	4	*Methanobacterium*	Mesophilic methanogens
			Methanothermaceae	1	*Methanothermus*	
	Methanococci	Methanococcales	Methanococcaceae	2	*Methanococcus*	
			Methanocaldococcaceae	2	*Methanocaldococcus*	
	Methanomicrobia	Methanomicrobiales	Methanomicrobiaceae	6	*Methanomicrobium*	
			Methanocorpusculaceae	1	*Methanocorpusculum*	
			Methanospirillaceae	1	*Methanospirillum*	
		Methanosarcinales	Methanosarcinaceae	8	*Methanosarcina*	
			Methanosaetaceae	1	*Methanosaeta*	
	Methanopyri	Methanopyrales	Methanopyraceae	1	*Methanopyrus*	Hyperthermophilic methanogens
	Halobacteria	Halobacteriales	Halobacteriaceae	28	*Halobacterium*	Halophiles
	Thermoplasmata	Thermoplasmatales	Ferroplasmaceae	1	*Ferroplasma*	Mesophilic acidophiles
			Picrophilaceae	1	*Picrophilus*	Thermophilic hyperacidophiles
			Thermoplasmataceae	1	*Thermoplasma*	Cell wall-less thermoacidophiles
	Thermococci	Thermococcales	Thermococcaceae	3	*Thermococcus*	Hyperthermophilic neutrophiles
	Archaeoglobi	Archaeoglobales	Archaeoglobaceae	3	*Archaeoglobus*	

[a] Based on information from NCBI Entrez Taxonomy Browser, http://www.ncbi.nlm.nih.gov/entrez/query.fcgi?db=Taxonomy.

The phylum Euryarchaeota consists of physiologically diverse species. Euryarchaeota is further subdivided into eight classes and nine orders (Table 41.2). Four of those classes — namely, Methanobacteria, Methanococci, Methanopyri, and Methanomicrobia — are composed of methanogenic Archaea. The class Halobacteria consists of halophilic species that require high salt concentrations for growth. Thermoplasmata include thermoacidophilic species, and the classes Thermococci and Archaeoglobi are comprised of hyperthermoacidophiles that require temperatures above 80°C for growth.

In recent years, two additional phyla were proposed in the Archaeal phylogenetic tree: (1) the Korarchaeota,[11] composed of several uncultured species; and (2) Nanoarchaeota,[12] which consists of a single species, *Nanoarchaeum equitans*. However, using 16S rRNA analysis, Korarchaeal species have since been placed firmly inside of the phylum Crenarchaeota.[13, 14] Similarly, while the lineage of Nanoarchaea remains uncertain, recent studies have demonstrated that Nanoarchaea represent a fast-evolving, deep branch of Euryarchaeota.[13, 14]

PHYLUM CRENARCHAEOTA

The phylum Crenarchaeota is composed of a single class, Thermoprotei, which is subdivided into five orders: Thermoproteales, Desulfococcales, Sulfolobales, Caldisphaerales, and Cenarchaeales. Cultured crenarchaeal species are morphologically diverse, including rods, cocci, filamentous, and disk-shaped cells. Some genera are motile. Almost all cultured species are obligate thermophiles, with optimal growth temperatures ranging from 70 to 113°C. Many cultured genera are also acidophiles, and most can metabolize sulfur.

Hyperthermophilic crenarchaeal species inhabit environments reminiscent of the conditions of early Earth, and have been studied as models of the original forms of life. These habitats include solfataras, which are hot, volcanic areas that give off sulfurous gases, as well as hydrothemal vents found at the bottom of the ocean, also known as "black smokers."

Interestingly, cultured species of Crenarchaea lack histone proteins, which are found in all euryarchaeal species studied. This observation led to a proposal that the eukaryotic nucleus comes from a euryarchaeal origin after the divergence of crenarchaea.[15] However, recent evidence from uncultured species indicates that histones are not absent from all crenarchaeal species.[16]

Hyperthermophilic and Thermophilic Crenarchaea: Thermoproteales, Sulfolobales, Desulfurcoccales, and Caldisphaerales

The vast majority of cultured crenarchaeotes are thermophilic, with optimal growth between 45 and 80°C, or hyperthermophilic, growing optimally at temperatures above 80°C. Most are sulfur metabolizers, and many produce sulfuric acid. Nearly all cultured crenarchaeal species are acidophilic to varying degrees.

The order Thermoproteales is composed of rod-shaped cells that are 0.1 to 0.5 μm in diameter and almost 100 μm in length.[17] Species in this order are hyperthermophilic, with optimal growth temperatures ranging from 75 to 100°C, and are widely distributed in solfotara hot springs and hydrothermal vents around the world.[17] Some thermoprotealates are chemolithoautotrophs, using carbon dioxide as a carbon source and gaining energy by the conversion of hydrogen and elemental sulfur to hydrogen sulfide. Others acquire energy by sulfur respiration of various organic substrates, yielding carbon dioxide and hydrogen sulfide. Some genera are able to gain energy by respiration using oxygen, nitrate, or nitrite as electron acceptors.[17]

Thermoproteus is the best described genus in this order. Morphologically, thermoproteal cells are long, thin rods that may be bent or branched. Species within this genus are organotrophic obligate anaerobes, and are hyperthermoacidophiles, with optimal pH values ranging from 1.7 to 6.5 and temperatures up to 100°C.[17] Species of *Thermoproteus* have been identified in acidic hot springs and water holes in Iceland, Italy, North America, New Zealand, and Indonesia.[17]

Genera of the order Sulfolobales are hyperthermoacidophiles, with optimal pH values around 2, and temperature optima ranging from 60 to 90°C. Sulfolobalate cells are coccoid and irregularly lobed, and may be aerobic, facultatively anaerobic, or obligately anaerobic.[17] All sulfolobalates are either facultatively or obligately chemolithoautotrophic sulfur metabolizers, and most release sulfuric acid as a byproduct of metabolism. Sulfolobalates have been isolated from acidic continental solfotara fields around the world, as well as on the surface of boiling mudholes.[17] *Sulfolobus* strains can grow either lithoautotrophically by oxidizing sulfur, or chemoheterotrophically using sulfur to oxidize simple reduced carbon compounds.[17] Heterotrophic growth has only been observed, however, in the presence of oxygen. *Sulfolobales* species are known for unusual tetraether lipids: in Sulfolobales, the ether-linked lipids are joined covalently across the "bilayer," making tetraethers. Technically, therefore, the tetraethers form a monolayer, not a bilayer. The tetraethers help *Sulfolobus* species survive extreme acid as well as high temperature.[18]

The three best-characterized species within this order are *Sulfolobus solfataricus*, *S. acidocaldarius,* and *S. tokodaii*. The genomes of these three species have been sequenced.[19–21] Sequencing analysis has demonstrated that *S. acidocaldarius* maintains a very stable genome organization, whereas *S. solfataricus* and *S. tokodaii* both possess extensive networks of autonomous and nonautonomous mobile elements.[20] *Sulfolobus* is now used as a model to study the molecular mechanisms of DNA replication in Archaea.[20] Additionally, certain species of *Sulfolobus* are of industrial significance. For example, *S. tokodaii* strain 7 is known to oxidize hydrogen sulfide to sulfate intracellularly, which has been used to treat industrial wastewater.[22]

Species in the order Desulfurococcales are hyperthermophilic, with optimal temperature ranges from 85 to 95°C, and maximal temperatures up to 102°C. An intron has been found in the 23S ribosomal RNA gene of *Desulfurococcus mobilis*, making it the first known instance of an intron within the ribosomal RNA of a prokaryote.[23] The cells of *Desulfurococcus* are regular to irregular coccoid to disk-shaped, and are 0.2 to 5 μm in diameter.[17] Their cell envelopes are composed of S-layers of protein or glycoprotein subunits.[24] *Desulfurococcus* species can be found in many hyperthermophilic environments, ranging from underwater thermal black smokers to freshwater hot springs in Iceland and the United States. Desulfurococcal species are strictly anaerobic, except for *Aeropyrum pernix,*[25] and are neutrophilic or weakly acidophilic, growing optimally at pH 5.5 to 7.5.[26] Two families, Desulfurococcaceae and Pyrodictiaceae, are known in the order.[17]

A novel genus of Crenarchaeota was recently identified that phylogenetically represents an independent lineage related to the order Desulfurococcales.[27] Known species of this newly defined order, *Caldisphaerales*, inhabit terrestrial hot springs. The type species for this order, *Caldisphaera lagunensis,* is strictly anaerobic and organotrophic, and uses sulfur as an electron acceptor. Most cells are regular cocci, 0.8 to 1.1 μm in width, and occur singly or in pairs. The cells are nonmotile and thermoacidophilic, growing optimally at 70 to 75°C and pH 3.5 to 4.0.[27]

Mesophilic and Psychrophilic Crenarchaea: Cenarchaeales

For many years it was believed that Crenarchaeal species were all thermophilic or hyperthermophilic. However, 16S rRNA analysis of marine samples revealed that there are several species of Crenarchaeota that can be found in both mesophilic ("normal" temperature range) and psychrophilic (cold temperature) environments. Very few of these species have been cultured thus far, but based on sequence information they have been placed into a single order, Cenarchaeales.

Planktonic Crenarchaeota of the order Cenarchaeales are now recognized to comprise a significant component of marine microbial biomass, roughly 10^{28} cells in today's oceans.[28] The recent isolation of *Nitrosopumilus maritimus*, the first cultivated nonthermophilic crenarchaeon, demonstrated that bicarbonate and ammonia can serve as sole carbon and energy sources for at least some members of this lineage.[29] This marine crenarchaeote grows chemolithoautotrophically by aerobically oxidizing ammonia to nitrite, the first observation of nitrification in the Archaea.[29] Efforts continue to attempt to culture more of these psychrophilic Crenarchaeal species, which may be of commercial significance to industry and biotechnology.

PHYLUM EURYARCHAEOTA

As mentioned above, Euryarchaeota is a physiologically diverse phylum. The following subsections address classes and orders of Euryarchaeota grouped together by their individual physiological and metabolic characteristics.

HALOPHILIC EURYARCHAEOTA: CLASS HALOBACTERIA

Halobacteria, sometimes referred to as haloarchaea, is a class of Archaea composed of 27 genera that grow under extreme salinity. Salt requirements of these species range from 1.5 to 5.2 M NaCl, although most strains grow best at 3.5 to 4.5 M NaCl, at or near the saturation point of salt (36% (wt/vol) salts). To maintain the osmolarity of these cells in their high-salt environment, halobacterial species accumulate up to 5 M intracellular levels of KCl to counterbalance the high extracellular salt concentration. As a result, the entire intracellular machinery, including enzymes and structural proteins, must be adapted to high salt levels, although these mechanisms are not entirely understood. The proteins of all haloarchaeal species have a very low isoelectric point and the genomes contain high GC contents that are well above 60%.[30]

Some species of halobacteria are motile by means of tufts of flagella, although many species are nonmotile.[17] Halobacteria are largely aerobic or facultative anaerobes. Halobacteria come in a wide variety of shapes, including rods, cocci, and a multitude of pleomorphic forms.[17, 31] The lack of turgor pressure within haloarchaeal cells enables the cells to tolerate the formation of corners and, as such, some species are triangular or square.[17, 31] Cell envelopes of coccoid haloarchaea are stable in the absence of salt but noncoccoid species maintain their integrity only in the presence of high concentrations of NaCl or KCl.[17] The surface of the cell envelope of noncoccoid species has a hexagonal pattern due to the regular packing of glycoprotein subunits that are held together only in the presence of salt.[17]

Species of Halobacteria are the primary inhabitants of salt lakes, inland seas, and evaporating ponds of seawater, such as the Dead Sea and solar salterns. They often tint the water column and sediments bright colors due to the presence of retinal-based pigments. Some of these pigments are capable of the light-mediated translocation of ions across cell membranes. The best-known halobacterial pigment is bacteriorhodopsin, which is an outwardly directed proton pump. Bacteriorhodopsin is involved in energy conservation, and is the only non-chlorophyll-mediated light energy transducing system known to date.[32]

Other retinal-based pigments found in Halobacteria include halorhodopsin, which is an inward chloride pump involved in osmotic homeostatis, and sensory rhodopsin I and II (SRI and SRII, respectively). SRI and SRII can mediate positive and/or negative phototaxis.[32] Surprisingly, rhodopsins have been revealed to be widely distributed among (eu)bacteria, including proteorhodopsin, which has been identified in bacterioplankton, and sensory rhodopsins similar to SRI and SRII found in cyanobacteria.[32]

METHANOGENIC EURYARCHAEOTA: METHANOBACTERIA, METHANOCOCCI, METHANOPYRI, AND METHANOMICROBIA

Methanogens are euryarchaeal species that are capable of producing methane, using a process referred to as methanogenesis. As opposed to Haloarchaea, methanogens are obligate anaerobes, requiring strict anoxic techniques to culture them. Habitats of methanogens include anoxic sediments, such as marshes and swamps; lake sediments, such as rice paddy fields; animal digestive tracts, including large intestines of monogastic animals, the rumen of ruminant animals, and the hindgut of cellulolytic insect (e.g., termites); hydrothermal vents, which function as sources for hydrogen and carbon dioxide; and artificial biodegradation facilities such as sewage sludge digesters.[3]

Some methanogens, described as hydrotropic, use carbon dioxide as a source of carbon and hydrogen as a source of energy. Some of the carbon dioxide reacts with, and is reduced by, hydrogen to produce methane. The methane, in turn, gives rise to a proton motive force across a membrane that is used to generate ATP. Other methanogens, called acetotrophic, use acetate (CH_3COO-) as a source of both carbon and energy. Still other methanogens exploit methylated compounds such as methylamines, methanol, and methanethiol. There are 11 different substrates for methanogenesis identified to date; they fall under the three categories described above: CO_2-type substrates (e.g., CO_2, formate [HCOO-]); methyl substrates (e.g., methanol [CH_3OH]); and acetotrophic substrates (e.g., acetate, pyruvate).

There are four different classes of methanogens. The class Methanobacteria consists of five genera that include rod-shaped, lancet-shaped, or coccoid methanogens that reduce CO_2 or methyl compounds with H_2, formate, or secondary alcohols as electron donors. They are nonmotile and contain cell walls made of psuedopeptidoglycan. Methanobacteria are widely distributed in nature, and are found in anaerobic habitats such as aquatic sediments, soil, anaerobic sewage digesters, and the gastrointestinal tracts of animals.[17]

As the name implies, the class Methanococci includes cells that are cocci or coccoid in shape, and contain protein cell walls.[17] All species of Methanococci are strict anaerobes, and obtain energy by the reduction of CO_2 to methane. Genera range from mesophilic (e.g., *Methanococcus*) to thermophilic (e.g., *Methanothermococcus*) to hyperthermophilic (e.g., *Methanocaldococcus*), and some species are mobile due to tufts of flagella.

Members of the class Methanomicrobia include a variety of cell shapes, including cocci, coccoid, rods, and sheathed rods. Most cells in this class have cells walls made of protein, and some cells are surrounded by a sheath. Most Methanomicrobia form methane by the reduction of carbon dioxide using a variety of electron donors, and all species are obligate anaerobes. Methanomicrobia can be found in a variety of habitats, including aquatic sediments, anaerobic sewage digesters, and the gastrointestinal tracts of animals.[17]

The class Methanopyri consists of a single genus, *Methanopyrus*. Cells of this genus are rod-shaped and contain cell walls made of psuedopeptidoglycan.[17] *Methanopyrus* are hyperthermophilic and grow between 84 and 110°C, with optimal growth at 98°C. These cells grow chemoautotrophically by the conversion of CO_2 and H_2 to methane.[17] Phylogenetic 16S rRNA studies have shown that *Methanopyrus kandleri* represents a very deep branch-off within Euryarchaeota, and is seemingly unrelated to any other methanogen.[17]

Methanogenic Phylogeny

There are several species of methanogens identified to date that are phylogenetically classified into four classes and five orders. The phylogenetic origin of methanogens and methanogenesis is currently under debate. Using 16S rRNA phylogenetic analysis, it appears that there are two distantly related groups of methanogens (called Class I and Class II) that are separated from each other by nonmethanogenic archaeans, including Thermoplasmata, Archaeoglobi, and Halobacteria.[33] However, analysis of enzymes required for methanogenesis reveals that methanogenesis evolution occurred only once (i.e., the two classes of methanogens did not develop the ability to produce methane by convergent evolution).[33] As a result, it has been proposed that methanogenesis was acquired early in the development of Euryarchaeota, but that the genes necessary for methanogenesis were subsequently lost in Thermoplasmates, Archaeoglobi, and Halobacteria.[33]

This scenario, in essence, proposes that the common ancestor of different physiologically and metabolically distinct groups within Euryarchaeota was a methanogen and this capability was independently lost in all other lineages. A recent study of phylogenomics, which analyzed protein sequences from across several archaean species, revealed that virtually all methanogenic species, even those that by 16S rRNA study were placed in distant classes, share at least 31 proteins that are not present in any other archaeal species.[13] This lends strong support for a monophyletic origin of all methanogenic species, and indicates that the branching patterns of various clades in phylogenetic trees may have been misleading and that they may have been affected by factors such as the long branch attraction effect.[13]

THEMOPHILIC EURYARCHAEOTA: THERMOPLASMATA

This class consists of one order (Thermoplasmatales) and three families (Thermoplasma, Picrophilus, and Ferroplasma). Thermoplasma and Ferroplasma are the only representatives of Archaea that do not contain a cell wall.[17] Species of Thermoplasma are obligate thermoacidophiles, with optimal growth achieved at 60°C and pH 2.[17] Thermoplasma are facultative anaerobes, and obligate heterotrophs, using elemental sulfur for respiration. Species may be found in self-heating coal refuse piles and in acidic solfatara fields.[17]

Members of the family Picrophilus are the most acidophilic organisms known thus far.[34, 35] Cells in this family are irregular cocci 1 to 1.5 μm in diameter and contain S-layer cell wall.[17] Picrophilus are thermophilic and hyperacidophilic, growing at temperatures between 47 and 60°C and pH ranges from below 0 to 3.5.[17] Their ability to grow at pH values at and below zero and at high temperatures has shifted the physico-chemical boundaries at which life was considered to exist.

Ferroplasma is the only member of this class that is not thermophilic, and has been identified at Iron Mountain in northern California.[36] It can grow where the temperature is between 15°C and 47°C, with an optimum temperature of 35°C, and where the pH is between 1.3 and 2.2, with an optimum pH of 1.7.[36] The cells are pleomorphic and lack a cell wall. Unlike other families in this class, *Ferroplasma* are autotrophic, oxidizing ferrous iron as the sole energy source and fixing inorganic carbon as the sole carbon source.[36]

HYPERTHERMOPHILIC EURYARCHAEOTA: ARCHAEOGLOBI AND THERMOCOCCI

The class Archaeoglobi is composed of a single family and three genera: *Archaeoglobus*, *Ferroglobus*, and *Geoglobus*.[37] Archaeoglobi are regular to irregular cocci occurring singly or in pairs.[17] Species of this class are strictly anaerobic and hyperthermophilic, growing optimally at 80°C and at neutral pH. These cells exhibit blue-green fluorescence at 420 nm. Organisms in the genus *Archaeoglobus* are autotrophic and/or organotrophic and reduce sulfate or sulfite for respiration.[17] Species of *Ferroglobus* grow by oxidation of Fe(II), S^{2-}, and H_2,[17] whereas *Geoglobus* grow anaerobically in the presence of acetate and ferric iron.[37]

Members of the class Thermococci are spherical and sometimes pleomorphic, and are about 1 μm in diameter. The cells often occur as diplococci or as clusters of up to 30 cells, and contain an S-layer cell wall. Thermococci are strictly anaerobic hyperthermophilic heterotrophs that generally perform sulfur respiration.[17] Optimal growth temperatures for the type genus *Thermococcus* range from 75 to 88°C.[17] *Thermococcus* releases strong-smelling sulfur-based products such as mercaptans.[17] Species of *Thermococcus* have been isolated from submarine solfataras, including deep-sea hydrothermal vents.[17]

One thermococcal species, *Pyrococcus furiosus,* is a hyperthermophilic anaerobic archaeon that grows optimally near 100°C using carbohydrates and peptides as carbon and energy sources.[17] This organism is commonly found in hydrothermal vents on the sea floor near volcanoes.[17] Its ability to grow to high cell densities under laboratory conditions without the need of elemental sulfur, and thus without producing toxic hydrogen sulfide, has made *P. furiosus* a useful model organism to study thermostable enzymes and adaptations to high-temperature environments.[38]

Parasitic Euryarchaea: Nanoarchaea

Karl Stetter and others recently described a novel archaeal species named *Nanoarchaeum equitans* that represents the smallest known living cell.[12] This tiny hyperthermophile grows and divides on the surface of crenarchaeal *Ignicoccus* species and cannot be cultivated independently, indicating a potential parasitic lifestyle.[39] Sequencing of the *N. equitans* genome revealed the smallest cellular genome presently known (480 kb) and raised fascinating questions regarding the origin and evolution of this archaeon.[39] Indeed, compared to typical genomes from parasitic/symbiotic microbes, the genome of *N. equitans* does not show any evidence of decaying genes and contains a full complement of tightly packed genes encoding informational proteins.[39] This suggests that the establishment of the dependence-relationship between *N. equitans* and *Ignicoccus* is probably very ancient.[39]

As stated previously, the lineage of *Nanoarchaeum equitans* is unclear. Based on SSU rRNA analysis, it appeared that *N. equitans* is the sole known member of a very ancient archaeal phylum, the Nanoarchaeota.[12,39] However, recent phylogenomic evidence places *N. equitans* within the phylum Euryarchaeota, although its exact location within that phylum is still under debate.[13] More research is underway to elucidate the origins of this unique archaeal species.

REFERENCES

1. Woese, C.R., Magrum, L.J., and Fox, G.E., Archaebacteria, *J. Mol. Evol.*, 11 (3), 245–251, 1978.
2. Woese, C.R., Kandler, O., and Wheelis, M.L., Towards a natural system of organisms: proposal for the domains Archaea, Bacteria, and Eucarya, *Proc. Natl. Acad. Sci. U.S.A.*, 87(12), 4576–4579, 1990.
3. Chaban, B., Ng, S.Y., and Jarrell, K.F., Archaeal habitats—from the extreme to the ordinary, *Can. J. Microbiol.*, 52(2), 73–116, 2006.
4. Duggin, I.G. and Bell, S.D., The chromosome replication machinery of the archaeon *Sulfolobus solfataricus, J. Biol. Chem.*, 281(22), 15029–15032, 2006.
5. Bartlett, M.S., Determinants of transcription initiation by archaeal RNA polymerase, *Curr. Opin. Microbiol.*, 8(6), 677–684, 2005.
6. Yoshinari, S., Itoh, T., Hallam, S.J., DeLong, E.F., Yokobori, S., Yamagishi, A., Oshima, T., Kita, K., and Watanabe, Y., Archaeal pre-mRNA splicing: a connection to hetero-oligomeric splicing endonuclease, *Biochem. Biophys. Res. Commun.*, 346(3), 1024–1032, 2006.
7. Krzycki, J.A., The direct genetic encoding of pyrrolysine, *Curr. Opin. Microbiol.*, 8(6), 706–712, 2005.
8. Koga, Y. and Morii, H., Recent advances in structural research on ether lipids from archaea including comparative and physiological aspects, *Biosci. Biotechnol. Biochem.*, 69(11), 2019–2034, 2005.
9. Makarova, K.S., Aravind, L., Galperin, M.Y., Grishin, N.V., Tatusov, R.L., Wolf, Y.I., and Koonin, E.V., Comparative genomics of the Archaea (Euryarchaeota): evolution of conserved protein families, the stable core, and the variable shell, *Genome Res.*, 9(7), 608–628, 1999.
10. Burggraf, S., Huber, H., and Stetter, K.O., Reclassification of the crenarchael orders and families in accordance with 16S rRNA sequence data, *Int. J. Syst. Bacteriol.*, 47(3), 657–660, 1997.
11. Barns, S.M., Delwiche, C.F., Palmer, J.D., and Pace, N.R., Perspectives on archaeal diversity, thermophily and monophyly from environmental rRNA sequences, *Proc. Natl. Acad. Sci. U.S.A.*, 93(17), 9188–9193, 1996.
12. Huber, H., Hohn, M.J., Rachel, R., Fuchs, T., Wimmer, V.C., and Stetter, K.O., A new phylum of Archaea represented by a nanosized hyperthermophilic symbiont, *Nature*, 417(6884), 63–67, 2002.
13. Gao, B. and Gupta, R.S., Phylogenomic analysis of proteins that are distinctive of Archaea and its main subgroups and the origin of methanogenesis, *BMC Genomics*, 8, 86, 2007.
14. Gribaldo, S. and Brochier-Armanet, C., The origin and evolution of Archaea: a state of the art, *Philos. Trans. R. Soc. Lond. B. Biol. Sci.*, 361(1470), 1007–1022, 2006.
15. Martin, W. and Muller, M., The hydrogen hypothesis for the first eukaryote, *Nature*, 392(6671), 37–41, 1998.
16. Cubonova, L., Sandman, K., Hallam, S.J., Delong, E.F., and Reeve, J.N., Histones in crenarchaea, *J. Bacteriol.*, 187(15), 5482–5485, 2005.
17. Boone, D.R., Castenholz, R.W., Garrity, G.M., *Bergey's Manual of Systematic Bacteriology*, 2nd ed. Springer-Verlag, New York, 2001, p. 721.

18. Chang, E.L., Unusual thermal stability of liposomes made from bipolar tetraether lipids, *Biochem. Biophys. Res. Commun.*, 202(2), 673–679, 1994.
19. Kawarabayasi, Y., Hino, Y., Horikawa, H., Jin-no, K., Takahashi, M., Sekine, M., Baba, S., Ankai, A., Kosugi, H., Hosoyama, A., Fukui, S., Nagai, Y., Nishijima, K., Otsuka, R., Nakazawa, H., Takamiya, M., Kato, Y., Yoshizawa, T., Tanaka, T., Kudoh, Y., Yamazaki, J., Kushida, N., Oguchi, A., Aoki, K., Masuda, S., Yanagii, M., Nishimura, M., Yamagishi, A., Oshima, T., and Kikuchi, H., Complete genome sequence of an aerobic thermoacidophilic crenarchaeon, Sulfolobus tokodaii strain7, *DNA Res.*, 8(4), 123–140, 2001.
20. Chen, L., Brugger, K., Skovgaard, M., Redder, P., She, Q., Torarinsson, E., Greve, B., Awayez, M., Zibat, A., Klenk, H.P., and Garrett, R.A., The genome of *Sulfolobus acidocaldarius*, a model organism of the Crenarchaeota, *J. Bacteriol.,* 187(14), 4992–4999, 2005.
21. She, Q., Singh, R.K., Confalonieri, F., Zivanovic, Y., Allard, G., Awayez, M.J., Chan-Weiher, C.C., Clausen, I.G., Curtis, B.A., De Moors, A., Erauso, G., Fletcher, C., Gordon, P.M., Heikamp-de Jong, I., Jeffries, A.C., Kozera, C.J., Medina, N., Peng, X., Thi-Ngoc, H.P., Redder, P., Schenk, M.E., Theriault, C., Tolstrup, N., Charlebois, R.L., Doolittle, W.F., Duguet, M., Gaasterland, T., Garrett, R.A., Ragan, M.A., Sensen, C.W., and Van der Oost, J., The complete genome of the crenarchaeon Sulfolobus solfataricus P2, *Proc. Natl. Acad. Sci. U.S.A.*, 98(14), 7835–7840, 2001.
22. Kletzin, A., Urich, T., Muller, F., Bandeiras, T.M., and Gomes, C.M., Dissimilatory oxidation and reduction of elemental sulfur in thermophilic archaea, *J. Bioenerg. Biomembr.*, 36(1), 77–91, 2004.
23. Silva, G.H., Dalgaard, J.Z., Belfort, M., and Van Roey, P., Crystal structure of the thermostable archaeal intron-encoded endonuclease I-DmoI, *J. Mol. Biol.*, 286(4), 1123–1136, 1999.
24. Watanabe, Y., Yokobori, S., Inaba, T., Yamagishi, A., Oshima, T., Kawarabayasi, Y., Kikuchi, H., and Kita, K., Introns in protein-coding genes in Archaea, *FEBS Lett.*, 510(1–2), 27–30, 2002.
25. Sako, Y., Nomura, N., Uchida, A., Ishida, Y., Morii, H., Koga, Y., Hoaki, T., and Maruyama, T., *Aeropyrum pernix* gen. nov., sp. nov., a novel aerobic hyperthermophilic archaeon growing at temperatures up to 100 degrees C, *Int. J. Syst. Bacteriol.*, 46(4), 1070–1077, 1996.
26. Huber, R. and Stetter, K.O., Discovery of hyperthermophilic microorganisms, *Methods Enzymol.*, 330, 11–24, 2001.
27. Itoh, T., Suzuki, K., Sanchez, P.C., and Nakase, T., *Caldisphaera lagunensis* gen. nov., sp. nov., a novel thermoacidophilic crenarchaeote isolated from a hot spring at Mt Maquiling, Philippines, *Int. J. Syst. Evol. Microbiol.*, 53(Pt. 4), 1149–1154, 2003.
28. Hallam, S.J., Konstantinidis, K.T., Putnam, N., Schleper, C., Watanabe, Y., Sugahara, J., Preston, C., de la Torre, J., Richardson, P.M., and DeLong, E.F., Genomic analysis of the uncultivated marine crenarchaeote *Cenarchaeum symbiosum*, *Proc. Natl. Acad. Sci. U.S.A.*, 103(48), 18296–18301, 2006.
29. Konneke, M., Bernhard, A. E., de la Torre, J. R., Walker, C. B., Waterbury, J.B. and Stahl, D.A., Isolation of an autotrophic ammonia-oxidizing marine archaeon, *Nature*, 437(7058), 543–546, 2005.
30. Soppa, J., From genomes to function: haloarchaea as model organisms, *Microbiology*, 152(Pt. 3), 585–590, 2006.
31. Walsby, A.E., Archaea with square cells, *Trends Microbiol.*, 13(5), 193–195, 2005.
32. Soppa, J., From replication to cultivation: hot news from Haloarchaea, *Curr. Opin. Microbiol.*, 8(6), 737–744, 2005.
33. Bapteste, E., Brochier, C., and Boucher, Y., Higher-level classification of the Archaea: evolution of methanogenesis and methanogens, *Archaea*, 1(5), 353–363, 2005.
34. Schleper, C., Puehler, G., Holz, I., Gambacorta, A., Janekovic, D., Santarius, U., Klenk, H.P., and Zillig, W., *Picrophilus* gen. nov., fam. nov.: a novel aerobic, heterotrophic, thermoacidophilic genus and family comprising archaea capable of growth around pH 0, *J. Bacteriol.*, 177(24), 7050–7059, 1995.
35. Schleper, C., Puhler, G., Kuhlmorgen, B., and Zillig, W., Life at extremely low pH, *Nature*, 375(6534), 741–742, 1995.
36. Edwards, K.J., Bond, P.L., Gihring, T.M., and Banfield, J.F., An archaeal iron-oxidizing extreme acidophile important in acid mine drainage, *Science*, 287(5459), 1796–1799, 2000.
37. Miroshnichenko, M.L. and Bonch-Osmolovskaya, E.A., Recent developments in the thermophilic microbiology of deep-sea hydrothermal vents, *Extremophiles*, 10(2), 85–96, 2006.
38. Poole, F.L., 2nd, Gerwe, B.A., Hopkins, R.C., Schut, G.J., Weinberg, M.V., Jenney, F.E., Jr., and Adams, M.W., Defining genes in the genome of the hyperthermophilic archaeon *Pyrococcus furiosus*: implications for all microbial genomes, *J. Bacteriol.,* 187(21), 7325–7332, 2005.

39. Waters, E., Hohn, M.J., Ahel, I., Graham, D.E., Adams, M.D., Barnstead, M., Beeson, K.Y., Bibbs, L., Bolanos, R., Keller, M., Kretz, K., Lin, X., Mathur, E., Ni, J., Podar, M., Richardson, T., Sutton, G.G., Simon, M., Soll, D., Stetter, K.O., Short, J.M., and Noordewier, M., The genome of *Nanoarchaeum equitans*: insights into early archaeal evolution and derived parasitism, *Proc. Natl. Acad., Sci. U.S.A.*, 100(22), 12984–12988, 2003.

42 Overview of Biofilms and Some Key Methods for Their Study

Irvin N. Hirshfield, Subit Barua, and Paramita Basu

CONTENTS

INTRODUCTION

Perhaps we can say that microbiology began in the late 17th century through the curiosity of Antony van Leeuwenhoek, who although an amateur had a passion for microscopy. He was the first to describe accurately the creatures of the "invisible world," which he called animalcules. We now appreciate that among these animalcules were bacteria and eukaryotic microbes. However, it should be realized that there were predecessors to van Leeuwenhoek. Jansen is credited with the development of early compound microscopes; and as far back as the mid-16th century, Girolamo Francastoro, among others, is hailed as suggesting that unseen organisms could cause disease [1, 2].

Travel to the last quarter of the 19th century. It was a period of ferment and discovery that led to modern microbiology. The giant figure of that era, Louis Pasteur, disproved spontaneous generation through his creatively designed long-necked flasks and also developed vaccines. Christian Gram developed a stain that is still used extensively, and divides bacteria into two major groups that bear his name: Gram-positive and Gram-negative. Moreover, in that period of time, an instrument for sterilization, the autoclave, was developed, that is of central importance in microbiology labs even now [1].

However, it can be argued that the most influential laboratory of that era was that of the German physician/scientist Robert Koch. It was in his lab that the Petri dish was introduced, agar as a solidifying agent was developed thanks to Fannie Hesse, culture media were invented, and of course there were Koch's postulates. During this time, a number of pathogens such as *Mycobacterium tuberculosis, Streptococcus pneumoniae,* and *Clostridium botulinum,* as well as the workhorse of genetics and molecular biology, *Escherichia coli,* were discovered through the use of techniques developed above and pure culture techniques. In that era, the growth of bacteria and their study were predicated on the concept that bacteria are free-living, planktonic cells, and this view dominated the study of bacteria for decades [3].

As frequently happens with the advent of new technology, new ideas are spawned. These ideas usually germinate in a small group of individuals, and like a newly inoculated culture can have a long lag time. However, once accepted, they flourish. This kind of history appears to apply to biofilms, which have multiple definitions, but generally refer to communities of microbes (prokaryotic and eukaryotic) that adhere to a surface and are enclosed in an extracellular polysaccharide matrix [4, 5].

Microscopy, especially light microcopy [6, 7], and transmission and scanning electron microscopy [3, 6, 8, 9] played a pivotal role in the realization of the biofilm concept. Through the application of transmission electron microscopy, morphological evidence was found for an attachment of *Escherichia coli* to the intestines of newborn calves. Subsequently, it was shown that biofilms form in aquatic ecosystems under high-shear conditions [10]. Another major advance occurred when scanning confocal laser microscopy was applied to the study of biofilms [11, 12]. Major advantages of this technology are that the samples are living and hydrated. Consequently, it was discovered that there is a complex organization to biofilms with mushroom- and tower-like structures enclosed in a matrix with intervening water channels [13, 14].

Numerous investigations have shown that biofilms have properties, probably derived from their organization, that are of both biological and clinical importance. It has been found consistently that biofilms are far more resistant to static and cidal agents than planktonic cells [15–17]. This applies to a broad spectrum of pathogens, including *Legionella, Listeria, Mycobacteria, Pseudomonas,* and *Vibrio.* There can be as much as a 100 to 1000 times greater resistance of the biofilms compared to their planktonic counterparts [15, 18, 19]. Treating biofilms of *P. aeruginosa* is of medical importance because of its pathogenesis in cystic fibrosis patients, and this has proved a difficult task [17–21]. The increased resistance of biofilms to antimicrobials is not the result of mutation or horizontal gene transfer because, upon removal of the cells from biofilms, and their subsequent growth as planktonic cells, the sensitivity to the agents re-emerges [3, 15, 21].

The resistance phenomenon also applies to fungal biofilms and has been documented with *Candida albicans* [22–24]. *Candida* is cited as being the fourth most-common cause of bloodstream infections [25], and is also a major clinical problem because of its ability to form biofilms on indwelling medical devices [22].

Multiple explanations have been advanced for the high resistance of biofilms to antimicrobials [15]. These include (1) depletion of the antimicrobial agent through its interaction and neutralization by the biofilm; (2) slow penetration of the antimicrobial into the biofilm [17]; (3) the presence of slow-growing and/or non-growing cells (e.g., due to oxygen, pH, or nutrient gradients in the biofilm) [13, 26]; (4) stress responses, induced or constitutive; and more controversially, (5) persister cells. The persister cells, if they exist, constitute only a small percentage of the population but represent a cell state that is impregnable to antimicrobial assault [27].

Due to the ability of microbial cells to adhere to surfaces, both biotic and abiotic, a major problem has developed in medicine with the colonization of indwelling medical devices and nosocomial infections. As the technology of medicine has advanced, and concomitantly the population of the aged is increasing, there has been increased utilization of indwelling medical devices. One major reason for this is that there has been a reduction in the length of hospital stays and a subsequent reliance on ambulatory care. For people of all ages, there are catheters to deliver medications or fluids. In addition, there are devices for those with heart defects and protheses for joint replacements, particularly hips and knees [19, 28].

A downside of these advances is that biofilms, whether bacterial or fungal, can form on these indwelling medical devices [29, 30]. These biofilms are difficult to treat and can lead to infections, which leads to a serious economic impact. Often, the only recourse is to remove the device and replace it. As an example, millions of intravascular catheters are commonly used in today's medicine in the United States. However, there are a huge number of bloodstream infections that arise from this use, which result in raising costs of treatment billions of dollars per annum [31].

This brief introduction gives a taste of the importance of biofilms, and why methods are needed to assess their formation as well as to test their response to antimicrobials.

METHODS TO ASSESS BIOFILM FORMATION

There are various strategies used to study biofilm formation, although they generally employ stationary phase cells. These modes of study use static systems, which are better for the study of early events in biofilm formation, and continuous flow or chemostat systems, which are preferable for the examination of mature biofilms [32].

Microtiter Plate Biofilm Assay

A popular static method that has the advantage of high throughput is the microtiter plate biofilm assay. This method is based on its development and modification by Christensen et al. [33], Mack et al. [34], and O'Toole and Kolter [35].

Procedure

1. The bacterial strain is grown in batch culture overnight at the desired temperature to stationary phase in the desired medium, such as minimal medium or Luria-Bertani medium. A 10-mL culture or less will suffice.
2. Dilute the cells 1:100 with the desired growth medium into the wells of a 96-well polystyrene microtiter plate. It is recommended that each strain be assayed in quadruplicate. Some wells should be filled with sterile water, and the plate should be covered with a lid and/or parafilm to prevent dehydration.
3. Incubate the cells in the plate at the appropriate temperature, 25 to 37°C is typically used [32], and the length of the incubation should be between 2 and 48 hr, depending on the strain, the medium, and the growth temperature.
4. After the incubation, the planktonic bacteria should be separated from those cells that have formed the biofilm. First, invert the plate and shake firmly over a wastewater tray. Alternatively, the cells can be washed using a water bottle that gently delivers the liquid. The wells should then be washed three more times, either by immersing them into water washes or gently applying water. Finally, invert the plates and tap them to remove any remaining water.

5. Air dry the wells for a few minutes at room temperature, and then add 0.1% crystal violet solution, 20 to 200 μL, to each well. Allow the staining to proceed for 10 to 20 min at room temperature.
6. Wash the wells with water to remove crystal violet that has not adhered to the cells. Repeat three more times and air dry the microtiter plate at room temperature.
7. Add 200 μL of 95% ethanol to each well, and incubate for 10 to 15 min at room temperature to solubilize the crystal violet. Mix with a pipette. Other solvents such as dimethylsulfoxide have been used instead of ethanol [36].
8. Transfer 100 to 150 μL of crystal violet solution to corresponding wells of an optically clear, flat-bottom 96-well plate, and measure the absorbance at 500 to 600 nm with a plate reader.
9. Alternatively, the crystal violet stain can be quantified by another means [35]:
   a. Add 200 μL of 95% ethanol twice to each microtiter plate well and mix.
   b. Transfer the ethanol to a 1.5-mL microcentrifuge tube.
   c. Bring the volume to 1 mL with deionized water.
   d. Transfer to a cuvette and read spectrophotometrically at 540 nm.

An *In Vitro* Cell Culture Method for Assessing Biofilm Formation [37]

This is a batch culture method, and batch culture is commonly used in studying biofilms because it is easy to grow the cells in a variety of media and vessels, and the cost is low [6]. In this method, the biofilm is grown on a coverslip that has been inserted into the culture medium.

Procedure

1. Cells are grown overnight in the appropriate medium. The bacteria can be grown aerobically or anaerobically, depending on the organism and the objective of the study.
2. Add 2.5 mL medium to individual sterile culture dishes.
3. To each dish, add a sterile 18-mm-diameter glass microscope coverslip and cover the dish.
4. Add the requisite amount of overnight culture to the dish and incubate overnight.
5. Remove the coverslips and rinse briefly with water. These can be viewed by phase-contrast microscopy.
6. For fluorescence microscopy, use the following approach:
   a. Remove the coverslip from the culture and transfer to a dry dish.
   b. Add 20 μL monoclonal antibody solution and incubate 30 min at room temperature [38].
   c. Add 5 μL fluorescein-conjugated goat secondary antibody; incubate at room temperature for 10 min.
   d. Briefly rinse the coverslips with water to remove unbound antibodies and cells, and then use immediately for microscopy to demonstrate biofilm development.

A Static Method for Assessment of *Staphylococcus* Biofilms

It has been well established that indwelling medical devices form excellent surfaces for biofilm formation [22]. *Staphylococcus* spp. — in particular *Staphylococcus epidermidis* and *Staphylococcus aureus* as well as coagulase-negative staphylococci — can produce a mucoid exopolysaccharide (referred to as slime) that allows them to form biofilms on the surfaces of these devices [39–41]. It is these staphylococci that are frequently involved in troublesome infections because of the difficulty in treating biofilms with antibiotics [15, 18, 19]. Biofilm formation can be assessed by a microtiter plate method [42] similar to that described above [33, 35]. It was found that adding 1% glucose to trypticase soy broth or 2% sucrose to brain heart infusion broth markedly increased the number of strains that formed biofilms [42].

As an alternative, a method that requires simpler equipment and detects the slime production is the tube method [41, 42]. This is considered a qualitative method, but Mathur et al. [42] have shown that it is equivalent to the plate method, at least for strong biofilm formers.

Tube Method Procedure

1. Grow staphylococcus on culture plate overnight.
2. Inoculate 10 mL TSB containing 1% glucose with a loopful of cells from the culture plate and incubate statically for 18 to 24 hr at 37°C in a glass or polystyrene test tube.
3. Decant the tube and wash with phosphate buffered saline (pH 7.3).
4. Air dry up to 10 min.
5. Stain with 0.1% crystal violet for 5 to 10 min at room temperature.
6. Wash with deionized water to remove the excess stain.
7. Invert the tube and allow it to air dry.
8. In assessing the results, the formation of a ring at the liquid interface is not considered a positive result. What should be observed is a visible film that lines the wall and the bottom of the tube [41, 42].

Flow Cell Method

Microscopic studies have been an important tool in the study of biofilm development. A route to the microscopic examination of developing biofilms is via the flow cell technique. This method was developed to allow direct, on-line observation of biofilm structure by light microscopy and scanning confocal laser microscopy [6, 43, 44]. Initially, light microscopy can be used; but as the biofilm thickens, SCLM (scanning confocal laser microscopy) is required to obtain interpretable images.

The system consists of five units. A container of medium is connected to a flow pump that is connected to a bubble trap; the medium then flows into a flow cell (chamber), and the effluent then goes to a waste bottle [43, 44]. The flow cell has four parallel flow channels (for details of its construction, see Wolfaardt et al. [43] or Christensen et al. [44]). The bubble chamber is present to prevent air bubbles from reaching the biofilm and disrupting its structure. The tubing between the bubble chamber and the flow cell must be of sufficient length to allow the flow cell to be mounted on a microscope. A critical component of the apparatus is a microscope glass coverslip that serves as a substratum for biofilm development as well as for microscopic observation. According to Christensen et al. [44], the coverslip is turned upward to prevent the sedimentation of flocs detached from other parts of the biofilm.

Procedure [44]

1. Sterilize flow cell with 0.5% (vol/vol) hypochlorite overnight, and then rinse with distilled water.
2. Using a syringe, 250 μL of an overnight culture diluted to an A_{450} of 0.05 is injected into the flow cell while the pump is turned off.
3. The flow cell is turned upside-down.
4. The influent line is clamped to prevent back-growth.
5. After 1 hr, the flow chamber is righted and medium is pumped through the system at a rate of 0.2 mm/sec.
6. The development of the biofilm can be followed by microscopy.

The Modified Robbins Device

A motivation for the development of the Robbins device derived from the problem of biofouling in industry [45–47]. This instrument, to a large degree, has been supplanted by the modified Robbins device [6, 18]. With the modified device, materials such as catheter surfaces can be employed to monitor the growth of biofilms [18]. Consequently, depending on the material chosen, the device can be used for industrial purposes, or for clinical or laboratory investigations [6, 18, 48]. It has also been used for the determination of the cidal effects of antibiotics on biofilms [18]. A key aspect in the design of this instrument is that the sampling ports have disks of the test surface attached at the bottom and in contact with the medium, but the flow of the medium does not disturb the biofilm. Because this is a multiport system, samples can be removed at various times or in multiples. The biofilms that grow can be examined by microscopy, or cells can be scraped off the surface of the disk material for enumeration [18]. The modified Robbins device can be used with continuous culture systems as well as with batch culture [49]. If there is an interest in purchasing this instrument, a source is Tyler Research Corporation, Edmonton, Alberta, Canada.

Procedure

The experimental design is that of Nickel et al. (1985) [18], which is the seminal publication with respect to the modified Robbins device. In this work, a batch culture was used. The disks attached to the sampling ports were derived from a urinary catheter, and the system was organized with a flask that contained the medium fitted with a line containing a valve going to a peristaltic pump. From the pump, medium flowed through tubing into the modified Robbins device and the effluent was collected in an appropriate vessel.

1. The entire experimental apparatus was sterilized using ethylene oxide.
2. Cells (*Pseudomonas aeruginosa*) were added to the medium so that they would be in log phase for the duration of the experiment.
3. The pump rate was 60 mL hr^{-1}.
4. During the course of the experiment, up to 8 hr, the disks were aseptically removed and examined by scanning electron microscopy or cells were scraped off the disks into medium for enumeration.

Microscopy

Microscopy has been an invaluable tool in studying the development of biofilms [3, 6–9]. Two major modes of microscopy used are scanning electron microscopy (SEM) and scanning confocal laser microscopy (SCLM). The major advantage of SEM is its high resolution; but because of fixation and dehydration of the samples in the preparative phases, it gives a distorted image of biofilm structure. In contrast, SCLM yields a lower resolution; but because the samples are more representative of their natural state, the image obtained is more accurate and has allowed for a much-improved understanding of biofilm structure [3].

Preparation of Samples for Scanning Electron Microscopy [50–54]

1. Place the sample in 2.5 to 5% glutaraldehyde in 0.05 to 0.1 M cacodylate buffer (pH 6.2 to 7.0) for 1 to 24 hr at 20°C. The mixture often contains 0.15% ruthenium red to stain the exopolysaccharide matrix.
2. If desired, the sample can be washed 2 to 3X with the buffer for 1 hr each.
3. Treat the sample with 1% osmium tetroxide [55, 56].
4. Dehydrate in an ethanol series from 25 to 100%. Each ethanol wash is for 10 min at room temperature.
5. The samples are critical-point dried.
6. The samples are coated with gold or platinum-palladium.

A variation on this procedure, which was used for the study of *Acinetobacter baumanii,* was published more recently by Tomaras et al. [54]. In this study, the bacteria were grown in Petri dishes without shaking.

1. After the appropriate time period, the dishes were flooded with 2.5% glutaraldehyde in 0.05 M cacodylate buffer.
2. This was incubated at room temperature for 2 hr.
3. The glutaraldehyde was removed and the cells were washed with distilled water.
4. Samples of 1 × 1 cm were cut from the Petri dish and rinsed gently with distilled water.
5. The preparation was treated with an ethanol series of washes as above.
6. The samples were then CO_2–critical-point dried for 60 to 120 min.
7. The samples were then gold-coated.

Scanning Confocal Laser Microscopy (SCLM) [57–61]

With SCLM it is possible to view live samples that have not been distorted by fixation. This has allowed for a much better understanding of the three-dimensional organization of biofilms. In addition, SCLM has been an effective experimental tool in conjunction with *in situ* hybridization [62, 63].

Sample Preparation and Analysis

Treatment of the sample in order to view it using SCLM depends on the type of instrument used:

1. **Upright microscope:** With this type of microscope, the samples are mounted with a material such as acid-free silicone glue.
   a. Grow biofilm overnight in the appropriate medium on a 35-mm, collagen-coated, glass coverslip that is sealed to the bottom of a culture dish [61]. An alternative is to use a poly-L-lysine-coated coverslip [60] glued to an appropriate vessel.
   b. Remove the medium and wash once with water or 0.85% saline.
   c. Mount the coverslip.
2. **Inverted microscope:**
   a. Grow biofilm on an appropriate substratum. Flat surfaces are better because they are easily mounted and stained. Flow cells and the disks of the modified Robbins device are well suited for this.
   b. Wash the biofilm in 0.85% saline or the medium of growth.
   c. The use of an inverted microscope dictates that the growth chamber is mounted upside-down using a coverslip (sealed around with nail polish) at the bottom [58]. This way, the laser light penetrates the coverslip from the bottom of the biofilm to the top, producing an inverted image [61].
3. **Staining:**
   a. There are a variety of fluors for nucleic acids. Positive stains include SYTO dyes, acridine orange, and the DAPI stain and propidium iodide [57]. Fluorescein can be used for either negative or positive staining, depending on the pH [43]. Depending on the experimental system, the amount of fluor added must be empirically determined.
   b. When more than one fluor stain is used, factors such as relative intensity, emission characteristics, photostability, interference, additive effects, and quenching effects must be considered [57].

4. **The mountant:** The primary purposes served by the mountant are to preserve the sample and prevent photobleaching. Dimethylsulfoxide (DMSO) is often used with bacterial samples, and mixed with the fluor before its addition to the sample.
   a. The mountant should have a pH that is compatible with the optimum pH for the fluor.
   b. The mountant must have an additive such as DMSO to prevent photobleaching.
   c. The refractive index of the mountant should be as close to that of the sample as possible. For bacteria, this is 1.517 [58].
5. **Magnification:** This is much less than with SEM and is 40X or 100X with an oil immersion lens. Another option is a 63X water immersion objective lens [61]. The sample (with the fluor) is then excited with light of appropriate wavelength, causing it to emit light (fluorescence).

6. **Imaging:**
   a. After locating suitable fields (usually with phase contrast or epi-fluorescence), the appropriate excitation and emission filters are set before scanning the sample.
   b. Either a single thin section (xy plane) is scanned or a series of xy optically thin sections through the biofilm is scanned. This is called a z series or section. It is also referred to as sagittal imaging [43]. The distance between the z sections depends on the thickness of the biofilm. For example, if the biofilm is approximately 30 to 40 μ thick, then z-slices are obtained every 2 μ [61]. Another option is to collect a single or a series of xz vertical sections through the specimen [57].
   c. These two-dimensional (2D) cross-sectional images are stacked using the SCLM software (provided by the manufacturer) to get 3D information. With the quantitative extraction of information contained in each image, a three-dimensional reconstruction of the principal biological events can be achieved [59].
7. **Image processing and analysis:**
   a. The SCLM images may require basic processing and enhancement before analysis. These include histogram analysis, gray-level transformation, normalization, contrast enhancement, application of median or lowpass filters or image addition, or multiplication. These are done using specialized digital image processing and analysis software. The processing applied also depends on the information desired. For example, when two different fluors are used (e.g., green and red), images of the biofilms can be converted to Adobe Photoshop files [59]. A region near the center of the biofilm is selected, a histogram of the region obtained, and the mean intensity value of the green pixels is divided by the mean intensity of the red pixels to obtain a semiquantitative measure of the accumulation of the fluorophore in the biofilm. Ratios of green/red intensity values obtained for the same region at different time points can be used to calculate the rate of fluorophore transport.
   b. Alternatively, images can be generated interactively by calling appropriate command-line scripts using the dialog and menu facilities of the SCLM software (provided by the manufacturer). Then digital image processing and analysis are performed with analytical packages such as the QUANTIMET 570 (Leica, Cambridge, United Kingdom) computer system, which enhances images, discards unwanted details, and accelerates gray-scale processing [57].
   c. Computerized image analysis comprises a sequence of operations that depend on the specimen being studied. Before the automated image processing is activated, pixel calibration is done through the QUANTIMET program. The size and position of the area of interest for measurement are specified, and the computer is then programmed to perform a sequence of automated operations such as image acquisition, gray image analysis, detection, measurements of signal intensity, volume integration, and data recording and display [57, 59].

d. Deconvolution can also be applied to SCLM images to sharpen the image by mathematical removal of "out-of-focus" information. All these functions help in smoothing the image, reducing the noise, and sharpening the image, thus defining the objects more accurately.

8. **Image output:** After obtaining the images, they are stored as TIFF, GIFF, PICT, or JPEG, depending on the degree of image fidelity and the degree of image compression required.

Methods for Examining the Effect of Antibiotics or Biocides on Biofilms

It is now accepted that bacterial growth in natural settings is most commonly in the form of biofilms. From a clinical perspective this presents a problem because biofilms grow on indwelling medical devices, particularly central venous catheters [29], and may be involved in the persistence of chronic infections [19]. A common experience is that the biofilms are considerably more resistant to antibiotics and biocides than planktonic cells [18, 64, 65]. Consequently, methods have been developed to examine the inhibition or killing of biofilms by antimicrobials, which has led to the term "minimal biofilm eradication concentrations" (MBECs) [64, 65]. These include the use of the modified Robbins device, the Calgary Biofilm Device, and simpler methods.

The Modified Robbins Device [18, 66]

As detailed above, the modified Robbins device was designed primarily to measure biofilm formation. However, it can be used to determine the ability of antibiotics to inhibit the growth or kill cells in a biofilm [18, 66].

Procedure

Grow cells in the modified Robbins device to form a biofilm as described above.

1. After the biofilm has formed, open the valve of a parallel reservoir containing antibiotic in the same medium.
2. After the appropriate period of antibiotic exposure, aseptically scrape the cells on a disk (urinary catheter) from a specimen plug into a tube containing sterile phosphate-buffered saline.
3. Subject the cells along with the disk to low-output sonication to disperse the cells.
4. Dilute the cells as needed and plate on nutrient agar for plate counting.

The Calgary Biofilm Device [6, 64, 65, 67]

Although the modified Robins device could be used to determine the susceptibility of biofilms to antibiotics, it was deemed less than optimal for rapid antibiotic susceptibility testing for clinical purposes [64]. The Calgary biofilm device is relatively simple in that it consists of two parts that are composed of plastic. One is the bottom, which contains 96 troughs (channels) for bacterial growth, and the other is a lid into which pegs are placed to fit into the troughs. The biofilms grow on the pegs. The pegs will also fit into standard 96-well polystyrene microtiter plates [65]. Ceri et al. [64, 65] demonstrated that the device can produce 96 equivalent biofilms and is highly reproducible. Advantages of this system are that it requires no pumps or tubing, thereby reducing problems of contamination; allows for multiple testing; has a platform that uses standard 96-well plates; and is considerably faster than the modified Robbins device [64, 65]. This system is also referred to as the MBEC (minimal biofilm eradication concentration) Assay System, and can be obtained from MBEC Biofilms Technologies Limited, Calgary, Alberta, Canada.

Procedure

1. Establish biofilm conditions:
   a. Add a standardized inoculum to the channels in the bottom plate using TSB as the growth medium.
   b. Attach the lid and place on a rocking table. The rocking action creates shear forces that promote biofilm formation. For aerobic cultures, the temperature is usually 35°C with 95% relative humidity.
   c. At selected times, remove duplicate pegs, place in 200 μL sterile buffer, and sonicate for 5 min using a sonic cleaning system.
   d. Plate on an appropriate medium to determine viable counts.
2. Susceptibility testing:
   a. After generating a standardized biofilm, then its susceptibility to antibiotics or biocides can be tested.
   b. Prepare working solutions of antibiotics in cation-adjusted Mueller–Hinton broth (CAMHB) to a concentration of 1024 μg mL^{-1}.
   c. Serially dilute the antibiotics in CAMHB and add to the wells of a 96-well plate.
   d. Wash the pegs (in the lid) on which biofilms have grown in a wash tray containing sterile buffer.
   e. Place the lid (with the pegs) over the wells to expose the pegs with the biofilms to the antibiotics. The assembly is usually incubated overnight.
   f. Determine the MBEC value (the minimal concentration of antibiotic that prevents growth of the biofilm).
3. Take the lid with the pegs and wash as above.
4. Then place the lid over a 96-well plate in which the wells contain recovery medium.
5. Sonicate as above to disperse the cells into the recovery medium.
6. Determine the viable cell number by plating onto the medium of choice.
7. Alternatively, measure the turbidity in the wells using a 96-well plate reader.

Other Methods for Assessing the Effects of Antibiotics or Biocides on Biofilm Viability [21, 68, 69]

Method 1

Stone et al. [68] examined the effect of tetracycline on biofilms of uropathogenic *Escherichia coli* grown in Plexiglas flow cells.

1. For enumeration, first aseptically scrape the biofilm into 400 μL of 0.1 M phosphate buffer at pH 7.0.
2. Vigorously vortex the cells.
3. Serially dilute the suspensions and plate on LB agar for enumeration.

Method 2

Anderl et al. [69] examined the effect of ampicillin and ciprofloxacin on *Klebsiella pneumoniae*. In this study the biofilms were propagated on polycarbonate filters.

1. To enumerate the biofilm, first immerse the membrane in 9 mL phosphate-buffered water.
2. Vigorously vortex the membrane to suspend the cells.
3. Plate serial dilutions on R2A agar by the drop plating method [70].

Method 3

Elkins et al. [21] studied the killing of planktonic cells and biofilms of *Pseudomonas aeruginosa* by hydrogen peroxide. The biofilms were generated using a drip-flow reactor with stainless steel slides as the substratum.

1. To enumerate the biofilms, remove the slides and scrape aseptically into 50 mL phosphate-buffered saline (pH 7.2) containing 0.2% sodium thiosulfate to neutralize the hydrogen peroxide.
2. To suspend the cells, homogenize the above mixture in a PT 10/35 Brinkman homogenizer for 15 sec at setting 4.
3. Appropriately dilute the homogenate and plate on R2A agar medium.

REFERENCES

1. Prescott, L.M., Harley, J.P., and Klein D.A., *Microbiology*, 6th ed., McGraw-Hill, New York, 2005, chap. 1.
2. Enge, R.S., Early developments in medical microbiology, *SCIEH Weekly Report*, 34, 7, 2000.
3. Costerton, J.W., A short history of the development of the biofilm concept, in *Microbial Biofilms*, Ghannoum, M. and O'Toole, G.A., Eds., ASM Press, Washington, DC, 2004, chap. 1.
4. Costerton, J.W. et al. [Authors are Costerton, Cheng, Geesy, Ladd, Nickel, Dasgupta, and Marrie], Bacterial biofilms in nature and disease, *Annu. Rev. Microbiol.,* 41, 435, 1987.
5. Costerton, J.W. et al. [Authors are Costerton, Lewandowski, De Beer, Caldwell, Korber, and James], Biofilms, the customized microniche, *J. Bacteriol.*, 176, 2137, 1994.
6. McLean, R.J.C. et al. [Authors are McLean, Bates, Barnes, McGowin, and Aron], Methods of studying biofilms, in *Microbial Biofilms,* Ghannoum, M. and O'Toole, G.A., Eds., ASM Press, Washington, DC, 2004, chap. 20.
7. Loeb, G.I., Measurement of microbial marine fouling films by light section microscopy, *Mar. Technol. Soc. J.*, 14, 14, 1980.
8. Marrie, T., Nelligan, J., and Costerton, J.W., A scanning and transmission electon microscopic study of an infected endocardial pacemaker lead, *Circulation*, 66, 1339, 1982.
9. Kinner, N.E., Balkwell, D.L., and Bishop, P.L., Light and electron microscopic studies of microorganisms growing in rotating biological contactor biofilms, *Appl. Environ. Microbiol.*, 45, 1659, 1983.
10. Geesy, G.G. et al. [Authors are Geesy, Richardson, Yeomans, Irvin, and Costerton], Microscopic examination of natural sessile bacterial populations from an alpine stream, *Can. J. Microbiol.,* 23, 1733, 1977.
11. Lawrence, J.R. et al. [Authors are Lawrence, Korber, Hoyle, Costerton, and Caldwell], Optical sectioning of microbial biofilms, *J. Bacteriol.*, 173, 6558, 1991.
12. Andersen, J.B. et al. [Authors are Andersen, Sternberg, Poulson, Bjorn, Givskov, and Molin], New unstable variants of green fluorescent protein for studies of transient gene expression in bacteria, *Appl. Environ. Microbiol.*, 64, 2420, 1998.
13. de Beer, D., Stoodley, P., and Lewandowski, Z., Effects of biofilm structure on oxygen distribution and mass transport, *Biotech. Bioeng.*, 44, 636, 1994.
14. Potera, C., Studying slime, *Environ. Health Perspect.*, 106, A604, 1998.
15. Stewart. P.S., Murherjee, P.K. and Ghannoum, M.A., Biofilm antimicrobial resistance, in *Microbial Biofilms*, Ghannoum, M. and O'Toole, G.A., Eds., ASM Press, Washington, DC, 2004.
16. Zheng, Z. and Stewart, P.S., Penetration of rifampin through *Staphylococcus epidermis* biofilms, *Antimicrob. Agents Chemother.*, 46, 900, 2002.
17. Walters, M.C. et al. [Authors are Walters, Roe, Bugnicourt, Franklin, and Stewart], Contributions of antibiotic penetration, oxygen limitation, and low metabolic activity to the tolerance of *Pseudomonas aeruginosa* biofilms to ciprofloxacin and tobramycin, *Antimicrob. Agents Chemother.*, 47, 317, 2003.
18. Nickel, J.C. et al. [Authors are Nickel, Ruseska, Wright, and Costerton], Tobramycin resistance of cells of *Pseudomonas aeruginosa* growing as a biofilm on urinary catheter material, *Antimicrob. Agents Chemother.*, 27, 619, 1985.
19. Donlan, R.M. and Costerton, J.W., Biofilms: survival mechanisms of clinically relevant microorganisms, *Clin. Microbiol. Rev.*, 15, 167, 2002.

20. Costerton, J.W., Stewart, P.S., and Greenberg, E.P., Bacterial biofilms: a common cause of persistent infections, *Science*, 284, 1318, 1999.
21. Elkins, J.G. et al. [Authors are Elkins, Hassett, Stewart, Schweizer, and McDermott], Protective role of catalase in *Pseudomonas aeruginosa* biofilm resistance to hydrogen peroxide, *Appl. Environ. Microbiol.*, 65, 4594, 1999.
22. Hawser S.P and Douglas, L.J., Resistance of *Candida albicans* biofilms to antifungal agents *in vitro*, *Antimicrob. Agents Chemother.*, 39, 2128, 1995.
23. Chandra, J.P. et al. [Authors are Chandra, Mukherjee, Leidich, Faddoul, Hoyer, Douglas, and Ghannoum], Antifungal resistance of *Candida* biofilms formed on denture acrylic *in vitro*, *J. Dent. Res.*, 80, 903, 2001.
24. Kuhn, D.M. et al. [Authors are Kuhn,George, Chandra, Mukherjee, and Gannoum], Antifungal susceptibility of *Candida* biofilms: unique efficacy of amphotericin B lipid formulations and echinocandins, *Antimicrob. Agents Chemother.*, 46, 1773, 2002.
25. Banerjee, S.N. et al. [Authors are Banerjee, Emori, Culver, Gaynes, Jarvis, Horan, Edwards, Tolson, Henderson, and Martone], Secular trends in nosocomial primary bloodstream infections in the United States 1980–1989: National Nosocomial Infections Surveillance Systems, *Am. J. Med.*, 91, 86S, 1991.
26. Zhang, T.C., Fu, Y.-C., and Bishop, P.L, Competition for substrate and space in biofilms, *Water Environ. Res.*, 67, 992, 1995.
27. Spoering, A.L., and Lewis, K., Biofilms and planktonic cells of *Pseudomonas aeruginosa* have similar resistance to killing by antimicrobials, *J. Bacteriol.*, 183, 6746, 2001.
28. Thomas, J.G., Ramage, G., and Lopez-Ribot, J.L., Biofilms and implant infections, in *Microbial Biofilms*, Gannoum, M and O'Toole, G.A., Eds., ASM Press, Washington, DC, 2004, chap. 15.
29. Maki, D.G., Infections caused by intravascular devices used for infusion therapy: pathogenesis, prevention, and management, in *Infections Associated with Indwelling Medical Devices, 2nd edition*, Bisno, A.L. and Waldovogel, F.A., Eds., ASM Press, Washington, DC, 1994, p. 155.
30. Nguyen, M.H. et al. [Authors are Nguyen, Peacock, Tanner, Morris, Nguyen, Snydman, Wagener, and Yu], Therapeutic approaches in patients with candidemia: evaluation in a multicenter prospective observational study, *Arch. Intern. Med.*, 155, 2429, 1995.
31. Raad, I.I., Intravascular-catheter-related infections. *Lancet*, 351, 893, 1998.
32. Merritt, J.H., Kadori, D.E. and O'Toole, G.A., Growing and analyzing static biofilms, *Current Protocols in Microbiology*, John Wiley & Sons, Inc., New York, 205, p1B-1.1, 2005.
33. Christensen, G.D., et al. [Authors are Christensen, Simpson, Younger, Badour, Barrett, Melton, and Beachy], Adherence of coagulase negative staphylococci to plastic tissue culture plates: a quantitative model for the adherence of staphylococci to medical devices, *J. Clin. Microbiol.*, 22, 996, 1985.
34. Mack, D. et al. [Authors are Mack, Nedelmann, Krokotsch, Schwarzkopf, Heesemann, and Laufs], Characterization of transposon mutants of biofilm-producing *Staphylococcus epidermidis* impaired in the accumulative phase of biofilm production: Genetic identification of a hexosamine-containing polysaccharide intracellular adhesion, *Infect. Immun.*, 62, 3244, 1994.
35. O'Toole, G.A. and Kolter, R., Initiation of biofilm formation in *Pseudomonas fluorescens* WCS365 proceeds via multiple, convergent signaling pathways: a genetic analysis, *Mol. Microbiol.*, 28, 449, 1998.
36. Danhorn, R. et al. [Authors are Danhorn, Hentzer,. Givskov, Parsek, and Fuqua], Phosphorous limitation enhances biofilm formation of the plant pathogen *Agrobacterium tumefaciens* through the PhoR-PhoB regulatory system, *J. Bacteriol.*, 186, 4492, 2004.
37. Merritt, J., et al. [Authors are Merritt, Qi, Goodman, Anderson, and Shi], Mutation of *luxS* affects biofilm formation in *Streptococcus mutans*, *Infect. Immun.*, 71, 1972, 2003.
38. Shi, W., A. Jewett, A., and Hume, W.R., Rapid and quantitative detection of *Streptococcus mutans* with species specific monoclonal antibodies, *Hybridoma*, 17, 365, 1998.
39. Raad, I., Management of intravascular catheter-related infection, *J. Antimicrob. Chemother.*, 45, 267, 2000.
40. Donlan, R.M., Biofilms and device associated infection, *Emerg. Infect. Dis.*, 7, 277, 2001.
41. Christensen, G.D. et al. [Authors are Christensen, Simpson, Bisno, and Beachey], Adherence of slime-producing strains of *Staphylococcus epidermidis* to smooth surfaces, *Infect. Immun.*, 37, 318, 1982.
42. Mathur, T. et al. [Authors are Mathur, Singhal, Khan, Upadhyay, Fatma, and Ratta], Detection of biofilm formation among the clinical isolates of staphylococci: an evaluation of three different screening methods, *Indian J. Med. Microbiol.*, 24, 25, 2006.
43. Wolfaardt, G.M. et al. [Authors are Wolfaardt, Lawrence, Robarts, Caldwell, and Caldwell], Molecular organization in a degradative biofilm community., *Appl. Environ. Microbiol.*, 60, 434, 1994.

44. Christensen, B.B. et al. [Authors are Christensen, Sternberg, Andersen, Palmer, Nielsen, Givskov, and Molin], Molecular tools for study of biofilm physiology. *Methods Enzymol.*, 310, 20, 1999.
45. McCoy, W.F. et al. [Authors are McCoy, Bryers, Robbins, and Costerton], Observations of fouling biofilm formations, *Can. J. Microbiol.*, 27, 910, 1981.
46. McCoy, W.F. and Costerton, J.W., Fouling biofilm development in tubular flow systems, *Dev. Indust. Microbiol.*, 23, 551, 1982.
47. Costerton, J.W. and Lashen, E.S., Influence of biofilm on efficacy of biocides on corrosion-causing bacteria, *Mater. Perform.*, 23, 34, 1984.
48. Domingue, G. et al. [Authors are Domingue, Ellis, Dasgupta, and Costerton], Testing antimicrobial susceptibilities of adherent bacteria by a method that incorporates guidelines in the national committee for clinical laboratory standards, *J. Clin. Microbiol.*, 32, 2564, 1994.
49. Millar, M.R., Linton, C.J., and Sherriff, A., Use of a continuous culture system linked to a modified Robbins device or flow cells to study attachment of bacteria to surfaces, *Methods Enzymol.*, 337, 43, 2001.
50. Marrie, T. J. and Costerton, J.W., Scanning electron microscopic study of uropathogen adherence to a plastic surface, *Appl. Environ.Microbiol.*, 45, 1018, 1983.
51. Marrie, J.T. and Costerton, J.W., Scanning and transmission electron microscopy of *in situ* bacterial colonization of intravenous and intraarterial catheters, *J. Clin. Microbiol.*, 19, 687, 1984.
52. Ladd, T.I. et al. [Authors are Ladd, Schmiel, Nickel, and Costerton], Rapid method for detection of adherent bacteria on Foley urinary catheters, *J. Clin. Microbiol.*, 21, 1004, 1985.
53. Kumo, H. et al. [Authors are Kumo, Ono, Iida, and Nickel], Combination effect of fosfomycin and ofloxacin against *Pseudomonas aeruginosa* growing in a biofilm, *Antimicrob. Agents Chemother.*, 39, 1038, 1995.
54. Tomaras, A.P. et al. [Authors are Tomaras, Dorsey, Edelmann, and Actis], Attachment to and biofilm formation on abiotic surfaces by *Acinetobacter baumanii*: involvement of a novel chaperone-usher pili assembly system, *Microbiology*, 149, 3473, 2003.
55. Malik, L.E. and Wilson, B.W., Modified thiocarbohydrazide procedure for scanning electron microscopy: routine use for normal, pathological or experimental tissues, *Stain Tech.*, 50, 265, 1975.
56. Kumon, H., T. Kaneshige, and H. Omori., T_4-labeling technique and its application with particular reference to blood group antigens in bladder tumors, *Scanning Electron Microsc.,* II, 939, 1983.
57. Lawrence, J.R. and Neu, T.R., Confocal laser scanning microscopy for analysis of microbial biofilms, *Methods Enzymol.*, 310, 131, 1999.
58. http://depts.washington.edu/keck/intro.htm, Confocal introduction.
59. Kuehn, M. et al. [Authors are Kuehn, Hausner, Bungartz, Wagner, Wilderer, and Wuertz], Automated confocal laser scanning microscopy and semiautomated image processing for the analysis of biofilms, *Appl. Environ. Microbiol.*, 64, 4115, 1998.
60. Cowan, E.S. et al. [Authors are Cowan, Gilbert, Khlebnikov, and Keasling], Dual labeling with green fluorescent proteins for confocal microscopy, *Appl. Environ. Microbiol.*, 66, 4113, 2000.
61. Jefferson, K.K., Goldmann, D.A., and Pier, G.B., Use of confocal microscopy to analyze the rate of vancomycin penetration through *Staphylococcus aureus* biofilms, *Antimicrob. Agents Chemother.*, 49, 2467, 2005.
62. Amann, R. et al. [Authors are Amann, Snaid, Wagner, Ludwig, and Schleifer], *In situ* visualization of high genetic diversity in a natural microbial community, *J. Bacteriol.*, 178, 3496, 1996.
63. Miller, S. et al. [Authors are Miller, Pedersen, Poulsen, Carstensen, and Molin], Activity and three-dimensional distribution of toluene-degrading *Pseudomonas putida*, in a multispecies biofilm assessed by quantitative in situ hybridization and scanning confocal laser microscopy, *Appl. Environ. Microbiol.*, 62, 4632, 1996.
64. Ceri, H. et al. [Authors are Ceri, Olson, Stremick, Read, Morck, and Buret], The Calgary biofilm device: new technology for rapid determination of antibiotic susceptibilities of bacterial biofilms, *J. Clin. Microbiol.*, 37, 1771, 1999.
65. Ceri, H. et al. [Authors are Ceri, Olson, Morck, Storey, Read, Buret, and Olson], The MBEC assay system: multiple equivalent biofilms for antibiotic and biocide susceptibility testing, *Methods Enzymol.*, 337, 377, 2001.
66. Raad, I. et al. [Authors are Raad, Darouiche, Hachem, Sacilowski, and Bodey], Antibiotics and prevention of microbial colonization of catheters, *Antimicrob. Agents Chemother.*, 39, 2397, 1995.

67. Bardouniotis, E. et al. [Authors are Bardouniotis, Huddleston, Ceri, and Olson], Characterization of biofilm growth and biocide susceptibility testing of *Mycobacterium phlei* using the MBEC® assay system, *FEMS Microbiol. Lett.*, 203, 263, 2001.
68. Stone, G. et al. [Authors are Stone, Wood, Dixon, Keyhan, and Matin], Tetracycline rapidly reaches all the constituent cells of uropathogenic *Escherichia coli* biofilms, *Antimicrob. Agents Chemother.*, 46, 2458, 2002.
69. Anderl, J.N., Franklin, M.J., and Stewart, P.S., Role of antibiotic penetration limitation in *Klebsiella pneumoniae* biofilm resistance to ampicillin and ciprofloxacin, *Antimicrob. Agents Chemother.*, 44, 1818, 2000.
70. Hoben, H.J. and Somasegaran, P., Comparison of the pour, spread, and drop plate methods for enumeration of *Rhizobium* spp. in inoculants made from presterilized peat, *Appl. Environ. Microbiol.*, 44, 1246, 1982.

43 Introduction to Bacteriophages*

Elizabeth Kutter and Emanuel Goldman

CONTENTS

INTRODUCTION

The Nature of Bacteriophages

Bacteriophages are viruses that only infect bacteria. They are like complex spaceships (Figure 43. 1), each carrying its genome from one susceptible bacterial cell to another in which it can direct the production of more phage. Each phage particle (virion) contains its nucleic acid genome (DNA or RNA) enclosed in a protein or lipoprotein coat, or capsid; the combined nucleic acid and capsid form the nucleocapsid. The target host for each phage is a specific group of bacteria. This group is often some subset of one species, but several related species can sometimes be infected by the same phage.

* Much of this text is excerpted from Chapters 3 and 7 in *Bacteriophages: Biology and Applications.*[1, 2]

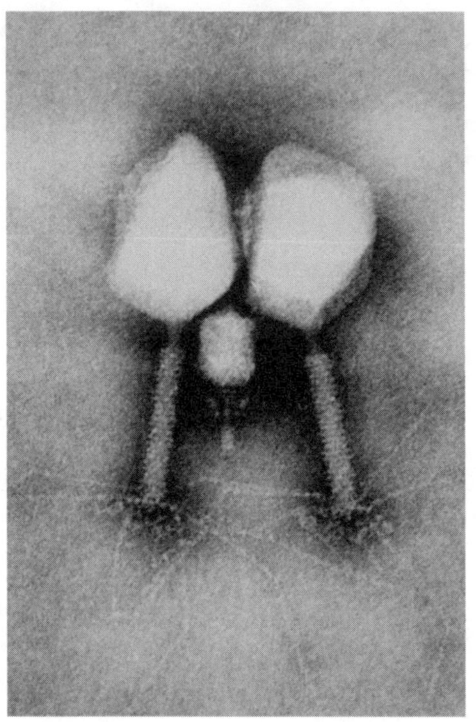

FIGURE 43.1 A "family portrait": bacteriophages φ29 and T2. (Electron micrograph provided by Dwight Anderson.)

Phages, like all viruses, are absolute parasites. Although they carry all the information to direct their own reproduction in an appropriate host, they have no machinery for generating energy and no ribosomes for making proteins. They are the most abundant living entities on Earth, found in very large numbers wherever their hosts live — in sewage and feces, in the soil, in deep thermal vents, and in natural bodies of water. Their high level of specificity, long-term survivability, and ability to reproduce rapidly in appropriate hosts contribute to their maintenance of a dynamic balance among the wide variety of bacterial species in any natural ecosystem. When no appropriate hosts are present, many phages can maintain their ability to infect for decades, unless damaged by external agents.

Some phages have only a few thousand bases in their genome, whereas phage G, the largest sequenced to date, has 498,000 basepairs — as much as an average bacterium, though still lacking the genes for such essential bacterial machinery as ribosomes. Over 95% of the phages described in the literature to date belong to the Caudovirales (tailed phages). Their virions are approximately half double-stranded DNA and half protein by mass, with icosahedral heads assembled from many copies of a specific protein or two. The corners are generally made up of pentamers of a protein, and the rest of each side is made up of hexamers of the same or a similar protein. The three families are defined by their very distinct tail morphologies: (1) 60% of the characterized phages are Siphoviridae, with long, flexible tails; (2) 25% are Myoviridae, with double-layered, contractile tails; and (3) 15% are Podoviridae, with short, stubby tails. The latter may have some key infection proteins enclosed inside the head that can form a sort of extensible tail upon contact with the host, as shown most clearly for coliphage T7.[3] Archaea have their own set of infecting viruses, often called archaephages. Many of these have unusual, often pleiomorphic shapes that are unique to the Archaea. However, many viruses identified to date for the Crenarchaeota kingdom of Archaea look like typical tailed bacteriophages.[4] See Chapter 41 for more information on Archaea.

The ten families of tailless phages described to date each have very few members. They are differentiated by shape (rods, spherical, lemon-shaped, or pleiomorphic); by whether they are enveloped in a lipid coat; by having double- or single-stranded DNA or RNA genomes, segmented or not; and by whether they are released by lysis of their host cell or are continually extruded from the cell surface. Their general structures, sizes, nucleic acids, adsorption sites, and modes of release are described in detail in recent books,[1, 5] as well as the basic infection processes of the Inoviridae, Leviviridae, Microviridae, and Tectiviridae. Relatively little is known about most of the others, which generally have been isolated under extremes of pH, temperature, or salinity and have only been observed in Archaea.

Phages can also be divided into two classes based on lifestyle: virulent or temperate. Virulent phages can only multiply by means of a lytic cycle; the phage virion adsorbs to the surface of a host cell and injects its genome, which takes over much of host metabolism and sets up molecular machinery for making more phages. The host cell then lyses minutes or hours later, liberating many new phages. Temperate phages, in contrast, have a choice of reproductive modes when they infect a new host cell. Sometimes the infecting phage initiates a lytic cycle, resulting in lysis of the cell and release of new phage, as just described. The infecting phage may alternatively initiate a lysogenic cycle; instead of replicating, the phage genome assumes a quiescent state called a prophage, generally integrated into the host genome but sometimes maintained as a plasmid. It remains in this condition indefinitely, being replicated as its host cell reproduces to make a clone of cells all containing prophages; these cells are said to be lysogenized or lysogenic (i.e., capable of producing lysis) because one of these prophages occasionally comes out of its quiescent condition and enters the lytic cycle. The factors affecting the choice to lysogenize or to reenter into a lytic cycle are described below. As discussed by Levin and Lenski,[6] the lysogenic state is highly evolved, requiring co-evolution of virus and host that presumably reflects various advantages to both. Temperate phages can help protect their hosts from infection by other phages and can lead to significant changes in the properties of their hosts, including restriction systems and resistance to antibiotics and other environmental insults. As discussed later in this chapter, they may even convert the host to a pathogenic phenotype, as in diphtheria or enterohemorrhagic *Escherichia coli* (EHEC) strains. Bacteriophages lambda (λ), P1, mu, and various dairy phages are among the best-studied temperate phages. (Note that mutation of certain genes can create virulent derivatives of temperate phages; these are still considered members of their temperate phage families, especially since they can pick up various genes through recombination with their integrated relatives.)

The larger virulent phages generally encode many host-lethal proteins. Some of them disrupt host replication, transcription, or translation; they may also degrade the host genome, destroy or redirect certain host enzymes, or alter the bacterial membrane. The temperate phages, in contrast, generally do much less restructuring of the host, and they carry few if any host-lethal proteins that would need to be kept under tight control during long-term lysogeny. They always encode a repressor protein, which acts at a few operator sites to block transcription of other phage genes. This repressor may be the only phage-encoded protein produced during the lysogenic state, but often a few other genes that may be beneficial to host survival are also expressed from the prophages. The repressor also blocks lytic infection by other phages of the same immunity group — that is, other phages whose genes can be regulated by the same repressor. In this way, a temperate phage generally protects its host bacterium from infection by several kinds of phages.

Antigenic Properties of Phages

Phages, like other protein structures, can elicit an antibody response, and antiphage antibodies have been used in a variety of ways since early phage researchers first showed that rabbits injected with phage lysates produced phage-neutralizing antibody; in 1926 d'Herelle,[7] and in 1959 Adams,[8] reviewed early studies in the field. For most phages, the neutralization follows first-order kinetics to an inactivation of about 99%, with the survivors then generally being more resistant and often having smaller plaque size. The activity of a given preparation at a standard temperature and salt concentration is denoted by a velocity constant *K*, which is related to the log of the percent of phage still infectious after a given exposure time *t*. As emphasized by Adams, such factors as salt concentration are very important in titering antibody activity, with monovalent cations above 10^{-2} M and divalent cations above 10^{-3} M being inhibitory. He cites an example of anti-T4 serum with an inactivation constant (*K* value) of over 105 min^{-1} under optimal conditions but only 600 min^{-1} in broth. Also, somewhat surprisingly, different assay hosts may give different estimates of the fraction of non-neutralized phage. Adams suggested this may be due to minor differences in the receptors recognized on various hosts, leading to differences in bonding strength. We also know now that some phages can use totally different receptors on different hosts, as described for T4 in the section below entitled "Adsorption." Once determined for a given set of conditions, the *K* value can be used to calculate the appropriate dilution for any given experiment.

With the aid of noted Russian electron microscopist Tamara Tikhonenko,[9] such techniques as immunoelectronmicroscopy were used to carry out particularly detailed studies of the process of inactivation by antibodies.[10] Implications of phage antigenicity for human therapy are discussed in Chapter 44.

Susceptibility of Phages to Chemical and Physical Agents

Phages vary greatly in their sensitivity to various chemical and physical agents, in ways that are generally unpredictable and need to be determined experimentally in each case. For example, no one has yet explained the stability of phage T1 to drying, a stability that has led to much grief in the biotech industry as well as in phage labs from contamination by ubiquitous T1-like phages. There are, however, certain general principles. For example, all phages are very susceptible to UV light in the range of 260 nm as well as in the far-UV; in addition to the general effects of sunlight, there are multiple reports of phage collections being lost to fluorescent-lighted refrigerators. Other factors potentially affecting phage viability include pH, ascorbic acid, urea, urethane, detergents, chelating agents, mustard gas, alcohols, and heat inactivation. The most extensive summary of the known effects of chemical and physical agents on phages is found in Adams;[8] there is also very good information in Ackermann and Dubow.[11]

Phages generally are stable at pH 5 to 8, and many are stable down to pH 3 or 4; each phage needs to be specifically characterized. Phages are often quite sensitive to protein-denaturing agents such as urea and urethane, but the level of inactivation depends on both concentration and temperature, and differs for different phages. Not surprisingly, detergents generally have far less effect on phages than they do on bacteria, although the few phages that are enveloped in membranes are quite susceptible, while chelating agents have strong effects on some phages but not on others; this apparently depends mainly on cofactor requirements for adsorption. Chloroform has little or no effect on non-enveloped phages; in fact, one needs to be very careful not to get phage contaminants into the chloroform being used to produce phage lysis. However, contaminants toxic to phages in high concentrations are found in some commercial chloroform. Mutagenic agents such as mustard gas, nitric oxide, and UV light inactivate phages and can induce the lytic cycle in many lysogens. Lytic phages can generally still infect cells recently inactivated with UV or other mutagens, because they do not require ongoing synthesis of host proteins after infection. In fact, UV inactivation of the host just prior to infection is sometimes used to virtually eliminate residual synthesis of host proteins for experiments using pulse labeling to examine patterns of phage protein synthesis.

INTERACTION OF PHAGE WITH THE HOST CELL

Virtually every structure on the outer surface of a bacterial cell is likely to serve as a receptor for some phage. See Chapter 13 for a discussion of bacterial cell walls.

The properties of the distinctive outer membrane of Gram-negative cells have been reviewed in detail by Nikaido[12, 13] and Ghuysen,[14] and are shown in Figure 43.2a. The inner face of the outer membrane is attached the peptidoglycan layer and has a phospholipid composition similar to that of the cytoplasmic membrane. In contrast, the lipid part of the outer face is mainly a unique substance called lipopolysaccharide (LPS), seen nowhere else in nature. LPS is composed of three parts: a hydrophobic lipid A membrane anchor, connected by a core polysaccharide to a complex distal polysaccharide O-antigen. The O-antigens play important roles in bacterial interactions with mammalian hosts and in virulence; they are indicated in strain names, such as *Escherichia coli* O157. They can also serve as efficient phage receptors; obviously, such phages are likely to have very narrow host ranges. Many common lab strains, including *E. coli* K-12, lack any O-antigen; *E. coli* B, used extensively for phage work, even lacks the more distal part of the LPS core, but they are found in most clinical strains. The outer membrane also contains several families of general porins — proteins that form large β-barrel channels with charged central restrictions that support non-

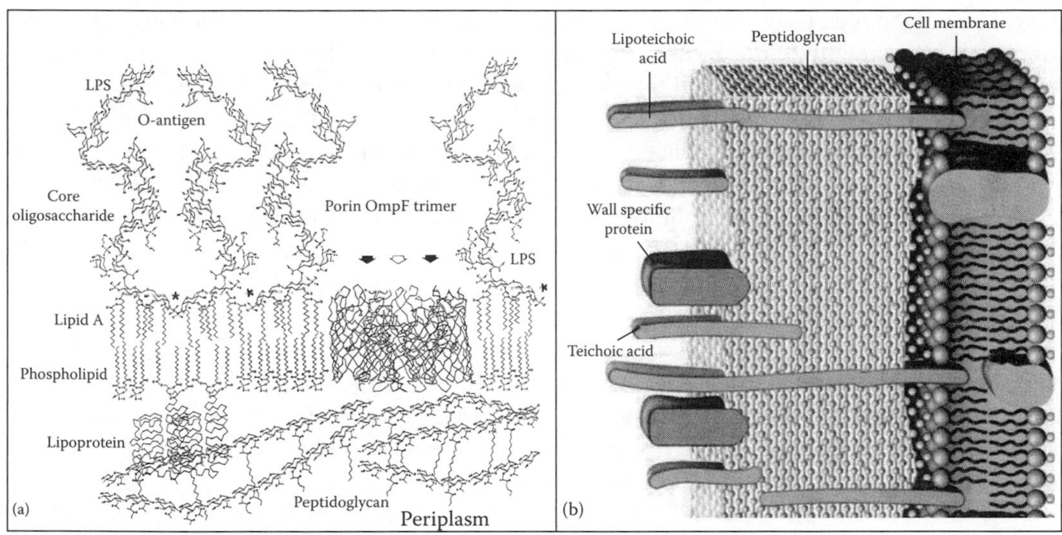

FIGURE 43.2 (a) Details of the outer membrane structures of Gram-negative bacteria. (With permission from Hancock, R.E.W., Karunaratne, D.N., and Bernegger-Egli, C. Molecular Organization and Structural Role of Outer Membrane Macromolecules, *The Bacterial Cell Wall*, J.M. Ghuysen, Hackenbeck, R., Eds., Elsevier, Amsterdam, 1994, 263–279.) (b) General structure of the cell wall of Gram-positive bacteria. (From Kutter et al.[2]) Recent work indicates that there is actually an inner wall zone between the peptidoglycan and the cell membrane in *S. aureus.*[64]

specific rapid passage of small hydrophilic molecules but exclude large and lipophilic molecules. It also contains various high-affinity receptor proteins that catalyze specific transport of solutes such as vitamin B12, catechols, fatty acids, and different iron derivatives. There are about 3 million molecules of LPS on the surface of each cell, plus 700,000 molecules of lipoproteins and 200,000 molecules of the porins. The concentrations of the specific other outer membrane proteins used as receptors vary considerably, depending on environmental conditions and the general need for the particular compound each one transports. Sex pili and other such structures are also used as sites of initial attachment by some phages, leading, for example, to phages that are male-specific. Some filamentous phages actually inject through the pili.

Appropriate positioning of the phage tail on the cell surface triggers irreversible events leading to DNA delivery into the host. In Gram-negative cells, this includes getting its DNA safely across the periplasmic space (periplasm), which appears to be a sort of viscous gel that contains many nucleases and proteases, in addition to transport across the inner membrane. There are few phages where much is known about the mechanism or energetics of this process.

Proteins TonB and TolA, which are anchored to the inner membrane but span the periplasmic space and are particularly important to various uptake systems, are also crucial to infection by such phages as T5. A detailed review by Letellier et al.[15] focuses primarily on three coliphages that have been best studied in this regard: myovirus T4, siphovirus T5, and podovirus T7. For each, a very long polyanionic genome — 50 times the cell length for T4 — must penetrate unscathed across two hydrophobic barriers, the outer and inner membranes, as well as the peptidoglycan sheath and the nuclease-containing periplasmic space. As they discuss, the rate of transfer can be as high as 3000 to 4000 basepairs/sec, in contrast to the 100 basepairs/sec seen for conjugation and natural transformation. Furthermore, the very high efficiency of infection for phages like T4 indicates that the process almost always occurs without significant damage to the infecting DNA molecule. The mechanisms are highly varied; often, ATP, a membrane potential, or enzyme action is involved, but some phage, such as T5, can enter cells in the absence of metabolic energy sources. It is clear that the volume of the rigid phage capsids does not change as the DNA is injected into the cell, and the simple release of energy from metastable packing in the particle does not provide sufficient energy to transport a DNA molecule several micrometers long through the very narrow channel inside the tail tube. For *Bacillus subtilis* phage SP82G, McAllister[16] showed that the rate of DNA transfer into the cell is constant over the whole genome and is highly temperature dependent, and the second-step transfer of the major part of coliphage T5 DNA proceeds normally even in experiments where the DNA has already been released from its capsid.[15] Though the myoviruses have contractile tails, it does not appear that their tail tubes actually pierce the inner membrane and a potential gradient is still required for DNA entry into the cell. All in all, as discussed by Molineux,[3] it is clear that the widely invoked "hypodermic syringe" injection metaphor is generally inaccurate and the mechanisms providing energy for the transfer differ from phage to phage. In the case of T7, for example, the energetics for transfer is largely supplied by transcription of the incoming DNA, starting from a promoter near the beginning of the molecule and depending on the fact that all of the transcription is in the same direction. The relatively slow speed of this process allows time for T7-encoded mechanisms to block damage to the DNA by host nucleases.

The interactions of phage with Gram-positive hosts have been much less well studied. Here, the phages recognize parts of the peptidoglycan layer and/or other molecules embedded in it (Figure 43.2b); specific interactions with the cell membrane may also be involved in the final, irreversible step.

Little is yet known about the specific adhesins of phage infecting Gram-positive bacteria or about the various receptors to which they bind. Phage specificity may relate to the substantial variations in the amino acids used for the cross-linking peptides in the exposed thick peptidoglycan structure; there are also variations in the teichoic acids protruding from it. A variety of cell-wall-associated proteins are also bound to the cell, either via a specific C-terminal anchor sequence or as N-terminal lipoproteins, and interact in various ways with the environment; staph protein A, which binds the constant region of IgG and has been implicated in pathogenesis, is the most famous of these.

Duplessis and Moineau[17] published the first identification and characterization of a phage gene involved in recognition of Gram-positive bacteria, for the sequenced phage DT1 and six related virulent phages with different host ranges on *Streptococcus thermophilus*. They confirmed that *orf18* encodes the adhesin, which has a conserved N-terminal domain of nearly 500 amino acids, a collagen-like sequence, and a largely conserved C-terminal domain with an internal 145 amino acid variable region (VR2). Using DT1 to infect a host in which MD4 genes *17–19* had been cloned, they generated five recombinant phages that now had the MD4 host range; all five contained the MD4 version of the VR2 region, but were otherwise largely like DT1.

Ravin et al.[18] showed that the specificity of adsorption of the isometric-headed phage LL-H and the prolate-headed JCL1032 to their host *Lactobacillus delbrueckii* involves a conserved C-terminal region in LL-H Gp71 and JCL 1032 ORF 474. Stuer-Lauridsen et al.[19] showed that four virulent prolate-headed phages (c2 species) of *Lactococcus lactis* with different host ranges first reversibly recognize a specific combination of carbohydrates in the outer cell wall. They then bind irreversibly with a host surface protein called PIP (phage infection protein), with no other known function, triggering DNA release from the capsid. Using the fully sequenced phages φc2 and φbIL67 and the sequenced least-conserved late regions of the other phages, the authors showed that the host-range-determining element is in the central 462-bp segment of a gene designated *115, 35*, and *2* in three different phages, with 65 to 71% pairwise similarity between them. While most genes in this region show no homology with unrelated phages, parts of this host-range-determining element resemble genes from phages of *Streptococcus thermophilus* and *Lactococcus lactis* that have also been implicated in host range determination processes. The product of a neighboring gene, *110/31/5*, with 83 to 95% similarity, had earlier been shown to bind to the tail spike and apparently to be involved in the infection process; they could not clone this second gene in their shuttle vector, perhaps because it is the one responsible for binding to PIP and helping transfer the DNA into the cell.

The phage infection process is very much dependent on the host metabolic machinery, so in most cases it is highly affected by what the host was experiencing shortly prior to infection, as well as by the energetic state and what nutrients and other conditions are present during the infection process itself.[20] Note that the phages that efficiently turn off host gene function, such as coliphages T4 and T5 and *Bacillus subtilis* phage SPO1, generally render their hosts incapable of responding to substantial environmental changes after infection. The energetic state of the cell also generally has significant effects on the probability of establishing lysogeny and, in some cases, on the reactivation of phages from lysogens. Most bacteria used in phage studies use two basic kinds of metabolic pathways to get the energy they need: fermentation and respiration. In both cases, electrons are withdrawn from an oxidizable molecule, freeing energy which is then stored as ATP, NADH, or similar coenzymes. These electrons, now at a lower potential, must in turn be removed by reducing some final acceptor. In fermentation, electrons removed from substrates such as glucose are transferred to breakdown products of the glucose to produce wastes such as ethanol or lactic acid, and only a small fraction of the energy in the glucose can be used. In respiration, the electrons are eventually passed through a membrane-bound electron transport system (ETS) and are carried away by reducing oxygen to water (in ordinary aerobic respiration), or, in anaerobic respiration, by reducing ferric to ferrous iron, sulfate to sulfide, fumarate to succinate, or nitrate to nitrite or N_2. The electron transport system is oriented in the cell membrane so as to transport protons, H^+, to the outside of the cell membrane. This builds up an electrochemical potential, known as the proton motive force (PMF), across the cell membrane. This PMF is critical for infection by most phages. As discussed by Goldberg,[21] the PMF is separable into two parts: a pH gradient and a membrane potential that is due to the general separation of charges. For many (but not all) phages, DNA transfer into the cell requires the membrane potential.

THE INFECTION PROCESS

Overview

The general lytic infection process for tailed phages is presented here, with some examples from model phages. More detail on these subjects can be found in Abedon and Calendar's *Bacteriophages*,[5] or in Granoff and Webster's *Encyclopedia of Virology*,[22] as well as in books and articles on the individual phages.

Since the early studies of d'Herelle, the details of phage-host interactions have been studied by use of the single-step growth curve (Figure 43.3a), as systematized in 1939 by Ellis and Delbrück.[23] Phages are mixed with appropriate host bacteria at a low multiplicity of infection. After a few minutes for adsorption, the infected cells are diluted (to avoid attachment of released phage to uninfected cells or bacterial debris) and samples are plated at various times to determine infective centers. An infective center is either a single phage particle or an infected cell that bursts on the plate to produce a single plaque (see Chapter 7 for a discussion of bacteriophage plaque assays). The number of plaques generally remains constant at the number of infected cells for a characteristic time, the latent period, and then rises sharply, leveling off at many times its initial value as each cell lyses and liberates the completed phage. The ratio between the numbers of plaques obtained before and after lysis is called the burst size. Both the burst size and the latent period are characteristic of each phage strain under particular conditions, but are affected by the host used, medium, and temperature. As first shown by Doermann,[24] if the infected cells are broken open at various times after infection, the phages seem to have all disappeared for a certain period. This eclipse period was a mystery until the nature of the phage particle and of the infection process was determined and it was realized that during this period only naked phage DNA is present in the cell. Using chloroform-induced lysis followed by plating, both the eclipse period and the subsequent rate of intracellular synthesis of viable phage particles are now routinely measured. Infecting at a high multiplicity (5 to 10 phages/cell) allows one to also measure the effectiveness of killing (i.e., the number of bacterial

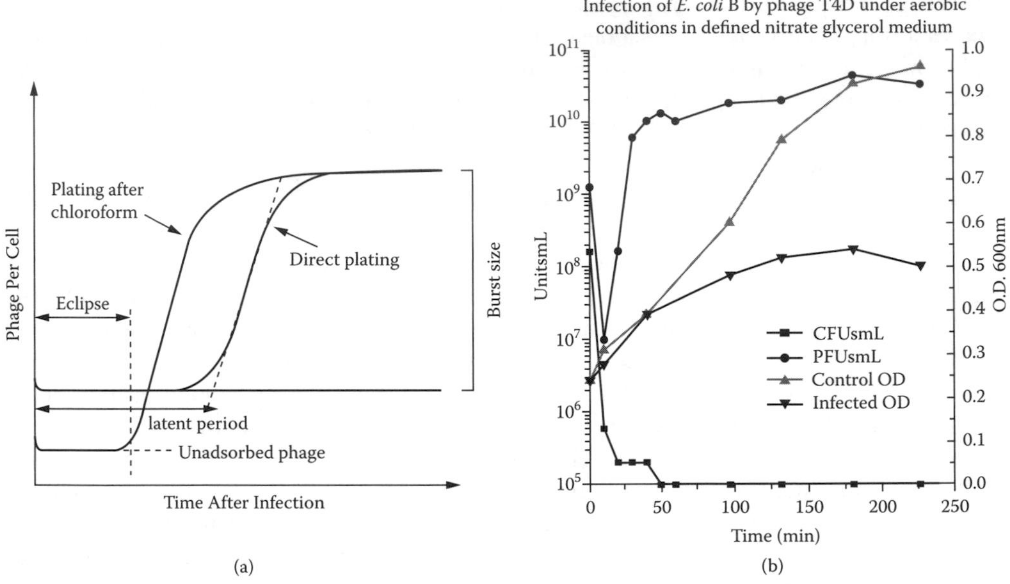

FIGURE 43.3 (a) Classical single-step growth analysis of phage infection; (b) infection of *E. coli* B by T4D at an MOI of 9 in a minimal medium. Note the very rapid initial drop in both surviving bacteria and unadsorbed phage, the doubling of the cell mass even after infection, as measured by absorbance (O.D. 600), and the fact that there is a long delay before lysis actually occurs under these conditions; this phenomenon of "lysis inhibition" is discussed in the text.

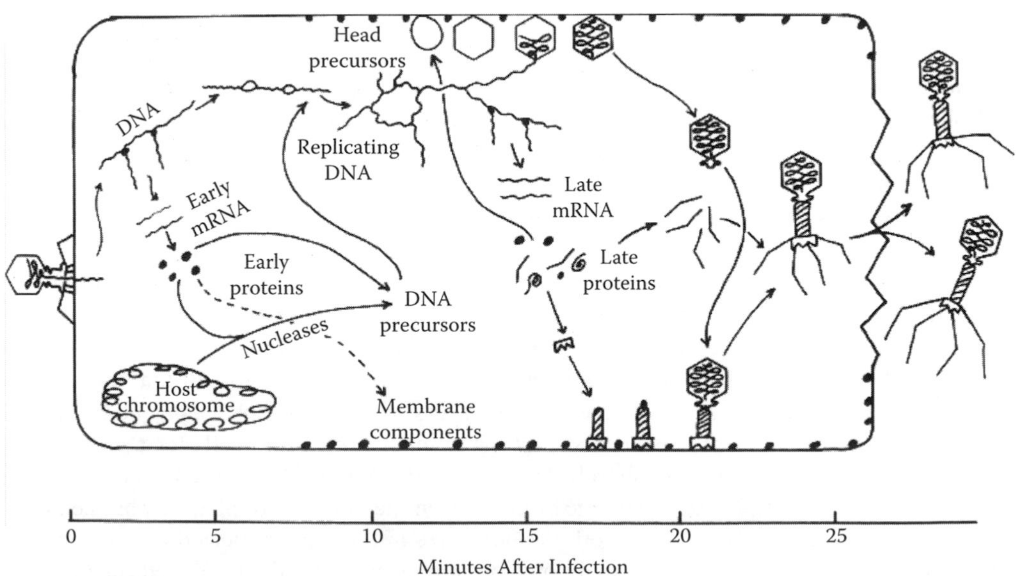

FIGURE 43.4 An overview of the T4 infection cycle. (From Mathews, C.K., Kutter, E., Mosig, G., and Berget, P.B., Eds., *Bacteriophage T4*. American Society for Microbiology, Washington, DC, 1983, as reproduced in Kutter et al.[1]).

survivors) and the impact of the phage infection in terms of continued expansion of cell mass and eventual cell lysis, as reflected in the absorbance or optical density (Figure 43.3b).

The infection process involves a number of tightly programmed steps, as diagramed in Figure 43.4. The efficiency, timing, and other aspects of the process may be very much affected by the host metabolic state, and in many cases the host can no longer adapt to major metabolic changes once the phage has taken over the cell. Thus, for example, studies on anaerobic infection can only be carried out in host cells that were themselves grown anaerobically.

Adsorption

Infection by tailed phages starts when specialized adsorption structures, such as fibers or spikes, bind to specific surface molecules or capsules on their target bacteria. In Gram-negative bacteria, virtually any of the proteins, oligosaccharides, and lipopolysaccharides can act as receptors for some phage. The more complex murein of Gram-positive bacteria offers a very different set of potential binding sites. Many phages require clusters of one specific kind of molecule that is present in high concentration to properly position the phage tail for surface penetration. However, coliphage N4 manages to use a receptor, called NfrA, which is only present in a few copies per cell. The attachment of some phages involves two separate stages and two different receptors. T4-like phages, for instance, must bind at least three of their six long tail fibers to primary receptor molecules to trigger rearrangements of baseplate components, which then bind irreversibly to a second receptor. Different members of the family use very different primary receptors, but they seem to all use the heptose residue of the LPS inner core for the secondary receptor.

Adsorption velocity and efficiency are important parameters that may vary for a given phage-host system, depending on external factors and host physiological state. For example, the lambda receptor is only expressed in the presence of the sugar maltose. Many phages require specific cofactors, such as Ca^{2+}, Mg^{2+}, or simply any divalent cation. T4B strains (but not T4D strains) require tryptophan for binding, but can bypass this requirement for a short "nascent" period when they are first released from the prior host and still attached to the inside of the membrane by their baseplates.[25–27] A nascent-phage period with broader adsorption properties was also observed for *Staphylococcus* phage in early work, and described for a *Streptococcus* phage by Evans,[28] but the mechanism(s) of nascent-phage interaction to give the extended host range is not known. Bacteria commonly develop resistance to a particular phage through mutational loss or alteration of receptors used by that phage. However, losing some receptors offers no protection against the many other kinds of phage that use different cell-surface molecules as receptors. Furthermore, in most cases, the phage can acquire a compensating adaptation through appropriate host range mutations, which alter the tail fibers so they can recognize the altered cell-surface protein or bind to a different receptor. This is likely less efficient than bacteria acquiring resistance, since phage adaptation to new receptors requires establishment of a new functional interaction, while development of host resistance is a "negative" event, requiring loss of function.[29] Still, some such mutants seem to be present in every substantial phage population. Furthermore, some phages, such as P1 and Mu, encode multiple versions of the tail fibers. Others can recognize multiple receptors; for example, T4 tail fibers bind efficiently to an *Escherichia coli* B-specific lipopolysaccharide and to the outer membrane protein OmpC found on K strains[30] The adhesion regions of the tail fibers are highly variable between related phages, with high rates of recombination that facilitate the formation of new, chimeric adhesins. Not surprisingly, there is much interest in engineering new receptor-recognition elements into the tail fibers of well-characterized phages so they can infect taxonomically distant hosts.

The physiological state of a cell can substantially change the concentration of particular cell-surface molecules and thus the efficiency of infection by certain phages. Many of the surface molecules that particular phages use as receptors are crucial to the bacterial cell, at least under some environmental conditions, so resistance may lead to loss of important functions and reduction in competitiveness.

Penetration

After irreversible attachment, the phage genome passes through the tail into the host cell. This is not actually an "injection" process, as has often been depicted, but involves mechanisms of DNA transfer specific for each phage. In general, the tail tip has an enzymatic mechanism for penetrating the peptidoglycan layer and then touching or penetrating the inner membrane to release the DNA directly into the cell; the binding of the tail also releases a mechanism that has been blocking exit of the DNA from the capsid until properly positioned on a potential host. The DNA is then drawn into the cell by processes that generally depend on cellular energetics, but are poorly understood except for a few phages; for example, as noted earlier, in T7, the entry of the DNA is mediated by the process of its transcription.

Once inside the cell, the phage DNA is potentially susceptible to host exonucleases and restriction enzymes. Therefore, many phage circularize their DNA rapidly by means of sticky ends or terminal redundancies, or have the linear ends protected. Many also have methods to inhibit host nucleases (T7, T4) or use an odd nucleotide in their DNA such as hydroxymethyldeoxyuridine (hmdU: SPO1) or hydroxymethyldeoxycytidine (hmdC: T4) for protection. In other cases, their genomes have been selected over evolutionary time to eliminate sites that would be recognized by the restriction enzymes present in their common hosts (staphylococcal phage Sb-1, coliphage N4).

The Transition from Host to Phage-Directed Metabolism

The initial step generally involves recognition by the host RNA polymerase of very strong phage promoters, leading to the transcription of immediate early genes. The products of these genes may protect the phage genome and restructure the host appropriately for the needs of the phage; they may inactivate host proteases and block restriction enzymes, directly terminate various host macromolecular biosyntheses, or destroy some host proteins. The large virulent phages such as T4, SPO1, and Sb-1 encode many proteins that are lethal to the host even when cloned individually and that appear to participate in this process of host takeover. A set of middle genes is often then transcribed, producing products that synthesize the new phage DNA, followed by a set of late genes that encode the components of the phage particle. For some phages, these transitions involve the synthesis of new sigma factors or DNA-binding proteins to reprogram the host RNA polymerase; other phages encode their own RNA polymerase. Degradation of host DNA, inhibition of host replication, and inhibition of transcription and translation of host mRNAs are other mechanisms that can contribute to reprogramming the cell for the synthesis of a new phage.

Morphogenesis

The DNA is packaged into preassembled icosahedral protein shells called procapsids. In most phages, their assembly involves complex interactions between specific scaffolding proteins and the major head structural proteins, followed by proteolytic cleavage of both the scaffolding and the N-terminus of the main head proteins. Before or during packaging, the head expands and becomes more stable, with increased internal volume for the DNA. Located at one vertex of the head is a portal complex that serves as the starting point for head assembly, the docking site for the DNA packaging enzymes, a conduit for the passage of DNA, and, for myoviruses and siphoviruses, a binding site for the phage tail, which is assembled separately. The assembly of key model phages provides the best-understood examples in biology of morphogenesis at the molecular level. It also has provided the models and components for several key developments in nanotechnology.

Cell Lysis

The final step — lysis of the host cell — is a precipitous event whose timing is tightly controlled.[31] If lysis happens too quickly, too few new phages will have been made to effectively carry on the cycle;

if lysis is delayed too long, opportunities for infection and a new explosive cycle of reproduction will have been lost.[32, 33] As discussed in Chapter 13, the shape-determining rigid wall that protects virtually all bacteria from rupture in media of low osmolarity and must be ruptured to release phage is actually one enormous molecule, forming a peptidoglycan or murein sac that completely encloses the cell — a sac that grows and divides without losing its structural integrity, and is the site of attack of antibiotics like penicillin (see Chapter 11). The glycan warp of this sack consists of chains of alternating sugars — *N*-acetylglucosamine (NAG) and *N*-acetyl muramic acid (NAM) — wrapped around the axis of the cylinder for rod-shaped bacteria. These chains are cross-linked by short peptides to make a two-dimensional sheet. The tailed phages all use two components for lysis: a lysin — an enzyme capable of cleaving one of the key bonds in the peptidoglycan matrix — and a holin — a protein that assembles pores in the inner membrane at the appropriate time to allow the lysin to reach the peptidoglycan layer and precipitate lysis. The timing is affected by growth conditions and genetics; mutants with altered lysis times can be selected. The T4-like phages often have an additional mechanism for adapting to host availability. If other phages of the same family try to enter the cell more than a few minutes after T4 infection, the infected cell interprets that as a sign that there are more phages around than available hosts, and substantially delays lysis. This phenomenon, termed "lysis inhibition," may last many hours; the precise length is characteristic of the specific phage involved and the environmental conditions.[32] For T4 grown aerobically at 37°C, it typically lasts about 6 hours and can be instigated even when the superinfection by the secondary phages occurs just a minute or two before the normal time of lysis. The general mechanisms and wide variety of holins and lysins are discussed by Young and Wang,[34] and by Kutter et al.[2]

LYSOGENY

The concept of lysogeny has had a checkered history. Early phage investigators in the 1920s and 1930s claimed to find phages irregularly associated with their bacterial stocks and believed that bacteria were able to spontaneously generate phage, which were thus long considered by many to be some sort of "ferment" or enzyme rather than a living virus. When Max Delbrück and colleagues began their work, they confined themselves to the classical set of coliphages designated T1–T7, none of which showed this property, and Delbrück attributed the earlier reports to methodological sloppiness. However, the phenomenon could no longer be denied after the careful work of Lwoff and Gutmann;[35] through microscopic observation of individual cells of *Bacillus megaterium* in microdrops, they demonstrated that cells could continue to divide in phage-free media with no sign of phage production, but that an occasional cell would lyse spontaneously and liberate phage. Lwoff named the hypothetical intracellular state of the phage genome a prophage and showed later that by treating lysogenic cells with agents such as ultraviolet light, the prophage could be uniformly induced to come out of its quiescent state and initiate lytic growth. The prophage carried by a bacterial strain is given in parentheses; thus, K(P1) means bacterial strain K carrying prophage P1.

Esther Lederberg[36] showed that strains of *Escherichia coli* K-12 carried such a phage, which she named lambda (λ). Meanwhile, Jacob and Wollman[37] had been investigating the phenomenon of conjugation between donor (Hfr) and recipient (F^-) strains of *E. coli*. They found that matings of F^- strains carrying λ and non-lysogenic Hfrs proceeded normally, but that reciprocal matings yielded no recombinants and, in fact, produced a burst of lambda phage. A mating of Hfr(λ) with non-lysogenic F^- would proceed normally if it were stopped before transfer of the *gal* (galactose metabolism) genes; but if conjugation proceeded long enough for the *gal* genes to enter the recipient cell, the prophage would be induced (zygotic induction). These experiments indicated that the λ prophage occupies a specific location near the *gal* genes; that a lysogenic cell maintains the prophage state by expression of one λ gene, encoding a specific repressor protein, which represses expression of all other λ genes; and that if the *gal* genes — and thus the λ prophage — are transferred into a non-lysogenic cell during mating, the prophage finds itself in a cytoplasm lacking repressor and therefore expresses its other genes and enters the lytic cycle.

Typical plaques made by phage λ are turbid, due to lysogenization of some bacteria within the plaque. Mutants of λ producing clear plaques are unable to lysogenize. Analysis of these mutants revealed three genes, designated *cI*, *cII*, and *cIII*, whose products are required for lysogeny. The *cI* gene encodes the repressor protein. Allen Campbell demonstrated that the sequence of genes in the prophage is a circular permutation of their sequence in the phage genome. He therefore postulated that the prophage is physically inserted into the host genome by circularization of the infecting genome followed by crossing-over between this genome and the (circular) bacterial genome.[101] We now know that the genome in a λ virion has short, complementary single-stranded ends; as one step in lysogenization, it circularizes through internal binding of these ends and is then integrated (by means of a specific integrase) at a point between the *gal* and *bio* (biotin biosynthesis) genes.

The mechanism of the molecular decision between lytic and lysogenic growth has been worked out in most detail for phage λ,[38] but Dodd and Egan[39] found great similarities for the unrelated temperate phage 186. In each case, lysogeny is governed by a repressor protein, CI, which binds to a set of operators in the lysogenic state and represses the expression of all genes except its own (Figure 43.5). Both phages have critical promoter regions, one promoting lysogeny and the other lysis, which are closely associated, and lysogeny is promoted by binding of the CI protein to these sites in such a way as to inhibit lytic growth. The CI proteins of both phages are strongly cooperative, forming tetramers or octomers. In the λ case, CI is involved in a molecular competition with another protein, Cro, which promotes the lytic cycle. The transition toward lysogeny in λ is also promoted by two other proteins, CII and CIII, which bind to critical promoters and stimulate transcription of the *cI* gene and others. The stability of CII is determined by factors that measure the cell's energy level. A cell with sufficient energy has little cyclic AMP (cAMP); when the cell is energy-starved (leading to low intracellular glucose), the cAMP concentration is high. A high level of cAMP promotes CII stabilization and thus lysogeny. It is clearly adaptive for a phage genome entering a new cell to sense whether there is sufficient energy to make a large burst of phage or whether the energy level is low, so its best strategy for survival is to go into a prophage state.

Notice that the critical event in establishing lysogeny is regulation of the phage genes. Integration of the phage genome into the host is secondary, and so it is understandable that other temperate phages, such as P1, can establish lysogeny with their prophages functioning as plasmids in the cytoplasm. Many interesting variations on lysogeny and temperate phages have been discovered.

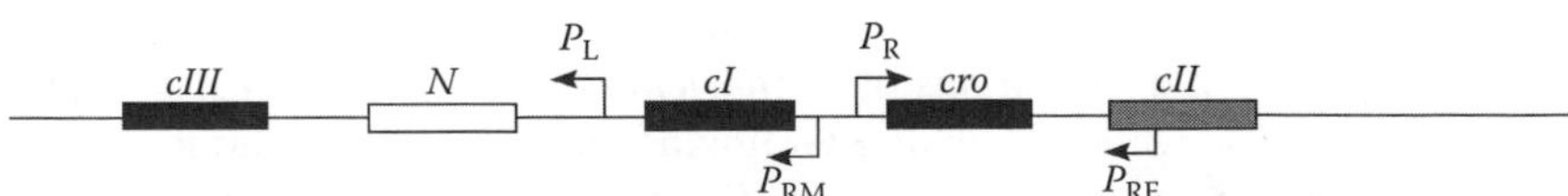

FIGURE 43.5 The regulatory region of the phage λ genome where the decision is made between lysis and lysogeny. The heart of the matter is competition between two proteins, CI and Cro. Initially, lysogeny is promoted by the combined action of the CII and CIII proteins, which act at P_{RE} (promoter right for establishment) and promote transcription of the *cI* gene; the CI protein then binds to various operator/promoter sites and represses them. In particular, CI prevents transcription of the *N* gene whose product acts at other sites in the genome to promote expression of genes required for lytic growth. CI also binds to P_{RM} (promoter right for maintenance) and promotes further transcription of its own gene. These lysogeny-directed processes are in competition with transcription from P_R of the *cro* gene and other genes to the right needed for lytic growth. Cro and CI compete for binding sites in the complex promoter/operator region that includes both P_{RM} and P_R where the lysis/lysogeny decision is ultimately made.

TRANSDUCTION

Generalized Transduction

As noted above, bacteriophages package their genomes into phage capsids. Also noted above, the host's bacterial DNA is often subject to degradation during the course of the phage infection. The consequence of these two phenomena is that capsids of many phages occasionally package pieces of host DNA instead of phage DNA, making a transducing particle (i.e., a vehicle that transmits a piece of DNA from one bacterium to another). This is because the phage particle containing the piece of bacterial DNA can attach to an uninfected bacterium and have its DNA contents taken up by the recipient cell. If the introduced donor DNA recombines with the chromosome of the recipient cell, then genes carried on the donor DNA will be incorporated into that recipient.

Originally discovered in 1952 by Zinder and Lederberg,[40] this process is called generalized transduction, because any portion of the genome can, in principle, be packaged and transferred to a recipient cell. In practice, some phages are much better than others in carrying out transduction, which is a consequence of the characteristics of the specific phage life cycle. Phage T4, for example, is not a very good transducing phage because it is too effective in degrading host DNA, whereas some integrity of host DNA is required for effective transduction, especially by a phage designed to package as long a DNA molecule as T4 does (about 180 kb). By contrast, phage P1 is a workhorse of transduction, very widely used by researchers, because it generates capsid-size chunks of host DNA and is capable of effectively packaging them in transducing particles. The frequency of P1 transduction of any given gene is generally taken to be about 10^{-6}. Multiplicity of infection (MOI) needs to be considered in transduction experiments, because too high an MOI will make it unlikely that a recipient cell will only receive a transducing particle and not be killed by a co-infecting virulent phage particle as well. An illustration of the principle of generalized transduction is shown in Figure 43.6.

Specialized Transduction

Temperate phages have the alternate lifestyle choice of lysogenizing the hosts they infect, integrating into the host chromosome, or, like phage P1, forming a plasmid and keeping their genomes largely dormant. But when adverse conditions activate a lysogen, the phages undergo a lytic cycle. Rarely, excision of temperate phage DNA from the chromosome is imprecise, resulting in an excised piece of DNA that contains most of the phage genome but also part of the host chromosome that

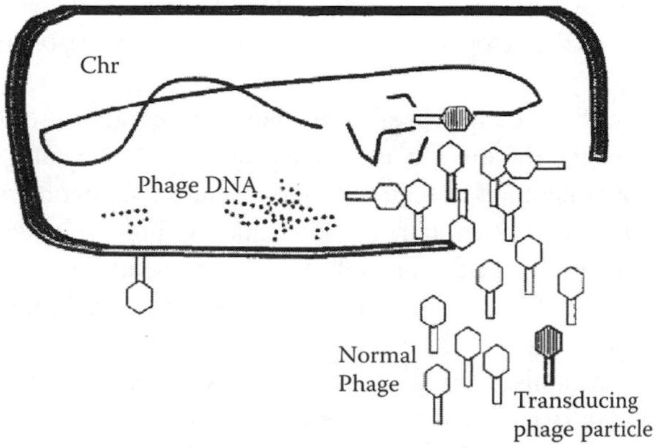

FIGURE 43.6 Generalized transduction. During the course of a lytic phage infection, host chromosomal DNA is degraded. Occasionally, a fragment of the host chromosome is inadvertently packaged into a phage capsid. This "transducing phage particle" can transfer the packaged part of the donor chromosome to a recipient cell. (Illustration courtesy of M. Zafri Humayun.)

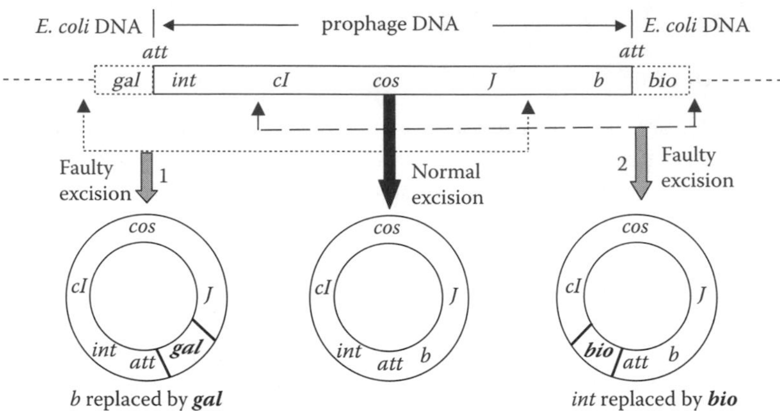

FIGURE 43.7 Specialized transduction results from faulty excision of the lambda prophage. The linear lambda phage DNA (not shown) has cohesive ends (*cos*) that lead to formation of circular lambda DNA. The attachment site (*att*) in the circular lambda DNA recombines with the *E. coli* chromosome between the *gal* and *bio* genes, integrating the lambda prophage DNA to make a lysogen. When the lysogen is induced to a lytic cycle, prophage DNA is excised and the restored circular lambda DNA will act as a template for virus growth. As a rare event, excision of the prophage is faulty, generating circular lambda DNA containing either the adjacent *E. coli gal* gene (excision 1 in the figure) or the adjacent *E. coli bio* gene (excision 2), in place of lambda genes. TransferS of these *E. coli* genes to recipients in subsequent infections constitute examples of specialized transduction. Representative lambda genes: *int*, integrase; *cI*, repressor; *J*, tail fiber; *b*, region of lambda chromosome that can be deleted with phage retaining viability. (Based in part on ideas described by Campbell.[101])

was physically adjacent to the prophage DNA on the chromosome (Figure 43.7). When this excised piece of DNA is packaged in a phage capsid, it becomes a vehicle for transmitting these adjacent host gene(s) to a recipient cell. Because of the imprecise excision, the partial phage genome is usually defective, so the recipient cell will not be subject to a lytic cycle of the phage; instead, the defective phage DNA, carrying the adjacent host DNA, will recombine with the recipient cell's chromosome and thereby transfer genetic material from donor to recipient.

First characterized in 1956 by Morse et al.,[41] this process is called specialized transduction because it can transfer only genes that were physically adjacent to the prophage when it resided in the chromosome of the donor (prior to induction). The host genes transferred can be from either side of the prophage DNA in the lysogen and, depending on the extent of imprecise excision, can include one or a few genes.

Many temperate phages have their own specific sites of integration, so the genes they transfer are characteristic of the phage and are a consequence of the specific integration sites. However, some phages, like Mu (for "mutator"), integrate randomly into the genome and then always carry some neighboring host DNA with them during the packaging process when they are induced, so they are able to effect a generalized transduction of short segments from throughout the genome, and the transducing phage is still fully capable of carrying out a normal infection cycle.

Transduction plays important roles in bacterial genetics, both in the lab and in the wild, and is one of the three major mechanisms for gene transfer between bacteria (the other two being conjugation and transformation). Briani et al.[42] have emphasized the importance of temperate phages as "replicons endowed with horizontal transfer capabilities," and they have probably been major factors in bacterial evolution by moving segments of genomes into new organisms. Broudy and Fischetti[43] have reported *in vivo* lysogenic conversion of Tox$^-$ *Streptococcus pyogenes* to Tox$^+$ with lysogenic streptococci or free phage in a mammalian host. Thus, this process presumably goes on all the time in nature.

ROLE OF BACTERIOPHAGE IN CAUSING PATHOGENESIS

Although most phage genes are dormant in a lysogen, particularly genes responsible for lytic growth of the phage, there are a few genes that continue to be expressed from the integrated phage genome. These include, of course, genes like the lambda repressor, which prevents expression of the lytic cycle. However, other phage-specific genes may also be expressed that direct synthesis of one or more products to change the cell's phenotype; the new product(s) did not exist in the cell prior to lysogeny and are often beneficial to host survival in some environments, giving the lysogenized host a selective advantage. This phenomenon is known as phage conversion.

Bacterial toxins, such as diphtheria, cholera, and Strep A toxin, and new antigens, such as *Salmonella* O antigen, are medically important examples of phage conversion, which led the famous bacterial geneticist William Hayes (1918–1994) to remark, "We incriminate the bacteria for the sins of its viruses." Table 43.1 lists a few prominent examples of phage-encoded pathogenesis factors. As an introductory chapter to the field, we will only briefly describe three of the very many known instances of phage-determined virulence. Extensive coverage of this topic may be found in recent publications.[44, 45]

Diphtheria Toxin

For an excellent review on diphtheria toxin (DT), see Holmes.[46] When *Corynebacterium diphtheriae* are lysogenized by phage β, the prophage expresses the *tox* gene, leading to the production of a precursor form of DT, which is then processed to the mature toxin. DT inhibits eukaryotic protein synthesis by inactivating elongation factor EF-2. It accomplishes this by covalent attachment of an ADP-ribosyl group to a unique residue in EF-2, a modified form of histidine called diphthamide.

Expression of the *tox* gene is controlled by an iron-regulated repressor encoded by the *dtxR* gene on the *Corynebacterium diphtheriae* chromosome. In the presence of iron, the DtxR protein is active as a repressor and prevents transcription of the *tox* gene. In the absence of iron, the repressor is inactive and there is no repression of *tox*, leading to production of DT (Figure 43.8). Because iron concentrations tend to be low in mammalian hosts, pathogenesis occurs when β-phage-carrying *C. diphtheriae* colonize the throat of an affected individual. See Chapter 26 of this volume for a full discussion of *Corynebacteria*.

TABLE 43.1
Examples of Bacteriophage-Encoded Pathogenesis Factors

Pathogenesis Factor	Bacteria	Phage
O-antigen[64–68] (immune system evasion)	*Salmonella* species	**ε15, ε34**
SopE[69] (invasion, type III secretion)	*Salmonella typhimurium*	SopEφ
SodC[70] (superoxide dismutase)	*Salmonella typhimurium*	Gifsy-2
Cholera toxin[47]	*Vibrio cholerae*	CTXφ
Diphtheria toxin[71]	*Corynebacterium diphtheriae*	Converting β-phage
Shiga-like toxin-I and -II[72, 73]	Enterotoxigenic *Escherichia coli*	Stx converting phage
Serum resistance[74]	*Escherichia coli*	λ
Enterotoxin A and staphylokinase[75]	*Staphylococcus aureus*	φ13
Streptococcal pyrogenic exotoxins A&C[76]	Group A *Streptococci*	T12 and 3GL16

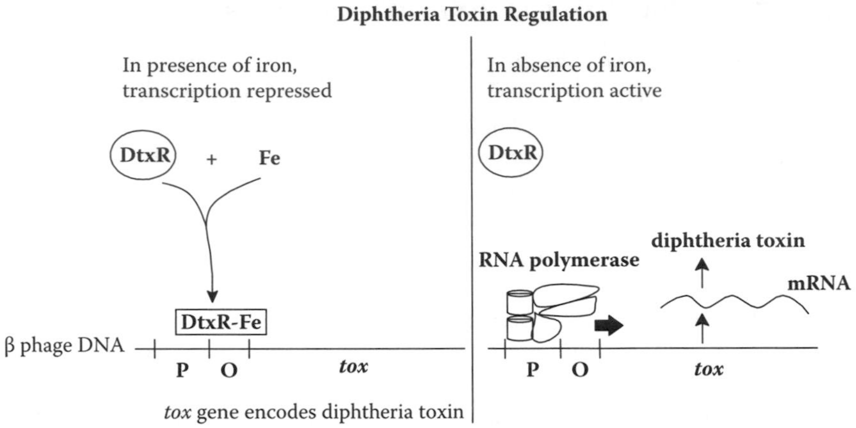

FIGURE 43.8 Negative control of diphtheria toxin by a repressor protein. The toxin is encoded by a phage gene in a lysogen. The DtxR repressor protein is encoded in the *C. diphtheria* genome. P, promoter; O, operator; Fe, iron.

Cholera Toxin

Recognition that cholera toxin (CT) is encoded by a lysogenic phage was relatively recent,[47] perhaps in part because the phage involved, CTXφ, is a filamentous phage that does not kill the host bacterial cell during productive growth. CT, comprised of one 'A' subunit and five 'B' subunits, inactivates a GTP-binding protein needed to control cellular cyclic-AMP synthesis, causing overproduction of cAMP and massive fluid and electrolyte loss from intestinal cells.[48]

Vibrio cholerae ordinarily occupy a salt-water habitat, which is not the environment of a mammalian gut. Thus, one can argue that a bacterial sensor system of a low-salt environment activates synthesis of CT, and the profuse resultant diarrhea facilitates the release and spread of the bacteria from the affected individual. It is known that environmental conditions affect *V. cholerae* virulence.[49] Refer to Chapter 21 of this volume for a full discussion of *Vibrio* species.

The sensor system for activation of CT synthesis is complex, with many cellular components[50] as well as a recently discovered lysogenic satellite phage, RS1.[51] The regulator ToxT activates transcription of the *ctxA* and *ctxB* toxin genes from a single promoter; there is much greater synthesis of the B subunit compared to the A subunit, which is probably controlled at the translational level. The *toxT* gene, as well as other virulence genes activated by ToxT (toxin co-regulated pilus [tcp] and accessory colonization factors [acf]), are located on the vibrio pathogenicity island (VPI) region of the *Vibrio cholerae* chromosome. This region was also thought to represent a lysogenic phage genome, but evidence for this was not compelling and this view is generally no longer current.[52]

Synthesis of ToxT itself is activated by other *Vibrio cholerae* genes, *toxR, toxS, tcpP,* and *tcpH*. The products of these genes are located in the cell membrane, where they may play a role in sensing a low salt level in the environment, activating ToxT synthesis, which in turn, activates CT production from the lysogen.

Recently, the satellite phage RS1 was shown to be another lysogen of *Vibrio cholerae* and to play a role in CT production. RS1 itself is defective and only grows during growth of CTXφ. The *ctxAB* genes can also be transcribed in the lysogen from an upstream promoter for phage growth, which is repressed by the RstR repressor. Phage RS1 encodes an anti-repressor, RstC, which inactivates repression by RstR, leading to increased synthesis of the CT genes. RstC itself is also repressed by the RstR repressor; therefore induction of the prophage by DNA damage (the "SOS" response), which inactivates RstR, is necessary to obtain production of RstC.[52] Thus, there is also a second mechanism available, mediated by satellite phage RS1, to stimulate synthesis of CT.

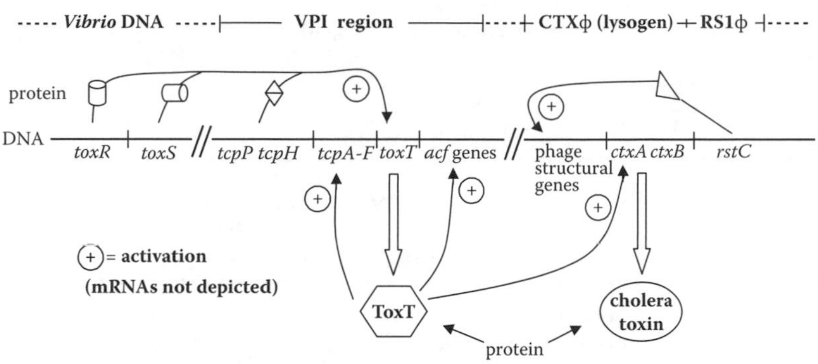

FIGURE 43.9 Regulation of virulence gene expression in *Vibrio cholerae*. TcpP/TcpH and ToxR/ToxS are located in the cell membrane and cooperate in activating transcription of *toxT*. ToxT activates transcription of *tcpA-F, ctxAB*, and *acf* genes. Genes for ToxR/ToxS are encoded on the ancestral *V. cholerae* chromosome and also regulate genes other than those on CTXφ and VPI (formerly thought to be a lysogen). Satellite phage RS1φ encodes the RstC anti-repressor, which inactivates the RstR repressor (not shown) of CTXφ growth. Inactivating the RstR repressor also leads to transcription of the *ctxAB* genes. Synthesis of RstC follows an SOS response.

A schematic diagram of the regulatory circuits for production of CT is shown in Figure 43.9. For simplicity, the effect of RstC is shown only as activation of transcription upstream of the *ctxAB* genes (whereas the actual mechanism involves the intermediate step of inactivation of the RstR repressor).

Shiga-Like Toxins

Escherichia coli, ordinarily a commensal universally found in the gut, can become highly pathogenic following lysogeny with Stx-converting phages. Pathogenicity is generally associated with the serotype O157:H7, resulting in hemorrhagic colitis, hemolytic uremic syndrome, and kidney failure. See Chapter 17 of this volume for a discussion of Enterobacteriaceae, including *Escherichia coli.*

The Shiga-like toxins, so-named due to similarity to toxins from *Shigella dysenteriae*, are also comprised of one ‘A’ subunit and five ‘B’ subunits. They inhibit protein synthesis by attacking eukaryotic ribosomal RNA.

The phages responsible for this are of the Lambda family, and carry genes, *stxA* and *stxB*, that encode the two toxin subunits. Of the two prominent toxin types, type 1 is regulated by iron, while type 2 is not. Type 2 is repressed in the lysogen, and expression follows induction of the lysogen to a lytic cycle.[53] Expression of Type 1 in the lysogen, under conditions of low iron concentration, is enhanced following induction of the lysogen to a lytic cycle.[54]

A simplified schematic diagram illustrating regulation of expression of Shiga toxin-type 2 is shown in Figure 43.10.

PRACTICAL INFORMATION ABOUT SELECTED PHAGES

Tailed Phages

Table 43.2 lists, in alphabetical order, selected tailed phages that have been extensively studied. Except for λ, pathogenesis-causing phages listed in Table 43.1 are not included. All of these phages have linear double-stranded DNA genomes. Many exhibit terminal redundancy with (e.g., P1, P22,

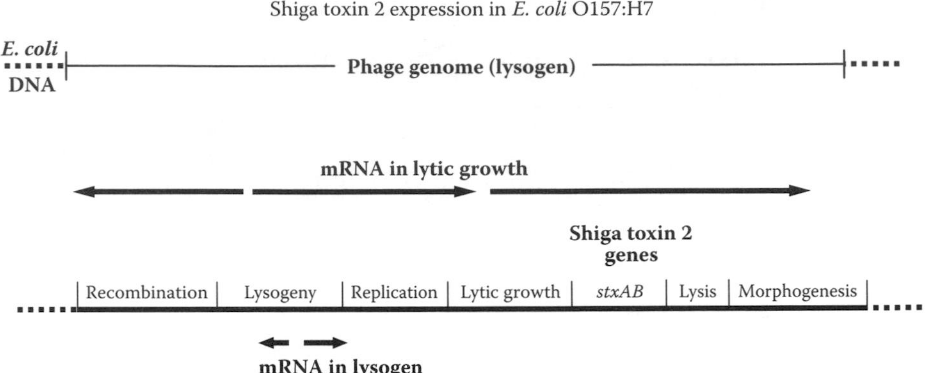

FIGURE 43.10 Shiga toxin 2 expressed following λ-family prophage induction in *E. coli* O157:H7. Putative genome organization (not to scale) and transcription of Stx2-encoding phage are based on known lambda-family phages. Shown are relevant functional genes clusters, as indicated. *stxAB* represents the genes encoding the pathogenic Shiga toxin 2. Below is shown the pattern of transcription initiating at the early promoters in the lysogen. Above is shown the patterns of transcription initiating at the early and late promoters during lytic growth.

T4) or without (e.g., λ, P2, P4, T5, T7) circular permutation.[55] Some key characteristics of the phages are presented, and the "comments" column highlights especially interesting or unique features.

More in-depth general resources include Granoff and Webster's *Encyclopedia of Virology,*[22] both editions of *The Bacteriophages,*[56, 57] Ackermann and DuBow,[11] and Adams.[8] The complete genomic sequences of a number of phages have now been determined and are available in the bacteriophage section of the genome sites at the National Center for Biotechnology Information <http://www.ncbi.nlm.nih.gov>, as are the sequences of some of the prophages determined in the course of microbial genome projects.

Phages of *Escherichia coli* and *B. subtilis* have generally been the best studied with regard to genetics and the physiology of the infection process, taking advantage of — and contributing to — the broad knowledge about these two key model organisms. Note that in general the designation of phage names has been quite arbitrary, giving no information about phage properties or relationships.

Non-tailed Phages

phiX174 (φX174): the prototype member of the *Microviridae*, with a 5.4 kb circular single-stranded DNA genome[58] and an icosahedral virion. The DNA is a positive (+) strand (i.e., sequence is in the same orientation as its mRNA) with overlapping genes. Infects Enterobacteriaceae such as *Escherichia coli, Salmonella,* and *Shigella,* and has been extensively used in studies of DNA replication.

M13: the prototype member of the *Inoviridae*, with a 6.4 kb circular positive single-stranded DNA genome,[59] and an amorphous, filamentous virion. Infects male *Escherichia coli* by adsorbing to the tip of the pilus; non-lytic, but turbid plaques can be discerned due to slower growth of infected bacteria. Has been extensively used for cloning and DNA sequencing, and in "phage display" technology, described in detail in Chapter 9.

MS2: the prototype member of the *Leviviridae,* with a 3.6 kb linear single-stranded highly structured RNA genome[60] and an icosahedral virion. Infects male *Escherichia coli* by adsorbing to the sides of the pilus; lytic virus, but does not shut off host metabolism. Plus strand RNA genome functions as an mRNA. High affinity of coat protein to specific sequences in RNA has been exploited in biotechnology.

phi6 (φ6): the prototype member of the *Cystoviridae,* with a genome of three double-stranded RNAs of 6.4, 4.1, and 2.9 kb (reviewed in Mindich[61]) and an enveloped virion. Infects the plant

TABLE 43.2
Survey of Selected DNA-Containing Tailed Phages (Caudovirales)*

Phage	Family[a]	Host(s)	Size[b]	Temperate[c]	Transduction	Comments
		Gram-negative bacteria				
Lambda (λ)	Sipho	*Escherichia coli*	48.5[77]	+	specialized	Genetic engineering vector
Mu	Myo	*E. coli, Salmonella,* others	36.7[78]	+	generalized	Random integration in host DNA
N4	Podo	*E. coli*	70.2[79]			Two unusual RNA polymerases
N15	Sipho	*E. coli*	46.3[80]	+		Prophage a linear plasmid
P1	Myo	broad host range	93.6[81]	+	generalized	Prophage a circular plasmid
P2	Myo	*E. coli*	33.6[82]	+		DNA packaged from monomeric circles
P4	Myo	*E. coli*	11.6[83]	+		Defective; satellite of phage P2
P22	Podo	*Salmonella typhimurium*	41.7[84]	+	generalized	Transduction via lytic cycle; modifies O antigen
phiKZ(φKZ)	Myo	*Pseudomonas*	280[85]		generalized	Transduction at very low frequency
T1	Sipho	*E. coli*	48.8[86]			Viable even after dried out
T4[d]	Myo	*E. coli*	169[87]			5-hmdC replaces C in DNA
T5	Sipho	*E. coli*	122[88]			Encodes tRNAs for all 20 amino acids
T7[e]	Podo	*E. coli* (F^- only)	39.9[89]			Many biotech applications
		Gram-positive bacteria				
C_1	Podo	group C *Streptococci*	16.7[90]			Potential medical use of "lysin"
G	Myo	*Bacillus megaterium*	498[91]			Largest characterized virus
L5	Sipho	*Mycobacteria*	52.3[92]	+	specialized	Useful for *Mycobacterial* genetics
MM1	Sipho	*Streptococcus pneumoniae*	40.2[93]	+		NOT coliphage MM1; many unknown genes
phi11(φ11)	Sipho	*Staphylococcus aureus*	43.6[94]	+	generalized	Closely related to pathogenic phage φ13
phi29(φ29)	Podo	*Bacillus subtilis*	19.3[95]			Biotech use of DNA polymerase
phi105(φ105)	Sipho	*B. subtilis*	39.3[96]	+	specialized	Similar to lambda
Sb-1	Myo	*S. aureus*	120[97]			Resistance rare; therapeutic use
SPβ	Sipho	*B. subtilis, B. globigii*	134[98]	+	specialized	Secretes betacin, kills nearby cells
SPO1	Myo	*B. subtilis*	140[99]			Transcriptional sigma cascade; hmdU replaces T

* For more detailed discussion of the phages in this table (and related phages), see Chapter 3 in *Bacteriophages: Biology and Applications.*[1]

[a] Virus type[100]: Myoviridae: viruses with contractile tails; Siphoviridae: viruses with long, noncontractile tails; and Podoviridae: viruses with short noncontractile tails. [b] Genome size in kbp. [c] + = known to be capable of lysogeny. [d] T2 and T6 very closely related to T4. [e] T3 very closely related to T7.

pathogen *Pseudomonas syringae* by initially adsorbing to the pilus, though other family members do not use the pilus for entry.

PRD1: the prototype member of the *Tectiviridae,* with a 14.9 kb linear double-stranded DNA genome,[62] and an icosahedral virion coating an internal lipid membrane. Infects Gram-negative bacteria including Enterobacteriaceae (*Escherichia coli, Salmonella*, etc.) and *Pseudomonas*. Certain conjugative plasmids required for phage adsorption to the cell wall (not pili).[63] Has garnered attention as a model for human adenovirus.

REFERENCES

1. Guttman, B., Raya, R., and Kutter, E. Basic phage biology (Chapter 3), *Bacteriophages: Biology and Applications,* Kutter, E. and Sulakvelidze A., Eds., CRC Press, Boca Raton, FL, 2005, 29–66.
2. Kutter, E., Raya, R., and Carlson, K. Molecular mechanisms of phage infection (Chapter 7), in *Bacteriophages: Biology and Applications,* Kutter, E. and Sulakvelidze, A., Eds., CRC Press, Boca Raton, FL, 2005, 165–222.
3. Molineux, I..J. No syringes please, ejection of phage T7 DNA from the virion is enzyme driven. *Mol. Microbiol.,* 40, 1–8, 2001.
4. Prangishvili, D. Evolutionary insights from studies on viruses of hyperthermophilic archaea. *Res. Microbiol.,* 154, 289–294, 2003.
5. Abedon, S.T. and Calendar, R.L. (Eds.). *The Bacteriophages*, second edition, Oxford University Press, New York, 2005, Part III, p. 129–210.
6. Levin, B.R. and Lenski, R.E. Bacteria and phage: a model system for the study of the ecology and co-evolution of hosts and parasites. *Ecology and Genetics of Host-Parasite Interactions*, The Linnean Society of London, 1985, 227–241.
7. d'Herelle, F. *The Bacteriophage and Its Behavior.* The Williams and Wilkins Company, Baltimore, MD, 1926.
8. Adams, M.H. *Bacteriophages*. Interscience Publishers, New York, 1959.
9. Tikhonenko, A.S., Gachechiladze, K.K., Bespalova, I.A., Kretova, A.F., and Chanishvili, T.G. Electron-microscopic study of the serological affinity between the antigenic components of phages T4 and DDVI. *Mol. Biol.* (Mosk), 10, 667–673, 1976.
10. Gachechiladze, K. Antigenicity of phages. *Bacteriophages: Biology and Applications,* Kutter, E. and Sulakvelidze A., Eds., CRC Press, Boca Raton, FL, 2005, 33–37.
11. Ackermann, H.W. and Dubow, M. *Viruses of Prokaryotes*. CRC Press, Boca Raton, FL, 1987.
12. Nikaido, H. Outer Membrane. *Escherichia coli and Salmonella*, Neidhardt, F.C. et al., Eds., American Society for Microbiology, Washington, DC, 1996, 29–47.
13. Nikaido, H., Molecular basis of bacterial outer membrane permeability revisited. *Microbiol. Mol. Biol. Rev.,* 67, 593–656, 2003.
14. Ghuysen, J.M. and Hackenbeck, R. *Bacterial Cell Wall*. Elsevier, Amsterdam, 1994.
15. Letellier, L., Boulanger, P., Plancon, L., Jacquot, P., and Santamaria, M. Main features on tailed phage, host recognition and DNA uptake. *Front. Biosci*., 9, 1228–1339, 2004.
16. McAllister, W.T. Bacteriophage infection: which end of the SP82G genome goes in first? *J. Virol*., 5, 194–198, 1970.
17. Duplessis, M. and Moineau, S. Identification of a genetic determinant responsible for host specificity in *Streptococcus thermophilus* bacteriophages. *Mol. Microbiol*., 41, 325–336, 2001.
18. Ravin, V., Raisanen, L., and Alatossava, T. A conserved C-terminal region in Gp71 of the small isometric-head phage LL-H and ORF474 of the prolate-head phage JCL1032 is implicated in specificity of adsorption of phage to its host, *Lactobacillus delbrueckii. J. Bacteriol*., 184, 2455–2459, 2002.
19. Stuer-Lauridsen, B., Janzen, T., Schnabl, J., and Johansen, E. Identification of the host determinant of two prolate-headed phages infecting *Lactococcus lactis. Virology,* 309, 10–17, 2003.
20. Kutter, E., Stidham, T., Guttman, B., Kutter, E., Batts, D., Peterson, S., Djavakhishvili, T., et al. Genomic map of bacteriophage T4. *Molecular Biology of Bacteriophage T4*, J. Karam, J.W. Drake, K.N. Kreuzer, G. Mosig, D.H. Hall, F.A. Eiserling, L.W. Black, et al., Eds., American Society for Microbiology, Washington, DC, 1994, 491–519.
21. Goldberg, E., Grinius, L., and Letellier, L. Recognition, attachment, and injection. *Molecular Biology of Bacteriophage T4*, J.D. Karam, J.W. Drake, K.N. Kreuzer, G. Mosig, D.H. Hall, F.A. Eiserling, L.W. Black, et al., Eds., American Society for Microbiology, Washington, DC, 1994, 347–356.

22. Granoff, A. and Webster, R.G., Eds.). *Encyclopedia of Virology*, Academic Press, New York, 1999.
23. Ellis, E.L. and Delbrück, M. The growth of bacteriophage. *J. Gen. Physiol.*, 22, 365–384, 1939.
24. Doermann, A.H. The vegetative state in the life cycle of bacteriophage: evidence for its occurrence and its genetic characterization. *Cold Spring Harbor Symp. Quant. Biol.*, 18, 3–11, 1953.
25. Wollman, E.L. and Stent, G.S. Studies on activation of T4 bacteriophage by cofactor. IV. Nascent activity. *Biochem. Biophys. Acta*, 9, 538–550, 1952.
26. Brown, D.T. and Anderson, T.F. Effect of host cell wall material on the adsorbability of cofactor-requiring T4. *J. Virol.*, 4, 94–108, 1969.
27. Simon, L.D. The infection of *Escherichia coli* by T2 and T4 bacteriophages as seen in the electron microscope. III. Membrane-associated intracellular bacteriophages. *Virology*, 38, 285–296, 1969.
28. Evans, A.C. The potency of nascent *Streptococcus* bacteriophage B. *J. Bacteriol.*, 39, 597–604, 1940.
29. Lenski, R.E. Coevolution of bacteria and phage: are there endless cycles of bacterial defenses and phage counterdefenses? *J. Theor. Biol.*, 108, 319–325, 1984.
30. Montag, D., Hashemolhosseini, S. and Heming, U. Receptor-recognizing proteins of T-even type bacteriophages. The receptor-recognizing area of proteins 37 of phages T4 TuIa and TuIb *J. Mol. Biol.* 216, 327–334, 1990.
31. Wang, I.N., Deaton, J., and Young, R. Sizing the holin lesion with an endolysin-betagalactosidase fusion. *J. Bacteriol.*, 185, 779–787, 2003.
32. Abedon, S.T. Selection for lysis inhibition in bacteriophage. *J. Theor. Biol.*, 146, 501–511, 1990.
33. Abedon, S.T. Phage Ecology (Chapter 5). *The Bacteriophages*, second edition, Abedon, S.T. and Calendar, R.L., Eds., Oxford University Press, New York, 2005, 37–48.
34. Young, R. and Wang, I.N. Phage Lysis (Chapter 10). *The Bacteriophages*, second edition, Abedon, S.T. and Calendar, R.L., Eds., Oxford University Press, New York, 2005, 104–128.
35. Lwoff, A. and Gutmann, A. Recherches sur un *Bacillus megaterium* lysogene. *Ann. Inst. Pasteur,* 78, 711–739, 1950.
36. Lederberg, E.M. and Lederberg, J. Genetic studies of lysogenicity in *Escherichia coli*. *Genetics*, 38, 51–64, 1953.
37. Jacob, F. and Wollman, E.L. *Sexuality and the Genetics of Bacteria*, Academic Press, New York, 1961.
38. Little, J.W. Gene Regulatory Circuitry of Phage λ (Chapter 8), *The Bacteriophages*, Abedon, S.T. and Calendar, R.L., Eds., Oxford University Press, New York, 2005, 74–82.
39. Dodd, I.B. and Egan, J.B. Action at a distance in CI repressor regulation of the bacteriophage 186 genetic switch. *Mol. Microbiol.*, 45, 697–710, 2002.
40. Zinder, N.D. and Lederberg, J. Genetic exchange in *Salmonella*. *J. Bacteriol.*, 64, 679–699, 1952.
41. Morse, M.L., Lederberg, E.M., and Lederberg, J. Transduction in *Escherichia coli* K-12. *Genetics*, 41, 142–156, 1956.
42. Briani, F., Deho, G., Forti, F., and Ghisotti, D. The plasmid status of satellite bacteriophage P4. *Plasmid,* 45, 1–17, 2001.
43. Broudy, T.B. and Fischetti, V.A. *In vivo* lysogenic conversion of Tox(–) *Streptococcus pyogenes* to Tox(+) with lysogenic Streptococci or free phage. *Infect. Immun.*, 71, 3782–3786, 2003.
44. Waldor, M.K., Friedman, D.I., and Adhya, S.L., Eds., *Phages: Their Role in Bacterial Pathogenesis and Biotechnology*. ASM Press, Washington DC, 2005.
45. Boyd, E.F. and Brüssow, H. Common themes among bacteriophage-encoded virulence factors and diversity among the bacteriophages involved. *Trends Microbiol.*, 10, 521–529, 2002.
46. Holmes, R.K. Biology and molecular epidemiology of diphtheria toxin and the *tox* gene, *J. Infect. Dis.*, 181(Suppl. 1), S156–167, 2000.
47. Waldor, M.K. and Mekalanos, J.J. Lysogenic conversion by a filamentous phage encoding cholera toxin, *Science*, 272, 1910–1914, 1996.
48. Spangler, B.D. Structure and function of cholera toxin and the related *Escherichia coli* heat-labile enterotoxin, *Microbiol. Rev.*, 56, 622–647, 1992.
49. Skorupski, K. and Taylor, R.K. Control of the ToxR virulence regulon in *Vibrio cholerae* by environmental stimuli, *Mol. Microbiol.*, 25, 1003–1009, 1997.
50. Cotter, P.A. and DiRita, V.J. Bacterial virulence gene regulation: an evolutionary perspective. *Annu. Rev. Microbiol.*, 54, 519–565, 2000.
51. Davis, B.M., Kimsey, H.H., Kane, A.V., and Waldor, M.K. A satellite phage-encoded antirepressor induces repressor aggregation and cholera toxin gene transfer. *EMBO J.*, 21, 4240–4249, 2002.

52. Davis, B.M. and Waldor, M.K. Filamentous phages linked to virulence of *Vibrio cholerae*, *Curr. Opin. Microbiol.*, 6, 35–42, 2003.
53. Wagner, P.L., Neely, M.N., Zhang, X., Acheson, D.W., Waldor, M.K., and Friedman, D.I. Role for a phage promoter in Shiga toxin 2 expression from a pathogenic *Escherichia coli* strain, *J. Bacteriol.*, 183, 2081–2085, 2001.
54. Wagner, P.L., Livny, J., Neely, M.N., Acheson, D.W., Friedman, D.I., and Waldor, M.K. Bacteriophage control of Shiga toxin 1 production and release by *Escherichia coli*, *Mol. Microbiol.*, 44, 957–970, 2002.
55. Fujisawa, H. and Morita, M. Phage DNA packaging. *Genes Cells*, 2, 537–545, 1997.
56. Calendar, R., Ed. *The Bacteriophages*, first edition, Plenum Press, New York, 1988.
57. Abedon, S.T. and Calendar, R.L., Eds. *The Bacteriophages*, second edition, Oxford University Press, New York, 2005.
58. Sanger, F., Coulson, A.R., Friedmann, C.T., Air, G.M., Barrell, B.G., Brown, N.L., Fiddes, J.C., Hutchison III, C.A., Slocombe, P.M., and Smith, M. The nucleotide sequence of bacteriophage phi X174. *J. Mol. Biol.* 125, 225–246, 1978.
59. van Wezenbeek, P.M., Hulsebos, T.J., and Schoenmakers, J.G. Nucleotide sequence of the filamentous bacteriophage M13 DNA genome: comparison with phage fd. *Gene*, 11, 129–148, 1980.
60. Fiers, W., Contreras, R., Duerinck, F., Haegeman, G., Iserentant, D., Merregaert, J., Min Jou, W., Molemans, F., Raeymaekers, A., Van den Berghe, A., Volckaert, G., and Ysebaert, M. Complete nucleotide sequence of bacteriophage MS2 RNA: primary and secondary structure of the replicase gene. *Nature*, 260, 500–507, 1976.
61. Mindich, L. Precise packaging of the three genomic segments of the double-stranded-RNA bacteriophage phi6. *Microbiol. Mol. Biol. Rev.*, 63, 149–160, 1999.
62. Bamford, J.K.H., Hänninen, A.-L., Pakula, T.M., Ojala, P.M., Kalkkinen, N., Frilander, M., and Bamford, D.H. Genome organization of membrane-containing bacteriophage PRD1. *Virology*, 183, 658–676, 1991.
63. Olsen, R.H., June-Sang Siak, J.-S., and Gray, R.H. Characteristics of PRD1, a plasmid-dependent broad host range DNA bacteriophage. *J. Virol.*, 14, 689–699, 1974.
64. Matias, V.R. and Beveridge, T.J. Native cell wall organization shown by cryo-electron microscopy confirms the existence of a periplasmic space in *Staphylococcus aureus*. *J. Bacteriol.*, 188, 1011–1021, 2006.
65. Robbins, P.W. and Uchida, T. Determinants of specificity in *Salmonella*: changes in antigenic structure mediated by bacteriophage. *Fed. Proc.*, 21, 702–720, 1962.
66. Losick, R. Isolation of a trypsin-sensitive inhibitor of O-antigen synthesis involved in lysogenic conversion by bacteriophage epsilon-15. *J. Mol. Biol.*, 42, 237–246, 1969.
67. Wright, A. Mechanism of conversion of the *salmonella* O antigen by bacteriophage epsilon 34. *J. Bacteriol.*, 105, 927–936, 1971.
68. Smith, H.W. and Parsell, Z. The effect of virulence of converting the O antigen of *Salmonella cholerae-suis* from 627 to 617 by phage. *J. Gen. Microbiol.*, 81, 217–224, 1974.
69. Mirold, S., Rabsch, W., Rohde, M., Stender, S., Tschape, H., Russmann, H., Igwe, E., and Hardt, W.D. Isolation of a temperate bacteriophage encoding the type III effector protein SopE from an epidemic *Salmonella typhimurium* strain. *Proc. Natl. Acad. Sci. U.S.A.*, 96, 9845–9850, 1999.
70. Figueroa-Bossi, N. and Bossi, L. Inducible prophages contribute to *Salmonella* virulence in mice. *Mol. Microbiol.*, 33, 167–176, 1999.
71. Groman, N.B. Conversion by corynephages and its role in the natural history of diphtheria. *J. Hyg. (London)*. 93, 405–417, 1984.
72. Muniesa, M., de Simon, M., Prats, G., Ferrer, D., Panella, H., and Jofre, J. Shiga toxin 2-converting bacteriophages associated with clonal variability in *Escherichia coli* O157:H7 strains of human origin isolated from a single outbreak. *Infect. Immun.*, 71, 4554–4562, 2003.
73. Huang, A., Friesen, J., and Brunton, J.L. Characterization of a bacteriophage that carries the genes for production of Shiga-like toxin 1 in *Escherichia coli*. *J. Bacteriol.*, 169, 4308–4312, 1987.
74. Barondess, J.J. and Beckwith, J. *bor* gene of phage λ, involved in serum resistance, encodes a widely conserved outer membrane lipoprotein. *J. Bacteriol.* 177, 1247–1253, 1995.
75. Coleman, D.C., Sullivan, D.J., Russell, R.J., Arbuthnott, J.P., Carey, B.F., and Pomeroy, H.M. *Staphylococcus aureus* bacteriophages mediating the simultaneous lysogenic conversion of beta-lysin, staphylokinase and enterotoxin A: molecular mechanism of triple conversion. *J. Gen. Microbiol.*, 135, 1679–1697, 1989.

76. Johnson, L.P., Schlievert, P.M., and Watson, D.W. Transfer of group A streptococcal pyrogenic exotoxin production to nontoxigenic strains of lysogenic conversion. *Infect. Immun.*, 28, 254–257, 1980.
77. Sanger, F., Coulson, A.R., Hong, G.F., Hill, D. F. and Petersen, G.B., Nucleotide sequence of bacteriophage lambda DNA. *J. Mol. Biol.,* 162: 729–773, 1982.
78. Morgan, G.J., Hatfull, G.F., Casjens, S., and Hendrix, R.W. Bacteriophage Mu genome sequence: analysis and comparison with Mu-like prophages in Haemophilus, Neisseria and Deinococcus. *J. Mol. Biol.,* 317, 337–359, 2002.
79. GenBank ACCESSION # NC_008720.
80. Ravin, V., Ravin, N., Casjens, S., Ford, M.E., Hatfull, G.F., and Hendrix, R.W. Genomic sequence and analysis of the atypical temperate bacteriophage N15. *J. Mol. Biol.*, 299, 53–73, 2000.
81. Lobocka, M.B., Rose, D.J., Plunkett, G., 3rd, Rusin, M., Samojedny, A., Lehnherr, H., Yarmolinsky, M.B., and Blattner, F.R. Genome of bacteriophage P1. *J. Bacteriol.,* 186, 7032–7068, 2004.
82. GenBank ACCESSION # NC_001895.
83. Halling, C., Calendar, R., Christie, G.E., Dale, E.C., Dehò, G., Finkel, S., Flensburg, J., Ghisotti, D., Kahn, M.L., and Lane, K.B. DNA sequence of satellite bacteriophage P4. *Nucleic Acids Res.,* 18, 1649, 1990.
84. Pedulla, M.L., Ford, M.E., Karthikeyan, T., Houtz, J.M., Hendrix, R.W., Hatfull, G.F., Poteete, A.R., Gilcrease, E.B., Winn-Stapley, D.A., and Casjens, S.R. Corrected sequence of the bacteriophage p22 genome. *J. Bacteriol.* , 185, 1475–1477, 2003.
85. Mesyanzhinov, V.V., Robben, J., Grymonprez, B., Kostyuchenko, V.A., Bourkaltseva, M.V., Sykilinda, N.N., Krylov, V.N., et al., The genome of bacteriophage ϕKZ of *Pseudomonas aeruginosa*. *J. Mol. Biol.*, 317: 1–19, 2002.
86. Roberts, M.D., Martin, N.L., and Kropinski, A.M. The genome and proteome of coliphage T1. *Virology*, 318, 245–266, 2004.
87. Miller, E.S., Kutter, E., Mosig, G., Arisaka, F., Kunisawa, T., and Ruger, W. Bacteriophage T4 genome. *Microbiol. Mol. Biol. Rev.*, 67, 86–156, 2003.
88. GenBank ACCESSION # NC_005859.
89. Dunn, J.J. and Studier, F.W. Complete nucleotide sequence of bacteriophage T7 DNA and the locations of T7 genetic elements. *J. Mol. Biol.*, 166, 477–535, 1983.
90. Nelson, D., Schuch, R., Zhu, S., Tscherne, M., and Fischetti, V.A., Genomic sequence of C1, the streptococcal phage. *J. Bacteriol.,* 185: 3325–3332, 2003.
91. Dore, E., Frontali, C., and Grignoli, M. The molecular complexity of G DNA. *Virology*, 79, 442–445, 1977.
92. Hatfull, G.F. and Sarkis, G.J. DNA sequence, structure and gene expression of mycobacteriophage L5: a phage system for mycobacterial genetics. *Mol. Microbiol.*, 7, 395–405, 1993.
93. Obregon, V., Garcia, J.L., Garcia, E., Lopez, R., and Garcia, P. Genome organization and molecular analysis of the temperate bacteriophage MM1 of *Streptococcus pneumoniae*. *J. Bacteriol.*, 185: 2362–2368, 2003.
94. Iandolo, J.J., Worrell, V., Groicher, K.H., Qian, Y. Tian, R., Kenton, S., Dorman, A., Ji, H., Lin, S., Loh, P., Qi, S., Zhu, H., and Roe, B. A. Comparative analysis of the genomes of the temperate bacteriophages phi 11, phi 12 and phi 13 of *Staphylococcus aureus* 8325. *Gene*, 289, 109–118, 2002.
95. Vlcek, C. and Paces, V. Nucleotide sequence of the late region of *Bacillus* phage phi 29 completes the 19,285-bp sequence of phi 29 genome. Comparison with the homologous sequence of phage PZA. *Gene*, 46, 215–225, 1986.
96. GenBank ACCESSION # AB016282.
97. Rezo Adamia, personal communication.
98. GenBank ACCESSION # AF020713.
99. Parker, M.L., Ralston, E.J., and Eiserling, F.A. Bacteriophage SPO1 structure and morphogenesis. II. Head structure and DNA size. *J. Virol.,* 46, 250–259, 1983.
100. Maniloff, J. and Ackermann, H.W. Taxonomy of bacterial viruses: establishment of tailed virus genera and the order Caudovirales. *Arch Virol.*, 143, 2051–2063, 1998.
101. Campbell, A. Phage integration and chromosome structure. A personal history. *Annu. Rev. Genet.*, 41, 1–11, 2007.

44 Phage Therapy
Bacteriophages as Natural, Self-Replicating Antimicrobials

Elizabeth Kutter

CONTENTS

INTRODUCTION

Bacteriophages — specific kinds of viruses that can only replicate in bacteria — have been discussed in much detail in the previous chapter. The art of using these phages to kill pathogenic microorganisms was first developed early in the last century; but because chemical antibiotics became available in the 1940s, phage therapy has been little used in the West. Today, however, the growing incidence of bacteria that are resistant to most or all available antibiotics is leading to widespread, renewed interest in the possibilities of phage therapy.[1–11] This overview is designed to put phage therapy into historical and ecological context; to briefly explore some of the most interesting and extensive applications carried out in eastern Europe, conducted primarily in a clinical rather than controlled-research mode; and to look at the original decline and current renaissance of phage therapy work in the West. Ironically, this renaissance takes full advantage of the advances in molecular biology that were themselves made possible by fundamental work with bacteriophages beginning in the 1940s under the leadership of Max Delbrück. The rapidly increasing emergence of multiantibiotic-resistant pathogenic bacteria has rekindled the interest of the Western scientific community, industry, and the general public in this almost century-old approach — as documented by the number of new phage therapy-related publications in the Western peer-reviewed and popular literature and the formation of new biotechnology firms commercializing phage-based technology in the West. Detailed discussions of phage therapy for both human and agricultural applications can be found in Kutter and Sulakvelidze,[7] along with chapters on related technologies, such as the use by Vincent Fischetti, Dan Nelson, and co-workers of phage lysin to kill *Streptococcus pneumoniae*, which holds promise for application in accessible places such as nasal passages. Recent work also suggests potential for phage lysin use in at least some systemic infections and even in biofilms.[12]

Biomedical technology today is very different from what it was in the early days of phage therapy research, and our understanding of biological properties of phages and the basic mechanisms of phage-bacterial host interaction has improved dramatically since the days of early therapeutic uses of bacteriophages. These advances can have a profound impact on the development of safe therapeutic phage preparations having optimal efficacy against their specific bacterial hosts and on designing science-based strategies for integrating phage therapy into our arsenal of tools for preventing and treating bacterial infections.

A major benchmark was reached in 2006 with the approval by the European Union of a listeria phage developed by Exponential Biotherapies Food Safety in Holland for the treatment of cheeses and by the U.S. FDA of an Intralytix phage cocktail against listeria in packaged meats and other food cocktails; the FDA also later gave GRAS (generally regarded as safe) status to the Exponential Biotherapies product. Intralytix also supplied the cocktail of eight sequenced phages being used in the first FDA-approved clinical trials in the United States, a physician-initiated test of phage against staphylococcus, pseudomonas, and *Escherichia coli* in leg ulcers begun in 2006 by Randy Wolcott in his Wound Care Center in Lubbock, Texas. In England, 2006 saw the advent of a clinical trial of phage against *Pseudomonas aeruginosa* in human ear infections, led by James Soothill and building on a small successful trial in dog-ear infections. In Poland, a phage-therapy clinic approved under European Union experimental-therapies guidelines has been set up by Andrzej Gorski, director of the Institute of Immunology and Experimental Therapy in Wroclaw, long a center of phage therapy research, as discussed below.

EARLY HISTORY

Edward Twort and Felix d'Herelle independently reported isolating these filterable entities that could lyse bacterial cultures; they are jointly given credit for the discovery. It was d'Herelle, a Canadian working at the Pasteur Institute in Paris, who gave them the name "bacteriophages" — using the suffix *phage* "not in its strict sense of *to eat*, but in that of *developing at the expense of*."[13] From the time he first discovered phages, d'Herelle was excited about their relationship to disease and their potential as therapeutic agents. Their discovery came while he was doing clinical work at the Pasteur Institute and exploring why enteric bacteria are only sometimes pathogenic. He was called to investigate an outbreak of bacillary dysentery in a group of French soldiers and examined their filtered stool samples for signs of invisible viruses that might alter the pathogenicity of the bacteria from the dysentery patients. He indeed observed clear, round spots in the confluent bacterial cultures covering his agar slants, and determined that the responsible agent multiplied indefinitely as long as appropriate living cells were present, and that cell lysis was required for multiplication.

Shortly after publishing his first paper describing bacteriophages (1917), d'Herelle used specially selected phages to treat avian typhoid in chickens and then human bacterial dysentery. The latter studies were conducted in 1919 with Victor Hutinel, chief of pediatrics at the Hôpital des Enfants-Malades in Paris.[6] A dose 100-fold higher than the therapeutic dose was first ingested by d'Herelle, Hutinel, and several interns, and when none of the volunteers showed any side effects a day later, Hutinel approved the first known human therapeutic use of phage. The first patient, a 12-year-old boy with severe dysentery (ten to twelve bloody stools per day), was given 2 mL phage orally after taking samples for microbiological analysis. The patient's condition rapidly improved after phage ingestion; he passed three more bloody stools that afternoon and one non-bloody stool during the night, and all symptoms disappeared by the next morning. Three brothers, ages 3, 7, and 10 years, were treated the following month; they were admitted in grave condition after their sister died of dysentery at home and each was given one dose of phage. All three started to recover within 24 hours.

Before attempting further clinical phage trials, d'Herelle turned to basic research, working out the details of phage infection of different bacterial hosts under a variety of environmental conditions. His 300-page 1922 book *The Bacteriophage*[13] includes excellent descriptions of plaque formation and composition, infective centers, the lysis process, host specificity of adsorption and multiplication, the dependence of phage production on the precise state of the host, isolation of phages from sources of infectious bacteria, and the factors controlling stability of the free phage. He quickly became fascinated with the apparent role of phages in the natural control of microbial infections. He noted, for example, the frequent specificities of the phages isolated from recuperating patients for disease organisms infecting them and the rather rapid variations over time of the phage populations. However, others did not wait for a better understanding of phage biology before applying them clinically. The first known published report of successful phage therapy came from Bruynoghe and Maisin,[14] who used phage to treat staphylococcal skin infections.

Throughout his life, d'Herelle worked to develop the therapeutic potential of properly selected phages against the most devastating health problems of the day. After travel to study epidemics in Latin America and a year at the Pasteur branch in Saigon, d'Herelle left the Pasteur Institute in 1922. He worked in Holland and then in Alexandria, Egypt, as a health officer for the League of Nations, applying phage therapy and sanitation measures to deal with major outbreaks of infectious disease throughout the Middle East and India. In 1928, he was invited to Stanford to give the prestigious Lane lectures, which were published as the monograph, *The Bacteriophage and its Clinical Applications*.[15] He accepted a regular faculty position at Yale but continued to spend summers in Paris working with the successful phage company he had established there, run by his son-in-law. He returned permanently to France in 1933, with excursions to Tbilisi, Georgia, to help establish phage work there. George Eliava, director of the Georgian Institute of Microbiology, saw bacteriocidal action of the water of the Koura river in Tbilisi, which he could not explain until he became familiar with d'Herelle's work while spending 1920–1921 at the Pasteur Institute. He became a very early collaborator of d'Herelle's.[13] The two developed the dream of founding an Institute of Bacteriophage Research in Tbilisi — to be a world center of phage therapy for infectious disease, including scientific and industrial facilities and supplied with its own experimental clinics. The dream quickly became a reality due to the support of Sergo Orjonikidze, the People's Commissar of Heavy Industry, despite KGB opposition to this "foreign project." A large campus on the river Mtkvari was allotted for the project in 1926. D'Herelle sent supplies, equipment, and library materials.

In 1934 and 1935, d'Herelle visited Tbilisi for 6 months and wrote a book on *The Bacteriophage and the Phenomenon of Recovery*.[16] He intended to move to Georgia; a cottage built for his use still stands on the institute's grounds. However, in 1937, Eliava was arrested as a "people's enemy" by Beria, then head of the KGB in Georgia and soon to direct the Soviet KGB. Eliava was executed, sharing the tragic fate of many Georgian and Russian progressive intellectuals of the time, and d'Herelle, disillusioned, never returned to Georgia. However, their institute survived and is still functioning at its original site on the Mtkvari (which it now shares with the more modern Institutes of Molecular Biology & Biophysics and of Animal Physiology). In 1951, it was formally transferred to the All-Union Ministry of Health set of Institutes of Vaccine and Sera, taking on the leadership role in providing bacteriophages for therapy and bacterial typing throughout the former Soviet Union. Under orders from the Ministry of Health, hundreds of thousands of samples of pathogenic bacteria were sent to the institute from throughout the Soviet Union to isolate more effective phage strains and to better characterize their usefulness. The Institute's industrial branch made up to 2 tons of phage products twice a week, 80% of it for the Soviet Army.[17] The challenges they faced following the break-up of the Soviet Union, loss of their main markets and forced privatization of the industrial arm, and more recent evolution into the Eliava Institute of Bacteriophage, Microbiology and Virology are further discussed below.

Early Attempts at Commercialization

From the beginning, a major medically important commercial use of phages has been for bacterial identification — this *phage typing* uses patterns of sensitivity to a specific battery of phages with particularly narrow host ranges to precisely identify microbial strains. This technique takes advantage of the fine specificity of many phages for their hosts and is still in common use around the world for epidemiological and diagnostic purposes, using various approved, carefully defined sets of commercially prepared phages.

Therapeutic administration of phages was explored extensively in many parts of the world, with successes being reported for a variety of diseases, including dysentery, typhoid and paratyphoid fevers, pyogenic and urinary tract infections, and cholera. Phages have been given orally, through colon infusion, as aerosols, and poured directly on lesions. They have also been given as injections: intradermal, intravascular, intramuscular, intraduodenal, intraperitoneal, and even into the lung, carotid artery, and pericardium. The early, strong interest in phage therapy is reflected in some 800 articles published on the topic between 1917 and 1956, many of which have been reviewed by Ackermann and DuBow,[18] and by Kutter and Sulakvelidze.[7] The reported results were quite variable. Many of the early physicians and entrepreneurs who became so excited by the potential clinical implications jumped into applications with very little understanding of phages, microbiology, or basic scientific process. Thus, many of the studies were anecdotal and/or poorly controlled; many of the failures were predictable, and some of the reported successes did not make much scientific sense. Often, uncharacterized phages, at unknown concentrations, were given to patients without specific bacteriological diagnosis, with no mention of follow-up, controls, or placebos.

Much of the understanding gained by d'Herelle was ignored in this early work, and inappropriate methods of preparation, "preservatives," and storage procedures were often used. On one occasion, d'Herelle reported testing 20 preparations from various companies and finding that none of them contained active phages. On another occasion, a preparation was advertised as containing a number of different phages, but the technician responsible had decided it was easiest to grow them in one large batch and one phage had out-competed all the others, so this was not, in fact, a polyvalent preparation. In general, there was no quality control except in a few research centers. Large clinical studies were rare, and the results of those that were carried out were largely inaccessible outside eastern Europe.

Specific Problems of Early Phage Therapy Work

Many still believe (erroneously) that phage therapy was proven not to work; however, it simply was never adequately researched, and the work that was done well is not widely enough known. It is thus important to carefully consider the reasons for the early problems and the question of efficacy. These included:

- Paucity of understanding the heterogeneity and ecology of either the phages or the bacteria involved
- Lack of availability or reliability of bacterial laboratories for carefully identifying the pathogens involved (necessitated by the relative specificity of phage therapy)
- Failure to select phages of high virulence against the target bacteria and test them *in vitro* before using them in patients
- Use of single phages in infections that involved mixtures of different bacterial species and strains
- Emergence of resistant bacterial strains (especially if only one phage strain is used against a particular bacterium); this can also happen by lysogenization if one uses temperate phages, discussed in the previous chapter
- Failure to neutralize gastric pH prior to oral phage administration

- Inactivation of phages by both specific and non-specific factors in bodily fluids, especially if phages are used intravenously

Key technical developments that helped clarify the general nature and properties of bacteriophages included:

- The concentration and purification of some large phages by means of very high-speed centrifugation and the demonstration that they contained equal amounts of DNA and protein[19]
- Visualization of phages by means of the electron microscope (EM)[20, 21]

A much better understanding of the interactions between lytic phages and bacteria began with one-step growth curve experiments of Ellis and Delbrück,[22] and the studies of Doermann.[23] These demonstrated an eclipse period during which the DNA began replicating, and there were no free phages in the cell; a period of accumulation of intracellular phages; and a lysis process that released the phages to go in search of new hosts, as discussed in the previous chapter. While this same pattern had been described by d'Herelle,[13] few people were aware of that work.

In 1943, an event occurred that was to have a major impact on the orientation of phage research in the United States and much of western Europe, strongly shifting the emphasis from practical applications to basic science. Physicist-turned-phage biologist Max Delbrück met with Alfred Hershey and Salvador Luria and formed the "Phage Group," which rapidly expanded through the influence of the summer "Phage Course" and phage meetings at Cold Spring Harbor, Long Island in 1945. Their influence on the origins of molecular biology has been well documented.[24, 25] A major element of the successes of phages as model systems for working out fundamental biological principles was that Delbrück convinced most phage biologists in the United States to focus on one bacterial host (*Escherichia coli* B) and seven of its most lytic phages, which he renamed T(type)1 through T7. As it turned out, T2, T4, and T6, originally isolated for potential therapeutic applications, were quite similar to each other, defining a family now called the "T-even phages." These phages were key in demonstrating that DNA is the genetic material, that viruses can encode enzymes, that gene expression is mediated through special copies in the form of "messenger RNA," that the genetic code is triplet in nature, and many other fundamental concepts. The negative side of this strong focus on a few phages growing under rich laboratory conditions, however, was that there was very little study or awareness of the ranges, roles, and properties of bacteriophages in the natural environment, or of potential applications.

PHAGES AND THE IMMUNE SYSTEM

A number of early experiments involving the injection of phages into animals led to the widespread impression that phage therapy could not in fact succeed because the phage were too rapidly cleared by the immune system; the remarks in this regard by Gunter Stent[26] had a particularly strong impact on the phage community. For example, two early experiments involving rabbits showed rapid disappearance of the particular phages used from the blood and organs, but long-term survival in the spleen.[27, 28] Subsequent experiments in rats and mice also showed rapid loss from the circulation. When Nungester and Watrous in 1934 injected 10^9 PFU of a staph phage intravenously into albino rats, a blood concentration of only 10^5 PFU mL^{-1} was seen after 5 min, and this dropped to 40 PFU mL^{-1} by 2 hr.[29, 30]

The pessimistic conclusions broadly reached based on this research, however, had two serious flaws. First, the experiments were done in the absence of host bacteria in which the phage could multiply and find protection. Furthermore, they were carried out by the very unnatural mode of intravenous injection, exposing the phage almost immediately to the reticulo-endothelial system. Many more recent results have made it clear that phages are often seen in the mammalian circulatory system; however, this generally occurs under conditions where they are entering the circulatory system from some sort of reservoir in other tissues and where the mammal is dealing with an infection by a bacterium that the phage can infect — precisely the sort of situation seen in gene therapy as currently practiced in eastern Europe.

One of the best early sets of experiments was published in 1943 by the noted Harvard bacteriologist René Dubos.[31] He injected white mice intracerebrally with a dose of a smooth *Shigella dysenteriae* strain that was sufficient to kill greater than 95% of the mice in 2 to 5 days and treated them by intraperitoneal injection of a phage mixture isolated from New York City sewage, grown in the same bacteria and purified only by sterile filtration. With no treatment or when treated with filtrates of uninfected bacterial cultures or with heat-killed phage, only 3/84 (3.6%) survived, whereas 46/64 (72%) of those given 10^7 to 10^9 phages survived. Dubos also carried out pharmacokinetic studies. When phages were given to uninfected mice, they appeared in the bloodstream almost immediately, but the levels started to drop within hours and very few were seen in the brain. In contrast, in the infected animals, brain levels quickly greatly exceeded blood levels; around 10^7 to 10^9 phages per gram were often seen between 8 and 114 hr, starting to drop anywhere between 75 and 138 hr. After the first 18 hr, the levels in the blood were far lower than in the brain, but they were still present at 10^4 to 10^5 per milliliter in those cases where the brain levels were still greater than 10^9 per gram. This clearly established that the phages themselves were responsible, not something in the lysate that just stimulated normal immune mechanisms; that phage could rapidly find and multiply in foci of infection anywhere in the body; and that phage could be maintained in the circulation as long as there was a privileged reservoir of infection where phages were continually being produced. Without providing data or pharmacokinetics, they mention that the mice also were rescued by subcutaneous or intravenous administration of phages, but not by stomach tube or in the water.

Carefully controlled experiments carried out from 1943 to 1945 by Henry Morton's group at the University of Pennsylvania[32] supported those of Dubos. They further showed the lack of any protection when lysates of phage with inappropriate host specificities were used. A final review authorized by the Council on Pharmacy and Chemistry discussed the major advantages of phages, such as the ability to replicate into problem areas and treat localized infections that are relatively inaccessible via the circulatory system and the fact that their high specificity greatly aided in reducing later resistance problems. It also emphasized that almost all the earlier research had been so poorly conceived and/or carried out that it offered no proof either for or against the promise of phage as antibiotics.

United States work with dysentery phages largely ended in 1944 when penicillin was made available to the general public. The military secrecy, the end of the war emergency funding, the rapid rise in antibiotic availability and their broad spectra, and Max Delbrück's success in convincing the phage community to shift its focus to basic mechanistic research involving a few model systems probably all contributed to the fact that there was little U.S. follow-up to these interesting and successful results and few even knew about them — or about two successful subsequent human applications.

Penicillin only worked against some kinds of bacterial infections. Typhoid, for example, was not treatable, and some excellent phage work was carried out in the interim. It was known that the strains of *Salmonella typhi* that created the main pathogenicity problems were those carrying one particular antigen, named Vi (for "virulence"). In 1936, a pair of Canadians identified phages specific against the Vi antigen. In the early 1940s, Walter Ward, of the Los Angeles County Hospital, was trying to deal with repeated serious outbreaks of typhoid that were killing 20% of those afflicted.[33] He tested the Vi-specific phages against mouse typhoid, and found that the death rate fell to 6%, versus 93% in the controls. Some of his colleagues then used these phages to treat typhoid patients; only 3/56 of their treated patients died, versus the 20% mortality they were seeing with the treatments available at the time.[34] Most impressively, the rest rapidly went from being largely comatose to full of vigor, with renewed appetite, in 24 to 48 hr. In 1948 and 1949, near Montreal, Desranleau treated nearly 100 dysentery patients with a *cocktail* of six Vi-specific phages, and the deaths dropped from 20 to 2%.[35] However, by 1947, chloramphenicol had been shown to work well against typhoid, and it was much easier for pharmaceutical companies to deal with, so that seems to have been the end of phage clinical trials in the Western hemisphere.

CLINICAL APPLICATION: PHAGE THERAPY WORK IN THE AGE OF ANTIBIOTICS

The strong understanding of phage biology has the potential to facilitate more rational thinking about the therapeutic process and the selection of therapeutic phages. During the evolution of molecular biology, there was little interaction between those who were so effectively developing the field using phage as tools to understand molecular mechanisms and structures and those working on phage ecology and therapeutic applications. As discussed extensively in the books by Kutter and Sulakvelidze,[7] Häusler,[11] and McGrath and van Sinderen,[36] the latter fields have grown greatly in recent years, spurred on by an increasing awareness of phage variety and roles in maintaining microbial balance in every ecosystem, and by concern about the increasing incidence of nosocomial infections and of bacteria resistant against most or all known antibiotics, as well as by the fact that phages are far more effective than antibiotics in areas where the circulation is bad, and in not disrupting normal flora. This strong sense of the potential importance of phages was seen particularly early on in Poland, France, Switzerland, and the former Soviet Union, where use of therapeutic phages never fully died out and where there has been some ongoing research and clinical experience. In France, Dr. Jean-François Vieu led the therapeutic phage efforts until his retirement some 15 years ago. He worked in the *Service des Entirobactiries* of the Pasteur Institute in Paris and, for example, prepared *Pseudomonas* phages on a case-by-case basis for patients. The experience there is discussed in Vieu[37] and Vieu et al.[38] In Vevey, Switzerland, the small pharmaceutical firm Saphal made "Coliphagine," "Intestiphagine," "Pyophagine," and "Staphagine" in drinkable and injectable forms, salves and sprays into the 1960s.[11] The preparations were officially approved and were paid for by insurance there.

Phage therapy was used extensively in many parts of eastern Europe, and several companies in Russia are reportedly making phages for this purpose. However, most of the research and much of the phage preparation came through key centers in Tbilisi, Georgia, and Wroclaw, Poland. In both cases, the close interactions between research scientists and physicians play an important role in the high degree of success obtained — just as appears to have been the case for d'Herelle's early work.

INSTITUTE OF IMMUNOLOGY AND EXPERIMENTAL MEDICINE, POLISH ACADEMY OF SCIENCES

The most detailed publications documenting phage therapy have come from the group of Stefan Slopek, for many years the director of the the Institute of Immunology and Experimental Medicine, Polish Academy of Sciences, Wroclaw. They published a series of extensive papers describing work carried out from 1981 to 1986 with 550 patients.[39–41] This set of studies involved ten Polish medical centers, including the Wroclaw Medical Academy Institute of Surgery Cardiosurgery Clinic, Children's Surgery Clinic and Orthopedic Clinic, the Institute of Internal Diseases Nephrology Clinic, and Clinic of Pulmonary Diseases. The patients ranged in age from 1 week to 86 years. In 518 of the cases, phage use followed unsuccessful treatment with all available other antibiotics. The major categories of infections treated were:

- Long-persisting suppurative fistulas
- Septicemia
- Abscesses
- Respiratory tract suppurative infections and bronchopneumonia
- Purulent peritonitis
- Furunculosis

In a final summary article, the authors carefully analyzed the results with regard to such factors as the nature and severity of the infection and monoinfection vs. infection with multiple bacteria.[41] Rates of success ranged from 75 to 100% (92% overall), as measured by marked improvement, wound healing, and disappearance of titratable bacteria; 84% demonstrated full elimination of the suppurative process and healing of local wounds. Infants and children did particularly well. Not surprisingly, the poorest results came with the elderly and those in the final stages of extended serious illnesses, two groups with weakened immune systems and generally poor resistance.

The bacteriophages all came from the extensive collection of the Bacteriophage Laboratory of the Institute of Immunology and Experimental Therapy, Polish Academy of Sciences, Wroclaw. In the later studies, some of the specific phages were named. All were virulent, capable of completely lysing the bacteria being treated. In the first study alone, 259 different phages were tested (116 for *Staphylococcus*, 42 for *Klebsiella*, 11 for *Proteus*, 39 for *Escherichia*, 30 for *Shigella*, 20 for *Pseudomonas*, and 1 for *Salmonella*); 40% of them were selected to be used directly for therapy. All of the treatment was in a research mode, with the phage prepared at the institute by standard methods and tested for sterility. Treatment generally involved 10 mL of sterile phage lysate orally half an hour before each meal, with gastric juices neutralized by taking (basic) Vichy water, baking soda, or gelatum. In addition, phage-soaked compresses were generally applied three times a day where dictated by localized infection. Treatment ran for 1.5 to 14 weeks, with an average of 5.3 weeks. For intestinal problems, short treatment sufficed, while long-term use was necessary for such problems as pneumonia with pleural fistula and pyogenic arthritis. Bacterial levels and phage sensitivity were continually monitored, and the phage(s) were changed if the bacteria lost their sensitivity. Therapy was generally continued for 2 weeks beyond the last positive test for the bacteria.

Few side effects were observed; those that were seen seemed to be directly associated with the therapeutic process. On about days 3 to 5, pain in the liver area was often reported, lasting several hours. The authors suggested that this might be related to the extensive liberation of endotoxins as the phage is destroying the bacteria most effectively. In severe cases with sepsis, patients often ran a fever for 24 hr on about days 7 or 8.[39]

Various methods of administration were successfully used, including oral, aerosols and infusion rectally, or in surgical wounds. Intravenous administration was not recommended for fear of possible toxic shock from bacterial debris in the lysates.[39] However, it was clear that the phages readily entered the body from the digestive tract and multiplied internally wherever appropriate bacteria were present, as measured by their presence in blood and urine as well as by therapeutic effects.[42] This interesting and rather unexpected finding has been replicated in many studies and systems.[43–46]

Detailed notes were kept throughout on each patient. The final evaluating therapist also completed a special inquiry form that was sent to the Polish Academy of Science research team along with the notes. The Computer Center at Wroclaw Technical University carried out extensive analyses of the data. They used the categories established in the WHO (1977) International Classification of Diseases in assessing results. They also looked at the effects of age, severity of initial condition, type(s) of bacteria involved, length of treatment, and other concomitant treatments. The articles included many specific details on individual patients, which helped give some insight into the ways phage therapy was used, as well as an in-depth analysis of difficult cases.

After Slopek's retirement, Beatta Weber-Dabrowska carried on with the treatment work and published a summary in English of the results with the next 1600 patients.[47] In 1998, immunologist A. Górski took over as institute director and revived a strong focus on phage therapy work, with special emphasis on the immunological consequences of phage treatment.[10, 48] He now has passed on the directorship of the institute and focuses all his energy on the Laboratory of Bacteriophages, which he heads (http://surfer.iitd.pan.wroc.pl/phages/phages.html). Under the experimental therapeutics rules of the European Union, to which Poland now belongs, they have opened their own clinic at the institute and are treating both Polish and occasional international patients there while collecting data in a carefully controlled fashion.[49] They are also now working with the basic phage group of M. Lobocka in Warsaw to sequence and further characterize key phages — another important step in eventually making them available to the outside world.

ELIAVA INSTITUTE OF BACTERIOPHAGE, MICROBIOLOGY AND VIROLOGY, TBILISI, GEORGIA

The most extensive work on phage therapy was carried out under the auspices of the Bacteriophage Institute at Tbilisi, in the former Soviet republic of Georgia, as discussed above; much more detail is given by Kutter and Sulakvelidze.[7] There, phage therapy is often part of the general standard of care, especially in pediatric, burn, and surgical hospital settings. Phage preparation was carried out on an industrial scale, employing 700 people in the factory and several hundred more in the research arm of the institute just before the break-up of the Soviet Union, and many tons of a variety of products were regularly shipped throughout the former Soviet Union. They were available both over the counter and through physicians. The largest use was in hospitals, to treat both primary and nosocomial infections, alone or in conjunction with other antibiotics and particularly when antibiotic-resistant organisms were found. The Georgian military is still one of the strongest supporters of phage therapy research and development because they have proven so useful for wound and burn infections as well as for preventing debilitating gastrointestinal epidemics among the troops. The International Science and Technology Centers (ISTC) program, set up jointly by the United States, Europe, and Japan to give constructive opportunities to scientists formerly working with Soviet military projects, is now also a strong supporter of basic and applied research in this area in Tbilisi, as are the Civilian Research and Defense Fund (CRDF) and the Science & Technology Center in Ukraine (STCU), with similar goals.

From the institute's inception, the industrial part was run on a self-supporting basis, while its scientific branch was government supported. The latter included the electron microscope facility; permanent strain collection; laboratories studying phages of the enterobacteria, staphylococci, and pseudomonads, and formulating new phage cocktails; and groups involved in immunology, vaccine production, *Lactobacillus* work, and other therapeutic approaches. It also carried out the very extensive studies needed for approval by the Ministry of Health in Moscow of each new strain, therapeutic cocktail, and means of delivery. This careful study of the host range, lytic spectrum, cross-resistance, and other fundamental properties of the phages being used was a major factor in the reported successes of the phage therapy work carried out through the institute, as was their method for initially selecting highly virulent phages from among the myriad potentially available against any given host. All the phages used for therapy are lytic, avoiding the problems engendered by lysogeny. The problems of bacterial resistance were largely solved by the use of well-chosen mixtures of phages with different receptor specificities against each type of bacterium, as well as of phages against the various bacteria likely to be causing the problem in multiple infections. The situation was further improved whenever the clinicians typed the pathogenic bacteria and monitored their phage sensitivity. Where necessary, new cocktails were then prepared to which the given bacteria were sensitive. Not infrequently, using a phage in conjunction with other carefully chosen antibiotics was shown to give better results than either the phage or the antibiotic alone.

A great deal of work went into developing and providing the documentation for Ministry of Health approval of specialized new delivery systems, such as a spray for use in respiratory tract infections, in treating the incision area before surgery, and in sanitation of hospital problem areas such as operating rooms. An enteric-coated pill was also developed, using phage strains that could survive the drying process, and accounted for the bulk of the shipments to other parts of the former Soviet Union.

The depth and extent of the work involved are very impressive. For example, from 1983 to 1985 alone, the institute's Laboratory of Morphology and Biology of Bacteriophages carried out studies of growth, biochemical features, and phage sensitivity on 2038 strains of *Staphylococcus*, 1128 of *Streptococcus*, 328 of *Proteus*, 373 of *P. aeruginosa*, and 622 of *Clostridium* received from clinics and hospitals in towns across the former Soviet Union. New, broader-acting phage strains were isolated using these and other institute cultures and were included in a reformulation of their extensively used Piophage preparation. In the years since, the formulation has continued to be improved based on further studies, and phages against *Klebsiella* and *Acinetobacter* have also been isolated and developed into therapeutic preparations. The other major product is *IntestiPhage*, used very extensively by the military, pediatric centers, and in regions with extensive diarrheal problems, which includes phages, often prepared in tablet form, active against a range of enteric bacteria.

Much of the focus in the past 30 years has been on combating nosocomial infections, where multi-drug-resistant organisms have become a particularly lethal problem and where it is also easier to carry out proper long-term research. Special mixtures were developed for dealing with strains causing nosocomial infections in various hospitals, and they were very effectively used in sanitizing operating rooms and equipment, water taps, and other sources of spread of the infections (most of them involving predominantly *Staphylococcus*). The number of sites testing positive for the problem bacteria decreased by orders of magnitude over the several months of the trial at each site. Clinical and prophylactic studies in collaboration with institutions such as the Leningrad (St. Petersburg) Intensive Burn Therapy Center, the Academy of Military Medicine in Leningrad, the Karan Trauma Center, and the Kemerovo Maternity Hospital as well as in Tbilisi at the Pediatric Hospital, the Burn Center, the Center for Sepsis and the Institute for Surgery, were used to further develop treatment protocols and phage cocktails, but Western-style clinical trials are still needed.

An exciting product completed the Georgian approval process and was licensed in 2000 by the Georgian Ministry of Health. "PhagoBioDerm™" is a biodegradable, nontoxic polymer composite that is impregnated with the Pyophage cocktail of phages, along with other antimicrobial agents.[50] Markoishvili et al.[51] reported the results of a study of PhagoBioDerm™ involving 107 patients with ulcers that had failed to respond to conventional therapy — systemic antibiotics, antibiotic-containing ointments, and various phlebotonic and vascular-protecting agents. They were treated with PhagoBioDerm™ alone or in combination with other interventions during 1999 and 2000. The wounds or ulcers healed completely in 70% of the 96 patients for whom there was follow-up data. In the 22 cases for which complete microbiologic analyses were available, healing was associated with the concomitant elimination or very marked reduction of the pathogenic bacteria in the ulcers. Other versions of the product are in the final developmental stage.

Guram Gvasalia, head surgeon at the medical school, started a Surgical Infections and Phage Therapy training program in 2005 with the support of the PhageBiotics Foundation to help prepare young Georgian surgeons for treatment of major wound infections and for carrying out serious research on phage therapy. He has now been made director of a new, very modern joint military-civilian hospital in Gori and is working together with Rezo Adamia, new director of the Eliava Institute, to develop and implement a research program that will meet Western standards.

The extensive observations there confirm that phages have many potential advantages:

- They are self-replicating but also self-limiting because they will multiply only as long as sensitive bacteria are present.
- They can be targeted far more specifically than most antibiotics to the problem bacteria, causing much less damage to the normal microbial balance in the gut. Particular resultant problems of antibiotic treatment include *Pseudomonads*, which are especially difficult to treat, and *Clostridium difficile*, the cause of serious diarrhea and pseudomembranous colitis.[52]
- Phages can often be targeted to receptors on the bacterial surface that are involved in pathogenesis, so any resistant mutants are attenuated in virulence.
- Few side effects have been observed, and none of them were serious.
- Phage therapy is particularly useful for people with allergies to antibiotics.
- Appropriately selected phages can easily be used prophylactically to help prevent bacterial disease at times of exposure or to sanitize hospitals and to help protect against hospital-acquired (nosocomial) infections.
- Especially for external applications, phages can be prepared fairly inexpensively and locally, facilitating their potential applications to underserved populations.
- Phages can be used either independently or in conjunction with antibiotics used intravenously and help reduce the development of bacterial resistance.

Recent Work in the West

Levin and Bull,[1] and Barrow and Soothill[3] provide good reviews of much of the animal research carried out in Britain and the United States since interest in the possibilities of phage therapy began to resurface in the early 1980s. The results, in general, are in very good agreement with the clinical work described above in terms of efficacy, safety, and importance of appropriate attention to the biology of the host-phage interactions, reinforcing trust in the reported extensive eastern European results.

In Britain, Smith and Huggins[43,44] carried out a series of excellent, well-controlled studies on the use of phages in systemic *Escherichia coli* infections in mice and then in diarrhetic disease in young calves and pigs. For example, they found that injecting 10^6 colony-forming units (CFUs) of a particular pathogenic strain intramuscularly killed ten out of ten mice, but none died if they simultaneously injected 10^4 plaque-forming units (PFUs) of a phage selected against the K1 capsule antigen of that bacterial strain. This phage treatment was far more effective than using such antibiotics as tetracycline, streptomycin, ampicillin, or trimethoprim/sulfafurazole. Furthermore, the resistant bacteria that emerged had lost their capsule and were far less virulent. In calves, they also found very high and specific levels of protection. They had to isolate different phages for each of their pathogenic bacterial strains, because they were focusing on high specificity against the pathogens and did not succeed in isolating phages specific for more general pathogenicity-related surface receptors such as the K88 or K99 adhesive fimbriae, which play key roles in attachment to the small intestine. Still, the phage treatment was able to reduce the number of bacteria bound there by many orders of magnitude and to virtually stop the fluid loss. The results were particularly effective if the phages were present before or at the time of bacterial presentation, and if multiple phages with different attachment specificities were used. Furthermore, the phage could be transferred from animal to animal, supporting the possibility of prophylactic use in a herd. If the phages were given only after the development of diarrhea, the severity of the infection was still substantially reduced, and none of the animals died.[46]

Levin and Bull[1] carried out a detailed analysis of the population dynamics and tissue phage distribution of the 1982 Smith and Huggins[43] study, which can be helpful in assessing the parameters involved in successful phage therapy and its apparent superiority to antibiotics. They have gone on to do interesting animal studies of their own and conclude that phage therapy is at least well worth further study.[1]

Barrow and Soothill[3] carried out a series of studies preparatory to using phages for infections of burn patients. Using guinea pigs, they showed that skin graft rejection could be prevented by prior treatment with phages against *Pseudomonas aeruginosa*. They also saw excellent protection of mice against systemic infections with both *Pseudomonas* and *Acinetobacter* when appropriate phages were used.[3] In the latter case, they reported that as few as 100 phages protected against infection with 10^8 bacteria — several times the LD_{50}!

Merrill et al.[53] have carried out a series of experiments designed to better understand the interactions of phages with the human immune system, and helped Richard Carlton start a company called Exponential Therapeutics to explore the possibilities of phage therapy. Their initial work was with lytic derivatives of the lysogenic phages lambda and P22 — a poor choice for therapeutic use, as discussed above and below — but they gathered some very interesting data about factors affecting interactions between these phages and the innate immune system and patented a process for isolating longer-circulating phages. They published successful animal studies with phages against vancomycin-resistant *Enterococcus* and Exponential Biotherapies completed a successful small, stage-one clinical trial with these phages, but focused more on work in the area of food safety. In the spring of 2006, a spin-off Dutch branch became the first company to receive governmental approval of phage as a food additive — a phage to destroy Listeria on cheeses — which was also granted GRAS status by the U.S. FDA that fall. Also in 2006, the FDA had approved a listeria-phage *cocktail* for use on packaged meats made by Baltimore-based Intralytix, after 4 years of very thorough and careful review.

PHAGE AND BACTERIAL PATHOGENICITY

Most bacteria are not pathogenic; in fact, they play crucial roles in the ecological balance in the digestive system, mucous membranes, and all body surfaces. They often actually help protect against pathogens. This is one reason why broad-spectrum antibiotics have such a large range of side effects and why more narrowly targeted bacteriocidal agents would be highly advantageous. Interestingly, most of the serious pathogens are close relatives of nonpathogenic strains.

Studies clarifying the mechanisms of pathogenesis at the molecular level have progressed remarkably in recent years, crowned by the determination of the complete sequence of many bacteria, and extensive cloning and sequencing of pathogenicity determinants. In general, a number of genes are involved, and many of these are clustered in so-called "pathogenicity islands," or Pais, which may be 50,000 to 200,000 basepairs long. They generally have some unique properties indicating that the bacterium itself probably acquired them as a sort of "infectious disease" at some time in the past, and then kept them because they helped the bacterium infect new ecological niches where there was less competition. Many of these Pais are carried on small extrachromosomal circles of DNA called *plasmids*, which can also be carriers of drug-resistance genes. Others reside in the chromosome where they are often found embedded in defective lysogenic prophages that have lost some key genes in the process and cannot be induced to form phage particles. However, they can sometimes recombine with related infecting phages. Therefore, it makes sense to avoid using lysogenic phages or their lytic derivatives for phage therapy to avoid any chance of picking up such pathogenicity islands.

For bacteria in the human gut, pathogenicity involves two main factors:

1. The production of toxin molecules, such as shiga toxin (from *Shigella* and some pathogenic *Escherichia coli*) or cholera toxin; these toxins modify proteins in the target host cells and thereby cause the problem.
2. The acquisition of cell-surface adhesions, which allow the bacterium to bind to specific receptor sites in the small intestine, rather than just moving on through to the colon.

They also all contain the components of so-called type III secretion machinery, related to structures involved in the assembly of flagella (for motility) and of filamentous phages, and instrumental in many plant pathogens. For all the pathogenic enteric bacteria, the infection process triggers changes in the neighboring intestinal cells. These include degeneration of the microvilli; formation of individual "pedestals" cupping each bacterium above the cell surface; and, in the case of *Salmonella* and *Shigella*, induction of cell-signaling molecules that trigger engulfment of the bacterium and its subsequent growth inside the cell.

Recently, *Escherichia coli* O157:H7 has been the subject of much concern, with contamination of such products as hamburgers and unpasteurized fruit juices leading to serious problems. Particularly in young children and the elderly, there is a high probability of death when O157 infections evolve into hemorrhagic colitis (bloody diarrhea) and hemolytic-uremic syndrome, where the kidneys are affected. Antibiotic therapy has shown no benefit; it is actually generally contraindicated because it leads to increased toxin release.[54] We have isolated phages that look promising for reducing O157 load in livestock from sheep naturally resistant to inoculation with *E. coli* O157:H7 at the USDA in College Station, Texas. A major job there is to eliminate this human pathogen that is found in the normal flora of 1/3 of the cattle brought to slaughter in the United States, contaminating water sources and creating massive recalls for industries from meat packers to Odwalla juices to packaged spinach. The first phage so isolated turns out to also be T4-like, and to infect virtually all tested O157 strains. Studies are suggesting that this and additional phages that we have isolated from sheep are good candidates for controlling O157 in the gastrointestinal tracts of ruminants.[55, 56]

PHAGE THERAPY WITH WELL-CHARACTERIZED, PROFESSIONALLY LYTIC PHAGES

A substantial fraction of the phages in various therapeutic mixes for Gram-negative bacteria also turn out to be relatives of bacteriophage T4, which has played such a key role in the development of molecular biology. This family is often called the T-even phages, an historical accident reflecting the fact that T2, T4, and T6 from the original collection of Delbrück's Phage Group all turned out to be related.

T4-like phages are found infecting all the enteric bacteria and their relatives.[57] Large sets have been isolated from all over the world: Long Island sewage treatment plants, animals in the Denver zoo, African lagoons, and dysentery patients in eastern Europe (the latter using *Shigella* as host). The laboratory of Harald Bruessow at Nestle, Inc., in Lausanne, Switzerland, has been exploring the possibility of using phages to treat infant diarrhea in underdeveloped countries — a long-standing concern of theirs, and no other approaches have worked for the 27% that involve coliform bacteria. They isolated a number of phages from patients at the world-famous Diarrheal Center in Bangladesh; all the broad-spectrum phages are T4-like.[58, 59] They have selected a group of therapeutic candidates and carried out various *in vitro* animal and human safety studies in preparation for human trials, now being initiated in Dacca.

Most of the T-even phages targeting enteric bacteria use 5-hydroxymethylcytosine instead of cytosine in their DNA, which protects them against most of the restriction enzymes that bacteria make to protect themselves against invaders; this gives the phage a much more effective host range. T4's entire DNA sequence is known, and we know a great deal about its infection process in standard laboratory conditions and about the methods it uses to target bacteria so effectively.[60, 61] We can potentially use this knowledge to develop more targeted approaches to phage therapy, particularly as more is learned about the similarities and differences in its extended family.[62, 63] We know that different members of the T-even family use different outer membrane proteins and oligosaccharides as their receptors, and we understand the tail-fiber structures involved well enough to potentially predict which phages will work on given bacteria and to engineer phages with new specificities.[64, 65]

There have still been far too few studies of T4 ecology and its behavior under conditions more closely approaching the natural environment and the circumstances it will encounter in phage therapy — often anaerobic and/or with frequent periods of starvation. The limited available information in that regard was summarized by Kutter et al.[60, 61] A variety of studies are shedding light on the ability of these highly virulent phages to coexist in balance with their hosts in nature. For example, they can reproduce in the absence of oxygen as long as their bacterial host has been growing anaerobically for several generations.

The T-even bacteriophages share a unique ability that contributes significantly to their widespread occurrence in nature and to their competitive advantage. They are able to control the timing of lysis in response to the relative availability of bacterial hosts in their environment. When *Escherichia coli* are singly infected with T4, they lyse after 25 to 30 min at body temperature in rich media, releasing about 100 to 200 phages per cell. However, when additional T-even phages attack the cell more than 4 min after the initial infection, the cell does not lyse at the normal time. Instead, it continues to make phages for as long as 6 hr.[66, 67] We have found that they can also survive for a period of time in a hibernation-like state inside starved cells, allowing their host to readapt when nutrients are again supplied, and produce a few additional phages. This is particularly interesting and important because bacteria undergo many drastic changes to survive periods of starvation, which increase their resistance to a variety of environmental insults.[68]

Thus, for many reasons, the T4-like phages make excellent candidates for therapeutic and prophylactic use against enteric and other Gram-negative bacteria, and widespread studies of their ecology and infection properties are now being carried out worldwide with these goals in mind.

For Gram-positive bacteria, the family of phages related to the well-characterized, recently sequenced *Bacillus subtilis* phage SBO1 appears to have similarly broad advantages. The Listeria phages recently approved for food-safety use belong to this family, as do Staphylococcus phages Twort and K, also one of the oldest staph phages, sequenced and extensively studied in Cork and shown to have a very broad host range,[69] the main staph phages in the Georgian cocktails (Kutateladze, personal communication), and some of the dairy phages.

SAFETY ISSUES

From a clinical standpoint, phages appear very safe, as discussed extensively in Chapter 14 of Kutter and Sulakvelidze.[7] This is not surprising, given that humans are exposed to phages from birth. Bergh et al. reported in 1989 that nonpolluted water contains about 2×10^8 phages mL^{-1}.[70] They are normally found in the gastrointestinal tract, skin, urine, and mouth, where they are harbored in saliva and dental plaque.[71–73] They also were shown to be unintentional contaminants of sera and thence of commercially available vaccines,[74–77] which were given dispensation to be sold despite this discovery, due to the general consensus that phages are safe for humans.

Extensive preclinical animal testing was required for approving new phage formulations in the former Soviet Union, but few of these studies were published. Bogovazova et al.[78, 79] evaluated the safety and efficacy of *Klebsiella* phages produced by the Russian company Immunopreparat. Pharmacokinetic and toxicological studies were carried out in mice and guinea pigs using intramuscular, intraperitoneal, or intravenous administration of phages. They found no signs of acute toxicity or gross or histological changes, even using a dose per gram that was 3500-fold higher than the projected human dose. They then evaluated the safety and efficacy of the phages in treating 109 patients infected with *Klebsiella*. The phage preparation was reported to be nontoxic for humans and effective in treating *Klebsiella* infections, as manifested by marked clinical improvements and bacterial clearance in the phage-treated patients.

Side effects such as occasional liver pain and fever reported in the early days of Western phage therapy may have been due to bacterial byproducts contaminating phage preparations used intravenously.[80–82] Concern for this possibility is a major reason why the Polish phage therapy group never administers their phage intravenously. The same is true for almost all the therapeutic work carried out in Tbilisi, and probably helps explain their virtually total lack of significant problems. Because the phages readily enter the bloodstream after infusion in or near wounds and other sites of localized infection and then travel to sites of infection throughout the body, as discussed above in the work by DuBow, there generally seems to be no particular reason for undergoing the extra risks of intravenous administration.

No negative effects on the efficacy or safety of other drugs have been reported as a result of phage administration in the long history of work in eastern Europe. No systematic studies have been carried out in this regard but phage are so specific in their actions that it is difficult to see where such interactions might be predicted to occur. On the other hand, at least some antibiotics would tend to interfere with phage treatment of localized infections in areas with poor circulation by killing off the most accessible of the bacteria in which the phage need to multiply as they work their way deeper into the lesion; this would be a particular problem in cases where the phages can still attach and infect but not complete their replication cycle. (Many of the Georgian physicians feel that antibiotics should never be used topically for wounds and deep-seated infections, because the decrease in antibiotic concentration below the surface provides a strong selection for antibiotic resistance, while this problem does not occur with phages.)

CONCLUSION

Clearly the time has come to look more carefully at the potential of phage therapy, both by strongly supporting new research and by looking carefully at the research already available,[5] such as the very interesting human anti-typhoid phage research carried out in this country in the 1940s that has recently come to light (Knouf et al.[34]), as well as the earlier European work and the very extensive applications in the former Soviet Union.

As Barrow and Soothill[3] conclude:

> "Phage therapy can be very effective in certain conditions and has some unique advantages over antibiotics. With the increasing incidence of antibiotic-resistant bacteria and a deficit in the development of new classes of antibiotics to counteract them, there is a need to investigate the use of phage in a range of infections."

Phages are quite specific as to the bacteria they attack, and the stipulations of Ackermann and DuBow[18] are important here. The specificity of phages means that:

> "Phages have to be tested against the patient's bacteria just as antibiotics [should be], and the indications have to be right, but this holds everywhere in medicine. However, phage therapy requires the creation of phage banks and a close collaboration between the clinician and the laboratory. Phages have at least one advantage. ... While the concentration of antibiotics decreases from the moment of application, phage numbers should increase. Another advantage is that phages are able to spread and thus prevent disease. Nonetheless, much research remains to be done ... on the stability of therapeutic preparations; clearance of phages from blood and tissues; their multiplication in the human body; inactivation by antibodies, serum or pus; and the release of bacterial endotoxins by lysis."

While it is clearly premature to introduce injectable phage preparations without further extensive research, the carefully implemented use of phages in external applications and for a variety of agricultural purposes could potentially help reduce the emergence of antibiotic-resistant strains and deal with problems we have difficulty in handling today. Furthermore, compassionate use of appropriate phages seems warranted in cases where bacteria resistant to all available antibiotics are causing life-threatening illness. They are especially useful in dealing with recalcitrant nosocomial infections, where large numbers of particularly vulnerable people are being exposed to the same strains of bacteria in a closed hospital setting. In this case, the environment as well as the patients can be effectively treated.

With the exploding possibilities and decreasing costs of genomic analysis, it is now possible to do at least partial genomic sequencing of phages to be included in general cocktails to know more about the phage families involved and exclude phages from temperate families and those likely to carry or acquire genes related to pathogenicity or toxin production; this is now being done standardly for phages being developed in the West. Such modern techniques now also are being applied to some of the Georgian phages with help from grants from the International Science and Technology Centers (ISTC) program and collaborations with scientists in a variety of countries, including the Sanger Pathogen Sequencing Center. This is an important step in considering their importation for topical and surgical use in the Western world, taking advantage of the very extensive clinical experience with these phages. The availability of a cocktail of eight sequenced Intralytix phages clearly played an important role in FDA approval of the physician-initiated clinical trial in Lubbock of phages targeting bacteria in leg ulcers — an important beginning.

REFERENCES

1. Levin, B., Bull, J.J., Phage therapy revisited: the population biology of a bacterial infection and its treatment with bacteriophage and antibiotics, *Am. Nat.,* 147: 881–898, 1996.
2. Radetsky, P., The good virus, *Discover,* 17: 50–58, 1996.
3. Barrow, P.A. and Soothill, J.S., Bacteriophage therapy and prophylaxis: rediscovery and renewed assessment of the potential, *Trends Microbiol.*, 5: 268–271, 1997.
4. Alisky, J., Iczkonski, K., Rapoport, A., and Troitsky, N., Bacteriophage shows promise as antimicrobial agents, *J. Infection*, 36: 5–13, 1998.

5. Sulakvelidze, A., Alavidze, Z., and Morris, J., Bacteriophage therapy, *Antimicrob. Agents Chemother.*, 45: 649–659, 2001.
6. Summers, W.C., *Felix d'Herelle and the Origins of Molecular Biology,* New Haven, CT: Yale University Press, 1999.
7. Kutter, E. and Sulakvelidze, A., *Bacteriophages: Biology and Applications.* Boca Raton, FL: CRC Press, 2005.
8. Merril, C.R., Scholl, D., and Adhya, S.L., The prospect for bacteriophage therapy in Western medicine, *Nat. Rev. Drug Discov.*, 2: 489–497, 2003.
9. Bruessow, H., Phage therapy: the Western perspective, in *Bacteriophage Genetics and Molecular Biology.* S. McGrath and D. van Sinderen, Eds., Caister Academic Press, Norfolk, UK, 2007, 159–192.
10. Gorski, A., Borysowski, J., Miedzybrodzki, R., and Weber-Dabrowska, B., Bacteriophages in medicine, in *Bacteriophage Genetics and Molecular Biology.* S. McGrath and D. van Sinderen, Eds., Caister Academic Press, Norfolk, UK, 2007, 125–158.
11. Hausler T., *Viruses vs. Superbugs: A Solution to the Antibiotics Crisis?* (originally published in German: Häusler T., Gesund durch Viren. Piper, Munchen, Germany, 2003), 2006.
12. Entenza, J.M., Loeffler, J.M., Grandgirard, D., Fischetti, V.A., and Moreillon, P., Therapeutic effects of bacteriophage CPL-1 lysin against *Streptococcus pneumoniae* endocarditis in rats, *Antimicrob. Agents Chemother.*, 49: 4789–4792, 2005.
13. D'Herelle, F. (Smith, G.H., translated), *The Bacteriophage: Its Role in Immunity.* Williams and Wilkins Company, Baltimore, MD, 1922.
14. Bruynoghe, R. and Maisin, J., Essais de therapeutique au moyen du bacteriophage du staphylocoque, *C.R. Soc. Biol.*, 85: 1120–1121, 1921.
15. D'Herelle, F., *The Bacteriophage and Its Clinical Applications.* Springfield, IL: CC Thomas, 1930.
16. D'Herrelle, F. (Eliava G., translated), *Bakteriofag i Fenomen Vyzdorovileniya.* Tbilisi, Georgia: Tbilisi National University Publications, 1935.
17. Saunders, M.E., Bacteriophages in industrial fermentations, in *Encyclopedia of Virology.* Webster, R.G. and Granoff, A., Eds., San Diego: Academic Press, 1994, p. 116–121.
18. Ackermann, H.W. and DuBow, M.S., Viruses of prokaryotes I: general properties of bacteriophages, in *Practical Applications of Bacteriophages.* Boca Raton, FL: CRC Press, 1987, chap. 7.
19. Schlesinger M., Reindarstellung eines Bakteriophagen in mit freiem Auge sichtbaren Mengen, *Biochem. Z.*, 264: 6, 1933.
20. Ruska H., Die Sichtbarmachung der Bakteriophagen Lyse im Übermikroskop, *Naturwissenschaften*, 28: 45–46, 1940.
21. Pfankuch, E. and Kausche, G., Isolierung U. Üebermikroskopische abbildung eines Bakteriophagen, *Naturwissenschaften*, 28: 46, 1940.
22. Ellis, E.L. and Delbrück, M., The growth of bacteriophage. *J. Gen. Physiol.*, 22: 365–384, 1939.
23. Doermann, A.D., The intracellular growth of bacteriophages. I. Liberation of intracellular bacteriophage T4 by premature lysis with another phage or with cyanide, *J. Gen. Physiol.*, 35: 645–356, 1952.
24. Cairns, J., Stent, G.S., and Watson, J.D., *Phage and the Origins of Molecular Biology.* Cold Spring Harbor, NY: Cold Spring Harbor Laboratory Press, 1966.
25. Fischer, E. and Lipson, C., *Thinking about Science: Max Delbrück and the Origins of Molecular Biology.* New York: Norton, 1988.
26. Stent, G., *Molecular Biology of Bacterial Viruses.* San Francisco: W.H. Freeman, 1963, p. 8.
27. Appelmans, R., Le bacteriophage dans l'organisme, *C. R. Seances Soc. Biol. Fil.*, 85: 722–724, 1921.
28. Evans, A.C., Inactivation of antistreptococcus bacteriophage by animal fluids. Public Health Reports, 48: 411–446, 1933. Cited by Merril, C. and Adhya, S., Phage Therapy, in *The Bacteriophages.* R. Calendar, Oxford University Press, 2006, p. 725–741.
29 Nungester, W.J. and Watrous, R.M., Accumulation of bacteriophage in spleen and liver following its intravenous inoculation, *Proc. Soc. Exp. Biol. Med.*, 31: 901–905, 1934.
30. Geier, M.R., Trigg, M.E., and Merril, C.R., Fate of bacteriophage lambda in non-immune germ-free mice, *Nature*, 246: 221–223, 1973.
31. Dubos, R.J., Straus, J.H., and Pierce, C., The multiplication of bacteriophage *in vivo* and its protective effects against an experimental infection with *Shigella dysenteriae*, *J. Exp. Med.*, 78: 161–168, 1943.
32. Morton, H.E. and Engely, F.B., Dysentery bacteriophage: review of the literature on its prophylactic and therapeutic uses in man and in experimental infections in animals, *J. Am. Med. Assoc.*, 127: 584–591, 1945.

33. Ward, W.E., Protective action of VI bacteriophage in *Eberthella typhi* infections in mice, *J. Infect. Dis.*, 72: 172–176, 1943.
34. Knouf, E.G., Ward, W.E., Reichle, P.A., Bower, A.G., and Hamilton, P.M., Treatment of typhoid fever with type specific bacteriophage, *J. Am. Med. Assoc.*, 132: 134–138, 1946.
35. Desranleu, J.-M., Progress in the treatment of typhoid fever with VI phages, *Can. J. Public Health*, 473-478, 1949.
36. McGrath, S. and van Sinderen, D., Eds., *Bacteriophage Genetics and Molecular Biology*. Norfolk, U.K.: Caister Academic Press, 2007.
37. Vieu, J.F., Les bacteriophages. In Fabre J, Ed. *Traite de Therapeutique,* Vol. *Serums et Vaccins*. Paris: Flammarion, p. 337–340b, 1975.
38. Vieu, J.F., Guillermet, F., Minck, R., and Nicolle, P., Données actuelles sur les applications therapeutiques des bacteriophages. *Bull. Acad. Natl. Med.*, 163: 61–66, 1979.
39. Slopek, S., Durlakova, I., Weber-Dabrowska, B., Kucharewicz-Krukowska, A., Dabrowski, M., and Bisikiewicz, R., Results of bacteriophage treatment of suppurative bacterial infections. I. General evaluation of the results, *Arch. Immunol. Ther. Exp.*, 31: 267–291, 1983.
40. Slopek, S., Kucharewica-Krukowska, A., Weber-Dabrowska, B., and Dabrowski, M., Results of bacteriophage treatment of suppurative bacterial infections. VI. Analysis of treatment of suppurative staphylococcal infections, *Arch. Immunol. Ther. Exp.*, 33: 261–273, 1985.
41. Slopek, S., Weber-Dabrowska, B., Dabrowski, M., and Kucharewica-Krukowska, A., Results of bacteriophage treatment of suppurative bacterial infections in the years 1981–1986, *Arch. Immunol. Ther. Exp.*, 35: 569–583, 1987.
42. Weber-Dabrowska, B., Dabrowski, M., and Slopek, S., Studies on bacteriophage penetration in patients subjected to phage therapy, *Arch. Immunol. Ther. Exp. (Warsz),* 35: 363–368, 1987.
43. Smith, H.W. and Huggins, M.B., Successful treatment of experimental *E. coli* infections in mice using phage: its general superiority over antibiotics, *J. Gen. Microbiol.*, 128: 307–318, 1982.
44. Smith, H.W. and Huggins, M.B., Effectiveness of phages in treating experimental *E. coli* diarrhea in calves, piglets and lambs. *J. Gen. Microbiol.*, 129: 2659–2675, 1983.
45. Smith, H.W. and Huggins, M.B., The control of experimental *E. coli* diarrhea in calves by means of bacteriophage, *J. Gen. Microbiol.,* 133: 1111–1126, 1987.
46. Smith, H.W., Huggins, M.B., and Shaw, K.M., Factors influencing the survival and multiplication of bacteriophages in calves and in their environment, *J. Gen. Microbiol.*, 133: 1127–1135, 1987.
47. Weber-Dabrowska, B., Mulczyk, M., and Gorski, A., Bacteriophage therapy of bacterial infections: an update of our institute's experience, *Arch. Immunol. Ther. Exp.*, 48: 547–551, 2000.
48. Weber-Dabrowska, B., Zimecki, M., Mulczyk, M., and Gorski, A., Effect of phage therapy on the turnover and function of peripheral neutrophils, *FEMS Immunol. Med. Microbiol.*, 34: 135–138, 2002.
49. Miedzybrodzki, R., Fortuna, W., Weber-Dabrowska, B., Górski, A., Phage therapy of staphylococcal infections (including MRSA) may be less expensive than antibiotic treatment, *Postepy Hig. Med. Dosw. (online)*, 61: 461–465, 2007, www.phmd.pl e-ISSN 1732-2693.
50. Katsarava, R., Beridze, V., Arabuli, N., Kharadze, D., Chu, C.C., and Won, C.Y., Amino acid-based bioanalogous polymers. Synthesis, and study of regular poly(ester amide)s based on bis(α-amino acid) α,β-alkylene diesters, and aliphatic dicarboxylic acids, *J. Polymer Sci.*, 37: 391–407, 1999.
51. Markoishvili, K., Tsitlanadze, G., Katsarava, R., Morris, J.G. Jr., and Sulakvelidze, A., A novel sustained-release matrix based on biodegradable poly(ester amide)s and impregnated with bacteriophages and an antibiotic shows promise in management of infected venous stasis ulcers and other poorly healing wounds, *Int. J. Dermatol.*, 41: 453–458, 2002.
52. Fekety, R. Antibiotic-associated diarrhea and colitis, *Curr. Opin. Infect. Dis.*, 8: 391–397, 1995.
53. Merril, C.R., Biswis, B., Carlton, R., Jensen, N.C., Creed, G.J., Zullo, S., and Adhya, S., Long-circulating bacteriophages as antibacterial agents, *Proc. Natl. Acad. Sci. U.S.A.,* 93: 3188–3192, 1996.
54. Greenwald, D. and Brandt, L., Recognizing *E. coli* O157:H7 infection, *Hosp. Pract. (Off. Ed.)*, 32: 123–126, 129–130, 133, 1997.
55. Raya, R.R., Varey, P., Oot, R.A., Dyen, M.R., Callaway, T.R., Edrington, T.S., Kutter, E.M., and Brabban, A.D., Isolation and characterization of a new T-even bacteriophage, CEV1, and determination of its potential to reduce *Escherichia coli* O157:H7 levels in sheep, *Appl. Environ. Microbiol.,* 72: 6405–6410, 2006.
56. Brabban, A.D., Nelson, D.A., Kutter, E., Edrington, T.S., and Callaway, T.R., Approaches to controlling *Escherichia coli* O157:H7, a food-borne pathogen and an emerging environmental hazard, *Environ. Pract.*, 6: 208–229, 2004.

57. Ackermann, H. and Krisch, H., A catalogue of T4-type bacteriophages, *Arch. Virol.*, 142: 2329–2345, 1997.
58. Chibani-Chennoufi, S., Sidoti, J., Bruttin, A., et al., Isolation of T4-like bacteriophages from the stool of pediatric diarrhea patients in Bangladesh, *J. Bacteriol.*, 186: 8287–8294, 2004.
59. Chibani-Chennoufi, S., Sidoti, J., Bruttin, A., et al., *In vitro* and *in vivo* bacteriolytic activities of *Escherichia coli* phages: implications for phage therapy, *Antimicrob. Agents Chemother.*, 48: 2558–2569, 2004.
60. Kutter, E., Kellenberger, E., Carlson, K., et al., Effects of bacterial growth conditions and physiology on T4 infection, in Karam, J.D. and Drake, J., Eds., *Molecular Biology of Bacteriophage T4.* Washington, DC: American Society for Microbiology, 1994, p. 406–420.
61. Kutter, E., Stidham, T., Guttman, B., et al., Genomic map of bacteriophage T4, in Karam, J.D. and Drake, J., Eds. *Molecular Biology of Bacteriophage T4.* Washington, D.C.: American Society for Microbiology, 1994, p. 491–519.
62. Nolan, J.M., Petrov, V., Bertrand, C., Krisch, H.M., and Karam, J.D. Genetic diversity among five T4-like bacteriophages, *Virol. J.*, 3: 30, 2006.
63. Kutter, E., Gachechiladze, K., Poglazov, A., Marusich, E., Aronsson, P., Napuli, A., et al., Evolution of T4-related phages, *Virus Genes*, 11: 285–297, 1995.
64. Henning, U. and Hashemolhosseini, S., Receptor recognition by T-even-type coliphages, in Karam, J.D. and Drake, J., Eds., *Molecular Biology of Bacteriophage T*4. Washington, DC: American Society for Microbiology, 1994, p. 291–298.
65. Tetart, F., Desplats, C., and Krisch, H.M., Genome plasticity in the distal tail fiber locus of the T-even bacteriophage: recombination between conserved motifs swaps adhesin specificity, *J. Mol. Biol.*, 282: 543–556, 1998.
66. Doermann, A.H., Lysis and lysis inhibition with *E. coli* bacteriophage, *J. Bacteriol.*, 55: 257–275, 1948.
67. Abedon, S., Lysis and the interaction between free phages and infected cells, in Karam, J.D. and Drake, J., Eds., *Molecular Biology of Bacteriophage T4.* Washington, DC: American Society for Microbiology, 1994, p. 397–405.
68. Kolter, R., Life and death in stationary phase, *ASM News*, 58: 75–79, 1992.
69. O'Flaherty, S., Coffey, A., Edwards, R., Meaney, W., Fitzgerald, G.F., and Ross, R.P., Genome of staphylococcal phage K: a new lineage of Myoviridae infecting Gram-positive bacteria with a low G+C content, *J. Bacteriol.*, 186: 2862–2871, 2004.
70. Bergh, O., Borsheim, K.Y., Bratbak, G., et al., High abundance of viruses found in aquatic environments, *Nature,* 340: 467–468, 1989.
71. Caldwell, J.A., Bacteriologic and bacteriophagic study of infected urines, *J. Infect. Dis.*, 43: 353–362, 1928.
72. Yeung, M.K. and Kozelsky, C.S. Transfection of *Actinomyces* spp. by genomic DNA of bacteriophages from human dental plaque, *Plasmid*, 37: 141–153, 1997.
73. Bachrach, G., Leizerovici-Zigmond, M., Zlotkin, A., et al., Bacteriophage isolation from human saliva, *Lett. Appl. Microbiol.*, 36: 50–53, 2003.
74. Merril, C.R., Friedman, T.B., Attallah, A.F., et al., Isolation of bacteriophages from commercial sera, *In Vitro,* 8: 91–93, 1972.
75. Geier, M.R., Attallah, A.F., and Merril, C.R., Characterization of *Escherichia coli* bacterial viruses in commercial sera, *In Vitro,* 11: 55–58, 1975.
76. Milch, H. and Fornosi, F., Bacteriophage contamination in live poliovirus vaccine, *J. Biol. Stand.*, 3: 307–310, 1975.
77. Moody, E.E., Trousdale, M.D., Jorgensen, J.H., et al., Bacteriophages and endotoxin in licensed live-virus vaccines, *J. Infect. Dis.*, 131: 588–591, 1975.
78. Bogovazova, G.G., Voroshilova, N.N., and Bondarenko, V.M., The efficacy of *Klebsiella pneumoniae* bacteriophage in the therapy of experimental *Klebsiella* infection, *Zh. Mikrobiol. Epidemiol. Immunobiol.*, 1991, p. 5–8.
79. Bogovazova, G.G., Voroshilova, N.N., Bondarenko, V.M., et al., Immunobiological properties and therapeutic effectiveness of preparations from *Klebsiella* bacteriophages, *Zh. Mikrobiol. Epidemiol. Immunobiol.*, 1992, p. 30–33.
80. Larkum, N.W., Bacteriophage as a substitute for typhoid vaccine, *J. Bacteriol.*, 17: 42, 1929.
81. Larkum, N.W. Bacteriophage from Public Health standpoint, *Am. J. Pub. Health,* 19: 31–36, 1929.
82. King, W.E., Boyd, D.A., and Conlin, J.H., The cause of local reactions following the administration of *Staphylococcus* bacteriophage, *Am. J. Clin. Pathol.*, 4: 336–345, 1934.

45 Introduction to Parasites

Frederick L. Schuster

CONTENTS

INTRODUCTION

The focus of this chapter is on human parasites of both cellular and multicellular organization — and particularly those that have achieved a high degree of prominence for the diseases they cause. In size, they range from protists* of microscopic dimensions to tapeworms that can attain a length of 20 to 30 feet. Parasitology is a broad discipline filling many tomes dedicated solely to the topic, as well as focused chapters in medical and veterinary texts. The reader is encouraged to consult these more extensive treatments for a fuller understanding of the human-parasite associations. In many ways, human parasitology as a science was a beneficent spin-off of colonialism, with imperialist nations founding institutes devoted to tropical medicine and seeking ways of protecting their nationals from diseases while serving abroad.

This chapter describes a limited number of parasitic diseases as representative of the major groups of parasitic organisms. Tables are used in lieu of prose to present comparative information. Names of larval stages and intermediate hosts have been kept to a minimum in descriptions of parasite life cycles. High-profile parasitic diseases (e.g., malaria and trypanosomiasis), because of their importance, are reviewed frequently in the literature. Areas such as drug therapy, diagnostic testing, and emerging pathogens are dynamic and subject to rapid change, which may require the reader to turn to the dozens of journals or the Internet in which current information can be found. In including references to the research literature, a number of review articles have been cited. This was done in an effort to limit the total number of reference citations at the end of the chapter and offer readers more comprehensive sources of background material on specific topics. The reader will find *Diagnostic Medical Parasitology*, 5th edition, by Garcia, [1] an up-to-date basic reference

* The terms "protist/protistan" are used throughout the chapter in place of protozoa/protozoan to describe "unicellular" parasites (see later).

text covering all aspects of the field in detail: biology, life cycles, diagnostic testing, etc. The major groups of parasites included in the chapter are the protists and helminths. Insects (mosquitoes, flies, etc.) and acarids (ticks), although undeniably human parasites, are dealt with only in the context of their role as disease vectors.

An aura of exoticism has been attached to parasites, supplemented by textbook pictures of indigenous peoples with elephantiasis, guinea worms being slowly extracted from legs and feet, and faces disfigured by mucocutaneous leishmanial infections. Laboratorians were trained to identify protistan cysts and helminth ova in human feces, trypanosomes and malaria parasites in blood films, and to distinguish between the different species of *Plasmodium*. The likelihood of ever encountering any of these parasites in the laboratory seemed remote except, perhaps, for those working in large metropolitan areas with diverse populations.

Global Dissemination of Parasites

Figuratively, the world has shrunk over the past several decades. From the 16th to the 19th centuries, the slave trade was instrumental in spreading a number of parasitic diseases (e.g., onchocerciasis, schistosomiasis [*S. mansoni*], leishmaniasis, hookworm [*Necator*], trypanosomiasis) from Africa to the New World [2]. Travel by ship or overland was slow, and sickly travelers were likely to die before reaching their destinations. Travel by jet aircraft is rapid and relatively inexpensive, making journeys between developing and developed areas of the world fast and affordable; currently, in-transit travelers are unlikely to die of a parasitic disease. Parasitic diseases that were once a rarity or were restricted to tropical or subtropical climates have become differential diagnoses for clinicians in the developed world. About 100 cases of malaria in airport personnel have been associated with import of mosquitoes from malarious regions via airplane cabins or possibly travelers' baggage [3]. Parasites limited to Southeast Asia can turn up in the midwestern United States owing to extensive patterns of immigration. Tourism, more than ever, via airplane or cruise ship has brought naive Westerners accustomed to safe water and food into areas of the world where quality standards are either ineffective or practically nonexistent; a case of amoebic dysentery may become a souvenir of a vacation resulting from brushing one's teeth with local tap water. In Switzerland, travelers including both immigrants returning from a visit to their homelands and traditional tourists were monitored over a 17-month period for various diseases. About 22 (5%) contracted malaria, mostly from visits to sub-Saharan Africa [4].

Poverty and lack of education about good health principles and practices help spread parasitic diseases. In countries with a public health infrastructure, sophisticated medical care, and accessible pharmaceuticals, parasitic diseases are not a serious problem. In the United States, diseases such as cysticercosis (infection with pork tapeworm larvae) and toxoplasmosis correlate with poor education and foreign birth. Cultural mores, lack of safe drinking water, and reluctance to change from a centuries-old way of life can also clash with programs designed to eliminate parasites. Practices such as eating raw flesh (fish, snakes, frogs), or using freshly killed animals as poultices in treating ailments, and spreading human waste as fertilizer perpetuate disease. In developing nations where parasitic diseases may be endemic, the rationale is that fuel for cooking can be scarce, the cost of drugs prevents their use in treating diseases, and sewage treatment facilities are unavailable.

Civil wars and political oppression have triggered mass movements of people fleeing for their lives, or brought them together in refugee camps meant to protect them from the elements, starvation, and marauders. Such concentrations of refugees also facilitate the spread of parasitic diseases. The appearance of HIV/AIDS about 25 years ago taught the public what it meant to be immunocompromised and what an opportunistic infection is. One-time parasitic diseases of little consequence became significant contributors to the morbidity and mortality of persons with impaired immune responses: *Pneumocystis* pneumonia, microsporidial infections. "New" or emerging diseases result from a sequence of interconnected events: spread of vectors and/or etiologic agents to new areas, opening of new ecological niches through deforestation, or building of dams resulting in changes in flora and fauna, as well as improved diagnostic techniques that can detect once-overlooked disease agents.

The numbers of people suffering and dying from parasitic diseases is staggering: 300 million with malaria, 900 million with hookworm, 1 million with African sleeping sickness, etc. These diseases have mortalities ranging from the thousands to millions annually, with children representing a large number of deaths. They occur mostly in developing nations but developed nations are no longer exempt because of the changing world.

Parasitism as a Form of Symbiosis

Symbiosis is a spectrum of intimate associations between two different species, ranging from mutualism, in which the relationship is beneficial to both partners, to parasitism, in which one partner benefits and the other is harmed. For the parasites dealt with in this chapter, the relationship is almost always obligatory for the parasite. Some free-living organisms, upon encountering a host with a compromised immune system, may become opportunistic (or facultative) parasites.

ADAPTATIONS FOR PARASITISM AS A WAY OF LIFE

Characteristics of Parasites

Protistan and animal parasites of humans and other animals are a varied collection of living forms, with representatives ranging from protists at the cellular level of organization to multicellular forms with distinct tissues and organs. Parasites are neither primordial creatures frozen at an early stage on the road to a free-living existence, nor are they degenerate representatives of their ancestry. They have undergone their own evolution. Amitochondriate organisms such as *Entamoeba histolytica*, an anaerobic amoeba, has relict "mitochondria" in its cytoplasm, as do trichomonads (hydrogenosomes), both of which are reported to have a common origin with eukaryotic mitochondria. Parasitism as a complex ecological niche requires many evolutionary adaptations for success, and parasites are subject to the same evolutionary pressures as free-living species. Some of these adaptations are listed in Table 45.1. Loss of synthetic capacity is typical of many parasites, making them dependent on the host for survival. Many parasites live in microaerophilic habitats such as the mammalian digestive tract, and rely on relatively energy-inefficient fermentative metabolism to satisfy their energy requirements (Table 45.1). Other parasites may require one or more growth factors necessary for survival and reproductive maturity. Parasites whose transmissive stages exit from the host into a hostile environment have thick, protective walls (amoebic cysts, helminth ova). Those that are transmitted by an insect vector (e.g., malaria, sleeping sickness) and are protected from direct exposure to the environment lack this protective covering. Baer has compared the host to an island on which the parasite is stranded, restricted from leaving by its dependence on the host [4]. Not that leaving the host would benefit the parasite — the longer it remains in residence, the longer the period of transmitting of progeny lasts.

The parasite load, or the number of parasites supported by the host, is another consideration in the parasite-host relationship. Given its dependence on the host and the inability of the adult parasite to move elsewhere, equilibrium develops between the host and the parasite. Some parasites may appear to have a benevolent association with their hosts; they may remain *in situ* for years or even decades, suggesting a stable relationship between the two organisms. In the host, it is to the advantage of the parasite to limit its numbers, as in many other ecological niches where density-dependent factors limit population size. The presence of a tapeworm in the host's intestine will prevent other members of the same species from establishing themselves and producing superinfection.

TABLE 45.1

Common Adaptations for Parasitism Although Not All Parasites Exhibit All Adaptations Listed

Adaptation	Advantage	Examples
Complex life cycles	Facilitates host-to-host transfer and wider dispersal	*Paragonimus westermani:* human, to snail, to crab, and back to human
Unisexual or hermaphroditic individuals	Hermaphroditism (helminths) assures production of progeny if an isolated worm is present in the host	Most tapeworms and flukes (trematodes) are hermaphroditic (monoecious). Schistosome worms are unisexual (dioecious); some hermaphroditic worms preferentially cross-fertilize with other individuals. Most nematodes are dioecious
Integration into the host's food chain	Exploitation of the host's food chain as a means of host-to-host transfer	*Clonorchis sinensis* larvae in fish muscle eaten raw or undercooked. *Trichinella spiralis* larvae in pork muscle. *Dracunculis medinensis* larvae in copepods
Low virulence for host	Parasites that rapidly kill the host limit their own ability to survive and reproduce	Most parasites attain equilibrium with host and can remain for years (e. g., tapeworms). Highly virulent parasites that rapidly kill the host are poorly adapted for parasitism
Protective, enzyme-impervious covering (cuticle)	Surfaces that resist host digestive fluids; particularly important in parasites of the gastrointestinal tract	Tapeworms (*Taenia* spp.), nematodes (*Ascaris lumbricoides*)
Holdfast structures	Suckers and/or hooks to prevent being flushed from the body	Tapeworm scolex or "head" (*Taenia* spp.) within the intestinal tract. Flatworms with oral and ventral suckers
Environmental transmissive stages within protective walls	Limits vulnerability of ova or cystic stages to desiccation, sunlight, and other environmental stresses	*Entamoeba histolytica, Giardia, Acanthamoeba spp.* with cyst walls. *Cryptosporidium and Toxoplasma* with thick-walled oöcysts. Ova of most species of cestodes, trematodes (flukes) and nematodes
Evading the host's immune response	Presence of the parasite within host cells prevents recognition by antibodies and/or destruction by phagocytic cells (see also antigenic variation)	Intracytoplasmic location: *Plasmodium* spp. and *Babesia* spp. in erythrocytes; also microsporidia *Leishmania*, *Toxoplasma*. Prevention of lysosomal fusion with the phagosome membrane; prevention of proinflammatory cytokine release; penetration directly into the cytoplasm of host cell, thus avoiding lysosomal action
Antigenic variation	Shedding and replacement of surface antigens to escape or delay recognition by host antibodies.; the longer a parasite remains in the host, the longer the period of transmission to other hosts	*Trypanosoma* spp., *Plasmodium* spp., *Pneumocystis jirovecii, Entamoeba histolytica, schistosome worms*
Host specificity	Adaptation to a particular host	*Trypanosoma brucei brucei* causes African sleeping sickness in game animals and livestock. *T. b. rhodesiensie* and *T. b. gambiense* infect humans
Fecundity	Large numbers of helminth offspring can arise from a single fertilized ovum; larval stages are able to undergo repeated rounds of asexual reproduction in intermediate hosts. Important because small percentage of larvae will never reach their definitive host	*Schistosoma japonicum* female can release up to 3500 ova/day. Uterus of *Dracunculus* may contain from 1 to 6 million larvae

Parasite Life Cycles: Definitions

The *definitive* host of a parasite is the one in which the parasite becomes sexually mature. Thus, for the malaria parasite, the mosquito is the definitive host, and the human or other animals (mammals, birds, reptiles) are intermediate hosts. *Intermediate* hosts (e.g., snails, fish) harbor developmental stages in the parasite life cycle and serve as food sources for the definitive host, thus completing the cycle from human to human. Some parasites may have a primary, secondary, and even a tertiary intermediate host. The *vector* is the means by which a parasite is spread. In African sleeping sickness, the tsetse fly is the vector and, in its quest for blood, transmits the parasite from animal to animal, animal to human, or from human to human. Water may also serve as a vector, transmitting water-borne infections. A *mechanical* vector is one that can transmit a parasite but in which no parasite development occurs. Flies and roaches as purveyors of filth may serve as *mechanical* vectors by transmitting disease organisms from fecal waste to food waiting to be consumed.

Mechanisms of Pathogenesis

A parasite inflicts damage upon the host. In people living in resource-poor areas of the world, the damage may lead to death. Competition with the parasite for nutrients may lead to malnutrition, slowed development of the host, or vulnerability to other infections. In well-fed, otherwise healthy individuals, parasites may cause morbidity but only rarely death. A tapeworm may even have an advantageous effect on its host. It has been suggested that a benign tapeworm can, by its presence in the intestine, present a target for the immune system, making it less likely that lymphocytes (see later) would cause autoimmune diseases [6]. In a mixed malarial infection, the presence of *Plasmodium malariae* may moderate symptoms caused by *P. falciparum* [7]. Parasites living in the intestinal tract, such as tapeworms, compete with the host for nutrients passing through the gut. Physical damage can result from attachment by hooks to the intestinal wall (some tapeworms) or migration of helminth larvae through body tissues and organs. Some parasitic worms (e.g., *Ascaris*) can, by their numbers, cause blockage of the intestine. Destruction of cells or tissues can be a consequence of feeding by the parasite. The malarial parasite develops within the host's erythrocytes, causing them to lyse. Toxins or waste products of the parasite can also cause damage to the host, as well as activating an intense immune response from the host that can have harmful consequences, akin to what occurs in toxic shock syndrome. The presence of a parasite or its secretions may cause an allergic response. Parasites produce a variety of enzymes, including proteases, peptidases, hydrolases, lipases, and phospholipases that can damage the integrity of host tissues and cell membranes, providing nutrients and shelter for the parasite and avenues for parasite expansion.

The course of parasitic infections can be influenced by factors such as the infective dose of the parasite, the virulence of the particular strain or species, the ability of the parasite to circumvent the host's immune system, and the immunostatus and/or health of the host. Immunosuppressed individuals, the elderly, and the young are more vulnerable to many diseases, parasitic or otherwise.

Parasite Genomes

Protists

Because of their impact on human existence, a number of parasites have undergone or are undergoing gene sequencing, particularly those that are significant agents of disease. *Plasmodium falciparum* has been a major target of genomic sequencing because it is the most virulent of the four *Plasmodium* species that cause malaria in humans, and the one most likely to cause death [8–10]. The kinetoplastid (mitochondrion-like organelle in trypanosomes associated with the kinetosome) flagellates have also been targeted for genomic sequencing, including agents of African sleeping sickness and Chagas' disease [11]: *Trypanosoma brucei* [12], *Trypanosoma cruzi* [13], and *Leish-mania major* [14]. Identification of the gene functions of parasites will enable tailoring drug therapy to specific gene products of the parasites. As a result of sequencing, vaccines can be developed to gene products that may offer broad protection against parasites in the various phases of their life cycles (e.g., sequential stages of development of the malaria parasite in the host). Recently, the genome of *Trichomonas vaginalis*, a sexually transmitted protist and the causal agent of vaginitis in females and urethritis in males, has been sequenced [15]. Other pathogenic protists that have been or are in the process of being sequenced are *Babesia* spp., *Entamoeba histolytica*, *Plasmodium vivax*, *Toxoplasma gondii,* and the microsporidia (www.sanger.ac.uk/Projects/Protists). Several other parasitic protists not covered in this chapter are also in various stages of being sequenced.

Helminths: Cestodes, Trematodes, and Nematodes

A number of helminth parasites included in this chapter either have been sequenced or are in the process of being sequenced, as are others not described here. These include *Schistosoma mansoni* and *S. haematobium* (schistosomiasis), *Echinococcus* spp. (hydatid cyst in the human brain), *Taenia solium (*pork tapeworm), *Ascaris lumbricoides*, *Ancylostoma,* and *Necator* (hookworms), *Onchocerca volvulus* (filarial worms causing river blindness), *Trichuris trichiura* (pinworms), and *Wucheraria bancrofti* (filariasis) (www.sanger.ac.uk/Projects/Helminths).

HOST FACTORS

The Immune Response

The Humoral Pathway

There are two major immune responses of the host. One is the *humoral* pathway resulting in antibody production by B-lymphocytes specifically targeted at the invading parasite, also called adaptive or inducible immunity. Immunity can be conferred to a neonate by transplacental transfer of maternal antibodies and provide protection for several months after birth before it begins to fade and before the child's own immune system becomes functionally mature. This offers a window of opportunity for parasites to infect and cause disease or death. Falciparum malaria has its greatest impact on children between 1 and 5 years of age. Antibodies produced during parasitic disease remain for varying lengths of time even after the disease has been cleared, conferring weak or strong immune protection against reinfection. Immunoglobulin M (IgM) is associated with the initial stage of infection, and IgG with post-acute to convalescent or later stages of infection, protection against recurrent infection, and placental transfer. IgA is a secretory antibody released from mucosal surfaces (intestinal, pulmonary and urogenital tracts, corneal surface), and has a role in antiparasite activity. IgE is linked to allergic responses and parasitic infections and causes degranulation of eosinophils, with release of hydrolytic enzymes, peroxidases, and a core cationic protein granule. Eosinophils can adhere via antibodies to the surface of intestinal helminths and release granule contents that aid in attacking the worm's surface, allowing other immune molecules to penetrate and destroy the worm.

Cell-Mediated Immunity

The second type is the *cell-mediated* immune response (innate or nonspecific immunity) leading to the activation of T-lymphocytes, activated macrophages, killer T-cells, and other cell types that are functionally equipped to destroy the parasite, particularly those parasites that are intracellular and sheltered from contact with antibodies. The cell-mediated response can activate the *complement system* (alternative and classical pathways) that may or may not damage and/or lyse the parasite. Cell-cell communication between immune response components is the function of interleukins, cytokines, and other molecules such as tumor necrosis factor-α and interferon-γ. *Plasmodium* toxins stimulate T-cell proliferation, which leads to overproduction of interferon-γ and tumor necrosis factor-α. An excess of these proinflammatory cytokines can be as damaging to the host as the disease itself, affecting erythrocyte formation and a possible cause of malaria-associated anemia [16]. Other parasites have evolved mechanisms to neutralize these activators. Tapeworms of various species can inhibit production of interferon-γ and interleukins, thus muting the immune response [6].

Parasite Evasion of the Immune Response

Parasites have intrinsic and extrinsic ways of avoiding the host's immune response. Chief among these is antigenic variation that is well developed in several protistan pathogens. Trypanosomes are particularly capable of shedding their antigenic coat (variant surface glycoprotein), to which the host has developed antibody sensitivity, and replacing it with a new coat requiring a build-up of new antibody by the host. It is estimated that about 10% of the *Trypanosoma* genome is dedicated to antigenic variation. This delay enables the parasites to remain within the host and prolong periods of transmission. *Leishmania* spp. and *Trypanosoma cruzi* develop safely within the host's macrophages, activist components of the antiparasite response. Other parasites (e.g., schistosome worms) cloak themselves with host molecules to escape detection as foreign antigen.

Components of the complement pathway can be neutralized by some parasites by preventing complement activation and subsequent lysis. Still other parasites may shed complement activation components as they bind to the parasite surface, or produce substances that activate and deplete complement, but at a safe distance from the surface of the parasite.

The Immunocompromised Host

Since the advent of HIV/AIDS, a number of parasitic diseases that were once of minor importance have come into prominence. The more prominent of these include the protist diseases *Pneumocystis* pneumonia, *Acanthamoeba* encephalitis, microsporidiosis, cryptosporidiosis, and cyclosporiasis. These disease agents are termed "opportunistic" parasites because they rarely attack hosts that can mobilize an effective immune response. Young children with immature immune systems, or elderly individuals with a reduced immune response are also subject to opportunistic infections. In still other individuals, immunosuppression can result from concurrent disease, alcoholism, diabetes, and cancer chemotherapy, making them susceptible to the same diseases that affect HIV/AIDS patients.

Steroids, which are effective as anti-inflammatory agents, may mute the immune response with fatal results [17]. While pain and other aspects of inflammation are lessened following corticosteroid use, steroids may make it possible for parasites to gain a foothold in a host receiving steroid therapy.

HUMAN PARASITIC DISEASES

Protists and helminths are two major groups of organisms that are parasites of humans (Tables 45.2 and 45.3). The insects and ticks include human parasites but they will be treated in this review with regard to their role as vectors of parasitic diseases (Table 45.4). These tables are not comprehensive and include the more familiar human parasites as well as those responsible for the more devastating diseases, causing both morbidity and mortality in large numbers of individuals. There are less than 400 species of protists and helminths that are parasitic in humans: approximately 300 species of worms and greater than 70 protist species [2].

Protists

Disease agents belonging to this group are responsible for some of the most widespread and fatal diseases of humans and other mammals. Malaria is at the top of the list of nonbacterial, nonviral infectious diseases for sheer numbers of infections and deaths annually. The deaths occur mostly in children between 1 and 5 years old, that is, after maternal transplacental immunity fades and before the child's own immune system has matured to the point at which it can respond to the infection. African sleeping sickness or trypanosomiasis is another major killer disease, not only of humans but also cattle and wild game — which serve as reservoirs of disease — preventing the raising and grazing of cattle in vast areas of Africa.

The taxonomy of the "Protozoa" has undergone a recent revision that is reflected in the supra-generic group headings included in Table 45.2: Amoebozoa (amoebae), Opisthokonta (protists with fungal affinities/origins), Chromalveolata (apicomplexan protists [intracellular parasites with an apical organelle involved in attachment to and penetration of host cells] and ciliates), and Excavata (flagellate or amoebo-flagellate protists) [17]. Some former Protozoa (*Pneumocystis*, microsporidia) are now known to have fungal origins. The term "protist" has been adopted for these organisms in place of the more familiar term "protozoa." "Protist" is an inclusive term, encompassing organisms with protozoan, algal, and fungal affinities. All protistan representatives have a cellular level of organization or, if multicellular, show no evidence of cellular differentiation into tissues. Protists can be intra- or extracellular parasites, and those found within host cells are able to avoid digestion by host cell enzymes. Intracellular parasites include *Plasmodium, Babesia, Leishmania, Toxoplasma,* and the microsporidia. Except for the microsporidia and *Babesia*, the other intracellular parasites are enclosed within a parasitophorous vacuole in which they are protected from lysosomal activity of the host cell, but are still able to take up nutrients. The microsporidia are "injected" directly into the cytoplasm of the host cell through a polar filament or tube (Figure 45.1).

African Sleeping Sickness

The subspecies *Trypanosoma brucei gambiense* is an etiologic agent of African trypanosomiasis, also known as Gambian sleeping sickness in West and Central Africa. The tsetse fly, *Glossina* spp., is the vector of the disease and, once infected, remains infected for its life. The fly acquires the parasite while obtaining blood from a human and passes it on to another host after the transmissive stage of the parasite enters the fly's salivary glands. In both the fly and the human, the parasite passes through several pleomorphic stages, ranging from a slender swimming form with an elongate undulating membrane and free flagellum (Figure 45.2) to a shorter stage lacking the free flagellum that is found in the salivary glands of the tsetse fly. *T. b. gambiense* causes a chronic form of sleeping sickness affecting the central nervous system, while a related subspecies, *T. b. rhodesiense*, causes a similar but fulminant infection in humans. Of diagnostic importance is elevated protein (and IgM) and leukocyte levels in the central nervous system (CNS). *T. b. rhodesiense* occurs in eastern and central Africa and is also found in wild game animals that serve as reservoirs for the parasite. *T. brucei brucei* infects cattle and wild animals (nagana), but not humans. Sleeping sickness is biphasic: an early blood phase and a later CNS phase, and different drugs are used for the different stages (Table 45.5).

American trypanosomiasis is caused by *Trypanosoma cruzi*, and transmitted by a triatomid ("kissing") bug. Interestingly, transmission from insect to human occurs when the bug defecates on the skin while feeding. The deposited fecal material contains the infective parasites and is usually rubbed into the break in the skin caused by the bite or into the eyes.

TABLE 45.2
Protistan Parasites of Humans

Parasite Genus	Species	Disease	Area Affected	Geographic Distribution	Transmission	Diagnosis
Amoebozoa (Amoebae)						
Acanthamoeba	Several species: *A. castelllanii* *A. culbertsoni* *A. healyi* *A. polyphaga* *A. hatchetti* et al.	Granulomatous amoebic encephalitis	Cutaneous lesions, CNS	Cosmopolitan	Contact with soil, water, air-blown cysts, immunocompromised status	IFA, brain/skin biopsy, culture, PCR
		Amoebic keratitis	Cornea		Corneal trauma, contact lens use	Corneal scraping, serology, culture, PCR
Balamuthia	*B. mandrillaris*	Granulomatous amoebic encephalitis	Cutaneous lesions, CNS	Cosmopolitan	Contact with soil, water, air-blown cysts	Brain/skin biopsy; culture, serology, PCR
Entamoeba	*E. histolytica*	Amoebiasis; amoebic dysentery	Colon, liver, brain	Cosmopolitan	Fecal-oral route, sewage-contaminated uncooked foods	Stool samples for trophic amoebae or cysts, immunoassay, serology for tissue stages
Opisthokonta (Protists with fungal affinities/origins)						
Microsporidia	*Encephalitozoon cuniculi* *Enterocytozoon bienusi*	Microsporidiosis	Intestinal tract, brain	Cosmopolitan	Fecal-oral transmission; immunocompromised status	Tissue biopsy; H&E, trichrome, and IFA staining; ELISA
Pneumocystis	*P. jirovecii*	*Pneumocystis (carinii) jirovecii* pneumonia (PCP)	Lungs	Cosmopolitan	Airborne transmission; immunocompromised status	Bronchoalveolar lavage
Chromalveolata (Protists with apicoplasts; ciliates)						
Babesia	*B. divergens* *B. microti*	Babesiosis	Blood	Europe, United States, Africa, Asia	Zoonotic disease; insect transmission (tick); blood transfusions	Blood films; PCR; ELISA; animal inoculation
Balantidium	*B. coli*	Balantidiosis	Intestine; mostly in pigs	Tropical to subtropical	Fecal-oral route	Trophic ciliates or cysts in stool sample
Cryptosporidium	*C. hominis*	Cryptosporidiosis	Small intestine	Cosmopolitan	Fecal-oral route, water, contaminated foods; immunocompromised status	Stool, sputum, urine samples for oocysts; modified acid-fast [positive]; IFA; PCR
Cyclospora	*C. cayatenensis*	Cyclosporiasis	Small intestine	Cosmopolitan	As for *Cryptosporidium*	As for *Cryptosporidium*

Plasmodium	*P. falciparum*	Malignant tertian malaria (48-hr periodicity)	Blood, brain	Africa, Asia, South America	Mosquitoes (*Anopholes* spp.)	Blood films; PCR; IFA; ELISA
	P. vivax	Benign tertian malaria	Blood	Asia, North Africa		Blood films; PCR; IFA
	P. malariae	Quartan malaria (72 periodicity)	Blood	SE Asia, Central America, Pacific area		Blood films; PCR; IFA
	P. ovale	Mild tertian malaria	Blood	West Africa, Pacific area, India		Blood films; PCR; IFA
Toxoplasma	*T. gondii*	Toxoplasmosis	Fetus	Cosmopolitan	Fecal-oral route, infected meat, cat feces; immunocompromised status	ELISA, PCR, IFA, culture; commercial kits available
Excavata (Flagellate or amoebo-flagellate protists)						
Giardia	*G. duodenalis*	Giardiasis	Small intestine	Cosmopolitan	Fecal-oral route	Stool sample; ELISA; IFA; commercial kits available
Leishmania	*L. tropica*	Old World leishmaniasis (Oriental sore)	Cutaneous ulcerated lesions	Africa, Middle East, Asia Minor, India	Sand fly (*Phelbotomus* et al.)	Presence of Leishman-Donovan bodies in aspirates (non-flagellate [amastigote] stages in tissue aspirates); ELISA; PCR; clinical ID
	L. major					
	L. donovani	Visceral leishmaniasis (Kala azar)	Enlarged liver and spleen cutaneous lesions	India, Africa, Meditteranean area		
	L. braziliensis	Mucocutaneous leishmaniasis (Espundia)	Cutaneous lesions	Latin America north to Mexico		
	L. mexicana	New World leishmaniasis (Chiclero ulcer)	Cutaneous lesions	South and Central Americas		
Naegleria	*N. fowleri*	Primary amoebic meningoencephalitis	CNS	Cosmopolitan	Swimming, water sports (fresh water)	Microscopic examination of CSF; culture; PCR
Trichomonas	*T. vaginalis*	Trichomoniasis	Urogenital tract of females and males	Cosmopolitan	Sexually transmitted disease	Vaginal discharge (wet-mount); commercial kits, PCR.
Trypanosoma	*T. brucei rhodesiense,* *T. brucei gambiese*	East and West African sleeping sickness	CNS	Africa	Tsetse fly (*Glossina* spp.)	Blood films; CSF; culture; lymph node aspirate; ELISA; card agglutination test
	T. cruzi	American trypanosomiasis (Chagas' disease)	Cardiovascular, gastrointestinal systems, brain	Argentina north to Mexican border	Reduviid bug (*Triatoma* spp.); blood transfusions	Blood films; IFA; culture; clinical ID; PCR; xenodiagnosis (in triatomid bug)

Note: Abbreviations: CSF cerebrospinal fluid; ELISA enzyme-linked immunosorbent assay; H&E hematoxylin-eosin stain (tissue); IFA immunofluorescent antibody staining; IIF indirect immunofluorescent staining (tissue); PCR polymerase chain reaction.

TABLE 45.3
Helminth Parasites of Humans

Parasite Genus	Disease	Body Area Affected	Distribution	Transmission
		Trematodes (Flukes)		
Clonorchis sinensis	Clonorchiasis	Liver; bile duct	Asia	Eating raw or undercooked fish
Schistosoma haematobium	Schistosomiasis (bilharziasis)	Bladder	Africa, Middle East	Aquatic larvae (cercariae) penetrate through skin
S. japonicum		Veins draining small intestine	Japan, China, SE Asia	
S. mansoni		Veins draining large intestine	Middle East, Africa	
		Cestodes (Tapeworms)		
Diphyllobothrium latum	Diphyllobothriasis (fish tapeworm)	Intestine	United States, Europe	Eating raw or undercooked fish
Echinococcus granulosus	Hydatid cyst (larval stage)	Brain	Cosmopolitan	Acquiring ova from infected dogs
Taenia saginatus	Taeniasis (beef tapeworm)	Intestine	Cosmopolitan	Eating raw or undercooked beef
Taenia solium	Taeniasis (pork tapeworm)	Intestine	Cosmopolitan	
	Cysticercosis (larval stage)	Brain		
		Nematodes (Round Worms)		
Ancylostoma duodenale	Hookworm	Small intestine	SE Asia, northern Africa, Europe	Terrestrial larvae penetrate through skin
Ascarisl umbricoides	Ascariasis	Intestine	Cosmopolitan	Contact with worm ova
Baylisascaris procyonis	Baylisascariasis	Migrating larvae	United States	Transmitted via raccoon feces (geophagia)
Dracunculis medinensis	Dracunculiasis (guinea worm)	Females release ova in subcutaneous sites (legs, feet)	Middle East	Intermediate host (crustacean) present in drinking water
Enterobius vermicularis	Enterobiasis	Colon; perianal area	Cosmopolitan	Contact with worm ova
Gnathostoma spinigerum	Gnathostomiasis	Migrating larvae	Asia, SW Pacific	Eating raw or undercooked fish
Necator americanus	Hookworm	Small intestine	SE Asia, United States, SW Pacific	Terrestrial larvae penetrate through skin
Onchocerca volvulus	Onchocerciasis (river blindness)	Eyes	Africa, Central and South Americas	Transmitted by black fly (*Simulium*)
Trichinella spiralis	Trichinellosis	Muscle tissues; migrating larvae	Cosmopolitan	Eating raw or undercooked pork
Wucheraria bancrofti	Filariasis (elephantiasis)	Lymphatic system	SE Asia, Pacific area, Central Africa	Transmitted by mosquitoes (Anopholes Culex, Aedes)

TABLE 45.4
Insects and Acarids Important as Vectors of Human Parasitic Disease

Insect	Genus	Disease	Transmission
Black fly	*Simulium*	Onchocerciasis	Injection from salivary glands
Cockroach	All species	None	Mechanical transmission
House fly	*Musca domestica*	None	Mechanical transmission
Mosquito	*Anopholes*	Malaria	Injection from salivary glands
Sand fly	*Phlebotomus*	Leishmaniasis	Injection
Tick	*Ixodes*	Babesiosis	Injection
Triatomid bug (kissing bug)	*Triatoma*	Chagas' disease	Defecation at site of insect bite
Tsetse fly	*Glossina*	Typanosomiasis	Injection

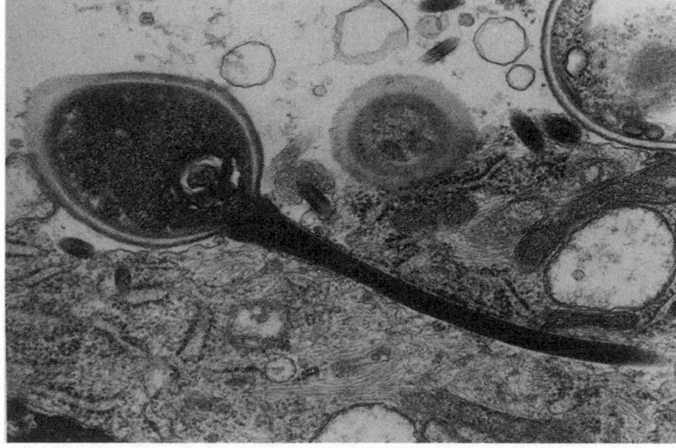

FIGURE 45.1 Transmission electron micrograph of a microsporidian with an extended polar tube projecting into a host cell. The parasite is injected directly into the host cell's cytoplasm, thus avoiding any contact with lysososomes or phagolysosomes. (Courtesy of Dr. Massimo Scaglia and DPDx Parasite Image Library at the Centers for Disease Control and Prevention.)

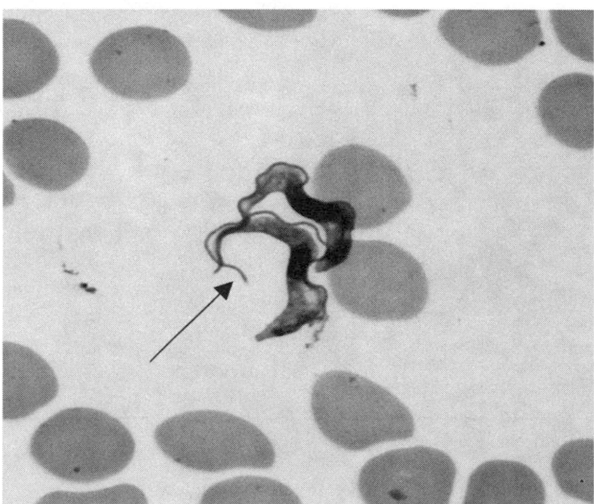

FIGURE 45.2 Trypanosomes seen in a stained blood film. The two organisms (trypomastigote stage) have a wavy outline due to undulating membrane that runs the entire length of the body. A free flagellum projects beyond the undulating membrane. Giemsa stain. (Courtesy of DPDx Parasite Image Library at the Centers for Disease Control and Prevention.)

TABLE 45.5
Drug Therapy for Protistan and Helminthic Infections

Disease	Recommended Drug Therapy
Protistan Diseases	
Acanthamoebiasis	Keratitis: polyhexamethylene biguanide, chlorhexidine gluconate with or without Brolene® Amoebic encephalitis: empirical therapy — pentamidine isethionate, sulfadiazine, flucytosine, fluconazole, or itraconazole
Balamuthiasis	Encephalitis: empirical therapy — pentamidine isethionate, sulfadiazine, clarithromycin (or azilthromycin), fluconazole, flucytosine
African sleeping sickness, west (*Trypanosoma b. gambiense*)	Human infection Early stage: pentamidine isethionate or suramin Late stage (CNS involvement): melarsoprol, eflornithine
African sleeping sickness, east (*Trypanosoma b. rhodesiense*)	Human infection Early stage: suramin Late stage: melarsoprol
Entamoebiasis (amoebic dysentery)	Iodoquinol or metronidazole or tinidazole
Chagas' disease	Benznidazole or nifurtimox
Giardiasis	Metronidazole or tinidazole
Leishmaniasis	Cutaneous, mucosal, and visceral disease: sodium stibogluconate
Microsporidiosis	Albendazole plus fumagillin
Naegleriasis (primary amoebic meningoencephalitis)	Amphotericin B plus miconazole
Plasmodium falciparum (malaria)	Malaria acquired in chloroquine-resistant area: atovaquone-proguanil or quinine sulfate plus doxycycline or mefloquine or artesunate plus mefloquine
Plasmodium vivax (malaria)	Malaria acquired in chloroquine-resistant area: quinine sulfate plus doxycycline or mefloquine
Plasmodium spp. (other human malarias)	Malaria acquired in chloroquine-sensitive area: chloroquine phosphate or quinidine gluconate (oral), quinidine dihydrochloride (parenteral) or artemether For prevention of relapse: primaquine phosphate General prevention: chloroquine-sensitive areas — mefloquine, doxycycline, chloroquine-resistant areas — chloroquine phosphate or atovaquone proguanil
Pneumocystis pneumonia (PCP)	Trimethoprim-sulfamethoxazole
Toxoplasmosis	Pyrimethamine plus sulfadiazine
Trichomoniasis (*Trichomonas vaginalis*)	Metronidazole or tinidazole

Protist Infections

1. *Amoebic keratitis and granulomatous encephalitis. Acanthamoeba* spp., free-living amoebae, are responsible for a painful corneal infection that can lead to blindness [19]. It occurs mainly in soft contact lens users and is often due to unhygienic practices in maintaining and cleaning of lenses and lens cases. Corneal trauma can also cause amoebic keratitis by allowing amoebae access to the damaged corneal surface. (Figure 45.3) Prevention: use sterile lens cleaning solutions (not prepared from tap water), keep lens case clean, do not wear lenses while swimming or relaxing in a hot tub or spa, and comply with manufacturer's instructions for cleaning and disinfecting lenses.

   Acanthamoeba encephalitis occurs mainly in immunocompromised hosts, in contrast to keratitis that occurs in immunocompetent hosts. It may start as a skin lesion but ultimately the amoebae are carried hematogenously to the CNS. The disease is usually fatal and antimicrobial successes have been based on empirical rather than optimal drug therapy. The amoeba is found virtually everywhere as trophozoites or as cysts (soil, water, in household water supplies and taps, potted plants, humidifiers, etc.) so that avoidance is all but impossible. Keratitis is the more common of the two disease manifestations; more than 5000 cases of keratitis in the United States compared to approximately 150 cases of amoebic encephalitis. The amoeba (in air conditioning cooling towers) may also carry *Legionella pneumophila* bacteria as endosymbionts, which may be a factor in outbreaks of legionellosis and the sick-building syndrome.
2. *Babesiosis.* This is a zoonotic blood infection transmitted by the bite of an infected tick, and similar to malaria in intraerythrocytic location and malaria-like symptoms (fever, chills, aches, and anemia). Two species are involved in human infections: *Babesia microti* in North America and *B. divergens* in Europe, with the latter species being the more virulent (5% vs. 42% mortality, respectively) [20, 21]. Co-infected ticks can spread both babesiosis and Lyme borreliosis disease (caused by the bacterium *Borrelia burgdorferi*) simultaneously. Splenectomized and immunosuppressed individuals are at high risk for developing fulminating disease if infected. Rodents, deer, elk, and cattle serve as reservoir hosts. Diagnosis can be difficult, and hamster inoculation (xenodiagnosis) is useful to boost parasitemia to make a diagnosis. *In vitro* cultivation is possible but exacting [22]. Prevention: wearing protective clothing (long sleeves, full-length pants) and applying tick

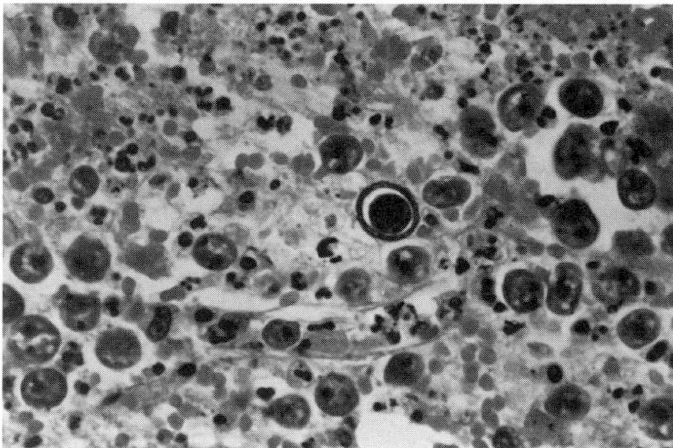

FIGURE 45.3 Section through brain with *Acanthamoeba* encephalitis. Numerous trophic amoebae are seen in the brain parenchyma in the vicinity of a blood vessel. A prominent encysted amoeba is also seen with its typical heavy wall. Hematoxylin-eosin stain. (Courtesy of Dr. G.S. Visvesvara.)

repellent when working or recreating in areas where ticks are abundant (particularly in the Northeast, upper Midwest, West, and Northwest coast of the United States).

3. *Balantidiasis. Balantidium coli* is the only ciliated protist that is parasitic in humans. Although the disease rarely occurs in developed nations, it is found in rural populations in tropical and subtropical areas (e.g., endemic in the Philippines) where pigs have access to human excrement containing *Balantidium* cysts. Malnourishment is a predisposing factor in susceptibility to balantidiasis. The parasite causes no morbidity in swine but is responsible for dysentery in humans. It is also a problem disease of institutionalized populations (prisons, mental hospitals). The ciliates are found in flask-shaped craters in the colon of humans and produce cysts that are shed in the stool; disease spread is by fecal-oral transmission. Prevention: avoidance of possibly contaminated drinking water and uncooked vegetables and fruits in *Balantidium*-endemic regions.
4. *Cryptosporidiosis and cyclosporiasis.* Caused in humans by *Cryptosporidium hominis, C. parvum,* and *Cyclospora cayetenensis*; spread by fecal-oral contamination, usually of drinking water but also food (raw shellfish, uncooked vegetables, salad greens). Person-to-person spread can occur among children (nurseries and day care) and among the elderly in nursing facilities. Infective oocysts are shed in stools; they are visible with modified acid-fast staining. Oocysts of *Cryptosporidium* are smaller (~5 μm) than those of *Cyclospora* (~9 μm) [1]. A major outbreak of cryptosporidiosis occurred in Milwaukee, Wisconsin, in 1993 that resulted in more than 400,000 infections was caused by a contaminated water supply.

   Cyclospora outbreaks have been traced to imported fruits (raspberries). Symptoms are typical of gastrointestinal infection and can be particularly serious in HIV/AIDS patients (severe watery diarrhea and extraintestinal [pulmonary] infections). Diarrhea is self-limiting (~2 weeks) in immunocompetent individuals. Antibiotics are not helpful in treatment except, perhaps, in HIV/AIDS patients. Prevention: avoid possibly contaminated drinking water (bottled water for HIV/AIDS patients), uncooked fruits and vegetables, and contact with farm animals.
5. *Falciparum malaria.* An abundant literature on falciparum malaria exists and this capsule summary is insufficient to convey all that is known of the disease. *Plasmodium falciparum* is the most lethal of the human malarial parasites, and the only one to affect the CNS, causing cerebral malaria [23]. The disease is actually two infections: malaria as the blood disease and as a fulminant cerebral infection. Cerebral malaria is more likely to develop in naive children than in adults with some immunity to the disease. Following injection of the infective stage (sporozoite) from *Anopholes* spp., the organism invades hepatic cells where asexual reproduction occurs. Infected liver cells release merozoites, which penetrate erythrocytes, reproduce asexually, and cause their hemolysis at approximately 48-hr intervals. The cycle of infection and lysis is synchronous. At lysis, there is release of toxins and wastes from the erythrocytes that causes the chills and fever typifying the disease. Infected erythrocytes become sequestered in the microvasculature of the brain. They bind to the endothelial walls of the blood vessels by parasite-induced knobs on the erythrocyte surface, and form rosettes of erythrocytes facilitating vessel occlusion [23]. The consequences of this blockage are oxygen insufficiency, necrosis, edema, hemorrhage, and inflammation in the brain. Overproduction of proinflammatory cytokines stimulated by *Plasmodium* antigens may aggravate the condition [24, 25]. Prevention: screens, netting, mosquito repellent, and prophylactic drugs if traveling in malaria-endemic areas.
6. Toxoplasmosis. Caused by *Toxoplasma gondii* and transmitted to humans in undercooked meat (horse meat, pork, lamb, beef) and by cat feces containing oocysts [26]. The domestic cat is the definitive host. In humans, the main pathway of infection is via meat, and the cat is unnecessary in transmission of *Toxoplasma* to humans. Symptoms are flu-like and pass quickly, but congenital infections (leading to hydrocephalus, microcephaly, chorioretinitis,

cerebral calcification, and mental retardation in the neonate and juveniles) can develop in pregnant women who are not immune from earlier exposure; the risk of fetal infection increases with the gestation period. The organism is also an opportunistic pathogen of HIV/AIDS patients and other immunosuppressed individuals. Reactivation of latent infections can occur in individuals who secondarily become immunosuppressed. Toxoplasmosis is more common abroad (20 to 40% of adult Americans are seropositive for *Toxoplasma* antibodies vs. approximately 70% in France and about 70 to 90% in Latin America) and in immigrants to the United States [27]. Prevention: avoid raw or undercooked meats that may transmit parasite. Pregnant women, in particular, should avoid eating undercooked meats or contact with cat litter boxes.

Helminths: Cestodes

The helminths are a group of parasitic organisms that includes the flatworms (cestodes and trematodes) with dorsoventrally flattened bodies and the round worms (nematodes). The cestodes are the tapeworms, best known as intestinal parasites but, as larval stages, can penetrate the intestinal wall and migrate through tissues and organs. Damage by the migration of the larvae and the final site of the larvae can cause life-threatening conditions. In a "normal" infection of the human, the larval stage (in beef, pork, or fish) is digested free in the intestine, grows, and reaches sexual maturity.

The body plan of the tapeworm includes a head, neck, and a series of segments (proglottids) making up the strobila. There is no intestinal tract, and nutrients are absorbed across the cuticular surface. The head has suckers and/or hooks to attach to the host intestine (Figure 45.4). The neck is the generative region that produces the proglottids. Cestodes are hermaphroditic. Each proglottid contains male and female reproductive organs, and self- or cross-fertilization can occur within a single segment or between segments of the same worm. As new proglottids are produced at the neck, the older ones move rearward toward the tail end of the strobila. Ultimately, the proglottid fills

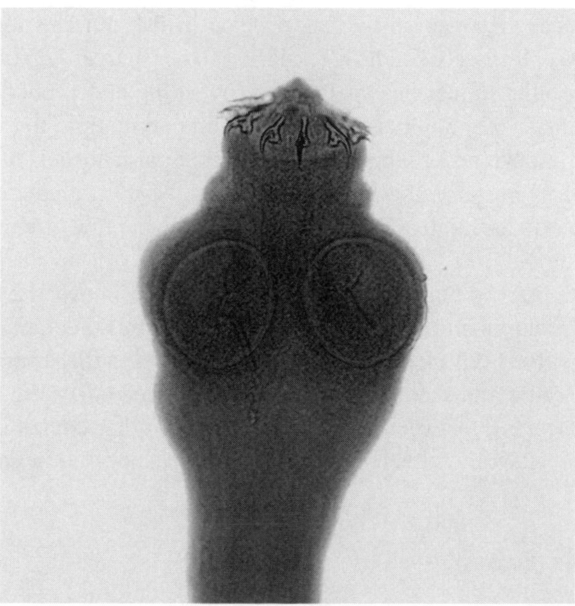

FIGURE 45.4. Scolex ("head") of an adult tapeworm, *Taenia solium*. The scolex has a ring of hooks (rostellum) and below that are four suckers (acetabula). The scolex tapers to form the neck, which is the generative region where proglottids are formed. (Courtesy of DPDx Parasite Image Library at the Centers for Disease Control and Prevention.)

with fertilized eggs to become a gravid proglottid that is released from the strobila and passes out of the host in the stool. In the stool, the eggs may remain viable for prolonged periods of time.

Pork Tapeworm

The human, as the definitive host, contracts the disease by eating undercooked pork. The pig, as the intermediate host, becomes infected with ova when it has access to human fecal matter containing proglottids and/or eggs. This occurs in areas of the world where human waste is used as fertilizer or in rural areas where pigs forage on refuse. In the pig intestine, the eggs hatch and produce larvae that penetrate the intestinal wall and migrate to muscle tissue where they become encapsulated. Pork flesh containing these larvae (bladder worms), which can be seen as lustrous beads, is called measly pork and can be identified with the naked eye. In the human intestine, the bladder worm is digested free from its capsule, attaches to the intestinal wall, and develops into the adult worm. The human can inadvertently become a dead-end intermediate host for the larva if the tapeworm egg itself is ingested. The freed larva (cysticercus) then migrates through host tissue, coming to rest in muscle or organs and causing a condition known as cysticercosis, or neurocysticercosis if the traveling larva ends its journey in the brain. The latter condition can be fatal. In the United States, most cases of cysticercosis occur in foreign-born Hispanics (Mexico and Latin America), while it is relatively rare in native-born citizens [28]. An outbreak of cysticercosis among Orthodox Jews — proscribed from eating pork or pork products — was traced to Hispanic household employees. Prevention: avoid undercooked pork products, particularly in developing countries where swine have access to garbage and other waste materials. Emphasize importance of hand washing to food handlers.

Cestode Infections

1. *Diphyllobothriasis.* Caused by a tapeworm found in fish, *Diphyllobothrium latum.* Larvae, present in copepods ingested by fish, develop in fish muscles and are infective for humans eating raw or undercooked fish. The larval stage (sparganum) of *Diphyllobothrium* spp. can also develop in humans as sparganosis, by eating undercooked fish, amphibians, snakes etc., or using such a poultice. The sparganum can migrate to the eye (ocular sparganosis) or brain, or develop as subcuticular nodules. Sparganosis is more common in the Orient than elsewhere because of consumption of raw fish, etc., and occasional application of raw meat to the skin as curative therapy. Prevention: avoid raw or uncooked fish, snakes, and frogs.
2. *Echinococcosis.* Caused by human ingestion of dog tapeworm ova; the human becomes an accidental intermediate host. Larval stage is a cyst (hydatid) that can develop in the liver or other portions of the body (kidney, bones, CNS) as a fluid-filled sphere. The tapeworm is endemic where cattle and sheep, the normal intermediate hosts, are raised. The disease is a result of close contact between humans and dogs or other canids in endemic areas of the world. Prevention: avoid intimate contact with dogs, especially in areas where livestock are raised.

Helminths: Trematodes (Flukes)

The fluke body is unsegmented. The adult worm may have one or two suckers, one ventrally located and the second a sucker that surrounds the mouth leading to a sac-like gut but lacking an anal opening. Most flukes are monoecious (have male and female sex organs in the same individual) although *Schistosoma* spp. are dioecious (the sexes in separate individuals). Intestinal flukes can be identified by the shapes of their ova as seen in stool samples (Figure 45.5)

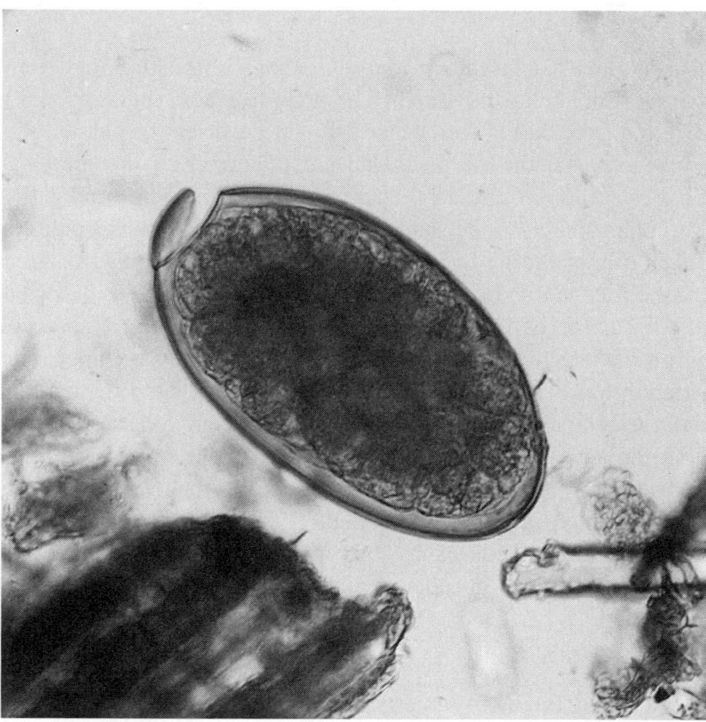

FIGURE 45.5 Ovum of a flatworm (*Fasciola hepatica*, liver fluke). The eggs are shed in the stool and identifying them is probably the chief way in which diagnosis is made. The egg has a distinctive operculum at one end through which the larva (miracidium) will emerge. (Courtesy of DPDx Parasite Image Library at the Centers for Disease Control and Prevention.)

Schistosomiasis

Caused by *Schistosoma* spp. The schistosomes are found in the circulatory system of humans while other flukes are found in the intestinal tract. The female is harbored within a groove of the male worm, ensuring close contact and fertilization of ova. Three of the medically important species recognized in this chapter are *S. haematobium;* which is found in veins associated with the urinary bladder and lower abdomen; *S. japonicum;* which is found in veins about the small intestine; and *S. mansoni,* which is found in veins associated with the large intestine. The ova of *S. haematobium,* and *S. mansoni* have projecting spines, located terminally and laterally, respectively; *S. japonicum,* has a laterally placed although rudimentary spine. Depending on the site of the adults in the body, eggs can be found in the urine or in the stool. Ova that do not pass out of the body can penetrate the venule walls, causing granuloma formation. Once in water, the egg hatches to release a ciliated miracidium that swims until it locates a snail as secondary host. The snail host is specific for each of the three different species. After undergoing development in the snail, a cercarial stage is released. The aquatic cercariae, using their tails as a propellant, swim to a human host and penetrate the skin. They then enter the circulatory system and are carried to the lungs and liver, and finally to their specific sites in the venous circulation. Humans may occasionally develop swimmer's itch, a condition caused by cercariae not adapted to humans (e.g., normally found in birds), penetrating the skin and initiating a strong immune reaction. Prevention: avoid streams, lakes, and ponds in endemic areas for washing and swimming because of cercariae in the water.

Trematode Infections

1. *Clonorchiasis.* A disease caused by *Clonorchis sinensis* (the Oriental liver fluke) endemic in parts of Asia (China, Japan, Korea). The definitive host (humans and other animals) acquires infection from eating freshwater fish — raw, dried, or pickled — from endemic areas. The fluke locates in the bile ducts, causing inflammation, thickening and, if infection is protracted, obstruction of bile flow. The result may be jaundice and gall stones. Ova are shed in the stool and hatch in the snail intermediate host following ingestion. Prevention: avoid uncooked fish, including frozen, pickled, or dried, from endemic areas.
2. *Paragonimiasis.* A zoonotic disease caused by *Paragonimus westermani*, endemic in parts of Asia and Africa. The definitive hosts include felines (cat, tiger) and canids; snails and crustaceans are the intermediate hosts. The fluke is found in the lungs, releases eggs that are transported upward from the lungs by the action of ciliated epithelial cells, pass into the intestinal tract, and are eliminated in the stool. The egg releases a ciliated miracidium that seeks out the appropriate snail species. After development in and release from the snail, the larval stage penetrates a crustacean and, ultimately, enters a human when the crab or crayfish is eaten raw or undercooked. The freed larva penetrates the wall of the intestinal tract, enters the body cavity, and migrates to the lungs. Prevention: thorough cooking of crabs and crayfish in endemic areas will protect against infection.

Helminth Infections: Nematodes

Nematode worms are among the most numerous organisms in the world. The group includes many free-living forms as well as medically significant parasites. The body is elongate with a complete mouth-to-anus intestinal tract. A fluid-filled coelom-like body cavity (pseudocoelom) is present. The body is protected from the host's digestive juices by a protective cuticle. Subcuticular musculature of the worms is longitudinally arranged, giving the worms a predominately sinuous type of movement. Sexes are separate in parasitic and most free-living species.

Onchocerciasis

The disease, also called river blindness, is caused by the filarial worm *Onchocerca volvulus*, and is transmitted by the black fly (*Simulium* spp.). The fly becomes infected with microfilariae when it feeds on the blood of an individual with onchocerciasis. The larvae (microfilariae) migrate from the fly's stomach to its proboscis. Once inoculated into the human host, the male and female worms accumulate in cutaneous nodules. Following mating, the larvae appear in the circulatory system. The disease is found in Central and South America, Mexico, the Middle East, and Africa, with distribution to the Americas a consequence of the slave trade. Microfilariae can also enter the eye where damage is done in large part by the host's immune response to the larvae. Prevention: water filtration in endemic areas, insecticide use to eliminate black flies.

NEMATODE INFECTIONS

1. *Ascariasis.* One of the most common nematode infections, it is caused by *Ascaris lumbricoides*. The adult worm is found in the human intestine and eggs are shed into the feces (Figure 45.6). The thick-walled eggs are highly resistant to environmental extremes and can remain viable for months after being shed. When a human ingests a mature egg, it hatches in the intestine and the larva begins a circuitous journey through the body (to the heart via the hepatic portal circulation, then to the lungs, to the pharynx where larvae are swallowed and returned to the intestine). Large numbers of worms can cause blockage of the intestinal lumen, requiring surgical removal. Ascariasis is more likely to occur in rural

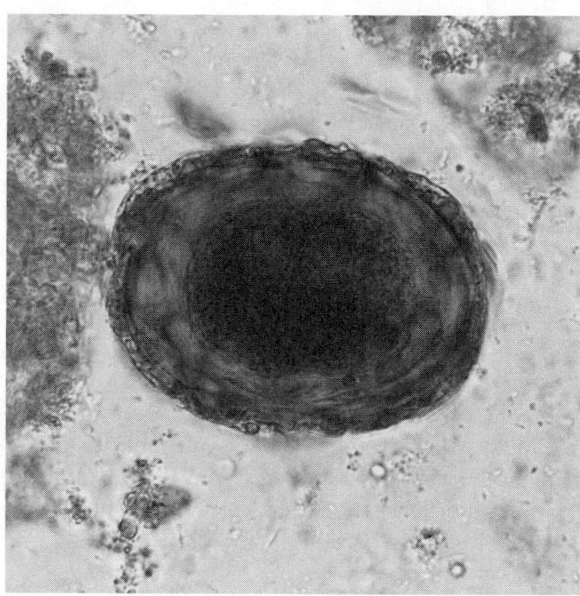

FIGURE 45.6 Ovum of the nematode worm *Ascaris lumbricoides*. The egg with an embryo inside has a thick, protective, and mammilate wall that can withstand long-term environmental exposure. (Courtesy of DPDx Parasite Image Library at the Centers for Disease Control and Prevention.)

farming areas where there is close contact between humans and pigs, both of which share the worm infection (*A. lumbricoides* in humans and *A. suum* in swine). Prevention: avoidance of soil containing human or swine excrement, particularly for children who are more likely to transfer eggs on their fingers after playing in or eating soil (geophagia).

2. *Baylisascariasis*. An emerging infection in humans caused by the ascarid worm *Baylisascaris procyonis* [29]. The worm is a parasite of raccoons, with humans as inadvertent intermediate hosts. Ova are discharged in raccoon feces. Children ingest eggs when playing in sandboxes or dirt around where raccoons defecate ("raccoon latrines"). The larvae migrate through tissues, especially the CNS, causing eosinophilic meningitis; there may also be ocular involvement. The disease is found in the United States (Northeast, West Coast). Effective drug therapy is not available, although steroids may be helpful in reducing inflammation [30, 31]. Prevention: keep children away from areas frequented by raccoons, keep garbage (which attracts raccoons) in tightly closed containers, and prevent children from eating soil.
3. *Dracunculiasis*. A disease reaching back to Biblical times (the "fiery serpent" that afflicted the Israelites) caused by the guinea worm *Dracunculis medinensis*. The disease is endemic to Africa, India, and parts of the Middle East. Humans are the definitive host and the copepod *Cyclops* (and related genera) the intermediate hosts [32]. The female worm is one of the largest nematodes, measuring approximately 100 cm in length, and is little more than an egg-filled uterus with compressed gut. The female, gravid with fertilized ova, migrates to the cuticular surface of the host's legs or feet, causing formation of a "fiery" painful blister. When the limb contacts cold water, the eggs are discharged from the worm's uterus. Ova are ingested by crustaceans, which are swallowed by humans when drinking water from a stream or pond. Filtration of water before drinking can prevent human infection. Concentrating people around a scarce water source, as during droughts, helps spread the disease. Since antiquity, the worm has been extracted by winding it around a twig or stick over several weeks. Although crude, it remains the treatment of choice, but accompanying

drug therapy may facilitate worm extraction [30, 31]. Aspirin is reportedly as effective as drug treatment in reducing inflammation [32]. Secondary bacterial infections with abscess formation are possible through the hole in the skin made by the female worm, particularly if the worm retracts, carrying bacteria subcutaneously. Prevention: water filtration and keeping water sources free of parasite ova can eliminate the disease.

4. *Filariasis. Wucheraria bancrofti*, a nematode worm that can cause elephantiasis as it increases in numbers sufficient to block lymph flow, resulting in lymph accumulation in tissues. Elephantiasis is a consequence of long-term infection. The disease is transmitted by the bite of a mosquito, and is endemic in tropical and subtropical areas (e.g., South Pacific islands). Prevention: use of netting and mosquito repellant.
5. *Gnathostomiasis.* A disease found mostly in the Orient and Southwest Pacific caused by *Gnathostoma spinigerum*. A consequence of ingesting undercooked definitive host (fish, snakes, frogs, chickens). Cutaneous larval migration can cause creeping eruption in humans without further complication, or eosinophilic meningitis, characterized by eosinophilia in the cerebrospinal fluid, if the larvae migrate to the CNS. Ocular involvement is possible. Drug treatment is problematic; dying worms may cause an aggravated immune response. Prevention: avoid raw or undercooked delicacies in endemic areas.
6. *Trichinellosis.* Caused by *Trichinella spirallis* present in pigs (the definitive host); the human is an inadvertent intermediate host. Transmission occurs when undercooked pork is ingested. The larvae that are digested free in the human intestine then penetrate the intestinal wall and migrate to muscle tissue (Figure 45.7), where they will eventually become encapsulated and calcified. The period of pathogenesis is during the migration of the larvae, at which time a heavy dose of larvae can damage vital organs with fatal results. Prevention: avoid undercooked pork products; pigs should not have access to human excrement.

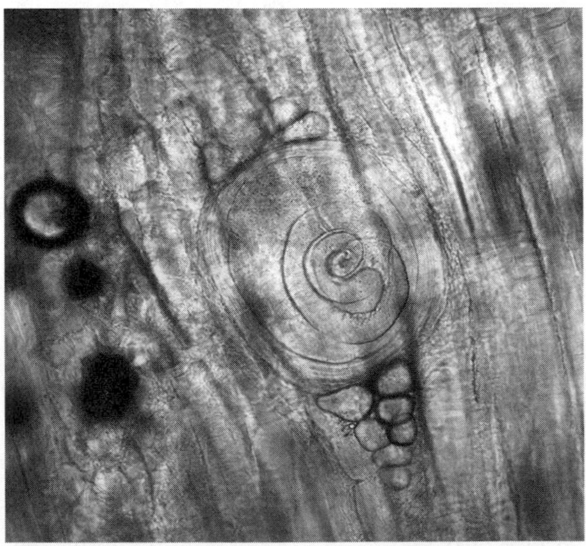

FIGURE 45.7 Section through muscle tissue showing an encysted larva of *Trichinella spiralis*. If not killed by cooking, the larvae are digested free of their capsules and develop into sexually mature adults. Larvae produced in the lumen penetrate the intestinal wall to encyst in muscle tissues. (Courtesy of DPDx Parasite Image Library at the Centers for Disease Control and Prevention.)

Insects/Acarids

The most familiar human parasites are the insects, found globally (Table 45.4). They are parasites because of their feeding on human blood. More importantly, they serve as vectors for parasitic protists and helminths: the *Anopheles* mosquito as transmitter of *Plasmodium*, the tsetse fly of trypanosomiasis, the triatomid bug of Chagas' disease, and the black fly of onchocerciasis. The tick, responsible for spreading babesiosis, is an acarid and not an insect. Insects such as the housefly and cockroach carry filth from excreta, human or otherwise [33].

PARASITE EPIDEMIOLOGY

Epidemiology: Ways that Parasites Spread

Parasite epidemiology attempts to explain how diseases are spread from person to person. Epidemiology has moved well beyond its beginnings in hand washing and the fundamental principles of hygiene. Modern epidemiology focuses more on the parasite genome, enabling identification of parasite strains and the presence of genes that confer drug resistance. The tools of the trade include evolutionary analyses and construction of phylogenetic trees, mapping the genomic composition of parasite populations, and, not forgetting the vectors, the genomes of the insects that transmit disease [34]. Molecular studies of parasite distribution and spread are the counterpart of studies tracing outbreaks of bacterial infections through plasmids and antimicrobial resistance genes that have revolutionized prokaryote epidemiology. In this chapter, however, more attention is given to the classic approach to epidemiology. Categories recognized below are not mutually exclusive and there much is overlap between them.

Insect Transmission

Elimination of insects as vectors — a desirable goal but not always possible — can reduce incidence of disease. Controlling populations of *Plasmodium*-carrying mosquitoes using insecticides or protecting susceptible children in malarious areas from being bitten by use of insecticide-impregnated bed netting, can reduce deaths due to malaria. Mosquitoes, however, have developed resistance to some insecticides (e.g., DDT), reducing their efficacy in controlling insect populations. As noted earlier, insects such as roaches and flies can transmit protistan cysts and helminth ova on mouthparts, legs, and feet as well as within their digestive tract. Regurgitation of gut contents prior to feeding and fecal deposition are other ways in which infectious material can be transferred [33].

Fecal-Oral Transmission

Human or animal feces transmit many parasitic diseases by parasite ova and protistan cysts. Such transmission can occur where feces are used as fertilizer ("night soil") or contaminate water used for bathing, drinking or irrigation. This is particularly the case with cysts of *Entamoeba histolytica, Giardia duodenalis,* and *Balantidium coli* that can contaminate the water supply in countries lacking water treatment standards. Feces-contaminated vegetables, eaten without cooking, are another result of fecal-oral transmission. This has become increasingly common in recent years as such foods, sometimes from abroad but also from "safe" farms, come on the market.

Direct Blood Transmission

This route is perhaps a consequence of human ingenuity that allows for blood collection for transfusions. Although the blood supply is monitored for HIV and prion diseases (bovine spongiform encephalopathy), etiologic agents of parasitic diseases are usually not looked for. Malaria and babesiosis, both blood infections, have been transmitted by human blood transfusions, as has Chagas' disease and toxoplasmosis.

Fomite Transmission

Inanimate objects can serve as vectors of some parasitic diseases: towels, tableware, and other common household articles and surfaces. Enterobiasis or pinworm infection can be readily spread through a household in which one member is infected. Female worms lay eggs in the perianal area at night and, as a result of scratching, can be transferred to fingers and then to household objects used by the host. Careful hand washing is essential in preventing the spread of worm eggs.

Zoonotic Transmission

A zoonotic disease is one that is spread from animal to human. Some of the zoonotic diseases include babesiosis (from rodents, cattle, deer, elk, and sheep) to humans via ticks; trypanosomiasis (from game animals and cattle via the tsetse fly); toxoplasmosis (from cats, pigs, lambs, and horses to humans), ascariasis, and echinococcosis. The transition from a mobile hunter-gatherer society to a settled life on farms, and ultimately communities, encouraged domestication of cattle, pigs, and other animals, offering opportunities for animal parasites to jump from one host to a new and virgin host. Nomadic people leave their wastes behind them but as communities increased in numbers, human and animal excreta would accumulate. Some was undoubtedly used as fertilizer as in the Orient today, but the cumulative build-up of waste would also contaminate water supplies, affording a means of disease transmission.

Sexual Transmission

Two protistan parasites can be transmitted via sexual contact. *Trichomonas vaginalis* can be spread as a sexually transmitted disease by intercourse, with transmission from male to female or female to male. *Entamoeba histolytica* has been spread by anal intercourse among male homosexuals.

Person-to-Person Transmission

Most parasitic diseases require an intermediate host before cycling to the human host. Pinworms, highly resilient intestinal parasites, can be spread directly from an infected child to other children and adults as a result of poor sanitary habits (careful hand washing). *Cryptosporidium* can be spread directly in high-risk groups such as children in daycare centers and institutionalized individuals. Some, such as *Pneumocystis*, the agent of *Pneumocystis* pneumonia (PCP), are transmitted directly in a pattern that is not completely understood (see later).

Airborne Transmission

Although airborne transmission of parasites is far less likely than spread of fungal spores, bacteria, and viruses, it does occur. *Pneumocystis* is found at one time or another in the lungs of healthy humans as transient colonizers without producing any ill effect, even expressing surface antigen variation to elude the immune system and prolong the period of residence [35]. But in the immunocompromised patient, lung infection results in PCP. In air samples collected in patients' hospital rooms and homes, *Pneumocystis* DNA was detected in greater than 50% of rooms of infected patients as compared to approximately 30% of rooms of noninfected patients, and not at all in storage and administrative areas in the hospital or in homes. The cysts of pathogenic species or strains of free-living amoebae that cause granulomatous and primary amoebic encephalitis (Table 45.2) can be wind-borne. *Naegleria fowleri*, a pathogenic species that has been isolated from the nostrils of healthy individuals, and *Acanthamoeba* spp. have been associated with sinus infections and amoebic keratitis in humans. Helminth ova, because of their size, are less likely to be spread by airborne transmission.

Food and Food Handler Transmission

Food may serve as a vehicle for transmission of many parasites, including those found in feces. Food contaminated with human or animal waste, particularly if not cooked (e.g., salad greens, fruits), can transmit *Entamoeba histolytica*, *Cryptosporidium,* and *Cyclospora.* Food handlers indifferent to washing hands after toileting have also been implicated in transmission.

Undercooked pork, beef, or fish can carry larval stages of tapeworms or *Trichinella* (pork). In Asia, intermediate hosts of some parasites (e.g., fish, crabs) may be eaten raw or lightly cooked and pass on larval stages to humans and other mammals as definitive hosts.

Water Transmission

Perhaps the simplest and most common vehicle for disease transmission is water, either directly or indirectly. Contaminated drinking water can transmit amoebic dysentery and giardiasis (see above: fecal-oral transmission). Numerous efforts have been made to provide water to water-scarce areas of Africa for drinking, irrigation, and hydroelectric power through the construction of dams. The resultant lakes and irrigation canals, however, have also attracted snails, in particular those that are intermediate hosts for schistosome worms, resulting in a sharp increase in schistosomiasis [36]. The water also provides a breeding ground for mosquitoes and black flies that spread malaria and river blindness, respectively. River blindness is transmitted by *Dracunculis* ova released into water. Eggs are eaten by aquatic copepods (*Cyclops*), which humans ingest in untreated water.

DIAGNOSTIC TECHNIQUES

As with all infectious diseases, a critical factor in initiating appropriate therapy is early identification of the etiologic agent. Cultivation of protist parasites from a clinical sample is helpful in some diagnoses. Helminth parasites, however, are not cultured.

After decades of testing, the protocols for detection of a number of parasitic disease agents are well established. For others, diagnosis is beyond the ability of the hospital or, at times, even the reference laboratory. Parasitic disease contracted during overseas travel or carried by immigrants may not be readily recognized by clinicians or local diagnostic facilities, and may require that clinical samples be sent abroad for identification to areas of the world where the disease is endemic. A definitive diagnosis may require invasive procedures that might not be feasible. It is thus noteworthy that diagnoses of some parasitic diseases are made postmortem.

Clinical Diagnosis

In developed countries, some parasitic infections can be recognized based on the patient's symptoms, recent travel history, occupation, contact with domestic and wild animals, unusual diets, immune status, insect bites, and place of birth (i.e., immigrant or native-born). Amoebic dysentery would be a logical diagnosis for a patient with diarrhea who had recently returned from a vacation in a developing country. Examination of a stool sample for amoebic cysts or trophozoites could verify the diagnosis and treatment could be started. Visceral leishmaniasis (kala-azar) would be a differential diagnosis for a child with an enlarged spleen and liver who emigrated from the Middle East (Table 45.2). Many other parasitic diseases have vague symptoms, are difficult to identify, and may require extensive testing. The clinician also has information from the standard panel of diagnostic tests, including the complete blood count (CBC); analysis of CSF (protein, white blood cells, and glucose in the CSF); and radiographic imaging (CT scans, MRIs). Neurocysticercosis, in which the larval stage of the pork tapeworm develops in the brain and eventually becomes encapsulated, will show up in an MRI. Other parasitic infections where calcification facilitates diagnosis are dracunculiasis, balamuthiasis, and baylisacariasis. Infection by parasites can lead to elevated IgE levels and eosinophilia. If the blood sample is counted electronically instead of using a stained blood film, eosinophils may not be differentiated from other leukocytes and might be missed. In malaria infections, a drop in erythrocyte numbers and anemia are to be expected and, coupled with a stained blood film, can confirm a diagnosis.

Laboratory Diagnosis

Where testing is based on established criteria, such as cysts in a stool sample or *Plasmodium*-infected erythrocytes in a blood film, a rapid diagnosis is possible. Many parasitic diseases, however, are more difficult to identify. Symptoms may be vague or non-specific and may mimic a number of other conditions.

Safety in the Laboratory

Laboratory personnel should be aware that parasitic protists and helminth ova could be just as dangerous as pathogenic bacteria, and should be treated and handled accordingly [37]. Laboratory infections can occur as a result of aerosol inhalation, needlestick injuries while doing animal inoculations, and oral and corneal infections resulting from splashes and failure to wash hands after handling potentially infectious materials and wearing protective goggles.

Work with moderately hazardous parasitic organisms should be done in a biosafety level-2 (BSL-2) facility. A laboratory coat or gown and gloves should be worn. There should be no mouth pipetting, eating, or drinking in the laboratory, and materials used in laboratory (pipettes, slides, culture vessels, etc.) as well as work surfaces should be decontaminated. In working with helminth ova, it should be remembered that these can remain viable even after formalin fixation. Additional information is available at the following Internet address: bmbl.od.nih.gov/sect3bsl2.htm.

Clinical Samples

The most common samples for testing include whole blood (malaria, trypanosomiasis), serum for serological testing (toxoplasmosis), stool samples for gastrointestinal parasites (worm ova), cysts of amoebae (dysentery), tissue samples for hematoxylin-eosin or other types of staining (Figures 45.6 and 45.7), sputum for lung parasites (*Paragonimus, Pneumocystis*), vaginal smears (trichomoniasis), and sigmoidoscopic (pinworms), or lymphatic aspirates (trypanosomes). Based on the type of specimen, special handling may be required. Although serum samples for antibody testing can be frozen and stored for prolonged periods of time, whole blood for blood film staining must be processed immediately. With other types of specimens, the integrity of the parasite might be damaged should the sample be kept at or near 37°C for prolonged time periods or, at the other extreme, placed in a refrigerator or freezer. Delayed transit of a sample to the laboratory can also reduce the chance of detecting more fragile parasites. Figure 45.8 presents an algorithm for the testing of clinical samples.

SPECIFIC TECHNIQUES

Direct and Indirect Testing for Parasites

Testing for the presence of parasites covers a broad range of techniques. Direct testing aims to visualize the parasite with a microscope in a blood film, stool, urine, or bronchoalveolar sample or biopsied tissue. Even with these long-established techniques, however, the parasite can elude the technician. For many parasites, multiple samples must be taken over several days because of periodicity in the presence of parasite stages in the clinical sample. The names characterizing the different malarial diseases (Table 45.2) denote differences in the time intervals at which the acute phase of disease occurs (i.e., approximately 48 hr in tertiary malaria; approximately 72 hr in quartan malaria). Early erythrocytic ("ring") stages of *Plasmodium falciparum* are infrequently seen

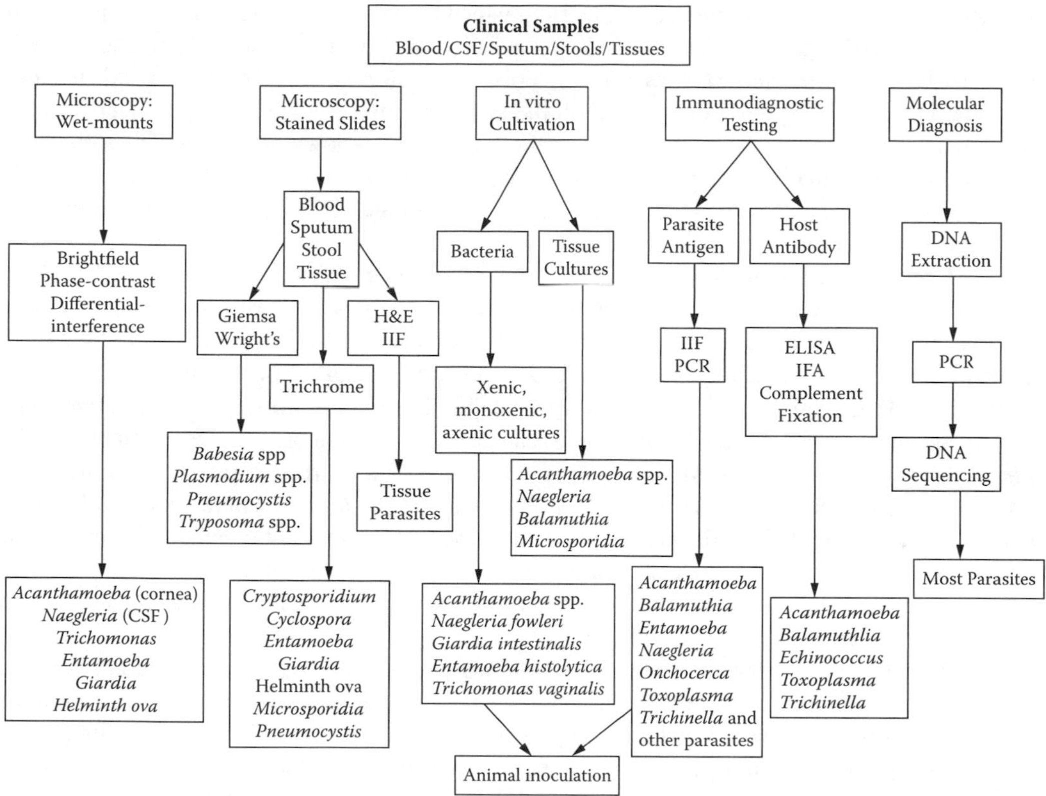

FIGURE 45.8 An algorithm based on the types of clinical samples and the various techniques that can be applied in identification. The same organisms can appear in several pathways, indicating that there is more than one way for identification of the parasite. It is assumed in this algorithm that most parasitic organisms can be identified by the PCR technique.

in circulating blood or blood films because these infected erythrocytes are likely to bind to the walls of blood vessels in the brain. The filarial worm *Wucheraria bancrofti* (cause of elephantiasis) appears in circulating blood at night, corresponding to the time that mosquito vectors would be most actively feeding.

In indirect testing, where testing for antibodies is involved, samples from the acute as well as the convalescent phases of the disease are needed to detect changes in antibody titer or in types of antibodies present (IgM vs. IgG). Antibodies in the patients' blood samples may cross-react with the "wrong" antigens, leading to misdiagnosis and unnecessary or inappropriate therapy. Some immunoserology tests may give problematic results (e.g., for neurocysticercosis) and may need corroboration by some other technique to get a definitive diagnosis.

Cultivation of Parasites

In vitro cultivation of parasites is useful in confirming identification, obtaining yields of parasites for molecular and biochemical studies, for assessing virulence, for electron microscopy, and for testing antimicrobial efficacy in treating infections. A number of protistan parasites can be cultured, although often requiring complex media, demanding growth conditions, and inconvenient transfer schedules to keep cultures from dying out. Some media are available commercially and the American Type Culture Collection (www.ATCC.org), in addition to having active cultures or frozen strains of parasitic protists available for purchase, can also provide media formulations and prepared media. Bacteria generally accompany protists established in culture from clinical samples, the number depending on the site in the body from which they were taken. CSF, for example, should be sterile and anything isolated (e.g., *Naegleria fowleri*) from it would be significant. Cultures with undefined bacteria are referred to as xenic or polyxenic cultures (from *xenos* [Greek], meaning stranger). The next step would be to establish the protist in monoxenic (single "stranger") culture with a defined bacterial species using antibiotics to eliminate unknown or unwanted microbes. The ultimate goal is eliminating all bacteria to obtain an axenic culture (without "strangers," or bacteria-free). Antibiotics frequently used for eliminating bacteria are penicillin-streptomycin, gentamycin, and amphotericin B for fungal contaminants.

Naegleria fowleri and *Acanthamoeba* can be cultured by transfer of clinical samples (*Naegleria*: brain tissue, CSF; *Acanthamoeba* spp.: brain or cutaneous tissues, corneal scrapings for amoebic keratitis) to non-nutrient agar plates with a film of bacteria as a food source (any of the following: *Escherichia coli*, *Klebsiella pneumoniae*, *Enterobacter* spp.) spread over the agar surface and incubated at 30 to 37°C. Amoebae will grow out within 24 to 72 hr [38]. *Balamuthia*, another free-living amoeba causing cutaneous and CNS infections, can be cultured from brain tissue (often obtained at autopsy) on tissue culture cells. Monkey kidney cells are a good source for growth, but other tissue culture lines have been used [39]. *Entamoeba histolytica*, *Giardia duodenalis*, and *Trichomonas vaginalis* require complex media and are often accompanied by a mixed population of indigenous bacteria that can overgrow in the enriched medium [40]. An *Entamoeba histolytica* look-alike, *E. dispar*, may also be present in cultured stool [41], as well as several other commensal amoebae from the intestine (*Entamoeba coli, Endolimax nana*).

The hemoflagellates (trypanosomes, *Leishmania*) are medically important protists that can be isolated from blood and tissues samples, and needle aspirates of cutaneous lesions. The media used for these organisms are varied and complex [42]. *Plasmodium* and *Babesia*, both intraerythrocytic blood parasites, can be grown in culture but are especially challenging in terms of media and growth conditions, requiring skills that are usually restricted to research laboratories [43, 44]. Microsporidia, intracellular parasites that can be isolated from urine, corneal scrapings, and bronchoalveolar lavage, can be maintained in a variety of tissue culture lines [45]. *Toxoplasma* can be identified by inoculation of clinical samples (e.g., CSF) into tissue cultures. *Pneumocystis jirovecii*, although an important opportunistic parasite, has yet to be cultured *in vitro*, beyond mere maintenance.

Animal Inoculations (Xenodiagnosis)

Mice, rats, hamsters, or guinea pigs can be inoculated with infectious material from a clinical sample or with isolated etiologic agents themselves if in culture, and observed for signs of disease. An advantage of this technique is that if parasitemia is low in the host and diagnosis is uncertain (as in *Babesia* infections), animal inoculation can allow for an increase under *in vivo* conditions and aid in diagnosis. The disadvantage with the technique is that it may take days or even weeks for results to develop. Some isolates such as amoebae (*Balamuthia mandrillaris, Acanthamoeba* spp.) from culture can be inoculated in mice via an intranasal or intracranial route and cause death of the animal within a week [46]. Insect vectors for parasitic diseases (*Anopheles* mosquitoes, triatomid bugs, etc.) can be inoculated by allowing them to feed on an infected human or animal. Some pathogenic strains lose virulence with prolonged *in vitro* culture and may require animal passage to maintain or up-regulate their virulence.

Immunofluorescent Testing for Antibodies (IFA)

Patient serum is prepared in a series of dilutions. Antibodies in the serum bind to the antigen (the parasite) that is fixed and dried on a slide. This, in turn, is treated with anti-human fluorochrome-conjugated antibody raised in goats or rabbits. Fluorescence resulting from binding of the fluorescent antibody to the human anti-parasite antibody can be visualized using a fluorescence microscope. The antibody titer is that serum dilution (e.g., 1:64), beyond which fluorescence can no longer be detected. The technique is useful for determining antibody titers at different stages of disease.

Immunofluorescent Staining for Parasites in Tissue (IIF)

IIF (indirect immunofluorescence) is a qualitative test for the presence of the parasite in tissue. Biopsy or necropsy tissue sections on slides are exposed to goat or rabbit anti-parasite serum, followed by fluorochrome-conjugated anti-rabbit or anti-goat serum. Parasites in tissue will fluoresce when the slide is examined with a fluorescence microscope, but not surrounding host tissue (Figure 45.9). The immunostaining is highly specific, making this a particularly reliable diagnostic test.

Enzyme-Linked Immunosorbent Assay (ELISA)

Enzyme immunoassay for antibody is widely used for diagnosis of many parasitic diseases. Commercial ELISA kits are available, precluding the do-it-yourself, time-consuming preparation of plates and assembling necessary reagents, thus simplifying the procedure. Because of considerations such as cost of the kits, the number of tests being performed in the laboratory, and the shelf life of the kit, it may not be feasible to perform ELISA testing, but rather to send samples to state or national reference laboratories.

Microscopic Diagnosis

For many diagnostic procedures, the microscope remains an important and affordable tool, especially in laboratories where more sophisticated testing and equipment are not available. A major use is in the examination of blood smears in conjunction with staining, using Giemsa or Wright's stains for blood smears. Stained blood films are useful in determining parasitemia, the percentage of infected erythrocytes. Changes in parasitemia can be used to check response to drug therapy or determine if a particular therapy is effective. Stained films are also effective in distinguishing between infection with *Plasmodium* or *Babesia* (pear-shaped in erythrocytes).

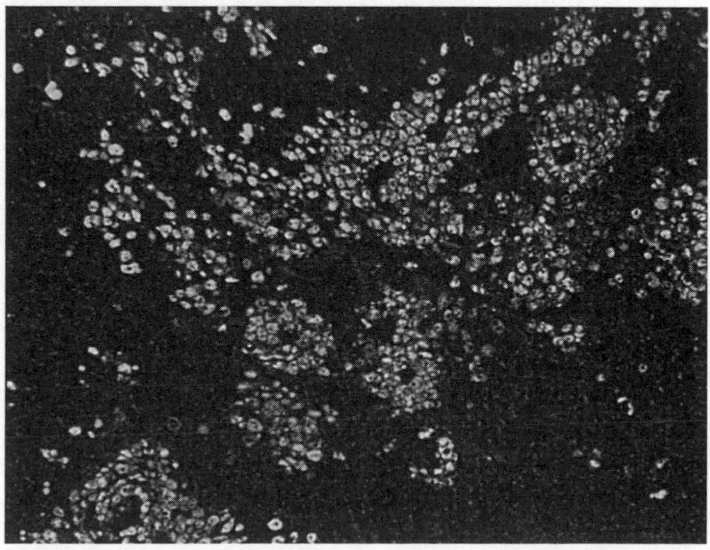

FIGURE 45.9 Section through brain stained with immunofluorescent antibody specific for *Balamuthia mandrillaris*. The amoebae fluoresce brightly against unstained parenchymal background. (Courtesy of Dr. G. S. Visvesvara.)

The microscope is used with a variety of clinical samples: fixed and stained with hematoxylin-eosin (tissue sections), or iron hematoxylin or trichrome for stool samples. Lugol's iodine, used in moderation, can also be a useful stain, enhancing the appearance of ova in stools and the presence of flagella on protists, although more conventional staining should be used to corroborate observations. A compendium of stains, formulas for preparation, and directions for their use can be found in Garcia [1]. In some instances, unstained material can be examined directly (e.g., CSF, lymph node aspirates) as wet-mounts.

A phase-contrast or differential-inference microscope can enhance the appearance of unstained specimens in wet-mounts and facilitate parasite identification. For extended viewing of wet-mounts, a combination of petroleum jelly and paraffin can be used to seal the coverslip to the slide, thus slowing drying. Fresh, unfixed wet-mounts are particularly useful for observing protist motility. Examination of CSF wet-mounts for the presence of active, rapidly moving amoebae is essential in the diagnosis of amoebic meningoencephalitis caused by *Naegleria fowleri*. In old or refrigerated CSF samples, leukocytes may be mistaken for amoebae, causing misdiagnosis. Flagellated trichomonads from vaginal discharge in a wet-mount typically show a wobbly movement, similar to that of a spinning top that is about to topple.

Concentrating the clinical sample is often useful for increasing the chance of detecting parasites or parasitic stages that may be sparse in samples. Thick vs. thin blood films can be helpful in detecting and identifying malaria parasites. A thick film offers a better chance of seeing parasites in the blood film, while a thin film is better able to identify the different species of *Plasmodium*. Helminth ova in stools may be obscured by fecal debris, and separating them using flotation (zinc floatation) or sedimentation (formal-ethyl acetate), the former being more reliable and easier to use than the latter, is frequently done [1].

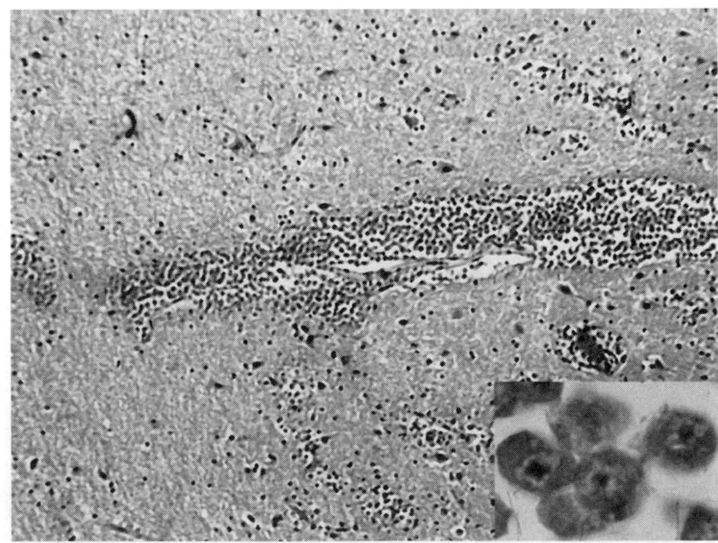

FIGURE 45.10 Section through brain with primary amoebic meningoencephalitis caused by *Naegleria fowleri*. A heavy concentration of amoeba is seen in the perivascular space. Hematoxylin-eosin stain. Inset: high-magnification image of *Naegleria* amoebae showing nucleus with large centrally located nucleolus in each amoeba. (Courtesy of Dr. G.S. Visvesvara.)

Histopathology

Sectioned tissue from biopsies, whether brain, liver, or intestine, stained with hematoxylin-eosin stain can provide definitive evidence for the identification of parasites (Figure 45.10 *Acanthamoeba* in brain tissue).

Molecular Diagnostic Techniques: PCR

Once a highly specialized technique used only in research and reference laboratories, the polymerase chain reaction (PCR) is now available to a wider audience of diagnosticians. Its basis is the use of primer sets containing basepairs unique for strands of parasite nuclear or mitochondrial DNA, giving an amplified product that can be visualized on a gel. The resulting bands identify the DNA as coming from a particular parasite whose DNA is present in host tissues or body fluids. Many, if not most, parasites can be identified by the PCR. In conventional PCR, the amplified DNA extract from a clinical sample is run on a gel and identified by the number of basepairs in the amplification product and by comparison with positive and negative controls. The identification can be further confirmed by DNA sequencing. Nested PCR is a modification of the conventional PCR technique in which sequential primer sets are used in producing an amplification product. Real-time PCR is an alternative technique that is more sensitive than conventional PCR and has additional advantages. The amplification and detection (fluorescent probe) steps occur in a closed tube, reducing the chance of contamination and environmental spread of DNA [47]. Additionally, the products are analyzed in real-time using a thermal cycler. The technique is faster (hours vs. overnight) and requires no post-amplification handling. Real-time PCR can identify DNA from different organisms simultaneously as, for example, *Naegleria fowleri*, *Acanthamoeba*, and *Balamuthia mandrillaris,* free-living amoebae responsible for amoebic encephalitis [48]. It has also been used for differentiating between species of *Plasmodium* more effectively than stained blood films and has shown evidence of mixed human infections [49].

PCR cannot distinguish between DNA from living vs. dead organisms, which can make a difference in deciding to treat or not treat a disease with drugs. PCR can be useful in retrospective studies of disease by analysis of archival slides and formalin-preserved tissues, but is less reliable in identifying DNA from parasite tissue samples because of fragmentation of the DNA caused by formalin fixation.

DRUG THERAPY

As with all infectious diseases, the ideal drug for use against an etiologic agent is one that will specifically target a particular function or metabolic pathway in the parasite that is either absent from the host's cells or less sensitive to the drug. Drug treatment is dynamic; new drugs come on the market, new variations of existing drugs become available, the U.S. Food and Drug Administration may approve drugs that were previously unavailable for use in the United States, or resistance to a particular drug therapy (e.g., use of chloroquine in areas where chloroquine resistance is common) may preclude its use in treatment and the cost of the drug, particularly when considering widespread use in endemic regions. Additionally, there are other considerations that will affect drug choice and dosage: dosage adjusted for age, oral vs. parenteral administration, stage of disease, drug use for prophylaxis, drug allergies or negative interactions with other medications, drug safety during pregnancy, disseminated or localized infections, and patient compliance with the recommended drug therapy. A drug that has unpleasant side effects, is relatively expensive, and has to be administered over prolonged time periods is less likely to be taken as directed by a patient. For these reasons, specifics of drug therapy, such as dosage, are not presented in the chapter other than to name current drugs that are recommended for treatment (Table 45.5). *The Sanford Guide to Antimicrobial Therapy* [30] and *The Medical Letter on Drugs and Therapeutics. Drugs for Parasitic Infections*

[31] provide updated information on currently recommended drugs of choice (and alternates) for parasitic diseases, including dosages for adults and children. Specific information on mechanisms of drug action can be found in specialized textbooks [50].

A consequence of drug therapy is death of parasites. Although this is a desired effect, massive release of antigens from disintegrating parasites may present more of a problem to the immune system than the presence of the living organisms. Large-scale death of helminths in tissue can elicit an allergic reaction from the host caused by an immune response to the dying parasites. Antihistamines and corticosteroids can mute the drug-initiated immune reaction, but in some instances palliative measures may be the preferred course of action. Table 45.5 presents general information about drugs for protistan and helminthic diseases. The reader, however, should consult current guidelines [30, 31] for drug use against specific diseases and/or seek medical advice.

Mechanisms of Action of Selected Drugs

Several drugs that are used against a number of different parasites are examined for their mechanisms of action. The general effects include interference with parasite metabolic pathways, inhibition of microtubule assembly preventing mitosis or disorganizing cell structure, alteration of permeability of the parasite tegument, and damage to parasite DNA.

Selected Drugs Effective in Treating Protistan Diseases

1. *Artemisinin and derivatives.* These antimalarial compounds are derived from the medicinal herb *Artemisia annua* that has been used for centuries in China as a treatment for malaria [51]. Derivatives of the compound include artemether and artesunate. The drugs themselves are effective for relatively short time periods and, to extend their duration of efficacy, they have been coupled with other drugs (e.g., artesunate-mefloquine). At the time of this writing, the U.S. Food and Drug Administration has not approved the drugs for use in the United States.
2. *Chloroquine.* Chloroquine was used successfully for many years in the treatment of malaria until drug resistance began to show up in malarial parasites. Now its use is restricted to parts of the world where malarial strains remain sensitive to the drug. The drug is a weak base and is taken up into acidic food vacuoles of the intracellular parasite. Several mechanisms of action have been reported for the drug: prevention of detoxification of internalized heme (from host erythrocyte hemoglobin) accumulated by the malarial parasite, formation of drug-heme complexes, or interference with nucleic acid biosyntheses. Chloroquine-sensitive strains accumulate the drug, while chloroquine-resistant strains do not.
3. *Metronidazole (Flagyl®) and tinidazole (Tindamax®; Fasigyn®).* Metronidazole is a synthetic nitroimidazole antiprotistan drug. When taken up, it is reduced by the parasite's electron transport system, resulting in production of free radicals. A gradient is established across the parasite membrane between the reduced molecule within the parasite and its normal oxidized molecular state outside the membrane, enhancing uptake of the drug. It has also been shown to damage the parasite's DNA and is a mutagenic agent (in rodents). It is particularly useful in treating infections caused by anaerobic or microaerophilic protists such as *Entamoeba histolytica, Giardia intestinalis,* and *Trichomonas vaginalis.* Tinidazole is similar to metronidazole in its action and the organisms affected, but is reportedly faster acting.
4. *Miltefosine (hexadecylphosphocholine).* Miltefosine is an alkylphosphocholine developed as an anticancer drug but has been shown effective in the treatment of leishmaniasis, including visceral [53], cutaneous, and mucocutaneous [54] forms. It is available as a pill

for oral administration and has limited side effects (diarrhea, vomiting) in patients. The drug produces an apoptosis-like effect on the parasite, resulting in membrane breakdown and lysis.

Selected Drugs Effective in Treating Helminth Diseases

1. *Praziquantel (Biltricide®).* Praziquantel alters the intracellular calcium levels of sensitive flat worms [52]. Among its effects on helminths, at low concentrations it causes spastic paralysis leading to expulsion of the worm from the host. At high concentrations, it leads to damage to the worm's tegument, alters permeability, and causes its disintegration, opening the tegument to the host's immune system. Both these events are triggered by an increase in calcium ions. It has also been used in treating neurocysticercosis. Although praziquantel is used in treating many fluke and cestode diseases, nematodes do not respond to treatment with this drug.
2. *Albendazole (Albenza®) and Mebendazole (Vermox®).* Both drugs are benzimidazole broad-spectrum antihelminthics. Their effects include preventing glucose absorption across the worm tegument, interfering with mitochondrial activity, and inhibiting microtubule formation by binding to tubulin. A concentration differential exists between sensitivity of nematode tubulin and host tubulin to benzimidazoles, the former being more sensitive to lower drug concentrations. Their use is primarily against nematodes: pinworm, hookworm, and mixed infections. They are used in the treatment of inoperable neurocysticercosis in conjunction with corticosteroids, the latter to prevent an inflammatory response to the disintegrating cysticercoid larvae.

CONCLUSIONS AND PERSPECTIVES

Parasites are well adapted to their ecologic niche, and in many diseases have achieved détente with their hosts. The parasites have had millennia to perfect their way of life; the host has decades to adjust. As the result of living in a disease-endemic region, the host may develop some immunity against the parasite (e.g., falciparum malaria in individuals older than 5 years), allowing survival.

Although major efforts are underway to eliminate specific parasitic diseases, problems remain:

1. Prohibitive cost of medication in economically depressed developing countries with endemic parasites
2. Convincing people to take necessary antiparasite medications despite unpleasant side effects or prolonged use
3. The difficulty in getting some people living in malarious areas to use insecticide-impregnated bed nets at night to protect children from mosquito bites
4. Development of resistance to drugs of choice
5. Prevention of re-infection in endemic areas

Progress on vaccines against the most devastating of the parasitic diseases, malaria, has been slow and the successes once envisioned have yet to be realized, with antigenic variation as a major obstacle to overcome; any vaccine, to be successful, has to protect against all stages of the parasite life cycle. With continued sequencing studies of parasite DNA and identification of gene products, new drugs aimed at specific functions in parasites may prove more effective in combating diseases than those in the current pharmacopoeia.

Looking ahead to the generally accepted effects of global climate change, conditions may become more favorable for survival and dissemination of human parasites. Developed nations will be able to adjust to increased incidence of parasitic diseases among travelers and immigrants.

Despite the subtropical temperatures, availability of the species of mosquito and suitable breeding sites in the United States, malaria, once a public health threat in this country, has been eradicated. Miasmas arising from the swamps around Rome were once connected in people's minds with the summertime occurrence of malaria ("bad air"); drainage removed the breeding sites for the vectors. Elimination of malaria was a product of mosquito abatement programs, testing, and the availability of antimalarial drugs. Resource-starved developing nations will likely bear the burden of increased parasite loads. As a result of global warming, water may become scarcer in areas of the world, further concentrating peoples around water sources leading to their contamination with enhanced development of larval stages and vectors (e.g., dracunculiasis).

REFERENCES

1. Garcia, L.S., *Diagnostic Medical Parasitology*, 5th ed., ASM Press, Washington, DC, 2007.
2. Cox, F.E.G., History of human parasitology, *Clin. Microbiol. Rev.*, 15, 595, 2002.
3. Thang, H.D., Elsas, R.M., and Veenstra, J., Airport malaria: report of a case and a brief review of the literature, *Neth. J. Med.*, 60, 441, 2002.
4. Fenner, L., Weber, R., Staffen, R., and Schllagenhauf, P., Imported infectious disease and purpose of travel in Switzerland, *Emerg. Infect. Dis.*, 13, 217, 2007.
5. Baer, J.G., *Ecology of Animal Parasites*, University of Illinois Press, Urbana, 1951, chap. 13.
6. McKay, D.M. and Webb, R.A., Cestode infection: immunological considerations from host and tapeworm perspectives, in *Parasitic Flatworms. Molecular Biology, Biochemistry, Immunology and Physiology*, Maule, A.G. and Marks, N.J., Eds., Cabi, Oxfordshire, United Kingdom, 2006, chap. 9.
7. Black, J, Hommel, M., Snounou, G., and Pinder, M., Mixed infections with *Plasmodium falciparum* and *P. malariae* and fever in malaria, *Lancet* 343, 1095, 1994.
8. Chiang, P.K., Bujnicki, J.M., Su, X., and Lanar, D.E., Malaria: therapy, genes and vaccines, *Curr. Mol. Med.,* 6, 309, 2006.
9. Gardiner, M.J., Hall, N., Fung, E., et al., Genome sequence of the human malaria parasite *Plasmodium falciparum*, *Nature,* 419, 498, 2002.
10. Volkman, S.K., Sabeti, P.C., DeCaprio, D., et al., A genome-wide map of diversity in *Plasmodium falciparum*, *Nat. Genet.*, 39, 113, 2007.
11. El-Sayed, N.M., Nyler, P.J., Blandin, G., et al., Comparative genomics of trypanosomatid parasitic protozoa, *Science*, 309, 404, 2005.
12. Berriman, M., Ghedin, E., Hertz-Fowler, C., et al., The genome of the African trypanosome *Trypanosoma brucei*, *Science*, 309, 416, 2005.
13. El-Sayed, N.M., Myler, P.J., Bartholomeu, D.C., et al., The genome sequence of *Trypanosoma cruzi*, etiologic agent of Chagas disease, *Science*, 309, 409, 2005.
14. Ivens, A.C., Peacock, C.S., Worthey, E.A., et al., The genome of the kinetoplastid parasite *Leishmania major*, *Science*, 309, 436, 2005.
15. Carlton, J.M., Hirt, R.P., Silva, J.C., et al., Draft genome sequence of the sexually transmitted pathogen *Trichomonas vaginalis*, *Science,* 315, 207, 2007.
16. Miller, L.H., Good, M.F., and Milon, G., Malaria pathogenesis, *Science,* 264, 1878, 1994.
17. Bloch, K. and Schuster, F.L., Inability to make a premortem diagnosis of *Acanthamoeba* species infetion in a patient with fatal granulomatous encephalitis, *J. Clin. Microbiol.*, 443, 3003, 2003.
18. Adl, S.M. Simpson, A.G.B., Farmer, M.A., et al., The new higher level classification of Eukaryotes with emphasis on the taxonomy of the Protists, *J. Eukaryot. Microbiol.,* 52, 399, 2005.
19. Khan N.A., *Acanthamoeba*: biology and increasing importance in human health, *FEMS Microbiol. Rev.*, 30, 564, 2006.
20. Homer, M.J., Aguilar-Delfin, I., Teleford, S.R. III, et al., Babesiosis, *Clin. Microbiol. Rev.,* 13, 451, 2000.
21. Zintl, A., Mulcahy, G., Skerrett, H. E., et al., *Babesia divergens*, a bovine blood parasite of veterinary and zoonotic importance. *Clin. Microbiol. Rev.*, 16, 622, 2003.
22. Schuster, F.L., Cultivation of *Babesia* and *Babesia*-like blood parasites: agents of an emerging zoonotic disease, *Clin. Microbiol. Rev.*, 15, 365, 2002.
23. Chen, Q.Q., Schlichtherle, M., and Wahlgren, M., Molecular aspects of severe malaria, *Clin. Microbiol. Rev.*, 13, 439, 2000.
24. Birbeck, G.L., Cerebral malaria, *Curr. Treat. Options Neurol.*, 6, 125, 2004.

25. Clark, I.A., Alleva, L.M., Mills, A.C., and Cowden, W.B., Pathogenesis of malaria and clinically similar conditions, *Clin. Microbiol. Rev.*, 17, 509, 2004.
26. Montoya, J.G. and Liesenfeld, O., Toxoplasmosis, *Lancet*, 363, 1965, 2004.
27. Jones, J.L., Kruszon-Moran, D., Wilson, M., et al., *Toxoplasma gondii* infection in the United States: seroprevalence and risk factors, *Am. J. Epidemiol.*, 154, 357, 2001.
28. Sorvillo, F.J., DiGiorgio, C., and Waterman, S.H., Deaths from cysticercosis, United States, *Emerg. Inf. Dis.*, 13, 230, 2007.
29. Gavin, P.J., Kazacos, K.R., and Shulman, S.T., Baylisacariasis, *Clin. Microbiol. Rev.*, 18, 703, 2005.
30. Moellering, R.C., Elliopoulos, G.M., and Sande, M.A. Eds, *The Sanford Guide to Antimicrobial Therapy*, Gilbert, D.N., Antimicrobial Therapy, Inc., Sperryville, VA, 2006.
31. Abramowicz, M. and Zuccotti, G., Eds., *Medical Letter on Drugs and Therapeutics. Drugs for Parasitic Infections*, The Medical Letter, Inc., New Rochelle, NY, 2004.
32. Cairncross, S., Muller, R., and Zagaria, N., Dracunculiasis (guinea worm disease) and the eradication initiative, *Clin. Microbiol. Rev.*, 15, 223, 2002.
33. Graczyk, T.K., Knight, R., and Tamang, L., Mechanical transmission of human protozoan parasites by insects, *Clin. Microbiol. Rev.*, 18, 128, 2005.
34. Conway, D.J., Molecular epidemiology of malaria. *Clin. Microbiol. Rev.*, 20, 188, 2007.
35. Stringer, J.R., Antigenic variation in *Pneumocystis*, *J. Eukaryot. Microbiol.*, 54, 8, 2007.
36. Fenwick A. Waterborne infectious diseases — could they be consigned to history? *Science*, 313, 1077, 2006.
37. Herwaldt, B.L., Laboratory-acquired parasitic infections from accidental exposures, *Clin. Microbiol. Rev.*, 14, 659, 2001.
38. Schuster, F.L, Cultivation of pathogenic and opportunistic free-living amebas, *Clin. Microbiol. Rev.*, 15, 342, 2002.
39. Visvesvara, G.S., Moura, H., and Schuster, F.L., Pathogenic and opportunistic free-living amoebae: *Acanthamoeba* spp., *Balamuthia mandrillaris*, *Naegleria fowleri*, and *Sappinia diploidea*, *FEMS Immunol Med. Microbiol.*, 50, 1, 2007.
40. Clark, C.G. and Diamond, L.S., Methods for cultivation of luminal protists of clinical importance, *Clin. Microbiol. Rev.,* 15, 329, 2002.
41. Diamond, L.S. and Clark, C.G., *Enatmaoeba histolytica*, Schaudinn 1903 (emended Walker, 1911) separating it from *Entamoeba dispar*, Brumpt 1925, *J. Eukaryot. Microbiol.*, 40, 340, 1993.
42. Schuster, F.L. and Sullivan, J.J., Cultivation of clinically significant hemoflagellates, *Clin. Microbiol. Rev.*, 15, 374, 2002.
43. Schuster, F.L., Cultivation of *Plasmodium* spp., *Clin. Microbiol. Rev.*, 15, 355, 2002.
44. Schuster, F.L., Cultivation of *Babesia* and *Babesia*-like blood parasites: agents of an emerging zoonotic disease, *Clin. Microbiol. Rev.*, 15, 365. 2002.
45. Visvesvara, G.S., *In vitro* cultivation of microsporidia of clinical importance, *Clin. Microbiol. Rev.*, 15, 401, 2002.
46. Visvesvara, G.S., Schuster, F.L., and Martinez, A.J., *Balamuthia mandrillaris*, N.G., N. Sp., agent of meningoencephalitis in humans and other animals. *J. Eukaryot. Microbiol.*, 40, 504, 1993.
47. Espy, M.J., Uhl, J.R., Sloan, M., et al., Real-time PCR in clinical microbiology: applications for routine laboratory testing, *Clin, Microbiol. Rev.*, 19, 165, 2006.
48. Qvarnstrom, Y., Visvesvara, G.S., Sriram, R., and da Silva, A.J., A multiplex real-time PCR assay for simultaneous detection of *Acanthamoeba* spp., *Balamuthia mandrillaris* and *Naegleria fowleri*, *J. Clin. Microbiol.*, 44, 3589, 2006.
49. Perandin, F., Manca, A., Calderaro, A., et al., Development of a real-time PCR assay for detection of *Plasmodium falciparum*, *Plasmodium vivax*, and *Plasmodium ovale* for routine clinical diagnosis, *J. Clin. Microbiol.*, 42, 1214, 2004.
50. Tracey, J.W. and Webster, L.T., Jr., Chemotherapy of parasitic infections, in *Goodman and Gilman's The Pharmacological Basis of Therapeutics*, Harmand, J.G., Limbird, L.E., Molinoff, R.B., and Ruddon, R.W., Eds., McGraw-Hill, New York, 1996, Sect. VIII.
51. Adjuik, M., Agnamey, P. Babiker, A., et al., The International Artemisinin Study Group. Artesunate combinations for treatment of malaria: meta-analysis, *Lancet*, 363, 9, 2004.
52. Greenberg, R.M., Praziquantel: mechanism of action, in *Parasitic Flatworms. Molecular Biology, Biochemistry, Immunology and Physiology.* Maule, A.G. and Marks, N.J., Eds., Cabi, Oxfordshire, United Kingdom, 2006, chap. 14.

53. Sundar, S., Jha, T.K., Thakur, C.P., et al., Oral miltefosine for Indian visceral leishmaniasis, *N. Eng. J. Med.*, 347, 1739, 2002.
54. Soto, J. and Soto, P., Miltefosine: oral treatment of leishmaniasis. *Expert Rev. Anti Infect. Ther.*, 4, 177, 2006.

46 Introduction to Yeasts

David H. Pincus

CONTENTS

INTRODUCTION

The original treatise by Macmillan and Phaff[1] has been modified significantly but portions were retained in this chapter because some of the information is timeless and still quite relevant. This update takes into account new information appearing in taxonomic texts and other literature, including a plethora of taxonomic changes due to the discovery of new genera and species as well as reassignment of previously known species into newly described genera or into other already-existing genera. The data derived from molecular studies (e.g., DNA hybridization, ribosomal RNA gene sequencing, etc.) have contributed to a more elaborate understanding of phylogeny and the relationships between species and genera of yeasts. Mating studies conducted on numerous members of anamorphic genera have also resulted in the discovery of many more teleomorphic states.

The term "yeast" is not easily defined taxonomically although it is used frequently in the scientific literature. It is originally derived from the Dutch word *gist,* which represents the foam in a brewing fermentation. Other words referring to yeast, such as the French word *levure,* are related to the role of yeast in the leavening (rising) of bread. The first concepts of yeast characteristics were based on those observed in the 19th century with yeasts such as *Saccharomyces cerevisiae,* the most well-known yeast to man due to its critical fermentative action in the production of bread and alcoholic beverages. At that time, yeasts were described as single-celled spherical to oval budding microorganisms that were hyaline (colorless) and later discovered to have the ability to form ascospores.[2]

The anaerobic conditions required for the fermentative process elaborated by Pasteur contributed to the theory that all yeasts could grow in the absence of air. Although many yeasts are facultative and able to ferment, it was later found that some yeasts such as *Rhodotorula* are non-fermentative and strictly aerobic. Other physiologic and morphologic characteristics were also expanded to accommodate the term "yeasts." Yeasts were originally assumed to only be hyaline, whereas later this was changed to include the yellow, orange, or red carotenoid pigments that could be formed by genera such as *Cryptococcus* and *Rhodotorula.*

Most yeasts can reproduce by forming blastoconidia (budding cells) while some genera are characterized by the absence of blastoconidia (e.g., *Schizosaccharomyces*, which divides by fission, or *Geotrichum*, which multiplies by disarticulation of mycelia that generates arthroconidia). Species of many anamorphic (asporogenous) and teleomorphic (sporogenous) genera (e.g., *Candida* and *Rhodosporidium*, respectively) are able to form true mycelia. Many genera are quite diverse with respect to their morphologic character, such as *Candida*, which includes species (e.g., *C. glabrata*) previously designated as members of the genus *Torulopsis*. They include species typically unable to form any filaments (pseudohyphae or true hyphae) or those that are only able to produce rudimentary pseudohyphae (short non-branching chains of cells) while other species of *Candida* can produce well-developed pseudomycelia and/or true mycelia such as *C. tropicalis* that typically produce both types of filaments.

Anamorphic yeasts differ from teleomorphic yeasts by their inability to form sexual spores (ascospores or basidiospores) and are therefore assigned to the Deuteromycetes (Fungi Imperfecti). While it was difficult previously to discern the relationship of asporogenous genera to either the Ascomycetes or Basidiomycetes, the Diazonium Blue B (DBB) test reaction is used today to group taxa into what appears to be a uniform relationship to either of the teleomorphic families. Anamorphs that are DBB-negative are considered ascomycetous, whereas those that are DBB-positive are considered basidiomycetous in their phylogenetic relationship.

Some of the basidiomycetous yeasts are able to form conidia that are forcibly ejected from sterigma (stalks). These conidia are known as ballistoconidia and resemble basidiospores although a sexual process is not involved in their formation. The teleomorphic yeasts that variably form ballistoconidia are also able to form sexual spores called basidiospores.

Characterization of the teleomorphic ascomycetous genera is sometimes facilitated by the properties of the ascospores, including shape, number, and whether they are retained in the ascus or liberated (evanescent). This is especially true of those genera or species that exhibit unusual shapes (e.g., *Metschnikowia* and needle-shaped ascospores without appendage) or unusually high numbers of spores (e.g., *Vanderwaltozyma polyspora* and 100 ascospores per ascus).

With all these variations and more, it is not surprising that a precise definition of yeasts on the basis of morphologic and physiologic characters is quite difficult although it appears to become easier with evolution of the taxonomy. Lodder and Kreger-van-Rij found it "most painful to have to state" the impossibility of giving a satisfactory definition to the term "yeasts" especially after authoring the first edition of the authoritative tome entitled *The Yeasts — A Taxonomic Study*. First published in 1952, it included the study of 1307 strains classified into 26 genera comprising 164 species.[2]

In its second edition edited by Lodder and published in 1970, the study of more than 4300 strains led to classification of 39 genera containing 349 species. Here, Lodder defined yeasts as "microorganisms in which the unicellular form is conspicuous and which belong to the fungi." This definition is still valid after nearly four decades and may be the best possible in view of the heterogeneity of these fungi.[3]

The third edition edited by Kreger-van-Rij was published in 1984 and included 60 genera containing 500 species.[4] Most recently, the fourth edition edited by Kurtzman and Fell was published in 1998 and included 100 genera and more than 700 species.[5]

The yeast database of the Centraalbureau voor Schimmelcultures (CBS) can be accessed online at http://www.cbs.knaw.nl/yeast/BioloMICS.aspx.[6] This website contains data on approximately 6500 strains available from the CBS collection as well as up to 900 species descriptions, and is updated regularly. The CBS website is also a valuable resource in checking taxonomic designations and synonyms. The tables in this chapter showing key morphologic and physiologic characteristics of the yeast genera were derived using data from the CBS website in conjunction with the newer taxonomic texts.[7] Limited data were available for the genera *Ascobotryozyma*, *Botryozyma*, *Cryptotrichosporon*, *Komagataella*, *Mastigobasidium*, *Ogataea*, and *Symbiotaphrina* and so additional publications were consulted.[8–15]

TABLE 46.1
Evolution of Yeast Taxonomy

Source	Year	Ref.	No. Genera	% Increase	No. Species	% Increase
Lodder and Kreger-van-Rij	1952	2	26	—	164	—
Lodder	1970	3	39	50.0	349	112.8
Kreger-van-Rij	1984	4	60	53.8	500	30.2
Kurtzman and Fell	1998	5	100	66.7	>700	>40.0
CBS	2007	6	122	22.0	1183	<69.0

As there appears to be no master list to which one can refer, it is possible that some genera may have been missed in checking the above sources. At last count, there were 122 genera comprising 1183 species that could be found at the CBS website. As one can see from Table 46.1, yeast taxonomy has evolved continually. While the growth of the number of genera seems to be slowing, the number of species is increasing at a still impressive rate.

Various types of taxonomic changes occur. There are new genera and species discovered as scientists explore previously uncharted locales and sample sources. New relationships are determined from more advanced classification tools (e.g., molecular techniques), and species may be reassigned to new genera (e.g., species of *Hansenula* reassigned to the genus *Pichia*). Sometimes, but more rarely, multiple species are grouped into one based on a better understanding of phylogenetic relationships. Examples of such "lumping" have occurred, with several earlier species of *Kluyveromyces* being grouped as synonyms of *K. marxianus* and multiple species of *Saccharomyces* grouped under *S. cerevisiae*. Such changes are difficult to follow for many scientists except taxonomists. It becomes very confusing to follow such changes in the literature, and subsequently one must survey all prior and current names in order to have a complete picture of available information. The same confusion holds true for some of the data found in genetic sequence databases such as GenBank because the nomenclature used may be outdated as it is a function of the sequence contributor.

Following the division of groups of yeasts in Kurtzman and Fell[5], there are four designated groups of yeasts, as follows:

1. Teleomorphic ascomycetous yeasts: yeasts capable of forming ascospores in asci and considered to be primitive ascomycetes: 58 genera (see Tables 46.2 and 46.6).
2. Anamorphic ascomycetous yeasts: yeasts most closely related to the teleomorphic ascomycetous yeasts but incapable of or not observed forming ascospores in asci: 16 genera (see Tables 46.3 and 46.7).
3. Teleomorphic basidiomycetous yeasts: yeasts capable of forming basidiospores and having life cycles similar to those of the order Ustilaginales of the Basidiomycetes: 22 genera (see Tables 46.4 and 46.8), of which six forcibly discharge ballistoconidia by a drop excretion mechanism; ballistoconidia resemble basidiospores but are considered an asexual rather than sexual means of reproduction.
4. Anamorphic basidiomycetous yeasts: yeasts most closely related to the teleomorphic basidiomycetous yeasts but incapable of or not observed forming basidiospores: 26 genera (see Tables 46.5 and 46.9), of which eight forcibly discharge ballistoconidia.

YEAST MORPHOLOGY

While many morphologic structures observed with yeasts are more confirmatory than diagnostic, there are certain structures or cell shapes that are very characteristic of certain genera and can aid in narrowing the choices in identification. Most morphologic structures are directly related to the

TABLE 46.2
Morphologic Characteristics of Teleomorphic Ascomycetous Yeast Genera

Genus	Number of Species	Colony Color[a, b]	Red Diffusible Pigment[b]	Cell Shape[c]	Budding[d]	Pseudohyphae[b]	True Hyphae[b]	Arthroconidia[b]	Ascospore Shape[c]	Ascospore Number	Other[b]
Ambrosiozyma	4	W, C, Y	–	A, C, K	ML	V	+	–	E	1–4	
Arthroascus	3	W, C, T	–	A, B, C, K	BP, ML	V	+	–	A, C	1–4	
Ascobotryozyma	2	C	–	C, K	ML	+	–	–	Q	4	
Ascoidea	4	C	–	B, D	V	V	+	–	C, E	16–160	V sympodial blastoconidia
Babjevia	1	W, C	–	A, C, K	BP, ML	V	–	–	A, B, C	1–4	
Cephaloascus	2	W, C	–	B, D	-	+	+	–	E	4	Sympodial blastoconidia
Citeromyces	2	W, C	–	A, B, C	ML	–	–	–	A, C	1–2	
Clavispora	2	W, C	–	B, C, K	ML	V	–	–	D, H	1–4	
Cyniclomyces	1	W	–	C, K	ML	+	–	–	C, H, K	1–4	Difficult to cultivate
Debaryomyces	19	W, C, Y	–	A, C, K	ML	V	–	–	A, B, C, (F, U)	1–4	
Dekkera	2	W, C	–	A, B, C, J, K	ML	+	V	–	E	1–4	
Dipodascopsis	2	W	–	A	ML	–	+	–	B, R	32–128	Sympodial blastoconidia
Dipodascus	14	W, C	–	K	–	–/(+)	+	+	A, B, C	1–30(128)	
Endomyces	1	W	–	D	–	–	+	–	B, E	2	Sympodial blastoconidia
Eremothecium	5	W, C	–	A, B, C, K	ML	V	V	–	I, X	2–32	
Galactomyces	3	W	–	K	–	–	+	+	A, B, C	1–2	
Hanseniaspora	10	W, C	–	C, K, M	BP	V	–	–	A, C, E, F, N	1–4	
Issatchenkia	5	W, C	–	A, B, C, K	ML	+	–	–	A, C	1–4	
Kawasakia	1	C	–	A, C	ML	–	–	–	A, C	1–16	
Kazachstania	29	W, C, T	–	A, C, K	ML	V	–	–	A, C, H, R	1–16	
Kluyveromyces	6	W, C, P	V	A, B, C, K	ML	V	–	–	A, B, C, R	1–6	
Kodamaea	4	W, C	–	C, K	ML	V	V	–	A, C, E	1–4	
Komagataella	3	W, C	–	A, C	ML	–	–	–	E	2–4	
Kregervanrija	3	W, C	–	C, K	ML	+	–	–	E, F	1–4	
Kuraishia	2	W, C	–	A, C	ML	–	–	–	E	1–2	
Lachancea	6	W, C	–	A, C	ML	V	–	–	A, C	1–4	
Lipomyces	11	W, C, Br	–	A, B, C	ML	–	–	–	A, B, C	1–20	
Lodderomyces	1	W, C	–	A, C, K	ML	+	–	–	A, B, C, H	1–2	
Metschnikowia	33	W, C, Y, (P)	–/(+)	A, C, K, (Q, V)	ML	V	V	–	W	1–2	
Nadsonia	2	W, C	–	C, K, M	BP	V	–	–	A, C	1–2	

Nakaseomyces	2	W, C	–	C	ML	–	–	–	C, H, R	1–8	
Naumovia	2	W, C	–	A, C	ML	–	–	–	A, C	1–4	
Ogataea	1	W, C	–	A, C, K	ML	–	–	–	A, E, F	1–4	
Pachysolen	1	W, C	–	A, B, C	ML	V	–	–	E, P	1–4	
Phaffomyces	3	W, C	–	A, C, K	ML	V	–	–	E	1–4	
Pichia	99	W, C	–	A, C, K, (T)	ML	V	V	–	A, E, F, P	1–4	
Protomyces	4	T, O	–	B, C, K	BP	–	–	–	B	50–200	Ascospores only on infected plants
Saccharomyces	7	W, C	–	A, B, C, K	ML	V	–	–	A, B, C	1–4	
Saccharomycodes	1	C, Y	–	C, K, M	BP	+	–	–	A, F	4	
Saccharomycopsis	8	W, C	–	A, C, K	ML	V	+	–	A, C, E, F, R	1–8	
Saturnispora	7	W, C, T	–	A, B, C, K	ML	V	–	–	A, C, E, F	1–8	
Schizosaccharomyces	3	W, C, T, Y	–	A, B, C, K	–	–	V	V	A, B, C	2–8	Fission
Smithiozyma	1	C	–	A, C	ML	–	–	–	A, C	1–4	
Sporopachydermia	3	W, C	–	B, C, K	ML	–	–	–	A, B, C	1–4	
Starmera	2	W, C	–	C, K	ML	V	V	–	E	1–4	
Stephanoascus	3	W, C	–	A, C	ML	–	V	–	E	1–4	Sympodial blastoconidia
Taphrina[e]	32	Y, P	–		BP	–	+	–			
Tetrapisispora	7	W, C	–	A, C, K	ML	V	–	–	A, C, R, S	1–12	
Torulaspora	5	W, C	–	A, C	ML	V	–	–	A, C	1–4	
Vanderwaltozyma	2	C	–	A, C, K	ML	V	–	–	A, C, R, S	1–100	
Wickerhamia	1	W, C, G, Y	–	C, K, M	BP	V	–	–	Z	1–2	
Wickerhamiella	5	W, C	–	A, C	ML, MP	–	–	–	A, E, L, S	1–2	
Williopsis	6	W, C	–	A, C, K	BP, ML	V	–	–	F	1–4	
Yarrowia	1	W, C	–	A, C, K	ML	+	+	–	A, C, E, F, P	1–4	
Zygoascus	2	W, C	–	C, K	ML	V	V	–	E, P	1–4	
Zygosaccharomyces	7	W, C	–	A, C, K	ML	V	–	–	A, B, C	1–4	
Zygotorulaspora	2	W, C	–	A, C, K	ML	V	–	–	A, C	1–4	
Zygozyma	3	W, C	–	A, C	ML	–	–	–	A, C, H, S	1–12	

[a] W = white; C = cream; T = tan; Y = yellow; O = orange; P = pink; R = red; Bu = buff; Br = brown; G = gray; OBr = olivaceous brown; OBl = olivaceous black; OG = olivaceous green

[b] + = positive for all species; – = negative for all species; V = variable; W = weak; D = delayed; ( ) = rarely seen.

[c] A = spheroidal or globose; B = ellipsoidal; C = ovoidal; D = clavate; E = hat-shaped; F = Saturn-shaped; G = dumbbell-shaped; H = oblong; I = fusiform; J = ogival; K = cylindroidal; L = elongate; M = apiculate or lemon-shaped; N = helmet-shaped; O = walnut-shaped; P = hemispheroidal; Q = crescentiform; R = reniform; S = allantoid (sausage-shaped); T = triangular; U = lenticular; V = pyriform; W = needle-shaped without appendage; X = spindle-shaped with appendage; Y = hemispheroidal with narrow ledge; Z = cap-shaped.

[d] BP = bipolar; ML = multilateral; MP = monopolar; – = no budding (exclusively arthroconidia or fission).

[e] Dimorphic plant pathogen with hyphal parasitic phase and yeast saprobic phase that shares some properties (e.g., DBB reaction) with basidiomycetous yeasts.

TABLE 46.3
Morphologic Characteristics of Anamorphic Ascomycetous Yeast Genera

Genus	Number of Species	Colony Color[a, b]	Red Diffusible Pigment[b]	Cell Shape[b, c]	Budding[d]	Pseudohyphae[b]	True Hyphae[b]	Arthroconidia[b]	Endoconidia[b]	Other[b]
Aciculoconidium	1	W, C	–	B, C	ML	–	+	–	–	Terminal needle-shaped conidia on hyphae
Arxula	2	W, C, Y	–	A, C, K	ML	V	+	+	+	
Blastobotrys	7	W, C	–	A, C, V	V	V	+	–	–	V sympodial blastoconidia
Botryozyma	3	C	–	C, K	ML	+	–	–	–	End cells transformed into appressoria
Brettanomyces	3	W, C	–	A, B, C, J, K, M, V	BP, ML	V	–	–	–	
Candida	282	W, C, (Bu)	–	A, C, K, L, (I, J, S, Q, T)	BP, ML, MP	V	V	–	–	
Geotrichum	14	W, C	–	K	–	–	+	+	V	
Kloeckera	1	W, C	–	C, L, M	BP	–	–	–	–	
Myxozyma	11	W, C	–	A, B, C	ML	–	–	–	–	
Oosporidium	1	O, P	–	A, C, K	ML	+	–	–	+	Fission
Saitoella	1	Y, O, R	–	C, K	ML	V	V	–	–	
Schizoblastosporion	2	W, C	–	B, C, K	BP	V	V	–	–	
Symbiotaphrina	2	C, P	–	A, C, V	ML	–	–	–	–	
Sympodiomyces	3	C	–	A, C	ML	–	+	–	–	Sympodial
Trichosporiella	3	C	V	A, C	ML	V	+	–	–	
Trigonopsis	1	W, C	–	B, T	ML, MP	–	–	–	–	

[a] W = white; C = cream; T = tan; Y = yellow; O = orange; P = pink; R = red; Bu = buff; Br = brown; G = gray; OBr = olivaceous brown; OBl = olivaceous black; OG = olivaceous green

[b] + = positive for all species; – = negative for all species; V = variable; W = weak; D = delayed; ( ) = rarely seen.

[c] A = spheroidal or globose; B = ellipsoidal; C = ovoidal; D = clavate; E = hat-shaped; F = Saturn-shaped; G = dumbbell-shaped; H = oblong; I = fusiform; J = ogival; K = cylindroidal; L = elongate; M = apiculate or lemon-shaped; N = helmet-shaped; O = walnut-shaped; P = hemispheroidal; Q = crescentiform; R = reniform; S = allantoid (sausage-shaped); T = triangular; U = lenticular; V = pyriform; W = needle-shaped without appendage; X = spindle-shaped with appendage; Y = hemispheroidal with narrow ledge; Z = cap-shaped.

[d] BP = bipolar; ML = multilateral; MP = monopolar; – = no budding (exclusively arthroconidia or fission).

TABLE 46.4
Morphologic Characteristics of Teleomorphic Basidiomycetous Yeast Genera

Genus	Number of Species	Colony Color[a,b]	Red Diffusible Pigment[b]	Cell Shape[b,c]	Budding[d]	Pseudohyphae[b]	True Hyphae[b]	Arthroconidia[b]	Endoconidia[b]	Ballistoconidia[b]	Other[b]
Agaricostilbum	1	C, Y	–	C, K	BP	V	V	–	–	–	
Bulleromyces	1	C	–	A, C, K	BP	V	V	–	–	+	Symmetrical ballistoconidia, sympodial blastoconidia
Chionosphaera	4	C	–	C, S	BP	V	V	–	–	–	
Cystofilobasidium	4	W, Y, O, P, R	–	A, C, G, K	BP	V	V	–	V	–	
Erythrobasidium	1	O, P	–	A, C, K	BP, ML	–	+	–	–	–	
Fibulobasidium	2	W, C, Y, G, Bu	–	A, C	BP	–	–	–	–	+	Symmetrical ballistoconidia
Filobasidiella	4	W, C, Y	–	A, C, M	BP, ML	–	V	–	–	–	
Filobasidium	5	W, C, P	–	A, B, C, K	BP, MP	V	V	–	–	–	
Holtermannia	1	W, C, Y	–	B, C, K	BP	–	+	–	–	–	
Kondoa	3	W, C, P	–	A, C, K	BP	V	V	–	–	–	V sympodial blastoconidia
Leucosporidium	5	W, C	–	C, K	BP	V	V	–	V	–	V sympodial blastoconidia
Mastigobasidium	1	W, C	–	A, C, K	BP	+	V	–	–	+	Asymmetrical ballistoconidia, sympodial blastoconidia
Mrakia	2	W, C, Y	–	C, K	BP	V	V	–	–	–	Sympodial blastoconidia
Rhodosporidium	9	Y, O, P, R, (Br)	–	A, C, K, (I)	BP	V	V	–	–	–	V sympodial blastoconidia
Sakaguchia	1	O, P, R	–	C	BP	V	V	–	–	–	
Sirobasidium	2	W, C, Y	–	A, C	BP, ML	V	–	–	–	+	V sympodial blastoconidia
Sporidiobolus	6	C, O, P, R	–	A, C, K, S	BP	V	V	–	–	+	Asymmetrical ballistoconidia, V sympodial blastoconidia
Sterigmatosporidium	1	W, C	–	C, K	ML	V	V	–	–	–	Blastoconidia on stalks
Tilletiaria	1	G-Br			–	–	+	–	–	+	
Tremella	19	W, C, Bu	–	A, B, C, K	BP, ML	–/(+)	–/(+)	–	–	–	V sympodial blastoconidia
Trimorphomyces	1	Y, P, Br	–	A, B, K	BP, ML	–	–	–	–	–	
Xanthophyllomyces	1	O, P, R	–	A, C	ML	V	–	–	–	–	

[a] W = white; C = cream; T = tan; Y = yellow; O = orange; P = pink; R = red; Bu = buff; Br = brown; G = gray; OBr = olivaceous brown; OBl = olivaceous black; OG = olivaceous green

[b] + = positive for all species; – = negative for all species; V = variable; W = weak; D = delayed; ( ) = rarely seen.

[c] A = spheroidal or globose; B = ellipsoidal; C = ovoidal; D = clavate; E = hat-shaped; F = Saturn-shaped; G = dumbbell-shaped; H = oblong; I = fusiform; J = ogival; K = cylindroidal; L = elongate; M = apiculate or lemon-shaped; N = helmet-shaped; O = walnut-shaped; P = hemispheroidal; Q = crescentiform; R = reniform; S = allantoid (sausage-shaped); T = triangular; U = lenticular; V = pyriform; W = needle-shaped without appendage; X = spindle-shaped with appendage; Y = hemispheroidal with narrow ledge; Z = cap-shaped.

[d] BP = bipolar; ML = multilateral; MP = monopolar; – = no budding (exclusively arthroconidia or fission).

TABLE 46.5
Morphologic Characteristics of Anamorphic Basidiomycetous Yeast Genera

Genus	Number of Species	Colony Color[a, b]	Red Diffusible Pigment[b]	Cell Shape[b, c]	Budding[d]	Pseudohyphae[b]	True Hyphae[b]	Arthroconidia[b]	Endoconidia[b]	Ballistoconidia[b]	Other[b]
Bensingtonia	13	W, C, Y, P, Br	–	A, B, C, K	BP, ML	V	V	–	–	+	V sympodial blastoconidia
Bullera	32	W, C, Y, P	–	A, B, C, K	BP, ML	V	V	–	–	+	V sympodial blastoconidia
Cryptococcus	84	W, C, Y, O, P, R	–	A, C, K, (G, S, M)	BP, ML, MP	V	V	–	–	–	V sympodial blastoconidia, fission in *C. statzelliae*
Cryptotrichosporon	1	Y, Bu	–	B, M	BP	V	–	–	–	–	
Dioszegia	9	C, O, R	–	A, C, K	BP, ML	V	V	V	–	+	V sympodial blastoconidia
Fellomyces	12	W, C, Y, O	–	A, B, C	ML	V	V	–	–	–	Blastoconidia on stalks, V sympodial blastoconidia
Hyalodendron	2	W, C, T	–	B, C	BP	V	+	+	–	–	
Itersonilia	1	C, Bu	–	A, C, V	BP, ML	V	+	–	–	+	
Kockovaella	11	C, Y, O	–	A, B, C, R	ML	V	–	–	–	+	Blastoconidia on stalks
Kurtzmanomyces	3	C, O, P	–	A, C, K	ML	–	V	–	–	–	Blastoconidia on stalks, V sympodial blastoconidia
Leucosporidiella	4	W, C, Y, Bu	–	A, C, K	BP	V	–	–	–	–	
Malassezia	11	C, T, Y, Bu	–	A, C, K	MP	–	–	–	–	–	
Moniliella	3	W, G, OBr, OBl, OG	–	B	BP, ML	V	+	+	–	–	
Pseudozyma	15	W, C, Y, P	–	B, C, I, K	BP	V	+	–	–	–	V sympodial blastoconidia
Reniforma	1	W, C	–	R, S	BP	–	–	–	–	–	
Rhodotorula	50	(W), C, Y, O, P, R, Bu	–	A, C, K, (I, T)	BP, ML, (MP)	V	V	–	–	–	V sympodial blastoconidia
Sporobolomyces	49	(W), C, Y, O, P, R, Bu	–	A, C, I, K, (S)	BP, ML	V	V	–	–	+	V sympodial blastoconidia
Sterigmatomyces	2	W, C	–	A, C	ML	V	–	–	–	–	Blastoconidia on stalks
Sympodiomycopsis	1	W, C	–	A, C, K	BP, ML	–	V	–	–	–	Sympodial blastoconidia
Tausonia	1	W, C, T	–		–	–	+	+	–	–	
Tilletiopsis	9	W, C, Y, P, Bu, Br	V[e]	B, L	BP	–	+	–	–	+	

Trichosporon	42	W, C, Y	–	A, C, D, K, S	BP, ML	–/(+)	+	+	V	–	
Trichosporonoides	5	C, OBr	–	A, B, C	ML	+	+	+	–	–	Sympodial blastoconidia
Tsuchiyaea	1	W, C	–	A, C, K, M	BP, ML	–	–	–	–	–	Blastoconidia on stalks
Udeniomyces	5	W, C, Y, P	–	A, B, C, I, K, S	BP	V	V	–	–	+	V sympodial blastoconidia
Zymoxenogloea	1	C, T	–	A, C, K	ML	–	–	–	–	–	

[a] W = white; C = cream; T = tan; Y = yellow; O = orange; P = pink; R = red; Bu = buff; Br = brown; G = gray; OBr = olivaceous brown; OBl = olivaceous black; OG = olivaceous green

[b] + = positive for all species; – = negative for all species; V = variable; W = weak; D = delayed; ( ) = rarely seen.

[c] A = spheroidal or globose; B = ellipsoidal; C = ovoidal; D = clavate; E = hat-shaped; F = Saturn-shaped; G = dumbbell-shaped; H = oblong; I = fusiform; J = ogival; K = cylindroidal; L = elongate; M = apiculate or lemon-shaped; N = helmet-shaped; O = walnut-shaped; P = hemispheroidal; Q = crescentiform; R = reniform; S = allantoid (sausage-shaped); T = triangular; U = lenticular; V = pyriform; W = needle-shaped without appendage; X = spindle-shaped with appendage; Y = hemispheroidal with narrow ledge; Z = cap-shaped.

[d] BP = bipolar; ML = multilateral; MP = monopolar; – = no budding (exclusively arthroconidia or fission).

[e] Brown or reddish-brown pigment may be exuded into the agar.

TABLE 46.6
Physiologic Characteristics of Teleomorphic Ascomycetous Yeast Genera

Genus	Number of Species	Fermentation[a]	Nitrate Utilization[a]	Cycloheximide 0.1%[a]	Growth in 10% NaCl[a]	Inositol Assimilation[a]	Urea Hydrolysis[a]	Diazonium Blue B Reaction[a]	Growth on 50% Glucose[a]	Growth at 37°C[a]	Other
Ambrosiozyma	4	V	V	+/D	–	–	–	–	V	V	
Arthroascus	3	V	–	V	–	–	–	–	–	–	
Ascobotryozyma	2	–	–	+/W	–	–	–	–	–	V	
Ascoidea	4	–	V	–	V	–/W	–/W	–/W		–	
Babjevia	1	–	–	V	–	–	–	–	–	–	
Cephaloascus	2	–/W	–		V	–		–		–	
Citeromyces	2	V	V	–	+	–	–	–	+	V	
Clavispora	2	+	–	–	V	–	–	–	V	+	
Cyniclomyces	1	+/W	–	–	–	–				+	Difficult to cultivate
Debaryomyces	19	V	–	V	+	–	–	–	V	V	
Dekkera	2	+/W	V	V	–	–	–	–	–	V	
Dipodascopsis	2	–	–	+		+		–	+	V	
Dipodascus	14	V	–	V	–	–	–	–	–	V	
Endomyces	1										
Eremothecium	5	V	–	V	–	–	–	–	–	V	
Galactomyces	3	V	–			–	–	–		V	
Hanseniaspora	10	+/W/D	–	V	V	–	–	–	V	V	
Issatchenkia	5	+	–	–	V	–	–	–	V	V	
Kawasakia	1	–	–	+	–	–	–	–	–	–	
Kazachstania	29	V	–	V	V	–	–	–	V	V	
Kluyveromyces	6	V	–	V	V	–	–	–	V	V	
Kodamaea	4	+	–	–	+	–	–	–	V	+	
Komagataella	3	+	–	V	–	–	–	–	–	V	
Kregervanrija	3	W/D	–	–	V	–	–	–	–	–	
Kuraishia	2	+	+	–	V	–	–	–	–	+	
Lachancea	6	+	–	V	V	–	–	–	V	V	
Lipomyces	11	–	–	+/(W/D)	–	V	–	–	–	V	
Lodderomyces	1	+	–	+	+	–	–	–	+	V	

Metschnikowia	33	V	–	–/(+)	V	–	–	–	V	V
Nadsonia	2	V	–	–	–	–	–	–	–	–
Nakaseomyces	2	+	–	–	+	–	–	–	+/D	+
Naumovia	2	+	–	–	V	–	–	–	–	V
Ogataea	1	V	–	+/D	–	–	–	–	–	V
Pachysolen	1	+	+	–	D	–	–	–	–	+
Phaffomyces	3	–	–	–	–	–	–	–	–	V
Pichia	99	V	V	V	V	–/(+)	–	–	V	V
Protomyces	4	–	V	–	–	–		–		–
Saccharomyces	7	+	–	–	V	–	–	–	V	V
Saccharomycodes	1	+	–	–	–	–	–	–	–	V
Saccharomycopsis	8	V	–	V	V	V	–	–	V	V
Saturnispora	7	+	–	–	–	–	–	–	–	–
Schizosaccharomyces	3	+	–	–	V	–	V	–	V	V
Smithiozyma	1	–	–	+	–	–		–	–	–
Sporopachydermia	3	–/W/D	–	+	–	+	–	–	–	+
Starmera	2	V	–	–	–	–	–	–	–	+
Stephanoascus	3	V	V	+	V	+/W/D	–	–	V	V
Taphrina[b]	32	–	V	–		+	V	V		–
Tetrapisispora	7	+	–	–	–/D	–	–	–	–	–
Torulaspora	5	+	–	V	V	–	–	–	V	V
Vanderwaltozyma	2	+	–	–	–	–	–	–	–	–
Wickerhamia	1	+	–	+	+	–	–	–	+	–
Wickerhamiella	5	–	V	V	+/W/D	–	–	–	+	–
Williopsis	6	+/W/D	V	V	V	–	–	–	V	V
Yarrowia	1	–	–	+	V		V	–	V	–
Zygoascus	2	+/D	–	+	+/W/D	+	–	–	V	+
Zygosaccharomyces	7	+/W/D	–	–	V	–	–	–	+/W/D	V
Zygotorulaspora	2	+	–	+	V	–	–	–	V	–
Zygozyma	3	–	–	+	–	V	–	–	–	–

[a] + = positive for all species; – = negative for all species; V = variable; W = weak; D = delayed; ( ) = rarely seen.

[b] Dimorphic plant pathogen with hyphal parasitic phase and yeast saprobic phase that shares some properties (e.g., DBB reaction) with basidiomycetous yeasts.

TABLE 46.7
Physiologic Characteristics of Anamorphic Ascomycetous Yeast Genera

Genus	Number of Species	Fermentation[a]	Nitrate Utilization[a]	Cycloheximide 0.1%[a]	Growth in 10% NaCl[a]	Inositol Assimilation[a]	Urea Hydrolysis[a]	Diazonium Blue B Reaction[a]	Growth on 50% Glucose[a]	Growth at 37°C[a]	Other
Aciculoconidium	1	V	–	+	–	–	–	–	–	–	
Arxula	2	V	+	+	+	+	–	–	+	+	
Blastobotrys	7	V	–			V				V	
Botryozyma	3	–	–	+/W/D	–	–	–	–	–	V	
Brettanomyces	3	+/W/D	–	V	–	–	–	–	–	V	
Candida	282	V	V	V	V	V	–	–	V	V	
Geotrichum	14	V	–	+		–	–	–	–		
Kloeckera	1	+	–	+	–	–	–	–	–	–	
Myxozyma	11	–	–	+	–	V	–	–	–	V	
Oosporidium	1	–	+			V	–	–	–	–	
Saitoella	1	–	+	–	–	–	+	–	–	–	
Schizoblastosporion	2	–	–	V	–	–	–	–	–	–	
Symbiotaphrina	2	–	–			–	+	–			
Sympodiomyces	3	–	–	+	D	+/W/D	–	–	–	–	
Trichosporiella	3	W/D	–	D		–	–	–	–	+	
Trigonopsis	1	–	–	+	D	–	–	–	–	+	

[a] + = positive for all species; – = negative for all species; V = variable; W = weak; D = delayed; ( ) = rarely seen.

TABLE 46.8
Physiologic Characteristics of Teleomorphic Basidiomycetous Yeast Genera

Genus	Number of Species	Fermentation[a]	Nitrate Utilization[a]	Cycloheximide 0.1%[a]	Growth in 10% NaCl[a]	Inositol Assimilation[a]	Urea Hydrolysis[a]	Diazonium Blue B Reaction[a]	Growth on 50% Glucose[a]	Growth at 37°C[a]	Other
Agaricostilbum	1	–	–	–	+	–		+	+	–	
Bulleromyces	1	–	–	V	–	+	+	+	V	–	
Chionosphaera	4	–	–	–	–	–	+		–	–	
Cystofilobasidium	4	–	+	V	V	+	+	+	V	–	
Erythrobasidium	1	–	+	–	–	–	+	+	–	–	
Fibulobasidium	2	–	V	–	–	V	+	+	–	–	
Filobasidiella	4	–	–	–	V	V	+	+	V	V	
Filobasidium	5	V	V	–	V	V	+	+	V	–	
Holtermannia	1	–	–	–		+	+	+	–	–	
Kondoa	3	–	V	–	V	–	V	V	–	V	
Leucosporidium	5	–	+	V	V	–	+	+	V	–	
Mastigobasidium	1	–	–	–	–	–	+	+	–	–	
Mrakia	2	V	+	–	–	V	V	+	–	–	
Rhodosporidium	9	–	V	V	V	V	+	+	V	V	
Sakaguchia	1	–	–	–	V	–	+	+	–	–	
Sirobasidium	2	–	–	–	V	+/D	+	+	–	–	
Sporidiobolus	6	–	V	V	V	–	+	+	V	V	
Sterigmatosporidium	1	–	–	–/W/D	–	+	+	+	–	–	
Tilletiaria	1	–	–			–	+		–	–	
Tremella	19	–	–	V	V	V	+/(–)	V	–/(W/D)	–/(+)	
Trimorphomyces	1	–	–			V	+	+		–	
Xanthophyllomyces	1	W/D	–	–	–	–	+	+	–/W/D	–	

[a] + = positive for all species; – = negative for all species; V = variable; W = weak; D = delayed; ( ) = rarely seen.

TABLE 46.9
Physiologic Characteristics of Anamorphic Basidiomycetous Yeast Genera

Genus	Number of Species	Fermentation[a]	Nitrate Utilization[a]	Cycloheximide 0.1 %[a]	Growth in 10% NaCl[a]	Inositol Assimilation[a]	Urea Hydrolysis[a]	Diazonium Blue B Reaction[a]	Growth on 50% Glucose[a]	Growth at 37°C[a]	Other
Bensingtonia	13	–	V	V	–	–	+	+	V	–	
Bullera	32	–	V	V	V	V	+	+	V	–	
Cryptococcus	84	V	V	V	V	V	V	V	V	V	
Cryptotrichosporon	1	–	–	–	–	V	+	+		–/D	
Dioszegia	9	–	–	–	–/W	V	+	+	V	–	
Fellomyces	12	–	–	+/(D)/(−)	V	V	V	+	–/(+)	–	
Hyalodendron	2	–					+	+			
Itersonilia	1	–	V			V	+	+	–	–	
Kockovaella	11	–	–	V	–/(W)	V	+	+	–	–	
Kurtzmanomyces	3	–	+	–	V	–	+	+	–	–	
Leucosporidiella	4	–	+	–	V	–	+	+	–	–	
Malassezia	11	–	–				+	+		V	Most species require fatty acid supplement
Moniliella	3	+	+			–	+	+		V	
Pseudozyma	15	–	+	V	V	+	+	+	–	V	
Reniforma	1	–	–	–	–	–	+	+	–	–	
Rhodotorula	50	–	V	V	V	V	+/(−)	+/(−)	V	V	
Sporobolomyces	49	–	V	V	V	–/(+/W)	+/(−)	+/(−)	V	V	
Sterigmatomyces	2	–	V	–	+	–	+	+	+	V	
Sympodiomycopsis	1	–	+	–	+	+	W	+	+	–	
Tausonia	1	–	+	–	–	D		+	–	–	
Tilletiopsis	9	–	+			V	+	+	–	–	
Trichosporon	42	–	–/(+)	V	V	V	+	+	V	V	
Trichosporonoides	5	+	+			–	+	+		V	
Tsuchiyaea	1	–	–	–	–	+	+	+	+	–	
Udeniomyces	5	–	+/W/D	–	V	V	+	+	–/W/D	–	
Zymoxenogloea	1	–	+	–	–	–	+?	+?	–	–	

[a] + = positive for all species; – = negative for all species; V = variable; W = weak; D = delayed; ( ) = rarely seen.

mode of asexual reproduction. Various types of asexual or vegetative reproduction are illustrated in Figure 46.1. Most yeasts reproduce by budding (i.e., blastoconidia), but fission occurs in some genera; and in others there is a combination (i.e., bud-fission) of the two processes. The following types of budding are referred to in morphologic Tables 46.2 through 46.5:

Multilateral: buds occur at various sites around the surface of the cell
Monopolar: buds occur only at one pole of the cell
Bipolar: buds occur only at the two opposite poles of a cell
Bud-fission: a combination of budding and fission, in which a broad based bud usually occurs at the pole of a cell and later a cross wall forms by centripetal growth across the base of the bud

Certain shapes of yeast cells can be very useful for identification as these can be typical of an entire genus. For example, bottle- or flask-shaped cells are characteristic of the monopolar budding exhibited by species of *Malassezia*. Apiculate or lemon-shaped cells are characteristic of bipolar budding seen with genera such as *Hanseniaspora*, *Nadsonia*, *Saccharomycodes*, and *Wickerhamia*. The triangular cells of *Trigonopsis* are very characteristic of this particular genus. The ogival cells seen with *Brettanomyces* and *Dekkera* are very helpful to narrow the field. Other shapes such as spheroidal, ellipsoidal, ovoidal, cylindroidal, and elongate are easily understood but not useful in genus-level identification. However, these more common cell shapes can be useful in species-level differentiation. For example, two common shapes can be used to differentiate yeasts that are otherwise very similar physiologically, such as *Debaryomyces hansenii* (anamorphic synonym = *Candida famata*, which has mostly spheroidal cells) and *Pichia guilliermondii* (anamorphic synonym = *Candida guilliermondii*, which has primarily ovoidal cells). This observation would be useful when the latter species fails to produce pseudohyphae, as can occur in many strains. Examples of various cell shapes are shown in Figure 46.2.

Members of several genera such as *Dipodascus*, *Eremothecium*, *Galactomyces*, *Geotrichum*, *Saccharomycopsis*, and *Trichosporon* produce true hyphae. The hyphae of some genera such as *Dipodascus*, *Galactomyces*, *Geotrichum*, and *Trichosporon* fragment into cells called arthroco-

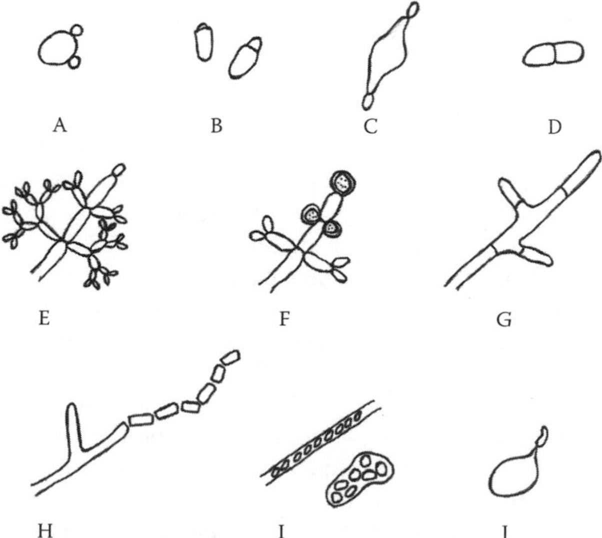

FIGURE 46.1 Types of asexual or vegetative reproduction in yeasts: A = multilateral blastoconidia; B = monopolar blastoconidia; C = bipolar blastoconidia; D = fission; E = pseudohyphae; F = chlamydospores; G = true hyphae; H = arthroconidia; I = endoconidia; J = ballistoconidia.

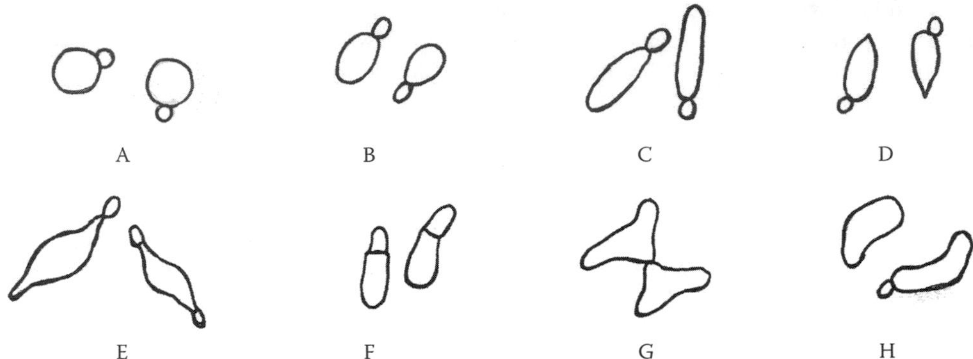

FIGURE 46.2 Examples of various cell shapes in yeasts: A = spheroidal; B = ovoidal or ellipsoidal; C = cylindroidal or elongate; D = ogival; E = apiculate or lemon-shaped; F = flask- or bottle-shaped; G = triangular; H = allantoid or sausage-shaped.

nidia. Some arthroconidial yeast genera, such as *Dipodascus*, *Galactomyces*, and *Geotrichum*, do not produce blastoconidia but their other characteristics resemble those of yeasts sufficiently to be included in this heterogeneous group. Other genera can produce both arthroconidia and blastoconidia, such as *Arxula* and *Trichosporon*.

In many budding yeasts, the buds remain attached and often elongate, forming long chains of cells that may or may not branch; the resulting filaments resemble those of true mycelia and are referred to as pseudohyphae or pseudomycelia. Pseudohyphae can be difficult to discern from true hyphae but there are a few ways to tell them apart:

1. True hyphae have septa (cross-walls) that are perpendicular to the walls of the main filament; whereas in pseudohyphae, the walls between two cells are typically curved.
2. Side branches of true hyphae typically have septa some distance away from the main filament where the wall in a pseudohyphal branch is at the juncture of the main filament.
3. Tip cells of true hyphae are typically continuous, whereas tip cells of pseudohyphae are typically ending with a smaller bud.

Refer to Figure 46.3, which shows a diagram of these differences.

Another very useful and specific morphologic characteristic is the production of ballistoconidia. These conidia are forcibly discharged from stalks called sterigma that develop from yeast cells or from hyphae in some of the basidiomycetous genera (e.g., *Bullera*, *Sporidiobolus*, and *Sporobolomyces*). The presence of ballistoconidia can be noted from the formation of satellite colonies. These are smaller colonies that develop in a random pattern outside the streak lines on an agar plate incubated for several days. Because ballistoconidia are forcibly discharged, one can invert a fresh agar plate over a ballistoconidiogenous yeast culture and, after a few days of incubation, a mirror image of the original culture will form.

Other specialized morphologic characters are useful in recognizing specific genera. Some genera such as *Sterigmatomyces* and *Sterigmatosporidium* reproduce asexually by the formation of conidia at the ends of sterigma. These conidia are different from ballistoconidia in that they are not forcibly discharged from the sterigma. The presence of terminal needle-shaped conidia on hyphae is specifically characteristic of one genus called *Aciculoconidium*. The development of sympodial conidia is a characteristic of genera such as *Mrakia* and *Sympodiomyces*. Conidiophores and new conidia remain attached and develop repetitively in a sympodial formation of several successive conidia.

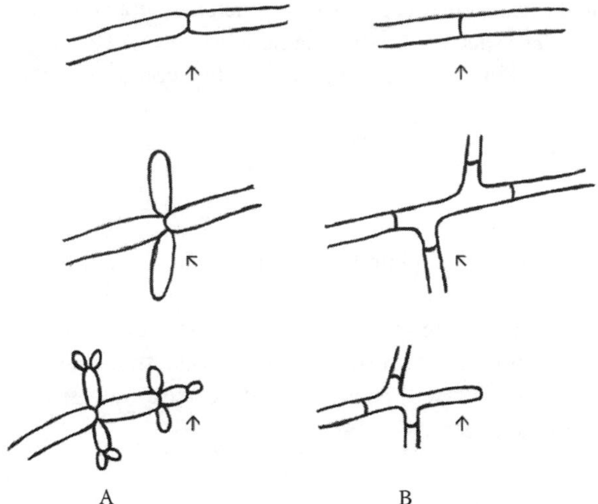

FIGURE 46.3 Differences between pseudohyphae (A) and true hyphae (B): Top = walls between pseudohyphal cells are typically curved while true hyphae have septa perpendicular to walls of the main filament; Middle = pseudohyphal branch is at the juncture of the main filament while true hyphae typically have septa some distance from the main filament; Bottom = tip cells of pseudohyphae typically end with a smaller bud while tip cells of true hyphae are typically continuous.

Chlamydospores are produced by some yeast species, such as members of the genera *Candida* and *Metschnikowia.* Although not thought to be spores, they are thick-walled, nondeciduous, intercalary or terminal asexual resting cells that form singly or in clusters. The presence of chlamydospores in a clinical yeast isolate is highly suggestive of either *Candida albicans* or *C. dubliniensis.* The latter species is thought to have more abundant production of chlamydospores occurring in clusters, but a more reliable way to differentiate the two species is by rapid or slow assimilation, respectively, of D-xylose, α-methyl-D-glucoside, or D-trehalose.

Finally, endocondial formation is only observed in older cultures of *Arxula, Cystofilobasidium, Geotrichum, Leucosporidium, Oosporidium,* and *Trichosporon.* These endoconidia are vegetative and produced within other cells or hyphae.

CULTURAL CHARACTERISTICS

Culture morphology is sometimes sufficiently specific to allow for recognition of certain genera. For example, the appearance of satellite colonies (smaller colonies appearing in a random pattern outside the streak lines) indicates the presence of ballistoconidia and limits the scope to 14 possible genera (e.g., *Bullera, Bulleromyces, Sporobolomyces, Sporidiobolus,* etc.). Likewise, cultures producing carotenoid pigments point to a relatively small group of less than 30 genera, such as *Rhodotorula, Sporobolomyces,* etc. The combination of carotenoid pigment and satellite colonies is specific to an even more limited set of about eight genera, including *Sporobolomyces* and *Sporidiobolus.* Some species in only three genera (*Kluyveromyces, Metschnikowia,* and *Trichosporiella*) can produce a diffusible red pigment due to pulcherrimin or a pulcherrimin-like substance.

Colony texture can be a useful and telling characteristic. Colonies displaying a more filamentous appearance can also be key in pointing to a limited set of genera, such as several of the arthroconidial genera (e.g., *Dipodascus, Galactomyces,* and *Geotrichum*). Some yeast species typically form ceribriform colonies with age, such as *Yarrowia lipolytica* and some species of *Trichosporon.* Farinose (chalky white or powdery) colonies are typical of certain species such as *Pichia farinosa.*

The odor of certain species is also very specific, such as the acidic odor generated by colonies of *Brettanomyces* and *Dekkera* species. As these genera produce acetic acid, cultures are short-lived unless cultivated on a neutralizing agar medium such as one containing calcium carbonate.

SEXUAL CHARACTERISTICS

TELEOMORPHIC YEASTS

Yeasts that produce sexual spores are related to either ascomycetes or basidiomycetes. It is thought that all anamorphic yeasts represent teleomorphs in which the sexual state has not yet been observed. This can be due to lack of appropriate mating strain combinations or inappropriate culture conditions (e.g., growth medium) in order to stimulate spore production. It is likely with sufficient time and studies that anamorphic genera will cease to exist because their species will be reassigned to their appropriate teleomorphic genera.

ASCOMYCETOUS YEASTS

The ascomycetous yeasts produce ascospores, which are a useful taxonomic characteristic. However, for practicality in routine yeast identification, laboratorians do not have the luxury of time or material resources to look for spore production. For those who are able to look for ascospores, the true identity of yeast is more readily discerned when sexual characteristics are taken into account. When ignored, one will typically identify yeasts using synonymous nomenclature for anamorphic states, which are more common in the scientific literature. For example, it would be rare to find a report of infection or antifungal resistance due to *Issatchenkia orientalis,* but it would be commonplace for one to find a similar report due to *Candida krusei* (its anamorphic obligate synonym). Similarly, it is unlikely to see reports of amphotericin B-resistant *Clavispora lustianiae* but one will find *Candida lustianiae*. For this reason, one must remain practical and less rigid in the application of teleomorphs and their associated synonymous anamorphic states. Routinely, ascospores are only observed by accident rather than by cultivation on special media over extended incubation periods. For the less routine application, much can be learned from the characteristics of ascospores.

The shape and number of ascospores per ascus are very useful features. Hat- or Saturn-shaped ascospores are found in less than half of the ascomycetous yeast genera. Needle-shaped ascospores without an appendage are only known to occur in *Metschnikowia,* while *Eremothecium* specifically forms spindle-shaped ascospores with an appendage. Other less-common shapes such as cap-shaped, crescentiform, and reniform are found in a few genera (e.g., *Wickerhamia*, *Ascobotryozyma*, and *Kluyveromyces*, respectively). Genera having numerous (100 or more) ascospores per ascus are rare and such a feature is restricted to only five genera (*Ascoidea*, *Dipodascopsis*, *Dipodascus*, *Protomyces*, and *Vanderwaltozyma*).

In addition, the manner of ascus formation, presence or absence of conjugation tubes, presence of protuberances resembling conjugation tubes, and whether ascospores are retained in the ascus or released from an evanescent ascus are all key features useful in the delineation of a yeast's identity. Ascospore-producing yeasts are either homothallic or heterothallic. "Homothallic" refers to the manner in which sexual reproduction takes place via identical nuclei undergoing fusion. "Heterothallic" refers to fusing nuclei that are not identical and originate from opposite mating types. Of course, it is not very practical to expect someone to routinely test opposite mating types of heterothallic yeasts such as *Kodamaea ohmeri* prior to their definitive identification.

BASIDIOMYCETOUS YEASTS

Because many basidiomycetous yeasts are heterothallic and require mixing opposite mating strains to observe the sexual state, it is of limited practicality to discuss these at length. For detailed information, the reader is referred to the taxonomic text of Kurtzman and Fell.[5]

There are certain features commonly associated with basidiomycetous yeasts that allow their recognition more readily. These features include positive reactions for the DBB test and urease production. Production of starch, encapsulated cells, and mucoid colonies are all characteristics found in the basidiomycetous yeasts. Red carotenoid pigments are more likely produced and, in addition, the production of ballistoconidia is restricted to this group.

In the second edition of *The Yeasts — A Taxonomic Study*, only two teleomorphic basidiomycetes were described (i.e., *Leucosporidium* and *Rhodosporidium*). These genera have complex life cycles that result in the formation of teliospores and basidiospores. In addition, clamp connections are observed in the mycelia. These are outgrowths from one hyphal cell joining the previous neighboring cell that allows nuclear migration and reestablishment of a dikaryotic state after formation of a new septum. While clamp connections are only observed in basidiomycetous yeasts, not all of them produce clamp connections.

In subsequent years, numerous teleomorphic basidiomycetes were described, including *Filobasidiella*, the genus containing the teleomorphic state of what is more commonly known as *Cryptococcus neoformans*. Today, there are more than 20 basidiomycetous genera that exhibit teleomorphic states.

Anamorphic Yeasts

As mentioned above, anamorphic yeasts are not known to produce sexual spores. This is due to either true inability or that appropriate conditions (i.e., mating studies, culture medium, and culture conditions) have not yet been discovered. In light of the impressive increase in knowledge and numbers of teleomorphic genera over the past several decades, it is thought that all anamorphic genera will eventually find their correct placement within teleomorphic genera.

Although sexual spores are not observed, other characteristics are useful to group anamorphic genera as they relate to either ascomycetes or basidiomycetes. The most clear-cut method to separate the two is by the DBB test being negative and positive, respectively. Other features are useful to determine basidiomycetous affinity, including production of urease, extracellular polysaccharide capsule (i.e., mucoid colonies), ballistoconidia, and red carotenoid pigments.

Most anamorphic genera are able to reproduce by budding, fission, or a combination of the two. Other types of propagules can be found, including arthroconidia, ballistoconidia, endoconidia, and conidia developing either at the ends of sterigma or ones developing sympodially from either yeast cells or hyphae.

Some anamorphic genera are nearly identical to ascomycetous genera, except for the absence of ascospores such as *Brettanomyces* and *Kloeckera*, which are the same as asporogenous *Dekkera* and *Hanseniaspora*, respectively. These are prime examples of the opportunity to discover conditions required to form ascospores.

PHYSIOLOGICAL CHARACTERISTICS

Certain physiological characteristics are quite useful to point to certain genera or species. Some of the characters with a high value are shown in Tables 46.6 through 46.9. Ascosporogenous yeasts that are nitrate-positive (utilizing potassium nitrate as the sole source of nitrogen) are likely to be a species of *Pichia*. Previously, these nitrate-positive *Pichia* species were included in the genus *Hansenula* but in 1984, the molecular studies of Kurtzman demonstrated that the genus *Hansenula* was synonymous with *Pichia*.[16] In the past, genera producing red carotenoid pigments such as *Rhodotorula* were known for their inability to ferment sugars. However, the more recently discovered genus of *Xanthophyllomyces* (teleomorphic basidiomycete and perfect state of *Phaffia*) possesses both carotenoid pigment and fermentative ability.

Conventional identification of yeasts relies on the combination of morphologic and physiologic properties. The classical broth methods were originally developed by Wickerham for assimilation and fermentation testing of yeasts.[17–19] In physiologic characterization, one tests for the ability of a yeast to utilize various carbon substrates (maltose, sucrose, lactic acid, etc.) as the sole source of carbon by employing a basal medium such as yeast nitrogen base (Difco) that contains ammonium sulfate (universal nitrogen source) and all vitamins, amino acids, and trace elements required for growth of yeasts. For testing various nitrogen substrates (potassium nitrate, sodium nitrite, amino acids, etc.) as the sole source of nitrogen, one can use yeast carbon base (Difco) that contains glucose (universal carbon source) and the vitamins, amino acids, and trace elements required for growth. To test fermentative abilities, a different basal medium (e.g., peptone and yeast extract) is employed. Normally, a 2% concentration of the sugar is added to the basal medium in a test tube that also contains an inverted Durham tube in order to observe production of CO_2 as this and ethanol are the byproducts of sugar fermentation. Yeasts able to ferment a particular sugar are also able to assimilate the same sugar; however, the reverse is not always true. As most yeasts are also able to assimilate ethanol as the sole source of carbon, it is necessary to incubate assimilation tests separately from fermentation tests because ethanol vapor produced by fermentation can dissolve in assimilation tests and cause false-positive results.

This method is a powerful tool for definitive characterization and taxonomy of yeasts. As most Wickerham media are unavailable commercially, they must be homemade. This is a long, tedious process that only very specialized laboratories can employ. Basal media and various substrates are prepared, sterilized, and dispensed prior to inoculation. Because many yeasts can "carry-over" nutrients from the isolation medium, one must run negative controls for each test type and organism. For up to 4 weeks, assimilation tests are read for turbidity, and fermentation tests are read for gas production. The "gold standard" Wickerham method is inappropriate for use in a routine laboratory and has been replaced by more practical methods available today.

Other key physiologic tests include temperature tolerance as certain genera and/or species are characteristically more or less tolerant of high temperatures. For example, the psychrophilic genus *Leucosporidium* generally only tolerates temperatures below 25°C. Temperature tolerance is also useful to differentiate otherwise quite similar species such as growth at 40°C, which is positive and negative, respectively, for *Issatchenkia orientalis* (syn. *Candida krusei*) and *Pichia fermentans* (syn. *Candida lambica*). Various other useful tests include growth in high concentrations of NaCl or glucose, vitamin requirements, growth in the presence of cycloheximide (actidione), starch formation, acetic acid production, urea hydrolysis, and the DBB test reaction. Unfortunately, with few exceptions, most of the classical tests are not practical for routine use. This has led to development of other methods, which enable rapid identification of the more commonly isolated yeasts. This is especially important for yeasts found causing infection because their rapid identification can aid in appropriate therapy.

A more rapid method for determining carbon assimilations is the auxanographic method. Here, a modified assimilation medium employs the use of agar with or without pH indicator (e.g., bromocresol purple) to test multiple substrates. The incorporation of agar speeds up the incubation. The use of a pH indicator can facilitate the observation of positive reactions. The organism is suspended in a molten basal medium (either yeast nitrogen base or yeast carbon base for carbon or nitrogen utilization, respectively), poured into a Petri dish, and allowed to solidify. Then, appropriate substrates or impregnated disks are placed on the surface of the seeded medium. After incubation of one to several days, positive reactions are noted by either turbidity or a color change surrounding the substrates.

YEAST IDENTIFICATION

Over the past three decades, commercial yeast identification systems have come to fruition, enabling even more rapid identification of commonly isolated yeasts. Although these systems focus mainly on clinically relevant species, there is certainly an overlap with yeast species that are relevant to food spoilage. See Figure 46.4 and Tables 46.10 and 46.11, which illustrate this overlap. Many of these commercial identification products utilize enzymatic substrates to detect enzymes such as

1a	Colonies cream, tan, or yellow; monopolar budding only	2
b	Colonies yellow, pink, orange, or red; urease positive	3
c	Colonies white or cream	7
2a	Lipid or fatty acid supplement required for growth	*Malassezia furfur*
b	Lipid or fatty acid supplement not required for growth	*Malassezia pachydermatis*
3a	Inositol assimilated	4
b	Inositol not assimilated	6
4a	Nitrate assimilated	*Cryptococcus albidus*
b	Nitrate not assimilated	5
5a	Growth at 37°C	*Cryptococcus neoformans*
b	No growth at 37°C	*Cryptococcus laurentii*
6a	Nitrate assimilated; ballistoconidia absent	*Rhodotorula glutinis*
b	Nitrate not assimilated; ballistoconidia absent	*Rhodotorula minuta* *Rhodotorula mucilaginosa*
c	Ballistoconidia present	*Sporobolomyces* spp.
7a	Arthroconidia present; blastoconidia absent	8
b	Arthroconidia present; blastoconidia present	*Trichosporon asahii* *Trichosporon inkin* *Trichosporon mucoides*
c	Arthroconidia absent	9
8a	Arthroconidia developing from side branches of the hyphae	*Geotrichum capitatum*[a] *Geotrichum klebahnii*
b	Arthroconidia developing from main filaments of hyphae	*Geotrichum candidum*[a]
9a	Sporangia present	*Prototheca* spp.
b	Sporangia absent	10
10a	True hyphae present; inositol assimilated	*Stephanoascus ciferrii*
b	True hyphae present; inositol not assimilated	*Candida albicans* *Candida dubliniensis* *Candida lipolytica*[a] *Candida tropicalis*
c	True hyphae absent	11
11a	Bipolar budding (apiculate cells)	*Kloeckera* spp.[a]
b	Multilateral budding; pseudohyphae absent	*Candida colliculosa*[a] *Candida famata*[a] *Candida glabrata* *Candida guilliermondii*[a] *Candida haemulonii* *Candida inconspicua* *Candida lusitaniae*[a] *Candida magnoliae* *Candida sphaerica*[a]
d	Multilateral budding; pseudohyphae present	*Candida boidinii* *Candida catenulata* *Candida freyschussii* *Candida guilliermondii*[a] *Candida intermedia* *Candida kefyr*[a] *Candida krusei*[a] *Candida lambica*[a] *Candida lusitaniae*[a] *Candida norvegensis*[a] *Candida parapsilosis* *Candida pelliculosa*[a] *Candida pulcherrima*[a] *Candida rugosa* *Candida sake* *Candida utilis*[a] *Candida zeylanoides*

[a]Imperfect (anamorphic) states of common ascosporogenous taxa

FIGURE 46.4 Key to identification of the most common clinically isolated yeasts.

TABLE 46.10
Yeast Pathogens of Man

Anamorphic or Common Name	Teleomorphic Name
Candida albicans	Unknown
Candida ciferrii	*Stephanoascus ciferrii*
Candida dubliniensis	Unknown
Candida famata	*Debaryomyces hansenii*
Candida glabrata	Unknown
Candida guilliermondii	*Pichia guilliermondii*
Candida haemulonii	Unknown
Candida inconspicua	Unknown
Candida kefyr	*Kluyveromyces marxianus*
Candida krusei	*Issatchenkia orientalis*
Candida lambica	*Pichia fermentans*
Candida lipolytica	*Yarrowia lipolytica*
Candida lusitaniae	*Clavispora lusitaniae*
Candida norvegensis	*Pichia norvegensis*
Candida parapsilosis	Unknown
Candida pelliculosa	*Pichia anomala*
Candida pulcherrima	*Metschnikowia pulcherrima*
Candida rugosa	Unknown
Candida tropicalis	Unknown
Candida utilis	*Pichia jadinii*
Candida zeylanoides	Unknown
Cryptococcus humicola	Unknown
Cryptococcus neoformans	*Filobasidiella neoformans*
Geotrichum candidum	*Galactomyces geotrichum*
Geotrichum capitatum	*Dipodascus capitatus*
N/A[a]	*Kodamaea ohmeri*
Malassezia furfur	Unknown
Malassezia pachydermatis	Unknown
N/A	*Pichia farinosa*
Prototheca wickerhamii[b]	N/A
Rhodotorula glutinis	*Rhodosporidium toruloides* (?)
Rhodotorula mucilaginosa	Unknown
N/A	*Saccharomyces cerevisiae*
Sporobolomyces salmonicolor	*Sporidiobolus salmonicolor*
Trichosporon asahii	Unknown
Trichosporon inkin	Unknown
Trichosporon mucoides	Unknown
Trichosporon ovoides	Unknown

[a] Not applicable.
[b] Colorless yeast-like alga.

precursors to carbon source utilization (i.e., glycosidases). There is also differential value in other enzymatic substrates, such as those that detect aminopeptidase, phosphatase, and esterase activities. Nitrate reductase is another enzymatic alternative to a conventional test (e.g., assimilation of potassium nitrate).[20]

As routine laboratories tend to have less resources (time and materials) and less skilled mycologists, the trend is to move away from morphology and have a definitive identification based solely on physiology. Of course, there is risk in this approach, depending on the spectrum of species found in a given environment.

TABLE 46.11
Most Common Yeasts in Food

Species	Food Source
Candida albicans	Beef, shellfish
Candida boidinii	Soft drinks, wine, cider
Candida catenulata	Beef, sausage, poultry, shellfish
Candida diddensiae	Beef, fish, shellfish
Candida etchellsii	Soft drinks, pickles, olives
Candida glabrata	Must, wine, poultry, fish, shellfish
Candida inconspicua	Soft drinks, juices, must, wine, beer, beef, fish
Candida intermedia	Beef, sausage, poultry, fish, shellfish
Candida magnoliae	Fruits, pickles, olives
Candida norvegica	Cider, winery, beef, shellfish
Candida parapsilosis	Must, wine, pickles, salads, bread, cheese, yogurt, butter, beef, sausage, poultry, fish, shellfish
Candida rugosa	Wine, sake, beer, pickles, olives, cheese, yogurt beef, sausage, poultry
Candida sake	Fruits, soft drinks, must, wine, winery, beer, salads, beef, shellfish
Candida stellata	Fruits, soft drinks, juices, must, wine
Candida tropicalis	Fruits, soft drinks, juices, wine, beer, brine, flour, dough, cheese, yogurt, beef, fish, shellfish
Candida versatilis	Pickles, beef
Candida zeylanoides	Fruits, wine, beef, sausage, lamb, fish
Cryptococcus albidus	Fruits, flour, beef, sausage, seafood
Cryptococcus humicola	Fruits, beef, seafood
Cryptococcus laurentii	Fruits, must, wine, winery, flour, cheese, butter, beef, lamb, ham, seafood
Cystofilobasidium infirmo-miniatum	Beef, lamb, seafood
Debaryomyces etchellsii	Wine
Debaryomyces hansenii (syn. *Candida famata*)	Fruits, juices, must, wine, winery, beer, brine, olives, salads, dough, bread, milk, yogurt, cheese, ice cream, beef, sausage, ham, fish, shellfish
Dekkera anomala	Soft drinks, wine, cider, beer, lambic beer
Dekkera bruxellensis	Soft drinks, wine, cider, beer, lambic beer
Galactomyces geotrichum (syn. *Geotrichum candidum*)	Milk, yogurt, cheese, sausage, ham
Hanseniaspora guilliermondii	Fruits, must, wine, winery
Hanseniaspora uvarum	Fruits, soft drinks, juices, must, wine, winery, cider, beer
Issatchenkia orientalis (syn. *Candida krusei*)	Fruits, soft drinks, juices, must, wine, beer, pickles
Kazachstania exigua	Must, wine, beer, brine, salads, dough, bread, beef, poultry, shellfish
Kluyveromyces lactis	Cheese, yogurt, cream
Kluyveromyces marxianus (syn. *Candida kefyr*)	Must, wine, beer, dough, cheese, yogurt, kefyr
Kregervanrija fluxuum	Fruits, wine, winery, beer, dough
Lachancea fermentati	Soft drinks
Lachancea kluyveri	Soft drinks, juices, wine
Lachancea thermotolerans	Soft drinks, juices
Lodderomyces elongisporus	Soft drinks, juices
Metschnikowia pulcherrima (syn. *Candida pulcherrima*)	Fruits, must, wine, winery, cider
Naumovia dairenensis	Salads
Pichia angusta	Pickles
Pichia anomala (syn. *Candida pelliculosa*)	Fruits, soft drinks, juices, must, winery, beer, pickles, olives, flour, dough, cheese, yogurt, sausage, ham, fish, shellfish
Pichia burtonii	Flour, bread, cheese, dairy, beef, ham, seafood
Pichia farinosa	Wine
Pichia fermentans (syn. *Candida lambica*)	Fruits, must, wine, beer, milk, cheese
Pichia guilliermondii (syn. *Candida guilliermondii*)	Fruits, soft drinks, juices, wine, cider, pickles, brine, olives, dough, sausage, seafood, shellfish

TABLE 46.11 (continued)
Most Common Yeasts in Food

Species	Food Source
Pichia jadinii (syn. *Candida utilis*)	Must, wine, beer
Pichia membranifaciens	Fruits, must, wine, cider, beer, brine, olives, salads, flour, bread, cheese, beef, fish
Pichia subpelliculosa	Must, winery, pickles
Rhodotorula glutinis	Fruits, must, winery, beer, flour, dough, beef, poultry, fish, seafood, shellfish
Rhodotorula minuta	Beef, sausage, poultry, fish
Rhodotorula mucilaginosa	Fruits, wine, winery, beef, sausage, ham, fish, shellfish
Saccharomyces bayanus	Must, wine, beer
Saccharomyces cerevisiae	Fruits, soft drinks, juices, must, wine, cider, sake, beer, olives, salads, flour, dough, bread, yogurt, kefyr, cheese, ice cream
Saccharomyces pastorianus	Beer, lambic beer
Saccharomycodes ludwigii	Must, wine, cider, beer
Saccharomycopsis fibuligera	Wine, beer, dough, bread
Schizosaccharomyces pombe	Fruits, must, wine
Sporobolomyces roseus	Fruits
Torulaspora delbrueckii	Fruits, soft drinks, juices, must, wine, winery, dough, bread, cheese, yogurt
Torulaspora microellipsoides	Soft drinks, fruit juices, flour
Trichosporon moniliiforme	Fruits, salads, flour, milk, butter, cheese, beef, sausage, poultry, fish
Trichosporon pullulans	Salads, beef, sausage, lamb, poultry, fish
Yarrowia lipolytica (syn. *Candida lipolytica*)	Olives, salads, milk, cheese, yogurt, beef, sausage, lamb, seafood
Zygosaccharomyces bailii	Fruits, soft drinks, juices, must, wine, cider, pickles, salads, bread
Zygosaccharomyces rouxii	Juices, pickles, ice cream

TABLE 46.12
Differential Characteristics of *Prototheca* Species

		Assimilation of:		
Species	Capsule	Fructose	Galactose	Trehalose
P. moriformis	+	+	–	–
P. stagnora	+	+	+	–
P. ulmea	+	–	–	–
P. wickerhamii	–	+	+	+
P. zopfii	–	+	–	–

YEAST-LIKE ORGANISMS

The fourth edition of *The Yeasts — A Taxonomic Study* edited by Kurtzman and Fell includes the yeast-like algal genus *Prototheca*. *Prototheca* bears close resemblance to the algal genus *Chlorella* except that *Prototheca* lacks chlorophyll. It grows well on Sabouraud dextrose agar and has a yeast-like appearance. Colonies have a distinctly foul odor rather than the more commonly pleasant odor associated with most yeasts. Because clinical labs need to be able to identify *Prototheca*, it is included in the key (Figure 46.4) for identification of the most common clinically isolated yeasts, which incorporates some of the features found in Tables 46.2 through 46.9. See Table 46.12 for characteristics useful in differentiating several *Prototheca* species.

There are other yeast-like genera that are excluded from taxonomic texts, such as the "black yeasts" that produce olivaceous, brown, or black pigments. Examples of genera include but are not limited to *Aureobasidium*, *Exophiala*, *Hormonema*, *Phaeoannellomyces*, and *Phaeococcomyces*. Genera that are more easily confused with the yeasts (such as *Cryptococcus* and *Trichosporon*) are those (*Aureobasidium* and *Hormonema*) with early growth being hyaline (usually cream to light pink) and only after several days to weeks developing dematiaceous (dark) pigment.

YEASTS AND MAN

Throughout recorded history, man has benefited from biochemical products of yeasts. Yeasts provide us with the metabolic processes needed to make bread, beer, wine, and other alcoholic beverages. In addition, yeasts provide us with many other products, including enzymes, coenzymes, and vitamins. Yeasts can be useful in breaking down waste products into nontoxic and/or usable products.

The main genera of commercial significance are *Saccharomyces* and *Candida*. Frequently, species were named on the basis of the fermentation they were associated with — for example, *Saccharomyces sake* and *Saccharomyces vini*. The morphological and physiological properties of many of these species are very similar, and separate names have not been justified from a taxonomic point of view. Special characteristics of certain yeast strains are extremely desirable for various industrial processes; therefore, strain differentiation may be more important than species classification.

Under certain circumstances, yeasts can cause infections. Factors such as prolonged broad-spectrum antibacterial therapy, immunosuppressive therapies (e.g., corticosteroids), indwelling catheters, underlying disorders (e.g., diabetes, leukemia, etc.), and nutritional status can all contribute to opportunism of yeasts. Table 46.10 provides a list of species pathogenic in man and some of the equivalent names found in the literature.[21, 22] Pathogenesis ranges from benign, superficial, mucocutaneous infections to deep-seated systemic infections that can invade internal organs and are associated with high levels of morbidity and mortality. Frequently, the source of infection is endogenous and results from an imbalance due to overgrowth of yeast flora otherwise held in check by other microbial flora (e.g., bacteria) and/or the host's immunocompetent state.

Yeasts can also be problematic by causing food spoilage. They do not typically compete well in mixed populations with bacteria, but their growth is favored by more acidic conditions (at pH 5 or lower), which allow for their outgrowth. Table 46.11 shows a compilation of the most common yeasts found in various food types. For a more thorough discussion on this subject, the reader is referred to the text published by Deak and Beuchat.[23]

REFERENCES

1. Macmillan, J.D. and Phaff, H.J. Yeasts, in O'Leary, W.M., Ed. *The Practical Handbook of Microbiology*, 1989, 251.
2. Lodder, J. and Kreger-van-Rij, N.J.W., Eds. *The Yeasts — A Taxonomic Study*, 1st ed. North-Holland Publishing Co., Amsterdam, 1952.
3. Lodder, J., Ed. *The Yeasts — A Taxonomic Study*, 2nd ed. North-Holland Publishing Co., Amsterdam, 1970.

4. Kreger-van Rij, N.J.W., Ed. *The Yeasts — A Taxonomic Study*, 3rd ed. Elsevier Science Publishers B.V., Amsterdam, 1984.
5. Kurtzman, C.P. and Fell, J.W., Eds. *The Yeasts — A Taxonomic Study*, 4th ed. Elsevier Science B.V., Amsterdam, 1998.
6. BioloMICS — Yeast species database of the Centraalbureau voor Schimmelcultures. http://www.cbs.knaw.nl/yeast/BioloMICS.aspx.
7. Barnett, J.A., Payne, R.W., and Yarrow, D., Eds. *Yeasts: Characteristics and Identification*, 3rd ed. Cambridge University Press, Cambridge, 2000.
8. Kerrigan, J. et al. *Ascobotryozyma americana* gen. nov. et sp. nov. and its anamorph *Botryozyma americana*, an unusual yeast from the surface of nematodes. *Antonie van Leeuwenhoek*, 79, 7, 2001.
9. Kerrigan, J. et al. *Ascobotryozyma cognata* sp. nov., a new ascomycetous yeast associated with nematodes from wood-boring beetle galleries. *Mycol. Res*., 107, 1110, 2003.
10. Kerrigan, J. et al. *Botryozyma mucatilis* sp. nov., an anamorphic ascomycetous yeast associated with nematodes in poplar slime flux. *FEMS Yeast Res.,* 4, 849, 2004.
11. Okoli, I. et al. *Cryptotrichosporon anacardii* gen. nov., sp. nov., a new trichosporonoid capsulate basidiomycetous yeast from Nigeria that is able to form melanin on niger seed agar. *FEMS Yeast Res*., 7, 339, 2007.
12. Kurtzman, C.P. Description of *Komagataella phaffii* sp. nov. and the transfer of *Pichia pseudopastoris* to the methylotrophic yeast genus *Komagataella. Int. J. Syst. Evol. Microbiol*., 55, 973, 2005.
13. Golubev, W.I.. *Mastigobasidium*, a new teleomorphic genus for the perfect state of ballistosporous yeast *Bensingtonia intermedia. Int. J. Syst. Bacteriol.,* 49, 1301, 1999.
14. Morais, P.B. et al. *Ogataea falcaomoraisii* sp. nov., a sporogenous methylotrophic yeast from tree exudates. *FEMS Yeast Res*., 5, 81, 2004.
15. Noda, H. and Kodama, K. Phylogenetic position of yeastlike endosymbionts of anobiid beetles. *Appl. Environ. Microbiol.,* 62, 162, 1996.
16. Kurtzman, C.P. Synonomy of the yeast genera *Hansenula* and *Pichia* demonstrated through comparisons of deoxyribonucleic acid relatedness. *Antonie van Leeuwenhoek*, 50, 209, 1984.
17. Wickerham, L.J. A simple technique for the detection of melibiose-fermenting yeasts. *J. Bacteriol*., 46, 501, 1943.
18. Wickerham, L.J. A critical evaluation of the nitrogen assimilation tests commonly used in the classification of yeasts. *J. Bacteriol*., 52, 293, 1946.
19. Wickerham, L.J. and Burton K.A. Carbon assimilation tests for the classification of yeasts. *J. Bacteriol*., 56, 363, 1948.
20. Pincus, D.H., Orenga, S., and Chatellier, S. Yeast Identification — Past, Present, and Future Methods. *Med. Mycol,* 45, 97, 2007.
21. Hazen, K.C. New and emerging yeast pathogens. *Clin. Microbiol. Rev.,* 8, 462, 1995.
22. Guého, E. et al. *Malassezia* and *Trichosporon*: two emerging pathogenic basidiomycetous yeast-like fungi. *J. Med. Vet. Mycol*., 32(Suppl. 1), 367, 1994.
23. Deak, T. and Beuchat, L.R., Eds. *Handbook of Food Spoilage Yeasts*, CRC Press, Boca Raton, FL, 1996.

47 Introduction to Virology

Ken S. Rosenthal

CONTENTS

INTRODUCTION

Viruses are the simplest infectious agents and yet, as they parasitize our bodies, they can cause devastating disease. Viral diseases range from the common cold and diarrhea to life-threatening encephalitis, hemorrhagic fever, and smallpox. Several different viruses may cause a single disease, as determined by the tissue or organ that is affected, or one virus may cause a unique disease due to the nature of its interaction with the body. Ultimately, the nature of the disease caused by the virus is a function of the tissue that is infected, the nature of the virus, and the host immune response to the virus. Interestingly, each of these parameters is determined by the structure and mode of replication of the virus. This chapter introduces the basic viral structures and modes of replication, and presents the consequences of these properties for the different virus families. *In many cases, knowing the structure of the virus allows prediction of many of the other properties of the virus.* This introduction also includes a discussion of general mechanisms of viral pathogenesis and disease production, epidemiology, antiviral drugs, and viral detection before a synopsis of each of the virus families is presented. Other discussion and more detailed descriptions of the viruses can be obtained from the referenced materials and the Internet, including *The Bad Bug Book at the FDA* (www. FDA.gov) and the *A to Z Index of the CDC* (www.cdc.gov) websites.

DEFINITION OF A VIRUS

Unlike bacteria, fungi, or parasites, a virus does not replicate by binary fission but rather by synthesis and assembly. Viruses neither have organelles nor can they make energy. Viruses are obligate intracellular parasites that depend on the biochemical machinery of the infected cell for replication. As obligate parasites, viruses must continue to find new host cells and individuals, and be able to use the biochemical machinery of the cell to replicate or the virus will disappear. The virus uses different molecular tricks to manipulate the cell and its host to replicate and spread.

STRUCTURE

Although we may think of a virus as having characteristics of a living microbe, it is not alive. The virion particle is a very large biochemical complex consisting of a nucleic acid genome protected and carried within either a protein shell, termed a capsid, or within a membrane envelope. (Although prions were considered atypical viruses at one time, they are rogue, infectious proteins and not viruses). The basic virion particle consists of either an RNA or DNA genome (but only one or the other) that is packaged for protection and delivery into a protein capsid or a membrane envelope. Additional proteins may be present to facilitate replication in the host.

The viral genome is the core of the virus and acts like a computer memory to store the genetic information for the virus. Like a computer memory (e.g., a hard drive, floppy disk, or CD-ROM), the viral genome can take different forms. It can consist of DNA — either linear, circular, single or double stranded — or single- or double-stranded RNA. DNA genomes can establish themselves in the nucleus and utilize the nuclear machinery for mRNA synthesis and replication of the genome, whereas RNA genomes are more temporary. Most RNA viral genomes replicate in the cytoplasm (where RNA belongs) and must provide the enzymes to make mRNA.

The outer structures of the virus are a package that protects the genome, delivers it from host to host, and provides the means for interacting with and entering the appropriate target cell (Figure 47.1). Capsids are protein shells usually in symmetrical icosahedral or icosadeltahedral shapes (soccer balls). Like hardboiled eggs, the protein of the capsid forms a shell that is impervious to drying, acids, and detergents. In contrast, enveloped viruses are surrounded by a membrane consisting of phospholipids and proteins, which must stay wet and can be disrupted by detergents and harsh environments, including the acids and bile of the gut. Of course there are exceptions to the generalizations that will be made and they will be identified throughout the text.

The nature of the viral disease, the means by which it spreads, the type of immune response necessary to control the viral infection, and the types of antiviral drugs that can be used to combat the infection are determined to a large extent by the viral structure and mechanism of replication (see below). The impervious shell of capsid viruses allows them to be transmitted by the fecal-oral route, resist drying, and also disinfection with detergents and solvents. This is not so for enveloped viruses. Rhinovirus (common cold) is an exception and is susceptible to acid pH. This is the basis for incorporating citric acid into tissues to kill cold viruses. As discussed later, antibody is usually sufficient to control most naked capsid viruses. This is why inactivated virus vaccines, which generate primarily an antibody response, are usually effective for capsid viruses but may not be effective for enveloped viruses. In contrast, enveloped viruses cannot withstand the harsh conditions of the gastrointestinal (GI) tract; they are readily disinfected by soap or solvents and are more sensitive to heat and dryness. Enveloped viruses are transmitted in fluids (e.g., aerosols, saliva, blood, semen, and tissue transplants). The exceptions are coronaviruses and hepatitis B virus, which can withstand the GI tract. Cell-mediated and antibody immune responses are important to control enveloped viruses.

Viruses range in size and complexity from the tiny Norwalk virus (22 nm), which causes outbreaks of diarrhea, to the poxviruses, which are almost visible under a light microscope (Tables 47.1 and 47.2). They may be as simple as the rabies virus, which has only 5 proteins, or as complex as the cytomegalovirus, which has approximately 110 proteins. The larger viruses have more space to hold a larger genome and encode more proteins. These "extra" proteins may facilitate the replication of the virus; allow replication in difficult cells, such as neurons; or facilitate escape from host protective innate and immune responses. Larger viruses also require larger aerosol droplets for transmission through the air.

REPLICATION AND CONSEQUENCES

Every virus must replicate and produce more viral progeny or disappear. In many respects, a virus utilizes host cells as a factory that provides the raw materials, energy, and most of the synthetic machinery to produce the genome and structural components of new viruses. For every infecting

TABLE 47.1
DNA Viruses

Family	Genome	Virion	Virion Size (nm)	Example(s)
Parvoviridae	ss, linear	Capsid	18–26	B19
Papillomaviridae	ss, circular	Capsid	45–55	Human papilloma
Polyomaviridae	ss, circular	Capsid	45–55	JC, BK
Hepadnaviridae	ds, circular	Enveloped	42	Hepatitis B
Adenoviridae	ds, linear	Capsid with fibers	70–90	Adenovirus
Herpesviridae	ds, linear	Enveloped	120–130	HSV, EBV, CMV
Poxviridae	ds, linear	Enveloped, brick	300 × 200 × 100	Smallpox

TABLE 47.2
RNA Viruses

Family	Genome	Virion	Virion Size (nm)	Example(s)
Caliciviridae	ss, positive	Capsid	35–40	Norwalk, HEV
Picornaviridae	ss, positive	Capsid	25–30	Polio, rhino, HAV
Coronaviridae	ss, positive	Enveloped	80–130	Corona, SARS
Paramyxoviridae	ss, negative	Enveloped	150–300	Measles, mumps
Orthomyxoviridae	ss, negative	Enveloped	80–120	Influenza A, B, C
Rhabdoviridae	ss, negative	Enveloped	180 × 75 bullet	Rabies
Filoviridae	ss, negative	Enveloped	800 × 80 fiber	Ebola
Reoviridae	Double stranded	Double capsid	60–80	Rota
Togaviridae	ss, positive	Enveloped	60–70	Rubella, EEV
Flaviviridae	ss, positive	Enveloped	40–50	Yellow fever
Bunyaviridae	ss, negative	Enveloped	90–100	Ca. Encelphalitis
Arenaviridae	ss, ambisense	Enveloped	50–300	Lassa fever
Retroviridae	ss, positive	Enveloped	80–110	HIV, HTLV

virus, the cell can produce 100 to 100,000 new infectious virions and many times more noninfectious (defective) particles. Viruses are notorious for making mistakes but when there are a million siblings, even 10% infectious virus is a lot. The virus can kill the cell quickly, slowly, or not at all. The outcomes depend on the type of virus and how it interacts with the cell.

All viruses follow a similar set of steps to generate new viruses (Figure 47.2). The basic steps for replicating the virus include:

1. Recognition and attachment
2. Entry into the cell
3. Uncoating
4. Synthesis of the macromolecular components, including the genome and the structural parts of the virion particle
5. Assembly of the new virion particles
6. Exit from the cell

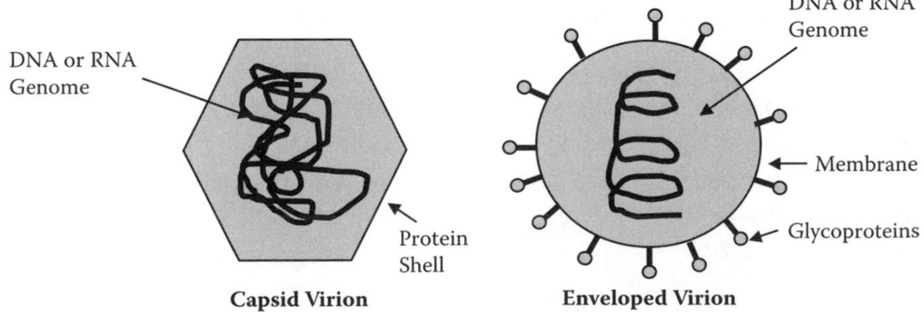

FIGURE 47.1 Virion structure. The virion particle consists of a DNA or RNA genome surrounded by either a protein shell (called a capsid) or a cell-like membrane (called an envelope). The virion may also contain other structures and proteins that will facilitate the replication of the virus within the cell. The outside structures of the capsid or the glycoproteins of the envelope mediate the interaction of the virus with the cell and are the primary targets of protective antibodies.

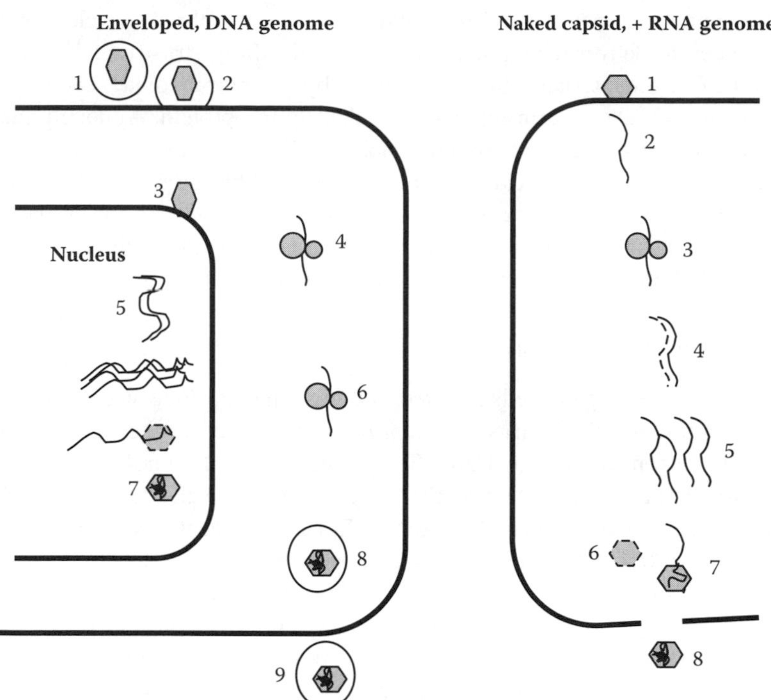

FIGURE 47.2 Virus replication. Virus replication proceeds in an ordered, sequential fashion. The steps in virus replication for an enveloped virus with a DNA genome, like a herpes virus, are (1) binding, (2) fusion and entry, (3) delivery of genome to nucleus, (4) immediate early and early mRNA and protein synthesis, (5) replication of the genome, (6) late protein synthesis, (7) capsid assembly and filling, (8) envelopment of genome-filled capsid, and (9) exit of cell by exocytosis or cell lysis. The steps in virus replication for a capsid virus with a positive-sense RNA genome, like a picornavirus (polio or rhino), are (1) binding, (2) entry and uncoating, (3) protein synthesis, (4) production of an RNA template, (5) genome replication, (6) capsid assembly, (7) filling capsid with genome, and (8) cell lysis and release.

Each of these steps is performed differently by different viruses, depending on the structure of the viral genome and whether the virion is a naked capsid or an envelope.

Recognition, Attachment, and Entry

Finding and binding of the virus to the appropriate cell is mediated by the surface structures of the naked capsid virus or by the glycoproteins that decorate the envelope. These viral attachment proteins (VAP) are the keys that interact with cell surface receptors and open the doors of entry into the cell. Some of these structures are like skeleton keys and bind to many different cell types, whereas others are very specific for certain cell types. A canyon on the surface of the rhino or polio virus (picornaviruses) acts as a keyhole that can be filled by specific receptors from the cell which are members of the immunoglobulin family of glycoproteins. The glycoproteins on the Epstein Barr virus bind to the receptor for the C3d component of complement (CD21), identifying the B lymphocyte as the principal target for the virus. The HIV gp120 glycoprotein picks out the CD4 molecule to target T cells and myeloid cells (dendritic cells and macrophages) but can switch target cells by mutating and changing the portion of the gp120 that binds to either a chemokine receptor present on myeloid cells and activated T cells (CCR5) or one that is present on other T cells (CXCR4).

The mechanism of virus entry into the cell depends on the virion structure. Binding of the picornavirus receptor to the canyon on the virion surface opens the capsid structure and creates a hydrophobic portal for injection of the viral genome into the cytoplasm. Some enveloped viruses, including herpes, retro-, and paramyxoviruses, will fuse their membrane envelope with the plasma membrane of the cell to deliver their genome into the cell. Many viruses bind to receptors on the cell surface that trigger the internalization of the virus by receptor-mediated endocytosis, the same pathway that cells use to take up transferrin or low-density lipoprotein. Although one would think that the normal acidification of the endosomal vesicle would be detrimental to a virus, they use this as a trigger to enter the cytoplasm. Acidification activates the fiber protein of the adenovirus capsid to lyse the vesicle and break into the cytoplasm. Acidification changes the conformation of the viral attachment glycoproteins of enveloped viruses, such as influenza or rabies virus, to promote fusion of their membrane with the vesicular membrane of the cell to deliver their genome into the cytoplasm.

Viral mRNA and Protein Synthesis

Once inside the cell, the viral genome is inserted into the cell's machinery to promote the synthesis of viral mRNA and protein. DNA viruses, except poxviruses, deliver their genome to the nucleus where it will be mistaken for cellular DNA. The sequences of the viral genes must contain the same genetic passwords as cellular DNA so that they can use the cell's machinery for making and processing mRNA. RNA will be made by the host DNA-dependent RNA polymerase and modified to make the viral mRNAs look and act like cellular mRNAs. The RNA will be processed into mRNA by removal of introns (if present), become stabilized by the addition of a 3′-polyadenosine tail, and pick up the recognition structure for attachment to the ribosome and protein synthesis with the addition of a 5′-methylated (7-methylguanosine) cap. Poxviruses are the DNA virus exception because they replicate in the cytoplasm. The poxviruses pay the price for being more independent of the nucleus by having to provide their own molecular tools for mRNA synthesis.

Most DNA viruses divide protein synthesis into early and late phases. The proteins made during the early phase convert the cell into a more efficient factory for replicating virus. Once the cell has been subverted and new viral genomes are being produced, then the late genes are turned on and the cell churns out large amounts of viral structural proteins and glycoproteins to make the capsid or envelope. Activation of the late genes and synthesis of the structural proteins of a DNA virus is a death sentence for the cell.

Most RNA viruses replicate in the cytoplasm, which provides easy access to the ribosome for protein synthesis. However, using RNA as the genetic information carrier is foreign to the cell because the cell has no polymerases to copy RNA from RNA, and cannot make mRNA or replicate the genome from an RNA template. Hence, most RNA viruses must encode and/or carry their own RNA-dependent RNA polymerases within the virion particle. In addition, these viruses must develop a way to acquire a 5′ cap or an alternative way to bind to the ribosome for protein synthesis.

The viruses with a genome that looks like mRNA (positive-sense) (except retroviruses) can immediately bind to ribosomes and initiate protein synthesis. The resulting protein, called a polyprotein, cleaves itself or is cleaved by other proteases into the individual viral proteins. Viruses with a negative-sense RNA (like a photographic negative) cannot be deciphered by the cell's ribosome. The genome of these viruses is decorated by enzymes (RNA-dependent RNA polymerase) that will convert the genome into individual mRNAs for each of the viral proteins when released from the virion particle. The orthomyxoviruses (influenza) are exceptional because their proteins are encoded by individual segments of negative-stranded RNA, and these segments are converted into mRNA in the nucleus. Influenza appropriates the 5′ cap from cellular mRNA, and the 3′-polyadenosine is attached before leaving the nucleus so that their viral mRNA will look like cellular messenger RNA. The reo- and rotaviruses have a segmented double-stranded RNA genome surrounded by components of the polymerase (an inner capsid) that will convert the negative strands of the genome into mRNA.

Although the retroviruses, including HIV, have a positive sense RNA genome, they carry a reverse transcriptase enzyme that converts the RNA genome into DNA and then another viral protein promotes the integration of the viral DNA into the host chromosome. The integrated retroviral

DNA becomes a very active gene within the host chromosome and viral mRNAs and even new copies of the virus are transcribed just like any other cellular gene.

Once the virus has manipulated the cell into making mRNA, the viral proteins are synthesized by the cell's ribosome, and glycoproteins are processed by the same machinery as cellular glycoproteins. Different viruses have different ways to regulate when early and late proteins are made. For the herpesviruses (DNA viruses), the early genes are activated by viral and cellular proteins and the late genes are activated by viral proteins. The early proteins of the togaviruses (positive RNA genome) are translated directly from the infecting genome while the mRNA for the late proteins is made from a negative RNA template. Especially during the late phase of protein synthesis, the cellular machinery is likely to be devoted to production of viral proteins to the exclusion of cellular proteins.

Genome Replication

The new genome of DNA viruses is replicated like cellular DNA, in a semiconservative manner. However, the amounts of deoxyribonucleotides and DNA polymerase in a resting cell are limited and this can prevent or retard the replication of a virus. The smaller DNA viruses get around this problem by replicating in growing cells or by removing or blocking the action of inherent cellular growth inhibitors (p53 or RB growth suppressor proteins). The larger viruses encode their own tools for viral replication, a viral DNA polymerase and even scavenging enzymes for deoxyribonucleotides, which provide more independence from the cell and speed up the replication of the genome.

Most RNA viruses replicate their genome using an RNA-dependent RNA polymerase that is carried into the cell or encoded by the virus and made soon after infection. This polymerase makes an RNA template from the genome and uses it to synthesize positive- or negative-sense RNA genomes. During some point during the production of an mRNA or a new genome, a double-stranded RNA structure will be generated between the template and the new copy. The double-stranded RNA replicative intermediate is the most powerful activator of Type 1 interferons and also activates other innate responses through Toll-like (TLR) and other receptors.

The retroviruses and hepatitis B virus utilize both RNA and DNA in their replication. During the early steps in retrovirus infection, the incoming RNA genome is converted into DNA by the viral reverse transcriptase enzyme (RNA-dependent DNA polymerase) and another viral enzyme facilitates the integration of the genome into the host chromosome. Viral mRNA and new genomes are produced by the cell's DNA-dependent RNA polymerase as if it were making mRNA. For hepatitis B virus, the cell makes a longer than full genomic length RNA copy from the DNA genome. Using a viral enzyme, reverse transcription produces a DNA genome.

The viral polymerase is the major target for antiviral drugs. Nucleotide analogs inhibit virus replication or promote hypermutation.

Virus Assembly

Ultimately, all parts of the virus must be assembled into the virion particle so that it can move to new target cells and hosts. At first thought, the process seems analogous to assembling a jigsaw puzzle by shaking the parts in a box. However, the process is not as random as the shaken puzzle box. For some capsid viruses, individual proteins associate into larger and larger complexes that eventually become empty capsid shells. The large number of viral proteins within the cell promotes enough collisions between the proteins to drive the process to completion. Many of these structures are built on scaffolds made of cellular membranes or other viral proteins. The genome can be inserted into an assembled capsid shell, or the capsid proteins can assemble around or associate with the genome. For enveloped viruses, the protein-coated genomes (nucleocapsid) associate with proteins on the inside of the cellular membrane and the structure wraps itself in the membrane to bud from the cell.

Virus Release

Most capsid viruses accumulate in the cytoplasm and must cause the cell to lyse in order to be released. Lysis of the cell may be a consequence of toxic viral proteins, the build-up of viral material, an inability to rebuild cellular structures, or induction of apoptosis or necrosis. Most enveloped viruses are more subtle and acquire their envelope from the plasma membrane as they push their way out of the cell without killing the cell. Herpes viruses are released by vesicular transport mechanisms similar to those used to release proteins.

Upon release, the viruses can spread to other cells and expand the infection. Some viruses (e.g., herpes simplex virus) can move between cells (intercellular spaces) to infect neighboring cells without exposing themselves to antibody detection. Other viruses (HSV, varicella zoster, measles and other paramyxoviruses, and HIV) recruit large numbers of cells into their factory by fusing them together into multinucleated giant cells (syncytia). Syncytia formation is also a diagnostic characteristic of these viruses.

HOST PROTECTIVE RESPONSES

Antiviral immune responses are a part of viral disease. They are essential for control of infection and yet they contribute to the pathology of the disease. They are such a threat that most viruses encode proteins and mechanisms to escape or shut off the immune response. Antiviral host protective responses are based on the following military-like rules:

1. Keep them out and they can cause no harm.
2. If they acquire a toehold, then prevent them from spreading.
3. Eliminate the invaders and their factories at all cost.

The initial antiviral protection consists of barriers such as skin and mucus; unfavorable conditions such as the acid and detergent-like bile found in the gut; and antimicrobial peptides such as the defensins, which are found in many parts of the body and especially the mouth. Body temperature and fever are deterrents to many viruses; for example, the rhinoviruses are restricted to the upper respiratory tract because they cannot grow at body temperature in the lower lung.

Innate responses react quickly in response to microbial components. These responses include the type 1 interferons, natural killer (NK) cells, macrophages, and dendritic cells. Neutrophils and complement have less important roles in antiviral responses than antibacterial responses.

Type 1 interferons consist of more than 20 different interferon alphas, also known as leukocyte interferon, and a single type of interferon beta, also known as fibroblast interferon. They received their name because they "interfere" with virus replication. The type 1 interferons are produced in response to viral DNA, RNA, and especially the double-stranded RNA replicative intermediate generated during RNA virus replication. Type 1 interferons are distinguished from interferon-γ (INF-γ), the only type 2 interferon, which is produced by NK and T cells in response to activation of immune responses (discussed later).

Interferon produced and released from a virus-infected cell acts as an early warning system to surrounding cells. The interferon binds to cell surface receptors and induces the production of antiviral proteins, which become activated upon virus infection of the cell. These antiviral proteins are part of the "antiviral state" and include a protein kinase (PKR), which when activated by double-stranded RNA will phosphorylate a key controlling protein in the ribosome assembly to prevent protein synthesis. Another protein activates an RNase in response to virus infection. Both of these interferon-induced and virus-activated responses work together to prevent protein synthesis and hence block virus replication. Interferon also acts as an early warning system and activates NK cells and other body defenses. Unfortunately, these protections cause the flu-like symptoms associated with many viral infections.

NK cells can recognize and kill viral-infected cells and also produce INF-γ. INF-γ acts as a cytokine to activate macrophages and dendritic cells. Activated macrophages also have killer cell activity. Activated dendritic cells initiate the antigen-specific antiviral immune response.

The goal of the protective innate and immune responses is to completely eliminate the virus (resolve the infection) by eliminating infectious virus and virus-producing cells. Antibody-mediated protection is primarily directed against the proteins and structures that bind to the cell surface receptor, the viral attachment proteins or structures. The antibody coats the virus and prevents it from being able to bind and infect cells and also promotes the uptake and destruction of the virus by macrophages (opsonization). T-cell responses are directed against viral peptides that decorate the major histocompatibility antigens of the infected cell. T lymphocytes can act as killer cells and make INF-γ to activate macrophages. Some viruses can be controlled by antibody, whereas other viruses require both antibody and cell-mediated responses. The rule is that if the virus kills the cellular factory as part of virus replication, then antibody is sufficient to resolve the infection. If the virus does not kill the infected cell, then cell-mediated immune responses must kill the cell to resolve the infection. Cell-mediated responses usually cause inflammation and tissue damage. More detail and a description of the antimicrobial immune response can be obtained in Ref. 2 through 4.

Mobilization of the antigen-specific immune response takes longer to activate than the local militia of innate responses, especially if it is the first challenge with this type of virus. Maturation of antigen-specific immune responses takes at least 5 to 7 days, sufficient time for the virus to spread, reach the target tissue for the disease (e.g., for polio virus, to the brain), and cause damage. As a result, a much stronger immune response is necessary to resolve the infection and this may result in greater immunopathology (see below). Prior immunization, either by natural infection or vaccination, shortens the time period to approximately 2 days and the infection is usually stopped before signs of disease begin.

PATHOGENESIS AND DISEASE

Viral disease is caused by a combination of viral pathogenesis and immunopathogenesis. For some viruses, such as polio, virus replication kills the target cell. Poliovirus killing of motor neurons of the anterior horn and brainstem is the primary cause of disease. A combination of pathogenesis caused by virus replication and the subsequent immune response causes the signs of disease for most viruses, from the rhinoviruses (common cold) to the more serious influenza and herpes viral diseases. Immunopathogenesis is the primary cause of disease for measles and mumps viruses.

Viral Pathogenesis

Viral pathogenesis is caused by changes in the cell that are induced by viral replication and viral replication-induced cell death. The type and extent of viral pathogenesis is determined by the nature of the virus-cell interaction. Many viruses, especially the larger viruses, encode proteins that can manipulate the cell and the host to promote virus replication. Only half of the approximately 80 genes encoded by HSV are necessary to replicate the virus in tissue culture; the other genes support virus replication in difficult cells, like neurons, and help the virus spread or escape the immune response.

Viral disease is primarily determined by the cell type and tissue (tropism) that are infected by the virus and the consequences of the virus-cell interaction (e.g., hepatitis, encephalitis). The expression of virus receptor molecules on the target cell is the initial and most important determinant of the nature of the viral disease.

Once the virus has picked, attached, and entered the target cell, there are several different possible outcomes: replication and viral-induced cell death, replication without cell death (chronic infection), latency, and transformation. Most naked capsid viruses are cytolytic because they must kill the infected cell to be released and spread to other cells, whereas many enveloped viruses leave the cell without killing the cell as they acquire their envelope from the cell membrane and bud away from the cell. For a DNA virus, the outcome can be determined by the ability of the cell to make the viral proteins. For example, HSV establishes a latent infection in neurons because only one RNA (latency associated transcript (LAT)) is transcribed in these cells, and this RNA does not encode a detectable protein. For simian virus 40 (SV40), the early gene, T antigen, stimulates cell growth, which facilitates replication of the DNA viral genome; but if the late genes for the capsid proteins cannot be transcribed, as in rodent cells, the stimulation of cell growth becomes uncontrolled, leading to oncogenic transformation. The T antigen protein prevents the function of the p53 and RB cellular growth suppressors. SV40 replicates and kills monkey cells, a situation that is inconsistent with oncogenic transformation. Similarly, the E6 and E7 proteins of the human papilloma virus Types 16 and 18 also prevent the function of the p53 and RB, and the virus stimulates cell growth. When the HPV DNA genome becomes circularized or integrated into the host chromatin, a gene important for virus replication is inactivated and the cell grows out of control. HPV 16 and 18 are associated with most human cervical carcinomas.

Although the virus may not kill the cell, it may cause characteristic changes to cellular structures within the cell. These changes may be seen as histological staining aberrations, including inclusion bodies, or syncytia formation (multinucleated giant cells). For example, herpes simplex virus causes syncytia formation, margination of the chromatin, and a Cowdry Type A intranuclear inclusion body. More examples are presented in Table 47.3.

Immunopathogenesis

Although the immune response is essential to control the virus infection, it is also a major contributor to the disease signs and symptoms. It takes energy to combat a viral infection, and many protective immune responses also cause cell damage. Interferon and cytokines not only mobilize the body's defenses against the virus, but also cause the flu-like symptoms of fever, malaise, headache, muscle ache, and lack of appetite. Inflammation, rashes, and tissue destruction can be the consequence of cell-mediated immune responses causing peripheral damage as they kill infected cells to combat the virus. Immune complexes between viral antigen and antibody can lead to hypersensitivity reactions as with the arthritis that accompanies adult infection by the B19 parvovirus or the rubella (German measles) virus. The large amount of viral antigen particles (HBsAg) during hepatitis B infection can cause glomerulonephritis. The benefits and costs of the anti-microbial immune responses are discussed further in Ref. 2, 3, and 4.

TABLE 47.3
Visual Examples of Viral Pathogenesis

Example	Virus
Cytolysis	Most viruses, especially herpes, picorna, adeno, pox
Syncytia formation	HSV, VZV, paramyxoviruses, HIV
Inclusion bodies:	
Cowdry type A intranuclear	HSV
Owl's eye intranuclear	CMV
Intranuclear basophilic	Adenovirus
Intracytoplasmic acidophilic	Poxvirus
Negri bodies intracytoplasmic	Rabies
Perinuclear cytoplasmic acidophilic	Reoviruses

LABORATORY DETECTION

This section is not a replacement for a chapter on the subject but provides just a brief summary.

The patient's symptoms and the source and nature of the clinical sample are the first clues to virus identification. Virus isolation has been the gold standard for detection of viruses but this is not possible in all cases. Growth of the virus isolate in appropriate cell cultures and the cytopathological effects induced by virus replication (see Table 47.3) can provide important clues to the trained eye. Due to the expense, difficulty, and risk, most laboratories have minimized the use of virus isolation and are identifying viruses by detection of viral components using immunological or molecular biological techniques, or by serology. ELISA, immunofluorescence, and related tests are useful for detecting viral antigen in serum or other sample and in infected tissue culture cells. *In situ* hybridization, polymerase chain reaction (PCR), reverse transcriptase PCR (RT-PCR), and related approaches can detect and quantitate the viral DNA or RNA genome. Serology is still very useful for identifying viruses that cause disease of long duration, such as Epstein Barr virus, hepatitis, or HIV-AIDS.

Serological analysis provides a history of a patient's infections. The presence of IgM (immunoglobulin M) to viral antigen indicates ongoing disease, and a fourfold increase in antibody titer between acute and convalescent serum indicates recent infection. Analysis of antibody responses to specific viral antigens can also be used to follow the course of some viral diseases. Detection of antibodies to the EBNA (Epstein Barr nuclear antigen) of Epstein Barr virus indicate convalescence from infectious mononucleosis or prior infection. In addition to the ELISA and immunofluorescence tests mentioned earlier, antibody neutralization of virus infection and inhibition of hemagglutination (for those viruses that bind to erythrocyte structures and agglutinate the cells), termed hemagglutination inhibition (HI), can be used to evaluate serum for antiviral activity.

EPIDEMIOLOGY

Similar to infectious disease, which studies the consequence of viral infection of the body and its control by the immune response, epidemiology studies the viral infection of the population and the control of its spread by immunization of the populace. Specifically, viral epidemiology studies the mechanisms and modes of transmission of the virus; the risk factors; the at-risk group; seasonal, geographical, and environmental considerations for infection; as well as modes of controlling virus spread.

TRANSMISSION

The most common means of virus transmission are the respiratory and fecal-oral routes. Other transmission routes include contact with fomites carrying the virus (e.g., tissues), contaminated secretions (e.g., saliva, semen) or by sexual contact, birth, blood transfusion, organ transplant, and zoonoses (insect and animal vectors).

Some viruses are zoonoses, capable of infecting insects or animals other than humans. A bird, rodent, or monkey may be a reservoir for maintaining the virus and an arthropod, such as a mosquito or tick, may be the vector for transmitting the virus. The arthropod would acquire the virus while taking a blood meal and transmit the virus to its next host. Humans are primarily at risk for infection when they encroach on the environment of the reservoir or vector.

The route of transmission is determined by several factors, most importantly the tissue that replicates the virus and the virion structure. As indicated earlier, capsid viruses can be transmitted by most routes but enveloped viruses must remain wet during transmission to be functional, and most enveloped viruses are destroyed in the GI tract. Viruses released by respiratory secretion replicate in the lung. For fecal-oral transmission, the virus can replicate in mucosal epithelium or lymphoid tissue of the intestines or the pharynx and then swallowed; most of these viruses are capsid viruses. Transmission in blood, by a mosquito or in a transfusion, requires that the virus infect blood cells or establish a sufficient viremia and be in the bloodstream when the Red Cross or the arthropod takes its blood.

Sexual transmission requires an additional criterion. Sexually transmitted viruses replicate on the genitalia or within secretory cells and virus transmission must occur prior to, or in the absence of symptoms. Aches, pains, lesions, or other symptoms are a deterrent to the transmission of STDs (sexually transmitted diseases).

Geography and Season

Many of the vector-borne viral diseases are endemic to the habitat of the insect or animal vector. This is especially true for many of the mosquito-borne encephalitis and hemorrhagic fever viruses. Unfortunately, the geographic limitations to virus spread have dissolved due to global transportation. Passengers and freight can carry a virus or an infected mosquito to new geographic regions within a day. A perfect example of this was the outbreak of severe acute respiratory syndrome (SARS) in 2002. SARS spread from Vietnam to China to Hong Kong and then to Toronto, Canada. Influenza spreads throughout the world in infected birds as well as infected people.

The reasons for the seasonal occurrence of viral infections are sometimes obvious — for example, the correlation between outbreaks of arboviral (arthropod-borne virus) encephalitis and the life cycle of the mosquito vector. The seasonal nature of influenza outbreaks may be due to the increased proximity of people during the fall and winter, or that the temperature and humidity of the air may stabilize or promote transmission of the virus-loaded aerosol.

Who Is at Risk

There are many different risk factors, in addition to geography, that determine whether an individual is likely to come into contact with a virus and the severity of the resulting infection. These include the age, health, and immune status of the individual, as well as occupation, travel history, lifestyle, sexual activity, and interactions with children, especially daycare facilities.

VACCINES AND ANTIVIRAL DRUGS

Vaccines

Vaccination is probably the most beneficial treatment that a physician can provide because the immune response is the only therapy for most viral diseases (more on vaccines can be found in References 3 and 5). Passive immunization can be used as a temporary therapy to prevent or ameliorate the course of viral disease. For example, rabies immunoglobulin is injected around the site of an animal bite to neutralize any virus that may have been injected. The use of human antibodies is preferable to avoid the possibility of developing serum sickness upon a second administration of animal antibody. Varicella zoster immunoglobulin (VZIG) can be given to children with leukemia because they are at risk of serious outcomes of infection.

Active immunization with an attenuated (non-disease causing) virus, inactivated, or subunit vaccine is preferable to passive immunization because it initiates a longer-term immune response in the individual (Table 47.4). Live, attenuated vaccines utilize virus variants that cause mild or no disease in humans because they cannot grow efficiently at human body temperature (37°C), they have been adapted to animal or tissue culture cells rather than human cells, or they are a related animal virus that does not grow well in humans but shares antigens with the human virus (e.g., smallpox and rotavirus vaccines). These vaccines elicit a natural, long-lived immunity with effective cellular and antibody responses. However, there is a minor risk of disease with live vaccines. Inactivated vaccines consist of virion particles or viral components that are purified and then

TABLE 47.4
Most Common Viral Vaccines

Vaccine	Type of Vaccine	Who Should Receive Vaccine
Measles	Live, attenuated	2-year-olds, booster for 12-year-olds
Mumps	Live, attenuated	2-year-olds
Rubella	Live, attenuated	2-year-olds
Varicella zoster	Live, attenuated	2-year-olds and >60-year-olds (zoster)
Polio	Killed, intramuscular	2-, 4-, and 6-month-olds
Polio	Live, attenuated, oral	2-month-olds (not recommended)
Influenza	Inactivated	Adults, annually
Influenza	Live, attenuated, aerosol	Teens and adults
Hepatitis A	Inactivated	Travelers and at-risk population
Hepatitis B	Virus-like particle	2-, 4-, and 8-month-olds and at-risk population
Rabies	Inactivated	Infected individuals and veterinarians
Rotavirus	Live, attenuated reassortant, oral	6-, 12-, and 18-week-olds
Human papillomavirus	Virus-like particle	Teen and adult women

heat, radiation, or chemically treated to destroy the infectivity of the virus. The hepatitis B virus and the new human papilloma virus vaccines are produced by genetic engineering and consist of virus-like particles (VLPs) containing one of the viral proteins. Inactivated vaccines produce a more short-lived and mainly antibody-, not cell-mediated immunity. Adjuvants, larger amounts, and booster immunizations of the inactivated vaccines are required to elicit protections with inactivated vaccines.

Immunization programs have led to the elimination of smallpox throughout the world; natural polio in most of the world; and measles, mumps, and rubella except in the Third World. Routine immunization programs are available for infants and for individuals at risk of infection due to their vocation or travel into an endemic region.

Antiviral Drugs

The best antiviral drug targets are the DNA polymerase of herpes viruses, the reverse transcriptase polymerase of retroviruses and hepatitis B virus, and other polymerases. Other targets include the protease of HIV, the neuraminidase of influenza, and the M2 protein of influenza A (Table 47.5). Inhibitors of viral polymerases are nucleoside analogs such as acyclovir, penciclovir, and azidothymidine (AZT). The anti-HSV and varicella zoster virus drugs (acyclovir and penciclovir) and their prodrugs (valacyclovir and famciclovir) are very safe drugs because they require activation by a viral enzyme called thymidine kinase. There are also non-nucleoside inhibitors for the HIV reverse transcriptase. The HIV protease is another good target for antiviral drugs because it is essential for the proper assembly of an infectious virion. Inhibitors of the influenza neuraminidase, such as zanamivir and oseltamivir, cause the virus to clump and this prevents virus release. Blockage of the M2 protein channel of influenza A by amantadine and rimantidine prevent the release of the genome into the cytoplasm. Ribavirin is a guanosine analog that induces hypermutation and inactivation of viral genomes as well as inhibiting other important processes.

TABLE 47.5
Selected Antiviral Drugs

Virus	Antiviral Drug	Trade Name
Human immunodeficiency virus		
RT inhibitor: nucleoside analog	Azidothymidine	Retrovir
	Dideoxycytidine	Hivid
	Dideoxyinosine	Videx
	Lamivudine	Viramune
	Stavudine	Zerit
RT inhibitor: non-nucleoside	Delavirdine	Rescriptor
	Nevirapine	Viramune
Protease inhibitors	Indinavir	Crixivan
	Nelfinavir	Viracept
	Ritanavir	Norvir
	Sasquinavir	Invirase
Fusion inhibitor	Enfuvirtide	Fuzeon
Herpes simplex and Varicella zoster	Acyclovir	Zovirax
	Valacyclovir	Valtrex
	Penciclovir	Denavir
	Famciclovir	Famvir
Cytomegalovirus	Ganciclovir	Cytovene
	Valganciclovir	Valcyte
	Cidofovir	Vistide
	Phosphoformate	Foscavir
Respiratory syncytial virus	Ribavirin	Virazole
Hepatitis C virus	Ribavirin and interferon-alpha	
Influenza A virus	Amantadine, rimantadine	Symmetrel, flumadine
Influenza A, B virus neuraminidase inhibitors	Zanamivir	Relenza
	Oseltamivir	Tamiflu
Hepatitis B virus	Lamivudine	Epivir
	Adefovir dipivoxil	Hepsera

DESCRIPTION OF VIRUS FAMILIES

DNA Viruses

Parvoviruses

The parvoviruses are the smallest DNA viruses and include one human pathogen, the B19 virus. Parvoviruses have a single-stranded DNA genome contained in a naked capsid. They require growing cells to provide DNA nucleotides and polymerase for replication. B19 targets erythroid precursor cells. During replication, the single-stranded genome is converted into double-stranded DNA.

B19 causes erythema infectiousum (fifth disease — one of the five childhood exanthems) in children and polyarthritis without rash in adults. In children, a period of high fever is followed several days later by a maculopapular rash caused by immune complexes. B19's predilection for erythroid precursor cells can cause an aplastic crisis in a person with chronic hemolytic anemia (e.g., sickle cell anemia) and neonatal infection may be fatal due to hydrops fetalis. There are neither vaccines nor antiviral drugs.

Papillomaviruses

There are at least 100 types of the wart-causing human papillomavirus (HPV) that infect cutaneous or mucosal tissues. HPVs are naked capsid viruses with a double-stranded DNA genome. These viruses can stimulate the cell to grow but utilize cellular polymerases for replication of the genome. In addition to the benign common and other warts of the skin, other HPV types can cause growths on mucosal epithelial surfaces of the throat, conjunctiva, and anogenital regions. Cervical infection by HPV 16, 18, and some other types are common sexually transmitted diseases that can lead to dysplasia and cervical carcinoma. Genome detection methods such as PCR and *in situ* DNA analysis are used to identify the virus. Surgical removal of the wart or growth is the most common therapy. A nucleotide analog, cidofovir, and an immunomodulator, imiquimod, are FDA-approved therapies. A vaccine against HPV 6, 11, 16, and 18 has recently been approved for young women.

Polyomaviruses

The polyomaviruses are close cousins of the papillomaviruses. The two human polyomaviruses, JC and BK, are ubiquitous but only cause disease in immunocompromised individuals. The JC virus causes progressive multifocal leukoencephalopathy, a disease of multiple neurologic symptoms; and the BK virus causes hemorrhagic cystitis. There are neither vaccines nor antiviral drugs.

Adenoviruses

The adenoviruses are larger double-stranded DNA viruses. Their naked icosadeltahedral capsid has a fiber at each of the 12 vertices that serves as the viral attachment protein. Respiratory diseases and pharyngoconjunctival fever (pink eye with a sore throat) are most common but adenovirus can cause other syndromes. There are live attenuated vaccines to adenovirus types 4 and 7 for military but not civilian use. These adenovirus types are also used as gene delivery vehicles for gene replacement therapy.

Herpesviruses

The herpesviruses have a double-stranded DNA genome packaged in an icosadeltahedral capsid that is surrounded by an envelope. Cell-mediated immune responses are essential for control of the herpesviruses. Herpesvirus replication progresses through immediate early, early, and late phases of mRNA and protein expression but transition through these phases depends on the type and state of the infected cell. The herpesviruses establish latent infection within specific cell types of the host, different for the different viruses. All the herpesviruses can reactivate upon immunosuppression and HSV reactivates upon stress — emotional as well as physical.

There are eight different human herpesviruses, most of which are ubiquitous. They include herpes simplex virus (HSV) types 1 and 2, which cause oral and genital herpes, encephalitis, keratoconjunctivitis, and severe neonatal infections; varicella zoster virus (VZV), which causes chicken pox and shingles; Epstein Barr virus (EBV), which causes heterophile-positive infectious mononucleosis; cytomegalovirus (CMV), which causes several different opportunistic diseases in immunocompromised individuals and is the most common cause of congenital viral disease; HHV6 and 7, which cause exanthema subitum (roseola), a rash on children; and HHV8, which causes Kaposi's sarcoma.

HSV and CMV can be isolated in tissue culture but these methods are being replaced by immunological and genome detection methods that are used for the other herpesviruses. Serology is commonly used to detect and follow the progression of EBV-associated infectious mononucleosis. Acyclovir, valaciclovir, penciclovir, and famciclovir are available for treatment of HSV and VZV. Ganciclovir, valganciclovir, cidofovir, and foscarnet are available for CMV. Passive immune protection with VZIG and prophylactic protection with a live vaccine are available for VZV.

Poxviruses

Poxviruses are the largest viruses that cause human disease. Poxviruses have a double-stranded DNA genome that is enclosed in an ovoid to brick-shaped enveloped virion with a complex structure. The poxviruses are the exception to most of the rules for virus replication because they are DNA viruses that replicate and assemble in the cytoplasm. As such, their genome encodes the polymerase and other enzymes necessary to transcribe mRNA as well as replicate its genome.

This virus family includes the deadly smallpox, animal pox viruses that cause similar vesicular diseases, and molluscum contagiosum. Through a very effective vaccination program and careful epidemiology, the smallpox virus was eradicated from all parts of the Earth except for the biowarfare factories of the former U.S.S.R. and the deep freezers of the CDC in the United States. Molluscum contagiosum is a wart-like skin disease. The smallpox vaccine consists of a related virus, vaccinia, which shares immunogenicity with smallpox but usually causes benign disease in immunocompetent humans.

RNA Viruses

Caliciviruses

The caliciviruses are naked capsid viruses with a positive-sense RNA genome. The most well-known of this genus are the Norwalk viruses. Most caliciviruses cause outbreaks of gastroenteritis. Hepatitis E virus (HEV) causes acute hepatitis similar to hepatitis A virus (HAV) but the mortality rate is 1 to 2% (ten times greater than for HAV) and up to 20% for pregnant women. HEV is detected by ELISA.

Picornaviruses

The picornaviruses (pico-small, RNA viruses) are also naked capsid viruses with a positive-sense RNA genome. This family includes the enteroviruses, which cause all kinds of different diseases except gastroenteritis, and the rhinoviruses, which cause the common cold. A canyon on the surface of these viruses binds to a receptor on the target cell and the RNA genome is injected into the cell. The positive-sense RNA genome binds to a ribosome and a polyprotein is made that becomes proteolyzed into an RNA-dependent RNA polymerase, capsid subunits, and other proteins. A negative-sense RNA template is made, from which more mRNA and new genomes are produced by the viral-encoded polymerase. An empty procapsid shell is filled with the RNA genome, and the virus kills and then leaves the cell.

The most famous member of the enteroviruses is poliovirus, infamous for causing paralytic disease. Fortunately, effective immunization programs with the attenuated-oral or the inactivated vaccines have led to the elimination of natural polio in most parts of the world. Other members of this family include the Coxsackie A viruses, which cause hand, foot and mouth disease, herpangina, common cold-like, polio-like, and other diseases; Coxsackie B virus (B is for body), which causes pleurodynia and myocardial infections; and echoviruses, which normally are benign but can cause viral meningitis in children younger than 1 year. Hepatitis A virus, which causes acute hepatitis, is also an enterovirus. All these viruses are transmitted by the fecal-oral route but some are also transmitted in aerosols and on fomites.

The rhinoviruses cause the common cold and are distinguished from the enteroviruses because they are disrupted by acids, cannot be transmitted by the fecal-oral route, and cannot grow efficiently at temperatures above 34°C.

Live attenuated and killed polio vaccines are composed of the three poliovirus types. Although the live vaccine is cheaper, easier to administer, and has other benefits, the killed vaccine is safer and currently preferred. The hepatitis A virus vaccine used in most parts of the world is also inactivated, but a live vaccine is used in China. Pleconaril is an antiviral drug that binds to the cell receptor binding canyon on the virion surface and prevents uncoating of the virus.

Coronaviruses

Coronaviruses are enveloped viruses with a positive-sense RNA genome. The virus derives its name from the "corona" seen in electron micrographs around the virion created by the viral glycoproteins. This structure also protects the virion from harsh conditions and allows this enveloped virus to be transmitted by the fecal-oral route in addition to the respiratory route. Most of these viruses cause the common cold. The SARS coronavirus (SARS-CoV) caused an outbreak of deadly severe acute respiratory syndrome, which started in Vietnam and China and spread to Toronto, Canada. There are neither vaccines nor antiviral drugs for coronaviruses but careful quarantine limited the spread and disease caused by SARS-CoV.

Paramyxoviruses

The paramyxoviruses include the well-known measles and mumps viruses, as well as very common parainfluenza, respiratory syncytial, and metapneumon respiratory disease viruses. The paramyxoviruses are enveloped viruses with a negative-sense RNA genome. As negative-strand RNA viruses, they carry their polymerase in the virion wrapped around the genome. Individual mRNAs and a genomic-length positive-sense RNA are transcribed and the latter form of RNA is used as a template to make new genomes. The paramyxoviruses express a fusion protein that allows them to enter the cell by fusion of the envelope with the plasma membrane, and the fusion protein also promotes cell-cell fusion into multinucleated giant cells (syncytia).

As respiratory viruses, the paramyxoviruses are very contagious. Parainfluenza and respiratory syncytial viruses (RSVs) are ubiquitous, causing anything from a serious cold to bronchiolitis or pneumonia. In young children, parainfluenza causes croup and RSV causes a serious pneumonia. The severity of RSV infections prompted the development of antiviral therapy with aerosolized ribavirin or treatment with anti-RSV immunoglobulin. The classic symptoms of measles are cough, coryza (runny nose), conjunctivitis, rash, and Koplik's spots (small vesicular lesions in the mouth). The more serious outcomes include pneumonia, encephalitis, and post-infectious encephalitis. Most of the disease symptoms are due to immune pathogenesis. Mumps infects glandular tissue and the immune response causes swelling resulting in parotitis, orchitis, and meningoencephalitis. Fortunately, there are very effective live, attenuated vaccines for protection against measles and mumps that are administered to children at 2 years of age and again at approximately 14 years of age.

Orthomyxoviruses

The orthomyxoviruses include influenza A, B, and C. These are enveloped viruses that have a segmented, negative-strand RNA genome. Most of the RNA genome segments encode individual viral proteins. The virions are decorated with hemagglutinin and neuraminidase glycoproteins. The hemagglutinin (HA or H) binds to a sialic acid cellular receptor, which is also on erythrocytes. The neuraminidase cleaves sialic acid from virion glycoproteins and mucus to prevent the virus from binding to itself and facilitates exit from the infected cell and spread. Having a segmented genome allows the influenza to reassort their segments when a cell is infected with more than one strain of influenza to create hybrid viruses.

Influenza A is a zoonose, a virus that infects animals and humans. Reassortment of a mixed animal-human influenza virus infection or adaptation of the animal virus to humans can create new viruses that can spread worldwide. These greatly feared pandemics usually originate in the Far East. Influenza viruses are named for their type, place of original isolation, date of original isolation, and type of hemagglutinin and neuraminidase (for example, A/Bangkok/1/79(H3N2) virus). Influenza B can also change, but less drastically than Influenza A. Influenza C causes mild disease. Influenza can be assayed by its ability to agglutinate erythrocytes but detection of antigen by ELISA tests or the genome by RT-PCR are preferred procedures.

Amantadine and rimantidine are antiviral drugs that inhibit a channel formed by the M2 protein of influenza A (not B or C) and this prevents the release of the genome from the virion particle within the cell to block replication. Oseltamivir (tamiflu) and zanamivir (relenza) inhibit the neuraminidase of influenza A and B. An annual vaccine program utilizes a mixture of inactivated viruses or their subunits comprising the influenza A and B types that are predicted to occur during that year. A live attenuated inhaled vaccine is also available.

Rhabdoviruses, Filoviruses, and Bornaviruses

These viruses have similar structure and were once considered to be in the same viral family. These are enveloped viruses with single-stranded, negative-sense RNA genomes. The rhabdoviruses have a characteristic bullet shape, whereas the filo- and borna-viruses are filamentous. The most famous rhabdovirus is rabies. Reservoirs of this virus are present in raccoons, foxes, skunks, other forest wildlife, bats, and even farm animals. Infection of humans is always fatal unless the individual is treated with anti-rabies immunoglobulin and immunized with an inactivated vaccine prepared in diploid fibroblasts. Rabies takes a long time to progress from the site of the bite to the brain and does not cause signs of disease until it reaches the brain. The characteristic fear of drinking water (hydrophobia) is due to the pain upon swallowing. Infection of the highly innervated salivary glands and mental changes in animals (e.g., mad dog) provides the source of virus and the attitude (madness) for transmitting the virus. Analysis of brain tissue from an infected animal will show Negri inclusion bodies in the cytoplasm but virus infection is usually detected by immunofluorescence or genome detection. As mentioned, there is an inactivated vaccine that is administered to people at risk for infection, such as veterinarians, and to people who have been bitten by an animal that is a possible carrier of rabies. A hybrid vaccine consisting of the rabies virus glycoprotein gene genetically inserted into the vaccinia virus is airdropped into the forest to immunize wild animals.

The Marburg and Ebola filoviruses cause severe or fatal hemorrhagic fevers and are endemic in regions of Africa. These viruses are so virulent that they often burn out their susceptible population by killing their human cohort before they can spread to another community.

Reoviruses

The reoviruses derive their name as the acronym for "respiratory enteric orphan" viruses because no disease was associated with them until the rotaviruses were identified. The Colorado tick fever virus is also a reovirus. Reoviruses have a double capsid and a double-stranded segmented RNA genome. Partial protease digestion exposes the viral attachment proteins and allows the virion to disassemble when inside the target cells of the GI tract. Like the orthomyxoviruses, the individual negative-strand segments of the genome are transcribed into positive-sense mRNA. An early set of proteins are made that corral the positive-sense segments into an inner capsid, late proteins are made, the double-stranded RNA is recreated, the outer virion capsid acquired, and the virus is released.

Rotaviruses are the most common cause of infantile diarrhea and a major cause of death due to dehydration in underdeveloped countries. New, live, attenuated, reassorted vaccines have been developed and approved to prevent this disease.

Togaviruses

The togaviruses consist of the alphaviruses, which are transmitted by mosquitoes and considered arboviruses (arthropod-borne viruses), and rubella virus, which is transmitted by aerosols. The togaviruses have an envelope that is tightly wrapped around a capsid containing a positive-strand RNA genome. After entering the cell by receptor-mediated endocytosis, the genome is released, binds to ribosomes, and a polymerase and other early proteins are made as part of the polysome. After a negative-strand RNA template is made for new genomes, smaller mRNAs are also made that encode a polyprotein for the capsid and envelope glycoproteins.

The alphaviruses are zoonoses with a bird or rodent reservoir and a mosquito vector. These viruses, which include Western, Eastern, and Venezuelan equine encephalitis viruses, cause either systemic flu-like symptoms or encephalitis in humans. Rubella virus causes German measles, a childhood rash with possible arthritic symptoms due to immune pathogenesis. Rubella was a major cause of congenital defects (deafness, mental retardation, and cataracts) until vaccine programs limited the spread of and infection by this virus. The vaccine is live and attenuated, and administered with the measles and mumps vaccines (MMR).

Flaviviruses

Most flaviviruses are arboviruses, except the hepatitis C virus (HCV). These viruses are enveloped with a positive-strand RNA genome that replicates like the picornaviruses, except with an envelope instead of a capsid. The arbo-flaviviruses cause mild systemic flu-like symptoms, hemorrhagic disease (dengue, yellow fever), or encephalitis (West Nile, Japanese encephalitis viruses). Vaccines are available for the yellow fever and Japanese encephalitis viruses. Hepatitis C (HCV) virus is transmitted in blood, semen, and vaginal fluids. HCV usually causes chronic hepatitis (70%), which can progress to cirrhosis (20%), liver failure (6%), or hepatocellular carcinoma (4%). HCV disease is treated with a combination of ribavirin and interferon alpha.

Bunyaviruses

The bunyaviruses are enveloped viruses with a negative-strand RNA genome. There are many different bunyaviruses with exotic names that are indicative of where in the world they were discovered. Most bunyaviruses are transmitted by arthropods, including mosquitoes, flies, or ticks, depending on the virus. The hantaviruses infect rodents and are spread in their urine and stool. The California encephalitis viruses and hantaviruses occur in the United States. These viruses cause encephalitis and hemorrhagic disease, respectively.

Arenaviruses

The arenaviruses (arena-gr. sandy) get their name because they incorporate ribosomes into the virion particle and this gives them a sandy appearance in electron micrographs. These enveloped viruses contain two circular RNA genome segments; the larger segment is a negative-sense RNA while the other is ambisense. Early mRNA from the ambisense segment is copied from the entering genomic negative-sense RNA. Later, when positive RNA templates are made to replicate the genome, another mRNA is also made for the late structural proteins. The arenaviruses are transmitted in urine from mice and other rodents and lassa fever is also transmitted in secretions and fluids from infected humans. Lassa fever causes a deadly hemorrhagic fever and is treated with immunoglobulin and ribavirin.

Retroviruses

The retroviruses include two human disease viruses — human immunodeficiency virus (HIV) and human T-leukemia virus (HTLV) — as well as oncogenic animal viruses, some non-pathogenic viruses, and many endogenous viruses that are carried in our chromosomes. The retrovirus envelope encloses a capsid that contains two copies of the positive-sense RNA, an RNA-dependent RNA polymerase (reverse transcriptase), and other enzymes. The basic retrovirus genome consists of three genes: the gag (capsid proteins), pol (polymerase, integrase and protease), and env (glycoprotein) genes. HTLV and HIV encode other genes to facilitate the replication and pathogenesis of the virus. HIV binds to CD4 molecules on T-cells and other cells and a chemokine receptor as a co-receptor. It then fuses its envelope with the cell membrane to enter the cell. The positive-strand RNA is converted into a negative-strand cDNA and then a circular double-stranded DNA by the reverse transcriptase enzyme carried in the virion. The HIV polymerase is extremely error prone and readily generates new versions of the virus. The genome integrates into the host chromosome and several mRNAs and a genomic-length, positive-sense RNA are transcribed by the cell as if from very active genes. A polyprotein, consisting of the capsid and enzymatic proteins, associates with viral glycoprotein modified plasma membranes. The genome then associates with the protein and the virion buds out of the cell. After leaving the cell, the virion protease cleaves the polypeptide into functional proteins. Eventually, the cell will die. HTLV also binds to CD4 molecules on T cells, its cDNA gets integrated into the host chromosome but the virus stimulates the growth of the cell.

Many books have been written about HIV and its disease, acquired immunodeficiency syndrome (AIDS). In summary, HIV infects the controlling cells of the immune response, the CD4 expressing macrophage, dendritic cell, and T-cell, to eventually deplete the person of immunity and cause the individual to die of opportunistic infections. Virus infection is detected by ELISA and then confirmed by Western blot analysis using antibodies to the viral components. More recently, the number of genomes is quantitated by real-time reverse transcriptase PCR analysis or other genome detection system. The high mutation rate of the virus promotes escape from antibody control and single antiviral drug treatment. A cocktail of antiviral drugs (highly active antiretroviral therapy (HAART)) consists of nucleotide and/or non-nucleotide analog inhibitors of the reverse transcriptase and a protease inhibitor.

HTLV infection is usually asymptomatic but can progress to adult acute T-cell lymphocytic leukemia (ATLL). HTLV can also cause tropical spastic paraparesis, a non-oncogenic neurologic disease.

Hepadnaviruses and Hepatitis D Virus

Hepatitis B virus (HBV) is the only member of the hepadnaviruses that infects man. The hepadnaviruses are enveloped DNA viruses but, like the retroviruses, encode a reverse transcriptase and replicate through an RNA intermediate. HBV replicates in hepatocytes without killing the cell. The DNA genome of the virus is transcribed into individual mRNAs and is larger than the genome RNA to be used as a template for making the genomic DNA. The negative DNA strand is synthesized from the template RNA by the viral encoded reverse transcriptase and as the positive DNA strand is being made, the RNA is degraded and the genome is enclosed in a capsid structure that causes synthesis to stop prior to completion. The virus acquires its envelope by budding from the plasma membrane without killing the cell. Large amounts of the glycoproteins of the virus are produced, and these form particles (HBsAg) that are present in the blood of infected individuals.

Hepatitis B, also known as serum hepatitis, has an insidious onset and usually will resolve, but it can also cause chronic disease in 10% of patients. Interestingly, the symptoms of disease are caused by the immune resolution of the infection and a mild case of HBV is likely to become chronic. This is especially true for children. The course of the disease is followed by detection of viral proteins and serology. The presence of the HBsAg and another viral antigen, HBeAg are indicative of active disease. Detection of antibodies to HBsAg indicate a vaccinated individual or previous infection. HBV is transmitted in blood, semen, vaginal secretions, and saliva. Antiviral drugs target the viral polymerase and include lamivudine (also effective against HIV), famciclovir (also effective against HSV), and adefovir dipivoxil. The HBV vaccine consists of virus-like particles that form from the

HBsAg. Originally, the vaccine was purified from the blood of chronic carriers of the disease but is now prepared by genetic engineering.

Hepatitis D virus (HDV) is somewhat of a parasite of hepatitis B virus. HDV can only replicate in HBV-infected cells and the HDV RNA genome is packaged in HBV envelopes. HDV exacerbates HBV disease, promoting fulminant hepatitis and cirrhosis.

PRIONS

Prions, by definition, are infectious proteins and not really viruses. The prion protein, called PrP^{SC}, is a mutant and/or structural variant of a normal and natural cell surface protein. Binding of the prion protein to the natural analog changes its configuration to resemble the prion and become a prion. Prions are very resistant to inactivation by detergents, disinfectants, heat, autoclaving, and irradiation. Prions cause spongiform encephalopathies, including Creutzfeld Jakob disease (CJD), Kuru, Gersmann-Stassler-Scheinker syndrome, and fatal familial insomnia, which are inheritable prion diseases. CJD was considered an atypical slow viral disease with an incubation period that could last for decades prior to the onset of a progressive degenerative neurologic disease. A variant CJD (vCJD) has a more rapid onset. The spongiform encephalopathies are characterized by a loss of muscle control, shivering, myoclonic jerks, and tremors, rapidly progressive dementia, and death. There are no vaccine treatments for these diseases.

REFERENCES

1. Murray, P., K.S. Rosenthal, and Pfaller M. *Medical Microbiology,* 5th ed., Mosby-Elsevier, St. Louis, MO, 2005.
2. Rosenthal, K.S. Are microbial symptoms “self-inflicted”? The consequences of immunopathology. *Infect. Dis. Clin. Prac.*, 13, 306, 2005.
3. Rosenthal, K.S. Vaccines make good immune theater: immunization as described in a three act play. *Infect. Dis. Clin. Prac.*. 14, 35, 2006.
4. Rosenthal, K.S. Viruses: microbial spies and saboteurs. *Infect. Dis. Clin. Prac.*, 14, 97, 2006.
5. Rosenthal, K.S. and D.H. Zimmerman. Vaccines: all things considered. *Clin. Vaccine Immunol.*. 13, 821, 2006.

Appendix

Survey of Selected Clinical, Commercial, and Research-Model Eubacterial Species

Compiled by Emanuel Goldman

Organism	Growth	Staining	Habitat	Pathogenesis (in Humans)	Comments
Acetobacter spp.	Aerobic	Gram– rod	Plants and soil	Nonpathogenic	Converts ethanol to acetic acid
Acidithiobacillus ferrooxidans	Aerobic	Gram– rod	Soil, water	Nonpathogenic	Thermo- & acidophillic; mine waste recovery
Acinetobacter spp.	Aerobic	Gram– rod	Soil and other environments	Lung, blood, wound infections	Nosocomial in debilitated patients
Actinomyces israelii	Anaerobic[a]	Gram+ rod	Normal mouth/ intestine flora	Lung, abdomen, cervicofacial abscess	Opportunistic pathogen, sulfur-like granules
Actinomyces naeslundii	Facultative anaerobe	Gram+ rod	Normal mouth flora	Periodontal disease, dental caries	Networked filaments, sulfur-like granules
Aeromonas spp.	Facultative	Gram– rod	Fresh & slightly salted water	Gastroenteritis, wound infections	Systemic disease in immunocompromised
Agrobacterium tumefaciens	Facultative	Gram– rod	Plants	Rarely pathogenic (catheters)	Plant tumors; plant genetic engineering
Alcaligenes spp.	Aerobic	Gram– rod	Soil and water	Lung infections in cystic fibrosis	Used to produce non-standard amino acids
Anabaena spp.	Facultative aerobe	Gram– oval	Water, soil, plankton	Nonpathogenic but some make toxin	Heterocysts; photosynthetic & nitrogen-fixing
Anaplasma spp.	Intracellular	Gram– rod	Ticks, animals	Pathogenic primarily in animals	Transmission by insect vectors
Arthrobacter spp.	Aerobic	Gram+ variable[b]	Soil	Nonpathogenic	Cleans some environmental pollutants
Bacillus spp.	Facultative & aerobes	Gram+ rod	Soil	Nonpathogenic	Forms spores (all *Bacillus*)
Bacillus anthracis	Facultative aerobe	Gram+ rod	Soil	Anthrax	Found on animal skins
Bacillus cereus	Facultative aerobe	Gram+ rod	Soil	Gastroenteritis	Food poisoning both by intoxication & growth
Bacillus megaterium	Facultative aerobe	Gram+ rod	Soil	Nonpathogenic	Large organism; soil inoculant & biotech uses
Bacillus subtilis	Facultative aerobe	Gram+ rod	Soil	Nonpathogenic	Model organism for research
Bacillus thuringiensis	Facultative aerobe	Gram+ rod	Soil	Nonpathogenic	Produces insecticide, agricultural use
Bacteroides fragilis	Anaerobic	Gram– rod	Normal intestinal flora	Peritonitis, abdominal & other sites	Causes disease in inappropriate sites
Bartonella spp.	Intracellular	Gram– rod	Ticks, insects, animals	Cat scratch disease, trench fever, carditis	Transmission to humans by insect vectors
Bdellovibrio bacteriovorus	Aerobic	Gram– rod	Fresh water and soil	Nonpathogenic	Kills other Gram-negative bacteria, in periplasm
Beggiatoa spp.	Facultative anaerobe	Gram– rod	Fresh water	Nonpathogenic	Filaments; sulfur granules; pollution indicator

Organism	Growth	Staining	Habitat	Pathogenesis (in Humans)	Comments
Beijerinckia spp.	Aerobic	Gram– rod	Soil	Nonpathogenic	Nitrogen-fixing
Bifidobacterium spp.	Anaerobic	Gram+ rod	Normal intestinal flora	Nonpathogenic	Symbiotic to humans; used as a probiotic
Bordetella pertussis	Aerobic	Gram– coccobacilli	Humans	Whooping cough	Effective vaccine has reduced incidence
Borrelia burgdorferi	Microaerophile	Gram– spirochete	Animals, insects	Lyme disease	Transmission to humans by ticks
Brucella melitensis	Facultative aerobe	Gram– rod	Animals	Brucellosis	Zoonotic disease in humans
Burkholderia spp.	Aerobic	Gram– rod	Soil, water, animals	Melioidosis; lung in cystic fibrosis	Degrades polychlorinated biphenyls
Campylobacter jejuni	Microaerophile	Gram– rod	Animals, chickens	Gastroenteritis	Widespread cause of food-poisoning
Caulobacter spp.	Aerobic	Gram– rod	Fresh water	Nonpathogenic	Asymmetric division; stalks; research model
Chlamydia pneumoniae	Intracellular	Gram– rod	Animals, insects, protozoa	Pneumonia	Elementary and reticulate body forms
Chlamydia trachomatis	Intracellular	Gram– rod	Humans	Chlamydia	STD[c] that can cause infertility
Chlorobium spp.	Facultative anaerobe	Gram– coccobacilli	Warm water	Nonpathogenic	Photosynthetic
Citrobacter spp.	Facultative	Gram– rod	Normal intestinal flora, soil	Rarely pathogenic; some UTIs[d]	Uses citrate as sole carbon source
Clostridium botulinum	Anaerobic	Gram+ rod	Soil	Botulism (food-poisoning)	Forms spores; toxin used cosmetically
Clostridium difficile	Anaerobic	Gram+ rod	Endogenous intestinal flora	Pseudomembraneous colitis	Forms spores; antibiotic-induced disease
Clostridium perfringens	Anaerobic	Gram+ rod	Soil and other environments	Gas gangrene; food-poisoning	Forms spores; also normal intestinal flora
Clostridium tetani	Anaerobic	Gram+ rod	Soil	Tetanus	Spores; wound infection; effective vaccine
Corynebacterium diphtheriae	Facultative anaerobe	Gram+ rod	Mammals	Diphtheria (upper respiratory disease)	Pathogenicity due to phage; effective vaccine
Corynebacterium spp.	Facultative anaerobe	Gram+ rod	Normal skin flora	Most are nonpathogenic	Can cause disease in debilitated patients
Coxiella burnetii	Intracellular	Gram– rod-like	Animals	Q-fever	Forms spores; transmitted by inhalation
Cyanobacteria genera & spp.	Facultative	Gram– mixed	Water and soil	A few species produce toxins	The "blue-green algae"; some used in food
Cytophaga spp.	Aerobic	Gram– rod	Water and soil	Nonpathogenic	Gliders; degrade cellulose; make antibiotics

Organism	Growth	Staining	Habitat	Pathogenesis (in Humans)	Comments
Desulfovibrio spp.	Anaerobic	Gram– rod	Water and soil	Nonpathogenic	Sulfate-reducing; pollution control
Ehrlichia spp.	Intracellular	Gram– coccus	Ticks, animals	Ehrlichiosis; blood disease	Transmission by insect vectors; dog pathogen
Enterobacter spp.	Facultative	Gram– rod	Normal intestinal flora	Skin and tissue	Opportunistic; nosocomial; debilitated patient
Enterococcus faecalis	Facultative	Gram+ coccus	Normal intestinal flora	UTIs,[d] endocarditis	Causes disease in inappropriate sites
Enterococcus faecium	Facultative	Gram+ coccus	Normal intestinal flora	UTIs,[d] endocarditis	Causes disease in inappropriate sites
Erwinia spp.	Facultative	Gram– rod	Soil, water, insects, plants	Nonpathogenic	Plant pathogen affects fruits & vegetables
Erysipelothrix rhusiopathiae	Facultative anaerobe	Gram+ rod	Animals, soil	Skin infections	Primarily an animal pathogen
Escherichia coli	Facultative	Gram– rod	Normal intestinal flora	Gastroenteritis, UTI,[d] septicemia	Fast grower; diseases at inappropriate sites
Escherichia coli O157:H7	Facultative	Gram– rod	Cattle	HUS,[e] hemorrhagic colitis	Food-poisoning: beef, manure runoff in water
Eubacterium spp.	Anaerobic	Gram+ rod	Endogenous mouth flora	Periodontal disease	Historically genus of convenience; diverse
Flavobacterium spp.	Aerobic	Gram– rod	Fresh water, soil	Nonpathogenic	Serious pathogen of fish
Francisella tularensis	Aerobic	Gram– coccobacilli	Animals; rodents, rabbits	Tularemia	Transmitted by insects, animals, food, air
Frankia spp.	Microaerophile	Gram+ rod	Soil, plant roots	Nonpathogenic	Nitrogen-fixing filamentous plant symbiont; spores
Fusobacterium spp.	Anaerobic	Gram– rod	Endogenous mouth flora	Periodontal disease, part of ANUG[f]	Biofilms; can cause disease at other sites
Gallionella ferruginea	Aerobic	No peptidoglycan	Fresh and salt water	Nonpathogenic	Chemolithotrophic; iron-fixing; biofilms
Gardnerella vaginalis	Facultatative anaerobe	Gram variable rod	Endogenous vaginal flora	Vaginosis	Not exclusively an STD[c]
Haemophilus ducreyi	Facultative aerobe	Gram– coccobacilli	Humans	Chancroid (genital lesions)	STD[c] facilitates HIV transmission
Haemophilus influenzae	Facultative aerobe	Gram– coccobacilli	Normal nasopharynx flora	Lung, ear, eye, meningitis in children	Effective vaccine for children
Haemophilus parainfluenzae	Facultative aerobe	Gram– coccobacilli	Normal oral flora	Endocarditis, meningitis, lung	Infrequent disease, secondary to trauma
Helicobacter pylori	Microaerophile	Gram– helical	Stomach and duodenum	Gastric ulcers	Survives in high acidity; carcinogen

Organism	Growth	Staining	Habitat	Pathogenesis (in Humans)	Comments
Klebsiella granulomatis[g]	Aerobic	Gram– rod	Humans, tropical/subtropical	Granuloma inguinale	STD[c] forms intracellular Donovan bodies
Klebsiella pneumoniae	Facultative	Gram– rod	Normal gut flora; water, soil	Pneumonia, UTIs,[d] septicemia	Often nosocomial; model research organism
Lactobacillus acidophilus	Facultative anaerobe	Gram+ rod	Normal flora gut, vagina etc.	Nonpathogenic	Probiotic, control of yeast infections
Lactobacillus spp.	Facultative anaerobe	Gram+ rod	Flora of humans & animals	Dental caries	Used in production of various foods
Lactococcus lactis	Facultative	Gram+ coccus	Plants, human gut flora	Nonpathogenic	Probiotic used in dairy food production
Legionella pneumophila	Facultative aerobe	Gram– rod	Fresh water; in amoeba	Legionnaire's disease (pneumonia)	Transmitted by inhalation of aerosols
Leptospirillum ferrooxidans	Aerobic	Gram– spiral	Fresh water, acidic	Nonpathogenic	Chemolithotrophic; iron-fixing; industrial use
Leptospira spp.	Aerobic	Gram– spirochete	Animals, water	Leptospirosis	Transmitted by contaminated water or food
Listeria monocytogenes	Facultative aerobe	Gram+ rod	Soil, animals	Listeriosis; food-poisoning; meningitis	Danger to pregnancy; grows in cold temperatures.
Methylococcus capsulatus	Aerobic	Gram– coccus	Soil, animals	Nonpathogenic	Methanotroph may help control pollution
Methylomonas spp.	Aerobic	Gram– rod	Soil, water	Nonpathogenic	Methane primary carbon and energy source
Micrococcus spp.	Aerobic	Gram+ coccus	Normal skin flora, soil ,water	Rarely pathogenic	Pathogen in immunocompromised hosts
Moraxella catarrhalis	Aerobic	Gram– coccus	Endogenous to nasopharynx	Respiratory tract infections	Other Moraxella spp. are Gram– rods
Morganella morganii	Facultative	Gram– rod	Human gut, animals, water	UTIs[d] and other infections	Nosocomial, opportunist
Mycobacterium avium	Aerobic	Gram+ acid-fast rod	Soil, water, animals, birds	Disseminated disease (wasting, etc.)	Immunocompromised patients at risk
Mycobacterium bovis	Aerobic	Gram+ acid-fast rod	Cattle	Tuberculosis (infrequent)	Source of the BCG vaccination; TB in cattle
Mycobacterium intracellulare	Aerobic	Gram+ acid-fast rod	Soil, water, animals, birds	Lung & disseminated disease	In the *Mycobacterium avium complex* (MAC)

Organism	Growth	Staining	Habitat	Pathogenesis (in Humans)	Comments
Mycobacterium leprae	Aerobic	Gram+ acid-fast rod	Soil, humans, animals	Leprosy (Hansen's disease)	Intracellular infection of many tissues
Mycobacterium marinum	Aerobic	Gram+ acid-fast rod	Fresh and salt water	Wound infections	Swimming/fish tank granuloma; fish pathogen
Mycobacterium smegmatis	Aerobic	Gram+ acid-fast rod	Water, soil, plants, biofilms	Nonpathogenic	Fast growth enables research model for genus
Mycobacterium tuberculosis	Aerobic	Gram+ acid-fast rod	Humans	Tuberculosis ("TB")	Slow growth; drug resistance a problem
Mycoplasma capricolum	Aerobic	No cell wall, round	Goats	Nonpathogenic	Goat pathogen; needs cholesterol for growth
Mycoplasma pneumoniae	Aerobic	No cell wall, rod	Humans	Atypical pneumonia	Small; common cell culture contaminant
Myxococcus xanthus	Aerobic	Gram− rod[h]	Soil, plants, fresh water	Nonpathogenic	Developmental cycle, fruiting bodies, spores
Neisseria gonorrhoeae	Aerobic	Gram− coccus	Humans	Gonorrhea	STD[c]; intracellular; antigenic variation
Neisseria meningitides	Aerobic	Gram− coccus	Humans	Meningitis	Entry to blood via nasopharyngeal mucosa
Nitrobacter spp.	Facultative	Gram− rod	Soil, fresh water	Nonpathogenic	Nitrite-oxidizing; wastewater treatment
Nitrosomonas spp.	Aerobic	Gram− rod	Soil, water	Nonpathogenic	Oxidizes ammonia to nitrite; treats wastewater
Nocardia asteroides	Aerobic	Gram+ acid-fast rod	Soil, water	Lung infection	Infection by inhalation, not person-to-person
Pantoea spp.	Facultative	Gram− rod	Plants, water, animal guts	Skin, blood, endocarditis	Plant pathogen, opportunistic in humans
Pasteurella multocida	Facultative anaerobe	Gram− coccobacilli	Animals	Inflammation, possible systemic	Zoonotic disease
Peptostreptococcus spp.	Anaerobic	Gram+ coccus	Endogenous mouth, gut flora	Multiple sites: skin, mouth, blood, etc.	Also contributes to caries
Photobacterium spp.	Facultative	Gram− rod	Salt water, marine organisms	Mostly nonpathogenic	Luminescent fish symbiont, some are pathogenic
Porphyromonas gingivalis	Anaerobic	Gram− rod	Endogenous mouth flora	Periodontal disease, gingivitis	Biofilms; black pigment from iron
Prevotella spp.	Anaerobic	Gram− rod	Mouth flora, cattle	Periodontal disease, caries	Used to treat acidosis in cattle
Propionibacterium acnes	Anaerobic	Gram+ rod	Skin, gut	Acne and other infections	As the name implies, produces propionic acid
Proteus mirabilis	Facultative	Gram− rod	Soil, water, humans	Kidney stones, UTIs[d]	Swarming motility, biofilms; urease
Proteus vulgaris	Facultative	Gram− rod	Normal intestinal flora	UTIs[d], wound infections	Swarming motility, putrefies meat

Organism	Growth	Staining	Habitat	Pathogenesis (in Humans)	Comments
Providencia spp.	Facultative	Gram– rod	Humans, animals, water	UTIs[d] from catheters, burn wounds	Opportunist can also cause gastroenteritis
Pseudomonas aeruginosa	Facultative[i] aerobe	Gram– rod	Soil, water, humans, plants	Lung and many other sites	Opportunist harmful for cystic fibrosis, others
Rhizobium spp.	Facultative aerobe	Gram– rod	Soil, plant nodules	Nonpathogenic	Major terrestrial nitrogen-fixing plant symbiont
Rhodopseudomonas spp.	Facultative aerobe	Gram– rod	Water, soil	Nonpathogenic	Photosynthetic; cleans environmental waste
Rhodospirillum spp.	Facultative	Gram– spiral	Marine environments, soil	Nonpathogenic	Photosynthetic; nitrogen-fixing
Rickettsia prowazekii	Intracellular	Gram– rod	Lice, humans	Epidemic typhus	Brill-Zinsser disease a relapse long afterward
Rickettsia rickettsii	Intracellular	Gram– rod	Ticks, humans	Rocky mountain spotted fever	Zoonotic disease (applies to all *Rickettsia*)
Rickettsia typhi	Intracellular	Gram– rod	Fleas, rodents, cats	Murine (endemic) typhus	Obligate intracellular parasite (all *Rickettsia*)
Salmonella enterica	Facultative	Gram– rod	Animals, birds, water	Mostly gastrointestinal diseases	>2000 subspecies including 5 serovars below
Salmonella choleraesuis[j]	Facultative	Gram– rod	Swine	Septicemia	Mycotic aneurysm a dangerous complication
Salmonella enteritidis	Facultative	Gram– rod	Animals, birds, water	Gastroenteritis	Ovarian transmission from chicken to eggs
Salmonella paratyphi	Facultative	Gram– rod	Animals, birds, water	Paratyphoid enteric fever	Fecal-oral transmission (all *Salmonella*)
Salmonella typhi	Facultative	Gram– rod	Animals, birds, water	Typhoid fever	Asymptomatic carriers in gall bladder
Salmonella typhimurium	Facultative	Gram– rod	Animals, birds, water	Gastroenteritis	Grows fast (all *Salmonella*); research-model
Sarcina spp.	Anaerobic	Gram+ coccus	Soil, water, humans	Dental caries, periodontal disease	Ferment carbohydrates; one sp. in stomach
Serratia marcescens	Facultative	Gram– rod	Soil, water, humans	UTIs,[d] wound infections, other sites	Nosocomial; red pigment
Shigella boydii	Facultative	Gram– rod	Humans, primates, water	Dysentery, shigellosis	Fecal-oral transmission (all *Shigella*)
Shigella dysenteriae	Facultative	Gram– rod	Humans, primates, water	Dysentery, shigellosis, HUS[e]	Makes shiga toxin, inhibits protein synthesis

Organism	Growth	Staining	Habitat	Pathogenesis (in Humans)	Comments
Shigella flexneri	Facultative	Gram– rod	Humans, primates, water	Dysentery, shigellosis	Reiter's syndrome sequel to primary disease
Shigella sonnei	Facultative	Gram– rod	Humans, primates, water	Dysentery, shigellosis	10 cells enough to cause disease (all *Shigella*)
Sphaerotilus natans	Aerobic	Gram– rod	Flowing and waste water	Nonpathogenic	Filamentous; sheathed; pipe-clogging
Spirillum volutans	Microaerophile	Gram– helical	Stagnant water	Nonpathogenic	Use in motility assay due to active flagella
Spiroplasma spp.	Intracellular	No cell wall, helix	Insects, plants	Nonpathogenic[k]	Insect and plant pathogen
Spirulina spp.	Facultative aerobe	Gram– spiral	Water, soil, plankton	Nonpathogenic	Food supplement with reputed health benefits
Staphylococcus aureus	Facultative	Gram+ coccus	Endogenous human flora	Skin infection, food-poisoning, TSS[l]	Other diseases as well; MRSA[m] a major problem
Staphylococcus epidermidis	Facultative	Gram+ coccus	Normal skin flora	Septicemia, endocarditis	Nosocomial from catheters & implants
Staphylococcus saprophyticus	Facultative	Gram+ coccus	Animals, human urinary tract	UTIs[d]	Primarily affects females
Streptobacillus moniliformis	Facultative	Gram– rod	Rodents	Rat-bite fever, Haverhill fever	Also transmitted by food; develops L-forms[n]
Streptococcus agalactiae	Facultative	Gram+ coccus	Gut & urogenital flora	Neonatal meningitis	CAMP[o] hemolysis test for identification
Streptococcus bovis	Facultative	Gram+ coccus	Human gut, cows, sheep	Endocarditis	bacteremia associated with colon cancer
Streptococcus gordonii	Facultative	Gram+ coccus	Human mouth flora	Dental plaque; endocarditis	Member of the *S. viridans* group of spp.
Streptococcus mutans	Facultative	Gram+ coccus	Human mouth flora	Dental caries	Member of the *S. viridans* group of spp.
Streptococcus pneumoniae	Facultative	Gram+ coccus	Human nasopharynx flora	Pneumonia, meningitis, others	Phage lysin has potential to eradicate species
Streptococcus pyogenes	Facultative	Gram+ coccus	Endogenous human flora	Strep throat; toxic shock; flesh eating	Rheumatic fever & glomerulonephritis sequelae
Streptococcus viridans (spp.)	Facultative	Gram+ coccus	Normal human flora	Generally nonpathogenic	Large group mostly α-hemolytic Streptococci
Streptomyces spp.	Aerobic	Gram+ rod	Soil, water	Nonpathogenic (a few exceptions)	Fungi-like filaments, mycelium, spores

Organism	Growth	Staining	Habitat	Pathogenesis (in Humans)	Comments
Streptomyces avermitilis	Aerobic	Gram+ rod	Soil	Nonpathogenic	Producer of anti-parasitic agent avermectin
Streptomyces coelicolor	Aerobic	Gram+ rod	Soil, water	Nonpathogenic	Major producer of antibiotics (all *Streptomyces*)
Streptomyces somaliensis	Aerobic	Gram+ rod	Soil	Bacterial mycetoma (Madura Foot)	Other bacterial & fungal causes of this disease
Streptomyces verticillus	Aerobic	Gram+ rod	Soil	Nonpathogenic	Produces anti-cancer drug bleomycin
Thiobacillus denitrificans	Facultative anaerobe	Gram– rod	Water, soil	Nonpathogenic	Autotrophic denitrification; sulfur oxidation
Thiothrix spp.	Aerobic	Gram– rod	Water	Nonpathogenic	Filaments; oxidizes sulfur; wastewater problem
Treponema denticola	Anaerobic	Gram– spirochete	Endogenous mouth flora	Periodontal disease, part of ANUG[f]	Forms biofilms with other organisms
Treponema pallidum	Microaerophile	Gram– spirochete	Humans	Syphilis	Slow growth; obligate intracellular parasite
Ureaplasma spp.	Facultative	No cell wall, oval	Human urogenital tract	Urethritis, neonatal illness, others	Hydrolyzes urea; opportunist; debilitated hosts
Veillonella spp.	Anaerobic	Gram– coccus	Mouth, gut, vaginal flora	Oral, bone, & other infections	Dental plaque biofilm; infrequent pathogen
Vibrio cholerae	Facultative	Gram– rod	Salt water	Cholera	Transmission by seafood & oral-fecal
Vibrio fischeri	Facultative	Gram– rod	Salt water, marine animals	Nonpathogenic	Bioluminescent; symbiotic with fish
Vibrio harveyi	Facultative	Gram– rod	Salt water	Nonpathogenic	Bioluminescent; research-model organism
Vibrio parahaemolyticus	Facultative	Gram– rod	Salt water	Gastroenteritis	From undercooked seafood, shellfish
Vibrio vulnificus	Facultative	Gram– rod	Salt water	Gastroenteritis, wounds, septicemia	Contaminated seafood; debilitated patients
Xanthomonas spp.	Facultative aerobe	Gram– rod	Plants	Nonpathogenic	Plant pathogen
Yersinia enterocolitica	Facultative anaerobe	Gram– coccobacilli	Rodents, animals, water, soil	Gastroenteritis	Poisoning from contaminated animal foods
Yersinia pestis	Facultative	Gram– coccobacilli	Rodents, animals, insects	Bubonic/pneumonic plague	Transmitted from rodents via fleas
Yersinia pseudotuberculosis	Facultative	Gram– coccobacilli	Rodents, animals	Disseminated disease from the gut	Fecal-oral transmission; intracellular in host

Organism	Growth	Staining	Habitat	Pathogenesis (in Humans)	Comments
Zoogloea spp.	Aerobic	Gram– rod	Wastewater	Nonpathogenic	Form flocs (gels); sewage & water treatment

[a] Some research papers describe *Actinomyces israelii* as a facultative anaerobe.
[b] Rods in log phase, cocci in stationary phase.
[c] STD, sexually transmitted disease.
[d] UTI, urinary tract infection.
[e] HUS, hemolytic uremic syndrome.
[f] ANUG, acute necrotizing ulcerative gingivitis.
[g] Formerly named *Calymmatobacterium granulomatis*.
[h] The rod shape of *Myxococcus xanthus* refers to the vegetative form. The "coccus" in the species name refers to the shape of the myxospore.
[i] Usually described as an obligate aerobe, *Pseudomonas aeruginosa* can also grow anaerobically, but does not have the ability to conduct fermentation.
[j] *S. choleraesuis* is the former species name now replaced with *S. enterica*. However, there also remains a subspecies serovar named *S. choleraesuis*.
[k] Some controversy as there is evidence from some workers that *Spiroplasma* spp. can cause Transmissible Spongiform Encephalopathies.
[l] TSS, toxic shock syndrome.
[m] MRSA, methicillin-resistant *Staphylococcus aureus*.
[n] L-forms are cell-wall deficient derivatives that render the bacteria resistant to penicillin.
[o] CAMP, acronym for Christie, Atkins, and Munch-Petersen, who discovered the basis for this test in 1944.

Index

A

E

F

G

H

M

S

T